NK CELLS
AND
OTHER NATURAL EFFECTOR CELLS

Academic Press Rapid Manuscript Reproduction

NK CELLS
AND
OTHER NATURAL EFFECTOR CELLS

Edited by

Ronald B. Herberman

National Cancer Institute
Division of Cancer Treatment
Biological Response Modifiers Program
Frederick Cancer Research Center
Frederick, Maryland

1982

ACADEMIC PRESS
A Subsidiary of Harcourt Brace Jovanovich, Publishers

New York London
Paris San Diego San Francisco São Paulo Sydney Tokyo Toronto

ACADEMIC PRESS, INC.
111 Fifth Avenue, New York, New York 10003

United Kingdom Edition published by
ACADEMIC PRESS, INC. (LONDON) LTD.
24/28 Oval Road, London NW1 7DX

Library of Congress Cataloging in Publication Data
Main entry under title:

NK cells and other naturel effector cells.

Includes index.
1. Killer cells. 2. Immunocompetent cells
I. Herberman, Ronald B., Date.
QR185.8.K54N48 1982 616.07'9 82-11406
ISBN 0-12-341360-5

PRINTED IN THE UNITED STATES OF AMERICA

82 83 84 85 9 8 7 6 5 4 3 2 1

GUSTAVO CUDKOWICZ

(1927–1982)

Who was a pioneer in studies of natural cell-mediated resistance in mice

Contents

Contributors *xxiii*
Preface *xxxix*

I. CHARACTERISTICS OF NK CELLS

A. LARGE GRANULAR LYMPHOCYTES

1. *HUMAN*

Morphology and Cytochemistry of Human Large Granular Lymphocytes 1
Carlo Enrico Grossi and Manlio Ferrarini
Analysis of Natural Killer Activity of Human Large Granular Lymphocytes at a Single Cell Level 9
Tuomo Timonen, John R. Ortaldo, and Ronald B. Herberman

2. *OTHER SPECIES*

Identification and Characterization of the Natural Killer (NK) Cells in Rats 17
Craig W. Reynolds, Robert Rees, Tuomo Timonen, and Ronald B. Herberman
Large Granular Lymphocytes as Effector Cells of Natural Killer Activity in the Mouse 25
Aldo Tagliabue, Diana Boraschi, Saverio Alberti, and Walter Luini

B. CHARACTERIZATION BY SURFACE MARKERS

1. *HUMAN*

Characterization of Human NK Cells Identified by the Monoclonal HNK-1 (Leu-7) Antibody 31
Toru Abo, Max D. Cooper, and Charles M. Balch
Phenotypic Characterization of Human Natural Killer and Antibody-Dependent Killer Cells as an Homogeneous and Discrete Cell Subset 39
Bice Perussia, Virginia Fanning, Giorgio Trinchieri

Analysis of Human NK Cells by Monoclonal Antibodies against Myelomonocytic and Lymphocytic Antigens 47
Helmut Rumpold, Gerhard Obexer, and Dietrich Kraft
Unique Combination of Surface Markers on Human NK Cells: A Phenotypic Compromise at Last 53
Gert Riethmüller, Jürgen Lohmeyer, Ernst Peter Rieber, Helmut Feucht, Günter Schlimok, and Eckardt Thiel
Phenotypes of Human NK Cells as Determined by Reduction of Cytotoxic Activity with Monoclonal Antibodies and Complement 59
Joyce M. Zarling
Analysis of Human Natural Killer Cells by Monoclonal Antibodies 67
Chris D. Platsoucas
Studies of Cell Surface Markers on NK Cells: Evidence for Heterogeneity of Human NK Effector Cells 73
Patricia A. Fitzgerald, Dahlia Kirkpatrick, and Carlos Lopez
Phenotypic and Functional Characterization of Natural Killer Cells by Monoclonal Antibodies 79
Marc G. Golightly, Barton F. Haynes, C. Phillip Brandt, and Hillel S. Koren
Natural Cytotoxic Activity of Purified Human T or Null Cells. Further Analysis with Monoclonal Antibodies 85
Jean Caraux, Bernard Serrou, and William O. Weigle
Human Blood Lymphocyte Subsets Characterized by Monoclonal Reagents 91
Giuseppe Masucci, Maria G. Masucci, Maria T. Bejarano, and Eva Klein
Size Distribution and β-Microglobulin Content of Human Natural Killer Cells as Analyzed by Fluorescence Activated Cell Sorting (FACS) 99
Marianne Hokland, Iver Heron, Kurt Berg, and Peter Hokland

2. *OTHER SPECIES*

Alloantigens Specific for Natural Killer Cells 105
Robert C. Burton, Scott P. Bartlett, and Henry J. Winn
Anti-NK 2.1: An Activity of NZB Anti-Balb/c Serum 113
Sylvia B. Pollack and Sandra L. Emmons
Current Status of Surface Antigenic Markers on (A) Interferon Induced Murine Natural Killer Cells and (B) Murine Antibody Dependent Cellular Cytotoxicity Cells 119
G. C. Koo, J. R. Peppard, A. Hatzfeld, and C. Colmenares
A Differentiation Antigen of Murine Natural Killer Cells 125
Masataka Kasai and Ko Okumura
Heterogeneity and Regulation of the Natural Killer Cell System 131
Nagahiro Minato and Barry R. Bloom
Cell Surface Antigenic Characteristics of Rat Large Granular Lymphocytes (LGL) 139
Craig W. Reynolds, John R. Ortaldo, Alfred C. Denn III, Teresa Barlozzari, Ronald B. Herberman, Susan O. Sharrow, Keith M. Ramsey, Ko Okumura, and Sonoko Habu

II. CHARACTERISTICS OF OTHER NATURAL EFFECTOR CELLS

A. HUMAN

Natural Killer Cells Are Distinct from Lectin-Dependent Effector Cells in Man as Determined by the Two-Target Conjugate Single Cell Assay 145
Thomas P. Bradley and Benjamin Bonavida

Virus Dependent Natural Cytotoxicity (VDCC) of Human Lymphocytes 153
Peter Perlmann, Abdulrazzak Alsheikhly, Bengt Härfast, Torbjörn Andersson, Claes Örvell, and Erling Norrby
Natural Cytotoxicity of Human Monocytes 159
Dina G. Fischer, Marc G. Golightly, and Hillel S. Koren
Natural Cytotoxicity of Human Mononuclear Phagocytes: Role of Contaminating Endotoxin and Expression by Different Macrophage Populations 165
Andrea Biondi, Barbara Bottazzi, Giuseppe Peri, Nadia Polentarutti, Giorgio Caspani, and Alberto Mantovani

B. MOUSE

Natural Killer Cell Subsets 173
Robert C. Burton, Scott P. Bartlett, John A. Lust, and Vinay Kumar
Solid Tumors Are Killed by NK and NC Cell Populations: Interferon Inducing Agents Do Not Augment NC Cell Activity under NK Activating Conditions 179
Edmund C. Lattime, Gene A. Pecoraro, Michael Cuttito, and Osias Stutmån
NC Cells Do Not Express NK-Associated Cell Surface Antigens and Are Not Culture Activated NK Cells 187
Edmund C. Lattime, Sally T. Ishizaka, Gene A. Pecoraro, Gloria Koo, and Osias Stutman
Enrichment and Characterization of Effector Pupulations Mediating NK, NC, ADCC, and Spontaneous Macrophage Cytotoxicity in Murine Peritoneal Exudate Cell Preparations 193
Edmund C. Lattime, Louis M. Pelus, and Osias Stutman
Further Studies on the Cytostatic Activity Mediated by Murine Splenocytes 201
Rachel Ehrlich, Margalit Efrati, and Isaac P. Witz

III. CELL LINEAGE OF NK AND RELATED EFFECTOR CELLS

A. EVIDENCE FOR OR AGAINST T CELL LINEAGE

Cell Lineage of NK Cell: Evidence for T Cell Lineage 209
Sonoko Habu and Ko Okumura
Immunoregulation of Mouse NK Activity by the Serum Thymic Factor (FTS) 215
Dominique Kaiserlian, Pierre Bardos, and Jean-Francois Bach
Distinctive Characteristics between Splenic Natural Killer Cells and Prothymocytes 225
G. C. Koo, Y. Cayre, and L. R. Mittl
Hypothesis on the Development of Natural Killer Cells and Their Relationship to T Cells 229
Zvi Grossman and Ronald B. Herberman
Comment on the Nomenclature of Lymphocyte Mediated Cytotoxic Effects 239
Eva Klein

B. EVIDENCE FOR OR AGAINST MACROPHAGE LINEAGE

Cells with Natural Killer Activity Are Eliminated by Treatment with Monoclonal Specific Anti-Macrophage Antibody plus Complement 243
Deming Sun and Marie-Luise Lohmann-Matthes

Effect of Anti-HLA and Anti-Beta-2 Microglobulin Antisera on the Natural Cytotoxic Activity of Human NK Cells and Monocytes 251
Rajiv K. Saxena, Queen B. Saxena, Robert S. Pyle, and William H. Adler

C. EVIDENCE FOR OR AGAINST OTHER OR SEPARATE LINEAGE

Could Human Large Granular Lymphocytes Represent a New Cell Lineage? 257
Manlio Ferrarini and Carlo Enrico Grossi

Natural Killer Cells: A Separate Lineage? 265
John R. Ortaldo

Natural Killer Activity in Mast Cell-Deficient W/W^{V} Mice 273
Aldo Tagliabue, Dean Befus, and John Bienenstock

IV. GENETICS OF NATURAL RESISTANCE

A. MOUSE

NK-Activity Against YAC-1 Is Regulated by Two H-2 Associated Genes 275
Gunnar O. Klein

Genetic Control of NC Activity in the Mouse: Three Genes Located in Chromosome 17 281
Osias Stutman and Michael J. Cuttito

Approaches to the Genetic Analysis of Natural Resistance *in Vivo* 291
George A. Carlson and Arnold Greenberg

The Beige (*bg*) Gene Influences the Development of Autoimmune Disease in SB/Le Male Mice 301
Edward A. Clark, John B. Roths, Edwin D. Murphy, Jeffrey A. Ledbetter, and James A. Clagett

NK Cell Activation in SM/J Mice 307
Nancy T. Windsor and Edward A. Clark

Chromosome 1 Locus: A Major Regulator of Natural Resistance to Intracellular Pathogens 313
Emil Skamene, Adrien Forget, Philippe Gros, and Patricia A. L. Kongshavn

B. RAT

Genetic Variation in Natural Killer (NK) Activity in the Rat 319
Craig W. Reynolds and Howard T. Holden

V. REGULATION OF CYTOTOXIC ACTIVITY

A. ONTOGENY OF NK CELLS

Ontogeny of Nk-1^{+} Natural Killer Cells 325
G. C. Koo, J. R. Peppard, A. Hatzfeld, and Y. Cayre

Marrow Dependence of Natural Killer Cells 329
Vinay Kumar, P. F. Mellen, and Michael Bennett

Age-Independent Natural Killer Cell Activity in Murine Peripheral Blood 335
Emanuela Lanza and Julie Y. Djeu

Ontogenic Development of Porcine NK and K Cells 341
Yoon Berm Kim, Nam Doll Huh, Hillel S. Koren, and D. Bernard Amos

B. AUGMENTATION

1. *INTERFERON OR INTERFERON INDUCERS*

Augmentation of Human Natural Killer Cells with Human Leukocyte and Human Recombinant Leukocyte Interferon 349
John Ortaldo, Ronald Herberman, and Sidney Pestka

Augmentation of Human Natural Killer Cytotoxicity by Alpha-Interferon and Inducers of Gamma-Interferon. Analysis by Monoclonal Antibodies 355
Chris D. Platsoucas

Human NK Cell Activation with Interferon and with Target Cell-Specific IgG 361
Måns Ullberg and Mikael Jondal

Interferons and Natural Killer Cells: Interacting Systems of Non-Specific Host Defense 369
Giorgio Trinchieri and Bice Perussia

Pleiotropic Effects of Interferon (IFN) on the Augmentation of Rat Natural Killer (NK) Cell Activity 375
Craig W. Reynolds, Tuomo Timonen, and Ronald B. Herberman

Differential Effects of Interferons on Porcine NK and K-Cell Activities 361
Tae June Chung, Nam Doll Huh, and Yoon Berm Kim

Role of Interferon in Natural Kill of Herpesvirus Infected Fibroblasts 387
Patricia A. Fitzgerald, Carlos Lopez, and Frederick P. Siegal

In Vitro Modulation of NK Activity by Adjuvants: Role of Interferon 395
Rosemonde Mandeville, Normand Rocheleau, and Jean-Marie Dupuy

Modulation of Macrophage Cytolysis by Interferon 401
Diana Boraschi, Elena Pasqualetto, Pietro Ghezzi, Mario Salmona, Domenico Rotilio, Maria Benedetta Donati, and Aldo Tagliabue

2. *OTHER AUGMENTING AGENTS*

The Relation between Human Natural Killer Cells and Interleukin-2 409
Wolfgang Domzig and Bedu M. Stadler

Modulation of Natural Cell Mediated Cytotoxicity of T-Colonies and PBL by Interleukin-2 415
Susanna Cunningham-Rundles, Richard S. Bockman, and Berish Y. Rubin

In Vivo Interleukin-2 Induced Augmentation of Natural Killer Cell Activity 421
Steven H. Hefeneider, Christopher S. Henney, and Steven Gillis

Augmentation of Natural Killer Activity by Retinoic Acid 427
Ronald H. Goldfarb and Ronald B. Herberman

Augmentation of Human Natural Killing Activity by OK432 431
Atsushi Uchida and Michael Micksche

Differences between Phorbol Esters and Interferon Induced Activations of Mouse NK Cells 437
Anna Senik and Jean Pierre Kolb

Sodium Diethyldithiocarbamate (DTC)-Induced Modifications of NK Activity in the Mouse 443
Gérard Renoux, Pierre Bardos, Danielle Degenne, and Mircea Musset

Augmentation of Mouse Natural Killer Cell Activity by Alloantibody: A Reverse ADCC Reaction 449
Rajiv K. Saxena

Specific and Nonspecific Modulation of NCMC by Serum Components 455
Mitsuo Takasugi and Eda T. Bloom

Phytohaemagglutinin Induced Susceptibility of Autologous Allogenic and Xenogenic NK Insensitive Target Cells to Lysis by Natural and Activated Killer Cells 463
Reinder Bolhuis

Effect of IFN and PHA on the Cytotoxic Activity of Blood Lymphocyte Subsets 475
Maria G. Masucci, Giuseppe Masucci, Maria T. Bejarano, and Eva Klein

Lectin-Dependent Natural Killer Cellular Cytotoxicity in Mice (NK-LDCC): A New Subpopulation of NK-Like Cytotoxic Cells 483
Benjamin Bonavida

Proliferation and Role of Natural Killer Cells during Viral Infection 493
Christine A. Biron and Raymond M. Welsh

Increased Anti K-562 Cytotoxicity of Blood Mononuclear Cells after Yellow Fever Vaccination, Probably the Function of Activated T Cells 499
Eva Klein, Astrid Fagraeus, Anneka Ehrnst, and Manuel Patarroyo

Modulation of Natural Killer Cell Activity during Murine Infection with *Trypanosoma cruzi* 503
Frank Hatcher and Raymond E. Kuhn

Analysis of Macrophage Activation and Biological Response Modifier Effects by Use of Objective Markers to Characterize the Stages of Activation 511
D. O. Adams and J. H. Dean

C. INHIBITION OR SUPPRESSION OF ACTIVITY

1. *SUPPRESSOR CELLS*

C. Parvum-Induced Suppressor Cells for Mouse NK Activity 519
Angela Santoni, Carlo Riccardi, Teresa Barlozzari, and Ronald B. Herberman

Natural Suppressor Cells for Murine NK Activity 527
Angela Santoni, Carlo Riccardi, Teresa Barlozzari, and Ronald B. Herberman

Suppression of Murine Natural Killer Cell Activity by Normal Peritoneal Macrophages 535
Michael J. Brunda, Donatella Taramelli, Howard T. Holden, and Luigi Varesio

Regulation of NK Reactivity by Suppressor Cells 541
M. Zöller, S. Matzku, G. Andrighetto, and H. Wigzell

Regulation of *in Vivo* Reactivity of Natural Killer (NK) Cells 549
C. Riccardi, T. Barlozzari, A. Santoni, C. Cesarini, and R. B. Herberman

Lack of Suppressor Cell Activity for Natural Killer Cells in Infant, Aged and a Low Responder Strain of Mice 557
Anthony G. Nasrallah, Michael T. Gallagher, Surjit K. Datta, Elizabeth L. Priest, and John J. Trentin

Suppressor Cells Active against NK-B But Not NK-A Cells in Mice Treated with Radioactive Strontium 563
Vinay Kumar, Paul F. Mellen, John A. Lust, and Michael Bennett

Does a Marrow Dependent Cell Regulate Suppressor Cell Activity? 569
Elinor M. Levy, Vinay Kumar, and Michael Bennett

Suppressor Lymphocytes of Human NK Cell Activity 575
Jussi Tarkkanen and Eero Saksela

Modulation of NK Activity by Human Mononuclear Phagocytes: Suppressive Activity of Broncho-Alveolar Macrophages 581
Claudio Bordignon, Paola Allavena, Martino Introna, Andrea Biondi, Barbara Bottazzi, and Alberto Mantovani

Suppression of NK Cell Activity by Adherent Cells from Malignant Pleural Effusions of Cancer Patients 589
Atsushi Uchida and Michael Micksche

2. OTHER MECHANISMS

Inhibition of Natural Killer Cell Cytotoxic Reactivity by Tumor Promoters and Cholera Toxin 595
Ronald H. Goldfarb and Ronald B. Herberman

Tumor-Promoting Diterpene Esters Induce Macrophage Differentiation, but Prevent Activation for Tumoricidal Activity of Macrophage and NK Cells 601
Robert Keller

Increase in Intra-Cellular Levels of Cyclic AMP Inhibits Target Cell Recognition by Human NK Cells 607
Måns Ullberg, Mikael Jondal, Gunnar Klein, Fred Lanefeldt, and Bertil B. Fredholm

Regulation of Cytotoxic Reactivity of NK Cells by Interferon and PGE_2 615
Kam H. Leung and Hillel S. Koren

Negative Regulation of Human NK Activity by Monomeric IgG 621
Andrei Sulica, Maria Gherman, Moiara Manciulea, Cecilia Galatiuc, and Ronald Herberman

Presence of FcR for IgG and IgM on Human NK Cells: The Role of Immune Complexes in the Regulation of NK Cell Cytotoxicity 631
Jean E. Merrill, Sidney Golub, Mikael Jondal, Fred Lanefeldt, and Bertil Fredholm

NK Activity in Mice Is Controlled by the Brain Neocortex 639
Gérard Renoux, Katleen Bizière, Pierre Bardos, Danielle Degenne, and Micheline Renoux

Decline of Murine Natural Killer Activity in Response to Starvation, Hypophysectomy, Tumor Growth, and Beige Mutation: A Comparative Study 645
Rajiv K. Saxena, Queen B. Saxena, and William H. Adler

Ethanol and Natural Killer Activity 651
Queen B. Saxena, Rajiv K. Saxena, and William H. Adler

D. ACCESSORY CELLS FOR CYTOTOXIC EFFECTOR CELLS

Regulation of Human NK Activity against Adherent Tumor Target Cells by Monocyte Subpopulations, Interleukin-1, and Interferons 657
Jan E. de Vries, Carl G. Figdor, and Hergen Spits

E. PRODUCTION OF INTERFERON BY NK CELLS

Production of Interferon by Human Natural Killer Cells in Response to Mitogens, Viruses and Bacteria 669
Julie Y. Djeu, Tuomo Timonen, and Ronald B. Herberman

Interferon Production and Natural Cytotoxicity by Human Lymphocytes Fractionated by Percoll Density Gradient Centrifugation 675
Eda T. Bloom and Mitsuo Takasugi

VI. SPECIFICITY OF NATURAL EFFECTOR CELLS

A. NATURE OF TARGET CELL STRUCTURES

Recognition Structures for Natural Killer Cells on Human Lymphocytes: A Panel Study 683
Miklós Benczur, Tamás Laskay, and Győső Petrányi

Specificity of Fresh and Activated Human Cytotoxic Lymphocytes 691
Farkas Vánky and Eva Klein

Fractionation of Natural Killer Cells on Target Cell Monolayers 699
Pamela J. Jensen, Patricia A. Weston, and Hillel S. Koren

Non-NK Leukocytes Demonstrate NK-Patterned Binding 705
Gerald E. Piontek, Rolf Kiessling, Alvar Grönberg, and Lars Ährlund-Richter

Target Cell Recognition by Natural Killer and Natural Cytotoxic Cells 713
Edmund C. Lattime, Gene A. Pecoraro, and Osias Stutman

The Relationship between NK and Natural Antibody Target Structures 719
A. H. Greenberg and B. Pohajdak

Effect of Sodium Butyrate and Hemin on NK Sensivity of K562 Cells 729
Marie-Christine Dokhélar and Thomas Tursz

Analysis of Differentiation Events Causing Changes in NK Cell Tumor-Target Sensitivity 733
Magnus Gidlund, Masato Nose, Inger Axberg, Hans Wigzell, Thomas Tötterman, and Kenneth Nilsson

Specificity of Natural Killer (NK) Cells: Nature of Target Cell Structures 743
Jerome A. Werkmeister, Stephen A. Helfand, Tina Haliotis, Hugh Pross, and John C. Roder

Effects of Interferon and Tumor Promoter, 12-0-Tetradecanoylphorbol-13-acetate, on the Sensitivity of Trophoblast Cells to Natural Killer Cell Activity 751
Kenneth S. S. Chang and Kenichi Tanaka

Serological Approaches to the Elucidation of NK Target Structures 757
David L. Urdal, Ichiro Kawase, and Christopher S. Henney

Surface Sialic Acid of Tumor Cells Inversely Correlates with Susceptibility to Natural Killer Cell Mediated Lysis 765
Ganesa Yogeeswaran, Raymond M. Welsh, Alvar Gronberg, Rolf Kiessling, Manuel Patarroyo, George Klein, Magnus Gidlund, Hans Wigzell, and Kenneth Nilsson

Inhibition of Spontaneous and Antibody-Dependent Cellular Cytotoxicity Using Mono- and Oligosaccharides 771
Peter Kaudewitz, Hans Werner (Löms) Ziegler, Gerd R. Pape, and Gert Riethmüller

B. NATURE OF RECOGNITION RECEPTORS ON EFFECTOR CELLS

The Use of a Monoclonal Antibody to Analyse Human NK Cell Function 777
Walter Newman

VII. CYTOTOXICITY BY CULTURED LYMPHOID CELLS

A. SHORT TERM CULTURES

Heterogeneity of MLC-Generated NK-Like Cells 785
Paula J. D'Amore, Marc G. Golightly, and Sidney H. Golub

Stimulation of Lymphocytes with Allogeneic Normal Cells and Autologous Lymphoblastoid Cell Lines: Distinction between NK-Like Cells and Cytotoxic T Lymphocytes by Monoclonal Antibodies 791
Joyce M. Zarling, Fritz H. Bach, and Patrick C. Kung

Generation of Lytic Potential in Mixed Cultures and Its Modification by IFN-α 799
Shmuel Argov and Eva Klein

Cultured Natural Killer Cells 807
Scott P. Bartlett and Robert C. Burton

Spontaneous Monocyte Mediated Cytotoxicity in Man: Evidence for T Helper Activity 815
Andrew V. Muchmore and Eugenie S. Kleinerman

B. CONTINUED CULTURES AND CLONES

Cultures of Purified Human Natural Killer Cells 821
Tuomo Timonen, John Ortaldo, and Ronald Herberman

Lytic Effect of T-Cell Cultures against Tumor Biopsy Cells and K562 829
Farkas Vánky and Eva Klein

Natural Killer Cell-Like Cytotoxicity Mediated by Herpesvirus Transformed Marmoset T Cell Lines 835
Donald R. Johnson, Mikael Jondal, and Peter Biberfeldt

Cloned Cell Lines with Natural Killer Activity 843
Gunther Dennert

Characterization of Cloned Murine Cell Lines Having High Cytolytic Activity against YAC-I Targets 851
Colin G. Brooks, Kagemasa Kuribayashi, Susana Olabuenaga, Mei-fu Feng, and Christopher S. Henney

Induction of NK-Like Anti-Tumor Reactivity *in Vitro* and *in Vivo* by IL-2 859
Eli Kedar and Ronald B. Herberman

Cloned Lines of Mouse Natural Killer Cells 873
Carlo Riccardi, Paola Allavena, John R. Ortaldo, and Ronald B. Herberman

Clonal Analysis of Human Natural Killer Cells 879
B. M. Vose, C. Riccardi, R. J. Marchmont, and G. D. Bonnard

Continuous Culture of Human NK-T Cell Clones 887
Daniel Zagury and Doris Morgan

Permanently Growing Murine Cell Clones with NK-Like Activities 893
Hans Hengartner, Hans Acha-Orbea, Rosmarie Lang, Lothar Stitz, Kendall L. Rosenthal, Peter Groscurth, and Robert Keller

Natural Killer Activity in Cloned IL-2 Dependent Allospecific Lymphoid Populations 903
John R. Neefe and Robin Carpenter

Regulation by Interferon and T Cells of IL-2-Dependent Growth of NK Progenitor Cells: A Limiting Dilution Analysis 909
Carlo Riccardi, B. M. Vose, and R. B. Herberman

Natural Cytotoxic Activity of Mouse Spleen Cell Cultures Maintained with Interleukin-3 917
Julie Y. Djeu, Emanuela Lanza, Andrew J. Hapel, and James N. Ihle

VIII. MECHANISMS OF CYTOTOXICITY

A. ROLE OF PROTEASES

Evidence for Proteases with Specificity of Cleavage at Aromatic Amino Acids in Human Natural Cell-Mediated Cytotoxicity 923
Dorothy Hudig, Doug Redelman, and Lory Minning

The Role of Neutral Serine Proteases in the Mechanism of Tumor Cell Lysis by Human Natural Killer Cells 931
Ronald H. Goldfarb, Tuomo T. Timonen, and Ronald B. Herberman

The Role of Surface Associated Proteases in Human NK Cell Mediated Cytotoxicity. Evidence Suggesting a Mechanism by which Concealed Surface Enzymes Become Exposed during Cytolysis 939
Gad Lavie

The Relationship between Secretion of a Novel Cytolytic Protease and Macrophage-Mediated Tumor Cytotoxicity 949
William J. Johnson, James E. Weiel, and Dolph O. Adams

B. ROLE OF PHOSPHOLIPIDS

Phospholipid Metabolism during NK Cell Activity: Possible Role for Transmethylation and Phospholipase A_2 Activation in Recognition and Lysis 955
Thomas Hoffman, Philippe Bougnoux, Toshio Hattori, Zong-liang Chang, and Ronald B. Herberman

C. ROLE OF SOLUBLE MEDIATORS

Role of Natural Killer Cytotoxic Factors (NKCF) in the Mechanism of NK Cell Mediated Cytotoxicity 961
Susan C. Wright and Benjamin Bonavida
Role of Lymphotoxins in Natural Cytotoxicity 969
Robert S. Yamamoto, Monica L. Weitzen, Karen M. Miner, James J. Devlin, and Gale A. Granger
Carbohydrate Receptors in Natural Cell-Mediated Cytotoxicity 977
James T. Forbes and Thomas N. Oeltmann

D. OTHER MECHANISMS OR INFORMATION ON MECHANISMS

Cellular Secretion Associated with Human Natural Killer Cell Activity 983
Eero Saksela, Olli Carpén, and Ismo Virtanen
Mechanism of NK Cell Lysis 989
Phuc-Canh Quan, Teruko Ishizaka, and Barry R. Bloom
Mechanisms of NK Activation: Models to Study Induction of Lysis and Enhancement of Lytic Efficiency 995
Stephan R. Targan
Augmented Binding of Tumor Cells by Activated Murine Macrophages and Its Relevance to Tumor Cytotoxicity 1003
Scott D. Somers and Dolph O. Adams
The Role of Free Oxygen Radicals in the Activation of the NK Cytolytic Pathway 1011
Stephen L. Helfand, Jerome Werkmeister, and John C. Roder
Cell Surface Thiols in Human Natural Cell-Mediated Cytotoxicity 1021
Dorothy Hudig, Doug Redelman, and Lory Minning
Activated Macrophage Mediated Cytotoxicity for Transformed Target Cells 1027
John B. Hibbs, Jr., and Donald L. Granger
The Uropod as an Integral and Specialized Structure of Large Granular Lymphocytes 1035
Kenneth E. Muse and Hillel S. Koren
Morphological Characteristics of Lymphocyte–Target Cell Interactions and Their Relations to Cytolytic Activity in the Human NK System 1041
Győző G. Petrányi, Miklós Benczur, and Miklós Varga

IX. NATURAL CELL-MEDIATED REACTIVITY AGAINST PRIMARY TUMOR CELLS AND AGAINST NON-TUMOR TARGETS

A. PRIMARY TUMOR CELLS AS TARGETS

Cytotoxic and Cytostatic Activity of Human Large Granular Lymphocytes against Allogeneic Tumor Biopsy Cells and Autologous EBV Infected B Lymphocytes 1047
Maria G. Masucci, Maria T. Bejarano, Farkas Vánky, and Eva Klein
Association of Human Natural Killer Cell Activity against Human Primary Tumors with Large Granular Lymphocytes 1055
Susana A. Serrate, Brent M. Vose, Tuomo Timonen, John R. Ortaldo, and Ronald B. Herberman
Auto-Tumor Lytic Potential of Lymphocytes Separated from Human Solid Tumors 1061
Farkas Vánky and Eva Klein
Natural Cell-Mediated Cytotoxicity against Spontaneous Mouse Mammary Tumors 1069
Susana A. Serrate and Ronald B. Herberman

B. NORMAL CELLS AS TARGETS

Natural Killing of Hematopoietic Cells 1077
Mona Hansson, Rolf Kiessling, and Miroslav Beran
Human Bone Marrow Mononuclear Cells as Effectors and Targets of Natural Killing 1085
Sudhir Gupta and Gabriel Fernandes

C. PARASITES AS TARGETS

Natural Killer (NK) Cell Activity against Extracellular Forms of *Trypanosoma cruzi* 1091
Frank M. Hatcher and Raymond E. Kuhn
Mechanisms of Natural Macrophage Cytotoxicity against Protozoa 1099
Santo Landolfo, Giovanna Martinotti, and Pancrazio Martinetto

D. FUNGI AS TARGETS

In Vitro Effects of Natural Killer (NK) Cells on *Cryptococcus neoformans* 1105
Juneann W. Murphy and D. Olga McDaniel

X. POSSIBLE NK CELL TUMORS OR PRESENCE OF NK CELLS AT SITE OF TUMOR GROWTH

A. HUMAN TUMORS

NK Activity of Tumor Infiltrating and Lymph Node Lymphocytes in Human Pulmonary Tumors 1113
Sidney H. Golub, Masayuki Niitsuma, Norihiko Kawate, Alistair J. Cochran, and E. Carmack Holmes
Natural Killer Activity in Human Ovarian Tumors 1119
Martino Introna, Paola Allavena, Raffaella Acero, Nicoletta Colombo, Pierangela Molina, and Alberto Mantovani
Natural Cytotoxic Effectors in Human Tumours and Tumour Draining Nodes 1127
B. M. Vose and M. Moore

Control of Natural Cytotoxicity in the Regional Lymph Node in Breast Cancer 1133
Susanna Cunningham-Rundles
Leukemic Blasts as Effectors in Natural Killing and Antibody Dependent Cytotoxicity Assay 1141
Sudhir Gupta and Gabriel Fernandes

B. OTHER SPECIES

Natural Killer Cell Activity Associated with Spontaneous and Transplanted Reticulum Cell Neoplasms 1147
Kenneth S. S. Chang and Richard Kubota
Reticulum Cell Sarcomas of SJL/J Mice: Pre-B Cell Lymphoma with Apparent Natural Killer Cell Function 1153
Nicholas M. Ponzio
Identification and Characterization of Large Granular Lymphocyte (LGL) Leukemias in F344 Rats 1161
Craig W. Reynolds, Jerrold M. Ward, Alfred C. Denn III, and E. William Bere, Jr.

XI. CLINICAL STUDIES WITH NATURAL EFFECTOR CELLS

A. ALTERATIONS WITH VARIOUS DISEASES

1. *CANCER*

Tumor Related Changes and Prognostic Significance of Natural Killer Cell Activity in Melanoma Patients 1167
Peter Hersey, Anne Edwards, William McCarthy, and Gerald Milton
The Assessment of Natural Killer Cell Activity in Cancer Patients 1175
Hugh F. Pross, Peter Rubin, and Malcolm G. Baines
NK Cell Activity in Patients with High Risk for Tumors and in Patients with Cancer 1183
Marc Lipinski, Marie-Christine Dokhélar, and Thomas Tursz
NK and K Cell Activity in Mammary and Cervix Carcinoma Patients in Relation to Radiation Therapy and the Course of Disease 1189
Tamás Garam, Tamás Pulay, Tibor Bakács, Egon Svastits, Gábor Ringwald, Klára Tótpal, and Győző Petrányi
Deficient NK and ADCC Mediated by Purified E-Rosette Positive and E-Rosette Negative Cells from Patients with B-Cell Chronic Lymphocytic Leukemia. Augmentation by *in Vitro* Treatment with Human Leukocyte Interferon 1195
Chris D. Platsoucas, Sudhir Gupta, Robert A. Good, and Gabriel Fernandes
Natural Killing in Patients with Hodgkin's Disease 1201
Sudhir Gupta and Gabriel Fernandes
Natural Killer Cells in Hamsters and Their Early Augmentation and Late Suppression during Tumor Growth 1207
Surjit K. Datta, John J. Trentin, and Takanobu Kurashige

2. *OTHER DISEASES*

NK Deficiency in X-Linked Lymphoproliferative Syndrome 1211
Janet K. Seeley, Thomas Bechtold, David T. Purtillo, and Tullia Lindsten

Mechanisms of Natural Killer Cell Depression in Multiple Sclerosis 1219
Ronald C. McGarry, J. C. Roder, and D. Brunet
Impaired Natural Killer Cell Function in Multiple Sclerosis and Association with the HLA System 1227
Miklós Benczur, Éva Gyodi, Győző Petrányi, Susan R. Hollán, Gyorgy Pálffy, Margarita Talás, Ivana Stőger, and István Fóldes
Interferon (IFN) Production and Natural Killer (NK) Cell Activity in Patients with Multiple Sclerosis: Influence of Genetic Factors Assessed by Studies of Monozygotic Twins 1233
Jochen Abb, Peter Kaudewitz, Helmut Zander, Hans-Werner Löms Ziegler, Friedrich Deinhardt, and Gert Riethmüller
Studies of Human NK Cell Function in Chronic Diseases 1241
P. Andrew Neighbour, Elizabeth Reinitz, Arthur I. Grayzel, Aaron E. Miller, and Barry R. Bloom
Peripheral Blood Natural Cell-Mediated Cytotoxicity in Patients with Atopic Dermatitis 1249
Nuha T. Kusaimi and John J. Trentin
Recovery of NK Cell Activity after Bone-Marrow Transplantation 1253
Marie-Christine Dokhélar, Marc Lipinski, and Thomas Tursz

B. AUGMENTATION OF REACTIVITY BY BIOLOGICAL RESPONSE MODIFIERS

1. INTERFERON

Enhanced NK Activity in Patients Treated by Interferon-α. Relation to Clinical Response 1259
Stefan Einhorn, Anders Ahre, Henric Blomgren, Bo Johansson, Håkan Mellstedt, and Hans Strander
NK Cytotoxicity in Interferon Treated Melanoma Patients 1265
Sidney H. Golub
Phase I Trial of Immunomodulatory Activities of Human Leukocyte Interferon in Advanced Cancer Patients 1273
Jerry A. Bash, James N. Woody, and John R. Neefe
Modulation of NK Activity by Recombinant Leukocyte Interferon in Advanced Cancer Patients 1279
Annette E. Maluish, John R. Ortaldo, and Ronald B. Herberman
In Vivo Effects of *Corynebacterium parvum* on Natural Cell-Mediated Immunity in Acute Myeloid Leukemia Patients 1285
Peter Hokland and Jørgen Ellegaard

2. OTHER AGENTS

Modification of Human Natural Cell-Mediated Cytotoxicity by MVE-2 1291
James T. Forbes, Anne Luck, and F. Anthony Greco
Interaction between Interferon and Natural Killer Cells in Humans after Administration of Immunomodulating Agents 1297
Tadao Aoki, Hideo Miyakoshi, Yoh Horikawa, Akira Shibata, Yoshitaka Aoyagi, and Mikio Mizukoshi
Augmentation of NK Cell Activity in Cancer Patients by OK432: Activation of NK Cells and Reduction of Suppressor Cells 1303
Atsushi Uchida and Michael Micksche

In Vivo Effects of Biological Response Modifiers and Chemotherapeutic Agents on NK Activity in Cancer Patients 1309
Didier Cupissol, Francois Favier, Augustin Rey, Bernard Longhi, Carine Favier, and Bernard Serrou

C. DEPRESSION OF REACTIVITY BY TREATMENTS

Effect of Diethylstilbestrol and Estramustine Phosphate (Estracyt®) on Natural Killer Activity in Patients with Carcinoma of the Prostate 1317
Sven Haukaas and Terje Kalland

XII. EVIDENCE FOR *IN VIVO* REACTIVITY OF NATURAL EFFECTOR CELLS

A. AGAINST TUMORS

1. *TRANSPLANTABLE TUMORS*

Evidence for *in Vivo* Reactivity: Against Transplantable and Primary Tumors 1323
Sonoko Habu and Ko Okumura

Acceleration of Metastatic Growth in Anti-Asialo GM1-Treated Mice 1331
E. Gorelik, R. Wiltrout, K. Okumura, S. Habu, and R. B. Herberman

The Effect of Selective NK Cell Depletion on the Growth of NK Sensitive and Resistant Lymphoma Cell Variants 1339
Ichiro Kawase, David L. Urdal, Colin G. Brooks, and Christopher S. Henney

Direct Evidence for Anti-Tumor Activity by NK Cells *in Vivo*: Growth of B16 Melanoma in Anti-NK 1.1-Treated Mice 1347
Sylvia B. Pollack

Involvement of Spontaneous Rather Than Induced Antitumor Mechanisms in Resistance to Primary Fibrosarcoma Implantation and Its Secondary Spread 1353
Robert Keller

Natural Killer Cell-Mediated Cytotoxicity against Solid Tumors: *In Vitro* and *in Vivo* Studies 1359
Nabil Hanna

The Beige Model in Studies of Natural Resistance to Syngeneic, Semisyngeneic, and Primary Tumors 1369
Klas Kärre, Gunnar O. Klein, Rolf Kiessling, Shmuel Argov, and George Klein

NK Cell and NAb Anti-Tumor Activity *in Vivo* 1379
Donna A. Chow, Garth W. Brown, and Arnold H. Greenberg

The *in Vivo* Effects of Interferon (IFN) Suppression of the NK Target Structure 1387
A. H. Greenberg, V. Miller, B. Pohajdak, and T. Jablonski

Interrelationship between NK Activity and T Cell-Mediated Immunity in Syngeneic Tumor Rejection 1393
James Urban and Hans Schreiber

2. *AGAINST PRIMARY TUMORS AND POSSIBLE ROLE IN SURVEILLANCE*

Evidence for *in Vivo* NK Reactivity against Primary Tumors 1399
Tina Haliotis, John Roder, and David Dexter

Transfer of NK Activity and Lymphoma Resistance to AKR Mice by Marrow from High NK, Lymphoma-Resistant (C57xAKR)F_1 Mice 1405
John J. Trentin, Surjit K. Datta, Elizabeth L. Priest, Michael T. Gallagher, and Anthony G. Nasrallah
Possible Role of NK Cells in Radiation Leukemogenesis: Adoptive Repair of NK Deficit of Fractionally Irradiated Mice by Marrow Transfusion 1411
Surjit K. Datta, Elizabeth L. Priest, and John J. Trentin
Role of Natural-Cell-Mediated Immunity in Urethan-Induced Lung Carcinogenesis 1415
Elieser Gorelik and Ronald B. Herberman
Depression of NK Reactivity in Mice by Leukemogenic Doses of Irradiation 1423
E. Gorelik, B. Rosen, and R. B. Herberman
The Suppression of NK Activity by the Chemical Carcinogen Dimethylbenzanthracene and Its Restoration by Interferon or Poly I: C 1431
Rachel Ehrlich, Elinor Malatzky, Margalit Efrati, and Isaac P. Witz
Effect of Depression of NK Activity by Neonatal Exposure to Diethylstilbestrol on Susceptibility to Transplanted and Primary Carcinogen-Induced Tumors 1437
Terje Kalland

B. AGAINST MICROBIAL AGENTS

The Role of NK(HSV-1) Effector Cells in Resistance to Herpesvirus Infections in Man 1445
Carlos Lopez, Dahlia Kirkpatrick, and Patricia Fitzgerald
The Role of Natural Killer Cells and Interferon in Resistance to Murine Cytomegalovirus 1451
Geoffrey R. Shellam, Jane E. Grundy (Chalmer), and Jane E. Allan
Loss of Genetic Resistance to Murine Cytomegalovirus Infection in C3H Mice Treated with 89Sr 1459
Aoi Masuda and Michael Bennett
Studies of NK Cell Activation and of Interferon Induction after Injection of Mouse Hepatitis Virus Type 3 (MHV3) 1465
Holger Kirchner and Liesel Schindler
Interferon, Natural Killer Cells, and Genetically Determined Resistance of Mice against Herpes Simplex Virus 1471
Holger Kirchner, Helmut Engler, and Rainer Zawatzky
Natural Resistance against Mouse Hepatitis Virus 3 1477
Jean-Marie Dupuy
Induction of Natural Killer Cells and Interferon Production during Infection of Mice with *Babesia microti* of Human Origin 1483
Mary J. Ruebush and Donald E. Burgess
Natural Cell-Mediated Immunity and Interferon in Malaria and Babesia Infections 1491
Elsie M. Eugui and Anthony C. Allison
Natural Cell-Mediated Resistance in Cryptococcosis 1503
Juneann Murphy
Natural Cell-Mediated Immunity against Bacteria 1513
Emil Skamene, Mary M. Stevenson, and Patricia A. L. Kongshavn
Genetic and Cellular Mechanisms of Natural Resistance to Intracellular Bacteria 1521
Christina Cheers and Jenny Macgeorge

C. RESISTANCE TO TRANSPLANTS OF BONE MARROW

The Role of Asialo GM1$^+$ (GA1$^+$) Cells in the Resistance to Transplants of Bone Marrow or Other Tissues 1527
Ko Okumura, Sonoko Habu, and Kazuo Shimamura

Direct Evidence for the Involvement of Natural Killer Cells in Bone Marrow Transplantation 1535
Eva Lotzová, S. B. Pollack, and C. A. Savary

D. POSSIBLE ROLE IN GRAFT VERSUS HOST DISEASE

In Vivo Role of NK(HSV-1) in the Induction of Graft versus Host Disease in Bone Marrow Transplant Recipients 1541
Carlos Lopez, Patricia Fitzgerald, and Dahlia Kirkpatrick

Index *1547*

Contributors

Numbers in parentheses indicate the pages on which the authors' contributions begin.

Jochen Abb (1233), *Max von Pettenkofer-Institut, Munich, West Germany*

Toru Abo (31), *Cellular Immunobiology Unit of the Tumor Institute, Departments of Surgery, Microbiology, and Pediatrics, University of Alabama in Birmingham, Birmingham, Alabama 35294*

Raffaella Acero (1119), *Istituto di Ricerche Farmacologiche, 'Mario Negri', Milan, Italy*

Hans Acha-Orbea (893), *Department of Experimental Pathology, University of Hospital Zürich, Zürich, Switzerland*

Dolph O. Adams (511, 949, 1003), *Departments of Pathology and Microbiology-Immunology, Duke University Medical Center, Durham, North Carolina 27710*

William H. Adler (251, 645, 651), *Gerontology Research Center, National Institute on Aging, National Institutes of Health, Baltimore, Maryland 2224*

Lars Ährlund-Richter (705), *Department of Tumor Biology, Karolinska Institute, Stockholm, Sweden*

Anders Ahre (1259, *Radiumhemmet, Karolinska Hospital, Stockholm, Sweden*

Saverio Alberti (25), *IRFMN, Milan, Italy*

Jane E. Allan (1451), *Department of Microbiology, University of Western Australia, Nedlands, Western Australia, Australia*

Paola Allavena (581, 873, 1119), *Istituto di Ricerche Farmacologiche, 'Mario Negri', Milan, Italy*

Anthony C. Allison (1491), *Institute of Biological Sciences, Syntex Research, Palo Alto, California 94303*

Abdulrazzak Alsheikhly (153), *Department of Immunology, University of Stockholm, Stockholm, Sweden*

D. Bernard Amos (341), *Division of Immunology, Department of Microbiology and Immunology, Duke University Medical Center, Durham, North Carolina 27710*

Torbjörn Andersson (153), *Departments of Infectious Diseases and Virology, Karolinska Institute Medical School, Stockholm, Sweden*

G. Andrighetto (540), *Istituto Immunopatologia, University of Padua, Verona, Italy*

Tadao Aoki (1297), *Research Division, Shinrakuen Hospital, Niigata, Japan*

Yoshitaka Aoyagi (1297), *1st Department of Internal Medicine, Niigata University of School of Medicine, Niigata, Japan*

Shmuel Argov (799, 1369), *Department of Tumor Biology, Karolinska Institute, Stockholm, Sweden*

Inger Axberg (733), *Department of Immunology, Uppsala University, Uppsala, Sweden*
Fritz H. Bach (791), *Immunobiology Research Center, Departments of Laboratory Medicine/Pathology and Surgery, University of Minnesota, Minneapolis, Minnesota 55455*
Jean-Francois Bach (215), *INSERM U 25, Hôpital Necker, Paris, France*
Malcolm G. Baines (1175), *Department of Microbiology and Immunology, McGill University, Montreal, Quebec, Canada*
Tibor Bakács (1189), *National Institute of Oncology, Budapest, Hungary*
Charles M. Balch (31), *Cellular Immunobiology Unit of the Tumor Institute, Departments of Surgery, Microbiology, and Pediatrics, University of Alabama in Birmingham, Birmingham, Alabama 35294*
Pierre Bardos (215, 443, 639), *Laboratoire d'Immunologie, Faculté de Médecine, Tours, France and INSERM U 25, Hôpital Necker, Paris, France*
Teresa Barlozzari (139, 519, 527, 549), *Biological Research and Therapy Branch, National Cancer Institute, Frederick, Maryland 21701*
Scott P. Bartlett (105, 173, 807), *Transplantation Unit, Massachusetts General Hospital, Boston, Massachusetts 02114*
Jerry A. Bash (1273), *Divisions of Immunologic Oncology and Medical Oncology, Lombardi Cancer Center, Georgetown University, Washington, D.C. 20007*
Thomas Bechtold (1211), *Department of Pathology and Laboratory Medicine, University of Nebraska Medical Center, Omaha, Nebraska 68105*
Dean Befus (273), *McMaster University, Hamilton, Ontario, Canada*
Maria T. Bejarano (91, 475, 1047), *Department of Tumor Biology, Karolinska Institute, Stockholm, Sweden*
Miklós Benczur (683, 1041, 1227), *National Institute of Haematology and Blood Transfusion, Budapest, Hungary*
Michael Bennett[1] (329, 563, 569, 1459), *Departments of Microbiology and Pathology, Boston University School of Medicine, Boston, Massachusetts 02215*
Miroslav Beran (1077), *Department of Radio biology, Karolinska Institute, Stockholm, Sweden*
E. William Bere, Jr. (1161), *Biological Research and Therapy Branch, Laboratory of Comparative Carcinogenesis, National Cancer Institute, Frederick, Maryland 21701*
Kurt Berg (99), *Institute of Medical Microbiology, University of Aarhus, Aarhus, Denmark*
Peter Biberfeldt (835), *Department of Pathology, Karolinska Institute, Stockholm, Sweden*
John Bienenstock (273), *McMaster University, Hamilton, Ontario, Canada*
Andrea Biondi (165, 581), *Istituto di Ricerche Farmacologiche, 'Mario Negri', Milan, Italy*
Christine A. Biron (493), *Department of Pathology, University of Massachusetts Medical School, Worcester, Massachusetts 01605*
Katleen Bizière[2] (639), *Laboratoiré d'Immunologie, Faculté de Médecine, Tours, France*
Henric Blomgren (1259), *Radiumhemmet, Karolinska Hospital, Stockholm, Sweden*
Barry R. Bloom (131, 989, 1241), *Departments of Pathology, Microbiology and Immunology, Medicine and Neurology, Albert Einstein College of Medicine and Montefiore Hospital and Medical Center, Bronx, New York 10461*
Eda T. Bloom (455, 675), *Geriatric Research, Education and Clinical Center, VA Wadsworth Medical Center and Department of Medicine, University of California, Los Angeles, California 90024*
Richard S. Bockman (415), *Laboratories of Clinical Immunology, Calcium Metabolism, and Interferon, Memorial Sloan-Kettering Cancer Center, New York, New York 10021*
Reinder Bolhuis (463), *Department of Immunology, Rotterdam Radio-Therapy Institute, Rotterdam; and Radiobiological Institute TNO, Rijswijk, The Netherlands*

[1]Present address: *Department of Pathology, Southwestern Medical School, Dallas, Texas 75235.*

[2]Present address: *Centre de Recherches Clin-Midy, 34082 Montpellier Cedex, France*

Benjamin Bonavida (145, 483, 961), *Department of Microbiology and Immunology, UCLA School of Medicine, Los Angeles, California 90024*

G. D. Bonnard (879), *Laboratory of Immunodiagnosis, National Cancer Institute, Bethesda, Maryland 20205*

Diana Boraschi (25, 401), *Sclavo Research Center, Siena, Italy*

Claudio Bordignon (581), *Clinica Medica Generale, Ospedale 'L. Sacco', Università di Milano, Milan, Italy*

Barbara Bottazzi (165, 581), *Istituto di Ricerche Farmacologiche, 'Mario Negri', Milan, Italy*

Philippe Bougnoux (955), *Biological Research and Therapy Branch, National Cancer Institute, Frederick, Maryland 21701*

Thomas P. Bradley (145), *Department of Microbiology and Immunology, UCLA School of Medicine, Los Angeles, California 90024*

C. Phillip Brandt (79), *Division of Immunology and Division of Rheumatic and Genetic Diseases, Duke University Medical Center, Durham, North Carolina 27710*

Colin G. Brooks (851, 1339), *Program in Basic Immunology, Fred Hutchinson Cancer Research Center, Seattle, Washington 98104*

Garth W. Brown (1379), *Manitoba Institute of Cell Biology, University of Manitoba, Winnipeg, Manitoba, Canada*

Michael J. Brunda (535), *Biological Development Branch, Biological Response Modifiers Program, National Cancer Institute, Frederick, Maryland 21701*

D. Brunet (1219), *Department of Neurology, Queen's University, Kingston, Ontario, Canada*

Donald E. Burgess[3] (1483), *Department of Microbiology and Immunology, Bowman Gray School of Medicine, Winston-Salem, North Carolina 27103*

Robert C. Burton (105, 173, 807), *Transplantation Unit, Massachusetts General Hospital, Boston, Massachusetts 02114*

Jean Caraux (85), *Department of Immunopathology, Scripps Clinic and Research Foundation, La Jolla, California 92037*

George A. Carlson (291), *The Jackson Laboratory, Bar Harbor, Maine 04609*

Olli Carpén (983), *Department of Pathology, University of Helsinki, Helsinki, Finland*

Robin Carpenter (903), *Divisions of Immunologic Oncology and Medical Oncology, Lombardi Cancer Center, Georgetown University, Washington, D. C. 20007*

Giorgio Caspani (165), *Istituto di Ricerche Farmacologiche, 'Mario Negri', Milan, Italy*

Y. Cayre[4] (225, 325), *Memorial Sloan-Kettering Cancer Center, New York, New York 10021*

C. Cesarini (549), *Institute of Pharmacology, University of Perugia, Perugia, Italy*

Kenneth S. S. Chang (751, 1147), *Laboratory of Cell Biology, National Cancer Institute, Bethesda, Maryland 20205*

Zong-liang Chang (955), *Biological Research and Therapy Branch, National Cancer Institute, Frederick, Maryland 20205*

Christina Cheers (1521), *Department of Microbiology, University of Melbourne, Parkville, Victoria, Australia*

Donna A. Chow (1379), *Manitoba Institute of Cell Biology, University of Manitoba, Winnipeg, Manitoba, Canada*

Tae June Chung (381), *Laboratory of Ontogeny of the Immune System, Sloan-Kettering Institute for Cancer Research, Rye, New York 10580*

James A. Clagett (301), *Department of Periodontics, University of Washington, Seattle, Washington 98195*

Edward A. Clark (301, 307), *Department of Genetics, University of Washington, and Genetics Systems Corporation, Seattle, Washington 98195*

[3] Present address: *Department of Immunology, Walter Reed Army Institute of Research, Washington, D.C. 20012.*

[4] Present address: *Department of Genie Genetique, Institut Pasteur, Paris, France.*

Alistair J. Cochran (1113), *Department of Surgery, Division of Oncology, UCLA School of Medicine, Los Angeles, California 90024*

C. Colmenares (119), *Memorial Sloan-Kettering Cancer Center, New York, New York 10021*

Nicoletta Colombo (1119), *Istituto di Ricerche Farmacologiche, 'Mario Negri', Milan, Italy*

Max D. Cooper (31), *Cellular Immunobiology Unit of the Tumor Institute, Departments of Surgery, Microbiology, and Pediatrics, University of Alabama in Birmingham, Birmingham, Alabama 35294*

Susanna Cunningham-Rundles (415, 1133), *Laboratories of Clinical Immunology, Calcium Metabolism, and Interferon, Memorial Sloan-Kettering Cancer Center, New York, New York 10021*

Didier Cupissol (1309), *Department of Chemo-Immunotherapy, and Laboratoire d'Immunopharmacologie des Tumeurs, INSERM U-236 and ERA-CNRS n 844, Centre Paul Lamarque, Hôpital St-Eloi, Montpellier, France*

Michael Cuttito (179, 281), *Cellular Immunology Section, Memorial Sloan-Kettering Cancer Center, New York, New York 10021*

Paula J. D'Amore (785), *Departments of Pathology and Surgery (Division of Oncology), UCLA School of Medicine, Los Angeles, California 90024*

Surjit K. Datta (557, 1207, 1405, 1411), *Division of Experimental Biology, Baylor College of Medicine, Houston, Texas 77030*

Jan E. de Vries (657), *Division of Immunology, The Netherlands Cancer Institute, Amsterdam, The Netherlands*

J. H. Dean (511), *National Toxicology Program, National Institute of Environmental Health Sciences, Research Triangle Park, North Carolina 27709*

Danielle Degenne (443, 639), *Laboratoire d'Immunologie, Faculté de Médecine, Tours, France*

Friedrich Deinhardt (1233), *Max von Pettenkofer-Institut, Munich, West Germany*

Alfred C. Denn, III (139, 1161), *Biological Research and Therapy Branch, National Cancer Institute, Frederick, Maryland 21701*

Gunther Dennert (843), *Department of Cancer Biology, The Salk Institute for Biological Studies, San Diego, California 92138*

James J. Devlin (969), *Department of Molecular Biology and Biochemistry, University of California, Irvine, Irvine, California 92717*

David Dexter (1399), *Department of Pathology, Queen's University, Kingston, Ontario, Canada*

Julie Y. Djeu (335, 669, 917), *Division of Virology, FDA Bureau of Biologics, Bethesda, Maryland 20205*

Marie-Christine Dokhélar (729, 1183, 1253), *Laboratoire d'Immunologie Clinique, Institut Gustave Roussy, Villejuif, France*

Wolfgang Domzig[5] (409), *Biological Research and Therapy Branch, National Cancer Institute-FCRF, Frederick, Maryland 21701*

Maria Benedetta Donati (401), *IRFMN, Milan, Italy*

Jean-Marie Dupuy (395, 1477), *Immunology Research Center, Armand-Frappier Institute, University of Quebec, Laval-des-Rapides, Quebec, Canada*

Anne Edwards (1167), *Medical Research Department, Kanematsu Memorial Institute and Melanoma Unit, Department of Surgery, Sydney Hospital, Sydney, N.S.W., Australia*

Margalit Efrati (201, 1431), *Department of Microbiology, The George S. Wise Faculty of Life Sciences, Tel Aviv University, Tel Aviv, Israel*

Rachel Ehrlich (201, 1431), *Department of Microbiology, The George S. Wise Faculty of Life Sciences, Tel Aviv University, Tel Aviv, Israel*

Anneka Ehrnst (499), *Department of Immunology, National Bacteriological Laboratory, Stockholm, Sweden*

Stefan Einhorn (1259), *Radiumhemmet, Karolinska Hospital, Stockholm, Sweden*

[5] Present address: *Max-Planck-Institute for Immunology, Freiburg/Br., West Germany.*

Jørgen Ellegaard (1285), *University Department of Medicine and Haematology, Aarhus Amtssygehus, Aarhus, Denmark*

Sandra L. Emmons (113), *Department of Biological Structure, University of Washington School of Medicine, Seattle, Washington 98195*

Helmut Engler (1471), *Institute of Virus Research, German Cancer Research Center, Heidelberg, West Germany*

Elsie M. Eugui (1491), *Institute of Biological Sciences, Syntex Research, Palo Alto, California 94303*

István Földes (1227), *Microbiological Research Group, National Institute of Health and Hygiene, Budapest, Hungary*

Astrid Fagraeus (499), *Department of Immunology, National Bacteriological Laboratory, Stockholm, Sweden*

Virginia Fanning (39), *The Wistar Institute of Anatomy and Biology, Philadelphia, Pennsylvania 19104*

Carine Favier (1309), *Department of Chemo-Immunotherapy, and Laboratoire d'Immunopharmacologie des Tumeurs, INSERM U-236 and ERA-CNRS n 844, Centre Paul Lamarque, Hôpital St-Eloi, Montpellier, France*

François Favier (1309), *Department of Chemo-Immunotherapy, and Laboratoire d'Immunopharmacologie des Tumeurs, INSERM U-236 and ERA-CNRS n 844, Centre Paul Lamarque, Hôpital St-Eloi, Montpellier, France*

Mei-fu Feng (851), *Program in Basic Immunology, Fred Hutchinson Cancer Research Center, Seattle, Washington 98104*

Gabriel Fernandes[6] (1085, 1141, 1195, 1201), *Memorial Sloan-Kettering Cancer Center, New York, New York 10021*

Manlio Ferrarini (1, 257), *Department of Clinical Immunology, University of Genoa, Genoa, Italy*

Helmut Feucht (53), *Institute of Immunology, University of Munich, Munich, West Germany*

Carl G. Figdor (657), *Division of Immunology, The Netherlands Cancer Institute, Amsterdam, The Netherlands*

Dina G. Fischer[7] (159), *Division of Immunology, Duke University Medical Center, Durham, North Carolina 27710*

Patricia Fitzgerald (73, 387, 1445, 1541), *Sloan-Kettering Institute for Cancer Research, New York, New York 10021*

James T. Forbes (977, 1291), *Division of Oncology, Department of Medicine, Vanderbilt University, Nashville, Tennessee 37232*

Adrien Forget (313), *Department of Microbiology and Immunology, University of Montreal, Montreal, Quebec, Canada*

Bertil Fredholm (607, 631), *Department of Pharmacology, Karolinska Institute, Stockholm, Sweden*

Cecilia Galatiuc (621), *Department of Immunology, Victor Babes Institute, Bucharest, Romania*

Michael T. Gallagher[8] (557, 1405), *Division of Experimental Biology, Baylor College of Medicine, Houston, Texas 77030*

Tamás Garam (1189), *National Institute of Haematology and Blood Transfusion, Budapest, Hungary*

Maria Gherman (621), *Department of Immunology, Victor Babes Institute, Bucharest, Romania*

Pietro Ghezzi (401), *IRFMN, Milan, Italy*

Magnus Gidlund (733, 765), *Departments of Immunology and Pathology, Uppsala University, Uppsala, Sweden*

[6] Present address: *Division of Clinical Immunology, University of Texas Health Science Center, San Antonio, Texas 78284.*

[7] Present address: *Department of Virology, Weizmann Institute of Science, Rehovot, Israel.*

[8] Present address: *City of Hope National Medical Center, Duarte, California 91010.*

Steven Gillis (421), *Program in Basic Immunology, Fred Hutchinson Cancer Research Center, Seattle, Washington 98104*

Ronald H. Goldfarb[9] (427, 595, 931), *Laboratory of Immunodiagnosis, NCI, National Institutes of Health, Bethesda, Maryland 20205*

Marc G. Golightly[10] (79, 159, 785), *Division of Immunology, Duke University Medical Center, Durham, North Carolina 27710*

Sidney Golub (631, 785, 1113, 1265), *Departments of Surgery (Division of Oncology), and Microbiology and Immunology, UCLA School of Medicine, Los Angeles, California 90024*

Robert A. Good (1195), *Memorial Sloan-Kettering Cancer Center, New York, New York 10021*

E. Gorelik[11] (1331, 1415, 1423), *Laboratory of Immunodiagnosis, National Cancer Institute, Bethesda, Maryland 20205*

Donald L. Granger[12] (1027), *Veterans Adminstration Medical Center and Department of Medicine, Division of Infectious Diseases, University of Utah School of Medicine, Salt Lake City, Utah 84148*

Gale A. Granger (969), *Department of Molecular Biology and Biochemistry, University of California, Irvine, Irvine, California 92717*

Arthur I. Grayzel (1241), *Departments of Pathology, Microbiology and Immunology, Medicine and Neurology, Albert Einstein College of Medicine and Montefiore Hospital and Medical Center, Bronx, New York 10461*

F. Anthony Greco (1291), *Division of Oncology, Department of Medicine, Vanderbilt University, Nashville, Tennessee 37232*

Arnold H. Greenberg (291, 719, 1379, 1387), *Manitoba Institute of Cell Biology, University of Manitoba, Winnipeg, Manitoba, Canada*

Alvar Grönberg (705, 765), *Department of Tumor Biology, Karolinska Institute, Stockholm, Sweden*

Philippe Gros (313), *Montreal General Hospital Research Institute, McGill University, Montreal, Quebec, Canada*

Peter Groscurth (893), *Institute of Anatomy, University of Zürich, Zürich, Switzerland*

Carlo Enrico Grossi (1, 257), *Department of Human Anatomy, University of Genoa, Genoa, Italy*

Zvi Grossman (229), *Mathematical Institute, Tel Aviv University, Tel Aviv, Israel*

Jane E. Grundy[13] (Chalmer) (1451), *Department of Microbiology, University of Western Australia, Nedlands, Western Australia, Australia*

Sudhir Gupta[14] (1085, 1141, 1195, 1201), *Memorial Sloan-Kettering Cancer Center, New York, New York 10021*

Éva Gyódi (1227), *National Institute of Haematology and Blood Transfusion, Budapest, Hungary*

Bengt Härfast (153), *Department of Immunology, University of Stockholm, Stockholm, Sweden*

Sonoko Habu (139, 209, 1323, 1331, 1527), *Department of Pathology, Takai University School of Medicine, Isehara, Kanagawa, Japan*

Tina Haliotis (743, 1399), *Departments of Microbiology and Immunology, and Radiation Oncology, Queen's University, Kingston, Ontario, Canada*

[9] Present address: *Cancer Metastasis Research Group, Department of Immunology and Infectious Disease, Pfizer Central Research, Pfizer Inc., Groton, Connecticut 06340.*

[10] Present Address: *Departments of Surgery (Division of Oncology), and Microbiology and Immunology, UCLA School of Medicine, Los Angeles, California 90024*

[11] Present address: *Surgery Branch, National Cancer Institute, Bethesda, Maryland 20205.*

[12] Present address: *Department of Physiological Chemistry, Johns Hopkins University, School of Medicine, Baltimore, Maryland 21218.*

[13] Present address: *National Cancer Institute, National Institutes of Health, Bethesda, Maryland 20205.*

[14] Present address: *Division of Basic and Clinical Immunology, University of California, Irvine, Irvine, California 92717.*

Nabil Hanna[15] **(1359),** *Cancer Metastasis and Treatment Laboratory, National Cander Institute, Frederick, Maryland 21701*

Mona Hansson (1077), *Department of Tumor biology, Karolinska Institute, Stockholm, Sweden*

Andrew J. Hapel (917), *Biological Carcinogenesis Program, National Cancer Institute, Frederick, Maryland 21701*

Frank Hatcher[16] **(503, 1091),** *Department of Biology, Wake Forest University, Winston-Salem, North Carolina 27109*

Toshio Hattori (955), *Biological Research and Therapy Branch, National Cancer Institute, Frederick, Maryland 21701*

A. Hatzfeld[17] **(119, 325),** *Memorial Sloan-Kettering Cancer Center, New York, New York 10021*

Sven Haukaas[18] **(1317),** *Institute of Anatomy, University of Bergen, Bergen, Norway*

Barton F. Haynes (79), *Division of Immunology and Division of Rheumatic and Genetic Diseases, Duke University Medical Center, Durham, North Carolina 27710*

Steven H. Hefeneider (421), *Program in Basic Immunology, Fred Hutchinson Cancer Research Center, Seattle, Washington 98104*

Stephen Helfand[19] **(743, 1011),** *Departments of Microbiology and Immunology, and Radiation Oncology, Queen's University, Kingston, Ontario, Canada*

Hans Hengartner (893), *Department of Experimental Pathology, University Hospital Zürich, Zürich, Switzerland*

Christopher S. Henney (421, 757, 851, 1339), *Program in Basic Immunology, Fred Hutchinson Cancer Research Center, Seattle, Washington 98104*

R. B. Herberman[20] **(9, 17, 139, 229, 349, 375, 427, 519, 527, 549, 595, 621, 669, 821, 859, 873, 909, 931, 955, 1055, 1069, 1279, 1331, 1415, 1423),** *Laboratory of Immunodiagnosis, National Cancer Institute, Bethesda, Maryland 20205*

Iver Heron (99), *Institute of Medical Microbiology, University of Aarhus, Aarhus, Denmark*

Peter Hersey (1167), *Medical Research Department, Kanematsu Memorial Institute and Melanoma Unit, Department of Surgery, Sydney Hospital, Sydney, N.S.W., Australia*

John B. Hibbs, Jr. (1027), *Veterans Adminstration Medical Center and Division of Infectious Diseases, Department of Medicine, University of Utah School of Medicine, Salt Lake City, Utah 84148*

Thomas Hoffman (955), *Biological Research and Therapy Branch, National Cancer Institute, Frederick, Maryland 21701*

Marianne Hokland[21] **(99),** *Institute of Medical Microbiology, University of Aarhus, Aarhus, Denmark*

Peter Hokland[22] **(99, 1285),** *University Department of Medicine and Haematology, Aarhus Amtssygehus, Aarhus, Denmark*

Howard T. Holden (319, 535), *Biological Development Branch, Biological Response Modifiers Program, National Cancer Institute, Frederick, Maryland 21701*

[15] Present address: *Smith Kline and French Laboratories, Department of Immunology, Philadelphia, Pennsylvania 19101.*

[16] Present address: *Department of Microbiology, Meharry Medical College, Nashville, Tennessee 37208.*

[17] Present address: *Institut de Pathologie Cellular, Kremlin Bicetre, France.*

[18] Present address: *Department of Anatomy, Lund, Sweden.*

[19] Present address: *Department of Neurology, Harvard Medical School, Boston, Massachusetts 02115.*

[20] Present address: *Biological Research and Therapy Branch, National Cancer Institute, Frederick, Maryland 21701.*

[21] Present address: *Division of Tumor Immunology, Sidney Farber Cancer Institute, Boston, Massachusetts 02115.*

[22] Present address: *Division of Tumor Immunology, Sidney Farber Cancer Institute, Boston, Massachusetts 02115.*

Susan R. Hollán (1227), *National Institute of Haematology and Blood Transfusion, Budapest, Hungary*

E. Carmack Holmes (1113), *Department of Surgery, Division of Oncology, UCLA School of Medicine, Los Angeles, California 90024*

Yoh Horikawa (1297), *Research Division, Shinrakuen Hospital, Niigata, Japan*

Dorothy Hudig (923, 1021), *Department of Medicine, University of California, San Diego, La Jolla, California 92093*

Nam Doll Huh[23] (341, 381), *Laboratory of Ontogeny of the Immune System, Sloan-Kettering Institute for Cancer Research, Rye, New York 10580*

James N. Ihle (917), *Biological Carcinogenesis Program, National Cancer Institute, Frederick, Maryland 21701*

Martino Introna (581, 1119), *Istituto di Ricerche Farmacologiche, 'Mario Negri', Milan, Italy*

Sally T. Ishizaka (187), *Cellular Immunology and Immunochemistry Sections, Memorial Sloan-Kettering Cancer Center, New York, New York 10021*

Teruko Ishizaka (989), *Department of Microbiology and Immunology, Albert Einstein College of Medicine, Bronx, New York 10461*

T. Jablonski (1387), *Manitoba Institute of Cell Biology, University of Manitoba, Winnipeg, Manitoba, Canada*

Pamela J. Jensen[24] (699), *Division of Immunology, Duke University Medical Center, Durham, North Carolina 27710*

Bo Johansson (1259), *Radiumhemmet, Karolinska Hospital, Stockholm, Sweden*

Donald R. Johnson (835), *Department of Pathology and Laboratory Medicine, University of Nebraska, Omaha, Nebraska 68105*

William J. Johnson (949), *Departments of Pathology and Microbiology-Immunology, Duke University Medical Center, Durham, North Carolina 27710*

Mikael Jondal (361, 607, 631, 835), *Department of Tumor Biology, Karolinska Institute, Stockholm, Sweden*

Klas Kärre (1369), *Department of Tumor Biology, Karolinska Institute, Stockholm, Sweden*

Dominique Kaiserlian (215), *INSERM U 25, Hôpital Necker, Paris, France*

Terje Kalland[25] (1317, 1437), *Institute of Anatomy, University of Bergen, Bergen, Norway*

Masataka Kasai (125), *Department of Tuberculosis, National Institute of Health, Tokyo, Japan*

Peter Kaudewitz (771, 1233), *Institute for Immunology, University of Munich, Munich, West Germany*

Ichiro Kawase (757, 1339), *Program in Basic Immunology, Fred Hutchinson Cancer Research Center, Seattle, Washington 98104*

Norihiko Kawate (1113), *Department of Surgery, Division of Oncology, UCLA School of Medicine, Los Angeles, California 90024*

Eli Kedar[26] (859), *Laboratory of Immunodiagnosis, National Cancer Institute, Bethesda, Maryland 20205*

Robert Keller (601, 893, 1353), *Immunobiology Research Group, Institute of Immunology and Virology, University of Zürich, Zürich, Switzerland*

Rolf Kiessling (705, 765, 1077, 1369), *Department of Tumor Biology, Karolinska Institute, and Department of Tumor Biology, Karolinska Institute Stockholm, Sweden*

Yoon Berm Kim (341, 381), *Laboratory of Ontogeny of the Immune System, Sloan-Kettering Institute for Cancer Research, Rye, New York 10580*

[23] Present address: *Department of Biochemistry, School of Hygiene and Public Health, Johns Hopkins University, Baltimore, Maryland 21218.*

[24] Present address: *National Cancer Institute, Frederick, Maryland 21701.*

[25] Present address: *Department of Anatomy, Lund, Sweden.*

[26] Present address: *The Lautenberg Center for General and Tumor Immunology, Hebrew University-Hadassah Medical School, Jerusalem , Israel.*

Holger Kirchner (1465, 1471), *Institute of Virus Research, German Cancer Research Center, Heidelber, West Germany*

Dahlia Kirkpatrick (73, 1445, 1541), *Sloan-Kettering Institute for Cancer Research, New York, New York 10021*

Eva Klein (91, 239, 475, 499, 691, 799, 829, 1047, 1061), *Radiumhemmet, Karolinska Hospital, Stockholm, Sweden*

George Klein (765, 1369), *Department of Tumor Biology, Karolinska Institute, Stockholm, Sweden*

Gunnar O. Klein (275, 607, 1369), *Department of Tumor Biology, Karolinska Institute, Stockholm, Sweden*

Eugenie S. Kleinerman (815), *Biological Response Modifiers Program, National Institutes of Health, Frederick, Maryland 21701*

Jean Pierre Kolb (437), *Laboratoire d'Immunologie Cellulaire, Institut de Recherches Scientifiques sur le Cancer, Villejuif, France*

Patricia A. L. Kongshavn (313, 1513), *Montreal General Hospital Research Institute, Montreal, Quebec, Canada*

G. C. Koo (119, 187, 225, 325), *Cellular Immunology and Immunochemistry Sections, Memorial Sloan-Kettering Cancer Center, New York, New York 10021*

Hillel S. Koren (79, 159, 341, 615, 699, 1035), *Division of Immunology and Division of Rheumatic and Genetic Diseases, Duke University Medical Center, Durham, North Carolina 27710*

Dietrich Kraft (47), *Institute of General and Experimental Pathology, University of Vienna, Vienna, Austria*

Richard Kubota (1147), *Laboratory of Cell Biology, National Cancer Institute, Bethesda, Maryland 20205*

Raymond E. Kuhn (503, 1091), *Department of Biology, Wake Forest University, Winston-Salem, North Carolina 27109*

Vinay Kumar[27] (173, 329, 563, 569), *Department of Pathology, Boston University School of Medicine, Boston Massachusetts 02215*

Patrick C. Kung (791), *Centocor, Inc., Malvern, Pennsylvania 19355*

Takanobu Kurashige[28] (1207), *Division of Experimental Biology, Baylor College of Medicine, Houston, Texas 77030*

Kagemasa Kuribayashi (851), *Program in Basic Immunology, Fred Hutchinson Cancer Research Center, Seattle, Washington 98104*

Nuha T. Kusaimi[29] (1249), *Division of Experimental Biology, Baylor College of Medicine, Houston, Texas 77030*

Santo Landolfo (1099), *Institute of Microbiology, University of Torino, Torino, Italy*

Fred Lanefeldt (607, 631), *Department of Pharmacology, Karolinska Institute, Stockholm, Sweden*

Rosmarie Lang (893), *Department of Experimental Pathology, Univeristy of Hospital Zürich, Zürich, Switzerland*

Emanuela Lanza (335, 917), *Division of Virology, FDA-Bureau of Biologics, Bethesda, Maryland 20205*

Tamás Laskay (683), *National Institute of Haematology and Blood Transfusion, Budapest, Hungary*

Edmund C. Lattime (179, 187, 193, 713), *Cellular Immunology and Immunochemistry Sections, Memorial Sloan-Kettering Cancer Center, New York, New York 10021*

Gad Lavie (939), *Division of Hematology and Center of Transfusion, The Beilison Medical Center, Petah Tiqva, Israel*

Jeffrey A. Ledbetter (301), *Genetics Systems Corporation, Seattle, Washington 98195*

[27] Present address: *Department of Pathology, Southwestern Medical School, Dallas, Texas 75235.*

[28] Present address: *Department of Pediatrics, Kochi Medical School, Kochi, Japan*

[29] Present address: *Department of Medicine, College of Medicine, University of Mosul, Iraq.*

Kam H. Leung (615), *Division of Immunology, Duke University Medical Center, Durham, North Carolina 27710*

Elinor M. Levy (569), *Departments of Microbiology and Pathology, Boston University School of Medicine, Boston, Massachusetts 02215*

Tullia Lindsten (1211), *Department of Tumor Biology, Karolinska Institute, Stockholm, Sweden*

Marc Lipinski (1183, 1253), *Laboratoire d'Immunologie Clinique, Institut Gustave Roussy, Villejuif, France*

Marie-Luise Lohmann-Matthes (243), *Max-Planck-Institut für Immunbiologie, Freiburg, West Germany*

Jürgen Lohmeyer (53), *Institute of Immunology, University of Munich, Munich, West Germany*

Bernard Longhi (1309), *Department of Chemo-Immunotherapy, and Laboratoire d'Immunopharmacologie des Tumeurs, INSERM U-236 and ERA-CNRS n 844, Centre Paul Lamarque, Hôpital St-Eloi, Montpellier, France*

Carlos Lopez (73, 387, 1445, 1541), *Sloan-Kettering Institute for Cancer Research, New York, New York 10021*

Eva Lotzová (1535), *M. D. Anderson Hospital and Tumor Institute, Houston, Texas 770301*

Anne Luck (1291), *Division of Oncology, Department of Medicine, Vanderbilt University, Nashville, Tennessee 37232*

Walter Luini (25), *IRFMN, Milan, Italy*

John A. Lust (173, 563), *Southwestern Medical School, Dallas, Texas 75235*

Jenny Macgeorge[30] (1521), *Department of Microbiology, University of Melbourne, Parkville, Victoria, Australia*

Elinor Malatzky (1431), *Department of Microbiology, The George S. Wise Faculty of Life Sciences, Tel Aviv University, Tel Aviv, Israel*

Annette E. Maluish (1279), *Biological Research and Therapy Branch, Biological Response Modifiers Program, National Cancer Institutes, Frederick, Maryland 21701*

Moiara Manciulea (621), *Department of Immunology, Victor Babes Institute, Bucharest, Romania*

Rosemonde Mandeville (395), *Immunology Research Center, Armand-Frappier Institute, Laval-des-Rapides, Canada*

Alberto Mantovani (165, 581, 1119), *Istituto di Ricerche Farmacologiche, 'Mario Negri', Milan, Italy*

R. J. Marchmont (879), *Department of Immunology, Paterson Laboratories, Christie Hospital and Holt Radium Institute, Manchester, England*

Pancrazio Martinetto (1099), *Institute of Microbiology, University of Torino, Torino, Italy*

Giovanna Martinotti (1099), *Institute of Microbiology, University of Torino, Torino, Italy*

Giuseppe Masucci (91, 475), *Department of Tumor Biology, Karolinska Institute, Stockholm, Sweden*

Maria G. Masucci (91, 475, 1047), *Department of Tumor Biology, Karolinska Institute, Stockholm, Sweden*

Aoi Masuda[31] (1459), *Department of Pathology, Boston University School of Medicine, Boston, Massachusetts 02215*

S. Matzku (540), *Institute of Nuclear Medicine, German Cancer Research Center, Heidelberg, West Germany*

William McCarthy (1167), *Medical Research Department, Kanematsu Memorial Institute and Melanoma Unit, Department of Surgery, Sydney Hospital, Sydney, N.S.W., Australia*

D. Olga McDaniel (1105), *Department of Microbiology and Public Health, University of Alabama, Birmingham, Alabama 35294*

[30] Present address: *Walter and Eliza Hall Institute, Parkville, Victoria, Australia.*

[31] Present address: *Department of Parasitology, Faculty of Medicine of Ribeirao Preto, Sao Paulo, S. P. Brazil.*

Ronald C. McGarry (1219), *Department of Microbiology and Immunology, Queen's University, Kingston, Ontario, Canda*

P. F. Mellen (329, 563), *Department of Pathology, Boston University School of Medicine, Boston, Massachusetts 02215*

Håkan Mellstedt (1259), *Radiumhemmet, Karolinska Hospital, Stockholm, Sweden*

Jean E. Merrill (631), *Department of Neurology, UCLA Medical School, Los Angeles, California 90024*

Michael Micksche (431, 589, 1303), *Institute for Cancer Research, University of Vienna, Vienna, Austria*

Aaron E. Miller (1241), *Departments of Pathology, Microbiology and Immunology, Medicine and Neurology, Albert Einstein College of Medicine and Montefiore Hospital and Medical Center, Bronx, New York 10461*

V. Miller (1387), *Manitoba Institute of Cell Biology, University of Manitoba, Winnipeg, Manitoba, Canada*

Gerald Milton (1167), *Medical Research Department, Kanematsu Memorial Institute and Melanoma Unit, Department of Surgery, Sydney Hospital, Sydney, N.S.W., Australia*

Nagahiro Minato (131), *Department of Medicine, Jichi Medical School, Jichi, Japan*

Karen M. Miner (969), *Department of Molecular Biology and Biochemistry, University of California, Irvine, Irvine, California 92717*

Lory Minning (923, 1021), *Department of Medicine, University of California, San Diego, La Jolla, California 92093*

L. R. Mittl (225), *Memorial Sloan-Kettering Cancer Center, New York, New York 10021*

Hideo Miyakoshi (1297), *Research Division Shinrakuen Hospital, Niigata , Japan*

Mikio Mizukoshi (1297), *Immunology Division, Research Center, Fujizoki Pharmaceutical Co., Ltd., Hachiohji , Japan*

Pierangela Molina (1119), *Istituto di Ricerche Farmacologiche, 'Mario Negri', Milan, Italy*

M. Moore (1127), *Department of Immunology, Paterson Laboratories, Christie Hospital and Holt Radium Institute, Manchester, England*

Doris Morgan (887), *Université Paris VI, Paris, France*

Andrew V. Muchmore (815), *Metabolism Branch, National Institutes of Health, Bethesda, Maryland 20205*

Edwin D. Murphy (301), *The Jackson Laboratory, Bar Harbor, Maine 04609*

Juneann Murphy (1105, 1503), *Department of Botany-Microbiology, University of Oklahoma, Norman, Oklahoma 73019*

Kenneth E. Muse (1035), *Division of Immunology, Duke University Medical Center, Durham, North Carolina 27710*

Mircea Musset (443), *Laboratoire d'Immunologie Virale, Institut M9rieux, Lyon, France*

Anthony G. Nasrallah[32] (557, 1405), *Division of Experimental Biology, Baylor College of Medicine, Houston, Texas 77030*

John R. Neefe (903, 1273), *Divisions of Immunologic Oncology and Medical Oncology, Lombardi Cancer Center, Georgetown University, Washington, D.C. 20007*

P. Andrew Neighbour (1241), *Departments of Pathology, Microbiology and Immunology, Medicine and Neurology, Albert Einstein College of Medicine and Montefiore Hospital and Medical Center, Bronx, New York 10461*

Walter Newman (777), *Program in Basic Immunology, Fred Hutchinson Cancer Research Center and Department of Microbiology and Immunology, University of Washington, Seattle, Washington 98195*

Masayuki Niitsuma (1113), *Department of Surgery (Division of Oncology), UCLA School of Medicine, Los Angeles, California 90024*

Kennith Nilsson (733, 765), *Department of Pathology, Wallenberg Laboratory, Uppsala, Sweden*

[32] Present address: *Cook Children's Hospital, Fort Worth, Texas 76102.*

Erling Norrby (153), *Departments of Infectious Diseases and Virology, Karolinska Institute Medical School, Stockholm, Sweden*

Masato Nose (733) *Department of Immunology, Uppsala University, Uppsala, Sweden*

Gerhard Obexer (47), *Institute of General and Exerimental Pathology, University of Vienna, Vienna, Austria*

Thomas N. Oeltmann (977), *Division of Oncology, Department of Medicine, Vanderbilt University, Nashville, Tennessee 37232*

Ko Okumura[33] (125, 139, 209, 1323, 1331, 1527), *Program in Basic Immunology, Fred Hutchinson Cancer Research Center, Seattle, Washington 98104*

Susana Olabuenaga (851), *Program in Basic Immunology, Fred Hutchinson Cancer Research Center, Seattle, Washington 98104*

Claes Örvell (153), *Departments of Infectious Diseases and Virology, Karolinska Institute Medical School, Stockholm, Sweden*

John R. Ortaldo (9, 139, 265, 349, 821, 873, 1055, 1279), *Biological Research and Therapy Branch, Biological Response Modifiers Program, National Cancer Institutes, Frederick, Maryland 21701*

György Pálffy (1227), *Department of Neurology and Psychiatry Medical School, Pécs, Hungary*

Gerd R. Pape (771), *Institute for Immunology, and Medical Clinic Grosshadern, University of Munich, Munich, West Germany*

Elena Pasqualetto (401), *IRFMN, Milan, Italy*

Manuel Patarroyo (499, 765), *Department of Immunology, National Bacteriological Laboratory, and Department of Tumor Biology, Korolinska Institute, Stockholm, Sweden*

Gene A. Pecoraro (179, 187, 713), *Cellular Immunology and Immunochemistry Sections, Memorial Sloan-Kettering Cancer Center, New York, New York 10021*

Louis M. Pelus (193), *Cellular Immunology and Developmental Hematopoiesis Sections, Memorial Sloan-Kettering Cancer Center, New York, New York 10021*

J. R. Peppard (119, 325), *Memorial Sloan-Kettering Cancer Center, New York, New York 10021*

Giuseppe Peri (165), *Istituto di Ricerche Farmacologiche, 'Mario Negri', Milan, Italy*

Peter Perlmann (153), *Department of Immunology, University of Stockholm, Stockholm, Sweden*

Bice Perussia (39, 369), *The Wistar Institute of Anatomy and Biology, Philadelphia, Pennsylvania 19104*

Sidney Pestka (349), *Roche Institute of Molecular Biology, Nutley, New Jersey 07110*

Győső Petrányi (683, 1041, 1189, 1227), *National Institute of Haematology and Blood Transfusion, Budapest, Hungary*

Gerald E. Piontek (705), *Department of Tumor Biology, Karolinska Institute, Stockholm, Sweden*

Chris D. Platsoucas (67, 355, 1195), *Memorial Sloan-Kettering Cancer Center, New York, New York 10021*

B. Pohajdak (719, 1387), *Manitoba Institute of Cell Biology, University of Manitoba, Winnipeg, Manitoba, Canada*

Nadia Polentarutti (165), *Istituto di Ricerche Farmacologiche, 'Mario Negri,' Milan, Italy*

Sylvia B. Pollack (113, 1347, 1535), *Department of Biological Structure, University of Washington School of Medicine, Seattle, Washington 98195*

Nicholas M. Ponzio (1153), *Department of Pathology, CMDNJ-New Jersey Medical School, Newark, New Jersey 07103*

Elizabeth L. Priest (557, 1405, 1411), *Division of Experimental Biology, Baylor College of Medicine, Houston, Texas 77030*

Hugh Pross (743, 1175), *Departments of Radiation Oncology, and Microbiology and Immunology, Queen's University, Kingston, Ontario, Canada*

[33] Present address: *Department of Immunology, Faculty of Medicine, University of Tokyo, Tokyo, Japan*

Tamás Pulay (1189), *Semmelweis University, Clinic of Gynecologic and Obstetrics, Budapest, Hungary*

David T. Purtillo (1211), *Department of Pathology and Laboratory Medicine, University of Nebraska Medical Center, Omaha, Nebraska 68105*

Robert S. Pyle (251), *Gerontology Research Center, National Institute on Aging, National Institutes of Health, Baltimore, Maryland 21224*

Phuc-Canh Quan (989), *Department of Microbiology and Immunology, Albert Einstein College of Medicine, Bronx, New York 10461*

Keith M. Ramsey (139), *FDA-Bureau of Biologics, Bethesda, Maryland 20205*

Doug Redelman (923, 1021), *Department of Medicine, University of California, San Diego, La Jolla, California 92093*

Robert Rees[34] (17), *Biological Research and Therapy Branch, National Cancer Institute, Frederick, Maryland 21701*

Elizabeth Reinitz (1241), *Departments of Pathology, Microbiology and Immunology, Medicine and Neurology, Albert Einstein College of Medicine and Montefiore Hospital and Medical Center, Bronx, New York 10461*

Gérard Renoux (443, 639), *Laboratoiré d'Immunologie, Faculté de Médecine, Tours, France*

Micheline Renoux (639), *Laboratoiré d'Immunologie, Faculté de Médecine, Tours, France*

Augustin Rey (1309), *Department of Chemo-Immunotherapy, and Laboratoire d'Immunopharmacologie des Tumeurs, INSERM U-236 and ERA-CNRS n 844, Centre Paul Lamarque, Hôpital St-Eloi, Montpellier, France*

Craig W. Reynolds (17, 139, 319, 375, 1161), *Biological Response Modifiers Program, National Cancer Institute, Frederick, Maryland 21701*

Carlo Riccardi (519, 527, 549, 873, 879, 909), *Biological Research and Therapy Branch, National Cancer Institute, Frederick, Maryland 20701, and Institute of Pharmacology, University of Perugia, Perugia, Italy*

Ernst Peter Rieber (53), *Institute of Immunology, University of Munich, Munich, West Germany*

Gert Riethmüller (53, 771, 1233), *Institute of Immunology, University of Munich, Munich, West Germany*

Gábor Ringwald (1189), *National Institute of Oncology, Budapest, Hungary*

Normand Rocheleau (395), *Immunology Research Center, Armand-Frappier Institute, Laval-des-Rapides, Canada*

John C. Roder (743, 1011, 1219, 1399), *Department of Microbiology and Immunology, Queen's University, Kingston, Canada*

B. Rosen[35] (1423), *Laboratory of Immunodiagnosis, National Cancer Institute, Bethesda, Maryland 20205*

Kendall L. Rosenthal (893), *Department of Experimental Pathology, Univeristy Hospital Zürich, Zürich, Switzerland*

John B. Roths (301), *The Jackson Laboratory, Bar Harbor, Maine 04609*

Domenico Rotilio (401), *IRFMN, Milan, Italy*

Berish Y. Rubin (415), *Laboratories of Clinical Immunology, Calcium Metabolism, and Interferon, Memorial Sloan-Kettering Cancer Center, New York, New York 10021*

Peter Rubin (1175), *Departments of Radiation Oncology and Microbiology and Immunology, Queen's University, Kingston, Ontario, Canada*

[34] Present address: *Department of Virology, University of Sheffield Medical School, Sheffield, England.*

[35] Present address: *Medical School, Beer-Sheba University, Beer-Sheba, Israel.*

Mary J. Ruebush[36] (1483), *Department of Microbiology and Immunology, Bowman Gray School of Medicine, Winston-Salem, North Carolina 27103*

Helmut Rumpold (47), *Institute of General and Experimental Pathology, University of Vienna, Vienna, Austria*

Eero Saksela (575, 983), *Department of Pathology, University of Helsinki, Helsinki, Finland*

Mario Salmona (401), *IRFMN, Milan, Italy*

Angela Santoni (519, 527, 549), *Biological Research and Therapy Branch, National Cancer Institute, Frederick, Maryland 21701; and Institute of Pharmacology, University of Perugia, Perugia, Italy*

C. A. Savary (1535), *M. D. Anderson Hospital and Tumor Institute, Houston, Texas 770301*

Queen B. Saxena (251, 645, 651), *Gerontology Research Center, National Institute on Aging, National Institutes of Health, Baltimore, Maryland 2224*

Rajiv K. Saxena (251, 449, 645, 651), *Gerontology Research Center, National Institute on Aging, National Institutes of Health, Baltimore, Maryland 2224*

Liesel Schindler (1465), *Institute of Virus Research, German Cancer Research Center, Heidelberg, West Germany*

Günter Schlimok (53), *Medical Clinic of Augsburg, Augsburg, West Germany*

Hans Schreiber (1393), *La Rabida-University of Chicago Institute, Committee on Immunology and Department of Pathology, University of Chicago, Chicago, Illinois 60649*

Janet K. Seeley (1211), *Department of Pediatrics, University of Connecticut Health Center, Farmington, Connecticut 06032*

Anna Senik (437), *Laboratoire d'Immunologie Cellulaire, Institut de Recherches Scientifiques sur le Cancer, Villejuif, France*

Susana A. Serrate (1055, 1069), *Biological Research and Therapy Branch, National Cancer Institute, Frederick, Maryland 21701*

Bernard Serrou (85, 1309), *Department of Tumor Immunopharmacology, Centre Paul Lamarque, Montpellier, France*

Susan O. Sharrow (139), *Immunology Branch, National Cancer Institute, Bethesda, Maryland 20205*

Geoffrey R. Shellam (1451), *Department of Microbiology, University of Western Australia, Nedlands, Western Australia, Australia*

Akira Shibata (1297), *1st Department of Internal Medicine, Niigata University of School of Medicine, Niigata , Japan*

Kazuo Shimamura (1527), *Department of Pathology, Tokai University of School of Medicine, Isehara, Kanagawa, Japan*

Frederick P. Siegal (387), *Mount Sinai Medical Center, New York, New York 10029*

Emil Skamene (313, 1513), *Montreal General Hospital Research Institute, Montreal, Quebec, Canada*

Scott D. Somers (1003), *Departments of Pathology and Microbiology-Immunology, Duke University Medical Center, Durham, North Carolina 27710*

Hergen Spits (657), *Division of Immunology, The Netherlands Cancer Institute, Amsterdam, The Netherlands*

Ivana Stoger (1227), *Microbiological Research Group, National Institute of Health and Hygiene, Budapest, Hungary*

Beda M. Stadler[37] (409), *Laboratory of Microbiology and Immunology, National Institute of Dental Research, Bethesda, Maryland 20205*

Mary M. Stevenson (1513), *Montreal General Hospital Research Institute, Montreal, Quebec, Canada*

[36] Present address: *Department of Preventive Medicine and Biometrics, Uniformed Services University of the Health Sciences, Bethesda, Maryland 20205.*

[37] Present address: *Institute of Clinical Immunobiology, University of Bern, Bern, Switzerland.*

Lothar Stitz (893), *Department of Experimental Pathology, Univeristy Hospital Zürich, Zürich, Switzerland*
Hans Strander (1259), *Radiumhemmet, Karolinska Hospital, Stockholm, Sweden*
Osias Stutman (179, 187, 193, 281, 713), *Cellular Immunology, and Immunochemistry Sections, Memorial Sloan-Kettering Cancer Center, New York, New York 10021*
Andrei Sulica (621), *Department of Immunology, Victor Babes Institute, Bucharest, Romania*
Deming Sun (243), *Max-Planck-Institut für Immunbiologie, Freiburg, West Germany*
Egon Svastits (1189), *National Institute of Oncology, Budapest, Hungary*
Aldo Tagliabue (25, 273, 401), *Sclavo Research Center, Siena, Italy*
Mitsuo Takasugi (455, 675), *Department of Surgery, University of California, Los Angeles, California 90024*
Margarita Tálas (1227), *Microbiological Research Group, National Institute of Health and Hygiene, Budapest, Hungary*
Kenichi Tanaka (751), *Laboratory of Cell Biology, National Cancer Institute, Bethesda, Maryland 20205*
Donatella Taramelli (535), *Biological Development Branch, Biological Response Modifiers Program, National Cancer Institute, Frederick, Maryland 21701*
Stephan R. Targan[38] (995), *Geriatric Research and Education Center, Medical and Research Services, Wadsworth VA Medical Center, Los Angeles, California 90024*
Jussi Tarkkanen (575), *Department of Pathology, University of Helsinki, Helsinki, Finland*
Eckardt Thiel (53), *Institute of Hematology, GSF, Munich, West Germany*
Tuomo Timonen[39] (9, 17, 375, 669, 821, 931, 1055), *Biological Research and Therapy Branch, and Laboratory of Immunodiagnosis, National Cancer Institute, Frederick, Maryland 21701*
Klára Tótpál (1189), *National Institute of Haematology and Blood Transfusion, Budapest, Hungary*
Thomas Tötterman (733), *Department of Pathology, Wallenberg Laboratory, Uppsala, Sweden*
John J. Trentin (557, 1207, 1249, 1405, 1411), *Division of Experimental Biology, Baylor College of Medicine, Houston, Texas 77030*
Giorgio Trinchieri (39, 369), *The Wistar Institute of Anatomy and Biology, Philadelphia, Pennsylvania 19104*
Thomas Tursz (729, 1183, 1253), *Laboratoire d'Immunologie Clinique, Institut Gustave Roussy, Villejuif, France*
Àtsushi Uchida (431, 589, 1303), *Institute for Cancer Research, University of Vienna, Vienna, Austria*
Måns Ullberg (361, 607), *Department of Tumor Biology, Karolinska Institute, Stockholm, Sweden*
James Urban (1393), *La Rabida-University of Chicago Institute, Committee on Immunology and Department of Pathology, University of Chicago, Chicago, Illinois 60649*
David L. Urdal (757, 1339), *Program in Basic Immunology, Fred Hutchinson Cancer Research Center, Seattle, Washington 98104*
Farkas Vánky (691, 829, 1047, 1061), *Department of Tumor Biology, Karolinska Institute, Stockholm, Sweden*
Luigi Varesio (535), *Biological Development Branch, Biological Response Modifiers Program, National Cancer Institute, Frederick, Maryland 21701*
Miklòs Varga (1041), *National Institute of Haematology and Blood Transfusion, Budapest, Hungary*
Ismo Virtanen (983), *Department of Pathology, University of Helsinki, Helsinki, Finland*

[38] Present address: *Department of Medicine, UCLA Center for the Health Sciences, Los Angeles, California 90024.*

[39] Present address: *Department of Pathology, University of Helsinki, Helsinki, Finland.*

B. M. Vose (879, 909, 1055, 1127), *Department of Immunology, Paterson Laboratories, Christie Hospital, and Holt Radium Institute, Manchester, England*

Jerrold M. Ward (1161), *Tumor Pathology and Pathogenesis Section, Laboratory of Comparative Carcinogenesis, National Cancer Institute, Frederick, Maryland 21701*

James E. Weiel (949), *Departments of Pathology and Microbiology-Immunology, Duke University Medical Center, Durham, North Carolina 27710*

William O. Weigle (85), *Department of Immunopathology, Scripps Clinic and Research Foundation, La Jolla, California 92037*

Monica L. Weitzen (969), *Department of Molecular Biology and Biochemistry, University of California, Irvine, Irvine, California 92717*

Raymond M. Welsh (493, 765), *Department of Pathology, University of Massachusetts Medical School, Worcester, Massachusetts 01605*

Jerome A. Werkmeister (743, 1011), *Departments of Microbiology and Immunology, and Radiation Oncology, Queen's University, Kingston, Ontario, Canada*

Patricia A. Weston[40] (699), *Division of Immunology, Duke University Medical Center, Durham, North Carolina 27710*

H. Wigzell (540, 733, 765), *Departments of Immunology and Pathology, University of Uppsala, Uppsala, Sweden*

R. Wiltrout (1331), *Laboratory of Immunodiagnosis, National Cancer Institute, Bethesda, Maryland 20205*

Nancy T. Windsor (307), *Department of Genetics and Regional Primate Research Center, University of Washington, Seattle, Washington 98195*

Henry J. Winn (105), *Transplantation Unit, Massachusetts General Hospital, Boston, Massachusetts 02114*

Isaac P. Witz (201, 1431), *Department of Microbiology, The George S. Wise Faculty of Life Sciences, Tel Aviv University, Tel Aviv, Israel*

James N. Woody (1273), *Divisions of Immunologic Oncology and Medical Oncology, Lombardi Cancer Center, Georgetown University, Washington, D.C. 20007*

Susan C. Wright (961), *Department of Microbiology and Immunology, UCLA School of Medicine, Los Angeles, California 90024*

Robert S. Yamamoto (969), *Department of Molecular Biology and Biochemsitry, University of California, Irvine, Irvine, California 92717*

Ganesa Yogeeswaran (765), *Department of Microbiology, Boston University School of Medicine, Boston, Massachusetts 02215*

M. Zöller (540), *Department of Immunology, University of Uppsala, Uppsala, Sweden*

Daniel Zagury (887), *Université Paris VI, Paris, France*

Helmut Zander (1233), *Labor für Immungenetik, Kinderpoliklinik, Munich, West Germany*

Joyce M. Zarling (59, 791), *Immunobiology Research Center, Departments of Laboratory Medicine/Pathology and Surgery, University of Minnesota, Minneapolis, Minnesota 55455*

Rainer Zawatzky (1471), *Institute of Virus Research, German Cancer Research Center, Heidelberg, West Germany*

Hans-Werner Loems Ziegler (771, 1233), *Institute for Immunology, University of Munich, Munich, West Germany*

[40] Present address: *Department of Pharmacology, Duke University Medical Center, Durham, North Carolina 27710.*

Preface

Within the last two years, the pace, scope, and even the character of research on effectors of natural immunity have undergone profound changes. That natural cell-mediated immunity plays an important role in protecting the host against threats from transformed cells and from the environment has attained wide acceptance. The present volume represents and reports the new directions from laboratories leading the current research, as amplified below.

The recent rapid advances in this field reflect the entry of many investigators from other disciplines and subspecialties, who have brought in new ways of dealing with problems peculiar to the study of natural effectors and have helped elucidate the characterization of the effector cells. Among these advances are the association of cells with a particular morphology, namely, large granular lymphocytes (LGL) with NK activity; the development of monoclonal antibodies against lymphoid cell populations; the purification of interferon and T-cell growth factor, and the production of recombinant species of some of these lymphokines; the findings of LGL tumors in man and in rats; and especially the new-found capability to grow NK cells *in vitro* when T-cell growth factor is supplied.

There has been a considerable shift in the focus of studies on NK cells and other natural effector mechanisms. With the association of LGL with NK cells, the definition of various cell surface antigens on NK cells, and the new ability to culture and to clone these effector cells, emphasis is now focused on detailed characterization of the effector cells and a comparison of the natural effectors with other, better known types of lymphoid cells. Whereas the early studies on the activities of natural effector cells demonstrated their antitumor effects, recent studies have repeatedly demonstrated the roles of these mechanisms in resistance to microbial infection, in bone marrow transplantation, and possibly also in graft-vs-host disease. Alterations in levels of activity have been recently noted in a variety of disease states. There is increased awareness of noncytotoxic roles for NK cells and other natural effectors, including the production of interferon. Even in the recent studies on host resistance to tumor growth, the emphasis has been shifting from the effects of NK cells on the local growth of transplantable tumors to the role of NK cells and other natural effectors in coping with primary carcinogenesis. Such studies have aroused interest in these cells as significant mediators of immune surveillance and of host resistance to the metastatic spread of tumors.

In this volume, all these recent and current developments, including previously unpublished data, have been given particular emphasis. Accordingly, there is essentially no overlap with an earlier volume, "Natural Cell-Mediated Immunity against Tumors" (Academic Press, 1980). Our coverage of this lively field of immune effector systems ranges from basic studies on their nature, regulation, and mechanisms of action to important practical issues such as their role in host resistance, their modulation by therapeutic intervention, and alterations of their activity in disease.

The special attention given the assemblage and organization of this voluminous material will provide the reader with a uniquely comprehensive overview and prospects for this dynamic multidisciplinary field of modern immunobiology. By offset-printing the text direct from the authors' typewritten manuscript the publisher's production time has been cut to only several months after final manuscript submission, thus ensuring the publication of fresh material. Furthermore, this process, eliminating typesetting costs for this very large book, has enabled the publisher to hold the price to about half of what it would have been otherwise.

MORPHOLOGY AND CYTOCHEMISTRY OF HUMAN LARGE GRANULAR LYMPHOCYTES[1]

Carlo Enrico Grossi
Department of Human Anatomy
University of Genoa
Genoa, Italy

Manlio Ferrarini
Department of Clinical Immunology
University of Genoa
Genoa, Italy

INTRODUCTION

Human peripheral blood cells with a lymphoid morphology and intracytoplasmic azurophilic granules have been described long ago and termed "monocytoid" or "leukocytoid" lymphocytes (Pappenheim and Ferrata, 1911). However, only recently have they been recognized as a distinct subset of mononuclear cells (Grossi et al., 1978; Ferrarini et al., 1980) with specialized functions, such as antibody-dependent cell cytotoxicity (ADCC) and natural killer (NK) activity (Timonen et al., 1981). These cells are usually referred to as Large Granular Lymphocytes (LGL) or NK cells.

LGL represent the bulk of two mononuclear cell subsets of the human peripheral blood, namely, the T_G cells (i.e., cells forming rosettes with sheep erythrocytes and expressing avid receptors for the Fc portion of IgG) and the third populations cells (TPC) (Ferrarini et al., 1980). TPC are defined as those "null" cells (i.e., cells not forming rosettes with sheep erythrocytes and not expressing surface immunoglobulin) with avid Fc receptors (Winchester et al., 1975).

Here we shall describe the morphological and cytochemical features of LGL, with a special emphasis on the granules and on the mechanism of their formation. These studies revealed possible stages of LGL maturation and this approach seems further exploitable for the search of LGL precursors.

[1]Supported by grants from the Italian CNR.

ISBN 0-12-341360-5

ULTRASTRUCTURE OF PERIPHERAL BLOOD LGL

Highly enriched LGL preparations have been used for morphological studies. Such enrichment has been obtained with three different procedures: 1) isolation of T_G cells (Grossi et al., 1978); 2) separation of TPC (Ferrarini et al., 1980); and 3) purification of LGL on Percoll gradients (Timonen et al., 1981). No differences in the results were obtained with these various preparations.

LGL appear as medium-sized lymphocytes with round or indented nuclei, clumped chromatin and usually prominent nucleoli. The cytoplasm is abundant and, besides rare mitochondria and isolated profiles of the rough endoplasmic reticulum (RER), displays the following distinctive features (Grossi et al., 1982): a) an extended Golgi apparatus which buds smooth vesicles surrounding stacks of parallel Golgi tubules and saccules; b) numerous "coated" vesicles which are sparse in the cytoplasm or, more often, concentrated in the nuclear notch; c) unit membrane-bound granules containing a matric of variable density; these organelles are likely to correspond to the azurophilic granules of the Giemsa-stained cells; d) the remarkable absence of phagolysosomes (Fig. 1).

Recently it has been suggested that parallel tubular arrays (PTA) which may be common in LGL (Costello et al., 1980; Payne and Glasser, 1981) could correspond to the azurophilic granules. However, in our experience, PTA are observed in a minority of cells only and therefore we believe that (i) PTA cannot be used as a marker for LGL identification and (ii) PTA do not correspond to the azurophilic granules (Fig. 1).

THE LYSOSOMAL SYSTEM OF PERIPHERAL BLOOD LGL

LGL contain several acid hydrolases, which are a specific marker of the lysosomal system. They are alpha-naphthyl acetate (ANAE) and -butirate (ANBE) esterase, acid phosphatase (AP) and beta-glucuronidase (B-Gluc) (Grossi et al., 1978, Saksela et al., 1979; Ferrarini et al., 1980; Costello et al., 1980). With azo-dye technique at the light microscope, these enzymes give a positive reaction in the paranuclear region, with a granular or more often diffuse pattern of staining. With a Gomori-type of substrate for AP, only a granular reaction is obtained with the positivity manily concetrated in the paranuclear area (Grossi et al., 1982). These findings suggest that the acid hydrolases could be in the granules and, possibly, in the "coated" vesicles which are more numerous in the paranuclear area. EM studies have confirmed this hypothesis by revealing the presence of AP in the granules and in the "coated" vesicles in addition to the perinuclear cisterna and

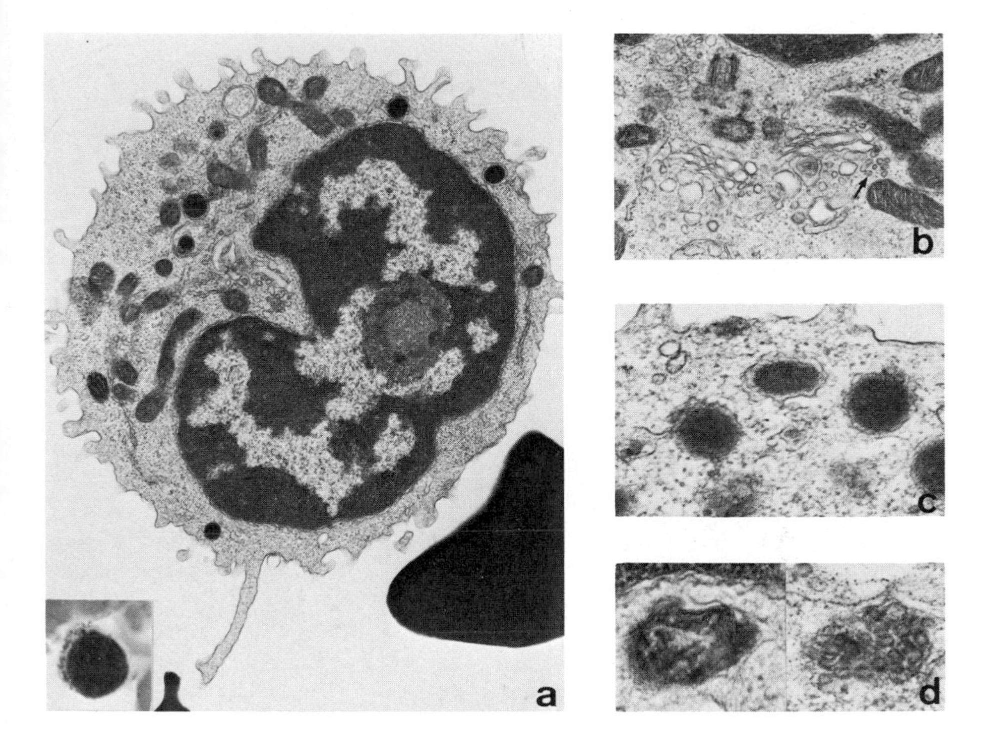

FIGURE 1. Ultrastructure characteristics of human peripheral blood LGL. a) Section of a whole LGL illustrating the overall features of this cell type. The inset shows a Giemsa-stained LGL containing numerous azurophilic granules. b) The Golgi region of a LGL. The arrow points to a group of "coated" vesicles. c) Ultrastructural characteristics of the mature LGL granules. d) Ultrastructure of two PTA. The organelle seen on the left is bounded by a unit membrane whereas the array of transversely-sectioned tubules on the rights is free in the cytoplasm.

some strands of the RER. By contrast AP is not detected in the Golgi apparatus, a finding that is of particular relevance to the understanding of the mechanism of the granulogenesis (Figs. 2 and 4).

Although identified as lysosomes, the granules or the "coated" vesicles never display peroxidase activity in contrast to that observed for cells of the granulocytic-monocytic lineage (Ferrarini et al., 1980). In addition, both the granules and the vesicles are not involved in the process of phagocytosis, since LGL fail to ingest latex particles or opsonized red cells (Grossi et al., 1982).

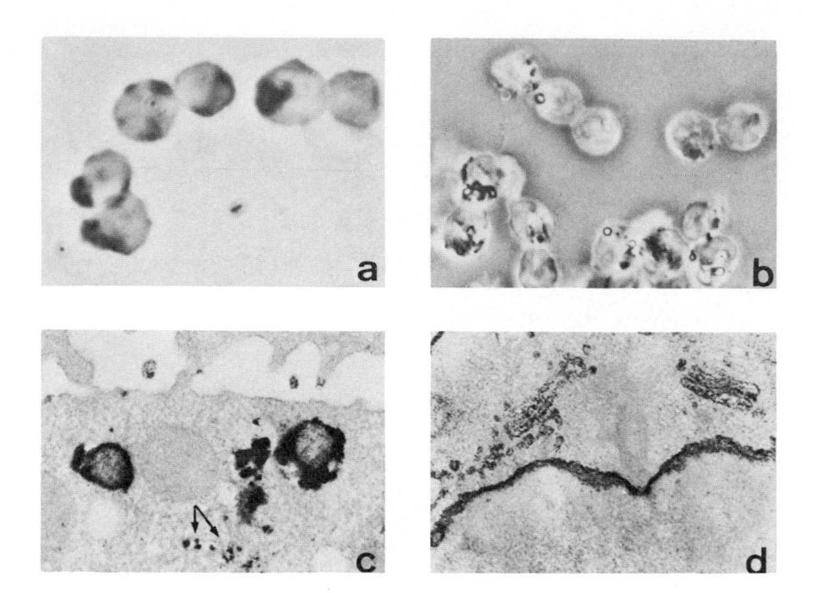

FIGURE 2. Cytochemistry of human peripheral blood LGL. a) Staining for ANAE activity reveals a paranuclear accumulation of the reaction product. b) Staining for AP with a Gomori-type of substrate shows granular deposits of the reaction product in the majority of the cells. c) Ultrastructural localization of AP activity within granules and "coated" vesicles (arrows). d) Ultrastructural localization of TPP in a portion of a LGL. The reaction product is found within the perinuclear cisterna, the stacks of Golgi-tubules and the Golgi-derived vesicles. Granules, as well as "coated" vesicles are negative.

PATTERNS OF GRANULE FORMATION IN LGL

In principle, the "coated" vesicles and the electron dense granules of LGL could represent distinct lysosomal types. Alternatively, the "coated" vesicles could subserve the function of transporting acid hydrolases from the RER to the mature granules. The latter hypothesis seems supported by the observations that "coated" vesicles fusing with immature granules (i.e., organelles containing a scarce matrix of low electron-opacity) are sometimes observed in mature LGL. This fusion occurs generally in the proximity of the Golgi apparatus where the immature granules and the "coated" vesicles are also more numerous indicating that the Golgi apparatus may be the site where the granules are packaged (Grossi et al., 1982).

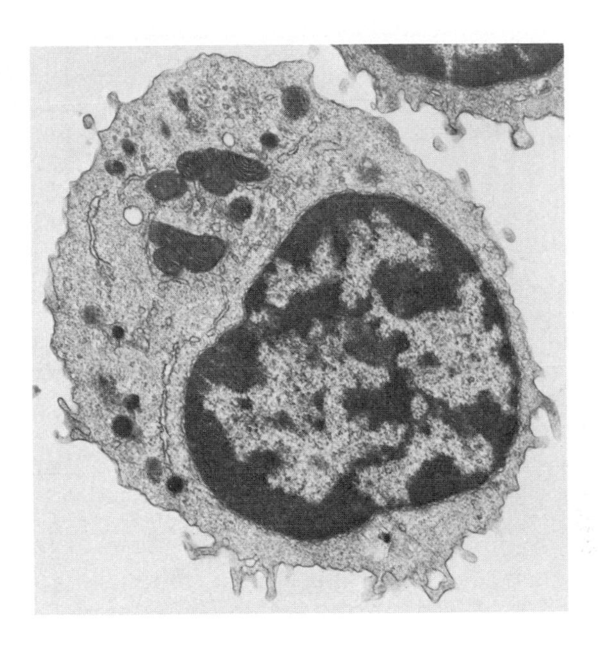

FIGURE 3. Ultrastructure of a LGL from a patient with an abnormally expanded LGL population. It is of note that numerous multivesicular bodies (arrow) were found in the cytoplasm of these cells. These organelles are rarely seen in normal peripheral blood LGL.

The opportunity of investigating further the mechanism of granule formation has been offered by the availability of immature LGL populations. These were the cells from a patient with an abnormally expanded LGL population, possibly of lalignant origin (Ferrarini et al., submitted) (Fig. 3) and the LGL isolated from normal (i.e., nonneoplastic) bone marrow aspirates (Grossi et al., 1981). Cells from both sources had very few mature granules and displayed a number of features indicating active granulogenesis such as: 1) an expanded Golgi apparatus budding numerous smooth vesicles; 2) some AP-positive "coated" vesicles, fusing with the smooth Golgi-derived vesicles to form 3) AP-positive multivesicular bodies containing "coated" vesicles surrounded by the Golgi-derived smooth membranes; 4) multivesicular bodies developing into granules through the accumulation of an electron-dense matrix. That the smooth vesicles, forming the limiting membrane of the multivesicular bodies were Golgi-derived, was shown by their positivity for a specific marker of the Golgi membranes, namcly, thiamine-pyrophosphatase (TPP) (unpublished observations)(Fig. 2).

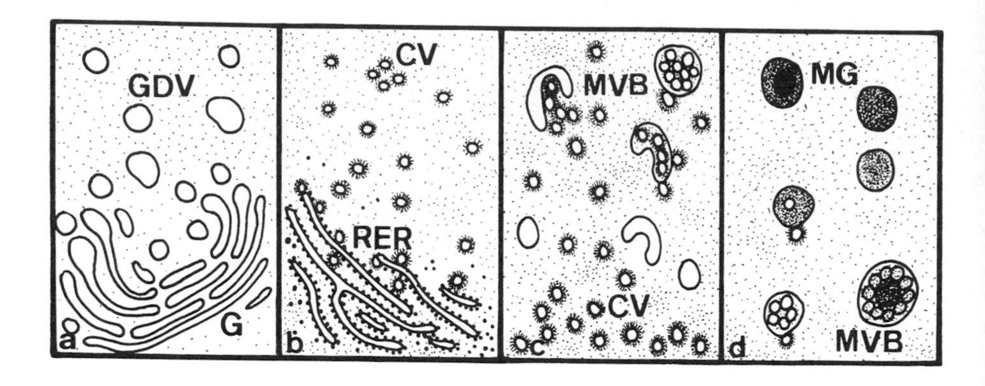

FIGURE 4. A schematic representation of the sequence which may lead to granule formation in LGL. This sequence has been deduced by ultrastructural and cytochemical analysis of LGL from normal bone marrow or from patients with abnormally expanded LGL populations. a) Smooth vesicles (GDV) budding from the Golgi apparatus (G). b) "Coated" vesicles (CV) derived from the rough endoplasmic reticulum (RER). c) "Coated" vesicles (CV) fusing with the smooth vesicles to form multivesicular bodies (MVB). d) Multivesicular bodies showing initial deposition of an electron-dense matrix (MVB) and maturing granules (MG).

Taken together, these observations suggest that acid hydrolases, synthesized within the RER, are transported by "coated" vesicles to the proximity of the Golgi apparatus where the "coated" vesicles fuse with Golgi-derived smooth vesicles to form multivesicular bodies which are the precursors of mature granules (Ferrarini et al., submitted) (Fig. 4).

CONCLUSIONS

Morphology and cytochemistry provide unique criteria for LGL identification, since other phenotypic characteristics so far investigated, such as the classical membrane markers or antigenic specificities detectable by monoclonal reagents, seem largely shared with cells of toher lineages (Reinherz et al., 1980; Abo and Balch, 1981; Ortaldo et al., in press). A possible exception is represented by the recently described HNK-1 monoclonal marker (Abo and Balch, 1981). The prominent morphological and cytochemical feature of LGL is constituted by a lysosomal system with different properties from those of other hemic granular cells (i.e., granulocytes and monocytes). The lysosomal system of LGL is clearly nonoperating in phagocytic functions. Since LGL possess cytotoxic activities, the possibility should be considered that the granules participate in the mechanism of killing, as would be supported by two observations, namely: 1) although capable of binding to the target, immature LGL equipped with a low number of mature granules are inefficient in killing (Ferrarini et al., submitted); 2) Chediak-Higashi patients have an impaired NK function (Haliotis et al., 1980) and their LGL display the same type of lysosomal defect detected in other cells (Abo, Balch, and Cooper, 1981, personal communication).

Granule formation does not normally occur in peripheral blood LGL but is observed in expanded, possibly malignant, LGL populations or in normal bone marrow cells. Since in other hemic lineages, granulogenesis takes place during cell maturation (Cohn and Benson, 1965; Bainton and Farquhar, 1966), it is likely that this is also the case of LGL. Granulogenesis might, therefore, represent a marker of LGL maturation.

REFERENCES

Abo, T., and Balch, C. M. (1981). J. Immunol. 127:1024.

Bainton, D. F., and Benson, B. (1965). J. Exp. Med. 121:135.

Costello, C., Catovsky, D., O'Brien, M., Morilla, R., and Varardi, S. (1980). Leuk. Res. 5:463.

Ferrarini, M., Cadoni, A., Franzi, A. T., Ghigliotti, C., Leprini, A., Zicca, A., and Grossi, C. E. (1980). Eur. J. Immunol. 10:562.

Ferrarini, M., Romagnani, S., Montesoro, E., Zicca, A., Del Prete, G. F., Nocera, A., Maggi, E., Leprini, A., and Grossi, C. E. (1981) submitted.

Grossi, C. E., Webb, S. R., Zicca, A., Lydyard, P. M., Moretta, L., Mingari, M. C., and Cooper, M. D. (1978). J. Exp. Med. 147:1405.

Grossi, C. E. Burgio, V. L., and Ferrarini, M. (1981). Haematologica 66 (suppl.):159.

Grossi, C. E., Cadoni, A., Zicca, A., Leprini, A., and Ferrarini, M. (1982). Blood, in press.

Haliotis, R., Roder, J., Klein, M., Ortaldo, J. R., Fauci, A. S., and Herberman, R. B. (1980). J. Exp. Med. 151:1039.

Ortaldo, J. R., Sharrow, S. O., Timonen, T., Herberman, R. B. (1981). J. Immunol., in press.

Pappenheim, A., and Ferrata, A. (1911). Folia Haemat. 10:78.

Payne, C. M., and Glasser, M. (1981). Blood 57:567.

Reinherz, E. L., Moretta, L., Roper, M., Breard, J. M., Mingari, M. C., Cooper, M. D., and Schlossman, S. F. (1980). J. Exp. Med. 151:969.

Saksela, E., Timonen, T., Ranki, A., and Hayry, P. (1979). Immunol. Rev. 44:71.

Timonen, T., Ortaldo, J. R., and Herberman, R. B. (1981). J. Exp. Med. 153:569.

Winchester, R. J., Fu, S. M., Hoffman, T., and Kunkel, H. G. (1975). J. Immunoo. 114:1210.

ANALYSIS OF NATURAL KILLER ACTIVITY OF HUMAN LARGE GRANULAR LYMPHOCYTES AT A SINGLE CELL LEVEL

Tuomo Timonen[1], John R. Ortaldo and Ronald B. Herberman

Biological Research and Therapy Branch
National Cancer Institute
Frederick, Maryland

INTRODUCTION

Recent evidence has shown an association of human natural killer (NK) activity with a morphological subpopulation of peripheral blood mononuclear cells, called large granular lymphocytes (LGL) (Timonen *et al.*, 1979; Timonen *et al.*, 1981). LGL comprise approximately 10% of peripheral blood lymphoid cells and they can be enriched up to $\geq 90\%$ purity by discontinuous density gradient centrifugation and subsequent depletion of high affinity sheep erythrocyte (E) rosette-forming cells from low density, LGL-enriched populations (Timonen *et al.*, 1981). The availability of this purified population has facilitated analyses of ultrastructure (Carpen *et al.*, 1982), surface antigenic phenotype (Ortaldo *et al.*, 1981), proliferative capacity (Timonen *et al.*, submitted for publication), lytic machinery (Carpen *et al.*, 1981) and target cell selectivity (de Landazuri *et al.*, 1981) of NK cells. Although most, if not all, peripheral blood NK activity is exerted by LGL, little information exists as to what proportion of LGL can actually function as NK cells. This question is directly relevant to the reliability of LGL morphology as a marker for NK cells. We have used the single cell cytotoxicity assay in agarose (Grimm *et al.*, 1979) to estimate the frequency of LGL capable of binding and lysing NK-sensitive target cells. The results indicate that up to 70% of LGL are NK cells. An additional parameter which affects the lytic potential of killer cells is their recycling capacity. We demonstrate here, by using

---

[1]Present address: Department of Pathology, University of Helsinki, Finland.

ISBN 0-12-341360-5

density gradient-purified LGL-target cell conjugates, that recycling is involved in NK lysis. Furthermore, we show that interferon (IFN) can augment all of the above mentioned phases of NK cell-mediated cytotoxicity.

Binding of LGL and Other Lymphoid Cells on K562 Target Cells and Estimation of NK Cell Frequencies Among LGL

The evaluation of the frequency of killer cells among lymphoid cells has recently been facilitated by the development of the single cell cytotoxicity assay in agarose (Grimm *et al.*, 1979). When various fractions, obtained by discontinuous density gradient centrifugation, were tested for NK cell frequencies by this method, using K562 (the cell line derived from a patient with chronic myelogenous leukemia) as a target cell, the following was seen (Table 1): a) the distribution of LGL and NK cells is similar, b) the number of LGL in each fraction is higher than the number of NK cells, c) among the purest LGL populations (fractions 2 and

Table 1. Binding and Lytic Characteristics of Human Peripheral Blood Lymphoid Cells Fractionated by Discontinuous Density Gradient Centrifugation on Percoll

Fract.	%Percoll[a]	%LGL	%Lysis[b] (E:T 20:1)		%Cells binding K562[c]		%LGL among binding cells		%Killer cells among binding cells[c]	
			-IFN	+IFN	-IFN	+IFN	-IFN	+IFN	-IFN	+IFN
1	42.5	50	38	48	40	40	83	85	40	72
2	45.0	86	73	85	50	48	>95	>95	50	80
3	47.5	48	44	53	30	28	80	81	36	68
4	50.0	27	28	35	22	23	64	67	20	43
5	52.5	8	5	8	6	5	50	49	24	38
6	55.0	<1	<1	<1	4	6	<1	<1	<1	<1
7	66.6	<1	<1	<1	6	5	<1	<1	<1	<1

a) Adjusted to 285 mOsm/kg H_2O (Timonen *et al.*, 1981).
b) In 4 hour chromium^{51}-release assay.
c) Analyzed by a 12 hour single cell cytotoxicity assay. (Grimm *et al.*, 1979). A 12 hour assay was used. Cells were preincubated for 3 hours in medium (-IFN) or with 800 IU/ml of fibroblast IFN (+IFN) (Timonen *et al.*, 1982).

3), only LGL bound to the target cells and a high cytotoxic activity was detected, d) among pure non-LGL populations (fractions 6 and 7), some lymphocytes bound to the target cells, but no cytotoxicity could be detected, e) preincubation of effector cells with IFN increased the number of NK cells among binding cells, but the number of killer cells never exceeded the frequency of LGL among binding cells. Taken together, the data indicate that LGL are the principal NK cells in peripheral blood. They also suggest that not all LGL are NK cells, since only about half of them bound to K562 after a single centrifugation of effector cells and target cells together. However, it was possible that the conjugation assay was not sensitive enough to detect all the conjugate-forming cells, or that for some reason, not all LGL are capable of binding target cells simultaneously. To test these possibilities, LGL purified by discontinuous density gradient centrifugation (fractions 2 and 3) and subsequent depletion of high affinity E-rosette forming cells, and target cells (either K562 or the anchorage-dependent mammary carcinoma line G-11) were centrifuged together at a 1:1 ratio, and the resulting conjugates and unbound target cells were separated from the rest of LGL by one step discontinuous density gradient centrifugation on 10% Percoll (Timonen _et al_., 1979; Carpen _et al_., 1981). The unbound LGL were then tested for their conjugation capacity with fresh target cells (Table 2). Some of the remaining LGL were shown to be capable of binding and lysing K562 and G-11. Even after a second depletion of conjugate-forming cells some unbound LGL were lytic. Thus, either the sensitivity of the binding assay is not sufficient to detect all the binders simultaneously, or LGL are not capable of synchronized binding. When the cumulative frequency of NK cells cytotoxic against K562 was calculated from the sequential conjugation experiments, about 50% of non-IFN-pretreated LGL and about 70% of IFN-pretreated LGL were shown to be NK cells. It is of interest that the number of NK cells against the anchorage-dependent G-11 was much less than against K562, and that IFN could increase the number of binding cells against G-11, whereas the number of binding cells did not change in the K562 system as a result of IFN-pretreatment of LGL. We have shown elsewhere that this dissociation between anchorage-dependent and suspension grown target cells is detectable by using various other target cells also (Carpen _et al_., 1981). The results suggest that there is some target cell selectivity in the binding capacity of LGL, and that IFN may augment NK activity by different mechanisms.

Table 2. Binding and Killing of K562 and G-11 by Initially Nonadherent LGL After Removal of Primary LGL-Target Cell Conjugates[a)]

Target cell	Binding cycle	%Binding cells		%Killer cells among binding cells		Estimated cumulative frequency of NK cells	
		-IFN	+IFN	-IFN	+IFN	-IFN	+IFN
K562	I	37	35	70	86	26	30
	II	41	55	86	92	48	63
	III	29	45	55	52	54	70
G-11	I	13	21	32	57	4	12
	II	16	11	38	43	9	16
	III	3	6	40	47	11	18

a) Conjugates, produced by mixing effector cells and target cells at 1:1 ratio, were centrifuged at 120g for 10 minutes. After a gentle resuspension, the conjugates were layered on a 10% Percoll gradient and spun at 40g for 8 minutes. Target-nonadherent cells were tested for conjugation capacity against fresh target cells (Binding cycles II and III) (Timonen *et al.*, 1979; Timonen *et al.*, 1982).

The Effect of Neuraminidase on the Binding and Lytic Capacities of LGL

Regardless of whether the detected partial binding capacity of LGL is due to de novo activation, reactivation, maturation, or low affinity binding capacity of some LGL, sialic acid appears to be involved in the refractory, non-binding state of LGL. If LGL and/or K562 were pretreated with 1 IU/ml of neuraminidase (VCN, Behringwerke AG, Marburg, Germany), increased binding and increased percentage of lytic cells among binders were detected (Table 3). The results indicate that practically all LGL estimated by sequential adsorption experiments to possess lytic potential could be induced to lyse simultaneously. This suggests that at least the insensitivity of the binding assay is not the major reason for the incomplete detection of NK cells among LGL by a single centrifugation of effector cells and target cells.

Recycling

A further dimension in the lytic capacity of lymphoid cells, in addition to binding and subsequent induction of

Table 3. The Effect of Neuraminidase (VCN) Treatment on the Binding and Lytic Capacity of LGL[a)]

Effector cells	Target cells	%Target binding cells	%Killer cells among binding cells	%Cytotoxicity (E:T 2:1)
LGL	K562	44	54	30
LGL(VCN)	K562	53	62	36
LGL	K562(VCN)	52	63	37
LGL(VCN)	K562(VCN)	68	78	53

a) K562 and/or LGL were pretreated for 1 hour at 37°C in serum-free conditions with 1 U/ml VCN, washed, and used in the single cell cytotoxicity assay. %Cytotoxicity measured with a 4 hour chromium^{51}-release assay.

lytic machinery, is their ability to sequentially lyse several target cells. This phenomenon, called recycling, has been suggested by an indirect method of analysis, to be involved also in NK activity (Ullberg *et al.*, 1981). We have directly studied the recycling capacity of purified LGL, by isolating LGL-K562 conjugates by one-step discontinuous density gradient centrifugation on Percoll as in Table 2. Both the rate of spontaneous dissociation of the conjugates and the capacity of the dissociated LGL to rebind K562 were then measured, along with the single cell cytotoxicity assay in agarose for the detection of NK cells among reconjugated LGL. As shown in Table 4, the majority of LGL spontaneously dissociated rapidly after the initial conjugate formation. If the LGL were not pretreated with IFN, there was a refractory period, and then rebinding approximately after 120 min. The results from the single cell assay indicate that the recycling is mainly regulated at the binding level, since during the whole experiment all those LGL that bound target cells lysed them with a similar efficiency. If the effector cells were pretreated with IFN, the rebinding and lysis could take place continuously, indicating that IFN can increase the recycling capacity of NK cells.

CONCLUSIONS

The morphological and density analyses of the NK cells in human peripheral blood strongly suggest that most, if not all, NK cells are morphologically LGL. Furthermore, we conclude that there is a high percentage of actual NK cells among LGL. Although other groups have reported NK activity in lymphoid populations apparently devoid of LGL (Hokland *et al.*, 1981), we have not been able to do so, even by using

Table 4. Lytic Capacity of Recycling LGL[a)]

Time (minutes)	%Binding cells				%Lytic cells among binding cells	
	Without centrifugation		With centrifugation		With centrifugation	
	-IFN	+IFN	-IFN	+IFN	-IFN	+IFN
0	65	68	72	70	52	80
60	25	33	32	81	47	82
120	33	39	63	72	54	78
210	20	29	33	78	55	79
240	23	33	68	80	60	83

a) Enriched LGL (after preincubation with or without IFN)-K562 conjugates were incubated for various time periods at 37°C and the spontaneous dissociation (without centrifugation) or the rebinding capacity after centrifugation at 120g for 10 minutes (with centrifugation) were monitored.

numerous other target cells in addition to K562 (Carpen *et al.*, 1981, de Landazuri *et al.*, 1981).

Our data do not indicate that LGL are a homogenous subpopulation of lymphoid cells. On the contrary, about 20-30% of them have no detectable NK activity, there is a clear size and density heterogeneity among LGL, and they are antigenically heterogenous (Ortaldo *et al.*, 1981). It is possible that the LGL inactive against K562 are cytotoxic to another target cell, since polyclonality of NK cells has been suggested (Phillips *et al.*, 1980). However, using seven different target cells we have shown that polyclonality can maximally account for an additional 10% LGL as NK cells, i.e., NK cell frequencies among LGL appear to be maximally around 80%. It is of interest that LGL reacting with the monoclonal antibody OKT8 do not seem to account for an appreciable proportion of NK activity (Ortaldo, *et al.*, 1981; Timonen *et al.*, submitted for publication). It will be important to study whether OKT8+ LGL would be K cells responsible for antibody-dependent cell-mediated cytotoxicity, some immature forms of NK cells not readily inducible to NK activity with IFN, or totally unrelated to NK activity, maybe exerting the suppressor cell activity known to be associated with the reactivity with OKT8 (Reinherz *et al.*, 1980).

The net result detected in the ^{51}chromium-release assays from the cytolytic activity of LGL can be attributed to the binding, lytic and recycling capacities of LGL. Therefore, mere calculation of the frequency of LGL among effector cells is not a sufficient measure of NK activity, since individual variations in these three parameters occur (Ullberg et al., 1981; our published results). However, the analyses of LGL frequencies among lymphoid cell populations should be helpful in deciding whether altered NK activity is due to major changes in the frequency of LGL, or to altered function.

Although most of the NK activity among fresh unprimed lymphocytes is exerted by LGL, not much is known about the cellular origin of NK-like effector cells that are induced during antigenic stimulation, for example in mixed lymphocyte cultures (Seeley et al., 1980), or in fetal calf serum-containing cultures (Ortaldo et al., 1980). The availability of the purified populations of both LGL and conventional T-cells will certainly facilitate the analysis of the cellular basis of these, as well as other forms of immune response.

ACKNOWLEDGMENT

This work has been supported in part by Grant No. 1 R01 CA 23809-01 from the National Cancer Institute, National Institutes of Health, Bethesda, Maryland.

REFERENCES

Carpen, O., Virtanen, I. and Saksela, E. (1981). Cell Immunol 58:97.

Carpen, O., Virtanen, I. and Saksela, E. (1982). J Immunol, in press.

Grimm, E. and Bonavida, B. (1979). J Immunol 123:2861.

Hokland, M., Heron, I. and Berg, K. (1981). Antiviral Research Abstr 1, 1:83.

de Landazuri, M.O., Lopez-Botet, M., Timonen, T., Ortaldo, J.R. and Herberman, R.B. (1981). J Immunol 127:1380.

Ortaldo, J. and Herberman, R.B. (1980). In: "Natural cell-mediated immunity against tumors (R.B. Herberman, ed.) Academic Press, New York, p. 465.

Ortaldo, J.R., Sharrow, S.O., Timonen, T. and Herberman, R.B. (1981). J Immunol 127:2401.

Phillips, W.H., Ortaldo, J.R. and Herberman, R.B. (1980). J Immunol 125:2322.

Reinherz, E.L., Kung, P.C., Goldstein, G. and Schlossman, S.F. (1980). J Immunol 124:1301.

Seeley, J.K. and Karre, K. (1980). In: "Natural cell-mediated immunity against tumors (R.B. Herberman, ed.) Academic Press, New York, p. 477.

Timonen, T., Ranki, A., Saksela, E. and Hayry, P. (1979). Cell Immunol 48:121.

Timonen, T., Ortaldo, J.R. and Herberman, R.B. (1981). J Exp Med 153:569.

Timonen, T., Ortaldo, J.R., Stadler, B.M., Bonnard, G.D., and Herberman, R.B. (1982). Submitted for publication.
Ullberg, M. and Jondal, M. (1981). J Exp Med 153:615.

IDENTIFICATION AND CHARACTERIZATION OF THE NATURAL KILLER (NK) CELLS IN RATS

Craig W. Reynolds
Robert Rees[1]
Tuomo Timonen[2]
Ronald B. Herberman

Biological Research and Therapy Branch
National Cancer Institute
Frederick, Maryland

I. INTRODUCTION

Recent evidence has suggested that natural killer (NK) cells are important mediators of a number of *in vitro* immunological phenomena. Little evidence, however, has been produced that directly examines the role of the NK cell in the *in vivo* mediation of these functions. A major reason for this deficiency is the lack of a suitable animal model that would permit the purification, transfer, and direct *in vivo* testing of NK cell function. Recently, however, we have been able to identify and isolate NK cells from rat spleen and blood (Reynolds *et al*, 1981a; Reynolds *et al*, 1981b). The present manuscript will review the morphology, organ distribution, frequency and target specificity of these rat LGL. The findings suggest that rats may provide a very useful animal model for detailed studies on the ontogeny, regulation and *in vivo* relevance of NK cells.

---

[1]Present address: Department of Virology, University of Sheffield Medical School, Sheffield, England.
[2]Present address: Department of Pathology, University of Helsinki Central Hospital, Helsinki, Finland.

ISBN 0-12-341360-5

II. MATERIALS AND METHODS

Animals. All experiments were performed with Wistar-Furth (WF) or Fischer (F344) rats.

Target Cells. The rat G_1-TC, the mouse YAC-1 and the human K562 leukemias were maintained *in vitro* as suspension cultures (Oehler *et al*, 1978). In addition, the following rat cell lines were maintained *in vitro* as monolayer cultures: W/Fu-T and ErTH/V-G rat sarcomas, the R35 rat mammary adenocarcinoma and the F2304 rat embryo cell (Nunn *et al*, 1976).

Preparation of Lymphoid Cells. Spleen, thymus, lymph node, bone marrow (BM), peritoneal exudate cells (PEC) and peripheral blood leukocytes (PBL) were prepared in Hanks' balanced salt solution (HBSS) as previously described (Ortiz de Landazuri & Herberman, 1972; Oehler *et al*, 1977). The removal of adherent cells was done by incubating leukocyte preparations on columns of nylon-wool (NW) at 37°C.

Percoll Fractionation - Evaluation of Cell Morphology. NW-nonadherent lymphocyte preparations were separated by centrifugation on discontinuous density gradients of Percoll (Pharmacia Chemicals, Uppsala, Sweden), as previously described (Reynolds *et al*, 1981a). For the morphological evaluation of lymphocyte preparations, air-dried cytocentrafuge preparations were fixed in methanol and stained with 10% Giemsa (pH 7.4).

^{51}Cr-release Cytoxicity Assay. Two-fold serial dilutions of effector cells were mixed with ^{51}Cr-labeled target cells and incubated at 37°C for 4 hours, as previously described (Reynolds & Herberman, 1981). Results are expressed as either percent cytotoxicity or as lytic units (LU) per 10^7 effector cells (Herberman *et al*, In Press).

III. RESULTS AND DISCUSSION

Figure 1A demonstrates that there are cells in rat peripheral blood with the morphological features of LGL. These cells have previously been described in the human (Timonen *et al*, 1979a; Timonen *et al*, 1979b; Timonen *et al*, 1981; Timonen and Saksela, 1980; Saksela *et al*, 1979) and characteristically have a high cytoplasmic to nuclear ratio, azurophilic granules in the cytoplasm and, generally, a kidney-shaped nucleus. Small lymphocytes and a monocyte can also be seen in Figure 1A, demonstrating the intermediate size of rat LGL. The importance of LGL in NK activity was first suggested by

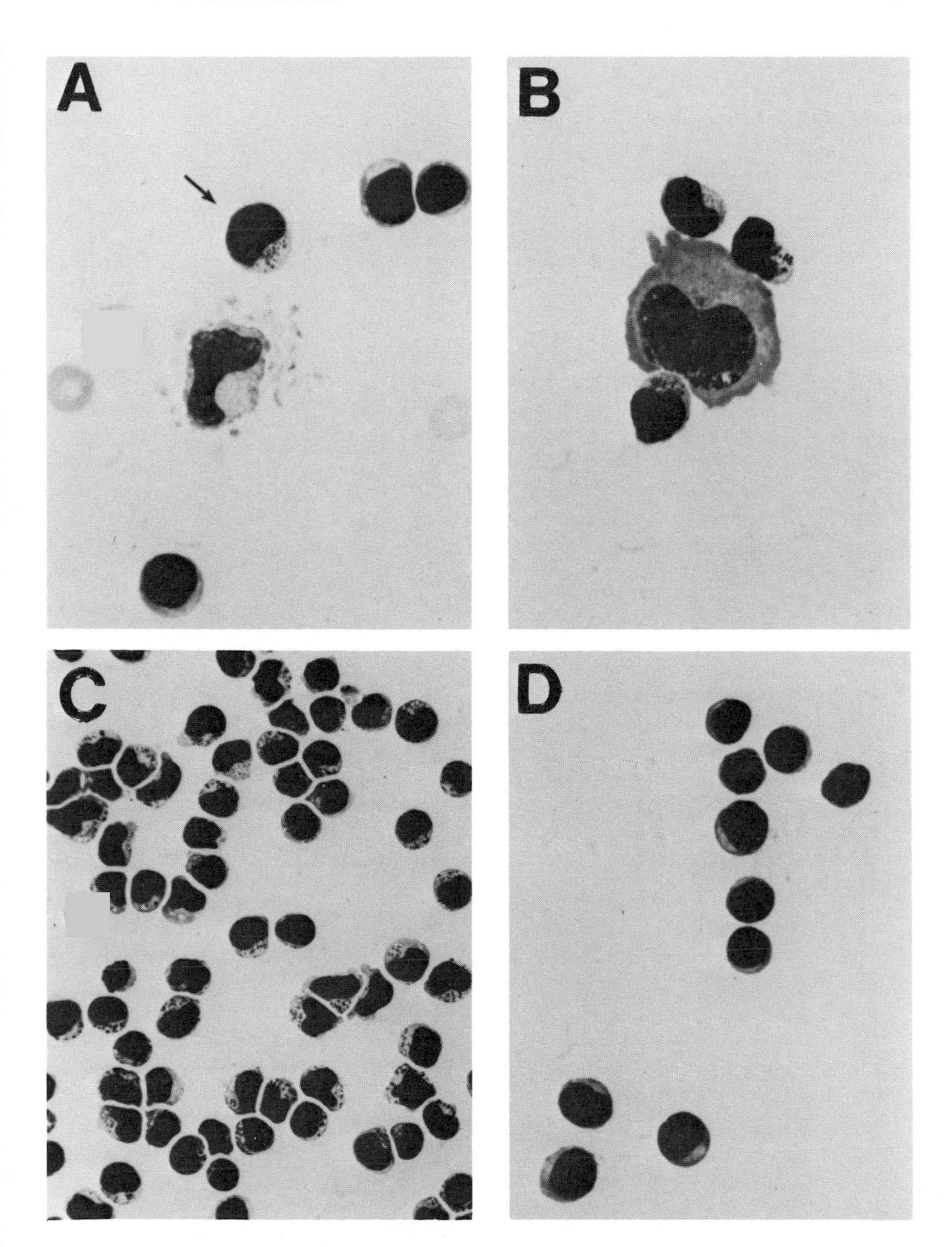

FIGURE 1. Morphology of rat peripheral blood lymphoid cells: A = Ficoll-Hypaque separated population (LGL indicated by arrow). B = LGL conjugates with G_1-TC target cell. C = Percoll isolated LGL population (>80% LGL). D = Percoll isolated small lymphocytes (<5% LGL).

their binding to NK sensitive target cells (Figure 1B). Subsequent studies have clearly demonstrated that these lymphocytes are also capable of killing these tumor cells (Reynolds et al, 1981a). Discontinuous density gradients of Percoll can also be used to isolate LGL and T cells from nylon wool passed blood. The morphology of these isolated populations can be seen in Figure 1C-D.

To determine if other organs contained LGL we screened a number of sites which had previously been shown to contain NK activity (Oehler et al, 1978; Nunn et al, 1976; Zoller & Matzku, 1980). The results in Table 1 clearly show that LGL are not confined to the blood. These cells are also found in significant numbers in the spleen, peritoneal cavity, and lungs. A small but significant number of LGL was always observed in lymph nodes but few, if any, LGL were found in the thymus and bone marrow. In addition, the passage of leukocyte preparations over nylon-wool columns always enriched for LGL but rarely enriched for NK activity. With the exception of the lungs, the LGL frequency pattern from various organs was similar to the distribution of NK activity (PBL > spleen > peritoneal exudate > lymph node > thymus = bone marrow). Interestingly, even the preincubation of lung-LGL with interferon (IFN) preparations did not greatly augment their cytotoxic potential (data not shown). These results are in general agreement with the hypothesis that rat NK cells are LGL, but also demonstrate that not all LGL have cytolytic activity. Mere measurement of LGL frequency would not appear to be an adequate means for predicting the NK activity of an effector cell population.

Table 1. LGL Frequency and Organ Distribution of NK Activity in W/Fu Rats

Organ	Median % Cytotoxicity	Median % LGL
PBL	28.3	7.0
-NW Passed	16.5	14.0
Spleen	20.5	4.0
-NW Passed	12.6	6.5
PEC	6.4	3.0
-NW Passed	2.7	8.0
Lymph node	1.1	<1.0
-NW Passed	1.0	<1.0
Lungs	0.0	7.0
Thymus	0.0	<0.5
Bone Marrow	0.0	<0.5

[a]E/T = 100:1 with G_1-TC target.

To directly examine the association of rat LGL with NK activity, Percoll density gradients were run using nylon-wool passed lymphocyte preparations (Figure 2).

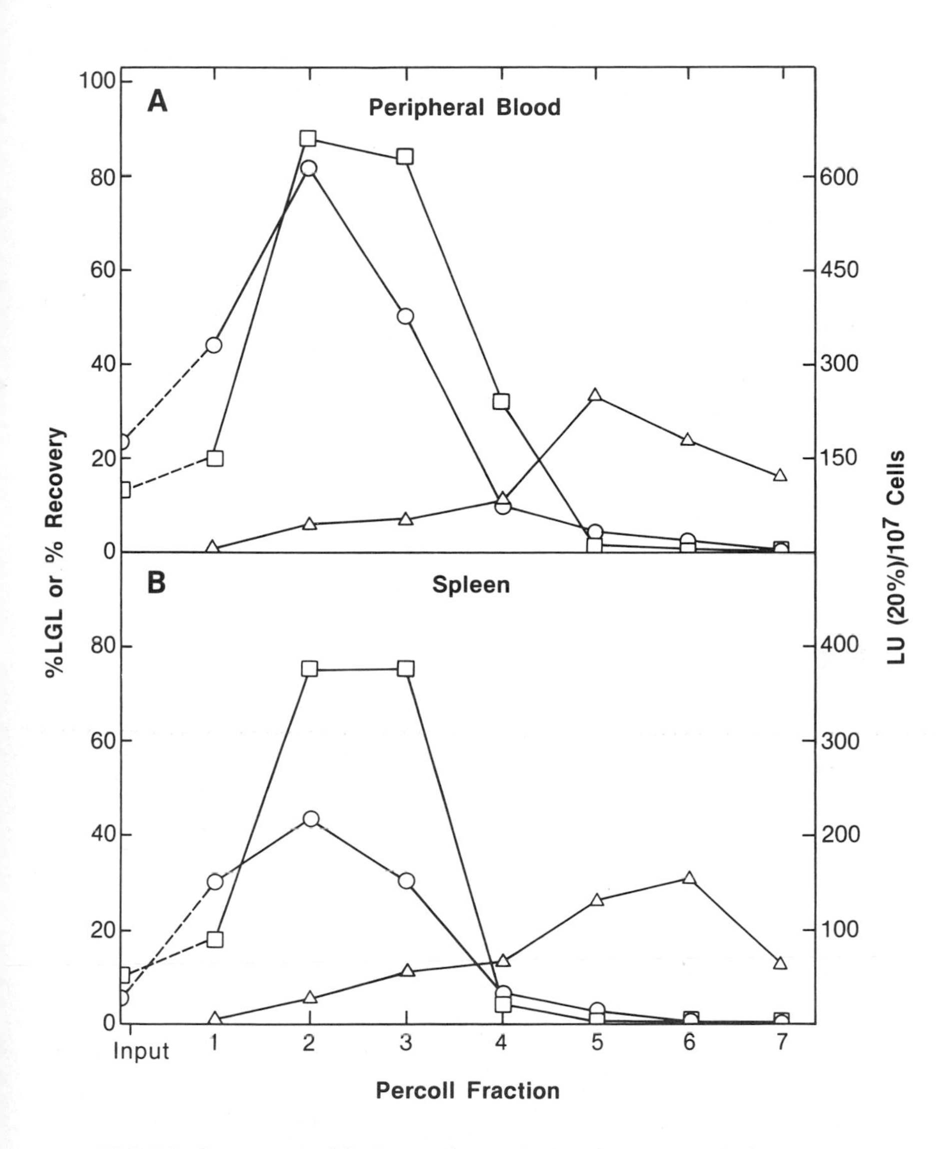

FIGURE 2. Percoll fractionation of rat peripheral blood (A) and spleen (B). O = lytic units/10^7 cells, △ = % of cells recovered, and □ = % of LGL.

These experiments demonstrated that most of the NK activity in both spleen and blood was contained in a minor population of cells which consisted primarily of LGL. These cells were highly enriched in fractions 2 and 3 with purities of 80-95% obtainable from the blood and 60-80% from the spleen. The enrichment of NK activity in the LGL fractions, the absence of NK in the LGL-depleted fractions, and the binding and lysis of NK susceptible target cells (data not shown) strongly indicate that rat LGL are the NK effector cell in this species.

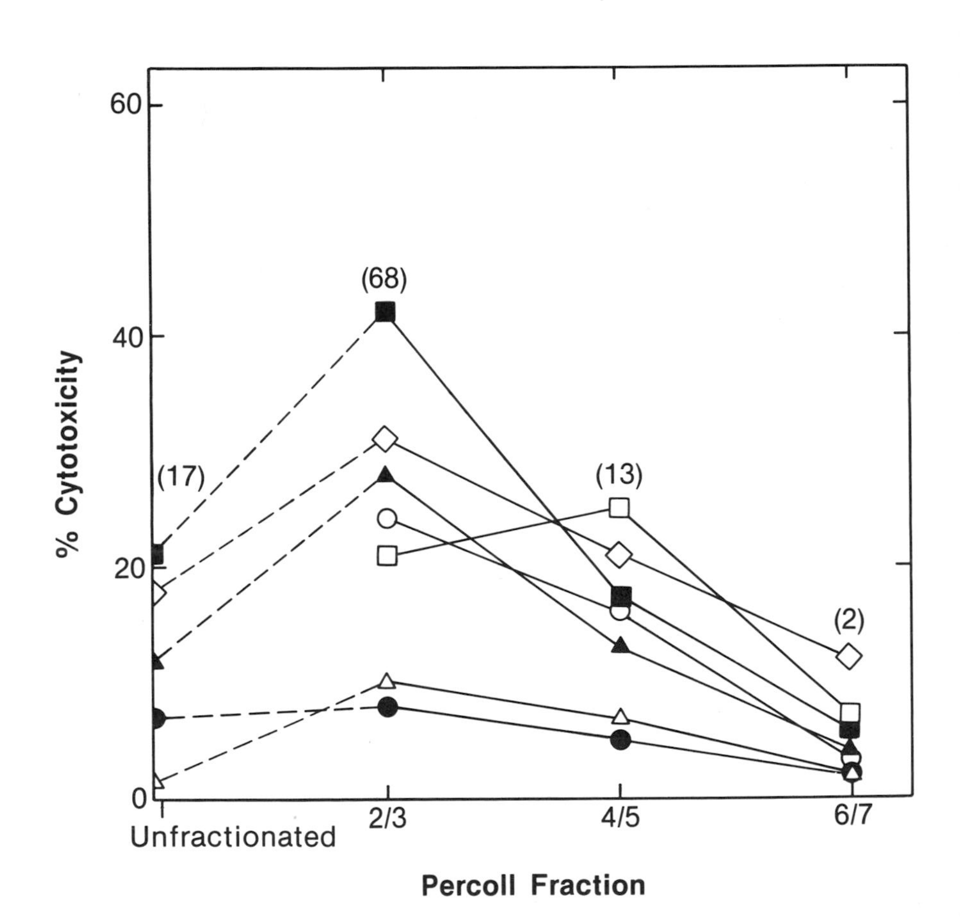

FIGURE 3. Cytotoxicity of Percoll fractions from PBL against: ● = R35, △ = ErTH/V-G, □ = W/Fu-T, ○ = W/Fu-P21, ▲ = G_1-TC, ■ = YAC-1 and ◇ = K562. E/T ratio = 50:1. ( ) = Average % LGL in each fraction.

Since the data in Figure 2 was obtained using a single lymphoma target (YAC-1), we were interested to know whether the cytotoxic cell for other targets was also an LGL. The results in Figure 3 demonstrated an enrichment in NK activity in fraction 2/3 with a corresponding decrease in both LGL and NK activity in later fractions. These results suggest that the natural cytotoxicity which is seen against syngeneic lymphomas (G_1-TC) syngeneic embryo cells (F2304), allogeneic sarcomas (ErTH/V-G, W/FU-T) and adenocarcinomas (R35), and xenogeneic leukemias (YAC-1, K562) is primarily mediated by LGL. These data imply that any heterogeneity in natural effectors for lymphoma targets and monolayer targets that may exist is within the small LGL subset.

REFERENCES

Herberman, R. B., Ortaldo, J. R., and Timonen, T. Methods in Enzymol (in press).
Nunn, M. E., et al., J. Natl. Cancer Inst. 56: 393 (1976).
Oehler, J. R., et al., Cell Immunol 29:238 (1977).
Oehler, J. R., et al., Int. J. Cancer 21:204 (1978).
Ortiz de Landazuri, M., and Herberman, R. B., J. Natl. Cancer Inst. 49:147 (1972).
Reynolds, C. W, Timonen, T., and Herberman, R. B., J. Immunol. 127:282 (1981a).
Reynolds, C. W., et al., J. Immunol. 127:2204 (1981b).
Reynolds, C. W., and Herberman, R. B., J. Immunol. 126:1581 (1981).
Saksela, E., et al., Immunol. Rev. 44:71 (1979).
Timonen, T., et al., Cell Immunol. 48:121 (1979a).
Timonen, T., et al., Cell Immunol. 48:133 (1979b).
Timonen, T., Saksela, E., J. Immunol. Methods 36:285 (1980).
Timonen, T., Ortaldo, J. R., and Herberman, R. B., J. Exp. Med. 153:569 (1981).
Zoller, M., and Matzku, S., J. Immunol. 124:1683 (1980).

LARGE GRANULAR LYMPHOCYTES AS EFFECTOR CELLS OF NATURAL KILLER ACTIVITY IN THE MOUSE

Aldo Tagliabue[1], Diana Boraschi[1], Saverio Alberti[2] and Walter Luini[2]
[1]Sclavo Research Center, Siena, Italy and [2]IRFMN, Milan, Italy.

I. INTRODUCTION

The possibility of identifying the effector cells of natural killer (NK) cytotoxicity by their morphology is a relatively new acquisition. In fact, it is only in the last two years that a few studies have shown that lymphocytes with a high cytoplasmic:nuclear ratio and characteristic azurophilic granules in the cytoplasm (large granular lymphocytes, LGL) are associated with the *in vitro* expression of NK activity in humans (1,2). We have recently investigated the possible relationship between LGL and mouse NK activity. Two main approaches were used: a) the distribution of LGL in different lymphoid and non-lymphoid organs was studied (3), with a focus on mouse peripheral blood in analogy with the human studies. b) We investigated the hypothesis that the large lymphocytes with cytoplasmatic granules, which are known to be present in high proportions in the epithelium of the small intestine of mammals, might exert NK activity at that site (4). Here we will summarize the results obtained, indicating that LGL are also effector cells of NK activity in mice, and we will report our most recent observations about murine LGL.

II. EXPERIMENTAL PROCEDURES

NK activity was assessed by a 6h ^{51}Cr release assay employing susceptible tumor target cells such as YAC-1 lymphoma (3,4) or non-susceptible tumor targets such as MN/MCA-1 fibrosarcoma (4). Lymphoid cell populations were obtained by procedures previously described in detail (3,4,5). LGL were stained after cytocentrifugation with Diff-Quik (Harleco, Gibbstown, NJ). The antisera employed were: anti-Thy 1.2 from New England Nuclear (Boston, Mass.), Lyt 1.1 and Lyt 2.1 from Cedarlane Laboratories (Hornby, Canada), anti-NK 1.2 from Dr. R.C. Burton, Harvard Medical School, Mass., anti-asialo GM 1 from Dr. K.

ISBN 0-12-341360-5

Okumura, Tokyo University. The complement dependent cytotoxic test (5) was performed employing low toxicity rabbit complement (Cedarlane Lab.).

III. RELATIONSHIP BETWEEN LGL AND NK ACTIVITY

Lymphoid cell populations from different organs were first examined to determine the frequency of LGL (Table I). LGL were always detectable in organs positive for NK activity (3), whereas no LGL were found where NK activity was absent. The positive correlation between % LGL and NK activity was further demonstrated in studies with lymphocytes from the peripheral blood of mice of different ages and strains. The highest % LGL were observed in C3H/HeN mice at 4-8 weeks, i.e. at the peak

TABLE I. Comparison between LGL and NK activity

Organ distribution[1]	% LGL	NK activity[4]
Peripheral blood	11	high
Intestinal mucosa	35	high
Lung	6	high
Spleen	3	high
Bone marrow	1	low
Peyer's patches	0	none
Peripheral lymphnodes	0	none
Thymus	0	none
Strain distribution[2]		
C3H/HeN	11	high
BALB/c	6	intermediate
BALB/c nu/nu	16	very high
Effect of age[3]		
4 weeks	13	high
8 weeks	8	high
12 weeks	2	intermediate
25 weeks	0-1	very low

[1] in C3H/HeN mice of 8 weeks of age
[2] tested employing peripheral blood lymphocytes
[3] tested employing peripheral blood lymphocytes from C3H/HeN mice
[4] tested in a 6h ^{51}Cr release assay against YAC-1 tumor cells.

of the NK activity. Moreover, nude BALB/c mice had higher % LGL and NK activity when compared to normal littermates. In addition, it was possible to obtain lymphoid populations enriched in LGL by separation of mononuclear cells from peripheral blood on discontinuous Percoll density gradients devised for mouse cells (3). In fact, in the fraction of the gradient containing 55% of Percoll (fraction 55), it was possible to obtain 40-85% of LGL when starting from an input population which contained 5-10% LGL. Table II shows a representative experiment which indicates how an increase in LGL in fraction 55 was parallel to an increase in NK activity. Interestingly, the fraction enriched in LGL was more cytotoxic against YAC-1 cells, but still failed to lyse fibrosarcoma cells, previously shown not to be susceptible to splenic NK activity (4).

TABLE II. Distribution of LGL and NK cytotoxicity among fractions of C3H/HeN peripheral blood obtained by discontinuous Percoll density gradient centrifugation

	%	% specific cytotoxicity against YAC-1		MN/MCA-1	
	LGL	70:1[a]	35:1	70:1	35:1
Peripheral blood:					
Input	8	30.1	18.7	1.3	-1.8
Percoll 55	38[b]	64.7[b]	54.0[b]	2.3	2.8
Percoll 60	6	19.2	13.3	0.3	3.2
Spleen	4	17.2	8.9	2.4	0.9

[a] Attacker to target ratio
[b] $P \lesssim 0.05$ versus input population.

IV. NK ACTIVITY OF MURINE GUT MUCOSAL LYMPHOID CELLS

Lymphocytes with a morphology similar to the LGL have been known for a long time to be present in the intestinal mucosa of mammals (6). Thus, we tested purified populations of lymphoid cells from the mouse intestine for their capacity to exert NK activity. Table III shows that intraepithelial lymphocytes from the gut (IEL) had NK activity comparable to the splenocytes and that interferon (IFN-β) boosted this activity.

TABLE III. NK activity of lymphocytes from different organs and its boosting by IFN-β

Organs	IFN-β	% specific cytotoxicity		
		50:1	25:1	12:1
Spleen	-	18.7	13.5	10.5
	+	33.9[a]	25.1[a]	20.6[a]
Intestinal mucosa	-	25.9	22.3	15.7
	+	46.7[a]	35.5[a]	22.6[a]
Peyer's patches	-	3.5	2.9	3.0
	+	0.1	4.3	3.9

[a] Increase above spontaneous level $P \leq 0.05$; for experimental details see ref. 3.

In contrast, no cytotoxicity was observed employing lymphocytes from the Peyer's patches with or without IFN-β. The characteristics of intestinal NK activity were similar to those of splenic NK cells (4). Intestinal NK activity was not affected by depletion of adherent cells. Moreover, the age dependency of NK activity was similar for IEL and splenocytes. Finally, NK-insensitive target cells were not lysed by IEL (4).

V. LGL AS THE NK EFFECTOR CELL IN THE INTESTINAL MUCOSA

In an attempt to investigate whether the LGL were the NK cells in the mouse small intestine, we studied their capacity to form conjugates (2) with YAC-1 cells. As shown in Table IV,

TABLE IV. NK activity and frequency of conjugation with YAC-1 cells of intraepithelial lymphocytes fractionated on Percoll gradients[a]

Source	% LGL	% specific cytotoxicity			% conjugates with LGL	% conjugates with small lymphocytes
		100:1	50:1	25:1		
Fraction 55	61	32.7	19.9	12.5	58	9
Fraction 70	2	1.4	0.6	0.5	0	6

[a] Cells were obtained from 8 week old CBA/J mice.

LGL were capable of preferentially forming conjugates with the tumor cells. We employed the Percoll gradients with the intestinal cells and in this case, too, it was possible to obtain populations enriched in LGL and NK activity in fraction 55, whereas fraction 70 contained very low percentages of LGL and lacked NK activity (Table IV).

VI. SEROLOGICAL CHARACTERIZATION OF THE INTESTINAL NK EFFECTOR CELL

In order to better characterize the effector cells of NK activity in the murine intestinal epithelium, IEL from fraction 55 were treated with a panel of antisera and complement prior to the cytolysis assays. Table V summarizes the results obtained. It appears that the intestinal NK cell is not identical to the splenic one. In fact, the gut NK activity was greatly reduced by anti-Thy 1.2, but only marginally affected by anti-asialo-GM 1 or anti-NK 1 (5).

TABLE V. Effect of treatment with various antisera plus complement on NK activity of splenocytes and IEL from fraction 55

	Percent reduction of cytotoxicity	
	Spleen	IEL 55[a]
Anti Thy 1.2	38[b]	75[c]
Anti Ly 1.1	11	+16
Anti Ly 2.1	17	5
Anti NK 1	83[b]	33
Anti asialo-GM 1	83[c]	31

[a] From 8 week old CBA/J mice
[b] Significant reduction from untreated or complement controls, $P \leq 0.05$
[c] Significant reduction from untreated or complement controls, $P \leq 0.001$.

VII. CONCLUSIONS AND SPECULATIONS

Taken all together, these results demonstrate that also in the mouse LGL are effector cells of NK activity. Several lines of evidence suggest that NK cells are a heterogeneous population. Thus it remains to be elucidated whether the LGL represents a subpopulation of NK cells. Interestingly, the serological characterization of the NK effector cells in the mouse small intestine, where the cytotoxicity is mainly exerted by LGL, revealed strong similarities between these cells and the NK_T subset recently described by Minato et al. (7). The possibility of purifying LGL with Percoll gradients also from other organs will allow a further analysis of the NK subsets.

Lymphocytes with cytoplasmatic granules have now been described in the epithelium of other organs besides the intestine, such as the mouse lung (3), the rabbit oviductal fimbria and endocervices (8), the male reproductive tract of rats and monkeys (9). Thus, it is tempting to speculate that LGL play a major role in immunosurveillance at the mucosal level.

ACKNOWLEDGEMENTS

We would like to acknowledge the collaboration of Prof. J. Bienenstock and Drs. A.D. Befus and D.A. Clark of the McMaster University, Hamilton, Ontario where part of this study was performed by A. Tagliabue, supported by a fellowship of the "Associazione Italiana per la Ricerca sul Cancro", Milan, Italy.

REFERENCES

1. Timonen, T., Saksela, E., Ranki, A., and Häyry, P., Cell. Immunol. 48: 133 (1979).
2. Timonen, T., Ortaldo, J.R., and Herberman, R.B., J. Exp. Med. 153:569 (1981).
3. Luini, W., Boraschi, D., Alberti, S., Aleotti, A., and Tagliabue, A., Immunol. 43:663 (1981).
4. Tagliabue, A., Luini, W., Soldateschi, D., and Boraschi, D., Eur. J. Immunol., 11:919 (1981).
5. Tagliabue, A., Befus, A.D., Clark, D.A., and Bienenstock, J., submitted for publication.
6. Rudzik, O., and Bienenstock, J., Lab. Invest. 30:260 (1974).
7. Minato, N., Reid, L., and Bloom, B.R., J. Exp. Med. 154: 750 (1981).
8. Odor, D.L., Fertil. and Steril. 25:1047 (1974).
9. Dym, M., and Romrell, L.J., J. Reprod. Fert. 42:1 (1975).

CHARACTERIZATION OF HUMAN NK CELLS IDENTIFIED BY THE MONOCLONAL HNK-1 (Leu-7) ANTIBODY[1]

Toru Abo, Max D. Cooper, Charles M. Balch

Cellular Immunobiology Unit of the Tumor Institute
Departments of Surgery, Microbiology, and Pediatrics,
University of Alabama in Birmingham

Natural killer (NK) cells and antibody-dependent killer (K) cells have been defined primarily by their functional properties (1,2). Partial purification of NK cells from lymphocyte populations has been possible because of their relatively lower density (3). While several types of surface markers have been identified on human NK cells, none have been precise markers for these cells because their distribution has not been restricted to NK cells (4,5).

We have recently produced a monoclonal antibody, HNK-1, that reacts with a differentiation antigen on human NK and K cells (5-8). We describe herein some characteristics of human NK cells identified by the HNK-1 (Leu-7) antibody, and suggest a possible differentiation scheme for NK cells.

I. PRODUCTION OF THE HNK-1 MONOCLONAL ANTIBODY

The monoclonal antibody, HNK-1, was produced against the cultured cell line, HSB-2, using the hybridoma technique introduced by Köhler and Milstein (5). Although HSB-2 cells have been considered to be of T cell origin, they lack T cell-associated antigens identified by the monoclonal antibodies, OKT1, T3, T4, T5 and T8 (5,9). In an earlier study, Kaplan *et*

[1] *Studies done in our laboratories have been supported by Grants CA 16673, CA 13148, CA 27197 and CA 03013, awarded by the National Cancer Institute.*

ISBN 0-12-341360-5

al. reported that a rabbit antiserum produced against HSB-2 reacted with human NK and K cells as well as with T cells (10).

When the HNK-1 antibody was tested against cultured lymphoid cell lines, it reacted with some T cell lines (HSB-2 and MOLT-4) but not with any other T cell line (MOLT-3), with any B cell lines or with cultured phagocytic cells (5). When human blood mononuclear cells were examined by immunofluorescence, the HNK-1 antibody reacted solely with a morphologically distinct population of granular lymphocytes with NK and K cell function.

II. NATURE OF THE HNK-1 ANTIGEN

The HNK-1 antigen was identified both on the surface membrane and in the cytoplasm of granular lymphocytes by both direct and indirect immunofluorescence (5,6).

Although the HNK-1 antigen was exclusively expressed on functional NK and K cells, it does not appear to be a functional receptor (5). Thus, neither NK nor K cell function was decreased when mononuclear cells were pretreated with the HNK-1 antibody or when the antibody was added to the culture medium for NK and K cell functional assays. The HNK-1 antigen expression on the cell surface was also resistant to treatment with 0.1% pronase, a concentration that inhibited NK and K cell functions.

III. CHARACTERISTICS OF HUMAN NK CELLS IDENTIFIED BY THE HNK-1 ANTIBODY

A. *Morphology and Cytotoxic Function of HNK-1$^+$ Cells*

The HNK-1 antibody reacted with a minority of peripheral blood lymphocytes (15 ± 7%) from normal adult donors (5). The HNK-1$^+$ cells, purified with a fluorescence-activated cell sorter, were a homogeneous population of medium-sized lymphocytes with abundant neutrophilic cytoplasm. This morphological appearance is similar to that previously described for NK cells isolated by direct adherence to target cells by Saksela *et al.* (11). On the other hand, the HNK-1$^-$ cells displayed the classical features of small lymphocytes with a narrow rim of cytoplasm and no granules.

The sorted HNK-1$^+$ and HNK-1$^-$ cells were also examined for NK and K cell function against K562 target cells and antibody-coated chicken erythrocytes. Almost all of the cytotoxcity resided in the HNK-1$^+$ cell population (5).

B. *Expression of Other Surface Antigens on HNK-1$^+$ Cells*

In spite of homogeneity of the morphology, cytotoxic function and most surface markers, HNK-1$^+$ cells are heterogeneous with respect to E-rosette receptor expression. HNK-1$^+$ cells distribute in both null cell ($ER^-sIg^-Fc\gamma R^+$) and the Tγ cell ($ER^+Fc\gamma R^+$) fractions (5).

The presence of T cell-associated antigens on HNK-1$^+$ cells was examined further using other monoclonal antibodies and two-color immunofluorescence (7). The HNK-1$^+$ subpopulation of cells showed variable expression of T cell antigens. The pan-T cell antigens, 3A-1, L17F12, T1 and T3, were co-expressed on 20 to 50% of HNK-1$^+$ cells. Ten to 20% of HNK-1$^+$ cells were reactive with T4, T5 and T8 antibodies.

In addition to the expression of T cell antigens by a minority of HNK-1$^+$ cells, a majority (60 to 70%) of HNK-1$^+$ cells expressed a myeloid antigen defined by the monoclonal M1 antibody. About 20% of HNK-1$^+$ cells also bore the HLA-DR determinant. While the monoclonal M1 antibody has previously been shown to react with NK and K cells (12,13), the M1 antigen expression is probably not restricted to cells of myeloid origin. Furthermore several macrophage/granulocyte antigens defined by monoclonal antibodies are not present on null cells and the Tγ cells (14,15), that are known to express the HNK-1$^+$ antigen.

C. *Tissue Distribution*

The proportion of HNK-1$^+$ cells in different lymphoid tissues was examined with the HNK-1 antibody (Table I). The spleen had the highest proportion of HNK-1$^+$ cells (10.4%), whereas few HNK-1$^+$ cells were demonstrated in the tonsil (3.7%), lymph nodes, bone marrow and thymus (<1.0%). In earlier studies, cell suspensions from these tissues exhibited a low level of NK and K cell function (1,2). The present study demonstrated that the low functional level in some lymphoid tissues was associated with a numerical deficiency of NK cells.

TABLE I Distributions of HNK-1^{+} T3^{+} T4^{+} and T8^{+} Cells in Human Lymphoid Tissues

Tissues	No tested	Percentages of positive cells			
		HNK-1^{+}	T3^{+}	T4^{+}	T8^{+}
Blood	6	15.0	55.4	32.3	19.7
Spleen	6	10.4	31.9	18.2	20.8
Tonsil	2	3.7	44.0	34.6	5.4
Lymph node	6	0.4	75.7	66.1	7.5
Bone marrow	6	0.7	4.1	0.9	1 2
Thymus	6	0.3	43.4	>90 0	77.5

D. Postnatal Expansion of NK Cell Population

When HNK-1^{+} cells were enumerated among blood mononuclear cells from healthy individuals, the proportion was extremely variable among individuals. We found that this variation was related to age and sex (6). Thus, the frequency of HNK-1^{+} cells was very low in the newborn (<1.0%) and increased progressively through childhood and adult life. The postnatal expansion of the human NK cell population was quite similar to that observed previously in rats, mice and piglets (1,2,16,17). It has also been observed that NK cell activity declines in older animals (1,2,17). To the contrary, we did not notice any age-related decline of HNK-1^{+} cells in humans.

We also noted a correlation of HNK-1^{+} cell levels with the sex of the donors (14). Males had a slightly higher proportion of HNK-1^{+} cells than females (16.9 vs. 13.1%, $P<0.05$). This result is in agreement with the findings of other investigators using functional assays (12,18,19). It is possible that hormones, including estrogen (20), may influence the development of NK cells.

E. Abnormality of NK Cells

It has been demonstrated that the patients with Chediak-Higashi (CH) syndrome exhibit depressed NK cell function (21). We have used the HNK-1 antibody to determine whether the depressed NK cell function is due to impaired function of NK cells or low numbers of this cell population. We found an abundance of HNK-1^+ cells in all CH patients tested (8 patients) and that HNK-1^+ cells from these patients had a unique morphological appearance (8). The HNK-1^+ cells in CH patients contained a single giant granule in the cytoplasm rather than multiple azurophilic granules seen in normal NK cells. The morphologically abnormal HNK-1^+ cells, isolated with a fluorescence activated cell sorter, lacked both NK and K cell function.

IV. DISTINCTION OF NK CELLS FROM ITS ANALOGUES

The expression of receptors for sheep erythrocytes and of T cell-associated antigens by some HNK-1^+ cells might suggest a T cell lineage. However, HNK-1^+ cells exhibited little or no response to mitogens (PHA, Con A and PWM) or alloantigens (7). Therefore, HNK-1^+ cells lack an important function ascribed to T cells.

The HNK-1^- cell fraction, in contrast, responded efficiently to both mitogens and alloantigens (7). Although fresh HNK-1^- cells lack cytotoxic properties, the stimulated HNK-1^- cells acquire a potent cytotoxic function. However, these cells never express the HNK-1 antigens even after incubation for up to 30 days. Recently, several types of effector cells have been described that require prior activation with alloantigens and cultured cell lines before they can spontaneously kill certain target cells (22). These analogues of NK cells in humans can be identified as activated HNK-1^- cells (7). Thus, activated cytolytic lymphocytes can be distinguished from the classically defined NK cells according to their expression of the HNK-1 antigen.

V. DIFFERENTIATION PATHWAY OF HUMAN NK CELLS

We have examined possible differentiation pathways of NK cells by correlating functional capability with surface antigen expression. HNK-1^+ cells were stained by two-color immunofluorescence, fractioned by cell sorter into HNK^+T3^- and HNK^+T3^+ cells, and then analyzed for their morphology and function (23). HNK^+T3^- cells contained multiple cytoplasmic granules and exhibited a high level of NK activity against K562 target cells. On the other hand, HNK^+T3^+ cells had low NK cell functional activity and a paucity of cytoplasmic granules. When these cell fractions were further analyzed, a reciprocal relationship between the T3 antigen and the M1 myeloid antigen expression was identified.

These results suggest that the HNK^+T3^+ cell may be a less mature NK cell. This hypothesis is strengthened by the observation that the majority (70%) of the $HNK\text{-}1^+$ cell population in the peripheral blood were $HNK^+T3^-M1^+$, whereas almost 95% of $HNK\text{-}1^+$ cell population in the bone marrow had the $HNK^+T3^+M1^-$ phenotype and a low level of NK activity. $HNK^+T3^+M1^-$ cells and $HNK^+T3^-M1^+$ cells therefore may represent different stages of NK cell maturation.

A survey of $HNK\text{-}1^+$ cell phenotypes was then performed in fetuses 13 to 15 weeks of gestational age. In fetal lymphohemopoietic tissues, only 0.01 to 0.05% of nucleated cells were $HNK\text{-}1^+$. However, a different phenotype was demonstrated that may represent a very immature form of NK cells. Sixty to 90% of $HNK\text{-}1^+$ cells in the liver, bone marrow and spleen derived from the fetuses did not express either T3 or M1 antigens. The remaining 10-40% of $HNK\text{-}1^+$ cells exhibited the same phenotype as that seen in adult bone marrow ($HNK^+T3^+M1^-$), while none of the cells were $HNK^+T3^-M1^+$.

Ortaldo *et al.*,(24) demonstrated that the T10 antigen was expressed on the majority of blood lymphocytes contained in a low density fraction of a Percoll density gradient (enriched for granular lymphocytes). Although these results suggested that the T10 antigen was expressed on NK cells, we observed no significant overlapping of $HNK\text{-}1^+$ cells with $T10^+$ cells by two-color immunofluorescence, either in the adult peripheral blood and fetal lymphoid tissues.

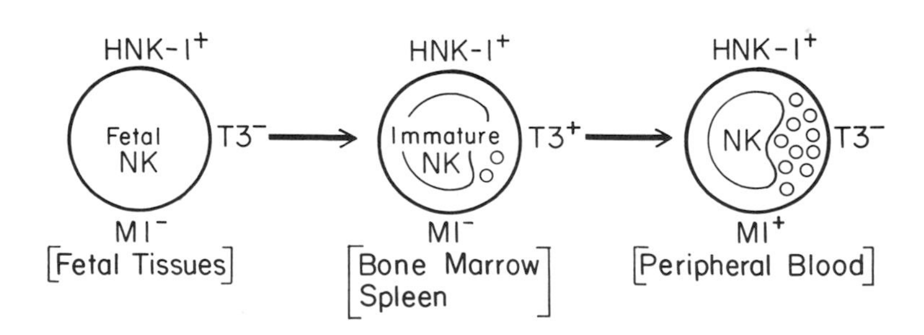

FIGURE 1. A possible differentiation pathway of human NK cells.

The results of our studies of adult and fetal lymphoid compartments suggest that the three distinct phenotypes may represent different stages in differentiation of HNK-1^+ cells. We propose a hypothetical model of NK cell differentiation in Figure 1.

The cell lineage of human NK and K cells is controversial. Many previous studies have suggested a relationship between NK cells and T cells (1,2). On the other hand, some myeloid antigens have been demontrated on NK cells (17,25), in addition to evidence for spontaneous cytotoxic function of promonocytes (26). Our results are more consistent with the idea that NK cells may differentiate along a distinct differentiation pathway separate from those followed by T cells, B cells and macrophages.

VI. SUMMARY

A differentiation antigen identified by the monoclonal HNK-1 antibody is exclusively expressed on human NK and K cells. HNK-1^+ cells are a relatively homogeneous population of granular lymphocytes that exhibit a unique postnatal expansion. The HNK-1 antibody also distinguishes the classically-defined NK cell from its analogues. Three distinct phenotypes of HNK-1^+ cells can be identified ($HNK^+T3^-M1^-$, $HNK^+T3^+M1^-$ and $HNK^+T3^-M1^+$). Only the latter subset of HNK-1^+ cells exhibits very efficient NK cell function, and predominate in adult blood. We propose a model in which these three cell phenotypes represent stages of a separate pathway of differentiation followed by the HNK cell.

REFERENCES

1. Herberman, R. B., and Holden, H. T., *Adv. Cancer Res. 27*, 305 (1978).
2. Roder, J. C., Kane, K., and Kiessling, R., *Prog. Allergy 28*, 66 (1981).
3. Timonen, T., and Saksela, E., *J. Immunol. Methods 36*, 285 (1980).
4. Cantor, H., Kasai, M., Shen, F., W., Leclerc, J. C., and Glimcher, L., *Immunol. Rev. 44*, 3 (1979).
5. Abo, T., and Balch, C. M., *J. Immunol. 127*, 1024 (1981).
6. Abo, T., Cooper, M. D., and Balch, C. M., *J. Exp. Med. 155*, 321 (1982).

7. Abo, T., Cooper, M. D., and Balch, C. M., *J. Immunol.* (Submitted).
8. Abo, T., Roder, J. C., Abo, W., Cooper, M. D., and Balch, C. M., *J. Clin. Invest.* (Submitted).
9. Reinherz, E. L., Kung, P. C., Goldstein, G., Levey, R. H., and Schlossman, S. F., *Proc. Natl. Acad. Sci. USA 77*, 1588 (1980).
10. Kaplan, J., Callewaert, D. M., and Peterson, W. D. Jr., *J. Immunol. 121*, 1366 (1978).
11. Saksela, E., Timonen, T., Ranki, A., and Hayry, P., *Immunol. Rev. 44*, 71 (1979).
12. Breard, J., Reinherz, E. L., Kung, P. C., Goldstein, G., and Schlossman, S. F., *J. Immunol. 124*, 1943 (1980).
13. Kay, H. D., and Horwitz, D. A., *J. Clin. Invest. 66*, 847 (1980).
14. Rosenberg, S. A., Ligler, F. S., Ugolini, V., and Lipsky, P. E., *J. Immunol. 126*, 1473 (1981).
15. Todd, R. F. III, Nadler, L. M., and Schlossman, S. F., *J. Immunol. 126*, 1435 (1981).
16. Kiessling, R., Klein, E., Pross, H. and Wigzell, H., *Eur. J. Immunol. 5*, 117 (1975).
17. Kim, Y. B., Huh, N. D., Koren, H. S., and Amos, D. B., *J. Immunol. 125*, 755 (1980).
18. Santoli, D., Trinchieri, G., Zmijemski, C. M., and Koprowski, H., *J. Immunol. 117*, 765 (1976).
19. Liburd, E. M., Pazderka, V., Russell, A. S., and Dossector, J. B., *Immunology 31*, 767 (1976).
20. Seaman, W. E., Merigan, T. C., and Talal, N., *J. Immunol. 123*, 2903 (1979).
21. Roder, J. C., Haliotis, T., Klein, M., Korec, S., Jet, J. R., Ortaldo, J., Herberman, R. B., Katz, P., and Fauci, A. S., *Nature 284*, 553 (1980).
22. Seeley, J. K., and Golub, S. H., *J. Immunol. 120*, 1415 (1978).
23. Abo, T., Gartland, G. L., Cooper, M. D., and Balch, C. M., *Fed. Proc.* (In press).
24. Ortaldo, J. R., Sharrow, S. D., Timonen, T., and Herberman, R. B., *J. Immunol. 127*, 2401 (1981).
25. Ault, K. A., and Springer, T. A., *J. Immunol. 126*, 359 (1981).
26. Lohmann-Matthes, M. L., Domzig, W., and Taskov, H., *J. Immunol. 123*, 1883 (1979).

PHENOTYPIC CHARACTERIZATION OF HUMAN NATURAL KILLER AND ANTIBODY-DEPENDENT KILLER CELLS AS AN HOMOGENEOUS AND DISCRETE CELL SUBSET

Bice Perussia[1]
Virginia Fanning
Giorgio Trinchieri

The Wistar Institute of Anatomy and Biology
Philadelphia, Pennsylvania

Natural Killer (NK) cells have been identified until now along with "null" lymphocytes (non-B, non-T) characterized by the presence of surface receptors for the Fc fragment of IgG (IgG-FcR) and, on a proportion of them, of receptors for sheep erythrocytes (1). The NK cell can be morphologically identified as a large granular lymphocyte (LGL) (2). Antibody-dependent killer (K) cells bear the same surface receptors and have the same morphology as NK cells (3). We used a panel of monoclonal antibodies (Table 1) to study the antigenic characteristics of NK and K cells. The reactivity of monoclonal antibodies with NK and K cells was analyzed using three methods.

Complement-Dependent Lysis. The number of lytic units (LU) per 10^7 original cell number was evaluated in effector cell populations, treated with antibody and selected non-toxic rabbit complement.

Indirect Rosetting. The effector cells were incubated sequentially with monoclonal antibodies and sheep erythrocytes coupled by the $CrCl_3$ method to affinity-purified goat $F(ab')_2$ anti-mouse immunoglobulins. Rosetting cells were

[1] The experimental work described in this paper was supported by NIH grants CA 10815 and CA 20833.

ISBN 0-12-341360-5

TABLE I. Monoclonal Antibodies Used for the Characterization of NK and K Cells

Antibody	Class	Antigen m.w. x 10^{-3}	Ref.	Reported specificity
OKT11A	IgG2a	46	6	T cells (SRBC receptor)
B67.1	IgG2a	46	N.R.[a]	as OKT11A
OKT3	IgG2a	20	7	T cells
L17F12	IgG2a	67	8	T cells
B36.1	IgG2b	67	5	as L17F12
OKT4	IgG2a	62	9	helper/inducer T cells
B53.4	IgM	N.R.	N.R.	as OKT4
OKT5	IgG1	33	10	cytotoxic/suppressor T
OKT8	IgG2a	33	11	cytotoxic/suppressor T
B33.1	IgG2a	28 and 32	4	HLA-DR
OKM1	IgG2a	N.R.	17	granulo-monocytes, LGL
B43.4	IgG2a	N.R.	5	as OKM1
B13.4	IgM	N.R.	4	granulo-monocytes
B34.3	IgG1	N.R.	5	granulo-monocytes
B44.1	IgM	N.R.	5	monocytes
B52.1	IgM	N.R.	N.R.	monocytes
B37.2	IgM	N.R.	5	granulocytes
B40.8	IgM	N.R.	5	as B37.2

[a]Not reported.

scored and separated from non-rosetting cells on a Ficoll-Hypaque density gradient. LU for both populations were computed.

Complement-Dependent Lysis After Single-Cell Cytotoxic Assay in Semi-Solid Medium. The single-cell assay in agarose was performed on microscope slides which were incubated for 3 hr at 37°C to allow the NK cells in the effector-target cell conjugates to lyse their target cell. A few drops of antibody and complement were then added and the slides were incubated for 60 min. The target cells lysed by the NK cells and the effector cells lysed by antibody plus complement were then identified by trypan blue exclusion.

The first two methods allow a very accurate determination of the cytotoxic activity in the purified populations. The last method allows the direct count of individual NK cells that are positive or negative for the antigen recognized by the monoclonal antibody. The results obtained by these three methods are, in most cases, superimposable.

I. REACTIVITY OF UNSTIMULATED NK AND K CELLS WITH ANTIBODIES SPECIFIC FOR T-CELLS OR T-CELL SUBSETS

About 80% of the cytotoxic activity in both NK and K cell populations is contained in the cells that react with monoclonal antibody OKT11A (6) or B67.1. The proportion of NK cells able to form spontaneous rosettes with sheep erythrocytes, however, had been reported previously as 50%. Serological data show, however, that OKT11A reacts with a slightly larger cell population than the one forming spontaneous rosettes. OKT11A-negative populations not only showed very low NK and K cytotoxic activities, but also a very decreased ability to form conjugates with K562 target cells. Single cell analysis of the cells lysing K562 target cells demonstrated that more than 80% of these cytotoxic cells were OKT11A-positive. When purified subpopulations of cells forming spontaneous rosettes with sheep erythrocytes (E-RFC) were treated with OKT11A and complement, NK cytotoxicity was completely abolished, whereas antibody-treated non-E-RFC cells retained all or most of their cytotoxic activity.

In contrast to OKT11A, six antibodies reacting with T cells [OKT3 (7), B36.1, L17F12 (8), B53.4 and OKT4 (9), OKT5 (10)] failed to detect NK and K cells. Unlike OKT5, the OKT8 (11) antibody, which is considered to recognize the same cytotoxic/suppressor T cell subset and precipitate an antigen of identical MW as OKT5 (12), reacts with about 40% of the NK cells. Antibody OKT8 recognizes a proportion of both E-RFC and non-E-RFC cytotoxic cells.

By depleting E-RFC of the B36.1-positive cells, a population of cells almost completely composed of cells with LGL morphology, with NK and K activity, and bearing IgG-FcR, is obtained. About 50% of these cells express the antigen recognized by the OKT8 antibody, although at a lower level than the OKT8-positive cells included in the E-RFC, IgG-FcR(-) population. The same cells, which correspond to the so-called T.G (13) cells [E-RFC, IgG-FcR(+)], react with OKT11A or with antibodies of analogous specificity with an intensity similar to or higher than that of B36.1- or OKT3-positive T cells. This contrasts with a previous report describing the low avidity of the receptor for sheep erythrocytes on NK or on T.G cells (14). It is, however, possible that the inability to form rosettes at 29°C, a criterion of low avidity receptors, actually reflects differences in the membrane properties of those cells and not differences in receptor density or antigenic characteristics.

Ortaldo _et al_. (15) found that 30% of the LGL were OKT8-positive, but they, as well as Zarling _et al_. (16) did not

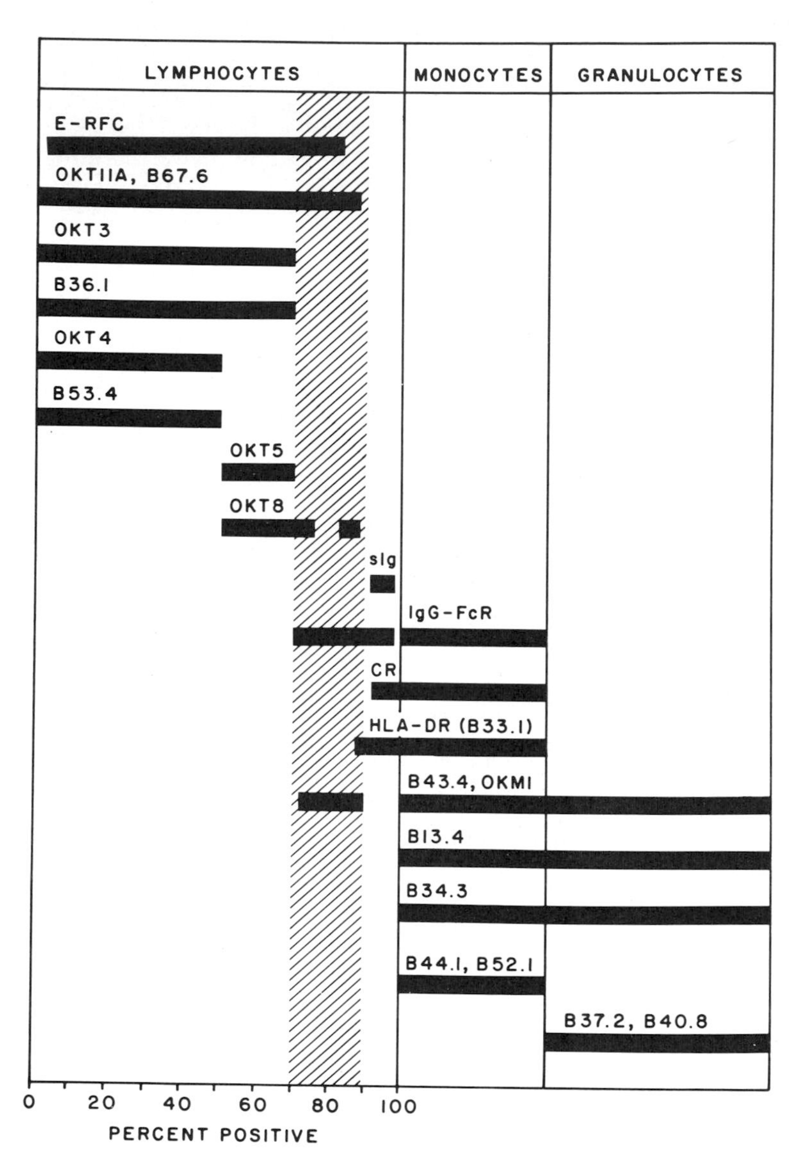

FIGURE 1. Distribution of the classical surface makers and of the antigens recognized by monoclonal antibodies on human peripheral blood nucleated cells. Each horizontal line identifies one surface marker. The shaded area indicates the cell types in which NK and K activity can be demonstrated. It is, however, possible that not all the cells with the phenotype included in the shaded area have cytotoxic activity (sIg: surface immunoglobulins; CR: complement receptors).

detect the OKT8 antigen on NK cells. However, Zarling *et al.* (16) eliminated the dead cells after complement-dependent lysis, and tested only the depleted live cells at the same cell concentrations as the control preparations. In this case, if an antibody reduces both cell number and cytotoxic activity by 30%, as OKT8 does, no difference will be seen in the total population because the negative cells are still equally active on a per cell basis. In our experimental protocols, the reduction in activity is determined independently of the number of cells depleted.

II. REACTIVITY OF UNSTIMULATED NK AND K CELLS WITH ANTI-MYELOMONOCYTIC ANTIBODIES

Our series of myelomonocytic antibodies reacts with seven different antigenic determinants that are all expressed on mature cells (granulocytes, monocytes or both), but that appear at different stages during differentiation (4, 5). NK cells bear only 1 of the seven antigens expressed on myelomonocytic cells. Antibodies B43.4 and OKM1 (17), which have similar specificity, recognized about 80% of the NK and K cells. It has, however, to be noted that the antigen recognized by the two antibodies is expressed on a larger proportion of lymphocytes than previously reported (17) (up to 60% of the lymphocytes from some donors are positive, without a proportionate decrease in other lymphocyte subsets), and NK and K cells are within the population of strongly reactive cells. These results argue against a possible myeloid origin of the NK cells. If, in fact, NK cells belong to the myelomonocytic lineage, they probably derive from an earlier precursor than the common monocyte-granulocyte precursor cell.

HLA-DR antigens, which are present on monocytes and B lymphocytes, are also expressed on about 10% of NK and K cells, in agreement with a report by other authors (18).

Preliminary data show that depletion of both OKT11A- and B43.4-positive cells from non-adherent mononuclear cells leaves a population of cells that is morphologically promonocytic. These cells lack NK activity even after treatment with IFN. This finding represents additional evidence against the hypothesis that cytotoxic cells are immature myeloid cells.

III. SURFACE MARKERS OF CYTOTOXIC CELLS WITH NK ACTIVITY ACTIVATED BY IFN OR ANTIGENIC STIMULATION

We have used monoclonal antibodies to determine: a) the ability of IFN to enhance the cytotoxic activity of subpopulations of lymphocytes separated on the basis of their surface antigens and b) the phenotypic characterization of the cytotoxic cells activated by IFN as already described for the unstimulated lymphocytes. The cytotoxic cells affected by IFN displayed surface markers very similar to those of the spontaneously occurring NK cells present in peripheral blood. Only the proportion of HLA-DR-positive cells was slightly increased in the NK cells stimulated by IFN, and the activity of the HLA-DR-positive NK cells was enhanced by IFN slightly more efficiently than that of HLA-DR-negative cells. Thus, IFN acts on cells of the same lymphocyte subpopulation that contains the spontaneous NK cells and does not induce any significant alteration in the surface markers identified by the monoclonal antibodies used in this study.

Allogeneic stimulation of B36.1-positive T cells (completely devoid of NK cell activity, even after IFN treatment) in mixed lymphocyte culture results in a completely different effector population that displays NK-like cytotoxic activity, and bears the B36.1 antigen but not the B43.4 antigen.

IV. CONCLUSIONS

Eighty to ninety percent of the cytotoxic effector cells express the antigen recognized by either OKT11A or OKM1/B43.4, and the majority of them express both antigens. The lack of one of the two antigens in some cytotoxic cells is more likely due to a quantitative heterogeneity in the distribution of the markers in a single subpopulation, than to the existence of NK cell subsets. There is virtually no NK activity in the OKT11A- and OKM1-negative lymphocytes. Thus, NK and K cells appear to be a relatively homogeneous cell type, with surface markers different from B or T lymphocytes, granulocytes or monocytes. Our data show that NK cells are not mature or immature myelomonocytic cells, but do not exclude the assignment of NK cells to a distinct lymphocyte subset related to T cells (bearing OKT11A and OKT8 antigen, both present on relatively immature cells of the T-lineage). The phenotypic characterization of NK cells by the use of monoclonal antibodies, even those possibly specific for differentiated NK cells, as one recently described

(19) will probably not define the NK cell lineage unambiguously. The question of whether NK cells and large granular lymphocytes are related to one of the classical hematopoietic lineages or represent a new separate lineage will be resolved only upon future analysis of NK cell precursors.

REFERENCES

1. Kay, H. D., Bonnard, G. D., West, W. H., and Herberman, R. B., J. Immunol. 118:2058 (1977).
2. Timonen, T., Ortaldo, J. R., and Herberman, R. B., J. Exp. Med. 153:569 (1981).
3. Perussia, B., Trinchieri, G., and Cerottini, J.-C., Transplant. Proc. 11:793 (1979).
4. Perussia, B., Lebman, D., Ip, S. H., Rovera, G., and Trinchieri, G., Blood 58:836 (1981).
5. Perussia, B., Trinchieri, G., Lebman, D., Janckiewicz, J., Lange, B., and Rovera, G., Blood, in press (1982).
6. Kung, P. C., and Goldstein, G., Vox Sang. 39:121 (1981).
7. Kung, P. C., Goldstein, G., Reinherz, E. L., and Schlossman, S. F., Science 206:347 (1979).
8. Wang, C. Y., Good, R. A., Ammirati, P., Dymbort, G., and Evans, R. L., J. Exp. Med. 151:1539 (1980).
9. Reinherz, E. L., Kung, P. C., Goldstein, G., and Schlossman, S. F., J. Immunol. 123:2894 (1979).
10. Terhorst, C., Van Agthoven, A., Reinherz, E., and Schlossman, S. F., Science 209:520 (1980).
11. Reinherz, E. L., Kung, P. C., Goldstein, G., and Schlossman, S. E., J. Immunol. 124:1301 (1979).
12. Reinherz, E. L., Hussey, R. E., Fitzgerald, K., Snow, P., Terhorst, C., and Schlossman, S. F., Nature 294:168 (1981).
13. Moretta, L., Webb, S. R., Grossi, C. E., Lydyard, P. M., and Cooper, H. D., J. Exp. Med. 146:184 (1977).
14. West, W. H., Cannon, G. B., Kay, H. D., Bonnard, G. D., and Herberman, R. B., J. Immunol. 118:355 (1977).
15. Ortaldo, J. R., Sharrow, S. O., Timonen, T., and Herberman, R. B., J. Immunol. 127:2401 (1981).
16. Zarling, J. M., Clouse, K. A., Biddison, W. E., and Kung, P. C., J. Immunol. 127:2575 (1981).
17. Breard, J., Reinherz, E. L., Kung, P. C., Goldstein, G., and Schlossman, S. F., J. Immunol. 124:1943 (1980).
18. Ng, A.-K., Indiveri, F., Pellegrino, M. A., Molinaro, G. A., Quaranta, V., and Ferrone, S., J. Immunol. 124:2336 (1980).
19. Abo, T., and Balch, C. M., J. Immunol. 127:1024 (1981).

ANALYSIS OF HUMAN NK CELLS BY MONOCLONAL ANTIBODIES AGAINST MYELOMONOCYTIC AND LYMPHOCYTIC ANTIGENS

Helmut Rumpold
Gerhard Obexer
Dietrich Kraft

Institute of General and Experimental Pathology,
University of Vienna, Austria

I. INTRODUCTION

NK cells are considered to play an important role in the body's defence mechanisms against cancer and viral infections and therefore tremendous efforts have been undertaken to characterize this cell during the last decade. Despite the fact that NK cell function could be attributed to a cell with a typical morphology, namely the large granular lymphocyte (LGL) (Timonen et al., 1979), there is still a controversy about the origin of this cell regarding T-, B- or myelomonocytic lineage. In particular, the fact that bone marrow-derived promonocytes exhibit NK cell activity against a variety of target cells similar to that observed using peripheral blood lymphocytes has led to the speculation that NK cells are of myelomonocytic origin (Lohmann-Matthes et al., 1979). To further approach this question highly purified human LGL (Timonen et al., 1981) were tested for reactivity with anti-human myelomonocytic monoclonal antibodies. In

---

1 supported in part by the Austrian Science Research Fund, Project No. 4568.

ISBN 0-12-341360-5

addition the expression of T cell associated antigens on human NK cells was studied using commercially available monoclonal anti-T cells antibodies.

Since most NK cell studies have so far been done *in vitro*, another important question is their *in vivo* significance. To elucidate this problem, monoclonal antibodies against NK cells would be helpful, especially to answer the question whether NK cells really do infiltrate tumors and metastases, since this is not easily determined by conventional histopathology. Mice were therefore immunized with highly enriched LGL and hybridomas established. Two monoclonal antibodies have so far been obtained which show reactivity predominantly with human NK cells.

II. INVESTIGATIONS OF LGL USING ANTI-MYELOMONOCYTIC AND ANTI-LYMPHOCYTIC MONOCLONAL ANTIBODIES

HL-60 is a promyelocyte cell line with ~30% myeloblasts and known to be capable to undergoing differentiation either to granulocytes or monocytes under appropriate stimuli *in vitro*. For this reason it was considered to be useful for induction of monoclonal antibodies against myelomonocytic differention antigens. VEP8 and VEP9 monoclonal antibodies were derived from a fusion experiment using HL-60 cells for immunization of mice.

The reactivity of these antibodies with various normal human cells is given in Table I. While VEP8 antibody reacts only with granulocytes and 70-80% bone marrow cells, VEP9 antibody reacts also with monocytes. Therefore we concluded that VEP9 antibody is a pan-myelomonocytic reagent reacting with mature granulocytes, monocytes and at least those bone marrow precursors which are in the differentiation stage of promyelocytes. When these antibodies were tested for reactivity with human LGL by means of an indirect membrane immunofluorescence test both of them showed no reactivity at all. When LGL were tested for OKM1 antigen expression a high percentage of cells was found to be positive, an observation which was also reported by other investigators (Kay and Horwitz, 1980; Zarling and Kung, 1980; Breard et al., 1981). However, this finding

TABLE I. Reactivity of VEP antibodies with normal human cells*

	VEP8	VEP9	VEP10	VEP13
PBL	-	-	8±2	24±8
T-cells	-	-	9±6	15±5
B cells	-	-	-	-
Thymocytes	-	-	>95	-
Bone marrow cells	77±4	69±5	40±7	6±4
Granulocytes	>95	>95	-	>95
Monocytes	-	>95	-	-
Thrombocytes	-	-	-	-
Erythrocytes	-	-	-	-

* Results are given as percentage of positive cells stained by means of indirect membrane immunofluorescence using FITC-labelled rabbit anti-mouse Ig $F(ab')_2$ (mean values from 3 or more experiments).

should not be interpreted as an argument for a myelomonocytic origin of human NK cells, since it was shown that OKM1 antibody, in addition to reacting with granulocytes and monocytes, also reacts with some T cells (Cauppens et al., 1981).

When various anti-T lymphocyte monoclonal antibodies were tested for reactivity with LGL a considerable percentage of positive cells was obtained with anti-human Lyt 3 antibody. This antibody is directed against the E receptor of T cells and therefore our finding supports previous results that human NK cells bear low avidity E receptors (West et al., 1977).

While no expression of T3, Leu 1, T4, Leu 3a, and T6 antigens was found on LGL, a considerable reactivity with OKT8 and Leu 2a antibodies was observed. Both antibodies are directed against a determinant known to be expressed on cytotoxic/ suppressor T cells.

TABLE II. Antigen expression of LGL enriched cell preparations* as defined by monoclonal antibodies against:

T3	13±11	Leu1	13±6	VEP8	<1
T4	<1	Leu2a	59±8	VEP9	<1
T6	<1	Leu3a	4±5	VEP10	55±9
T8	35±8	huLyt3	55±4	VEP13	75±11
M1	81±11				

*LGL were enriched by means of Percoll density gradient centrifugation (Timonen et al., 1981), yielding 76±15% as determined in cytocentrifuge preparations stained according to Pappenheim and Wright-Giemsa. Mean values from 3 or more experiments are given as percentage of cells stained by indirect membrane immunofluorescence using FITC-labelled rabbit anti-mouse Ig $F(ab')_2$.

III. MONOCLONAL ANTIBODIES OBTAINED FROM MICE IMMUNIZED WITH LGL

In an attempt to produce monoclonal antibodies against LGL and to get more insight into the antigens expressed on NK cells, mice were immunized with highly purified LGL and hybridomas were established. From these experiments two monoclonal antibodies have been obtained so far: VEP10 and VEP13. Both of them reacted with LGL but also with various other cells (Table I). VEP10 antibody detects an antigen which, in addition to expression on LGL, is also expressed on thymocytes, bone marrow cells, some T cell lines (Molt 4, Yurkat), B cell lines (Radji, Daudi) and a null cell line (Reh-6), but not on myelomonocytic cell lines (HL 60, U937, K562). VEP13 antibody showed strong reactivity with mature granulocytes in addition to its reactivity with LGL. When peripheral blood lymphocytes were treated with these antibodies plus complement, a strong reduction of NK activity against a variety of target cells was observed (Fig.1) thus indicating that VEP10 and VEP13 antigens are indeed expressed on functional NK cells. To approach the question whether VEP10 and VEP13 antigens are also expressed on human K cells, peripheral blood lym-

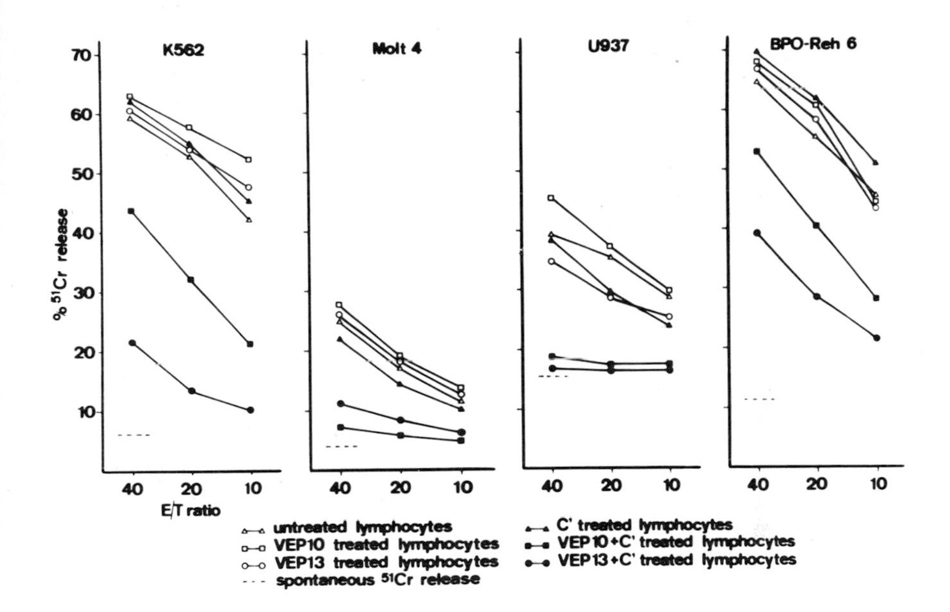

FIGURE 1. Depletion of NK and K cell activity by monoclonal antibody and C' treatment of peripheral blood lymphocytes.

phocytes were treated with VEP10 or VEP13 antibody and complement and tested for K cell activity, using hapten-coated Reh-6 cells and an affinity chromatography purified anti-hapten IgG antibody. In previous experiments this cell line was shown to be resistant to NK cell lysis in a 4 hr ^{51}Cr release assay and therefore only K cell activity is measured in this system. Again both antibodies were able to strongly reduce K-cell activity, thus supporting the concept that NK and K cell acitivty are mediated by the same cell type.

IV. CONCLUSIONS

Monoclonal antibodies specific for myelomonocytic antigens showed no reactivity with LGL. When monoclonal antibodies against T cell associated antigens were tested for reactivity with LGL, only those against cytotoxic/suppressor T cells and anti- human Lyt3 were found to be reactive. By immunizing mice with LGL one monoclonal antibody (VEP10) was obtained showing reactivity with LGL, thymocytes and 20% of bone marrow cells and another one (VEP13) reacting with LGL and granulocytes.

Taking these findings together, two groups of antibodies reacting with LGL can be distinguished: those reacting with LGL and other lymphocytes (i.e. Leu2a, OKT8, anti-human Lyt 3 and VEP10 antibody) and those reacting with LGL, lymphocytes and cells of the myelomonocytic lineage (OKM1, VEP13). The latter finding is not a surprising one since LGL are known to be of bone marrow origin and thus the expression of antigens common to lymphocytes and myelomonocytic cells on LGL is expected. However, the observation that when LGL are tested with monoclonal antibodies, which discriminate between lymphocytes and myelomonocytic cells only those against lymphocyte antigens react also with LGL, favours the concept that LGL are of lymphoid origin. Since antigens common to mature T cells (OKT3, Leu1) are not expresssed on LGL the question whether LGL are immature T cells or represent a third lymphocyte subtype or are precursor cells for both lymphoid as well as myelomonocytic cells still awaits resolution.

REFERENCES

Breard, J., Reinherz. E.L., O'Brien, C. and Schlossman, S.F., Clin. Immunol. Immunopathol. 18:145 (1981).

Cauppens, J.L., Goodwin, S.S. and Searls, R.P., Cell Immunol. 64:292 (1981).

Kay, H.D., and Horwitz, D.A., J. Clin. Invest. 66:847 (1980).

Lohmann-Matthes, M.-L., Domzig, W., and Roder, J., J. Immunol. 123:1883 (1979).

Timonen, T., Ortaldo, J.R., and Herberman, R.B., J. Exp. Med. 153:569 (1981).

Timonen, T., Ranki, A., Saksela, E., and Häyry, P., Cell Immunol. 48:121 (1979).

West, W.H., Cannon, G.B., and Herberman, R.B., J. Immunol. 118:355 (1977).

Zarling, J.M., and Kung, P.C., Nature 288:394 (1980).

UNIQUE COMBINATION OF SURFACE MARKERS ON HUMAN NK CELLS: A PHENOTYPIC COMPROMISE AT LAST

Gert Riethmüller[*]
Jürgen Lohmeyer[*]
Ernst Peter Rieber[*]
Helmut Feucht[*]
Günter Schlimok[**]
Eckardt Thiel[***]

[*] Institute of Immunology
University of Munich
[**] Medical Clinic of Augsburg
[***] Institute of Hematology, GSF, Munich

I. INTRODUCTION

A favourite topic in the current discussion on NK cells concerns their heterogeneity which, however, is more frequently invoked than adequately demonstrated. Recent serologic, genetic, and morphologic studies related to the genealogy of NK cells promote the view of a unique cellular entity of those non-adherent and non-phagocytic cells capable of cytolysis without preceding specific induction (Herberman et al.1980). Nevertheless, because of various phenotypic and functional overlaps with the monocyte/macrophage lineage as well as with thymus determined lymphocytes, the picture of the NK-cells is still blurred. In view of the key role that malignant aberrations, e.g. plasmocytomas and insulinomas, have played in the identification of the normal counterpart cells, a neoplastic NK cell would seem to be a desirable object for the identification of these elusive killer cells. We have screened therefore various malignant tumor cells, particularly leukemias, for NK activity. As anticipated, the detection of an unique malignant cell helped us to identify its physiologic counterpart and to characterize it.

II. THE MALIGNANT T_{NK} CELL

After having screened several leukemic cells for NK activity, we came across a rather remarkable chronic lymphatic leukemia afflicting a 71 year old patient. The patient had a leukocytosis of 480.000 leukocytes/mm^3, 99 % of the cells being medium to large granular lymphoid like-cells with the histochemical characteristics of the large granular lymphocytes described by Timonen et al (Timonen et al.1979). In spite of a di-

ISBN 0-12-341360-5

stinct splenomegaly, the patient had no lymphnode enlargement. Rather unusual was a polyclonal hyperglobulinemia with 25 g/l IgG, 5,2 g/l IgA and 1.3 g/l IgM. IgD and IgE were not significantly elevated. The isolated leukemic cells were cytotoxic against MOLT 4 cells and they showed NK activity as well as antibody-dependent cytotoxicity against Mel-Wei tumor cells (Table 1). The T cell nature seemed to be undisputable because of E rosette formation, reactivity with heterologous rabbit anti human thymic antigen (HuTLA of Dr. Rodt) and positive staining with T3 and T8 monoclonal antibodies of the OKT series. The leukemic cells expressed uniformly the Fcγ-receptor, while the receptors for $C3_b$ and $C3_d$ as defined by rosette-formation were undetectable. The leukemic cells had no effect on B cell differentiation in coculture experiments in which B lymphocytes were stimulated by pokeweed mitogen to transform into Ig secreting cells (Schlimok et al. 1981).

TABLE 1 NK and ADCC Activity of T-CLL Cells (Si) against non-Adherent and Adherent Tumor Cells

	Target MOLT 4			Target Mel-Wei		
	4h ^{51}Cr release			40 h 3H-Proline release		
E/T	100:1	50:1	25:1	100:1	50:1	25:1
NK	62*	42	43	57	38	29
ADCC	-	-	-	32(23)§	21(18)	13(11)

* % cytotoxicity

§ ADCC expressed as increment over spontaneous cytotoxicity using human anti melanoma serum Utz 1:10. In brackets control effector cells of normal donor.

III. THE SEARCH FOR THE PHYSIOLOGIC COUNTERPART

By density centrifugation, the T-CLL cells could be easily isolated to high purity from the patient's blood. Because of their remarkable homogeneity, they lend themselves as most suitable immunogens for monoclonal antibody induction. After immunization of BALB/c mice with two injections of 2 x 10^7 leukemic T cells hybridomas were produced by fusion of spleen cells with P3 x 63-Ag8.653 myeloma cells (Rieber et al. 1981, Lohmeyer et al. 1981). Three IgG_1 secreting clones were first selected from more than 100 Ig-secreting hybrids. While reacting with > 90% of the T-CLL cells, each of the three monoclonals exhibited a distinct staining pattern with a panel of normal lymphocytes and other target cells (table 2).

TABLE 2 Reactivity of Monoclonal Antibodies Induced With T-CLL Cells of Patient Si.

Cells tested	T411	T811	M522
T_{γ}-CLL Si	96*	98	95
PBL normal donors (n=10)	62 ± 8	20 ± 10	35 ± 8
E-RFC	94 ± 4	25 ± 9	5 ± 2
Adherent cells	1	1	80 ± 10
Granulocytes	1	1	85 ± 8
$S\text{-}Ig^{+}$ cells	1	1	1
Thymus cells (n=3)	15 ± 10	90 ± 5	1
T_{μ} CLL	90	1	1
B-CLL (n=5)	1	1	1

* Percent cells positive in indirect immunofluorescence.

The antigens recognized by the monoclonal antibodies T411,T811, M522 were characterized by immunoprecipitation and SDS-polyacrylamide electrophoresis. The T811 antibody apparently recognizes the same functional T cell subpopulation and a surface molecule of similar size as the OKT8 antibody (Rieber et al, 1981). The distribution of the M522 antigen seemed to be most remarkable since it was present on the T-CLL cells as well as on macrophages and granulocytes. Its absence from B-CLL cells and its failure to block the Fc_{γ} receptor up to a concentration of 1 mg/ml on monocytes support the view that the monoclonal antibody is not directed to the Fc-receptor itself. The apparent molecular weight of M522(110 kd and 180 kd) also speaks against the possibility of the antigen being the Fc-receptor itself.
In view of the coexpression of antigens M522, T411 and T811 and OKM1, OKT3 and OKT8 respectively on the T-CLL cells, the effort was warranted to search for such double staining cells in normal lymphocyte populations. Indeed, using either direct double fluorescence with green or red fluorescent microbeads (Covaspheres^{R}) or reversed rosetting together with immunoautoradiography the simultaneous occurence of T811 and M522 antigens could be demonstrated on 4±2 % of non-adherent lymphocytes. In addition, the microbead technique allowed a triple marker analysis demonstrating either the E receptor or the Fc receptor on the double stained cells. According to these rosette assays, the $T811^{+}/M522^{+}$ subset consisted of 85 ± 5 % E rosette-forming cells and 91±2% Fc receptor-bearing cells (Lohmeyer et al. 1981).

After having identified this hybrid phenotype in peripheral blood of normal donors, the T-CLL cells of patient Si were re-examined with another anti-monocyte monoclonal reagent. The analysis clearly showed that in addition to the T3 and T8 antigens of the OKT-series the T-CLL cells also carried the M1 antigen while the T4 antigen associated with the helper/inducer T cells was absent.

IV. NATURAL CYTOTOXICITY ASSOCIATED WITH A CELL COEXPRESSING THYMIC AS WELL AS MONOCYTIC/GRANULOCYTIC ANTIGENS

Since it has been found that cytotoxic T cells as well as their precursors belonged to the $T811^+$ subset (Rieber et al.1981) it was of interest to know to which degree the $T811^+$ subset excerted NK function. Furthermore, the question arose whether the $T811^+$ population characterized as T cytotoxic/suppressor subset could be subdivided into a fraction carrying the M522 antigen and effectuating natural cytotoxicity. To answer these questions we developed a density fractionation technique based on reverse rosette formation. In its direct or indirect form, the technique allowed enrichment as well as depletion of lymphocyte subsets with the help of monoclonal antibodies. The details of the technique have been outlined previously (Lohmeyer et al. 1981). By this technique a purity of more than 90 % could be obtained for the relevant lymphocyte subpopulations. As demonstrated in Table 3, NK activity was assessed in various isolated fraction prepared either from non-adherent lymphocytes or from E-rosette-forming cells. The total NK activity of both non-adherent lymphocytes and E^+ lymphocytes could be separated into the $M522^+$ fraction, a confirmation of the data of Zarling and Kung (1980).However, when the E^+ lymphocytes were separated into the $M522^+$ or $T811^+$ subsets, respectively, the total NK activity was found in each of these fractions. The corresponding $M522^-$ or $T811^-$ populations were devoid of any lytic activity. Thus, it was clearly demonstrated that NK activity associated with E^+ lymphocytes was effectuated by the $M522^+$/$T811^+$ subset. In further experiments, consecutive reverse rosette separations with M522 and T811 monoclonal antibodies were carried out. Nearly 90 % of the isolated double marker cells belonged to the medium to large granular lymphocytes described by Timonen et al (1979). The $M522^+$/$T811^+$ subset when isolated by consecutive rosetting was distinctly active in NK activity. As demonstrated by the controls, the reverse rosetting procedure, the density gradient centrifugation and dissociation did not substantially affect the cytolytic capacity.

TABLE 3 Anti K562 Cytotoxicity of Lymphocyte Subsets Separated by Monoclonal Antibodies

Effector cell Population	% Recovery of cells	Cytotoxicity % CTX25:1	$LU_{33}/10^7$
NAL	100	29.0	75.8
NAL_c a)	80	19.0	51.2
$M522^-NAL$	60	3.8	0.1
$M522^+NAL$	12	59.0	227.0
$T811^-NAL$	62	17.0	33.0
$T811^+NAL$	15	30.0	66.0
E^+	72	25.8	39.8
E^+_C a)	53	17.0	30.9
$M522^-E^+$	50	1.0	0.1
$M522^+E^+$	6	34.0	90.0
$T811^-E^+$	30	3.9	0.1
$T811^+E^+$	10	30.8	74.0

a) NAL_c = non-adherent lymphocytes separation controls: Lymphocytes are incubated with monoclonal antibodies and reverse rosettes are formed and subsequently dissociated. Ox erythrocytes are separated by Percoll[R] gradient centrifugation.

CONCLUDING REMARKS

The simultaneous expression on one cell of two usually discordant antigens, the T cell antigen T811 and the myeloid antigen M522 clearly provides a compromise for the two contending views of the NK cell nature, the one being most strongly supported by data of Lohmann-Matthes (Lohmann-Matthes et al. 1979) and Zarling and Kung (Zarling & Kung. 1980) who favor a monocyte lineage, the other view, stressing the T cell lineage, had its proponents from the early days of the NK science (see review Herberman et al.(1980). The malignant cell described here and the corresponding double marker cell can be interpreted in a way that a transitional status between the cell lineages in early maturation stages occurs and that this is correlated with NK function. However, the M522 and OKM1 antigens on the T811 cytotoxic/suppressor T subset need to be further analyzed particularly with regard to biosynthesis and modulation of of MHC restricted cytotoxicity.

The presence of the OKM1 on T cells in general had been noted by others (Breard et al. 1980, Fast et al. 1981, Kay & Horwitz 1980). The functional role of the defined surface molecules, particularly their involvment in the various forms of cellular cytotoxicity, becomes now amenable to direct examination, which may solve the puzzle of NK heterogeneity.

ACKNOWLEDGEMENTS

We thank Sybille Brodmann, Gerti Rank and Simone Heydeke for technical assistance. Renate von Carnap helped us to cope with the intricacies of camera-ready proof typing.

REFERENCES

Breard, J., Reinherz, E.L., Kung, P.C., Goldstein, G. and Schlossman, S. J. Immunol. 124:1943 (1980)

Fast, D.L., Hansen, J.A. and Newman, W. J. Immunol. 127: 448 (1981)

Herberman, R.B., Timonen, T., Ortaldo, J.R., Bonnard,G.D. and Gorelik, E. in "Progress in Immunology IV" (M. Fougereau and J. Dausset, eds.) p. 691 Academic Press, London, 1980.

Kay, H.D. and Horwitz, D.A. J. Clin.Invest. 66:847 (1980)

Lohmeyer, J., Rieber, E.P., Feucht, H., Johnson, J.P., Hadam, M.R. and Riethmüller, G. Europ.J. Immunol. 11:997 (1981)

Rieber, E.P., Lohmeyer, J., Schendel, D.J. and Riethmüller, G. Hybridoma 1: 59 (1981)

Schlimok, G., Thiel, E., Rieber, E.P., Huhn, D., Feucht, H., Lohmeyer, J. and Riethmüller, G. manuscript submitted.

Timonen, T., Saksela, E., Ranki, A. and Häyry, P. Cell. Immunol. 48: 133 (1979)

Zarling, J.M. and Kung, P.C. Nature 288:394 (1980)

PHENOTYPES OF HUMAN NK CELLS AS DETERMINED BY REDUCTION OF CYTOTOXIC ACTIVITY WITH MONOCLONAL ANTIBODIES AND COMPLEMENT[1]

Joyce M. Zarling[2]

Immunobiology Research Center
Department of Laboratory Medicine/Pathology
University of Minnesota
Minneapolis, Minnesota

I. INTRODUCTION

As discussed also in other chapters in this volume, studies have been undertaken to determine the morphology and cell surface markers of human NK cells in attempts to elucidate the cell lineage of these spontaneously cytotoxic cells and to determine whether cells with NK activity have characteristics which distinguish them from other mononuclear cell subpopulations.

Although originally considered to be "null" cells, it is apparent that at least some NK cells share with T cells the ability to form rosettes with sheep red blood cells (E) (West et al., 1977) and to react with xenogeneic anti-human T cell serum (Kaplan and Callewaert, 1978). Further, most human NK cells express receptors for the Fc portion of IgG (West et al., 1977; Eremin et al., 1978) and appear to be morphologically large granular lymphocytes (LGL) (Timonen et al., 1981). In collaboration with Dr. Patrick Kung, who

---

[1]This work was supported by NIH grant 26738. This is paper #284 from the Immunobiology Research Center, University of Minnesota, Minneapolis, MN.

[2]J.M. Zarling is a Scholar of the Leukemia Society of America.

ISBN 0-12-341360-5

developed several monoclonal antibodies reactive with various human mononuclear cell subpopulations (for review, see Kung et al., 1980) studies were undertaken to determine which, if any of these antibodies in the presence of complement (C) would reduce human NK cell activity against human leukemia cell lines. Results of our recent studies (Zarling and Kung, 1980; Zarling et al., 1981) are reviewed herein.

II. METHODS AND MATERIALS

A. Monoclonal Antibodies

The monoclonal antibodies used in these studies were produced as reviewed (Kung et al., 1980). OKT11A reacts with 100% human thymocytes and peripheral E rosetting cells (Verbi et al., in press); OKT3 reacts with approximately 90% of peripheral E rosetting cells (Kung and Goldstein, 1980); OKT4 reacts predominantly with helper/inducer T cells (Reinherz et al., 1979); OKT8 reacts with cytotoxic/suppressor T cells (Kung et al., 1980); OKM1 reacts with monocytes and a small proportion of non-adherent mononuclear cells (Breard et al., 1980).

B. Preparation of Effector Cells and Antibody Treatments

Ficoll-hypaque isolated peripheral blood mononuclear cells were depleted of monocytes by two cycles of plastic adherence and were stimulated for 16 hrs. with or without polyinosinic: polycytidylic acid (poly I:C, 125μg/ml) or were cultured for 6-7 days with x-irradiated (2500R) allogeneic normal cells as detailed (Zarling and Kung, 1980). Effector cells were treated with the antibodies at a final dilution of 1:100 or, in the case of OKM1, at a final dilution of 1:20 for 1 hr. at 4°C followed by two treatments with non-toxic rabbit C as detailed (Zarling et al., 1981). This double C treatment was necessary to completely eliminate certain cell subpopulations. Dead cells were eliminated by Ficoll gradient centrifugation and the viable cells were washed and tested for cytotoxic activity in 5-7 hr ^{51}Cr release assays.

Table I. Reduction of CTL activity by OKT3 or OKT8 plus C [a]

% Specific ^{51}Cr release from allogeneic normal cells ± S.D.		
Treatment of allo-stimulated effector cells		
Control ascites + C	OKT8 + C	OKT3 + C
56.8 ± 8.9	3.7 ± 3.9	7.0 ± 4.3

[a]Effector cells, stimulated for 7 days with allogeneic normal cells, were treated with monoclonal antibodies and C as detailed in the Methods section. E:T ratio = 30:1.

III. EFFECTS OF OKT3 AND OKT8 ON CTL AND NK ACTIVITIES

The results shown in Table I indicate that treatment of lymphocytes stimulated with allogeneic normal cells in mixed leukocyte culture (MLC), with OKT3 or OKT8 plus C, virtually eliminates cytotoxic T lymphocyte (CTL) activity against ^{51}Cr labeled allogeneic normal lymphocytes. This finding agrees with observations obtained by positive selection that CTLs are OKT3+, OKT8+ (or OKT5+), OKT4- (for review, see Kung and Goldstein, 1980). In contrast, results in Figure 1a show

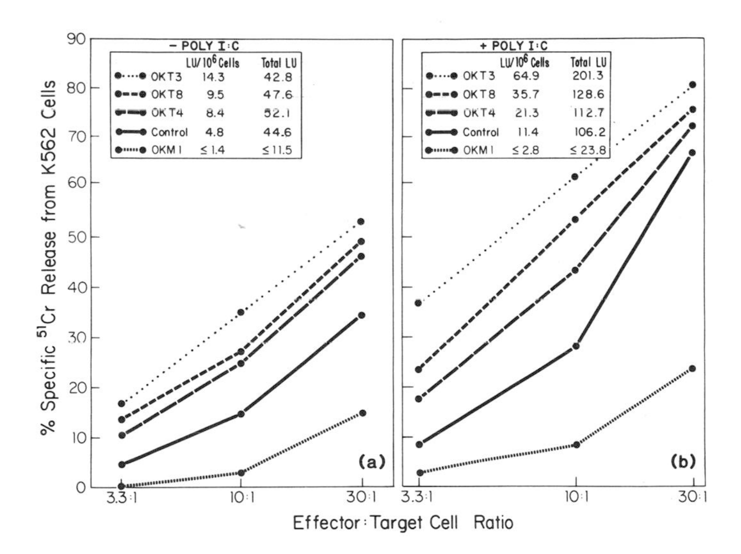

FIGURE 1. Effect of Antibodies Plus C on NK Activity. Effectors were treated as detailed in the Methods section and lytic units (LU) recovered after treating 1 x 10^7 cells were calculated as described (Zarling and Kung, 1980).

that NK activity is not reduced after treatment with OKT3, OKT8 or OKT4 plus C. Figure 1b indicates that NK cell activity is boosted by poly I:C as we and others have reported (Djeu et al., 1980; Zarling et al., 1980; Koren et al., 1981) and that the cytotoxic activity of poly I:C augmented NK cells is not reduced following treatment with monoclonal antibodies OKT3, OKT8 or OKT4.

IV. REDUCTION IN NK CELL ACTIVITY BY MONOCLONAL ANTIBODY OKM1

Results in Figure 1a indicate that treatment of fresh monocyte-depleted peripheral blood mononuclear cells with monoclonal antibody OKM1 plus C reduced NK activity against K562 cells approximately 4 fold. In a total of more than 15 experiments, we have found that NK activity against K562 cells was reduced 2-4.5 fold following treatment with OKM1 and C. Similarly, treatment of poly I:C activated NK cells with OKM1 and C reduced the lytic activity by approximately 4 fold in terms of total lytic units. Findings that NK cells can express the determinant against which monoclonal antibody OKM1 is directed is in agreement with observations of Kay and Horwitz (1980) and also of Ortaldo et al. (1981) who recently reported that a high proportion of the LGL-enriched populations is OKM1+.

V. REDUCTION IN NK CELL ACTIVITY BY MONOCLONAL ANTIBODY OKT11A

Since NK cell activity was reduced but not eliminated following treatment with OKM1 plus C, we hypothesized that some NK cells may have too low a density of the OKM1 antigen to be lysed by the antibody plus C or, alternatively, there may be an OKM1- NK cell population (Zarling and Kung, 1980). Because NK cell activity can reside within the E rosetting cell population, yet NK cell activity was not decreased following treatment with OKT3 which reacts with 90-95% of E rosetting cells, it seemed apparent that NK cells may reside within the small proportion of E rosetting cells which is OKT3-. Indeed we found that treatment with OKT11A, which reacts with 100% of peripheral E rosetting cells (Verbi et al., in press) also reduces NK cell activity (Table II). Additional evidence that NK cells reside within the OKT11A+ OKT3- populations derives from our observations that treatment of OKT3 depleted cells with OKT11A plus C resulted in a

TABLE II. Reduction in NK cell activity against K562 cells following treatment with monoclonal antibodies OKM1 and OKT11A plus C

Effector cell treatment	% specific ^{51}Cr release E:T ratio	
	30:1	10:1
C alone	60.3	32.3
OKM1 + C	40.8	16.2
OKT11A + C	38.7	20.0
OKM1 + OKT11A + C	9.7	4.1

marked reduction of cytotoxic activity (Zarling et al., 1981). Treatment with OKM1 together with OKT11A and C was found to be more efficacious in reducing NK activity against K562 than was treatment with either antibody alone plus C (Table II). Treatment of monocyte-depleted mononuclear cells with either OKM1 or OKT11A also reduced NK cell activity against 2 T cell leukemia lines, MOLT-4 and HSB-2, and treatment with both antibodies together with C resulted in virtually no detectable NK cell activity against these targets (Zarling et al., 1981).

VI. EVIDENCE FOR PHENOTYPIC HETEROGENEITY OF NK CELLS

Why does treatment with OKM1 together with OKT11A reduce human NK cell activity more so then does treatment with either antibody alone with C? We considered the possibility that NK cells may express both OKM1 and OKT11A antigens but perhaps the density of each antigen is so low on some NK cells that treatment with both antibodies together with C is required to kill this weakly antigen positive population of cells. Alternatively it could be that some NK cells are OKM1+ OKT11A- and others are OKM1- OKT11A+. Evidence to support the latter contention derives from two findings. First, following treatment with either OKM1 or OKT11A plus C, residual NK cell activity was detected yet we have found that our method for depleting the relevant antigen-bearing cells (i.e., treatment with antibody and then twice with C) is efficacious in that <1% OKT11A+ cells was detected following treatment with OKT11A and C and <1% OKM1+ cells was detected following treatment with OKM1 plus C. Thus, residual NK activity after treatment with OKM1 or OKT11A plus C seems unlikely to be explained on the basis of ineffective depletion of the antigen positive cell populations. Second, we asked whether, within the OKT3- cell population which had been monocyte depleted, there are cells which are OKM1+ OKT11A- and vice versa. Our collaborator, Dr. William Biddison, by using two-color flow microfluorometric analysis of the OKT3 depleted population, found that approximately one-half of the OKM1+ cells were OKT11A- and one-half of the OKT11A+ cells were OKM1- (Zarling et al., 1981).

VII. SUMMARY AND CONCLUDING REMARKS

In summary, antibodies reactive with mature T cells including the OKT3+, OKT4+ or OKT8+ subsets do not detectably reduce NK cell activity following the addition of C. Although negative selection with OKT3 or OKT8 and C is clearly a means to distinguish most CTLs from NK cells, it might be inaccurate to conclude that there exists not even a very small proportion of NK cells which expresses OKT3 and/or OKT8 markers. Treatment with OKM1 or OKT11A plus C markedly reduced NK cell activity; treatment with both antibodies together with C virtually eradicated NK cell activity against three different NK-sensitive leukemia cell lines. These results, together with others discussed above, enabled us to conclude that there are at least 2 populations of NK cells lytic for NK-sensitive leukemia cells. It is also possible that some NK cells may reside within the population of OKT3- cells which expresses both OKM1 and OKT11A markers. Double positive selection would be required to verify the existence of an OKM1+ OKT11A+ NK cell population. It should be noted that our experiments have dealt with NK cells lytic for human leukemia cell lines and we have not as yet determined the phenotype of the effectors which are lytic for other tumor cells or virus-infected target cells that are susceptible to NK-mediated lysis.

ACKNOWLEDGMENTS

The author is indebted to Drs. Patrick Kung and William Biddison for their collaboration, M.S. Dierckins, K.A. Clouse and E.A. Seveniçh for carrying out many of the experiments, and J. Ritter and J. Gfrerer for secretarial assistance.

REFERENCES

Breard, J., Reinherz, E.L., Kung, P.C., Goldstein, G., and Schlossman, S.F. (1980). J. Immunol. 124:1943.
Djeu, J.Y., Heinbaugh, J.A., Holden, H.T., and Herberman, R.B. (1979). J. Immunol. 122:175.
Eremin, O., Ashby, J., and Plumb, D. (1978). J. Immunol. Methods 24:257.
Kaplan, J., and Callewaert, D.M. (1978). J. Natn. Cancer Inst. 60:961.
Kay, H.D., and Horwitz, D.A. (1980). J. Clin. Invest. 66:847.
Koren, H.S., Anderson, S.J., Fischer, D.G., Copeland, C.S., and Jensen, P.J. (1981). J. Immunol. 127:2007.
Kung, P.C., and Goldstein, G. (1980). Vox Sanguinis 39:121.
Kung, P.C., Talle, M.A., DeMaria, M.E., Butler, M.S., Lifter, J., and Goldstein, G. (1980). Transplant. Proc. 12:141.
Ortaldo, J.R., Sharrow, S.O., Timonen, R., and Herberman, R.B. (1981). J. Immunol. 127:2401
Reinherz, E.L., Kung, P.C., Goldstein, G., and Schlossman, S.F. (1979). Proc. Natl. Acad. Sci. 76:4061.
Timonen, T., Ortaldo, J.R., and Herberman, R.B. (1981). J. Exp. Med. 153:569.
Verbi, W., Greaves, M.S., Schneider, C., Koubek, K., Janossy, G., Stein, H., Kung, P.C., and Goldstein, G. Europ. J. Immunol. In press.
West, W.H., Cannon, G.B., Kay, H.D., Bonnard, G.D., and Herberman, R.B. (1977). J. Immunol. 118:355.
Zarling,.J.M., Schlais, J., Eskra, L., Greene, J.J., Ts'o, P.O.P., and Carter, W.A. (1980). J. Immunol. 124:1852.
Zarling, J.M., and Kung, P.C. (1980). Nature 288:394.
Zarling, J.M., Clouse, K.A., Biddison, W.E., and Kung, P.C. (1981). J. Immunol. 127:2575.

ANALYSIS OF HUMAN NATURAL KILLER CELLS BY MONOCLONAL ANTIBODIES

Chris D. Platsoucas[1]

Memorial Sloan-Kettering Cancer Center
New York, New York

I. INTRODUCTION

Natural killer (NK) or spontaneous cytotoxicity is mediated by a nonadherent, nonphagocytic population of cells that possess Fc receptors for IgG and reside within the population of large granular lymphocytes (Kay et al, 1977; West et al, 1977; Timonen et al, 1981; Herberman, 1980). Both E-rosette positive and E-rosette negative cells mediate NK cytotoxicity (Kay et al, 1977; West et al, 1977; Kall and Koren, 1978; Gupta et al, 1978). Recently, the development of monoclonal antibodies recognizing cell surface differentiation antigens expressed on distinct population of hemopoietic cells (Reinherz and Schlossman, 1980; Kung and Goldstein, 1980) permits further analysis of the heterogeneity and functional properties of natural killer cells in man.

II. METHODS

OKT3, OKT4, OKT8, OKM1, OKT11a and OKT10 monoclonal antibodies were obtained from Ortho Pharmaceuticals Corporation, Raritan, N.J. Clone 9.6 and αIa (Clone 7.2) monoclonal antibodies were obtained from New England Nuclear, Boston, MA. Monocyte-depleted human peripheral blood lymphocytes (Platsoucas et al, 1980) were used in these studies. Cells to be separated using the fluorescence-activated cell sorter were incubated with the appropriate monoclonal antibody (0.25 μg/$1x10^6$ cells) for 30 min at 4°C, and labeled with an FITC-conjugated $F(ab^1)_2$ fragment of goat antiserum to mouse Ig.

[1]Supported by grant CH-151 from the American Cancer Society and NIH grant CA 8748.

ISBN 0-12-341360-5

TABLE 1

NATURAL KILLER CYTOTOXICITY OF 9.6-POSITIVE, OKT11a-POSITIVE AND OKM1-POSITIVE HUMAN PERIPHERAL BLOOD LYMPHOCYTES ISOLATED BY THE FLUORESCENCE ACTIVATED CELL SORTER.

Population	% CYTOTOXICITY	
	100:1	50:1
Monocyte-depleted peripheral blood lymphocytes	44%	26%
Clone 9.6-positive cells	29%	21%
OKT11A-positive cells	40%	20%
OKM1-positive cells	35%	17%

The cells then were separated in the FACS IV (Becton-Dickinson). Antibody and complement treatment (rabbit complement, Pel Freeze, Little Rock, Ark.) was carried out by a two step protocol as previously described (Platsoucas and Good, 1981). In certain experiments, treatment with complement (C) was repeated for a second time. Lysis by the OKT10 monoclonal antibody (IgG1) was achieved after addition of rabbit anti-mouse immunoglobulin (RαM). RαM immunoglobulin and C treatment alone did not affect the NK cytotoxicity. Immunofluorescence analysis, after removal of dead cells by centrifugation on Ficoll/Hypaque, revealed that antibody and complement treatment under the conditions employed here resulted in complete depletion of the appropriate cell population. NK cytotoxicity against the K562 and Daudi targets was determined as previously described (Platsoucas *et al*, 1980). Cell numbers after treatment with antibody and complement were not adjusted for dead cells. Experiments to investigate the ability of monoclonal antibodies to block NK cytotoxicity in the absence of complement were carried out by adding the antibodies to the effector cells and then mixing with the ^{51}Cr-labelled targets (Platsoucas and Good, 1981).

III. RESULTS

A representative experiment of the natural killer cytotoxicity mediated by 9.6-positive, OKT11a-positive or OKM1-positive cells, purified by the fluorescence activated cell sorter, is shown in Table 1. Statistically significant differences in the NK cytotoxicity mediated by purified 9.6-positive and OKT11a-positive cells were not observed. In a previous report (Platsoucas *et al*., 1981), the OKT11a mono-

TABLE 2

CELL SURFACE PHENOTYPES OF HUMAN NATURAL KILLER CELLS.*

Population	% of control (mean ±s.d.)	Range
Monocyte-depleted peripheral blood lymphocytes	100%	
OKT 10	75±20%	45-95 (n=5)**
OKT 11a	64±16%	30-83 (n=9)
Clone 9.6	67± 9%	60-81 (n=4)
OKM 1	55± 7%	32-80 (n=7)
OKIa	28±16%	13-45 (n=4)
OKT 8	14±12%	0-29 (n=5)
OKT 4	<5%	0- 5 (n=4)
OKT 3	<10%	0-10 (n=4)

* NK cytotoxicity was determined against K 562 target cells
** number of experiments

clonal antibody was erroneously designated as OKT11. Other experiments using in vitro treatment with monoclonal antibodies and complement resulted in similar findings. These results are summarized in Table 2. Percent of control values were calculated by the formula [100 - (% Cr-51 release of Ab+C treated cells / % Cr-51 release of C-treated cells alone) X 100]. Natural killer cells from the peripheral blood express the following antigens which have been reported to be present on cells from various hemopoietic lineages: T11a (the receptor for sheep erythrocytes); T10 (present on early T cells, thymocytes, activated T cells, some bone marrow cells, etc.); and M_1 (present on monocytes and granulocytes). All NK cells appear to be within the OKT3-negative cell population. Furthermore, a proportion of NK cells may express Ia antigens on their surface membrane. Similarly, natural killer-like cytotoxic cells generated in MLC were OKT3-negative, OKT4-negative and OKT8-negative (data not shown) (Platsoucas and Good, 1981). These results are in agreement with reports in the literature (Kay and Horowitz, 1980; Zarling and Kung, 1980; Lohmeyer et al, 1981; Platsoucas, 1981; Fast et al, 1981; Ortaldo et al, 1981; Zarling et al, 1981).

In other experiments (Table 3), the effect of in vitro treatment on natural killer cytotoxicity with the following combinations of monoclonal antibodies and complement was investigated: (a) OKT11a and OKM1; (b) OKT11a and OKT10 (in the presence of RαM); and (c) OKM1 and OKT10 (in the presence of RαM). In all cases, higher reduction of NK cytotoxicity was observed by simultaneous treatment with two monoclonal

TABLE 3

EFFECT OF IN VITRO TREATMENT WITH THE OKT11a, OKM1 AND/OR OKT10 MONOCLONAL ANTIBODIES AND COMPLEMENT ON THE NATURAL KILLER CYTOTOXICITY OF MONOCYTE-DEPLETED PERIPHERAL BLOOD LYMPHOCYTES.

Antibody and Complement	% CYTOTOXICITY Targets: K 562 Cells 50:1	25:1	12.5:1	Targets: Daudi Cells 50:1	25:1	12.5:1
C alone	56%	33%	20%	30%	22%	16%
OKT11a+C	41%	24%	15%	13%	10%	6%
OKM1+C	45%	28%	19%	13%	6%	6%
OKT11a+OKM1+C	13%	11%	7%	5%	4%	1%
OKT10+RαM[b]+C	31%	21%	13%	11%	8%	5%
OKT10+RαM+OKT11A+C	17%	9%	4%	5%	2%	1%
OKT10+RαM+OKM1+C	17%	11%	7%		ND	

[a] Effector to target ratio; [b] Rabbit anti-mouse IgG.

TABLE 4

INABILITY OF THE OKT11a, OKM1, OKT8, OKT3, OKIa and αIa MONOCLONAL ANTIBODIES TO INHIBIT NATURAL KILLER CYTOTOXICITY IN THE ABSENCE OF ADDED COMPLEMENT.

Monoclonal Antibody	% Cytotoxicity[a] Donor 1	Donor 2	Donor 3	Donor 4	% Inhibition
None	49%	44%	60%	65%	--
OKT11a	44%	30%	60%	68%	11 (0-33)[b]
OKM1	50%	42%	62%	60%	1 (0-6)
OKT8	48%	41%	ND	ND	5 (2-8)
OKT3	ND	ND	55%	60%	4 (0-8)
OKIa	43%	39%	ND	ND	9 (6-12)
αIa (clone 7.2)	45%	42%	ND	ND	8 (4-12)

[a] Effector to target ratio, 50:1; [b] range of % inhibition.

antibodies than with a single antibody. These results suggest that certain non-overlapping or partially overlapping populations bearing these antigens (T11a, M1 and T10) may be, in part, responsible for natural killer cytotoxicity. Experiments are in progress using cell separation methods permitting positive selection of cells and double labeling techniques to clarify these issues. Recently, Zarling _et al._ (1981) reported the existence of at least two populations of NK cells, one of which was found to be OKM1-positive and the other OKT11a-positive. Similarly, Lohmeyer _et al_ (1981) reported the presence of partially non-overlapping populations of NK cells bearing a monocytic antigen and/or an early thymocyte antigen.

The OKT3, OKT4, OKT8, OKT11a, OKM1, OKIa, αIa (Clone 7.2), anti-Leu1, anti-Leu2a and anti-Leu3a monoclonal antibodies did not inhibit the natural killer or natural killer-like cytotoxicity (Platsoucas and Good, 1981) in the absence of added complement. In contrast, we reported elsewhere that the OKT3 and anti-Leu2a monoclonal antibodies (and in certain cases the OKT8 monoclonal antibody), inhibit significantly the specific T cell mediated cytotoxicity against allogeneic targets in the absence of added complement (Platsoucas and Good, 1981; Evans _et al_, 1981).

IV. CONCLUDING REMARKS

Natural killer cells in man constitute a complex and heterogeneous system. The development of monoclonal antibodies recognizing cell surface antigens expressed on distinct populations of hemopoietic cells provides a new tool for the analysis of this complex system. Natural killer cells express cell surface antigens (such as M1, T10, T11a) present on cells from various lineages. The use of monoclonal antibodies has provided evidence suggesting the existence of subpopulations of NK cells. It is possible that these subpopulations belong to different cell lineages. The lysis of various target cells, such as cells from tumor cell lines, virus-infected cells or fresh tumor cells, by these populations of NK cells remains to be investigated, and their specificity to be determined. These populations of NK cells may respond differently to various biological response modifiers.

REFERENCES

Evans R.L., Wall D.W., Platsoucas C.D., Siegal F.P., Fikrig S.M., Testa C.M. and Good R.A. PNAS 78:544 (1981).

Fast L.D., Hansen J.A. and Newman W. J. Immunol. 127:448 (1981).

Gupta S., Fernandes G., Nair M. and Good R.A. PNAS 75:5137 (1978).

Herberman R.B. (ed.). Natural Cell-Mediated Immunity Against Tumors, Academic Press, New York, 1980.

Kall M.A. and Koren H. Cell. Immunol. 40:58 (1978).

Kay H.D., Bonnard G.C., West W.H., Herberman R.B. J. Immunol. 118:2058 (1977).

Kay H.D. and Horowitz D.A., J. Clin. Invest. 66:847 (1980).

Kung P.C. and Goldstein G. Vox Sang. 39:121 (1980).

Lohmeyer J., Rieber P., Feucht H., Hadam M. and Reithmuller G. In: Mechanisms of Lymphocyte Activation (K. Resch and H. Kirchner, eds.), Elsevier/North Holland Biomedical Press, p. 282, (1981).

Ortaldo J.R., Sharrow S.O., Timonen T. and Herberman R.B. J. Immunol. 127:2401 (1981).

Platsoucas C.D. In: Mechanisms of Lymphocyte Activation (K. Resch and H. Kirchner, eds.), Elsevier/North Holland Biomedical Press,p. 290, (1981).

Platsoucas C.D. and Good R.A. PNAS 78:4500 (1981).

Platsoucas C.D., Fernandes G., Gupta S.L. Kempin S., Clarkson B., Good R.A. and Gupta S. J. Immunol. 125:1216 (1980).

Reinherz E.L. and Schlossman S.F. Cell 19:821 (1980).

Timonen T., Ortaldo J.R. and Herberman R.B. J. Exp. Med. 153:569 (1981).

West W.H., Cannon G.B., Kay H.D., Bonnard G.D. and Herberman R.B. J. Immunol. 118:35 (1977).

Zarling J.M. and Kung P.C. Nature 288:394 (1980).

Zarling J.M., Clouse K.A., Biddison W.E. and Kung P.C. J. Immunol. 127:2575 (1981).

STUDIES OF CELL SURFACE MARKERS ON NK CELLS: EVIDENCE FOR HETEROGENEITY OF HUMAN NK EFFECTOR CELLS

Patricia A. Fitzgerald
Dahlia Kirkpatrick
Carlos Lopez

Sloan-Kettering Institute for Cancer Research
New York, New York

I. INTRODUCTION

Natural killer (NK) cells have received much attention during the past several years as potential effectors of early anti-tumor and anti-viral immunity and as potential mediators of hemopoietic homeostasis. Our interest has been in the role of NK cells in the resistance to herpesvirus infections because of studies of a mouse model which suggested that these effector cells probably are an important aspect of genetic resistance to HSV-1 (1). An NK assay, using HSV-1 infected fibroblasts as targets [NK(HSV-1)], was developed to determine whether these KN cells played an important role in resistance to herpesvirus infections in man. In another chapter of this volume, we present evidence of a strong correlation between low NK(HSV-1) and an increased susceptibility to herpesvirus infections (3, Lopez et al., this volume) as well as an impressive association between normal pretransplant NK(HSV-1) responses and the development of graft versus host disease (GvHD) after engraftment with bone marrow from a matched sibling donor (4, Lopez et al., this volume). In contrast to this latter finding, Livnat et al (5) found no correlation between pretransplant NK of K562 erythroleukemia targets, the most commonly used cells for evaluating human NK, and the development of GvHD in the post-transplant period. These studies were carried out at two different institutions using somewhat different patient groups

---

Supported in part by NIH grants CA-23766, CA-08748, and CA-17404; NIH Training grant CA-09149; grant IM-303 and Faculty Research Award #193 (to CL) from American Cancer Society.

ISBN 0-12-341360-5

but the major possibility for these differences was that the two NK assays evaluated different effector cell subpopulations (6).

In murine systems, considerable evidence has been developed demonstrating that NK cells are indeed heterogeneous, and that the different populations can be distinguished by biological and serological techniques (7,8). The studies summarized here were undertaken to determine whether human NK effector cells demonstrate similar heterogeneity. We found that the effector cells which lyse HSV-1 infected fibroblasts differ from the cells which kill K562 targets, although it is possible that an overlapping population may be responsible for some kill of both targets.

II. MATERIAL AND METHODS

A. NK Assays

Natural killer cell assays were performed as previously described (2). Briefly, effector cells from human peripheral blood were isolated on a Ficoll Hypaque shelf. Targets were ^{51}Cr-labeled human foreskin fibroblasts infected for two hours with HSV-1, uninfected fibroblasts or K562 cells. Effector and target cells were mixed at effector:target (E:T) ratios of 100:1 to 3:1 (2 fold dilutions) and incubated for 14 hours at 37°C. Supernatants were harvested for determination of ^{51}Cr release and lysis at E:T of 50:1 was estimated from the linear regression obtained after application of the Von Krogh Equation (9) to the data.

B. Complement Dependent Cytotoxicity Assays

Peripheral blood mononuclear cells were incubated with monoclonal antibodies for 45 minutes at 4°C, then with a 1:4 dilution of rabbit complement for 45 min. at 37°C. The antibody and complement cycles were repeated, then viable cell recovery was determined by trypan blue exclusion.

C. "Panning"

Positive and negative selection of lymphocyte subpopulations was performed using a modification of the procedure described by Wycsocki and Sato (10). Briefly, nylon wool passaged mononuclear cells were coated at 4°C with mouse monoclonal antibodies, then incubated on petri dishes which were coated with sheep anti-mouse IgG. After incubation at 4°C for 45 minutes, non-adherent or "negative" cells were rinsed off the petri dish. Adherent or "positive" cells were recovered by gently scraping the petri dishes with a rubber policeman. Purity of selected cell populations was determined by flow cytometry and was generally $\geq$ 90%. As a control, cells not coated with monoclonal antibody were incubated on sheep anti-mouse IgG plates. Less than 5% of the cells adhered non-specifically to the plates, leading to no depletion of the NK in the negative population. The control "positive" population contained little or no kill.

III. RESULTS

A. Patient Studies

Studies in our laboratory have shown a strong correlation between normal pre-transplant NK(HSV-1) and subsequent development of graft-versus-host disease (GvHD) after engraftment with allogeneic bone marrow (4). In similar studies using K562 effector cells, no correlation was found between NK(K562) and subsequent GvHD (6). Because of these differing observations, NK(HSV-1) and NK(K562) activity were simultaneously studied in both control and patient groups. Although NK(HSV-1) and NK(K562) levels were most often either both normal or both low, a number of individuals have been found to have low NK (HSV-1) but normal NK(K562). A representative group showing this dichotomy, adult patients with aplastic anemia, is shown in Table I.

TABLE I. NK in Adult Patients with Aplastic Anemia*

Patient	NK(HSV-1)	NK(K562)
1	8.4	65.1
2	11.2	41.0
3	9.5	35.6
4	8.1	40.8

*Patients were studied in 14 hr. NK assays.
Normal range (mean 2 S.D.) for NK(HSV-1) was 21-69% and for NK(K562) was 29.8-98.8%.

B. Studies with Monoclonal Antibodies

A further definition of the effector cells mediating NK was achieved using monoclonal antibodies to define cell surface phenotypes. Nylon wool-passaged peripheral blood lymphocytes were coated with antibodies to Leu-1, Leu-4 (pan T-cell markers), Leu-2 (on suppressor cells), or Leu-3 (on helper cells) and the positive and negative populations were separated using the panning technique. NK(HSV-1) and NK(Fs) were enriched in all of the negative populations and depeleted from the positively selected cells. Most of the NK(K562) activity was found in the negative populations but a small fraction could be recovered in the Leu-1 and Leu-4 positive populations.

Monoclonal antibodies to Lyt-3 (clone 9.6, recognizing the sheep red blood cell rosette receptor) and Ia (clone 7.2) are complement-fixing and were used in both the panning assay and the complement-dependent elimination assay. Monoclonal

Lyt-3 required a facilitating rabbit anti-mouse IgG for full expression of activity. As shown in Table II, treatment of effector cells with anti-Lyt-3 plus complement reduced NK (K562) activity by about 50% but had no effect on NK(HSV-1) or NK(Fs). Conversely, two cycles of anti-Ia plus complement reduced NK(HSV-1) and NK(Fs) dramatically but left NK(K562) unaffected.

Table II. NK After Complement Elimination

Treatment		Target % ^{51}Cr-Release	
	Fs	HSV-Fs	K562
Exp. 1. C'control	11.5	50.6	30.0
Lyt-3 + C'	13.5	52.6	15.5
Exp. 2. C'control	37.0	78.4	32.3
Anti Ia + C'	14.3	23.9	33.8

Results similar to those obtained with complement elimination were obtained by panning with antibodies to Lyt-3 and Ia: NK(K562) activity was found in both Lyt-3 positive and negative fractions but was not found in Ia positive selected cells. NK(Fs) and NK(HSV-1) activity were found in Ia positive, Ia negative, and Lyt-3 negatively selected populations. In addition, OKMI, an antigen found on macrophages and some null cells, was found on the effector cells for all three targets using the panning procedure; although OKMI positive cells had greater activity against K562 than Fs or HSV-1 Fs targets (Fitzgerald _et al_., in preparation).

IV. DISCUSSION

The studies outlined here provide strong evidence for heterogeneity within human NK populations. A summary of our findings comparing NK(FS), NK(HSV-1) and NK(K562) is presentend in Table III. The patient studies demonstrate that NK (HSV-1) and NK(K562) can be independent of one another. The studies with monoclonal antibodies also suggest that NK(K562) and NK(HSV-1) can be mediated by different subpopulations, but it is possible that an overlapping subpopulation exists, perhaps having the Ia(-), 9.6 (-) phenotype. Results from competition experiments are consistent with the possibility that there exists both separate and overlapping cell populations as effectors of NK(HSV-1) and NK(FS). Varying numbers of unlabeled target cells were mixed with effector cells and ^{51}Cr-labeled target cells at fixed E:T ratios. K562 cells competed well with ^{51}Cr-labeled K562 cells but only partially with FS or HSV-FS cells. Conversely, unlabeled FS or HSV-FS cells competed well with labeled FS or HSV-FS but only partially with K562 (Fitzgerald _et al_., in preparation).

Our finding that treatment with either Lyt-3 or OKMI reduces only a fraction of NK(K562) activity is somewhat different than that of Fast *et al.*, (11) who reported that treatment with either of these monoclonal antibodies reduced most of the NK(K562) activity. The complement eliminations in that study, however, were performed after the effector cells had been incubated overnight at 37°C, whereas we have studied freshly isolated cells. We have found that following 24 hours incubation, a significant portion of the NK(K562) activity was found in Leu-1 or Leu-4 positive population, whereas nearly all of the NK(K562) activity of freshly isolated cells was found in Leu-1 and Leu-4 negative populations (12). In recent studies, Zarling *et al.*, (13), have suggested that OKMI and an antibody similar to anti-Lyt-3 recognize different populations of NK(K562) effector cells. Our data would tend to support this hypothesis.

TABLE III. Summary of Characteristics of NK(FS), NK(HSV-1) and NK(K562)

Characteristic	NK(FS)	NK(HSV-1)	NK(K562)
Enriched in large granular fractions of Percoll gradients?	Yes	Yes	Yes
Competed for by unlabeled:			
FS	Yes	Yes	Partially
HSV-FS	Yes	Yes	Partially
K562	Partially	Partially	Yes
Predictive of GvHD	No	Yes	No
Cell Surface Markers (0--++++)			
Leu 1	0	0	0/+
Leu 2	0	0	0
Leu 3	0	0	0
Leu 4	0	0	0/+
Lyt 3	0	0	++
Ia	+	+	0
OKMI	+	+	++

Timonen *et al* (14,15) have reported that the NK activity against K562 cells and several other targets is mediated by cells which band in the low density fractions of Percoll density gradients, fractions which are highly enriched for "large granular lymphocytes" (LGL). We similarly find the NK activity against FS, HSV-FS and K562 cells in LGL enriched fractions of Percoll gradients (Fitzgerald *et al.*, in preparation). Thus although we are able to distinguish between NK (K562) and NK(HSV-1) effector cells in patient studies, competition studies and serologically, it appears that both types of NK are mediated by cells with a similar morphology. It remains to be determined whether the different subpopulations are simply different maturational stages within the same cell lineage, or whether they actually represent totally independent lineages.

In summary, we have presented evidence for heterogeneity of human NK cells. This heterogeneity may explain discrepancies in NK data being obtained in different laboratories (e.g. the correlation of NK(HSV-1) but not NK(K562) with GvHD in bone marrow transplant recipients (4,6) and emphasizes the importance of choosing appropriate target cells for human NK studies.

REFERENCES

1. Lopez, C., Ryshke, R., and Bennett, M. Infect. Immun. 28:1208 (1981).
2. Ching, C. and Lopez, C. Infect. Immun. 26:48 (1978).
3. Lopez, C., Kirkpatrick, D., Read, S., Fitzgerald, P., Pitt, J., Pahwa, S., Ching, C., and Smithwick, E. Submitted.
4. Lopez, C., Kirkpatrick, D., Sorell, M., O'Reilly, R. and Ching, C. Lancet 2:1103 (1979).
5. Livnat, S.. Seigneuret, M., Storb, R., and Prentice, R.L. J. Immunol. 124:481 (1980).
6. Lopez, C.. Kirkpatrick, D., Livnat, S., and Storb, R. Lancet 2:1025 (1980).
7. Stutman, O., Lattime, E., and Figarella, E. Fed. Proc. 50:2699 (1981).
8. Lust, J., Kumar, V., Burton, R., Bartlett, S., and Bennett, M. J. Exp. Med. 154:306 (1981).
9. Pross, H., Barneo, M., Ruban, P., Shragge, P., and Patterson, M. J. Clin. Immunol. 1:51 (1981).
10. Wycsocki, L., and Sato, V. Proc. Nat. Acad. Sci. USA 75:2844 (1978).
11. Fast, L., Hansen, J., and Newman, W. J. Immunol. 127:448 (1981).
12. Fitzgerald, P., Evans, R., and Lopez, C. in "Mechanisms of lymphocyte activation", (K. Resch and H. Kirchner, eds). p. 595, Elsevier/North Holland, N.Y. (1981).
13. Zarling, H., Cluse, K., Biddison, W., and Kung, P. J. Immunol. 127:2575 (1981).
14. Timonen, T., and Saksela, E. J. Immunol. Methods. 36: 285 (1980).
15. Timonen, T., Ortaldo, E.J., and Herberman, R. J. Exp. Med. 154:569 (1981).

Phenotypic and Functional Characterization of Natural Killer Cells by Monoclonal Antibodies.[1]

Marc G. Golightly, Barton F. Haynes, C. Phillip Brandt, and Hillel S. Koren.[2]

Division of Immunology and
Division of Rheumatic and Genetic Diseases
Duke University Medical Center
Durham, North Carolina

I. INTRODUCTION

Large granular lymphocytes (LGL), which can be purified by Percoll gradient centrifugation, have recently been described to be responsible for natural killer (NK) cell activity (Timonen et al., 1979). The lineage of these NK cells continues, however, to be an unresolved question (Herberman and Ortaldo, 1981) despite the fact that highly purified NK cells can be obtained. Recently, Ortaldo et al. (1981) using monoclonal antibodies have demonstrated that these highly purified cells shared certain surface determinants with T cells, monocytes, and myeloid cells. Using peripheral blood lymphocytes (PBL), it has been demonstrated that NK activity can be reduced using selected monoclonal antibodies when employed in conjunction with complement (Zarling and Kung, 1980; Zarling et. al., 1981).

The aim of our study was to employ a selected panel of monoclonal antibodies in order to characterize the phenotypes of highly purified endogenous and interferon (IFN)-activated LGLs. We also examined the effect of these reagents on the functional properties of LGL at the single cell level.

The results presented in this chapter expand and confirm the information regarding the repertoire of cell surface determinants present on LGL and demonstrate that the OKT10 antibody was capable of interfering with the killing of K562 tumor target cells at the single cell level.

[1]Supported by NCI grants CA 09058, CA 23354, and CA 28936.
[2]Recipient of Research Cancer Development Award CA 00581 from the National Cancer Institute.

ISBN 0-12-341360-5

II. METHODS

A. LGL Preparation

Human PBL obtained by Ficoll Hypaque were subjected to plastic and nylon wool adherence procedures as previously described (Kall and Koren, 1978). These cells were then separated on a seven step discontinuous Percoll density gradient as described elsewhere (Timonen et al., 1981). The most highly purified and active fraction, as determined by Giesma stain and cytotoxicity against ^{51}Cr labeled K562 targets, was used as the LGL source. In some cases these cells were further purified by preparative high-affinity (29°C) E-rosetting (Ortaldo et al., 1981).

B. Monoclonal Fluorescent Assay

The fluorescent assay was performed as previously described (Haynes et.al, 1981a). Briefly, the LGLs were incubated with each monclonal antibody (45 min @ 4°C) followed by washing and incubation (45 min @ 4°C) with affinity purified FL-α-mouse IgG (F/P=6). A minimum of 200 cells were scored for each monoclonal antibody.

C. Single Cell Assay

The ability of various monoclonal antibodies to interfere with the binding or killing of K562 target cells was determined at the single cell level using the conjugate/agarose assay described for monocytes (Fischer et al., 1982). However, the assay time was shortened to one hour and the LGL were incubated 45 min. at 22°C in the presence of the appropriate monoclonal antibodies prior to using in the assay.

III. RESULTS

A. Phenotypic Characterization of Endogenous and IGN-Activated LGL

The panel of monoclonal antibodies used to examine the phenotypic characteristics of endogenous LGL is listed in Table 1. The data show that all T cell-directed monclonal antibodies reproducibly stained various proportions of the LGL preparation except for NA1/34 which detects human cortical thymocytes but did not stain the LGLs. The majority of the LGLs were stained with the monocyte (4F2) and monocyte/myeloid (OKM1) antibodies. It is of interest that clone 61D3 (BRL Inc., Gaithersburg, MD.) specific for monocytes was notdetected

TABLE I. Phenotypic Characterization of Endogenous and IFN-Activated LGL[1]

Monoclonal[2] Antibody	General Specificity	Fluorescent Positive Cells (%)				
		Exp1	Exp2		Exp3	
		IFN -	IFN -	IFN[4] +	IFN -	IFN[4] +
3A1	Pan T	94	95	93	95	95
17F12	Pan T	-	24	34	19	23
10.2 *	Pan T	58	-	-	-	-
LEU4	Pan T	-	32	18	24	30
OKT3 *	Pan T	47	-	-	-	-
OKT4	Inducer T cell	28	19	18	21	21
9.3	Inducer T cell	69	-	-	-	-
OKT8	Cytotox/Suppressor	25	33	42	37	29
OKT10	Thymocoytes	-	63	69	59	64
NA1/34	Human Cortical thymocytes Langerhans Cells	0	0	0	0	1
4F2	Monocyte	-	98	95	98	100
OKM1[3]	Monocyte, Myeloid	48	73	80	85	86
L243	Ia-like	5	-	-	-	-
OKI1*	Ia-like	-	2	7	5	4
5E9	Transferrin Receptor	4	0	0	0	0
3F10	non-polymorphic HLA	100	99	100	100	100

[1]Typical LGL morphology by Giemsa Staining of the LGL preparation ranged from 53% to 60% .
[2]Grouping of monoclonal antibodies indicated by an (*) recognize the same molecule. Supernant from P3 clone (an Ig control) was always performed and was less than 1% positive.
[3]Average of two determinations for each experiment.
[4]Cells activated for 18 hrs with 100u/ml IFNα

on the LGL (data not shown). Most LGL do not express Ia-like molecules as detected by L243 and OKI1 but do express HLA antigens. Interestingly, overnight activation of LGL with IFNα did not change the phenotypic expression of those cells; neither did the profile of labeling of control cells incubated for 18 hrs change. The transferrin receptor felt to be an "activation" marker as detected by the 5E9 monoclonal antibody (Haynes, et.al.) was absent on both endogenous and IFN-activated NK cells. That activation did occur was always verified by examining the NK activity against ^{51}Cr labeled K562 targets.

To further increase the purity of the LGL, they were rosetted under conditions which enrich for low-affinity E rosetting cells (see methods section). Most of the monoclonal antibodies stained similar proportions of the low -affinity LGLs with the exception of several T cell markers: 17F12, Leu4 OKT3), 9.3 and Leu3A (OKT4). These decreased to 5-6, 2-4, 0.5-2, and 1-2%, respectively, In contrast the level of staining with OKT10 increased to 85%.

B. Functional Inhibition of LGL by Monoclonal Antibodies at the Single Cell Level.

A select group of monoclonal antibodies were examined for their ability to inhibit binding or killing in the conjugate agarose assay. The data presented in Fig 1 show that several of the monclonal antibodies including those that did not stain LGL (Table I -NA1/34&5E9) inhibited binding between 0 to 35%, probably in a nonspecific fashion. However, OKT10 (present on 59 to 69% of LGL-Table I) was the only antibody which inhibited killing profoundly in 3/3 experiments (57,75, and 90% inhibition). 4F2 which detects nearly 100% of the LGL (Table I) caused neither inhibition of binding nor of killing.

IV. DISCUSSION

In spite of the intensive research and numerous attempts aimed at understanding the lineage of NK cells, we still have not resolved this intriguing question. Recently, an extensive study using monoclonal antibodies directed against a variety of determinants associated with various lymphoid subpopulations, established that the LGL is the predominant cell type in the perpherial blood detected by OKT10 monoclonal antibody (Ortaldo et al., 1981).

The major goals of our study were to compare the phenotypes of endogenous LGL with IFN-activated LGL by employing a broad spectrum of monoclonal antibodies and to test their ability to inhibit binding and/or killing of tumor target cells. Our data have shown a very similar distribution of the majority of the monoclonal antibodies previously described (Ortaldo et al., 1981). In addition, we have tested several monoclonal antibodies [4F2, 17F12 (10.2), 9.3, NA1/34, 5E9, 3F10, and OKI1 (L243)] not previously examined with LGL. The fact that 4 of 7 T cell monoclonal antibodies [17F12, Leu4 (OKT3), 9.3 and Leu 3A(OKT4)] detected extremely low percentages of positive cells after depleting high-affinity E rosett-

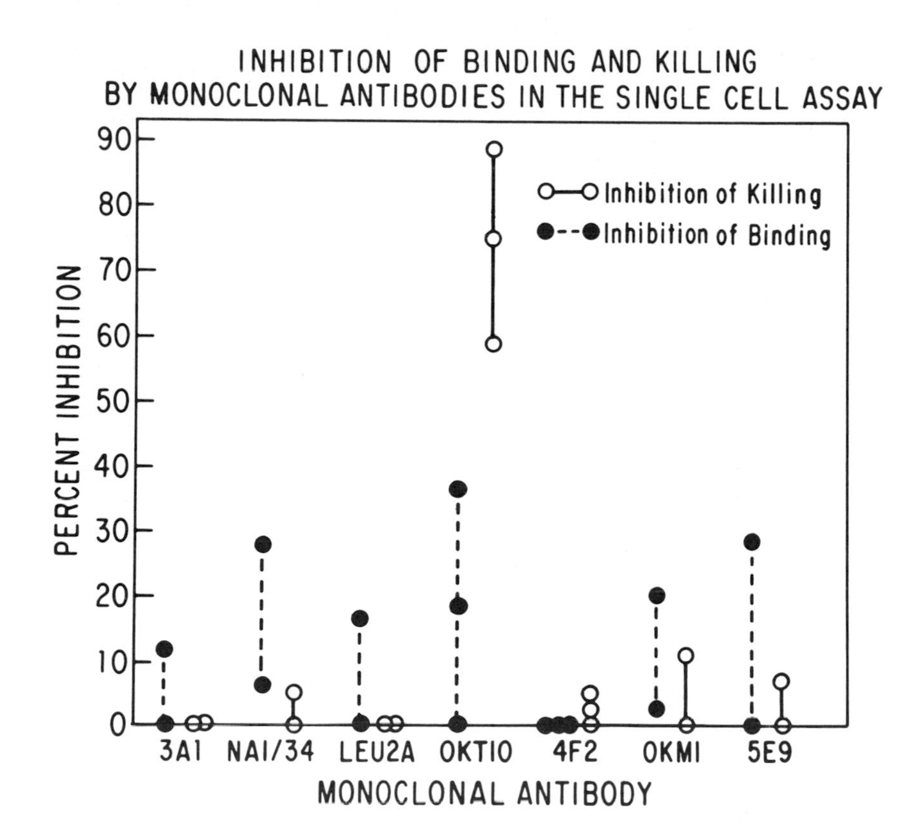

Figure 1- optimal dilution of each monoclonal antibody as determined by fluorescense (3A1 = 1:1000, NA1/34 = 1:100, Leu 2a (OKT8) = 1:50, OKT10 + 1:500, 4F2 = 1:500, OKM1 = 1:20, 5E9 = 1:1000). When a 4-fold dilution of the optimal concentration was used similar results were obtained (data not shown). The lines represent the range of percent inhibitory activity by each of the monclonal antibodies tested in 2 separate experiments, except for OKT10 and 4F2 where the range represnets 3 experiments performed with 2 different individuals. The binding and killing in the conjugate/agarose assay without monoclonal antibody ranged between 25-38% and 19-37%, respectively.

ing cells from the LGL preparation may be the result of a higher degree of purity in this preparation or due to heterogeneity of the LGL with respect to some of the T cell determinants.

One of the features of IFN is its ability to modulate the expression of cell surface determinants (Gresser, 1977). Our results with IFNα activated cells, failed to show major detectable changes in the expression of the determinants recognized by the monoclonal antibodies employed, including the transferrin receptor often associated with activated cells (Haynes, et al., 1981b). However, it is possible that minor antigenic changes were not detected by the fluorescent assay.

The fact that only OKT10 inhibited killing at the single cell level significantly and reproducibly is novel and intriguing. To date, inhibition of natural killing was only demonstrated by pretreating PBL with OKM1 or OKT11A +C as detected in a ^{51}Cr release assay (Zarling et al., 1981; Zarling and Kung, 1980). In the single cell conjugate/agarose assay inhibition of binding to K562 targets was observed with several monoclonal antibodies. This inhibition, however, was most likely nonspecific since at least 2 of them (NA1/34 and 5E9) do not have determinants on LGL (table I), moreover, killing was not affected by any of these monoclonal antibodies. The ability of OKT10 to inhibit killing dramatically suggests that the LGL determinants recognized by the OKT10 antibody may be involved in the lytic step of the LGL-tumor cell interaction.

ACKNOWLEDGMENTS

The authors thank the following individuals for exchange of reagents; Drs Edgar Engelman, Robert Evans, Giedon Goldstein, John Hanson, Ron Levy, Andrew McMichael, and Robert Mittler.

REFERENCES

Fischer, D.G., Golightly, M.G., and Koren, H.S., Chapter in this book.
Gresser, I., Cell. Immunol. 34:406 (1977).
Haynes, B., Metzgar, R.S., Minna, J.D., and Bunn, P.A., N. Engl. J. Med., 304:1319 (1981).
Haynes, B.F., Hemler, M., Contner, T., Mann, D.L., Eisenbarth, G.S., Strominger, L., and Fauci, A., J. Immunol. 127:347 (1981)
Herberman, R.B., and Ortaldo, J.R., Science 214:24 (1981).
Kall, M., and Koren, H.S., Cell. Immunol, 40:58 (1978).
Ortaldo, J.R., Sharrow, S.O., Timonen, T., and Herberman, R.B. J. Immunol. 127:2401.
Timonen, T., Ortaldo, J.R., and Herberman R.B., J. Exp. Med. 153:569 (1981).
Timonen, T., Sakesela, E., Ranki, R., and Häyry, P., Cell. Immunol. 48:133 (1979).
Zarling, J.M., and Kung, P.C., Nature 288:394 (1980).
Zarling, J.M., Clouse, K.A., Biddison, W.E., and Kung, P.C., J. Immunol. 127:2575 (1981).

NATURAL CYTOTOXIC ACTIVITY OF PURIFIED HUMAN T OR NULL CELLS. FURTHER ANALYSIS WITH MONOCLONAL ANTIBODIES.[1]

Jean Caraux[2]
Bernard Serrou
William O. Weigle

Department of Immunopathology
Scripps Clinic and Research Foundation
La Jolla, California, USA

and

Department of Tumor Immunopharmacology
Centre Paul Lamarque
Montpellier, France.34.

I. INTRODUCTION

Human natural killer (NK) cells have been characterized by a variety of conventional functional and surface markers of lymphoid cells (1). These studies indicate that they are non-phagocytic non-adherent cells with surface Fc receptor for IgG. NK activity is enriched in the null cell compartment but also present in the E-rosetting population of lymphocytes (T cells) (2-4). More recently, their phenotype was also studied using monoclonal antibody reagents against surface determinants of human mononuclear cells. These NK cells have been described to be OKT3$^-$, OKT8$^-$ (T cytotoxic/suppressor subset), OKT4$^-$ (helper/

1. Work supported by funds from the CNRS, INSERM, DGRST and Ligue Nationale Francaise contre le Cancer.
2. Present address: Scripps Clinic and Research Foundation.

ISBN 0-12-341360-5

inducer) and partially OKM1$^+$ (monocytes, null cells) (5). The purpose of the present study was to quantitate the NK cytotoxic potential of purified T cells or purified null cells after treatment with monoclonal antibodies and C' to remove T or null cells bearing the corresponding determinants.

II. METHODS

Isolation of enriched human T cells or enriched null cells has already been described elsewhere (6) as well as their surface markers (E, EA, SmIg, Peroxidase)(6), ADCC activity (6), or T cell colony generation potential (7). Briefly, monocytes were removed from peripheral blood mononuclear cells preparations by phagocytosis of iron particles and adherence to Dextran. B cells were removed by anti-F(ab')2 immunoadsorbtion and the resulting T and null cell preparation was further split into enriched T cells and enriched null cells by rosetting with sheep red blood cells and sedimentation of the rosetting cells on Ficoll Hypaque. NK activity was assayed in micro-well dishes against ^{51}Cr labeled target cells (the human JM Tcell line). These targets were chosen for their high and reproductive sensitivity to NK cell lysis, allowing precise titration of the lytic capacity of individual cell populations. In order to remove cells bearing a given monoclonal antibody defined determinant, enriched T cells or enriched null cells were exposed to a dilution of the monoclonal antibody and subsequently to C'. Cell death was estimated by the Trypan Blue dye exclusion technique, in Terasaki micro-wells. Monoclonal antibody defined phenotype was estimated by microcytotoxicity as well as by surface immunofluorescence using FITC labeled anti-mouse immunoglobulin.

III. RESULTS AND DISCUSSION

As shown in Figure 1, NK activity against the JM cells was enriched after monocyte and B cells removal (T and null population,▼) as compared to unfractionated mononuclear cells (●). It was highly concentrated in the null cell fraction (△) and also present , though to a much lesser extent into the E-rosetting T cell fraction (□). This confirmed previous reports which used different human tumor target cells. It is also worth noting that this pattern of cytotoxic reactivity was strikingly similar to that of K cell activity which we previously described (6) using a similar approach.

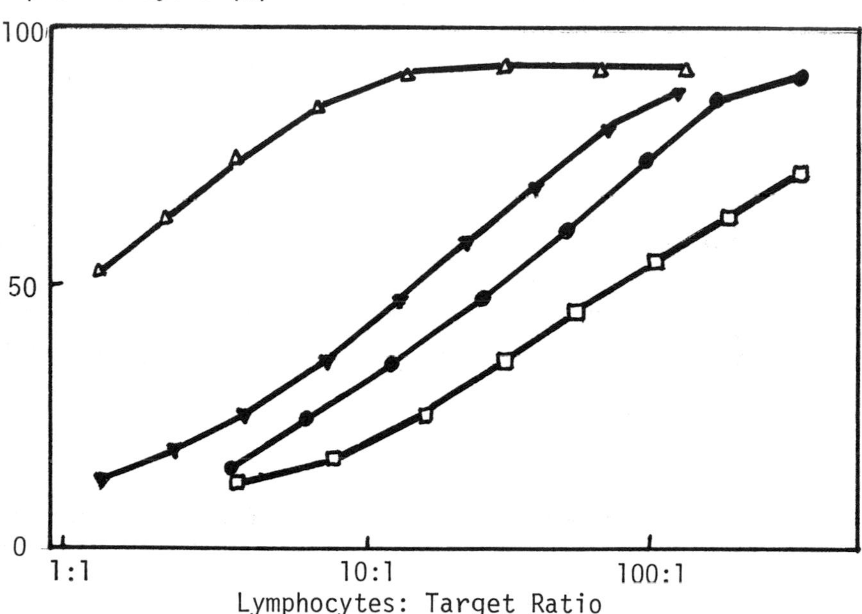

Fig. 1. Titration of NK cell activity in human mononuclear cells (●), monocytes and B cells depleted T and null cells (▼), enriched T cells (◘) or enriched null cells (▲). ^{51}Cr labeled JM target cells, 10 hours assay.

We then proceeded to elimination of OKM1+ cells or OKT3+ cells in purified null cells or T cells by subsequent treatment with monoclonal antibody and complement. Titration of NK cell activity against JM cells was then assayed on the remaining depleted T cells or depleted null cells.

In the null cell population (Figure 2), treatment with OKT3 and C' did not produce any major change in the high cytolytic capacity of the Null cells. This is not surprising since under 5% of the null cells were OKT3 positive as estimated by indirect fluorescence. In contrast, exposure to OKM1 and C' lysed 65% of the Null cells and resulted in complete abrogation of their NK activity. We could thus conclude that null-type NK cells were OKM1+ and OKT3$^-$.

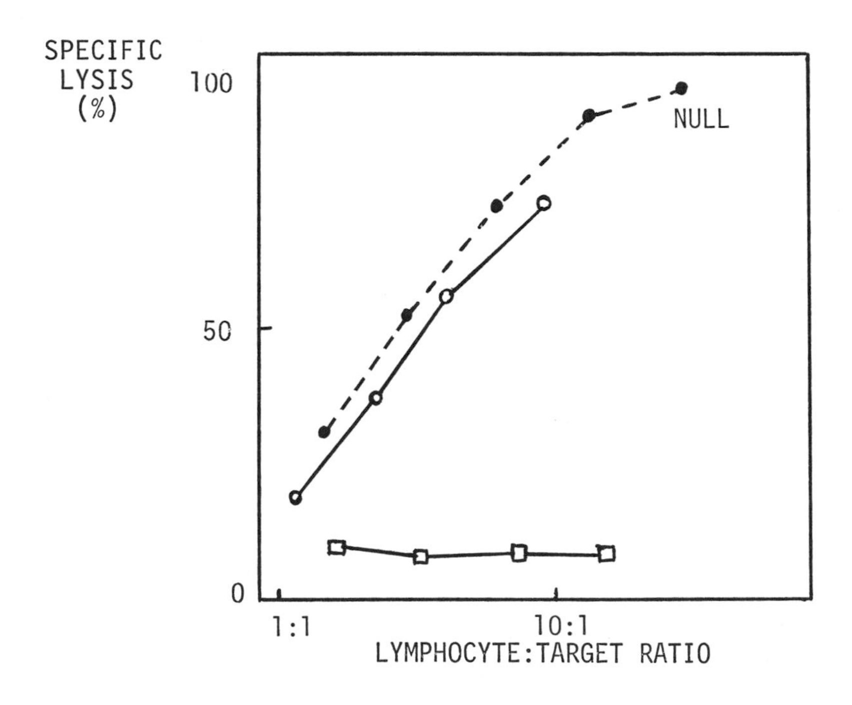

Figure 2. Titration of NK cell activity in human null cells (●--●) after treatment with OKT3 + C' (○) or OKM1 + C' (□). 51 Cr labeled human JM target cells.

In the T cell population, treatment with OKT3 and C' depleted 88% of the E-rosetting cells but produced a striking increase in the NK capacity of the remaining T cell subset. This could have been accounted for by selection of some $OKM1^+$ null cells in the OKT3 depleted E-rosetting fraction. However, simultaneous treatment of the E-rosetting T cells with OKM1 and OKT3 also resulted in an increased NK activity. This indicates that NK effector cells are present in the E-rosetting T cells but are different from their null cell counterparts in that they are $OKM1^-$. (Figure 3).

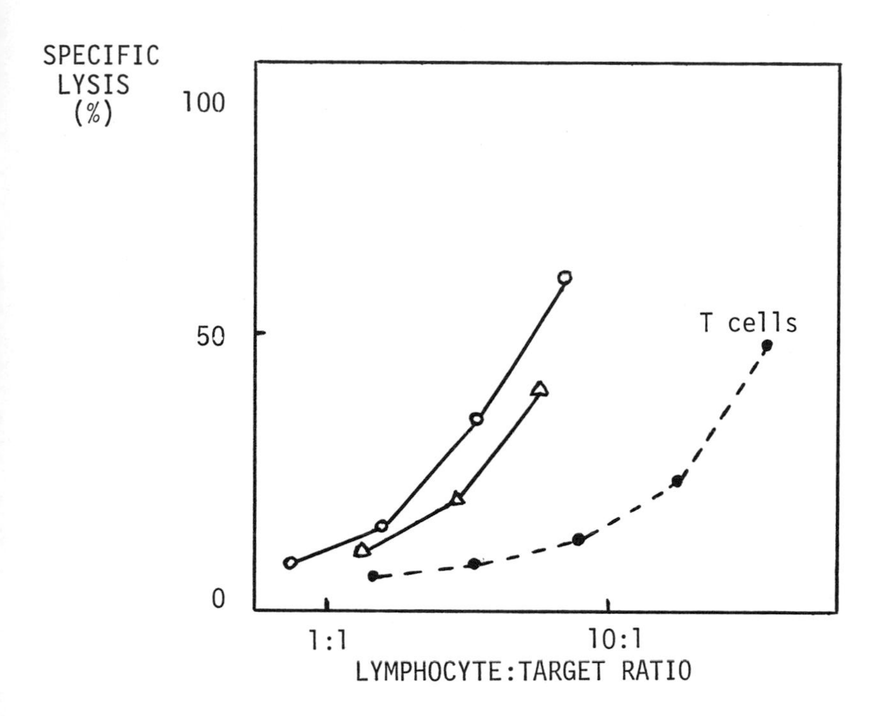

Figure 3. Titration of NK cell activity in human T cells (---) after treatment with OKT3 and C' (○) or OKT3 and OKM1 and C' (□). ^{51}Cr labeled JM target cells.

Previous studies using depletion of specific human peripheral blood mononuclear cell subsets with monoclonal antibody and C' have allready demonstrated that treatment of lymphocytes with OKM1 + C' resulted in diminution but not complete abrogation of the NK activity. Our present studies further indicate that this is due to depletion of null cell type effectors which can thus be defined as E-rosette$^-$, OKT3$^-$, OKM1$^+$ NK cells.

On the other hand, we can also define a different subset of cells containing NK effectors. This subset segregates with E-rosetting T cells but is not lysed by OKT3 or OKM1. It can thus be defined as E-rosette$^+$, OKT3$^-$, OKM1$^-$ NK cells.

Attribution of NK cytotoxic capacity to T cells or null cells was previously a matter of controversy. This was due to the enrichment of NK effectors in the small null cell subset while E-rosetting NK cells were diluted in the much larger T cell population. From the present studies it appears that NK effectors can indeed be found in both T or null cells. T or null cell type NK cells can be differentiated not only on the basis of E-rosetting properties but also of OKM1 reactivity. It remains to be known whether these two phenotypes correspond to different functional states of the NK lineage, one being acquired upon activation, or whether it delineates two different pathways of NK cell differentiation.

REFERENCES

1. Herberman, R.B., ed. Natural Cell-Mediated Immunity against Tumors. Academic Press, New-York. (1980).
2. Kay, H.D., Bonnard, G.D., West, W.H. and Herberman, R.B. J.Immunol. 118,2058 (1977).
3. West, W.H., Cannon, G.B., Kay, H.D., Bonnard, G.D. and Herberman R.B., J.Immunol. 118,355 (1977).
4. Kay, H.D., Horwitz, D.A., J.Clin.Invest. 66,847 (1980).
5. Zarling, J.M., Kung, P.C. Nature 288, 384 (1980).
6. Caraux, J., Thierry, C., and Serrou, B. Eur. J. Immunol. 8, 806 (1978).
7. Klein, B., Caraux, J., Thierry, C.,Gauci, L., Causse, A. and Serrou, B. Immunology 43, 39 (1981).

HUMAN BLOOD LYMPHOCYTE SUBSETS CHARACTERIZED BY MONOCLONAL REAGENTS

Giuseppe Masucci
Maria G. Masucci
Maria T. Bejarano
Eva Klein

Department of Tumor Biology
Karolinska Institute
Stockholm, Sweden

I. INTRODUCTION

The abundant literature which deals with the characterization of lymphocyte subset with NK effect reflects its phenotypic heterogeneity. The lymphocytes have been studied for cell membrane markers such as nylon adherence, expression of SRBC, Fc and lectin receptors, morphological features and reactivity with monoclonal reagents. Taking in consideration these features the activity of the human blood lymphocytes can be enriched by separating nylon wool passed, large, granular, Fc gamma receptor positive cells not expressing high avidity E receptors and reacting with the monoclonal antibody OKM1 (Breard et al., 1980).

Due to the quantitative variations in the expression of these markers, each differring independently from the other, it is not possible to separate <u>all</u> active cells and <u>all</u> inactive cells in clean populations.

Subsets separated on the basis of nylon adherence, expression of E receptors of different avidity and capacity to form EA rosettes, differ in the anti-K562 activity (Bakács et al., 1978; Masucci et al., 1980).

Supported by Grant No 5 R01 CA-25250-02 awarded by the National Cancer Institute DHEW and by the Swedish Cancer Society.

ISBN 0-12-341360-5

The recent development of monoclonal antibodies directed against surface moieties of white blood cells has introduced a new dimension in the characterization and separation of lymphocyte subsets. We have analyzed the subsets separated on the basis of nylon adherence E and EA rosetting capacity for their composition, with regard to reactivity with the monoclonal reagents: OKT3, OKT4, OKT8, OKM1 and OKIa1.

II. MATERIAL AND METHODS

Our aim was to obtain subpopulation of lymphocytes having E receptors of different avidity (E^-, E^+_{low}, E^+_{high}) with or without concomitant expression of Fcγ receptors (EA^+, EA^-).

The separation of lymphocyte subsets was carried out in four steps involving nylon column passage, E and EA rosetting (Masucci et al., 1980).

Reactivity with the monoclonal antibodies (OKT3, OKT4, OKT8, OKM1 and OKIa1, Ortho Pharmaceutical Co. Raritan, New Jersey, 08869, USA), was assessed by indirect immunofluorescence (Reinherz et al., 1979a; Reinherz et al., 1979b; Kung et al., 1979; Reinherz et al., 1979c; Reinherz et al., 1980).

Cell morphology was studied on cytospin centrifuge preparations stained with Giemsa.

The cytotoxicity assay was performed as described previosly (Masucci et al., 1980). Each fraction was tested with and without 2 hrs interferon pretreatment (1000 U/ml human IFN-α, specific activity 10^6U/ml).

III. RESULTS

The cell yield and surface marker characterization of the subsets are given in Table I.

A. E Rosette Negative Subsets

The nylon adherent E^- population contained the SIg positive B lymphocytes and a high proportion of cells reacted with OKM1 and OKIa1. The nylon passed E^- fraction had 27% OKT3, 45% OKM1 and 25% OKIa1 positive cells. The $OKM1^+$ cells expressed Fc receptors and could be selected out by EA rosetting. The EA^- cells contained the cells which reacted with the T cell defining reagents (OKT3, OKT4, OKT8). The sum of cells reacting with OKT3, OKM1 and OKIa1 did not reach 100%, suggesting that part of the population was not defined by the reagents used.

TABLE I. Characterization of the lymphocyte subsets[a]

	Yield[b]	% LGL	% Rosetting cells			SIg	Reactivity with monoclonal reagent[d]				
			E	E^c	EA		OKT3	OKT4	OKT8	OKM1	OKIa1
Total population		19	66	75	16	12	72	43	21	22	11
Adherent E^-	2	5		<1	n.d.	58	5	3	2	50	59
Non adherent E^-EA^-	3	40		<1	2	0	23	17	6	20	13
- " - E^-EA^+	2	61		<1	87	1	3	2	1	63	15
Adherent E^+	5	12		84	n.d.	10	67	42	23	13	23
Non adherent $E^+_{low}EA^-$	2	26			n.d.	2	16	8	6	25	12
- " - $E^+_{high}EA^-$	20	5	81	>99	n.d.	0	86	56	25	2	0
- " - $E^+_{low}EA^+$	2	67	0	70	n.d.	0	10	3	29	64	4
- " - $E_{high}EA^+$	12	61	81	>99	n.d.	0	86	52	23	3	1

[a]Mean of 9 experiments.
[b]Expressed as percentage of input.
[c]Rosetting in presence of 7% Dextran.
[d]Expressed as percentage of positive cells.

B. E Rosette Positive Subsets

The majority of the cells in the nylon adherent E^+ fraction reacted with OKT3, 23% reacted with OKIa1. It is likely that the latter reagent detects activated T cells. Compared to the total population, the nylon passed E^+ fraction was slightly depleted of $OKM1^+$ cells, which separated with the EA^+ part of this subset. Elimination of the E_{low} cells resulted in a relative increase in $OKT3^+$ and $OKT4^+$ and in a substantial depletion in $OKM1^+$ cells. Accordingly the $OKM1^+$ cells were found in the E_{low} subsets; the majority in the EA rosetting part. In spite of the separation on the basis of E rosetting, this subset did not react with OKT3 and OKT4. However a proportion of the cells were $OKT8^+$. There was an inverse relationship between the ratio $OKT4^+/OKT8^+$ (0.1) cells when compared to the total population (2.0). In the unfractionated population, 19% of the cells were large granular lymphocytes (LGL). All three EA^+ fractions, E^-, E_{low} and E_{high} were enriched in these cells. The proportion of LGL decreased progressively in the EA^- fractions, depending on their E rosetting capacity.

C. Recoveries of Cells Defined with The Monoclonal Reagents

The recoveries of each cell category relative to the input are shown in Table II. 44% of the cells which reacted with the T reagents were recovered after the separation and since this percentage was similar to the total recovery of lymphocytes, there was no selective loss.

A substantial proportion of $OKM1^+$ cells was lost. This occurred in part during nylon wool column passage, because after this step only 45% $OKM1^+$ cells were recovered while the total cell recovery was 68%. Only 27% of these cells were present in the final fractions.

The $OKT3^+$ cells were in the E_{high} fractions. The distribution of the various categories of T cells (OKT3, OKT4 and OKT8) was similar in all fractions, except in the $OKM1^+$ rich E_{low} EA^+ subset which contained 6% of the recovered $OKT8^+$ cells but much less of the $OKT3^+$ and $OKT4^+$ cells. A high proportion of $OKM1^+$ cells (48%) could be collected in the EA^+ fractions. The majority (42%) had no or low avidity E receptors. Since these cells represented only 12% of the recovered lymphocytes the enrichment was considerable. As expected the $OKIa1^+$ cells were in the nylon adherent fractions. 34% of these were E^+. These data are in accordance with the observation that activated T cells, which are known to express Ia antigen, adhere to nylon wool.

TABLE II. Cytotoxic Activity and Distribution of The Various Cell Types in The Subsets

	Distribution of recovery[a]						Cytotoxic activity lytic units (LU)/10^6 [b]		
	cells	OKT3	OKT4	OKT8	OKM1	OKIa1	K562		Daudi
							–	+IFN	+IFN
Total pop.							23	50	15
Adherent E^-	4.2	0.2	0.2	0.2	16.5	34.9			
Non adherent $E^- EA^-$	6.3	2.0	2.3	1.7	9.9	9.8	100	192	83
– " – $E^- EA^+$	4.2	0	0.2	0.2	20.9	8.8	125	325	8
Adherent E^+	10.4	10.4	10.3	11.6	10.6	33.9	27	80	71
Non adherent $E^+_{low} EA^-$	4.2	0.9	0.6	1.0	8.0	6.8	10	23	20
– " – $E^+_{high} EA^-$	41.6	53.9	55.9	51.2	6.6	0	0	0	0
– " – $E^+_{low} EA^+$	4.2	0.6	0.2	5.8	21.3	2.3	31	100	8
– " – $E^+_{high} EA^+$	25.0	31.7	30.1	28.0	5.8	3.2	1	15	1

[a]The figures are calculated from the data presented in Table I. % total recovery cells: 48, OKT3: 44.1, OKT4: 46.5, OKT8: 46.4, OKM1: 27.2, OKIa1: 30.6.

[b]Calculated for 40% lysis against K562 and 20% lysis against Daudi.

D. Cytotoxic Potential

The anti-K562 killer cells were enriched in the E^- and E_{low} fractions (Table II), which coincided with the enrichment of $OKM1^+$ cells. On a per cell basis, the E_{low} EA^+ subset was four times more efficient than the cells which remained in the interphase after EA rosette separation.

Both of the E^- subsets (EA^+ and EA^-) had similar strong anti-K562 activity, though they had a different profile of reactivity with the monoclonals.

Daudi was killed by the EA^- cells of the E^- and E_{low} subsets. The nylon adherent E^+ fraction which contained 23% $OKIa1^+$ cells had a high anti-Daudi activity.

In order to confirm this difference depending on Fc receptor we separated EA rosetting subsets from the cells collected at the 40% interface of Percoll density gradient. In addition total lymphocyte populations have been exposed to antigen-antibody complex monolayers (Table III). The recovered EA rosetting cells represented 23% of the low density subset and 20% of the lymphocyte attached to the oxEA monolayer. In both systems the cytotoxic activity against K562 was higher in the EA^+ fraction while against the Daudi the EA^- subsets were more active.

IV. CONCLUSIONS

This separation schedule confirms that the LGL population is heterogeneous with regard to cell surface markers. There seems to be a pattern which relates to their anti K562 lytic activity. Among the non-adherent Fc receptor positive subsets, the E^- was more active than the low avidity E^+ one, and the high avidity E and EA receptor positive subset, with 60% LGL, was not active without IFN stimulation.

Another marker which was proposed to define the killer cells was reactivity with the OKM1 monoclonal reagent (Zarling et al., 1980). The results in Table I show that the LGL-s are heterogeneous for this marker since the E_{high} EA^+ subset which contained 61% LGL had only 3% $OKM1^+$ cells. This subset reacted with the other monoclonals which define T cells (OKT3, OKT4 and OKT8) and it had no lytic effect. The lack of activity substantiates the proposal that only OKM1 positive cells are NK active. This seems thus to be valid for the high avidity E positive cells. However, even these markers, i.e. EA receptor expression, OKM1 reactivity and LGL morphology do not define the active cells without qualifications, since two EA^- subsets with similar constitution for these markers, the E^- and the E_{low}, differred in cytotoxic potential.

TABLE III. Cytotoxic Activity of Lymphocyte Subsets Isolated on The Basis of Fcγ Receptor Expression

	Percoll fr. 2[b]		EA_{ox} monolayer[c]	
	EA^-	EA^+	Non adherent($Fc\gamma^-$)	Adherent($Fc\gamma^+$)
Recovery[a]	50 ± 13	23 ± 17	46 ± 11	20 ± 12
K562[d]	33 ± 12	64 ± 28	28 ± 17	46 ± 18
Daudi[d]	22 ± 3	12 ± 7	17 ± 8	8 ± 5

[a]Percentage of input cells: Mean and SD from 3 experiments

[b]Performed according to Timonen et al. (1981). Lymphocytes from fraction 2 (interphase of 40% Percoll) were further separated by EA rosetting. This fraction contained 72% LGL.

[c]Fc receptor positive cells were depleted by adherence to monolayers of ox erythrocytes coated with rabbit IgG anti ox erythrocytes. The adherent cells have been removed from the monolayer by a rubber policeman.

[d]Percentage of specific ^{51}Cr release at 25:1 effector to target ratio. IFN treated effectors. Mean and SD of 3 experiments.

It was also proposed that only cells with Fc receptor are NK active (Santoli et al., 1978). (The consensus is that the receptor is only a marker and it is not participating in the killing function). However, our separation shows that a small proportion of cells which did not sediment as EA rosettes have NK function and these are within the non-adherent E^- population. It is important to note the great difference between the Fc receptor negative E^- and the E_{low} subsets in activity. Their proportion of $OKM1^+$ cells is similar but they differ in LGL content.

There was a high proportion of $OKIa1^+$ cells in the fractions cytotoxic for Daudi. This suggests that activated T cells may contribute to this function. OKIa1 reacts with Ia antigens present on B cells and on activated T cells (Reinherz et al., 1979c; Reinherz et al., 1980).

The presence of activated T cells in the nylon non-adherent E^- population is also suggested by the finding that this subset had higher spontaneous ^3H-thymidine incorporation values compared to the total population (Masucci et al., submitted). 29% of the cells in this fraction reacted with OKT3 which defines mature T cells in the blood. When Daudi was used as target, only the EA^- $OKM1^+_{poor}$ subpopulations expressed significant enrichment in cytotoxic activity.

The heterogeneity of NK active cells has been acknowledged by all investigators who studied the conventional cell surface markers. The recent development of monoclonal reagents and the assignment of the effect to cells with a certain morphology has led to their more precise characterization. Heterogeneity is however still a fact.

REFERENCES

Bakacs, T., Klein, E., Steinitz, M., and Gergely, Y.P. (1978-79). Haematologia 12(1-4):113.

Breard, J., Reinherz, E.L., Kung, P.C., Goldstein, G., and Schlossman, S.F. (1980). J. Immunol. 124:1943.

Kung, P.C., Goldstein, G., Reinherz, E.L., and Schlossman, S.F. (1979). Science 206:347.

Masucci, G., Masucci, M.G., and Klein, E. Submitted for publication.

Masucci, M.G., Masucci, G., Klein, E., and Berthold, W. (1980). Proc. Nat. Acad. Sci. USA 77:3620.

Reinherz, E.L., Kung, P.C., Goldstein, G., and Schlossman, S.F. (1979a). Proc. Nat. Acad. Sci. USA 76:4061.

Reinherz, E.L., Kung, P.C., Goldstein, G., and Schlossman, S.F. (1979b). J. Immunol. 123:1312.

Reinherz, E.L., Kung, P.C., Pesando, J.M., Ritz, J., Goldstein, G., and Schlossman, S.F. (1979c). J. Exp. Med. 150:472.

Reinherz, E.L., Kung, P.C., Breard, J.M., Goldstein, G., and Schlossman, S.F. (1980). J Immunol. 124:1883.

Santoli, D.G., Trinchieri, G., and Koprowski, H. (1978). J. Immunol. 121:532.

Timonen, T., Ortaldo, J.R., and Herberman, R.B. (1981). J. Exp. Med. 153:569.

Zarling, J.M., and Kung, P.C. (1980). Nature 288:394.

SIZE DISTRIBUTION AND β_2-MICROGLOBULIN CONTENT OF HUMAN NATURAL KILLER CELLS AS ANALYZED BY FLUORESCENCE ACTIVATED CELL SORTING (FACS)[1]

Marianne Hokland[2]
Iver Heron
Kurt Berg

Institute of Medical Microbiology
University of Aarhus
Aarhus, Denmark

Peter Hokland[2]

University Department of Haematology and Medicine
Aarhus Amtssygehus
Aarhus, Denmark

I. INTRODUCTION

Several approaches have been used in the characterization of human natural killer (NK) cells, and it now seems fairly well established that NK cells cannot readily be categorized to only one lineage of cells (Breard et al., 1981), though many of them appear in Giemsa-stained smears as large granular lymphocytes (LGL). Thus, one criterion for purifying NK cells could be their size, and we therefore decided to use the fluorescence activated cell sorter (FACS) for purifying mononuclear cell suspensions of different sizes. Prompted by our discovery that interferon (IFN) enhances the expression of MHC products on lymphoid cells, using β_2-Microglobulin (β_2-M)as a marker (Heron et al., 1978; Hokland et al., 1981) as well as boosting NK activity, we investigated the possible correlations between the amounts of β_2-M on different lymphocyte subsets and NK activity.

[1]Supported by grants from the Danish Cancer Society (M.H. and K.B.) and the Danish MRC (I.H. and P.H.)
[2]Present address: Division of Tumor Immunology, Sidney Farber Cancer Institute, 44 Binney St., Boston, MA USA 02115

ISBN 0-12-341360-5

II. NK ACTIVITY OF FACS PURIFIED CELL PREPARATIONS

Normal peripheral blood mononuclear cells were isolated by Isopaque-Ficoll method and stained with a FITC-conjugated rabbit anti-β_2-M antiserum (Dako, Copenhagen, Denmark). The cells were subsequently processed on a FACS II machine (Becton Dickinson, Mountain View, CA). In Fig. 1, the dot plot screen of the FACS II from a typical experiment is shown, each cell being represented by a dot giving its size (or forward light scatter, x-axis) and the fluorescence intensity (y-axis). As shown, both lymphocytes and monocytes (as verified by Giemsa-stained smears) can be delineated and lymphocytes can be further subdivided into 4 lymphocyte subsets according to their

FIGURE 1. Dot plot from the FACS II of mononuclear cells labelled with an anti-β_2-M antiserum. Each dot represents a cell with its size (x-axis) and fluorescence intensity (y-axis). c, all lymphocytes; d, small lymphocytes; e large lymphocytes; f, monocytes; h, cells with high amounts of β_2-M; l, cells with low amounts of β_2-M.

β_2-M content and size. The cell sorting function of the FACS II enabled us to purify the above outlined subpopulations and test them for NK activity against the sensitive erythromyeloid cell line K562. This was done after an overnight incubation with or without 500 units IFN/ml of the cultures.

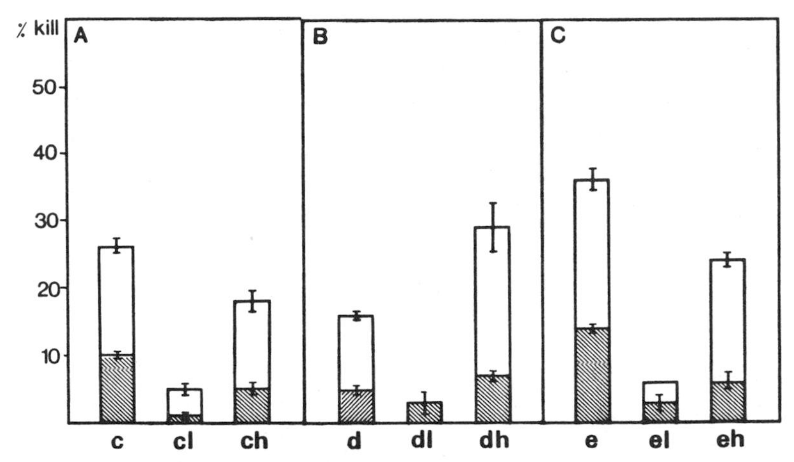

FIGURE 2. NK activity against K562 (E/T ratio, 10:1) of anti-β_2-M labelled PBL sorted according to size into small lymphocytes (panel B) and large lymphocytes(panel C). Unseparated cells run through the FACS were used as controls (panel A). Each population was subsequently separated into high and low fluorescence fractions. Hatched part of columns: NK without addition of IFN. Clear part of columns: total NK after addition of 500 units IFN.

Fig. 2 shows a representative experiment where lymphocytes were sorted, first according to size into small and large cells and then into high and low β_2-M containing cells, thus resulting in 9 cell fractions (PBL sorted only according to

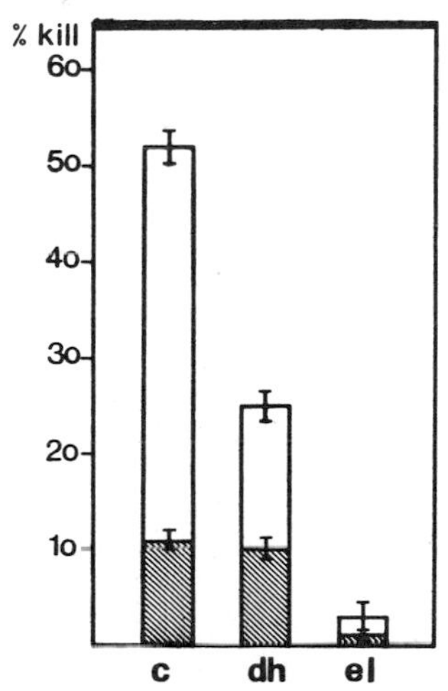

FIGURE 3. NK activity against K562 cells (E/T ratio, 10:1) of *dh* and *el* subsets sorted as indicated in Fig. 1. β_2-M labelled cells run through the FACS served as controls (left column).

β_2-M being included as controls). The FACS was set up to sort the 20% least and the 20% most fluoresceing cells, and it was found that cells with low β_2-M content, irrespective of size, always exhibited lower NK than the high β_2-M containing cells and that IFN inducible NK cells moreover were found in the β_2-M rich cell fractions, indicating that the association between β_2-M and NK might be stronger than that between NK and cell size. Direct evidence for this notion was obtained from experiments where large lymphocytes with low β_2-M content purified by one FACS sorting were compared in NK activity to small lymphocytes with high β_2-M content, since the high β_2-M containing cells exhibited much higher NK activity than those with low β_2-M contents (Fig. 3). The association between β_2-M and NK could also be observed in experiments where a portion of the lymphocytes were sorted according to their content of an antigen defined by a heterologous anti lymphocyte antiserum (ALS), their content of β_2-M and finally only according to their size. From Table 1 it appears that of the small lymphocytes, only those selected on the basis of β_2-M content showed appreciable NK activity.

TABLE 1

NK ACTIVITY OF PBL SORTED ACCORDING TO SIZE AND CONTENT OF A COMMON LYMPHOCYTE ANTIGEN OR β_2-M

			^{51}Cr release of cells sorted according to content of			
			β_2-M		ALKG	
	IFN[a]	Spec ^{51}Cr release[b]	low	high	low	high
Unfractionated PBL	-	3 ± 1	11 ± 2[c]		7 ± 2[c]	
	+	18 ± 3	52 ± 4		21 ± 3	
Small PBL	-	0 ± 1	0 ± 1	10 ± 2	NT	1 ± 1
	+	2 ± 1	1 ± 1	25 ± 3	NT	2 ± 2
Large PBL	-	9 ± 2	0 ± 1	12 ± 3	9 ± 3	NT
	+	42 ± 4	1 ± 1	35 ± 4	22 ± 4	NT

[a]Preincubation with or without IFN (500 iu/ml) for 16h at 37°C.

[b]4h cultures against K562 cells (E/T ratio, 50:1); mean ± SD of triplicate.

[c]Unfractionated cells labelled with antisera but not sorted according to antigen content.

III. CONCLUSIONS

The experiments described in this chapter have clearly demonstrated the advantage of the FACS for sorting cells with different antigen densities. In terms of definition of the human NK cell, our data have to a large extent corroborated the "LGL concept" of Timonen et al. (1981) in that most of the NK activity of PBL was found in the large lymphocyte fraction.

However, by measuring and sorting out cells according to their β_2-M content, we found an even stronger association between NK and β_2-M content (Fig. 3 and Table 1). The heterogeneity of human NK was furthermore clearly illustrated by the delineation of an NK subset of small lymphocytes with high β_2-M content. With respect to the NK cells which could be boosted by IFN, our data again to some extent support those of Timonen et al. (1981), who found this effect in the LGL fraction. On the other hand, lymphocytes with high β_2-M content were more prone to IFN boosting than were the large with small β_2-M content.

In experiments not included here, we were able to demonstrate that though β_2-M content and NK activity correlated closely, the β_2-M molecule (or a molecule with structural homologies to that) is not a part of the NK receptor, since $F(ab')_2$ fragments of the anti-β_2-M antiserum did not inhibit NK. This could in turn imply that β_2-M is co-expressed with the NK receptor, or that the amounts of β_2-M correlated with the activation state of the NK cell. In any case, since neither FACS sorting according to size alone or to content of a common lymphocyte antigen showed the same correlation to NK (Table 1), the β_2-M molecule seems to be selectively associated with NK function.

REFERENCES

Breard, J., Reinherz, E.L., O'Brien, C. and Schlossman, S.F. (1981). Clin. Immunol. Immunopathol. 18:145.
Heron, I., Hokland, M. and Berg, K. (1978). Proc. Natl. Acad. Sci. USA 75:6215.
Hokland, M., Heron, I. and Berg, K. (1981). J. Interferon Res. 4:483.
Timonen, T., Ortaldo, J.R. and Herberman, R.B. (1981). J. Exp. Med. 153:569.

ALLOANTIGENS SPECIFIC FOR NATURAL KILLER CELLS

Robert C. Burton[1]
Scott P. Bartlett
Henry J. Winn

Transplantation Unit
Massachusetts General Hospital
Boston, Massachusetts

I. INTRODUCTION

The best evidence for a subset of lymphohemapoietic cells is the expression of a specific cell surface marker that distinguishes cells of that subset from other cells with similar morphology and/or properties. So the discovery of the Lyt antigens of T cell subsets has been crucial to our understanding of the functions and lineal relations of these cells (1), and the discovery of the NK-1 specific cell surface alloantigen of murine natural killer (NK) cells which kill lymphoma cells _in vitro_ (NK_A cells), constitutes the best evidence for the existence of these cells as a separate sub-class of lymphohemapoietic cells (2,3).

The current phenotype of the NK_A cells: NK-1$^+$, Thy-1$^+$(50%), Lyt-1$^+$(25%), Lyt-2$^-$, Qa-1$^-$(50%), Qa-3$^+$, Qa-4$^+$, Qa-5$^+$, Ly-5$^+$, Ly-6$^+$, Ly-10$^+$, Ly-11$^+$, LyM-1$^-$ Asialo-Gm1$^+$, Mac-1$^+$ does not clearly fit it into

[1]This study was supported by Grants CA-17800 and CA-20044 awarded by the National Cancer Institute, Bethesda, Maryland and by Grants AM-07055 and HL-18646 from the National Institutes of Health, Bethesda, Maryland; and RCB was supported by the John Mitchell Crouch Fellowship of the Royal Australiasian College of Surgeons and a National Health and Medical Research Council of Australia Fellowship in Applied Health Sciences.

ISBN 0-12-341360-5

any known cell lineage (Review: 4). While the discovery of NK-1 has not resulted in the definition of the cell lineage of NK cells, it has been an important advance and has aided or enabled (i) study of ontogeny of these cells (4), (ii) experiments showing that cells bearing these antigens are effective both in preventing tumor growth in Winn assays (5) and development of experimental metastasis in adoptive transfer experiments (6), (iii) the development of NK-1$^+$ cell lines (7), (iv) direct estimation of the fre- quency of these cells in the lymphohemapoietic tissues of adult mice (8-10), and (v) the demonstration that at least 2 subsets of NK cells exist in the spleen of normal mice (11- 13).

This is not to deny that other less specific markers of NK cells which kill lymphomas in vitro have no value. Far from it, Asialo Gml and Qa-5 which are expressed by a proportion of T cells (14,15) and Ly-5 which is expressed by most lymphohemapoietic cells (16) have also proven to be of use, for instance, in studying NK cell mediated lysis, which is blocked by anti-Ly-5 antisera (2,17) , in establishing "NK cell" lines (7), in determining the effects of interferon on NK cells (18), and in studying heterogeneity of activated NK cells (19). It has also recently been shown that Asialo-Gml treated nude mice show enhanced growth of syngeneic and allogeneic lymphomas (20). Nonetheless the finding of specific markers of NK cells and the continued search for subset markers constitute one of the most promising avenues to a better understanding of the biology of these cells. Therefore, in this brief review we will concentrate on the current state of our research in this area.

II. STRATEGIES FOR DEFINING NK CELL SPECIFIC ALLOANTIGENS

A. Alloantigens of NK_A cells

In Table 1 are shown the sera made between different strains of mice of the same H-2 type that we have tested for complement (C) dependent anti-NK activity using normal murine spleen cells and the Moloney virus induced A strain lymphoma YAC. Many sera have some anti-NK activity, so we have concentrated on characterizing those with strong anti-NK activity using a strategy we have developed for identifying specific serological markers of very minor lymphohema poetic subpopulations (3,4). This involves genetic and functional tests which show (i) that the anti-NK activity is separate from any other cytotoxic antibody activity of serum, (ii) that it does not affect T or B cell function and is not directed against macrophages and polymorphs, and (iii) that it is directed against a minor (<10%) subpopulation of murine spleen cells. Backcross and F_2 experiments which indicate the number of loci involved and the relationship between different specific anti-NK alloantisera are also performed.

To date, only the anti-NK-1.2 sera (CE anti-CBA, (CE x NZB)F_1 anti-CBA, (CE x NZB)F_1 anti-BALB/c) have been extensively genetically characterized (3). Anti-NK-1.1 sera (C3H anti-CE, CBA anti-CE, (C3H x BALB/c)F_1 anti CE, C3D2 anti- CE) have not, and so the number of anti-NK cell antigens those sera react with are not known (2,21). Although (C3H x BALB/c)F_1 anti-CE sera do not react with T or B cells (21), data concerning their reactivity with monocytes/ macrophages has not been published.

The relationship between the alloantigens detected by these anti-NK-1.1 and anti-NK-1.2 sera is still under investigation, but it is now very likely that anti-NK-1.2 sera in fact detect not an allele of NK-1 but a new alloantigenic system NK-3. The reasons for provisionally designating anti- NK-1.2 sera as such have been published (3). Since that publication, simultaneous analysis of 55 (CE x CBA)F_2 mice with a CE anti-CBA and a (C3H x BALB/c)F_1 anti-CE antiserum for removal of splenic NK activity against YAC has identified 5 double negative mice. This is not compatible with these sera detecting allelic gene products, and progeny testing of appropriate matings for that F_2 analysis is in progress as a final confirmation of this finding.

The C3H anti-ST serum identifies a specific alloantigen of the NK cell, and the strain distribution of this antigen had suggested that is was not an allele of NK-1 (22). Simultaneous testing of the F_2 backcross mice referred to above with this serum as well, has indicated that it does identify a new NK cell specific alloantigen, NK-2.1 (23). Since all three types of sera react with the same NK cell, the NK cell which kills YAC *in vitro*, we presume that all three identify the NK_A cell. The strain distributions of all three sera are shown in Table II.

Of the other strong anti-NK alloantisera (Table I), the A.BY anti-B10 has been partially characterized and found to contain 2 anti-NK alloantibodies. One of these is shared with lymphocytes and the other is specific for NK cells. Absorption of this serum with B10 thymocytes, and a repeat of the strain analysis should reveal whether this is a new specific anti-NK antibody or not.

TABLE I. Survey of Antisera Made in H-2 Compatible Strain Combinationss for Anti-NK Activity YAC Target

Anti-NK activity		
Strong	Weak	Absent
CE anti-CBA	CBA anti-CE	(AKR x DBA/2)Fl Anti-CE
(CE x NZB)Fl Anti-CBA	(CE x B6)Fl Anti-CBA	(AKR x DBA/2)Fl Anti-BALB/c
C3H Anti-ST	B10 Anti-129	(C3H x DBA/2)Fl Anti-CBA
(C3H x DBA/2)Fl Anti-ST	(AKR x DBA/2)Fl Anti-A	CBA Anti C3H
(C3H x DBA/2)Fl Anti-CE	(AKR x DBA/2)Fl Anti-B10BR	C3H Anti-CBA
RF Anti-AKR	B10 Anti-A.BY	AKR Anti-CBA
B10.A Anti-A	(C3H x DBA/2)Fl Anti-A	(B6 x CE)Fl Anti-A.BY
ABY Anti-B10	C3H Anti-CE	(B6 x DBA/1)Fl Anti-C3H.SW
(CBA x SJL)Fl Anti-C58	(B6 x DBA/1)Fl-Anti-BUB	
(CE x NZB)Fl Anti-BALB/c		
(CE x BALB/c)Fl Anti-B10.D2		

[a] $\geq$50% removal of NK activity from spleen cell suspension of immunizing strain when tested at a 1:10 serum dilution.

[b] <50% removal of NK activity with significant difference ($p<0.05$) from the C alone control, when tested at a 1:10 serum dilution.

[c] Genetic analysis indicates that this serum contains specific anti-NK cell alloantibodies.

B. Alloantigens of NK_B Cells

When spleen cells of appropriate strains of normal mice are treated with CE anti-CBA or C3H anti-ST sera and C, NK activity against YAC is abolished, leaving NK activity against WEHI-164.1, a 3 methyl-cholanthrene induced fibrosarcoma largely intact (11,13). This activity - NK_B cell activity is however sensitive to anti-H-2 or rabbit anti-mouse serum and C treatment (13). To date, 7 sera from Table I have been tested, and one has been found which identifies an alloantigen of NK_B cells (Table III). It is the C3H anti-ST serum. However, the strain distribution of this reactivity indicates that it is quite separate to the anti-NK_A activity (compare Table II and Table III). This serum has yet to be tested with the FLD-3, Friend virus induced NK_B erythroleukemia target, and the systematic characterization necessary to show that it is or is not specific for NK_B cells has been initiated.

III. DISCUSSION

Natural killer cells were orginally defined as lymphohemapoietic cells from normal mice which sponntaneously lysed lymphoma tumor target cells *in vitro* (24). Since T cell, B cells, macrophages, polymorphonuclear leukocytes and platelets can all kill target cells *in vitro* given appropriate conditions, it has been apparent for some time that multiple cells types, perhaps of different lineages, could mediate NK like spontaneous killing of target cells *in vitro*.

This becomes even more apparent when cells activated by interferon and interferon inducers or by *in vitro* culture, under a variety of conditions are considered (9,19,25,26). We have identified 3 phenotypically distinct cells with NK activity in unstimulated cultures of murine spleen cells (26) and Minato et al have identified four using stimulated and unstimulated cells (19).

It seems clear that the source of the spleen cells: from normal, virus-infected, interferon or interferon-mediator treated, very young or old and/or nude mice; whether they are assayed fresh or after short or long term culture; the conditions under which they are cultured prior to assay; the length of the assay itself; and, in particular, the type of tumor target cell used, including whether it is or is not virus infected, all markedly influence the patterns and amounts of target cell killing. As these lytic activities are studied with specific anti-NK antibodies and other reagents directed against cell surface markers it

TABLE II. Strain Distribution of NK Alloantigens

Strains	NK-1 (C3HxBALB/c)F_1 Anti-CE[a]	CE Anti-CBA[b]	C3H Anti-ST[c]
NZB	+	–	+
Ma/My	+	–	+
CE	+	–	+
SJL	+	–	+
BUB	nt	–	nt
CBA	–	+	–
C3H	–	+	–
RF	–	+	+
BALB/c	–	+	+
DBA/2	–	+	+
129	–	+	+
AKR	–	+	–
A	–	+	–
AU	nt	+	nt
C57BL	+	+	+
C58	+	+	+
ST	+	+	+
DBA-1	–	–	+

[a]C3H Anti-CE, CBA Anti-CE, (BALB/c x C3H)F1 Anti-CE, C3D2F1 Anti-CE
[b]CE Anti-CBA, (NZBxCE)F1 Anti-CBA, (CE x CBA)F1 x CE Anti-CBA
[c]C3H Anti-ST, (C3HxST)F1 x C3H Anti-ST
nt, not tested

appears increasingly certain that "natural killing" will be seen to be an *in vitro* property of a number of different subsets of lymphohemapoietic cells, perhaps even of different differentiation lineages. Furthermore, it is possible that this *in vitro* lytic function will, for some of these subsets, have no *in vivo* counterpart, and merely be another detectable shared marker of these cells.

ACKNOWLEDGEMENTS

The authors wish to thank Gloria koo for her generous supply of anti-NK-1.1 antiserum.

TABLE III. Strain Distribution of NK_A and NK_B Alloantigens Reactive with C3H Anti-ST Serum

Strain	Abolition of Splenic NK Activity (C3H anti-ST + C)	
	YAC (NK_A)	WEHI-164.1 (NK_B)
ST/bJ	+	+
NZB/BINJ	+	–
DBA/2J	+	+
C57BL/6J	+	–
C58/J	+	–
Ma/MyJ	+	–
CBA/J	–	–
C3H/HeJ	–	–
BALB/cJ	–	–

DBA/2 Spleen	WEHI-164.1	(Percent Specific Lysis)[a]
	100:1	50:1
C	34±9	26±6
C3H anti-ST+C	14±7	8±4

[a]Mean ± s.e.m. 4 experiments

REFERENCES

1. Cantor, H., and Boyse, E. (1977) Immunol. Rev. 33:105.
2. Cantor, H., Kasai, M., Shen, F., Leclerc, J., Glimcher, L. (1979) Immunol. Rev. 44:1.
3. Burton, R., and Winn, H. (1980) J. Immunol 126:1985.
4. Koo, G., and Burton, R. In "Differentiation Antigens of Lymphocytes" (N. Warner, ed.), Academic Press, New York, (In press).
5. Kasai, M., Leclerc, J., McVay-Boudreau, L., Shen, F. and Cantor, H.(1979) J. Exp. Med. 149:1260.
6. Hanna, N., Burton, R. (1981) J. Immunol. 127:1754.
7. Nabel, N., Bucalo, L., Allard, J., Wigzell, H. and Cantor, H. (1981) J. Exp. Med. 153:1582.
8. Tam, M., Emmons, S., and Pollack, S. (1980) J. Immunol. 124:650.
9. Tai, A., Warner, N. (1980) In "Natural Cell Mediated Immunity Against Tumors" (R. Herberman, ed.) Academic Press, New York, P. 241.
10. Burton, R., Bartlett, S., Kumar, V., and Winn, H. (1981) Fed. Proc. 40:1006.

11. Burton, R., Bartlett, S., Kumar, V., and Winn, H. (1981)
Trans. Proc. 13:783.
12. Lust, J., Kumar, V., Burton, R., Bartlett, S., and Bennett, M. (1981) J. Exp. Med. 154:306.
13. Burton, R., Bartlett, S., Kumar, V., and Winn, H. (1981)
J. Immunol. 127:1864.
14. Stein, K., Schwarting, G., and Marcus, D. (1978) J. Immunol. 121:304.
15. Hammerling, G., Hammerling, U., Flaherty, L. (1979) J. Exp. Med. 150:108.
16. Scheid, M. and Triglia, D. (1979) Immunogenetics 9:423.
17. Pollack, S., Tam, M., Nowinski, R. and Emmons, S. (1979)
J. Immunol. 123:1818.
18. Minato, N., Reid, L., Cantor, H., Lengyel, P. and Bloom, B. (1980) J. Exp. Med. 152:124.
19. Minato, N. Reid, L. and Bloom, B. (1981) J. Exp. Med. 154:750.
20. Habu, S., Fukui, H., Shimamura, K., Kasai, M., Nagai, Y., Okumura, K. and Tamaoki, N. (1981) J. Immunol. 127:34.
21. Glimcher, L., Shen, R. and Cantor, H. (1977) J. Exp. Med. 145:1.
22. Burton, R. (1980) In "Natural Cell Mediated Immunity Against Tumors" (R. Herberman, Ed.) Academic Press. New York, P. 19.
23. Burton, R., Bartlett, S. and Winn, H. (1980) In "Genetic
Control of Natural Resistance to Infection and Malignancy" (E. Shamene, P. Kongshavn, and M. Landy, eds.) Academic Press, New York, p. 413.
24. Kiessling, R., Klein, E., Pross, H. and Wigzell, G. (1975) Eur. J. Immunol. 5:117.
25. Tai, A., Burton, R. and Warner, J. (1980) J. Immunol. 124:1705.
26. Bartlett, S. and Burton, R. J. Immunol. (In press).

ANTI-NK 2.1: AN ACTIVITY OF NZB ANTI-BALB/c SERUM

Sylvia B. Pollack
Sandra L. Emmons

Department of Biological Structure
University of Washington School of Medicine
Seattle, WA 98195

I. INTRODUCTION

We have immunized young NZB/J mice with BALB/c spleen cells (SC) in attempts to raise antisera to NK-associated alloantigens. This strain combination was chosen because both strains are H-2^d, Thy 1.2, Lyt 1.2, Lyt 2.2, Lyt 3.2 and Ly 5.1. NZB is positive for the NK-associated antigen NK 1.1 which was defined by BALB/c x C3H F_1 anti-CE (Glimcher _et al._, 1977; Cantor _et al._, 1979) whereas BALB/c is negative for NK 1.1.

Analysis of NZB anti-BALB/c serum revealed that antibodies to NK-associated antigens were indeed present but that the activity was not to an allele of NK 1. Data summarized here and presented in detail elsewhere (Pollack and Emmons, submitted) have led us to conclude that this antiserum detects an NK-associated antigen provisionally designated NK 2.1, which is neither an allele of nor linked to NK 1.

II. MATERIALS AND METHODS

A. Antisera Production and Testing

Anti-NK 1.1 was raised by immunizing BALB/c x C3H F_1 mice with CE spleen cells (SC) as described in detail previously

[1]Supported by USPHS CA-18647 and DOE Research Contract 2225 (AT 06-79EV10270).

ISBN 0-12-341360-5

TABLE I

C Dependent Effects of NK 1.1 and NK 2.1 Antisera on NK Activity of BALB/c and NZB SC

Treatment of SC[a]	% Lysis of YAC-1 by BALB/c			NZB/J		
	100:1[b]	50:1	25:1	100:1	50:1	25:1
Medium Control	40	29	19	51	28	19
C Control	41	27	18	44	29	18
Normal NZB Serum	49	29	17	43	28	19
Normal NZB Serum + C	40	26	18	42	30	19
α NK 1.1 Serum	44	29	21	36	26	17
α NK 1.1 + C	40	26	18	15**	10**	5**
α NK 2.1 Serum	43	28	19	NT	NT	NT
α NK 2.1 + C	20**	11**	9**	43	32	21

[a]SC were treated in a one-step lysis with serum and/or C (newborn rabbit serum) as listed.
[b]Effector to target ratio.
**Significant decrease in NK activity.

TABLE II

The NK Associated Antigen Detected by NZB Anti-BALB/c Serum Is Not Allelic to NK 1.1.

	No. BALB/c x NZB F_2 Progeny with Phenotype[a]			
	$NK1^+$ $NK2^-$	$NK1^+$ $NK2^+$	$NK1^-$ $NK2^+$	$NK1^-$ $NK2^-$
% of total (136)	28 21%	3 2%	36 26%	69 51%

[a]$NK1^+$ denotes a greater than 30% decrease in NK activity in SC treated with C and BALB/c x C3H F_1 anti-CE (anti-NK 1.1) serum. NK 2^+ denotes a greater than 25% decrease in NK activity in SC treated with C and NZB anti-BALB/c (anti-NK 2.1) serum.

(Pollack *et al.*, 1979). The same immunization schedule was used to raise NZB anti-BALB/c, i.e., NZB/B1NJ mice were injected intraperitoneally with progressively greater doses of BALB/c SC at 2-week intervals. After the fourth immunization, the NZB mice were bled individually and their sera tested for C dependent depletion of NK activity. BALB/c SC were treated with antiserum and C or antiserum alone in a one-step lysis (Pollack *et al.*, 1979) and then tested at 3 effector:target ratios on ^{51}Cr-labeled YAC-1 target cells. The sera were also screened for autoantibody by dye exclusion assay on BALB/c SC. Sera with autoantibody or low titers of anti-NK activity were discarded and the remaining sera pooled. Sera were subsequently obtained on days 4 and 7 after further immunizations. These antisera killed 5% or fewer of BALB/c SC in C dependent assays.

B. Mice

Mice were obtained from the Jackson Laboratories except for BALB/c, BALB/c x NZB F_1 and BALB/c x NZB F_2 which were raised in our laboratory.

III. ANTI-NK ACTIVITY OF NZB ANTI-BALB/c SERUM

A. Comparison of NK 1.1 and NK 2.1 on BALB/c and NZB SC

SC from BALB/c and from NZB mice were treated with antiserum and/or C as detailed in Table I and tested for NK activity on YAC-1 targets. As expected, NK 1.1 antiserum mediated a C-dependent decrease in NZB NK activity but had no effect on BALB/c. The NZB anti-BALB/c serum had a reciprocal effect. Whereas this result suggested that the anti-NK activity in the NZB anti-BALB/c serum was to NK 1.2, further characterization of the serum showed that the activity was not to an allele of NK 1.

B. Segregation Analysis of NK 1.1 and NK 2.1 Reactivity in BALB/c x NZB F_2

One hundred thirty-six BALB/c x NZB F_2 mice were tested for C dependent sensitivity of their splenic NK activity to NK 1.1 antiserum and to NK 2.1 antiserum. The expected segregation ratios, if the two sera detected allelic products, are one quarter 1^+2^-, one half 1^+2^+, one quarter 1^-2^+. No double negatives should occur. As shown in Table II, half of the F_2 mice were, in fact, 1^-2^- compared to the sensitivity of the parental strains. The presence of intermediate phenotypes suggested a possible gene dosage effect.

Linkage analysis of 63 F_2 mice for which coat color data were available suggested that NK 2 is linked to the agouti locus on chromosome 2. Linkage of NK 1 to chromosomes 2, 4 or 7 was excluded.

C. Strain Distribution of NK 1.1 and NK 2.1

SC from 18 strains of mice were typed for sensitivity to NK 1.1 and NK 2.1 antisera (Table III). Quantitative differences in sensitivity were observed, particularly with the NK 2.1 antiserum. C57Bl/6 (B6) was of special interest since it reacted strongly with NK 1.1 antiserum but also gave a borderline positive result with NK 2.1 serum in assays of C dependent depletion of NK activity. Analysis of B6 SC with the fluorescence activated cell sorter confirmed that they did bind NK 2.1 antiserum, although the profile of positive cells was lower and shifted to the left compared to BALB/c SC. B6 SC were sorted into NK 2.1^+ and NK 2.1^- populations. All of the NK activity was in the NK 2.1^+ cells (Table IV).

IV. CONCLUSIONS AND DISCUSSION

Several alloantisera which detect NK associated antigens have been described. Glimcher *et al.* (1977) reported that anti-Ly 1.2 serum raised in C3H mice immunized with CE lymphoid cells had anti NK activity. By immunizing (BALB/c x C3H) F_1 with CE, they obtained an NK specific serum which they called anti-NK 1.1. We subsequently used NK 1.1 antiserum and the FACS to select for B6 NK cells and were able to demonstrate enrichment of NK activity both *in vitro* and *in vivo* (Pollack *et al.*, 1979, Tam *et al.*, 1980). Recently we have developed a model for *in vivo* depletion of NK cells using NK 1.1 antiserum (Pollack and Hallenbeck, 1982; Pollack, this volume; Lotzova *et al.*, this volume).

Burton and Winn (1981) reported that CE anti-CBA sera detected an allele of NK 1 which they designated NK 1.2. Their conclusions were based on strain distribution of reactivity and segregation in (CE x CBA)F_1 x CE backcross progeny. The presence of strains which reacted both with NK 1.1 antiserum and with CE anti-CBA serum, e.g., C57Bl/6 and C58, raised some doubts about whether the two antisera in fact were to allelic products of the same locus, however.

Tai and Warner (1980) detected anti-NK activity in (NZB x CE)F_1 anti CBA serum. Unlike CE anti-CBA serum which kills significant numbers of SC other than NK cells, (NZB x CE)F_1 anti CBA serum was not significantly cytotoxic to SC, yet depleted NK activity. These investigators suggested that the activity might be to NK-associated antigen other

TABLE III

Strain Distribution of Sensitivity to Anti NK 1.1 and Anti-NK 2.1 in C Dependent Depletion Assays

Phenotype	Strains
NK 1.1^{++},NK2.1^{-}	B10.A, CE/J, C57Br/J, C57L/J, C58/J, MA/MY, NZB/J, SM/J
NK1.1^{++},NK2.1^{+}	C57Bl/6J, 129/J
NK1.1^{-},NK2.1^{++}	A/J, BALB/c, C3H/HeJ, DBA/2J
NK1.1^{-},NK2.1^{-}	DBA/1J, SJL/J

TABLE IV

Enrichment of B6 NK Activity in Sorted NK 2.1^{+} SC

Effector Cell Population*	% Specific Lysis at Effector:Target Ratios of:						
	25:1	12.5:1	10:1	6:1	5:1	3:1	2.5:1
Unseparated	5.2	1.9		1.7		1.0	
NK 2.1^{-}	1.0	0.0		0.0		0.2	
NK 2.1^{+}			10.8		6.2		3.9

[a]B6 SC were passed over nylon columns 2x and labeled sequentially with anti-NK 2.1 (1:20) and then FITC-goat anti-mouse Ig (1:25; Meloy) as described previously in studies with anti-NK 1.1 serum (Tam *et al.*, 1980). Labeled cells were sorted on a FACS-II (Becton-Dickinson) into "bright" NK 2.1^{+} and "dull" NK 2.1^{-} populations. Each population (unseparated, NK 2 positive and NK 2 negative) was tested for effector activity on ^{51}Cr-labeled YAC-1 targets.

than NK 1 but further analysis was not reported at that time.

We have now raised an NK specific antiserum in NZB mice immunized with BALB/c spleen cells. The strain distribution of reactivity with this antiserum is similar but not identical to that obtained with CE anti-CBA and (NZB x CE)F_1 anti-CBA sera.

Analysis of SC from 18 strains of mice and from F_1 and F_2 progeny of BALB/c x NZB that were tested concurrently for sensitivity to NK 1.1 and to NZB anti-BALB/c serum led us to conclude that the antisera detect two distinct NK-associated antigens which are neither allelic nor linked. Further analyses of the inheritance of NK 1 and NK 2 are in progress. Neither antigen appears to be inherited as a dominant. The expression of NK 1 and NK 2 may be influenced by regulatory genes that segregate independently from the structural genes which determine the cell-surface associated antigens detected by the antisera.

REFERENCES

Burton, R. C. and Winn, H. J. (1981). J. Immunol. 126:1985.

Cantor, H., Kasai, M., Shen, F. W., LeClerc, J. C. and Glimcher, L. (1979). Immunol. Rev. 44:1.

Glimcher, H., Shen, F. W. and Cantor, H. (1977). J. Exp. Med. 145:1.

Pollack, S. B., Tam, M. R., Nowinski, R. C. and Emmons, S. L. (1979). J. Immunol. 123:1818.

Tai, A. and Warner, N. L., in "Natural Cell-Mediated Immunity to Tumors" (R. B. Herberman, ed.), p. 241. Academic Press, New York, 1980.

Tam, M. R., Emmons, S. L. and Pollack, S. B. (1980). J. Immunol. 124:650.

CURRENT STATUS OF SURFACE ANTIGENIC MARKERS ON (A) INTERFERON INDUCED MURINE NATURAL KILLER CELLS AND (B) MURINE ANTIBODY DEPENDENT CELLULAR CYTOTOXICITY CELLS

G. C. Koo[1]
J. R. Peppard
A. Hatzfeld[2]
C. Colmenares

Memorial Sloan-Kettering Cancer Center
New York, New York

I. INTRODUCTION

Through the studies of antigenic phenotype of murine NK cells, it becomes evident that NK cells are heterogeneous with regard to their antigenic expression. At the same time antigenic markers have also been useful in categorizing these heterogeneous killer cells (Lust et al.,1981 and Minato et al., 1981). We are interested in examining two questions with regard to surface phenotype of NK cells: (a) whether interferon induced NK cells express antigens different from endogenous NK cells (b) whether murine ADCC cells could be distinguished from NK cells based on surface phenotype analysis.

II. SURFACE ANTIGENS OF INTERFERON (IF) INDUCED NK CELLS

We used BSA fractionated spleen cells from C57BL/6 mice and induced the cells for about 18 hrs with 10^3 units of partially purified IF from mouse fibroblasts (generous gift from Dr. W. Stewart). The cells were then treated with normal mouse serum (NMS) or various alloantisera + C in a microcytotoxicity assay as described by Burton and Winn (1981). Table I shows that both endogenous NK and IF induced NK cells expressed similar antigens with the exception of Ly-6 antigen. Other antigens that are not presented in this table that we have also examined are T-200, Lyt-1.1 and Lyt-2.2 in which we observed similar

[1]Supported by NIH grants CA-08748, CA 25416, ACS-FRA-167, NSF and NATO
[2]Present address: Institut de Pathologie Cellular, Kremlin Bicetre, France

ISBN 0-12-341360-5

antigenic expression on NK and IF-NK. (NK cells are positive for T-200, partially positive for Lyt-1 and negative for Lyt-2). Ly-6 antigen is found on T and B cells in lymphoid tissue and is expressed to a larger degree on activated T and B cells, such as mitogen stimulated blast cells and plaque forming cells (McKenzie et al.,1977 and Feeney, 1978). Recently a number of monoclonal antibodies have defined antigenic determinants closely linked to each other, genetically (Eckhardt and Herzenberg,1980;Kimura et al.,1980b, and Takei et al.,1980). We have tested three of these anti-mLy-6 antibodies and they eliminated NK activities at various extent. The fact that Ly-6 antigen is expressed more densely on IF-NK cells and the fact that there are variant forms of Ly-6 antigen(s) may account for the inconsistent findings reported earlier (Koo and Hatzfeld,1980;Pollack et al.,1980). It seems likely that at times the endogenous NK cells may constitute some activated NK cells as well, thereby become more susceptible to anti-Ly-6 and C. Ly-10 antigen was also reported to be negative on NK cells (Koo and Hatzfeld,1980) but with repeated testing, it was found to be positive on NK cells. Ly-10 antigen is found on both T and B cells (Kimura et al.,1980a).

TABLE I. Antigenic expression of NK and Interferon Induced NK cells

Treatment with C + antiserum	% Reduction of specific lysis[a] NK	IF-NK[b]
NMS[c]	0	0
anti-Nk-1	82	75
anti-mThy-1.2	60	62
anti-Qa-2,3	77	79
anti-mQa-4	60	62
anti-mQa-5	90	86
anti-mLy-6	54	76
anti-mLy-10	74	74

[a]% reduction of specific lysis = % reduction of lysis after treatment with antisera + C compared to treatment with NMS + C; target cells were YAC-1 cells. E:T ratio 100:1, mean of NK lysis was 16%, mean of IF-NK lysis was 24%.

[b]BSA fractionated spleen cells were induced with 10^3 IF for 18 hr before the 4 hr. ^{51}Cr release assay

[c]NMS: normal mouse serum; m designates monoclonal antibody These antisera have been previously described in Koo and Hatzfeld, 1980).

Presently the only antisera known to eliminate NK activity specifically are anti-Nk-1 (Glimcher et al.,1977;Koo and Hatzfeld 1980 and Pollack et al.,1980), anti-Nk-2 (Burton and Winn 1981), anti-Qa-5 (Koo and Hatzfeld 1980) and possibly anti-Asialo-Gm-1 (Kasai et al.,1980). Monoclonal anti-Nk-1 and Nk-2 are not yet available.

II. SURFACE ANTIGENS OF ADCC CELLS

It is the contention of most investigators that NK cells and ADCC cells are the same cells, based on tissue distributions, age dependency and other criteria (Ojo and Wigzell,1978) yet there are discrepancies that indicate NK and ADCC cells may be distinct populations (Koren and Williams 1981 and Huh et al.,1981). We have thus used both the Nk-1 and Qa-5 antisera to examine this issue and compared two types of ADCC killing, namely: (a) RL♂1-ADCC and (b) CRBC-ADCC. RL♂1-ADCC was assayed with ^{51}Cr labelled RL♂1 in the presence of goat anti-gp70 and CRBC-ADCC was assayed with ^{51}Cr labelled chicken RBC in the presence of rabbit anti-chicken RBC (CRBC).

TABLE II. Similarities and differences between NK and ADCC cells

	RL♂1-NK	RL♂1-ADCC	CRBC-ADCC
Nk-1 antigen	+[a]	+	+/-
Qa-5 antigen	+	N.T.	+/-
BSA gradient[b]	enriched	enriched	not enriched

[a]+ indicates 80-100% reduction of lytic activity to RLol in presence of C and antisera; +/- indicates partial elimination (∿50%) of lytic activity in presence of C + antisera. N.T. - not tested

[b]NK activity was enriched in A + B layers of BSA gradient fractionated spleen cells (Koo and Hatzfeld, 1980).

TABLE III. Splenic RL♂1-ADCC and CRBC-ADCC activities in mice with high and low RL♂1-NK activity[a]

	% Specific lysis			
	129-Gix⁻	BALB	129	B6
RL♂1-NK	1	2	6	13
RL♂1-ADCC	1	4	11	22
CRBC-ADCC	60	N.T.	65	50

[a]Spleen cells were fractionated on BSA gradient and cells from A + B layers are used for the assay. E:T ratio = 100:1.

Table II and III show that RL♂1-NK and RL♂1-ADCC cells are very similar in terms of antigenic expression, distribution in BSA gradient and strain distribution, suggesting that they are likely the same cells. Both RL♂1-NK and RL♂1-ADCC cells are in part distinct from CRBC-ADCC cells by these criteria. Both anti-Nk-1 and anti-Qa-5 eliminated 80-100% of NK activity but only eliminated about 50% of CRBC-ADCC activity. That there was a distinct CRBC-ADCC cell was particularly striking when 129-Gix⁻ mice showed good CRBC-ADCC killing and virtually no RL♂1 NK or RL♂1-ADCC killing (Table II). We conclude from these data that NK cells can act as RL♂1-ADCC and also CRBC-ADCC in the presence of appropriate antibodies, but a second non-NK population can also mediate CRBC-ADCC. These non-NK CRBC-ADCC cells are probably the monocytic cells implied by earlier studies (Kiessling et al.,1976). Conceptually, CRBC-ADCC cells are a broader population compared to NK cells (see figure below). We believe this scheme of killing mechanism may resolve the controversial issues on the two types of cytolytic cells.

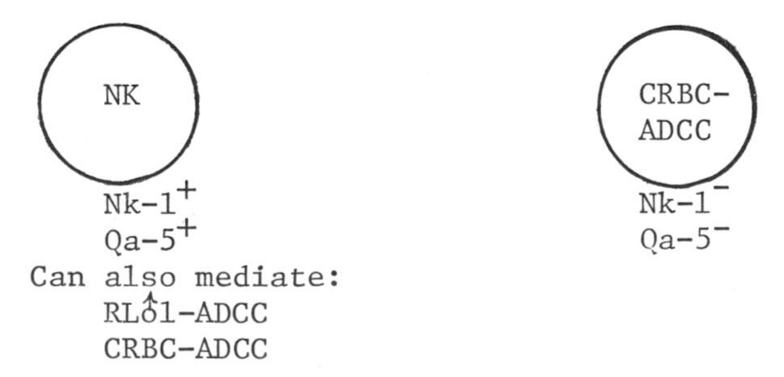

ACKNOWLEDGMENTS

We thank Drs. S. Kimura, N. Tada and U. Hammerling for generous supply of antisera, and Barbara LoFaso for preparation of this manuscript.

REFERENCES

Burton, R.C., and Winn, H.J. (1981). J. Immunol. 126:1985.
Eckhardt, L.A., and Herzenberg, L.A. (1980). Immunogenet. 11:275.
Feeney, A.J. (1978) Immunogenet. 7:537.
Glimcher, L., Shen, F.-W., and Cantor, H. (1977). J. Exp. Med. 145:1.
Huh, N.D., Kim, Y.B., and Amos, D.B. (1981). J. Immunol. 127: 2190.
Kasai, M., Iwamori, M., Nagai, Y., Okumura, K. and Tada, T. (1980). Eur. J. Immunol. 10:175.
Kiessling, R., Petranyi, G., Karre, K., Jondal, M., Tracey, D., and Wigzell, H. (1976). J. Exp. Med. 143:772.
Kimura, S., Tada, N., and Hammerling, U. (1980a). Immunogenet. 10:363.
Kimura, S., Tada, N., Nakayama, E., and Hammerling, U. (1980b). Immunogenet. 11:373.
Koo, G.C., and Hatzfeld, A. (1980). In "Natural cell-mediated immunity against tumors" (R.B. Herberman, ed.), p. 105. Academic Press, New York.
Koren, H.S. and Williams, M.S. (1981). J. Immunol. 121:1956.
Lust, J.A., Kumur, V., Burton, R.C., Bartlett, S.P., Bennett, M. (1981). J. Exp. Med. 154:306.
McKenzie, I.F.C., Cherry, M., and Snell, G.D. (1977). Immunogenet. 5:25.
Minato, N., Reid, L., and Bloom, B.R. (1981). J. Exp. Med. 154:750.
Ojo, E., and Wigzell, H. (1978). Scand J. Immunol. 7:297.
Pollack, S.B., Emmons, S.J., Hallenbeck, L.A. and Tam, M.R. (1980). In "Natural cell-mediated immunity against tumors" (R.B. Herberman, ed.), p. 139. Academic Press, New York
Takei, F., Galfre, G., Alderson, T., Lennox, E.S., and Milstein, C. (1980). Eur. J. Immunol. 10:241.
Young, W.W., Hakomori, S., Durdik, J.M., and Henney, C.S. (1980). J. Immunol. 124:199.

A DIFFERENTIATION ANTIGEN OF MURINE NATURAL KILLER CELLS

Masataka Kasai

Department of Tuberculosis
National Institute of Health
Tokyo, Japan

Ko Okumura

Department of Immunology
Faculty of Medicine
University of Tokyo
Tokyo, Japan

I. BRAIN ASSOCIATED ANTIGENS OF NATURAL KILLER (NK) CELL

During the past few years, a number of surface antigens of NK cells particularly those of mice, have been characterized (1-4). Antiserum to the neutral glycolipid, asialo GM1 (GA1), to be discussed in this paper, is a unique antiserum in terms of its strong anti-NK activity. Notably, this antibody specificity was found during the course of analysis of anti-NK activity in anti-BAT (brain-associated T cell antigen) antiserum. It was previously shown that anti-BAT antiserum contains various antibody activities such as anti-BAθ (brain-associated θ), anti-stem cell and anti-erythrocyte (5-7). However, these antibody activities appear to be quite different from anti-NK activity (8). In order to determine the membrane antigen detected on NK cells by anti-BAT, we focused our attention on a variety of glycolipids which are thought to be major components of brain tissue. When the reactivity of anti-BAT with various purified glycolipids was tested by microflocculation using lecithin-cholesterol micelles, the anti-BAT exhibited a strong reactivity with GA1. Much less activity was found with

ISBN 0-12-341360-5

micelles containing GM1 or GD1b, even though these gangliosides are the major glycolipid components in the brain tissue. Furthermore, the absorption of anti-BAT with GA1 effectively diminished its anti-NK activity. These results suggest that the majority of anti-NK activity contained in anti-BAT antiserum is attributable to anti-GA1 antibody (4).

The presence of anti-GA1 antibody in anti-BAT is somewhat surprising for us, since GA1 is a trace component of brain tissue and is barely detectable by means of quantitative analysis (9). During the course of this experiment, however, we became aware of the fact that GA1 was to be highly immunogenic when injected into rabbits with CFA (complete Freund's adjuvant). The anti-GA1 antiserum showed little cross-reactivity with structurally related glycolipid, e.g., GM1, GD1b and asialo GM2, in the microflocculation test. When lymphoid cells from spleen, lymph node and bone marrow of different inbred mouse strains were treated with anti-GA1 antiserum and complement, NK activity in these cell preparations was completely abolished. The cytotoxic titer of this antiserum against NK cells was extremely high, e.g., a high dilution of anti-GA1 which showed no detectable cytotoxic killing was capable of eliminating mouse NK activity in the presence of complement.

II. EFFECT OF ANTI-GA1 ON NK ACTIVITY *IN VIVO*

Considering the various biological roles of glycolipids such as receptor or recognition molecules in cellular interaction, we set up an experiment to determine the *in vivo* effect of anti-GA1 NK activity. *In vivo* administration of anti-GA1 to BALB/c nude mice resulted in a marked reduction in NK activity (10). The amount of anti-GA1 required for elimination of NK activity was much smaller than we expected. The dose response carve clearly indicated that even a few microliters of anti-GA1 could deplete virtually all of the NK activity and this depletion lasted from several days to 1 week after one injection of the antiserum. Therefore, anti-GA1 was used to study the biological role of NK *in vivo*. Intravenous injection of anti-GA1 into BALB/c nude mice was found to greatly enhance both the local growth and incidence of tumor take of the syngeneic lymphoma, RL♂1 (10,11). This strong correlation between the presence of NK function and resistance to tumor growth agrees with the previous result which indicates that positively selected NK cells can retard tumor growth (12). There are also remarkable similarities between the present result and the recent observation that the growth of transplanted tumor cells is enhanced in beige mice that have an impairment of NK activity (13,14).

As the reduction in NK activity was observed even 30 min after injection of anti-GA1, we set up an experiment to determine the mode of action of anti-GA1 on NK activity. As can be seen in Table I-B, incubation of BALB/c nu/nu spleen cells with anti-GA1 for 30 min at 37°C resulted in a strong reduction of NK activity only when normal mouse serum (NMS) was present. As was the case in vivo, NK activity was inhibited even with a minute amount of anti-GA1, whereas NMS alone was not effective. The effect of NMS was abolished by heating it for 30 min at 56°C. Because of the weak activity of mouse complement, it seems unlikely that abolition of NK activity is due to a cytolytic effect caused by microliter amounts of anti-GA1. NK activity was reduced equally in the presence of anti-GA1 and serum from AKR mice with a genetically determined deficiency of C5, or with serum from mouse strain possessing the complete spectrum of complement components. Serum from C4-deficient guinea pigs was also effective in removing NK activity in the presence of anti-GA1. These results indicate that a complement dependent cytoloytic membrane attack pathway is not neccessarily required for this effect. Therefore, this suggests that only the function of NK cells is seriously damaged by anti-GA1 antibodies and serum factors.

As interferon seems to have substantial influence on the NK function both in vivo and in vitro, we thought it appropriate to determine whether anti-GA1 would also alter the effect of interferon on NK cell activity. Table I-A shows that intraperitoneal injection of interferon markedly augmented the cytotoxic activity of NK cells in normal nude mice. However, interferon could not restore NK activity that was depressed by injection of anti-GA1. Similar results were also obtained in vitro (Table I-B). Based on the above experiments, we can postulate that the receptor able to bind interferon and, thus, trigger an augmentation of NK activity is expressed on NK cells at a differentiation stage when these cells are also reactive to anti-GA1. Other NK cell surface antigens, such as Ly-5 and Qa-5, were determined serologically and it was recently reported that interferon triggers the differentiation of NK precursor cells (Qa-5^+, Ly-5^-) into NK effector cells (Qa-5^+, Ly-5^+) (15). If this is true, the present results indicate that the phenotype of NK precursor cells is GA1^+, Qa-5^+, Ly-5^-. An attractive possibility is that GA1, the cell surface structure responsible for killing, and the receptor for interferon are closely inter-connected in the NK cell membrane. The results presented here might give us some clue to understanding the cytolytic mechanism of NK cells and the mode of action of interferon.

TABLE I.

A) Effect of Interferon after Treatment with Anti-GA1 *In Vivo*

Antisera injected	Interferon	% lysis of YAC-1
None	-	18.5
None	+	41.7
Anti-GA1	-	0.5
Anti-GA1	+	1.3

B) Effect of Interferon after Treatment with Anti-GA1 *In Vitro*

Cells treated with	Interferon	% lysis of YAC-1
NMS	-	17.9
NMS	+	38.6
NMS + anti-GA1	-	0.3
NMS + anti-GA1	+	1.2

A. BALB/c nu/nu mice were injected i.v. with 5 μl of anti-GA1 or with 500 U of interferon. Interferon was also injected 30 min after injection of anti-GA1. NK activity of spleen cells was determined 24 hr after administration of each reagent. Effector/target ratio was 50:1.

B. Spleen cells from BALB/c nu/nu mice were treated with NMS of the same mice or with NMS plus anti-GA1 (1:200) for 30 min at 37°C. After being washed, each group of spleen cells was tested for NK activity against YAC-1 cells. Interferon (500 U) was also added to the ^{51}Cr-release assay.

III. ANTI-GA1 REACTIVITY TO LEUKEMIC CELLS IN AKR MICE

The another interesting feature of anti-GA1 antibody is its reactivity to leukemic cells in AKR mice. In AKR mice, leukemia develops spontaneously in the majority of animals over 6 months of age. Since thymectomy in early life greatly reduces the incidence of leukemia, the thymus in AKR mice appears to have a central role in the pathogenesis of leukemia. With the onset of leukemia, the level of murine leukemia virus (MuLV)-related antigens on thymocytes sharply increase, whereas the differentiation antigens, such as Thy-1, are reduced (16). The majority of thymocytes of 2 month old AKR mice do not react with anti-GA1 by immunofluorescence, whereas the number of thymocytes from 6 months old preleukemic AKR mice stained with anti-GA1 markedly increased (17). This

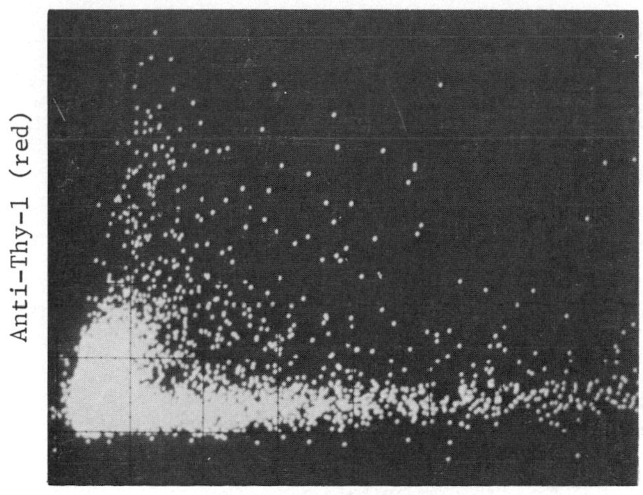

FIGURE 1. Two colour analysis of anti-GA1 and anti-Thy-1 stains of preleukemic AKR thymocytes.

increase in GA1+ cells is closely correlated with the increase in MuLV-antigen expression. By two color immunofluorescence analysis, preleukemic AKR thymocytes were stained with anti-GA1, but not with anti-Thy-1 antisera (Fig. 1). It is interesting to know whether or not preleukemic AKR thymocytes reactive to anti-GA1 is responsible for providing the leukemogenic signal.

The results presented here suggest that antigenic determinants recognized by anti-GA1 can be expressed on cell that have undergone some phase of differentiation in the thymic environment. Although it is well accepted that NK activity can develop without the thymus, when considering NK cell ontogeny, it is interesting to note that a number of the antigens so far determined to be on murine NK cells are also expressed during certain differentiation stages of T cells. Thus, the unique antiserum, found as one of the specificities of anti-BAT, will be valuable for studying natural defence mechanism as well as some aspects of the ontogeny of various lympho-hematopoietic cells.

REFERENCES

1. Cantor, H., Kasai, M., Shen, F.W., Leclerc, J.C., and Glimcher, L., Immunol. Rev. 44:1 (1978).
2. Kasai, M., Leclerc, J.C., Shen, F.W., and Cantor, H., Immuogentics 8:153 (1979).
3. Chun, M., Pasanen, U., Hammerling, U., Hammerling, G., and Hoffmann, M.K., J. Exp. Med. 150:426 (1979).
4. Kasai, M., Iwamori, M., Nagai, Y., Okumura, K., and Tada, T., Eur. J. Immunol. 10:175 (1980).
5. Golub, E.S., J. Immunol. 109:168 (1972).
6. Golub, E.S., J. Exp. Med. 136:369 (1972).
7. Golub, E.S., Exp. Hematol. (Copenhagen) 1:105 (1973).
8. Habu, S., Hayakawa, K., Okumura, K., and Tada, T., Eur. J. Immunol. 9:938 (1979).
9. O'Brien, J.S., in "The Metabolic Basis of Inherited Diseases" p. 841. McGraw Hill, New York (1978).
10. Kasai, M., Yoneda, T., Habu, S., Maruyama, Y., Okumura, K., and Tokunage, T., Nature 291:334 (1981).
11. Habu, S., Fukui, H., Shimamura, K., Kasai, M., Nagai, Y., Okumura, K., and Tamaoki, N., J. Immunol. 127:34 (1981).
12. Kasai, M., Leclerc, J.C., McVay-Boudreau, L., Shen, F.W., and Cantor, H., J. Exp. Med. 149:1260 (1979).
13. Talmage, J.E., Meyers, K.M., Prieur, D.J., and Starkey, J.R., Nature 284:622 (1980).
14. Karre, K., Klein, G.O., Kiessling, R., Klein, G., and Roder, J.C., Nature 284:624 (1980).
15. Minato, N., Reid, L., Cantor, H., Lengel, P., and Bloom, B.R., J. Exp. Med. 152:124 (1980).
16. Kawashima, K., Ikeda, H., Stockert, E., Takahashi, T., and Old, L.J., J. Exp. Med. 144:193 (1976).
17. Kasai, M., Habu, S., Tamaoki, N., Okumura, K., and Tada, T., Proceeding of 4th International Congress of Immunology, Paris, 10.1.31 (1980).

HETEROGENEITY AND REGULATION OF THE NATURAL KILLER CELL SYSTEM

Nagahiro Minato
Barry R. Bloom

Dept. of Medicine, Jichi Medical School, Japan
Department of Microbiology and Immunology, Albert Einstein College of Medicine, Bronx, N.Y.

Since the original description of natural killer (NK) cells, major efforts have been undertaken by a number of investigators to clarify three fundamental questions: 1) the phenotypic characteristics and cell lineage of NK effector cells; 2) the mode of physiological regulation of the NK system; and 3) the role of NK cells in host defense mechanisms <u>in vivo</u>. We summarize here briefly our studies attempting to clarify each of these questions.

I. HETEROGENEITY

Using serological analysis of cell surface phenotypes of murine NK cells, we were able to identify four distinct NK effector cell subsets (Minato et al, 1981) (Table 1). The NK_I and NK_T subsets, present in fresh and precultured spleens, were cytotoxic for HeLa-Ms (and YAC) but not P815 cells. TK cells were generated from cultured spleens and capable of killing allogeneic P815 cells but not HeLa-Ms cells. NK_M cells were generated in cultures of bone marrows and were capable of lysing HeLa-Ms but not P815 cells.

Table 1

Characteristics of Four Subsets of Spontaneously Cytotoxic Cells in Mice

	NK_I	NK_T	TK	NK_M
Serological phenotype	$Thy\text{-}1^-$, $Lyt\text{-}2^-$ $Qa\text{-}5^+$, $Ly\text{-}5^+$	$Thy\text{-}1^+$, $Lyt\text{-}2^-$ $Qa\text{-}5^+$, $Ly\text{-}5^+$	$Thy\text{-}1^+$, $Lyt\text{-}2^+$ $Qa\text{-}5^-$, $Lyt\text{-}5^+$	$Thy\text{-}1^-$, $Lyt\text{-}2^-$ $Qa\text{-}5^-$, $Lyt\text{-}5^+$
Target cell lysis				
HeLa-Ms (or YAC-1)	+	+	−	+
P815	−	−	+	−
Regulated by IFN	+	Not tested	−	−
Regulated by IL-2	−	+	+	−
Organ distribution				
Bone marrow	− (+) [−]	− (+) [+]	− (−) [−]	+ (−) [−]
Spleen	+ (+) [−]	+ (+) [+]	+ (−) [+]	− (−) [−]
	− (−) [−]	− (−) [−]	− (−) [−]	− (−) [−]
Activity in *be/be* mice	−	−	+	+
Activity in *nu/nu* mice	+	+	−	+

Parentheses activity augmented after treatment with IFN; Brackets indicate activity augmented after treatment with IL-2.

ISBN 0-12-341360-5

That these serologically distinguishable NK subsets represented independent populations was indicated by experiments using mutant mice of the C57Bl/6 series, namely nu/nu and bg/bg mice. Spleen cells from bg/bg mice, either fresh or cultured, failed to generate any detectable NK activity against HeLa-Ms cells and lacked the NK_T and NK_I subsets of NK cells,in contrast to +/+ or nu/nu mice. Bone marrow cells from bg/bg mice, however, generated cytotoxic activity comparable to cells from +/+ or nu/nu mice, and the effector phenotype was clearly that of the NK_M subset. These results establish that the NK_M cell belongs to an independent lineage from NK_I and NK_T effector cells. Of particular interest is the possibility that NK_M in our systems correspond to the NC cells described by Stutman and his collaborators (Lattime et al, 1981), which share an ability to kill solid tumor targets and appear to be unresponsive to IFN or IL-2. TK cells, which share the same phenotype as CTL, cannot be generated from bone marrow cells of +/+ mice or from spleen cells from nu/nu mice, indicating that TK activity is a function of post thymic T cells, whereas NK_M as well as NK_I and NK_T subsets are clearly thymus-independent. It remains unclear at present whether TK activity simply reflects polyclonal activation and expansion of preexisting alloreactive CTL precursors in culture, or a distinct subset of T cells exerting spontaneous cytotoxicity.

2. REGULATION OF NK ACTIVITY

A. INTERFERON: It has been widely established that IFN is a major regulatory factor in activating NK cells. We have previously shown that IFN can act on Qa 5^+, Ly 5^- NK precursor cells and induce them to become cytotoxic and express the Ly 5 surface antigen, thereby becoming NK_I or NK_T effector cells (Minato et al, 1980). Spontaneous and IFN-augmented splenic NK activity was totally mediated by Qa 5^+, Ly 5^+ cells, the only significant difference being a large increase in the proportion of Thy 1^+ NK_T effector cells in IFN-treated cultures. In bg/bg mice, however, even after the addition of IFN, cytotoxic activity was Thy 1^-, Qa 5^- (NK_M cells). Thus IFN failed to augment cytotoxic activity of bg/bg mice lacking NK_I and NK_T type effector cells.

It was consistently observed that the susceptibility of NK effector cells to anti-Thy 1 increased after exposure to IFN in both +/+ and nu/nu mice (Fig. 1). A small but significant generation of NK_T cell activity was also observed even after pretreatment of nu/nu spleen cells with anti-Thy1+C prior to exposure to IFN. Combined with previous observations that IFN can stimulate the cytotoxic activity in Thy 1 depleted populations, it appears that the conversion of NK_I to NK_T cells is mediated by IFN, and partially involved in the preferential generation of NK_T cell activity by IFN. The NK-augmenting activity of IFN was rapid but only transient, and

Figure 1

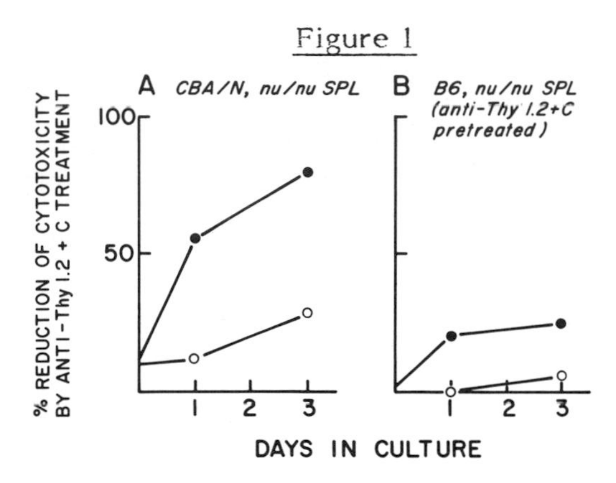

Spleen cells from normal CBA/N nu/nu mice (A), or B6 nu/nu mice which had been pretreated with anti-Thy 1.2+C (B) were cultured in the presence of (O) or absence (O) of Il-2 preparation for 1 or 3 days. The cells were then harvested and treated with C alone or anti-Thy 1.2+C just before the cytotoxicity assay against HeLa-Ms cells. % reduction of the cytotoxicity by the treatment with anti-Thy 1.2+C compared with that by C alone in each group is indicated.

usually followed by a depression of activity below control levels (Fig. 2). In contrast to the selective activation of the NK_I and NK_T subsets, the suppressive effect of IFN was observed as well on the generation of NK_M and TK activities, perhaps reflecting the known growth-inhibitory activities of IFN. The important point to bear in mind is that IFN can give diametrically opposed effects depending on the timing, dose and subset being examined.

Figure 2

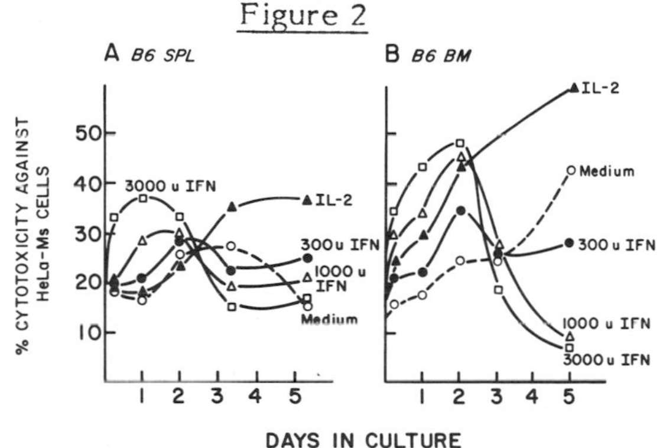

Spleen or bone marrow cells from normal B6 mice were cultured with medium alone or with varying doses of partially purified IFN or 10% of partially purified Il-2. Three hours, 1 day, 2 days, 3 days, or 5 days later, the cells were harvested and the cytotoxic activity was assayed against HeLa-Ms cells.

B. INTERLEUKIN-2 (IL-2): It is well known that NK activity can be strongly augmented in MLC. We found the effector phenotype to be the NK_T subset CTL, and their generation was absolutely dependent upon the presence of T cells. MLC supernatants also potentiated the generation of NK_T, and chromotographic analysis indicated that, in addition to IFNγ, fractions containing IL-2 augmented NK activity in vitro.(Minato et al, 1980).

Using purified IL-2, Henney et al. have established that IL-2 was responsible for NK augmenting activity in T cell-conditioned medium (Henney et al, 1981). In our experiments the regulation of NK activity by IL-2 differs in some respects from the that by IFN. In terms of subset selectivity, IL-2 markedly stimulated the generation of TK as well as NK_T activity, but failed to augment NK_M or NK_I activity. Indeed, although both IFN and IL-2 induced NK_T activity in spleen cell cultures, IL-2 failed to stimulate Thy 1-depleted populations, whereas IFN fully augmented activity of the same populations. (Table 2).

Table 2

Exp. no.	Organ	Additives in culture	% Cytotoxicity after the pretreatment with		
			C alone	a-Thyl.2+C	a-Qa5+C
I	B6 spleen	medium	8.5	9.0	3.2
		IFN(10^3 u/ml)	37.7	33.3	3.0
II	B6 spleen	medium	27.0	12.1	11.8
		Il-2	48.2	13.5	11.5

Normal B6 spleen cells were first treated with C alone or various antibodies + C, and then cultured with or without IFN or IFN-free Il-2 preparation for 2 days or 4 days respectively. The cells were then harvested and assayed for the cytotoxicity against HeLa-Ms cells.

Finally, in contrast to the transient action of IFN, IL-2 induced rather prolonged augmentation of NK activity in culture. These observations strongly suggest that the primary target cells of these two regulatory factors are different; Thy 1^- cells for IFN, and Thy 1^+ for IL-2. It is well established that IFN can effectively augment NK activity in newborn mice, which have only negligible endogenous NK activity. IL-2 preparations, on the other hand, can stimulate NK activity only after significant endogenous NK activity was detected in young mice (Fig. 3).

C. RELATIONSHIP BETWEEN IFN AND IL-2 REGULATION OF NK ACTIVITY: From the above, it seems very likely that the augmentation of NK activity by IFN and IL-2 involve different mechanisms. Clearly, IFN can act at multiple levels on NK activity, including recruitment of precursor cells, enhancement of lytic activity and facilitation of recycling of effector cells for IFN action. IL-2, on the other hand, was originally defined as a "growth factor" and stimulation or maintenance of proliferative

Figure 3

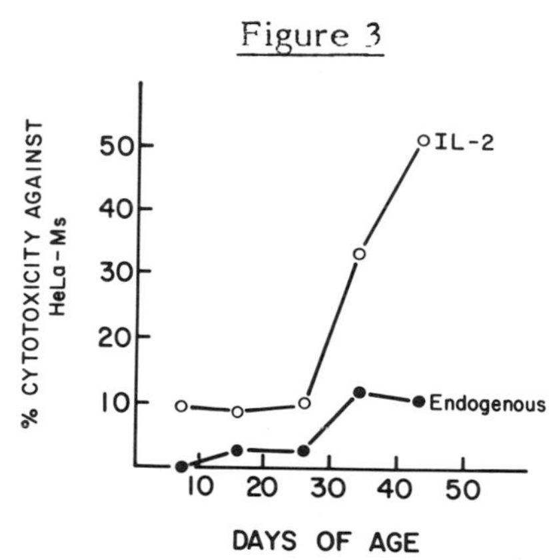

Spleen cells from various ages of Balb/c nu/nu mice (7 days old to 44 days old) were cultured in the presence or absence of Il-2 preparation for 3 days, and the cytotoxic activity against HeLa-Ms cells was assayed.

activity appears to be one of its major effects on T cells. IFN-free IL-2 can augment both NK activity and IFN production in vitro (Table 3).

Table 3

Organ	Additives in culture	Days in the culture: 3 days % Cytotoxicity	3 days IFN(u/ml)	5 days % Cytotoxicity	5 days IFN(u/ml)
B6 B.M.	medium	10.1	< 20	9.5	< 20
	IFN(300 u/ml)	20.8	1,280	2.9	1,280
	IFN-free Il-2	16.8	1,280	48.0	2,560
	IFN + Il-2	17.1	2,560	5.1	1,280

We have previously reported that IFN production in vitro induced by HeLa-Ms was predominantly IFNα, judging by acid stability and antiserum neutralization, thus establishing the NK_I and NK_T subsets as important contributors to IFN production (Minato et al, 1980). On the basis of these findings, two basic possible mechanisms of action of IL-2 on NK cells can be considered: 1) IL-2 stimulates NK activity indirectly through induction of IFN production; and 2) IL-2 acts directly on the NK_T or TK subsets independent of IFN. We favor the latter view, since NK augmentation by IL-2 could not be inhibited by anti-IFNα/β serum (IFNγ cannot be excluded) Since we have already provided evidence that IFN can act on the Thy 1^-, NK_I subset and induce conversion to the NK_T subset, and that the primary targets of IL-2 are NK_T cells, the possibility that IFN and IL-2 act synergistially and in sequential fashion is an appealing one. In this model, IFN would serve as an inductive signal, and IL-2 as a proliferative signal which serves to expand the effector population. Using Il-2 preparations, it was possible to enrich and maintain NK activity in longterm cultures, which could not be achieved with IFN alone (Fig.4).

Figure 4

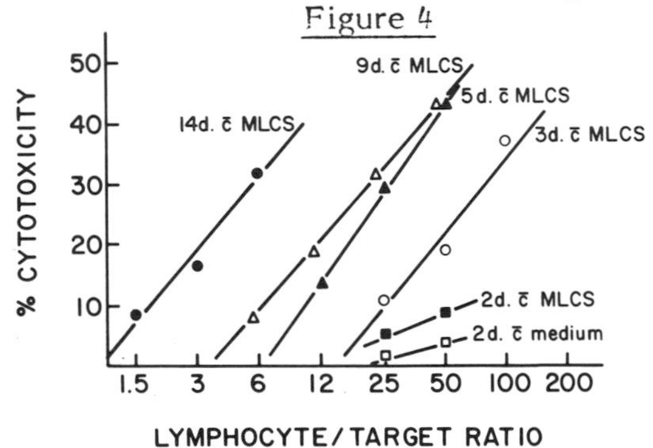

Bone marrow cells from normal B6 mice were cultured in the presence of conditioned medium of MLC(25%) with feedings of every 3 days/ Various days after the culture, the cells were harvested and the cytotoxic activity against HeLa-Ms cells was assayed with varying A/T ratio. We were able to maintain the activity over amonth this way, and clone them on the feeder layers of irradiated syngeneic bone marrow cells with conditioned medium.

III. BIOLOGICAL SIGNIFICANCE OF THE NK SYSTEM

Using a variety of virus persistently (PI) tumor cell lines, we showed a striking correlation between rejection of these tumor cells in vivo in athymic nude mice and susceptibility of the tumor cells to killing by NK cells in vitro (Reid et al, 1981). Serologically the phenotypes of effector cells in such mice corresponded to the NK_I and NK_T subsets. In nude mice inoculated with virus PI tumor cells, not only was NK activity markedly augmented in the spleen, but also the percentage of Thy 1^+ NK_T cells doubled over uninoculated control nude mice. Treatment of nude mice with anti-IFN serum (AIF) completely abrogated the augmentation of NK activity by HeLa-Ms cells without affecting endogenous NK activity (Fig. 5). In parallel, the induction of IFN production was similarly abolished by preinjection of mice with anti-IFN serum. These results strongly support the view that the IFN-mediated NK (NK_I and NK_T) feedback system described for in vitro experiments is indeed operating in vivo.

When nude mice treated with anti-IFN serum were inoculated with virus PI tumor cells that were unable to form tumors in normal nude mice, 100% of the mice developed aggressively growing tumors with marked local invasiveness, and a significant proportion developed remote metastases (Table 4). Since the xenogeneic tumor cells were refractory to the growth inhibitory effects of murine IFN,

Figure 5

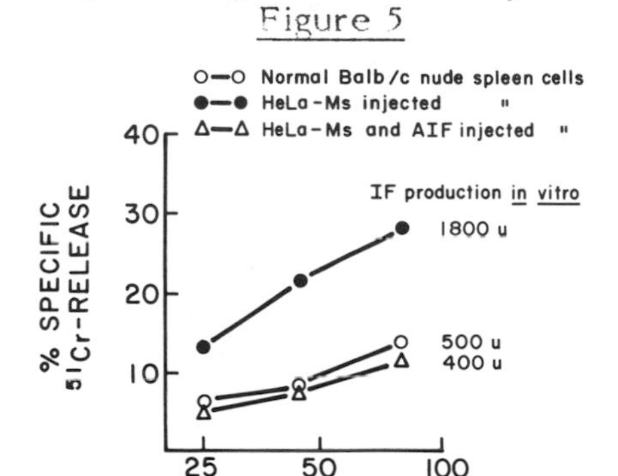

and since nude mice obviously cannot produce IL-2 endogenously, these experiments established that the IFN-NK system was primarily

Table 4

Effect of Anti-IFN Globulin on Tumorigenicity

Cell	Virus Infection	Treatment	Tumor Frequency	% Invasion	Metastatic
BHK	---	Control†	8/8	0	0
		Anti-IFN	8/8	100	37
BHK	Mumps	Control†	13/26	0	0
		Anti-IFN globulin	18/18	100	25

†Controls included treatment with PBS, normal sheep serum or sheep anti-contaminants in IFN preparation.

responsible for restriction of tumor growth and metastasis in this experimental system. Clearly in conventional animals, IL-2 as well as IFN may well be involved in augmenting activity of NK cells. These results, together with the findings from other laboratories showing that interferon can restrict the growth of tumors *in vivo* which themselves are resistant to growth inhibitory effects of IFN *in vitro* (Gresser et al, 1972; Shellekens et al, 1979) and the increased metastases of melanoma cells in *beige* mice, and their reduction by inducers of IFN (Talmadge et al, 1980), all suggest that the NK_I and NK_T subsets do have a functional role in resistance to tumors and virus infected cells *in vivo*.

1. Gresser, I., Maury, C. and Brouty-Boyé, D. Nature 239:167, 1972.
2. Henney, C.S., Kuribayashi, K., Kern, D.E., Gillis, S. Nature 291:235, 1981.
3. Lattime, E.C., Pecoraro, H.G. and Stutman, O. J. Immunol. 126:2011, 1981.

4. Minato, N., Reid, L., Cantor, H. and Bloom, B.R. J. Exp. Med. 152:124, 1980.
5. Minato,N., Reid, L. and Bloom, B.R. J. Exp. Med. 154:750, 1981.
6. Reid, L., Minato, N., Gresser, I., Holland, J. Kadish, A. and Bloom, B.R. P.N.A.S. 78:1171, 1981.
7. Shellekens, H., Weiman, W., Cantell, K. and Stitz, L. Nature 278:742, 1979.
8. Talmadge, J.E., Meyers, K.M., Prieur, D.J. and Starkey, J.R. Nature 284:622, 1980.
9. Ullberg, M. and Jondal M. J. Exp. Med. 153:615, 1981.

Supported by USPHS grants AI09807, AI10402, and NMSS 1006 and ACS BC301.

CELL SURFACE ANTIGENIC CHARACTERISTICS OF RAT LARGE GRANULAR LYMPHOCYTES (LGL)

Craig W. Reynolds[1], John R. Ortaldo[1], Alfred C. Denn III[1], Teresa Barlozzari[1], Ronald B. Herberman[1], Susan O. Sharrow[2] Keith M. Ramsey[3], Ko Okumura[4], Sonoko Habu[4]

[1]Biological Research and Therapy Branch
National Cancer Institute
Frederick, Maryland

[2]Immunology Branch, National Cancer Institute
Bethesda, Maryland

[3]FDA-Bureau of Biologics
Bethesda, Maryland

[4]Tokyo University
Tokyo, Japan

I. INTRODUCTION

Recent advances in cell separation techniques have facilitated the isolation of rat natural killer (NK) cells and their characterization as large granular lymphocytes (LGL) (Reynolds *et al*, 1981a; Reynolds *et al*, 1981b). LGL have distinct azurophilic granules in their cytoplasm and have been shown to be highly associated with *in vitro* NK activity. However, we still know very little about the detailed nature and ontogeny of these cells. Therefore, we have initiated studies with monoclonal antibodies and continuous flow microfluorometry, to examine the cell surface antigenic characteristics of highly enriched rat LGL, relative to those of T cells, monocytes and polymorphonuclear leukocytes (PMNs).

ISBN 0-12-341360-5

The results demonstrate that, like human LGL and mouse NK cells, rat LGL are an antigenically distinct subpopulation of lymphocytes which share some characteristics with both T cells and monocytes.

II. MATERIALS AND METHODS

Leukocyte preparations. All experiments were performed with peripheral blood leukocytes (PBL) prepared as previously described (Oehler et al, 1978; Nunn et al, 1976; Oehler et al, 1977) from W/Fu or rnu/+ rats. Monocytes were isolated from PBL by adherence on FBS coated tissue culture flasks (Reynolds et al, 1981b). LGL, T cells, and PMN were prepared from nylon wool (NW) nonadherent PBL by centrifugation on a discontinuous density gradient of Percoll (Reynolds et al., 1981a). Viability was over 95% as judged by trypan blue exclusion. Morphological analyses of these fractions are as follows: fraction 2 - LGL (75 to 90%); fraction 3 - T cells (>80%); fraction 4 PMN (60 to 90%).

Flow microfluorometry (FMF). FMF analysis was performed as previously described (Reynolds et al, 1981b; Miller et al, 1978; Ortaldo et al, 1981) using two fluorescence-activated cell sorters (FACS II-III) (Becton-Dickinson FACS Systems).

III. RESULTS AND DISCUSSION

With the recent advances in the separation of monocytes, T cells and NK cells (LGL) from rat blood and spleen, it has become possible to directly examine the cell surface antigens present on enriched populations of these cells. Studies with anti-rat monoclonal antibodies (Table 1) indicated that most rat LGL express the leukocyte common (L-C), W3/13 and T cell associated OX-8 antigens. A significant number of LGL from ART-1^a rats but not from ART-1^b animals were also reactive with the anti-T cell BC-84 antibody. Interestingly, this BC-84 alloantigen was also present on LGL from nude rats (data not shown). Antigens not found on LGL were the W3/25 (on helper T cells), Ia and sIg. These results clearly show that LGL share the L-C and W3/13 antigens with most rat leukocytes, but differ significantly from T cells in their lack of expression of W3/25 and from monocytes in their expression of OX-8.

Since asialo-GM1 (aGM1) (Ortaldo et al, 1981; Kasai et al, 1980; Durdik et al, 1980), thy 1 (Durdik et al, 1980; Mattes et al, 1979; Herberman et al, 1978), and M1/70 (Ault & Springer, 1981; Holmberg & Ault, unpublished) antibodies

TABLE 1. Reactivity of Anti-Rat Monoclonals with Rat Leukocytes

Antibody/ Antigen	Mean FU (% positive)[a]			
	LGL	Monocytes	T Cells	PMN
W3/13	30(83)	45(76)	20(88)	20(95)
W3/25	30(15)	15(23)	30(51)	<5
OX-1/(L-C)	40(98)	40(97)	35(99)	30(98)
OX-8	40(84)	7(4)	30(34)	<5
Ia(poly)	7(8)	10(12)	<5	<5
Ia(common)	<5	10(12)	<5	<5
BC-84/ (ART-1[a])	40(64)[b]	30(58)[b]	45(93)[b]	<5
sIg	<5(8)	<5(2)	<5	<5

[a]Control samples were used as baselines and were always <5 FU.
[b]Mean of ART-1[a] (+) animals only. Cells from ART-1[b] animals were always <5 FU.

have previously been shown to react with mouse or human NK cells, we also examined their reactivity with enriched populations of rat LGL, monocytes and T cells (Table 2). Only aGM1 was detected on the rat cells and this was present in the majority of cells in each subpopulation.

TABLE 2. Reactivity of Other Antibodies with Rat Leukocytes

Antibody	Mean FU (% positive)[a]			
	LGL	Monocytes	T Cells	PMN
Asialo GM1	300(78)	70(76)	150(59)	100(95)
Thy 1.1	10(14)	10(23)	7(14)	10(30)
M1/70	15(10)	10(5)	<5	NT

[a]Control samples were used as baseline and were always <5 FU.

Although this antiserum reacted strongly with rat LGL (300 FU) it also showed a lower but significant reactivity with monocytes, T cells and PMN. Antihuman monoclonal antibodies (OKM1, OKT3, OKT4 and OKT8) were not reactive with any of the rat populations (data not shown).

Previous studies in the mouse have shown that the *in vivo* injection of aGM1 antibody causes a marked reduction in NK activity (Kasai *et al*, 1981). Therefore we initiated similar studies in the rat using the aGM1 and OX-8 antibodies to determine if this loss of NK activity could be accounted for by a decrease in LGL. The results in Table 3 demonstrate that the *in vivo* injection of either of these antibodies produces a marked reduction in both splenic and PBL NK activity. The mechanism of this reduction, however, may be different for these two antibodies since a significant reduction in the LGL percentage can be seen with the asialo GM1 antibody, whereas no effect on the LGL percentage was seen after injection of OX-8. These results suggest that at least two mechanisms may be responsible for the *in vivo* reduction of NK activity after the injection of anti-LGL antibodies. These are: 1) the elimination or redistribution of LGL from the spleen and blood (aGM1) and 2) the inactivation or blocking of existing LGL without their elimination (OX-8). Additional antibodies are currently being used to further examine the mechanisms of NK reduction and the effect of this reduction on immune surveillance in rats. These results should provide much needed information regarding the developmental lineage and *in vivo* function of NK cells.

TABLE 3. Reduction in NK Activity by In Vivo Injection of Anti-LGL Antibody[a]

Experimental Group[b]	% Cytotoxicity[c]					
	G_1-TC		YAC-1		% LGL	
	Spleen	PBL	Spleen	PBL	Spleen	PBL
Normal Control	NT	NT	15.6	23.1	NT	11.0
Asialo GM1 (1:25)	NT	NT	5.9	7.3	NT	1.0
PBS Control	6.7	11.5	18.6	33.8	1.0	8.5
OX-8 (1:500)	0.0	3.4	11.0	11.9	1.5	9.5

[a]Antibody injected I.V. (0.25-0.50 ml), 4 days before assay.
[b]For OX-8, animals were rnu/+; for anti-asialo GM1, W/Fu were used.
[c]E/T ratio 50:1, 4 hour assay.

REFERENCES

Ault, K.A., and Springer, T.A. (1981). J. Immunol. 126:359.
Durdik, J.M., Beck, B.N., and Henney, C.S. (1980). In "Natural cell-mediated immunity against tumors" (R. Herberman, ed.), p. 37. Academic Press, New York.
Herberman, R.B., Nunn, M.E., and Holden, H.T. (1978). J. Immunol 121:304.
Holmberg, L., and Ault, K., unpublished observation.
Kasai, M., et al. (1980). Eur. J. Immunol. 10:175.
Kasai, M., et al. (1981). Nature 291:334.
Mattes, M.J., et al. (1979). J. Immunol. 123:2851.
Miller, M.H., et al. (1978). Rev. Sci. Instrum 49:1137.
Nunn, M.E., et al. (1976). J. Natl. Cancer Inst. 56:393.
Oehler, M.R., et al. (1977). Cell Immunol. 29:238.
Oehler, J.R., et al. (1978). Int. J. Cancer 21:204.
Ortaldo, J.R., et al. (1981). J. Immunol. 127:2401.
Reynolds, C.W., et al. (1981a). J. Immunol. 127:282.
Reynolds, C.W., et al. (1981b). J. Immunol. 127:2204.

NATURAL KILLER CELLS ARE DISTINCT FROM LECTIN-DEPENDENT EFFECTOR CELLS IN MAN AS DETERMINED BY THE TWO-TARGET CONJUGATE SINGLE CELL ASSAY[1]

Thomas P. Bradley and Benjamin Bonavida

Department of Microbiology and Immunology
UCLA School of Medicine
Los Angeles, California

I. INTRODUCTION

Normal peripheral blood lymphocytes (PBL) from non-immunized, disease-free hosts express various cytotoxic functions. These functions include natural killing (NK), antibody-dependent cellular cytotoxicity (ADCC), and lectin-dependent cellular cytotoxicity (LDCC). LDCC was first described in a murine system (1-3). Allospecific cytotoxic T lymphocytes were seen to lose their specificity and lyse nonspecific (syngeneic or xenogeneic) targets when the lectins concanavalin A (Con A) or phytohemagglutinin (PHA) were added to the CTL-target cell mixture. Non-immune, "normal" lymphoid populations were not able to mediate LDCC. Thus, in mice, LDCC appeared to be mediated by a primed CTL population rather than by a "naturally" occurring cytotoxic cell in the non-immune animal (4). LDCC in Man was first examined by Bonavida *et al*. (5). A major difference between the human and murine system was that normal non-immune *human* PBL, in the presence of lectin, expressed LDCC activity in a short-term (2-3 hour) assay, without priming or activation of the lymphocytes. The human effector cell of LDCC, then, appears to be different from the murine LDCC effector. Bonavida *et al*. described the human LDCC effector as a nonadherent, Fc receptor (FcR)-bearing cell found in both

---

[1]Supported by Grant CA-12800 from the National Cancer Institute, DHHS and Training Grant A1-7126 from the National Cancer Institute which was awarded to T.P. Bradley.

ISBN 0-12-341360-5

E rosette positive and negative fractions. Since both NK and ADCC activities are also attributed to FcR-bearing cells (6-8) the question arose as to the relationship between these effector cells. Studies using whole populations of effector cells have claimed that NK and ADCC effectors are either the same as (9,10) or distinct from NK effector cells (11-13). However, no such studies have attempted to determine the relationship between the LDCC effector and the NK and ADCC effectors. With the recent introduction of the single cell assay in agarose (14,15) and the two-target conjugate assay (4), new tools are now available to directly determine the relationship between these effectors at the single cell level. The relationship between NK and ADCC effectors is the topic of another manuscript which has been submitted for publication. The present report is focused on the relationship between NK and LDCC effector cells. We use the discontinuous Percoll gradient method of Timonen et al. (16) to enrich for NK effectors, large granular lymphocytes (LGL) which, along with non-enriched PBL, are used as effectors. Using the two-target conjugate assay (4), we attempted to simultaneously bind a single lymphocyte to both an NK target and an LDCC target with lysis of one or more targets observed. If a single effector cell kills both NK and LDCC targets, then the single effector can be called both an NK and LDCC effector cell. Alternately, if a single effector kills only one or the other target (but not both), then, with the necessary controls, one can conclude that NK and LDCC effector cells are distinct. This latter hypothesis proved correct.

II. MATERIALS AND METHODS

Cell Preparation, Treatment and Assay: Heparinized fresh human blood is separated on Ficoll-hypaque (17), incubated on nylon wool columns (18), and purified on a seven-step discontinuous Percoll gradient by the method of Timonen, et al. (16). Up to 70% of giemsa-stained cells in pooled fractions 2 and 3 are of the large granular lymphocyte (LGL). Effector cells are suspended in RPMI-1640 supplemented with either 10% fetal bovine serum or 1% human AB serum. Target cells K562 and MOLT-4 are used as NK-sensitive targets and RAJI as an NK-insensitive target. Reagents and media are used as described in an earlier report (19). Lectin and IO_4 treatments and the ^{51}Cr-release assay are accomplished as previously described (4).

Single and Two-Target Conjugate Formation and Lysis: This procedure is described in detail in a recently published manuscript (4). To form two-target conjugates, effector to target ratios of 1:1:1 (effector:NK target:LDCC target) are used. A gentle means of resuspending the pellet (post-centrifugation of the effector-target cell mixture) is critical to obtaining a high frequency of two-target conjugates.

III. RESULTS AND DISCUSSION

Our initial experiments were designed to test whether LDCC activity is enriched in the NK-enriched LGL fraction. Using the ^{51}Cr-release assay, we found NK activity to be clearly enriched, as reported by Timonen et al. (16) (Table I).

TABLE I. Lysis of NK and LDCC Targets by PBL and LGL in a ^{51}Cr-release assay[a]

	Lytic Units[b]				
	Target Cell[d]				
	K562	RAJI	RAJI (Con A)	RAJI (PHA)	RAJI (IO_4)
PBL[c]:	125	13.3	160	182	191
LGL :	571	14.2	333	n.d.[e]	n.d.

[a]NK-enriched lymphocytes (LGL) and non-enriched PBL are used as effectors in a three-hour ^{51}Cr-release assay.

[b]Lytic units per 10^7 lymphocytes taken at 33% mean specific lysis.

[c]^{51}Cr-labeled targets ($5x10^3$ per microtitre well) and effectors are incubated at various effector to target ratios.

[d]^{51}Cr-labeled targets are either left untreated (K562 and RAJI) or treated with Con A (20 ug/ml final concentration) PHA (20ul stock solution/ml cells) or IO_4 (7.5 mM final concentration) at 30°C, then washed twice prior to use.

[e]Values not determined.

Lytic activity against K562 is increased from 125 lytic units (L.U.) per 10^7 lymphocytes at 33% lysis in non-enriched PBL to 571 L.U. in the LGL fraction (a 4.6 fold increase). LDCC is also enriched in the LGL fraction (160 L.U. in PBL to 333 L.U. in LGL) although this enrichment is not as marked as that for NK activity (only a 2 fold increase). These results indicate that enriching for NK activity also enriches for LDCC activity, and that both effectors can be found in the LGL fraction. Since a single population of LGL effectors mediates both NK and LDCC activity, this tentatively suggested that a single effector cell might be responsible for both NK and LDCC activities. After all, both LGL (20) and LDCC effector cells (5) have been shown to be FcR positive cells. However, conclusions concerning the relationship between the NK and LDCC effector cells using whole populations of effectors with the ^{51}Cr-release assay remained equivocal. The relationship could be directly resolved only at the single cell level. The frequency of single effector cell (PBL and LGL) binding and killing of NK and LDCC targets is shown in Table II. Against K562, the frequency of effector to target binding is increased 53% in the LGL population. Killing of lymphocyte-bound targets is increased 81% over that in the non-enriched PBL population. Binding and lysis of lectin or periodate treated RAJI is also enhanced in the LGL population. As seen with the ^{51}Cr-release assay (Table I), this increase is also not as marked as the increase seen against NK targets (for LDCC an increase of 23% binding and 26% killing in the LGL population). These differences in the LGL population between NK and LDCC in both binding and lysis might suggest that these effectors, although both in the LGL fraction, are distinct. Two controls for the LDCC reaction were performed. First, we ascertained that RAJI could not serve as a target in LDCC when not treated with lectin. Although binding of lymphocytes to untreated RAJI did occur, the binding did not lead to a significant amount of lysis (less than 7% in Table II). Furthermore, LGL did not show an enrichment over nonenriched PBL in binding or killing of untreated RAJI. Second, to rule out any artifact of lectin target treatment, we oxidatively modified RAJI with the chemical sodium periodate (IO_4). We have previously shown (4) that oxidation with IO_4 yields an LDCC-like reaction which we have termed oxidation dependent cellular cytotoxicity (ODCC).

The definitive answer to our question of whether an NK effector cell is also an LDCC effector cell is determined by using the two-target conjugate assay. It is first necessary to establish whether a single effector cell is able to bind to two different targets simultaneously. Using a gentle method of resuspension (see Materials and Methods), we found that up to 20-30% of all effector cells bound to targets are bound to more than a single target. Of these two-target conjugates, up to one-third are comprised of two different

TABLE II. Binding and Lysis of NK and LDCC Targets by PBL and LGL in the Single Cell Assay

	Effector Cell	Target Cell				
		K562	RAJI	RAJI(ConA)	RAJI(PHA)	RAJI(IO_4)
% Effectors conjugated to Targets[a]	PBL	16.3 $\pm$ 1.5[b]	11.0 $\pm$ 2.0	31.7 $\pm$ 2.1	27.0 $\pm$ 1.9	23.7 $\pm$ 2.3
	LGL	25.0 $\pm$ 4.4	13.1 $\pm$ 1.8	39.0 $\pm$ 1.4	n.d.	n.d.
% Conjugates with Killed Targets[a]	PBL	26.7 $\pm$ 4.5	6.6 $\pm$ 1.5	28.0 $\pm$ 0.9	31.8 $\pm$ 4.4	33.4 $\pm$ 3.8
	LGL	48.3 $\pm$ 6.1	5.3 $\pm$ 2.2	31.6 $\pm$ 4.3	n.d.	n.d.

[a] These calculations are described in a previous report (4).

[b] Each value represents the median value of three to five separate experiments.

targets (one K562 and one lectin or IO_4-treated RAJI). It was then necessary to establish that a single effector cell was capable of inflicting lethal hits to multiply-bound targets. In the murine system, Zagury et al. (21) had shown that a single effector cell was capable of sequentially killing up to four attached targets. Also in the mouse system, we had previously shown that an allospecific CTL could kill both a specific target as well as a lectin treated syngeneic target (4). The human cytotoxic system, however, had not been examined in this manner. Our results show that a single lymphocyte (PBL or LGL) can indeed kill two simultaneously attached targets. As seen in Table III, a single PBL or LGL can kill two K562, two MOLT-4, two Con A-treated, PHA-treated, or IO_4-treated RAJI.

TABLE III. Lysis of Two Targets Bound Simultaneously to a Single Effector Cell

	Two-Target Conjugates of:[a]	Percent of All Two-Target Conjugates With:[b] Both Targets Killed:	One Target Killed:	No Targets Killed:
Identical Targets	K562-PBL-K562	18.0[c]	36.8	45.1
	K562-LGL-K562	24.7	48.2	27.1
	MOLT-PBL-MOLT	22.3	35.1	42.6
	RAJI(ConA)-PBL-RAJI(ConA)	23.8	42.9	33.3
	RAJI(ConA)-LGL-RAJI(ConA)	20.1	57.2	22.7
	RAJI(PHA)-PBL-RAJI(PHA)	21.6	25.2	53.2
	RAJI(IO_4)-PBL-RAJI(IO_4)	19.2	28.7	52.1
Different Targets	K562-PBL-MOLT	25.4	37.5	37.1
	K562-PBL-RAJI(ConA)	0.0	53.3	46.7
	K562-LGL-RAJI(ConA)	0.0	58.3	41.6
	K562-PBL-RAJI(PHA)	0.0	51.4	48.6
	K562-PBL-RAJI(IO_4)	0.0	62.6	37.4

[a]Two-target conjugates consisting of two-identical or two-different targets are incubated for four hours. Incubations of up to 16 hours did not significantly increase two-target killing over background lysis.

[b]At least 53 two-target conjugates are counted per experimental point.

[c]Data representative of at least four separate experiments.

A single effector can also kill both MOLT-4 and K562 targets when simultaneously bound, indicating that killing was not restricted to the killing of two identical targets. However, it is clearly shown that a single effector cell is not capable of killing two targets when one was an NK target and one an LDCC or ODCC target. This result directly shows that the NK and LDCC effector cells are distinct and thereby constitute different subsets of the LGL population. The failure of an NK effector cell to lyse an LDCC target (as well as the failure of the LDCC effector to lyse an NK target) may be due to the effector cells' lack of "lytic" receptors for the appropriate target or lectin. Presumably, the lectins are not able to accomodate a lytic trigger on the NK effector cells, enabling the NK effector cells to kill NK-insensitive, lectin-treated targets. This result is somewhat unexpected since the murine CTL which has specific receptors for the sensitizing antigen can be triggered to nonspecifically lyse lectin-treated xenogeneic and even syngeneic targets in an LDCC reaction (1). Since binding to two different targets is seen without lysis of the two targets, it is inferred that binding and lysis are distinct events.

The present study offers evidence that cells with lytic potential are present within an NK-enriched LGL population that are distinct from NK effector cells. The study also shows that lysis of one target does not facilitate the lysis of a proximally located second target. The system presented here allows further investigation into the mechanism of lysis and populations involved in NK and LDCC.

REFERENCES

1. Bonavida, B., and Bradley. T.P. (1976). Transplantation 21:94.
2. Forman, J., and Möller, G. (1973). J. Exp. Med. 138:672.
3. Bevan, M.J., and Cohn, M. (1975). J. Immunol. 114:559.
4. Bradley, T.P., and Bonavida, B. (1981). J. Immunol. 126:208.
5. Bonavida, B., Robins, A., and Saxon, A. (1977). Transplantation 23:261.
6. West, W.H., Cannon, G.B., Kay, H.D., Bonnard, G.D., and Herberman, R.B. (1977). J. Immunol. 118:355.
7. Kay, H.D., Bonnard, G.D., West, W.H., and Herberman, R.B. (1977). J. Immunol. 118:2058.
8. Perlmann, P., Perlmann, H., and Wigzell, H. (1972). Transplant. Rev. 13:91.

9. Ojo, E., and Wigzell, H. (1978). Scand. J. Immunol. 7:297.
10. Landazuri, M.O., Silva, A., Alvarez, J., and Herberman, R.B. (1979). J. Immunol. 123:252.
11. Koren, H.S., and Williams, M.S. (1978). J. Immunol. 121:1956.
12. Bolhuis, R.L.H., Schuit, H.R.E., Nooyen, A.M., and Rontelap, C.P.M. (1978). Eur. J. Immunol. 8:731.
13. Neville, M.E. (1980). J. Immunol. 125:2604.
14. Grimm, E.A., and Bonavida, B. (1979). J. Immunol. 123:2861
15. Bonavida, B., Bradley, T.P., and Grimm, E.A. (1982). Methods in Enzymology, vol.82.
16. Timonen, T., Ortaldo, J.R., and Herberman, R.B. (1981). J. Exp. Med. 153:569.
17. Boyum, A. (1968). Scan. J. Clin. Lab. Investig. 21 (Suppl.97):1.
18. Julius, M., Simpson, H.E., and Herzenberg, L.A. (1973). Eur. J. Immunol. 3:645.
19. Bradley, T.P., and Bonavida, B. (1978). Transplantation 26:212.
20. Herberman, R.B., Timonen, T., Reynolds, C., and Ortaldo, J.R. (1980). In "Natural Cell-Mediated Immunity Against Tumors," ed. R.B. Herberman, Academic Press, p.89.
21. Zagury, D. *et al*. (1979). J. Immunol. 123:1604.

VIRUS DEPENDENT NATURAL CYTOTOXICITY (VDCC) OF HUMAN LYMPHOCYTES

Peter Perlmann
Abdulrazzak Alsheikhly
Bengt Härfast

Department of Immunology
University of Stockholm
Sweden

Torbjörn Andersson[1]
Claes Ørvell
Erling Norrby

Departments of Infectious Diseases
and Virology
Karolinska Institute Medical School
Stockholm, Sweden

Treatment of lymphocytes with live or UV-inactivated virus (Parotis or Sendai) enhances their cytotoxicity for a large variety of target cells. The effector cells appear to be heterogeneous. Which effector cells predominate in a given reaction depends on the type of target cells used. With some target cells, T-cells lacking Fc receptors for IgG are highly cytotoxic. Neither specific recognition of viral antigen nor mitogenic activation are required for VDCC induction. Virus particles freed of their surface glycoproteins do not cause VDCC. However, VDCC is easily induced by the peplomer carrying hemagglutinin and neuraminidase activity. Although this protein probably serves as recognition factor, establishing the close effector cell/target cell contact necessary for cytotoxicity, its aggregating effect is not sufficient for VDCC-induction.

[1]Supported by SMRC grant 16X-04973

ISBN 0-12-341360-5

Lymphocytes from the peripheral blood of normal human donors exhibit an enhanced natural cytotoxicity to a variety of virus infected target cells. This cytotoxicity is neither virus specific nor HLA-restricted and is antibody independent (Andersson et al., 1975; Härfast et al., 1978; Santoli et al., 1978). A similar enhanced lymphocyte cytotoxicity can be induced by exposing peripheral blood lymphocytes (PBL) to small amounts of either live or UV-inactivated virus before incubation with target cells (Härfast et al., 1980a). This cytotoxicity affects a large number of non-infected target cells, comprising both established tumor cell lines, allogeneic or autologous fibroblasts (Casali et al., 1981) or PHA-transformed lymphoblastoid cells. Here, we will summarize some of our recent work with Parotis or Sendai virus on the nature of the effector cells and on mechanisms of induction.

When characterized by means of surface markers, the effector cells in VDCC are heterogenous, comprising both T- and "non-T" lymphocytes. Originally, it appeared that the effector cells in general had high avidity receptors for IgG (Härfast et al., 1978; 1980a). Recent experiments with a larger panel of human tissue culture cell lines indicated that purified T cells, thoroughly depleted of Tγ-cells, frequently exhibited a strong activity in VDCC. The cytolytic potential of the Tγ-depleted T cell fractions depended primarily on the type of the target cell (Table I). At present, we cannot exclude that the effector cells in the Tγ-depleted T cell fractions were cells with low avidity Fc receptors for IgG, or aquired such receptors during incubation (Perlmann and Cerottini, 1979). In any event, such experiments show that the target cells may select the effector cells which will predominate in a given VDCC-system. This also implies that different mechanisms of target cell recognition and/or cytotoxicity induction may prevail with different types of target cells. Similar findings have previously been made in other systems (Saksela et al., 1979; Troye et al., 1980; Klein and Vanky, 1981).

The fact that T-cells were active in VDCC raised a number of important questions. Thus, it could not be excluded that cytotoxicity was in many instances induced by the specific interaction of viral antigen with the corresponding antigen receptors on some memory T-cells, present in the circulation after a previous mumps infection. Therefore, we studied the VDCC activity of cord blood lymphocytes against a number of human tumor cell lines. In general, in the samples (∿20) tested, VDCC was very high, regardless of whether the experiments were performed by pretreating the lymphocytes with UV-inactivated virus or by adding them to target cells acutely infected with virus. With the target cells used in the typical experiment of Table II, strong VDCC was again seen in both the total

TABLE I. NK and VDCC of lymphocyte fractions to different target cells

Donor	Lymphocytes[b]	^{51}Cr-Release[a]							
		T 24		Mel-1		Chang		Vero	
		NK	VDCC	NK	VDCC	NK	VDCC	NK	VDCC
1	Unfractioned	9	56	4	36	7	18	17	45
	"Non-T"	20	59	nd	nd	12	34	36	48
	Total T	7	45	4	33	2	15	8	18
	Tγ-depleted T	5	46	7	36	0	5	2	5
		(14)		(29)		(13)		(16)	
2	Unfractioned	9	45	6	65	21	56	52	80
	"Non-T"	20	57	4	51	31	70	61	83
	Total T	4	42	3	46	9	27	27	34
	Tγ-depleted T	1	37	1	48	1	15	0	7
		(18)		(32)		(15)		(18)	

[a]From three human (T 24, Mel-1, Chang) and one monkey (Vero) cell lines, spontanous release (in paranthesis) subtracted; 18h, effector to target ratio 50:1. -
[b]Unfractionated or fractionated by rosetting and column fractionation. - nd: not done.

T-cell fraction and in the Tγ-depleted T-cells, consisting almost entirely of lymphocytes binding the monoclonal anti-T cell antibodies OKT3 or Leu1 (Reinherz et al., 1980; Wang et al., 1980). Whether the strong cytotoxic activity of the "non--T" fraction was due to its high content of OKT3 - Leu1 binding lymphocytes is unknown. In any event, these cord blood experiments indicate that VDCC is not the result of a previous immunization of the lymphocyte donor with mumps antigen.

As interactions between antigen sensitive T-cells and viral antigens were not responsible for VDCC induction, it may be asked if the virions induce cytotoxicity by polyclonal T-cell activation (Clark, 1975; Waterfield et al., 1976). Lymphocytes from either healthy donors or from cord blood were treated with UV-inactivated mumps virions. After incubation for up to 7 days, VDCC and stimulation of DNA synthesis were measured in parallel. Although VDCC was seen in all cases, DNA synthesis of the lymphocytes was not significantly elevated over that of the untreated controls. Thus, there is presently no evidence for VDCC beeing due to mitogenicity of the inducing virus preparation.

To further elucidate the mechanism of VDCC-induction,

TABLE II. Surface markers and cytotoxicity of cord blood lymphocytes

Lymphocytes	Surface Markers[a]				^{51}Cr-Release[b]			
	E^+	EAG^+	$T3^+$	$Leu1^+$	NK		VDCC	
					30:1	60:1	30:1	60:1
Unfractionated	87	14	73	82	5	8	34	50
"Non-T"	33	43	nd	32	9	15	38	54
Total T	91	12	94	98	1	2	31	43
Tγ-depleted T	94	0	97	99	4	2	36	47

[a]Percent lymphocytes forming rosettes with sheep erythrocytes (E^+) or with IgG coated bovine erythrocytes (EAG^+), or stained with monoclonal anti T cell antibodies OKT3 or Leu1 by indirect immunofluorescence. -

[b]From human carcinoma line T 24, spontaneous release (26%) subtracted; 18h, ratios 30:1 and 60:1.

Parotis virions were treated with proteolytic enzymes, thereby stripping them of their two major surface glycoproteins (peplomers), carrying either hemagglutinin and neuraminidase activities (HN) or hemolytic (HL) and cell fusion (F) activities (Örvell, 1978). The capacity of such naked virions to enhance lymphocyte cytotoxicity above the NK background was strongly reduced (Härfast et al., 1980b). In contrast VDCC was induced when the lymphocytes were treated with an extract containing a mixture of the HN- and F-peplomers in soluble form. Similar results with measles virus have recently been reported by others (Casali et al., 1981). These experiments showed that the viral surface glycoproteins were essential for VDCC induction. This was further supported by inhibition experiments with rabbit antibodies specific for either the HN- or the F-protein.

In these experiments, no definite answer as to the relative importance of the individual glycoproteins was reached. For this reason, we recently performed similar experiments with Sendai virus. When Sendai virions are treated with chymotrypsin, the HN-peplomer is removed selectively while treatment with trypsin removes the F-protein. Removal of HN abolished VDCC, but removal of F had no effect. Conversely, treatment of the lymphocytes with the HN glycoprotein in solution but not with the F-protein resulted in VDCC induction (Alsheikhly et al., 1982).

TABLE III. Lymphocyte cytotoxicity to chicken-erythrocytes(E_c)

Lymphocytes	^{51}Cr-Release[a]		
	No Antibody		Antibody[b]
	40:1	80:1	2:1
Untreated	1	2	4
Virus-treated[c]	2	2	48

[a]From ^{51}Cr-E_c, spontaneous release (2%) subtracted, 18h, lymphocyte: E_c ratios as indicated. - [b]Rabbit anti-E_c, final dilution 10^{-5}.-[c]10 μg UV-inactivated Parotis virus /$4x10^{-6}$ lymphocytes, 30 min. 37°C, then 2 washes.

Since the HN peplomer appeared to be the only viral protein involved in VDCC it could be assumed that cytotoxicity is brought about by the agglutinating effect of this protein, resulting in close effector cell - target cell contacts. However, although such a contact is a necessary prerequisite for full expression of VDCC, it is probably not sufficient to bring about target cell lysis. Not all lymphocytes which bind to virus infected target cells are cytotoxic and treatment with neuraminidase which reduces binding does not reduce cytotoxicity (Härfast et al., 1977). Human PBL, treated with UV--inactivated mumps virus also form aggregates with chicken erythrocytes. This aggregation strongly enhances ADCC but does not result in lysis of these target cells without antibodies (Table III). These results imply that, in addition to binding, at least one second step of target cell recognition appears to be required for VDCC expression.

Incubation of lymphocytes with virus infected target cells results in production of interferon which enhances lymphocyte cytotoxicity (Trinchieri et al., 1980). Thus far, we have been unable to detect interferon in the supernatants from virus treated lymphocytes incubated with non-infected target cells. Further experiments are, however, needed to establish if VDCC induced in this way is interferon independent.

In conclusion, these experiments show that certain viral glycoproteins are efficient inducers of lymphocyte cytotoxicity, apparently not involving conventional immune recognition by humoral antibodies or specific T-cell receptors. These latter mechanisms are believed to affect the course of many virus infections by limiting the spread of the virus. Since VDCC is generated earlier than the conventional immune response (Welsh, 1982) it may have an important protective function at the beginning of a virus infection. At the same time, because of its low selectivity, comprising both infected and non-infected targets, VDCC may also be a harmful tissue damaging factor, appearing in the course of a persisting infection.

REFERENCES

Alsheikhly, A., Örvell, C., Härfast, B., Andersson, T., Norrby, E., and Perlmann, P. Scand. J. Immunol.(in press).
Andersson, T., Stejskal, V., and Härfast, B. (1975). J. Immunol. 114:237.
Casali, P., Sissons, J. G., Buchmeier, M. J., and Oldstone, M. B. A. (1981). J. Exp. Med. 154:840.
Clark, W. R. (1975). Cell Immunol. 17:505.
Härfast, B., Andersson, T., Stejskal, V., and Perlmann, P. (1977). J. Immunol. 118:1132.
Härfast, B., Andersson, T., and Perlmann, P. (1978). J. Immunol. 121:755.
Härfast, B., Andersson, T., Alsheikhly, A., and Perlmann, P. (1980a). Scand. J. Immunol. 11:357.
Härfast, B., Örvell, C., Alsheikhly, A., Andersson, T., Perlmann, P., and Norrby, E. (1980b).Scand. J. Immunol. 11:391.
Klein, E., Vãnky, F. (1981). Cancer Immunol. Immunother. 11:183.
Örvell, C. (1978). J. Gen. Virol. 41:527.
Perlmann, P., and Cerottini, J. C. (1979). In "The Antigens" (M. Sela, ed.), p. 173. Academic Press, New York.
Reinherz, E. L., Hussey, R. E., and Schlossman, S. F. (1980). Eur. J. Immunol. 10:758.
Saksela, E., Timonen, T., Ranki, A., and Häyrÿ, P. (1979). Immunol. Rev. 44:71.
Santoli, D., Trinchieri, G., and Lief, F. (1978). J. Immunol. 121:156.
Trinchieri, G., Perussia, B., and Santoli, D. (1980). In "Natural Cell-mediated Immunity against Tumors" (R. B. Herberman, ed.), p. 655. Academic Press, New York.
Troye, M., Vilien, M., Pape, G. R., and Perlmann, P. (1980). Int. J. Cancer. 25:33.
Welsh, R. M. (1981). In "Current Topics in Microbiology and Immunology" (O. Haller, ed.), p. 83. Springer Verlag, Berlin.
Wang, C. Y., Good, R. A., Ammirati, P., Dysenhart, G., and Evans, R. L. (1980). J. Exp. Med. 151:1539.
Waterfield, G. D., Waterfield, E. M., Anaclerio, A., and Möller, G. (1976). Transplant Rev. 29:277.

NATURAL CYTOTOXICITY OF HUMAN MONOCYTES[1]

Dina G. Fischer[2]
Marc G. Golightly
Hillel S. Koren[3]

Division of Immunology
Duke University, Medical Center
Durham, North Carolina

I. INTRODUCTION

The ability of freshly isolated human peripheral blood monocytes to lyse a variety of tumor target cells was demonstrated *in vitro* by several investigators (Mantovani et al., 1979; Hammerstrom, 1979; Horwitz, 1979; Cameron and Churchill, 1979; Rinehart et al., 1979). In most of these studies long-term isotope release assays (18-72 hrs) were used. In some cases (Cameron and Churchill, Rinehart et al. 1979), the monocytes were cultured for several days, before the addition of target cells, resulting in their maturation to macrophages. Thus, in most of the investigations the cell population studied was presumably the mature macrophage.

Since recruitment of mononuclear phagocytes to various tissues in response to inflammatory stimuli involves primarily circulating monocytes (Volkman and Gowans 1965; Spector and Willoughby 1963), it was important to study these cells after only minimal *in vitro* manipulation. Using a novel isolation technique we were able to obtain highly purified human peripheral blood monocytes which were capable of lysing a variety of tumor cells in a 3 hr ^{51}Cr-release assay (Fischer et al. 1981). In this chapter we will summarize results of

---

[1]Supported by N.C.I. grants CA 23354, CA 29589
[2]Present address: Department of Virology, Weizmann Institute of Science, Rehovot, Israel.
[3]Recipient of Research Cancer Development Award CA 00581 fron the National Cancer Institute.

ISBN 0-12-341360-5

our studies on the cytotoxic activity of the monocytes, and its potentiation by lymphokines and interferon. The interaction of the monocytes with tumor target cells was examined by both the ^{51}Cr-release assay, and at the single cell level with the modified conjugate/agarose technique (Bradley and Bonavida, 1977).

II. EXPERIMENTAL PROCEDURES

The novel technique for the isolation of monocytes from peripheral blood has been previously described (Fischer et al., 1981). The unique step in the adherence procedure involves a pretreatment of the plastic tissue culture plates with serum autologous to the monocyte donor. After 15 min at 37°C, the serum was removed and mononuclear cells, separated on Ficoll-Hypaque density gradient, were added in RPMI-1640 medium supplemented with 20% fetal bovine serum (FBS). Following 1 hr incubation at 37°C, the nonadherent cells were removed and the adherent monocytes were physically detached from the plates after exposure to Versene for 15 min.

Cytolytic activity of the monocytes was measured by ^{51}Cr-release from prelabelled tumor target cells after 3-4 hrs of interaction with effector cells (Fischer et al.,1981). In addition, the cytolytic activity was studied at the single cell level with the conjugate/agarose technique using a modification of the method described for cytotoxic T lymphocytes (Bradley and Bonavida, 1977). Briefly, monocytes and tumor target cells at 1:1 ratio were incubated in a pellet for 10 min at 30°C and 20 min at 4°C. The cells were then gently resuspended in molten 0.5% agarose at 39°C and quickly spread on a microscope slide. After an incubation for 3-4 hrs at 37°C the slides were incubated with trypan blue to stain the dead cells. Each slide was scored for the percent bound monocytes and the percent dead conjugated target cells, and the number of cytotoxic monocytes was calculated.

III. RESULTS AND DISCUSSION

A. Natural Cytotoxicity of Peripheral Blood Monocytes

The ability of a monocytes population to lyse tumor target cells seems to be dependent on the procedure used to isolate the effectors. We have found that minor variations in the technique, such as subsituting one type of serum with another one, greatly affected the activity of the cell population obtained (Fischer et al.,1981).

The various human suspension cell lines used as targets

differed in their sensitivity to lysis, with K562, a myeloid cell line, being the most sensitive one. The cytotoxic activity of the monocytes varied both among donors and among different preparations. When the cytotoxic activity was measured in a 3 hr ^{51}Cr-release assay, at a 20:1 effector-target ratio, with K562 as the target, percent specific lysis ranged from 11% to 49% (20 experiments). Using the T cell lines CEM, MOLT-4 or HSB as targets the highest specific lysis observed was 28% when tested under the same assay conditions.

Several lines of evidence suggest that the cytotoxic activity of the monocytes preparations could not be attributed to contaminating natural killer (NK) lymphocytes. Most of the preparations had less than 2% lymphocytes, and their presence could not explain the observed activity, particularly, since monocytes obtained from several donors were able to lyse CEM target cells more efficiently than the lymphocytes from the same donor (Fischer et al., 1981). In addition, using the conjugate/agarose technique, we have studied the interaction between the monocytes and the tumor target cells on the single cell level. Morphological observations combined with peroxidase stain further demonstrated that the cells bound to the tumor targets were indeed monocytes (Golightly et al., 1982)

The binding of monocytes to the target cells based on studies carried out with the conjugate/agarose technique was found to be rapid and temperture dependent. Maximal binding was reached after incubation for 5 min at 30°C. To detect lysis, the slides had to be incubated for longer periods of time (1-4 hrs) at 37°C. Interestingly, while 30-50% of the monocytes were capable of binding K562 targets, only 15% of the total monocytes were capable of killing. These observations demonstrate that binding and killing are distinct and separable events. Comparing the single cell and ^{51}Cr-release assays revealed a similar time course (Golightly et al., 1982). However, while lysis in agarose plateaued after 3-4 hrs, ^{51}Cr-release continued to increase beyond this time period, most likely due to the ability to recycle (i.e: to kill more than one target cell). Microcinematographic studies have indeed confirmed the ability of monocytes to recycle (K. Muse, personal communication).

B. Potentiation Of The Cytolytic Activity By Lymphokines And Interferon

Monocytes incubated overnight or longer in microtiter wells gradually lost their cytolytic activity. The loss was more rapid in the presence of human serum than with FBS. When incubated in tubes under conditions that prevented the adherence of the monocytes their cytotoxic activity was maintained

(Fischer et al., 1982). Peripheral blood monocytes have previously been shown to undergo changes associated with maturation as a result of *in vitro* culturing (Johnson et al., 1977). The maturation process was found to be accelerated by human serum (Unsgaard, 1979) and to require adherence of the monocytes (Dr. R.A. Musson, personal communication). It is therefore possible that the ability of the monocytes to rapidly lyse tumor cells is associated with their maturation.

While the cytotoxic activity of the monocytes declined after incubation in control medium, overnight incubation in the presence of lymphokines (supernatants of mononuclear cells stimulated by phytohemagglutinin) or partially purified fibroblast interferon resulted in several-fold increase in cytolytic activity (Table IA). Studies of the effector-target interaction at the single cell level revealed that activation by either lymphokines or interferon did not increase the fraction of the monocytes which were capable of binding and lysing tumor target cells (Table IB), suggesting that the stimulants potentiated the efficiency of already active cells but did not induce cytotoxic activity in cells that lacked this function.

Potentiation of the killing efficiency seems to be due to an increase both in the rate of lysis and in the ability of the effectors to recycle, as suggested by killing kinetics of ^{51}Cr labeled target cells. While tumor cell killing by control monocytes reached almost maximal levels after 6 hrs, the killing by lymphokine-treated monocytes was not only rapid during the initial 6 hrs, but also continued at a steeper slope during the additional incubation period (Fischer et al., 1982). The ability of stimulated monocytes to kill neoplastic cells more efficiently than the unstimulated ones might therefore be related to increased recycling.

IV. CONCLUSIONS AND SPECULATIONS

A fraction (5-10%) of the adherent peripheral blood monocytes was found to be capable of rapidly binding and lysing tumor target cells. Several lines of evidence support the possibility that the capacity to kill neoplastic cells resides in the immature population of circulating monocytes. These cells can respond to stimuli (e.g. lymphokines and interferon) which enhance their killing efficiency and which might interfer with their maturation. Under conditions favorable for maturation (e.g.: adherence or autologous serum) these monocytes lose not only their cytotoxic activity, but also their responsiveness to stimulation.

Monocyte cytotoxicity may play an important role in resistance against tumor growth (Adams and Snyderman, 1979; Evans and Alexander, 1976). Since a wide variety of drugs, includ-

TABLE I. Effect of lymphokines and interferon on the cytotoxic activity of monocytes

A. ^{51}Cr Release — Target Cells[a]

	Monocytes	Stimuli	K562	SB	HSB
1.	day 0[b]	none	8.3±0.8	<0.1	3.9±0.5
	day 1[c]	none	6.5±0.6	0.2±0.2	3.9±0.5
		Lymphokines	23.7±0.9	4.9±0.6	16.4±1.2
		Interferon	25.6±0.7	11.4±0.7	15.3±3.1
2.	day 0	none	20.5±1.0		
	day 1	none	6.8±0.9		
		Lymphokines	36.1±2.3		
		Interferon	30.4±0.8		

B. ^{51}Cr Release and conjugates

	Monocytes	Stimuli	% specific[a] lysis±SE	%cytotoxic[d] monocytes
1.	day 0	none	16.0±0.3	5.28
	day 1	none	16.2±0.8	5.70
		Lymphokines	27.7±0.9	5.89
2.	day 0	none	9.7±0.5	6.32
	day 1	none	4.4±0.6	4.30
		Interferon	24.5±1.0	5.72

[a] Activity was determined in a 3 hr ^{51}Cr-release assay at effector: target ratio of 20:1 and expressed as percent specific lysis ±SE.
[b] Monocytes tested as effectors immediately after their isolation.
[c] Monocytes incubated overnight in microtiter wells in the presence or absence of 25% lymphokines or 10^3 U/well of β interferon.
[d] Calculation based on the conjugate/agarose assay (=percent bound monocytes x percent dead conjugated targets).

ing biological response modifers (e.g. lymphokines and interferon) are available for the treatment of cancer patients, considerations such as their effect on the maturation of monocytes should play a role in selecting the most efficacious drug. Based on our hypothesis, it might be important to select agents known to delay maturation and to recruit immature monocytes from the bone marrow in order to achieve maximal resistance to tumor growth.

REFERENCES

Adams, D.O., and Synderman, R.J.,Nat. Cancer Inst. 62: 1341 (1979).

Bradley, L.P., and Bonavida, B., J. Immunol. 126:208 (1977).

Cameron, D.J., and Churchill, W.H., J. Clin. Invest. 63: 977 (1979).

Evans, R., and Alexander, P., in "Immunobiology of the Macrophage" (D.S. Nelson, ed.) p. 535, Academic Press, New York, 1976.

Fischer, D.G., Hubbard, W.J., and Koren, H.S., Cell Immunol. 58:426 (1981).

Fischer, D.G., Golightly, M.G. and Koren, H.S., Submitted for publication.

Golightly, M.G., Fischer, D.G.,, and Koren, H.S., Submitted for publication.

Hammerstrøm, J., Scand, J. Immunol. 10:575 (1979).

Horwitz, D.A., Knight, N., Temple, A., and Allison, A.C., Immunology 36:221 (1979).

Johnson, W. D., Jr., Mei, B., and Cohn, Z.A. J. Exp. Med. 146:1613 (1977).

Mantovani, A., Jerrells, T.R., Dean, J.H., and Herberman, R.B. Inst. J. Cancer 23:18 (1979).

Rinehart, J.J., Vessella, R., Lange, P., Kaplan, M.E., and Gormus, B.J., J. Lab. Clin. Med. 93:361 (1979).

Spector, W.G., and Willoughby, D.A.,Bact. Rev. 27:117 (1963).

Unsgaard, G., Acta Path. Microbiol Scand. Sect . C. 87:141 (1979).

Volkman, A., and Gowand, J.L., Brit. H. Expt. Path. 46:50 (1965).

NATURAL CYTOTOXICITY OF HUMAN MONONUCLEAR PHAGOCYTES: ROLE OF CONTAMINATING ENDOTOXIN AND EXPRESSION BY DIFFERENT MACROPHAGE POPULATIONS[1]

Andrea Biondi
Barbara Bottazzi
Giuseppe Peri
Nadia Polentarutti
Giorgio Caspani
Alberto Mantovani

Istituto di Ricerche Farmacologiche
"Mario Negri"
Milan, Italy

I. INTRODUCTION

Human blood monocytes and macrophages from different anatomical sites have appreciable levels of spontaneous cytotoxicity against susceptible target cells, in the absence of deliberate stimulation (1-7). Expression of cytotoxicity is a function of the maturation stage and anatomical site of origin of mononuclear phagocytes and can be modulated by exposure to interferons (IFN) and lymphokines (8-10).

Spontaneous cytotoxicity of macrophages has also been reported in mice and rats (11,12), but evidence suggested that is may be related to endotoxin contamination of tissue culture reagents (13) or to the microbial status of donor animals (14). The experiments reported herein were designed to analyse the role of endotoxin contamination in the expression of natural human monocyte cytotoxicity. We have also investigated the expression of antibody-dependent and independent cytotoxicity by various mononuclear phagocyte populations and found that

[1] Supported by CNR (No. 80.01579.96) and by grant 1R01 CA 26824 from National Cancer Institute

ISBN 0-12-341360-5

these two effector functions can be dissociated in different macrophage populations.

II. EXPERIMENTAL PROCEDURES

Cytotoxicity was measures in a long term as ^{3}H-thymidine release (1,2). Mononuclear phagocytes were obtained as described (1,2,15). The murine TU5 line was used as target for direct cytotoxicity. The murine TLX9, coated with rabbit antibody (resistant to NK activity and to natural monocyte cytotoxicity) was used for antibody-dependent cellular cytotoxicity (ADCC) experiments (16,17). Endotoxin contamination of tissue culture reagents was measured by the limulus amebocyte lysate (LAL) assay (No. L09081 and No. L17481, sensitivity 0.27-0.36 ng/ml - M.A. Bioproducts). The following LAL-negative reagents, were used thoughout: Ficoll-Hypaque (Lot No. 136723, MSL Eurobio, Paris); pyrogen-free saline for clinical use (Lots. No. F046, F133, F184, E045, Bieffe, Italy); pyrogen-free distilled water for clinical use (Lot No. 2248/3, Bieffe, Italy); RPMI 1640 medium (10 x concentrated, Lot. No. L708304, U703002, Gibco Europe); glutamine (Lot. 1115 Eurobio Paris and Lot L701704, Gibco Europe); penicillin and streptomycin for clinical use (Lot No. 1=0002, Farmitalia; Lot No. 8072143/3, Squibb); fetal bovine serum (FBS Lot 100222 Sterile Systems Inc.); human cord serum (HCS), aseptically collected by one of us. By way of comparison LAL+FBS(Lot No. 95919 M.B.A. and Lot No. 101011 SeraLab) was used. Lymphokine supernatants were obtained with PHA (HA16 Lot K0275, Wellcome R.L.) and tested with LAL before use. We have tested several human leukocyte or fibroblast IFN preparations and all of them were LAL^{+}. Therefore, we attempted removal of contaminating endotoxin from fibroblast IFN by taking advantage of its interaction with the LAL. For this purpose 10^{5}U/ml solution of partially purified fibroblast IFN was added to a vial of lysate from the amebocyte of the horseshoe crab, Limulus poliphemus, and the mixture was incubated for 2 hours at 37°C. High-speed (12.000 r.p.m.) supernatants of the resulting gel failed to induce gelation when added to fresh lysate.

III. NATURAL CYTOTOXICITY OF HUMAN MONOCYTES IN ENDOTOXIN-FREE CONDITIONS

Table 1 shows a typical experiment in which we measured the cytotoxic capacity of human blood monocytes in various sources

TABLE I. Natural Cytotoxicity of Human Monocytes under LAL-Condition. Influence of Various Sources of Serum (LAL^+FBS; LAL^-FBS and LAL^-HCS) at Different Attacker to Target Ratios

Expt.No.	Serum	Specific Lysis %			
		10:1	20:1	40:1	80:1
1	LAL^+FBS	5.1±1.6	19.8±3.3	51.9±1.4	49.5±3.0
	LAL^-FBS	4.3±1.3	17.8±7.4	35.5±3.1	56.6±4.3
	LAL^-HCS	4.0±2.0	43.5±1.7	65.6±2.2	53.5±7.2

of serum (LAL^+FBS, LAL^-FBS and LAL^-HCS) using a range of attacker to target (A:T) ratios. Cytotoxicity was consistently observed with the 3 sera. In all experiments except 2 cytolysis levels were higher with LAL^+FBS than with LAL^-FBS, but cytotoxicity was maximal with LAL^-HCS. Killing of TU5 target cells was also observed when adult human AB serum (LAL^-) was used, but the tumor cells showed little growth under these conditions and therefore these data are not presented. In an effort to assess the role of minute amounts of endotoxin below the sensitivity of the LAL assay, in a series of experiments Polymyxin B was added to the cultures. This antibiotic binds to endotoxin and inhibits several of its biological activities, including activation for cytotoxicity of murine macrophages (18,19). Polymyxin B did not appreciably affect the expression of natural cytotoxicity of human monocytes.

We previously observed that in vitro exposure to lymphokine supernatants or to IFN augmented the cytotoxicity of human blood monocytes and macrophages (8-10). In mice it has been shown that endotoxin can act synergistically with lymphokine in activating macrophages for cytotoxicity (18) and, at least in some systems, the presence of minute amounts of endotoxin is required for the expression of tumoricidal activity by macrophages exposed to lymphokines (18,20) but not to IFN (21). IFN or lymphokine augmented cytotoxicity was expressed by human monocytes under LAL^- conditions and Polymyxin B did not affect the tumoricidal activity.

IV. FUNCTIONAL HETEROGENEITY IN THE EXPRESSION OF TUMORICIDAL ACTIVITY

Mononuclear phagocytes express cytolytic activity against tumor cells not only directly but also via antibody. In the presence of specific antibody purified human blood monocytes or *in vitro* cultured monocytes were shown to be efficient effectors of antibody dependent cellular cytotoxicity (ADCC) against tumor cells (22-24). No information is available concerning the ADCC effector capacity of human tissue macrophages nor it is known whether the antibody-dependent and independent effector capacities, two mechanistically distinct functions, are dissociated in different mononuclear phagocyte populations in humans.

To further characterize the capacity of human mononuclear phagocyte populations from diverse anatomical sites to mediate killing of tumor cells via antibody or directly and to evaluate functional heterogeneity in the response to stimuli (IFN and lymphokines) we examined direct cytotoxicity and ADCC of monocytes, milk macrophages and peritoneal macrophages. Only 6 out of 10 milk macrophage preparations caused significant lysis of TU5, as compared to 13/15 monocyte and 11/12 peritoneal macrophage preparations. In our previous study we tested only 3 pooled milk samples and found spontaneous cytotoxicity

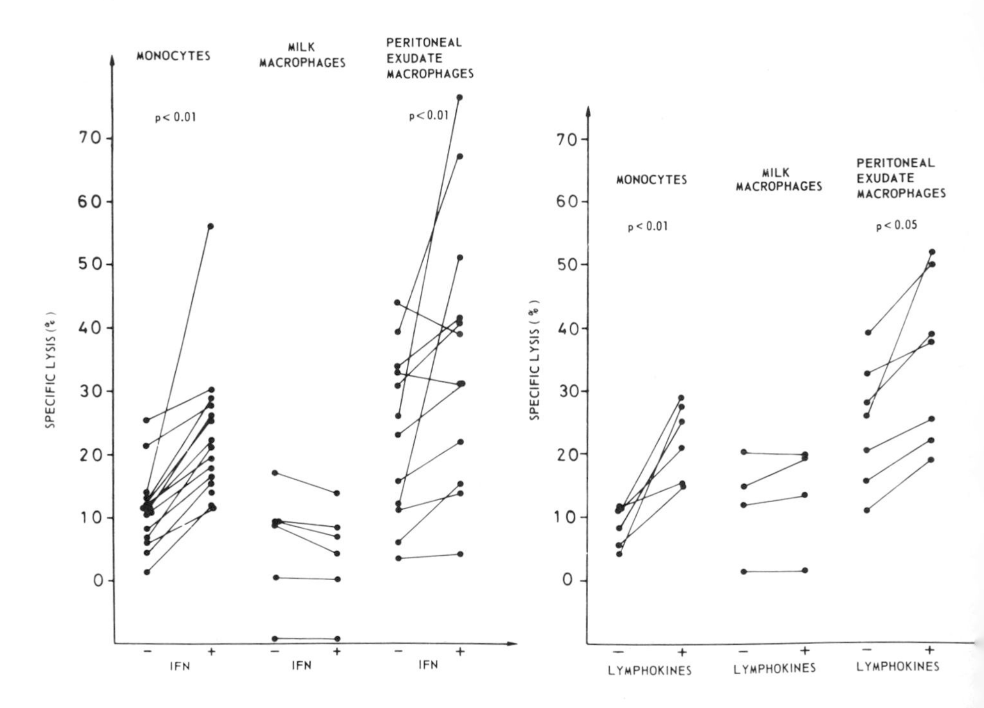

FIGURE 1. Effect of partially purified IFN (1000 units/ml) and lymphokines (1/3 diluted) on tumoricidal activity of human monocytes, milk macrophages and peritoneal exudate macrophages. Horizontal bars indicate the median.

comparable to mononuclear phagocytes from other sites (2): the use of milk pooled from a large number of donors may have obscured the defective cytotoxicity of some samples. Fig. 1 summarizes the effect of in vitro exposure to IFN or to lymphokine supernatants (1000 units/ml of IFN and 1/3 diluted lymphokines). Unlike blood monocytes and peritoneal macrophages, milk macrophages consistently failed to respond to these stimuli in terms of enhanced tumoricidal capacity. In spite of this defective stimulation, milk macrophages proved efficient effectors of ADCC against tumor cells like monocytes and peritoneal macrophages, as shown in Fig. 2. Therefore capacity to mediate antibody dependent and independent tumoricidal activity can be dissociated to some extent in human mononuclear phagocyte populations from diverse anatomical sites.

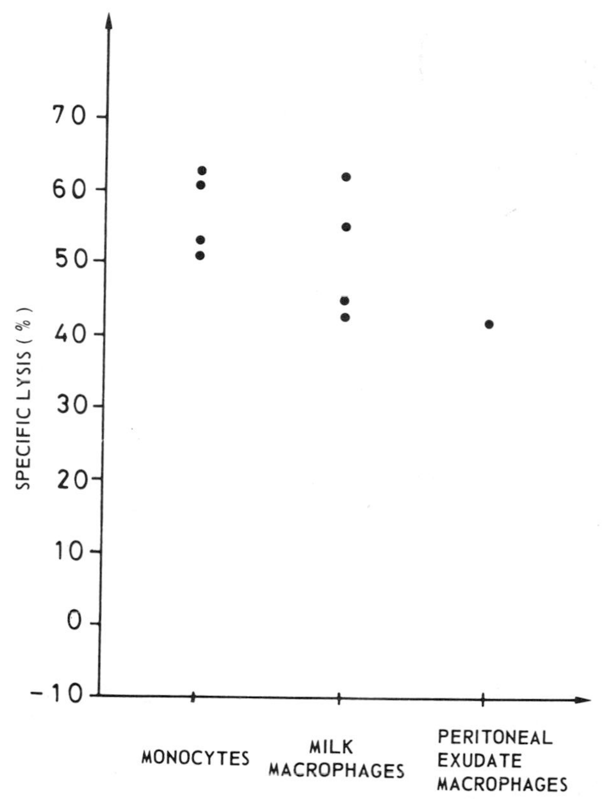

FIGURE 2. ADCC effector capacities of human monocytes, milk macrophages and peritoneal exudate macrophages. ADCC effector capacity was measured against the murine TLX9 line in the presence of 1/10,000 diluted rabbit antiserum. Horizontal bars indicate the median.

V. CONCLUDING REMARKS

The results presented here strongly suggest that endotoxin contamination of tissue culture reagents, which can account for spontaneous killing by rodent macrophages (13), is not responsable for the appreciable levels of spontaneous cytotoxicity of human monocytes and macrophages observed in a 48-72 h assay with susceptible (TU5) target cells. The evidence that, in this and other monocyte cytotoxicity system, effector cells belong to the monocyte-macrophage lineage, was discussed elsewhere (17). Studies in mice suggest that natural cytotoxicity of peritoneal macrophages can be related to the microbial status of the host, in the absence of overt infection (14). Whether this mechanism plays a role in the spontaneous cytotoxicity of human monocytes and macrophages remains a matter of speculation.

Human macrophage populations from diverse anatomical sites are considerably heterogeneous in their capacity to mediate natural cytotoxicity and to respond to IFN and lymphokines and Table VI summarizes the characteristics of macrophage populations studied so far by our group (1,2,25). As previouly reported human lung alveolar macrophages were unresponsive to IFN but showed augmentation of cytolytic and cytostatic activity when exposed to lymphokine supernatants (25). Therefore the total absence of responsiveness to both IFN and lymphokines appears to distinguish the defect of milk macrophages from that of lung bronchoalveolar macrophages. The availability of human mononuclear phagocyte populations with defective responsiveness to stimuli which augment direct tumor cytotoxicity to stimuli which defective responsiveness

TABLE VI. Expression of Direct Tumoricidal Activity and ADCC by Different Human Mononuclear Phagocyte Populations.

	ADCC	Direct cytotoxicity		
		natural	IFN-boosted	Lymphokine-boosted
Monocytes	+	+	+	+
Peritoneal macrophages	+	+	+	+
Alveolar macrophages	+	-	-	+
Milk macrophages	+	±	-	-

to stimuli which augment direct tumor cytotoxicity offers a useful tool for studying the regulation of macrophage-mediated tumoricidal activity in humans. It is of interest in this contest that in _situ_ macrophages isolated from human solid ovarian and renal (26 and unpublished data) carcinomas were unresponsive to IFN and lymphokine supernatants.

ACKNOWLEDGMENTS

We thank Prof. D. Fumarola for suggestions and criticisms.

REFERENCES

1. Mantovani, A., Jerrells, T.R., Dean, J.H., Herberman, R.B. (1979). Int. J. Cancer 23: 18.
2. Mantovani, A., Bar Shavit, Z., Peri, G., Polentarutti, N., Bordignon, C., Sessa, C., and Mangioni, C. (1980). Clin. Exp. Immunol. 39: 776.
3. Hammerström, J. (1980). Acta Pathol. Microbiol. Scand. sect.C. 88: 201.
4. Fischer, D.G., Hubbard, W.J., and Koren, H.S. (1981). Cell. Immunol. 58: 426.
5. Balkwill, F.R., and Hogg, N. (1979). J. Immunol. 123:1451.
6. Horwitz, D.A., Kight, N., Temple, A., and Allison, A.C. (1979). J. Immunol. 36:221.
7. Rinehart, J.E., Lange, P., Gormus, B.J., and Kaplan, M.E. (1978). Blood 52: 211.
8. Cameron, D.J., and Churchill, W.H. (1979). J. Clin. Invest. 63: 977.
9. Jett, J., Mantovani, A., and Herberman, R.B. (1980). Cell. Immunol. 54: 425.
10. Mantovani, A., Dean, J.H., Jerrells, T.R., and Herberman, R.B. (1980). Int. J. Cancer 25: 691.
11. Keller, R. (1978). Br. J. Cancer 37: 732.
12. Tagliabue, A., Mantovani, A., Kilgallen, M., Herberman, R.B., and McCoy, J.L. (1979). J. Immunol. 122: 2363.
13. Martin, F., Martin, M., Jeannin, J.F., and Lagneau, A. (1978). Eur. J. Immunol. 8: 607.
14. Lohmann-Matthes, M.-L., and Domzig, W. (1980). In"Natural Cell-Mediated Immunity against Tumors" (R.B. Herberman, ed.), p.117. Academic Press, New York.
15. Zanella, A., Mantovani, A., Mariani, M., Silvani, C., Peri, G., and Tedesco, F. (1981). J. Immunol. Methods 41: 279.

16. Mantovani, A., Polentarutti, N., Gritti, P., Bolis, G., Maggioni, A., and F. Spreafico (1979). Eur. J. Cancer 15: 797.
17. Mantovani, A., Peri, G., Polentarutti, N., Allavena, P., Bordignon, C., Sessa, C., and Mangioni, C. (1980). In "Natural Cell-Mediated Immunity against Tumors" (R.B. Herberman, ed.), p.1271. Academic Press, New York.
18. Weinberg, J.B., Chapman,H.A. Jr., and Hibbs, J.B. Jr. (1978). J. Immunol. 121: 72.
19. Johnson, W.J., and Balish, E. (1981). J. Reticuloendothel. Soc. 29: 369.
20. Pace J.L., and Russell, S.W. (1981). J. Immunol. 126: 1863.
21. Taramelli, D., and Varesio, L. (1981). J. Immunol. 127:58.
22. Mantovani, A., Caprioli, V., Gritti, P., and Spreafico, F. (1977). Transplantation 24: 291.
23. Shaw, G.M., Levy, P.C., and Lobuglio, A.F. (1978). J. Immunol. 121: 573.
24. Shaw, G.M., Levy, P.C., and Lobuglio, A.F. (1978). J. Clin. Invest. 62: 1172.
25. Bordignon, C., Avallone, R., Peri, G., Polentarutti, N., Mangioni, C., and Mantovani, A. (1980). Clin. Exp. Immunol. 41: 336.
26. Peri, G., Polentarutti, N., Sessa, C., Mangioni, C., and Mantovani, A. (1981). Int. J. Cancer 28: 143.

NATURAL KILLER CELLS SUBSETS

Robert C. Burton[1]
Scott P. Bartlett

Transplantation Unit
Massachusetts General Hospital
Boston, Massachusetts

John A. Lust
Vinay Kumar[2]
Southwestern Medical School
Dallas, Texas

I. INTRODUCTION

Elsewhere in this volume we have discussed the crucial role that antibodies specific for natural killer (NK) cell subsets have played in better defining the phenomenon of natural killing (Burton, Bartlett and Winn). Here we wish to summarize our current investigations into NK cell subsets

[1]This study was supported by Grants CA-17800 and CA-20044 awarded by the National Cancer Institute, Bethesda, Maryland and by Grants AM-07055 and HL-18646 from the National Institutes of Health, Bethesda, Maryland; and RCB was supported by the John Mitchell Crouch Fellowship of the Royal Australasian College of Surgeons and a National Health and Medical Research Council of Australia Fellowship in Applied Health Sciences.
[2]Supported by Grants CA-21401 and CA-15369 from the National Institutes of Health, Bethesda, Maryland; and VK was a Cancer Research Scholar of the American Cancer Society, Massachusetts Division during these studies.

ISBN 0-12-341360-5

which make use of a number of approaches in addition to the use of specific anti-NK alloantibodies. In these studies neonatal, aged, irradiated, ^{89}Sr and cyclophosphamide (CY) treated and beige mutant mice have been used. In addition mice were infected with LCM virus and treated with interferon inducers, and spleen cells from normal mice were "activated" in vitro by short term culture with interferon or interferon inducers. More detailed description of NK cell subsets identified in long term cultures can be found elsewhere herein (Bartlett and Burton).

A. NK Cell Subsets in Fresh Murine Spleen

The use of multiple tumor targets is an important factor in detecting NK cell subsets, and we have chosen the four tumors which, in our experience, best illustrate this (1). They are YAC-1, a Moloney leukemia virus-induced T lymphoma; EL-4, a chemically induced T lymphoma; WEHI-164.1, a cloned non adherent line of a 3-methyl cholanthrene-induced fibrosarcoma; and FLD-3, a Friend virus-induced erythroleukemia.

In Table I are summarized our findings when spleen cells from various types of mice were assayed in short (4-6 hours) and long (12-24 hours) assays on these 4 target cells (1-4). Included in the Table are the results of treating spleen cells from appropriate strains of mice with CE anti-CBA (anti-NK-1.2)(2) serum and monoclonal anti-Thy-1.2 antibodies and complement (C). The results are expressed as: +, significant target cell lysis; -, lack of target cell lysis; +/-, variable target cell lysis. From these findings one can deduce that 4 subsets of cells in murine spleen capable of spontaneous "NK like" activity against one or more of these tumor target cells can be discerned. NK-1.2$^+$ (NK_A) cells appear divisible into two subsets, one which is susceptible to Sr treatment of mice, i.e. marrow dependent, which lyses YAC and a second which lyses EL-4 which is marrow independent (5). Since both cells appear to express at least some NK-1.2 it is possible that they are in the same differentiation lineage but at different stages of maturation. It should be pointed out that several sublines of EL-4 exist. Some are totally resistant to NK cell mediated lysis (6), whereas others behave variously as NK_A or NK_B targets (5).

The finding that about half of this NK-1.2$^+$ activity is also susceptible to monoclonal anti-Thy-1.2 + C could mean either that all NK-1.2$^+$ cells express Thy-1.2, or a cross reacting antigen, but with a wide variation in density so that only a proportion would fix enough antibody to be

TABLE I. Heterogeneity of Murine Natural Killer Cells

Spleen Cell Source	Assay Duration	Target Cell Lysis YAC	EL-4	WEHI-164.1	FLD-3
Normal	short	+	-	-	-
	long	+	+	+	+
Neonatal	long	-	+	+	-
Aged Mice	long	decreased	+	+	+
Sr mice	long	-	+	+	+
Beige mutant	short	-	-	-	-
	long	+	+	+	+
CY treated mice (-3D)	long	-	-	+	-
Lethal XRT (-3D)	long	+	+	+	-
MC anti-Thy-1.2 + C	long	50% decrease	+	+	+
CE anti-CBA + C: CBA	long	-	+/-	+	+

TABLE II. Heterogeneity of NK Cells in Beige Mutant Mice

Spleen Cells Source	Mean Percent Specific Lysis[a] YAC 100:1	50:1	25:1	WEHI-164.1 100:1	50:1	25:1	5:1
6 Hour Assay							
C57BL/6J	24	16	8	13	9	4	
C57BL/6J-bg^J	4	3	1	9	5	1	
16 Hour Assay							
C57BL/6J	33	19	9	70	63	55	39
C57BL/6J-bg^J	26	14	5	72	66	57	39
16 Hour Assay							
C57BL/6J-bg^J C	18	10	5	65	60	54	44
Anti-NK-1.2+C	3	1	0	61	56	52	46

[a]Means of 3-6 experiments

eliminated by C mediated lysis, or that are two distinct NK-1.2$^+$ NK$_A$ cell subpopulations: one Thy-1$^+$ and the other Thy-1$^-$ (1). Studies using flow cytometry (7) and interferon or interleukin II induction of spleen cells treated with anti-Thy-1, anti-Ly5 and anti-Qa5 + C (8) strongly suggest that the latter is the case.

Studies with beige mutant mice have clearly confirmed the dichotomy between NK$_A$ and NK$_B$ cells revealed by use of anti-NK-1.2 sera. These are summarized in Table II. Here it can be seen that NK$_B$ activity is normal in C57BL/6 homozygous beige mutant mice and that NK-1.2$^+$ (NK$_A$) activity is relatively but not absolutely impaired (4).

The NK$_B$ (NK-1.2$^-$) subset also appears to be divisible into 2 subsets: one present in neonatal and lethally irradiated mice which lyses WEHI-164.1, and a second which is not present in these mice which lyses FLD-3 (Table I). Again, it is possible that these are cells in the same differentiation lineage but at different stages of differentiation. The newly discovered marker of NK$_B$ cells, described elsewhere in this volume (Burton, Bartlett and Winn) may be of use in the further study of these NK$_B$ subsets. The relationship of NK$_B$ cells to the natural cytotoxic (NC) cells described by Stutman is not resolved at present, since NK_B cells are susceptible to anti-H-2 antibodies and complement (1), whereas NC cells appear to be resistant (9).

B. Activated Natural Killer Cells

When spleen cells recovered from viral infected mice or mice treated with interferon inducers such as C. parvum or Poly-I - poly-C are assayed *in vitro*, enhanced levels of spontaneous lysis of tumor target cells are observed (3,10, 11). The cell lineage(s) of these activated NK cells and their relationship to the NK cell subsets found in normal murine spleen has not been resolved. However, there is some evidence that these activated NK cells are comprised of more than one discrete subset. The range of target cells that activated NK cells lyse *in vitro* is much broader than NK cells from normal murine spleen, and includes targets specifically resistant to NK_A and NK_B cells such as P-815 (8,12). When activated NK cells are separated on the basis of size, at least 2 subsets of cells with lytic activity, as defined by the different target cells they kill, can be discerned (13).

We have also found that activated NK cells which kill YAC are much more resistant to treatment with anti-NK-1.2 +C than NK_A cells (14). This could be interpreted as meaning that these are a homogeneous population of NK cells, expressing little NK-1.2, or that there are at least 2 subpopulations of NK cells present, a major NK-1.2$^-$ subset and a minor NK-1.2$^+$ subset. Our data do not distinguish between these alternatives, but Tai and Warner (13) were able to sort all NK activity from LCM infected mice with anti-NK-1.2 serum using a FACS II. This suggests that activated NK cells do express NK-1.2, but that it is less dense than on NK_A cells.

Studies of Minato et al (8,15) have indicated that interferon regulates the activity of Thy-1$^-$ NK cells found in normal spleen and that interleukin II affects the activity of the Thy-1$^+$ NK cells. It would be of interest to further phenotype these cells with anti-NK cell alloantisera.

II CONCLUSIONS

The original functional definition of NK cells clearly includes several discrete subsets of cells. Evidence from our own and other laboratories indicates that at least 4 such NK like cells can be discerned on the basis of cell surface markers, studies of various types of mice, and *in vivo* experiments with cultured cells. What is less certain is the lineage of these subsets and their relationships to each other.

It is likely that one or more of these apparently discrete cell types will be a precursor of another and it is still possible that a single differentiation lineage accounts for all the activity we measure as natural killing of spleen cells harvested directly from mice.

REFERENCES

1. Burton, R.C., Bartlett, S.P., Kumar, V., and Winn, H.J. (1981) J. Immunol. 127:1864.
2. Burton, R.C., and Winn, H.J. (1981) J. Immunol. 126: 1985.
3. Lust, J.A., Kumar, V., Burton, R.C., Bartlett, S.P., and Bennett, M. (1981) J. Exp. Med. 154:306.
4. Bartlett, S.P., and Burton, R.C. J. Immunol., (In press).
5. Kumar, V., Luevano, E., and Bennett, M. (1979) J. Exp. Med. 150:531.
6. Hamilton, M.S., Burton, R.C., and Winn, H.J. (1981) J. Trans. Proc. XIII:787.
7. Mattes, M.J., Sharrow, S.O., Herberman, R.B. and Holden, H.T. (1979) J. Immunol. 123:2851.
8. Minato, N., Reid, L., and Bloom, B.R. (1981) J. Exp. Med. 154:750.
9. Lattime, E.C., Pecoraro, G.A., and Stutman, O. (1981) J. Immunol. 126:2011.
10. Tai, A., Burton, R.C., and Warner, N.L. (1980) J. Immunol. 124:1705.
11. Djeu, J.Y., Heinbaugh, J.A., Holden, H.T., and Herberman, R.B. (1979) J. Immunol. 122:175.
12. MacFarlane, R.I., Cerdig, R., and White, D.O. (1979) Infect. Immun. 26:832.
13. Tai, A., and Warner, N.L. (1980) In "Natural Cell-Mediated Immunity Against Tumors" (R.B. Herberman, ed.) p. 241, Academic Press, New York.
14. Kumar, V., Barnes, M.C., Bennett, M., and Burton, R.C. J. Immunol., (In Press).
15. Minato, N., Reid, L., Cantor, H., Lengyel, P., and Bloom, B.R. (1981) J. Exp. Med. 152:124.

SOLID TUMORS ARE KILLED BY NK AND NC CELL POPULATIONS: INTERFERON INDUCING AGENTS DO NOT AUGMENT NC CELL ACTIVITY UNDER NK ACTIVATING CONDITIONS [1]

Edmund C. Lattime
Gene A. Pecoraro
Michael Cuttito
Osias Stutman

Cellular Immunology Section
Memorial Sloan-Kettering Cancer Center
New York, New York

I. INTRODUCTION

Our proposition that natural cell mediated cytotoxicity (NCMC) in mice is mediated by at least two distinct types of effector cells (1), has been confirmed and extended in other laboratories (2-4). Depending on the few targets used initially for screening, NCMC effector cells with different properties were described, the two main prototypes being natural killer (NK) cells (5) and natural cytotoxic (NC) cells (1). An assortment of in vitro activated cells capable of killing NK-susceptible targets have also been described (2,4,5). Based on a variety of criteria (type of assay, type of target, surface phenotype, genetic control as well as physical properties) we suggested that NK and NC cells were related effector populations but belonged to a heterogeneous NCMC system (1, 6-8). Both effector types share properties which are different from other cytotoxic mechanisms, especially T cells, such as pre-existence at high levels without priming, lack of detectable memory, lack of H2-restriction for killing and the ability to be blocked by simple sugars (8). In the present

[1] Supported by National Institute of Health Grants CA-08748, CA-17818, CA-25932 and American Cancer Society Grant IM-188.

ISBN 0-12-341360-5

paper, we show that: 1) NC cells are less susceptible than NK cells to boosting by interferon (IFN) or IFN inducing agents and 2) that target cells derived from solid tumors are susceptible to killing by both NK and NC populations. This latter point should be stressed since, as the NK and NC populations were characterized using "prototype" lymphoma (NK) and solid (NC) tumor targets this "solid vs. lymphoid" idea was perpetuated until it became an understood rule.

II. NCMC OF SOLID AND LYMPHOID TUMORS IS AUGMENTED BY POLY-IC

Following our description of the NC cell as a second NCMC effector population which lysed solid tumors (1), we examined the possible regulation of the NCMC populations by the interferon inducing agent poly-IC. Using a number of lymphoid (YAC-1, R1.1, RL♂1, L-5178 cl. 27v) and solid (Meth-A, Meth-E4, Meth-X, Meth-Z, CMS-4, CMS-11, UV-2237, WEHI-164.1) tumor targets, we noted a pattern in their susceptibility to lysis by effector populations from either untreated or boosted animals (Table I).

TABLE I. NCMC Against Solid and Lymphoid Tumors is Boosted by Poly IC [a]

Effector	Target (type)	% Cytotoxicity Untreated	% Cytotoxicity Poly-IC
C57BL/6J	YAC-1 (L)	30	64*
	R1.1 (L)	0	18*
	RL♂1 (L)	13	29*
	L-5178 cl.27v. (L)	6	25*
	UV-2237 (S)	1	18*
	Meth-X (S)	15	34*
	Meth-E4 (S)	37	53*
	Meth-Z (S)	6	18*
	CMS-4 (S)	27	50*
	CMS-11 (S)	0	20*
	Meth-A (S)	35	38
	WEHI-164.1 (S)	29	34

[a]Experiments were done using 18 hr. ^{51}Cr release assays and the indicated lymphoid (L) or solid (S) tumor target cell lines (5×10^3 labelled targets plus 2.5×10^5 whole spleen ef-

Confirming the findings of other laboratories (9,10), uniform augmentation of lysis of lymphoid targets was seen, however, on the basis of a number of experiments, the solid tumors were divisible into two groups: those which showed increased lysis by the boosted animals (Meth-E4, Meth-X, Meth-Z, CMS-4, CMS-11 and UV-2237) and 2) those targets whose lysis was not effected by the boosting regimen (Meth-A and WEHI-164.1). Although the magnitude of boosting varied from experiment to experiment, the pattern of susceptibility remained consistant.

We have previously reported that NC cells do not express a number of NK cell associated antigens (7). Among the antisera we have studied, we have found that NK cell activity is best removed by anti-Qa-5 + C, which removes 90-100% of activity, even in long term assays (11). Using NK elimination with anti-Qa-5 and NCMC boosting with poly-IC, studies showed that the poly-IC boosted component of solid tumor lysis was the function of a Qa-5+ (NK) population (Table II).

Using such a scheme, it was possible to separate solid and lymphoid tumor targets into three groups: 1) NK (Qa-5+) susceptible, (YAC-1, R1.1, Meth-Z), 2) NK (Qa-5+) and NC (Qa-5-) susceptible (Meth-E4, Meth-X, CMS-4) and 3) NC (Qa-5-) susceptible (Meth-A, WEHI-164.1). It is worth stressing that Meth-Z in group 1 is a chemically induced fibrosarcoma. The results also suggested, that in those cases where targets are lysed by both NC and NK populations, the NC cell activity is not boosted (no significant difference in cytotoxicity by Qa-5- boosted and Qa-5- untreated populations is seen). The finding that solid tumor lines are susceptible to lysis by both NK and NC cells is supported by our previous studies demonstrating that solid and lymphoid targets share recognition structures (7,12). Our studies to date have failed to identify a lymphoid target which is lysed by NC cells although such a case has been described (13).

These studies clearly demonstrate that the distinction between NK and NC cells based on lymphoid and solid tumor targets (1,2) is no longer valid since NK cells kill solid

Footnote for Table I continued:

fectors from either untreated or poly-IC (100 μg i.p. + 24 hr) treated mice (8-10 wks.). Effector populations were pooled from 3-6 mice. Untreated and treated populations were assayed on the same target in the same experiment. The target cells used have previously been described [Meth-A, E4, X, Z (1); WEHI-164.1 (2); R1.1, L-5178 (12); CMS-4 and CMS-11 are chemically induced sarcomas of BALB/c mice received from Dr. A. DeLeo, SKI] (* = $p < 0.05$ change from untreated control).

TABLE II. Poly-IC Augmented NCMC is a Function of a Qa-5+ Effector Poplulation [a]

Effector	Target[b]	% Cytotoxicity			
		Non-Boosted		Poly-IC	
		C	Qa-5 + C	C	Qa-5 + C
C57BL/6J	YAC-1 (L)	45	0	70*	1
	R1.1 (L)	0	0	24*	0
	Meth-Z (S)	3	0	23*	0
	Meth-E4 (S)	32	25	58*	28
	Meth-X (S)	20	16	31*	12
	CMS-4 (S)	27	13	55*	11
	Meth-A.N (S)	32	28	32	25
	WEHI-164 (S)	29	31	35	28

[a]See Table I for methods.

[b]For the description of the solid (S) and lymphoid (L) tumor cell lines see Table I. Aliquots of the same effector population were used with each target. Pretreatment cell numbers were not restored following Ab + C. * = $p < 0.05$ significant boosting from control.

tumors. Two possible interpretations can be raised in explaining the fact that the NK anti-solid tumor activity is primarily noted in boosted populations: 1) that the particular solid tumors studied are less susceptible to NK cell lysis than the lymphoid tumors, which would not be surprising since the lymphoid targets were chosen on the basis of NK susceptibility; or 2) that treatment with poly-IC results in the induction of an NK population with a broadened spectrum of recognition. Although the studies were not designed to answer this question, both the above data and our unpublished results suggest the former to be the case. When spleen cell populations from non-boosted animals were treated with anti-Qa-5, or antisera towards other NK-associated antigens + C and tested on the NK and NC susceptible targets, a varying degree of NCMC inhibition was seen, suggesting that, in the resting state, a small proportion of solid tumor lysis was due to NK cells (Table II and ref. 7). Significant variability in the magnitude of the Qa-5+ component was observed, likely due to different states of activation of the particular effector pool. If this is the case, as our data would suggest, then it adds a complicating factor to studies into the regulation of NCMC activity towards solid tumors.

III. NC CELL ACTIVITY IS NOT AUGMENTED BY REGIMENS WHICH AUGMENT NK CELL ACTIVITY

With the above studies in mind, we examined the ability of interferon inducing agents to augment NC cell activity. Using the prototype NK susceptible YAC-1 and NC susceptible Meth-A targets, we have studied regulation of the two activities using the interferon inducing agents poly-IC, Tilorone and C. parvum (9, 10). At doses and times at which significant NK cell augmentation was noted, no augmentation of NC cell activity was observed (Table III), thereby suggesting either that NC is not regulated by interferon inducing agents, or at least not under the same conditions (dose, time, etc.). Broader dosage and kinetic studies are underway to address this question.

TABLE III. NC Activity is Not Boosted Under NK Boosting Conditions [a]

Effector	Target	Agent	% Cytotoxicity	
			Control	Boosted
BALB/c	YAC-1	Poly-IC	16	61*
	Meth-A	in vivo	53	47
	YAC-1	Poly-IC	1	20*
	Meth-A	in vitro	60	55
	YAC-1	Tilorone	16	63*
	Meth-A		53	49
	YAC-1	C. parvum	5	29*
	Meth-A		36	28

[a]NK activity was measured in 18 hr ^{51}Cr release assays ($5x10^3$ targets + $2.5x10^5$ effectors). NC activity was measured in 18 hr. ^3H-proline assay (10^3 targets + $5x10^4$ effectors). Poly-IC in vivo (100 μg i.p. + 24 hrs.), in vitro (25 μg/ml + 24 hrs.), Tilorone (2 mg p.o. + 24 hrs.) and C. parvum (7 mg i.p. + 72 hrs.) were used as boosting agents. Effector cells are pooled spleen populations from 3-6 mice 8-10 weeks of age. * = $p < 0.05$ significant boosting from control populations.

IV. INTERFERON INDEPENDENCE OF NC CELLS?

Table IV shows an example of the minimal effects of alpha-interferon on NC cell activity, when added directly to the assay at dosage and incubation times which produced at least a 50-60% increase in NK cell activity (data not shown). In addition, we have shown that during the long term assay used for NC cell activity no detectable IFN is produced and that a potent anti-IFN antibody (kindly provided by Dr. B.R. Bloom), had no detectable effects on NC cell activity in vitro (data not shown). In those instances where "NC activity" has been boosted by in vivo treatment with the IFN inducing poly-IC, most of the increased activity was mediated by a serologically defined NK cell population. An IFN insensitive NK-like cell in bone marrow, termed NK^m has recently been described (4) which we feel is comparable to the NC cell in spleen and marrow (1, 6, 14).

TABLE IV. Effect of Interferon on NC Activity [a]

Effector	Target	Treatment	% Cytotoxicity	
			100:1	25:1
BALB/c	Meth-A	none	35	27
		IFN 10 IU	53	27
		IFN 100 IU	36	30
		IFN 1000 IU	37	35

[a] See Table III for methods. Mouse alpha IFN was added directly to the wells during the 18 hr. assay at a concentration of the indicated IU/ml. IFN provided by Dr. P. von Wussow, New York University, New York.

V. CONCLUSIONS

The studies described above show that: 1) depending on the target cell used, solid tumors are lysed by NK, NC or both effector populations, and 2) the augmentation of solid tumor lysis by interferon inducing agents is a function of boosted NK and not NC cell activity.

This overlapping tumor cell lysis by NK and NC populations adds to the complexity of studies into their characteristics since, up until now, it has been a working assumption that solid and lymphoid tumors were lysed by NC and NK cells respectively. Our demonstration that, depending on the state of activation of the effector cell pool and the target cell used, varying proportions of NC and NK cell activities may be involved in target cell lysis, may require a reassessment of prior studies into NCMC. For example our earlier demonstration of NC cell activation (14) may be the result of a boosted NK population

These studies show that certain solid tumors are clearly susceptible to lysis by both resting and boosted NK cells. By screening a relatively low number of solid tumors, we obtained examples of almost all of the possible permutations of NK and NC cell activities. The corollary of these observations is that the frequently quoted statement equating NK cells to an anti-lymphoma system and NC to an anti-solid tumor system, is probably incorrect, and reflects only the apparent properties of the limited number of targets used for the studies. It may well be that a certain degree of "preference" may exist in the NC and NK compartments, but the preference is dictated by the recognition structures on the targets and not by the target cell type. It is probable that both systems can recognize different surface determinants, as is suggested by differential blocking of NC and NK cell activities by monosaccharides (15) as well as similar surface determinants as evidenced by cross cold target inhibition studies (7, 12).

ACKNOWLEDGMENTS

The authors wish to thank Drs. C. Henney, A. DeLeo, M. Palladino, R. Burton, N. Hanna and the Salk Tissue Center for providing cell lines. We wish to thank Dr. U. Hammerling for anti-Qa-5 antisera, Ms. M.L. Devitt for expert technical assistance and Ms. L. Stevenson for preparing this manuscript.

REFERENCES

1. Stutman, O., Paige, C.J., and Feo Figarella, E., J. Immunol. 121: 1819 (1978).
2. Burton, R.C., in "Natural Cell-Mediated Immunity Against Tumors" (R.B Herberman, ed.), p. 19. Academic Press, New York, 1980.
3. Kumar, V., Luevano, E., and Bennett, M., J. Exp. Med. 150: 531 (1979).
4. Minato, N., Reid, L., and Bloom, R.B., J. Exp. Med. 154: 750 (1981).

5. Herberman, R.B., and Holden, H.T., Adv. Cancer Res. 27: 305 (1978).
6. Paige, C.J., Feo Figarella, E., Cuttito, M.J., Cahan, A., and Stutman, O., J. Immunol. 121: 1827 (1978).
7. Lattime, E.C., Pecoraro, G.A., and Stutman, O., J. Immunol. 126: 2011 (1981).
8. Stutman, O., Lattime, E.C., and Feo Figarella, E.F., Fed. Proc. 40: 2699 (1981).
9. Gidlund, M., Orn, A., Wigzell, H., Senik, A., and Gresser, I., Nature 273: 759 (1978).
10. Djeu, J., Y., Heinbaugh, J.A., Holden, H.T., and Herberman, R.B., J. Immunol. 122: 175 (1979).
11. Lattime, E.C., Ishizaka, S.T., Pecoraro, G.A., Koo, G., and Stutman, O. (this publication).
12. Lattime, E.C., Pecoraro, G.A., and Stutman, O. (this publication).
13. Lust, J.A., Kumar, V., Burton, R.C., Bartlett, S.B., and Bennett, M., J. Exp. Med. 154, 306 (1981).
14. Stutman, O., Feo Figarella, E., Paige, C.J., and Lattime, E.C., in "Natural Cell-Mediated Immunity Against Tumors" (R.B. Herberman, ed.), p. 187. Academic Press, New York, 1980.
15. Stutman, O., Dien, P., Wisun, R.E., and Lattime, E.C., Proc. Natl. Acad. Sci., U.S.A. 77: 2895 (1980).

NC CELLS DO NOT EXPRESS NK-ASSOCIATED CELL SURFACE ANTIGENS AND ARE NOT CULTURE ACTIVATED NK CELLS[1]

Edmund C. Lattime
Sally T. Ishizaka
Gene A. Pecoraro
Gloria Koo
Osias Stutman

Cellular Immunology and Immunochemistry Sections
Memorial Sloan-Kettering Cancer Center
New York, New York

I. INTRODUCTION

The original studies into natural cell-mediated cytotoxicity (NCMC) resulted in the description of the murine natural killer (NK) cell (1) which has been shown to lyse a number of tumor cell targets *in vitro* (2,3). Subsequent studies in our laboratory led us to describe a second NCMC effector cell population which we have termed the natural cytotoxic cell (NC) (4). The general description of NC cells and a comparison with NK cells has previously been reported (4,5).

The studies described in this section address two questions: 1) do NK and NC cells express similar cell surface antigens; and 2) if not, can differential antigen expression allow the separation of NK and NC functions and thus facilitate studies into the relationship of the two populations.

---

[1]Supported by National Institute of Health Grants CA-08748, CA-17818, CA-25932, CA-25416 and American Cancer Society Grants IM-188 and FRA-167.

ISBN 0-12-341360-5

II. NK AND NC SURFACE ANTIGEN EXPRESSION

NK cells have been shown to share a number of cell surface antigens with other lymphoid populations (6-9). NK cells express the T cell associated Qa-2,3 (6), Qa-5 (6), and Ly-11 (7) antigens. NK cells bear the neutral glycolipid asialo GM-1 (8), NK 1.1 (6) and NK 1.2 (9) antigens, and have been shown to express high levels of H-2 (9).

Studies which led us to describe the NC cell utilized the adherent fibrosarcoma cell lines Meth-A and Meth-113 in a long term (24 hr.) ^{3}H proline assay (4,5). Utilizing this system, we have recently reported (10) that the NC cell does not express a number of the above NK-associated antigens. The use of the adherent Meth-A tumor as an NC target required that we either utilize an adherent cell assay (^{3}H-proline) or trypsinize the target and use a ^{51}Cr release assay. Either option made a direct NC-NK cell comparison under the same assay conditions impossible. We have isolated a non-adherent variant of the Meth-A tumor cell line, which we have termed Meth-A.N (11). Using this cell line for NC activity and the YAC-1 lymphoma as a target for NK cell activity in long term (18 hr.) ^{51}Cr release assays, a direct comparison can be made between the antigen expression of the NK and NC cell populations. Table I lists the antisera used, its preparation, and source.

TABLE I. Antisera Used[a]

Antigen	Preparation	Ref.	Supplied by:
Asialo GM-1	Rabbit anti ASGM-1	3,8	C. Henney
Qa-2,3	B6.K1 anti-B6	6	L. Flaherty
Qa-5	Monoclonal	6	U. Hammerling
NK 1.1	(C3H X BALB) anti-CE	6	G. Koo
NK 1.2	CE anti-CBA	9	R. Burton
Ly-11	(A.A1 X BALB/B)anti-B10.A	7	D. Meruelo
H-2K^{b}	(B10.D2 X A)anti-B10.A(5R)		NIH Sera

a. All antisera were used as previously described (10). Rabbit C was used at a 1:20 dilution with all of the antisera except anti-asialo GM-1 which was used with guinea pig C at a 1:10 dilution.

Studies described in Table II demonstrate that NK and NC cells markedly differ in antigen expression. NC cell activity is not affected by treatment with antisera directed to the Qa-2,3, Qa-5, Ly-11, NK 1.2, or asialo GM-1 antigens, all of which deplete NK cell activity. The results would also suggest that the NC cell is low in H-2 expression. Studies using the NK susceptible L5178 cl. 27v (3) and NC susceptible WEHI-164.1 (9) targets showed a similar pattern of effector cell antigen expression (data not shown). These results confirm our earlier findings using the Meth-A target and the ^{3}H-proline assay.

TABLE II. NC Cells Do Not Express NK Associated Antigens[a]

Exp.	Effector	Treatment	% Cytotoxicity (% Reduction)	
			NK (YAC-1)	NC (Meth-A.N)
1	C57BL/6J	C	20	19
		Qa-2,3 + C	4 (80)	31 (0)
		Qa-5 + C	0 (100)	24 (0)
2	C57BL/6J	C	11	17
		Qa-2,3 + C	0 (100)	18 (0)
		Qa-5 + C	0 (100)	22 (0)
		$H\text{-}2K^{b}$ + C	0 (100)	17 (0)
3	C57BL/6J	C	17	32
		ASGM-1 + C	5 (60)	37 (0)
		Ly-11 + C	3 (82)	34 (0)
		Qa-5 + C	1 (95)	30 (7)
4	BALB/c	C	12	10
		NK 1.2 + C	0 (100)	10 (0)

a. All experiments were done using 18 hr. ^{51}Cr release assays and either YAC-1 (NK) or Meth-A.N (NC) targets. 5×10^{3} labelled targets were assayed with 2.5×10^{5} C or Ab + C treated whole spleen populations from either C57BL/6J or BALB/c mice (8 wks. of age). Aliquots of the same effector cell populations were used with each target. Pretreatment cell numbers were not restored.

Since none of the antisera tested significantly affected NC activity when used in a C-mediated elimination method, we examined the NC cell for the presence of NK 1.1 antigen through the use of a rosetting technique. C57BL/6 Fv1^{n}

spleen cells were first enriched for NK cells using a discontinuous BSA gradient (6). The NK enriched fractions (at the 10-23% and 23-26% interfaces) were incubated with anti-NK 1.1 antisera and the positive cells removed with protein A coated sheep red blood cells (6). The resultant NK 1.1^- population was examined for NK and NC cell activities. As seen in Table III, greater than 80% of the NK cell activity was removed by this treatment; however, NC cell activity in the same cell population was not significantly reduced, demonstrating that NC cells are NK 1.1^-.

TABLE III. NC Cells Do Not Express the NK 1.1 Antigen[a]

Effector	Treatment	% Cytotoxicity			
		NK (YAC-1)		NC (Meth A)	
		100:1	50:1	100:1	50:1
C57BL/6	NMS	53	36	55	45
	NK 1.1	10	5	49	43

a. NK cell activity was measured in a 4 hr ^{51}Cr release assay using $5x10^3$ YAC-1 target cells. NC activity was measured in a 24 hr ^3H-proline assay using 10^3 Meth-A target cells.

III. NC CELL ACTIVITY IS NOT THE FUNCTION OF A CULTURE ACTIVATED NK CELL POPULATION

The studies described in Table II show that when NK cells are removed from spleen cell populations by pretreatment with specific antibody + C, activity is not regenerated during the long term (18 hr.) assay period. Thus, NC cell activity is not a function of an NK cell population which is either not removed by the antisera treatment or regenerated during the assay. In order to further explore a possible NK-NC relationship, we have studied the two activities in the NK deficient beige mutant mouse. We have previously reported that the beige mutant mouse exhibits normal NC cell activity (12). Using ^{51}Cr release assays and the YAC-1 (NK) and Meth-A.N (NC) target cells, we have studied the relative activities and Qa-5 phenotypes of the effector cells in the mutant.

TABLE IV. NK Cells in Beige Mutant Mice are Qa-5+ [a]

Effector	Treatment	% Cytotoxicity			
		NK (YAC-1)		NC (Meth-A.N)	
		4 hrs	18 hrs	4 hrs	18 hrs
C57BL/6 bg/+	C	10	25	2	24
	Qa-5 + C	1	1	0	29
C57BL/6 bg/bg	C	0	11	0	37
	Qa-5 + C	0	0	0	42

a. $5x10^5$ effector spleen cells were assayed with $5x10^3$ ^{51}Cr labelled target cells for 18 hrs. Aliquots from the same effector populations were assayed with each target. Pretreatment cell numbers were not restored following Ab + C.

The studies shown in Table IV demonstrate 1) that while there is little or no NK cell activity in the beige mouse when measured in a 4 hr. assay, significant activity is seen in an 18 hr assay, and 2) although requiring a long term assay, as is the case in the normal animal, NC cell activity is normal in the mutant mouse. As we have described above in normal animals, lysis of the YAC-1 (NK) target in the beige animal is a function of a classical NK ($Qa-5^+$) cell while lysis of the Meth A.N (NC) target is the function of a NC ($Qa-5^-$) effector cell. The demonstration that resting NK cell activity in the beige mutant is a function of a $Qa-5^+$ NK population supports the earlier report that BCG boosted NK cell activity in the animal expresses the NK-associated asialo-$GM-1^+$ phenotype (8).

IV. CONCLUSIONS

The studies described above support previous studies from our (4,5,10-12) and other (9,13) laboratories which have shown that NCMC is not the function of a single effector cell population under different states of activation but of at least two distinct effector cells. NC cell activity can clearly be separated from NK cell activity through the use of specific antisera. The studies also show that when long term assays are used, significant NK cell activity ($Qa-5^+$) can be demonstrated in the beige mutant mouse, which is also distinct from NC cell activity ($Qa-5^-$).

ACKNOWLEDGMENTS

The authors wish to thank Drs. C. Henney, R. Burton, D. Meruelo, U. Hammerling and L. Flaherty for specific antisera, Ms. Mary Lou Devitt for technical assistance, and Ms. Linda Stevenson for the preparation of this manuscript.

REFERENCES

1. Herberman, R.B., and Holden, H.T., Adv. Cancer Res. 27: 305 (1978).
2. Herberman, R.B., Djeu, J.Y., Kay, D., Ortaldo, J.R., Riccardi, C., Bonnard, G.D., Holden, H.T., Fagnani, R., Santoni, A., and Puccetti, P., Immunol. Rev. 44: 43 (1979).
3. Durdik, J.M., Beck, B.N., Clark, E.A., and Henney, C.S., J. Immunol. 125; 683 (1980).
4. Stutman, O., Pagie, C.J., and Feo Figarella, E., J. Immunol. 121; 1819 (1978).
5. Paige, C.J., Feo Figarella, E., Cuttito, M.J., Cahan, A., and Stutman, O., J. Immunol. 121; 1827 (1978).
6. Koo, G., and Hatzbeld, A., in "Natural Cell-Mediated Immunity Against Tumors" (R.B. Herberman, ed.), p. 105. Academic Press, New York, 1980.
7. Meruelo, D., Paolino, A., Flieger, N., and Offr, M., J. Immunol. 125: 2713 (1980).
8. Durdik, J.M., Beck, B.N., and Henney, C.S., in "Natural Cell-Mediated Immunity Against Tumors" (R.B. Herberman, ed.), p. 37. Academic Press, New York, 1980.
9. Burton, R.C., in "Natural Cell-Mediated Immunity Against Tumors" (R.B. Herberman, ed.), p. 19. Academic Press, New York, 1980.
10. Lattime, E.C., Pecoraro, G.A., and Stutman, O., J. Immunol. 126: 2011 (1981).
11. Lattime, E.C., von Wussow, P., Pecoraro, G.A., and Stutman, O., (Submitted for publication).
12. Stutman, O., and Cuttito, M., Nature 290: 254 (1981).
13. Minato, N., Reid, L., and Bloom, R.B., J. Exp. Med. 154: 750 (1981).

ENRICHMENT AND CHARACTERIZATION OF EFFECTOR POPULATIONS MEDIATING NK, NC, ADCC, AND SPONTANEOUS MACROPHAGE CYTOTOXICITY IN MURINE PERITONEAL EXUDATE CELL PREPARATIONS[1]

Edmund C. Lattime
Louis M. Pelus
Osias Stutman

Cellular Immunology and Developmental Hematopoiesis Sections
Memorial Sloan-Kettering Cancer Center
New York, New York

I. INTRODUCTION

Primarily by virtue of the specific tumor target used, murine resident peritoneal cells (PEC) have been shown to manifest natural killer (NK), natural cytotoxic (NC), antibody dependent cellular cytotoxicity (ADCC) and spontaneous macrophage (SK) cytotoxicity (1-3). Because of the relatively low levels of cytotoxicity found when testing certain tumor targets, particularly the NK sensitive target cells, it has been difficult to determine whether the various activities are the function of a single or multiple effector populations (3).

Through the use of cell separation by sedimentation at unit gravity (velocity sedimentation), the studies described in this report address two questions: 1) are the lysis of the NK susceptible YAC-1 and the NC susceptible Meth-A tumor cell lines the function of the same population that manifests ADCC and SK killing and 2) is it possible to separate or enrich the various populations and thus facilitate their characterization.

---

[1]Supported by National Institute of Health Grants CA-08748, CA-17818, CA-25932, CA-28512; American Cancer Society Grant IM-188 and a grant from the Gar Reichman Foundation. L.M.P. is a special fellow of the Leukemia Society of America.

ISBN 0-12-341360-5

II. METHODS

Resident non-induced peritoneal cells were obtained from 8-12 wk. (C57BL/6 x DBA/2)F_1 mice by lavage (4). Viable nucleated cells were suspended in mouse tonicity phosphate buffered saline (320 mOsm) and separated by velocity sedimentation at unit gravity according to the method of Miller and Phillips (5). After sedimentation, 35 ml fractions were collected, washed, counted and appropriately diluted as described (6,7).

NK, NC and ADCC cell functions were tested as previously described (8,9). SK cytotoxicity was measured in a 48 hr. proline assay with 10^3 Meth-113 target cells and a 1:1 E:T ratio (8).

Morphological and functional characterization of separated cell populations was analysed as previously described (6,7).

The proportion of cells bearing the Qa-2,3, Ia^d and Thy-1.2 antigens was determined by complement mediated cytotoxicity and trypan blue exclusion (11).

III. SEPARATION OF EFFECTOR POPULATIONS

Following separation by velocity sedimentation, the various fractions were assayed for NK, NC, ADCC and SK cytotoxicity as described in methods above. The peak of each effector function was found in fractions sedimenting at different velocities (NC, 1-2 mm/hr.; NK, 3-4 mm/hr.; ADCC, 5-12 mm/hr. and SK, 8-12 mm/hr.) (Fig. 1). These results show that, although there may be overlapping activity of some effector populations with the various targets, this separation technique allowed both the isolation and enrichment of distinct populations mediating the different functions.

FIGURE 1. Analysis of effector cell function of unseparated and velocity sedimented cell fractions.

Cytolytic activity of each fraction was determined as described in the methods. Data are expressed as the percentage of maximum activity for each assay. Unseparated cell activity is likewise expressed as a percent of maximum in order to denote enrichment in effector cell activity obtained following separation. All data were obtained from one sedimentation analysis, and have been repeated in three other experiments with equivalent results.

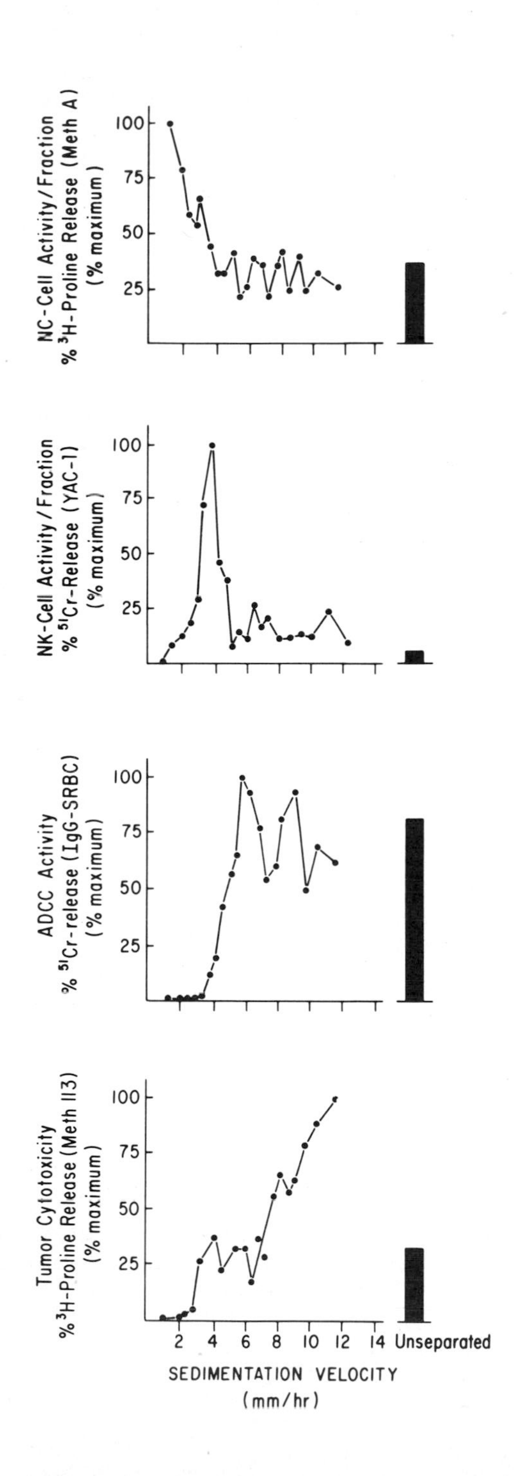
NC-Cell Activity / Fraction
% ^{3}H-Proline Release (Meth A)
(% maximum)
100
75
50
25
NK-Cell Activity / Fraction
% ^{51}Cr-Release (YAC-1)
(% maximum)
100
75
50
25
ADCC Activity
% ^{51}Cr-release (IgG-SRBC)
(% maximum)
100
75
50
25
Tumor Cytotoxicity
% ^{3}H-Proline Release (Meth 113)
(% maximum)
100
75
50
25
2 4 6 8 10 12 14 Unseparated
SEDIMENTATION VELOCITY
(mm/hr)

IV. CHARACTERIZATION OF EFFECTOR POPULATIONS

In order to further characterize the various effector enriched fractions, a number of studies were undertaken (Table 1 and Figure 2). The results obtained both support previous studies, and yield further information as to the nature of the effector populations. Although significant separation and enrichment has been possible, the cytotoxic effectors are not pure populations and therefore characteristics of a given fraction are not necessarily characteristics of the effector population. With this in mind, however, one is able to make statements regarding the type of cell involved in the various activities.

Peak NC cell activity was seen in the smallest, predominantly lymphoid, populations (1-3 mm/hr.), fractions characterized by non-phagocytic, Fc^-, $esterase^-$, $Qa\text{-}2,3^-$, Ia^-, $Thy\text{-}1.2^+$ cells. Peak NK cell activity was seen in fractions sedimenting at 2-5 mm/hr. which are primarily Fc^+, $esterase^+$, $Qa\text{-}2,3^+$, $Thy\text{-}1.2^-$, Ia^- and morphologically mixed lymphoid-monocytoid. Fractions high in ADCC and SK activities are morphologically macrophage/monocyte, phagocytic, Fc^+, $esterase^+$ and Ia^+. Although similar in their characteristics, the two activities are separable, with high ADCC activity seen in the SK low, 5-8 mm/hr. fractions.

V. CONCLUSIONS

The studies described above demonstrate that: 1) significant enrichment and separation of cytotoxic effector populations can be achieved using velocity sedimentation, and 2) velocity sedimentation may represent a valuable pre-enrichment method for secondary characterization studies.

ACKNOWLEDGMENTS

The authors wish to thank Drs. U. Hammerling, C. David and L. Flaherty for specific antisera, Dr. P. Ralph for ADCC reagents, Mr. G. Pecoraro for expert technical assistance and Ms. Linda Stevenson for preparing this manuscript.

TABLE 1. Characterization of Mononuclear Cell Populations

Sedimentation Velocity (mm/hr)	% Recovered Cells (Range; 4 Expts.)	NK-Cell Activity/Fraction % maximum $\bar{x} \pm$ S.D. (2 Expts.)	NC-Cell Activity/Fraction % maximum	Phagocytosis (%+/fraction)	Fc Receptor (%+/fraction)	Non-specific Esterase (%+/fraction)	$Qa_{2,3}$ (%+/fraction)	Thy 1.2 (%+/fraction)	Ia^d (%+/fraction)
11.0 - 12.0	0.1 - 0.5	7 ± 3	25	67	89	69	2	32	15
10.0 - 11.0	0.1 - 0.6	13 ± 6	31	73	87	77	0	36	12
9.0 - 10.0	0.8 - 2	8 ± 6	31	74	81	82	4	35	20
8.0 - 9.0	3 - 7	19 ± 7	32	84	72	95	3	20	32
7.0 - 8.0	6 - 9	20 ± 4	29	93	58	98	0	21	29
6.0 - 7.0	10 - 19	17 ± 6	37	88	59	99	2	10	14
5.0 - 6.0	17 - 23	14 ± 2	29	68	73	96	7	8	14
4.0 - 5.0	10 - 24	40 ± 9	31	25	82	87	15	3	10
3.0 - 4.0	18 - 31	100 ± 0	54	6	61	67	30	6	2
2.0 - 3.0	8 - 10	28 ± 4	63	2	32	28	14	25	0
1.0 - 2.0	0.2 - 1.5	4 ± 2	100	0	10	18	0	17	0
Unseparated		*	†	51	71	80	24	1	49

* NK-Cell activity by unseparated resident peritoneal cells was 3 ± 2% lysis of ^{51}Cr-labelled YAC-1 cells, and represents less than 10 % of the activity of the peak fraction.

† NC-Cell activity by unseparated cells was 21% lysis of ^{3}H-proline-labelled Meth A cells, and represents 30% of the activity of the peak separated cell fraction.

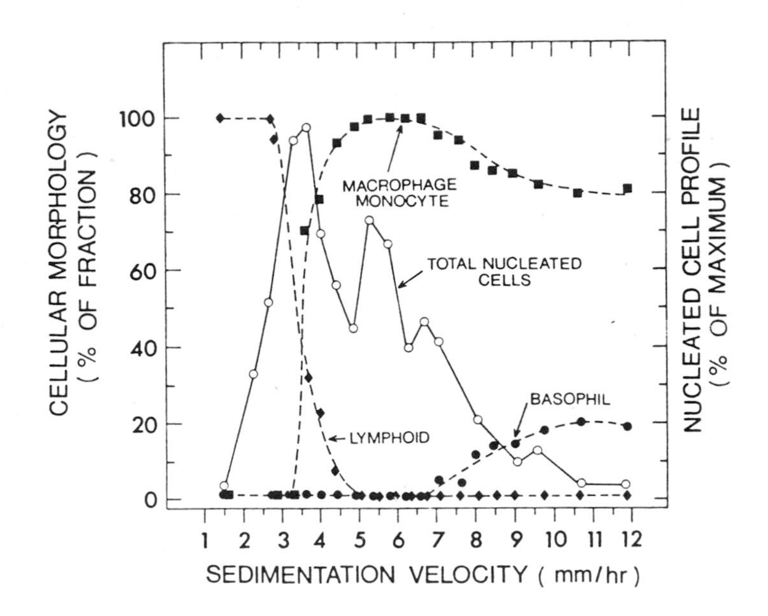

FIGURE 2. Velocity sedimentation analysis of unstimulated resident murine peritoneal cells.

The total nucleated cell profile is indicated, as is the morphological classification of individual cell fractions.

REFERENCES

1. Paige, C.J., Feo Figarella, E., Cuttito, M.J., Cahan, A., and Stutman, O., J. Immunol. 121: 1827 (1978).
2. Ralph, P., and Nakoinz, I., J. Immunol. 119: 950 (1977).
3. Tracey, D.E., Wolfe, S.A., Durdik, J.M., and Henney, C. S., J. Immunol. 119; 3 (1977).
4. Pelus, L.M., and Bockman, R.S., J. Immunol. 123: 2118 (1979).
5. Miller, R.G., and Phillips, R.A., J. Cell. Physiol. 73: 191 (1969).
6. Pelus, L.M., Broxmeyer, H.E., DeSousa, M., and Moore, M. A.S., J. Immunol. 126, 1016 (1981).
7. Pelus, L.M., and Moore, M.A.S., in "Methods for Studying Mononuclear Phagocytes" (D. Adams, ed.) Academic Press, New York (in press).

8. Lattime, E.C., Pecoraro, G.A., and Stutman, O., J. Immunol. 126: 2011 (1981).
9. Ralph, P., Nakoinz, I., Diamond, B., and Yalton, D., J. Immunol. 125, 1885 (1980).
10. Koski, I.R., Poplack, D.G., and Blaese, R.M., in "In Vitro Methods in Cell-Mediated Immunity" (B.R. Bloom, ed.) p. 359. Academic Press, New York, 1976.
11. Lemke, H., Hammerling, G.J., Hohmann, C., and Rajewsky, K., Nature 256: 495 (1978).

FURTHER STUDIES ON THE CYTOSTATIC ACTIVITY[1] MEDIATED BY MURINE SPLENOCYTES

Rachel Ehrlich
Margalit Efrati
Isaac P. Witz[2]

Department of Microbiology
The George S. Wise Faculty of Life Sciences
Tel Aviv University
Tel Aviv Israel

I. INTRODUCTION

Among immunocyte populations participating in the natural cellular reactivities against tumors are macrophages (1), NK (natural killer) cells (2), cells mediating antibody dependent cellular cytotoxicity (K cells) (3), NK (natural cytotoxic) (4) and natural cytostatic cells (5). These activities may be mediated either by different classes of cells differing in physical and biological characteristics, or by a single cell population (or a very limited number of populations) performing multiple functions.

In previous studies from this laboratory we demonstrated that splenocytes from normal mice have the capacity to prevent replication of adherent tumor cell lines (5,6). In these studies some important differences between NK and cytostatic activities were described; however, certain overlaps could not be ruled out. In the present study we describe several additional characteristics of the cytostatic activity and of cells mediating it.

---

[1]This investigation was supported by Contract N01-CB74134 awarded by the National Cancer Institute DHEW and by a grant of Concern Foundation in conjunction with the Cohen-Appelbaum-Feldman Families Cancer Research Fund.

[2]Incumbent: The David Furman Chair of Immunobiology of Cancer.

ISBN 0-12-341360-5

II. RESULTS

Cytostatic and Cytotoxic Activities Mediated by Splenocytes from Normal C57BL/6 Mice at Various Ages

Preliminary observations (6) revealed that the natural cytostatic activity (measured by the inhibition of ^{125}IUDR incorporation (^{125}IUDR I-I) into B16-F10 melanoma cells), unlike the NK activity does not decline with age. Further experiments (table I) confirmed this and showed that very young mice (10-12 days old) that have a low NK activity (even when assayed in an 18 h assay) have a fully expressed cytostatic activity similar to mature mice. These findings show that ontogenetically the cytostatic cell develops earlier than the NK cell and that aged animals exhibit a high cytostatic activity. From this aspect the cytostatic cell resembles the natural cytotoxic cell (NC) which was described by Stutman et al (4). These results also support the conclusion that the NK and the cytostatic activities are mediated largely by different cell populations.

The Effect of Carrageenan and Hydrocortisone Acetate on Natural Cytostatic and Cytotoxic Activities

Carrageenan and hydrocortisone acetate are known to suppress NK activity (7). The suppression is caused either by a direct toxic effect on the cells or by inducing a regulatory suppressive cell population, possibly macrophages. Table II shows that in contrast to the decrease in NK activity, the cytostatic activity was not inhibited at all, or even somewhat enhanced. This enhancement could be the result of an enrichment of cytostatic cells on account of other cells whose activity was affected by the drugs. However, the possibility is not excluded that cytostasis is mediated by cells which are activated by carrageenan and hydrocortisone acetate. Such cells were demonstrated by Cudkowicz and Hochman (7) to be phagocytic and to belong to the monocyte-macrophage lineage.

Augmentation of Cytostatic and NK Activities of C57BL/6, C3H/eB and C3H/HeJ Splenocytes by LPS

LPS derived from gram-negative bacteria is a polyclonal B cell activator known to boost NK activity (8). Figure 1 compares the NK activity and the cytostatic activity of splenocytes from LPS-treated C57BL/6 mice. LPS activated NK cells within one hour after its i.v. administration. LPS-

TABLE I. Cytostasis and NK Activity Mediated by Splenocytes from C57BL/6 Mice at Various Ages

Age of Mice	% Cytostasis (Mean±S.D.[a])	% NK activity (Mean±S.D.[a])
10 days(9)[b]	25.2±16.9	9.6±6.3
21 days(6)[b]	27.7±14.9	14.0±3.7
2-3 months(15)[b]	27.4±14.9	40.8±13.2
12 months(4)[b]	24.5±9.2	26.5±8.2

[a] S.D. = standard deviation

[b] Numbers in parentheses indicate the number of experiments performed. A pool of 1-3 spleens was used in each experiment. The % activity is that obtained at an effector to target ratio of 50:1.

TABLE II. Effect of Carrageenan and Hydrocortisone Acetate on Cytostatic and NK Activities

Treatment	% Cytostasis (Mean±S.D.[a])	% NK (Mean±S.D.[a])
saline[b]	32.7±19.5(3)[c]	31.0±1.3(12)[c]
carrageenan[b]	43.0±21.7(4)[c]	23.0±6.5(8)[c]
hydrocortisone[b]	37.3±6.4(3)[c]	20.8±3.7(9)[c][d]

[a] % activity levels are those obtained at effector to target ratio of 100:1. S.D. = standard deviation.

[b] Control C57BL/6 mice were injected intraperitoneally with saline, 1.6 mg lambda carrageenan (sigma) or 5 mg hydrocortisone acetate (sigma). NK activity was tested 24-48 hours later.

[c] Numbers in parentheses indicate the number of mice tested individually.

[d] The differences between the NK activity levels of control and treated mice were statistically different (students' T test - $p < 0.05$)

boosted NK activity reached peak levels 3 hours after treatment and declined gradually thereafter. Forty-eight hours after the treatment the NK activity in LPS treated and saline treated mice was identical. The cytostatic activity showed a different pattern of response to LPS. An activity peak was evident 48 hours after the treatment and increased cytostatic activity could still be measured 72 hours after treatment. The fact that 48 hours after LPS administration the cytostatic activity reached its peak while the NK activity decreased to control levels or lower, indicates, again, that these activities are mediated by different cell populations and are subjected to different patterns of activation.

C3HeB and C3H/HeJ mice differ in their response to LPS. Whereas the former mice are good responders, the latter are defective in their response to LPS. The LPS mutation is expressed in macrophages (9), B cells (10) and fibroblasts (11). Figure 2 shows that although there was no difference in the level of NK activity in C3HeB and C3H/HeJ mice, NK activity from the latter strain could not be boosted by LPS. These results indicate that the LPS defect is also expressed in NK cells of this strain or that these mice lack a helping function or a specific receptor which is needed for LPS-mediated NK stimulation.

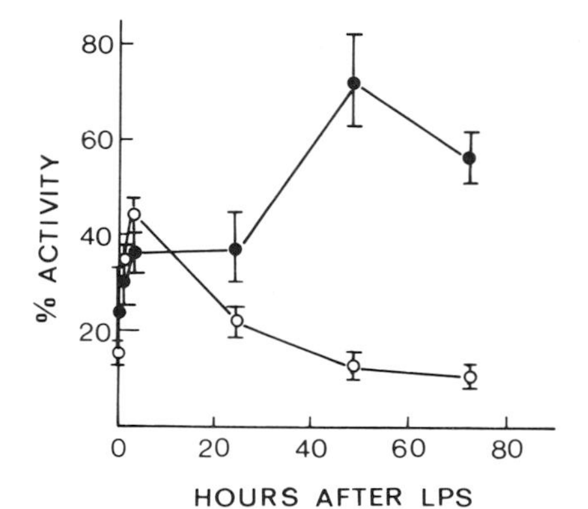

FIGURE 1. The effect of LPS on cytostasis and NK activity of C57BL/6 mice.

C57BL/6 mice were each injected i.v. with saline or 75 μg LPS derived from E-coli. Cytostasis (●) and NK activity (○) mediated by splenocytes from control and LPS treated mice were tested at different time intervals after the injection. The results show the mean ± S.D. activity of 3 mice at each time interval, at an effector to target ratio of 100:1.

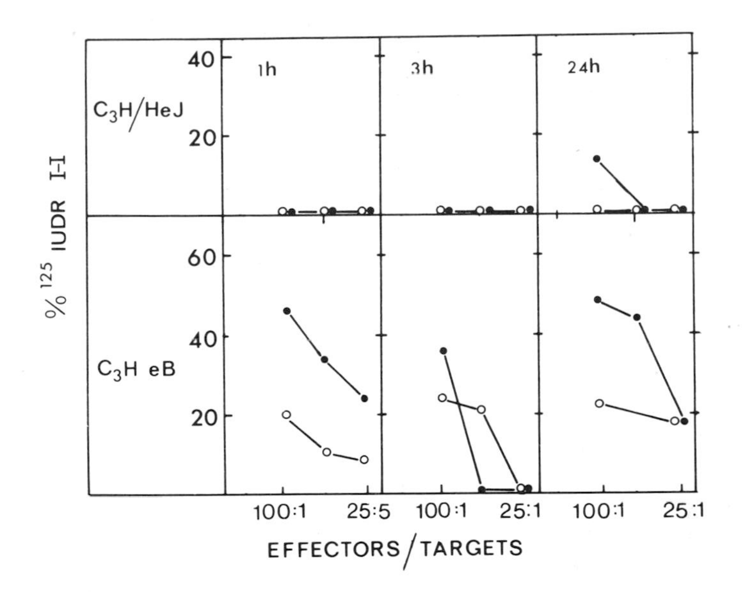

FIGURE 3. The effect of LPS on the cytostatic activity of C3HeB and C3H/HeJ mice.

For experimental details see legend of Figure 2. Cytostatic activity of saline (o) or LPS injected (•) mice was measured by the ^{125}IUDR I-I assay.

consisted of 63-72% monocytes, 18% polymorphonuclear cells and 7-18% lymphocytes.

III. DISCUSSION

The results of this study as well as those from previous ones (5,6) summarised in table IV show that NK activity and cytostasis are diverse biological activities. The latter activity is mediated by 2 cell populations differing in their adherence properties.

The surface markers associated with the cytostatic murine splenocytes seem to indicate that these cells are either early macrophage-monocytes or granulocytes or both. Both types of cells were shown to exert cytostatic activities against a wide variety of target cells (13).

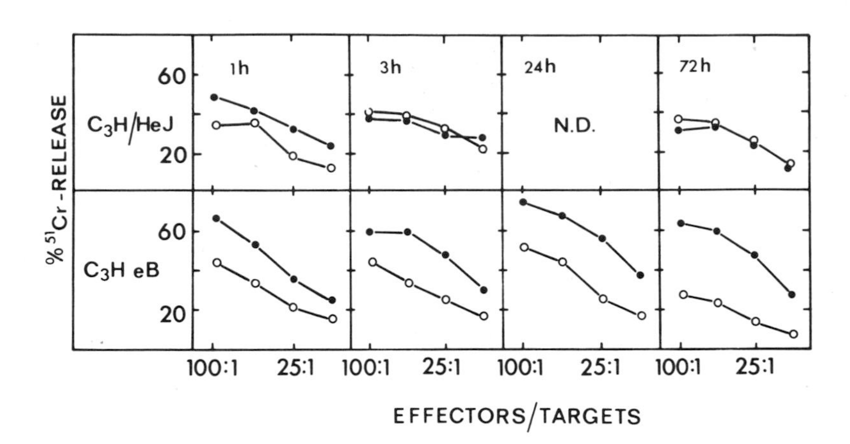

FIGURE 2. The effect of LPS on the NK activity of C3HeB and C3H/HeJ mice.

Five month old C3HeB or C3H/HeJ mice were each injected i.v. with saline (o) or 75 μg of LPS (●) derived from E-coli. NK activity was measured by the 18hr ^{51}Cr-release assay at various time intervals after the injection. Spleens from 3 control and 3 LPS-treated mice were pooled and assayed at each time interval.

In contrast to the normal NK activity of untreated C3H/HeJ mice there was essentially no measurable cytostatic activity in these mice and this activity could also not be boosted by LPS (Figure 3).

Further Characterisation of the Cytostatic Cells

Initial observations revealed that the cytostatic cell population is Thy-1 negative, Fc-receptor positive and non-phagocytic (5,6). The cytostatic cells are partially adherent. We enriched the adherent cytostatic population by passing splenocytes on Sephadex G-10 columns and eluting the adherent population by chemical or mechanical means. The results confirm those of others (12) that most of the NK cells were not adherent to columns, and demonstrate (Table III) that the cytostatic activity is mediated both by non adherent as well as adherent cells. The adherent population of splenocytes (eluted by mechanical means) was found to be highly enriched for cytostatic activity. A morphological examination of this fraction revealed that the non adherent fraction consisted mainly of lymphocytes (94%), the adherent fraction

TABLE III. Cytostasis Mediated by Splenocytes Fractionated According to Adherence

E/T Ratio	% Cytostasis[a] (Mean ± S.D.[b])			
	Un fractionated	Non[c] Adherent	Loosely[d] Adherent	Adherent[e]
100:1	22.5±13.1	13.0±12.4	12.5±10.6	51.5±21.0
50:1	8.2±7.5	4.2±6.2	12.7±16.3	29.4±20.4
25:1	3.8±8.5	0	11.0±13.3	17.4±13.8

a) Five experiments were performed. A pool of 3 spleens was used in each experiment.
b) S.D. = standard deviation.
c) Cells which did not adhere to Sephadex G-10 columns.
d) Cells which adhered to Sephadex G-10 columns and eluted by Alsever's solution.
e) Cells which adhered to Sephadex G-10 columns and were not eluted by Alsever's solution but were eluted by a strong mechanical agitation.

TABLE IV. Comparison Between Cytostasis and NK Activity

	Cytostasis	NK activity
Adherence	+ & -	-
Phagocytosis	-	-
Thy-1	-	-
Fc-receptor	+	-
Activity in:		
10 day old mice	fully expressed	lower than young adults
12 month " "	" "	" " "
Effect of:		
Incubation at 37°C	no effect	activity disappears
Hydrocortisone acetate (in vivo)	no effect (or enhancement)	decreased activity
Carrageenan	"	" "
LPS	relatively long-lived (longer than 5 days)	relatively short-lived (2 days)
Effect of primary tumor-bearing		
Urethan induced	decreased	no change
DMBA "	increased	decreased
Induced by forced-breeding	no change	"

Cytostasis may be of interest to those studying tumor-host immune relations for 2 major reasons. The first being that this activity is significantly boosted in mice bearing certain autochthonous tumors (14). The second reason is that cytostasis may be one of the major mechanisms responsible for tumor dormancy (15,16).

REFERENCES

1. Keller, R. In: "Natural Cell Mediated Immunity Against Tumors" (Herberman, R.B. ed.), p.1219, Academic Press, New York, 1980.
2. Herberman, R.B. In: "Natural Cell Mediated Immunity Against Tumors" (Herberman, R.B. ed.), p.973, Academic Press, New York, 1980.
3. Perussia, B., Santoli, D. and Trincieri, G. In: "Natural Cell Meidated Immunity Against Tumors" (Herberman, R.B. ed.), p.365, Academic Press, New York, 1980.
4. Stutman, O., Figarella, E.F., Paige, C.J. and Lattime, E.C. In: "Natural Cell Mediated Immunity Against Tumors" (Herberman, R.B. ed.), p.187, Academic Press, New York, 1980.
5. Ehrlich, R., Efrati, M. and Witz, I.P. J. Immunol. Meth. 40:193 (1981).
6. Ehrlich, R., Efrati, M. and Witz, I.P. In: "Natural Cell Mediated Immunity Against Tumors" (Herberman, R.B. ed.), p.47, Academic Press, New York, 1980.
7. Cudkowicz, G. and Hochman, P.S. Immunol. Rev. 44:13 (1979).
8. Djeu, J., Heinbaugh, J.A., Holden, H.T. and Herberman, R.B. J. Immunol. 122:175 (1979).
9. Glode, L.M., Jacques, A., Mergenhagen, S.E. and Rosenstreich, D.L. J. Immunol. 119:162 (1977).
10. Glode, L.M., Schere, I., Osborne, B. and Rosenstreich, D.L. J. Immunol. 116:454 (1976).
11. Ryan, J. and McAdam, K.P.W.J. Nature 269:153 (1977).
12. Herberman, R.B. In: "Cell Mediated Immunity Against Tumors" (Herberman, R.B. ed.), p.277, Academic Press, New York, 1980.
13. Korec, S., Herberman, R.B., Cannon, G.B., Reid, J. and Braatz, J.A. Int. J. Cancer 28:119 (1981).
14. Ehrlich, R., Efrati, M., Bar-Eyal, A., Wollberg, M., Schiby, G., Ran, M. and Witz, I.P. Int. J. Cancer 26: 315 (1980).
15. Haran-Ghera, N. J. Natl. Cancer Inst. 60:707 (1978).
16. Wheelock, E.F., Weinhold, K.J. and Levich, J. Adv. Cancer Res. 34:107 (1980).

CELL LINEAGE OF NK CELL: EVIDENCE FOR T CELL LINEAGE

Sonoko Habu

Department of Pathology
Tokai University School of Medicine
Isehara, Kanagawa, Japan

Ko Okumura

Department of Immunology
Faculty of Medicine
University of Tokyo
Tokyo, Japan

I. INTRODUCTION

For the past five years since the discovery of NK cells, intensive efforts to characterize these cells have been made. In earlier reports, the NK population was characterized as "null" cells, i.e. non-T, non-B cells and non-macrophages because of the lack of appropriate surface markers (1-3). Thus, NK cells were classified as a new cell population. However, there is increasing evidence that NK cells share certain surface antigens with T cells, such as Thy-1 (4,5), Ly-5 (6,7), Qa-2, Qa-4 and Qa-5 (8).

In this paper, we will discuss whether the membrane phenotype of NK cells indicates an independent cell lineage both with regard to ontogenical differentiation and function, or whether they belong to an already known cell population, especially the T cell lineage.

ISBN 0-12-341360-5

II. ANTI-ASIALO GM1 ANTISERUM IS SELECTIVE FOR NK CELLS

Since there were suggestive reports that NK cells belong to immature T cells, we examined the antisera against brain tissue which were known to react with T cells including mature and immature thymocytes (9,10). As mentioned elsewhere in this volume by Kasai and Okumura, anti-asialo GM1 (anti-GA1) reactive with NK cells was found in the antisera against brain tissue. As previously reported, the anti-GA1 antiserum in the presence of complement could selectively eliminate NK activity _in vitro_ as well as _in vivo_ (11-13), but had no effect on cytotoxic T cells (11), carrier specific helper T cells, concanavalin A (Con A) reactive T cells, sIg bearing B cells, or LPS reactive lymphocytes (12). Antigen presenting and accessory cells were not also eliminated by the antiserum (data not shown). These results indicate that as far as the mature effector cells are examined, GA1 antigen is preferentially expressed on NK cells in adult mice.

III. ASIALO GM1 (GA1) AS A MARKER OF IMMATURE CELLS OF T CELL LINEAGE

To determine the relationship of the GA1 positive cells ($GA1^+$ cells) to the other cell lineages, we examined the distribution of the cells reacting with anti-GA1 ($GA1^+$ cells) in adult and embryonic lymphoid tissues. Of particular interest, was the detection of $GA1^+$ cells in the embryonic, but not adult liver and thymus (Table I). At an early stage of gestation, embryonic thymocytes expressed GA1 but no Thy-1 antigen (14). The proportion of such cells rapidly declined thereafter, in contrast with the later increase of Thy-1^+, $GA1^-$ cells. This reciprocal correlation strongly suggests that GA1 is present on immature thymocytes which do not yet express Thy-1 antigen, but are committed to express it in the thymic environment. However, in order to assign $GA1^+$ cells to the immature cells of T cell lineage, further study is required: the relationship of $GA1^+$ cells in the fetal thymus and fetal liver must be determined. Then, we performed transfer experiments of fetal liver cells of CBA (Thy-1.2) into the irradiated (900Rd) AKR mice (Thy-1.1). If the transfered cells were pretreated with anti-GA1 and complement, Thy-1.2^+ cells of the CBA donor differentiated poorly in the recipients possessing Thy-1.1 (Table II). In contrast to the effect on T cell lineage, cytotoxic treatment of fetal liver cells with anti-GA1 did not affect B cell differentiation. Fetal liver cells from CSW (Igh^a), with or

TABLE I. Proportion of GA1+ Cells and Thy-1+ Cells in Lymphoid Tissues from CBA Mice[a]

Cell source	% GA1+ cells[b]	%Thy-1+ cells[c]
Thymocytes from		
11 days embryos	95	8
13 days embryos	71	39
15 days embryos	15	88
17 days embryos	11	95
19 days embryos	4	99
Thymocytes from infants (6 days)	1	99
Thymocytes from adults (4 wks)	1	99
Fetal liver from		
15 days embryos	18	0
18 days embryos	10	0
Spleen cells from adults (4 wks)	12	31
Lymph node cells from adults (4 wks)	7	63
Bone marrow cells from adults (4 wks)	7	3

a. Percent-positive cells were counted by fluorescence microscope.
b. Lymphoid cell suspensions were incubated with pepsin digested F(ab')2 fragments of anti-GA1 at 4°C for 30 min and were stained with fluoresceinated F(ab')2 fragments of goat anti-rabbit Ig.
c. Fluoersceinated monoclonal anti-Thy-1 was utilized for direct staining.

without anti-GA1 pretreatment were transfered into irradiated CWB mice (Igh^b) which are congenic mice for immunoglobulin (Ig) allotype. Equivalent amounts of donor type Igs were produced in recipients injected with the CWB cells untreated or treated with anti-GA1 (data not shown). In addition, it was confirmed by staining with fluoresceinated antisera that the cells possessing cytoplasmic and surface IgM were remained after cytotoxic treatment with anti-GA1 (data not shown). Reviewing these experimental results, we conclude that GA1 is expressed on immature cells of the T cell lineage prior to being influenced by the thymic environment, but not on cells of the B-cell lineage.

TABLE II. Generation of Thy-1+ Cells from CBA Fetal Liver Cells in AKR Thymus

Transfered cells		% Thy-1.2+ cells in recipient(AKR) thymus
Treatment	% GA1+ cells	
NRS	29.0	44.4
Anti-GA1 + Complement	0.5	22.5
Complement	25.1	40.1

a. 2 x 10^7 fetal liver cells from CBA (15 days of gestation were intravenously injected into each irradiated (900Rd) AKR which was maintained under specific pathogen free.
b. See foot note of Table I.
c. The cells obtained from AKR thymus were stained with fluoresceinated monoclonal anti-Thy-1.2. Percent positive cells represent arithmetic means of 20 mice.
d. p value compares of Thy-1.2+ cells in the thymocytes from AKR receiving CBA fetal liver cells treated with the antisera or normal rabbit serum (NRS) in the presence of rabbit complement ($p < 0.005$).

IV. ALLOANTISERA REACTIVE TO NK CELL

According to recent studies by Koo _et al_. and Minato _et al_. (15,16), antigens such as Thy-1, Ly-5, Qa-2, Qa-4 and Qa-5 are not always expressed on all NK cells. This indicates that the NK cell population can be divided into distinct subpopulations with or without expression of various alloantigens. Although some T cell associated surface antigens are expressed on some NK cells, it is unlikely that NK cells are a subset of T cells. This idea is supported by the fact that NK cells are not sensitive to T cell specific antisera, such as anti-Lyt-1 and anti-Lyt-2 and that the typical T-cell marker, Thy-1 antigen, is not expressed on all NK cells. Furthermore, anti-Thy-1, anti-Ly-5, anti-Qa-2, anti-Qa-4 and anti-Qa-5 antisera, are reported to react with cells other than T cells (17-20), implying that the cells bearing such antigens can belong to a cell lineage distinct from T cells. Among the alloantisera reactive to NK cells, anti-NK-1 is considered to be specific for NK cells, and unreactive to T cell or B cells(21-23). Therefore, alloantisera may distinguish subsets among NK cells that as a whole population express the common surface marker NK-1, which may be equivalent to the Thy-1 antigen for the T cell lineage. It is evident from our study, as well as those of others, that NK cells are not part of the B cell lineage because they lack universal B cell markers such as sIg (24,25), Lyb-2 (15), Ia and ThB (26).

In summary, although NK cells share certain antigens with T cells, they express unique antigen(s) absent on T cells. NK cells might be derived from T cell line ontogenically i.e., NK cells may share a common precursor with T cells but they may be committed prior to thymus migration and differentiate independently from the T cell developmental pathway. With this in mind, our tentative hypothetical schema is depicted below.

REFERENCES

1. Kiessling, R., Klein, E., and Wigzell, H., Eur. J. Immunol. 5:112 (1975).
2. Herberman, R.B., Nunn, M.E., Holden, H.T., and Larvin, D.H., Int. J. Cancer 16:230 (1975).
3. Sendo, F., Aoki, T., Boyse, E.A., and Buofo, C.K., J. Natl. Center Inst. 55:603 (1975).
4. Mattes, J., Sharrow, S., Herberman, R.B., and Holden, H.T., J. Immunol. 123:2851 (1975).
5. Herberman, R.B., Nunn, M.E., and Holden, H.T., J. Immunol. 121:304 (1978).
6. Cantor, H., Kasai, M., Shen, F.W., Leclerc, J.C., and Glimcher, L., Immunol. Rev. 44:1 (1979).
7. Kasai, M., Leclerc, J.C., Shen, F.W., and Cantor, H., Immunogenetics 8:153 (1979).
8. Koo, G.C., and Hatzfel, A., in "Natural Cell Madiated Immunity against Tumors" (R.B. Herberman, ed.), p. 105. Academic Press, New York (1980).
9. Golub, E.S., J. Immunol. 109:168 (1972).
10. Golub, E.S., J. Exp. Med. 136:369 (1972).
11. Habu, S., Hayakawa, K., Okumura, K., and Tada, T., Eur. J. Immunol. 9:938 (1979).
12. Kasai, M., Iwamori, M., Nagai, Y., Okumura, K., and Tada, T., Eur. J. Immunol. 10:175 (1980).
13. Habu, S., Fukui, H., Shimamura, K., Kasai, M., Nagai, Y., Okumura, K., and Tamaoki, N., J. Immunol. 127:34 (1981).
14. Kasai, M., Yoneda, T., Habu, S., Maruyama, Y., Okumura,

K., and Tokunaga, T., Nature 291:334 (1981).
15. Habu, S., Kasai, M., Nagai, Y., Tamaoki, N., Tada, T., Herzenberg, L.A., and Okumura, K., J. Immunol. 125:2284 (1980).
16. Koo, G.C., Jacobson, J.B., Hämmering, G., and Hämmering, U., J. Immunol. 125:1003 (1980).
17. Minato, N., Reid, L., and Bloom, B.R., J. Exp. Med. 159:750 (1981).
18. Thierefelder, S., Nature 296:691 (1977).
19. Williams, A.F., Contemp. Top. Mol. Immunol. 6:83 (1977).
20. Scheid, M.P., and Triglila, D., Immunogenetics 9:423 (1979).
21. Hämmering, G., Hämmering, U., and Flaherty, L., J. Exp. Med. 150:108 (1979).
22. Glimcher, L., Shen, F.W., and Cantor, H., J. Exp. Med. 145:1 (1977).
23. Pollack, S.B., Tam, M.R., Nowinski, R.C., and Emmons, S.L., J. Immunol. 123:1818 (1979).
24. Tam, M.R., Emmons, S.L., and Pollack, S.B., J. Immunol. 124:650 (1980).
25. Roder, J.C., Kiessling, R., Biberfeld, P., and Anderson, B., J. Immunol. 121:2509 (1978).
26. Gidlund, M., Ojo, E.A., Orn, A., Wigzell, H., and Murgita, R.A., Scand. J. Immunol. 9:167 (1979).

IMMUNOREGULATION OF MOUSE NK ACTIVITY BY THE SERUM THYMIC FACTOR (FTS)

Dominique Kaiserlian
Pierre Bardos
Jean-François Bach

INSERM U 25
Hôpital Necker
Paris - France

I. ABSTRACT

Immunoregulation of mouse NK activity can be achieved by a serum thymic factor (FTS) both in vivo and in vitro. We have observed a decrease or an increase of the NK cytotoxicity according to the mouse strain and the age of the animals. Treatment of thymectomized mice with FTS returned the augmented NK activity to normal levels, indicating that NK cell activity is under thymic influence. The role of macrophages and T cells in FTS-induced regulation of NK cytotoxicity has been studied. Different mechanisms of action of FTS on NK cells are discussed.

II. INTRODUCTION

Although NK cells do not express the functional or antigenic characteristics of typical macrophages, T or B cells(1) indirect arguments (such as the elimination of NK activity by repeated treatment with anti-Thy 1 antibodies and complement (1)) suggested that NK cells could belong to the T cell lineage and were likely to be pre-T cells. In addition, it has been found that NK cells were submitted to both positive and negative regulation (2). Hence the idea of studying regulation of NK cell activity by T cell modulating agents.

The thymic serum factor (FTS) is a nonapeptide secreted by the thymic epithelium, which induces both intra- and extrathymic maturation of T lymphocytes. Beside its major role in T cell differentiation, FTS has also been reported to be a potent immunoregulator of T-dependent responses in thymectomized

ISBN 0-12-341360-5

and in normal animals. It appeared, therefore, of great interest to examine the effect of FTS both in vivo and in vitro on NK activity and determine the relationship between thymus and NK cells.

III. METHODS

FTS Treatment

In vivo treatment was performed by injecting young mice (3-4 weeks of age) i.p. 5 times weekly for variable periods (24 hours to 12 weeks) at doses of 0.1 or 1 ug synthetic FTS per day. Control mice received saline.

In vitro treatment was performed by incubating splenocytes, thymocytes and bone marrow cells from old CBA mice (5-7 months) for 2 hours at 37°C without fetal calf serum with graded concentrations of FTS (from 10 ng to 10^{-2}ng/ 3 x 10^6 cells).

Assay for NK activity

The cytotoxic activity of NK cells was measured in a direct ^{51}Cr release assay using the YAC-1 cell line, as previously described (3).

Assay for suppressor or helper activity

The induction of helper or suppressor activity regulating the NK reactivity of normal spleen cells was determined by a coculture assay. The cytotoxicity of graded numbers of cells (2.10^5 to 8.10^5) from control and FTS-treated animals was first evaluated separately. The sum of the two activities was calculated for a constant total cell number of 10^6 (calculated cytotoxicity) and compared to the values obtained when using mixtures of spleen cells from control and FTS-treated animals (experimental cytotoxicity). When experimental cytotoxicity was higher than calculated cytotoxicity, we concluded that FTS had had a helper effect, and conversely, a suppressor effect when experimental cytotoxicity was lower than calculated cytotoxicity.

Depletion experiments

Macrophage-depleted populations were obtained after passage on Sephadex G-10 columns. Depletion of Thy-1+ cells was performed by treating spleen cells twice with monoclonal anti-Thy 1.2 antibodies in the presence of complement. Removal of plastic adherent and carbonyl-iron phagocytic cells was performed as previously described (4).

Thymectomy and thymus grafting

Thymectomy was performed by suction at the age of 3-4 weeks. Syngeneic thymuses from neonatal donor mice were grafted 6 weeks after thymectomy. NK activity was measured 6 weeks later. The results were compared to those of control mice and of a group of thymectomized mice treated with FTS for 6 weeks.

IV. RESULTS

Effect of FTS on NK activity of normal mice

When administered *in vivo* to normal adult mice, FTS increased or decreased the splenic NK activity, depending upon the mouse strains and the duration of treatment. In high NK responder strains (CBA, NZB), FTS treatment resulted in a marked diminution of the NK activity, whereas it stimulates NK reactivity of low NK responder strains (C57BL/6, DBA/2, A/J)(Fig. 1).

MODULATION OF NK ACTIVITY IN MICE BY FTS

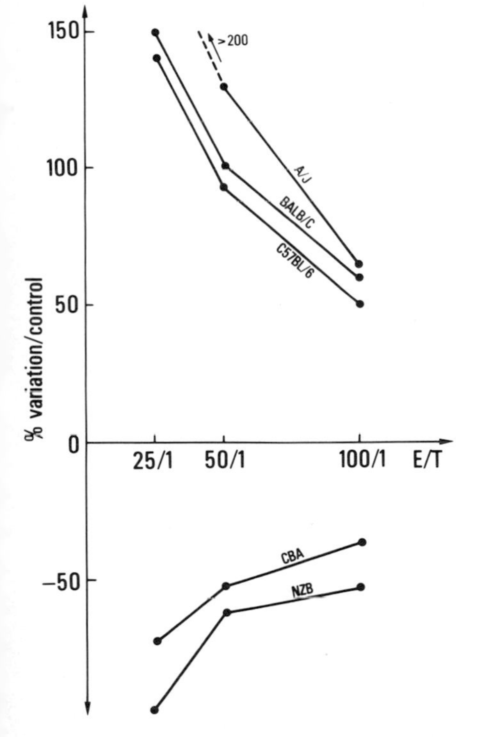

FIGURE 1. In vivo effect of FTS on mouse NK activity in various strains of mice

3-4 week old mice were treated with 10 ng of FTS 3 times a week for 8 weeks. Results are expressed as the percent variation of control mice injected with NaCl (8 mice per group).

$$\frac{\text{\% cytotox. in FTS-treated mice} - \text{\% cytotox. in control mice}}{\text{\% cytotox. in control mice}} \times 100$$

The study of the in vitro effect of FTS on the NK activity of mouse lymphoid cells from various origins showed that a 2-hour incubation with FTS resulted in a significant increase of NK cytotoxicity, not only in spleen cells but also in thymocytes and bone marrow cells of old CBA mice (Fig.2). This immunostimulating activity of FTS was shown to be specific for the active conformation of the peptide, since an inactive structural analog of FTS $(DSer^4)$-FTS, was totally ineffective on NK cells, whereas an active analog (Har^3)-FTS enhanced the NK activity of splenocytes and thymocytes to the same extent as FTS itself. It is interesting to note that FTS decreased in vitro the NK response of adult NZB mice (between 2 and 6 months of age but did not modify that of young (4 weeks) or old (7-10 months) NZB mice (Fig.3).

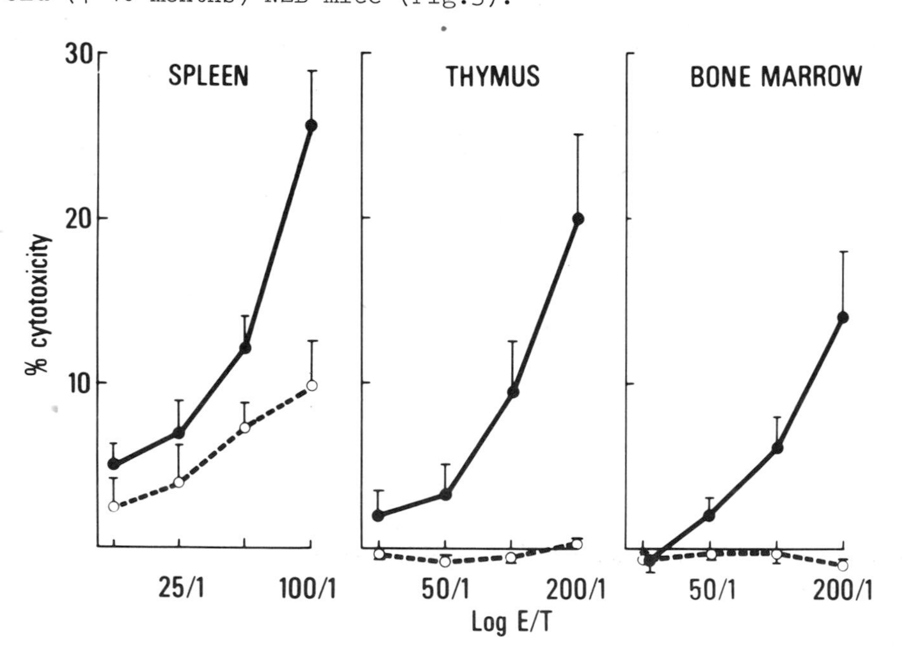

FIGURE 2, In vitro effect of FTS on mouse NK activity

Spleen cells, thymocytes or bone marrow cells from 5 month old CBA mice were incubated for 2 hours at 37°C + 5% CO_2 with FTS (10 ng/3.10^6 cells) (●—●) or with medium alone (○- - -○). The NK activity of FTS-treated and control cultures was then tested in a 4 hour ^{51}Cr release assay against YAC-1.

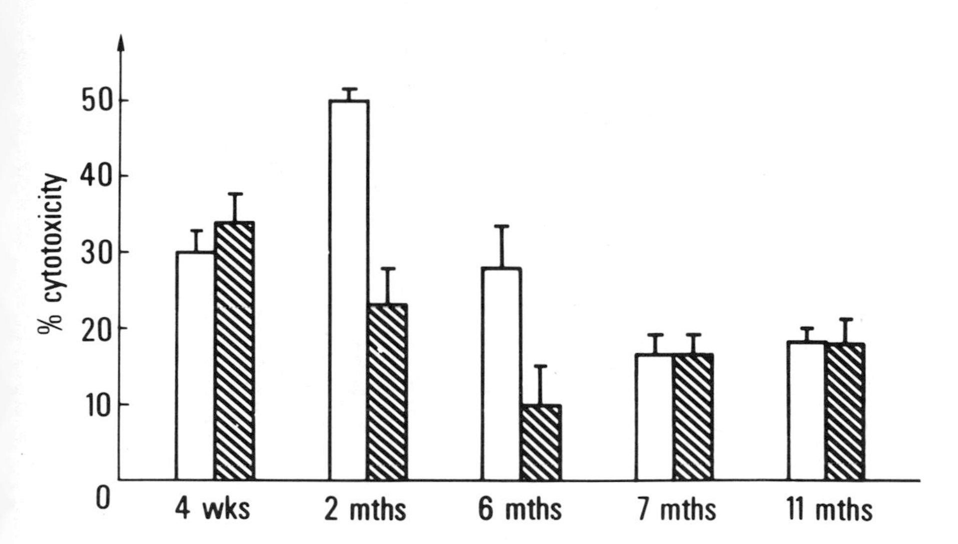

FIGURE 3, In vitro inhibition of the splenic NK activity of NZB mice

Spleen cells from NZB male mice of various ages were incubated for 2 hours at 37°C + CO_2 with FTS (10 ng/3.10^6 cells) ▧, or with medium alone □. Assay for NK cytotoxicity was performed in a further 4-hour incubation of treated and control cells with ^{51}Cr-labeled YAC-1.

There is a good correlation between in vivo and in vitro effects of FTS on the NK activity of NZB mice. In the CBA strain, we found that treatment of young adult mice with FTS reduced the splenic NK activity while in vitro, incubation of FTS with splenocytes or thymocytes from old CBA mice increased the initially low level of NK reactivity. These findings indicate that FTS-induced modulation of NK cytotoxicity in the CBA strain varies according to the initial level of NK activity.

Effect of FTS on NK activity of thymectomized mice

Thymectomy performed at weaning (3-4 weeks) led to an increase of NK activity (Fig.4). Syngeneic thymus grafts or FTS treatment in vivo returned NK activity to normal levels, indicating that NK cell activity is under thymic influence.

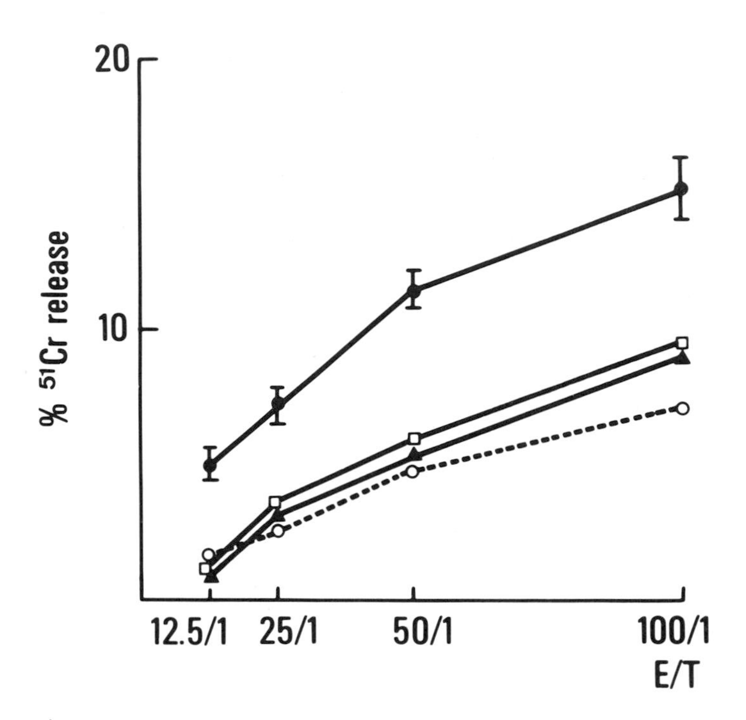

FIGURE 4, Effect of FTS treatment or thymus grafts on the NK activity of thymectomized mice

Two days following thymectomy at weaning, C57BL/6 mice were either grafted with syngeneic thymus or treated with 100 ng FTS (5 days a week for 6 weeks). The splenic NK cytotoxicity was tested 6 weeks after thymectomy in thymectomized mice, either untreated (●—●) or grafted with syngeneic thymus (▲—▲) or treated with FTS (□—□), and was compared to that of age-matched normal control mice (○- - -○).

Mechanisms of FTS-mediated regulation of NK activity

1. Suppressor or helper effect of spleen cells from FTS-treated mice on NK activity. In the co-culture system presented above, spleen cells from FTS-treated NZB mice suppressed the NK activity of untreated syngeneic splenocytes. Conversely, splenocytes from FTS-treated C57BL/6 mice enhanced the NK cytotoxicity of syngeneic splenocytes. These data suggest that FTS may exert its in vivo action on NK activity via subsets of cells regulating NK cells. Furthermore, preliminary data suggest that soluble mediators recovered from the supernatant of FTS-treated spleen cells might be implicated.

Role of regulatory cells in FTS-mediated regulation of mouse NK activity. In vivo FTS treatment increased NK activity in C57BL/6 mice. This stimulating effect of FTS disappeared after depletion of adherent and phagocytic splenocytes, but remained unchanged after depletion of adherent cells only, indicating a helper effect of phagocytic cells. Conversely, in NZB mice, depletion of adherent and phagocytic cells increased the NK activity of FTS-treated NZB mice and the same effect was obtained after depletion of adherent splenocytes.

In contrast with these in vivo data, we have observed that the stimulating effect of FTS in vitro on CBA mouse splenocytes was not affected by depletion of macrophages. In addition, the stimulatory effect of FTS on NK activity remained unchanged after treatment of spleen cells with monoclonal anti-Thy 1.2 antibody on two consecutive occasions (Fig.5).

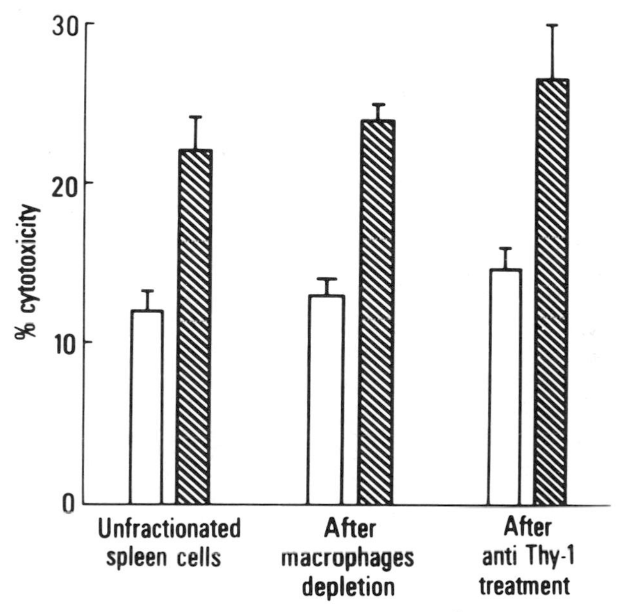

FIGURE 5, Effect of macrophage and $Thy-1^+$ cell depletion on FTS-induced stimulation of the NK activity

Spleen cells from 5 month old CBA mice were either depleted of macrophages (after passage on Sephadex G-10 column), or depleted of $Thy-1^+$cells after 2 consecutive treatments with monoclonal anti-Thy 1.2 antibodies and rabbit complement. The suspensions were then incubated for 2 hours at 37°C + 5% CO_2 in air with FTS (10 ng/3.10^6 cells) (hatched bar) or with medium alone (open bar) and tested for NK cytotoxicity in a 4-hour release assay.

V. DISCUSSION

The effect of FTS on NK activity can be explained by either a direct action on NK cells, or an indirect action mediated by regulatory cells.

The indirect action could be mediated by a regulatory cell, possibly a macrophage, whose role in modulating NK activity has already been shown (5), indicating that NK activity was increased by poly-IC. Other authors (6) have reported a decrease in NK activity after hydrocortisone treatment. Our experiments indicate that adherent and/or phagocytic cells play a role in the in vivo regulation of NK activity by FTS. FTS could also act on cells which produce soluble factors which could either activate (e.g. interferon) or inhibit (e.g. prostaglandins E_2) NK activity. Such an hypothesis would be in keeping with another recent report (7) showing that incubation of mononuclear cells with a thymic factor (TP_1) induces an appreciable increase of interferon production. Moreover, we have observed that two peptidic fragments of thymopoietin (TP5 and TP13) (both containing the minimal active structure of thymopoietin) can stimulate in vitro the NK activity of old CBA mice (data not shown). Experiments are now under progress to clarify the putative relationship between FTS-induced interferon production and stimulation of NK activity.

A direct action of FTS on NK activity might also operate. It could be explained by two non-mutually exclusive hypotheses 1/ recruitment and/or maturation of pre-NK cells into active NK cells ; 2/ increase of the cytotoxic potential of already mature and immunocompetent NK cells. The maturation of NK precursor cells (Lyt 5^-) into functional NK cells (Ly 5^+) has been reported to be the consequence of interferon stimulation (8). Such indication of a cellular maturation process could be affected by FTS, if one considers that NK cells are indeed pre-T cells, and that FTS is able to induce in vitro differentiation of medullary precursors of human T cells into mature T cells (9). The hypothesis of FTS action on the differentiation pathway from pre-NK to T cells would explain the increased NK activity observed after thymectomy (by accumulation of pre-T cells) and its decrease following thymus grafting or FTS treatment in thymectomized mice (maturation of pre-T cells into T cells).

The in vitro effect of FTS on NK activity merits separate discussion. Only a rapid mechanism, independent of cell division, can account for the in vitro modulation of NK cells by FTS, which excludes the hypothesis of recruitment and/or differentiation of NK precursors into active NK cells. On the contrary, the hypothesis of an increase in the lytic potential of mature NK cells seems more likely. According to this assumption, the in vitro effect of FTS on NK activity could be either due 1/ to a direct action of the peptide on NK cells, which could eventually be mediated by interferon, 2/ or to an indirect action mediated by a Thy-1 negative cell or a subset of immature T cells expressing low density of Thy-1 antigen (e.g. the post-thymic precursors (10).

In conclusion, our studies indicate that the thymus, and more precisely FTS, may play an important role in the physiological regulation of NK cell activity. Our data are thus in favor of the relationship between the T cell lineage and NK cells. The relevance of FTS effects to the therapeutic use of this peptide in cancer patients remains open to speculation. One should note, however, that recent reports indicate that FTS restores normal level of NK activity in cancer patients in whom NK reactivity was decreased.

REFERENCES

1. Herberman,R.B., Djeu,J.Y., Kay,H.D., Ortaldo,J.R., Riccardi C., Bonnard,G.D., Holden,H.T., Fagnani,R., Santoni,A. and Puccetti,P. in "Immunological Review" (G.Möller,ed.) vol.44, p.43-70, Munksgaard, Copenhagen, 1979.
2. Herberman,R.B., Bonnard,G.D., Brunda,M., Domzig,W., Fagnani R., Goldfarb,R.H., Holden,H.T., Ortaldo,J.R., Reynolds,C. Riccardi,C., Santoni,A., Taramelli,D., Timonen,T. and Varesio,L. in "The Biological Significance of Immune Regulation" (M.E. Gershwin and Ruben L.N. eds.) (in press) New York, Marcel Dekker.
3. Bardos,P., Altman,J., Guillou,P. and Carnaud,C. Clin.Exp. Immunol. 43: 534 (1981)
4. Bardos, P. and Bach,J.F. Scand.J.Immunol. (Submitted to Publ)
5. Djeu,J.Y., Heinbaugh,J.A., Holden,H.T. and Herberman,R.B. J.Immunol. 122: 182 (1979)
6. Hochman,P.S. and Cudkowicz,G. J.Immunol. 123: 968 (1979)
7. Shoham,Y., Eshel, Y., Aboud,M. and Salzberg,S. J.Immunol. 125:54 (1980)
8. Minato,M., Reid,L., Cantor, H., Lengyel,P. and Bloom,B.R. J.Exp.Med. 152:124 (1980)
9. Incefy,G.S., Nertelsmann,R., Yata,K., Dardenne,M., Bach,J.F. and Good,R.A. Clin.Exp.Immunol. 40: 396 (1980)
10. Stutman,O. in "Contem.Topics in Immunobiology" 7:1 (1977).

DISTINCTIVE CHARACTERISTICS BETWEEN SPLENIC NATURAL KILLER CELLS AND PROTHYMOCYTES

Koo, G.C.[1]
Cayre, Y.[2]
Mittl, L.R.

Memorial Sloan-Kettering Cancer Center
New York, New York

I. INTRODUCTION

Recent studies have implied that NK cells may be prothymocytes or pre-T cells (Herberman *et al.*, 1980) because nude mice have high NK activity, and because NK cells express low levels of Thy-1 antigen. We have approached this issue by investigating whether NK cells fulfill the criteria of the definition of prothymocytes (Koo *et al.*, 1981). In this context, prothymocytes are defined as the subset of lymphoid cells that are negative for Thy-1 and other T-cell differentiation antigens but can be induced by thymopoietin to express Thy-1 antigen (Komuro and Boyse 1973). Prothymocytes are also found to have receptors for peanut agglutinin (Haran-Ghera et al., 1978), a lectin highly specific for galactosyl residues binding preferentially with Gal-β-(1,3)-Gal-NAc segment (Pereira et al., 1976). The results of the comparative study of NK cells and prothymocytes are summarized below.

II. DEPLETION OF $Nk\text{-}1^{+}$ CELLS DOES NOT DEPLETE PROTHYMOCYTES

Both splenic prothymocytes and $Nk\text{-}1^{+}$ cells are enriched in A and B layers of BSA fractionated spleen cells (Komuro and Boyse, 1973 and Koo *et al*. 1980). Cells from A + B layers could be induced to express Thy-1 antigen by Tp-5 (a synthetic

[1]Supported by NIH grants CA-08748, CA-25416, ACS(FRA-167) and the French Affairs Ministry
[2]Present Address: Department of Genie Genetique, Institut Pasteur, Paris, France

ISBN 0-12-341360-5

TABLE I. Effect of TP-5 on splenic NK activity and proportion of Nk-1+ and Thy-1+ cells after depletion of Nk-1+ cells[a]

Depletion of[b] Nk-1+ cells	Induction with Tp-5	NK activity (% lysis)	Nk-1+ (% PA-SRBC rosettes)[e]	Thy-1+ cells (% PA-SRBC rosettes)[e]
-	-	17 ± 3	21 ± 4	22 ± 3
-	+	18 ± 2	22 ± 3	27 ± 1
+	-	4 ± 3	3 ± 3	10 ± 2
+	+	3 ± 3	3 ± 1	17 ± 4

[a]Average ± S.D. of 3 experiments

[b]Spleen cells from A + B layers of BSA gradient were sensitized with anti-Nk-1.1 serum and PA-SRBC; the rosettes were depleted by ficoll-isopaque gradient

[c]Cells were induced with 10 ug/ml of Tp-5 for 2 1/2 hrs.

[d]E:T ratio = 100:1

[e]% PA-SRBC rosettes =

$$\frac{\text{\% rosettes of test - \% rosettes of reagent control}}{\text{100 - \% rosettes of reagent control without serum}} \times 100$$

Nk-1+ spleen cells were assessed by using anti-Nk-1.1 serum (C3HxBALB) anti-CE spleen.
Thy-1+ cells were assessed by using monoclonal anti-Thy 1.2 serum.

pentapeptide proved to be an analog of thymopoietin). When these cells were depleted of Nk-1+ cells, inducibility by Tp-5 was not affected, implying that Nk-1 antigen is not on prothymocytes (Table I). Furthermore, Tp-5 has no effect on Nk-1+ cells or NK activity.

III. Nk-1+ CELLS AND NK ACTIVITY ARE IN SPLENIC PNA− FRACTION AND PROTHYMOCYTES ARE IN PNA+ FRACTION

BSA enriched spleen cells were fractionated into PNA+ and PNA− populations. Nearly all the Nk-1+ cells and NK activity were in the PNA− fraction, whereas prothymocytes (cells inducible for Thy-1 antigen) were in the PNA+ fraction (Table II).

We concluded from these results that splenic Nk-1+ cells

are distinct from prothymocytes, at least at this stage of T-cell development. There is at present no indication that $Nk\text{-}1^+$ or NK cells will develop into T cells. $Nk\text{-}1^+$ cells are likely "end" cells that belong to a distinct lineage from T cells.

TABLE II. Proportion of $Nk\text{-}1^+$ and $Thy\text{-}1^+$ cells in whole spleen and A + B layers of BSA gradient before and after PNA fractionation

	% PA-SRBC rosettes[a]			% specific lysis[b]
	$Nk\text{-}1^+$ cells	$Thy\text{-}1^+$ cells	$Thy\text{-}1^+$ cells after Tp-5 induction[c]	
whole spleen	7 ± 3	22 ± 3	-- --	10
A + B	18 ± 7	21 ± 3	27 ± 1	26
PNA^+ of A+B[d]	8 ± 1	6 ± 5	32 ± 5	10
PNA^- of A+B	27 ± 4	27 ± 5	26 ± 5	40

[a]Average ± S.D. of three experiments; % PA-SRBC, see footnore in Table 1
[b]E:T ratio = 100:1, targets were YAC cells
[c]Cells induced with 5 ug/ml of Tp-5 for 2 1/2 hr
[d]Purity of each fraction was 95-97% as determined by fluorescein-labeled PNA

ACKNOWLEDGEMENTS

We thank Barbara LoFaso for preparation of the manuscript.

REFERENCES

1. Haran-Ghera, N., Rubio, N., Leef, F., and Goldstein, G. Cell. Immunol. 37:308 (1978).

2. Herberman, R.B., Timonen, T., Reynolds, C., and Ortaldo, J.R. in "Natural Cell mediated immunity against tumors" (R.B. Herberman, ed), p. 89, Academic Press, New York, 1980.

3. Komuro, K., and Boyse, E.A. J. Exp. Med. 138:479, (1973).

4. Koo, G.C., Jacobson, J.B., Hammerling, G.J., and Hammerling, U. J. Immunol. 125:1003 (1980).

5. Koo, G.C., Cayre, Y., and Mittl, L.R. Cellular Immunol. 62:164 (1981).

6. Pereira, M.E., Kabat, E.A., Lotan, R., and Sharon, N. Carbohydr. Res. 51:107 (1976).

HYPOTHESIS ON THE DEVELOPMENT OF NATURAL KILLER CELLS AND THEIR RELATIONSHIP TO T CELLS

Zvi Grossman[1] and Ronald B. Herberman[2]

[1]Mathematical Institute, Tel Aviv University
Tel Aviv, Israel

[2]Biological Research and Therapy Branch
National Cancer Institute
Frederick, Maryland

There are a series of major issues about natural killer (NK) cells which to date are poorly understood: (a) the possible relation of NK cells to the T cell lineage; (b) the nature and specificities of the recognition receptors on NK cells and of the antigens on the target cells; and (c) the mechanisms which regulate the number and activity of NK cells reactive against a given target. In the absence of sufficient direct evidence, we have invoked theoretical considerations meant to indicate a direction in which the answers might be sought. The proposed scheme is consistent with, and provides a unifying interpretation of, the main characteristics of NK cells which have been observed; and it makes some testable predictions.

The Origin and Specificity of NK Cells

The characteristics of NK cells can be incorporated into a recently proposed theoretical approach to the immune system (Grossman and Cohen, 1980). Clones of lymphocytes compete in an essentially Darwinian sense for prominence through processes of growth, differentiation, and feedback suppression. The consequence of competition depends on the relative growth rates of individual clones . The assessment of clonal growth capacities in terms of affinities and antigen concentrations is defined by the concept of "balanced growth." Accordingly, the number of divisions which an activated lymphocyte

ISBN 0-12-341360-5

undergoes depends on regulatory factors outside the cell. Clonal growth and differentiation are mutually antagonistic processes (though not necessarily exclusive). A large antigen dose favours the expansion of low-affinity clones, and a low dose promotes the development of high affinity clones of T as well as B cells. An expansion of low affinity clones without maturation to the effector cell level is possible. During the generation of the T cell repertoire, in particular, T cells which interact with certain self antigens (notably, MHC products) only "proliferatively" are dynamically selected, and this constraint ensures tolerance to self. MHC restriction and alloreactivity are essentially expressions of "heteroclicity" in this selection, i.e., interaction ("priming") with one group of antigens leads to the generation of a T cell memory directed primarily at related, but different, antigens. This scheme requires that the affinity threshold for maturation be higher than that for proliferation, and that the threshold for killing be even higher. Following Mitchison and Lake (1980), we make a distinction between "weak" and "strong" antigens with respect to their ability to elicit certain types of immune responses. Operationally, certain cell-surface antigens, including tumor antigens, are "weak antigens" if they fail to stimulate a significant primary response.

It has been suggested that MHC products, as presented by certain cells, are strong antigens because they can induce extensive proliferation of clones with a wide range of affinities (Grossman, 1982). Several other cell-surface antigens are weak antigens (Knight et al, 1979), not necessarily because they are recognized with low affinity; rather, those precursor T cells which recognize such antigens with a sufficiently high affinity to differentiate and exhibit effector activity also have low probability of regenerative proliferation.

As a further step, we propose that NK cells and their precursors are T cells with the latter characteristics. Clones of such cells would be susceptible to competition and suppression by other proliferating, cross-reactive clones, especially at higher antigen doses. Cross-reactivity between weak and strong antigen-reactive clones could be due to the presentation of weak and strong antigens by the same target cell; or perhaps by the uptake of molecules, presented as weak antigens on the original target cells, by other autologous cells, where they are "strongly" presented in association with MHC-products (Clark et al, 1979). The best immunization regimen for NK precursor cells would be a prolonged exposure to a low dose of the antigen (e.g., a small number of target cells): This would reduce differentiation rates and trigger high affinity clones to a slow proliferation; and in addition, the effect of competition and cross-suppression would be limited under these conditions (Grossman, 1982). As is the case with MHC-restricted cells, we envisage the selection and development of the NK cell population as driven by the self-environment. The processes of MHC-restricted and nonrestricted T cell selection should take place at different sites, because otherwise the dominance of the MHC-restricted cells would lead to their exclusive expression.

Our hypothesis can be stated as follows: (a) prethymic T cell precursors expressing receptors specific for various (non-MHC) cell surface antigens are the origin of NK cells; (b) NK cells which can be driven by interacting with the respective self antigens into a relatively prolonged proliferation are positively selected. Those presumably include clones with low affinities to abundant self antigens on the surface of cells [in analogy to the MHC-restricted clones which are selected (Grossman, 1982)] and clones with high affinities to certain infrequently expressed or transitory antigens (e.g., differentiation antigens); (c) the selection of NK cells is mostly extra-thymic, since the conditions in the thymus are selective against weak antigen-reactive clones; (d) because of the nature of recognition of weak antigens as defined above, many cell-surface antigen-reactive clones end up as terminally differentiated effector cells, or close to that state, prior to challenge with foreign or modified cell-surface antigens; (e) further clonal expansion of NK precursors is favoured by a continuous low-level challenge with the antigen.

T cell proliferation is often associated with recognition of (modified) MHC Ia-antigens on antigen-presenting cells. Such T cells proliferate and can help other B and T cells, such as CTL, to proliferate. The generation of B and T cell growth factors (BCGF and TCGF) appears to be involved. The "weakness" of surface antigens recognized by NK cells may stem from the fact that they are seen not in association with Ia antigens: potential helper cells do not expand and do not produce appreciable amounts of TCGF. Direct differentiation of NK precursor cells into effector cells is then promoted. The helper cells could either be ordinary T cells, or alternatively a helper analogue of NK cells, i.e. "natural helper" (NH) cells, could be postulated. The involvement of another subpopulation of cells in the expression of NK effector activity, or at least in the development of the NK cell population (by clonal expansion), is a question open to experimentation. Recent evidence on "helper activity" of large granular lymphocytes (LGL) in vitro (Ferrarini and Grossi, 1982) supports such a functional heterogeneity within the population of "naturally reactive" cells. Further, recent experiments have indicated that LGL may produce TCGF, but at considerably lower levels than T cells (Domzig and Stadler, this volume). Finally we note that if it is indeed the lack of self-induced growth factor which limits the proliferative response of NK cell precursors, then the possibility of cooperative interaction of clones directed at strong and weak antigens on the same target cell, respectively, can be envisioned: the growth factor generated by MHC-restricted, or certain allo-reactive, proliferating clones can drive antigenically-activated NK precursors into proliferation as well. Selective blocking of strong determinants, or suppression of the respective clones, may result in the selective expression of a conventional NK cell activity against the same target.

Interpretation of the Similarities and Differences Between NK Cells and CTL

The hypothesis that the receptors on NK cells may be similar to those on CTL has been raised by a number of investigators (e.g., Kaplan and Callewaert, 1982). The apparent clonal distribution of the recognition structures with the population of NK cells, the fact that they share several T cell-associated markers, and their ability to grow in response to TCGF and T cell mitogens all indicate a possible close relationship of NK cells to the T cell lineage. However the nature of recognition by NK cells appears to be substantially different from that of CTL: (a) There are disparities in specificity: antigenic specificities recognized by NK cells are more widely distributed, and include syngeneic, allogeneic and also some xenogeneic targets; in contrast to CTL, NK cells do not appear to recognize cell membrane structures in association with MHC-products. (b) NK cells are not thymic-dependent. (c) Induction of CTL is usually dependent on sensitization with a given antigen, the resultant CTL have specific reactivity limited mainly to the immunogen, and later re-exposure to the same antigen evokes a rapid and specific memory response. In contrast, functional NK cells arise without known antigenic stimulation and as soon as they are detectable, have reactivity against the full array of specificities that are associated with this effector system. In addition, NK cells have not been demonstrated to undergo a specific memory response but rather, upon stimulation with cells bearing NK-related antigens, or with interferon, rapidly develop, without a requirement for proliferation, augmented reactivity against a wide range of specificities.

The hypothesis presented in the previous section can reconcile these differences with the assumption of a common T cell lineage, without having to postulate qualitatively different rules for NK cells and CTL. Some of the differences with respect to specificity may result directly from the identity of the antigens which are recognized: MHC products are not recognized by NK cells and thus are not restricting. The cell-surface antigens for NK cells possibly represent a wider spectrum of recognizable structures as compared to modified MHC-products, and hence the wider distribution of such NK cell targets. It should be noted that the relevant cell-surface structures could also be manifested as changes in structure, pattern or density of molecules on normal, or transformed self, cell membranes.

The other differences in specificity and function between NK cells and CTL can be attributed to the quantitatively different nature of the interaction of T cells with "weak" and "strong" antigens, respectively. The spontaneous effector activity of NK cells is related to the advanced state of differentiation induced earlier by interacting with the autologous cellular environment. Also, interaction of foreign, modified or transformed cell-surface antigens with NK cell precursors readily leads to the expression of effector function with no, or little, proliferation. Consequently, the response is prompt, does not require sensitization and usually does not involve memory generation (amplification). Moreover, the fact that a considerable proportion of the NK cell population can bind to, and a smaller but significant proportion can kill, a given target cell may reflect, in part, the underlying cross-reactivity at the level of antigen recognition by individual cells of the T cell lineage. It is the competition among expanding clones during priming, and the resulting dynamic selection, which accounts for the exquisite specificity of a conventional T cell mediated response at the effector cell level (Grossman, 1982). In contrast, for NK cells, which do not proliferate extensively in response to antigens, and therefore are not subjected to much competition, a wider range of clones with various specificities are co-expressed. This, and the possible existence of several cell-surface antigens common to many target cells, can explain the apparent limited degree of specificity exhibited by NK cells.

Proliferation, competition and selection play a role in the pre-selection of the NK cell repertoire. Among clones interacting with self-antigens during the same period and located in the same region, those possessing higher proliferative capacities would be positively selected, and the other clones suppressed. In the thymus, MHC-specific clones dominate, and they do so in other sites as well. However, compartmentalization in some tissues may allow independent development of NK cells. The NK cell system is thus not only thymus-independent, but in athymic individuals, more NK precursors escape suppression and may develop into NK cells, as is indeed indicated by an increased NK cell frequency in nude rats (Reynolds and Holden, this volume). It may also be noted that recent evidence indicates that the thymus is not an exclusive site for the selection of the T cell repertoire in general (Wagner _et al_, 1981). Again, cells with a range of intermediate affinities to certain abundant self antigens, allowing for a certain degree of regeneration of precursor NK cells, will be selected. As in the case of the MHC-associated T cell recognition, the fact that self-recognition tends to trigger proliferation rather than effector reactivity ensures "tolerance to self", and at the same time allows for the expression of effector reactivity to "modified self." With respect to modified or transformed cells, NK reactivity is expected to be suppressed at high antigenic loads, which trigger proliferation of low affinity T cell clones and exhaust high affinity NK cells and their precursors. On the other hand, a small number of transformed cells, even if not initially eliminated by an excess of NK activity, may trigger effective NK clones to proliferate and maintain themselves, and also to differentiate, at a level sufficient to limit the growth of the target cell population. Such a steady state for a tumor-NK system might keep the tumor "under control" (i.e., benign, latent and possibly undetectable). The inverse correlation noticed between the size of tumors and in situ NK reactivity (Gerson, 1980) may thus be explained as a two-directional effect, where the tumor growth is not, or not only, a result of the absence of NK cells, but may be the cause for this absence.

Finally, some of the morphologic and cell-surface antigenic differences between NK cells and typical T cells may just reflect the different states of differentiation along a common pathway (with NK cells functionally more mature or activated), or alternatively may be due to the maturation outside the thymus and/or to skipping of intermediate differentiation steps, or the rapidity of differentiation, by NK cell precursors. In support of this possibility, D. Zagury (personal communication) has recently noted that cloned CTL, after binding to their target cells, develop a large number of cytoplasmic granules, similar to those seen in LGL.

Surveillance, Self Recognition and Growth Control

We have conjectured above that natural reactivity arises from, and is regulated by, previous interactions with the cellular self-environment. Even if the selected population is primarily heteroclitic in its effector reactivity, one might predict that cytotoxicity against autologous cells is more likely to be manifested by NK cells than by MHC-selected T cells, since we propose that the former cells are more differentiated.

NK cells have been shown to react against some autologous or syngeneic normal cells (e.g., Hansson et al, this volume) and some authors have indeed suggested that NK cells may be involved in the growth control of certain autologous, normal cell populations. Within our model, the development in some compartments of a NK precursor population directed against cells related to, but not identical with, the stimulating or "driving" cells, is quite possible. Thus the driving cells might be maturing hemopoietic cells in the bone-marrow, while the principal target cells might be stem cells. The target antigens might be differentiation antigens, as suggested recently by several authors (Dokhelar & Turz and Gidlund et al, this volume), and then exhibited heteroclicity might be associated with differences in structure or density of the relevant antigens on the cell membrane. Another possibility is that the postulated binding affinity of NK cells for the driving cells is sufficient in order for them to exert cytostatic effects and control of the growth of these cells, without their elimination by cytolysis. A similar proposition has recently been made with regard to the possible role of self anti-idiotypic interactions among lymphocytes in the control of the immune system (Grossman, 1982). The possible cytostatic activity of relatively mature, non-proliferating NK cells against their own progenitors, which possess growth potential, may be a factor in the control of growth of the NK cells themselves.

Another mode of feedback growth control is the induction of differentiation of proliferating cells by their progeny. It is tempting to speculate that interferon, which can be secreted by mature NK cells (Djeu et al, this volume) as well as T cells, may regulate simultaneously in this way both cell growth and the expression of effector reactivity. A unifying conjecture is that NK cells, like other T cells, require specific antigenic activation which then renders them sensitive to the influences of various growth factors that regulate proliferation and differentiation. Furthermore, the induction of subpopulations of NK cells ("natural helper cells") to secrete interferon (Djeu et al, this volume) and TCGF (Domzig and Stadler, this volume) may be analogous to that of other lymphokine-secreting T cells, and also require antigenic stimulation.

Experimental Implications and Conclusions

The hypothesis on the origin and specificity of NK cells admittedly involves some unsubstantiated speculations. Yet, it is consistent with the presently known facts, within the limits of the prevailing uncertainties, and it also suggests certain experimental procedures to test it or to clarify some questions which it invokes: (a) Based on our assumption of some clonality of NK cells, it should be possible to selectively inhibit the activity of various NK clones by separate cell-surface molecules. Indeed, it has already been possible to demonstrate some non-cross reactive cold-target inhibition of cytotoxicity by mouse NK clones (Kedar and Herberman, this volume). (b) Since we propose that NK cells are a subpopulation of T cells that react with non-MHC self antigens, the procedures effective in priming for a T cell response against minor histocompatibility antigens (Mitchison and Lake, 1980) might also prepare "NK cell memory." In particular, prolonged exposure to a small number of target cells may induce a slow proliferation of NK cell precursors. Simultaneously, local suppression of potentially competing MHC-restricted responses, e.g. by anti-MHC antibody, may promote such a process. More generally, blocking of strong MHC antigens on target cells, especially that of Ia-antigens, could "unmask" NK activity either by removing competition or alternatively by inhibiting the production of growth factor which can induce proliferation non-specifically (but locally). This second mode would lead to a transient expression of NK cell function rather than to a prolonged and stable elevation of their activity. (c) Allogeneic cells, presenting both MHC and non-MHC surface antigens, would be expected to trigger both NK cell and T cell-restricted responses. The NK response should peak earlier, and disappear concomitantly with the proliferation of conventional T cell clones. In fact, the "anomalous killers" (AK) described by Seeley and Karre (1980) can be interpreted as representing such an early NK response. (d) The intriguing possibility that conventional CTL could reach an end state with the morphologic and functional properties of NK cells should be relatively easy to test, since NK cells are recognizable as LGL (Timonen _et al_, 1981). This possibility is suggested by our assumption of an overlap between NK cells and terminally differentiated T cells. NK precursors, resembling morphologically ordinary T cell precursors, may be cloned and expanded by TCGF and then induced to differentiate and switch into LGL. Similarly, cloned CTL could be induced to express a high degree of effector activity, and a morphological switch into LGL for a proportion of cells might be directly detected. Indeed, as mentioned earlier, Zaquary has recently noted an apparent terminal differentiation of cloned CTL into heavily granulated cells. Further, it has already been reported that apparent clones of CTL, with specific anti-MHC reactivity, developed simultaneously (Neefe and Carpenter, this volume) or after loss of specific reactivity (Hengartner *et al.*), cytotoxic activity against typical NK-susceptible cells. (e) Our interpretation of the extrathymic origin of NK cells is based on a competitive relationship between CTL and NK cells. This suggests that the same suppressor cells could be effective against both populations. This could be tested in vitro and in adoptive transfer systems. (f) Antigen dose and affinity are oppositely related in our model. During the growth of an induced or transferred tumor, an early phase of NK cell cytotoxic activity is expected to be replaced by a transient increase in CTL and in nonlytic NK cell binding activity. Moreover, concomitant NK activity outside the tumor site, for a localized tumor at an early phase, is expected even when such activity is suppressed at the tumor site itself. (In analogy to the phenomenon of "concomitant immunity".) In fact, such a disparity between *in situ* and systemic NK activity has been generally observed in cancer patients (e.g., Gerson, 1980; Vose, 1980). (g) A straight-forward generalization of the concept that natural killers are differentiated cell-surface antigen-specific cytotoxic T cells predicts the existence of similar natural helper (NH) cells. These would interact with NK precursors, via cell contact or production of factors, to induce proliferation and/or terminal differentiation. Limiting dilution and log-dose/log-response measurements (Coppelson and Michie, 1966) could be carried out in order to assess the number of cooperating cell populations required for the expression of effector responses. In addition, the phenotypic composition of cell clusters formed at the site of NK-target interaction might provide information about this question. (h) Genetic control of the level of NK activity to certain target cells might be traced to genes coding for the expression of surface antigens on these cells, or to genes coding for non-MHC self antigens affecting NK cells pre-selection, or to both. (i) The NK cell receptor should be homologous to that part of the T cell recognition structure directed at the same non-MHC antigens. It is of much interest in this regard that Okumura (1982) has recently reported that monoclonal antibodies to allotypic specificities, apparently linked to the immunoglobulin V_H region, reacted with mouse NK cells. As a corollary, NK cells and antibody directed at common target molecules would be expected to share idiotypic determinants.

REFERENCES

Clark, E., Lake, P., Mitchison, N.A., and Nakashima, I. (1979). In "Natural and Induced Cell-Mediated Cytotoxicity" Academic Press, New York.
Coppelson, L.W., Michie, D. (1966). Proc. Roy. Soc. London (Biol). B163:555.
Ferrarini, M. and Grossi, C.E. (1982). In "Natural Cell-Mediated Immunity, Vol. 2" (R.B. Herberman, ed.), Academic Press, in press.
Gerson, J.M. (1980). In "Natural Cell-Mediated Immunity Against Tumors" (R.B. Herberman, ed.), p. 1047. Academic Press, New York.
Grossman, Z., and Cohen, I.R. (1980). Eur. J. Immunol. 10:633.
Grossman, Z. (1982). Submitted for publication.
Kaplan, J. and Callewaert, D.M. (1980). In "Natural Cell-Mediated Immunity Against Tumors" (R.B. Herberman, ed.), p. 893. Academic Press, New York.
Knight, J., Knight, A., and Mitchison, A.N. (1979). In "The Molecular Basis of Immune Cell Function" (J.G. Kaplan, ed.), p. 139. Elsevier/North-Holland, Amsterdam.
Mitchison, N.A., and Lake, P. (1980). In "Immune System: Genetics and Regulation" (E. Sercarz, L.A. Herzenberg, and C.E. Fox, eds.), p. 555. Academic Press, New York.
Okumura, K. (1982). In "Macrophage and NK Cell Regulation and Function. Proceedings of 9th International RES Congress" (E. Sorkin and S. Normann, eds.), in press.
Seeley, J.K. and Karre, K. (1980). In "Natural Cell-Mediated Immunity Against Tumors (R.B. Herberman, ed.), p. 477. Academic Press, New York.
Timonen, T., Ortaldo, J.R., and Herberman, R.B. (1981). J. Exp. Med. 153:569.
Wagner, H., Hardt, T.C., Stockinger, H., Pfizenmaier, K., Bartlett, R., and Rollinghoff, M. (1981). Immunol. Rev. 58:95.

COMMENT ON THE NOMENCLATURE OF LYMPHOCYTE MEDIATED CYTOTOXIC EFFECTS

Eva Klein

Department of Tumor Biology
Karolinska Institutet
Stockholm, Sweden

In the past few years several important discoveries emerged in the field of in vitro cytotoxicities which influence the interpretations and provide new methodological possibilities. Such are: 1. the demonstration that histocompatibility antigens are involved in the lytic interaction between effectors and targets when haptens, viral antigens, products of minor histocompatibility antigens are recognized; 2. the natural killer potential of unmanipulated lymphocyte populations; 3. the possibility to detect distinct subsets acting on different targets in the same effector populations by the cold target competition assay; 4. the characterization of lymphocyte subsets with monoclonal reagents determining well defined functional entities; 5. the demonstration of lymphokines, produced when lymphocytes encounter antigens; 6. the possibility to enlarge lymphocyte clones in vitro.

The present view distinguishes cytotoxic T lymphocytes - CTL - and natural killer cells - NK. The basis for the distinction is the in vivo and/or in vitro sensitization event in the generation of the former and its absence in the latter effect. Plasma membrane marker characteristics of the effectors and specificity patterns of the lytic effects are also considered in the designation of various systems.

The outcome of the short term cytotoxic assays - performed either with freshly harvested effectors or with lymphocytes kept in culture with various stimuli - depends on several factors, such as the immunological history of the lymphocyte donors, the activation profile of the lymphocyte population, the treatment of the effectors, the characteristics of the target, and the species or allo-relationship of the effectors and target.

ISBN 0-12-341360-5

It is likely therefore that the basis for the lymphocyte target interaction is not identical in all systems.

Because of the heterogeneity of the lymphocyte population and the various cell membrane moieties which can serve as targets, even in one experiment the recorded lysis is the sum of concommittant events in which the details of the effector and target interaction may differ. Due to generalizations on the basis of results with various systems, the literature on the characteristics of cell mediated cytotoxicity is not as clear as the present state of knowledge could permit. A more precise nomenclature would help to sort out the different phenomena and thereby provide a clearer picture.

The analysis of the natural killer effect with regard to phenotypic characteristics of the active subset and its selectivity, at least in certain systems, provides evidence that it is in part a T cell function. We proposed therefore previously, that the distinction between CTL and NK cells is not as sharp as initially suggested.

In view of the heterogeneity of the lymphocyte population with regard of antigen recognizing receptors the definition "specificity" ought to be modified: specificity may still be the case on the level of functioning subsets even if the total population acts on several non cross-reactive targets. Therefore specificity on the population and on the acting subset level ought to be distinguished.

The designation "natural" is operational. Consequently, lytic activity of cultured, stimulated lymphocyte population does not comply with the name "natural" even if they are assayed against NK sensitive targets. In several papers these cells are referred to as "activated" killers. The different designation for the two types of effect is well motivated because the sensitive target panel and the surface marker characteristics of the effector populations which function in the natural and activated systems differ.

Results obtained in cold target competition assays and with cloned killer lines suggest that at least a proportion of lytic cells act on different targets with different mechanisms. With one mechanism lysis is based on the recognition of cell surface epitope, it is performed by effectors derived from clones of the relevant receptor carrying lymphocytes, the other may be triggered by an as yet unknown type of interaction between the cell surfaces. This effect is brought about therefore by a population which includes lymphocytes with receptors to a variety of antigens i.e. it is polyclonal.

The following hypothesis may be proposed: In a certain differentiation state T cells are capable to lyse targets in either way. Other cytotoxic subsets may exert only one or the other action. The lymphocytes which have only the non-discriminative effects may not express the antigen receptors. Another subset functions only if triggered through the interaction between antigen and its specific receptor on the plasma membrane.

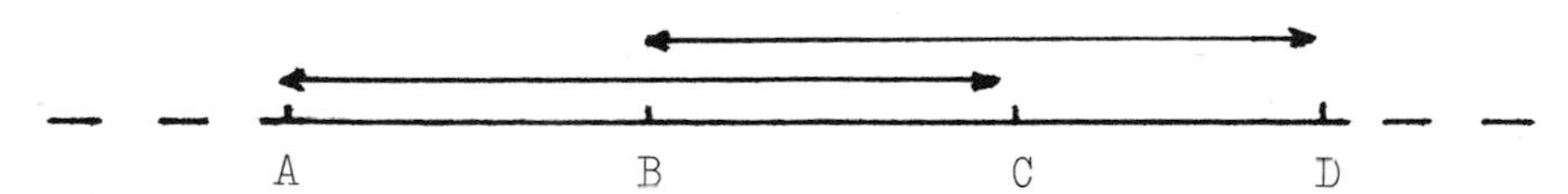

The figure depicts this assumption. A-D represents the lymphocytes with lytic potential. Heterogeneity in this population is at least on two levels: 1. the activity profile, with regard to lytic potential, 2. expression of the repertoire of receptors for antigens. A-C can lyse targets without involvement of the antigen-receptor interaction, provided these express certain membrane properties. Lymphocytes in the subset B-D can lyse targets on the basis of epitope recognition. C-D can only exert the lytic effect on the latter basis. B-C can function in both ways.

The relative representation of the subsets differs individually and also in lymphocyte populations of variour sources (blood, spleen, lymph node) and can be shifted temporarily both in vivo and in vitro. Shifts may be caused as a consequence of direct interactions between different lymphocyte subsets or soluble factors.

The main distinction between population A-C acting on the NK sensitive prototype targets and B-D acting through recognition of antigens on the target cell membrane is that the former is polyclonal while the latter is clonal.

Another lymphocyte subset does not respond with lytic function but is triggered to enter into the mitotic cycle after the encounter with antigens or mitogens. Among these cells killer precursors are present.

With regard to the lytic effects, introduction of the clonality aspects in the designations could help the evaluation and comparison of results. Lytic effects whether natural or induced, based on antigen recognition are likely to follow the rules laid down for cellular immunology, involving specificity and memory. On the other hand, the polyclonal effects detectable with certain targets, such as K562, reflects the activation or maturation profile of the population and belongs to the topic of lymphocyte differentiation.

The differentiation profile of antigen stimulated lymphocyte cultures has been shown to vary depending on the experimental conditions. This variation determines the relative strength of the lytic effect against the stimulator cell and the "non specific" target. It is also a common experience that shifts of the specificity pattern occurs in the cytotoxic action of clonal lines.

Distinction should be made between lytic events:

1. Natural and activated
   a. antigen restricted, clonal
   b. non-restricted, polyclonal
2. Immune
   a. sensitization specific, restricted, clonal
   b. transactivated, restricted, clonal
   c. non-restricted, polyclonal

Category 2b is the result of amplification of the activation event through humoral mediators which recruits T cells devoid of receptors to the antigen stimulus. Since these lymphocytes represent the antigen recognizing repertoire the population can exert antigen restricted lytic effects. The activated lymphocytes may also lyse targets sensitive to the cytotoxic function even without the involvement of antigen-receptor interaction. Thus 2c represents the action of cells including a collection of clones. The type of reactivity is then determined by the characteristics of the target presented to the effector population.

It is always difficult to change a nomenclature which is widely used. NK, CTL and CTL-P (precursor) can be retained, however presently it is used in a way which mainly distinguishes non restricted and restricted lytic events. Since both NK and CTL may have these two characteristics, the unqualified use of these terms may create some confusion. Since the time these designations have been coined the knowledge about T cell differentiation and the properties of the various lytic systems has increased. It may be therefore timely to give a thought whether the nomenclature is still appropriate.

Natural, antigen restricted killer effects are most likely to occur in systems in which alloantigens are involved. It is known that the number of lymphocytes which recognize the products of MHC is high. It is a general rule that allogeneic grafts are rejected by non-immunized recipients. It is therefore possible that target cell lysis on the basis of alloantigen recognition can be brought about by lymphocytes derived from non-immunized donors.

Cells with natural killer activity are eliminated by treatment with monoclonal specific anti-macrophage antibody plus complement

Sun, Deming, and Lohmann-Matthes, Marie-Luise
Max-Planck-Institut für Immunbiologie, Freiburg, FRG

Introduction

In the last edition of "natural resistance against tumors" we described that in cultures of bone-marrow cells of mice in the presence of L-cell conditioned medium cells can be generated, which have macrophage surface characteristics and mature to typical macrophages (1, 2). These macrophage precursor cells, which are non-adherent and non-phagocytic, exhibited strong NK-activity against YAC-1 target cells. Later we had difficulties to reproducibly generate these macrophage precursor cells in our bone-marrow cultures, presumably because of changed health conditions of the bone-marrow donor mice. Because of this difficulty, we approached the question of the nature of the NK-cell with other means. For that purpose we raised monoclonal antibodies against mouse-macrophages by immunizing rats with thioglycolate-, corynebacterium parvum-induced and bone-marrow cultured macrophages. We used these macrophage mixtures in order to obtain many different antimacrophage specificities. Out of several fusion experiments we selected four clones, which produced antibodies reacting specifically with mouse-macrophages. These antibodies were selected according to their property to react only with a small proportion of a given macrophage population, since we were mainly interested to identify macrophage subpopulations. One of these four antibodies was very effective in eliminating NK-activity from spleen cells of nu/nu-mice and from peritoneal exudate cells from corynebacterium parvum pretreated mice (3).

Materials and Methods

Animals

All animals used are from our own breeding colony at the age of 8 - 12 weeks.

Culture of bone marrow macrophages

Bone marrow-derived macrophages were cultured as described by Meerpohl et al. (4).

Corynebacterium parvum (C. P.) induced peritoneal macrophages and Thioglycollate induced peritoneal macrophages were obtained as described (3).

Labeling of macrophages and tumor cells

ISBN 0-12-341360-5

Labeling of macrophages

10 - 20 x 10^6 macrophages of different origin were incubated for 40 min in 1.0 ml medium (Eagle's medium, Dulbeccos modification, DMEM) plus 10 % FCS containing 500 uCi ^{51}Cr. The cells were then washed three times with PBS and adjusted to 5 - 8 x 10^6 cells/ml for use.

Labeling of tumor cells

2 x 10^6 tumor cells were incubated with 250 uCi ^{51}Cr or 0.1 uCi ^{125}IUDR for 30 min or 3 h. After 3 washings with PBS the cells were ready for use.

Cell fusion and cloning of hybrids was done as described by Sun et al. (3)

Indirect radioimmunoassay was done as described by Sun et al. (3)

Complement-mediated cytotoxicity assay

^{51}Cr-labeled macrophages were used as target in the cytotoxicity assay. 50 ul of Eagle's medium containing 3 x 10^5 labeled macrophages per well were incubated with 100 ul supernatant of hybrids. The control groups were incubated with normal rat IgG. After 30 min, the cells were washed twice and mixed with 50 ul 1:30 diluted rabbit anti rat Ig (as facilitating antibody) for 5 min. After two more washings, 100 ul of an appropriately selected rabbit complement (1:6) were added and incubated for 60 min. One half of the cell-free supernatants was counted in the gamma-counter. Percentage cytotoxicity was:

$$\frac{\text{cpm supernatant}}{\text{maximal cpm supernatant after detergent lysis}} \times 100$$

Test of NK activity and inhibition by monoclonal antibodies

Natural killing activity was measured by a standard 6 h ^{51}Cr release assay using the cell line YAC-1 as target. The effector cells of the test groups were treated with the supernatants of the hybrids, facilitating antibody (in the case of M 57 and M 102) and rabbit complement. The control groups were treated with normal rat IgG, facilitating antibody and C'. The NK activity was calculated according to the formula: % cytotoxicity =

$$\frac{\text{cpm supernatant}}{\text{cpm sediment + cpm supernatant}} \times 100$$

Results

1. Detection of rat anti mouse monoclonal antibodies with the indirect RIA

^{125}I labeled goat IgG anti-rat IgG was used a second stage antibody. Mature bone marrow cultured macrophages from 7 day old cultures were used as indicator cells. Goat anti-rat antibody was chosen as second antibody since it turned out that goat serum has much less naturally occuring antibodies against mouse cell surface determinants as rabbit serum. Clonal supernatants were tested against mouse macrophages, T lymphoma BW 5147, B lymphoma SP 2/0 Ag14, T lymphocytes and B lymphocytes from mouse spleen, granulocytes, L 929 fibroblasts and rat macrophages. Only those clones were selected which did not show any cross-reactivity with other cells than macrophages (Table 1).

Several controls were done to test the possibility that nonspecific binding to Fc-receptors of macrophages could mimic antigen-specific binding to macrophages (3).

2. Complement dependent lysis of different macrophage populations with the monoclonal antibodies

All 4 clonal supernatants were tested for direct lysis with many different rabbit complements. Only clone 43 and 143 performed direct lysis of macrophages in the presence of C'. The two other clonal supernatants were however, cytotoxic in the presence of facilitating antibody (rabbit anti rat IgG). Table 2 shows lysis of ^{51}Cr labeled 7 day bone-marrow macrophages by clonal supernatants with or without facilitating antibody and with C'. All controls are included. The data clearly show that none of the clones performs complete lysis of the macrophage populations. These results present evidence that we may be dealing with antibodies directed against macrophage subpopulations.
In order to further clarify, whether monoclonal antibodies indeed recognized subpopulations of macrophages, we did sequential treatment with either the same or different antibodies plus C. A clear-cut additive lysis was only observed when different antibodies were used (3).

3. Inhibition of NK-activity of nu/nu-spleen cells by treatment with clone M 102 + C

All 4 clonal supernatants were tested for their capacity to influence NK-activity of spleen cells of nu/nu-mice. Whereas 3 clonal supernatants had no effect on NK-activity, the supernatant of clone M 102 plus C' stably inhibited the NK-activity of nu/nu spleen cells. The degree of inhibition varied between 50 - 90 % from experiment to experiment (Table 3).

4. Inhibition of the NK-activity of corynebacterium parvum induced peritoneal cells

Similar experiments were performed using as a source for NK-cells peritoneal exudate from corynebacterium parvum pretreated mice.

Again only clone M 102 plus C' reduced NK-activity as shown in table 3. The effector NK-cells had to be lysed in order to obtain the inhibitory effect on NK-activity. The mere addition of M 102 without complement did not reduce NK-activity.

Table 4 summarizes the functional properties of the four monoclonal anti-macrophage antibodies as known so far. Whereas M 102 reacts only with NK-cells and does not affect any other macrophage function tested, other clonal supernatants react with e. g. lymphokine-activated macrophages (M 43) or antibody-dependent cytotoxicity effector macrophages (M 43 and M 57).

Discussion

The first rat anti mouse monoclonal antibody (MAC-1) was described by Springer et al. (5). Recently Ault and Springer (6) reported that this MAC-1 antibody reacts with human NK-cells. Similar results have been reported by Breard et al. (7) using a human monoclonal anti monocytic antibody.

There has been some arguing about the specificity of these anti-macrophage antibodies since they crossreact e. g. with granulocytes. Such a cross-reaction is, in fact, not very astonishing since macrophages and granulocytes have a common precursor cell in the bone-marrow. Our data presented now demonstrate that the supernatant of clone 102, directed specifically against mouse-macrophages, also reacts with mouse NK-cells and can eliminate them. The selectivity of the action of clone 102, which reacts with a certain subpopulation of macrophages and does not interfere with other so far tested macrophage functions tends us to suggest that mouse NK-cells represent a subpopulation of the monocytic lineage.

This observation fits with our previously described data, where we reported, that macrophage precursor cells cultured from mouse bone-marrow can perform NK-activity (1, 2). Either by modulating the in vitro culture conditions (8) or by modulating the in vivo health conditions of the mice (9) the composition of macrophages cultured from the bone-marrow of mice can be influenced. Since NK-cells appeared to be a subpopulation of cells of the monocytic lineage, the generation of these cells in bone-marrow cultures may vary according to the culture conditions involved.

Table 1 Binding of monoclonal antibodies to various target cells measured in the indirect radio-immuno-assay

targets	medium cpm	M 43 cpm	M 57 cpm	M 102 cpm	M 143 cpm	rat anti-mouse serum cpm
mouse 7 day cultured bone-marrow macrophages	223 ± 2	2370 ± 6	2200 ± 28	1500 ± 16	2400 ± 24	5800 ± 190
Corynebacterium parvum induced macrophages	196 ± 2	2260 ± 18	2050 ± 16	2220 ± 10	1000 ± 40	4580 ± 40
splenic T + B lymphocytes	600 ± 34	450 ± 11	390 ± 14	404 ± 10	510 ± 16	5800 ± 190
BW 5147 + SP 2/0	270 ± 12	286 ± 10	230 ± 5	237 ± 4	220 ± 18	1490 ± 19
L 929 fibroblasts	243 ± 9	121 ± 8	157 ± 12	138 ± 8	148 ± 10	4000 ± 158
Granulocytes	120 ± 32	114 ± 4	460 ± 21	120 ± 19	150 ± 5	1350 ± 120
rat bone-marrow macrophages	62 ± 2	73 ± 2	71 ± 6	69 ± 4	70 ± 3	-

Table 2 Cytotoxicity of anti-macrophage monoclonal supernatants + C' for macrophages of different origin

treatment of cells	% ^{51}Cr release	
	Bone-marrow-macrophages	Corynebacterium parvum induced macrophages
Medium + C'	23 ± 1	25 ± 2
rat anti-mouse serum + C'	80 ± 3	75 ± 2
Normal rat IgG + C'	23 ± 2	24 ± 1
M 43 + C'	55 ± 2	42 ± 2
M 57* + C'	44 ± 3	55 ± 3
M 102* + C'	41 ± 2	49 ± 2
M 143 + C'	37 ± 1	30 ± 1
rabbit anti rat + C'	22 ± 1	24 ± 1

*Facilitating antibody was added for 5' as described in materials and methods.

Macrophages from different sources were labeled with ^{51}Cr. 3×10^5 cells/well in 50 ul medium were incubated with 100 ul clonal supernatants. As control normal IgG (1:100) and rabbit anti rat (the facilitating antibody) were used. After 30' at $37^{o}C$ the cells were washed and 100 ul selected rabbit C' (1:6) were added. After 60' at 37° 50 ul supernatant were counted in the gamma-counter.

Table 3 Effect of pretreatment with antimacrophage monoclonals + C' on the NK activity of Corynebacterium parvum induced peritoneal cells and of nu/nu spleen cells

Pretreatment of effector cells	% ^{51}Cr release from YAC-cells	
	CP-PE effector cells[a)]	nu/nu spleen effector cells[b)]
Ø	82 ± 2	65 ± 2
Medium + C'	85 ± 1	64 ± 2
M 43 + C'	78 ± 3	66 ± 1
M 57* + C'	65 ± 1	58 ± 2
M 102* + C'	40 ± 2	24 ± 1
M 143 + C'	77 ± 2	63 ± 2
rat IgG* + C'	80 ± 3	64 ± 3
Rabbit anti-rat IgG + C'	82 ± 2	66 ± 1

a) Spont. release of YAC cells was 28 ± 2. Effector/target ratio was 10:1. Assay time was 8 h.

b) Spont. release from YAC cells was 21 ± 2. Effector/target ratio was 50:1. Assay time was 6 h.

*Facilitating antibody (rabbit anti rat) was added for 5' as described in material and methods.

Table 4 Characteristics of 4 rat anti-mouse macrophage monoclonal antibodies

clone	Sub-class	cross reaction with granulo-cytes in RIA	Cytotoxic for NK-cells	Cytotoxic for lymphokin-activated macrophages	Cytotoxic for effector-macrophages in ADCC
43	IgM	-	-	+	+
57	IgG 2b	+	-	-	+
102	IgG 1	-	+	-	+
143	IgG 2a	-	-	-	-

References

1. Lohmann-Matthes, M.-L. and Domzig, W., 1980, In: Natural resistance against tumors, ed. R. Herberman and M. Landy, Academic Press, p. 117.
2. M.-L. Lohmann-Matthes, Roder, J., and Domzig, W., 1979, J. of Immunol. 123, 1883.
3. Sun, D. M. and Lohmann-Matthes, M.-L., 1981, Eur. J. Immunol. in Press.
4. Meerpohl, H. G., Lohmann-Matthes, M.-L., and Fischer, H., 1976, Eur. J. Immunol. 9, 213.
5. Springer, T., Galfre, G., Secher, D. S., and Milstein, C., 1979, Eur. J. Immunol. 9, 301.
6. Ault, K. and Springer, T., 1981, I. Immunol. 126, 359.
7. Breard, J. E. L., Reinherz, P. C., Kung, G., Golgstein, G., and Schlossmann, S. F., 1980, J. Immunol. 124, 1943.
8. Lee, K. C. and Wong, M., 1980, J. of Immunol. 125, 86.
9. Bursuker, I., Goldman, R., Schade, U., and Lohmann-Matthes, M.-L., Cell. Immunol., in Press.

This work was supported by the Volkswagen Stiftung and by the German BMFT.

EFFECT OF ANTI-HLA AND ANTI-BETA-2 MICROGLOBULIN ANTISERA ON THE NATURAL CYTOTOXIC ACTIVITY OF HUMAN NK CELLS AND MONOCYTES

Rajiv K. Saxena
Queen B. Saxena
Robert S. Pyle
William H. Adler

Gerontology Research Center
National Institute on Aging
National Institutes of Health
Baltimore, Maryland

I. INTRODUCTION

Elsewhere in this volume, we have reviewed our work which indicates that the treatment of mouse spleen cells with specific anti H-2 antisera raises their natural killer (NK) activity against K562 target cells but not against YAC (1). An antibody dependent cell mediated cytotoxicity (ADCC) like mechanisms with a reversed polarity of the antibody bridge (reverse-ADCC) was proposed as a possible mechanism for allo-antiserum NK augmentation in mouse system (1,2). Using K562 and YAC target cell lines, we have extended our studies to determine the effect of anti HLA antisera on the natural killer activity of human peripheral blood leukocytes (HPBL) (3). Results obtained in this direction are summarized in this chapter.

II. NATURAL CYTOTOXIC ACTIVITY OF ANTIBODY COATED HPBL

NK activities of HPBL preparations were determined simultaneously against K562 and YAC target cells in a 4 hr ^{51}Cr

ISBN 0-12-341360-5

release assay (4). While HPBL preparations from healthy donors invariably had high levels of anti K562 lytic activity, their cytotoxic activity against YAC target cells was variable and generally poor. Results of a representative experiment in Figure 1A indicate that the anti-K562 lytic ability of HPBL was about 25 fold higher than their anti-YAC activity. When the HPBL preparation was pretreated with specific anti-HLA alloantiserum (Figure 1B) or with anti beta-2 microglobulin xenoantiserum (Figure 1C), their cytotoxic activity against YAC target cells was totally abolished whereas the anti-K562 lytic activity was not influenced in a similar manner. One reason for the selective inhibitory effect of membrane reactive antibody on the natural cytotoxic activity of HPBL could be that different populations of effector cells having

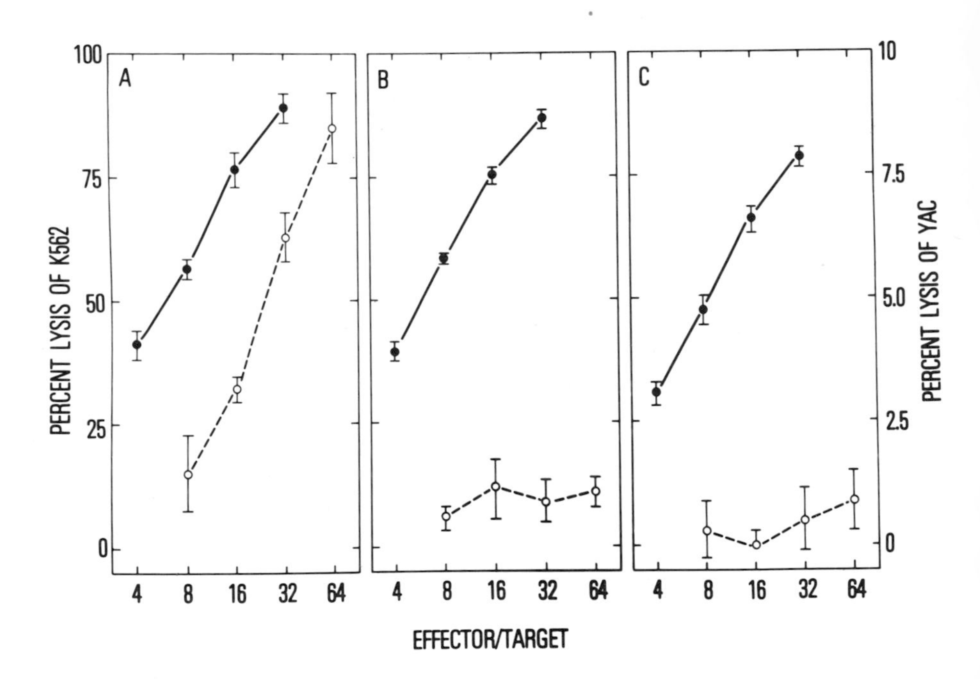

FIGURE 1. Anti-K562 (●) and anti-YAC (○) natural cytotoxicity of HPBL. Natural cytotoxicity of HPBL against K562 and YAC cells was measured in a 4 hr ^{51}Cr release assay (4). Cytotoxic activities of control (A), specific anti-HLA alloantiserum treated (B) and rabbit anti human beta-2 microglobulin treated (C) HPBL preparations are shown. For antiserum pretreatment, 10^7 HPBL/ml in RPMI + 10% FCS were incubated with 1/100 dilution of antisera for 30 min in ice and then washed.

different susceptibility to the two antibody preparations were responsible for lysing K562 and YAC cells. This proposal was examined by cell fractionation studies.

III. EVIDENCE FOR A DIFFERENT LINEAGE OF CELL TYPES IN HPBL WHICH LYSE K562 AND YAC CELLS

Nonadherent (lymphocyte enriched) and plastic adherent (monocyte enriched) populations of HPBL were prepared by previously described procedures (5) and their natural cytotoxic activity against YAC and K562 target cells was studied. Results of a typical experiment are given in Figure 2 and indicate that while the adherent cell preparation was enriched

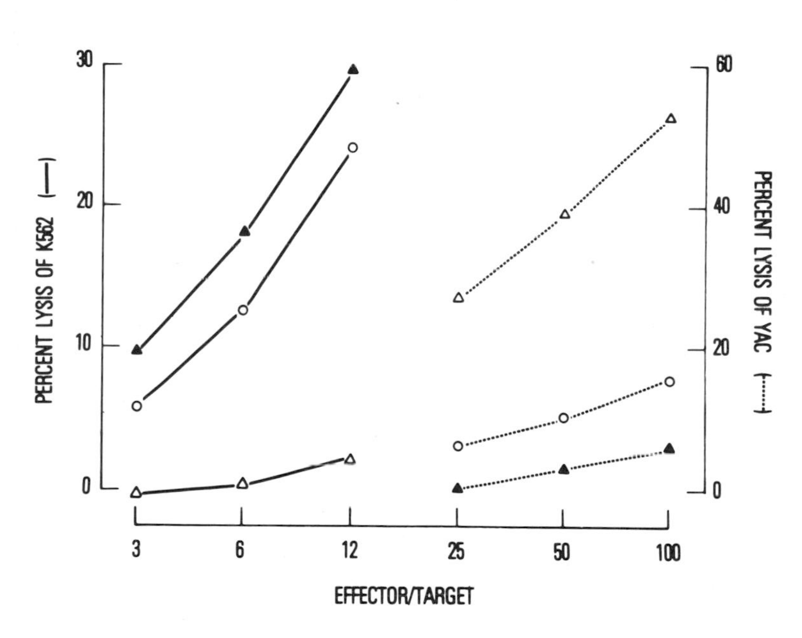

FIGURE 2. Anti-K562 and anti-YAC cytotoxic activity of monocyte enriched and lymphocyte enriched populations of HPBL. Monocyte and lymphocyte enriched populations were derived from HPBL on the basis of plastic adherence property (5) and their cytotoxic activities were measured against K562 and YAC target cells in a 4 hr ^{51}Cr release assay (4).

TABLE I. Inhibition of An i-YAC Cytotoxic Activity Of Human Monocytes by ABM Antiserum[a]

Addition to Cytotoxicity Assay Medium	% YAC Lyses at Monocyte: YAC Ratio of 100:1
None	38.7
ABM Antiserum 1:360 dilution	9.1
ABM Antiserum 1:120 dilution	6.6
ABM Antiserum 1:40 dilution	5.3

[a] Assays were conducted as described (3).

TABLE II. Lack of Effect of ABM Antiserum on the Binding of YAC Cells by Human Monocytes[a]

Addition to the Target Binding Assay Medium	% of Target Bound Effector Cells	
	Expt. 1	Expt. 2
None	10.3	12.2
ABM Antiserum (1:100 dilution)	12.5	10.6

[a] Target binding assay was done by using human monocytes (plastic adherent HPBL) and YAC target cells by the procedure described by Roder et al (7).

in its anti-YAC cytotoxic activity, it did not induce significant lysis of K562 cells. Lymphocyte enriched cell preparation on the other hand became enriched in its anti-K562 cytotoxic activity but lacked any detectable anti-YAC cytotoxic activity. These results indicate that the anti-YAC cytotoxic activity of HPBL may be mediated by monocytes which are enriched in the plastic adherent population of HPBL (about 90% of the cells in adherent HPBL preparations were monocytes as determined by staining with a monocyte specific monoclonal antibody preparation from Bethesda Research Laboratories, Bethesda, Maryland). The nonadherent nature of anti-K562 effector cells is in concurrence with the previous characterization of these effector cells as NK lymphocytes (6).

IV. EFFECT OF ANTI BETA-2 MICROGLOBULIN (ABM) ANTISERUM ON THE NATURAL CYTOTOXICITY OF MONOCYTES

Since human monocytes in HPBL mediate the cytotoxicity against YAC target cells, the inhibition of this reaction by anti-HLA and ABM antisera may represent a direct inhibition of

Table III. Lytic Ability of Control or Anti Beta-2 Microglobulin Antiserum Treated Human Monocytes Against Staphylococcus Aureus[a]

Effector Cells	Pretreatment to Effector Cells[b]	Residual Bacterial Plaques per µl of Incubation Medium After 90 min. Incubation
None	-	1324, 1280
HPBL-1	None	91, 72
HPBL-1	ABM	89, 54
HPBL-2	None	61, 87
HPBL-2	ABM	71, 64
HPBL-3	None	50, 94
HPBL-3	ABM	75, 51
HPBL-4	None	53, 43
HPBL-4	ABM	78, 88

[a] Bactericidal activity of four HPBL preparations (monocyte concentration 6 to 11%) were assayed against staphylococcus aureus as described before (8). Results are given in duplicate.

[b] HPBL (10^7/ml in plain RPMI medium) were incubated with 1/100 dilution of a rabbit anti human beta-2 microglobulin (ABM) antiserum for 30 min at 4° C, and washed twice with fresh medium.

monocyte cytotoxic activity. This proposal is supported by results in Table I, indicating that the anti-YAC cytotoxic activity of the adherent monocyte preparations derived from HPBL was markedly suppressed in the presence of ABM antiserum. Since the binding of monocytes with YAC cells was not influenced (Table II) ABM antiserum appears to inhibit some subsequent step(s) in the cytotoxic reaction leading to the lysis of YAC cells. Interestingly, though the anti-YAC cytotoxic activity of human monocytes is blocked by ABM antiserum, the effect is specific since in another functional assay, the lytic activity of human monocytes against Staphylococcus aureus is not influenced by the presence of ABM antiserum (Table III).

V. CONCLUSION

Anti-tumor cytotoxic activity of human monocytes has generally been studied in prolonged cytotoxicity assays (5). In the present chapter, we have shown that the anti-YAC lytic activity of HPBL is mediated by human monocytes and can be detected in a 4 hr ^{51}Cr release assay. Fresh human lymphocytes are unable to lyse YAC cells though lymphocytes cultured overnight may acquire this capacity (our unpublished data). These results point to an important difference in the human and mouse systems since in the latter K562 and YAC cells are lysed by the same subpopulation of NK cells (2). Another important difference was that unlike in the mouse system, anti-K562 lytic activity of HPBL could not be augmented by anti-HLA antisera. It should however be emphasized that the K562 target cells which are poor targets for mouse NK cells, are highly susceptible to lysis by human NK cells and consequently a further facilitation of human NK-cell and K562 target cell interaction may not occur through an antibody bridge as is proposed for the mouse system (1,2). Other target cell lines moderately susceptible to lysis by human NK cells and expressing Fc receptors, will have to be utilized to determine whether R-ADCC mechanisms of target lysis can operate in the human system.

ACKNOWLEDGMENTS

We wish to thank Ms. Eleanor Wielechowski for typing this manuscript and to Ms. Charlotte Adler for preparing the figures.

REFERENCES

1. Saxena, R. K., (in this volume).
2. Saxena, R. K., Saxena, Q. B., and Adler, W. H., Cell. Immunol. Vol. 65 (in press).
3. Saxena, R. K., Spees, E., and Adler, W. H., Indian J. Exp. Biol. 19:595 (1981).
4. Saxena, Q. B., Mezey, E., and Adler, W. H., Int. J. Cancer 26:413 (1980).
5. Mantovani, A., Tagliabue, A., Dean, J. H., Jerrells, T. R., and Herberman, R. B., Int. J. Cancer 23:28 (1979).
6. Kay, H. D., Bonnard, G. D., and Herberman, R. B., J. Immunol. 122:675 (1979).
7. Roder, J. C., Kiessling, R., Biberfeld, R., and Andersson, B., J. Immunol. 121:2509 (1978).
8. Kaplan, E. L., Laxdal, T., and Quie, P. G., Pediatrics 41:591 (1948).

COULD HUMAN LARGE GRANULAR LYMPHOCYTES REPRESENT A NEW CELL LINEAGE?

Manlio Ferrarini
Department of Clinical Immunology
University of Genova
Genova - Italy

Carlo Enrico Grossi
Department of Human Anatomy
University of Genova
Genova - Italy

INTRODUCTION

Amongst the cells with natural killer (NK) activity towards transformed or virus-infected cells, Large Granular Lymphocytes (LGL) represent a special subpopulation distinguishable through a number of morphological and cytochemical properties. In the human normal peripheral blood, NK activity is only detected within the LGL fraction (Timonen et al., 1981; Abo and Balch, 1981). However, activated T cells and cells of the monocytic/granulocytic lineages can, under special circumstances, exert this function (Rimm et al., 1981; Mantovani et al., 1980). On the other hand, there is no formal proof that the totality of LGL possess NK activity.

No conclusive evidence has been so far obtained demonstrating that LGL belong to any of the known hemic lineages (e.g., as discussed by Herberman, 1980). Here we shall consider the possibility that LGL constitute a new cell lineage. Support to this hypothesis may come from i) the lack of evidence that LGL belong to other hemic lineages and ii) the demonstration of a distinctive pattern of LGL maturation.

Supported by grants from the Italian CNR.

ISBN 0-12-341360-5

ESSENTIAL CRITERIA FOR HUMAN LGL IDENTIFICATION

Before discussing the above points, it is necessary to establish the essential features that a given cell should display in order to be defined as LGL. According to our experience, these features can be described as follows.

- LGL exhibit a lymphoid morphology and contain intracytoplasmic azurophilic (electron-dense) granules.
- LGL are positive for acid hydrolases and negative for peroxidase.
- LGL are non-adherent, non phagocytic cells.
- The large majority of LGL have avid receptors for IgG (Winchester et al., 1979).

Other "conventional" markers such as sheep erythrocyte receptors (Ferrarini et al., 1980), complement receptors (Nocera et al., 1982) and DR antigens (Greaves et al., 1979; Yu et al., 1980) are expressed only by a proportion, but not the totality, of LGL. Analysis with monoclonal reagents has proved further that LGL have a rather heterogeneous surface phenotype (Ortaldo et al., 1981), although one monoclonal antibody, seems to detect LGL specifically (Abo and Balch, 1981).

COULD HUMAN LGL BELONG TO THE MYELO-MONOCYTIC LINEAGE?

The main evidence in support of the above hypothesis, can be summarized as follows:

- LGL share surface markers with the cells of the myelo-monocytic lineage (i.e. avid Fc receptors, DR antigens, receptors for C3b and C3bi, the asialo-GM1 marker) and some antigens detected by monoclonal reagents (i.e. OKM1, Mac1) (Ortaldo et al., 1981).
- Like monocytes, LGL display surface membrane acid esterases inhibitable by NaF (Zucker-Franklin, 1981; Grossi et al., 1982). This marker is not detected on the surface of B or T cells.
- Recently, cell suspensions enriched for NK activity have been isolated from the human peripheral blood and induced to differentiate into monocytes in vitro. These data have been interpreted as an indication that NK cells are promonocytes (Lohmann-Matthes and Domzig, 1980).

In our opinion, no one of the above three points provides compelling evidence that LGL belong to the myelo-monocytic lineage. For example, in addition to "monocytic" markers, LGL also express determinants or receptors in common with the T or B lineages. The experiments of Lohman-Matthes and colleagues, although of interest, are open to a variety of interpretations, ranging from the possibility that the NK activity that they measure is not exerted by LGL, to the hypothesis that their suspensions contained both LGL and monocyte precursors. In the absence of information on the expression of peroxidase activity by these putative monocyte precursors, all of the above uncertainties cannot be solved. These comments reiterate the importance of the peroxidase activity for the distinction between LGL and the cells of the myelo-monocytic lineage (Ferrarini et al., 1980). This point is relevant because endogenous peroxidase is particularly abundant in early cells of the monocytic lineage and precedes the appearance of the phagocytic function (Cohn et al., 1966; Nichols et al., 1971).

COULD HUMAN LGL BELONG TO THE T-CELL LINEAGE?

This problem can be approached in two different ways, namely by discussing whether or not LGL constitute a mature T-cell subset, or by examining the possibility that LGL are immature (pre-thymic) T cells. The first hypothesis would be supported by the following findings.

- LGL share surface markers with mature T cells i.e. receptors for sheep red cells and some antigenic specificities detected by monoclonal reagents (Abo and Balch, 1981; Ortaldo et al., 1981).
- Cell suspensions highly enriched for LGL respond, although in a somewhat variable manner, to T-cell mitogens (Moretta et al., 1976; Lum et al., 1980).
- Cell suspensions highly enriched for LGL have been maintained and expanded in media containing IL2 (Timonen et al., 1981). This result has been interpreted as evidence for the capacity of LGL to respond to T-cell growth factors.

All of the above three points are open to alternative interpretations. The presence of common markers in two different cells does not necessarily indicate their common origin. For example, disparate cell types can express the same membrane structures (Metzgar et al., 1981). In the experiments with the mitogens the responsive cells could have been a minor fraction of contaminant T cells since those studies were carried out on purified T_G cell fractions which, although enriched in LGL (up to 80%), contain also as yet unidentified cell types (Ferrarini et al., 1980). Support to the above possibility comes from the observation that expanded LGL populations from patients with lymphoproliferative disorders, have consistently failed to respond to T-cell mitogens (Callard et al., 1981; Ferrarini et al., 1981). As for the experiments with IL2, although of great interest, there is so far no sufficient evidence that the responding cells met all of the requirements to be classified as LGL. An additional note of caution could be perhaps provided by the finding that in one patient with a lymphoproliferative disorder of the LGL, we have been unable to expand and maintain the LGL in media containing IL2.

The possibility that LGL could represent early (pre-thymic) T-cell precursors was raised on the ground of studies in the mouse showing that NK cells express cartain alloantigens of the immature T cells (as reviewed in Herberman, 1980). Since murine NK cells seem more heterogeneous than their human counterpart (Lust et al., 1981; Minao et al., 1981) it is difficult to extrapolate from the mouse to the human system. Furthermore, the surface marker analysis of circulating human LGL would not support the above hypothesis since, with the exception of the OKT10 marker, LGL do not express any of the early T-cell antigens so far identified (Abo and Balch, 1981; Ortaldo et al., 1981). Immature T-cell markers are also absent from expanded LGL populations of lymphoproliferative disorder patients, where LGL display other phenotypic characteristics of immaturity (i.e. ultrastructure and enzyme espression) (Ferrarini et al., 1981). Additional negative evidence that LGL could be pre-thymic cells comes from the observation that they are consistently absent from the thymus where a further maturational step of LGL could be expected (Moretta et al., 1979).

Observations on immunodeficiency patients have shown that NK cells do not belong to the B-lineage since X-linked agammaglobulinemia patients have normal NK activity and normal LGL percentages (Koren et al., 1978; Lipinsky et al., 1980; Sirianni et al., 1981; Nocera et al., 1982). The finding of a reduced NK activity on a limited number of severe combined immunodeficiency (SCID) patients has been interpreted as an indication that NK cells are pre-thymic cells. However an alternative explanation might be that SCID patients could have an impaired immune interferon production which is known to enhance NK activity (see Herberman, 1980). Patients with documented defective interferon production display a decreased NK activity (Lipinsky et al., 1980). Furthermore, some of the immunodeficiency patients may have circulating immune complexes. Recently a good correlation has been found between high levels of circulating immune-complexes and low NK activity in patients with systemic lupus erythematous (Hoffman, 1980) and IgA deficient patients (Lipinsky et.

al., 1980). These findings have suggested that contact with immune complexes may alter LGL distribution (with consequent decrease of the proportion of circulating cells) or may inhibit their cytotoxic activity.

EVIDENCE THAT LGL COULD BELONG TO A SEPARATE CELL LINEAGE

Previous observations from our laboratories pointed out that LGL from normal peripheral blood are rich in mature electron-dense granules and lack ultrastructural features indicating active granulogenesis (Grossi et al., 1982). Active granulogenesis is also absent from other hemic granular cells, such as circulating granulocytes and monocytes (Cohn and Benson, 1965; Bainton and Farquhar, 1966). These findings have been interpreted as an indication that circulating cells represent a final stage of maturation at which granulogenesis has already ceased. Accordingly, granulogenesis is considered as a marker of immaturity.

The availability of expanded LGL populations with immature features from lymphoproliferative disorders has permitted both the identification of the steps leading to the granule formation and of other phenotypic features attributable to immature LGL (Ferrarini et al., 1981). The same characteristics of immaturity have been observed in LGL from normal bone marrow (Grossi et al., 1981). These studies have indicated that i) the bone marrow may be the source of LGL and ii) both the granules and the process leading to their formation in LGL differ from those occurring in the other hemic series. The latter finding is supporting evidence that LGL are distinguishable from other myeloid cells already at their early stages of maturation.

Comparative studies on the surface phenotype of mature and immature LGL using monoclonal antibodies have shown that, whereas the former express preferentially antigenic determinants of the monocytic lineage, the latter are characterized by the presence of markers of the T-suppressor/cytotoxic subset. Thus, the heterogeneity observed for the surface phenotype of LGL, could be explained by changes occurring within this lineage during maturation (Ferrarini et al., 1981). Indication that LGL represent a distinct lineage may come also from observations on lymphoproliferative disorders and immunodeficiencies. Although patients with expanded LGL populations have been described, no direct evidence for their monoclonal origin has been provided. However, if some of these patients had indeed a malignancy, they could not be classified as typical granulocytic, monocytic or lymphocytic leukemias (see Editorial by Cooper, 1980). Studies on immunodeficiencies have been carried out on a relatively small number of patients so that they cannot provide definitive information on alterations of both LGL numbers and functions. However, a marked increase of circulating LGL has been reported in a number of severe combined immunodeficiency patients perhaps suggesting that these cells might have a compensatory function (Ault et al., 1981; and our unpublished data).

Phylogeny studies would support the hypothesis that LGL are a separate lineage. Cells with morphological features identical to those of mature LGL (lymphocytic coelomocytes) have been described in the earthworm, Lumbricus terrestris, which lacks a specific immune system, but possesses phagocytic cells (Linthicum et. al., 1977 a.). Lymphocytic coelomocytes play a predominant role in the process of allograft rejection (Linthicum et al., 1977 b). Therefore, LGL may be part of a primitive "recognition" system which has diverged from the granulocytic-monocytic lineage during phylogeny before the appearance of the immune system. The specific function of these cells may be that of a control or surveillance of cell growth. The recent reports that NK cells, beside their cytotoxic activity may inhibit be growth of other hemic lineages or even affect bone marrow engrafting (Broxmeier et al., 1979; Cudkowicz and Hochman, 1979; Bacigalupo et al., 1981), lend further support to a regulatory function of LGL. This function would not simply consist in the elimination of "aberrant" cells, but might also reside in a promoting activity on cell growth and maturation.

REFERENCES

Abo, T., and Balch, C. M. (1981). J. Immunol. 127:1024.

Ault, K. A., Smith, B. R., and Holmberg, L. A. (1980). New Engl. J. Med. 303:881.

Bacigalupo, A., Podestà, M., Mingari, M. C., Moretta, L., Piaggio, G., van Lint, M. T., Durando, A., and Marmont, A. M. (1981). Blood 57:491.

Broxmeyer, H. E., Ralph, P., Margolis, V. B., Nakoinz, I., Meyers, P., Kapoor, N., and Moore, M. A. S. (1979). Leuk. Res. 3:193.

Callard, R. E., Smith, C. M., Worman, C., Linch, D., Cawley, J. C., and Beverley, P. C. (1981). Clin. exp. Immunol. 43:497.

Cohn, Z. A., Fedorko, M. E., and Hirsh, J. G. (1966). J. exp. Med. 123:757.
Cooper, M. D. (1980). New Engl. J. Med. 302:964.
Cudkowicz, C., and Hochman, P. S. (1979). Immunol. Rev. 44:13.
Ferrarini, M., Cadoni, A, Nocera, A., Di Primio, R., Zicca, A., Leprini, A., Franzi, A. T., and Grossi, C. E. (1980). Eur. J. Immunol. 10:562.
Ferrarini, M., Romagnani, S., Montesoro, E., Zicca, A., Del Prete, G. F., Nocera, A., Maggi, E., Leprini, A., and Grossi, C. E. (1981) submitted.
Greaves, M. F., Verbi, W., Festentein, H., Papasteriadis, C., Jaraquemada, D., Hayward, A. (1979). Eur. J. Immunol. 10: 562.
Grossi, C. E., Burgio, V. L., and Ferrarini, M. (1981). Haematologica 66 (suppl.):159.
Grossi, C. E., Cadoni, A., Zicca, A., Leprini, A., and Ferrarini, M. (1982). Blood in press.
Herberman, R. B. (ed.) (1980). Natural cell-mediated Immunity against Tumors. Academic Press, New York.
Herberman, R. B. (1980). In "Natural Cell-mediated Immunity gainst Tumors". Herberman R. B. ed., Academic Press, New York, 277.
Hoffman, T. (1980). Arthritis and Rheumatism 23:80.
Koren, H. S., Amos, D. B., and Buckley, R. H. (1978). J. Immunol. 120:796.
Linthicum, D. S., Stein, E. A., Marks, D. H., and Cooper, E. L. (1977). Cell. Tiss. Res. 185:315.
Linthicum, D. S., Marks, D. H., Stein, E. A., and Cooper, E. L. (1977). Eur. J. Immunol. 7:871.
Lipinski, M., Virelizier, J. L., Tursz, T., and Griscelli, C. (1980). Eur. J. Immunol. 10:246.
Lohmann-Matthes, M. L., and Domzig, W. (1980). In "Natural Cell-Mediated Immunity against Tumors". In Herberman . R. B. ed., Academic Press, New York., 117.
Lum, L. G., Benveniste, E., and Blaese, R. M. (1980). J. Immunol. 124:702.
Lust, J. A., Kumas, V., Burton, R. C., Barlett, S. P., and Bennett, M. (1981). J. exp. Med. 154:306.
Mantovani, A., Peri, G., Polentarutti, N., Allavena, P., Bordignon, C., Sessa, A., and Mangioni, C. (1980). In "Natural Cell-mediated Immunity against Tumors". Herberman, R. B. ed,

Academic Press, New York, 1271.
Metzgar, R. S., Borowitz, M. J., Jones, N. H., and Dowell, B. L. (1981). J. exp. Med. 154:1249.
Minato, N., Reid, L., and Bloom, B. R. (1981). J. exp. Med. 154:750.
Moretta, L., Ferrarini, M., Mingari, M. C., Moretta, A., and Webb, R. S. (1976). J. Immunol. 117:2171.
Moretta, L., Ferrarini, M., and Cooper, M. D. (1978). Contemp. Top. Immunobiol., 8:19.
Nichols, B. A., Bainton, D. F., and Farquhar, M. G. (1971). J. Cell Biol. 50:498.
Nocera, A., Cadoni, A., Zicca, A., Di Primio, R., Leprini, A., and Ferrarini, M. (1982). Scand. J. Immunol. in press
Ortaldo, J. R., Sharrow, S. O., Timonen, T., and Herberman, R. B. (1981). J. Immunol. in press
Rimm, I. J., Schlossman, S. F., and Reinherz, E. L. (1981). Clin. Immunol. Immunopath. 21:134.
Sirianni, M. C., Fiorilli, M., Pandolfi, F., Quinti, I., and Aiuti, F. (1981). Clin. Immunol. Immunopath. 21:12.
Timonen, T., Ortaldo, J. R., Bonnard, G. D., and Herberman, R. B. (1981). Immunobiology 159:206.
Timonen, T., Ortaldo, J. R., and Herberman, R. B. (1981). J. exp. Med. 153:569.
Trinchieri, G., and Santoli, D. (1978). J. exp. Med. 147:1314.
Winchester, R. J., Hoffman, T., Ferrarini, M., Ross, G., and Kunkel, H. G. (1979). Clin. exp. Immunol. 37, 126.
Yu, D. T. Y., Winchester, R. J., Fu, S. M., Gibosky, A., Ko, H. S., and Kunkel, H. G. (1980). J. exp. Med. 151:91.
Zucker-Franklin, S., Lavie, G., and Franklin, E. C. (1981). J. Histochem. Cytochem. 29 (3A): 451.

NATURAL KILLER CELLS: A SEPARATE LINEAGE?

John R. Ortaldo

Biological Research and Therapy Branch
Biological Response Modifiers Program
National Cancer Institute - FCRF
Frederick, Maryland

I. INTRODUCTION

Over the past few years, extensive studies by many investigators have attempted to characterize natural killer (NK) cells. NK cells were shown to be Fc receptor-positive, nonadherent, nonphagocytic, cytotoxic effector cells. Until recently, it was not possible to find characteristics particularly associated with NK cells. However, the ability to fractionate human NK cells on Percoll gradients by the technique of Timonen *et al* (1980, 1981) demonstrated that cells with a characteristic morphology, i.e., large granular lymphocytes (LGL), have a high degree of association with NK cells. LGL account for all the human NK activity and studies of these cells provide an opportunity for direct characterization of NK cells. LGL have a high cytoplasmic:nuclear ratio and azurophilic granules in the cytoplasm, form rosettes with sheep erythrocytes, are positive for fluoride-resistant alpha-naphthyl-acetyl-esterase and tartrate-sensitive acid phosphatase, and are negative for lysozyme, peroxidase, and alpha-naphthyl-butyl-esterase.

II. STUDIES OF LGL WITH MONOCLONAL ANTIBODIES

It has been possible to study in detail the surface phenotype of LGL, using monoclonal antibodies and fluorescence flow cytometry (Ortaldo *et al*, 1981). The reactivity of LGL have been compared to that of T cells, monocytes, and

ISBN 0-12-341360-5

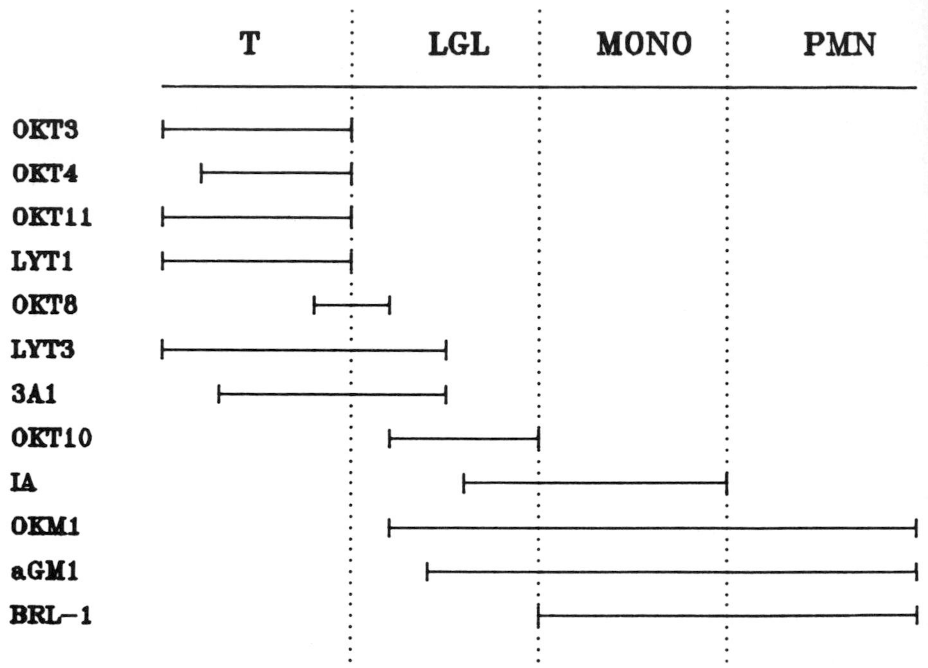

Figure 1. Diagrammatic representation of reactivity of monoclonal and other antibodies with various leukocyte subpopulations.

polymorphonuclear (PMN) leukocytes. Figure 1 summarizes the results with a variety of monoclonal antibodies against T cells or monocytes, or subpopulations thereof. Several monoclonal antibodies, i.e. OKT3, OKT4, OKT11, LyT1, LyT2, and T101, recognized only T cells. However, other monoclonal antibodies, i.e. OKT8, LyT3 (against the receptor for sheep), 3A1, and OKT10, recognized NK cells as well as T lineage cells. It should be noted that OKT10 is undetectable on peripheral T cells but present on early thymocytes and some bone marrow cells. Antigens such as Ia, OKM1, Mac-1 and asialo GM1 were found on both monocytes and NK cells, and some were also on PMNs. None of the monoclonal antibodies demonstrated conclusively whether NK cells are of either T cell or monocyte lineage.

In a recent study, mice were immunized repeatedly with highly purified populations of LGL isolated from peripheral blood mononuclear cells on Percoll gradients. As summarized in Figure 2, hybridomas prepared from the spleen of such mice produced antibodies that were evaluated by their reactivity against LGL, T cells and monocytes. Most prevalent were antibodies against leukocyte common antigens shared by all cell

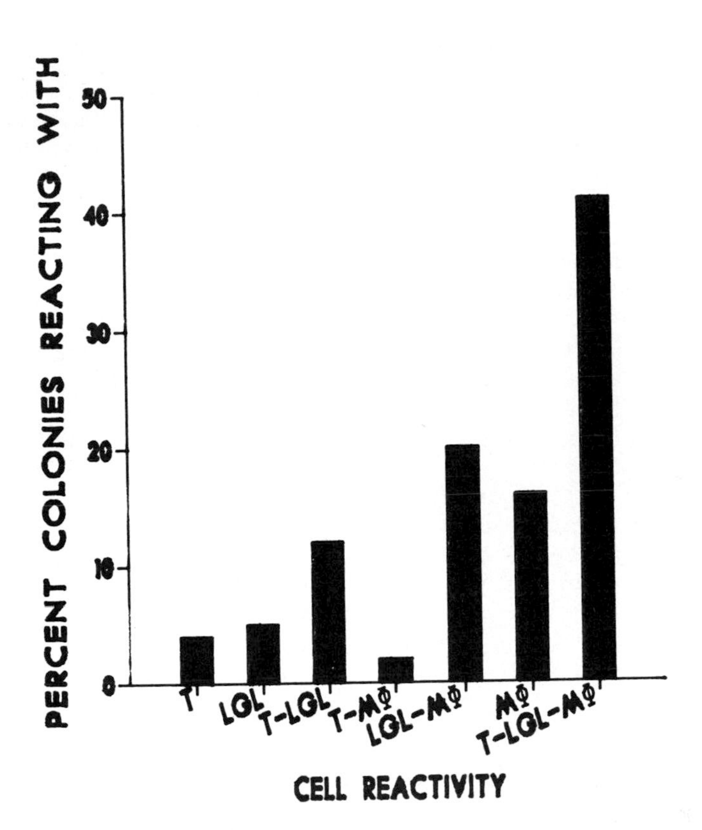

Figure 2. Summary of reactivities of monoclonal antibodies developed against purified LGL's. Results were obtained using radio-immunoassay, with the percent reactivity of all antibodyproducing populations shown.

populations. Only a very small percentage of hybridomas demonstrated significant reactivity against just LGL, with the majority reacting with LGL plus either T cells or monocytes. This is not unexpected, considering the reactivity patterns seen with the established monoclonal antibodies described above.

There have been a number of other monoclonal antibodies which have been used in attempts to elucidate the lineage of NK cells. Kraft et al (1981) and Trinchieri et al (this volume) recently reported a series of myelomonocytic antigens, which are on all monocytic precursor cells but were not found on human NK cells.

III. DISCUSSION

The lineage of NK cells is a quite controversial issue and several alternatives need to be considered: 1) NK cells may be in the T cell lineage; 2) NK cells may be in the monocyte lineage; or 3) NK cells may be in a separate lineage, with a precursor stem cell common to other PBL, i.e. T cells and monocytes.

The argument that NK cells are associated with the T cell lineage has been based mainly on a large body of evidence indicating the sharing of characteristics by both rodent and human NK cells with mature T cells. Human NK cells have been shown to form E-rosettes and express T cell-associated antigens. Several groups have recently reported the ability of both mouse and human NK cells to expand in response to supernatants containing interleukin 2 (T cell growth factor) (Ortaldo and Timonen, 1981). The apparent growth of NK cells in response to supernatants containing IL-2, although suggestive of some relationship to the T cell lineage, can not be taken as definitive evidence. Even the partially purified IL-2 preparations that have been used in some experiments (Ortaldo and Timonen, 1981) might contain other growth factors in addition to IL-2. Furthermore, although IL-2 has been considered entirely selective for the growth of T cells, it might in fact support the growth of NK lineage cells as well as T cells. In addition, NK cells have been shown to proliferate in response to T cell lectins (PHA, Con A) and during their growth on IL-2, LGL express mature T antigens, as indicated by reactivity with OKT3 (Ortaldo and Timonen, 1981).

In the mouse NK cells have been shown to express Thy-1, and in nude or neonatally thymectomized mice, the increased NK cell activity might be attributed to an accumulation of prethymic T cells or increased development of T lineage cells along a thymic-independent side pathway. Our demonstration of T10 expression on human NK cells similarly might be taken as an indication that these cells are at an early stage in the T lineage, before development of the antigens associated with more typical peripheral blood T cells.

Regarding the monocyte lineage of NK cells, we have been unable to demonstrate either the maturation or growth of NK cells on colony stimulating factor. In contrast, Lohmann-Matthes _et al_ (1979) have reported that mouse pro-monocytes from cultures of bone marrow exhibited NK activity. However, they failed to demonstrate that the cells which developed into mature monocytes were indeed the same cells which demonstrated the NK activity. Evidence certainly indicates that NK cells are derived from stem cells in the bone marrow (Haller, 1978), and such cells may have differentiated in the cultures of

Lohmann-Matthes et al, in addition to promonocytes.

Finally, the possibility that LGL are separate cell lineage derived from a common bone marrow precursor is consistent with the previous postulation of a pleuri-potent hemic stem cell (see Figure 3). Depending on the stimulation of such a stem cell, progeny may be induced or differentiate toward any of the major, mature hemotologic cell types. Particularly at the early stages of such differentiation, the various cell types might be expected to share a variety of antigenic and other characteristics. The use of monoclonal antibodies and other characteristics such as the various soluble factors have been used by other investigators in an attempt to place NK cells in either the T lineage or the monocyte lineage. However, the subpopulation reagents which react with the T or T lineage cells, and also NK cells, do not in fact react with cells of the myelomonocytic cell lineage. Conversely, reagents (such as OKM1, Mac 1, and asialo GM1) reacting with monocytes and NK cells, do not react with T lineage cells. The only possible exception to this is the presence of framework Ia antigens on activated T cells, B cells and presumably represent an activation antigen which may be present on any leukocyte. In fact, the only cell population which tends to have a high degree of overlap with T cells and monocytes is the NK cell. However, present evidence (Ortaldo et al, 1981; Dennert et al, 1981) indicates that long term cultures of NK cells do not develop either monocyte nor T cell activity, and indeed these cultured cells maintain their characteristic morphology and function, which would not be expected if one were dealing with a precursor cell. Because of the high degree of phenotypic overlap with T cells and monocytes, and because of the atypical expression of other markers (rosette-forming ability with SRBC, Thy 1 in rodent, etc., it is my contention that NK cells are in a third lineage. Such a separate lineage for LGL would explain the many shared functions and surface markers with T cells and monocytes. A common cell origin between monocyte and the PMN exists; however this cell differentiates into distinct mature functional and morphological cells in the peripheral blood. The NK cell could share a very early common stem cell origin with T cells, cells, PMNs and monocytes, and therefore share a number of characteristics. At present, conclusive evidence to place NK cells into a third lineage does not exist, and this remains only a speculation. However, because of the simultaneous presence of antigens such as M1 and T10 on NK cells, the possibility that NK cells are a third lineage should be considered as a real alternative to placement within either the T cell or monocyte lineages.

Figure 3 is a modification of a scheme of stem cell differentiation as postulated by McCullogh and Till (1971). Further investigations are needed to elucidate whether NK cells under the proper stimuli or proper culture conditions might mature into other cell types. A possible way to examine this question is to attempt to stimulate the maturation or differentiation of LGL. It would be of much interest if agents such as sodium butyrate or DMSO induced the differentiation of purified LGL toward either T or monocyte lineage cells. An alternative possibility would be the in vivo maturation of natural killer cells. At present, the ability in the rat system to isolate, label with tracers and re-inject LGL into syngeneic animals, in order to follow the migration, differentiation, and possible morphological changes of these cells, offers an excellent possibility to examine the question of whether LGL mature further in vivo.

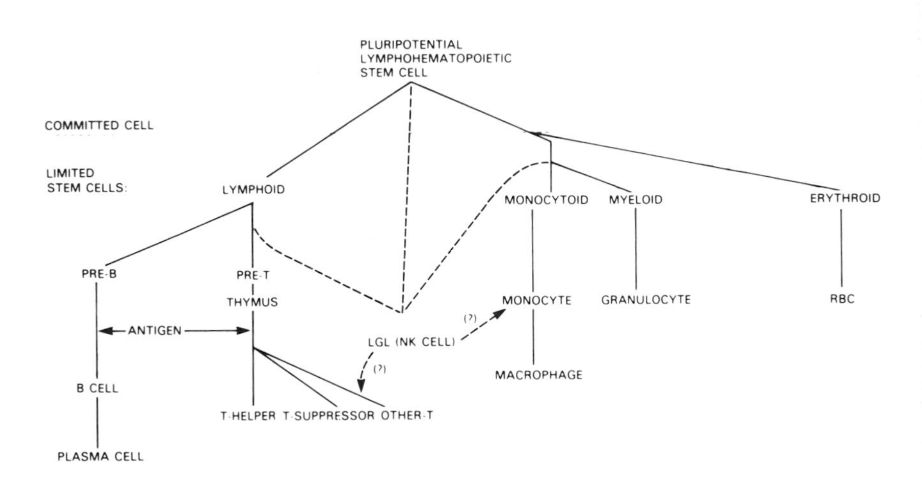

Figure 3. Possible lineages of human NK cells.

REFERENCES

Dennert, G., Yogeeswaren, G., and Yamagata, S. (1981) J. Exp. Med. 151:545.

Haller, O., Kiessling, R., Orn, A., and Wigzell, H. (1978) J. Exp. Med. 145:1411.

Herberman, R.B. (1980) "Natural Cell-Mediated Immunity Against Tumors." Academic Press, New York.

Kraft, D., Rumpold, H., Steiner, R., Radeklewicz, T., Swethy, R., and Wiederman, G. (1981) In "Mechanism of Lymphocyte Activation (K. Resch and H. Kirchner, eds.), p. 279, Elsevier/North Holland, New York.

Lohmann-Matthes, M.L., Domzig, W., and Roder, J. (1979) J. Immunol. 123:1993.

McCulloch, E.A. and Till, J.E. (1971) Am. J. Pathol. 65:601.

Ortaldo, J.R. and Timonen, T.T. (1981) In "Mechanisms of Lymphocyte Activation (K. Resch and H. Kirchner, eds.), p. 286, Elsevier/North Holland, New York.

Ortaldo, J.R., Sharrow, S.O., Timonen, T., and Herberman, R.B. (1981) J. Immunol. 127:2401.

Timonen, T. and Saksela, E. (1980) J. Immunol. 36:285.

Timonen, T., Ortaldo, J.R., and Herberman, R.B. (1981) J. Exp. Med. 153:569.

NATURAL KILLER ACTIVITY IN MAST CELL-DEFICIENT W/W^v MICE

Aldo Tagliabue
Sclavo Research Center
Siena, Italy

Dean Befus
John Bienenstock

McMaster University
Hamilton, Ontario, Canada

The cell lineage of NK cells is still a matter of debate. The recent finding that large granular lymphocytes (LGL) are NK cells (1,2) further complicates this issue. In fact, cells morphologically similar to LGL have been found in the epithelium of the mouse small intestine and it has been proposed that they are mast cell precursors with T-cell characteristics (3). Since it has been shown on the one hand that NK activity can be exerted by the intraepithelial lymphocytes from the small intestine (4) and that, on the other hand, mature peritoneal mast cells can be naturally cytotoxic for certain tumor cells (5), we decided to investigate the possible relationship between NK cells and mast cells. For this purpose we used W/W^v mice, which have a genetically determined macrocytic anaemia associated with a defect in the pluripotential haematopoietic stem cell that also causes a 99% reduction in mast cells (6). We tested the NK activity of splenocytes from 8-week old W/W^v mice and from their littermates recognizable by their coat colour (6). As shown in the figure, W/W^v mice displayed a low but detectable activity against YAC-1 tumor cells in a 6h ^{51}Cr release assay. This NK activity was not different from that observed in the littermates of W/W^v mice, although lower than the cytotoxicity observed in a high NK strain such as the CBA/J mice. Moreover, LGL were observed in cell suspensions obtained from spleen and intestinal epithelium of W/W^v mice. Thus, these

NK CELLS AND OTHER NATURAL EFFECTOR CELLS

ISBN 0-12-341360-5

results indicate that it is very unlikely that NK cells are mast cell precursors.

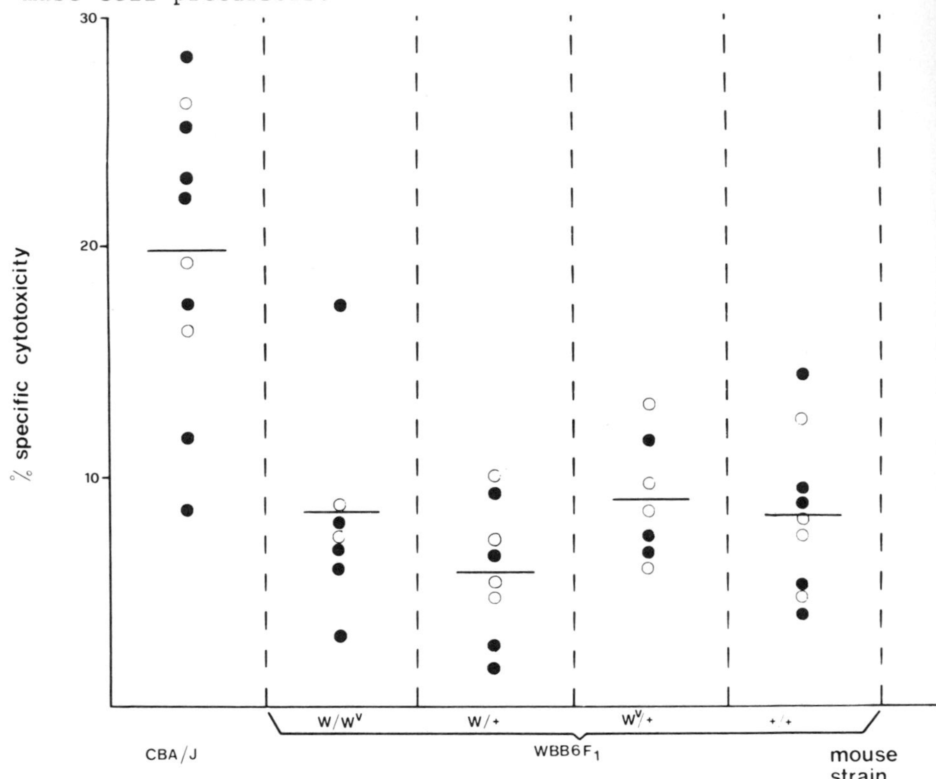

NK activity against YAC-1 tumor cells of spleen from CBA/J mice and from WBB6F$_1$/J littermates. Each circle represents one mouse. Results are from two experiments (black and white circles). Attacker to target ratio 50:1.

REFERENCES

1. Timonen, T., Ortaldo, J.R., and Herberman, R.B., J. Exp. Med. 153:569 (1981).
2. Luini, W., Boraschi, D., Alberti, S., Aleotti, A., and Tagliabue, A., Immunol. 43:663 (1981).
3. Guy-Grand, D., Griscelli, C., and Vassalli, P., J. Exp. Med. 148:1661 (1978).
4. Tagliabue, A., Luini, W., Soldateschi, D., and Boraschi, D., Eur. J. Immunol. 11:919 (1981).
5. Farram, E., and Nelson, D.S., Cell. Immunol. 55:294 (1980).
6. Kitamura, Y., Go, S., and Hatanaka, K., Blood 52:447 (1978).

NK-ACTIVITY AGAINST YAC-1 IS REGULATED BY TWO H-2 ASSOCIATED GENES

Gunnar O. Klein[1]

Dept. of Tumor Biology
Karolinska Institute
Stockholm, Sweden

I. INTRODUCTION

It has recently become clear that the NK-pehonemenon is heterogenous (1,2). The present study is concerned exclusively with one type of NK-activity that could perhaps be called classical with the YAC-1 lymphoma as the prototype target cell. This NK-activity of normal spleen cells against YAC-1 is henceforth referred to as "NK" alone, in the conviction that this is representative of a number of other lymphomas as well. NK-activity against certain other tumors will however probably be under a different genetic control.

The YAC tumor is of strain A/Sn origin but we have found that the same H-2 associated genes that will be discussed in this chapter operate against a number of C57BL/6 derived lymphomas as well (data not shown).

In the first studies of genetic influence on NK-activity it was found that high activity was dominant and that $H\text{-}2^{b/a}$ heterozygotes were significantly higher in NK than $H\text{-}2^{a/a}$ homozygotes in the (B6[2] x A) x A and (C57L x A) x A backcrosses (3). It was concluded that the gene associated with $H\text{-}2^b$ was located on chromosome 17 but outside H-2 as the A congenic $H\text{-}2^b$ carrying A.BY strain was as low as strain A.

[1]This work was supported by Grant No. 3 R01 CA 14054-0651, No. 5 R01 CA 25250-02 and No. 1 R01 CA 26782-02, NIH and the Swedish Cancer Society
[2]B6=C57BL/6

ISBN 0-12-341360-5

Harmon and Clark found that B10.A, B10.A(18R) and B10.D2 were higher in NK-activity than B10 against EL-4 (4). We have confirmed the importance of the D-end of the H-2 complex against YAC-1 in a larger series of B10 congenic strains (5).

It was however an apparent contradiction that $H-2^a$ when introduced to the relatively NK-high C57BL background gave high activity whereas it in the backcross studies described above, was associated with low activity. B10 and B6 have the same NK-activity (unpublished observation). This paradox could however be explained by the hypothesis that non-H-2 linked background genes modify the influence of the H-2 associated NK-activity genes. This would also explain the finding with the A.BY strain and is further supported by the fact that $AKR.H-2^b$ was on exactly the same low NK-level as AKR; far below that of B6 (5).

In an attempt to localize the H-2 D^d associated gene further, we have studied a set of strains on B6 background with crossovers between H-2 D and the Tla locus. We then found that the $B6.Tla^a$ strain was not higher in NK-activity than the B6 controls but in fact significantly lower. This has led to the conclusion that there are at least two genes on chromosome 17 which controls NK-activity in mice of C57BL background.

II. MATERIALS AND METHODS

Mice: The $B6.Tla^a$, B6.K1, B6.K2, $B6.H-2^k$, B6/By and $A.Tla^b$ strains were kindly provided by Dr E.A. Boyse, Memorial Sloan-Kettering Cancer Center. These strains and A/Sn were bred at our conventional animal facilities and used when 5-8 weeks of age. In each test however, the mice were agematched $\pm$ 3 days.

Cytotoxicity assay: Spleens of normal mice were tested individually in a 51-Cr release assay against the tissue culture line YAC-1 as described previously (5). To reduce the influence of the day to day variation of target cell sensitivity, all data were transformed with activity of B6 mice as a reference according to the following formula:

$$\% \text{ Transformed lysis} = \frac{\text{Activity in test}}{\text{Activity of B6}} \times 100$$

TL-typing: Segregating mice from the cross ($B6 \times B6.Tla^a$) x B6.K2 were typed for TL-type with an anti-TL antiserum obtained from Dr E.A. Boyse, ($B6 \times A.Tla^b$) immunized with the strain A leukemia ASL1. Thymocytes were tested in a complement dependent cytotoxicity assay with GPC 1:15 and anti-TL 1:400 as described previously (6).

III. RESULTS AND DISCUSSION

Figure 1 shows the origin of the relevant parts of chromosome 17 of the recombinant strains used in this study. B6.Tla^a is a recombinant between B6 and A with the whole T-region from the A strain. The strain B6.H-2^k has all of H-2 and the T-reg-region from AKR and the two lines B6.K1 and B6.K2 are recombinants between B6.Tla^a and B6.H-2^k with crossovers in different locations between H-2 D and Tla.

Assays of natural killer cell activity were performed with normal spleen cells of individual mice. B6 mice were included in each test to provide a reference as the sensitivity of the assay varied considerably. All values of percent lysis were therefore transformed according to the formula in Materials and Methods beofre pooling of data from different experiments. The actual levels of percent lysis were for B6 between 8.2 % and 52.8 %.

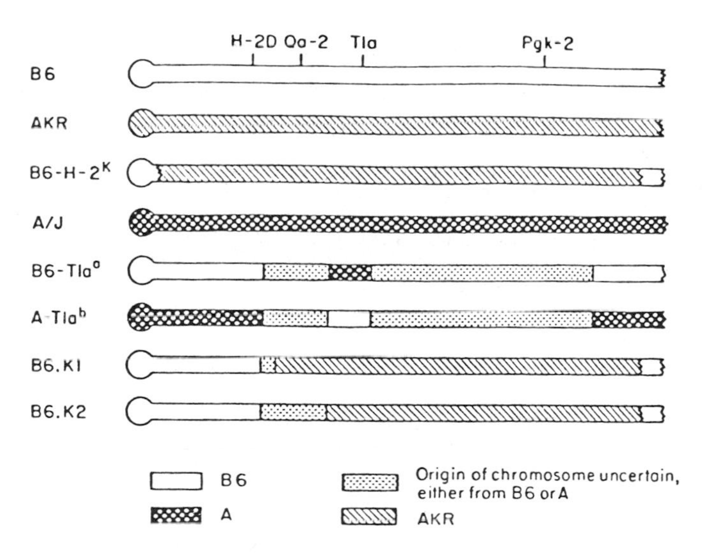

FIGURE 1. The origin of H-2 and the adjacent T-region. Qa-1 which is probably identical to the previously described H-2 T (Jan Klein personal communication) cannot be distinguished from the Tla locus. Modified from (7).

There are clearly two levels of NK-activity indicated with B6 and B6.H-2^k on the same relatively high level and B6.Tla^a together with B6.K1 and B6.K2 on a lower level with about 3/4 of the lytic capacity of B6 (figure 2). The fact that B6.Tla^a was significantly lower than B6 while B10.A was higher than B10, indicates that there is another gene influencing NK-activity (apart from the H-2 D^d associated one previously described). This gene is somewhere in the T-region with B6.H-2^k and B6 having the high allele and with B6.Tla^a having a low activity allele.

B6.K1 has all previously known loci in the T-region from B6.H-2^k (Qa-2, Qa-3, Qa-1, Tla etc.). In spite of this it seems to have received the NK-activity gene from B6.Tla^a. This argues for a localization of this gene, which we tentatively designated NKT, between H-2 D and Qa-2.

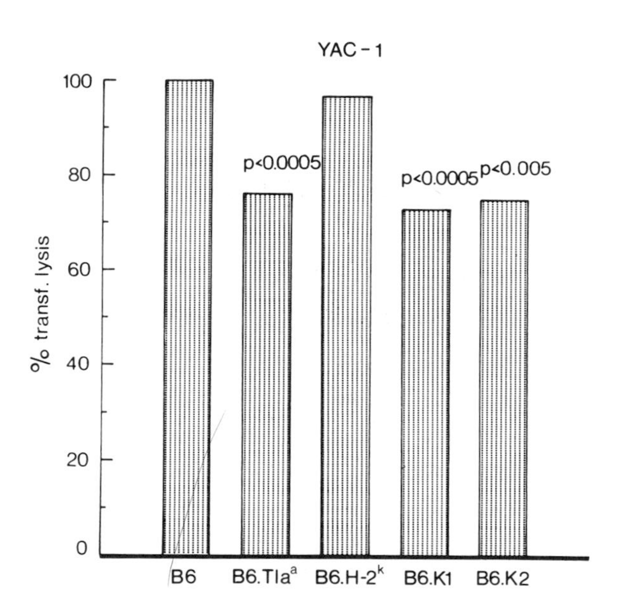

FIGURE 2. Mean tranformed % lysis against YAC-1. Pooled data from 8 experiments with effector to target cell ratio 100:1. The significance of the difference compared to B6 is indicated above each bar (Students T-test).

The effect of this gene was further shown in segregating mice from the cross (B6 x B6.Tla^{a}) x B6.K2 where 50 % of the mice were TL^{+} (originating from B6.Tla^{a}, the only TL positive of these strains). The Tla^{b} allele carried by both B6.K2 and B6 does not lead to expression of any TL antigens on normal thymocytes. These mice were tested for NK-activity of their spleens and TL-type of the thymocytes. The results are shown in figure 3. The TL^{-} (Tla^{b}/Tla^{b}) mice were significantly higher in NK-activity than the TL^{+} (Tla^{a}/Tla^{b}) heterozygotes.

We also tested the reciprocal recombinant, A.Tla^{b}. The origin of Qa-2 in this strain is uncertain as both B6 and A have Qa-2^{a}. As shown in Table I, A.Tla^{b} was exactly on the same NK-level as the A/Sn control. This could be due to the strain A background which seems to give low NK-activity irrespective of H-2 linked genes (c.f. A.BY which has the same activity as A/Sn). We therefore crossed A.Tla^{b} to B6.Tla^{a} to get the high B6 background without introducing the H-2^{b} associated NK-gene. This F$_{1}$-hybrid was however not different from the F$_{1}$-hybrid A/Sn x B6.Tla^{a} (Table I).

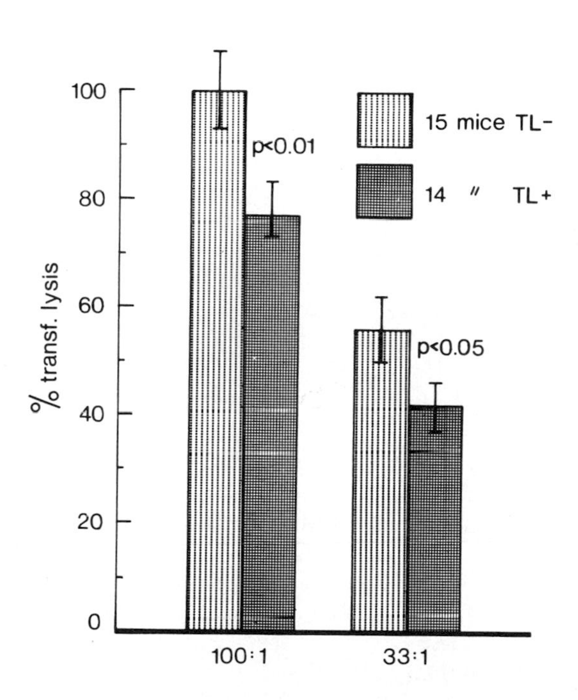

Figure 3. Mean percent transformed lysis of 29 segregating mice from the cross (B6 x B6.Tla^{a}) x B6.K2. The statistical significance of the difference between TL^{+} and TL^{-} mice is indicated in the figure (Students T-test).

TABLE I. NK-activty of A.Tla^b

Strain	Number of mice	Mean lysis
A/Sn	11	10.2 %
A.Tla^b	11	10.1 %
A/Sn x B6.Tla^a	17	10.1 %
A.Tla^b x B6.Tla^a	20	10.7 %

The most likely explanation for this is that the crossing-over in A.Tla^b has occured distally to the crossover in B6.Tla^a and that the NK-activity gene (NKT) therefore is not from B6 in A.Tla^b. This interpretation is consistent with the data with B6.K1 which also indicated that the NKT gene is between H-2 D and the other known loci in the T-region (Qa-2 etc.).

IV. CONCLUSIONS

We have thus shown that NK-activity against YAC-1 is influenced by two genes on chromosome 17. One of these genes which we tentatively designated NKD may at present not be separated from H-2 D/L by recombinants. It is in fact possible that this gene is identical to the gene coding for the H-2 D or L molecule. This is suggested by our preliminary experiments with two H-2 D^d mutants. These two (H-2^{dm1} and H-2^{dm2}) had both lower NK-activity than the corrseponding wild types.

The other gene regulating NK-activity is telomeric of H-2 D but centromeric of Qa-2 and thus defines a new locus - NKT.

REFERENCES

1. Lust, J. A., Kumar, V., Burton, R. C., Bartlett, S., P., and Bennett, M., J. Exp. Med. 154:306 (1981).
2. Minato, N., Reid, L., and Bloom, B. R., J. Exp. Med. 154:750 (1981).
3. Petranyi, G., Kiessling, R., Povey, S., Klein, G., Herzenberg, L., and Wigzell, H., Immunogenetics 3:15 (1976).
4. Harmon, R. C., Clark, E., O'Toole, C., and Wicker, L., Immunogenetics 4:601 (1977).
5. Klein, G. O., Kärre, K., Kiessling, R., and Klein, G., J. Immunogenetics 7:401 (1980).
6. Stanton, T.H., and Boyse, E.A. Immunogenetics 3:525 (1976).
7. Hämmerling, G. J., Hämmerling, U., and Flaherty, L., J. Exp. Med. 150:108 (1979).

GENETIC CONTROL OF NC ACTIVITY IN THE MOUSE: THREE GENES LOCATED IN CHROMOSOME 17 [1]

Osias Stutman
Michael J. Cuttito

Cellular Immunology Section
Memorial Sloan-Kettering Cancer Center
New York, New York

I. INTRODUCTION

Our earlier studies showed that the mouse strain distribution of high or low reactivity for NC (natural cytotoxic) cells was different from that of NK (natural killer) cells (1-4). Some mouse strains had normal high NC activity and low NK activity, including mice carrying the beige mutation (1-3). Although some variation of high-low reactivity within a strain has been observed with different targets for NC (1-4) and NK (5-7), the screening of different mouse strains, especially congenic pairs, against a single prototype target has proven informative, both in NK (5-7) as well as NC studies (2). In this paper we will present some of our studies on the genetic control of NC reactivity of different mouse strains and congenic partners against Meth 113 targets, a fibrosarcoma of BALB/c origin (1). Our results strongly suggest that at least 3 genes, probably all located in chromosome 17, distally from the D-end of the H2 region, determine the levels of NC reactivity in spleen against the Meth 113 target. None of the putative "NC genes" appear to affect NK activity.

---

[1] Supported by National Institutes of Health grants CA-08748, CA-15988, CA-17818 and American Cancer Society grant IM-188.

ISBN 0-12-341360-5

II. METHODS

The mice used were from our colony (or obtained from other sources but bred in our colony, indicated as SKI in Table I), the colony of Dr. E.A. Boyse at our Institute (indicated as Boy in the Tables) and from the Jackson Laboratories (Bar Harbor, ME; indicated as J in the Tables). Four to 8 week old animals of both sexes were tested (see ref. 1 for details on housing and animal care). Spleen cells were used at 100, 50 and 10:1 effector:target ratios against Meth 113 target cells pre-labelled with ^{3}H-proline, and percent cytotoxicity measured as target cell loss 18 hrs later (for further details on cell preparation and cytotoxicity assay see refs. 1 & 4). The NC reactivity of males and females was identical in both low (i.e. C57BL/6) or high (A/J reactors, thus the results presented are from pooled data for both sexes. Based on the total of more than 650 individual determinations, we decided that at the 100:1 effector:target ratio, a cytotoxicity value of 35% or more could be considered "high" and a value of 15% or less as "low". NK activity (see Table III) was measured against ^{51}Cr-pre-labelled YAC.1 targets in 4 or 18 hr. assays. For details on the methods see Lattime et al., in this volume.

III. RESULTS AND DISCUSSION

Strain distribution. Table I shows the results of testing the NC activity of spleen cells from different mouse strains and congenic partners against Meth 113 targets (for the sake of brevity, only 100:1 ratios are shown). As we previously reported, A/J and other sublines are high NC reactors, while C57BL/6 are low reactors. The breeding and housing of the BALB/c, C57BL/6 or A/J mice in our animal facilities did not significantly change the reactivity patterns of each strain (animals marked as SKI in Table I). In addition, BALB/cJ were also low reactors, including BALB carrying the nude trait (1). On the otherhand, BALB/cByJ were high NC reactors (1,2). Since these two BALB sublines differ only at the Qa 2,3 region of chromosome 17 (8), which is a region located between the D-end of H2 and Tla (8), and since a strong influence of genes in the vicinity of the D-end of H2 has been described for NK reactivity (5,9-12), we decided to explore the possible role of genes in that region on NC regulation. Especially using the congenic mice developed by

Dr. E.A. Boyse in which genes from the H2 region derived from A and AKR strains (both high NC reactors) had been introduced into the C57BL/6 low NC reactor background.

In addition to the BALB/cJ-BALB/cByJ pair, Table I shows that while the C57BL/6 strain is a low NC reactor (see the Boy subline), the C57BL/6.Tla^a congenic (B6.Tla^a) which contains the Tla region derived from A strain mice and the B6.H2K and B6.K1 with the whole H2 or the Qa 2,3 regions derived from AKR, are all high NC reactors. Both A and AKR (Table I) are high NC reactors. It should be stressed that in all these congenics in the C57BL/6 background, the only difference from the original strain is on variable segments of chromosome 17, including the H2 region and adjacent regions distal to the D-end of H2 (13).

Three genes. Before describing the meaning of the three genes in the genotype presented in Table I (as well as the meaning of the brackets), it is apparent from glancing at the results that with two exceptions (marked with asterisks), all the strains that are high NC reactors have at least two "positive" alleles at any of the three genes proposed, while the strains with only one "positive" gene or no "positive" genes, are low NC reactors. The three main prototypes of the high NC reactors are: 1) Strains which are positive for the "TLA" and "Qa 2,3" genes: such as BALB/cByJ, NZB and C58; 2) Strains which are positive for the "Tla" and the "A-AKR" genes: such as A/J, and other sublines and some of the appropriate congenics (A.BY and A.CA), some recombinant strains from complex crosses which included A derived material (A.AL, A.TH, A.TL) and some congenics of A in the C57BL/10 background [B10.A(3R), B10.A(5R)] and 3) Strains which are positive for the "Qa 2,3" and "A-AKR" genes: such as the AKR strain and some C57BL/6 congenics derived from AKR material (AKR/Boy, B6.H2K, B6.K1). The two main exceptions are: 1) the A.Tla^b congenic which has its Tla region derived from C57BL/6 and, in spite of being "positive" only for the "A-AKR" gene, is a high NC reactor and 2) the B6.K2 which is a congenic strain of C57BL/6 with AKR and is also a high NC reactor being positive only for the "A-AKR" gene.

A few words about the three genes. Tla is a gene located approximately 1.5 map units from the D-end of H2 (8). Of its different alleles (8), the b allele is associated with low NC reactivity (the b allele is the negative allele for expression of the Tla surface antigens, see refs. 8 and 13). On the other hand, all of the other alleles that we tested (a, c, & d) are associated with high NC reactivity, provided that the "positive" allele at "Qa 2,3" (as is the case in BALB/c ByJ) or at the "A-AKR" genes (as is the case for A strain and sublines) are also expressed. The reason for all the brac-

TABLE I. NC Reactivity of Spleen Cells from Different Mouse Strains Against Meth 113 Targets [a]

Strain	Genotype "Tla"	"Qa,2,3"	"A-AKR"	No.	Mean Cytotoxicity 100:1 (±SEM)	NC activity [b]
BALB/cJ	+	-	-	19	10.3 ± 1.4	L
BALB/cJ SKI	+	-	-	22	6.3 ± 0.8	L
BALB/cJ nu/nu	+	-	-	7	11.1 ± 1.5	L
BALB/cByJ	+	+	-	10	35.2 ± 1.8	H
BALB/cByJ SKI	+	+	-	16	38.1 ± 2.4	H
C57BL/6J	-	-	-	54	15.2 ± 1.9	L
C57BL/6J SKI	-	-	-	32	13.8 ± 1.7	L
C57BL/6J A^{vy}	-	-	-	4	11.8 ± 2.8	L
C57BL/6 Boy (B6)	-	-	-	28	14.3 ± 1.1	L
B6.Tla^{a} Boy	+	-	+	38	39.1 ± 1.1	H
A/J	+	-	+	22	37.9 ± 1.7	H
A/J SKI	+	-	+	12	36.9 ± 1.9	H
A/Boy	+	-	+	11	36.9 ± 1.1	H
A.Tla^{b}Boy	-	-	+	6	36.2 ± 1.6	H*
A/WySnJ	+	-	+	5	39.0 ± 2.1	H
A.By	+	-	+	15	39.9 ± 1.0	H
A.CA	+	+	+	11	43.6 ± 2.2	H
A.SW	-	-	+	19	13.3 ± 3.1	L
A.AL	+	?	+	3	35.6 (39,35,33)	H
A.TH	+	-	+	10	35.0 ± 3.3	H
A.TL	+	-	+	10	42.8 ± 2.2	H
C57BL/10J	-	-	-	10	14.4 ± 2.1	L
B10.$D2^{n}$	+	-	-	7	11.1 ± 1.5	L
B10.A	+	-	-	7	9.6 ± 1.5	L
B10.A (2R)	-	-	-?	4	7.8 ± 1.9	L
B10.A (4R)	-	-	-?	7	10.4 ± 1.5	L
B10.A (3R)	+	-	+?	7	24.4 ± 1.5	I
B10.A (5R)	+	-	+?	5	33.6 ± 2.4	I
AKR/Boy	-	+	+	8	45.2 ± 2.6	H
B6.H2k	-	+	+	9	37.6 ± 1.9	H
B6.K1	-	+	+	5	35.4 ± 0.9	H
B6.K2	-	-	+	20	25.4 ± 2.7	I*
DBA/2	+	-	-	7	6.5 ± 1.9	L
NZB	+	+	-	8	34.6 ± 3.5	H
C58	+	+	-	2	36 (36,36)	H
C3H/HeJ	-	+	-	5	21.4 ± 4.4	I
C3H/BiUmc	-	+	-	8	18.8 ± 3.7	I

[a]Spleen cells from 4 to 8 week old mice of indicated strains tested individually in an 18 hr. cytotoxicity assay using ^{3}H-proline prelabelled Meth 113 target cells (see ref. 1 for

kets used so far is that it is most probable that neither Tla nor Qa 2,3 products _per se_ are engaged in NC regulation, and that closely linked genes may be the "real" NC regulating genes. Both Tla and Qa 2,3 are genes encoding for surface antigens of lymphoid cells, and in the case of Tla, antigens expressed exclusively on thymocytes (see refs. 8 and 13). With Qa 2,3, the association of high NC reactivity is reversed, and it is the b allele (i.e. the negative allele) which is associated with high NC reactivity (see 8 for further discussion of Qa 2,3; see ref. 13 for serological typing of Tla and Qa 2,3 in all the strains described in Table I). It is worth noting that Qa 2 (and perhaps also 3) are expressed on NK cells and not detectable on NC cells (14). The "A-AKR" gene is a postulated gene which we described (see 2, in that reference it was termed the "A background" gene) to interpret some of the NC data, especially the A.Tla^b high reactivity (or the B10.A low reactivity). Later studies showed that the AKR strain could also provide for such a background gene (see Table I).

The patterns of NC reactivity in some congenics such as B6.Tla^a and B6.H2k, as well as the high NC reactivity of the A. AL, A.TH and A.TL recombinant group (see Table I and ref. 15,16 for origin of such strains), strongly indicates that the "A-AKR" gene is also located on chromosome 17. In some cases like A.TL, there is a suggestion that it may be located between D-end and Tla (based on the A.SW or A.CA components as defined by typing of Qa 1,2,3,4,5; H2T; Qed 1 and Tla; see refs. 15 for A.TL, 8 for Qa 1 to 3, 17 for Qa 4 & 5 and 13 for the rest). On the other hand, some congenics like B10.A (Table I), although carrying the whole H2 region of A origin, including Tla, are low responders and thus, do not have the "A-AKR" gene; suggesting that it may be located distally of Tla. Conversely, the B10.A (3R) and (5R) (Table I) are good responders, and may have obtained the gene of A origin through recombination together with the Tla^a region (18). In the case of the C3H sublines which are positive only for the "Qa 2,3" gene and show intermediate reactivity, it may be possible that they carry the "A-AKR" gene (which if true may

---

Footnote for Table I continued:

further details). For description of the 3 genes under "Genotype" see text. Cytotoxicity of 35% or more is considered "high" and cytotoxicity of 15% or less, "low". The (*) indicate exceptions to the "2 positive genes out of three" theory, see text.

[b]L = Low; H = High; I = Intermediate.

be called the "k" gene since it may relate to such haplotype, A being a KD recombinant). In the next paragraphs, while discussing gene complementation, it will become apparent that appropriate typing for these different genes can be done by gene complementation in F_1 hybrids. In summary, the two exceptions, A.Tla^b and B6.K2 may be possible candidates for being carriers of the "A-AKR" gene as well as any of the other two (i.e. the "Tla" or "Qa 2,3") which has been actually dissociated from its closely linked real Tla or Qa 2,3 genes.

Gene complementation. Table II shows some F_1 data with hybrids from high and low reactors (first three lines), two examples of gene complementation in which the parents are both low reactors which give high or intermediate reactive progeny (lines 4 & 5) and low reactor crosses which give low reactor progeny, as expected from the "two out of three" theory presented in the previous section. Probably, the most informative cross is the (B10.A x A.SW)F_1 (line 4, Table II) which gives a high reactive progeny in spite of the parents being both low reactors (Table I) and also gives credence to the proposed "A-AKR" gene. Thus, the case of gene complementation between the "Tla" and "A-AKR" genes provided by the B10.A and A.SW parent respectively, appears quite strong from the results of this cross. The fact that the (B6 x B6.Tla^a)F_1 (lines 1 & 2) which is a cross between one parent being negative for all genes and one parent being positive for two genes, as well as the gene complementation of "Tla" and "Qa 2,3" in the (B10.A x C3H/Bi)F_1 cross (line 5), which also produced intermediate progeny, may suggest that in addition to the appropriate allelism at two of the three genes, cis and trans positioning may also be of importance. Lines 1 and 2 in Table II also show that there is no evidence of "maternal effect" of the high reactor in the F_1 breeding. Finally, the studies with (B6 x B6.Tla^a)F_2 hybrids, still in progress, show that with the first 67 animals tested, the proportion of high and low reactors favors a two gene model (i.e. 28 of 67 animals were low reactors, which gives a 58.2% of high reactors; a one gene model predicts 75% and a 2 gene model predicts 56.2% high reactors in the F_2). It is obvious that a more refined genetic analysis of these strongly suggestive results is mandatory. It is also apparent that only after such a detailed study, can one begin thinking of proposing actual names for the "genes" described in this study.

No effect on NK activity. Table III shows that no major differences in NK reactivity in spleen were observed between the 3 pairs of prototype animals, when tested in 4 or 18 hr. assays against YAC targets (4 hr. data not shown).

TABLE II. NC Activity of Spleen Cells from Different F_1 Hybrids Against Meth 113 [a]

Hybrid	Genotype "Tla"	Genotype "Qa2,3"	Genotype "A-AKR"	No.	Mean Cytotoxicity 100:1(±SEM)	NC activity [b]
1. (B6 x B6.Tla^a)	+	-	+	28	23.5 ± 1.4	I
2. (B6.Tla^a x B6)	+	-	+	22	25.2 ± 1.8	I
3. (BALB/cJ x A/J)	+	-	+	8	20.4 ± 1.9	I
4. (B10.A x A.SW)	+	-	+	23	36.4 ± 1.9	H
5. (B10.A x C3H/Bi)	+	+	-	10	21.5 ± 2.6	I
6. (B10.A x B10)	+	-	-	16	12.8 ± 1.2	L
7. (B6 x DBA/2)	+	-	-	7	8.1 ± 1.1	L

[a]For details on the strains used, assay and experimental design, see Table I. Cytotoxicity values of 35% or more are considered as "high" and of 15% or less as "low". For details on the "genotype" see Text.

[b]L = Low; I = Intermediate; H = High.

TABLE III. NK Activity in Spleen Against YAC.1 Target Cells in Mice with High or Low NC Activity [a]

Strain	No. of animals	Mean % cytotox. ± SEM 100:1	Mean % cytotox. ± SEM 50:1	Nc activity
C57BL/6 Boy	17	36.0 ± 2.0	29.1 ± 2.1	Low
B6.Tla^a	12	30.6 ± 0.9	24.5 ± 1.9	High
BALB/cJ	7	13.3 ± 4.2	6.2 ± 1.7	Low
BALB/cByJ	7	13.0 ± 2.3	9.0 ± 2.9	High
A/Boy	7	10.0 ± 2.3	6.3 ± 1.6	High
A.Tla^b	7	10.3 ± 2.4	8.3 ± 3.5	High

[a]NK activity tested on ^{51}Cr-labelled YAC.1 cells in an 18 hr. assay (for further details see ref. 14). For details on strains used see footnote Table I. For details on "NC activity" and its definition see Table I.

IV. CONCLUSIONS

Three direct conclusions can be extracted from the present study: 1) the genetic control of NC activity is different from that of NK activity (i.e. the present putative 3 gene system affects only NC levels); 2) the genetic control of natural cell-mediated cytotoxicity is complex (see refs. 1-7 and 9-12 for further discussion) and 3) the region in chromosome 17 distal to the D-end of H2 appears quite active in the control of both NC (present paper, also 2) and NK activity (5,9-12). One additional statement can be added, and that is that in almost none of the genetic studies available, do we know the actual mechanisms of the genetic control of natural cell mediated cytotoxicity (for the possible exceptions see refs. 5-7).

ACKNOWLEDGMENTS

We would like to thank Drs. E.A. Boyse and Lorraine Flaherty for rather extensive discussions over these studies and Ms. Linda Stevenson for preparing the manuscript.

REFERENCES

1. Stutman, O., Paige, C.J., and Feo Figarella, E., J. Immunol. 121: 1827.(1978).
2. Stutman, O., and Cuttito, M.J., in "Natural Cell-Mediated Immunity Against Tumors" (R.B. Herberman, eds.), p. 431. Academic Press, New York, 1980.
3. Stutman, O., and Cuttito, M.J., Nature 290: 254 (1981).
4. Stutman, O., Feo Figarella, E., Pagie, C.J., and Lattime, E.C., in "Natural Cell-Mediated Immunity Against Tumors" (R.B. Herberman, ed.), p. 187, Academic Press, New York, 1980.
5. Clark, E.A., and Harmon, R.C., Adv. Cancer Res. 31: 227 (1980).
6. Clark, E.A., Windsor, N.T., Sturge, J.C., and Stanton, T.H., in "Natural Cell-Mediated Immunity Against Tumors" (R.B. Herberman, ed.), p. 417, Academic Press, New York, 1980.
7. Clark, E.A., Shultz, L.D., and Pollack, S.B., Immunogenetics 12: 601 (1981).
8. Flaherty, L., in "Origins of Inbred Mice" (H.C. Morse III, ed.) p. 409, Academic Press, New York, 1978.
9. Petranyi, G., Kiessling, R., and Klein, G., Immunogenetics 2: 53 (1975).
10. Petranyi, G., Kiessling, R., Povey, S., Klein, G., Herzenberg, L., and Wigzell, H., Immunogenetics 3: 15 (1976).
11. Klein, G.O., Klein, G., Kiessling, R., and Karre, K., Immunogenetics 6: 561 (1978).

12. Klein, G.O., Karre, K., Klein, G. and Kiessling, R., J. Immunogenetics 7: 401 (1980).
13. Klein, J., Flaherty, L., VandeBerg, J.L., and Shreffler, D.C., Immunogenetics 6: 489 (1978).
14. Lattime, E.C., Pecoraro, G.A., and Stutman, O., J. Immunol. 126: 2011 (1981).
15. Shreffler, D.C., and David, C.S., Tissue Antigens 2: 232 (1972).
16. David, C.S., and Shreffler, D.C., Tissue Antigens 2: 241 (1972)
17. Hammerling, G.J., Hammerling, U.,and Flaherty, L., J. Exp. Med. 150: 108 (1979).
18. Stimpfling, J.H., and Reichert, A.E., Transplant Proc. 2: 39 (1970).

APPROACHES TO THE GENETIC ANALYSIS OF NATURAL RESISTANCE IN VIVO[1]

George A. Carlson

The Jackson Laboratory
Bar Harbor, Maine

Arnold Greenberg

Manitoba Institute of Cell Biology
The University of Manitoba
Winnipeg, CANADA

I. INTRODUCTION

Natural resistance against tumour transplantation has been demonstrated in a variety of systems, and it has been clearly shown that NK cells can play a role in the nonadaptive rejection of tumour cells in vivo (Karre et al., 1980; Riccardi et al., 1980; Herberman and Ortaldo, 1981). Although positive correlation between in vivo resistance and natural killing has been obtained in many studies, it is evident that NK cells alone cannot account for all the manifestations of natural resistance (Carlson and Campbell, 1980; Gorelik and Herberman, 1981 Chow et al., 1981a,b). In vitro natural killing is due to at least three distinct subpopulations of cells (Lust et al., 1981; Minato et al., 1981), but the relative contributions of NK subtypes to resistance in vivo have yet to be determined. Reconstitution of natural resistance in vivo by transfer of NK-enriched cell populations has demonstrated the contribution of partially defined cell types (Riccardi et al., 1981; Hanna and Burton, 1981), and this approach should provide information on the efficacy of the various NK subpopulations in vivo.

[1]This work was supported by NIH grant number CA28231 from The National Cancer Institute, and by grants from the Medical Research Council and National Cancer Institute of Canada.

ISBN 0-12-341360-5

Genetic analysis is another powerful tool for resolution of the different natural resistance systems operating in vivo. Resistance to tumour transplantation is a complex phenomenon. Using tumour growth rate as the assay, net retardation of tumour growth may depend not only on a variety of nonadaptive defense mechanisms but also on classical immune responses. To avoid this difficulty, we monitored directly the death and metastatic distribution of ^{131}I or ^{125}I-IdUrd prelabelled tumour cells in vivo using the technique of Hofer et al. (1969). This method has proven valuable in the study of natural resistance and allows the early events due to nonadaptive rejection to be distinguished from the subsequent effects of the immune response (Karre et al., 1980; Riccardi et al., 1980; Gorelik and Herberman, 1981; Carlson and Wegmann, 1977; Carlson and Terres, 1978; Carlson et al. 1980). In this report we describe some of the approaches to the genetic analysis of natural resistance currently being employed in our laboratories.

II. THE RELATIONSHIP BETWEEN NK CELLS AND H-2 ASSOCIATED NATURAL RESISTANCE

It is often assumed that NK activity is the in vitro counterpart of natural resistance against normal or transformed cells of the haemopoietic system (Kiessling et al., 1977; Lotzova and Savary, 1981). For example, both NK activity and H-2 associated natural resistance are independent of normal thymic function, but can be abrogated by treatment with antimacrophage agents, ^{89}Sr, or cortisone acetate (Carlson et al., 1980; Lotzova and Savary, 1981; Shearer et al., 1976; Herberman et al., 1975; Haller and Wigzell, 1977; Cudkowicz and Hochman, 1980). However, nonadaptive resistance to bone marrow or tumour transplantation in irradiated and nonirradiated mice was most strongly expressed against H-2 non-identical cells (Carlson and Wegmann, 1977; Snell, 1958; Cudkowicz and Stimpfling, 1964). In contrast, with few exceptions (Harmon et al., 1977), NK activity is independent of H-2 non-identity between the effector and target cells, although the H-2 haplotype of the effector can influence the level of activity expressed (Petranyi et al., 1975; Klein et al., 1980).

We compared the role of H-2 in natural resistance with its influence on NK activity by assaying the two activities in H-2 typed CBA/J x (CBA/J x DBA/2J)F_1 backcross mice. This strain combination was chosen because CBA mice are high in NK activity while DBA/2 are lower; and secondly, the $H-2^d$ haplotype has been shown to increase NK activity in mice of the C57BL background over that seen with the H-2^k or H-2^b haplotype (Klein et al., 1980). These mice allowed us to test the postulate that NK cells are involved in natural resistance against cells which are apparently insensitive to NK cell-mediated lysis in vitro. Natural resistance was monitored against the H-2^d, NK-resistant leukaemia L20u.2 (a oubain-resistant clone of L1210). All mice were injected i.v. with 2 x 10^6 ^{125}I-IdUrd prelabelled cells and the amounts of radioactivity remaining in the spleens and lungs were determined 24 hours later. The results of one such experiment are shown in Figure 1.

Note that the H-2^k homozygous mice had fewer tumour cells in their spleens and lungs than did the H-$2^{k/d}$ heterozygotes. In a series of 24 backcross mice the greater resistance of the H-$2^{k/k}$ than H-$2^{k/d}$ animals was statistically significant ($P<10^{-7}$). This is consistent with our earlier findings that nonadpative resistance against H-2 incompatible cells was stronger than that against semi-syngeneic cells, and that H-2 antigens themselves may serve as the target or stimulus for natural resistance.

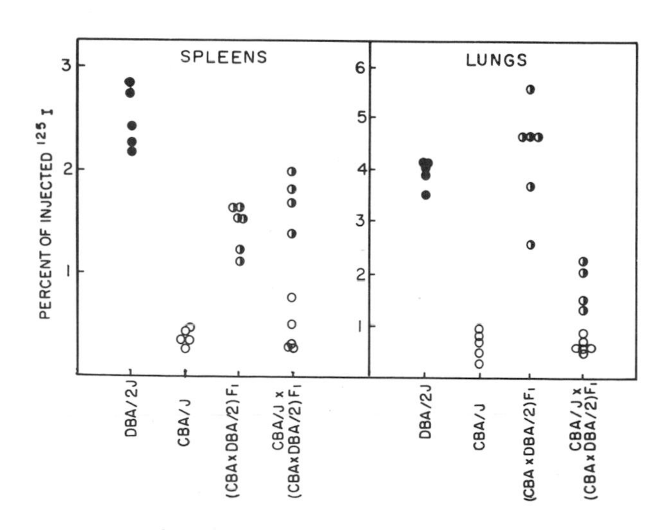

FIGURE 1. Natural resistance against L20u.2 in individual CBA/J, DBA/2J, F_1, and backcross mice. Closed circles represent H-2^d homozygotes, open circles H-2^k homozygotes, and half-closed circles H-$2^{k/d}$ heterozygotes.

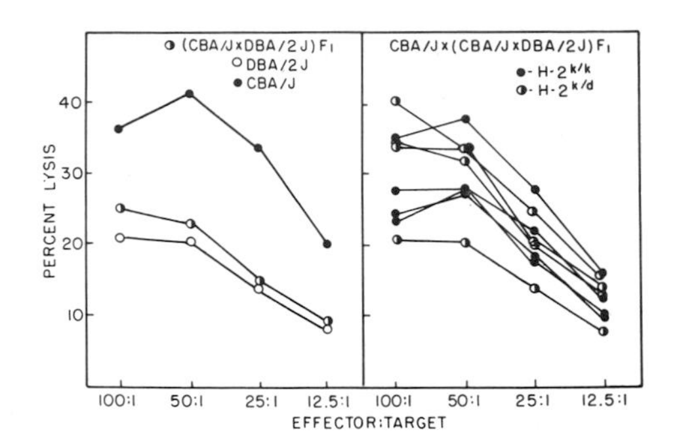

FIGURE 2. NK activity against YAC-1 cells in CBA/J x (CBA/J x DBA/2J)F_1 mice.

In contrast, when in vitro activity against NK-sensitive YAC-1 cells (H-2^a) was tested in these backcross mice no statistically significant association with H-2 type was seen. The results of one such experiment, shown in Figure 2, illustrate that the NK levels in the backcross mice ranged from intermediate to high, and that the H-2 type was not predictive of NK level. The NK level in the F_1 in this experiment was lower than usually obtained; in most experiments, intermediate levels were found.

These experiments indicate that, unlike the results obtained with mice of the C57BL background, H-2 haplotype does not influence NK activity in the CBA/J x DBA/2J combination; this suggests H-2-non-H-2 gene interaction in the control of NK level. It is also clear that NK cells alone cannot account for the phenomenon of H-2 associated natural resistance in vivo. That confrontation of the host with H-2 nonidentical cells serves as a trigger, or stimulus, for NK-mediated cell lysis has not been ruled out. One such indirect mechanism could be mediated via the production of interferon or other NK inducers after stimulating the host with H-2 nonidentical cells (Djeu et al., 1980). However, we were unable to detect any H-2 dependent augmentation of NK activity following i.v. injection of L20u.2 into compatible and incompatible strains (Carlson and Truesdale, unpublished results).

One intriguing possibility is that, even in H-2 associated natural resistance, tumour survival in the lung may give an indication of host NK levels. Note in Figure 1 that tumour survival in the lungs of the H-$2^{k/d}$ backcross mice, though greater than that in the $H-2^{k/k}$ homozygotes, is less than that in the F_1 mice. The possible association between NK activity and L20u.2 survival in the lung is further indicated by preliminary results obtained using recombinant inbred strains of mice. G. O. Klein of the Karolinska Institute and B. Taylor of the Jackson Laboratory have tested all 24 BXD recombinant inbred (RI) lines for in vitro NK activity against YAC-1 (personal communication). In collaboration with B. Taylor we are testing these lines for natural resistance against the $H-2^d$ leukaemia L20u.2. As expected from our previous results (Carlson and Wegmann, 1977; Carlson et al., 1980), the RI strains fell into two distinct groups when injected with L20u.2. Strains which were H-2 identical with the tumour showed splenic colonization by the leukaemia at levels similar to that in the DBA/2J parent, while the $H-2^b$ strains showed as little splenic colonization as the C57BL/6J parent (approximately 10 fold fewer cells surviving than in DBA/2J). Tumour survival in the lung showed an interesting pattern, however. Of the four $H-2^b$ BXD strains tested to date (there are 8 in all), three, BXD-2, 19, and 29, allowed significantly more tumour survival in the lung than did the B6 parent; these strains were low in NK activity. BXD-13, which had high NK activity, showed the same resistance in the lung as did B6. Conversely, although most of the $H-2^d$ lines (of the 13 tested to date) exhibited colonization of the lung by the $H-2^d$ leukaemia to the same extent as the D2 parent, BXD-30 and 31 had significantly less tumour in their lungs. BXD-30 was found to have high NK activity, while BXD-31 was intermediate. It must be noted, however, that not all the $H-2^d$ strains shown to be high in NK activity by Klein exhibited decreased L20u.2 survival in the lung.

III. NATURAL RESISTANCE AGAINST TUMOUR CLONES WITH DIFFERENTIAL SENSITIVITY TO IN VITRO EFFECTORS

We have used paired tumour clones with high and low relative tumour frequencies in syngeneic DBA/2 mice to demonstrate that tumor incidence after a threshold inoculation can correlate with their relative sensitivity to syngeneic natural effector mechanisms (Chow et al., 1981a). For example, L5178Y-F9 had a lower tumour frequency after a threshold inoculum than did L5178Y-1 and was more sensitive to syngeneic natural antibody; both clones showed low sensitivity to NK-mediated lysis. SL2-5 was less tumorigenic than SL2-9 correlating with its sensitivity to NK cells; both of these clones were sensitive to syngeneic natural antibody. Using ^{131}I-

IdUrd-prelabelled cells we found that differential elimination of these clones based on their sensitivity to in vitro effectors was dependent on the route of tumour inoculation (Chow et al., 1982). For example, natural antibody mediated systemic elimination of the tumour was most pronounced after i.v. inoculation (i.e. more L5178Y-F9 cells were eliminated than L5178Y-1). Although differential killing monitored by whole body counting between NK-sensitive SL2-5, and NK-resistant SL2-9 was most pronounced after s.c. injection of syngeneic animals, significantly fewer SL2-5 cells survived in the lung 4 hr after i.v. injection. This result was in keeping with the observations noted above. Given the value of these cell lines in dissecting natural resistance against syngeneic cells, we are now using these tumour clones in the genetic analysis of natural resistance.

DBA/2J, C57BL/6J, and B6D2F$_1$ mice were injected i.p. with 10^6 ^{131}I-labelled SL2-5 or SL2-9 cells. These same tumour clones were also tested for sensitivity to in vitro NK-med-

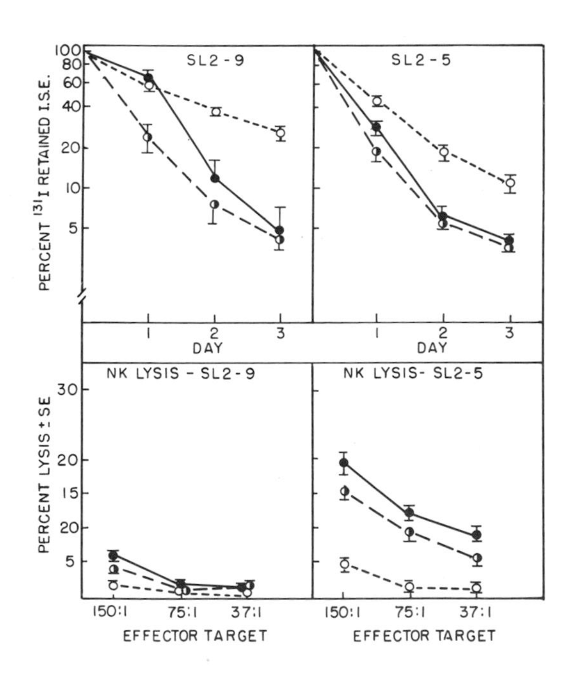

FIGURE 3. In vivo resistance and in vitro natural killing of NK-resistant SL2-9 and NK-sensitive SL2-5 by DBA/2J (o), C57BL/6J (●), and B6D2F$_1$ (◐) mice.

iated lysis by these same strains. The results of one experiment are shown in Figure 3. Note that although SL2-9 was 4-fold less susceptible to natural killing in vitro, allogeneic and hybrid resistance by C57BL/6 and B6D2F_1 mice were comparable to that seen with SL2-5. The F_1 and B6 eliminated SL2-5 at the same rate, while killing of the SL2-9 cells by B6 mice was consistently slower than by the F_1 hybrid. The NK-sensitive SL2-5, however, survived less well in the syngeneic DBA/2 host than did SL2-9.

IV. MUTATIONS INFLUENCING NATURAL RESISTANCE

Several mutations that affect the immune system have been shown to influence NK activity in vitro (Clark et al., 1981). The beige (bg) mutation exhibits a relatively selective defect in NK cells (Roder and Duwe, 1979) and has been used to show a correlation between tumour resistance, inhibition of metastasis, and NK activity (Karre et al., 1980; Talmadge et al., 1980). However, as shown by Gorelik and Herberman (1981), tumour resistance is not uniformly impaired in bg/bg mice. These investigators found that C57BL/6-bg/bg mice were at least as resistant to intrafootpad inoculation of NK-sensitive YAC-1 as were the normal controls.

We have tested resistance to i.v. injected L20u.2 in beige and motheaten (me) mice. Mice homozygous for me have been shown to be deficient in NK cells (Clark et al., 1981) and to exhibit a severe immunodeficiency and autoimmune syndrome, surviving only to 7 weeks of age (Shultz and Green, 1976; Sidman et al., 1978). Their NK deficiency is not due to suppressor cells and cannot be boosted by treatment with poly I:C or interferon (Carlson and Shultz, ms. in preparation).

Motheaten mice showed no natural resistance to L20u.2 inoculation, allowing even greater tumour survival than syngeneic DBA/2J. In contrast, bg/bg mice ultimately showed the same level of resistance as their littermates against the DBA/2J leukemia. However, the rejection process occurred more slowly in beige mice. This could indicate that NK cells are playing a role in resistance against this NK-resistant tumour, or, more likely, that macrophages or granulocytes whose function is also impaired by the bg mutation are involved.

V. CONCLUDING REMARKS

One broad category of nonadaptive defense operating in vivo can be termed NK-related. Under this aegis we include H-2 associated natural resistance against tumours, Hh-1 dependent resistance to haemopoietic stem cell transplantation, and resistance which has been shown to be affected by the beige mutation. All these phenomena have been shown to be mechanistically related to NK-cell mediated lysis in vitro, and the common involvement of a H-2D (Hh-1) associated gene (s) can be taken as evidence for similar genetic control (Klein et al., 1978). However, on closer genetic analysis it is clear that in vivo resistance is not simply a reflection of in vitro NK activity. It remains to be determined whether these differences in genetic control are revealing regulatory networks controlling NK cells in vivo, or are indicative of different classes of effector cells.

REFERENCES

Carlson, G. A., and Wegmann, T. G., (1977). J. Immunol. 118: 2130.

Carlson, G. A., and Terres, G. (1978). J. Immunol. 121:1752.

Carlson, G. A., and Campbell, G. (1980). in "Genetic Control of Natural Resistance to Infection and Malignancy" (E. Skamene, P. A. L. Kongshavn, and M. Landy, Eds.) p. 445.

Carlson, G. A., Melnychuk, D., and Meeker, M. J. (1980). Int. J. Cancer 25:111.

Chow, D. A., Wolosin, L. B., and Greenberg, A. H. (1981a). Int. J. Cancer 27:459.

Chow, D. A., Miller, V. E., Carlson, G. A., Pohajdak, B. and Greenberg, A. H. (1981b). Invasion and Metastasis in press.

Chow, D. A., Carlson, G. A, and Greenberg, A. H. (1982). Submitted for publication.

Clark, E. A., Shultz, L. D., and Pollack, S. B. (1981). Immunogenetics 12:601.

Cudkowicz, G., and Stimpfling, J. H. (1964). Immunology 7:219.

Cudkowicz, G., and Hochman, P. S. (1980). Cellular Immunol. 53:395.

Djeu, J. Y., Huang, K.-Y., and Herberman, R. B. (1980). J. Exp. Med. 151:781.

Gorelik, E., and Herberman, R. B. (1981). Int. J. Cancer 27:709.

Haller, O., and Wigzell, H. (1977). J. Immunol. 118:1503.
Hanna, N., and Burton, R. C. (1981). J. Immunol. 127:1754.
Harmon, R. C., Clark, E. A., O'Toole, C., and Wixler, L. S., (1977). Immunogenetics 4:601.
Herberman, R. B., Nunn, M. E., Holden, H. T., and Lavrin, D. H. (1975). Int. J. Cancer 16:230.
Herberman, R. B., and Ortaldo, J. R. (1981). Science 214:24.
Hofer, K. G., Prensky, W., and Hughes, W. L. (1980). J. Nat. Cancer Inst. 43:763.
Karre, K., Klein, G. O., Kiessling, R., Klein, G., and Roder, J. (1981). Nature 284:624.
Kiessling, R., Hochman, P. S., Haller, O., Shearer, G. M., Wigzell, H., and Cudkowicz, G. (1977). Eur. J. Immunol. 7:655.
Klein, G. O., Klein, G., Kiessling, R., and Karre, K. (1978). Immunogenetics 6:561.
Klein, G. O., Karre, K., Klein, G., and Kiessling, R. J. (1980). Immunogenetics 7:401.
Lotzova, E., and Savary, C. A. (1981). Exp. Hematol. 9:766.
Lust, J. A., Kumar, V., Burton, R. C., Bartlett, S. P., and Bennet, M. (1981). J. Exp. Med. 154:306.
Minato, N., Reid, L., and Bloom, B. R. (1981). J. Exp. Med. 154:750.
Petranyi, G., Kiessling, R., and Klein, G. (1975). Immunogenetics 2:53.
Riccardi, C., Santoni, A., Barlozzari, T., Puccetti, P., and Herberman, R. (1980). Int. J. Cancer 25:475.
Riccardi, C., Barlozzari, T., Santoni, A., Herberman, R. and Cesarini, C. (1981). J. Immunol. 126:1284.
Roder, J., and Duwe, A. K. (1979). Nature 278:451.
Shearer, G. M., Cudkowicz, G., Schmitt-Verhulst, A. M., Rehn, T. G., Waksal, H., and Evans, P. D. (1976). Cold Spr. Hbr. Symp. Quant. Biol., Vol. XLI, p. 511.
Shultz, L. D., and Green, M. C. (1976). J. Immunol. 116:936.
Sidman, C. L., Shultz, L. D., and Unanue, E. R. (1978). J. Immunol. 121:2399.
Snell, G. D. (1958). J. Natl. Cancer Inst. 21:843.
Talmadge, J. E., Meyers, K. M., Prieur, D. J., and Starkey, J. R. (1980). Nature 284:622.

THE BEIGE (bg) GENE INFLUENCES THE DEVELOPMENT OF AUTOIMMUNE DISEASE IN SB/Le MALE MICE

Edward A. Clark[1,4]
John B. Roths[3]
Edwin D. Murphy[3]
Jeffrey A. Ledbetter[4]
James A. Clagett[2]

Departments of Genetics[1] and Periodontics[2]
University of Washington, Seattle, Washington

[3]The Jackson Laboratory, Bar Harbor, Maine

[4]Genetic Systems Corporation, Seattle, Washington

INTRODUCTION

The recessive mutation beige (bg) produces a number of pleiotropic effects in mice including increased susceptibility to pneumonitis (1) and cytomegalovirus infections (2), defective granulocyte chemotaxis (3) and lysosomal enzyme secretion (4), and giant lysosomal granules in granule-containing cells (5). Roder and coworkers (6,7) reported that the beige mutant selectively impairs natural killer (NK) cell reactivity against lymphoid tumors. The frequency of NK cells in beige mice is normal (6), and, after exposure to interferon (IFN) or IFN inducers, NK cell activity in beige mice can be augmented to almost normal levels (8,9). This suggests that the bg defect may be affecting the activation or perhaps the lytic efficiency of NK cells.

The recombinant inbred strain BXSB was derived from offspring of the intercross (C57BL/6J ♀ x SB/Le ♂). A progressive fatal autoimmune disease develops spontaneously that is markedly

---

This work was supported by NIH grants CA26713 and RR00166 (E.A.C.), CA-22948 (E.D.M.), DE-02600 (J.A.C.), and Genetic Systems Corporation (J.A.L. and E.A.C.).

ISBN 0-12-341360-5

accelerated in male animals (10-12). The enhancement of this disease in BXSB males is determined by a factor linked to the Y chromosome. The Y chromosome of BXSB mice was contributed by the strain SB/Le in the original intercross (11). Diseased BXSB mice characteristically develop a progressive diffuse glomerulonephritis, moderate lymphoadenopathy, splenomegaly, and have high levels of serum immunoglobulins including erythrocyte, thymocyte, and DNA autoantibodies (11-13). Here we report that like the BXSB mice derived from them, male SB mice develop a fatal lupus-like syndrome and that the beige locus has a profound effect on the development of this autoimmune disease.

The segregating strain SB/LeMp- bg/+ and bg/bg (N13F18) was developed in a research colony at the Jackson laboratory, Bar Harbor, Maine (E.D.M. and J.B.R.). These mice were bred in a colony at the University of Washington (E.A.C.). Where possible only littermates were compared. The method used for measuring NK cell activity against YAC-1 target cells has been described (14). Lymphocyte proliferative responses to E. coli lipopolysaccharide (LPS) and concanavalin A (Con A) were performed as described earlier (15). Frozen kidney sections were prepared and treated with fluorescein-conjugated (FITC)-F(ab')$_2$ fragments of a goat anti-mouse IgM (μ chain specific) or anti-mouse C3 sera and kidneys scored (1^+ - 4^+) based on the frequency of glomerular deposits as well as fluorescence intensity. Cell subsets were quantified on an Ortho-fluorescence-activated cell sorter (FACS) using fluorescein-conjugated rat monoclonal antibodies specific for mouse differentiation antigens (16,17) or FITC-conjugated GAM.

RESULTS AND DISCUSSION

The germinal observation that stimulated this study was the discovery by one of us (J.B.R.) that the survival times of SB/Le-bg/+ and bg/bg male mice are dramatically different. A comparison of the longevity of SB/Le female bg/+ (682 $\pm$ 55) days and bg/bg (573 $\pm$ 35 days) mice showed no statistical differences. However, the difference in survival between male bg/+ (248 $\pm$ 42 days) and bg/bg (483 $\pm$ 37 days) animals was striking and highly significant ($p < .001$) with heterozygous mice dying usually more than 250 days before their homozygous counterparts. A plot and statistical analysis of the cumulative percent survival to be reported elsewhere (Clark *et al.* submitted) shows that in both bg/+ and bg/bg male mice, there is a phase where a rapid decrease in survival occurs, but that the onset of this phase in beige males occurs approximately 290 days after it does in bg/+ mice. However, the slopes and the curves during the rapid mortality phase in bg/+ and bg/bg groups are not statistically different.

Although homozygosity at the beige locus prevents early mortality, the beige gene does not appear to have any major effect on other phenotypic abnormalities including lymphadenopathy, splenomegaly, serum IgG_1 levels, and incidence of positive serum ds DNA autoantibodies.

The SB-bg/+ male mice invariably die early of a progressive diffuse glomerulonephritis similar to that seen in +/+ BXSB male mice. A comparison of the development of immune complex and complement deposition in the kidneys of bg/+ and bg/bg male mice (Fig. 1) revealed that significantly more bg/+ mice (15/21) have elevated levels of IgM deposition in their kidneys (>2+) than bg/bg littermate controls (5/18, $p<0.01$). C3 deposition is also more frequent and intense in heterozygotes. Surprisingly, IgM deposition is quite evident even in young SB bg/+ male mice. Therefore, we elected to examine in detail several immunologic parameters in this age group.

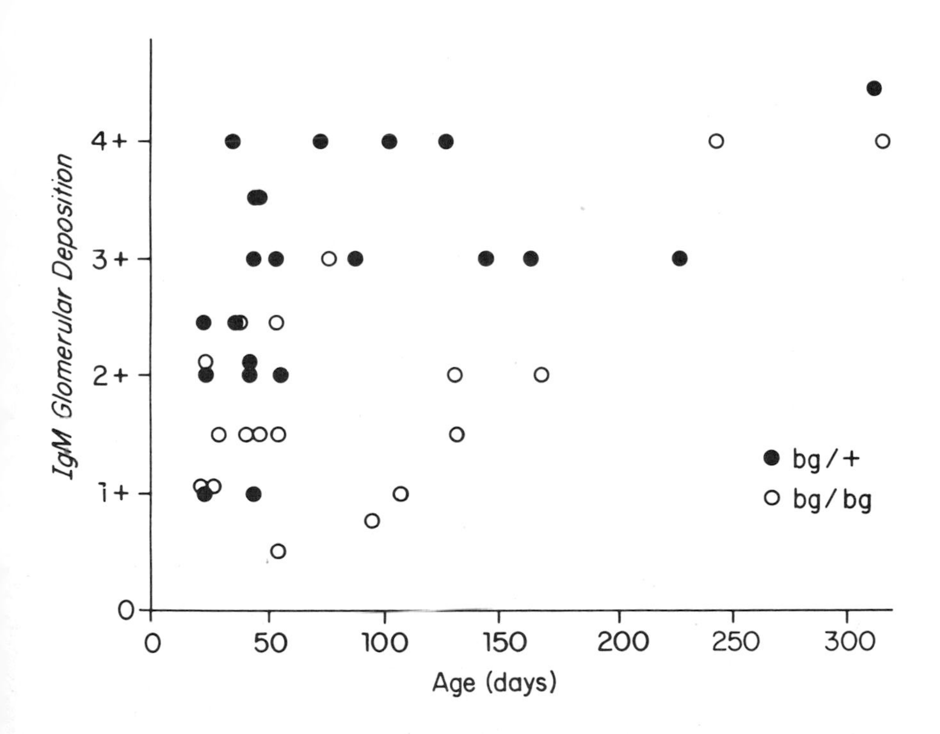

FIGURE 1. Ontogeny of IgM immune complex deposition in SB male mice. ●, bg/+; ○, bg/bg.

(1) NK cell activity. As expected from studies of other bg alleles (6,8,9), bg/bg male homozygotes had lower NK cell activity than bg/+ heterozygous counterparts (Fig. 2). This difference was most pronounced between 6-10 weeks of age. NK activity in the SB strain overall is low compared to most other strains.

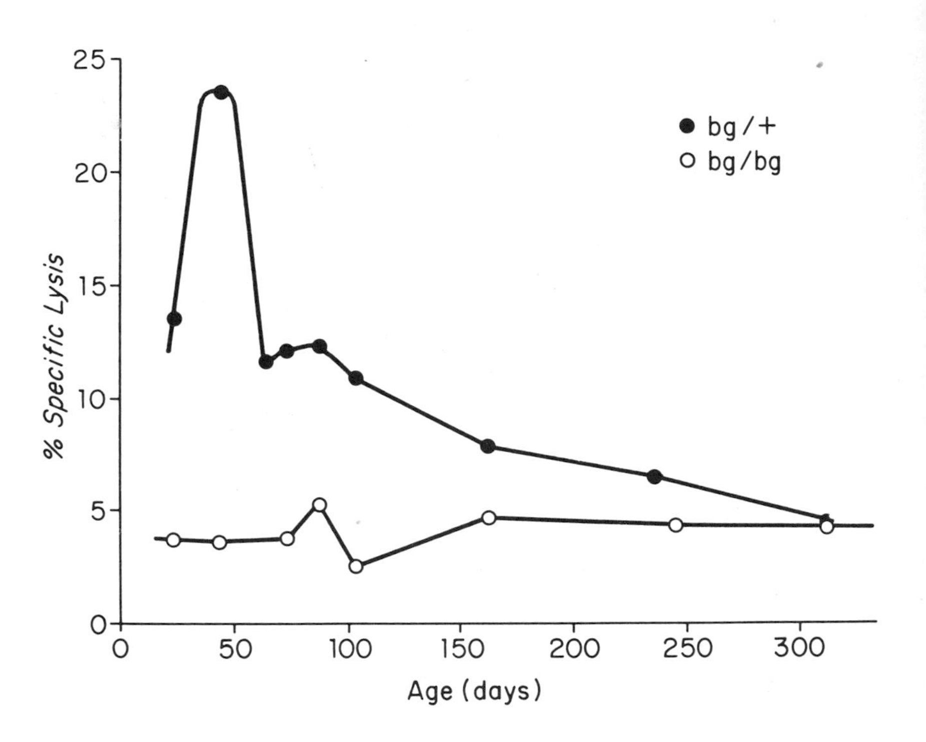

FIGURE 2. Ontogeny of NK activity in SB male mice. ●, bg/+; ○ bg/bg.

(2) Response to LPS and Con A. To our surprise, 8-week old SB bg/+ mice had somewhat higher responses to LPS mice than bg/bg homozygous littermates. An experiment comparing two groups of littermates is shown in Table I. LPS responses were higher in both male and female heterozygotes indicating that higher responsiveness to B cell mitogens in bg/+ alone cannot account for the accelerated autoimmune disease in male mice. Responses to the T cell mitogen Con A were not significantly different in the bg/+ and bg/bg groups.

TABLE I. Responses to LPS and Con A in SB/Le Mice

Litter	Genotype	Sex	Proliferative Resp. (Mean cpm±SD) LPS	Con A
1	bg/+		94,478 (3,163)	280,695 (14,777)
			95,774 (7,282)	182,271 (14,073)
	bg/bg		58,624 (1,286)	241,358 (18,552)
			45,178 (3,297)	216,305 (6,633)
2	bg/+		69,861 (3,360)	216,929 (8,646)
	bg/bg		42,489 (1,070)	176,730 (5,808)
			50,745 (1,741)	265,752 (9,778)
			33,943 (1,955)	244,828 (11,282)
	bg/+		110,283 (3,657)	230,220 (8,025)
			101,508 (1,362)	195,959 (5,575)
	bg/bg		40,038 (1,307)	57,710 (5,796)

Litter 1: 57 days old Litter 2: 58 days old

(3) Cell sorter analysis of the frequency of Thy-1, Lyt-1, Lyt-2, THB, and Ig-positive spleen cells in bg/+ and bg/bg mice revealed that the mean frequency of Lyt-2 (12.0% versus 10.5%), and Ig bearing cells was not clearly different in the two groups. However, bg/+ mice had somewhat more THB$^+$ cells than their bg/bg counterparts (26.7% versus 16.7% in the most distinctive experiment). The significance of this finding will be discussed in detail elsewhere (Clark *et al.*, submitted). Thy-1$^+$ and Lyt-1$^+$ cell levels were not significantly different in the two groups.

Thus, the beige mutation influences several immune parameters in SB mice including NK activity, LPS responsiveness, frequency of THB$^+$ cells, and onset of autoimmune disease. Others have reported that the beige gene affects macrophage activity (18). Clearly, the beige mutation does not selectively impair NK lymphoid cells and should not be evoked simply as a murine model of NK deficiency. To do so would be as simplistic as describing nude mice as having no T cell activity.

In NK cell defective, LPS low responsive, SB beige male mice, the onset of autoimmune disease is delayed. The mechanism responsible for this protective effect is not known. A compelling but at this time somewhat myopic interpretation would be that NK cells play a role in immunoregulation. One more likely possibility is that an inefficiency in handling interactions at the cell membrane (19) that normally triggers autoreactivity in SB mice is responsible for the delay in disease onset in beige homozygotes. The decrease in Ig deposition in the kidneys, LPS responsiveness, and frequency of THB$^+$ cells in beige homozygotes suggests that the protective effect of the beige gene may be acting directly on B cells.

REFERENCES

1. Lane, P. W., and Murphy, E. D., Genetics 72:451 (1972).
2. Shellam, G. R., Allan, J. E., Papadimitriou, J. M., and Bancroft, G. J., Proc. Nat. Acad. Sci. USA 78:5104 (1981).
3. Gallin, J. I., Bujak, J. S., Patten, E., and Wolff, S. M., Blood 43:201 (1974).
4. Brandt, E. J., Elliott, R., and Swank, R. T., J. Cell. Biol. 67:774 (1975).
5. Oliver, C., and Essner, E., J. Histochem. Cytochem. 21:218 (1973).
6. Roder, J. C., and Duwe, A. K., Nature 278:451 (1979).
7. Roder, J. C., Lohmann-Matthes, M. L., Domzig, W., and Wigzell, H., J. Immunol. 123:2174 (1979).
8. Talmadge, J. E., Meyers, K. M., Prieur, D. J., and Starkey, J. R., Nature 284:622 (1980).
9. Welsh, R. M., and Kiessling, R. W., Scand. J. Immunol. 11:363 (1980).
10. Murphy, E. D., and Roths, J. B., in "Genetic Control of Autoimmune Disease" (N. R. Rose, P. E. Bigazzi, N. L. Warner, eds.), p. 207. Elsevier North-Holland, New York, 1978.
11. Murphy, E. D., and Roths, J. B., Arth. Rheum. 22:1188 (1979).
12. Andrews, B. S., Eisenberg, R. A., Theofilopoulos, A. N., Izui, S., Wilson, C. B., McConohey, P. J., Murphy, E. D., Roths, J. B., and Dixon, F. J., J. Exp. Med. 148:1198 (1978).
13. Eisenberg, R. A., Theofilopoulos, A. N., Andrews, B. S., Peters, C. J., Thor, L., and Dixon, F. J., J. Immunol. 122:2272 (1979).
14. Clark, E. A., Russell, P. H., Egghart, M., and Horton, M. A., Int. J. Cancer 24:688 (1979).
15. Engel, D., Clark, E. A., Held, L., Kimball, H., and Clagett, J. A., J. Exp. Med. 154:726 (1981).
16. Ledbetter, J. A., and Herzenberg, L. A., Immunol. Rev. 47:362 (1979).
17. Eckhardt, L. A., and Herzenberg, L. A., Immunogenetics 11:275 (1980).
18. Mahoney, K. H., Morse, S. S., and Morhan, P. S., Cancer Res. 40:3934 (1980).
19. Oliver, J. M., Zurier, R. B., and Berlin, R. D., Nature 253:471 (1975).

NK CELL ACTIVATION IN SM/J MICE

Nancy T. Windsor
Edward A. Clark

Department of Genetics
and Regional Primate Research Center
University of Washington
Seattle, Washington

INTRODUCTION

The genetic control and the mechanisms for the regulation and activation of NK cells are still poorly understood. Minato et al. (1981) have recently suggested that several distinct in vitro cell populations are involved in natural cytotoxicity. In vivo genetic models for this complex phenotype would be most useful.

In particular, little is known about the genes that regulate NK cell activation. Recently, we reported that the SM/J inbred strain has both hyper NK and K cell activity (Clark et al., 1981), and hyperresponsiveness to B cell mitogens (Engel et al., 1981). NK activity was elevated even in older SM/J mice. Although target binding NK cell frequencies were not different between SM/J and C57BL/6 (B6) mice, SM/J mice had more actively lytic NK cells (Clark et al., 1981). We concluded that SM/J NK cells may be chronically activated and that the SM/J strain may provide a new genetic model for examining NK cell activation.

The elevation of NK activation in the SM/J strain is due to the chronic activation of a subset of $NK1^{+}$, $Qa\text{-}5^{-}$ NK cells. This responsiveness appears to be under polygenic, non H-2 control. In this study, three aspects of SM/J mice were examined: 1) Do SM/J NK cells differ in expression of markers other than Qa-5? 2) Can NK activity in SM/J mice be activated or triggered to even higher levels? 3) How does splenic interferon (IFN) activity in this hyper NK strain compare with IFN levels in other strains?

Supported by NIH grants CA26713 and RR00166 and by Genetic Systems Corporation, Seattle, Washington.

ISBN 0-12-341360-5

MATERIALS AND METHODS

Mice. The A/J, B6/J and SM/J mice used in this study were obtained from the Jackson Laboratory, Bar Harbor, ME and bred in our colony. The $H\text{-}2^v$ congenic B10.SM strain was kindly provided by Dr. Jack Stimpfling, McLaughlin Research Institute, Great Falls, MT.

Antisera. Monoclonal anti Qa-4 and Qa-5 sera were kindly donated by Dr. Gloria Koo, Sloan-Kettering Memorial Cancer Center (Koo et al., 1980). Monoclonal F7D5 Thy 1.2 antibody has been described previously (Lake et al., 1980). NK specific anti-NK 1 serum was the generous gift of Dr. Sylvia Pollack, Dept. of Biological Structures, University of Washington (Pollack et al., 1980).

Interferon Assay. Interferon was assayed by the inhibition of plaque-forming units (PFU) of vesicular stomatitis virus (VSV) on mouse L cells (Welsh and Kiessling, 1980). Briefly, each well of a 96-well flat bottom tissue culture plate was seeded with $2x10^4$ mouse L cells. After the monolayer was established, the supernatants were removed, and serial dilutions of spleen cells from various mouse strains were added in duplicate. In addition, to each plate were added dilutions of a control Mouse Interferon Standard (obtained from the National Cancer Institute, NIH, Bethesda, MD). Plates were incubated 14 to 18 hours at 37°C, then washed twice to remove the spleen cells. Forty PFU of VSV were added to each well and incubated for 90 min at 37°C. The wells were washed several times to remove all virus, and overlayed with MEM plus 5% methyl cellulose. After two days, the wells were stained with 2% crystal violet for 10 min, rinsed, and allowed to dry. Plaques were counted and IFN units were calculated by plotting PFU vs. Standard IFN units. The number of IFN units/spleen cell concentration was then extrapolated.

In Vivo Augmentation of NK Activity. $1.0x10^8$ cells of Bacillus Calmette-Guerin (BCG, Pasteur strain, Trudeau Institute, Saranac Lake, NY) were administered to mice i.p. in 0.5 ml of PBS. Peritoneal exudate cells and spleen cells were tested for NK activity at various times after treatment (Tracey et al., 1977).

NK Assay. The NK cytolytic assay has been described previously (Clark et al., 1979). Briefly, ^{51}Cr-labeled YAC-1 cells were used as targets in a 4-hour assay in round bottom microtiter plates. Percent specific lysis was determined by using the formula: 100 x (cpm test-cpm spontaneous)/(cpm detergent-cpm spontaneous). The spontaneous release of chromium was less than 10% for the data presented.

RESULTS AND DISCUSSION

In our previous study, we found that SM/J NK cells are NK-1$^+$ but Qa-5$^-$, unlike their H-2^v-identical congenic B10.SM counterparts, which have NK-1$^+$ Qa-5$^+$ cells (Clark et al., 1981). SM/J and B10.SM NK cells also differ in their expression of Qa-4 antigens (Table I); SM/J NK cells are Qa-4$^-$ while B10.SM NK cells are Qa-4$^+$. These differences may be because 1) SM/J NK cells are chronically activated, 2) the B10.SM strain has a recombination between H-2D and Qa-5 regions, or 3) non H-2 SM genes are affecting the expression of Qa antigens. Non Qa genes have been reported to influence Qa antigen expression (Stanton et al., 1981). We currently favor the third possibility based on preliminary experiments in progress.

TABLE I. Sensitivity of SM/J NK Activity to Various Antisera Plus Complement (C)

Antiserum + C (dilution)	Sensitivity of Splenic NK Activity of Strain		
	SM/J	B10.SM	B6
NK-1.1 (1:10)	+	+	+
Qa-5 (1:100)	-	+	+
Qa-4 (1:50)	-	±	+
Thy-1.2 (1:100)	±	±	+

+, >50% activity removed; ±, <50% activity removed; -, activity not removed.

The frequency of target binding NK cells is not different in SM/J, B10.SM, and B6 mice, but SM/J mice have a higher frequency of actively lytic NK cells (Clark et al., 1981). Thus, it is possible that SM/J NK cells are easily triggered or poorly regulated and/or that SM/J mice have elevated levels of IFN which chronically acivates their NK cells. To distinguish these possibilities, we compared the activation of NK cells with BCG in SM/J and B6 mice (Table II). NK activity could be boosted in SM/J mice and the NK response was greater and more rapid in the SM/J strain than in B6 mice.

Spleen cell IFN levels were also greater in SM/J mice than in other strains (Table III). The "NK high" SM/J strain consistently had three- to fivefold more IFN activity/10^6 spleen cells than did "NK low" A/J mice.

TABLE II. Rapid Activation in vivo of NK Activity in SM/J Mice

Time	Strain	Treatment	Percent Specific Lysis: Spleen cells[1]	PEC
3 hr	B6/J	none	25.2	0.7
		BCG[2]	25.6	0.6
	SM/J	none	36.5	0.4
		BCG	49.8	3.1
18 hr	B6/J	none	20.2	2.0
		BCG	25.1	7.4
	SM/J	none	55.2	3.4
		BCG	48.9	37.7
48 hr	B6/J	none	29.3	1.4
		BCG	28.8	11.8
	SM/J	none	57.7	0.7
		BCG	64.1	30.4

[1] Spleen cell E:T ratio 100:1; Peritoneal exudate cell (PEC) E:T ratio 10:1 except 48 hr, 25:1.

[2] $1x10^8$ BCG inoculated i.p.

TABLE III. Elevation of Spleen Cell IFN Levels in SM/J Mice

Experiment	Strain	Mean IFN Units/Well: $1x10^6$ [1]	$5x10^5$	$1x10^5$
1	SM/J	1125	700	160
	B10.SM	560	170	70
	B6/J	730	300	80
	A/J	375	90	26
2	SM/J	900	500	175
	B6/J	450	175	ND
	A/J	280	160	52

[1] Spleen cell dose added to duplicate microcultures.

Thus, SM/J NK cells can be activated rapidly in vivo and their splenic IFN levels are elevated. However, it is not known whether the SM/J NK cells themselves have a lower activation threshold than NK cells from other strains or if the hyperactivation of SM/J NK cells is mediated via IFN produced by another rapidly triggered cell. We currently favor the second possibility, since B cells are known to produce IFN (Weigent et al., 1981) and since SM/J B cells are hyperresponsive to polyclonal B cell activators (Engel et al., 1981). Experiments with our NK cell-associated monoclonal antibodies (Windsor et al., in preparation) should help to distinguish between these and other possibilities.

Using the cross-intercross method of Snell, we have begun to develop a congenic line A.SM derived from the "NK-low" A/J strain and the "NK high" SM/J strain. We are currently breeding generations 7 and 8. Intercross progeny mice were tested at generations 4 and 6 for NK activity and responsiveness to B cell mitogens. Mitogen responsiveness and NK activity segregated independently of each other. Progeny with elevated NK activity (30-45%) vs. A/J (5-15%) and SM/J (35-70%) have been selected for further breeding. Generation 2 (AxSM) F_2 mice were tested for NK and K cell activity concurrently, and these two phenotypes segregated closely together.

Recently stimulated NK cells have been reported to have broader reactivity than "endogenous" NK cells (e.g., Welsh et al., 1979). Therefore, we tested SM/J NK cells for reactivity against the NK insensitive lines P815 and L5178Y cl 27v and av (Durdik et al., 1980). Although YAC-1 targets were readily killed (51% specific lysis), P815 (0.6%) and cl aV (5.1%) were as insensitive to killing by SM/J NK cells as they were to B6 NK cells.

In summary, the activated NK cells in SM/J mice cannot readily be categorized into one of the subsets described by Minato et al. (1981). They are Qa-5$^-$ and do not lyse P-815, properties similar to the NK_M cell. However, SM/J NK cells can be activated and IFN levels are elevated in SM/J mice. NK cell activity is present in both the spleen and the bone marrow of SM/J mice (data not shown). Thus, more than one SM/J NK cell subset may be activated.

REFERENCES

Clark, E. A., Engel, D., and Windsor, N. T., J. Immunol. 127:2391 (1981).

Clark, E. A., Russell, P. H., Egghart, M., and Horton, M. A., Int. J. Cancer 24:688 (1979).

Durdik, J. M., Beck, B. N., Clark, E. A., and Henney, C. S., J. Immunol. 125:683 (1980).

Engel, D., Clark, E. A., Held, L., Kimball, H., and Clagett, J., J. Exp. Med., 154:726 (1981).

Koo, G. C., Jacobson, J. B., Hammerling, G. J., and Hammerling, U., J. Immunol. 125:1003 (1980).
Lake, P. E., Clark, E. A., Khorshidi, M., and Sunshine, G., Eur. J. Immunol. 9:875 (1979).
Minato, N., Reid, L., and Bloom, B. R., J. Exp. Med. 154:750 (1981).
Pollack, S. B., Tam, M. R., Nowinski, R. C., and Emmons, S. L., J. Immunol. 123:1818 (1979).
Stanton, T. H., Carbon, S., and Maynard, M., J. Immunol. 127:1640 (1981).
Tracey, D. E., Wolfe, S. A., Durdik, J. M., and Henney, C. S., J. Immunol. 119:1145 (1977).
Weigent, D. A., Langford, M. P., Smith, E. M., Blalock, J. E., and Stanton, G. J., Infect. Immun. 32:508 (1981).
Welsh, R. M., and Kiessling, R. W., Scand. J. Immunol. 11:363 (1980).
Welsh, R. M., Zinkernagel, R. M., and Hallenbeck, L. A., J. Immunol. 122:475 (1979).

CHROMOSOME 1 LOCUS : A MAJOR REGULATOR OF NATURAL RESISTANCE TO INTRACELLULAR PATHOGENS.

Emil Skamene
Adrien Forget*
Philippe Gros
Patricia A.L. Kongshavn

Montreal General Hospital Research Institute
McGill University and *Department of Microbiology and Immunology, University of Montreal, Montreal, Quebec, Canada

Natural resistance to infection with several intracellular pathogens - both bacterial and protozoal - has recently been demonstrated to be under genetic control. The ability of genetically-resistant mouse strains to prevent growth, in the reticuloendothelial tissues, of *Mycobacterium bovis* (BCG) is controlled by a single, dominant, autosomal gene designated **Bcg** (1). Similarly, resistance to the growth of *Salmonella typhimurium* is controlled by the **Ity** gene (2) and resistance to *Leishmania donovani* by the **Lsh** gene (3).

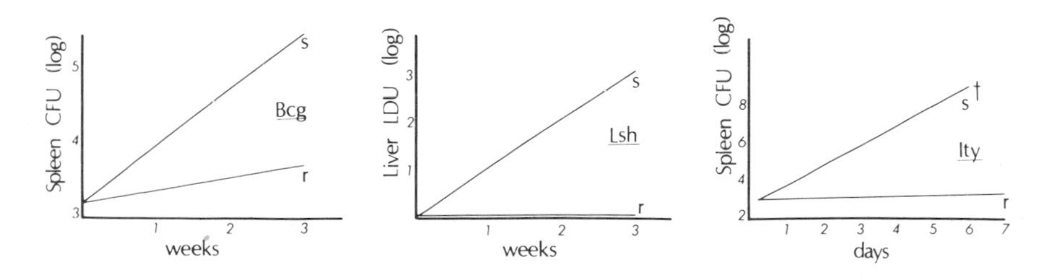

FIGURE 1. Course of infection with *M. bovis*, *L. donovani* and *S. typhimurium* in mice that carry either the resistant **(r)** or the susceptible **(s)** allele of **Bcg, Lsh** and **Ity** genes (adapted from Ref. 1, 2 and 3).

This communication advances the hypothesis that the natural resistance to all these pathogens (and most likely also to many others) is regulated out of the same locus that maps on Chromosome 1 of the murine genome.

INBRED STRAIN SURVEY FOR Bcg, Lsh AND Ity ALLELES

Mice of 17 inbred strains were infected intravenously with 10^4 CFU of *M. bovis* (BCG) and **Bcg** typed 3 weeks later

ISBN 0-12-341360-5

as either resistant **(r)** or susceptible **(s)** according to the level of bacterial burden in their spleens (Fig. 2). When compared with the previously established **(r)** and **(s)** types for **Lsh** and **Ity** , an identical distribution of **r** and **s** alleles for these three genes was found.

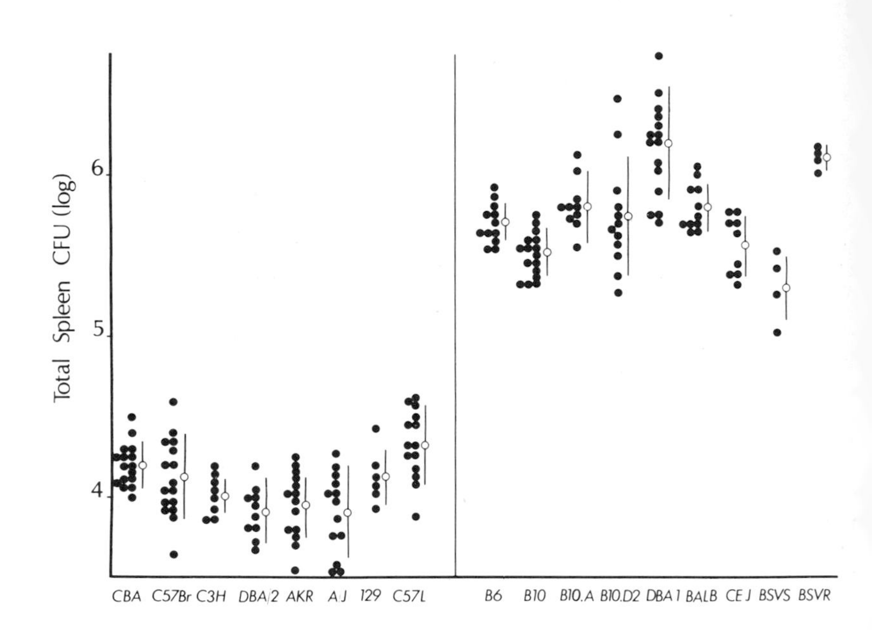

Bcg	r	r	r	r	r	r	r	r	s	s	s	s	s	s	s	s	s
Lsh	r	r	r	r	r	r	r	r	s	s	s	s	s	s	s	s	s
Ity	r	r	r	r	r	r	r	r	s	s	s	s	s	s	s	s	s

FIGURE 2. **Bcg** typing of inbred mouse strains. Closed circles: **Bcg** types of individual animals. Open circles: mean **Bcg** value of each strain ± S.E. **Lsh** and **Ity** types of the inbred strains taken from Ref. 2 and 3.

SURVEY OF RECOMBINANT INBRED STRAINS FOR **Bcg, Lsh** AND **Ity**

The typing method described in the previous paragraph was used to establish the **Bcg** type of BXD and BXH recombinant inbred strains that were derived from BCG-susceptible C57BL/6 (B) progenitor and BCG-resistant DBA/2 (D) or C3H/HeJ (H) progenitors, respectively (4). The strain distribution pattern of **Bcg**^{r} and **Bcg**^{s} alleles was compared with that established for **Lsh**^{r}/**Lsh**^{s} and **Ity**^{r}/**Ity**^{s} alleles (2,3,5,9) and found to be identical in 38 out of 38 recombinant inbred strains tested.

LINKAGE ANALYSIS

Since the strain survey and the analysis of recombinant inbred strains suggested a close linkage (or identity) of **Bcg, Lsh** and **Ity** genes, a formal proof was sought by the examination of individual animals obtained from a segregating population for the distribution of the three phenotypes in question. In these studies the A/J mice were used as the representative resistant **(Bcg^r, Lsh^r and Ity^r)** strain while B10.A mice were used as representative susceptible **(Bcg^s, Lsh^s and Ity^s)** parents (Fig. 2). Animals of the backcross (B10.A x A)F_1 x B10.A population were examined for **Bcg-Lsh** and **Bcg-Ity** linkage.

A. **Bcg-Lsh** Linkage

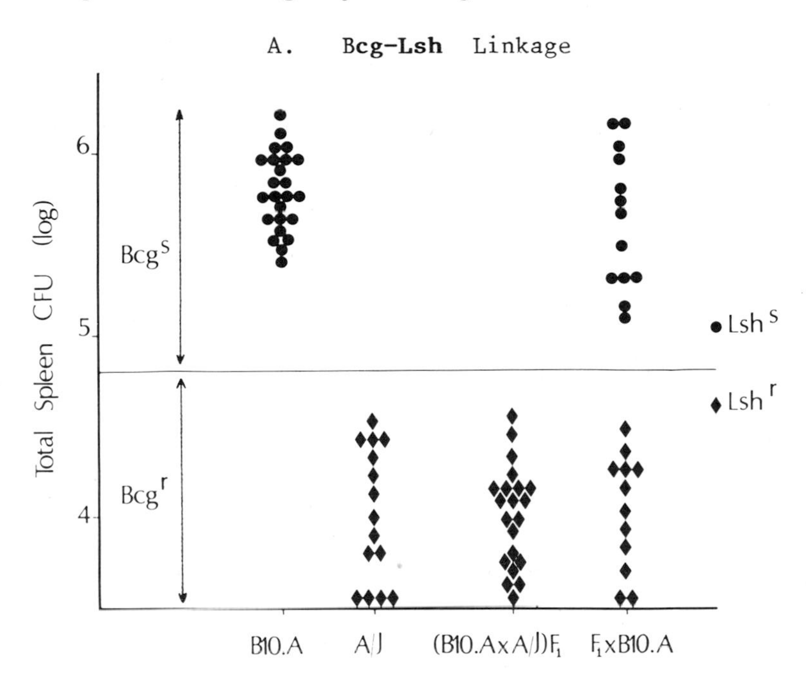

FIGURE 3. Backcross linkage analysis of **Bcg-Lsh** . Any individual animal of the backcross and control population that was typed as **Bcg^r** appears (with its **Bcg** value) below the horizontal line, any **Bcg^s** animal appears above the line. ● = **Lsh^s** , ♦ = **Lsh^r** .

Twenty-six backcross animals and the appropriate parental and F_1 hybrid controls were infected with BCG as described above and they were splenectomized 3 weeks later for **Bcg** typing. The same animals were then infected with *L. donovani* after a further 3 weeks and the **Lsh** type was established in their livers (6). Preliminary experiments confirmed that neither splenectomy nor previous infection with BCG influenced the **Lsh** typing. Twelve of the backcross animals were typed as $\mathbf{Bcg^r}$ and fourteen were typed as $\mathbf{Bcg^s}$. All twelve of the $\mathbf{Bcg^r}$ mice proved to be resistant to L. donovani ($\mathbf{Lsh^r}$) and all of the $\mathbf{Bcg^s}$ mice proved to be $\mathbf{Lsh^s}$ (Fig. 3).

B. **Bcg-Ity** Linkage.

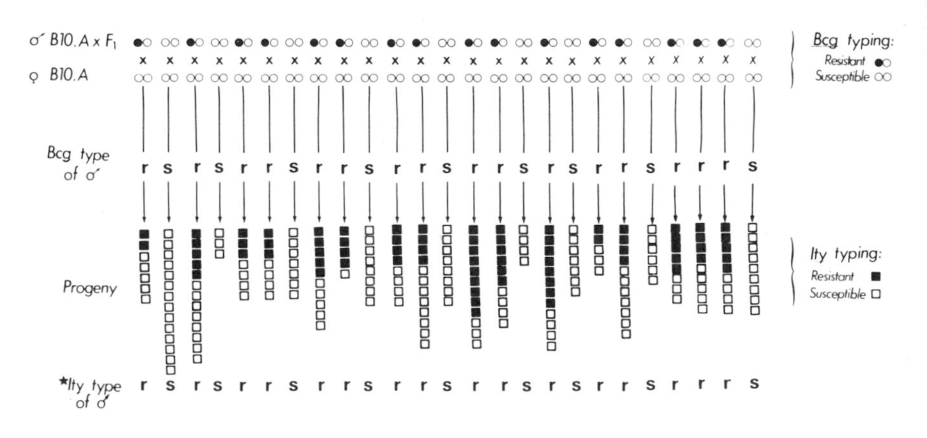

FIGURE 4. **Bcg-Ity** Linkage by Progeny Testing.

Since **Bcg-Ity** or **Ity-Bcg** typing of individual animals (by the double infection method) proved technically impossible, such linkage was sought by progeny testing. Twenty randomly chosen males of the backcross generation were mated with $\mathbf{Bcg^s/Ity^s}$ females of B10.A strain and their progeny were **Ity** typed at 7 days after intravenous infection with 10^4 S. typhimurium as $\mathbf{Ity^r}$ (survivors) or $\mathbf{Ity^s}$ (dead). The **Ity** type of each backcross male was then deduced from the survival pattern of the progeny, as either $\mathbf{Ity^s}$ (all progeny dead) or $\mathbf{Ity^r}$ (progeny segregating between survivors and dead). Subsequently, each of the backcross males was **Bcg** typed and the established $\mathbf{Bcg^r}$ or $\mathbf{Bcg^s}$ type was compared with the **Ity** type as deduced from the progeny typing. No conclusive **Bcg-Ity** discordance was found (Fig. 4).

MAPPING OF THE Bcg GENE

The close linkage (or, probably, an identity) in a

chromosomal locus of the **Bcg, Lsh** and **Ity** genes was confirmed by gene mapping. The strain distribution pattern of **Bcg^r** and **Bcg^s** alleles among the recombinant inbred strains was compatible with the gene being located on the chromosome 1 in the vicinity of **Idh-1** and **Pep-3** genes (coding for enzymes isocitrate dehydrogenase-1 and peptidase-3, respectively). The recombination frequency of **Bcg - Idh-1** and **Bcg - Pep-3**, which was established by the recombinant inbred strain analysis, places the **Bcg** gene between those two genetic markers (Fig. 5). Recent mapping of the **Ity** gene using the classical five-point backcross analysis is compatible with **Ity** being positioned within the same chromosomal region (7).

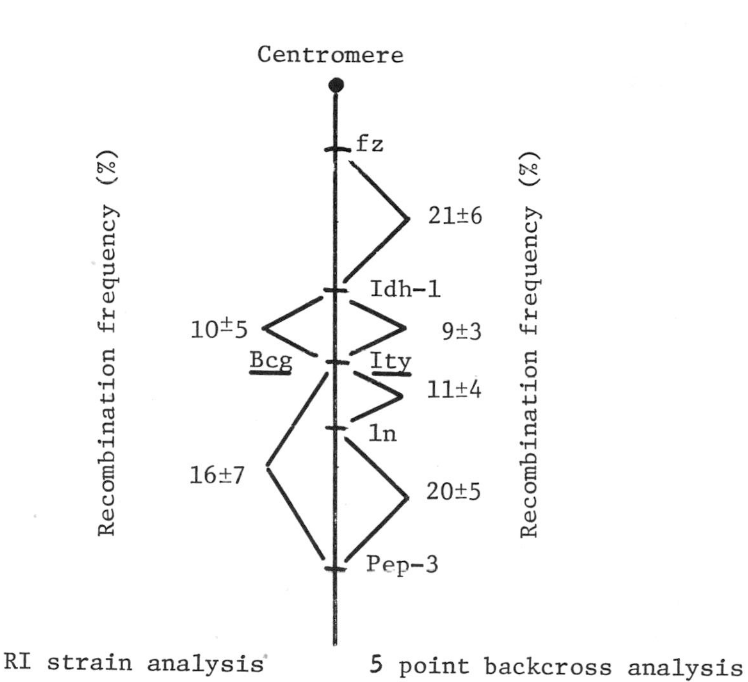

FIGURE 5. Mapping of **Bcg** and **Ity** loci on Chromosome 1.

CONCLUSION

The four lines of evidence presented in this paper (inbred strain survey, recombinant inbred strain study, linkage analysis and gene mapping) support the hypothesis that a single genetic locus on the murine Chromosome 1 controls resistance to (at least) three intracellular pathogens. The phenotypic expression of the **Bcg (Lsh, Ity)** gene, as it is known presently, is compatible with a cell-mediated natural resistance phenomenon. The genetic advantage of the resistant strain is apparent very early in the course of the host response to these infections, clearly before a specific immune response sets in. The fact that the three pathogens in question are antigenically and taxonomically unrelated also argues against specific acquired immunity being regulated by a single gene. Furthermore, the resistant phenotype is preserved in animals of the **Bcg^r** background which are depleted of (or genetically deficient in) the T and B lymphocyte populations. The resistance mechanism is silica-sensitive, and it is derived from hemopoietic bone-marrow precursors as demonstrated in radiation bone-marrow chimera experiments. Although no direct experimental proof is available as yet, all the indirect evidence points to the macrophage as being the cell expressing the genetic resistance. It seems likely that the cellular expression of the resistant allele is a microbiostatic (rather than microbicidal) effect of the intracellular milieu of the parasitized macrophage (8). The recent availability of mouse strains that are congenic except for the **Bcg (Lsh, Ity)** gene should greatly enhance the functional studies on the phenotypic expression of this gene.

REFERENCES

1. Gros, P., Skamene, E., and Forget, A. J. Immunol. 127:2417 (1981).
2. Plant, J., and Glynn, A.A. Clin. Exp. Immunol. 37:1 (1979).
3. Bradley, D.J., Taylor, B.A., Blackwell, J., Evans, E.P., and Freeman, J. Clin. Exp. Immunol. 37:7 (1979).
4. Taylor, B.A. in "Genetic Control of Natural Resistance to Infection and Malignancy". (E. Skamene, P.A.L. Kongshavn and M. Landy, eds.), p. 101. Academic Press, New York, 1980.
5. Plant, J.E., Blackwell, J.M., O'Brien, A.D., Bradley, D.J., and Glynn, A.A. Nature (1982), in print.
6. Actor, P. Exp. Parasit. 10:1 (1960).
7. Taylor, B.A., and O'Brien, A.D. Infect. Imm. (1982), in print.
8. Hormaeche, C.E. Immunology 41:973 (1980).
9. O'Brien, A.D., Rosenstreich, D.L., and Taylor, B.A. Nature 287:440 (1980).

GENETIC VARIATION IN NATURAL KILLER (NK) ACTIVITY IN THE RAT

Craig W. Reynolds
Howard T. Holden

Biological Response Modifiers Program
Division of Cancer Treatment
NCI - FCRF
Frederick, Maryland

I. INTRODUCTION

It has recently become clear that immunogenetic factors play an important role in determining the levels of *in vitro* and *in vivo* natural killer (NK) activity. However, the mechanism(s) responsible for these alterations in activity are not clear. Two main alternatives for these alterations might be considered: 1) an increased or decreased number of effector cells, or 2) an altered regulation in cytolytic activity. To investigate these possibilities, it would be helpful to be able to isolate, morphologically identify, enumerate and directly examine the interaction of effector cells with target cells. Recently we have been able to isolate and identify a specific subpopulation of rat lymphocytes which mediate NK activity (Reynolds *et al*, 1981a; Reynolds *et al*, 1981b). These lymphocytes are termed large granular lymphocytes (LGL) and can be distinguished by their high cytoplasm to nuclear ratio, eccentric kidney-shaped nucleus and characteristic azurophilic granules in the cytoplasm. This observation and the availability of a large number of inbred rat strains makes the rat a highly promising model for studying the role of immunogenetic factors in determining NK cell activity. The present study examines in detail the levels of NK activity in a number of these strains. The results demonstrate that a variety of genetic factors may be important in determining NK levels.

ISBN 0-12-341360-5

II. MATERIALS AND METHODS

Animals . Unless otherwise stated, all experiments were performed with 7-12 wk old male rats. A complete list of these animals can be found in Table 1.

Target cells. The rat G_1-TC and mouse YAC-1 lymphomas were maintained in vitro as suspension cultures (Oehler et al, 1978; Nunn et al, 1976). The rat pheochromoblastoma, SP3, was kindly provided by Dr. Margot Zoller and maintained in vitro as a monolayer culture.

Preparation of lymphoid cells. Spleen, thymus, lymph node, bone marrow (BM), peritoneal exudate cells (PEC) and peripheral blood leukocytes (PBL) were prepared as previously described (Ortiz de Landazuri & Herberman, 1972; Oehler et al, 1977). Adherent cells were removed on columns of nylon-wool (NW). For morphological evaluation of lymphocyte preparations, air dried cytocentrifuge preparations were stained with 10% Giemsa (pH. 7.4).

^{51}Cr-release cytotoxicity assay. Two-fold serial dilutions of effector cells were mixed with ^{51}Cr-labeled target cells and incubated at 37°C for 4 or 18 hours, as previously described (Reynolds & Herberman, 1981). Results are expressed as either percent cytotoxicity or lytic units (LU) per 10^7 effector cells (Herberman et al, 1982).

III. RESULTS AND DISCUSSION

Since previous studies have shown considerable variation in NK cell activity among various rat strains (Oehler et al, 1978; Zoller & Matzku, 1980), we investigated whether a simple correlation could be found between the major histocompatibility complex (MHC) in the rat and NK activity (Table 1). Little or no correlation was observed between the MHC and NK activity. However, it should be noted that only one $RT1^n$ strain was examined (BN) and it was found to be deficient in NK activity. Clearly, it will be important to determine if other $RT1^n$ strains or congenic $RT1^n$ stains may also have low activity. Studies are currently under way to investigate this possibility. The Rowlett nude rat, tentatively designated $RT1^c$ (Rolstad & Fossem, 1980) was found to have very high activity whereas all other $RT1^c$ stains were found to have intermediate cytotoxicity. These results suggest that although immunogenetic factors do play a role in determining NK levels in rats but the genetics of these control mechanisms are not clear.

Table 1. NK Activity of Athymic and Euthymic Rat Strains[a]

Strain	RT1	RT2	RT3	Cytotoxicity
Athymic				
rnu/rnu	c[b]	-[c]	-	High
Euthymic				
ACI	a	b	a	Intermediate
AUG	c	a	a	Intermediate
BD II	u	-	-	Intermdeiate
BD IX	d	-	-	Intermediate
BD X	d	-	-	Intermediate
BN	n	a	b	Low
BUF	b	a	b	Intermediate
COP	a	b	-	Intermediate
DA	a	b	a	Intermediate
F344	l	a	b	Intermediate
LA/N	b	b	-	Intermediate
LEW	l	a	a	Intermediate
M520	b	a	a	Intermediate
NBR	l	b	a	Intermediate
PVG	c	b	a	Intermediate
PVG-RT1[a]	a	b	a	Intermediate
WF	u	b	a	Intermediate

[a]Genetic designations from: The Laboratory Rat Volume 1, Edited by H. J. Baher, *et al.*, Academic Press, New York, New York (1979).
[b]Tentative designation.
[c]Designation unknown.

F_1 offspring between an intermediate and low cytotoxicity strain were next examined for NK activity against an allogeneic lymphoma (G_1-TC), a xenogeneic lymphoma (YAC-1) and an allogeneic adherent pheochromoblastoma (SP-3). The results (Table 2) demonstrate that in both a 4 and 18 hour assay that these F_1-offspring had NK levels between that of either parental strain. Although the interpretation of these results will be difficult until the mechanism for low NK activity in BN animals becomes clear, it does provide contrasting information to the mouse where intermediate or high NK seems to be a dominant trait (Petranyi *et al*, 1975; Petranyi *et al*, 1976; Kiessling *et al*, 1975a; Kiessling *et al*, 1975b).

In contrast to the BN, effect of the nude (rnu) gene on NK activity is much more clear. Heterozygous animals (rnu/+) have intermediate levels of both NK activity and LGL in all

Table 2. Cytotoxic Activity of F_1-Offspring from Intermediate and Low NK Strains

		% Cytotoxicity (E/T=25:1)			
		YAC-1	G_1-TC	SP3	
Strain[a]	Organ	4 hr	4 hr	4 hr	18 hr
LEW	Spleen	59.6	22.8	23.2	42.4
	PBL	71.9	27.6	21.6	35.8
BN	Spleen	9.4	1.1	1.6	0.0
	PBL	18.4	4.5	0.0	12.8
(LEW/BN)F_1	Spleen	30.7	10.9	10.1	20.4
	PBL	45.6	18.8	23.6	31.4

[a]Eight to twelve week old animals.

organs examined (Table 3). In this case, intermediate NK activity (and normal thymus development) is dominant to the nude mutation. Only homozygous nudes (rnu/rnu) were found to have high NK activity. The mechanism by which this high NK activity is achieved seems to be through an increase in

Table 3. Frequency and Organ Distribution of NK Activity in Athymic (rnu/rnu) and Euthymic (rnu/+) Rats[a]

Strain	Organ	Cytotoxicity LU (20%)/10^7 cells[b]	% LGL
rnu/rnu	PBL	93.4	22
	-NW Passed	62.0	53
	Spleen	71.8	4
	-NW Passed	167.2	35
	Lymph node	13.4	2
	PEC	0.9	1
	Bone marrow	<1.0	<1
rnu/+	PBL	18.2	8
	-NW passed	18.3	14
	Spleen	10.7	2
	-NW passed	41.1	5
	Lymph node	<1.0	<1
	PEC	<1.0	<1
	Bone marrow	<1.0	<1
	Thymus	<1.0	<1

[a]Six-twelve weeks of age
[b]4 hour assay against G_1-TC cells

the percentage of LGL effector cells. In general, a 2-7 fold increase in both NK activity and LGL percentage was observed in nude rats as compared to their euthymic littermates. These results are consistent with previous suggestions that NK cells are in the T cell lineage but thymic independent, and thus, in an athymic environment, more T cell progenitors may be available to develop along the NK pathway. Alternately these results also are consistent with a role for the thymus in the negative regulation of effector cell numbers.

We next examined high, intermediate and low NK strains for age variation in cytotoxic activity and LGL numbers (Table 4). With the exception of BN rats, there was very little decline in either NK numbers or activity with age. In contrast, the BN rats showed a marked age-related decline in

Table 4. Strain and Age Variation in LGL Frequency and NK Activity[a]

Strain	Age (wk)	Organ	% Cytotoxicity	% LGL
WF	8-10	Spleen	20.5	4.0
		PBL	22.1	7.0
	15-20	Spleen	15.8	3.5
		PBL	21.4	7.0
F344	8-10	Spleen	17.8	1.5
		PBL	29.6	6.0
	15-20	Spleen	14.0	NT[b]
		PBL	NT	NT
BN	8-10	Spleen	5.5	5.0
		PBL	20.2	8.5
	15-20	Spleen	0.0	1.0
		PBL	1.6	8.0
rnu/+	8-10	Spleen	16.4	2.0
		PBL	20.8	8.0
	15-20	Spleen	17.6	1.0
		PBL	27.0	5.0
rnu/rnu	8-10	Spleen	30.1	5.0
		PBL	30.6	22.0
	15-20	Spleen	26.1	4.5
		PBL	24.1	27.5

[a]4 hour assay against G_1-TC. E/T ratio euthymic 100:1 for spleen, 50:1 for PBL; rnu/rnu- 50:1 for spleen and 25:1 for PBL.

[b]Not tested.

peripheral blood NK activity. The spleens from these animals had low or undetectable levels of cytotoxicity at all ages tested. Interestingly, by 20 weeks of age a decline in LGL percentages was seen in the spleen of these BN rats but not in the blood. These results suggest that the low NK activity seen in BN rats may be due to multiple factors. At least one factor is the decrease in LGL numbers found in the spleens of older animals. This observation however can not account for the low activity seen in the blood of older animals, since normal levels of LGL are still present. Preliminary studies from our laboratory have shown that the NK activity of PBL in these older animals can be augmented by interferon-containing supernatants (data not shown), although the NK levels never reached that of intermediate strains. These results do suggest that there is no defect in the augmenting potential of LGL from BN rats. Another possible factor which could affect NK activity is the suppressor cell. Since these cells have been described previously, this must be considered as a possible explanation for loss of NK activity in BN rats.

The ability to identify NK cells (LGL) in the rat and the observed genetic variation in their cytotoxic activity should provide a sound basis for the further analysis of high and low NK strains and their genetic nature.

REFERENCES

Herberman, R. B., Ortaldo, J. R., and Timonen, T. Methods Enzymol., In Press (1982).
Kiessling, R., Klein, E., and Wigzell, H., Eur. J. Immunol 5:112 (1975a).
Kiessling, R., et al., Int. J. Cancer 15:933 (1975b).
Nunn, M. E., et al., J. Natl Cancer Inst. 56:393 (1976).
Oehler, J. R., et al., Cell Immunol 29:238 (1977).
Oehler, J. R., et al., Int. J. Cancer 21:204 (1978).
Ortiz de Landazuri, M., and Herberman, R. B., J. Natl. Cancer Inst. 49:147 (1972).
Petranyi, G. G., Kiessling, R., and Klein, G., Immunogenetics 2:53 (1975).
Petranyi, G. G., et al., Immunogenetics 3:15 (1976).
Reynolds, C. W., Timonen, T., and Herberman, R. B., J. Immunol. 127:282 (1981a).
Reynolds, C. W., et al, J. Immunol. 127:2204 (1981b).
Reynolds, C. W., and Herberman, R. B., J. Immunol. 126:1581 (1981).
Rolstad, B., and Fossem, S., Rat News Letter 8:14 (1980).
Zoller, M., and Matzku, S., J. Immunol. 124:1683 (1980).

ONTOGENY OF Nk-1+ NATURAL KILLER CELLS

Koo, G.C.[1]
Peppard, J.R.
Hatzfeld, A.[2]
Cayre, Y.[3]

Memorial Sloan-Kettering Cancer Center
New York, New York

Recent experiments have demonstrated Nk-1 antigen to be specific for murine NK cells capable of lysing lymphoma targets (Glimcher et al., 1977; Koo et al., 1980). We have attempted to follow the ontogenetic development of Nk-1+ cells in fetal, baby (1-2 wk) and old (1 yr) mice. We used the protein A-SRBC rosette assay to monitor the proportion of Nk-1+ cells in tissues of mice and further performed induction assays with interferon to look for precursors of Nk-1+ cells (Koo et al., 1982). Our results are summarized as follows: (a) About 18% of Nk-1+ cells were found in fetal thymus on day 14 of gestation but these Nk-1+ cells were transient and became undetectable on day 16. Instead, Nk-1+ cells (∿30%) were observed in spleen and liver until birth (Table I).

TABLE I. Reactivity of anti-Nk-1.1 serum with C57BL/6 fetal thymus, liver and spleen cells

	% PA-SRBC rosettes[a]		
Days of gestation	thymus	liver	spleen
14	18	14	NT
16	3	30	32
18	2	38	34
20	2	5	35

[a]% PA-SRBC rosettes were assessed with anti-Nk-1.1 serum at a dilution of 1:15.

[1]Supported by NIH grants CA-08748, CA-25416, ACS(FRA-167) NSF and NATO
[2]Present address: Institut de Pathologie Cellular, Kremlin Bicetre, France
[3]Present address: Department of Genie Genetique, Institut Pasteur, Paris, France

ISBN 0-12-341360-5

Contrary to Nk-1+ cells, Thy-1+ cells were continuously found in thymus from day 14 of gestation until birth. Very low numbers of Thy-1+ cells were detected in liver and spleen. Thus, Nk-1 antigen, similar to Asialo-Gm-1 (Habu et al.,1981) is an early hematopoietic antigen, but we do not support the contention that Nk-1+ cells are prothymocytes based on our earlier findings that Nk-1+ NK cells and prothymocytes appear to be distinct (Koo et al.,1981) (b) A large proportion (20-30%) of Nk-1+ cells are found in spleen of mice from birth to 2 wk. old. Very low lytic activity could be demonstrated to RL♂1 or YAC cells, even with an 18 hr ^{51}Cr release assay. These Nk-1+ cells do bind to targets and elimination of Nk-1+ cells abolished the target binding cells (Table II). Induction with IF for 18 hours resulted in an increase of Nk-1+ cells but little increment in NK activity. We concluded that baby Nk-1+ cells were likely immature physiologically and our data substantiated the suggestion that target binding cells in baby spleen compete for the targets when baby spleen cells were observed to "suppress" NK lysis (Cudkowicz and Hochman, 1979)

TABLE II. Target Binding Cells in Adult and Baby Spleens Are Nk-1+

Eliminated with[a]	Adult spleen cells[b] % Nk-1+ [c]	Adult spleen cells[b] % TBC[d]	Baby spleen cells % Nk-1+	Baby spleen cells % TBC
- -	17	18	20	27
NMS	13	13	18	18
anti-Nk-1.1	1	2	3	5

[a]After formation of PA-SRBC rosettes with anti-Nk-1.1 serum, Ficoll elimination was performed as described in Koo et al. (1980) NMS, normal mouse serum.

[b]adult (6-8 wk old) spleen cells were enriched by BSA gradient; baby (1-2 wk old) spleen cells were not enriched.

[c]% Nk-1+ cells were assessed by PA-SRBC test using anti-Nk-1.1 serum (dil. 1:10).

[d]TBC, target binding cells, were assessed by using YAC cells.

TABLE III. Effect of Interferon on the Proportion of $Nk\text{-}1^+$ Cells and NK activity in splenocytes of young (5d) and old (1 yr) C57BL/6 mice

	$Nk\text{-}1^+$ cells [a] (% rosettes)		NK activity [b] (% lysis)	
Age of mice	Without IF	With IF[c]	Without IF	With IF[c]
6-8 wk [d]	14 ± 1	26 ± 3	17 ± 2	26 ± 3
5 day[d]	23 ± 5	38 ± 3	1 ± 1	5 ± 3
12-14 months[e]	11 ± 4	24 ± 7	9 ± 1	23 ± 1

[a]Assessed by PA-SRBC rosette assay, using anti-Nk-1.1 antiserum: (C3H x BALB/c) anti CE spleen; spleen cells from 6-8 week old and 12-14 month old mice were enriched by BSA gradient

[b]NK activity to YAC cells. E:T ratio = 100:1

[c]Induction experiments with interferon: 10^7 cells in one ml were incubated with RPMI or 10^3-10^4 units IF for 24 hr. at 37°C and 5% CO_2; cells were washed 1x and assayed for $Nk\text{-}1^+$ cells and NK activity

[d]Mean ± S.D. of 3 representative tests from 14 experiments

[e]Mean ± S.D. of 3 experiments.

(c) The discordance between $Nk\text{-}1^+$ cells and NK activity was also observed in old (1 yr old) mice. Old mice had similar proportion of $Nk\text{-}1^+$ cells as young adult mice (6-8 wk old), although their lytic activities were generally lower. Induction with IF could however induce NK activity to YAC and these induced NK cells were positive for Nk-1 antigen (Table III).

From these studies, one can therefore distinguish two populations of $Nk\text{-}1^+$ cells: inactive $Nk\text{-}1^+$ and active $Nk\text{-}1^+$ NK cells. The increase of $Nk\text{-}1^+$ cells after IF induction suggests the existence of a precursor of $Nk\text{-}1^+$ ($Nk\text{-}1^-$ or pre-Nk-1) cells. Thus, using the serological marker, Nk-1 antigen we observed a scheme of NK differentiation, similar to that proposed by Targan and Dorey (1980):

$$\text{pre-Nk-1} \rightarrow \text{Nk-1}^+ \rightarrow \text{Nk-1}^+\ \text{NK}$$

INACTIVE INACTIVE ACTIVE

Presently we cannot conclude if Nk-1+ cells are the immediate precursors to Nk-1+ NK cells, and we also have not yet identified the antigenic phenotype of pre-Nk-1 cells. Based on our ontogenetic data presented here splenic Nk-1 antigen is detectable earlier than other NK associated antigens, namely Thy-1 (Spear et al.,1973), Qa 2,3 (Kincade et al. 1980), Qa 4,5 (Olson et al.,1981) antigens. Therefore, these antigens are probably not on the pre-Nk-1 cells. Recently, we have separated splenic population into fractions that are positive or negative for the receptors to the lectin, peanut agglutinin (PNA). We showed that Nk-1+ NK cells were negative for PNA receptor (Koo et al.,1981, also see Koo et al. under cell lineage in this book). We have also induced cells from splenic PNA+ and PNA− fractions with interferon to identify the putative pre-Nk-1 cells. We have observed an increase of Nk-1+ cells in the PNA+ fraction, suggesting that the precursors, pre-Nk-1 have receptors for PNA. In addition, there was no increase in Nk-1+ cells in the PNA− fraction after IF induction, only enhanced lytic activity was observed. These results further support our contention that IF can induce both a precursor pre-Nk-1 to become Nk-1+ and can also induce Nk-1+ to become active NK cells.

ACKNOWLEDGMENTS

We thank Barbara LoFaso for preparation of the manuscript.

REFERENCES

Cudkowicz, G. and Hochman, P.S. (1979). Immunol. Rev. 44:13.
Glimcher, L., Shen, F.W. and Cantor, H. (1977). J. Exp. Med. 145:1.
Habu, S., Kasai, S.M., Nagai, Y. (1981). J. Immunol. 125:2284.
Kincade, P.W., Flaherty, L., Lee, G., Watanabe, T. and Michaelson, J. (1980). J. Immunol. 124:2879.
Koo, G.C., Jacobson, J.B., Hammerling, G.J. and Hammerling, U. (1980). J. Immunol. 125:1003.
Koo, G.C., Cayre, Y., and Mittl, R. (1981). Cell. Immunol. 62:164.
Koo, G.C., Peppard, J.R., and Hatzfeld, A. (1982). Submitted for publication.
Olson, C.L., Hammerling, U., and Abbott, J. (1981). Cell. Immunol. 57:265
Spear, P.G., Wang, A.L., Rutishauser, U. and Edelman, G. (1973) J. Exp. Med. 138:557.
Targan, S., and Dorey, F. (1980). J. Immunol. 124:2157.

MARROW DEPENDENCE OF NATURAL KILLER CELLS[1]

Vinay Kumar[2]
P. F. Mellen
Michael Bennett[2]

Department of Pathology
Boston University School of Medicine
Boston, Massachusetts

I. INTRODUCTION AND RESULTS

In mice, elimination of functional bone marrow, either by irradiation with the bone-seeking isotope, ^{89}Sr, or by induction of osteopetrosis with estradiol, leads to marked impairment in NK cell activity against certain tumor cells (1,2). Thus NK cell activity (NK-A) against YAC-1, MPC-11, RL--1 and C1-18 tumor cells is extremely low in mice injected with ^{89}Sr, whereas NK cell activity (NK-B) against WEHI-164.1, FLD-3, EL-4 and Meth A target remains largely normal (1). We propose that an intact bone marrow microenvironment is critical for the differentiation of functional NK-A cells (1-3). Here we will present this hypothesis and discuss its validity in the light of recent evidence.

[1]Supported by NIH grants CA21401, CA31792 and AI18811.
[2]Present address: Department of Pathology, Southwestern Medical School, Dallas, Texas.

ISBN 0-12-341360-5

A. The Hypothesis: Testing the Predictions

The hypothesis in its simplest form is presented in Fig. 1. It states that bone marrow constitutes a "central lymphoid" organ for the differentiation of NK-A cells, just as the thymus and bursa-equivalent tissue constitutes the central lymphoid tissues for T and B cell differentiation. The effect of bone marrow destruction on NK-A cell function may thus be viewed as analogous to the effect of thymectomy on T cell functions. This hypothesis, which is based largely on the study of NK(YAC-1) activity in 89Sr-treated mice, leads to three testable predictions indicated in Figure 1. A corollary of this hypothesis is that NK(YAC-1) cells may not be in the direct lineage of T cells or macrophages.

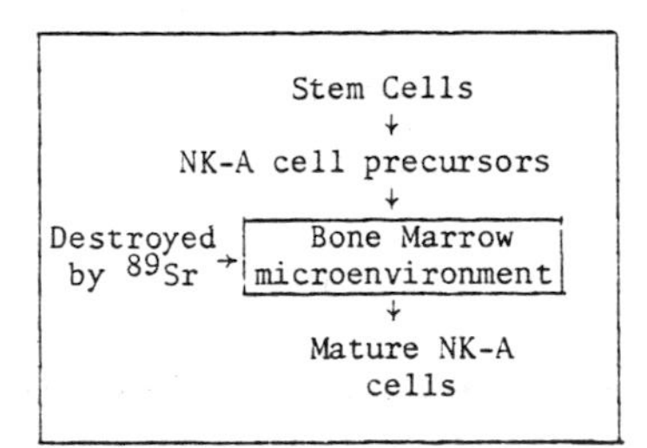

1. Spleen should have precursors of NK-A cells.
2. Grafting bone marrow cells should not restore NK-A activity.
3. Grafting bone marrow microenvironment should restore NK-A activity.

FIGURE 1. Hypothesis--predictions in mice treated with 89Sr.

1. Precursors of NK-A cells in 89Sr-treated B6D2F1 Mice. If the low NK(YAC-1) activity in 89Sr-treated mice is due to arrested differentiation of NK cell precursors, 89Sr-treated mice should possess precursors of NK(YAC-1) cells. Since the spleen takes over all of the stem cell functions in 89Sr-treated mice, we asked the queston whether spleens of 89Sr-treated mice possess transplantable precursors of NK(YAC-1) cells. Lethally (800R) irradiated mice were infused with 6 to 24 x 10^6 splenocytes from 89Sr-treated or control mice, and NK(YAC-1) activity generated in the recipients' spleens 4-8 weeks after cell transfer was measured. The NK(YAC-1) cells detected under these conditions would be derived from their precursors present in the donor inoculum. Both normal spleen cells (NSC) and spleen cells from 89Sr-treated mice (89SrSC) generated new NK(YAC-1) cells. 89SrSC reconstituted mice developed somewhat lower levels of NK(YAC-1) activity as compared with NSC. However, after injection of polyinosinic: polycytidylic acid or heat-killed Corynebacterium parvum organisms, spleen cells from both the groups of reconstituted mice showed equal and strong (YAC-1) NK activity (4). This experiment indicates that spleens of 89Sr-treated mice contain transplantable precursors of NK(YAC-1) cells.

2. Grafting Bone Marrow Cells into 89Sr-treated mice. According to the hypothesis presented (Fig. 1), one would expect that infusion of NK cell precursors into 89Sr-treated mice would not result in the generation of mature and functional NK(YAC-1) cells (just as infusion of T cell precursors in thymectomised recipients would fail to generate mature T cells). 89Sr-treated B6D2F1 mice (100 uCi, -5 weeks) were infused with 10^7bone marrow cells. Some of the 89Sr-treated mice were lethally irradiated (800R) prior to bone marrow cell infusion. Controls were age-matched, lethally irradiated B6D2F1 mice infused with the same pool of bone marrow cells. NK(YAC-1) activity was measured in the spleens 2 to 10 weeks after bone marrow cell transfer. Whereas injection of bone marrow cells generated new NK(YAC-1) cells within 2-3 weeks in lethally irradiated normal recipients, they failed to do so in unirradiated or irradiated 89Sr-treated mice (3). To exclude the possiblity that the bone marrow cells infused failed to differentiate due to some suppressive influence exerted by the spleen, the above experiments were repeated in 89Sr-treated mice which were splenectomised prior to bone marrow cell transfer. Once again, bone marrow cells failed to restore NK(YAC-1) activity in splenectomized 89Sr-treated mice (see details of these experiments elsewhere in this Volume, Kumar, et.al.) Together these experiments indicate that NK(YAC-1) cells cannot be generated from their precursors in the absence of an intact bone marrow.

3. Grafting Bone Marrow Microenvironment in 89Sr-treated Mice. From the above experiments, one could suggest that the low NK(YAC-1) activity in 89Sr-treated mice is due to lack of some component of the bone marrow, other than the marrow stem cells. As a preliminary test of this hypothesis, we grafted 4-5mm long segments of normal syngeneic tibias into the splenic pulp of 89Sr-treated mice. The bone segments were flushed prior to insertion into the spleen of 89Sr-treated mice. Four to six weeks after bone implantation, NK(YAC-1) activity of the recipient mice was tested. The bone implant was carefully dissected and removed from the spleen prior to making the cell suspensions. Under these conditions, there was significant restoration of the NK(YAC-1) activity in 89Sr-treated mice implanted with normal bones. The level of killing by the spleen cells of such mice was 50-60% of the controls (100% value=35% lysis).

Incomplete restoration was not entirely unexpected, since only a single bone segment could be implanted into the spleen. NK(YAC-1) activity remained low in sham-operated ^{89}Sr-treated donors. These experiments, although preliminary, support the hypothesis that NK(YAC-1) cell differentiation is dependent upon intact bone marrow microenvironment.

B. Relationship Between NK-A and Pre-T Cells

Several characteristics of NK cells have led to the suggestion that NK-A cells may be pre-T cells, i.e. those cells which are committed to the T cell pathway but have not yet differentiated into mature T cells. To explore the possible relationship between pre-T cells and NK cells, we performed three sets of experiments. (i) We investigated the effect of pre-T cell depletion from the bone marrow on its ability to repopulate thymocytes and NK(YAC-1) activity in lethally irradiated mice. B6D2F1 (H-$2^{d/b}$)bone marrow cells were incubated with various doses of thymopoietin (10^{-5} to 1ug) for 2 hours, followed by treatment with monoclonal anti-Thy-1.2 antibody (1:100) + C. Thymopoietin causes enhanced expression of Thy 1.2 antigen on pre-T cells, many of which are then lysed by anti Thy 1.2 + C(5). Treated and control bone marrow cells were injected into lethally irradiated BALB/c (H-2^{d}) mice (1 x 10^6/mouse). At several time points after bone marrow cell infusion, the repopulation of the thymus glands by donor type cells was assessed by serologic detection of H-2^b positive cells within the thymus. At the same time, the generation of NK-A cell activity was monitored by performing NK(YAC-1) assays with the spleen cells. Under these conditions, pre-T cell depleted bone marrow cells were significantly impaired in their ability to repopulate the thymus but the repopulation of NK(YAC-1) activity was equal to that of control bone marrow cells. (Table I. Exp. 1) This was true at all time points tested between 3 to 8 weeks after cell transfer. These experiments indicate that NK(YAC-1) cells and thymocytes do not share a common <u>thymopoietin sensitive</u> (pre-T cell) precursor. (ii) In the second set of experiments, we tested the effect of depleting NK 1.2 antigen positive cells (NK-A cells) on the repopulation of thymocytes. B6D2F1 (H-$2^{d/b}$) spleen cells were pretreated with anti NK-1.2 (CE anti CBA) serum + C, a procedure which removes 90% of NK(YAC-1) activity but lyses only 5-10% of the spleen cells. Fifteen x 10^6 treated and control cells were injected into lethally irradiated BALB/c (H-2^d) mice and 15-30 days later, donor type thymocytes were enumerated in the recipient mice (as described above). NK 1.2 + C treated spleen cells were as efficient as control spleen cells in repopulating the thymus. (Table I Exp 2)

Thus mature NK 1.2 positive NK-A cells seem not to be the precursors of most thymocytes. It cannot be ruled out, however, that NK 1.2 positive NK cells may be the precursors or a minor population or subset of T cells. (iii) The third approach utilized was to determine if incubation of spleen cells with thymopoietin leads either to a reduction of the splenic NK(YAC-1) activity or to an enhanced expression of Thy 1.2 antigen on NK(YAC-1) cells. These two would be expected if differentiation along NK(pre-T) → T cell pathway occurs. Following incubation with thymopoietin, NK(YAC-1) activity was completely unaffected and the degree of reduction caused by monoclonal anti Thy 1.2 + C was also unaltered. (Table I, Exp. 3)

TABLE I. DISSOCIATION OF PRE-T CELLS FROM NK-A CELLS.

Exp.	Number Cells Transferred($X10^6$)	Pretreatment of cells	Thymocytes % Donor-type[a]	NK(YAC-1) Activity[b]
1	1. marrow	Anti-Thy-1.2+C	>70	++
	1. marrow	Thymopoietin, then Anti-Thy-1.2 + C	30	++
2	15. spleen	C only	>80	ND
	15. spleen	Anti-NK-1.2+C	>80	ND
3	Spleen Cells treated but not transferred	Vehicle	--	++
		Thymopoietin		++
		Anti-Thy-1.2+C		+
		Thymopoietin, then Anti-Thy-1.2 + C		+

a) Determined by H-2 typing of cells (see text)
b) ++=30-50%, +=15-30% specific cytotoxicity; ND, not done

II. COMMENTS AND CONCLUSIONS

It is well established that mice treated with 89Sr have markedly lowered NK(YAC-1) activity. Two mechanisms other than the one proposed by us deserve comments. (i) Suppressor cells acting at the effector level have been suggested (6), but despite extensive attempts, we have been unable to detect suppressors of NK(YAC-1), either at the effector or precursor levels in 89Sr-treated mice. (See elsewhere in this Volume, Kumar, V., et.al.). (ii) Bone marrow may be a unique source of NK cell precursors, and these are depleted in 89Sr-treated mice. This seems unlikely, since injection of normal bone marrow cells, (even repeatedly) fails to correct the NK defect in 89Sr-treated mice, whereas similar treatment restores the low NK(YAC-1) in beige mutant mice, which are known to have defective NK stem cells (7). Furthermore, NK(YAC-1) cell precursors can be detected in spleens of 89Sr-treated mice (4).

It is interesting to note that the frequency of NK antigen positive cells is normal in 89Sr-treated mice (Kumar, V., Burton, R.C., and Koo, G., unpublished). Not only are the NK antigen positive cells unable to lyse YAC-1 cells, but they cannot be boosted significantly by interferon or its inducers (3). Based on these observations, we suggest that NK-A cell differentiation can be separated into an early marrow-independent and a later marrow-dependent phase. The former is characterized by the acquisition of NK antigens, ability to bind to the tumor targets (2) and insensitivity to IFN. The marrow-dependent phase which follows requires the influence (humoral and/or stromal) of an intact bone marrow. During this phase, the cells acquire lytic potential and responsiveness to IFN. It should be emphasized that all lymphohemopoietic cells are normally marrow-derived. However, T cells, B cells, and macrophages are not marrow dependent, since their differentiation seems to proceed normally in the absence of an intact bone marrow. The rather selective NK-A cell defect seen in 89Sr-treated mice suggests that these cells are not only marrow-derived, but also strictly marrow-dependent (M) cells.

The hypothesis that NK-A cells are marrow-dependent does not preclude some relationship with T cells or macrophages. Indeed, there is evidence that certain stages of T cell differentiation may require an intact bone marrow (8). We do believe, however, that the sharing of antigens between T cells and macrophages is not enough evidence to assign a particular lineage to NK cells. Many important questions about NK cell lineage remain unanswered, including the relationship if any, between NK-A and NK-B cells.

REFERENCES

1. Lust, J.A., Kumar, V., Burton, R.C., Bartlett, S.P. and Bennett, M. J. Exp. Med. 154:306, 1981.
2. Seaman, W.E., Gindhart, T.D., Greenspan, J.S., Blackman, M.A. and Talal, N. J. Immunol. 122:2541, 1979.
3. Kumar, V., Ben-Ezra, J., Bennett, M. and Sonnenfeld, G. J. Immunol 123:1832, 1979.
4. Levy, E.M., Kumar, V. and Bennett, M. J. Immunol. 127:1428, 1981.
5. Kommuro, K., Goldstein, G. and Boyse, E.A. J. Immunol. 115:195, 1975.
6. Cudkowicz, G. and Hochman, P.S., Immunol. Rev. Rev. 143:728, 1979.
7. Roder, J.C., J. Immunol. 123:2168, 1979.
8. Stutman, O. Ann. Immunol. (Inst. Pasteur) 127C:943, 1976.

AGE-INDEPENDENT NATURAL KILLER CELL ACTIVITY IN MURINE PERIPHERAL BLOOD

Emanuela Lanza
Julie Y. Djeu

Division of Virology
Bureau of Biologics
Bethesda, Maryland

I. INTRODUCTION

There is general agreement that rodent natural killer (NK) cells show an age-related development of activity (Kiessling *et al.*, 1975; Herberman *et al.*, 1975). This activity is absent in the newborn spleen, matures to adult levels between 4-8 weeks of age, and is then followed by a period of decline, often to undetectable levels by 12 weeks and thereafter. The in vivo kinetics of mouse NK cell development differs significantly from that of humans, which appears to maintain a steady state for years, with little fluctuation seen between young and elderly persons (Herberman and Holden, 1978). This apparent difference disappears, however, when comparable lymphoid organs are analyzed in the two species. Analysis of mouse peripheral blood lymphocytes (PBL) showed that NK activity matures around 4 weeks of age but is maintained in circulation throughout a major portion of adult life.

II. METHODS

Heparinized whole blood was collected from mice by intracardiac puncture, diluted 1:3 in balanced salt solution and layered over a Ficoll-Hypaque gradient that had been adjusted for optimal separation of mouse PBL (Parish and Hayward, 1974). Spleen cells from the same mice were pooled and the levels of cytotoxicity in the 2 sites were measured in a 4 hr ^{51}Cr release assay against YAC-1 tumor cells. Removal of Thy-1 positive cells or asialo GM_1 positive cells was by pretreatment with allogeneic AKR anti-C3H (Herberman *et al.*, 1973) or rabbit anti-asialo GM_1 serum (Kasai *et al.*, 1980) in the presence of rabbit complement.

ISBN 0-12-341360-5

Inactivation of macrophages was by preincubation with silica or carrageean (Djeu *et al.*, 1979b) for 3 hr at 37°C prior to testing for NK activity.

III. RESULTS AND CONCLUSIONS

The development of NK activity was followed in CBA/J and C3H/HeN mice from 2 weeks to more than one year (Table 1). The activity was absent in both the spleen and PBL of 2 week old mice but it started to rise shortly thereafter. Adult levels were attained between 4-8 weeks in the 2 organs but beyond this point the levels in the PBL and the spleen began to diverge. NK activity remained highly detectable in PBL at all ages tested, up to 68 weeks. Although PBL-NK activity showed a slight downward trend with age, it did not exhibit the age-dependent loss of activity characteristic of the spleen. The pattern in mouse PBL was therefore comparable to that observed for NK cells in human peripheral blood. Of considerable interest was the consistently higher activity observed in PBL than in the spleen of the same mice tested, even when these were examined in 6-8 week old mice at the height of splenic activity.

TABLE 1. NK Activity: Age Dependence in Spleen and PBL

Weeks of Age	% Cytotoxicity			
	CBA/J		C3H/HeN	
	Spleen	PBL	Spleen	PBL
	200:1[a]	200:1	200:1	200:1
2	-	-	0.9	1.2
4	20.5	35.0	-	-
6	27.2	45.3	22.3	40.7
8	30.0	44.2	10.2	19.8
12	15.3	34.4	7.8	19.8
24	10.1	30.0	11.4	22.6
32	3.0	31.4	5.2	26.3
40	3.2	26.2	5.0	23.8
52	5.5	29.6	4.8	13.6
68	2.5	31.5	-	-

[a]Effector/Target ratio. (a lower ratio, 50:1 was tested with comparable results)

Mouse strains have generally been classified as being high or low reactive, by tests of splenic effector cells against YAC-1 tumor and other target cells (Herberman *et al.*, 1978a; Kiessling *et al.*, 1975). In testing PBL, the same strain distribution was seen, suggesting a common origin with splenic NK cells (Table 2). Moreover, they appeared to be under the same genetic influence. CBA/J and C3H/HeN mice that had high splenic activity were also found to contain high PBL-NK activity while BALB/c had intermediate levels in both organs. On the other hand, SJL or beige mice (Roder, 1978), which have been shown to have low splenic NK function, also had low or barely detectable activity in their PBL. Again, it was noted that the PBL-NK activity was substantially higher than the splenic activity in all the strains tested.

TABLE 2. NK Activity of Various Spleen and PBL Strains

Strain[a]	% Cytotoxicty	
	Spleen	PBL
	200:1[b]	200:1
nu/nu	31.8	41.6
CBA/J	28.2	43.5
C3H/HeN	22.8	33.0
BALB/c	7.5	28:1
SJL	6.5	8.0
bg/bg	2.5	5.0

[a]All the mice used were 6 wk of age.

[b]Effector/Target ratio. (a lower ratio, 50:1 was also tested with comparable results

Analysis of surface markers showed that PBL-NK cells possessed asialo-GM_1 but low levels of Thy-1 antigens (Table 3), which are characteristic of splenic NK cells (Herberman *et al.*, 1978a; Djeu *et al.*, 1979a; Herberman *et al.*, 1978b; Durdik *et al.*, 1980). These cells were also non-phagocytic, and showed resistance to macrophage-toxic agents such as silica and carageenan (Djeu *et al.*, 1979a).

TABLE 3. Detection of NK Activity after Treatment with Antisera or Macrophage-Toxic Agents

Treatment	CBA/J		Nu/Nu	
	Spleen	PBL	Spleen	PBL
Media	++	++	++	++
Anti-Thy-1 + C	+	+	+	+
Anti Asialo GM1 + C	-	-	-	-
Silica[a]	++	++	N.D.	N.D.
Carrageenan[a]	++	++	N.D.	N.D.

[a]Dose 50 ug/ml.

Another property shared with splenic NK cells was the ability to lyse targets other than YAC-1. The rank order of lysis of the target cells were identical between PBL and splenic effector cells, with YAC-1 being the most sensitive, followed by RBL-5, RLo1, B16, MOLT-4 and the most resistant being K562 (Table 4). Of interest was the sensitivity of WEHI-164 to lysis by PBL effector cells, which extends previous observations of the presence of natural cytotoxic (NC) cells in the spleen of mice (Lust et al., 1981). Lability at 37°C was another common characteristic, although PBL-NK activity appeared more resistant to this type of decay. After 4 hr of incubation, only 30% of splenic NK activity remained, as compared to 75% in the PBL. Longer incubation, up to 18 hr, however, brought both activities to similar low levels.

These data indicated that the long-held assumption that mouse NK activity differs from the human counterpart in its maturation and maintenance, may be incorrect. The activity in the mouse develops and matures to peak levels between 4-8 weeks of age and is then maintained in the PBL throughout adult life, much like the occurrence in human PBL. The activity in PBL followed the same strain distribution as had been reported for the spleen (Kiessling et al., 1975; Roder, 1978), which suggests that the genetic factors regulating NK activity are the same at the 2 sites. Analysis of surface and biological properties also suggest that the 2 effector populations stem from the same source. The only major difference that has so far been observed is the lack of influence of age on PBL-NK cells.

TABLE 4. Natural Cell-Mediated Cytotoxicity Against Different Target Cells

Target Cells[b]	Species Origin	Tumor Type	% Cytotoxicity Spleen 200:1[a]	PBL 200:1
YAC-1	Mouse	Moloney-leukemia	36.5	46.8
WEH1-164	Mouse	Fibrosarcoma	29.3	32.7
RBL-5	Mouse	Rauscher-virus	18.2	38.9
RL♂ 1	Mouse	radiation-induced leukemia	13.6	21.8
B-16	Mouse	Lung melanoma	5.9	16.0
MOLT-4	Human	T cell leukemia	6.3	8.2
K562	Human	Myeloid leukemia	0.4	3.2

[a]Effector/Target ratio (a lower ratio, 50:1 was tested with comparable results)

[b]The incubation time was 4 hr with YAC-1, RBL-5, RL♂ 1, and MOLT-4 and K562; 18 hr with WEHI-164 and B-16.

The reason for the different age distribution of mouse NK activity in the spleen and PBL is unclear. It may represent an actual difference in the number of NK effector cells and/or a difference in negative regulatory factors such as the presence of a larger number of suppressor cells in older spleens as have been detected in the spleens of infant mice whose lymphoid cells lacked detectable cytotoxic activity (Herberman *et al.*, 1978b; Hochman and Cudkowicz, 1979). Another possibility could be the presence of positive regulatory factors, *e.g.*, more production in PBL than in the spleen of interferon, which is known to boost NK activity in mice (Djeu *et al.*, 1979a). An alternative explanation may be that NK cells circulate freely in the peripheral blood, rather than remain resident in the spleen. This would facilitate rapid homing to organs that are being invaded by virus and/or tumor cells. Elucidation of the mechanism of differential regulation of NK cells in the 2 organs should provide some insight into the role of these cells in host resistance.

REFERENCES

Djeu, J.Y., Heinbaugh, J.A., Holden, H.T., and Herberman, R.B. J. Immunol. 122, 141 (1979a).

Djeu, D.Y., Heinbaugh, J.A., Holden, H.T., and Herberman, R.B. J. Immunol. 122, 175 (1979b).

Durdik, J.M., Beck, B.M. and Henney, C.S. in Natural Cell-Mediated Immunity against Tumors (ed. Herberman, R.B.) (Academic, New York, 1980).

Herberman, R.B., Nunn, M.E., Larvin, D.H., and Asofky, R. J. Natl. Cancer Inst. 51, 1509 (1973).

Herberman, R.B., Nunn, M.E., and Lavrin, D.H. Int. J. Cancer 16, 216 (1975).

Herberman, R.B., and Holden, H.T. Adv. Cancer Res. 27, 305 (1978a).

Herberman, R.B., Nunn, M.E., and Holden, H.T. J. Immunol. 121, 306 (1978b).

Hochman, P.S., and Cudkowicz, G. J. Immunol. 123, 968 (1979).

Kasai, M., Iwamori, M., Nagai, T., Okumura, K., and Tada, T. Eur. J. Immunol. 10, 175 (1980).

Kiessling, R., Klein, E., and Wigzell, H. Eur. J. Immunol. 5, 112 (1975).

Kiessling, R., Klein, E., Pross, H., and Wigzel. Eur. J. Immunol. 5, 117 (1975).

Lust, J.A., Kumar, V., Burton, R.C., Bartlett, S.P., and Bennett, M. J. Exp. Med. 154, 306 (1981).

Parish, C.R., and Hayward, J.A. Proc. Roy. Soc. Lond. B. 187, 65 (1974).

Roder, J.C. in Natural Cell-Mediated Immunity against Tumors (ed. Herberman, R.B.) (Academic, New York, 1980).

ONTOGENIC DEVELOPMENT OF PORCINE NK AND K CELLS[1]

Yoon Berm Kim
Nam Doll Huh[2]

Laboratory of Ontogeny of the Immune System
Sloan-Kettering Institute for Cancer Research
Rye, New York

Hillel S. Koren
D. Bernard Amos

Division of Immunology
Department of Microbiology and Immunology
Duke University Medical Center
Durham, North Carolina

I. INTRODUCTION

In light of the potential importance of natural killer (NK) and killer (K) cell systems in the host defense mechanism (Herberman and Holden, 1978; Kiessling and Haller, 1978; Herberman, 1980; Roder et al., 1981), we have initiated studies to define the ontogenic development, tissue distribution, cellular nature and regulatory mechanisms for development and/or activation of NK/K cells in gnotobiotic miniature swine (Koren et al., 1978; Kim et al., 1980; Huh et al., 1981a; Huh et al., 1981b) by taking advantage of the

[1]This work was supported in part by grants No. IM-233 from the American Cancer Society, Inc. and CA-08748 and HD-12097 from the National Institutes of Health, USPHHS.

[2]Present address: Department of Biochemistry, Johns Hopkins University, School of Hygiene and Public Health, Baltimore, Maryland.

ISBN 0-12-341360-5

fact that immunologically "virgin" piglets would not be contaminated with "natural" antibodies *in vivo* (Kim, 1975; Scheffel and Kim, 1979).

We have established that target cell systems for our miniature swine effector cells in spontaneous or natural killing (e.g., K562, HSB) and antibody-dependent cellular cytotoxicity (ADCC) (e.g., TNP-SB) assays are the same as that for human cytolytic effector cells and the majority of porcine effector cells for NK and ADCC are non-T, non-B and nonadherent (nonphagocytic) cells (Koren et al., 1978; Kim et al., 1980). This paper will briefly summarize our studies on the ontogenic development of NK cells in specific pathogen-free (SPF) and germfree (GF) piglets, the relationship of NK and K cells, and discuss a proposed model of the cell lineage(s) of NK and K cells.

II. ONTOGENIC DEVELOPMENT OF NK AND ADCC

We have examined the ontogenic development of NK and ADCC in GF, colostrum-deprived piglets which were maintained in GF isolators and compared with that of SPF, naturally-farrowed, colostrum-fed piglets which were maintained in the SPF facility (Huh et al., 1981a). The K cells for ADCC developed before birth, but NK cells developed after the GF piglets were 3-4 weeks of age (ages of hysterectomy-derived GF piglets were adjusted according to full term of 114 days of gestation as 0-day old); and the SPF piglets developed NK activity two weeks after birth (Table I). There was little difference in the level of ADCC activity between GF and SPF animals at any age. In addition, GF, colostrum-deprived piglets were associated with anaerobic normal flora (*Lactobacillus* *sp*. and *Streptococcus* *sp*. isolated from SPF miniature swine) at 1 week of age and maintained in GF isolators while the development of NK activity was followed. As shown in Table II, 2-week old piglets developed significant NK activity comparable to SPF, colostrum-fed piglets of the same age, indicating that microflora is responsible for acceleration of maturation of NK activity during ontogeny. We believe that the microbial effect is mediated by interferon induction and the interferons then effect early maturation of NK activity (see Chung et al., this volume). However, the maintenance of NK activity after maturation does not depend upon the microbial or antigenic environment of the host. The prenatal development of K cells for ADCC and the fact that there are no differences in the levels of ADCC in GF and SPF piglets suggest that both ontogenic development and maintenance of K cells for ADCC are independent of microbial and environmental influences on the host.

TABLE I. Ontogenic Development of NK and ADCC in Germfree and Specific Pathogen-free Piglets

Tests	Age of animals	% Specific lysis (mean ± S.E.) Germfree	Specific pathogen-free
NK	E:T (PBL:K562) = 100:1		
	0 day	1.8 ± 0.6 (8)[a]	1.2 ± 0.6 (8)
	2 weeks	2.5 ± 1.0 (8)	14.3 ± 7.0 (8)
	3 weeks	6.6 ± 3.9 (8)	11.8 ± 7.6 (8)
	4 weeks	16.3 ± 4.8 (17)	17.0 ± 4.3 (6)
	5 weeks	12.9 ± 4.6 (7)	9.4 ± 3.1 (8)
	6 weeks	49.7 ± 12.0 (8)	43.8 ± 6.5 (8)
	Adults	43.1 ± 6.2 (17)	40.6 ± 4.4 (17)
ADCC	E:T (PBL:TNP-SB + anti-TNP) = 100:1		
	0 day	18.6 ± 5.2 (8)	36.0 ± 9.8 (8)
	2 weeks	39.1 ± 11.2 (8)	34.0 ± 7.0 (8)
	3 weeks	33.0 ± 8.4 (8)	35.1 ± 9.5 (8)
	4 weeks	30.5 ± 4.0 (17)	49.1 ± 5.1 (6)
	5 weeks	42.8 ± 9.5 (7)	46.0 ± 6.8 (8)
	6 weeks	42.0 ± 5.1 (8)	61.0 ± 4.3 (8)
	Adults	42.6 ± 4.0 (17)	47.5 ± 5.2 (17)

[a]Number of animals tested in parenthesis

III. RELATIONSHIP OF NK AND K CELLS

Table III summarizes the tissue distributions of effector cells for NK and ADCC and the effects of "natural" antibodies, enzymes, interferons and "anti-NK" sera on NK and ADCC activities.

1. Tissue Distribution of Effector Cells for NK and ADCC. The effector cells for NK in the SPF young adult miniature swine were present only in peripheral blood and not in spleen, lymph nodes, tonsils, thymus or bone marrow. In contrast the K cells for ADCC were present in spleen and bone marrow as well as in peripheral blood. In GF, colostrum-deprived

TABLE II. Effect of Microflora on Ontogenic Development of NK and ADCC in Gnotobiotic Piglets[a]

Age of animals	% Specific lysis (mean ± S.E.) NK E:T (PBL:K562) = 100:1	ADCC E:T (PBL:TNP-SB + anti-TNP) = 100:1
2 weeks	16.4 ± 6.7 (6)[b]	43.2 ± 10.3 (6)
3 weeks	12.9 ± 2.5 (6)	35.0 ± 4.8 (5)
4 weeks	8.6 ± 2.1 (6)	41.3 ± 7.5 (6)
6 weeks	23.1 ± 8.9 (2)	64.4 ± 16.4 (2)

[a]Germfree piglets associated with anaerobic flora (*Lactobacillus sp.* and *Streptococcus sp.*) at one week of age and maintained identical to germfree piglets.

[b]Number of animals tested in parenthesis

immunologically "virgin" piglets no NK activity was detected in any tissue examined, including peripheral blood, spleen, lymph nodes, thymus, bone marrow and liver. K cells for ADCC were present in bone marrow, liver and peripheral blood (Kim et al., 1980). This dissociation of tissue distribution of NK and ADCC would suggest that their effector cells may be distinct subpopulations or may represent different stages of development of a single lineage.

2. *Effect of "Natural" Antibodies on NK Activity*. To examine the possible involvement of *in vivo* armed "natural" antibodies in NK activity (Akira and Takasugi, 1977), newborn piglets were fed colostrum resulting in high levels of maternal "natural" antibodies in their circulation (Kim and Watson, 1966; Kim, 1975); and cells from various tissues including peripheral blood, spleen, lymph nodes, thymus, liver and bone marrow of these colostrum-fed piglets were examined. They uniformly failed to show any NK activity, even though K cells for ADCC were present in peripheral blood, spleen, liver and bone marrow. In addition, *in vitro* arming experiments demonstrated that cells incubated in the sera of NK-positive animals or in the culture fluid of NK-positive cells failed to show any NK activity (Kim et al., 1980), indicating that NK activity is not simply a function of arming with "natural" antibodies *in vivo*.

TABLE III. Differential Characteristics of Porcine NK and ADCC

	NK activity	ADCC activity
1. Tissue Distribution		
peripheral blood	++++	++++
spleen	(±)	+++
lymph nodes	-	-
tonsils	-	-
thymus	-	-
bone marrow	-	++++
2. "Natural" antibodies for NK		
in vivo arming	none	
in vitro arming	none	
3. Enzyme treatment		
0.1% pronase	susceptible	resistant
0.1% bromelase	susceptible	resistant
4. Interferon effect		
porcine IFN	augment	no effect
human IFN α	augment	no effect
IFN Lr-A	augment	no effect
IFN Lr-D	augment	no effect
5. Effect of "anti-NK" sera		
anti-NK + C'	reduction	no effect

3. *Differential Susceptibility to Enzyme Treatment on NK and ADCC*. When peripheral blood lymphoid cells (PBL) of SPF miniature swine were treated with increasing concentrations of pronase or bromelase at 37°C for 30 minutes, NK activity decreased, but ADCC activity increased. With 0.1% pronase treatment, NK activity was completely abolished while ADCC activity was enhanced (Kim et al., 1980).

4. *Differential Effects of Interferon on NK and ADCC*. Porcine interferon, human leukocyte interferon (HuIFN-α, Cantell's) and purified cloned recombinant human leukocyte IFN Lr-A and IFN Lr-D (Hoffmann-La Roche's) were tested; and all augmented NK activity with little effect on ADCC (see Chung et al., this volume).

5. Effect of "Anti-NK" Sera on NK and ADCC. To investigate further the relationship between the effector cells for NK and ADCC, the effect of "anti-NK" sera on NK and ADCC activities was examined. "Anti-NK" sera were prepared from rabbits injected with porcine PBL high in NK activity. The "anti-NK" sera dissociating NK and ADCC activities at a certain dilution were selected after screening many antisera. When PBL were incubated with "anti-NK" sera (1:800 dilution) for 30 minutes at 37°C and treated with guinea pig complement (preabsorbed with liver and kidney cells of immunologically "virgin" piglets and diluted 1:20), about 6% of the cells were lysed, and this resulted in a marked reduction of NK activity without affecting ADCC activity (Huh et al., 1981b), suggesting the presence of NK cells distinct from K cells for ADCC. This demonstration of the existence of cells for mediating only NK activities complements the prior observation of the presence of only K cells for ADCC activity from the cells of immunologically "virgin" piglets as well as from the cells of adult spleen and bone marrow (Kim et al., 1980). Monoclonal anti-NK antibody and anti-K antibody are being prepared by hybridoma methodology for further characterization of porcine NK and K cells.

IV. A PROPOSED CELL LINEAGE(S) OF NK AND K CELLS

Figure 1 shows two possible hypothetical models for cell lineage(s) of NK and K cells under investigation. Model A is a single sequential lineage model. Stem cells and progenitor NK/K cells in bone marrow express neither NK nor K antigens on the cell surface. As these cells differentiate into pre-K cells (in bone marrow and periphery), they begin to express K antigens and FcR on the cell surface and are target binding cells (TBC) for antibody-coated target cells but are unable to kill. Fully differentiated K cells express K antigens and FcR on the cell surface and the lytic machinery dependent on FcR modulation develops. Next, K/NK cells appear and express both K and NK antigens and FcR on the cell surfaces. A lytic machinery which is independent of FcR modulation develops in these cells. Fully differentiated NK cells lose K antigens and FcR while NK antigens and/or receptors and the lytic machinery become fully developed. Therefore, K cells and K/NK cells participate in ADCC and K/NK cells and NK cells participate in NK activity. On the other hand, Model B proposes a two distinct lineages model. Stem cells and progenitor NK/K cells in bone marrow express neither NK nor K antigens on the cell surface. These cells differentiate into pre-K cells (in bone marrow and periphery) which begin to express K antigens

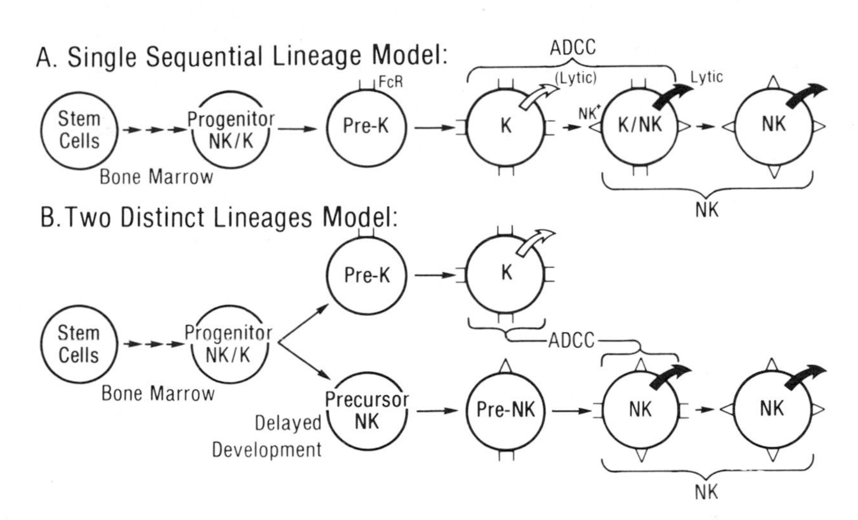

FIGURE 1. Cell lineage of NK/K cell system (hypothetical model)

and FcR on the cell surface and are TBC for antibody-coated target cells but are unable to kill. Finally, K cells fully express K antigens and FcR on the cell surface and the lytic machinery dependent on FcR modulation develops. Precursor NK cells may develop independently of the K-cell lineage later than K-cell development in bone marrow and periphery. Thus, precursor NK cells may begin to develop NK antigens on the cell surface but not FcR nor lytic machinery. These cells develop into pre-NK cells which express NK antigens and/or receptors and bind NK susceptible target cells and also express FcR on the cell surface but have no lytic machinery. Thus, pre-NK cells may be interferon-responsive TBC for NK. These cells further differentiate into early NK cells having both NK antigens and/or receptors and FcR and could serve as both effector cells for NK and ADCC but the lytic machinery need not require FcR modulation for activation as in the case of K cells. Finally, fully differentiated NK cells lose FcR and would serve as effectors for only NK. The FcR^+ NK cells could also explain the reports of one cell with two functions (NK and ADCC), but K cells and NK cells could be distinct lineages.

V. SUMMARY AND CONCLUSION

The differential onset of the ontogenic development of effector cells for NK and ADCC, the difference in tissue distributions and susceptibility to enzyme treatment, and the differential effects of interferons and "anti-NK" sera on NK and ADCC suggest that porcine NK and K cells are distinct subpopulations. This unique gnotobiotic miniature swine model may provide us with an excellent opportunity to further investigate the definition of the cellular nature of NK and K cells and to examine their cell lineages and relationships.

ACKNOWLEDGMENTS

The authors wish to acknowledge the assistance of Mr. Gene Monson, Mr. Donald Moody and Mr. Brian Hepburn in procuring animals for this study; Ms. Karen Becker for her excellent technical assistance, and Mrs. Rose Vecchiolla for the preparation of the manuscript.

REFERENCES

Akira, D., and Takasugi, M. (1977). Int. J. Cancer 19:747.

Herberman, R. D., ed. (1980). "Natural Cell-Mediated Immunity Against Tumors." Academic Press, New York.

Herberman, R. B., and Holden, M. T. (1978). in "Advances in Cancer Research," Vol. 27 (G. Klein and S. Weinhouse, eds.), p. 305. Academic Press, New York.

Huh, N. D., Kim, Y. B., Koren, H. S., and Amos, D. B. (1981a). Int. J. Cancer 28:175.

Huh, N. D., Kim, Y. B., and Amos, D. B. (1981b). J. Immunol. 127:2190.

Kiessling, R., and Haller, O. (1978). in "Contemporary Topics in Immunobiology, Vol. 8 (N. L. Warner and M. D. Cooper, eds.), p. 171. Plenum Press, New York.

Kim, Y. B. (1975). in "Immunodeficiency Diseases in Man and Animals" (D. Bergsma, R. A. Good, and J. Finnstad, eds.), p. 549. Sinauer Associates, Inc., Sunderland, Mass.

Kim, Y. B., and Watson, D. W. (1966). Ann. N.Y. Acad. Sci. 133:727.

Kim, Y. B., Huh, N. D., Koren, H. S., and Amos, D. B. (1980). J. Immunol. 125:755.

Koren, H. S., Amos, D. B., and Kim, Y. B. (1978). Proc. Natl. Acad. Sci. USA 75:5127.

Roder, J. C., Karre, K., and Kiessling, R. (1981). Prog. Allergy 28:66.

Scheffel, J. W., and Kim, Y. B. (1979). Infect. Immun. 26: 202.

AUGMENTATION OF HUMAN NATURAL KILLER CELLS WITH HUMAN LEUKOCYTE AND HUMAN RECOMBINANT LEUKOCYTE INTERFERON

John Ortaldo[1], Ronald Herberman[1], and Sidney Pestka[2]

[1]Biological Research and Therapy Branch
National Cancer Institute - FCRF
Frederick, Maryland
and
[2]Roche Institute of Molecular Biology
Nutley, New Jersey

INTRODUCTION

Recent research indicates that the antitumor effect of interferon may be mediated by at least two mechanisms (Epstein, 1977; Gresser et al., 1972): 1) the direct antiproliferative effects on tumor cells by the interferons (Evinger et al., 1981a, 1981b), and 2) the modulation of the host defense mechanisms. Among the various effects of interferon which have been studied in great detail is its ability to rapidly augment cell-mediated cytotoxic responses. One of these cytotoxic responses which is most sensitive to human leukocyte interferon is the natural killer (NK) activity. During the initial studies of the effects of interferon on cytotoxic reactivity, there was concern as to whether the interferon in the heterogeneous materials was actually responsible for the augmenting activity. However, we have recently shown that homogeneous leukocyte interferon exhibited NK and ADCC (antibody-dependent cell-mediated cytolysis) activities and was at least as active relative to the crude or partially purified interferons (Herberman et al., 1981). Because of the recent discovery of a large number of discrete molecular species of human leukocyte interferon (Pestka et al., 1980; Rubinstein et al., 1981; Hobbs et al., 1981), it was therefore important to determine all or some of the various homogeneous leukocyte interferons had regulatory activity as measured by their ability to augment NK cells. In addition, the recent attention which

NK CELLS AND OTHER NATURAL EFFECTOR CELLS

ISBN 0-12-341360-5

has been given to the possibility of obtaining large amounts of human interferon from bacteria by recombinant DNA technology raised the possibility of examining recombinant forms of human leukocyte interferon.

RESULTS

Ten species of human leukocyte interferon (Rubinstein *et al.*, 1981), eight purified to homogeneity, were tested for their ability to modulate cytolytic activity of NK cells. All of these interferons were tested multiple times against randomly selected normal donors in an *in vitro* 4 hour ^{51}Cr release assay by prior treatment of effector cells with interferon at 37°C for 1 hour. All interferons tested had a specific activity of 1 to 4 x 10^8 units/mg of protein, and all species demonstrated significant augmentation of NK activity when tested at high levels, i.e., >500 units of interferon/ml. However, as the dosage of the leukocyte interferon species was titrated, the ability to augment cytolytic activity varied considerably among the various species, although all preparations were tested at equivalent antiviral titers. A summary of the results with the various species of leukocyte interferon is shown in Table 1. The preparations could be subdivided according to augmenting ability relative to their antiviral potency. Several species of leukocyte interferon (α_1, β_2, and β_3) induced strong augmentation of NK activity at low antiviral doses, i.e., 10-50 units/ml; whereas other preparations were much less effective, only demonstrating significant augmentation at >100 units/ml. Therefore, it became apparent from these results that quite substantial differences (as much as 10-50 fold) in the number of antiviral units of various homogeneous species were necessary to achieve similar levels of NK augmentation.

It was of interest to examine the species specificity of homogeneous preparations of recombinant human leukocyte interferons (Staehelin *et al.*, 1981). This was of particular significance since a partially purified preparation of recombinant human leukocyte interferon had been previously reported to lack the typical species specificity reported in the antiviral systems (Masucci *et al.*, 1980). With the pure recombinant leukocyte A interferon (IFLrA), no significant augmentation of mouse NK activity was seen. In contrast, significant augmentation of activity from large granular lymphocytes (LGL) and human peripheral blood NK activity was seen at very low interferon titers. As expected, control preparations of human fibroblast interferon and mouse leukocyte interferon augmented the respective effector cells

Table 1. Effect of Natural and Recombinant Human Leukocyte Interferon Species on NK Activity

	Natural Leukocyte	Recombinant Leukocyte
High-level boosters[a]	IFL-β_2 [c]	IFLrA
	IFL-β_3	
	IFL-γ_1	
Low-level boosters[b]	IFL-α_1	IFLrB
	IFL-α_2	IFLrD
	IFL-β_1	
	IFL-γ_2	
	IFL-γ_3	
	IFL-γ_4	
	IFL-γ_5	

a) Significant augmentation of activity seen at <50 units/ml of interferon. All interferon preparations demonstrated boosting of activity at 100 units/ml or greater.
b) Require 100 units/ml of interferon or greater to demonstrate NK cytolytic activity.
c) See Pestka and Baron (1981) for discussion of interferon nomenclature.

of the same species but did not augment xenogenic effector cells. The evaluation of various recombinant interferons has led to similar conclusions obtained with the naturally occurring leukocyte interferon species. Two additional preparations of recombinant leukocyte interferon, B and D, were tested (D was homogeneous, B was highly purified) for their relative ability to augment human NK cells and both were level boosters in their ability to augment cytolytic activity relative to identical antiviral units of IFLrA. These results are summarized in Table 1.

DISCUSSION

Pure human leukocyte and recombinant leukocyte interferons were shown to be able to augment appreciably cytolytic activity of NK cells. Furthermore, dose titrations have demonstrated that as little as 10-100 units of antiviral activity is sufficient to induce significant augmentation of cytolytic activity. However, analysis of the multiple species of human leukocyte interferon has demonstrated significant differences in the ability of the various

Table 2. Species Specificity of Pure Recombinant Human Leukocyte A Interferon on NK Activity

Preparation	Cytolytic Function[a)] Human NK	Mouse NK
Control	+	+
IFLrA[b)]	+++	+
Partially purified human fibroblast interferon	+++	+
Partially purified mouse L-cell interferon	+	+++

a) Relative cytolytic function as measured in a 4 hour ^{51}Cr release assay.
b) Purified as described by Staehelin _et al._ (1981).

species to achieve levels of augmentation of cytolytic function. This is especially of interest since these preparations were tested at identical amounts of antiviral units. In a recent study comparing these same preparations for their ability to modulate and augment monocyte function, similar differences were seen. However, preparations which augmented cytolytic activity of human NK cells were not necessarily the most potent modulators of human monocyte function (data not shown; Ortaldo _et al._, 1982), indicating multiple actions of the interferon, dependent on the type of assay used. Since many of these interferons share common amino acid sequences, it appears that relatively minor differences in structure could considerably change their biological activity. More recently available recombinant interferons, fibroblast (Goeddel _et al._, 1980), and hybrid leukocyte interferons (Rehberg _et al._, 1982), might lend considerable information toward understanding of the biological biochemical requirements for activation of various cytolytic activities (which might include NK cell, T-cell, and monocyte functions). As has been previously reported with various types of leukocyte and fibroblast interferon preparations, the purified interferons demonstrated no significant augmentation of cytolytic activity of heterogeneous species.

The species-restricted effects found were in contrast to the results previously reported by Masucci _et al._ (1980) with partially purified materials, which could have been due to endotoxin or other bacterial contaminants which might have been in the preparations. However, IFLrD which exhibits antiviral activity on mouse cells and is closely related to the recombinant leukocyte interferon α, tested by Masucci et al. (1980), might increase cytolytic activity on mouse cells. The demonstration that the homogeneous leukocyte interferon preparations had potent effects on the NK effector functions provides optimism for the potential usefulness of these preparations in clinical therapy of cancer patients and other disease states. However, since the various species as well as the recombinant forms, at the same number of antiviral units, demonstrated large differences in their functional efficacy, it might be of clinical importance to utilize for therapeutic trials species or mixtures of species of interferon that have high potency for augmenting a variety of functions.

ACKNOWLEDGMENTS

The authors wish to thank Drs. Courtney McGregor, Joseph Tarnowski, Hsiang-fu Kung and Mr. Robert Bartlett for preparations of recombinant human leukocyte interferons IFLrA, B, and D; and Bruce Kelder and Linda Petervary for antiviral assays, and Mr. Richard Jaffee for technical assistance.

REFERENCES

Epstein, L.B. (1979). In "Biology of the Lymphokines" (S. Cohen, E. Pick and J.J. Oppenheim, eds.), p. 443, Academic Press, New York.

Evinger, M., Maeda, S., and Pestka, S. (1981a). J Biol Chem 256:2113.

Evinger, M., Rubinstein, M., and Pestka, S. (1981b). Arch Biochem Biophys 210:319.

Goeddel, D.V., Shepard, H.M., Yelverton, E., Leung, D., Crea, R., Sloma, A., and Pestka, S. (1980). Nucleic Acids Res 8:4057.

Gresser, I., Maury, C., and Brouty-Boye, D. (1972). Nature 239:167.

Herberman, R.B., Ortaldo, J.R., Rubinstein, M., and Pestka, S. (1981). J Clin Immunol 1:149.

Hobbs, D.S., Moschera, J., Levy, W.P., and Pestka, S. (1981). In "Methods in Enzymology (S. Pestka, ed.), Vol. 78, p. 472, Academic Press, New York.

Masucci, M.G., Szigetti, R., Klein, E., Klein, G., Gruest, J., Montagnier, L., Taira, H., Hall, A., Nagata, S., and Weissmann, C. (1980). Science 209:1431.

Ortaldo, J.R., Mantovani, A., Hobbs, D., Rubinstein, M., Pestka, S., and Herberman, R.B. (1982). J Biol Chem, in press.

Pestka, S. and Baron, S. (1981). In "Methods in Enzymology" (S. Pestka, ed.), Vol. 78, p. 3, Academic Press, New York.

Pestka, S., Evinger, M., McCandliss, R., Sloma, A., and Rubinstein, M. (1980). In "Polypeptide Hormones" (R.F. Beers, Jr. and E.G. Bassett, eds.), p. 33, Raven Press, New York.

Rehberg, E., Kelder, B., and Pestka, S. (1982). In preparation.

Rubinstein, M., Levy, W.P., Moschera, J.A., Lai, C.-Y., Hershberg, R.D., Bartlett, R.T., and Pestka, S. (1981). Arch Biochem Biophys 210:307.

Staehelin, T., Hobbs, D.S., Kung, H.-F., Lai, C.-Y., and Pestka, S. (1981). J Biol Chem 256:9750.

AUGMENTATION OF HUMAN NATURAL KILLER CYTOTOXICITY BY ALPHA-INTERFERON AND INDUCERS OF GAMMA-INTERFERON. ANALYSIS BY MONOCLONAL ANTIBODIES.

Chris D. Platsoucas[1]

Memorial Sloan-Kettering Cancer Center
New York, New York

I. INTRODUCTION

Human leukocyte interferon has been shown to augment both *in vitro* and *in vivo* the natural killer (NK) and antibody-dependent cell-mediated cytotoxicity (ADCC) of human peripheral blood mononuclear leukocytes (1-3). It is likely that augmentation of NK cells by IFN-α is one of the mechanisms of its biological action *in vivo*. Both purified E-rosette positive and E-rosette negative cells from the peripheral blood mediate NK cytotoxicity (4-6). In the majority (approximately 80%) of normal donors, the cells that are augmented by IFN-α to perform NK are E-rosette forming cells(7-9). Essentially similar observations were made with fibroblast interferon (10). We report here an analysis of the augmentation of human NK cells by IFN-α and inducers of gamma interferon (staphylococcal enterotoxin A and OKT3 monoclonal antibody), using monoclonal antibodies to lymphoid and monocytoid cell populations.

II. METHODS

Monocyte-depleted human peripheral blood lymphocytes were employed in these studies (8). OKT3, OKT4, OKT8, OKM1, OKT11a and OKT10 monoclonal antibodies were obtained from Ortho Corporation. Cell separation procedures using the fluorescence activated cell sorter or treatment of the cells

[1] Supported by grant CH-151 from the American Cancer Society and NIH grant CA 8748.

ISBN 0-12-341360-5

with monoclonal antibody and complement (11) were discussed in another paper of this volume. Treatment of the cells with monoclonal antibody and complement resulted in complete depletion of the population of interest, as judged by immunofluorescence analysis. Cell numbers employed for the NK determinations were not adjusted for the proportions of lysed cells by antibody and complement treatment. Human leukocyte interferon (IFN-α) had a specific antivity of $1x10^6$ IU/mg of protein and was provided by the Interferon Laboratories of this Center. Staphylococcal enterotoxin A (SEA) was obtained from FDA, Cincinnati, Ohio. Mixed lymphocytes cultures were set up as previously described (11). Natural killer cytotoxicity was determined against the K 562 targets (8).

III. RESULTS AND DISCUSSION

In vitro treatment of purified (by fluorescence-activated cell sorting) OKT11a-positive, clone 9.6-positive or OKM1-positive cells with human leukocyte interferon (14 hrs, 37°C) resulted in significant augmentation of their NK cytotoxicity. A representative experiment is shown in Table 1. In other experiments, monocyte-depleted peripheral blood lymphocytes were treated with various monoclonal antibodies and complement, washed and incubated for 14 hrs at 37°C with either 1000 U/ml of IFN-α or 0.05 μg/ml of SEA. SEA is a T cell mitogen and a powerful inducer of gamma interferon (12, 13). We have observed that *in vitro* treatment of monocyte-

TABLE 1

NATURAL KILLER CYTOTOXICITY MEDIATED BY 9.6-POSITIVE, OKT11a-POSITIVE AND OKM1-POSITIVE HUMAN PERIPHERAL BLOOD LYMPHOCYTES ISOLATED BY THE FACS. AUGMENTATION BY HUMAN LEUKOCYTE INTERFERON.

	% CYTOTOXICITY*	
SUBPOPULATION	100:1	50:1
Monocyte-depleted PBL	44%	26%
" , 1000 u/ml IFN-α	79%	42%
Clone 9.6⁺	29%	21%
Clone 9.6⁺, 1000 u/ml IFN-α	52%	22%
OKT11a⁺	40%	20%
OKT11a⁺, 1000 u/ml IFN-α	70%	27%
OKM1⁺	35%	17%
OKM1⁺, 1000 u/ml IFN	64%	32%

*determined against the K 562 target cells.

TABLE 2

EFFECT OF DEPLETION OF NK CELL POPULATIONS ON THE AUGMENTATION OF NK CYTOTOXICITY BY IFN-α OR SEA*

ANTIBODY AND COMPLEMENT	% CYTOTOXICITY**					
	IFN-α			SEA		
	Donor 1	2	3	Donor 1	2	3
C alone	29	57	38	46	66	62
OKT11a+C	19	31	ND	41	73	65
OKM1+C	37	54	48	49	73	60
OKT3+C	28	67	47	42	73	55
OKT8+C	28	51	44	46	70	62

*Monocyte-depleted PBL were treated with various monoclonal antibodies and complement or complement alone, washed and incubated for 14 hrs at 37°C with either 1000 u/ml IFN-α or 0.05μg/ml SEA. **NK was determined against K 562 targets. E/T ratio 50:1. Results on NK cytotoxicity of non-boosted lymphocytes are shown in another paper of this volume.

depleted lymphocytes with SEA resulted in highly significant augmentation of NK cytotoxicity (14). It is possible that this augmentation is mediated by gamma interferon (14). Treatment with the OKT11a monoclonal antibody and complement significanly decreased the augmentation of NK cytotoxicity by subsequent treatment with IFN-α. In contrast, treatment with the OKM1, OKT3 or OKT8 monoclonal antibodies did not significanly reduce the augmentation of NK by IFN-α (Table 2). Although NK mediated by purified M_1-positive cells is augmented by interferon (Table 1), depletion of these cells from monocyte-depleted lymphocytes (by OKM_1 antibody and C treatment) did not significantly affect the augmentation of NK by subsequent exposure to IFN-α (Table 2). It is possible that IFN-α induces the differentiation of pre-NK cells (present in monocyte-depleted lymphocytes), to NK effector cells. In addition, the augmentation (by IFN-α) of the NK cytotoxicity mediated by the remaining (M_1-negative) NK cell populations, may not permit detection, under these experimental conditions, of a decrease in NK due to removal of M_1-positive cells prior to boosting with interferon. Furthermore, treatment with the OKT11a or OKM1 monoclonal antibodies and complement did not prevent highly significant augmentation of NK cytotoxicity after treatment with SEA. Additional studies are in progress using combinations of monoclonal antibodies to identify the NK cell populations that are augmented by these modifiers.

TABLE 3

AUGMENTATION OF NATURAL KILLER CYTOTOXICITY OF HUMAN MONOCYTE-DEPLETED PERIPHERAL BLOOD LYMPHOCYTES BY IFN-α AND SEA. ANALYSIS BY MONOCLONAL ANTIBODIES.*

Treatment	Antibody and Complement	% Cytotoxicity**	% Inhibition	Level of Significance
None	C alone	35.5 ± 6.0	--	--
	OKM1 + C	14.7 ± 4.0	60%	$p < 0.01$
	OKT11a + C	10.1 ± 3.2	72%	$p < 0.005$
	OKT3 + C	34.5 ± 0.6	0.02%	NS
IFN-α 500 u/ml	C alone	55.0 ± 8.2	--	--
	OKM1 + C	20.7 ± 3.2	62%	$p < 0.005$
	OKT11a + C	7.8 ± 1.3	86%	$p < 0.001$
	OKT3 + C	50.0 ± 8.0	9%	NS
SEA 0.05 μg/ml	C alone	57.7 ± 7.6	--	--
	OKM1 + C	47.3 ± 7.0	18%	NS
	OKT11a + C	34.6 ± 6.0	40%	$p < 0.01$
	OKT3 + C	50.0 ± 0.7	13%	NS

*Monocyte-depleted human peripheral blood lymphocytes were incubated with IFN-α or SEA for 14 hrs at 37°C. The cells were washed and treated with the appropriate monoclonal antibody and complement. **Natural Killer cytotoxicity was determined against the K 562 targets. E/T ratio 50:1.

In other experiments, we investigated the cell surface phenotypes of NK cells after *in vitro* treatment with IFN-α or SEA. NK cytotoxicity after augmentation with IFN-α treatment was significantly reduced (by 70 to 90%) after removal of T10-positive cells (data not shown) or T11α-positive cells (by 60 to 90%). Removal of M1-positive cells by OKM1 antibody and complement treatment resulted in partial reduction of NK cytotoxicity, which was not affected by depletion of T3-positive cells (Table 3). NK cytotoxicity after augmentation with SEA treatment *in vitro* was only partially reduced after depletion of T10-positive cells (by 50%) (data not shown) or T11a-positive cells (by 50%) and was not affected significantly, on a statistical basis, by depletion of M1-positive cells or T3-positive cells.

In other studies, we observed that the OKT3 monoclonal antibody in the absence of complement augments the NK (Table 4) and ADCC of monocyte-depleted peripheral blood lymphocytes from certain donors (data not shown) (15). This monoclonal antibody is an inducer of gamma-interferon (16).

TABLE 4

AUGMENTATION OF NATURAL KILLER CYTOTOXICITY OF MONOCYTE-DEPLETED PERIPHERAL BLOOD LYMPHOCYTES BY *IN VITRO* TREATMENT (18 HRS, 37°C) WITH THE OKT3 MONOCLONAL ANTIBODY.

OKT3 (μg/ml)	% CYTOTOXICITY (AGAINST K 562 TARGETS) Donor 1	Donor 2
0.00	18%	40%
0.05	35%	48%
0.25	46%	46%
0.50	45%	44%

Studies on the augmentation of natural killer-like cytotoxicity generated in MLC by IFN-α and SEA is shown in Table 5. IFN-α addition to MLC on day 5 resulted in significant augmentation of the natural killer-like cytotoxicity against the K 562 targets on day 6, but had no effect on the specific T cell-mediated cytotoxicity. Similarly, addition of SEA on day 5 resulted in significant augmentation of NK-like cytotoxicity. Addition of interferon on day 0 had no effect on NK-like cytotoxicity on day 5, although it augmented significantly the specific cytotoxicity against allogeneic targets (data not shown). Addition of IFN-α on both days 0 and 5

TABLE 5

AUGMENTATION BY SEA AND α-INTERFERON OF THE NATURAL KILLER-LIKE CYTOTOXICITY (AGAINST THE K562 TARGETS) GENERATED IN MIXED LYMPHOCYTE CULTURE

IMMUNOMODULATOR	% CYTOTOXICITY* Experiment 1	Experiment 2
None	32%	41%
α-Interferon, 500 U/ml (added on day 5)	50%	79%
SEA, 0.1 μg/ml (added on day 5)	66%	88%
α-Interferon, 500 U/ml (added on day 0)	36%	54%
α-Interferon, 500 U/ml (added on day 0) 500 U/ml (added on day 5)	22%	25%
α-Interferon, 500 U/ml (added on day 0) SEA 0.1 μg/ml (added on day 5)	77%	87%

* Determined on day 6. Effector to target ratio 50:1.

failed to augment NK-like cytotoxicity on day 6. It appears that cells that have already been exposed to α-interferon fail to respond again to this agent within this short time interval. However, addition of SEA on day 5 to mixed lymphocyte culture, treated with IFN-α on day 0, resulted in highly significant augmentation of NK-like cytotoxicity.

In conclusion, we investigated the populations of human NK cells that are augmented by IFN-α and inducers of gamma-interferon (SEA and OKT3 monoclonal antibody), using monoclonal antibodies. These NK cell populations may respond differently to various biological response modifiers.

REFERENCES

1. Einhorn S., Blomgren H. and Strander H., Int. J. Cancer, 22:405 (1978).
2. Trinchieri G. and Santoli D., J. Exp. Med. 147:1314 (1978).
3. Herberman R.B., Ortaldo J.R. and Bonnard G.D., Nature 227:221 (1979).
4. Kay M.D., Bonnard G.D., West W.H. and Herberman R.B., J. Immunol. 118:2058 (1977).
5. Kall M.A. and Koren H., Cell. Immunol. 40:58 (1978).
6. Gupta S., Fernandes G., Nair M. and Good R.A., PNAS 75:5137 (1978).
7. Platsoucas C.D., Gupta S., Good R.A. and Fernandes G. Fed. Proc. 39:932 (1980).
8. Platsoucas C.D., Fernandes G., Gupta S.L., Kempin S., Clarkson B., Good R.A. and Gupta S., J. Immunol. 125:1216 (1980).
9. Platsoucas C.D., Fernandes G., Good R.A. and Gupta S., submitted.
10. Ortaldo J.R., Lang N.P. Timonen T. and Herberman R.B., J. Inter. Res. 1:253 (1981).
11. Platsoucas C.D. and Good R.A., PNAS 78:4500 (1981).
12. Johnson H.M., Stanton G.J. and Baron S., Proc. Soc. Exp. Biol. Med. 154:138 (1977).
13. Langford M.P., Stanton G.J. and Johnson H.M. Infection Immunity 22:62 (1978).
14. Platsoucas C.D., Stewart II W.E., Wiranowska-Stewart M. and Good R.A. The Biology of the Interferon System (E. DeMayer, G. Galasso and H. Schellenkens, eds.), p. 256, Amsterdam, Elsevier/North Holland (1981).
15. Platsoucas C.D., Von Wussow P. Unpublished observations.
16. Von Wussow P., Platsoucas C.D., Wiranowska-Stewart M. and Stewart II W.E., J. Immunol. 127:1197 (1981).

HUMAN NK CELL ACTIVATION WITH INTERFERON AND WITH TARGET CELL-SPECIFIC IgG

Måns Ullberg[1]
Mikael Jondal

Department of Tumor Biology
Karolinska Institutet
Stockholm, Sweden

I. INTRODUCTION

Interferon (IF) is a highly active molecule on a multitude of cellular functions (Paucker et al., 1962; Finter, 1973) among which are NK cell lysis (Trinchieri and Santoli, 1978; Herberman, 1980). Pre-treatment of effector cells leads to an overall increase in lytic efficiency, partly due to recruitment of "pre-NK" cells.

Cytotoxicity induced by target cell specific IgG has been called antibody-dependent cellular cytotoxicity (ADCC) and the effector cell, K cell (Perlmann, 1981). The exact relationship between NK and K cells is not clear although much experimental evidence indicate that these two functions are mediated by the same cell.

The present paper summarizes our recent experiments on the finer mechanisms behind interferon and target cell specific IgG enhancement of NK and K cell killing respectively (Ullberg et al., 1981; Ullberg and Jondal, manuscript in preparation) and the relationship between these two cell populations. We have used two different methods to detect cytotoxicity, the agarose single cell assay as described by Grimm et al. (1979) and a modified ^{51}Cr release assay for the estimation of the maximum number of target cells that 10^5 effector cells can lyse given optimal conditions (Vmax) (Ullberg and Jondal, 1981). By

[1]Supported by the Swedish Cancer Society.

ISBN 0-12-341360-5

using these two methods in combination, different steps involved in killing can be distinguished: recognition of target cells, the lytic event, target cell death and recycling of effector cells (Ullberg and Jondal, 1981).

Our results indicate that IF and target cell-specific IgG do not influence effector cell recognition but rather increase the efficiency of the lytic event and of the recycling capacity. Although the effects are similar, they add on top of each other. Consequently, with human hematopoietic cell lines as target cells, it appears as if the role of the IgG antibody in ADCC is to act as a "second signal" that will activate the lytic event of the NK cell against NK resistant target cells and also increase the recycling capacity. Thus, it is probably accurate to view ADCC (and also lectin dependent killing) as variants of NK lysis when present in normal peripheral blood.

II. METHODS

Nylon wool purified lymphocytes were used in all experiments to minimize non-specific binding. Interferon was obtained from Dr. Kari Cantell, Helsinki. For IF stimulation, 10^4 U/ml of IF was added and the cells were incubated for 1 hour. Chromium release assays were performed with quadruplicate wells in V-shaped microplates. A dilution series of target cells with six different concentrations (diluted 3:5, with 2×10^5 target cells as the highest concentration) was used together with a constant number of effector lymphocytes (10^5 cells). Cytotoxicity assays were run for 3 hours at 37°C. Percentage lysis was determined by using the formula:

$$100 \times \frac{(\text{test release}) - (\text{spontaneous release})}{(80\% \text{ of total label}) - (\text{spontaneous release})}$$

The number of killed target cells in each well is obtained by multiplying the percentage lysis by the initial number of target cells. The results from the chromium assays fit to the Michaelis-Menten equation expressed as:

$$V = \frac{V\,max \times T}{Km + T}$$

In this equation V represents the number of killed target cells at the end of the assay, T the initial number of target cells, Vmax the maximal lytic potential (when the number of target cells approaches infinity) and Km the Michaelis-Menten constant. The Vmax values are determined by using the Lineweaver-Burk plot. In this, the reciprocal values of V are plotted as a function of the reciprocal values of T (Ullberg and Jondal, 1981). Regression analysis is used to obtain a straight line and the Vmax values are obtained from the reciprocal of the Y intercept.

Cytotoxic assays in agarose were performed as described by Grimm *et al.* (1979) with slight modifications. Equal numbers of lymphocytes and target cells (2×10^5 in a total volume of 0.2 ml of medium) were spun together and incubated for 10 minutes. The pellet was subsequently resuspended by firm shaking for 4 sec on a whirlmixer. 0.5 ml of 0.5% agarose solution in RPMI 1640 kept at 48°C was precooled for 20 sec and added to the cells. The suspension of cells in agarose was mixed with a pasture pipet and the solution dripped from the height of 30 cm onto plastic Petri dishes (Nunc 60 mm) to form a thin layer. After solidifying, 6 ml of medium with 10% FCS was added. The dishes were incubated for 3 hours at 37°C. During this incubation the active killer cells bound to target cells would kill their targets. The dishes were subsequently stained with trypan blue, washed 3 times in PBS for 10 min and finally fixed with 1% formic aldehyde solution. The dishes were scored using an inverted microscope. The percentage TBC:s (target binding cells) was determined by counting the number of conjugates in 400 lymphocytes. The percentage dead conjugates was determined by counting the number of dead target cells in 100 conjugates. By multiplying these two parameters, the fraction of active NK cells of all lymphocytes was calculated.

The maximal recycling capacity (MRC) was estimated by combining data in the chromium and agarose assays. The Vmax values were divided with the number of active NK cells (fraction of active NK cells multiplied with the number of lymphocytes in the chromium assays, 10^5 cells). The calculated MRC values reflect the average recycling potential among the active NK cells in three hours given a saturated system.

III. RESULTS

A. Human NK cell activation by interferon.

IF treatment of NK cells give increased Vmax values for all lines tested. The killing of Molt-4 increased with a factor of 1.8 ($p < 0.0025$), against K-562 with 2.1 ($p < 0.01$)

and against BJAB with 3.2 ($p < 0.0025$). Thus, the highest increase occurs against the most resistant lines (Table I).

TABLE I. Human NK cell activation by interferon

Target cell[b]	Treatment	Parameters[a]				
		Vmax (x 10 3)	TBC:s (%)	Cytotoxic TBC (%)	Active NK cells (%)	MRC (cells/ 3h)
Molt-4(6)	-	9.5	6.8	52	3.5	2.8
Molt-4(6)	IF	16.8	7.0	49	3.5	4.9
K-562(6)	-	6.1	8.0	19	1.5	3.8
K-562(6)	IF	12.9	8.0	32	2.6	5.0
BJAB(5)	-	6.1	7.7	24	1.9	4.0
BJAB(5)	IF	19.6	7.3	44	3.2	6.8

[a]Parameters explained in detail in the Methods section.
[b]Number of performed experiments in parenthesis.

Neither interferon, nor target cell specific IgG influenced effector cell recognition as the TBC values remained basically identical (Table I and II).

An increase in the percentage TBC:s that kill their target cells reflect a recruitment of new effector cells into the system. Depending on the target cell used, recruitment will be seen or not (Table I). With cell lines such as Molt-4, that are highly susceptible to NK cell lysis and that have a large fraction of active killer cells in the TBC population (approx. 50%), no recruitment is seen. On the other hand with cell lines such as K-562 and BJAB, that have a smaller fraction of active TBC:s initially, a statistically significant recruitment is seen (K-562, $p < 0.001$ and BJAB, $p < 0.001$). By dividing the Vmax value with the number of active NK cells, a mean maximal recycling capacity MRC can be estimated (Table I). Interferon stimulation leads to increased MRC values against all cell lines tested (Molt-4, $P < 0.005$, K-562, $p < 0.05$ and BJAB, $p < 0.005$).

B. Human NK cell activation by interferon and target cell specific IgG.

Pre-coating Daudi target cells with rabbit IgG directed against human IgM did not alter the percentage of TBC:s,

TABLE II. Human NK cell activation by interferon and target cell specific IgG

Target cell[b]	Treatment	Parameters				
		Vmax (x 10^3)	TBC:s (%)	Cytotoxic TBC (%)	Active NK cells (%)	MRC (cells /3h)
Daudi	-	3.3	9.8	26	2.5	1.4
Daudi	IF	10.7	10.2	39	4.0	2.7
Daudi	anti-IgM	18.7	10.6	41	4.3	4.4
Daudi	IF and anti-IgM	26.6	10.1	46	4.6	5.9

[a]Parameters explained in detail in the Methods section. Values are means from three experiments.
[b]Daudi is a Burkitt lymphoma derived B cell line expressing high amounts of surface-bound IgM (Klein *et al.*, 1968).

thus the role of the antibodies is not primarily target cell recognition (Table II). Both interferon and target cell specific IgG increased the fraction of active TBC:s to the same degree (Table II). When interferon pre-treated cells were used against IgG coated target cells, no further increase in this fraction was seen although Vmax values increased significantly. Consequently, this increase was reflected in the MRC values, meaning that IF increased the recycling against un-coated target cells and that a still higher recycling was seen with IgG-caoted target cells (Table II).

IV. DISCUSSION

The fact that IF enhancement of NK activity occurred without any change in the size of the TBC fraction suggests that it is unrelated to target cell recognition as earlier published (Herberman, 1980; Ullberg *et al.*, 1981). An increased efficiency of the NK cell during the lytic event could be seen as faster killing in the single cell assay (data not shown) in combination with faster recycling.

Maybe slightly surprising was the finding that target cell specific IgG seemed to have an effect very similar to IF.

This is in contrast to earlier theories that the K cell would only be capable of killing once whereafter the Fc-IgG receptor would be modulated and non-functional (Perlmann and Cerrotini, 1981; Ziegler and Henney, 1975; 1977). This is obviously not the case at the optimal concentrations used. However, with large immune complexes, and with high IgG concentrations on the target cells, no induction of ADCC occurs, NK activity is switched off and the Fc-IgG receptor irreversibly lost in contrast to the NK function which reappears after overnight incubation (Merrill et al., 1981).

The present data also clearly demonstrates that the K cell, as earlier defined (Perlmann and Cerrotini, 1981), is part of the TBC fraction that has the capacity to spontaneously conjugate to human hematopoietic cell lines (Ullberg and Jondal, Clin. Lab. Immunol., in press) and that it overlaps with the NK cell fraction that kills after IF stimulation. This population is probably identical to the one that spontaneously kills the most susceptible NK target (i. e. Molt-4 cells in Table I). Thus the IgG antibody has the role of activating the NK cell after its initial binding to the target cell. Whether the comparatively high, optimal IgG concentrations used also have the capacity to increase the number of TBC:s against non-hematopoietic target cells which lack the surface structures responsible for the spontaneous TBC reaction is presently unclear. A biochemical analysis of the ADCC, with these target cells, activation step would be interesting as it could reveal mechanisms involved in the lytic event and in effector cell recycling. Especially, transmethylation of membrane phospholipids, activation of phospholipase A2 and formation of arachidonic acid would have to be investigated, as such reactions have been reported to occur during IF stimulation of human peripheral blood cells (Bougnoux et al., 1981).

ACKNOWLEDGEMENTS

The excellent technical assistance of Miss Maj-Britt Alter is gratefully acknowledged.

REFERENCES

Bournoux, P., Hirata, F., Timonen, T., and Hoffman, T. (1981). In "Proceedings of the 14th Int. Leucocyte Culture Conference", in press.

Finter, N. B. (1973). "Interferon and interferon inducers", North-Holland Publ. Co., Amsterdam
Grimm, E. A., Thoma, J., and Bonavida, B. (1979). J. Immunol. 113, 2870.
Herberman, R. B. (Editor) (1980). "Natural cell-mediated immunity against tumors". Academic Press, New York.
Klein, E., Klein, G., Nadkarni, J. S., Nadkarni, J. J., Wigzell, H., and Clifford, P. (1968) Cancer Res. 28: 1300.
Merrill, J., Ullberg, M., and Jondal, M. (1981). Eur. J. Immunol. 11:536.
Paucker, K., Cantell, K., and Henle, W. (1962) Virology 17:324.
Perlmann, P., and Cerrotini, J. C. (1981). "The antigens" (M. Sela, Ed.). Academic Press, New York.
Saksela, E., Timonen, T., and Cantell, K. (1979). Scand. J. Immunol. 10:257.
Targan, S., and Dorey, F. (1980). J. Immunol. 124:2157.
Trinchieri, G., and Santoli, D. (1978). J. Exp. Med. 147: 1314.
Ullberg, M., and Jondal, M. (1981). J. Exp. Med. 153: 615.
Ullberg, M., Merrill, J., and Jondal, M. (1981). Scand. J. Immunol., 14:285.
Ullberg, M., and Jondal, M. Clin. Lab. Immunol., in press.
Ullberg, M., and Jondal, M. Manuscript in preparation.
Ziegler, H. K., and Henney, C. S. (1975). J. Immunol. 115:1500.
Ziegler, H. K., and Henney, C. A. (1977). J. Immunol. 119:1010.

INTERFERONS AND NATURAL KILLER CELLS: INTERACTING SYSTEMS OF NON-SPECIFIC HOST DEFENSE

Giorgio Trinchieri[1]
Bice Perussia

The Wistar Institute of Anatomy and Biology
Philadelphia, Pennsylvania

Interferons (IFN) play a complex modulatory role in the regulation of the cytotoxic activity of natural killer (NK) cells. We and others originally described the strong enhancement of NK cell cytotoxic efficiency as a result of treatment with IFN and IFN inducing agents (1-5), as well as the protection from NK cell-mediated lysis of target cells treated with IFN (2, 6, 7). In addition, monocytes and lymphocytes, possibly including the NK cells, may produce IFN under conditions that would seem to require NK cell activity *in vivo*, e.g., virus infection and tumor invasion. The IFN, in turn, would regulate NK activity (2, 3, 8). Our findings on the effect of IFN on NK cells have been summarized in the preceding volume of this series (9-11). We briefly present here the results of our studies in the last two years and their possible significance for understanding the defense role of NK cells.

I. ACTIVATION AND INACTIVATION OF NK CELLS

IFN treatment of human peripheral blood lymphocytes (PBL) rapidly boosts their spontaneous cytotoxicity. The cytotoxic cells that are activated by IFN have the same surface markers as those spontaneously active in peripheral blood, as judged by the presence of classical surface markers (2, 3) or by the use of monoclonal antibodies (Perussia *et*

[1]The experimental work described in this paper was supported by NIH grant CA 10815 and CA 20833.

ISBN 0-12-341360-5

al., this volume). IFN treatment of lymphocytes determines an increase in the expression of the products of HLA A, B and C loci on all cells (12, 13), including NK cells, but this increase does not correlate with the increase in NK activity (13). HLA-DR expression, which occurs only on a very small proportion of NK cells, increases on monocytes but not on lymphocytes (our unpublished results and I. Heron, personal communication). The expression of the other surface markers of NK cells is not affected by IFN.

We performed experiments in which the spontaneous cytotoxicity of human PBL was abrogated either by mixing unlabeled target cells with PBL or by absorption-elution of PBL from K562 target cells (14). IFN is released in the supernatant of mixed cultures of lymphocytes and most tumor-derived cell lines, including K562 (1, 8) and the IFN released in the culture enhances the cytotoxicity of the NK cells cocultured with the cell line (2, 3). In most of these studies, therefore, we used a cloned K562 cell line that is unable to induce IFN production when co-cultured with lymphocytes. Our results (14) showed that NK cells are inactivated after direct interaction with target cells, and they are no longer able to kill target cells susceptible to spontaneous killing. The inactivation is complete after 3 to 4 hr at 37°C, whereas contact between effector and target cells at 4°C or 20°C does not induce any inactivation. The inactivation was not due to inhibiting factors or suppressor cells. These results do not exclude the possibility that each NK cell may lyse more than one target cell, but after one or more cycles of lysis in the absence of IFN, NK cell cytotoxic ability is exhausted. The NK cells inactivated after contact with K562 are also unable to lyse other sensitive target cells. Furthermore, cytotoxic activity of inactivated NK cells does not reappear during a 48-hr incubation but is almost completely restored when the inactivated NK cells are incubated in the presence of IFN for a few hours. These results (14) raise the possibility that IFN boosts the cytotoxicity of NK cells in part by enhancing their ability to be recycled and thus kill several target cells.

Interaction at 37°C between antibody-dependent killer (K) cells with various types of immune complexes, including antibody-coated target cells, results in an irreversible modulation of the lymphocyte surface receptor for the Fc fragment of IgG molecules and in the inactivation of K cells (15, 16). The interaction with immune complexes also results in a significant inhibition of NK cell cytotoxic activity (16). Our preliminary results show that IFN can reverse, in part, the inactivation of NK cells following interaction with immune complexes without an apparent effect on the expression of Fc-receptors or a consistent effect on the K cell activity of those cells.

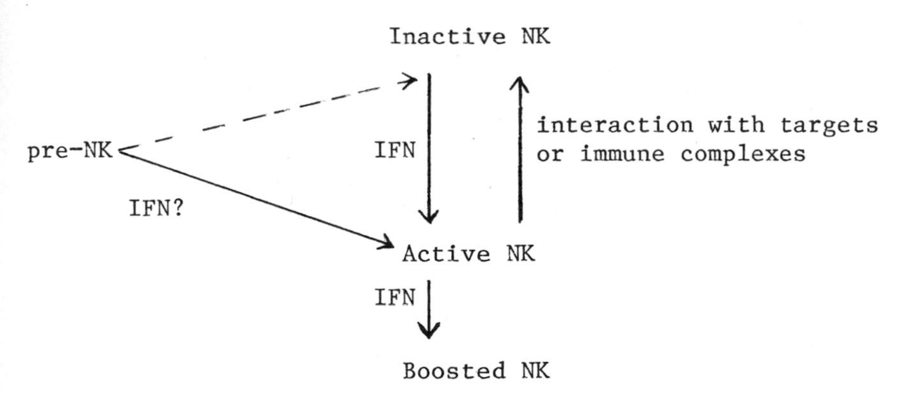

FIGURE 1. Model of pathways of NK cells maturation/activation.

Thus, it appears that IFN, in addition to increasing the cytotoxic efficiency of NK cells in lysing target cells and also possibly endowing inactive pre-NK cells with cytotoxic ability, can reverse the inactivation of NK cells due to interaction with target cells or immune complexes. These various mechanisms are represented schematically in Figure 1. As NK cells *in vivo* may continuously interact with potential target cells or immune complexes, it has been suggested (2, 17) that a subliminal level of IFN may be required to maintain NK cell baseline activity.

II. PRODUCTION OF IFN BY PERIPHERAL BLOOD MONONUCLEAR CELLS

Mononuclear cells from peripheral blood may produce two different types of IFN, α and γ. IFNγ is produced mostly by T lymphocytes in response to immune stimulation. An anamnestic response is characteristic for the production of IFNγ, e.g., only lymphocytes from immune donors respond with IFNγ production to viral antigens (18), and lymphocytes primed to alloantigens respond to the allogeneic cells with IFNγ production in a few hours, whereas unprimed lymphocytes respond with a lower production in several days (19). Both monocytes and lymphocytes can produce IFNα in response to virus infection, but only lymphocytes do so when the cells are co-cultured with tumor-derived cell lines (1, 8, 9) or with freshly explanted tumor cells (C.J. Stanton, personal communication). Although the type of lymphocyte able to produce IFNα has been the subject of extensive investigation in several laboratories, it has not been definitively identified (8-10). Our recent results showed that the IFNα-producing cells are of low buoyant density; they express IgG-Fc receptors but not the surface markers of classical T cells (E-rosetting and T antigens recognized by monoclonal antibodies); and they are predominantly HLA-DR-positive and OKM1-negative. These data, with minor variations depending on the viruses or the cell lines used for induction, restrict the choice of the IFN producing cells to B cells, to a proportion of NK cells (or large granular lymphocytes) or to an as yet unidentified cell type.

Regardless of the exact identity of the cell type producing IFN, it is well-established that blood mononuclear cells in contact with an IFN-inducing stimulus can produce IFN. The circulating NK cells are probably unable to damage normal tissue cells because of the relative resistance of the latter to spontaneous lysis and possibly also because of the continuous inactivation of the NK cells that interact with them. All types of IFN activates NK cells to destroy pathological cells, such as tumor cells and virus-infected cells, and also the normal cells. Thus, the non-selective enhancement of NK cells mediated by IFN to activate an efficient "defense" system becomes problematic unless the effects of IFN on the other cells of the organism are considered.

III. PROTECTIVE EFFECT OF IFN ON TARGET CELLS

Normal human fibroblasts that have been preincubated with IFN are not lysed by NK cells (2, 6, 7). The degree of protection is directly proportional to the dose of IFN and to the sensitivity of the target cells to IFN (2). Normal fibroblasts are almost completely protected by low doses of IFN, whereas virus-infected cells and some tumor-derived cell lines are not protected even by high doses of IFN (2). The induction of protection is dependent on RNA and protein synthesis in the treated cells and parallels (in IFN dosage, time of induction and recovery) the induction of the antiviral state (2, 6). The lack of protection of virus-infected cells is probably secondary to the block in host protein synthesis during virus replication. By various absorption studies and single-cell cytotoxic assays in agarose we have shown that NK cells bind to the IFN-treated target cells but are unable to lyse the target cells (6). This mechanism of protection is specific for NK cell cytotoxicity; resistance to antibody-dependent K cells, cytotoxic T cells or antibody and complement lysis does not increase (2, 6).

The effects of IFN on the target cells for cytotoxic effectors are therefore pleomorphic, but are all directed toward increasing the selectivity of the defense mechanisms against pathologically altered cells: a) virus-infected cells or some tumor cells are not protected against the non-specific cytotoxicity mediated by activated NK cells, and thus destroyed, but normal tissue cells are completely protected; b) IFN does not alter the susceptibility of target cells to cytotoxic mechanisms dependent on soluble antibody (both cell-mediated or complement-mediated) in which the specificity of destruction is ensured by the fine specificity of the antibodies; and c) IFN increases the expression of MHC products, rendering all cells more susceptible to MHC-specific or restricted cytotoxic T cells (7), but only cells with antigens recognized by the specific receptor of the T lymphocytes are destroyed.

The target cells treated with IFN not only become resistant to the cytotoxic effect of NK cells but also lose their ability to inactivate the NK cells that bind to them (6). Thus, the chance of the NK cells reaching the relevant non-protected target cells in a state of functional inactivation is decreased. This, in turn, contributes to the enhancing effect of IFN on the cytotoxicity of NK cells.

IV. CONCLUSIONS

The interplay between IFN and effector cells such as NK cells, monocytes/macrophages and cytotoxic T cells represents what is probably one of the most important regulatory mechanisms of the defense of the organism. Other factors (e.g., interleukin 1 and 2, colony-stimulating factor, lysozyme, prostaglandins, etc.) and other cell types almost certainly play a role in this regulation, but the action of IFN on the NK cell system may be the major or one of the major effects on the immune system during spontaneous IFN induction (e.g., virus infection) or during IFN therapy. Our knowledge is still very fragmentary, but studies from our laboratory and many others have allowed the placement of many tesserae of this complex mosaic. Several pieces are certainly misplaced, and others are probably _in vitro_ artifacts without relevant counterparts _in vivo_. However, our overall understanding of the system is slowly increasing, and several of the parameters of these mechanisms now can be accurately evaluated _in vitro_ (e.g., enhancement of NK activity, protection of the tumor cells, etc.). Thus, approaches to IFN therapy that are less empirical than those currently followed may soon be possible.

REFERENCES

1. Trinchieri, G., Santoli, D., Dee, R. R., and Knowles B., J. Exp. Med. 147:1299 (1978).
2. Trinchieri, G., and Santoli, D., J. Exp. Med. 147:1314 (1978).
3. Trinchieri, G., Santoli, D., and Koprowski, H., J. Immunol. 120:1849 (1978).
4. Herberman, R. B., Ortaldo, J. R., and Bonnard, G. D., Nature 277:221 (1979).
5. Gidlund, M., Orn, A., Wigzell, H., Senik, A., and Gresser, I., Nature 273:759 (1978).
6. Trinchieri, G., Granato, D., and Perussia, B., J. Immunol. 126:335 (1981).
7. Welsh, R. M., Karre, K., Hausson, M., Kunkel, L. A., and Kiessling, R. W., J. Immunol. 126:219 (1981).
8. Trinchieri, G., Santoli, D., and Knowles, B., Nature 270:611 (1977).
9. Trinchieri, G., Perussia, B., and Santoli, D., in "Natural Cell-Mediated Cytotoxicity Against Tumors" (R. Herberman, ed.), p. 665. Academic Press, New York, 1980.
10. Trinchieri, G., Perussia, B., and Santoli, D., in "Natural Cell-Mediated Cytotoxicity Against Tumors" (R. Herberman, ed.), p. 1199. Academic Press, New York, 1980.
11. Santoli, D., Perussia, B., and Trinchieri, G., in "Natural Cell-Mediated Cytotoxicity Against Tumors" (R. Herberman, ed.), p. 1171. Academic Press, New York, 1980.
12. Heron, I., Hokland, M., and Berg, K., Proc. Natl. Acad. Sci. USA 75:6215 (1978).
13. Perussia, B., Santoli, D., and Trinchieri, G., Ann. N.Y. Acad. Sci. 380:55 (1981).
14. Perussia, B., and Trinchieri, G., J. Immunol. 126:754 (1981).
15. Ziegler, H. K., and Henney, C. S., J. Immunol. 115:1500 (1975).
16. Perussia, B., Trinchieri, G., and Cerottini, J. C., J. Immunol. 123:681 (1979).
17. Herberman, R. B., Ortaldo, J. R., Djeu, J. Y., Holden, H. T., Jett, J., Lancy, N. P., Rubinstein, M., and Pestka, S., Ann. N.Y. Acad. Sci. 350:63 (1980).
18. Youngner, J. S., and Salvin, S. B., J. Immunol. 111:1914 (1973).
19. Perussia, B., Mangoni, L., Engers, H. D., and Trinchieri, G., J. Immunol. 125:1589 (1980).
20. Kirchner, H., Peter, H. H., Zawatzky, R., and Engler, H., Ann. N.Y. Acad. Sci. 350:587 (1980).

PLEIOTROPIC EFFECTS OF INTERFERON (IFN) ON THE AUGMENTATION OF RAT NATURAL KILLER (NK) CELL ACTIVITY

Craig W. Reynolds, Tuomo Timonen[1], Ronald B. Herberman

Biological Research and Therapy Branch
National Cancer Institute
Frederick, Maryland

I. INTRODUCTION

It has previously been shown by a number of laboratories that IFN inducers, IFN-containing supernatants and even purified IFN can augment NK activity. Recent studies using heterogeneous lymphocyte populations have indicated that this increase in activity may be due either to the recruitment of new effectors or to an activation of the lytic process (Targan & Dorey, 1980; Saksela *et al*, 1979). To examine in detail the mechanism(s) by which IFN augments rat NK activity, we investigated the effects of IFN on the ability of large granular lymphocytes (LGL) to bind to target cells and to subsequently lyse them. The results demonstrate that the effects of IFN on NK cells are pleiotropic. From these results we propose a model for augmentation which includes four stages of LGL activation.

II. MATERIALS AND METHODS

Animals. All experiments were performed with 7-12 wk old, male and female, specific pathogenic-free Rowlett nude (rnu/rnu) rats from the Mammalian Genetics Section, Division of Research Services, NIH.

[1]Present address: Department of Pathology, University Helsinki Central Hospital, Helsinki, Finland.

ISBN 0-12-341360-5

Target cells. The rat G_1-TC and mouse YAC-1 lymphoma cell lines were maintained *in vitro* as suspension cultures described (Oehler *et al*, 1978; Nunn *et al*, 1976).

Lymphocyte preparation. Peripheral blood leukocytes (PBL) were separated from heparinized whole blood as previously described (Oehler *et al*, 1977). Adherent cells were removed on columns of nylon wool (NW) at 37°C. Enriched LGL preparations were separated by discontinuous Percoll density gradients (Reynolds *et al*, 1981a; Reynolds *et al*, 1981b) and checked for purity by morphological examination of Giemsa stained preparations.

Interferon (IFN) pretreatment of effector cells. Purified LGL populations were incubated either alone or with IFN-containing supernatants for 1-2 hr at 37°C, washed once and then tested. The IFN-supernatants used in these studies (kindly provided by Dr. Robert Weisman, Johns Hopkins University, Baltimore, Md.) contained 200 units of rat IFN/ml and were produced by Newcastle disease virus infection of a normal rat kidney (NRK) tissue culture cell line.

Slide conjugate assay. The ability of different cell populations to bind various target cells was measured as previously reported (Reynolds & Herberman, 1981; Timonen & Saksela, 1980).

Single cell cytotoxicity assay in agarose. To analyze killer cell frequencies among LGL, the single cell cytotoxicity assay in agarose recently developed by Grimm & Bonavida (1979) was used.

III. RESULTS AND DISCUSSION

The results of these studies demonstrate that the vast majority of nylon-wool-passed nude blood and spleen cells which formed conjugates were LGL (Table 1). Incubation of the cells overnight at 4°C produced notably fewer conjugate forming cells than at 37°C. This effect was only transient, however, since this population of 4° cells recovered binding activity within 60-90 minutes and was highly active in a 4 hr cytotoxicity assay (data not shown).

Using the single-cell agarose assay we were able to demonstrate that treatment of LGL with IFN induced a significant augmentation in the frequency of binding to G_1-TC but not to YAC-1 (Table 2). This difference between targets may reflect the relative number of receptors or the affinity of binding required for each interaction. An additional effect of IFN was to increase the rate of killing by LGL which formed conjugates with either G_1-TC or YAC-1. This

Table 1. Effects of Overnight Incubation on the Conjugate Frequency of Nude Rat LGL

	% LGL in conjugates		% Other lymphocytes in conjugates	
Experimental Group	G_1-TC	YAC-1	G_1-TC	YAC-1
Spleen - 4°C	4	NT[a]	<1	<3
- 37°C	17	22	<1	4
PBL - 4°C	<3	7	<1	<3
- 37°C	13	15	<1	7

[a]Not tested.

Table 2. Effects of IFN Treatment on the Binding and Killing of Target Cells by Nude Rat LGL

Target	Incubation time (hr)[a]	% Binders[b] alone	% Binders[b] +IFN	% Killers alone	% Killers +IFN
G_1-TC	0	16	36	6	8
	1	12	30	25	45
	2	12	27	46	66
	4	13	27	60	72
	12	13	29	67	86
YAC-1	0	49	47	<5	8
	1	51	46	37	64
	2	50	43	60	78
	4	51	48	80	83
	12	52	50	100	99

[a]Total incubation time in agarose.
[b]Slide conjugate frequency with G_1-TC (alone = 15%, +IFN = 20%) and YAC-1 (alone = 52%, +IFN = 49%).

increase in cytotoxicity probably reflects an IFN-induced increase in lytic machinery and can be seen with virtually all target cells tested (Timonen et al, 1982). A third effect of IFN on LGL was the induction of a higher percentage of killers after 12 hr incubation. As noted for the effects on binding frequency, this difference was seen only with G_1-TC and not YAC-1. In additional experiments with G_1-TC, in which the assay period was extended to 20 hr, virtually all IFN-treated LGL that formed conjugates could be shown to cause lysis of the bound targets, whereas little if any increase in the percent killers was seen in the non-IFN-treated cells (data not shown). These results demonstrated that an appreciable number of G_1-TC-binding LGL were not lytic unless pretreated _in vitro_ with IFN. Either this subpopulation of conjugate-forming/nonlytic cells did not bind to YAC-1 or they were rapidly activated by YAC-1 and therefore were indistinguishable from conjugate-forming/lytic LGL. This latter possibility might result from the rapid production of IFN by conjugate-forming LGL and would be consistent with recent evidence that purified populations of human or rat LGL can produce IFN when cultured with tumor cells (Djeu _et al_, 1982; Timonen _et al_, 1980).

From the previous studies with human NK cells (Targan & Dorey, 1980; Timonen _et al_, 1982) and from the present data, we propose the following model of activation of NK cells by IFN (Figure 1). Included are four stages of LGL activation: A) nonbinding/nonlytic; B) binding/nonlytic; C) binding/lytic; and D) binding/highly lytic. At least three possible sites of IFN action are envisioned: 1) an increase in the number of conjugate-forming cells; 2) a shift of some binders from a nonlytic state to become killers; and 3) an increase in the rate of killing by LGL. Other intermediate forms of

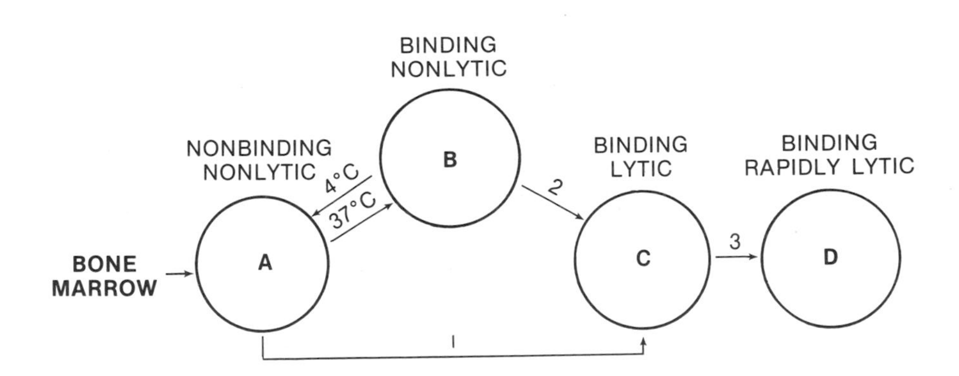

FIGURE 1. Model of LGL activation stages induced by IFN. Four functionally different LGL stages are presented: A) nonbinding/nonlytic; B) binding/nonlytic; C) binding/lytic; and D) binding/rapidly lytic. Three different effects of IFN are also shown: 1) Activation of non-binding/nonlytic cells into a binding/lytic stage; 2) activation of a binding/nonlytic cell to kill; and 3) the further activation of a lytic cell into more rapid lysis.

LGL and other sites of IFN action may also occur but have not been identified using our present techniques. For example, we have not yet examined the differentiation of NK-precursor cells from the bone marrow into LGL and the possible involvement of IFN in this process. In addition, we have not yet examined recycling of NK cells and the associated activation stages. The overall effect of IFN may be to activate or speed up many steps within the cycle. The use of the nude rat should provide some interesting answers to the above questions.

REFERENCES

Grimm, E., and B. Bonavida, J. Immunol. 13: 2861 (1979).
Djeu, J., Timonen, T., and Herberman, R., in Role of Natural Killer Cells, Macrophages and Antibody-Dependent Cellular Cytotoxicity in Tumor Rejection and as Mediators of Biological Response Modifier Activity (ed. M. Chirigos), in press, (1982).
Nunn, M. E., et al, J. Natl. Cancer Inst. 56:393 (1976).
Oehler, R. B., et al, Cell Immunol 29:238 (1977).
Oehler, J.R., et al, Int. J. Cancer 21: 204 (1978).
Reynolds, C. W., Timonen, T., and Herberman, R. B., J. Immunol. 127:282 (1981a).
Reynolds, C. W., et al, J. Immunol. 127:2204 (1981b).
Reynolds, C. W., and Herberman, R. B., J. Immunol. 126:1581 (1981).
Saksela, E., Timonen, T., and Cantell, K., Scand. J. Immunol 10:257 (1979).
Targan, S., and Dorey, F., J. Immunol. 124: 2157 (1980).
Timonen, T., et al. Eur. J. Immun. 10:422 (1980).
Timonen, T., Saksela, E., J. Immunol. Methods 36:285 (1980).
Timonen, T., Ortaldo, J.R., and Herberman, R. B., Submitted for publication (1982).

DIFFERENTIAL EFFECTS OF INTERFERONS ON PORCINE NK AND K-CELL ACTIVITIES[1]

Tae June Chung
Nam Doll Huh[2]
Yoon Berm Kim

Laboratory of Ontogeny of the Immune System
Sloan-Kettering Institute for Cancer Research
Rye, New York

I. INTRODUCTION

Interferon may be one of the most important regulatory substances for natural killer (NK) cells. Augmentation of NK activity by interferon has been demonstrated in both man and animals (Trinchieri et al., 1978; Herberman et al., 1979; Minato et al., 1980; Targan and Dorey, 1980; Djeu et al., 1981). Interferon might activate preexisting NK cells and/or induce pre-NK cells to become active NK effector cells (Targan and Dorey, 1980; Djeu et al., 1981). Whether interferon also augments killer (K) cell for antibody-dependent cellular cytotoxicity (ADCC) is controversial (Trinchieri et al., 1978; Herberman et al., 1979). We have begun to study the role of interferons in the regulation of NK cell and/or K-cell activity in the gnotobiotic miniature swine system (Kim et al., 1980; Huh and Kim, 1981).

---

[1]This work was supported in part by grants IM-233 from the American Cancer Society, Inc. and CA-08748 and HD-12097 from the National Institutes of Health, USPHHS.

[2]Present address: Department of Biochemistry, Johns Hopkins University School of Hygiene and Public Health, Baltimore, Maryland.

ISBN 0-12-341360-5

II. EFFECT OF INTERFERON INDUCER (POLY I:C) AND PORCINE ALVEOLAR MACROPHAGE INTERFERON

The effects of the well-known interferon inducer, double-stranded complex polynucleotides of polyinosinic-polycytidylic acid (poly I:C), has been tested. When specific pathogen-free (SPF) adult miniature swine peripheral blood lymphoid cells (PBL) were incubated with poly I:C (100 μg/ml) for 18 hr at 37°C, NK activity was enhanced two-to-threefold while there was no enhancement of ADCC activity (Table I). Note that the apparent enhancement of ADCC (TNP-SB plus anti-TNP) was due to the enhancement of spontaneous cytotoxicity against TNP-SB in the absence of added anti-TNP

TABLE I. Effect of Poly I:C Treatment on NK and ADCC Activities of SPF Adult Miniature Swine Peripheral Blood Lymphoid Cells

		% Specific Lysis[a]			
Animal No.	Poly I:C Treatment[b]	K562	TNP-SB	TNP-SB + Anti-TNP	Net ADCC[c]
SPF 3116-5	-	14.2±2.4	2.6±1.0	38.3±2.2	35.7
	+	35.4±3.6	15.9±0.8	51.9±1.3	36.0
SPF 4004-2	-	29.7±0.4	3.8±1.7	33.1±2.0	29.3
	+	51.0±2.8	13.6±4.0	43.1±3.8	29.5
SPF 4021-2	-	15.5±0.4	2.8±1.9	36.1±2.9	33.3
	+	48.3±2.9	27.1±3.3	53.2±1.9	26.1
SPF 4021-2-2[d]	-	7.6±0.8	5.8±2.4	27.3±1.2	21.5
	+	35.8±0.4	10.5±2.2	35.1±1.2	24.6

[a]Effector-to-target cell ratios were 100:1.

[b]Effector cells (PBL) were incubated without (-) or with (+) 100 μg poly I:C/5 x 10^6 cells/ml for 18 hr at 37°C and washed three times with cold BSS before assay.

[c]Net ADCC were calculated by subtracting % specific lysis of TNP-SB from that of TNP-SB plus anti-TNP.

[d]SPF 4021-2-2 was examined at the second time 8 months after the first test (SPF 4021-2).

antibody, therefore, net ADCC were calculated by subtracting percent specific lysis of TNP-SB from that of TNP-SB plus anti-TNP to determine the effects on ADCC only.

Porcine interferon was induced by incubating porcine alveolar macrophages with poly I:C (10 μg/ml) for 18 hr at 37°C. These culture supernatants contained 300 Units/ml of interferon activity, as measured by the cytopathic effect of vesicular stomatitis virus on MDBK cells using NIH Human Leukocyte Reference Interferon #6-023-901-527. As shown in Table II, when SPF porcine PBL were treated with porcine IFN, there was significant enhancement of NK activity but not ADCC activity. Controls with poly I:C (10 μg/ml) alone or with supernatants of alveolar macrophages cultured without poly I:C were negative. Human leukocyte interferon α was included as a positive control (see also Table III).

III. EFFECT OF HUMAN INTERFERONS ON PORCINE NK AND ADCC

We have examined effects of human leukocyte interferon (Cantell's HuIFN-α, specific activity of 10^6 Units/mg protein) on porcine NK and ADCC since there are many similarities between the human and porcine lymphoid system. As shown in Table III, we found that HuIFN-α is a good inducer and/or augmenter of porcine NK activity but not ADCC activity. To rule out a possibility that impurities in HuIFN-α preparations may cause enhancement of porcine NK activity, we have examined the effect of highly purified cloned recombinant human leukocyte interferon IFN-LrA and IFN-LrD (Hoffmann-La Roche, Inc.) (Goeddel et al., 1981) on porcine NK and ADCC. The PBL treated with IFN-LrA or IFN-LrD at concentrations as low as <0.1 ng/5 x 10^6 cells/ml had augmented NK activity (data not shown). Once again HuIFN-LrA and HuIFN-LrD enhance NK activity but not ADCC activity (Table IV).

IV. CONCLUSIONS

Interferon inducers or interferons have differential effects on porcine NK and ADCC activities; they enhance NK cell activity but not K-cell activity. Human interferons effectively augment porcine NK activity suggesting our gnotobiotic miniature swine system may serve as a valuable animal model for *in vivo* and *in vitro* studies of the effects of human interferons.

TABLE II. Effect of Porcine Interferon on NK and ADCC Activities of SPF Adult Miniature Swine Peripheral Blood Lymphoid Cells

		% Specific Lysis[a]			
Animal No.	Treatment[b]	K562	TNP-SB	TNP-SB + Anti-TNP	Net ADCC[c]
SPF 3098-5	None	16.8±2.9	21.7±1.5	60.6±3.3	38.9
	A-MØ IFN[d]	39.3±1.6	35.7±2.0	71.9±1.2	36.2
	A-MØ Super[e]	17.8±0.6	22.5±0.7	61.1±1.5	38.6
	Poly I:C[f]	25.3±0.6	29.0±3.4	58.4±1.2	28.5
	Human IFN-α[g]	35.1±1.2	38.9±1.1	73.7±3.4	34.8
SPF 4026-2	None	20.7±2.7	4.6±0.6	39.1±3.0	34.5
	A-MØ IFN[d]	39.6±1.8	20.6±1.8	53.1±5.4	32.5
	A-MØ Super[e]	18.1±2.4	5.9±4.8	36.9±0.8	31.0
	Poly I:C[f]	20.3±2.5	7.5±1.5	43.9±4.8	36.4
	Human IFN-α[g]	37.6±3.5	10.7±1.4	48.4±2.1	37.7
SPF 4078-3	None	17.7±1.8	11.5±2.1	54.8±2.9	43.3
	A-MØ IFN[d]	36.7±0.9	26.6±5.2	62.7±6.4	36.1
	A-MØ Super[e]	19.8±1.3	16.6±0.6	54.1±3.4	37.5
	Poly I:C[f]	25.1±1.0	17.6±2.4	53.1±4.7	35.5
	Human IFN-α[g]	35.0±1.5	22.4±5.9	57.3±2.8	34.9

[a]Effector-to-target cell ratios were 100:1.

[b]Effector cells (PBL 5 x 10^6 cells/ml) were treated with various reagents for 3 hr at 37°C and washed three times with cold BSS before assay.

[c]Net ADCC were calculated by subtracting % specific lysis of TNP-SB from that of TNP-SB plus anti-TNP.

[d]Porcine alveolar macrophages (A-MØ) were incubated with 10 μg poly I:C/2 x 5 cells/ml for 18 hr at 37°C and supernatants were collected. IFN activity of supernatant was determined to be 300 Units/ml.

[e]A-MØ (2 x 10^5 cells/ml) were incubated for 18 hr at 37°C and the supernatant was collected and used as A-MØ supernatant control.

[f]Poly I:C 10 μg/ml in media as control.

[g]Human leukocyte interferon IFN-α 500 Units/ml in media.

TABLE III. Effect of Human Leukocyte Interferon (HuIFN-α) on NK and ADCC Activities of SPF Adult Miniature Swine Peripheral Blood Lymphoid Cells

		% Specific Lysis[a]			
Animal No.	HuIFN-α Treatment[b]	K562	TNP-SB	TNP-SB + Anti-TNP	Net ADCC[c]
SPF 3152-9	-	15.3±1.6	11.1±1.9	36.2±4.0	25.1
	+	40.1±4.4	19.8±1.7	44.0±6.3	24.2
SPF 4078-3	-	17.7±1.8	11.5±2.1	54.8±2.9	43.3
	+	35.0±1.5	22.4±5.9	57.3±2.8	34.9
SPF 4108-3	-	24.2±2.3	19.8±2.5	69.0±1.7	49.2
	+	43.0±1.8	26.3±3.3	79.9±1.8	53.6
SPF 4102-6	-	16.8±4.7	10.1±1.1	42.1±2.1	32.0
	+	33.4±3.3	23.3±3.4	58.9±6.0	35.6

[a]Effector-to-target cell ratios were 100:1.

[b]Effector cells (PBL) were incubated without (-) or with (+) 500 Units of HuIFN-α/5 x 10^6 cells/ml for 3 hr at 37°C and washed three times with cold BSS before assay.

[c]Net ADCC were calculated by subtracting % specific lysis of TNP-SB from that of TNP-SB plus anti-TNP.

ACKNOWLEDGMENTS

The authors wish to acknowledge Dr. Richard O'Reilly for his sharing the sample of Cantell's human leukocyte interferon and Dr. Patrick W. Trown, Hoffmann-La Roche, Inc., Nutley, NJ for his gift of purified cloned recombinant human leukocyte interferons for this study; Ms. Michele Czajkowski for her excellent technical assistance, and Mrs. Rose Vecchiolla for the preparation of the manuscript.

TABLE IV. Effect of Recombinant Human Leukocyte Interferons on NK and ADCC Activities of SPF Adult Miniature Swine Peripheral Blood Lymphoid Cells

		% Specific Lysis[a]			
Animal No.	IFN Treatment[b]	K562	TNP-SB	TNP-SB + Anti-TNP	Net ADCC[c]
SPF 4075-3	-	17.7±1.2	5.1±1.2	59.7±2.4	54.6
	IFN-LrA	41.0±0.6	7.0±0.7	61.7±9.2	54.7
SPF 3114-8	-	7.0±1.3	7.7±0.6	70.3±2.3	62.6
	IFN-LrA	20.8±1.1	12.3±1.6	70.1±2.8	57.8
	IFN-LrD	21.9±1.7	12.8±1.0	54.1±2.4	41.3
SPF 4105-1	-	34.1±1.4	11.1±1.3	36.1±3.1	25.0
	IFN-LrA	50.8±1.2	19.1±3.0	50.2±1.9	31.1
	IFN-LrD	54.1±4.0	17.1±1.5	41.7±1.4	24.6

[a]Effector-to-target cell ratios were 100:1.

[b]Effector cells (PBL) were incubated without (-) or with 1.5 ng of IFN-LrA or IFN-LrD/5 x 10^6 cells/ml for 3 hr at 37°C and washed three times with cold BSS before assay.

[c]Net ADCC were calculated by subtracting % specific lysis of TNP-SB from that of TNP-SB plus anti-TNP.

REFERENCES

Djeu, J. Y., Stocks, N., Varesio, L., Holden, H. T., and Herberman, R. B. (1981). Cell. Immunol. 58:49.
Goeddel, D. V., Leung, D. W., Dull, T. J., Gross, M., Lawn, R. M., McCandliss, R., Seeburg, P. H., Ullrich, A., Yelverton, E., and Gray, P. W. (1981). Nature 290:20.
Herberman, R. B., Ortaldo, J. R., and Bonnard, G. D. (1979). Nature (London) 277:221.
Huh, N. D., and Kim, Y. B. (1981). Fed. Proc. 40:1127.
Kim, Y. B., Huh, N. D., Koren, H. S., and Amos, D. B. (1980). J. Immunol. 125:755.
Minato, N., Reid, L., Cantor, H., Lengyel, P., and Bloom, B. R. (1980). J. Exp. Med. 152:124.
Targan, S., and Dorey, F. (1980). J. Immunol. 124:2157.
Trinchieri, G., Santoli, D., and Koprowski, H. (1978). J. Immunol. 120:1849.

ROLE OF INTERFERON IN NATURAL KILL OF HERPESVIRUS INFECTED FIBROBLASTS

Patricia A. Fitzgerald
Carlos Lopez

Sloan-Kettering Institute for Cancer Research
New York, New York

Frederick P. Siegal

Mount Sinai Medical Center
New York, New York

I. INTRODUCTION

Natural killer (NK) cells active against tumor cells, virus-infected cells, and some normal cells have been described in a number of species. We have been interested in the natural kill of herpes simplex virus-type 1 (HSV-1) infected fibroblasts [NK(HSV-1)] because of earlier observations which indicated that these cells probably play an important role in genetic resistance to HSV-1 in the mouse. More recently, we have found that patients unusually susceptible to severe or persistent herpesvirus infections demonstrated low NK(HSV-1) responses (1, Lopez *et al*., in this volume). In addition, we found a correlation between normal NK(HSV-1) activity prior to bone marrow transplantation and the development of GvHD after engraftment (2, Lopez *et al*., in this volume). However, these biologic correlates have been developed using HSV-1 infected fibroblasts as targets whereas the correlation with GvHD could not be made with the more commonly used K562 targets (3). Other studies (4, Fitzgerald *et al*., in this volume) also clearly indicated that the effectors of NK(HSV-1) differ from those of NK(K562).

Supported in part by NIH grants CA-23766 and CA-08748; NIH training grant CA-09149; grant IM-303 and Faculty Research Award #193(to C.L.) from the American Cancer Society.

ISBN 0-12-341360-5

Since the discovery that interferons (IFN) of all classes are potent stimulators of NK activity, considerable attention has been directed towards understanding IFN's role in the NK response. Trinchieri et al., (5,6) have observed that many NK targets induce IFN production when incubated with human effector cells. These observations lead to the suggestion that the preferential lysis of virus-infected cells is due to the induction of IFN by virus which in turn leads to non-specific augmentation of lysis of the virus-infected targets. In this chapter we summarize the results of our studies which indicate that NK(HSV-1), although usually associated with IFN production, is not dependent on its generation (7).

II. METHODS

NK(HSV-1) assays were performed as described earlier (8). Briefly, effector cells were freshly isolated peripheral blood mononuclear cells. Target cells of monolayers of human foreskin fibroblasts (FS) were infected or mock infected with strain 2931 HSV-1, then trypsinized, washed and labeled with ^{51}Cr. Effector and target cells were mixed to yield effector to target (E:T) ratios ranging from 100:1 to 3:1, and assays were incubated for 14 hours at 37°C. Supernatants were harvested from each well for determination of ^{51}Cr-release. Data for each NK assay were analyzed using the Von Krogh equation and a linear regression of these points was obtained (9). The lysis at E:T of 50:1 was estimated from the regression line and used for comparative purposes. Supernatants for interferon analysis were collected from microtiter wells at the termination of NK assays. A cytopathic effect inhibition assay was used for quantitation of the antiviral activity (10). Each assay included an international human leukocyte reference standard (NIAID #G-023-901-527). The sheep antisera to partially purified human leucocyte interferon(#G-026-502-568) and its control antisera (#G-027-501-568) were obtained from the NIAID. The anti-interferon has 99% or greater reactivity with IFN-α and less than 1% cross reactivity with IFN-β .

III. RESULTS

A. Characterization of IFN Produced During NK Assays

When mononuclear cells from normal individuals were incubated with HSV-1 infected fibroblasts, during 14 hour NK assays, IFN was generated in the supernatant. Infected or uninfected fibroblasts by themselves or mononuclear cells incubated with uninfected fibroblasts produced little or no detectable IFN. This interferon had the properties of IFN-α :

1) both human and to a lesser extent bovine cells were protected from cytopathic effects of vesicular stomatitis virus, 2) the anti-viral activity was pH 2 stable, 3) an antiserum to IFN-α but not to IFN-β neutralized both the antiviral and NK boosting activities of the supernatants.

B. Relationship between IFN Production and NK(HSV-1)

To determine the relationship between IFN production and cytotoxicity generated during 14 hour NK(HSV-1) assays, supernatants were harvested and tested for both IFN production at E:T 50:1 and ^{51}Cr-release. 19 normal controls produced between 60 and 3000 IU of IFN-α and NK(HSV-1) values of 19-87%. No correlation was found between the amount of target cell lysis and IFN produced (linear correlation co-efficient r=0.16).

Further evidence for the independence of *in vitro* IFN production and NK(HSV-1) activity comes from some of our patient studies. We have studied a number of young male homoxexuals with acquired immunodeficiency who show high susceptibility to opportunisitic infections, including herpesviruses. A number of these patients have been found to be deficient in NK(HSV-1) activity, as well as having deficiencies in cell mediated immunity (11). In subsequent studies of similar individuals, we have found that not all of these individuals are deficient in their NK(HSV-1) activity, but interestingly, all of them made little or no IFN-α during the NK(HSV-1) assay (Table 1). We therefore have further evidence that NK(HSV-1) is not dependent on IFN generated *in vivo*. Conversely, preliminary patient studies have suggested that certain immunodeficient individuals can show normal levels of IFN production during NK(HSV-1) assays but have highly deficient NK activity (not shown).

TABLE I. Discussion of NK(HSV-1) and IFN Production in Certain Patients with Acquired Immunodeficiency

Patient	NK(HSV-1)+	IFN-Production (IU)++
1	34.4	< 10
2	22.3	< 10
*3a	65.8	< 10
3b	38.6	10
4	26.7	10

* Individual 3 was studied on two occasions

\+ NK range for normal controls in these experiments 19-87%

++ Range of IFN production for normal controls in these experiments: 60-3000 IU.

C. Effect of Pretreatment with IFN on NK

Additional evidence for the independence of NK(HSV-1) and IFN production was derived from experiments where effector cells were pretreated with optimal quantities of IFN prior to the NK assays. If the generation of IFN due to the interaction of effector cells with the virus-infected target cells were responsible for the increased lysis of infected or uninfected cells, it would be expected that similar lysis of infected and uninfected fibroblasts by IFN-pretreated effectors would be seen. When pretreated for 3-4 hours with IFN-α (10^4 IU/ml) prior to the NK assay, effector cells from normal individuals gave significantly higher lysis of both infected and uninfected target cells, but the uninfected cells were still not lysed as readily as the infected cells. The mean NK responses for 17 individuals whose effector cells were pretreated with IFN were 40.5% and 60.5% against FS and HSV-FS targets respectively. Mean values for untreated cells were 16.1% and 40.5% for FS and HSV-FS targets.

D. Effect of anti-IFN-α on NK(HSV-1)

In a final series of experiments designed to directly test the role of IFN generation on NK levels during NK(HSV-1) assays, experiments were performed using antisera directed against IFN-α . Addition of anti-IFN-α sufficient to neutralize the anti-viral effects of 3000 IU of IFN-α fully abolished the NK-augmenting effects of 1000 or 3000 IU of exogenous IFN-α but not IFN-β and did not affect NK against uninfected fibroblasts.

TABLE 2. Effect of Anti-IFN-α on NK*

Exp.	Additions	^{51}Cr-Targets FS %Lysis	FS Supernatant IFN	HSV-FS % Lysis	HSV-FS Supernatant IFN
1.	Medium	15.2	<10	32.8	300
	Control-antisera	15.7	<10	30.8	600
	Anti-IFN-α	9.6	<10	32.2	<10
2	Medium	23.7	<10	62.3	3000
	Control-Antisera	23.5	<10	54.9	2000
	Anti-IFN-α	23.2	<10	58.1	<10
3	Medium	38.1	<10	76.4	3000
	Anti-IFN-α	42.0	<10	61.2	<10

*Anti-IFN-α sufficient to neutralize 3000 IU/ml of IFN-α , control antisera, or media were added to each well along with effectors and target cells at E:T ratios of 50:1, 25:1,12.5:1, and 6.2:1. The mixtures were incubated together for 14 hours, after which supernatants from each well were harvested for determination of % lysis and supernatant interferon. Results expressed as % lysis or IFN present at E:T 50:1.

We then added anti-IFN-α or a control antiserum to effector cells just prior to the addition of ^{51}Cr labeled infected or uninfected fibroblasts. Addition of anti-IFN-α had no effect on NK(HSV-1) in two of three experiments and only a small effect in the third. NK(FS) was unaffected by anti-IFN-α. Analysis of supernatants from NK(HSV-1) wells containing anti-IFN-α revealed no detectable IFN, while wells incubated with control antisera had generated high levels of IFN-α (Table 2).

IV. DISCUSSION

A number of sutides have shown IFN to be a potent inducer of NK activity, both _in vivo_ and _in vitro_. It is clear that pretreatment of effector cells leads to augmentation of lysis of target cells, including those which are poorly lysed by uninduced effectors (12). The studies presented here were undertaken to determine the role of interferon generated _in vitro_ during NK(HSV-1) assays. The IFN was found to be IFN-α and was produced by effector cells in Percoll gradient fractions enriched for large granular lymphocytes (Fitzgerald, _et al_., unpublished). Because IFN was produced during NK (HSV-1) assays, it was possible that the induction of IFN _in vitro_ led to the subsequent augmentation of NK and non-specifically increased lysis of infected over uninfected cells, as has been suggested (5,6). Several of our findings argue against this possibility: 1) We failed to find any correlation between levels of cytotoxicity and the amount of IFN generated by normal controls during NK(HSV-1) assays. Further, patients were studied who showed normal levels of NK(HSV-1) but were unable to produce IFN during the assay, again arguing against a strict interrelationship between cytotoxicity and IFN generated (Table 1). 2) Even after effector cell pretreatment with IFN, there remained a preferential lysis of infected over uninfected targets and 3) Addition of anti-IFN to NK assays neutralized all of the IFN produced but did not reduce the kill of infected fibroblasts to that of uninfected cells (Table 2). Together, these results suggest that this preferential lysis of virus infected targets can be independent of interferon generated _in vitro_.

Similar independence of NK activity and interferon production have been reported by other investigators (13,14), although IFN may eventually account for some of the kill late in long term assays (13).

Although the herpesvirus-infected targets were killed to a greater degree than uninfected FS cells, the mechanism for this increased lysis is still not well understood. Spontaneous release values were similar for both infected and uninfected targets arguing against an inherent fragility of virus infected targets. It is possible that viral glycoproteins specifically or nonspecifically induce NK in an interferon independent manner, as has been suggested by Casali *et al*., in a measles virus system (15).

Furthermore, although our findings rule out the necessity of *in vitro* generations of IFN for augmented lysis of HSV-1 infected targets, it is likely that IFN induced *in vivo* may lead to the recruitment and/or augmentation of NK activity. Such an *in vivo* recruitment of NK cells by IFN was suggested by our studies of a patient whose peripheral blood cells initially had very little NK activity and were unresponsive to IFN *in vitro* but who showed normal NK and clinical improvement after two courses of IFN administered *in vivo* (Kirkpatrick, *et al*., unpublished).

REFERENCES

1. Lopez, C., Kirkpatrick, D., Read, S., Fitzgerald, P., Pitt, J., Pahwa, S., Ching, C. and Smithwick, E. Submitted.
2. Lopez, C., Kirkpatrick, D., Sorell, M., O'Reilly, R., and Ching, C. Lancet 2:1103 (1979).
3. Lopez, C., Kirkpatrick, D., Livnat, S., and Storb, R. Lancet ii:1025 (1980).
4. Fitzgerald, P., Evans, R., and Lopez, C., in "Mechanism of Lymphocyte Activation" (K. Resch and H. Kirchner, eds) p.595, Elsevier/North Holland, New York (1981).
5. Trinchieri, G., Santoli, D., and Koprowski, H. J. Immunol. 120:1849 (1978).
6. Trinchieri, G., Perussia, B., and Santoli, D. in "Natural Cell Mediated Immunity against Tumors" (R. Herberman, ed.)p.655 (1980).
7. Fitzgerald, P., von Wussow, P., and Lopez, C. Submitted.
8. Ching, C., and Lopez, C. Infect. Immun. 26:48 (1979).
9. Pross, H., Barnes, M., Rubin, P. and Patterson, M.J. Clin. Immunol. 1:51 (1981).
10. Stewart, W.E. in "The Interferon System", Springer-Verlag (1981).
11. Siegal, F., Lopez, C., Hammer, G., Brown, A., Kornfeld, S., Gold., J., Hassett, J., Hirschman, S., Cunningham-Rundles, C. Adelborg, B., Parhem, D., Siegal, M., Cunningham-Rundles, S., and Armstrong, D. N. Engl. J. Med. 305:1439 (1981).
12. Welch, R., and Hallenbeck, L.J. Immunol 124:2491(1980).
13. Minato, N., Reid, L.M., Cantor, H., Lengyel, P., and Bloom, B.R. J. Exp. Med. 152:124 (1980).

14. Copeland, C., Koren, H., and Jensen, P. Cell Immunol. 62: 220 (1981).
15. Casali, P., Sisson, S.J., Buchmeier, M., and Oldstone, M. J. Exp. Med. 154:840 (1981).

IN VITRO MODULATION OF NK ACTIVITY BY ADJUVANTS: ROLE OF INTERFERON

Rosemonde Mandeville
Normand Rocheleau
Jean-Marie Dupuy

Immunology Research Center
Armand-Frappier Institute
Laval-des-Rapides, Canada

I. INTRODUCTION

Current theory suggests that natural cell-mediated cytotoxicity (NCMC) is an important effector mechanism in limiting tumour development in animals and man (reviewed by Herberman and Holden, 1978). This is also supported by the observation of deficient tumour rejection in NK deficient beige mice, and by the high incidence of spontaneous "lymphoma-like" disorders in patients with selective NK deficiencies (Chediak-Higashi syndrome) (Roder, 1980). The assay and manipulation of such a function may have considerable therapeutic potential in cancer patients, mainly in the selection of individuals who may benefit from an adjuvant immunotherapy. In man, NK activity can be augmented significantly above the spontaneous level by *in vivo* or *in vitro* treatment with interferon and interferon-inducers (Einhorn, 1980; Herberman *et al.*, 1979; Ortaldo *et al.*, 1980). Several studies have demonstrated that interferon is a major internal regulator of NK activity, but there is still the question of whether it is the sole positive regulator in this system. To investigate further the mechanism of regulation and to identify the metabolic requirements for stimulation of NK activity, we have studied the boosting effects of several adjuvants including interferon, BCG, *B. abortus*, *C. parvum* and influenza vaccine. These adjuvants were used in parallel on effector cells from the same individuals. The potential regulatory role of interferon in the induction of high levels of NCMC is also discussed.

ISBN 0-12-341360-5

II. MATERIALS AND METHODS

Peripheral blood lymphocytes (PBL) obtained from 35 normal healthy donors were treated with carbonyl-iron and isolated on a Ficoll-Hypaque gradient. They were washed three times in RPMI, and resuspended to the maximum concentration (generally 10^6 cells/ml) in RPMI + 10% fetal calf serum (FCS) and gentamycin (10 μg/ml). The cytotoxicity technique consisted of an overnight ^{51}Cr-release assay using K-562 erythroleukemic cells as already described by Pross *et al* (1981).

Addition of adjuvants directly to the test system was carried out in this study using freshly thawed material: (a) *Bacillus Calmette-Guerin* (BCG), Montreal substrain, from Institut Armand-Frappier, was used at concentrations ranging from 10^2 to 10^8 bacilli/ml (b) *Corynebacterium parvum* (*C.parvum*) strain CN 6134, a formalin-killed suspension from Burroughs Wellcome, (Beckenham) Kent, U.K.) was used in concentrations ranging from 0.1 μg to 100 μg/ml (c) human leucocyte interferon was a gift from Dr. Virelizier (Hôpital des Enfants Malades, Paris) and was used in concentrations ranging from 0.05 to 50 U/ml (d) *Brucella abortus* (*B.abortus*) vaccine preparation (Institut Armand-Frappier) was inactivated at 65°C for 1 hour and used in concentrations ranging from 10^2 - 10^8 bacilli/ml (e) influenza vaccine (Fluviral, Institut Armand-Frappier) is a polyvalent vaccine containing 15 μg HA/ml of each of 3 different strains of viruses A/Bangkok, B/Brazil and A/Singapore. It was used in concentrations ranging from 5×10^{-4} to 5×10^{-1} μg HA/ml.

Pretreatment of 3×10^6 cells for 10-12 hours with metabolic inhibitors was carried out using a concentration of 10 μg/ml of either actinomycin D or cycloheximide. At the end of the incubation period, cells were washed three times in medium without FCS, counted, checked for viability and resuspended in RPMI-FCS before being used in the cytotoxicity test.

III. RESULTS

A. Modulation of NCMC by Adjuvants

Incubation of effectors and targets together with adjuvants for 16-18 hours (Table I) markedly augmented the cytotoxicity level against K-562 target cells in only 71%, 25 out of 35 of the donors tested. This augmented activity could be demonstrated in repeated experiments in the same individuals. Its expression appeared to be regulated by inherent properties of the effector lymphocytes since (a) the dose of adjuvant required for maximal boosting varied considerably among individuals, in some donors, as little as

0.5 U/ml of interferon or 10^2 bacilli/ml of BCG, producing the most significant boosting; (b) treatment of mononuclear cells from the same individual with several adjuvants showed that, for each donor, the maximum level of boosting was approximately the same for all types of adjuvants including interferon (Table I); (c) the high doses of all bacterial adjuvants (10^8 for BCG and B.abortus and 100 µg/ml for C.parvum) produced an inhibition of NCMC level. This inhibition was more pronounced with BCG;

TABLE I. Effect of Treatment with Adjuvants on NK Activity

Treatment	Concentration	% cytotoxicity (E/T, 10:1) in donor			
		A	B	C	D
None	-	50 ± 1.5	40 ± 0.6	30 ± 0.1	32 ± 2.8
Interferon (U/ml)	.05	59 ± 0.6	46 ± 2.9	39 ± 0.7	49 ± 2.2
	.5	61 ± 2.4	52 ± 0.9	42 ± 0.9	51 ± 2.0
	5	64 ± 1.2	51 ± 1.1	41 ± 2.5	52 ± 0.7
	50	62 ± 4.6	55 ± 1.1	41 ± 0.7	58 ± 1.7
BCG (bacilli/ml)	10^2	61 ± 2.2	46 ± 2.0	41 ± 1.9	40 ± 3.4
	10^4	56 ± 0.5	45 ± 0.9	40 ± 0.2	42 ± 0.7
	10^6	55 ± 3.1	44 ± 1.1	44 ± 2.2	38 ± 0.7
	10^8	27 ± 1.6	36 ± 1.7	18 ± 1.2	32 ± 3.0
B.abortus (bacilli/ml)	10^2	57 ± 1.3	48 ± 0.8	41 ± 0.9	35 ± 0.6
	10^4	59 ± 0.1	42 ± 1.0	42 ± 0.3	36 ± 0.4
	10^6	54 ± 2.2	45 ± 1.7	40 ± 3.4	38 ± 0.1
	10^8	53 ± 0.7	37 ± 0.6	38 ± 2.0	32 ± 1.8
C.parvum (µg/ml)	.1	59 ± 1.7	47 ± 2.8	39 ± 1.1	38 ± 2.3
	1	51 ± 2.4	44 ± 2.7	41 ± 0.6	32 ± 0.7
	10	53 ± 7.1	42 ± 0.8	41 ± 0.6	39 ± 1.2
	100	50 ± 1.1	42 ± 2.9	37 ± 2.1	34 ± 1.3
Influenza (µg HA/ml)	10^{-4}	63 ± 0.9	47 ± 1.7	44 ± 1.3	44 ± 0.1
	10^{-3}	60 ± 0.1	48 ± 1.4	37 ± 1.3	40 ± 0.8
	10^{-2}	62 ± 1.5	53 ± 1.3	44 ± 0.9	41 ± 2.8
	10^{-1}	60 ± 1.7	53 ± 1.0	42 ± 0.3	43 ± 2.4

(d) in 10 out of 35 (29%) of these donors, no stimulation was demonstrated with any of the adjuvant used. This lack of response was demonstrated in repeated experiments in the same individual.

These findings strengthen the contention that boosting of NK activity by adjuvants is an inherent property of the effector lymphocytes since it can be demonstrated for a variety of bacterial and viral adjuvants. In addition, they show that the dose of adjuvant producing the highest level of boosting differs among individuals and that, like natural or spontaneous NK activity, some persons lack this property. Based on these observations, the property of effector lymphocytes to be boosted or not by in vitro treatment with one or several adjuvants could be of great clinical significance since it could be used in the identification of individuals most able to respond to immunotherapy.

B. Metabolic Requirements for Augmentation of NK Activity

To evaluate the metabolic requirements for RNA and protein synthesis, effector cells were pretreated with 10 μg/ml of either actinomycin D (an RNA synthesis inhibitor), or cycloheximide (a protein synthesis inhibitor).

Both normal and augmented NK activity (Table II) were completely abolished by the treatment with actinomycin D, demonstrating the absolute requirement for normal RNA synthesis for the maintenance of a normal NK activity as well as for the development of a boosting effect. However, when effector cells were pretreated with cycloheximide, no effects on normal NK levels could be identified (Table II), but, a complete block of boosting by all the adjuvants used (including interferon) was demonstrated.

These findings suggest that unlike normal NK activity the generation and maintenance of augmented activity are dependent on new protein and RNA synthesis, also illustrating that boosting is a dynamic process depending on the normal cell machinery for its expression.

IV. DISCUSSION

Treatment of human mononuclear cells with several adjuvants leads to substantial boosting of NK activity in 71% of a normal population. Our results suggest that this augmentation of activity is an inherent property of the effector cells and is independent on the adjuvant used. In mice, immunogenetic factors have been reported to play an important role in determining the NCMC level.

TABLE II. Effect of Pretreatment with Metabolic Inhibitors on Spontaneous and Adjuvant-boosted NK Activity

Donor	Treatment[b]	Pretreatment with inhibitors[a]		
		None	Cyclo-heximide	Actino-mycin D
A	None	37 ± 1.5	39 ± 2.0	6 ± 0.8
	Interferon	49 ± 2.4	36 ± 2.1	8 ± 0.9
	BCG	41 ± 0.8	38 ± 0.3	10 ± 0.3
	B.abortus	45 ± 2.1	42 ± 0.9	9 ± 0.9
	C.parvum	49 ± 1.0	40 ± 1.0	10 ± 0.6
	Influenza	49 ± 2.6	38 ± 3.1	8 ± 0.2
B	None	51 ± 3.5	33 ± 1.8	13 ± 0.9
	Interferon	62 ± 2.1	32 ± 0.9	14 ± 1.4
	BCG	59 ± 0.2	34 ± 1.0	17 ± 3.2
	B.abortus	60 ± 2.8	32 ± 1.7	11 ± 0.6
	C.parvum	59 ± 1.8	35 ± 0.6	13 ± 0.4
	Influenza	61 ± 3.4	37 ± 0.5	13 ± 0.9
C	None	47 ± 2.9	47 ± 2.1	25 ± 2.4
	Interferon	63 ± 1.2	51 ± 2.4	28 ± 1.1
	BCG	64 ± 5.7	51 ± 1.5	28 ± 0.3
	B.abortus	59 ± 1.5	46 ± 0.4	25 ± 0.3
	C.parvum	66 ± 4.0	50 ± 2.0	25 ± 0.7
	Influenza	75 ± 3.9	51 ± 2.5	25 ± 2.7

[a]Lymphocytes were incubated with inhibitors (10 μg/ml) for 10-12 hours. Cells were washed twice, readjusted to the desired concentration and added to K-562 cells in the presence or absence of adjuvants.

[b]Concentration used: interferon (5 U/ml), BCG and *B. abortus* (10^2 bacilli/ml), *C.parvum* (10 μg/ml) and influenza vaccine (10 μg HA/ml).

In man, only a few reports have described an association of high and low activity with HLA haplotypes (Petranyi *et al.*, 1974; Santoli *et al.*, 1976). More importantly, a genetic mutation, defined by an autosomal recessive CH gene, has been associated with low NK activity in Chediak-Higashi patients (Roder, 1980). The possibility that a genetic control could regulate the responsiveness of NK cells to adjuvants is presently under investigation in our laboratory.

It is of much interest that protein synthesis is required for the development of an augmented activity, while normal or spontaneous cytotoxicity is unaffected by protein synthesis inhibition. This observation points to the fact that both natural and augmented NK activity seem to have separate mechanisms of regulation and control. Further, since interferon production by cells was shown to be inhibited by cycloheximide and actinomycin D (Stewart, 1979), our results strongly suggest that the metabolic events involved in boosting of NK activity are very different from those previously described for induction of interferon. Moreover, even treatment with exogenous interferon does not seem to help restore this lost function.

These observations fail to support the hypothesis of interferon as the internal mediator of augmentation of NK activity, and illustrate the needs for reevaluation of the mechanisms regulating this inherent property of NK cells.

REFERENCES

Einhorn, S. (1980). In "Natural cell-mediated immunity against tumors" (R.B. Herberman, ed.), p. 529. Academic Press, New York.

Herberman, R.B., and Holden, H.T. (1978). Adv. Cancer Res. 27: 305.

Herberman, R.B., Ortaldo, J.R., and Bonnard, G.D. (1979). Nature 277: 221.

Ortaldo, J.R., Phillips, W., Wassserman, K., and Herberman, R.B. (1980). J. Immunol. 125: 1839.

Petranyi, G., Benczur, M., Onody, C.E., and Hollan, S.R. (1974). Lancet 1: 736.

Pross, H.F., Baines, M.G., Rubin, P., Shragge, P., and Patterson, M.S. (1981). J. Clin. Immunol. 1: 51.

Roder, J.C., and Haliotis, T. (1980). In "Natural cell-mediated immunity against tumors" (R.B. Herberman, ed.), p. 379. Academic Press, New York.

Santoli, D., Trinchieri, G., Zmijewski, C.M., and Koprowski, H. (1976). J. Immunol. 117: 765.

Stewart, W.E. (1979). In "The Interferon System" Springer-Verlag/Wien, New York.

MODULATION OF MACROPHAGE CYTOLYSIS BY INTERFERON

Diana Boraschi[1], Elena Pasqualetto[2], Pietro Ghezzi[2], Mario Salmona[2], Domenico Rotilio[2], Maria Benedetta Donati[2], and Aldo Tagliabue[1]

[1]Sclavo Research Center, Siena, Italy, and [2]IRFMN, Milan, Italy

I. INTRODUCTION

Natural resistance against tumors involves several cellular and humoral mechanisms. Among the cellular components thought to contribute to the immunosurveillance events, mononuclear phagocytes, together with "natural killer" (NK) cells, seem to play a major role in anti-tumor defence. Macrophages (Mø) with high nonspecific cytolytic activity against tumor cells can be recovered from the peritoneal cavities of mice chronically infected with several microorganisms, or induced by administration of a variety of agents (1). Resident peritoneal Mø and monocytes from normal mice can also express significant levels of tumoricidal activity, although lower than that of "activated" Mø (2,3). This "natural" cytolytic activity of normal mononuclear phagocytes can be increased _in vitro_ by exposure to products of activated lymphocytes (lymphokines), bacterial endotoxins, interferons, or a variety of other substances. Among the factors of potential _in vivo_ relevance in the modulation of natural cytotoxicity against tumor cells, interferons are of special interest because of their regulatory effects on both NK (4) and Mø cytotoxic activities (5,6). We will discuss here the effects of mouse fibroblast interferon (IFN-β) on tumoricidal activity and other functions of peritoneal Mø, and the possible relevance of these effects.

II. EXPERIMENTAL PROCEDURES

Cytolytic activity of peritoneal Mø was measured as ^{3}H-TdR release from prelabeled mKSA-TU5 tumor cells (at a ratio of 20:1) after 48 h of cocultivation. Specific lysis was calculated by subtracting the spontaneous release of tumor cells cultured in absence of Mø (20% in Table I and III). Mø suppressive activity was measured as inhibition of ^{3}H-TdR incorporation by spleen lymphocytes prepulsed with mitogen or specific antigen, as previously described (7). Superoxide

ISBN 0-12-341360-5

TABLE I. Enhancement of Mø cytolysis by IFN-β [a]

Macrophages exposed to	Tumor cytotoxicity (% specific lysis ± SEM)			
	0	1 U	10 U	100 U
medium	1.9±0.9			
Mouse IFN-β	-	12.1±1.4	20.4±1.8	54.7±4.1
Mouse Mock IFN-β	-	3.3±0.4	4.1±0.9	3.2±0.5
Human IFN-β	-	1.8±0.1	3.7±1.4	3.0±1.3

[a] $1x10^5$ Mø from PBS-injected mice were exposed to IFN for 4 h.

anion release from Mø exposed 90' to opsonized zymosan was measured as superoxide dismutase inhibitable reduction of ferricytochrome c. Hydrogen peroxide production was measured as scopoletin oxidation in the presence of horseradish peroxidase. Endogenous metabolism of arachidonic acid (AA) was evaluated by thin-layer chromatography of the 24h-supernatants of Mø cultures ($1x10^7$ cells/culture), pre-exposed to culture medium or to IFN-β (10^4 U) for 20 h and pulsed 4 h with ^{14}C-AA.

III. MODULATION OF CYTOLYSIS

Peritoneal Mø of C3H/HeN mice can develop high cytolytic activity against tumor cells when exposed in vitro to partially purified mouse IFN-β. In contrast, neither mouse mock IFN-β nor heterologous human IFN-β could significantly enhance Mø cytotoxicity (Table I). The enhancing effect of IFN-β was assayed on either resident Mø (all mature, peroxidase-negative cells) or PBS-induced, inflammatory Mø (which include about 50% young, monocyte-like, peroxidase-positive cells). In contrast with the lymphokine Mø activating factor (MAF) which could increase only the cytolytic activity of inflammatory Mø, IFN-β was equally active on both resident and inflammatory cells (Table II).

The kinetics of cytotoxicity enhancement also differed between IFN-β and MAF. In fact, IFN- β-induced increase of cytolysis was evident after 4h of incubation with Mø and persisted after 20h. In contrast, the enhancing activity of MAF strongly declined with prolonged exposure times (Table II). Thus, it appears that the modulation of Mø cytotoxicity by IFN-β follows mechanisms and acts on Mø subsets clearly

TABLE II. Modulation of Mϕ cytolysis by IFN-β and MAF[a]

Macrophages exposed to	Tumor cytotoxicity of Mϕ treated in vitro for 4 h		20 h	
	% release ± SEM	% release over natural cytolysis	% release ± SEM	% release over natural cytolysis
Resident Mϕ				
medium	24.1±1.2	0	11.4±0.2	0
MAF	29.3±0.4	5.2	13.4±1.0	2.0
IFN-β	43.1±0.6	19.0	27.1±0.6	15.7
Inflammatory Mϕ				
medium	30.1±1.0	0	14.8±0.7	0
MAF	58.3±1.5	28.2	15.5±0.8	0.7
IFN-β	56.7±2.3	25.6	36.8±0.8	22.0
No Mϕ	10.6±0.2		10.4±0.3	

[a] 1.5×10^5 Mϕ were exposed to culture medium, MAF (diluted 1/5) or IFN-β (1,000 U) for 4h or 20h before addition of target cells.

distinguishable from those of MAF. In fact, MAF would preferentially affect cytolytic activity of immature Mϕ, lacking any significant effect on resident Mϕ. On the other hand, IFN-β activity seems principally directed to mature Mϕ. The Mϕ effector cells did not appear to express surface Ia molecules.

TABLE III. Effect of Ia^+ Mϕ depletion on tumoricidal activity[a]

Macrophages exposed to	Tumor cytotoxicity (% specific lysis ± SEM) of Mϕ after treatment with			
	medium	C	ATHαATL	ATHαATL + C
medium	9.1±0.3	10.8±0.7	11.0±0.4	12.4±1.7
MAF	29.4±0.6	29.2±1.2	32.3±0.7	31.2±0.8
IFN-β	35.7±0.2	32.7±0.4	32.2±1.5	34.0±2.5

[a] 2×10^5 PBS-induced Mϕ were exposed to culture medium, MAF (diluted 1/3) or IFN-β (1,000 U), after being treated with ATHαATL serum and complement.

As shown in Table III, neither spontaneous cytolysis nor IFN-β- nor MAF-induced tumoricidal activity of Mϕ from C3H/HeN mice (H-2I^k) were affected by ATHαATL serum (specifically directed to Ia antigens encoded by the I^k subregion of the H-2 complex) and complement treatment.

IV. MODULATION OF SUPPRESSION

The ability of IFN-β to modulate *in vitro* another Mϕ activity, thought to be controlled by the same mechanisms of Mϕ cytolysis, was investigated. Both resident and inflammatory Mϕ are naturally suppressive for proliferation-dependent and independent lymphocyte activities (e.g. replication in response to mitogens or specific antigens and lymphokine production)(7, 8). Stimulation *in vivo* with agents able to increase Mϕ tumoricidal activity in parallel enhances Mϕ-mediated suppression of lymphoproliferation (8,9). We show in Table IV that it was however possible to distinguish between modulation of cytotoxocity and modulation of suppression by the use of IFN-β *in vitro*. In fact, incubation of Mϕ for 20h with IFN-β strongly reduced their suppressive activity, whereas MAF was without effect. In contrast with cytolysis data, neither IFN-β nor MAF had any effect on Mϕ suppression after 4h of incubation. Thus, Mϕ suppressive activity appears to be regulated by mechanisms quite distinct from those controlling cytotoxicity. Also, as shown in Table V, depletion of Ia$^+$ Mϕ was without effect on Mϕ natural suppressive activity, thus suggesting that the effector of suppression would be an Ia$^-$ Mϕ. However, the

TABLE IV. Modulation of Mϕ suppression by IFN-β [a]

Macrophages exposed to	Suppressive activity (% reduction lymphoproliferation ± SEM) by			
	Resident Mϕ		Inflammatory Mϕ	
	4 h	20 h	4 h	20 h
medium	73.7±2.2	48.8±5.0	58.6±5.2	43.0±5.6
MAF	74.7±3.2	45.6±3.9	65.0±2.0	36.9±3.7
IFN-β	77.3±2.5	9.6±5.6	57.0±5.6	2.0±1.2

[a] $5x10^4$ Mϕ were exposed to culture medium, MAF (diluted 1/3) or IFN-β (500 U) before addition of $5x10^5$ PPD-pulsed spleen lymphocytes from BCG-immune mice.

modulatory effect of IFN-β on Mø suppression was dramatically affected by depletion of Ia^+ Mø. Whereas IFN-β was able to decrease the suppressive activity of untreated resident Mø (which include about 10% Ia^+ cells; 10), its effect was only marginal on Mø depleted of the Ia^+ subset (Table V). Thus, the modulatory effects of IFN-β appear to be mediated by an Ia^+ regulatory cell which in turn affects the Ia^- suppressor Mø.

TABLE V. Effect of Ia^+ Mø depletion on suppressive activity[a]

In vitro treatment	Suppressive activity (% reduction lymphoproliferation ± SEM) of Mø exposed to medium	IFN-β
medium	46.6 ± 4.0	6.8 ± 0.3
ATHαATL + C	53.3 ± 2.2	40.7 ± 2.7

[a] $1x10^5$ resident Mø were treated with ATHαATL serum and complement, then exposed for 20h to culture medium or to IFN-β (250 U), before addition of $5x10^5$ ConA-pulsed spleen lymphocytes.

TABLE VI. Production of oxygen intermediates by Mø

Macrophages exposed to	Superoxide production (nmoles reduced cyt. c/ mg protein ± SEM)[a]		Hydrogen peroxide production (nmoles/ mg protein ± SEM)[b]	
	4 h	20 h	4 h	20 h
medium	126.5±14.4	222.9±15.5	167.2±4.6	193.1±5.3
MAF	119.0±7.0	263.1±21.0	160.8±2.8	215.4±12.5
IFN-β	123.0±7.5	77.5±2.7	146.0±9.5	33.0±11.4

[a] $1x10^6$ resident Mø were exposed to culture medium, MAF 1/4 or IFN-β (2,500 U).
[b] As in (a), but $3.5x10^5$ resident Mø were used.

V. MODULATION OF RELEASE OF OXYGEN INTERMEDIATES AND PROSTAGLANDINS

A variety of soluble factors released by Mø have been hypothesized as responsible for their cytotoxic activity.

Among them, of special interest are oxygen metabolites and prostaglandin E_2, since both may also be involved in the effector mechanism of Mϕ suppression (11). The dissociation observed between the modulation by IFN-β of Mϕ suppression and cytolysis prompted us to investigate the effect of IFN-β on the production of these cytotoxic and suppressive molecules by Mϕ. Table VI illustrates the release of superoxide anion and hydrogen peroxide by Mϕ exposed for either 4h or 20h to IFN-β. In correlation with the suppression data, the production of both oxygen intermediates was drastically depressed in Mϕ exposed to IFN-β for 20h, while it was unaffected in MAF-treated Mϕ. As for suppressive activity, no change in Mϕ release of oxygen metabolites was observed after 4h of exposure to either IFN-β or MAF. Preliminary data on arachidonic acid (AA) metabolism of Mϕ show that production of PGE_2 was also depressed in cells exposed to IFN-β for 20h (Fig. 1).

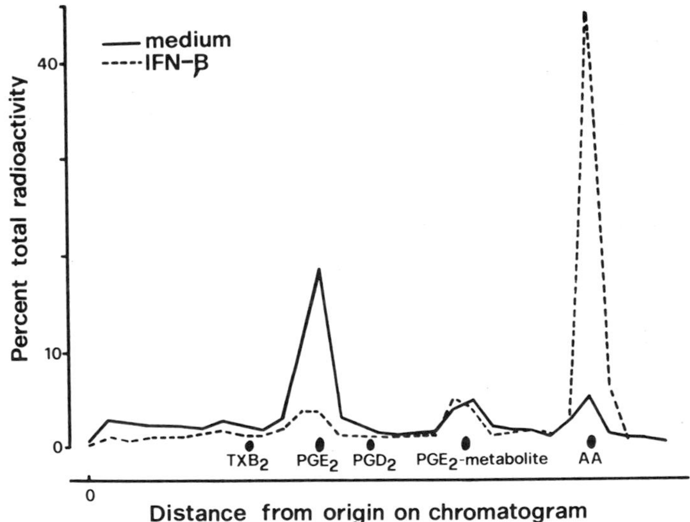

FIGURE 1. Modulation of arachidonic acid metabolism by IFN-β.

This suggests that IFN-β could decrease Mϕ suppressive capacity by depressing their ability to produce both oxygen metabolites and PGE_2. On the other hand, neither oxygen intermediates nor PGE_2 would seem to have a major role as effector molecules in the IFN-β-mediated augmentation of Mϕ cytolytic activity (12).

VI. CONCLUDING REMARKS

The observation that IFN-β could differentially modulate Mϕ functions leads to speculation about the importance of these events in the *in vivo* regulation of immune reactions in a viral infection. IFN-β is produced by virus-infected cells and can itself inhibit viral replication. Mø exposed to IFN-β exhibit enhanced cytocidal capacity, thus becoming more effective in destroying the virus-infected cells. As a third line of defence, T lymphocytes become able to react adequately to activation stimuli (provided by the infection events), since the natural suppression mechanism exerted by Mø is drastically reduced by IFN. When cells infected with the virus are all destroyed, the IFN production will stop. Thus, Mø would return to a lower level of cytotoxicity, since high cytolytic activity is no longer required. In parallel, suppressive capacity would return to high levels and Mø would be able to shut off, in a feed-back mechanism, lymphocyte activation.

ACKNOWLEDGEMENT

The ATHαATL serum was produced and screened in the laboratory of John E. Niederhuber, to whom we are greatly indebted.

REFERENCES

1. Hibbs, J.B., Jr., in "The Macrophage in Neoplasia" (M.A. Fink, ed.), p. 63. Academic Press, New York, 1976.
2. Keller, R., Brit. J. Cancer 37:732 (1978).
3. Tagliabue, A., Mantovani, A., Kilgallen, M., Herberman, R.B., and McCoy, J.L., J. Immunol. 122:2363 (1979).
4. Djeu, J.Y., Heinbaugh, J.A., Holden, H.T., and Herberman, R.B., J. Immunol. 122:175 (1979).
5. Schultz, R.M., Papamatheakis, J.D., and Chirigos, M.A., Science 197:674 (1977).
6. Boraschi, D., and Tagliabue, A., Eur. J. Immunol. 11:110 (1981).
7. Boraschi, D., Soldateschi, D., and Tagliabue, A., Eur. J. Immunol. in press (1982).
8. Baird, L.G., and Kaplan, A.M., Cell. Immunol. 28:22 (1977).
9. Kirchner, H., Fernbach, B.R., and Herberman, R.B., in "Mitogens in Immunobiology" (D.L. Rosenstreich and J.J. Oppenheim, eds.), p. 495. Academic Press, New York, 1976.
10. Cowing, C., Schwartz, B.D., and Dickler, H.B., J. Immunol. 120:378 (1978).
11. Metzger, Z., Hoffeld, J.T., and Oppenheim, J.J., J. Immunol. 124:983 (1980).
12. Boraschi, D., Ghezzi, P., Salmona, M., and Tagliabue, A., Immunology in press (1982).

THE RELATION BETWEEN HUMAN NATURAL KILLER CELLS AND INTERLEUKIN 2

Wolfgang Domzig[1,2]
Beda M. Stadler[3]

[2]Biological Research and Therapy Branch
National Cancer Institute-FCRF
Frederick, Maryland

[3]Laboratory of Microbiology and Immunology
National Institute of Dental Research
Bethesda, Maryland

The association of natural killer (NK) cells and their growth in interleukin 2 (IL-2)-containing medium (1), as well as the ability of this lymphokine to augment mouse NK cell activity (2), generated new questions about the relationship of NK cells and IL-2: 1) To what extent is IL-2 necessary for the NK cell activity? 2) Do NK cells not only respond to IL-2 but also produce IL-2 themselves? The following will summarize experiments which have been done to clarify this relation.

I. RELATION BETWEEN HUMAN NK CELL ACTIVITY AND IL-2

The purification of human IL-2 (3) and the generation of monoclonal antibodies against this lymphokine (4) made it possible for us to study the effect of IL-2 or monoclonal anti-IL-2 antibody on highly enriched NK cell populations (5), obtained from human peripheral blood. Enriched NK cell populations were collected from discontinuous density Percoll gradient and then further depleted of 29°C sheep erythrocyte-rosetting cells (6). The cytolytic activity of such highly enriched NK cell populations could easily be enhanced with the addition of low amounts of purified IL-2 preparations into the 4 hr test, as shown in Table 1. The boosting of human NK cell activity by IL-2 is similar to that reported in the murine system (2).

[1]Supported by DFG grant Do 244/2-2.
[2]Present address: Max-Planck-Institute for Immunology, Freiburg/Br., Germany.
[3]Present address: Institute of Clinical Immunobiology, University of Bern, Bern, Switzerland.

ISBN 0-12-341360-5

TABLE 1. Boosting of human NK cell activity by Purified IL-2

Treatment	NK cell activity (LU ± SD)[a]		
	Experiment 1	Experiment 2	Experiment 3
None	147 ± 20	317 ± 36	250 ± 50
IL-2 (30 U/ml)	833 ± 30	588 ± 18	500 ± 52

[a]The cytotoxicity test was done for 4 hr. The NK cell activity is expressed in lytic units per 1 x 10^7 cells.

Anti-IL-2, at concentrations sufficient to block growth of the IL-2 dependent cell line CT6, caused 50% inhibition of the cytotoxicity compared to cells treated with control antibodies. When NK cells were preincubated for 16 hr with anti-IL-2, the NK cell activity declined further, to about 10% of that of the control. This finding indicated the importance of IL-2 for maintenance of *in vivo* occurring spontaneous human NK cell activity. Upon addition of low amounts of IL-2 (<30 units/ml of purified or crude IL-2 preparations), partial restoration of the cytolytic activity could be seen. These experiments raised the question as to whether NK cells or cells associated with them can produce IL-2 and thereby account for the high levels of spontaneous NK cell activity.

II. RELATIONSHIP OF NK CELLS AND IL-2-PRODUCTION

Nonadherent cells from human PBL were separated in a seven step discontinuous Percoll gradient as described in (7). Each fraction was tested for NK cell activity as well as stimulated with PMA and Con A as described in (8). The supernatants were then tested for IL-2 content. The results (as the mean of eight experiments) showed peak NK cell activity in fractions 2 and 3, intermediate activity in fraction 4 and low NK cell activity in fractions 1 and 5, whereas IL-2 was highest in fraction 5 (500 ± 157 U/ml), intermediate in fraction 4 and 6 and less in fractions 1, 2, 3 and 7. Thus there was no direct correlation between NK cell activity and the possibility to stimulate the cells to produce IL-2. But cell populations associated with the highest NK cell activity in the gradient still produced significant levels of IL-2 after stimulation: fraction 2 (75 ± 21 U/ml) and fraction 3 (150 ± 49 U/ml). What was the source of IL-2-production in these NK cell populations? As these cell populations had not been further purified, it was of particular concern whether this was due to contaminating T cells.

III. PURIFICATION OF NK CELL POPULATIONS FROM T CELLS

Considering the hypothesis that T cells provided the IL-2 (9) which was found in populations with high NK cell activity after stimulation or which was necessary for the NK cell activity, further steps were taken to eliminate T cells from the enriched NK cell populations. First, NK cell rich fractions from the Percoll gradient were depleted of 29°C sheep erythrocyte-rosetting cells. For further purification an adsorption technique was chosen using the monoclonal antibody OKT3, purchased from Ortho Pharm. Corp., Raritan, New York. Antibody labeled cells were adsorbed on plastic which had been coated with purified $F(ab')_2$ fragments of sheep anti-mouse IgG.

TABLE 2. IL-2-Production after Stimulation of NK cell Populations of Varying Purity

Cells from Percoll gradient	Experiment 1		Experiment 2		Experiment 3	
	IL-2[a]	LU[b]	IL-2	LU	IL-2	LU
NK cells	85.9	n.d.	289.3	323	57.4	238
NK cells depleted of 29°C E^+ cells	37.1	n.d.	17.1	714	39.3	357
T cells from fraction 5	650.0	n.d.	512.0	-	n.d.	n.d.

[a] IL-2 content is expressed in units/ml as described in (8).
[b] NK cell activity was tested for 4 hr and is expressed in lytic units per 1×10^7 cells.

Table 2 shows the effect of the elimination of 29°C sheep erythrocyte-rosetting T cells from NK cell populations on their IL-2 production after PMA plus Con A stimulation, as well as on their cytolytic function. This procedure resulted in a further enrichment of NK cell activity whereas the IL-2 production declined but was not eliminated.
To examine whether the remaining IL-2 production was still due to some few T cells, we further removed OKT3 positive cells by adsorption. Table 3 shows that there was still IL-2 production after stimulation of such populations, but this treatment reduced the cytolytic activity. In a total of 15 experiments, we found that the adsorption of OKT3 positive cells from already highly enriched NK cell populations resulted in a mean of about 50% inhibition of NK cell activity, which we did not

see using an unrelated monoclonal antibody like OKT6.

TABLE 3. IL-2-Production after Stimulation and NK Cell Activity of Highly Enriched NK Cells after Elimination of OKT3 Positive Cells

	Experiment 1		Experiment 2		Experiment 3	
NK cells	IL-2[a]	LU[b]	IL-2	LU	IL-2	LU
After rosetting	42.1	476	78.6	400	130	286
After rosetting and adsorption of $OKT3^+$ cells	26.3	106	30.0	143	76	41

[a] IL-2 content is expressed in units/ml.
[b] NK cell activity was tested for 4 hr and is expressed in lytic units per 1 x 10^7 cells.

The adsorption of cells positive with OKT3, an antibody which is specific for T lymphocytes (10) and reported absent on NK cells (11), showed a similar effect on the NK cell activity as did the anti-IL-2 treatment in the short term assay. Thus, although there remained cells which were able to produce IL-2 after additional stimulation, we may have removed spontaneously IL-2 producing cells in this cell population, which resulted in the decline of the NK cell activity as if we removed the IL-2 with anti-IL-2 antiserum from the cytotoxicity test.

Beside the morphological assessment of T cell content on slides, which appeared not very accurate for low numbers of T cells, we tried to find a more reliable method to indicate any remaining T cells. One promising approach was to measure the proliferative response after activation with OKT3 (12). There was no difference between the ability to stimulate the proliferation with OKT3, as measured by ^{3}H-incorporation, using PBL which contain a very high amount of T cells, or enriched populations of NK cells taken from fractions of the Percoll gradient with a very small amount of T cells, or this same cell population after additional depletion of cells which rosette at 29°C sheep erythrocytes as seen in Table 4. The stimulation with OKT3 was specifically related to this antibody, since control ascites as well as ascites of monoclonal antibody against a virus did not affect these cells.

TABLE 4. Proliferation of NK Cells after OKT3 Stimulation

Treatment	cpm		
	PBL	Enriched NK cells	NK cells after rosetting
None	315	1,841	4,077
OKT3 (25 ug/ml)	44,141	46,231	35,455

IV. DISCUSSION

The experiments using the anti-IL-2 antibody lead to the suggestion that the cytolytic activity of NK cells is dependent on IL-2. NK cell activity, though present *in vivo* may in itself not be a spontaneous activity but rather NK cells may exist in an inactive stage, which has to be activated with a lymphokine. *In vivo* we would always find active NK cells, as the activating lymphokines are present. The existence of "binders" and "nonbinders" among NK cells which can be activated to bind and finally lyse the bound target support this suggestion, as do the reports of others (13,14). We show that IL-2 is a major factor for this activity. In addition, it is known that interferon (IFN) also plays an important role for the NK cell activity, probably a regulating role, as others have found for the murine system that IL-2 and IFN have a cooperative effect on NK cell activity (15).

Whether or not NK cells themselves are IL-2 producers can not be directly concluded from the experiments. As IL-2-production is cell cycle dependent (8), we may have a cell cycle-related separation in the Percoll gradient and thus we may be unable to monitor the total capability for IL-2-production. As we only measured the ability to produce IL-2 in response to PMA + Con A, it may be that this additional stimulus is inefficient for already IL-2 producing NK cells or a different stimulus may be necessary. Furthermore, we can only measure IL-2 released into the medium, and we cannot detect the factor which is taken up and utilized by the cells.

Two suggestions are provided by the experiments in which highly enriched NK cell populations were activated to proliferate by OKT3: 1) OKT3 may actually be present on NK cells. 2) In response to this stimulus, NK cells may produce IL-2 which causes their proliferation. If OKT3 is present on NK cells, it is suggested that this expression is stronger on active cells, since in each adsorption experiment the same number of cells with the morphology of LGLs (16) were used. Alternatively, it is possible that the response to OKT3 was by stem cells, as stem cells of various kinds would be expected in these fractions.

The experiments unexpectedly documented the difficulty to distinguish between NK cells and T cells using markers recognized by monoclonal antibodies. As NK cells exhibit surface markers associated with monocyte-macrophages as well as with T cells (11), one would like to find an NK-private marker unrelated to any other cell type.

To solve the question whether NK cells do produce IL-2, NK-clones may be most useful to answer this question. It will be of interest to determine whether they can be stimulated to produce IL-2 and whether their cytolytic activity can be blocked by anti-IL-2.

REFERENCES

1. Timonen, T., Ortaldo, J.R., Stadler, B.M., Bonnard, G.D., and Herberman, R.B., submitted 1982.
2. Henney, C.S., Kuribayashi, K., Kern, D.E., and Gillis, S., Nature 291:335 (1981).
3. Stadler, B.M., and Oppenheim, J.J., in "Lymphokines" Vol. 6 (E. Pick, ed.), Acadmeic Press, New York, in press.
4. Stadler, B.M., Berstein, E.M., Siraganian, R.P., and Oppenheim, J.J., in "Mechanisms of Lymphocyte Activation" (K. Resch and H. Kirchner, eds.), p. 570. Elsevier/North Holland (1981).
5. Timonen, T., and Saksela, E. J. Immunol. Meth. 36:285 (1980).
6. West, W.H., Cannon, G.B., Kay, H., Bonnard, G.D., and Herberman, R.B. J. Immunol. 118:355 (1977).
7. Domzig, W., Stadler, B.M., and Herberman, R.B., in preparation.
8. Stadler, B.M., Dougherty, S.F., Farrar, J.J., and Oppenheim, J.J. J. Immunol. 126:2321 (1981).
9. Smith, K.A. Immunol. Rev. 51:337 (1980).
10. Kung, P.C., Goldstein, G., Reinherz, E.L., and Schlossman, S.T. Science 206:347 (1979).
11. Ortaldo, J.R., Sharrow, S.O., Timonen, T., and Herberman, R.B. J. Immunol. 127:2401 (1981).
12. Van Wauwe, J.P., De Mey, J.R., and Goossens, J.G. J. Immunol. 124:2708 (1980).
13. Timonen, T., Ortaldo, J.R., and Herberman, R.B. J. Immunol., in press (1982).
14. Masucci, M.C., Klein, E., Argow, S. J. Immunol. 124:2458 (1980).
15. Kuribayashi, K., Gillis, S., Kern, D.E., and Henney, C.S. J. Immunol. 126:2321 (1981).
16. Timonen, T., Ortaldo, J.R., and Herberman, R.B. J. Exp. Med. 153:569 (1981).

MODULATION OF NATURAL CELL MEDIATED CYTOTOXICITY OF T-COLONIES AND PBL BY INTERLEUKIN-2

Susanna Cunningham-Rundles
Richard S. Bockman
Berish Y. Rubin

Laboratories of Clinical Immunology,
Calcium Metabolism, and Interferon
Memorial Sloan-Kettering Cancer Center
New York, New York

I. INTRODUCTION

Exponential proliferation of T-cell lines developed from mitogen-stimulated T lymphocyte clonal expansion has been found to be highly dependent upon the presence of T-cell growth factor, now designated Interleukin-2 (1,2). The possibility of investigating immune interactions using homogeneous populations of T lymphocytes has led to extensive use of IL-2-maintained T lymphocyte colonies.

Initiation of clonal expansion has been shown to be positively regulated by Interleukin-1 (IL-1) and negatively regulated by prostaglandin E_2 (PGE_2) which are both produced by monocyte/macrophages (3,4). Analysis of factors like IL-2 which may influence or direct colony formation is critical to understanding which immune functions or interactions may be studied by these means and how selection pressure associated with the use of growth promoters may affect colony phenotype. The studies with IL-2-developed T lymphocyte colonies and IL-2-stimulated normal peripheral blood lymphocytes (PBL) presented here indicate that natural cytotoxicity is induced by IL-2 and therefore that expression of this function would be expected to be both induced and conserved during colony establishment.

ISBN 0-12-341360-5

II. METHODS

The T cell colonies used in these experiments were obtained from peripheral blood mononuclear cells from healthy donors as described previously (4). The concentration of IL-2 used was 1.0 unit per ml. Natural cytotoxicity was carried out using a 4 hr ^{51}Cr release assay as described previously (5). Interferon was measured (6) as inhibition of cytopathic effect of vesicular stomatitis virus (VSV) on GM2767 cells with appropriate standards. Gamma Interferon (γ IFN) was prepared by one of us (B.Y.R.) and alpha Interferon (α IFN) was cloned IFN received from Hoffman-La Roche in collaboration with Dr. H.F. Oettgen (MSKCC). The experiments shown in Table IV were carried out in flasks during the preculture phase. Cells were harvested, counted and replated under standard conditions for assessment of cytotoxicity against K562. In these experiments normal PBL were plated and targets were added either immediately or after 18 hr of preculture. An additional 100 μl of medium was added and subsequently removed to protect cells against drying and substrate deprivation during the preculture phase.

III. RESULTS AND DISCUSSION

Individual T-lymphocyte colonies were removed from soft agar culture then placed into liquid medium containing 1.0 unit per ml IL-2 until a density of 7.0 X 10^5 lymphocytes per ml or greater was obtained. The cells were then harvested, washed, and resuspended in RPMI 1640 with 1.0 unit per ml IL-2 for functional studies. For proliferation analyses, cultures were incubated for 4 to 6 days without additional IL-2 or other supplementation. As shown in Table I, for three clones originating from one normal donor proliferative activity was observed. Significant IL-2 production was unlikely to have occurred in the activated cultures in the absence of adherent cells. The unstimulated cultures remained viable and demonstrated thymidine uptake comparable to cultured PBL. The data shown here reflect maximum activation by any concentration of stimulant. Analysis of concentration dependence of response showed that for PHA, the strongest response occurred at the highest concentration of PHA (286 μg/ml) in contrast to the strongest response of the normal control which occurred at 47 μg/ml (two intervening concentrations were also tested in these experiments). This same relationship held true for response to PWM but not to Con A where each colony had a

TABLE I. Proliferative Response of T-Colony-Derived Clones

Activator	Clones F-4	E-5	E-8	Control PBL
PHA	19.4[a]	15.3	14.6	144.0
Con A	25.4	22.7	15.7	82.0
PWM	12.2	21.6	17.1	70.9
E. coli	7.9	5.9	7.7	14.0
C. albicans	2.4[b]	6.9	3.9	66.0
HSV	2.3[b]	3.5	< 1.0[b]	14.0

[a]The data are given as stimulation ratio of response to activator divided by unstimulated response.
[b]negative.

TABLE II. T-Colony Anti-K562 Activity at Different Concentrations of IL-2

Colony	Concentration of IL-2 1.0 U/ml	0.5 U/ml	0.3 U/ml
E-8	32.9[a]	26.3	N.D.
E-5	26.9	N.D.	11.6
F-4	36.7	N.D.	N.D.

[a]Percent release ^{51}Cr at E-T ratio 10:1.

different point of maximum activation (8 concentrations tested). These results indicate that different colonies may have different binding affinities for the same stimulant. The responses to microbial activators and to Herpes Simplex virus suggest that functionally distinct colonies have been selected. In contrast, all three clones had strong anti-K562 activity as shown in Table II when an E-T ratio of 10:1 was used. This level of cytotoxicity is average for unseparated PBL at an E-T ratio of 100:1 and therefore indicates a selective enrichment of cytotoxic cells. Less than 1% of the starting population of mononuclear cells is recovered as clonally proliferating T lymphocyte colonies and selection pressure would be expected to be most intense during the first 24 to 48 hr of culture. Since PHA can activate pre-NK cells, it is likely that NK activity would be protected during the first phase of colony growth.

The presence of IL-2 during clonal proliferation may have a critical impact on functional expression and T-cell colony selection. The relationship between IL-2 concentration and the NK activity of T colonies in a 4 hr assay is shown in

TABLE III. Relative Effect of IL-2 and Interferons on PBL NK Activity During 4 Hr Assay

Culture condition	IFN/ml[a] (units)	% Cytotoxicity E-T: 100:1	E-T: 50:1
no addition	0.0	30.1	23.9
IL-2[b]	18.0	26.1	16.7
αIFN	1500.0	35.1	33.1

[a]Per 5 X 10^6 lymphocytes.
[b]IL-2, 1 Unit/ml.

TABLE IV. Effects of IL-2 and Other Augmenting Agents on NK Activity in 18 Hr Incubation

Culture condition	Units IFN preculture	Units IFN postculture	% release
no addition	0	0.0	5.6
IL-2[b]	15[c]	56.0	63.0
αIFN	60	150.0	39.2
OKT-3	< 6	< 4.5	24.1

[a]Per ml, per 10^6 lymphocytes.
[b]IL-2, approximately 1.0 Unit/ml.
[c]IFN contained is γIFN.

Table II. The amount of IL-2 present at each concentration was equivalent to one tenth the noted amount on a per cell basis compared to conditions required for colony growth. Significant cytotoxicity (20.1%) was also seen at an E-T ratio of 5:1 with 0.5 U/ml but not with 0.3 U/ml which represented negligible IL-2 per cell. This experiment does not rule out the possibility that different clones might have different intrinsic levels of endogenous NK activity separate from IL-2-related modulation. Reduction of IL-2 did not affect cell viability during the period of study. IL-2 at the level of maximum supplementation shown in Table II did not augment NK activity of normal PBL during a 4 hr incubation with labeled target cells, as shown in Table III.

In contrast, the results shown in Table IV indicate that IL-2 does cause significant augmentation of NK activity in an 18 hr incubation. The low endogenous NK activity seen here is typical of many normal donor PBL cultured at 37°C for 18 to 24 hr following mononuclear cell isolation. As shown, IL-2

TABLE V. Assessment of Pre-NK Pool Following 18 Hr Preculture of PBL with NK Augmenting Agents

Preculture condition	Assay condition	100:1	50:1	25:1
1. no preculture	- IFN	30.6	22.8	16.1
no preculture	+ IFN	45.0	43.2	32.8
2. 18 hr	- IFN	19.2	18.0	7.8
18 hr	+ IFN	27.1	24.5	12.6
3. 18 hr + αIFN[a]	- IFN	60.7	58.2	38.0
18 hr + αIFN	+ IFN	65.3	52.7	39.0
4. 18 hr + γIFN[b]	- IFN	39.3	38.6	24.1
18 hr + γIFN	+ IFN	66.1	46.8	29.3
5. 18 hr + IL-2[c]	- IFN	73.2	72.5	61.4
18 hr + IL-2	+ IFN	71.9	74.8	64.4
6. 18 hr + indomethacin[d]	- IFN	52.5	37.9	21.7
	+ IFN	66.2	44.1	32.3

[a]1500 units IFN per culture (10^6, 5 X 10^5, 7.5 X 10^5 PBL.
[b]7.5 units IFN per culture.
[c]4.0 units αIFN per culture, 0.2 units IL-2 per culture.
[d]4.4 X 10^{-6} M final concentration.

was shown to be the strongest NK augmenting agent in comparison with other augmenting agents. Further studies were carried out to assess the relative size of the pre-NK cell compartment as reflected in isolated mononuclear cells precultured with NK augmenting agents. Addition of αIFN during the 4 hr assay period permitted estimation of pre-NK cells by comparison with cytotoxicity without αIFN under identical culture conditions. This estimation is clearly relative to the amount and type of αIFN tested.

As shown in Table V, αIFN did augment endogenous anti-K562 activity in the absence of preculture (as shown in line 1). Following overnight incubation of plated mononuclear cells, endogenous NK activity dropped and this drop was accompanied by a lesser decrease in the observed increment of NK activity induced by αIFN in 4 hr when augmented NK activity was compared to the observed endogenous anti-K562 activity in both cases. In the presence of αIFN (see line 3) significant augmentation of anti-K562 activity relative to endogenous lytic capacity of fresh cells was observed. The results obtained with IL-2 addition clearly demonstrate this NK activator to be the strongest tested in this series. The amount of IFN present in the IL-2 preparation added was small (4.0 units per culture in cultures of 10^6, 5 X 10^5 and 2.5 X 10^5 lymphocytes, respectively). Since αIFN had no significant further effect on NK activity following IL-2 augmentation, it is tempting to speculate that effectively all of the pre-NK cells had become functionally active. In additional experiments, not shown, we have found that at least 86% ^{51}Cr release is technically possible in the system with the same level of endogenous NK activity as that shown here. The experiment with indomethacin shown in line 6 strongly suggests that prostaglandin may have been responsible for the depression of endogenous NK activity during preculture incubation. Further, we tentatively conclude that a principal effect of NK augmenting agents may be to offset negative control mediated by prostaglandins.

The effect of IL-2 in this system indicates that under standard conditions of T cell colony establishment, expression of natural cytotoxicity would be strongly promoted.

ACKNOWLEDGMENTS

The excellent technical assistance of Mrs. K.M. Smith is gratefully acknowledged. These studies were supported in part by USPHS grants CB 84228-34, CA 08748-17 and the Richard Molin Memorial Foundation for Cancer Research.

REFERENCES

1. Morgan, D.A., Ruscetti, F.W., and Gello, R. Science 193: 10007 (1976).
2. Aarden, L.A. J. Immunol. 123:2928 (1979).
3. Zeevi, A., Goldman, I., and Rozenszagn, L.A. Cell. Immunol. 28:235 (1977).
4. Bockman, R.S., and Rothschild, M. J. Clin. Invest. 64: 812 (1979).
5. Cunningham-Rundles, S., Filippa, D.A., Braun, D.W., Antonelli, P., and Shikari, H. J. Natl. Acad. Sci. 67: 585 (1981).
6. Stewart, W.E., II. The Interferon System, Springer Verlag Inc., New York (1979).

IN VIVO INTERLEUKIN-2 INDUCED AUGMENTATION OF NATURAL KILLER CELL ACTIVITY

Steven H. Hefeneider
Christopher S. Henney[1]
Steven Gillis[2]

Program in Basic Immunology
Fred Hutchinson Cancer Research Center
Seattle, Washington

I. INTRODUCTION

Interleukin-2 (IL-2), a class of T cell regulatory molecules, has been shown to trigger the clonal expansion of antigen or mitogen activated T cells (1, 2, 3). By doing so, the lymphokine will also provide that T cell help necessary for the generation of plaque forming cell responses from T cell deficient lymphoid populations (4), and promote the *in vitro* differentiation of antigen specific cytolytic effector T cells (CTL) from both nude mouse splenocytes (5) and immature thymocyte populations (6). Recently, Henney and co-workers (7, 8) demonstrated that exposure of both normal mouse and nude mouse splenocytes *in vitro* with IL-2 resulted in augmentation of natural killer (NK) cell activity. This potentiation of *in vitro* NK cell activity was IL-2 dependent, in that precipitation of potentiating IL-2 containing solutions with a murine monoclonal IgG antibody, directed against IL-2, abrogated the ability of the potentiating supernates to enhance NK cytolytic responses. The augmenting effects of IL-2 on NK activity were further shown to be distinct from

[1] Supported by NCI grant # CA24537, NIAID grant # AI15383, AI 15384 and grant IM-274 from the American Cancer Society.

[2] Special Fellow of the Leukemia Society of America, supported by NCI grant 28419 and grant I-724 from the National Foundation.

ISBN 0-12-341360-5

the known potentiating effects of interferon (IF).

Based on the ability of IL-2 to modulate NK cell activity in vitro and on the well-documented effects this lymphokine has on T cell responses, it appears that IL-2 may be a crucial component in the development of competent immune responses. While convincing data obtained from a variety of in vitro experiments support this contention, little evidence exists to document an in vivo regulatory function for this lymphokine. With the recent discovery of tumor cell lines which produce high concentrations of IL-2 upon mitogen activation, and biochemical protocols sufficient for the purification of this lymphokine to molecular homogeneity, we may now question directly the effect of in vivo administration of purified IL-2 on immune responses.

II. RESULTS

A. Effect of in vivo administration of purified IL-2 on endogenous NK activity

To examine the effect of in vivo administration of IL-2 on NK cell activity highly purified preparations of this lymphokine were prepared. In these studies, IL-2 was purified from tissue culture medium conditioned by 1% PHA stimulated LBRM-33-5A4 murine lymphoma cells. Purification protocols, including sequential ammonium sulfate precipitation, gel exclusion, ion exchange and hydrophobic affinity chromatography and preparative flatbed isoelectric focussing (IEF) have been previously described in detail (9). The IL-2 fractions used in these studies contained no detectable colony stimulating factor, interleukin-1, or interferon. The IL-2 titer of the material was determined in a standard bioassay, testing the capacity of the material to maintain cloned T cell proliferation as assessed by T cell line incorporation of tritiated thymidine (10). Bovine serum albumin (BSA) (0.5 mg/ml in saline) was added to the IL-2 preparation as carrier protein, to ensure stability of the lymphokine during in vitro manipulations required for injection.

A single intraperitoneal (IP) injection of 100 units of purified IL-2 resulted in a significant augmentation of NK cell activity observed in both splenic and peritoneal exudate cell (PEC) populations (Fig. 1). IL-2-BSA was administered to CBA/J mice 48 hrs prior to testing for NK cytotoxic activity against 51 Cr-labelled YAC-1 tumor cells using a standard 4 hr 51 Cr-release assay (11). Control animals received a single IP injection of saline containing 0.5 mg/ml BSA administered 48 hrs prior to assay for splenic/PEC NK cytotoxicity. As detailed in Figure 1, PEC obtained from saline-BSA treated mice mediated 5.7% lysis of YAC-1 target cells at an effector/target cell ratio of 100/1. In contrast, PEC obtained from IL-2-BSA treated animals mediated 52.8% lysis of YAC-1 cells at an identical effector/target cell ratio. As further evidence of the *in vivo* regulatory capacity of IL-2, spleen cells obtained from these same IL-2-BSA treated mice mediated 21.4% lysis of YAC-1 cells at an effector/target cell ratio of 100/1 as compared with saline-BSA treated control animals who mediated only 5.2% lysis. Clearly, administration of 100 units of highly purified IL-2 resulted in potentiation of cytolytic activity observed in both splenic and PEC populations of CBA/J mice.

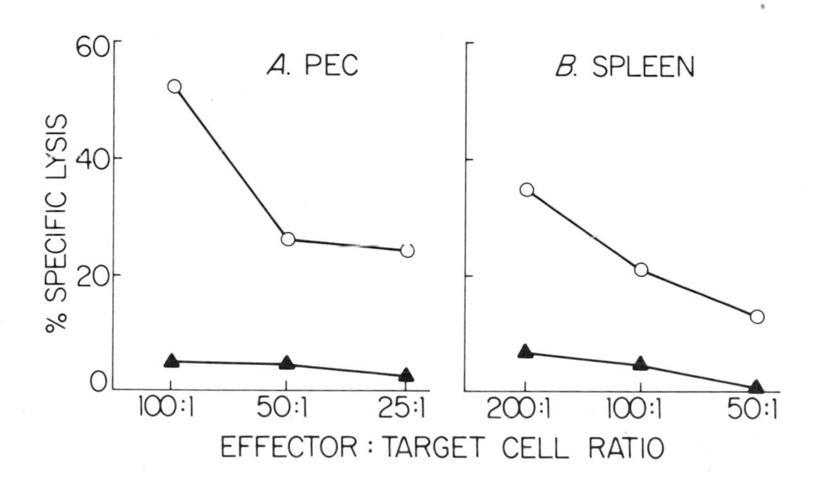

FIGURE 1. Ability of purified IL-2 to augment natural killer cell activity.

CBA/J mice received either saline-BSA (▲) or IL-2 in saline BSA (○), 48 hrs prior to evaluation of either PEC (A) or splenic (B) NK activity in a standard 4 hr 51 Cr-release assay using YAC-1 target cells.

While 51 Cr-labelled YAC-1 tumor cells are frequently used as indicator targets to assess NK cell killing, we felt it important to establish the specificity of the IL-2 induced cytolytic response. To address this issue, we examined the effect of IL-2 administration on splenic effector cell killing of two NK susceptible (YAC-1 and L5178Y clone 27v) and one NK insusceptible (a subclone of L5178Y termed clone 27av) tumor targets (12). Using the same treatment protocol as described above (100 units IL-2 BSA, administered IP 48 hrs prior to testing for NK activity), we observed that while administration of 100 units of purified IL-2-BSA resulted in substanial agumentation of resident CBA/J splenic effector cell activity against both NK susceptible targets (Table I), lymphokine treatment did not induce a lytic response against the NK cell insusceptible clone 27av cells. It is important to note that while clone 27av cells are not lysed by NK cells, these cells are readily killed by appropriately primed cytotoxic T lymphocytes (12). These data suggested that the augmentation of splenic cytolytic activity, observed after treatment with IL-2, was mediated by NK cells.

Table I. IL-2 dependent in vivo augmentation of NK cell activity.

Treatment[a]	Target[b]	% specific lysis Effector/target cell ratio 200/1	100/1	50/1
Saline-BSA	YAC-1	32.7	26.3	21.5
IL-2-BSA	YAC-1	54.0	46.0	32.9
Saline-BSA	Clone 27v	10.4	6.4	3.5
IL-2-BSA	Clone 27v	21.1	12.0	8.4
Saline-BSA	Clone av	0	0	0
IL-2-BSA	Clone av	0	0	0

[a]CBA/J mice (2 mice/group) received either saline-BSA or 100 units IL-2 in saline-BSA 48 hrs prior to evaluation of splenic cytolytic activity in a standard 4 hr 51 Cr-release assay.

[b]YAC-1 and clone 27 v of L5178Y are NK sensitive targets while clone av of L5178Y is NK insusceptible.

Having established the specificity of killing of the IL-2 induced augmentation of NK cell activity, subsequent studies focused on elucidation of optimal parameters for induction of this response. From these studies it was observed that a similar magnitude of IL-2 dependent potentiation of NK cell activity was seen after either intravenous (IV) or IP administration of the lymphokine (data not shown). In addition, we questioned what concentration of in vivo administered purified IL-2 resulted in optimal augmentation of splenic NK cell activity. As detailed in Table II, significant potentiation of splenic NK cell activity was noted after IP administration of either 50,100 or 150 units of IL-2-BSA. In contrast, treatment with 25 units of IL-2 or mock IL-2 (a fraction obtained from the IEF profile which lacked IL-2 activity, as assessed by its inability to support IL-2 dependent T cell line proliferation) did not result in an augmented NK cell response. These studies suggested that 100 units of IL-2-BSA, administered IP, 48 hrs prior to testing for splenic NK cell activity was optimal for potentiation of resident

NK cell activity in CBA/J mice.

Table II. Effect of IL-2 dose-response on augmentation of splenic NK cell activity.

	% specific lysis Effector/target cell ratio		
Treatment[a]	200/1	100/1	50/1
Experiment # 1			
Saline-BSA	18.4	14.8	10.3
Mock IL-2[b]	22.3	12.8	11.3
IL-2-BSA			
25 units	18.3	10.9	8.8
50 units	37.6	23.4	16.3
Experiment #2			
Saline-BSA	14.0	11.6	6.4
IL-2-BSA			
100 units	36.3	22.4	14.7
150 units	30.8	16.1	11.3

[a]CBA/J mice (2 mice/group) received a single ip injection of either saline-BSA or IL-2 in saline-BSA 48 hrs prior to testing of spleen cell cytotoxic activity in a standard 4 hr 51 Cr-release assay (YAC-1 target cells).

[b]An IEF fraction, obtained as a by-product during preparation of purified IL-2, which lacks IL-2 activity, as assessed by its inability to sustain growth of IL-2 dependent T-cells.

III. DISCUSSION

The ability of IL-2 to function in a variety of in vitro immune responses has been well established (1-6). The lymphokine is a T cell product and functions in vitro by binding to mitogen or antigen activated T cells, causing their proliferative expansion and/or differentiation (13). Recent observations by Henney and colleagues (7, 8) show that in addition to effecting T cell responses, in vitro exposure to IL-2 resulted in potentiation of NK cell activity. If, as some data suggest that NK cells are part of the T cell lineage, then the observation that IL-2 modulates NK cell activity would be consistent with the previously documented cellular specificity of this lymphokine. In support of this contention is the observation that IL-2 expanded NK cell clones express Thy 1 antigen, as well as the neutral glycolipid asialo GM1 on their surface (8). In sum, the evidence from in vitro studies supports the hypothesis that IL-2 is intimately involved in the development of competent immune responses and provides the basis to question an in vivo regulatory function for this molecule.

Armed with highly purified preparations of IL-2, our studies demonstrated that in vivo IL-2 treatment of CBA/J mice resulted in significant augmentation of NK cell activity. The enhanced NK cell cytolytic response appeared 48 hrs after treatment with a single IP injection of 100 units of IL-2. This potentiated NK cell response was noted in both splenic and PEC populations. The mechanism by which IL-2 potentiates in vivo NK cell activity is unknown. Henney et al. (7) were able to demonstrate that IL-2 induced augmentation of in vitro NK cell activity was independent of the known NK cell potentiating effects of interferon. Currently we cannot conclusively state whether IL-2 treatment in vivo augments NK cell activity directly or functions by activating another lymphokine (interferon?).

The results of these studies suggest that IL-2 can potentiate in vivo resident NK cell activity. If the in vivo regulatory potential of IL-2 can be established to function in parallel with its known in vitro regulatory capacity, the potential of IL-2 as an agent for the treatment of various immune deficiencies and T-cell disorders will be worthy of pursuit.

REFERENCES

1. Baker, P.E., S. Gillis, M.M. Ferm and K.A. Smith, J. Immunol. 121:2168 (1978).
2. Gillis, S. and K.A. Smith, Nature. 268:154 (1977).
3. Watson, J., J. Exp. Med. 150:1510 (1979)
4. Watson, J., S. Gillis, J. Marbrook, D. Mochizuki and K.A. Smith, J. Exp. Med. 150:849. (1979).
5. Gillis, S., N.A. Union, P.E. Baker, and K.A. Smith, J. Exp. Med. 149:1460 (1979).
6. Wagner, H., M. Rollinghoff, K. Pfizenmaier, C. Hardt and G. Johnscher, J. Immunol. 124:1058 (1980).
7. Henney, C.S., K. Kuribayashi, D.E. Kern, and S. Gillis, Nature. 291:335 (1981).
8. Kuribayashi, K., S. Gillis, D.E. Kern and C.S. Henney, J. Immunol. 126:2321 (1981).
9. Mochizuki, D., J. Watson and S. Gillis, J. Immunol. 125:2579 (1980).
10. Gillis, S., M.M. Ferm, W. Ou and K.A. Smith, J. Immunol. 120:2027 (1978).
11. Okada, M. and C.S. Henney, J. Immunol. 125:300 (1980).
12. Durdik, J., B.N. Beck, E.A. Clark and C.S. Henney, J. Immunol. 125:683 (1980).
13. Shaw, J., B. Caplan, V. Paetkau, P.L.M. Pilarski, T.L. Delovitch and I.F.C. McKenzie, J. Immunol. 124:2231 (1980).

AUGMENTATION OF NATURAL KILLER ACTIVITY BY RETINOIC ACID

Ronald H. Goldfarb[1,2]
Ronald B. Herberman

Laboratory of Immunodiagnosis, NCI,
NIH, Bethesda, MD

I. INTRODUCTION

Natural killer(NK) cells comprise a non-adherent, non-phagocytic lymphoid cell sub-population that is involved in anti-tumor cytotoxic reactivity(1). Recently, in our laboratory, we have isolated human NK cells to a high degree of purity(2), and have demonstrated that NK cells may be morphologically identified as large granular lymphocytes(LGL).

NK cells are present in unimmunized individuals, and unlike other lymphoid effector cells, do not require long time periods to be primed for cytolytic capacity; NK cells have therefore been considered to be an important component in immune defense against malignant disease. If NK cells are indeed involved in anti-tumor immune surveillance, then augmentation of NK reactivity may function to prevent the initiation or promotion of malignant disease; a full understanding of regulatory controls over the expression of NK cytotoxic reactivity shall be critical for the elucidation of various biochemical pathways by which NK cells can be augmented.

To date, the major positive regulators of NK cytotoxic reactivity, appear to be interferon(IFN) and inducers of IFN (1,2). Recently, we have examined a number of agents with well-established regulatory potential for their effects on the cytolytic function of human NK cells(LGL)(4). In this chapter, we discuss the role of retinoic acid as an augmenting agent for NK cytolytic reactivity.

---

[1]Supported by PHS Fellowship 1F32CA06681 to RHG by NCI.

[2]Present address: Cancer Metastasis Research Group, Dept. of Immunology & Infectious Disease, Pfizer Central Research, Pfizer Inc., Groton, CT 06340.

ISBN 0-12-341360-5

II. RETINOIC ACID AND ANTI-TUMOR IMMUNITY

Retinoic acid(RA) has been implicated in both the stimulation of cell-mediated immune responses, and in the enhancement of anti-tumor immune reactivity(5). For example, RA has been reported to stimulate host anti-tumor immunity in fibrosarcoma bearing mice(6). Furthermore, it has been reported that RA increases the induction and cytotoxic activity of murine killer T cells(7). RA has also been shown to increase specific tumor immunity against a tumor virus induced sarcoma (8) and to augment specific cell-mediated immunity against melonoma-bearing animals(9). Interestingly, it appears that retinoic acid-mediated enhanced cytolytic activity(7,8) was abrogated upon treatment of spleen cells with anti-theta serum plus complement, suggesting that RA acted on cytotoxic T cells or cells in the T cell lineage. It has been suggested that RA has the capacity to enhance the induction of human peripheral blood lymphocytes against Raji lymphoma cells *in vitro*(5). In addition, RA, an agent considered to be useful in the chemoprevention of cancer(10), is often antagonistic to the action of tumor promoters(5), and can inhibit tumor promoter-mediated effects on human lymphocytes(11,12).

III. AUGMENTATION OF NK ACTIVITY BY RETINOIC ACID

We have recently observed(4) that RA modulates NK activity mediated by human LGL. It appears that RA augments human NK activity against K562 cells(Table I). We have shown that the optimal concentration for RA-mediated NK cell augmentation was between 10^{-5} and 10^{-6}M(4). In addition, we have observed that retinoids other than RA also enhance human NK activity (R.H. Goldfarb and D.L. Slate, unpublished observations). Unlike the boosting capacity of IFN, it appears that RA boosts NK activity in only 60% of all individuals tested (4). For many donors, RA can stimulate NK activity to a level equivalent to, or greater than, that induced by IFN(4). It is of particular interest that treatment of NK cells(LGL) with optimal doses of IFN plus RA resulted in less augmentation than that stimulated by either agent alone(4); this finding suggests that RA might stimulate NK cells through a mechanism that shares similarities with the IFN-induced mechanism of boosting. However, RA has not been shown to be an IFN inducer in human fibroblasts(13) or lymphocytes(14).

It is of interest that both RA and IFN induce alterations in cell surface components(15) and increase the expression of cell surface receptors(16). Therefore, each of these membrane reactive agents may lead to similar cell surface alterations on NK cells, and thereby enhance NK cytotoxic reactivity; the question of whether RA and other retinoids function by an IFN dependent, or independent mechanism, is under study through the use of well-characterized and specific anti-IFN antisera (R.H. Goldfarb and D.L. Slate, unpublished observations). The resolution of this issue is of great importance since it has been shown that RA can partially reverse IFN protection against tumor growth in mice(17); it has therefore been suggested that a need for caution is warranted if RA is used in tumor-bearing individuals being treated with IFN(17).

TABLE I. Effect of Retinoic Acid on NK Cell(LGL) Lysis of K562 Cells

Incubation Conditions	Lytic Units/10^8cells* (-IFN)	(+IFN 1000 /ml)
Input(non-purified cells)	48	N.D.
Purified human NK cells(LGL)	104	156
LGL + retinoic acid(10^{-4}M)	158	138
LGL + retinoic acid(10^{-5}M)	204	140
LGL + retinoic acid(10^{-6}M)	172	138
LGL + retinoic acid(10^{-7}M)	132	118
LGL + retinoic acid(10^{-8}M)	108	98

*1 lytic unit is the number of effector cells required to cause 30% lysis of K562 cells.

A role for retinoids in augmentation of NK reactivity has recently been substantiated by in vivo experimentation; we have observed that NK activity is transiently boosted in response to retinolacetate(R.B. Herberman and M. Nunn-Hargrove, unpublished observatios). Further in vivo experimentation with other retinoids, in vivo, has been initiated.

IV. CONCLUSION

The capacity of tumor preventative agents, such as RA and other retinoids(10) to augment NK cell-mediated cytolysis (4), suggests that NK cells may play a significant role in protection against the promotion of malignant disease. Our results therefore suggest that optimal chemoprevention of cancer by retinoids, as well as other agents, may be dependent on development of better protocols for their augmentation of natural cell-mediated immunity against tumors.

1. Herberman, R.B., "Natural Cell-Mediated Immunity Against Tumors". Academic Press, New York, 1980.
2. Timonen, T., Ortaldo, J.R., and Herberman, R.B., J. Exp. Med. 153:569 (1981).
3. Goldfarb, R.H., and Herberman, R.B., in "Adv. in Inflammation Res." (G. Weissmann, Ed.) 4:45 (1982).

4. Goldfarb, R.H., and Herberman, R.B., J. Immunol., 126: 2129 (1981).
5. Lotan, R., Biochem. Biophys. Acta. 605:33 (1980).
6. Tannock, I.F., Suit, H.D., and Marshall, N., J. Natl. Cancer Inst., 48:731 (1972).
7. Dennert, G., Crowley, C., Kouba, J., and Lotan, R., J. Natl. Cancer Inst. 62:89 (1979).
8. Glaser, M. and Lotan, R., Cell. Immunol. 45:175 (1979).
9. Felix, E.L., Lloyd, G., and Cohen, M.H., Science, 189: 886 (1972).
10. Sporn, M.B., Newton, D.L., Roberts, A.B., DeLarco, J.E., and Todaro, G.J. in "Molecular Actions and Targets for Cancer Chemotherapeutic Agents." Academic Press, New York. In Press.
11. Skinnider, L., and Giesbrecht, K., Cancer Res., 39:3332 (1979).
12. Abb, J. and Deinhardt, F., Int. J. Cancer, 25:267 (1980).
13. Blalock, J.E., Tex. Rep. Biol. Med., 35:69 (1977).
14. Abb, J., and Deinhardt, F., Exp. Cell Biol. 48:169 (1980).
15. Bollag, W., Cancer Chemother. Pharmacol., 3:207 (1979).
16. Jetten, A.M., Nature, 284:626 (1980).
17. Baron, S., Kleyn, K.M., Russell, J.K., and Blalock, J.E., J. Natl. Cancer Inst., 67:95 (1981).

AUGMENTATION OF HUMAN NATURAL KILLING ACTIVITY BY OK432

Atsushi Uchida
Michael Micksche

Institute for Cancer Research
University of Vienna

Natural killer (NK) cells are thought to play an important role in host defence mechanisms against tumors (Herberman and Holden, 1978). It has been convincingly demonstrated that interferon (IFN) plays a central role in the augmentation or boosting of NK cells (Gidlund et al., 1978). Recently, another form of augmentation of NK activity has been observed by using alloantibodies (Brunda et al., 1981), interleukin 2 (Henney et al., 1981), retinoic acid (Goldfarb and Herberman, 1981), and prostaglandins (Kendal and Targan, 1980). On the other hand, OK432, a heat- and penicillin-treated lyophilized powder of a low virulent Su strain of streptococcus pyogenes A3, has been extensively used in cancer patients as an immunotherapeutic agent (Uchida and Hoshino, 1980a,b). The agent has been shown to activate macrophages in rats (Ishii et al., 1976) and to induce immune IFN in mice (Matsubara et al., 1979). We report here that OK432 augments NK activity in vitro independently of IFN induction, even in the presence of NK suppressor cells.

Blood lymphocytes were incubated alone or with 50 ug OK432 per ml for 20 hours at 37° C, then washed and tested for NK activity against K562 cells in a 4-hour chromium release assay. As previously reported (Uchida and Micksche, 1981a,b), lymphocytes incubated in medium showed no significant change in NK activity during 24 hours. OK432-treated lymphocytes showed a marked increase in NK activity, whereas the mere addition of OK432 to the assay caused no enhancement of NK activity (Table I). Lymphocytes from both normal controls and cancer patients were activated by OK432. OK432 was also able to augment cytotoxicity against both NK-susceptible and NK-resistant target cells. This augmentation was dose-dependent and could be detected at concentrations as low as 0.1 ug OK432/ml. reaching a maximum at 50 µg OK432/ml. Significant enhancement of NK

ISBN 0-12-341360-5

TABLE I. In Vitro Augmentation of NK Activity by OK432

OK432 treatment (50 ug/ml)	Cytotoxicity (LU/10^7)	
	Normal	Cancer
None	91 (46-174)	39 (2-127)
0 hour	82 (29-189)	33 (3-141)
20 hours	286 (85-978)	182 (44-673)

Cytotoxicity assays were done with several E:T ratios and the activity was expressed in lytic units (LU) per 10^7 cells, 1 LU being defined as the number required to produce 33% specific cytotoxicity. Results are expressed as mean (range) of 15 individuals.

activity was detectable after 4-hour preincubation with OK432 and peaked at 16-24 hours. OK432 has been reported to have a direct cytostatic or cytotoxic effect on certain tumor cells (Okamoto et al., 1967). However, it seems unlikely that OK432-augmented NK activity was due to a direct cytotoxic effect of OK432 remaining on the surface of effector cells, since K562 cells were not killed by OK432 (50 ug/ml) in a 4-hour assay.

The mechanism by which OK432 augments NK activity was examined. Pretreatment of lymphocytes with OK432 at 4° C failed to enhance NK activity, indicating that active cell metabolism of lymphocytes is necessary for the enhancement of NK activity by OK432. Then, the metabolic processes required for development of increased cytotoxicity following OK432 treatment were evaluated by using various metabolic inhibitors (Table II). To know whether the augmentation of NK activity by OK432 requires DNA synthesis, lymphocytes were treated with mitomycin C, an irreversible inhibitor of DNA synthesis, prior to OK432 treatment. While mitomycin-treated lymphocytes showed little DNA synthesis, they were able to develop increased NK activity in response to OK432. No difference was observed in NK activity of control lymphocytes treated with and without mitomycin C. Similarly, the enhancement of NK activity by IFN has been independent of cell proliferation (Einhorn et al., 1978). To evaluate the requirement for RNA synthesis in the augmentation of NK activity, lymphocytes were treated with actinomycin D, an irreversible inhibitor of RNA synthesis, prior to treatment with OK432. Actinomycin-treated lymphocytes were not able to mediate cytotoxicity even after OK432 treatment. The requirement for protein synthesis was examined by using puromycin, a reversible inhibitor of protein synthesis, before and during OK432 treatment. Puromycin-treated cells showed a drastic impairment of NK activity and were not activated by OK432. These

TABLE II. Effects of Metabolic Inhibitors on Enhancement of NK Activity by OK432

Treatment	Cytotoxicity (LU/10^7)	
	Control	OK432
None	110	364
Mitomycin C	95	303
Actinomycin D	1	1
Puromycin	1	1

Lymphocytes were pretreated for 1 hour with mitomycin C (50 µg/ml)m actinomycin D (10 µg/ml), or puromycin (10 µg/ml), washed (mitomycin, actinomycin), then incubated with OK432 (50 µg/ml) for 20 hours.
Results are expressed as mean of 6 experiments.

results indicate that RNA and protein synthesis are necessary for the manifestation of augmented NK activity by OK432. It might be possible that OK432 activates NK cells without induction of RNA and protein synthesis but the increased cytotoxic function is completely suppressed by RNA and protein synthesis inhibitors. Our findings are keeping with other observations that NK activity is highly susceptible to inhibitors of RNA or protein synthesis and IFN-induced augmentation of NK activity requires RNA and protein synthesis (Ortaldo et al., 1980).

Soluble factors produced by lymphocyte-tumor cell cultures (Koide and Takasugi, 1978) and antigen-stimulated lymphocyte cultures (Lazda and Baram, 1979) have augmented NK activity. IFN present in these supernatants are found to play a central role in boosting of NK cells (Trinchieri et al., 1978). The possible role of IFN in the augmentation of NK activity by OK-432 was considered. Lymphocytes pretreated for 20 hours with supernatants from 20-hour OK432-stimulated lymphocyte cultures showed no increase in NK activity (Table III). In addition, the supernatant did not contain any detectable amounts of IFN. Next, lymphocytes were pretreated for 20 hours with OK432 in the presence or absence of anti-α-IFN antibodies. Anti-IFN antibodies did not abrogate the augmentation of NK activity by OK432. In contrast, lymphocytes treated with IFN plus anti-IFN antibodies showed no increase in NK activity but remained at levels comparable to those of cells incubated in medium, while IFN-treated cells had a strong increase in NK activity. These results suggest that the augmentation of NK activity by OK432 is independent of IFN induction, though undetectable amounts of immune IFN might be involved in this enhancement. It seems likely that binding of OK432 on the surface of NK cells acti-

TABLE III. Effects of Supernatants and Anti-Interferon Antibodies on OK432-Augmented NK Activity

Pretreatment	Cytotoxicity ($LU/10^7$)*
Medium	
Supernatant**	
OK432	
OK432 + anti-IFN***	
IFN****	
IFN + anti-IFN	

* Mean of 6 experiments.
** Supernatants from 20-h OK432-stimulated lymphocyte culture.
*** 1 x 10^4 neutralizing units/ml.
**** Purified human leukocyte interferon (1,000 U/ml).

vates their lytic function. An alternative possibility is that OK432 stimulates the production of another lymphokine that is in turn responsible for the enhancement of NK activity.

It ia important to know whether the effect of OK432 is on already mature NK cells or pre-NK cells. Human NK activity has been reported to be exerted by a morphological subpopulation of lymphoid cells, large granular lymphocytes (LGL) (Timonen and Saksela, 1980). The LGL-enriched fraction isolated by discontinuous Percoll gradient centrifugation showed a substantial increase in NK activity above that of unseparated lymphocytes, while the LGL-depleted fraction had no NK activity (Table IV). When LGL-rich cells were treated with OK432, they had much higher levels of NK activity than untreated LGL-rich cells or OK432-activated whole lymphocytes. In contrast,

TABLE IV. Effects of OK432 on NK Activity and LGL/K562 Conjugates*

	OK432 treatment	Conjugate (%)	Cytotoxicity ($LU/10^7$)
Unseparated	-	16	94
	+	17	263
LGL-enriched	-	43	372
	+	45	1042
LGL-depleted	-	8	1
	+	1	1

* Percentage of LGL binding K562 after 2-hour incubation at 1:1 ratio. Results represent mean of 5 experiments.

TABLE V. Augmentation of NK Activity by OK432 in Presence of Suppressor Cells*

	Cytotoxicity ($LU/10^7$)			
	Control**	Medium	OK432	IFN
PBMC	79	87	261	247
PBMC + PEAC	75	31	243	35
PEMC	1	1	52	1

Blood and effusion mononuclear cells (PBMC, PEMC), and PBMC plus 10% effusion adherent cells (PEAC) were pre-incubated alone, or with OK432 or IFN for 20 h at 37°C.
* Mean of 6 experiments.
** Incubation at 4° C.

LGL-depleted cells did not develop cytotoxicity even after OK-432 treatment. The results indicate that LGL are involved in both spontaneous and OK432-augmented NK activity. Similarly, LGL have been shown to be the population that can be boosted by IFN (Timonen and Saksela, 1980). Furthermore, the frequency of LGL forming conjugates with target cells was not altered by OK432. IFN also has enhanced NK activity without altering the number of lymphocytes binding K562 cells (Silva et al., 1980). It is evident from the present study that cells responding to OK432 have a pre-existing ability to recognize their target cells. However, it could not be concluded whether OK432 activates mature NK cells or pre-NK cells. Our results can be explained either if OK432 augments the lytic function of mature NK cells or if OK432 activates pre-NK cells that are within the population of LGL and have the ability to recognize their target cells.

We have observed that carcinomatous pleural effusions of cancer patients contain adherent cells that suppress the maintenance and development of NK activity (Uchida and Micksche, 1981b, other chapters). To know whether OK432 can augment NK activity in the presence of suppressor cells, blood lymphocytes were invubated with OK432 in the presence or absence of adherent effusion cells at 37° C for 20 hours. OK432 enhanced NK activity regardless of the presence of adherent suppressor cells (Table V). In contrast, adherent suppressor cells inhibited spontaneous and IFN-augmented NK activity. Similarly, effusion lymphocytes were activated by OK432, but not by IFN. These results suggest that OK432 not only augments NK activity but also reduces NK suppressor activity. Indeed, treatment of adherent effusion cells with OK432 resulted in a reduction of suppressor activity. It may be possible that OK432 activates a lytic mechanism resistant to suppressor cells.

In conclusion, OK432 can rapidly augment NK activity. Although the mechanism of this enhancement is not clarified, our results strongly suggest that OK432 activates NK cells independently of interferon induction. The possibility that OK432 activates the motility and/or secretion of NK cells is under investigation.

REFERENCES

Brunda, M. J., Herberman, R. B., and Holden, H. T. (1981). Int. J. Cancer 27:205.

Einhorn, S., Blomgren, H., and Strander, H. (1978). Int. J. Cancer 24:244.

Gidlund, M., Örn, A., Wigzell, H., Senik, A., and Gresser, I. (1978). Nature 273:759

Goldfarb, R. H., and Herberman, R. B. (1981). J. Immunol. 126: 2129.

Henney, C. S., Kuribayashi, K., Ker, D. E., and Gillis, S. (1981). Nature 291:335.

Herberman, R. B., and Holden, H. T. (1978). Adv. Cancer Res. 27:305.

Ishii, Y., Yamaoka, H., Toh, K., and Kiuchi, K. (1976). Gann 67:115.

Kendal, R. A., and Targan, S. (1980). J. Immunol. 125:2770.

Koide, Y., and Takasugi, M. (1978). J. Immunol. 121:872.

Lazda, V. A., and Baram, P. (1979). J. Immunol. 122:2086.

Matsubara, S., Suzuki, F., and Ishida, H. (1979). Cancer Immunol. Immunother. 6:41.

Okamoto, H., Shoin, S., Koshimura, S., and Shimizu, R. (1967). Jpn. J. Microbiol. 11:323.

Ortaldo, J. R., Philips, W., Wasserman, K., and Herberman, R. B. (1980). J. Immunol. 125:1839.

Silva, A., Bonavida, B., and Targan, S. (1980). J. Immunol. 125:479.

Timonen, T., and Saksela, E. (1980). J. Immunol. Methods 36: 285.

Trinchieri, G., Santoli, D., Dee. R. R., and Knowles, B. B. (1978). J. Exp. Med. 147:1299.

Uchida, A., and Hoshino, T. (1980a). Cancer 45:476.

Uchida, A., and Hoshino, T. (1980b). Int. J. Cancer 26:401.

Uchida, A., and Micksche, M. (1981a). Int. J. Immunopharmac. 3:365.

Uchida, A., and Micksche, M. (1981b). Cancer Immunol. Immunother. 11:255.

DIFFERENCES BETWEEN PHORBOL ESTERS AND INTERFERON INDUCED ACTIVATIONS OF MOUSE NK CELLS

Anna SENIK[1]
Jean Pierre KOLB[1]

Laboratoire d'Immunologie Cellulaire[2]
Institut de Recherches Scientifiques sur le Cancer
Villejuif, France

I. INTRODUCTION

Compared with B and T lymphocytes which display exquisite target specificity and require antigenic stimuli to undergo proliferation and differentiation, NK cells display broad target recognition and are able to spontaneously destroy in vitro various target cells. This precommitment to lysis suggests that NK cells are maintained by some homeostatic mechanism (still unknown) at a basal level of activity which may act as a first line of defence of the host against viral infection or neoplasic growth. Interferon can significantly enhance NK activity above this basal level (1), and recent studies in our laboratory have attempted to investigate the mechanisms responsible for in vitro activation of mouse NK cells by interferon (2) or by the phorbol ester TPA, a second reagent recently shown by us to stimulate NK cells as strongly as interferon (3). Results of these studies demonstrate that NK cell activation can proceed according to two different pathways : one (interferon-induced) requires new RNA and protein synthesis, while the other (TPA-induced) does not require such synthesis. This report further extends the investigation of the TPA model, and when appropriate, compares it to the NK activation produced by interferon.

[1] I.R.S.C. B.P. n° 8 - 94802 - VILLEJUIF Cedex, France
[2] supported by INSERM grant 80-1008

ISBN 0-12-341360-5

II - RESULTS

A - Relationship between the Tumor Promoting Capacity of Phorbol Esters and their in vitro stimulating Effect on Spleen Cell Cytotoxicity against Yac-1 Target Cells

Previous experiments reported by Keller have demonstrated that 12-0-tetradecanoyl phorbol-13-acetate (TPA) has an inhibitory effect on rat NK activity (4). This finding was later confirmed in both mouse and man systems (5,6). However results in our laboratory, obtained in the meantime, reproducibly demonstrated an enhancing effect of this reagent upon mouse NK cell activity.

A possible role of suppressive macrophages in this system will be proposed in the discussion, as a tentative explanation to overcome this discrepancy. As illustrated in figure 1 a 3 hour pretreatment of nude mouse spleen cells with TPA, a potent tumor promoter, or with PDB (phorbol-12-13-dibenzoate), a weak tumor promoter, resulted in a pronounced increase of spleen cell cytotoxicity. The enhancing effects were dose dependent, the peak response being obtained with 10-20 ng/ml of TPA, and 100-200 ng/ml of PDB. In contrast 4-α-PDD (4-α-phorbol-12,13-didecanoate) another phorbol diester that is inactive as a tumor promoter, had no effect. This indicates that there is a positive correlation between the enhancing properties of the phorbol esters and their tumor promoting capacity but further experiments with a larger series of phorbol esters will be required to ascertain this assumption.

B - TPA-Stimulated Cytotoxic cells are Similar to Spontaneous and Interferon-Stimulated NK cells

Two series of experiments were undertaken. In the first one IFNα and TPA-activated spleen cells were tested for cytotoxicity on various tumor cells, and were shown to display the same target cell specificity. Thus each NK-sensitive target cell tested (mouse YAC-1 lymphoma cells, human K-562 and MOLT-4 tumor cells) was lysed to comparable extent by both kinds of effector cells, while NK-insensitive target cells (mouse L 1210 and YAC-222 tumor cells) resisted this lysis.

In the second set of experiments, the splenocytes used as effector cells were submitted to various cell fractionation procedures. After filtration of C57Bl/6 nude mouse

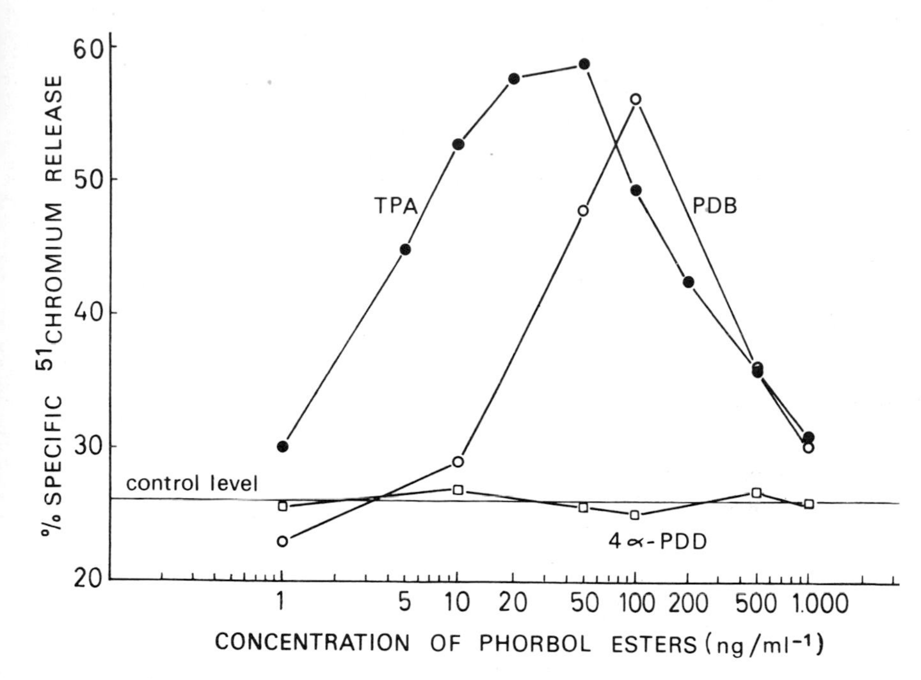

FIGURE 1. Effect of phorbol esters on NK activity - (Spleen cells from Balb/c nude mice preincubated for 3 hours with the reagents, washed and tested on YAC-1 targets at a 100:1 effector: target cell ratio).

spleen cells through a nylon wool column, the NK activity, either spontaneous or TPA or IFNα-induced, was recovered in the non adherent cell population ; in the same way the whole NK activity was abolished upon treatment with a monoclonal anti-Thy-1 antibody (clone F7D5, Olac laboratories) in the presence of complement.

C- Comparison between the Pathways of NK cell Activation Induced by TPA and IFNα

The above described experiments suggest that accessory cells are not involved in the stimulation of NK cells by TPA and IFNα since neither removal of T or B cells, nor even macrophages did prevent any of these two activation processes. In both cases a short time of contact (a few minutes) between spleen cells and the reagents was sufficient to trigger the cytolytic mechanism (data not shown). Furthermore, the number of target binding cells was unchanged after TPA-treatment (data not shown) as after interferon treatment (2), suggesting that the increased lytic activity is not due to newly recruited NK cells, but rather to an increase in the individual lytic capacity of precommited NK cells.

However, there is a major difference between the mechanisms of TPA and IFNα-induced NK activation. Treatment of the attacking cells with cycloheximide (50 μg/ml) or actinomycin D (25 μg/ml), inhibitors respectively of protein and RNA synthesis, prevented the IFNα-induced increase of NK cytotoxicity without affecting the TPA enhancing effect (or the cytotoxicity level of already committed NK cells). These results indicate that IFNα and TPA trigger two different activation pathways of NK cells, the former being dependent of a maturation process involving RNA and protein synthesis, the latter probably resulting from modifications of the cell membrane, as suggested by the short latent period of the TPA-induced cytotoxicity (not shown).

III - CONCLUDING REMARKS

Our interest for the effects of phorbol esters on NK cell activity was determined by three series of arguments : 1) TPA has been shown to enhance the in vitro production of interferon (7), which is a potent activator of NK cytotoxicity, 2) this reagent is able to act on the terminal differentiation of a variety of specialized cells including macrophages (8) and B cells (9), and 3) a part of its tumor promoting ability in vivo could be to modify the "immune surveillance" against nascent tumor cells which is believed to be exerted mainly by cytotoxic cells of the NK and macrophage type (4). Our results show that in vitro preincubation of mouse spleen cells with active phorbol esters leads to a marked increase in the lytic potential of NK cells. They also indicate that the enhancing effect of each phorbol ester parallels its in vitro tumor promoting ability. Under no circumstances interferon could be detected in the supernatants of TPA-treated spleen cells, indicating that the stimulation of NK activity was truly due to TPA. These results are at variance with those reported by others, showing on the contrary an inhibitory effect of TPA upon NK activity (4-5). We believe that macrophages may play an important regulatory role on the NK systems, as is the case of pyran copolymer and adriamycin-induced modulations of NK activity (10). In fact, according to Seaman et al., the TPA-induced inhibition of human NK cell activity was significantly reduced and even sometimes replaced by a stimulating effect when adherent cells were removed from the peripheral blood cells (6).

TABLE 1. Removal of macrophages improves TPA-induced activation.

Preincubation with	Percent specific ^{51}Cr release[a]	
	whole spleen cells	Sephadex G-10 non adherent cells
Medium	54.5	55.3
TPA (10ng/ml)	75.5	92.2

[a] Spleen cells from C57Bl/6 nude spleen cells were incubated for 2 hours with TPA, washed twice and tested for cytotoxicity against YAC-1 target cells. S.E. did not exceed ± 3 %

In our hands, also, removal of adherent cells by filtration through Sephadex G-10 resulted in a better increase of the lytic capacity of the TPA-treated effector cells (Table I). This increase was not simply due to an enrichment of the NK cells but rather to the removal of inhibitory macrophages. It is of value to note that we sometimes observed, but only for high concentration of TPA (100-1000 ng/ml) an inhibitory effect on NK activity. On the other hand, pretreatment of YAC-1 target cells with high doses of TPA led to a decreased susceptibility to lysis of these cells. Taken together, these data suggest that, at least, a part of the inhibitory effects exerted by TPA might be explained by membrane alterations of the target cells (due to residual TPA molecules present during the cytotoxicity assay). The mechanism involved in NK cell activation by TPA and interferon are quite different since TPA effect does not require RNA and protein synthesis. These results, together with those of Brunda et al. (11), using alloantibodies or lectins as a stimulatory agents, indicate at least two different activation pathways of NK cells. Recently, phospholipase A_2 has been reported to be involved in the lytic process of NK cells (12). Since TPA is able to induce in several cell types both phospholipase activity (13) and various products of the arachidonic acid metabolism (14 (PGE_2 and thromboxane TX B2) known as NK inhibitors, slight differences in the induction of one or the other of these sets of regulatory molecules could thus lead to either a stimulation or an inhibition of the natural killer cells.

REFERENCES

1. Gidlund, M., Orn, A., Wigzell, H., Senik, A., and Gresser, I., Nature (London) 273:759 (1978)

2. Senik, A., Kolb, J.P., Orn, A., and Gidlund, M., Scand. J. Immunol., 12:51 (1980)
3. Kolb, J.P., Senik, A., and Castagna, M., Cell Immunol., in press.
4. Keller, R., Nature (London) 282:279 (1979)
5. Goldfarb, R.H., and Herberman, R.B., J. Immunol. 126: 2129 (1981)
6. Seaman, W.E., Gindhardt, T.D., Blackman, M.A., Dalal, B., Talal, N., and Werb, Z., J. Clin. Invest. 67:1324 (1981)
7. Vilcek, J., Sulea, I.T., Volvovitz, F., and Yip, Y.K., in "Biochemical Characterization of lymphokines" (A.L. Weck, F. Kristensen, and M. Landu, Eds) p. 323, Academic Press, New York, 1980.
8. Mizel, S.B., Rosenstreich, D.L., and Oppenheim, J.J., Cell Immunol. 40:230 (1978)
9. Castagna, M., Yagello, M., Rabourdin-Combe, C., and Fridman, W.H., Cancer Lett. 8:365 (1980)
10. Santoni, A., Ricardi, C., Barlozzari, T., and Herberman, R.B., in "Natural-cell mediated immunity against tumors", (R.B. Herberman, Ed). p. 753, Academic Press, New York (1980)
11. Brunda, M.J., Herberman, R.B., and Holden, H.T. in "Natural cell mediated immunity against tumors" (R.B. Herberman, Ed.) p. 525, Academic Press, New York, 1980.
12. Hoffman, T., Hirata, F., Bougnoux, P., Fraser, B.A., Goldfarb, R.H., Herberman, R.B., and Axelrod, J. Proc. Natl. Acad. Sci. 78, 3839-3843 (1981)
13. Hamilton, J.A., Cancer Res. 40, 2273 (1980)
14. Chang, J., Wigley, F., and Newcombe, D., Proc. Natl. Acad. Sci. U.S.A., 77, 4736.

Sodium Diethyldithiocarbamate (DTC)-induced modifications of NK Activity in the mouse

Gérard Renoux[1]
Pierre Bardos
Danielle Degenne

Laboratoire d'Immunologie
Faculté de Médecine
Tours, France

Mircea Musset
Laboratoire d'Immunologie virale
Institut Mérieux
Lyon, France

I. INTRODUCTION

Sodium diethyldithiocarbamate (DTC) enhances T-cell-mediated events without affecting B cells (Renoux and Renoux 1979, 1981 a ; Renoux 1980). These effects are mediated by the increased production of selective inducers of prothymocytes, and even in *nu/nu* mice (Renoux et al. 1979 b, 1980 c ; Renoux and Renoux 1980, 1981 a), and controlled by the brain neocortex (Renoux 1980 ; Renoux et al. 1980 a, b). Current studies show that DTC induces double-peaked time-response curves, which would correspond to its mechanisms of action. DTC is a lipophilic agent and penetrates easily into cells to modify their enzymatic activities and, in turn, their protein synthesis. This mechanism is likely to be involved in the early production of an extra-thymic peptide able to recruit Thy-1^{+} cells from bone-marrow cells (Renoux and Renoux 1981 b), confirming the prediction of Stutman (1974) that prethymic

[1]Supported by a grant from Institut Mérieux.

ISBN 0-12-341360-5

commitment events are needed to explain the action of thymic hormones. Lately, and almost certainly through the influence of this peptide, DTC activates the thymus for an increased production of thymic hormones (Renoux and Pompidou, manuscript in preparation). Taken together, the data suggest that DTC could intervene in the T-cell differentiation pathway at early prethymic events as well as at post-thymic steps, by direct influences on cell metabolism and indirect, hormone-mediated mechanisms.

In this report, we present results which, although largely provisional, suggest that DTC could increase natural cytotoxic activity without association with augmented endogenous interferon levels.

MATERIALS AND METHODS

Female C3H/He mice were treated as indicated in Results by subcutaneous injection of DTC (Institut Mérieux, Lyon), or of pyrogen-free saline in controls.

Cytotoxicity Assay. Spleen lymphocyte suspensions as effector cells were added in microplate wells to ^{51}Cr labelled YAC-1 target cells at various effector to target, E/T, ratios (1 : 100 to 12.5 : 1) and the mixture in triplicate was incubated for 4 hrs at 37°C in a 5 % CO_2 atmosphere. The percentage of chromium release was plotted versus the log of the various E/T ratios to determine lytic units (LU), as described by Kay et al. (1977).

Interferon determination. Interferon levels were determined on serum aseptically sampled, by reduction of the cytopathic effect of vesicular stomatitis virus (VSV) on L cell monolayers (Gresser et al. 1974), and expressed as units per milliliter, standardized with MRC Research Standard B 69/19.

Statistical methods. The data were analyzed using a paired *t*-test at each time and dose point.

RESULTS

To evaluate the role of DTC as an interferon inducer, 8-week-old female C3H/He mice were s.c. inoculated with 25 mg/kg of DTC, or pyrogen-free saline in controls, and their spleen lymphocytes were tested for NK activity at different times after treatment. At sacrifice, blood was aseptically sampled for titration of serum interferon levels. The sera of 10 more mice of the same strain, sex and age, were also tested for interferon activity. As shown in Table I, DTC increased at 2 hr

Table I. In vivo modification of NK Activity and Interferon level in 8-week old C3H/He mice by DTC.

Treatment	LU per 10^7 cells	% of control LU	Serum Interferon titer
0 (saline)	158.3	—	2
2 hr	280.8	177*	2
6 hr	172.9	92	64
24 hr	124.6	79	2
48 hr	NT	—	2

*P < 0.01 (comparison of treated with untreated control by Student's t-test). Four mice per groups, and ten additional mice for interferon determination.

the spontaneous spleen cell cytotoxicity in 8-week-old mice, at the peak of the age-related changes in NK activity, and at a time where serum interferon level was below the sensitivity of the test. A transient, slightly elevated titer of early interferon occurred subsequently, at which time NK activity was not significantly (P > 0.05) different from that of control.

Table II. In vivo influence of DTC on NK Activity in 26-week-old C3H/He mice.

Treatment time	LU per 10^7 cells	% of control LU
0	26	—
6 h	9.5	38*
24 h	6.8	26*
2 d	28	107
7 d	48.6	187*

*P < 0.01 in comparison of treated with untreated (yime 0 = saline-treated mice) control by Student's t-test. Four mice per group.

Another approach was to determine the effects of DTC on older mice. A representative experiment (Table II) shows that a treatment with 25 mg/kg of DTC resulted in an early inhibition of the NK activity in C3H/He of 26 weeks of age, whose spontaneous NK activity has already declined to low levels. In contrast, natural cytotoxicity to YAC-1 cells was found significantly augmented beyond that of control when tested for at day 7, post-treatment. The early inhibition and the delayed increase of NK cytotoxicity induced by DTC in mice of the age of 6.5 months, appears as important differences with the known effects of interferons, and interferon inducers (Gidlund et al. 1978 ; Djeu et al. 1979 a ; Herberman et al. 1979).

The influence of age was also examined in the assays shown in Table III, in which spleens were tested for NK activity 4 days after treatment, a time interval where T-cell activities have been found maximally increased by DTC, likely through the enhanced production of thymic hormones (Renoux and Renoux 1981 a). A rise in natural cell-mediated cytotoxicity induced by 25 mg/kg of DTC, was observed in 14-week-old mice, whereas the treatment was inactive on younger or older mice, irrespective of the spontaneous levels of NK activity (Table III). However, as 2.5 mg/kg of DTC augmented above that of control, the NK activity in mice of 4 weeks of age, the data suggest that the influence of DTC might be modulated by epigenetic factors of the host, as already observed with levamisole (Renoux and Renoux, 1979).

Table III. In vivo modification of NK activity in C3H/He mice of different ages by DTC administered 4 days before test.

Treatment	Age (weeks) of mice					
	4		14		18	
	LU[a]	% of control	LU[a]	% of control	LU[a]	% of control
saline	71.2	—	53.1	—	14	—
DTC						
25 mg/kg	57.7	81	83.8	158*	12.4	88
2.5 mg/kg	145.2	254*	NT		NT	

[a]Lytic Units per 10^7 spleen lymphocytes.

*$P < 0.01$ in comparison with saline-treated controls ; other results, not significantly different. Four to eight mice per group.

DISCUSSION

It seems likely that the modifications in natural cell-mediated cytotoxicity produced by the administration of DTC, are not associated with interferon production. DTC can either activate or suppress NK activity depending on the mouse age, and at time periods differing from those where interferons or interferon inducers are active. The agent is neither an antigen nor a mitogen (Renoux and Renoux 1979, 1981 a), and the chances are little it could induce interferon-gamma. Immunostimulants of bacterial origin trigger in older mice high levels of NK activity at about 3 days that decline to baseline levels by 7 to 10 days, in contrast, with the late boosting efficacy of DTC.

An alternative to an interferon-like effect could be evolved from a comparison between the available data on the heterogeneity recognized within the populations of cells expressing NK activity and the immunopharmacologic activities of DTC.

As evidenced in this volume, the majority of cells which function as NK cells belongs to the T-cell population, mostly as pre-thymic subsets, yet a monocyte-macrophage population possesses also NK activity (Djeu et al., 1979 b ; Warner and Tai, 1981). The role for macrophages in the augmentation of NK activity by interferon inducers suggest that macrophages may be involved in maintaining the levels of NK activity in normal mice (Djeu et al., 1979 b).

So far, DTC has no direct effect on macrophages, yet induces differentiation steps in precursor cells, and even in bone-marrow cells, and activates already differentiated T cells. The bone marrow is the probable site for the murine natural killer cell (Haller and Wigzell, 1977). Also, thymocytes and T cells recognize cells of their own species, which effect could in some instances, e.g. when the target is the mouse line YAC-1, conduct to target lysis and particularly when the activation state of the lymphocytes is high (Klein, 1980). A T-cell recruiting agent, such as DTC could thus modify NK cytotoxicity against YAC-1 cells, depending upon the kinetics of its direct or indirect activity.

The present study suggests also complex relationships between features of the host and its ability to develop increased NK activity, as age of the animals, dose of the agent and treatment time modulate the influence of DTC.

It seems possible that detailed studies on the concentrations and schedules of administration of DTC alone, or in conjunction with interferon, will provide more insight on the cell subsets expressing NK activity, and result in improved immunotherapeutic regimen.

REFERENCES

Djeu, J.Y., Heinbaugh, J.A., Holden, H.T., and Herberman, R.B. (1979 a). J. Immunol. 122 : 175.
Djeu, J.Y., Heinbaugh, J.A., Holden, H.T., and Herberman, R.B. (1979 b). J. Immunol. 122 : 182.
Gidlund, M., Orn, A., Wigzell, H., Senik, A., and Gresser, I. (1978). Nature 273 : 759.
Gresser, I., Bandu, M.T., Bronty-Boye, D., and Tovey, M.G. (1974). Nature 251 : 543.
Haller, O., and Wigzell, H. (1977). J. Immunol. 118 : 1503.
Herberman, R.B., Djeu, J.Y., Kay, H.D., Ortaldo, J.R., Riccardi, C., Bonnard, G.D., Holden, H.T., Fagnani, R., Santoni, A., and Puccetti, P. (1979). Immunological Rev. 44 : 43.
Kay, H.D., Bonnard, G.D., West, W.H., and Herberman, R.B. (1977). J. Immunol. 118 : 2058.
Klein, E. (1980). Immunology today 1 : iv.
Renoux, G. (1980). In "New Trends in Human Immunology and Cancer Immunotherapy" (B. Serrou and C. Rosenfeld, eds.), p. 986. Doin, Paris.
Renoux, G., and Renoux, M. (1979). J. Immunopharmacol. 1 :247.
Renoux, G., and Renoux, M. (1980). Thymus 2 : 139.
Renoux, G., and Renoux, M. (1981 a). In "Augmenting Agents in Cancer Therapy" (E.M. Hersh, M.A. Chirigos, M.J. Mastrangelo, eds.), p. 427. Raven Press, New York.
Renoux, G., and Renoux, M. (1981 b). Abst. 8th Internat. Congress Pharmacol., p. 437.
Renoux, G., Renoux, M., and Guillaumin, J.M. (1979 a). Internat. J. Immunopharmacol. 1 : 43.
Renoux, G., Renoux, M., Guillaumin, J.M., and Gouzien, C. (1979 b). J. Immunopharmacol. 1 : 415.
Renoux, G., Bizière, K., Renoux, M., and Guillaumin, J.M. (1980 a). C.R. Acad. Sci. (Paris) 290 D : 719.
Renoux, G., Bizière, K., Renoux, M., Gyenes, L., Degenne, D., Guillaumin, J.M., Bardos, P., and Lebranchu, Y. (1980 b). Internat. J. Immunopharmacol. 2 : 156.
Renoux, G., Renoux, M., Gyenes, L., and Guillaumin, J.M. (1980 c). Internat. J. Immunopharmacol. 2 : 167.
Stutman, O. (1974). In "The Biological Activity of Thymic Hormones" (D.W. Van Bekkum, ed.), p. 87. Halsted Press, New York, London.
Warner, N.L., and Tai, A.S. (1981). Fed. Proceed. 40 : 2711.

AUGMENTATION OF MOUSE NATURAL KILLER CELL ACTIVITY BY ALLOANTIBODY: A REVERSE ADCC REACTION

Rajiv K. Saxena

Gerontology Research Center
National Institute on Aging
National Institutes of Health
Baltimore, Maryland

The human leukemia cell line K562 is only moderately susceptible to lysis by mouse spleen cells but the effector cells treated with specific anti H-2 antiserum, have considerably enhanced levels of anti-K562 natural killer (NK) activity. For the last few years, we have been studying this phenomenon and the present chapter is a summary review of this work (1-6).

I. CHARACTERISTICS OF ALLOANTISERUM NK AUGMENTATION

A. Specificity of Action

Alloantiserum NK augmentation can be demonstrated either by including the effector cell reactive alloantiserum in the cytotoxicity incubation medium or by using effector cell preparations pretreated with the alloantiserum (1,2). The component of alloantisera which is responsible for the NK activation is IgG, since the depletion of IgG from alloantiserum by passage through a protein-A column, abrogates its NK augmenting capacity. Specificity of the alloantibody NK augmentation has been tested by using a panel of alloantisera directed against single or multiple specificities in K,D or I regions of the H-2 complex. A given combination of alloantiserum and

ISBN 0-12-341360-5

the strain of spleen cell donor mouse always resulted in augmented anti-K562 NK activity if the alloantiserum recognized some antigen(s) on the effector cells (3). These results indicated that the binding of alloantibody to effector cells is necessary for NK activation. Interestingly, anti Ia antisera also augmented the anti-K562 NK activity of mouse spleen cells, which provides strong evidence in support of expression of Ia antigens on NK cells (3). An important feature of the alloantiserum effect is that the enhanced cytotoxicity is observed against certain target cell lines (K562, and many human lymphoblastoid cell lines) but not against several other cell lines (YAC, MOLT-4, EL-4) (3-5). In this regard we have found that in order to demonstrate an alloantibody NK augmentation, the expression of Fc receptors on target cells is a necessary but may not be a sufficient condition (5).

B. Direct Action of Alloantibody on NK Cells

Anti K562 NK activity of spleen cell preparations depleted of T-cells (by anti Thy-1 + C treatment), B-cells (by passage through nylon wool columns) or macrophages (by carbonyl iron and magnet treatment) can still be augmented by alloantisera, indicating that T-cells, B-cells and macrophages are not required in the NK activation process and that the alloantibody may exert its effect directly by binding to NK effector cells (2,3). This conclusion is also supported by our observation that an anti NK-1 antiserum which reacts with a non H-2 antigen selectively expressed on a non-T non-B population of lymphocytes, can also augment the anti-K562 NK activity of mouse spleen cells (5).

C. Evidence for the Modification of NK Effector Target Interaction by Alloantibody

Certain other NK activating agents like interferon, interferon inducers and mitogens require a lag period of several hours before optimal NK augmentation takes place and their effects can invariably be blocked by actinomycin-D (4, Table 1). In contrast, NK activation in response to alloantisera is immediate and is not abrogated by actinomycin-D (4, Table 1). A typical induction phase therefore may not be required for an alloantiserum effect. In addition, alloantiserum treatment does not raise the cytotoxicity of NK cells against all target cells, indicating that a true NK activation is not induced by alloantibody. It is therefore likely that the higher levels of target lysis induced by the alloantibody treated effector

Table I. Effect of Actinomycin-D on the Activation of NK Activity by Various Agents[a]

Activating Agent	% Lysis of K562	
	Control Spleen Cells	Actinomycin-D Treated Spleen Cells
None	15.2	9.8
Anti H-2[b] Antiserum	33.7 ($p < 0.01$)	17.8 ($p < 0.01$)
Interferon	36.9 ($p < 0.01$)	10.4
Pokeweed Mitogen	45.7 ($p < 0.01$)	7.9
Concanavalin A	40.1 ($p < 0.01$)	8.9

[a]Spleen cells pooled from C57BL/6 mice (2×10^6/ml) were treated with Actinomycin-D (50 μg/ml) for 45 min at 37° C followed by 3 washings with cold RPMI-1640 + 10% FCS. NK activity of control of Actinomycin-D treated spleen cell preparations were tested in a 20 hr ^{51}Cr release assay at an E:T of 100:1 in presence or absence of various activating agents as described before (4).

cells may result from a facilitated interaction of antibody coated effector cells and Fc receptor bearing target cells.

II. REVERSE ADCC HYPOTHESIS

A facilitated effector target interaction may be conceived in which the Fc region of effector cell bound antibody interacts with Fc receptors on target cells. If this proposition is true, interference of the interaction between the effector bound alloantibody Fc region and Fc receptor on target cells should block the NK activation. We have found that the protein-A which binds to the Fc region of IgG, can inhibit the alloantiserum augmented but not the basal NK activity of mouse spleen cells towards K562 target cells (5). Moreover, the $F(ab)_2$ fraction of the alloantibody is not effective in augmenting the anti-K562 NK activity of mouse spleen cells. These results clearly indicate the importance of IgG-Fc receptor interaction in NK augmentation (5). We have therefore proposed that the antibody bridge formed by effector cell bound alloantibody interacting with Fc receptor on target cells, may reinforce the basal effector-target interaction and thereby results in an enhanced target lysis. This proposal for explaining the alloantiserum NK augmentation is similar to an antibody dependent cell mediated cytotoxicity (ADCC) reaction with the important difference of a reversal of polarity of the antibody bridge between the effector and target cells. The phenomenon of alloantiserum NK augmentation has therefore been referred to as reverse-ADCC or R-ADCC reaction (5).

III. RELATIONSHIP BETWEEN NK, ADCC AND R-ADCC ACTIVITIES

Though the R-ADCC effector cells are distinct from T or B lymphocytes, it can not be said with certainty whether they are identical to NK and/or ADCC effector cells. We have recently obtained indirect evidence which supports a common effector hypothesis for all three types of cytotoxic activities in mouse spleen cells (6). (a) Levels of anti-K562 and anti-YAC NK activities, anti-K562 ADCC and anti-K562 R-ADCC were simultaneously determined in several mouse spleen cell preparations and a strong correlation was obtained between the levels of these cytotoxic activities (6). These results would be expected if a single cell population mediated NK, ADCC and R-ADCC activities. (b) If ADCC and R-ADCC activities were mediated by different effector cells their simultaneous elicitation should give an additive cytotoxicity response. We have however found that the two types of antibody dependent cytotoxic responses are not additive (6). ADCC and R-ADCC effector cells may therefore be identical. (c) YAC cells are highly susceptible to lysis by mouse spleen cells and are known to express target structures recognized by NK cells. If R-ADCC and ADCC effector cells are distinct from NK cells, the former should not by definition, have any cytotoxic activity in the absence of appropriate antibody and thus would not interact with NK sensitive YAC cells. Addition of YAC cells to the cytotoxic assay medium however specifically inhibited in a dose dependent manner, the anti K562 ADCC and R-ADCC reactions (6). These results further support the proposition that the NK, ADCC and R-ADCC effector cells may be identical.

IV. CONCLUDING REMARKS

Our presently held view on the mechanism of alloantiserum NK augmentation phenomenon and its relationship with NK and ADCC effector cells is given in Figure 1. According to this scheme, a basal NK cell-target cell interactions is reinforced by the formation of an antibody bridge which results in a higher level of target lysis. Depending upon the direction of the antibody bridge, the augmented target lysis would represent either ADCC or R-ADCC phenomenon. Antibody dependent cytotoxicities thus become special cases of basal NK interactions and in thus respect, the present view is in contrast to some previous suggestions which attempted to explain the basal NK activity as a form of ADCC reaction taking place in the presence of naturally occurring target reactive antibody (7,8).

The R-ADCC phenomenon represents a novel way in which antibody can influence the interaction between effector and target cells. It should however be emphasized that the scheme in Figure 1 is essentially built upon experiments utilizing K562 target cells and mouse spleen cells. Further work would be required to determine the extent to which this scheme can be generalized to other systems. The biological significance of the R-ADCC phenomenon if any is not known at present. It is however reasonable to speculate that R-ADCC mechanisms may play a role in auto-immune disorders in which an anti NK cell antibody may be present. Brunda et al (9) have recently shown an NK augmenting effect of monoclonal alloantibodies on mouse spleen cells. If noncytotoxic monoclonal antibodies specifically reactive to NK cells are available, it would be interesting to determine if such antibody preparations may be used to modulate the activity of NK cells in vivo.

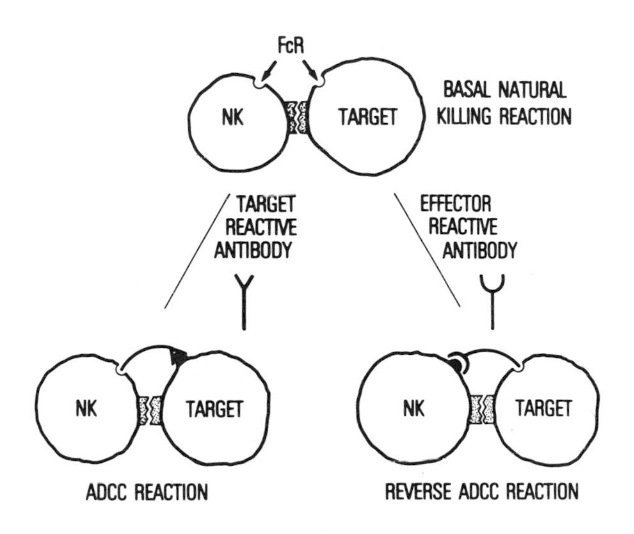

FIGURE 1

ACKNOWLEDGMENTS

I am grateful to Dr. William H. Adler for constant encouragement and for providing all laboratory facilities. I also wish to thank Ms. Eleanor Wielechowski for the painstaking job she has done in typing this chapter and to Ms. Charlotte Adler for preparing the illustration.

REFERENCES

1. Saxena, R. K., and Adler, W. H., Fed. Proc. 38:1278 (1979).
2. Saxena, R. K., and Adler, W. H., J. Immunol. 123:846 (1979).
3. Saxena, R. K., Saxena, Q. B. and Adler, W. H., Cell. Immunol. 59:89 (1980).
4. Saxena, R. K., Adler, W. H., and Nordin, A. A., Cell. Immunol. 63:28 (1981).
5. Saxena, R. K., Saxena, Q. B., and Adler, W. H., Cell. Immunol. Vol. 65 (1981) in press.
6. Saxena, R. K., Saxena, Q. B., and Adler, W. H., Immunology (in press).
7. Koide, Y., and Takasugi, J., J. Natl. Cancer Inst. 58:1009 (1977).
8. Troye, M., Perlmann, P., Pape, G. R., Spiegelberg, H. L., Naslund, I., and Gidlof, A., J. Immunol. 119:1061 (1977).
9. Brunda, M. J., Herberman, R. B., and Holden, H. T., Int. J. Cancer 27:205 (1981).

SPECIFIC AND NONSPECIFIC MODULATION OF NCMC BY SERUM COMPONENTS[1]

Mitsuo Takasugi

Department of Surgery
University of California
Los Angeles, California

Eda T. Bloom

Geriatric Research, Education and Clinical Center
VA Wadsworth Medical Center
and Department of Medicine
University of California
Los Angeles, California

Natural cell-mediated cytotoxicity (NCMC) is a polyspecific system involving an effector mechanism recognizing many antigens on target cells (Takasugi et al., 1977c; Greenberg and Takasugi, 1980). The terms, 'selective' and 'nonselective', have been quite useful in describing such reactions where the result is the sum total of many specific reactions (Klein, 1975; Takasugi and Mickey, 1976). The reaction between effector and target cell is not clearly positive or negative but relative, i.e., stronger or weaker. Nonselective reactions can be predicted mathematically by analyzing reactivities for a number of effector cells against several target cells. It does not provide any information on the specificity of the reaction. 'Selectivity', on the other hand, is used to describe a special interaction between the effector and target cells. It is the basis for the investigation of specificity in polyspecific systems.

NK cells which mediate NCMC are capable of recognizing a variety of antigens on cell surfaces of cultured target cells producing an apparent nonselective effect. Closer examination of the reactions, however, gives indications of some selectivity to NCMC. The clearest evidence supporting the existence of specificity in NCMC is derived from indirect tests as the

[1]This work was supported in part by contract NO1-CP43211 from the Biological Carcinogenesis Branch, Division of Cancer Cause and Prevention, National Cancer Institute.

ISBN 0-12-341360-5

cross-competition assay (CCA) (Takasugi et al., 1977c; Ortiz de Landazuri et al., 1973). In the CCA, NCMC by the same effector cells against several target cells is inhibited through competition with the same unlabeled target cells. The test is conducted in a two-dimensional array with all combinations of ^{51}Cr-labeled target cells and unlabeled competitor cells. Selective inhibition is usually observed for identical target cells, indicating the recognition of unique antigens on each target cell. The detection of specificity in a polyspecific system is facilitated through the use of indirect tests as competition and absorption. Through selective inhibition of apparent nonselective effects, unique antigens are recognized on each target cell with the same effector suspension, supporting our initial proposal that NCMC is a polyspecific system.

When polyspecific effector cells are tested directly on different target cells, some selectivity is observed, but for the most part, the reaction appears nonselective. We have used the interaction analysis (Takasugi and Mickey, 1976; Takasugi et al., 1977b) to distinguish between selective and nonselective components of the NCMC reaction. The average reactivities for the effector and target cells is used to calculate an expected nonselective cytotoxicity. The calculation employs the two-way analysis of variance as follows:

Nonselective cytotoxicity = $\bar{X}_e + \bar{X}_t - \bar{X}_{et}$, where:

$\bar{X}_e$ = mean reactivity of effector cells for all targets
$\bar{X}_t$ = mean reactivity of target cells with all effectors
$\bar{X}_{et}$ = mean reactivity for all tests.

Selective reactions are determined by variations from the expected score.

Selective cytotoxicity = X_o - nonselective cytotoxicity

Where X_o is the observed score.

The differentiation of selective from nonselective effects is the initial step in the investigation of specificity in polyspecific systems. The second step is the identification of a specificity through consistent detection of shared selective reactions on different target or effector cells. We can then identify the antigen(s) as one shared by two or more target cells or detected by given effector cells. Identification of an antigen is improved by using increased numbers of effector and target cells which recognize or bear the antigen. Finally, the antigen is associated with some element (e.g., viral, differentiation, histocompatibility antigens) common to the effector or target cells.

The nonselective portion of the total reaction provides no information on specificity since it cannot be distinguished from nonspecific effects. Physical conditions such as crowding or production of toxic products can increase ^{51}Cr-release above control levels but the effects are not selective for an effector-target combination and are obvious factors contributing to nonselective effects. Specific effects can also cause similar patterns. Specific reactivities against an antigen on all target cells produces nonselective patterns which cannot be distinguished from other nondiscriminating effects. A polyspecific effector system reacting against different antigens on each target cell also produces apparent nonselective effects. However, indirect testing and the CCA can be used advantageously in this situation to isolate reactions against these antigens. Without employing these approaches, nonselective effects cannot be differentiated from truly nonspecific effects and is not informative for studies of specificity in polyspecific systems.

In past studies on the modulation of NCMC with antiserum, we concluded that natural antibodies in serum had a significant role in determining the polyspecificity of the reaction (Takasugi et al., 1977a; Akira and Takasugi, 1977; Koide and Takasugi, 1977; Takasugi and Akira, 1979). Treatment of effector cell with trypsin decreased NCMC activity, and incubation of these cells in serum partially restored this activity. When the serum was absorbed with different target cells prior to incubation, the restored cytotoxicity was selectively less for the absorbing target cells. Moreover, incubation of trypsin-treated effector cells in the eluate imparted to these effector cells selective activity against the appropriate targets. The selectivity of these reactions led us to conclude that antibodies were important in NCMC.

The modulation of effector cells by incubation in serum is complex, and it is difficult to predict how NCMC or antibody dependent cell-mediated cytotoxicity (ADCC) will be affected

TABLE 1. Modulation of Effector Cells Following Treatment with Serum

	Target Cells									
	Daudi		CEM		K562		ALS-Lymph		Average	
	%cyt	Rank	%cyt	Rank	%cyt	Rank	%cyt	Rank	%cyt	Rank
Original effector cell	4.33		10.45		27.02		14.50		14.08	
Treated with serum										
A	9.34	1	16.60	1	25.61	1	21.00	1	18.14	1
B	4.30	2	11.00	2	21.61	2	7.83	3	11.19	2
C	4.05	4	7.80	3	19.42	4	12.50	2	10.94	3
D	4.13	3	6.85	4	19.53	3	6.00	4	9.13	4

TABLE 2. Testing of Serum for Interferon Activity

Serum	Interferon Units (I.U.) Test 1	Test 2	Average	Rank
A	415	357	386	1
B	145	117	131	2
C	77	121	99	4
D	118	116	117	3

without previous experience with the serum. Besides antibodies, most sera contain many factors which can enhance or inhibit NCMC and ADCC. Recently, interferon has received considerable attention for its ability to enhance the activity of effector cells for NCMC and ADCC (Trinchieri and Santoli, 1978; Trinchieri et al., 1978; Koide and Takasugi, 1978; Koide and Takasugi, 1980). In Table 1, a lymphocyte suspension was incubated in serum and tested against three target cells for NCMC and against antilymphocyte serum treated lymphocytes for ADCC. Sera were ranked according to the cytotoxicity produced by the preincubated effector cells. The ranking order of cytotoxicity for each target system varied slightly but agreed with the general order of effector activity. Thus, cytotoxicity by an aliquot of effector cells treated with Serum A was strongest followed by serum B. Serum C and D varied in NCMC according to the target system whereas the order of B and C was reversed in ADCC. The interferon levels of the sera were also tested through their ability to protect G-11 cells from the cytopathic effects of vesicular stomatitis virus. The results were converted to interferon units by comparison with a standard interferon reference from the Research Resources Branch, NIAID, NIH. Table 2 presents the ranking of interferon activity for the four sera shown in Table 1. The ranks correlate well with the modulating effect on cytotoxicity by the sera used to treat effector cells, although the final effect of a serum with its many possible factors could not always be attributed solely to interferon activity. Thus, serum-treated effector cells frequently showed weaker activity than the original untreated cells, indicating the presence in the serum of other modulating activity. The final result may also depend on the target cell used.

The effect we have just described of serum on effector cells which correlated with interferon levels is nonselective. The difference in cytotoxicity between the original and serum-treated effector cells was used as a measure of modulation by the serum and was analyzed for selectivity through application of the interaction analysis (Table 3). The observed modulation was remarkably similar to the nonselective expected

TABLE 3. Selective Modulation of NCMC by Serum Treatment of Effector Cells

Serum	Change	Change in % Cytotoxicity from Original			
		Daudi	CEM	K562	Average
A	Observed	5.01	6.15	-1.41	3.25
	Nonselective	5.79	4.77	-0.82	
	Selective	-0.78	1.38	-0.59	
B	Observed	-0.03	0.55	-5.41	-1.63
	Nonselective	0.91	-0.11	-5.70	
	Selective	-0.94	0.66	0.30	
C	Observed	-0.28	-2.65	-7.60	-2.10
	Nonselective	-0.97	-1.99	-7.58	
	Selective	0.69	-0.66	-0.02	
D	Observed	-0.20	-3.60	-7.49	-3.76
	Nonselective	-1.22	-2.24	-7.83	
	Selective	1.02	-1.36	0.34	
	Average	1.13	0.11	-5.48	-1.41

scores, and the strongest selective effect was a 1.38 change in score by serum A when tested on CEM. Therefore, modulation of effector cell activity in NCMC is mostly nonselective.

When effector cells are treated with serum, activity against target cells can be enhanced, inhibited, or unaffected, indicating the presence of other modulating factors besides interferon. If the serum is absorbed with target cells prior to the incubation of trypsin-treated lymphocytes in the serum, cytotoxic activity is selectively decreased against the absorbing target cell (Akira and Takasugi, 1977; Koide and Takasugi, 1977). Eluates from the absorbing cells can also redirect the same effector cells for selective activity against the absorbing target cells (Takasugi and Akira, 1979). Examples of these results are presented in Tables 4 and 5. It appears that serum from most individuals contain natural antibodies which recognize antigens on cultured target cells (Takasugi et al., 1979). These natural target antigens (NTA) include unique as well as shared antigens for each target cell. During absorption of the serum, antibodies to antigens unique for the absorbing cell and for antigens shared by all target cells are removed while antibodies to antigens unique for other target cells remain. The cytotoxic pattern achieved with effector cells incubated in absorbed serum is the result of selected antibodies and other modulating factors. Antibodies eluted from target cells redirect NCMC to

TABLE 4. Selective Modulation of NCMC by Incubation of Effector Cells in Absorbed Sera

		Percent cytotoxicity against target cells				
Effector		K562	CEM	NC37	8382	Average
Original		44.4	70.0	75.3	80.2	
Trypsin-treated		6.4	40.3	45.1	45.5	
Serum absorbed with	Cytotoxicity					
K562	Observed	1.9	44.6	49.7	42.6	34.7
	Nonselective	7.6	43.8	45.0	42.6	
	Selective	-5.7	0.8	4.7	0.0	
CEM	Observed	13.0	42.9	49.5	46.9	38.1
	Nonselective	11.0	47.2	48.4	46.0	
	Selective	2.0	-4.3	1.1	0.9	
NC37	Observed	12.8	48.2	46.4	45.7	38.3
	Nonselective	11.2	47.4	48.6	47.2	
	Selective	1.6	0.8	-2.2	-0.5	
8382	Observed	10.2	46.9	42.1	42.7	35.5
	Nonselective	8.4	44.6	45.8	43.4	
	Selective	1.8	2.3	-3.7	-0.7	
	Average	9.5	45.7	46.9	44.5	36.6

antigens unique for a target cell and those shared by all target cells. It is less likely that modulating factors as interferon are significant in eluates since they would be removed during washing of the absorbing cells. The imposition of selectivity to NK activity when effector cells are incubated in absorbed serum or eluates support a signficant role for antibodies in NCMC. It suggests that antibodies are important in selective activity and can also contribute to apparent nonselectivity. Much of the polyspecificity of NCMC is explained by natural antibodies.

The results achieved by rearming effector cells with antibodies selected by absorption and elution are presented in Tables 4 and 5. A selective negative cytotoxicity is observed along the descending diagonal of the test array in Table 4. Employment of the same order of target and absorbing cells along the vertical and horizontal axes in the test array defines the patterns of interactions between the reactants. The descending diagonal describes a relationship between identical cells which share the most NTA, whereas the

TABLE 5. Selective Modulation of NCMC by Incubation of Effector Cells in Eluted Antibodies

		Percent cytotoxicity against target cell				
Effector		K562	CEM	NC37	8382	Average
Original		29.0	14.1	13.8	46.6	
Trypsin -treated		10.0	3.7	4.2	5.8	
Serum absorbed with	Cytotoxicity					
K562	Observed	30.6	13.4	12.9	18.1	18.8
	Nonselective	25.0	14.8	16.4	19.1	
	Selective	5.6	-1.4	-3.5	-1.0	
CEM	Observed	15.2	13.6	13.3	13.2	13.8
	Nonselective	20.0	9.8	11.4	14.1	
	Selective	-4.8	3.8	1.9	-0.9	
NC37	Observed	21.3	7.8	14.1	14.7	14.5
	Nonselective	20.7	10.5	12.1	14.8	
	Selective	0.6	-2.7	2.0	-0.1	
8382	Observed	21.2	12.6	13.6	18.8	16.6
	Nonselective	22.8	12.6	14.2	16.9	
	Selective	-1.6	0.0	-0.6	1.9	
	Average	22.1	11.9	13.5	16.2	15.9

ascending diagonal describes the relationship between different cells expressing the most dissimilar NTA. Although the last selective score along the descending diagonal in Table 4 is only weakly negative, it is significant that all of the scores in the diagonal are negative. The analysis assumes that all scores in the test array are relative; each score influences and is influenced by other scores in the same row, column, and along the diagonal. A strong reaction along a diagonal reinforces other results from the same diagonal in the same direction (Takasugi and Mickey, 1976). The detection of four selectively negative scores along the descending diagonal and its absence in the ascending diagonal argues for the detection of specificity, of unique antigens on each target cell, and the role of natural antibodies in detecting these antigens. The same arguments can apply to the selectively positive scores along the descending diagonal in Table 5 testing effector cells reconstituted with antibodies eluted from target cells.

Just as arming effector cells with antibodies specific for a target cell increases cytotoxicity, effector cells can be redirected with antibodies against antigens not on the target cell resulting in a loss of activity. Thus, antibodies can contribute selectively to an increase or decrease in cytotoxic activity depending on their specificities and on the antigens on target cells. Due to the presence of antibodies against antigens shared among target cells and to the presence of multiple antibody specificities, antibodies also cause considerable nonselectivity. We know that NCMC varies extensively between individuals. It is probably regulated by the combined effects from several modulating factors in serum, including antibodies and interferon.

REFERENCES

Akira, D., and Takasugi, M., Int. J. Cancer 19:747 (1977).

Greenberg, J., and Takasugi, M., in "Natural Cell-Mediated Immunity against Tumors" (R. B. Herberman, ed.), p. 835. Academic Press, New York, 1980.

Klein, E., quoted in Bean, N. A., Bloom, B. R., Herberman, R. B., Old, L. J., Oettgen, H. F., Klein, G., and Terry, W. D., Cancer Research 35:2092 (1975).

Koide, Y., and Takasugi, M., J. Natl. Cancer Inst. 59:1099 (1977).

Koide, Y., and Takasugi, M., J. Immunol. 121:872 (1978).

Koide, Y., and Takasugi, M., in "Natural Cell-Mediated Immunity against Tumors" (R. B. Herberman, ed.), p. 537. Academic Press, New York, 1980.

Ortiz de Landazuri, M., and Herberman, R. B., Nature (London), New Biol. 238:18 (1973).

Takasugi, M., and Mickey, M. R., J. Natl. Cancer Inst. 57:255 (1976).

Takasugi, J., Koide, Y., and Takasugi, M., Eur. J. Immunol. 7:887 (1977a).

Takasugi, M., Akira, D., Takasugi, J., and Mickey, M. R., J. Natl. Cancer Inst. 59:69 (1977b).

Takasugi, M., Koide, Y., Akira, D., and Ramseyer, A., Int. J. Cancer 19:291 (1977c).

Takasugi, M., and Akira, D., in "Natural and Induced Cell-Mediated Cytotoxicity" (R. Reithmuller, P. Wernet, G. Cudkowicz, eds.), p. 83. Academic Press, New York, 1979.

Takasugi, M., Hardiwidjaja, M., McAllister, R., Peer, M., J. Natl. Cancer Inst. 63:1299 (1979).

Trinchieri, G., and Santoli, D., J. Exp. Med. 147:1314 (1978).

Trinchieri, G., Santoli, D., and Koprowski, H., J. Immunol. 120:1849 (1978).

PHYTOHAEMAGGLUTININ INDUCED SUSCEPTIBILITY OF AUTOLOGOUS ALLOGENIC AND XENOGENIC NK INSENSITIVE TARGET CELLS TO LYSIS BY NATURAL AND ACTIVATED KILLER CELLS

Reinder Bolhuis
Dept. of Immunology
Rotterdam Radio-Therapy Institute,
Rotterdam;
and
Radiobiological Institute TNO,
Rijswijk
The Netherlands

INTRODUCTION

It is well-known that the phenomenon of cellular recognition is of importance in many areas of biology. The immunologist, in particular, is faced with the problem of understanding the complex relationship between cells of the immune system and especially that between cytotoxic effector cells and their target cells. This involves the capacity of target cells to induce the production of soluble factors such as interferon as a result of their interaction with the effector cells and the capacity of effector cells to "read" gradients of these diffusable factors. Similarly, the effector cells can "read" the cell surface characteristics such as the glycoproteins of the major histocompatibility complex on their target cells.

Natural killer cells represent a subpopulation of lymphocytes in the peripheral blood of normal individuals. They cause lysis of syngenic, allogenic and xenogenic tumour cells and sometimes normal cells, among which are fetal thymocytes and fetal bone marrow cells. Suspensions can be enriched for these cells by employing density gradient centrifugation techniques on percoll (18). We have previously presented evidence that NK cells and mixed lymphocyte culture (MLC) activated killer (AK) cells are probably heterogeneous in nature and are activated *de novo*, i.e., the indi-

---

This work was supported by Grant RRTI 81-04 of the Dutch National Cancer League, the "Koningin Wilhelmina Fonds".

ISBN 0-12-341360-5

vidual killer cells may differ in the target cell repertoire which they are capable of lysing (7,8). The cell surface characteristics have been well documented (2, 3, 5, 6).

We also studied the target cell specificity range of MLC-AK cells and compared their characteristics with the cytotoxic T lymphocytes (CTL) which are simultaneously generated in MLC (7-9). Among other findings, we and others reported that MLC-AK cells are not likely to be identical with the MLC-generated CTL (7-9, 11, 16). This has been confirmed by Zarling et al. using cytotoxic monoclonal antibodies for the elimination of CTL after MLC and showing that MLC-AK cell activity was maintained (29). In the studies mentioned above (7, 9), we also observed lysis of autologous phytohaemagglutinin (PHA) lymphoblasts. Such autologous or "anomalous" lysis of lymphoblasts was inhibited by K-562 myeloid leukemia cells. When no "anomalous" or autologous cytotoxicity of PHA lymphoblasts was observed, no inhibition by K-562 cells was seen (9 and Table I). Thus the inhibition by was not due to nonspecific events. From these data, we concluded that the total percent lysis observed in CML was the sum of that caused by MLC-activated AK cells and that caused by MLC generated immune specific CTL. The cytolytic T cells

Table I

COLD TARGET CELL CROSS-INHIBITION ASSAYS

effector cell	inhibitor cell	inhibition to target cell ratio	per cent ^{51}Cr-release of target cells		
			A-PHA	B-PHA	K-562
B.Ax			35	21	
	A-PHA	50:1	10		
	B-PHA	50:1	21		
	K-562	12:1	15		
A.Bx			0	50	61
	A-PHA	30:1		40	62
	B-PHA	30:1		18	59
	K-562	10:1		40	43

Effector to target cell ratio 50:1. Spontaneous 51Cr-release from PHA lymphoblasts ranged from 9% to 19%. 51Cr-release from target cells in the presence of inhibitor cells was equal to medium control values. Technical details of the Materials and Methods are described in ref. 9.

have well defined specificities, the recognition structures to which NK, AK cells are directed are in contrast ill-defined. The two effector cell types, however, appeared to recognize different target structures on the same target cell (7, 9).

After observing the lysis of autologous lymphoblasts by the MLC activated killer cells, we decided to further study the role of PHA in this NK-like cytotoxicity for a number of reasons: carbohydrates are the characteristic molecules on cell surfaces and these are often associated with proteins and lipids. These glycoproteins as glycolipids, or their moieties, are involved in recognition phenomena. The main characteristic of lectins is their capacity to bind to specific sugars, to agglutinate cells and to stimulate lymphocytes, although some are nonmitogenic. The interaction of lectin-like structures and simple sugars or sugar moieties of complex glycoproteins or glycolipids present on cell surfaces are being studied extensively. Such interactions play a role in cell adhesion, aggregation and agglutination, cell differentiation, platelet activation and the binding of bacteria and viruses to cell surfaces.

The tumour target cell specificity repertoire of "fresh" NK and MLC-AK cells

As previously reported, the specificity repertoire of MLC-AK cells is broader than that of "fresh" NK cells. Representative experiments are presented in Table II. The data in this Table show that both homologous and xenogeneic tumour cells, which are NK cell-resistant, are lysed by MLC-AK-like cells.

Table II

THE TARGET CELL SPECIFICITY SPECTRUM OF NK CELLS AND MLC-AK CELLS

				per cent cytotoxicity to targets					
Exp. no	donor	effector* cells	day of testing	K-562	T-24	Mel-1	RPMI 7666	P-815	GRSL
1	A	NK **	0	45	2	1	1	0	0
		NK, 1 wk	7	2	0	0	0	0	0
		MLC-AK	7	60	50	45	52	54	35
2	B	NK	0	45	n.t.	n.t.	n.t.	1	3
		NK, 1 wk	7	4	n.t.	n.t.	n.t.	0	0
		MLC-AK	7	65	n.t.	n.t.	n.t.	51	28

* Effector to target cell ratio 25 : 1

**NK: "fresh" natural killer cells; NK, 1 wk: "fresh" NK cells, cultured for 1 week in RPMI-1640 medium plus 10% human AB serum without stimulator cells; MLC-AK: MLC-activated effector cells.

The lysis of autologous and allogeneic PHA-lymphoblasts by NK and MLC-AK cells and the enhancement of their activity by Interferon

As we reported previously (7, 9), PHA-lymphoblasts are lysed by their autologous MLC-AK cells. We extended these observations in the present experiments. From the data presented in Tables II and IV, it can be seen that the PHA lymphoblasts are lysed by both "fresh" NK and autologous MLC-AK cells. Moreover, the MLC-AK cell lytic activity against K-562 and against NK resistant mouse tumour target cells (P-815 and GRSL) can be considerably increased after pretreatment of the effector cells with interferon, which is known to be the key mediator of enhancement of "fresh" NK cell activity in vitro (Table III). It is noteworthy that the lysis of autologous PHA lymphoblasts is also increased after treatment of NK and MLC-AK cells with interferon (Table III), while such an effect is not observed for the CTL lysis of lymphoid cells (data not shown). The per cent

Table III

THE LYSIS OF PHA-LYMPHOBLASTS AND HOMOLOGOUS TUMOUR CELLS BY NK AND MLC-AK CELLS AND THE EFFECT OF INTERFERON ON THEIR LYTIC CAPACITY

		per cent cytotoxicity to targets			
donor	effector cell type	A-PHA blast*		B-PHA blast	
		control	IF^	control	IF
A.Bx	CTL; MLC-AK**	15	29	28	39
B.Ax	CTL; MLC-AK	44	52	13	28
C	NK	22	31	16	30
		K-562		T-24	
A.Bx	CTL; MLC-AK	64	65	36	57
B.Ax	CTL; MLC-AK	54	61	41	54
C	NK	40	53	5	12

* lymphocytes were cultured for 3 days and stimulated with PHA on day 3 and used as target cells on day 7.

**CTL: MLC generated immune specific cytotoxic cells;
MLC-AK: MLC-activated effector cells;
NK: "fresh" natural killer cells.

^ 10×10^6 effector cells were incubated in 2.5 ml RPMI-1640 in Falcon 2058 tubes at 37°C with 1×10^4 units leukocyte interferon for 1 h, washed and used.

Table IV

THE EFFECT OF PHA PRETREATMENT ON NK AND MLC-AK CELL LYSIS AND ON TARGET CELL SUSCEPTIBILITY TO LYSIS

effector° cells of donor	CTX at day	per cent cytotoxicity to target[x]			
		K-562	A-PHA** blast	A-PHA^ Ly	A.Bx-PHA
A	0	35		21	
A-PHA	0	23		3	
A.Bx	7	64	40	70	40
A.Bx-PHA	7	40	2		
			C-PHA blast	C-PHA Ly	C.Dx-PHA
C	0	34		23	
C-PHA	0	37		3	
C.Dx	7	61	53	60	39
C.Dx-PHA	7	34	5		
			D-PHA blast	D-PHA Ly	D.Ex-PHA
D.Ex	7	61	53	69	39
D.Ex-PHA	7	34	5		

* Frozen-stored lymphocytes were thawed and cultured 1 day before using them as target cells.

**Lymphocytes were cultured for 3 days and stimulated with PHA on day 3 and used as target cells on day 7.

^ Like *, but they were treated with PHA for 1 h and washed three times

° A: "fresh" NK cells; A-PHA: ibid, but pretreated (1 h) with PHA; A.Bx: MLC activated effector cells; A.Bx-PHA: ibid, but pretreated with PHA.

x Effector to target cell ratios tested were 50:1; 25:1; 12:1 and 6:1. The 50:1 ratio is presented.

N.B.: The present lysis of lymphoid targets and MLC-AK cells used as target cells was below 5%.

lysis of allogeneic PHA-blast cells thus represents the sum of lysis by immune specific CTL and MLC-AK cells (9); hence, the increase in per cent lysis of allogeneic PHA-(blast) cells is most likely due to the enhanced MLC-AK cell activity.

Induction of susceptibility of NK resistant tumour and autologous lymphoid cells to lysis by PHA pretreatment of these target cells.

A possible explanation for the results described so far is that the lymphoid cells used as effector cells were activated by the lectin during pretreatment. To investigate this possibility and to further explore the interaction of PHA with target cells, three approaches were taken. The first was to determine whether blast transformation of the lymphoid cells was a prerequisite for them to become susceptible to lysis by NK and MLC-AK cells. For this purpose, lymphocytes, PHA-lymphoblasts and lymphocytes pretreated with PHA for 1 hr were used as target cells. Aliquots of the effector cells (e.g., the NK or MLC-AK cells) containing populations of mononuclear cells with or without pretreatment with PHA for 1 hr were also used as target cells. The second approach was to determine whether PHA still induced this lysis when the effector cells rather than the target cells were pretreated with PHA. The third procedure was to determine whether NK resistant tumour cells would become susceptible to lysis by "fresh" NK cells after pretreatment with PHA. The criss-cross cytotoxicity assays performed were aimed at determining the specificity of the cytotoxicity patterns observed and the effect of treatment of the effector cells and/or target cells with PHA, respectively. The results are presented in Tables IV and V A and they demonstrate that: 1) untreated autologous lymphocytes are resistant to lysis by NK and MLC-AK cells; 2) PHA treated lymphocytes as well as NK resistant tumour cells are lysed by NK and MLC-AK cells; 3) pretreatment of the effector cells does not result in lysis of NK resistant untreated target cells; and 4) nonspecific agglutination of PHA pretreated effector cells making them incapable of reaching the target cells was excluded by demonstrating that the PHA-treated effector cells when ^{51}Cr-labeled and used as target cells were lysed and thus could be reached. Moreover, no increased ^{51}Cr-released was observed in the control ^{51}Cr-labeled PHA pretreated lymphoid cells incubated at effector cell concentration (data not shown).

Enrichment of NK cells on percoll gradients

We then enriched for NK cells using density gradient centrifugation of lymphoid cells on percoll (18) and tested the lymphoid cells of the different percoll fractions for their lytic activity employing both PHA pretreated NK resistant lymphoid (autologous) and tumour target cells and K-562 cells. Representative experiments are presented in Tables VI A and B and the results suggest that the effector cells that lyse the PHA pretreated target cells are contained in the same cell fraction which is responsible for lysing the K-562 target cells. This fraction contains the large granular lymphocytes (LGL) which are known to exert the NK cell cytotoxicity (18). The lower the percentage of LGL cells (in the lower fractions), the lower is the activity found against both K-562 and PHA pretreated target cells (Tables VI A and B).

TABLE V A

ANALYSIS OF CYTOTOXICITY OF DENSITY GRADIENT SEPARATED LYMPHOID CELLS ON PERCOLL, K-562 AND P815-PHA TARGET CELLS

effector cells of Percoll fractions	E:T	per cent cytotoxicity to target K-562	P-815-PHA	P-815	% LGL	% Ly
unfractionated	50:1	32	23	4		
	25:1	18	13			
	12:1	8	8			
1	50:1	3	3			
	25:1	1	1			
	12:1	0	0			
2	50:1	60	54		72	27
	25:1	54	39			
	12:1	38	25			
3	50:1	47	35		21	78
	25:1	29	22			
	12:1	19	13			
4	50:1	23	15		9	90.5
	25:1	15	10			
	12:1	7	6			
5	50:1	9	6		2	97
	25:1	4	4			
	12:1	2	0			
6	50:1	2	3		0.8	98
	25:1	1	1			
	12:1	0	1			
7	50:1	2	0		0.4	99.2
	25:1	1	0			
	12:1	1	0			

TABLE V B

ANALYSIS OF CYTOTOXICITY OF DENSITY GRADIENT SEPARATED LYMPHOID CELLS ON PERCOLL, K-562 AND P815-PHA TARGET CELLS

effector cells of Percoll fractions	E:T	per cent cytotoxicity to target K-562	Ly-PHA[a]	% LGL	% monocytes	% Ly
unfractionated	40:1	37	22	12	20	66
	20:1	25	15			
	10:1	15	13			
	5:1	9	9			
after nylon	40:1	55	24	16	0	84
	20:1	46	24			
	10:1	33	16			
	5:1	19	11			
1	40:1	58	28	50	0	50
	20:1	55	22			
	10:1	44	19			
	5:1	27	11			
2	40:1	50	26	38	0	62
	20:1	34	20			
	10:1	22	13			
	5:1	12	8			
3	40:1	42	19	9	0	91
	20:1	28	14			
	10:1	16	9			
	5:1	9	7			
4	30:1	8	8	1	0	99
	15:1	5	4			
	7:1	2	1			
	3:1	1	2			
5	30:1	2	3	1	0	99
	15:1	0	0			
	7:1	0	3			
	3:1	0	0			

[a] donor 1038

The fact that PHA lymphoblasts and PHA treated lymphoid cell and NK-resistant tumour cells were lysed by a NK cell-containing fraction and by MLC-AK cells can be explained in a number of ways. One can hypothesize that: 1) PHA forms bridges between effector and target cells and lysis occurs; or the NK and MLC-AK cells have receptors for lectin on their membranes and additional as yet undefined triggering steps are necessary for killing to occur (10, 12, 15); 2) the PHA treated target cells activate or increase the level of lytic activity of hitherto nonreactive cells, resulting in a lytic mechanism which is independent of antigen recognition; 3) PHA simply change the membrane properties of the target cell resulting in increased NK-sensitivity or increase in the level of lytic activity of the effector cells; and 4) antigen receptors are involved in the lysis of PHA treated target cells: all NK sensitive target cells have in part similar antigens which are rearranged or exposed on NK insensitive targets following interaction with PHA and thus can interact with antigen receptors on the NK and MLC-AK effector cells (see later).

The experiments in which the effector cells were used as both effectors and targets (Tables IV A and B) strongly suggest that PHA does not merely function as a bridge between effector and target cells since:

1) Lysis is unidirectional, i.e., only the PHA pretreated cell is lysed.
2) The unidirectional lysis of only PHA pretreated cells allows the conclusion that PHA does not activate the lytic machinery since the PHA pretreated effector cell is not activated.

PHA pretreatment of the effector cells (MLC-AK and CTL) does either not effect or reduce (K-562 target cells or allogeneic lymphoid target cells) or even completely abolish (PHA-treated target cells) their lytic capacity. This may be due to reduced cell mobility or to cross linking of receptors involved in the lytic process.

Lysis of PHA-treated target cells is not inhibited by monoclonal anti-CTL antisera

To further investigate the involvement of CTL allospecific receptors in the lysis of PHA pretreated target cells and tumour cells MLC-responder cells were tested for cytotoxicity against autologous PHA-pretreated autologous lymphocytes, stimulates lymphocytes and tumour cells either with or without anti-CTL monoclonal antisera present during the lytic assay. The anti-CTL monoclonal antisera were obtained from Drs. Mawas and Malissen (Marseille). It can be concluded from the data presented in Table VII that only the allo-immune reactivity is inhibited, setting the MLC-AK cell lytic mechanism apart from CTL lysis. This observation argues against the involvement of MHC specific receptors on the MLC-AK effector cells against PHA pretreated target cells and NK resistant tumour cells.

CONCLUDING REMARKS

Extensive research is performed to unravel the mechanism whereby effector cells bind to and kill target cells, but this issue has largely remained unsolved. Known are the facts that accomodation of the immune specific CTL binding receptors is a prerequisite for allospecific cytolysis to occur. The LDCC cells kill nonspecifically in the presence of PHA on the target cells. Thus it was suggested that lectins just bridge the effector cell and target cell and resulting in lysis of the target by the activated effector cell. Our study demonstrates that the target cell is the principal site of PHA action.

In part, similar data were obtained by Berke et al. in mice (4). They conclude that the lytic mechanism of LDCC and CTL is similar or even identical. Berke et al. used polyclonally activated T cells (Con A treatment) carrying a wide spectrum of MHC-specific receptors. The hypothesis they provide is that Con A distributes the self MHC proteins, now capable of high avidity interaction with a wide range of pre-CTL receptors with which they would ordinarily not react due to low affinity of these receptors. Thus, they view CTL and LDCC as essentially the same. Relating our experimental results (Table V A and B) we would have to assume that LGL are pre-CTL. That this is not the case was shown by Abo et al., who purified NK cells by cell sorting with the use of NK specific monoclonal antibodies. The NK cells did not differentiate into CTL during MLC (Abo,personal communication).

An alternative explanation for the PHA induction of susceptibility to lysis of NK resistant target cells may be offered: "Lectin-like" receptors have been described previously in the membrane of the mouse peritoneal macrophage (19) and are widely distributed amongst phagocytes of different types and sources. It is therefore possible that PHA pretreatment of NK resistant lymphoid or tumour cells reduce the fibronectin exposure on these cells with subsequent exposure of underlying sugars that can now bind to "lectin-like" receptors on NK and MLC-NK cells. It may still be possible that NK susceptible and PHA-treated targets are lysed by NK cells through different mechanisms and then that NK cells may have three different lytic mechanisms, i.e., NK, ADCC and LDCC.

At present we cannot exclude the possibility that there is an overlap between MLC-AK cells lysing autologous PHA-treated cells and allospecific CTL.

REFERENCES

1. Bach, M.L., Bach, F.H. and Zarling, J.M. Lancet 1, 20, 1978.
2. Bakacs, T., Gergely, P. and Klein, E. Cell. Immunol. 32, 317, 1977.
3. Bakacs, T., Klein, E., Yefenof, E., Gergely, P. and Steinitz, M. Z.Immun. Forsch. 154, 121, 1978.

4. Berke, G., Hu, V., McVey, E. and Clark, W.R. J. Immunol. 127, 776, 1981.
5. Bolhuis, R.L.H. Cancer Immunol. Immunother. 2, 245,1977.
6. Bolhuis, R.L.H., Schuit, H.R.E., Nooyen, A.M. and Ronteltap, C.P.M. Europ. J. Immunol. 8, 731, 1978.
7. Bolhuis, R.L.H. and Ronteltap, C.P.M. Immunol.Letters 1, 191, 1980.
8. Bolhuis, R.L.H. In: Herbermann, R.B. (ed.) Natural killer (NK) cells and related effector cells. Academic Press. New York, 1980, pp. 449-463.
9. Bolhuis, R.L.H. and Schellekens, H. Scand. J. Immunology 13, 1981, 401-412.
10. Bonavida, B. and Bradley, T.P. Transplantation 21, 94, 1976.
11. Callewaert,D.M., Lightbody,J.J., Kaplan,J., Joroszewski, J., Peterson, W.D. and Rosenberg, J.C. J. Immunol. 121, 81, 1978.
12. Gately, M.K. and Martz, E. J. Immunol. 119, 1711, 1977.
13. Malissen, B., Charmot, D., Liabeuf, A. and Mawas, C. J. Immunol. 123, 1781, 1979.
14. Ortaldo, J. Behring Inst. Mitt. 67, 258-264, 1981.
15. Rubens, R.P. and Henney, C.S. J. Immunol. 118, 180,1977.
16. Seeley, J.K., Masucci, G., Poros, A., Klein, E. and Golub, S.H. J. Immunol. 123, 303, 1979.
17. Timonen, T., Ortaldo, J.R. and Herberman, R.B. J. Exp. Med.153, 569, 1981.
18. Timonen, T. and Saksela, E. J. Immunol. Methods 36, 285, 1980.
19. Weir, D.M., Grahame, L.M., Ogmundsdottir, H.M. J. Clin. Lab.Immunol. 2, 51, 1979.
20. Zarling, J.M. and Kung, P.C. Nature 288, 394, 1980.

EFFECT OF IFN AND PHA ON THE CYTOTOXIC ACTIVITY OF BLOOD LYMPHOCYTE SUBSETS

Maria G. Masucci
Giuseppe Masucci
Maria T. Bejarano
Eva Klein

Department of Tumor Biology
Karolinska Institute
Stockholm, Sweden

I. INTRODUCTION

It is known that blood lymphocytes of individuals differ in spontaneous cytotoxic potential. The activity can be enhanced by short term IFN pretreatment of the effectors (interferon activated killing, IAK) and by addition of PHA to the short term assay (lectin-dependent cellular cytotoxicity LDCC). We have compared the effect of IFN and PHA on the lytic activity of lymphocyte subsets separated according to cell surface markers and cell density.

This investigation was in part supported by Grant no. 5R01 CA-25250-02 awarded by the National Cancer Institute DHEW and also by the Swedish Cancer Society.

Maria Grazia Masucci is recipient of a fellowship from the Foundation Blancefor-Boncompagni-Ludovisi född Bildt, Stockholm Sweden.

ISBN 0-12-341360-5

II. MATERIAL AND METHODS

A. IFN Preparations

Leukocyte and immune interferon (IFN-α, $10^{6,5}$ μg/mg protein and IFN-γ) from Dr. K. Cantell (Helsinki, Finland), lymphoblastoid interferon (1-IFN-α, 10^7 μg/mg protein) and fibroblast interferon (10^6 μg/mg protein) from Dr. W. Berthold (Biberach, Germany), recombinant interferon produced by E Coli: IFNα-1 (4,2 x 10^3 μg/mg protein) and IFNα-2 from Dr. Ch. Weissman (Zürich, Switzerland).

B. Fractionation of The Lymphocyte Population on The Basis of Surface Markers

Lymphocytes separated from blood were separated with the aim to obtain subsets that express E rosettes of different affinity, with and without concomitant Fcγ receptor (Bakacs et al., 1977; Masucci et al., 1980a).

C. Fractination of Lymphocytes on Percoll Density Gradients

The procedure described by Timonen and Saksela (1980), was used with minor modifications.

D. Morphological Characterization

2 x 10^5 lymphocytes in 0.2 ml medium were centrifuged for 4 min at 1000 rpm on microscope slides using a cytospin centrifuge. The preparations were air dried and fixed for 5 min in methanol and stained for 7 min in 10% Giemsa (E. Merck, Darmstadt, Germany). Morphological evaluation was done by inspection of the slides by oil immersion microscopy. At least 200 cells were scored.

E. Cytotoxicity Assay

The ^{51}Cr release assay was used as in our previous publications (Masucci et al., 1980a).

III. RESULTS

A. Effect of IFN on Spontaneous Cytotoxicity Against K562 and Daudi

Exposure of lymphocytes to IFN for 2-3 hours resulted in considerably enhanced cytotoxic activity (Table I). All IFN preparations had similar enhancing effects.

When tested in short term ^{51}Cr release assays, cell lines can be divided in three categories: 1) sensitive in short-term assays, such as K562; within the conventional range of the effector/target ratios used, considerable target cell damage occurs at low ratios, 2) sensitive only in long-term assays, such as Daudi, 3) insensitive.

IFN-pretreated effectors could kill the low sensitive targets with the kinetics shown for the high sensitive K562 (Masucci et al., 1980b; Berthold et al., 1981).

The short-term NK and IAK systems differ only in the technical aspects in that the latter involves pretreatment of the effector cells. The two systems overlap, however, because the

TABLE I. Effect of Various IFNs on The Cytotoxic Activity of Fresh Lymphocytes Against K562 and Daudi

IFN[a]	No. of experiments	Increase in efficiency[b]	
		K562	Daudi
IFN-α	10	2.5	8.7
1-IFN-α	20	2.1	8.5
IFN-β	5	2.3	8.0
IFN-γ	1	2.0	8.3
Coli IFNα-1	3	2.0	5.2
Coli IFNα-2	3	2.9	9.2

[a]For characterization of the IFN preparations see Material and Methods.

[b]Ratio between the number of effector cells required to obtain the same value of specific ^{51}Cr release by untreated and after IFN pretreatment.

level of cytotoxic activity exerted spontaneously in a strong NK donor is similar to the one induced by IFN in a donor with low spontaneous activity (Table II). Thus what can be achieved artifically with the lymphocytes of one individual can be a spontaneous level in another individual.

B. Effect of IFN and PHA on The Cytotoxic Activity of Lymphocyte Subsets

1. Lymphocytes Separated on The Basis of Surface Markers. The cytotoxic potential of unfractionated lymphocyte populations was enhanced by preincubation of the effectors with IFN or in the presence of PHA during the assay. The effects brought about by IFN and PHA were quantitatively similar. The extent of enhancement depended on the level of inherent cytotoxicity, with the treatments inducing only a slight increase in a strong reactive donor. Lymphocytes separated on the basis of nylon adherenence, or E and EA receptor expression by rosette separation, differred in spontaneous cytotoxic activity. The activity of the nylon passed subsets against K562 ranked as follows: E^-EA^+, E^-EA^-, $E^+_{low}EA^+$, $E^+_{low}EA^-$, $E^+_{high}EA^+$, E^+_{high} EA^-. IFN and PHA treatment influenced the activity of the subsets to different extent (Table III). The E^- and E^+_{low} fractions were enhanced more substantially by IFN than by PHA, while the opposite was true for the E^+_{high} cells. Among the

TABLE II. Effect of IFN Treatment on The Cytotoxic Activity of "High" and "Low" NK Donors

	Effector/target ratio	Cytotoxic activity[a] K562 -IFN	K562 +IFN[b]	Daudi -IFN	Daudi +IFN[b]
"high" NK	25:1	45	56	30	47
	8:1	27	42	20	32
	3:1	11	20	9	22
"low" NK	25:1	25	41	10	37
	8:1	14	29	5	25
	3:1	5	10	2	12

[a]Cytotoxic activity was measured in a 4h ^{51}Cr release assay.
[b]The effector cells were pretreated for 3h with 1000 U/ml lymphoblastoid IFN (1-IFN-α).

TABLE III. Enhancing Effect of IFN and PHA on The Cytotoxic Activity of Lymphocyte Subsets Separated on The Basis of E and EA Rosetting[a]

	Increase in anti K562 efficiency[b]	
	IFN	PHA
Unfractionated	2.1±0.5[b]	2.1±0.5[b]
E^- EA^+	4.5±1.0	1.8±0.6
E^- EA^+	3.8±1.2	1.6±0.6
E^+_{low} EA^+	4.6±1.3	1.7±1.2
E^+_{low} EA^-	3.5±0.9	1.5±0.2
E^+_{high} EA^+	2.8±0.8	4.5±1.5
E^+_{high} EA^-	1.4±0.5	6.8±1.6

[a]Mean and S.D. of 4 experiments.
[b]Increase in efficiency was calculated as in Table I.

E^+_{high} subpopulations, the EA^+ cells which exhibited spontaneous cytotoxic activity responded better to IFN than the EA^- cells. The E^+_{high} EA^- fraction contains the majority of mature resting T cells and has only low spontaneous anti-K562 activity with the majority of individuals.

2. Lymphocytes Separated on Percoll Density Gradients. In accordance with published data, only LGL-containing fractions were sensitive to the effect of IFN (Timonen et al., 1981). The strongest enhancement occurred with the fraction collected at the 40% Percoll interface, which represented 10% of the recovered cells (Table IV). In this subset, 55% of the cells had the LGL morphology compared to 10% in the unfractionated population. This correspond to a 5-fold enrichment in LGLs. Cells collected at the high density interfaces were deplted in LGL and had none or only very low spontaneous cytotoxic activity. This was not influenced by preincubation of the effector cells with 1000 U/ml IFN. Addition of PHA to the lytic assay induced cytotoxic activity also in the high density IFN insensitive fractions. The level of PHA induced activation was in these fractions higher than in the low density, IFN sensitive fractions suggesting that the cells sensitive to PHA and IFN activation are different.

TABLE IV. Effect of IFN and PHA on The Cytotoxic Activity of Lymphocyte Subpopulations Separated on Percoll Density Gradients

	% recovery of cells	% LGL	% lysis	Increase % lysis by effectors treated with: IFN	PHA
Unfractionated		10	26[a]	10	12
Fraction number					
2	10.0	55	44	22	14
3	37.3	30	22	14	24
4	32.8	4	15	0	23
5	10.7	2	8	-3	24
6	8.2	0	5	-3	20

[a]% specific ^{51}Cr release at 25.1 effector/target ratio.

In order to test whether the PHA-induced activation of the cytotoxic potential acts through induction of IFN production, antibodies against IFN-α were added to the test. The presence of anti-human IFN serum did not influence NK activity against K562 (Table V). As expected, the IFN-induced enhancement was abrogated by the antiserum even in the lowest concentration used. The PHA-induced increase of cytotoxicity was not influenced.

IV. CONCLUSIONS

It is known that blood lymphocytes of individuals differ in spontaneous cytotoxic potential. In the low NK donors, the activity can be enhanced by short term IFN pretreatment of the effectors (interferon activated killing, IAK) and by addition of PHA to the short term assay. Lymphocyte subpopulations differ in their response to these measures.

The effect of IFN-treated lymphocytes of low NK donors were similar in strength and subset ranking order to the spontaneous activity of the high NK donors. Therefore, the distinction between NK and IAK is only operational. Cells with high NK potential can be enriched in nylon-passed E^- and low avidity E^+ or the low density subsets. All these subsets can be triggered for enhanced activity by IFN.

TABLE V. Effect of Anti-IFN on The IFN and PHA-Induced Enhancement of Cytotoxic Activity Against K562[a]

Antiserum dilution	% lysis		
	Control	+IFN[b]	+PHA[c]
0	12	40	30
1:100	15	21	29
1:1000	14	25	28
1:10000	13	30	30

[a] 10^6 lymphocytes were incubated for 3h in 1 ml medium containing the indicated amount of anti IFN serum (goat anti-human IFN-α serum, 4.5×10^5 antiviral units/ml, a gift from Dr. K. Cantel Helsinki, Finland).

[b] 1000 U of lymphoblastoid IFN were added during the preincubation.

[c] PHA (Wellcome, 1:400 dilution of the working dilution) was added during the cytotoxicity test. ^{51}Cr release assay performed at 25:1 E/T ratio.

The inactive cells were in the high avidity E^+ EA^- subset or in the high density subset. These were not triggered by IFN but addition of PHA to the lytic assay induced activity. Thus there was a reciprocity between the IFN and PHA effects. In accordance with the difference, the PHA-imposed triggering event did not depend on IFN-α production because anti-IFN-α serum did not abrogate it.

The results showed thus that all non-B lymphocyte subsets separated on the basis of nylon wool adherence, SRBC and EA rosetting or density contain cells with lytic potential which can be induced after exposure to the appropriate stimulus.

The response to IFN is limited to a certain differentiation state of lymphocyte. It has been shown that the lytic activity of cells cultivated in the presence of IFN for 6 days cannot be enhanced even if it is low (Argov and Klein, 1982).

Taking in account that the IFN-induced potentiation of killer function acts on the lymphocyte subsets which exert the spontaneous NK activity (Trinchieri et.al., 1978; Masucci et al., 1980a) and is more obvious with low NK-sensitive targets, we have formulated the following hypothesis: The potentially lytic lymphocyte population is heterogeneous with regard to the strength of the lytic function. Various cell lines differ in the threshold of the lytic function which can inflict damage on them. The differences in the dose response curves with different targets reflect the proportion of effectors reaching these thresholds. We assume that IFN elevates the lytic function temporarily. Consequently IFN treated cells will also affect targets whose sensitivity threshold is relatively high and are thus damaged only by lymphocytes expressing strong activity.

Cells defined as prekillers in one target system may be killers when a more sensitive cell is used as target.

It seems that lymphocytes which express spontaneous lytic function are further potentiated by IFN, while in the resting, inactive lymphocytes, PHA can induce lytic activity.

REFERENCES

Argov, S., and Klein, E. (1982). Submitted for publication.

Bakács, T., Gergely, P., and Klein, E. (1977). Cell. Immunol. 32:137.

Berthold, W., Masucci, M.G., Klein, E., and Strander, H. (1981). Human Lymph. Diff. 1:1.

Masucci, M.G., Masucci, G., Klein, E., and Berthold, W. (1980a). Proc. Nat. Acad. Sci. 77:3620.

Masucci, M.G., Masucci, G., Klein, E., and Berthold, W. (1980b). In "International Symposium on New Trends in Human Immunology and Immunotherapy" (B. Serrou and C. Rosenfeld, eds), p. 887.

Timonen, T., and Saksela, E. (1980). J. Immunol. Meth. 36:285.

Timonen, T., Ortaldo, J., and Herberman, R. (1981). J. Exp. Med. 153:569.

Trinchieri, G., Santoli, D., Dee, R.R., and Knowles, B.B. (1978). J. Exp. Med. 147:1299.

LECTIN-DEPENDENT NATURAL KILLER CELLULAR CYTOTOXICITY IN MICE (NK-LDCC): A NEW SUBPOPULATION OF NK-LIKE CYTOTOXIC CELLS

Benjamin Bonavida

Department of Microbiology and Immunology
UCLA School of Medicine
University of California at Los Angeles
Los Angeles, California

I. INTRODUCTION

Several cytotoxic effector cell subpopulations have been described in recent years. These include the cytotoxic T cell (CTL)[1], the Fc-bearing K cell in ADCC, the natural killer cell (NK) and cytotoxic macrophages. In addition, cytotoxicity has been obtained upon the addition of lectin (LDCC) (1,2) or target modification by chemicals (IO_4) or enzymatic modification (NAGO) (3). In man, LDCC can be observed with normal PBL (1) whereas in mice, the effector cells must be presensitized to exert LDCC and ODCC (2,3).

In the mouse system, LDCC or ODCC can be achieved by already sensitized CTL as normal cells show no cytotoxic activity. Previous studies in our laboratory suggested that LDCC is mediated by the same antigen-specific CTL, although other studies suggested that LDCC is mediated by a separate subpopulation of cells (4,5). However, unequivocal evidence from our laboratory has shown at the single cell level that a CTL is capable of lysing both antigen-specific and antigen-

[1]Abbreviations used: CMC, cell-mediated cytotoxicity; CTL, cytotoxic T lymphocyte; LDCC, lectin-dependent cellular cytotoxicity; Con A, concanavalin A; PHA, phytohemagglutinin; WGA, wheat germ agglutinin; C', complement.

[2]Manuscript in preparation.

ISBN 0-12-341360-5

nonspecific targets in LDCC and ODCC (2). These results have suggested that the lytic step is nonspecific provided the initial recognition and activation steps take place. In no case was it possible to show LDCC with non-sensitized populations of murine lymphocytes against a battery of targets. However, similar studies were not examined with NK-sensitive targets.

The purpose of this study was to (1) determine whether normal mouse lymphocytes mediate LDCC against NK-sensitive targets, and (2) delineate whether the defect in NK activity in low NK strains of mice can be corrected by lectin.

II. MATERIALS AND METHODS

Effector Cells: Spleen cell suspensions were treated with H_2O for 15 seconds to remove red blood cells. The cells were fractionated on nylon wool columns for one hour at 37°C, and the eluted cells resuspended in RPMI in 5% FCS (6).

Reagents: Anti-Thy-1.2, hybridoma #H0314 was obtained from the Salk Institute Distribution Center and was used to produce monoclonal antibodies in ascites (7). Rabbit antiserum directed against asialo GM_1 was the generous gift of Dr. Okumura (8).

Cell-Mediated Cytotoxic Assay: Radiolabeling of targets with ^{51}Cr and the cytotoxic assay were done as previously described (6). Briefly, 50ul of effector cells, 50ul of target cells, and 100ul of lectin solution were added to microtiter wells. The plates were incubated for four hours at 37°C and then centrifuged and aliquots of the supernatant were counted for radioactivity.

III. RESULTS

A. Con A Mediated Augmentation of Cytotoxicity Against YAC-1 Target Cells by Normal Spleen Cells

The effect of Con A on NK-CMC against YAC-1 targets was tested in high and low NK strains of mice. The cytotoxic activity of the high NK strain CBA was not regularly enhanced by Con A, although in some experiments an increase was observed. In contrast, the CMC by the low NK strains SJL/J and Beige/Beige mice was significantly enhanced by the addition of Con A (Figure 1). The enhancement was abolished by the Con A sugar inhibitor, α-methyl-D-mannoside. These

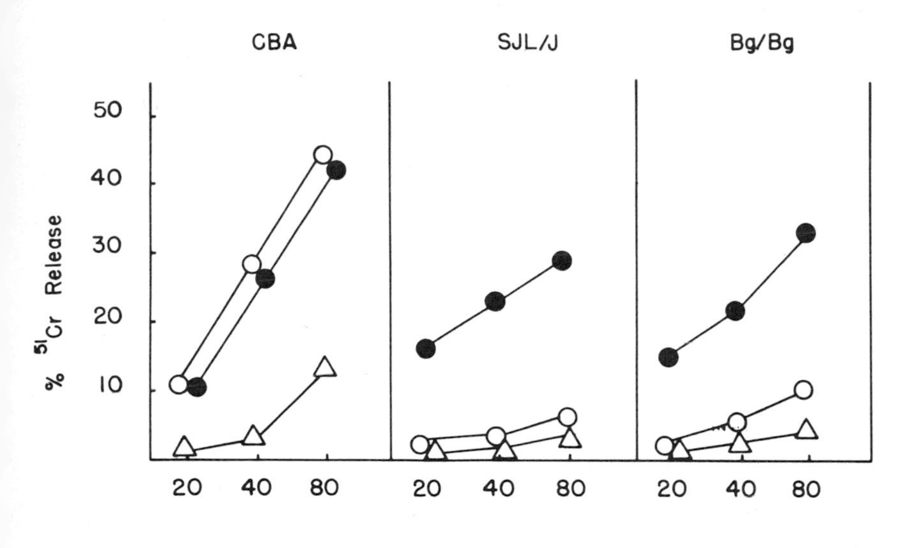

o—o No Con A; •—• Con A; △—△ Con A + alpha methyl-mannoside

Figure 1: NK-LDCC by spleen cells derived from different strains of mice. The Con A concentration used was 20 μg/ml. Cytotoxicty was done in a four-hour ^{51}Cr-release assay.

---

results showed that cytotoxicity to YAC-1 targets is significantly augmented by Con A in the low NK strains of mice.

B. Con A-Induced Augmentation of Cytotoxicity by Normal Spleen Cells is Restricted to NK-Sensitive Targets

Previous studies in the murine system have shown that normal spleen cells do not mediate CMC in the presence of Con A when tested against a battery of non-NK-sensitive target cells. Similar experiments were repeated using NK-sensitive and NK-resistant targets. The results showed that lysis in the presence of Con A is restricted to NK-sensitive targets (Table I). Thus, we will refer to the Con A induced cytotoxic activity as NK-LDCC.

C. Con A but not PHA or WGA Mediate NK-LDCC

Several lectins have been shown to mediate LDCC with mouse CTL. The effect of Con A, PHA, and WGA on NK-LDCC was tested (Figure 2). Clearly, all these lectins are effective in mediating LDCC with allosensitized CTL using a syngeneic non-NK EL4 target cell (Figure 2F). The lectin of Con A augmented

TABLE I. Target Specificity in NK-LDCC

	Cytotoxicity (% ^{51}Cr-release ± S.E.)				
Strains	E:T	Con A	YAC-1	EL-4	AKSL-2
CBA/J	20	-	11.7	1.9	5.6
		+	27.7	6.1	4.4
	40	-	21.9	8.4	6.2
		+	39.1	7.4	4.7
	80	-	37.3	6.2	9.1
		+	52.9	7.2	5.6
SJL/J	20	-	0.8	ND	0.2
		+	16.2	ND	5.8
	40	-	0.3	ND	1.8
		+	22.1	ND	5.6
	80	-	2.2	ND	1.1
		+	23.4	ND	9.8
Beige/Beige	20	-	0.3	0.3	2.0
		+	12.0	4.4	2.6
	40	-	0.4	0.6	1.3
		+	20.0	6.2	1.7
	80	-	0.2	0.9	0.9
		+	24.1	5.3	3.8
C57Bl/6 αP815 (CTL-LDCC)	2	-		0.1	0.3
		+		58.6	34.1
	6	-		0.3	1.2
		+		78.6	56.2

NK-LDCC in SJL/J, Bg/Bg, and C57Bl/6 mice, whereas there was no detectable augmentation in the high NK Balb/c nu/nu mice. However, in contrast to CTL-LDCC, neither PHA nor WGA augmented NK-LDCC. In fact, PHA reduced the NK activity of high NK strains DBA and Balb/c nu/nu. These results showed that there exists different lectin requirements for NK-LDCC and CTL-LDCC, and therefore NK-LDCC is different from CTL-LDCC.

D. Effector Cells Involved in NK-LDCC are Thy-1.2^{+} AGM_1^{-}

The above findings, showing that NK-CMC of Balb/c nu/nu mice are not affected by Con A, suggested that the effector cell may not be the classical NK effector cell. We tested the effect of cytotoxic anti-Thy 1.2 and anti-A-GM_1 treatment of

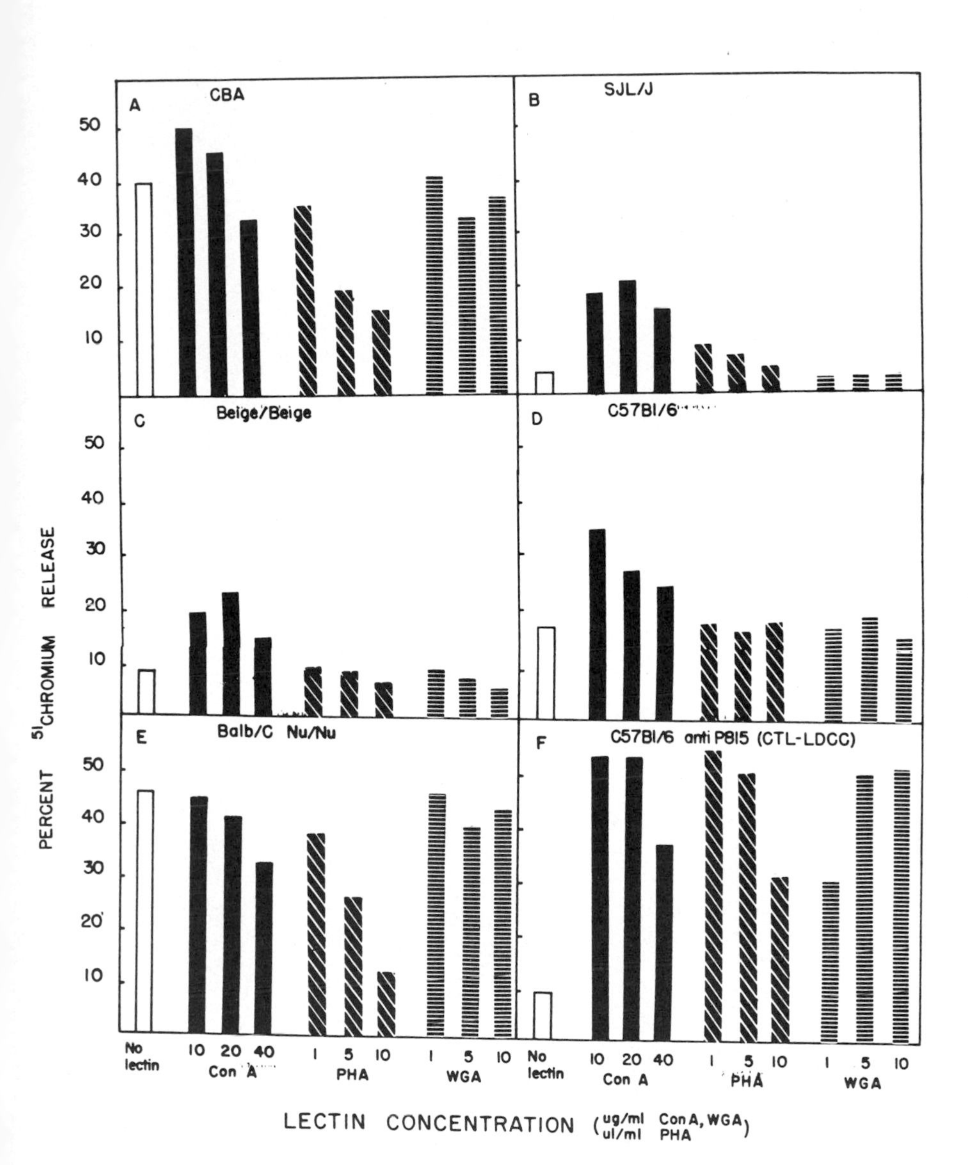

Figure 2: The effect of various lectins on NK-LDCC in different strains of mice were done. The ^{51}Cr-release assay was four hours.

No lectin Con A PHA WGA

effector cells in both NK-CMC and NK-LDCC. The results in Table II show that treatment with anti-Thy-1.2 plus C' abrogates the Con A-augmented effect without reducing the NK activity observed in the absence of Con A. This treatment, however, eliminates the CTL-LDCC as expected. In contrast, the treatment of cells with anti-A-GM_1 abrogates the NK-CMC but does not affect the NK-LDCC activity. The specificity of the anti-GM_1 was previously tested in CTL-LDCC whereby no reduction of cytotoxicity is observed (unpublished). These results show clearly that NK-LDCC is mediated by a cell which has the phenotype Thy-1.2^+, AGM_1^-.

TABLE II. Effect of Anti-Thy-1.2 and Anti-AGM1 Treatment on NK-LDCC[a]

		Cytotoxic Antibody					
		Anti-Thy-1.2			Anti-AGM1		
Strain	Con A	Sham	C'	Thy-1.2 +C	Sham	C'	Anti-AGM1 +C'
CBA	-	42.2	49.0	46.5			
	+	59.7	51.9	39.5			
Beige/Beige	-	0.6	1.3	19.0			
	+	13.5	10.6	21.2			
Beige/B6	-	11.6	13.3	26.2	10.7	8.4	4.0
	+	43.0	41.0	18.4	20.2	23.1	16.7
SJL/J	-	6.2	1.8	7.8	12.1	10.6	3.7
	+	18.1	13.2	8.6	23.2	21.0	15.3
C57Bl/6	-	13.8	12.8	14.2	21.9	15.9	3.6
	+	24.8	26.9	13.7	36.8	27.3	14.2
C57Bl/6αP815	-	6.2	2.1	0.8			
(LDCC)	+	31.1	23.7	0.3			

[a] The effector cells were treated with either complement (C') alone or with antibody and complement. The cytotoxic activity was performed at E:T of 80:1 except for C57Bl/6 anti-P815 (CTL) where the E:T was 10:1. The target for the CTL-LDCC was the syngeneic EL-4 target.

IV. DISCUSSION

Evidence is presented which defines a new subpopulation of murine spleen cells which mediate NKCMC in the presence of Con A. The surface phenotype of the effector cell is distinct from classical NK cells and is Thy-1.2$^+$ and AGM_1^-. First, the effector cell is not a classical T-LDCC effector cell since it can be distinguished by various criteria, namely, (1) the NK-LDCC effector cell is normal whereas T-LDCC is a sensitized lymphocyte; (2) the effector cells lyse only NK-sensitive targets whereas CTL-LDCC cells lyse both NK-sensitive and NK-insensitive targets; (3) lysis of NK-sensitive target cells by NK-LDCC effectors is mediated by Con A but not by PHA or WGA, whereas all three lectins are effective to mediate CTL-LDCC; (4) there seems to be gradation in the ability of certain strains of mice to mediate NK-LDCC. In CTL-LDCC, such variation is not usually seen with sensitized T cells. Therefore, the results suggest that NK-LDCC is mediated by a subpopulation of cells which is distinct from effector cells which mediate classical CTL-LDCC. Second, the effector cell in NK-LDCC can also be differentiated from classical NK cells by the following criteria: (1) The NK-LDCC cell is sensitive to treatment with anti-Thy-1.2 but not with anti-AGM_1, whereas the NK cell is insensitive to anti-Thy-1.2 and sensitive to anti-AGM_1; (2) High and low NK strains do not correlate with NK-LDCC. It appears that low NK strains exert a strong NK-LDCC, whereas high NK strains exhibit poor NK-LDCC.

The demonstration of a unique subpopulation of cells effective in NK-LDCC raises several questions relevant to the mechanism of NK-CMC. Such questions include the lineage of the effector cell, the mode of action of the lectin Con A, the strain distribution of NK-LDCC activity, the receptors involved, etc.

The nature of the effector cell in NK-LDCC classifies the cell as distinct from classical NK effector cells. It may be that the NK-LDCC cell is related to the NK_B effector cell (9, 10). This may be tested using anti-NK 1.2 serum and compared to the known characteristics of NK_B effectors. Studies by Dennert _et_. _al_. have shown that established murine NK lines show enhanced activity in the presence of Con A (11). Such cells are Thy-1$^+$ Lyt-2$^-$. Further characterization is needed to establish the relationship between these effector cells and the NK-LDCC cells. In addition, Minato _et al_. have reported several subpopulations of cells that are effective in NK-CMC, based on surface characteristics (12). One of the effector cells is Thy-1$^+$ Lyt-2$^-$ and Qa5$^+$ cell. This cell responds to

IL-2 but not to interferon (12). It is possible to speculate that the NK-LDCC effector cell is a premature NK cell which responds to IL-2 that is induced by Con A.

Brunda et al. (13) have reported that the addition of the lectin Helix pomatia to mixture of B6 nu/nu spleen cells and NK resistant K562 targets resulted in enhanced cytotoxicity. The enhancement correlated with the levels of NK activity against YAC-1 target cells. In addition, Saxena and Adler (14) have reported that anti-H-2 serum directed against normal spleen cells enhanced the cytotoxicity against NK resistant K562 target cells. It is not clear at the present time how these studies relate to the present findings in NK-LDCC. Clearly, the systems used are different and further studies are needed to sort out and define the effector cells involved in these systems.

The failure to detect NK-LDCC with certain lectins (PHA and WGA) shown to be highly effective in CTL-LDCC suggests that the effector cell either lacks receptors for these lectins or that the relevant activating signals are not induced. The role of Con A in NK-LDCC is not clear. It may be that the effector cell recognizes and binds to NK-sensitive targets but that the binding does not lead to triggering of the cell. Studies at the single cell level corroborate this question (unpublished[2]). Furthermore, it has been shown in low NK strains of mice (SJL/J and Bg/Bg) that binding takes place without killing (15,16). Thus, Con A may be inducing the triggering event leading to lysis. Alternatively, Con A may be modulating certain surface membrane structures on the effector cell important in NK-CMC.

Based on the findings reported here, a model is proposed to account for NK-LDCC. The NK-LDCC effector cell may be a precursor cell which bears receptors for NK target antigens and, when matured, becomes an NK effector cell. This maturation step proceeds normally in high NK strains but is defective in poor NK strains. It is considered that the defect is localized in the trigger or activation step of the cytolytic pathway. The addition of Con A corrects this defect and leads to lysis. This model is consistent with our findings showing that poor NK strains respond well to Con A.

REFERENCES

1. Bonavida, B., Robin, A., and Saxon, A. Transplantation 22:261, 1977.
2. Bonavida, B., and Bradley, T.P. Transplantation 21:94, 1976.

3. Bradley, T.P., and Bonavida, B. J. Immunol. 126:208, 1981.
4. Rubens, R., and Henney, C.S. J. Immunol. 118:180, 1977.
5. Laux, D.C., and Tunticharen, M. Cell. Immunol. 41:294, 1978.
6. Fan, J., Ahmed, A., and Bonavida, B. J. Immunol. 125:2444 1980.
7. Marshak-Rothstein, A., Fink, P., Gridley, T., Raulet, D.H. Bevan, M.J., and Gefter, M.L. J. Immunol. 122:2491, 1979
8. Kasai, M., Iwannow, M., Nogai, Y., Okumura, K., and Tada, T. Eur. J. Immunol. 10:175, 1980.
9. Paige, C.J., Figarella, E.F., Cuttito, M.J., Cahan, A., and Stutman, O. J. Immunol. 121:1827, 1978.
10. Burton, R.C. Transplantation Proceed. 13:783, 1981.
11. Dennert, G., Yogeeswaran, G., and Yamagata, S. J. Exp. Med. 153:545, 1981.
12. Minato, N., Reid, L., and Bloom, B.R. J. Exp. Med. 154:750, 1981.
13. Brunda, M.J., Herberman, R.B., and Holden, H.T. In "Natural Cell Mediated Immunity Against Tumors" R.B. Herberman, editor, Academic Press, N.Y., p.525, 1980.
14. Saxena, R.K., and Adler, W.H. J. Immunol. 123:846, 1979.

PROLIFERATION AND ROLE OF NATURAL KILLER CELLS DURING VIRAL INFECTION[1]

Christine A. Biron[2]
Raymond M. Welsh[3]

Department of Pathology
University of Massachusetts Medical School
Worcester, Massachusetts

I. INTRODUCTION

Natural killer (NK) cells are activated during viral infection to lyse a broad spectrum of target cells with high efficiency (Welsh, 1978). Although interferon (Welsh, 1978; Gidlund, et al., 1978; Trinchieri and Santoli, 1978) and viral glycoproteins (Casali, et al., 1981) directly activate NK cells, these treatments by themselves do not elicit the levels of activity generated *in vivo* during viral infections. Our laboratory has been studying the *in vivo* activation of NK cells to ask 1. what happens to this population during the course of infection and 2. what role do these cells play in the control of infection. Our results, presented here, indicate that blastogenesis of a NK cell population is induced during infection and that a natural defense mechanism preferentially eliminates virus infected cells *in vivo*.

II. PROLIFERATION OF NATURAL KILLER CELLS DURING VIRAL INFECTION

Endogenous NK cells isolated from murine spleens and fractionated on the basis of size by centrifugal elutriation are contained within a limited class of small-to-medium size cells (Table 1, fractions 2 and 3). In contrast, NK cells activated during lymphocytic choriomeningitis virus (LCMV) infection have a broad size distribution, and a proportion of these are found in the large, blast-size fractions (Table 1,

---

[1] Supported by USPHS Grant AI 17672
[2] Leukemia Society of America Fellow
[3] Recipient of USPHS RCDA AI 00432

ISBN 0-12-341360-5

TABLE I

NATURAL KILLER CELL BLASTOGENESIS DURING VIRUS INFECTION *IN VIVO*[a]

Characterization	Effector Cell[b]	Unseparated	Fraction[c] 1	2	3	4	5	6
% Lysis[d]	Control	6.4	2	10	4	2	<1	<1
	LCMV	20	7	6	32	42	48	18
% Blast[e]	Control	6	0.8	3	18	23	21	18
	LCMV	0.5	1	<0.5	4	19	13	17
% of Total Conjugates with Dead Target Cells[f]	Control	6	5	8	8	5	10	5
	LCMV	14	11	18	23	25	28	12
% Blast Cell Conjugates with Dead Target Cells	Control	0[g]	10[g]	0[g]	10	3	10	8
	LCMV	25[g]	0[g]	0[g]	44	27	41	21

[a] C3H/St mice, 4-to-8 weeks old, were infected with 10^4 PFU of LCMV i.p.
[b] Spleen cells were prepared from normal (control) or LCMV infected mice (1).
[c] Cells were fractionated on the basis of size by centrifugal elutriation. Small, medium, and large size cells were collected respectively in fractions 1 through 6 (5).
[d] Cytotoxicity was determined in a 4H ^{51}Cr-release assay on YAC-1 target cells.
[e] Cells incorporating ^{3}H-thymidine during a 1 H pulse were scored as blast by autoradiography.
[f] K562 target cells were used in a 4 H single-cell assay. Slides were then fixed, dried, and used for autoradiography.
[g] These numbers based on fewer then 10 conjugates.

fractions 4 and 5) (Kiessling, et al., 1980). The large cells express NK antigens; the activity in these fractions can be reduced >70% by treatment with anti-asialo GM1 or anti-NK 1.2. Although the activity in fractions 4 and 5 may be mediated by a population of large but non-proliferating cells, it is clear that these fractions are greatly enriched for dividing cells as judged by ^{3}H-thymidine uptake.

To directly visualize killing by dividing cells, we have combined the single-cell assay of Grimm and Bonavida (1979) with autoradiography using Kodak nuclear track emulsion. Spleen cells which had been pulsed for 1 H with ^{3}H-thymidine were used in a single-cell assay with K562 target cells. K562 cells are only marginally sensitive to endogenous but are easily lysed by activated NK cells. During a 4 H assay 25 and 28 % of the K562 cells in conjugates with effector cells from fractions 4 and 5, respectively, were killed (Table 1). Approximately 20% of this killing was mediated by cells which had incorporated ^{3}H-thymidine, and more of the blast cells in conjugates appeared to be lytic than the non-blast cells in conjugates.

Since it was possible to demonstrate killing by dividing cells, it was of interest to know if conditions which inhibit proliferation _in vivo_ also inhibit the generation of the large NK cells during viral infection. We examined the generation of activated NK cells in two such systems: 1. in irradiated animals and 2. in animals with high interferon levels. Spleen cells from mice exposed to 1000 RADS prior to virus infection had high NK levels in fraction 3, but reduced activity in the blast-size fractions 4 and 5 (Table II). A similar activity profile was observed with spleen cells from animals that had been infected with high doses of virus (Table II). These animals had spleen interferon concentrations at levels known to inhibit cellular proliferation (>10^4 units). Taken together, these data suggest that a population of NK cells is induced to proliferate during viral infection.

III. NATURAL IMMUNITY AGAINST VIRUS-INFECTED CELLS _IN VIVO_

The role of natural killer cells in the control of virus infection is being examined by following the clearance of ^{125}Iudr-labeled cells _in vivo_ after intravenous injection. Rejection measured by this technique strongly correlates with natural killer cell activity (Riccardi, et al., 1979). LCMV-infected L929 cells were rejected twice as rapidly as non-infected cells (Table III). The rejection was dependent on virus infection of the target cells since pre-treatment of cells with defective interfering (DI) virus protected the cells from standard virus infection and also inhibited _in vivo_

TABLE II

INHIBITION OF NK ACTIVITY IN LARGE-CELL FRACTIONS

	% Lysis[a]						
		Fraction					
Treatment	Unseparated	1	2	3	4	5	6
LCMV	6	2	5	25	24	28	10
γ-Irradiated + LCMV[b]	3	1	5	24	12	10	8
LCMV	13	4	7	16	14	16	8
LCMV (high dose)[c]	12	7	6	14	9	6	9

[a] Cytotoxicity was determined in a 4-to-9 H ^{51}Cr-release assay with YAC-1 target cells. Effector cells were fractionated by centrifugal elutriation as described in Table I. Activated spleen cells were isolated from LCMV-infected mice.

[b] C3H/St mice were exposed to 1000 RADS prior to virus infection. Spleen cells were harvested at 3 days post-treatments.

[c] Mice were infected with doses (10^7 PFU) of LCMV which induce >10^4 units of IFN/spleen. Low doses of virus induce about 10^3 units of IFN/spleen.

TABLE III
CLEARANCE OF VIRUS-INFECTED CELLS IN-VIVO

Treatment	% Counts Remaining[a]			
	Uninfected Cells	STD LCMV Infected Cells	DI + STD[b] LCMV Infected Cells	NK Activity (% lysis)[c]
Control	41	19	41	N.D.
Inhibition				
Control	35	18		10
Cortisone	66	58		4
Cyclophosphamide	63	63		2
NK Deficient				
Control Bg/+	25	15		6
Bg/Bg	41	33		<1
Complement Depleted				
Control	59	25		2
CVF[d]	40	25		4
Activation				
Control	62	42		8
Day 3 LCMV-infected	17	22		53
Day 1 Poly I:C	33	30		10

[a]Mice were injected i.v. with 5×10^5 ^{125}Iudr labeled cells. Lungs were harvested at 4-to-8 H post-injection.
[b]Cells were first infected with defective interfering (DI) virus and then with standard (STD) LCMV; <6% of the cells expressed viral antigens.
[c]NK activity was determined on YAC-1 target cells.
[d]Mice were complement-depleted by treatment with cobra venom factor.

rejection (Table III). The difference in the rejection was not simply a virus-induced lysis of the target cells, since treatments which depressed NK activity such as cortisone and cyclophosphamide also interfered with the in vivo rejection. Using the histocompatible MC57G cells, it was shown that the rejection of both infected cells and uninfected cells in NK deficient beige mice, C57BL, was reduced (Table III). These data, along with the observation that complement depletion does not affect the rejection, suggest that a nonimmune cellular mechanism causes rejection of virus-infected cells. This is a generalized phenomenon; cytomegalo, herpes simplex, and Sindbis virus-infected cells were all rejected more rapidly than the uninfected cells (Biron and Welsh, 1982). The differential rejection may involve a localized NK cell activation. This NK cell hypothesis is supported by the fact that rejection of both infected and uninfected target cells was similar and more rapid in animals whose NK cells had been activated by virus infection or by treatment with poly I:C (Table III).

IV. SUMMARY

Our laboratory is examining the activation of natural killer cells during infection and discerning their role in the control of infection. The data presented here indicate that the infection results in the activation of blast NK cells and suggests that this proliferation is tightly regulated. In addition, we have demonstrated a cytotoxic mechanism in non-immune mice which preferentially lyses virus infected cells in vivo.

REFERENCES

Biron, C.A., and Welsh, R.M. (1982). Med. Microb. Immunol., In Press.

Casali, P., Sissons, J.G.P., Buchmeier, M.J. and Oldstone, M.B.A. (1981). J. Exp. Med. 154: 840.

Gidlund, M., Orn, A., Wigzell, H., Senik, A., and Gresser, I. (1978). Nature 273: 759.

Grimm, E., and Bonavida, B. (1979). J. Immunol. 123: 286.

Kiessling, R., Eriksson, E., Hallenbeck, L.A. and Welsh, R.M. (1980). J. Immunol. 125: 1551.

Riccardi, C., Puccetti, P., Santoni, A., and Herberman, R.B. (1979). J. Nat'l Cancer Inst., 63: 1041.

Trinchieri, G. and Santoli, P. (1978). J. Exp. Med., 147: 1314.

Welsh, R.M. (1978). J. Exp. Med., 148: 163.

INCREASED ANTI K562 CYTOTOXICITY OF BLOOD MONONUCLEAR CELLS AFTER YELLOW FEVER VACCINATION, PROBABLY THE FUNCTION OF ACTIVATED T CELLS

Eva Klein
Astrid Fagraeus
Anneka Ehrnst
Manuel Patarroyo

Department of Tumor Biology
Karolinska Institute
Stockholm Sweden
and
Department of Immunology
National Bacteriological Laboratory
Stockholm, Sweden

I. INTRODUCTION

Vaccination in man with yellow fever has been found to lead to interferon production and increased DNA-synthesis in a proportion of E-rosetting blood lymphocytes (Wheelock and Sibley, 1965; Ehrnst et al., 1978). It has been shown that virus infection, IFN administration or IFN inducing substances elevate the NK potential. Since at least part of the lymphocytes with NK effect belong to the T subset and the function may be due to their activated state (Hersey et al., 1975;

Supported by the Swedish Medical Research Council B82-16X-04976-06; PHS Grant number 5R01 CA 25250-03, National Cancer Institute, DHHS and the Swedish Cancer Society.

ISBN 0-12-341360-5

West et al., 1977; Bakács et al., 1978; Klein, 1980), we have undertaken the present experiments in which the cytotoxic potential of blood lymphocytes of vaccinated individuals was tested and the cells which interact with K562 cells were characterized. The detailed results are in press (Fagraeus et al., 1982).

II. RESULTS

Subsequent to the vaccination the cytotoxic potential of the blood lymphocytes against K562 increased (Table I). The maximal cytotoxic effect was seen on the 8th-12th day. The cytotoxic potential could be further enhanced by 2 hours IFN treatment of the lymphocytes prior to the assay.

In accordance with earlier results ^{3}H-thymidine uptake by the mononuclear cells increased, with a peak on day 12 (Table II).

TABLE I. Anti K562 Effect of Blood Lymphocytes Assayed after Yellow Fever Vaccination[a]

	Days in relation to vaccination				
	-2	8	12	19	34
Expt. 1	40-50	48-62	57-73	32-50	46-50
Expt. 2	22-30	40-60	24-43	16-41	20-40

[a] ^{51}Cr release in a 4h assay at 50:1 effector target ratio. The second value is the effect of IFN-α pretreated lymphocytes.

TABLE II. ^{3}H-thymidine Uptake in Blood Lymphocytes Assayed after Yellow Fever Vaccination

Days in relation to vaccination	Expt. 1		Expt. 2	
	CPM[a]	%[b]	CPM	%
-5, -2	983	0.1	1075	0.2
+8	1461	0.2	1450	0.2
+12	8505	1.3	11042	1.6
+19 or 21	1437	0.1	1750	0.3

[a] Measured on 10^{6} lymphocytes.
[b] %positive cells scored in autoradiography.

A prerequisite for the lytic effect is a close contact between the target and the effector cells. The conjugate formation has been often used for the characterization of the events in cytotoxicity (Berke and Gabison, 1975). We have studied therefore the conjugate-forming cells with regard of reactivity with the monoclonal antibody Leu4, known to react with the majority of T cells, and with OKM1, which reacts with circulating monocytes and a proportion of lymphocytes (Breard et al., 1980).

Lymphocytes which attach to K562 were characterized in fixed preparations by indirect immunofluorescence or after staining with May-Grünwald-Giemsa. Mixtures of K562 and lymphocytes at 1:5 ratio were kept in RPMI with 8% fetal calf serum for four hours at 37°C and then overnight at +4°C. After washing they were smeared on slides, air-dried and fixed in acetone for 20 minutes at -20°C.

In the course of the studies we have discovered that the OKM1 reagent reacted with K562 cells. Viable tumor cells exposed to the reagent showed the same dotted fluorescent staining as the OKM1 positive lymphocytes. Neither tumor cells nor lymphocytes reacted with the OKM1 reagent after acetone and ether-etanol fixation, which indicates that the reactive cell surface moiety seems to be of lipid nature. Exposure of K562-lymphocyte mixture to the OKM1 reagent caused a coagglutination. Due to these two facts the proportion of OKM1 positive cells in the conjugates were not scored.

A high proportion of Leu4 positive cells were seen in the conjugates with K562 (Table III). On the 8th and 12th day after vaccination the proportion of Leu4 positive cells conju-

TABLE III. Blood Lymphocytes in Conjugates with K562 Assayed after Yellow Fever Vaccination

Day after vaccination	% of all lymphocytes seen in conjugates with K562 scored on May-Grünwald-Giemsa-stained smears		% of Leu4 positive lymphocytes in the conjugates	
	Expt. 1	Expt. 2	Expt. 1	Expt. 2
8	5.2	7.9	60	50
12	12.8	12.6	78	81
21	7.2	5.5	59	64
42	7.5	7.8	51	40

gating with K562 increased, representing the majority of attached cells. These K562-attached Leu4-positive cells were larger than the free, not attached cells, but some of the free Leu4-positive cells were also blasts. Some of the large cells had microvilli which could be seen with the Leu4 reagent, but they were more conspicuous after the anti-actin staining. The villous pattern is characteristic for activated lymphcytes such as seen after PHA-stimulation. On the 21st and 42nd day after vaccination the majority of Leu4-positive cells in the conjugates were small and had no microvilli.

III. CONCLUSION

The results show that activated T cells appear in the blood after yellow fever vaccination and the cytotoxic effect of the lymphocyte population is elevated. The properties of lymphocytes which were attaching to K562 cells suggest that activated T cells are responsible for the increase in cytotoxicity.

REFERENCES

Bakács, T., Klein, E., Yefenof, E., Gergely, P., and Steinitz, M. (1978). Z. Immunitätsforsch. Immunobiol. 123:1312.
Berke, G., and Gabison, D. (1975). Eur. J. Immunol. 5:671.
Breard, J., Reinherz, E.L., Kung, P.C., Goldstein, G., and Schlossman, S.F. (1980). J. Immunol. 124:1943.
Ehrnst, A., Lamberg, B., and Fagraeus, A. (1978). Scand. J. Immunol. 8:339.
Fagraeus, A., Ehrnst, A., Klein, E., Patarroyo, M., and Goldstein, G. (1982). Cell. Immunol., in press.
Hersey, P., Edwards, A., Edwards, J., Adams, E., Milton, G.W., and Nelson, D.S. (1975). Int. J. Cancer 16:173.
Klein, E. (1980). Immunol. Today 1, IV.
West, W.H., Cannon, G.B., Kay, H.D., Bonnard, G.D., and Herberman, R.B. (1977). J. Immunol. 48:355.
Wheelock, E.F., and Sibley, W.A. (1965). The New Engl. J. Med. 273:194.

MODULATION OF NATURAL KILLER CELL ACTIVITY DURING MURINE INFECTION WITH *TRYPANOSOMA CRUZI*[1]

Frank Hatcher[2]
Raymond E. Kuhn

Department of Biology
Wake Forest University
Winston-Salem, North Carolina

I. INTRODUCTION

The cellular and molecular events which regulate the expression of NK cell activity in the normal mouse are complex and reflect the physiology, environment, and genetic background of the mouse (Kiessling and Wigzell, 1979). Recent evidence indicates that the lytic activity of NK cells can be augmented by various pharmacologic and biologic agents (as reviewed in the first volume of this series), apparently as a result of the stimulation and subsequent action of interferon production (Djeu et al., 1979).

We have investigated the level and regulation of various murine effector cell responses during the course of experimental infection with the pathogenic protozoan parasite, *Trypanosoma cruzi* (Brazil strain). During these studies the depression of *in vitro* proliferative (Cunningham et al., 1981) and cytotoxic (Hatcher and Kuhn, 1981) T lymphocyte responses to alloantigens were detected in lymphoid tissue of infected mice. If spleen or peritoneal exudate cells from infected mice were tested for cytotoxic activity, a significant level of lysis was observed against YAC-1 or P-815 tumor cells above that in cells from age- and sex-matched control mice (Hatcher and Kuhn, 1981).

[1]Supported by NIH grant AI-14841
[2]Present address: Department of Microbiology, Meharry Medical College, Nashville, Tennessee

ISBN 0-12-341360-5

The results presented here are from an investigation of the cells which mediated increased cytotoxicity against allogeneic tumor cell lines during murine infection with _T. cruzi_. The data suggest the involvement of murine cells typical of NK cells found in normal mice and another population of cytotoxic cells with characteristics of NK cells but which are Thy 1+ and NK 1⁻.

II. RESULTS

The first phase of augmented cytotoxic activity against tumor cells following injection of _T. cruzi_ was detected 1-2 days later in cells from the spleen or peritoneal exudate (SC or PEC, respectively) in both susceptible (C3H(He) Dub) and resistant (C57Bl/6) strains of mice (Hatcher and Kuhn, 1981). Tumor cell lines susceptible to this activity were the YAC-1 lymphoma and the RL o 1 lymphoma cell lines, while P-815 tumor cells were not appreciably affected unless large numbers of _T. cruzi_ were injected (5×10^4 as compared to 10^3 trypomastigotes, i.p.). Subsequent to a partial resolution of this phase of cytotoxic activity, infected mice developed a second lytic phase in SC and PEC (and cells in the lymph nodes, unpublished) populations. This phase was most clearly distinguished from the first phase in mice given 10^3 _T. cruzi_, and was detected in assays using YAC-1 or P-815 tumor cells as targets (Hatcher and Kuhn, 1981). Other cell lines tested which were susceptible to this second phase of cytotoxic activity were the RL o 1, EL-4, and WEHI 164.1 cell lines (unpublished). Cytotoxic activity against these tumor cell lines during this period of infection was significantly above that in normal mice. Peak activity was detected 16-19 days post-infection and partially resolved by day 22. Evidence of a third phase has been detected in two of three experiments using B6 mice infected with _T. cruzi_ between days 30 and 40. Cytotoxic activity against these tumor cell lines by cells from mice infected for longer periods of time were not significantly above that in normal controls, although only YAC-1 and P-815 tumor cells have been tested thus far (unpublished). These increased cytotoxic responses during the course of acute experimental infection are not mediated by antibody specific for the tumor cells, nor have we been able to demonstrate any cross reacting antigens between _T. cruzi_ and YAC-1 or P-815 tumor cells (Hatcher and Kuhn, 1981). Thus, the induction of lytic activity against tumor cells in mice infected with this parasite were designated _T. cruzi_-induced lytic activity (TILA). The first two phases of TILA have been the most extensively characterized and have been tentatively termed TILA-1 and -2 based on the initial observation of differential susceptibility of target cells and the kinetics of the augmented activity. In the following text, we will present the results of experiments which suggest that additional differences exist between cells effecting the first two phases of TILA.

The cytotoxic activity present in recently infected mice that represents TILA-1 has been shown to be mediated by NK cells which closely resemble splenic NK cells in normal mice (Hatcher et al., 1981). Representative experiments which illustrate the similarity of murine NK cells with TILA-1 effector cells are presented in Table I. Cytotoxic activity against YAC-1 target cells were mediated by nonadherent cells in both the SC and PEC populations of mice infected for 2 days and were sensitive to pretreatment with anti-NK 1.2 plus complement (C'). TILA-1 was partially reduced by pretreatment with anti-Thy 1.2, but insensitive to anti-mouse immunoglobulin, plus C'. These findings are similar to that observed in assays testing normal splenic NK cells. TILA-1 was detectable in infected beige hybrid mice (F_1(B6 bg/bg X B6 +/+) but not in infected beige mutant mice (B6 bg/bg). The latter results are consistent with the genetic deficiency beige mutant mice have in the expression of NK cell activity (Roder and Duwe, 1979).

TABLE I. The effect of various pretreatment protocols and condition of infection on the cytotoxic activity against YAC-1 target cells by normal splenic NK cells, or by cells collected from infected mice during TILA-1 or TILA-2

		Percent cytotoxic activity in C57Bl/6 mice						
			Plastic		C' plus			
Cells tested[a]		Untreated	NA	Adh	NK 1.2	Thy 1.2	bg/bg	bg/+
Normal SC	Exp 1	15.4	19.5	2.5	-	-	2.3	21.1
	Exp 2	13.3	-	-	2.5	9.9		
Normal PEC		0.9	-	-	-	-	-0.5	1.2
TILA-1 PEC	Exp 1	17.5	24.3	1.9	-	-	1.1	23.3
	Exp 2	14.9	-	-	2.6	12.1		
TILA-2 PEC	Exp 1	29.9	26.7	7.8	-	-	25.3	34.6
	Exp 2	24.0	-	-	23.7	2.7		

[a]SC or PEC from normal, day 2 (TILA-1) or day 19 (TILA-2) B6 or beige mutant (bg/bg), or beige hybrid (bg/+) mice infected with 10^3 trypomastigote forms of _T. cruzi_ i.p.
[b]Four to five hour ^{51}Cr-release assays using 10^4 YAC-1 target cells and a 100:1 effector to target ratio.
[c]Not tested.

An investigation of the characteristics of the cells mediating TILA-2 in mice infected for 16-19 days revealed significant differences in effector cell phenotype from TILA-1. Results from several experiments are also presented in Table I illustrating our findings. First, although some cytotoxic activity against YAC-1 or P-815 tumor cells was recovered in the nonadherent cell populations. Second, no appreciable loss of TILA-2 activity was detected following pretreatment with antibodies specific for the NK 1.2 alloantigen, whereas anti-Thy 1.2 alloantiserum (plus C') almost completely abolished cytotoxic activity against either tumor cell line. Antibodies against mouse immunoglobulin when present during the cytotoxic assay were not effective in reducing TILA-2 mediated lysis (not shown). Third, TILA-2 responses similar in kinetics of infection and degree of lysis against YAC-1 or P-815 target cells was detected in either beige mutant or beige hybrid mice (results of assays using P-815 targets are not shown). These results were obtained using either SC or PEC as effectors in cytotoxicity assays.

Interestingly, the Thy 1^+ effectors active during TILA-2 are not typical cytotoxic T lymphocytes and share some properties with murine NK cells. TILA-2 against either target cell line was not reduced appreciably following pretreatment with either anti-Ly 1 or anti-Ly 2 alloantisera plus C', whereas alloantigen stimulated CTL from one way MLC were sensitive to the latter antiserum. Also, the mechanism by which the TILA-2 effector cells interact with the target cells

TABLE II: Inhibition of murine NK and TILA-2 by simple sugars but not by pretreatment with anti-Ly 1.2 or -Ly 2.2 plus C'.

Cells tested[a]	Mean percent cytotoxic activity[b]					
	A. Antisera pretreatment + C'			B. Sugar (100 mM)		
	C' alone	anti Ly 1.2	anti-Ly 2.2	None	d-Man	d-Gal
Normal SC	16.9	15.9	16.3	20.2	8.3	10.2
Anti-P-815 CTL	24.5	22.3	4.5	28.9	27.5	29.1
TILA-2 SC	26.1	22.5	24.1	38.5	12.0	18.7

[a]Spleen cell from normal B6 mice or mice infected for 18 days (10^3 T. cruzi i.p.) or from MLC using B6 spleen cells and P-815 stimulator cells (day 5 of culture) were used as effectors against YAC-1 target cells for normal and infected splenic activity and against P-815 target cells for cytotoxic T lymphocyte activity.

[b](100:1) effector to target

[c]Antisera pretreatment were done as described in Table I.

may not involve a specific recognition of target membrane components, since the addition of d-mannose or d-galactose (50-100 mM) inhibited cytotoxic activity. The absence of detectable expression of Ly 2 alloantigens on murine NK cells as well as the inhibition of NK cell activity by simple sugars are not characteristic of cytotoxic T lymphocytes responding to alloantigens (Glimcher et al., 1977; Stutman et al., 1980). Thus the Thy 1^+ cells mediating TILA-2 have several characteristics associated with murine NK cells even though they do not express, detectable NK 1 surface antigen.

We have subsequently examined the serum from mice infected for various periods of time for the presence of interferon. Serum from mice infected for 1,2,3,16 or 19 days (10^3 *T. cruzi*) did not demonstrate any increase in interferon levels (unpublished). These results are contrary to that observed by others (Rytel and Marsden, 1970) who detected an interferon-like substance in outbred mice infected with the Peru strain of *T. cruzi*. Our results do not exclude the possibility that interferon production is stimulated in the microenvironment of the spleen or peritoneal cavity of infected mice which could result in augmented NK cell activity.

In other experiments TILA-1-like responses could be detected in mice injected with heat-killed *T. cruzi* (Hatcher et al., 1981) but not with plasma from infected mice. Neither plasma from infected mice or heat-killed parasite preparations resulted in a cytotoxic response similar to TILA-2 in infected mice. Thus it would seem likely that the cytotoxic responses observed are not due to the passive transfer of factors which are present in the blood of infected mice in addition to the parasites themselves.

III. DISCUSSION

These results indicate that one of the consequences of murine infection with *T. cruzi* is the activation of NK cells in the spleen and peritoneal cavity. Further, the data from mice infected for longer periods of time indicate the absence of NK 1^+ cytotoxic cells but an increase in cytotoxic activity by Thy 1^+ effector cells. Thus, infection with *T. cruzi* appears to have both a quantitative and a qualitative effect on the expression of naturally cytotoxic effector cell activity in the murine model system. The role that TILA effector cell activity may have in terms of murine resistance to infection with this parasite is not known. However, evidence is presented in another chapter in this volume which indicates that augmented NK cells (poly I:C-or parasite-induced) have cytotoxic activity against extracellular forms of this parasite *in vitro*.

The cellular and molecular mechanisms involved in generating TILA against tumor cell lines is also not known at this time. Perhaps the most perplexing aspect of these observations are those suggesting a major change in phenotype of the TILA effector cell (NK 1^+ Thy $1^{\pm}$ to NK 1^- Thy 1^+) along with the detection of TILA-2 in infected beige mutant mice and the inhibition of TILA-2 activity *in vitro* by the addition of sugars. There are several possible mechanisms which might account for some of these results. The one that is most consistent with the experimental findings obtained thus far is that TILA-2 represents the activity of NK-like cells which have an altered surface antigen phenotype (either by selection or differentiation) from the NK cell system in normal mice.

The results with beige mutant mice infected with *T. cruzi* are not inconsistent with the possible involvement of NK-like cells during TILA-2. The genetic defect in beige mutant mice affecting NK cell lytic activity is apparently at the level of the lytic mechanism and not the number of target-binding cells (Roder, 1979), a defect which can apparently be overcome following the injection of certain agents which induce interferon and NK cell activity (Welsh and Kiessling, 1980b; Karre et al., 1980). The possible involvement of other natural cytotoxic cell subpopulations during murine infection with *T. cruzi* should also be determined since increased lysis of cells from the WEHI 164.1 fibrosarcoma line was also detected during TILA-2. Other investigations have determined that the murine effector cells mediating lysis of WEHI 164.1 are natural cytotoxic cells lacking the NK 1.2 surface antigen found on NK cells (Burton, 1980).

ACKNOWLEDGMENTS

The data from experiments testing for the presence of NK 1.2 surface antigens on TILA effector cell populations were done in collaboration with Dr. Robert C. Burton. The authors wish to acknowledge the expert technical assistance of M. Cerrone.

REFERENCES

Burton, R.C. (1980) *In* Natural Cell-Mediated Immunity Against Tumors. Ed. R.B. Herberman, Academic Press, N.Y., p. 37.

Cunningham, D.S., Kuhn, R.E., and Hatcher, F.M. (1981) Exp. Parasitol. 51: 141.

Djeu, J.Y., Heinbaugh, J.A., Holden, H.T., and Herberman, R.B. (1979) J. Immunol. 122: 175.

Glimcher, L., Shen, F., and Cantor, H. (1977) J. Exp. Med. 145: 1.

Hatcher, F.M., and Kuhn, R.E. (1981) J. Immunol. 126: 2436.

Hatcher, F.M., Kuhn, R.E., Cerrone, M.C., and Burton, R.C. (1981) J. Immunol. 127: 1126.

Karre, K., Klein, G.O., Kiessling, R., Klein, G., and Roder, J.C. (1980) Nature 284: 624.

Kiessling, R., and Wigzell, H. (1979) Immunol. Rev. 44: 165.

Roder, J.C. (1979) J. Immunol. 123: 2168.

Roder, J.C., and Duwe, A.K. (1979) Nature 278: 451.

Rytel, M.W., and Marsden, P.D. (1970) Amer. J. Trop. Med. Hyg. 19: 929.

Stutman, O., Dien, P., Wisun, R.E., and Lattime, E.C. (1980) Proc. Natl. Acad. Sci. 77: 2895.

Welsh, R.M., and Kiessling, R.W. (1980b) Scand. J. Immunol. 11: 363.

ANALYSIS OF MACROPHAGE ACTIVATION AND BIOLOGICAL RESPONSE MODIFIER EFFECTS BY USE OF OBJECTIVE MARKERS TO CHARACTERIZE THE STAGES OF ACTIVATION

D.O. Adams
and
J.H. Dean

Departments of Pathology and Microbiology-Immunology
Duke University Medical Center
Durham, North Carolina
and
National Toxicology Program
National Institute of Environmental Health Sciences
Research Triangle Park, N.C.

INTRODUCTION

The activation of mononuclear phagocytes is an important host defense against both microbes and neoplasms (Hibbs, Remington, and Stewart, 1980). Recent evidence indicates the activation of murine mononuclear phagocytes proceeds via a defined sequence of stages (Hibbs, et al., 1977; Meltzer, 1981). In brief, macrophages (Mϕ) from sites of inflammation, though not resident peritoneal Mϕ, respond effectively to lymphokines by gaining responsiveness to traces of endotoxin (Hibbs, et al., 1977; Meltzer, 1981). These latter Mϕ, which have been termed primed Mϕ and which are not activated for cytolytic destruction of tumor cells, become cytolytically activated when confronted with ng amounts of endotoxin (Hibbs, et al., 1977; Meltzer, 1981). These stages of development are also pertinent to activation for tumor cytotoxicity in vivo and to activation for microbicidal function (Russell, Doe, and McIntosh, 1977; Buchmuller and Mauel, 1979). Establishing the degree of activation reached by Mϕ taken from tumors, sites of microbial invasion or tissues after administration of a biological response modifier is thus important for understanding the host's response to various challenges and for rational planning of immunotherapeutic intervention against these challenges.

Analysis of the degree of activation expressed by a given population of mononuclear phagocytes, however, has been frustrating.

ISBN 0-12-341360-5

The various stages of activation are currently defined operationally by the signal(s) which must be applied to push the Mϕ to full activation. Since the assays for tumor cytolysis or microbicidal kill require several days for completion, the long interval between signal application and quantitation of effects further complicates analysis. One approach to this problem would be to define objective and quantitative markers, which would distinguish Mϕ in the various states of activation. Cohn and colleagues have very successfully exploited this approach to distinguish Mϕ resident in the peritoneal cavity from Mϕ taken from sites of inflammation (Cohn, 1978). Previous studies from this laboratory have indicated that activated Mϕ are characterized by two separate capacities: the capacity to complete augmented and selective binding of neoplastic cells and the capacity to secrete a novel cytolytic serine protease ($\sim$ 40,000 M_r), which we have termed CF (for review, see Adams, Johnson, and Marino, 1982). The lytic potential of CF is assayed by testing macrophage-free supernatants against tumor cells in the absence of serum.

We here review studies indicating these two capacities plus several other capacities of Mϕ can be used as objective markers to characterize Mϕ in various stages of activation and then show how these markers can be employed to analyze the effects of several biological response modifiers.

Objective Markers of the Stages of Activation

To approach this problem, we employed Mϕ in various stages of activation, produced by methods originally described by Meltzer (Meltzer, 1981). In brief, resident Mϕ were taken from the peritoneal cavity of unmanipulated C57BL/6J mice. Responsive Mϕ were taken from the peritoneal cavity 24 hours after injection of fetal calf serum (Meltzer, 1981). Primed Mϕ were obtained by treating these responsive Mϕ with lymphokine-enriched supernatants of Con-A stimulated spleen cells, while activated Mϕ were obtained by pulsing the primed Mϕ with traces of endotoxin (Meltzer, 1981). All the experiments were conducted in endotoxin-free conditions (< 0.13 ng of endotoxin/ml as measured by the Limulus Amebocyte lysate assay). The status of the resident, responsive, primed and activated Mϕ was verified by quantifying their cytolytic potential under appropriate circumstances (Meltzer, 1981).

Employing a panel of functions, we found objective markers which characterize each of these stages (Adams, Johnson, and Marino 1982). The relevant characteristics of each type of Mϕ can be catalogued (Table I). To summarize the salient observations, responsive Mϕ are clearly inflammatory Mϕ, since they display three markers characteristic of such Mϕ (Adams and Marino, 1981; Adams, Johnson, and Marino, 1982). Furthermore, they display two other

Table I

Objective Markers of the Stages of Activation

	RESIDENT MØ	RESPONSIVE MØ	PRIMED MØ	ACTIVATED MØ
Spreading (% cells spread/hr)	5	72	68	74
Secretion of Plasminogen Activator (EU/10^6 MØ/24 hr)	0	18	41	39
Phagocytosis of EIgG (CPM of ^{51}Cr-SRBC/10^6 MØ)	410	2600	3350	3200
Specific Binding of ^{125}I-Mannose-BSA (fmol/mg protein bound at saturation)	---	314	120	55
Secretion of Cytolytic Factor (LU/10^6 MØ/4 hr)	0	2	2	36
Secretion of Cytolytic Factor with LPS (10 ng/ml)	0	4	35	54
Binding of Tumor Cells (Targets bound/10^6 MØ x 10^3)	4	11	42	43
Cytolysis of Tumor Cells (% net cytolysis)	0	0	4	56
Cytolysis of Tumor Cells with LPS (10 ng/ml)	0	0	48	69

(Data from References 3,5,7,15,17)

such markers: decreased content of 5' nucleotidase (< 4.0 units/mg of Mϕ protein) and extensive phagocytosis via the C3 receptor (phagocytosis of > 300 SRBC/100 Mϕ) (compare with Cohn, 1978). The responsive Mϕ are not prepared to secrete CF, even when given a second signal such as ng of endotoxin. Primed Mϕ, which also express these markers, have gained the capacity for augmented binding of tumor cells (Adams, Johnson, and Marino, 1982). Primed macrophages do not secrete CF spontaneously but can do so when pulsed with a second signal such as ng amounts of endotoxin (Adams and Marino, 1981; Adams, Johnson, and Marino, 1982). By contrast, activated Mϕ, which bear the markers of inflammatory Mϕ and which can effect augmented binding, secrete cytolytic factor spontaneously without need of an additional signal (Adams, Johnson, and Marino, 1982; Marino and Adams, 1980; Adams, et al. 1980).

Analysis of the Effect of Biological Response Modifiers on Macrophage Activation

We have examined the effects of pyran copolymer (methyl-vinyl-ether copolymer, fraction 2; Adria Laboratories) on Mϕ under standard tissue culture conditions (Dean, et al., 1981) and more recently under endotoxin-free conditions (Adams et al., 1982). The pertinent characteristics of these Mϕ are summarized (Table II). By comparison with Table I, it is clear that the pyran-elicited Mϕ are in the primed state. This conclusion has been critically verified by examining their cytolytic potential -the pyran-elicited Mϕ are not spontaneously cytolytic, but become so when pulsed with only traces of endotoxin (Adams, et al., 1982).

We have also examined several other biological response modifiers for their effects on Mϕ activation. The results of our studies to date are summarized (Table III). We have now observed agents which can induce responsive Mϕ *in vivo*, induce primed Mϕ *in vivo*, push primed Mϕ to the fully activated state *in vitro* and induce fully activated Mϕ *in vivo*.

CONCLUSIONS

The data summarized in this chapter present a system for analyzing precisely the state of activation of various mononuclear phagocytes. The method of analysis is advantageous because the markers employed are both objective and quantitative. Furthermore, the markers can all be quantified in four hours or less, permitting close correlation between experimental manipulation of the macrophages *in vivo* or *in vitro* and completion of the analyses. It is worth emphasizing that the system of assessment applies to populations of

Table II

Characteristics of Pyran Elicited MØ[1]

Spreading (% cells spread/hr)	78 ± 2
Phagocytosis of EIgG (CPM of ^{51}Cr-SRBC/10^6 Mø)	2,759 ± 133
Specific Binding of ^{125}I-Mannose-BSA (fmol/mg protein bound at saturation)	120 ± 5
Secretion of Cytolytic Factor (LU/10^6 Mø/4 hr)	0
Secretion of Cytolytic Factor with LPS (10 ng/ml)	50 ± 2
Binding of Tumor Cells (Targets bound/10^6 Mø x 10^3)	48 ± 3
Cytolysis of Tumor Cells (% net cytolysis)	2 ± 1
Cytolysis of Tumor Cells with LPS (10 ng/ml)	65 ± 3

[1] Mø elicited by 100 µg of MVE-2 given intraperitoneally into C57Bl/6J mice six days before harvest of the macrophages. Data taken from References 7 and 15.

Table III

Effects of Various Biological Response Modifiers on Mϕ Activation

Agent	Effect	Reference
Bacillus Calmette-Guerin	Activates mϕ *in vivo*	5
C. Parvum	Activates mϕ *in vivo*	1
Pyran copolymer (MVE-2)	Primes mϕ *in vivo*	7
Fractions of *C. Parvum* cell walls	Primes mϕ *in vivo*	9
Muramyl Di-peptide (Liposome encapsulated)	Pushes primed mϕ to activated mϕ *in vitro*	8
Glucan	Elicits responsive mϕ *in vivo*	1

Mϕ and that heterogeneity in these populations can exist. Several of the markers, such as spreading, binding of tumor cells, or lysis of tumor cells under agarose, can be employed on single cells and, hence, used to assess heterogeneity.

We have already applied this system to analyzing several biological response modifiers in vivo and in vitro (see Table III). We are also applying this system to analyzing the effects of various immunotoxicants upon Mϕ function. The method is extremely useful in this latter regard, because the immunotoxicants tested to date have had pleitropic effects upon the mononuclear phagocyte system. A critical test of this system of analysis will be to see if it can screen potentionally active biological response modifiers and rationally predict effective methods for employing them in vivo.

The data are of interest in another regard. Development of the capacity for augmented binding of tumor cells and for secretion of CF correlate well with present concepts on how these two capacities are induced (for review, see Adams and Marino, 1982). Currently available evidence indicates that lymphokines alone, particularly Mϕ activating factor, can induce augmented binding capacity (Marino and Adams, 1982). By contrast, the secretion of CF, like many other Mϕ-derived products, is regulated in two stages (Adams, et al., 1980). An initial preparatory signal prepares the Mϕ for secretion, but a second triggering signal is needed to effect actual release of CF (Adams, et al., 1980). This implies the cascade of signals, which are required for the activation of murine mononuclear phagocytes, may work principally by inducing those capacities requisite for completion of Mϕ-mediated tumor cytotoxicity (for a review of how augmented binding and secretion of CF interact to complete cytolysis, see Adams, Johnson, and Marino, 1982). The concept of activation as the application of signals which induce capacities necessary for completion of a complex function is consistent with evidence from a variety of sources (for review, see Adams and Marino, 1982). Use of the markers described here may therefore help to define the regulatory physiology of Mϕ development.

ACKNOWLEDGEMENTS

The work described here was supported by USPHS Grants CA16784, CA14236, and CA29589.

REFERENCES

1. Adams, D.O. and J.H. Dean. 1981. Unpublished observations.
2. Adams, D.O., W.J. Johnson and P.A. Marino. 1982. (Fed. Proc., in press, April).

3. Adams, D.O., W.J. Johnson and P.A. Marino. 1982. (Manuscript in preparation).
4. Adams, D.O., K.J. Kao, R. Farb and S.V. Pizzo. 1980. J. Immunol. 124:293.
5. Adams, D.O. and P.A. Marino. 1981. J. Immunol. 126:981.
6. Adams, D.O. and P.A. Marino. 1982. Contemporary Topics Hematol/Oncology III. (In preparation).
7. Adams, D.O., W.J. Johnson, P.A. Marino, and J.H. Dean. 1982. (Manuscript in preparation).
8. Adams, D.O., W. Fogler and I.J. Fidler. 1981. Unpublished observations.
9. Adams, D.O. and R.L. Tuttle. 1981. Unpublished observations.
10. Buchmuller, Y. and J. Mauel. 1979. J. Exp. Med. 150:359.
11. Cohn, Z.A. 1978. J. Immunol. 121:813.
12. Dean, J.H., et al. 1981. p. 267. Augmenting Agents in Cancer Tumors. E. Hersch, Mustrangalo, and M. Chirigos, eds. Raven Press, New York.
13. Hibbs, J.B., J.S. Remington and C.C. Stewart. 1980. Pharmacol. Ther. 8:37.
14. Hibbs, J.B., R.R. Taintor, H.A. Chapman and J.B. Weinberg. 1977. Science 197:279.
15. Imber, M., S. Pizzo, W.J. Johnson and D.O. Adams. 1982. J. Biol. Chem. (In press).
16. Marino, P.A. and D.O. Adams. 1980. Cell Immunol. 54:11.
17. Marino, P.A. and D.O. Adams. 1982. (Submitted for publication).
18. Meltzer, M.S. 1981. Lymphokines 3:319.
19. Russell, S.W., W.F. Doe and A.T. McIntosh. 1977. J. Exp. Med. 146:1151.

C. PARVUM-INDUCED SUPPRESSOR CELLS FOR MOUSE NK ACTIVITY

Angela Santoni, Carlo Riccardi, Teresa Barlozzari,
and Ronald B. Herberman

Biological Research and Therapy Branch
National Cancer Institute
Frederick, Maryland

I. INTRODUCTION

The cytotoxic activity of natural killer (NK) cells is highly susceptible to immunopharmacological manipulations, with resultant potentiation or abrogation of the spontaneous levels of activity. A variety of agents have been found to produce both stimulatory and inhibitory effects, which vary with the circumstances. Such opposite effects by the same agent have been seen with pyran copolymer (Santoni _et al_, 1980a), BCG (Wolfe _et al_, 1976; Ito _et al_, 1980), _Corynebacterium parvum_ (Cp) (Ojo _et al_, 1978), glucan (Lotzova and Gutterman, 1979) and adriamycin (Santoni _et al_, 1980a). Augmentation tends to be an early event and correlates with induction of interferon production. Depression tends to occur later and has been frequently associated with the activation of suppressor cells. The cell lineage of the suppressor cells has been difficult to define, but macrophages (Santoni _et al_, 1980b) as well as nonadherent cells, lacking easily detectable markers (Cudkowicz and Hochman, 1979; Ito _et al_, 1980), have usually been found to be involved.

Cp, a potent interferon inducer and RES stimulator, has been found to induce both stimulatory and inhibitory effects on NK activity. Three days after Cp injection (0.7 or 2.1 mg/mouse, ip), substantial boosting of NK activity can be observed in the spleen and peritoneal cavity, where it persists for 10-15 days. In contrast, during this period, cytotoxicity by spleen cells is pro foundly depressed. Such

ISBN 0-12 341360-5

a decline in reactivity might be attributed to direct inhibitory effects of the agent on NK cells, inactivation of accessory cells, or to induction of suppressor cells. We summarize here our studies to better understand the mechanisms underlying the depression of NK activity induced by Cp.

SUPPRESSION OF NK ACTIVITY BY CP-INDUCED ADHERENT SPLEEN CELLS

Cp has been shown to cause profound inhibitory effects on T cell-mediated immune functions (Kirchner *et al*, 1975; Lichtenstein *et al*, 1981) and many of these negative effects have been attributed to the presence of activated macrophages. To analyze whether macrophage-like cells might account for the depressed levels of NK activity observed 7-14 days after Cp injection, mixture experiments were performed as previously described by Cudkowicz and Hochman (1979) and by us (Santoni *et al*, 1980b). Figure 1 summarizes the results of a representative experiment with Cp-treated mice. As already reported (Ojo *et al*, 1978; Savary and Lotzova, 1978), the spleen cells from mice treated 10 days earlier with Cp, had depressed NK activity when compared to that of age-matched untreated mice. Mixture of splenocytes from Cp-treated mice with those of normal high NK-reactive mice resulted in a substantial inhibition of activity. This inhibitory activity was associated with adherent cells and was not affected by treatment with anti-Thy 1.2 antiserum plus complement, whereas removal of phagocytic cells by treatment with iron and magnet removed most of it (data not shown), suggesting that the suppressor cells were macrophages.

EFFECT ON NK ACTIVITY OF REMOVAL OF ADHERENT CELLS FROM THE SPLEENS OF CP-TREATED MICE

To determine whether the inhibitory activity exerted by adherent spleen cells from Cp-treated mice could explain, at least in part, the low levels of NK activity of their spleens, we examined whether removal of adherent suppressor cells by passage through a nylon column, could reverse the depressed NK activity. Although we removed almost all macrophages, as judged by latex particle ingestion, the levels of cytotoxic activity remained substantially low,

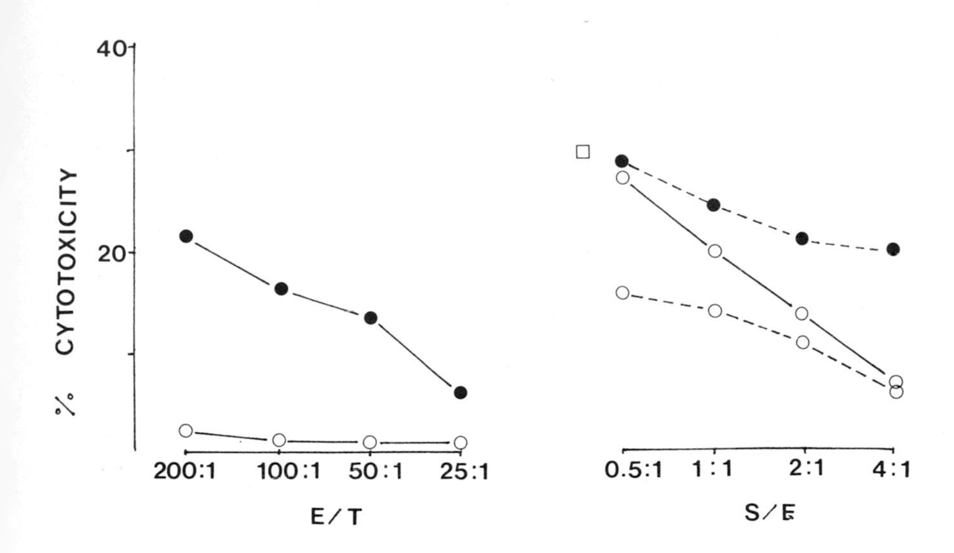

Figure 1. Supression of NK activity by adherent spleen cells from normal and Cp-treated 9-wk old C57Bl/6 mice. NK activity of normal and Cp-treated mice (left panel) and test for suppression of NK activity by mixture of normal 8-wk old CBA spleen cells (□) (E/T ratio 50:1) and spleen cells from Cp-treated mice (right panel). Spleen cells of untreated mice (●——●); cells of mice treated with Cp (2.1 mg/mouse, ip, day-10) (O——O); plastic adherent spleen cells of untreated mice (●---●); plastic adherent spleen cell from Cp-treated mice (O---O).

(4.1% lysis by unseparated, 7.0% by column-passed Cp-spleen cells cf 25.2% by untreated; 200:1 E/T vs. YAC-1), suggesting that the presence of adherent suppressor cells is not the only mechanism responsible for Cp-induced depression of NK activity.

SUPPRESSION OF NK ACTIVITY BY CP-INDUCED NONADHERENT SPLEEN CELLS

The failure to reverse the low levels of NK activity in the spleen of Cp-treated mice, after removal of adherent cells by passage through a nylon column, indicates that Cp

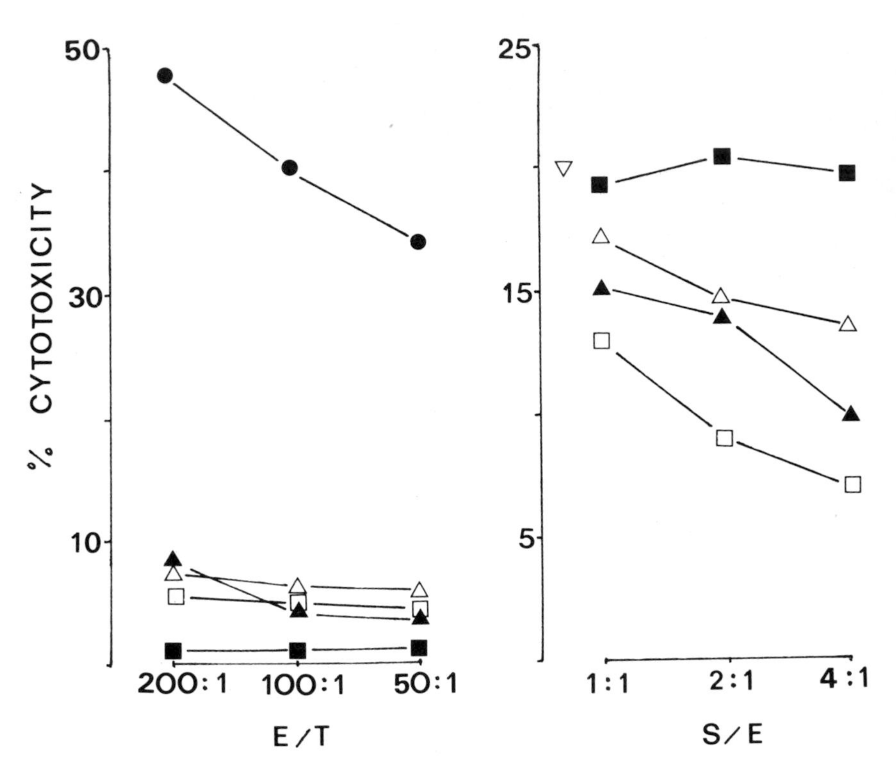

Figure 2. Inhibition of NK activity by Cp-induced non-adherent suppressor cells: effect of treatment of suppressor cells with anti-thy 1.2 plus complement. NK activity of normal and Cp-treated 8-wk old CBA/J mice (left panel) and suppression of NK activity by mixture of normal 7-wk old CBA spleen cells (▽) (E/T ratio 50:1) and spleen cells from Cp-treated mice (right panel). Spleen cells of untreated mice (●——●); nylon nonadherent spleen cells from Cp-treated mice (2.1 mg/mouse, ip, day-10) (△——△); nylon nonadherent spleen cells from Cp-treated mice, after treatment with anti-Thy 1.2 plus complement (□——□) or with complement alone (▲——▲); thymus cells of untreated mice (■——■).

either has a direct toxic effect on NK cells or it activates nonadherent suppressor cells, which cannot be physically separated from NK cells by column passage. To test the latter hypothesis, nylon or plastic nonadherent spleen cells from Cp-treated mice were tested for their ability to inhibit in vitro the lytic phase of NK activity (Figure 2). Nylon and plastic-nonadherent spleen cells from Cp-treated mice were found to be capable of inhibiting the effector phase of natural cytotoxicity. The suppressor cells were resistant to treatment with anti-Thy 1.2 (Fig. 2) or anti-asialo GM-1 antisera (not shown) plus complement, indicating that they were neither mature T cells nor mature, inactive NK cells. Further support for the conclusion that the suppressor cells were not mature NK cells was given by the finding of nonadherent suppressor cells in the spleen of Cp-treated beige mice, which are considered genetically deficient in NK activity.

FAILURE TO INDUCE SUPPRESSOR CELLS BY CP IN THE SPLEEN OF ATHYMIC NUDE MICE

Some of the biological effects caused by Cp have been found to depend upon T lymphocytes (Sljivic and Watson, 1977). To analyze whether the inhibitory effects on NK activity exerted by Cp-induced suppressor cells were also dependent on T cells, we examined the presence of Cp-induced suppressor cells in the spleen of athymic nude mice as well as in that of their heterozygous littermates. In contrast to the depression caused in euthymic mice, the decreased levels of NK activity in the spleen of Cp-treated nudes were not associated with detectable adherent suppressor cells (Figure 3). Adherent spleen cells from Cp-treated nude mice were not found to be capable of inhibiting NK activity, indicating that activation of such suppressor cells might depend on the presence of mature T cells. Moreover, removal of nylon- adherent cells substantially increased the low levels of NK activity exhibited by splenocytes of Cp-treated nudes (data not shown), ruling out a direct toxic effect of Cp on NK cells from nude mice. However, in preliminary experiments there has been some evidence for the presence of nonadherent suppressor cells in the spleens of Cp-treated nudes, which could be revealed only by depleting much of the NK activity by pretreatment with anti-Thy 1.2 + C.

MECHANISMS FOR INHIBITION OF NK ACTIVITY BY Cp-INDUCED SUPPRESSOR CELLS

The mechanisms by which suppressor cells for NK activity inhibit the lytic function of NK cells have not been yet determined, but pyran (Santoni _et al_, 1980a) as well as hydrocortisone (Hochman and Cudkowicz, 1979) and i-carra-

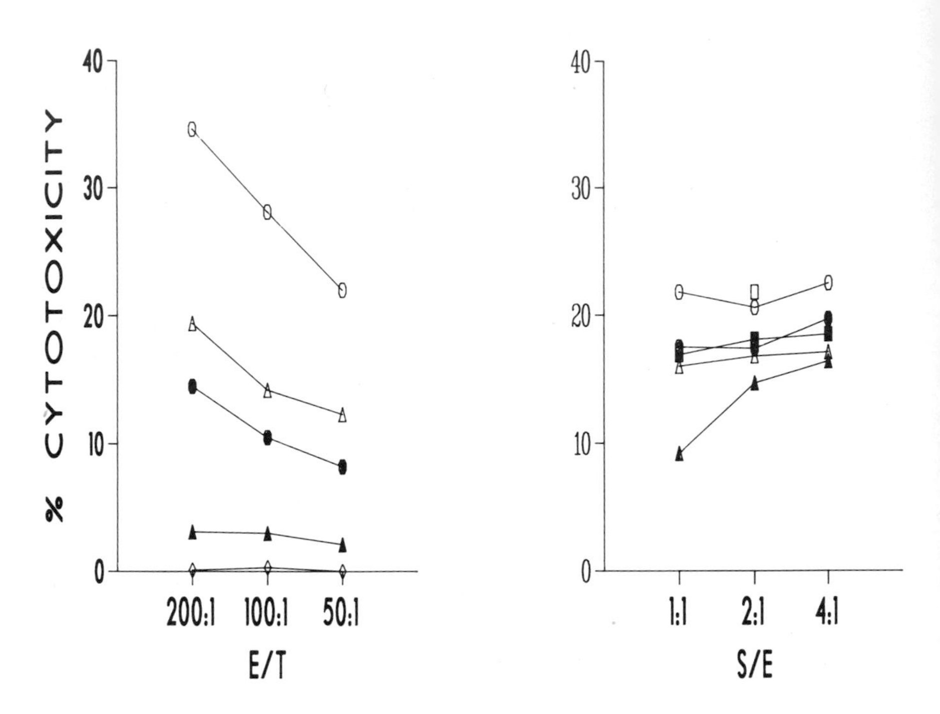

Figure 3. Failure of Cp to induce adherent suppressor cells in nude mice. NK activity of normal and Cp-treated mice (left panel): Spleen cells of untreated nu/+ BALB (△); from Cp-treated nu/+ (▲); from untreated nu/nu (O); and from Cp-treated nu/nu (●). Thymus cells from nu/+ (◇). Suppressor activity by mixture of normal 8-wk old CBA spleen cells, E/T ratio of 50:1 (□) and adherent spleen cells from normal and Cp-treated mice (right panel): Splenic adherent cells from untreated nu/+ (△); from Cp-treated nu/+ (▲); from untreated nu/nu (O); and from Cp-treated nu/nu (●). Thymus cells from nu/+ (■).

geenan (Hochman et al, 1981)-induced suppressor cells appear to act by release of undefined soluble mediators. Cell-free supernatants capable of inhibiting NK activity were also obtained by culturing spleen cells from Cp-treated mice at 37°C for 4 hrs. (data not shown). The nature of the inhibitory soluble factors remains to be determined. Prostaglandins are good candidates as mediators of the observed suppression although the addition of indomethacin during the cytotoxicity assay did not reverse the *in vitro* inhibition of NK activity by splenocytes of Cp-treated mice.

CP-INDUCED AUGMENTATION OF NK ACTIVITY IN THE PERITONEAL CAVITY

At the time when depression of NK activity occurred in the spleen of Cp-treated mice, increased levels of cytotoxicity were seen in the peritoneal cavity. Since adherent suppressor cells for NK activity were found to naturally occur in the peritoneal cavity of normal mice (Santoni *et al*; Brunda *et al*, this book), we analyzed whether their suppressive activity could be modified by Cp treatment. As reported elsewhere in this book, adherent suppressor cells from Cp-treated mice showed a reduced capacity to inhibit the effector phase of NK cells, suggesting that the inhibitory activity of naturally occurring peritoneal suppressor cells could be depressed by Cp.

CONCLUSION

The results of our studies indicate that the decreased natural cytotoxicity observed in the spleen, 7-14 days after Cp injection, could be attributed, at least in part, to the presence of two types of suppressor cells capable of inhibiting the lytic phase of NK activity: a) macrophage-like cells with adherent and phagocytic properties; b) null cells, i.e. nylon and plastic nonadherent, lacking Thy 1.2 and asialo GM-1 markers. Moreover the development of adherent suppressive activity appears to be thymic-dependent, since adherent suppressor cells could not be detected in the spleen of Cp-treated athymic nude mice. The augmentation of NK activity observed in the peritoneal cavity at the time when depression occurred in the spleen is consistent with previous findings (Ojo *et al*, 1978) that this agent can cause both stimulatory and inhibitory effects. The augmented levels of NK activity observed in the peritoneal cavity were correlated with a reduction of inhibitory activity mediated by adherent suppressor cells, that naturally occur in the peritoneal cavity of normal mice. Our results indicate that *in vivo* modulation of NK activity by Cp is a complex phenomenon, involving various components of the immune system. This agent might exert some of its effects by acting on cells with suppressor capacity, either activating them or reducing their activity. The development or maintenance of inhibitory activity might also depend on interactions of suppressor cells with other cells, such as T lymphocytes.

REFERENCES

Cudowicz, G. and Hochman, P.S. (1979). Immunol. Rev. 44:13.
Hochman, P.S. and Cudowicz, G. (1979). J. Immunol. 123:968.
Hochman, P.S., Cudkowicz, G. and Evans, P.D. (1981). Cell Immunol. 61:200.
Ito, M., Ralph, P. and Moore, M.A.S. (1980). Clin. Immunol. Immunopathol. 16:30.
Kirchner, H., Holden H.T. and Herberman, R.B. (1975). J. Immunol. 115:1212.
Lichtenstein, A., Murahata, R., Sugasawara, R. and Zighelboim, J. (1981). Cell. Immunol. 58:257.
Lotzova, E. and Gutterman J.U. (1979). J. Immunol. 123:607.
Ojo, E., Haller, O., Kimura, A. and Wigzell, H. (1978). Int. J. Cancer 21:446.
Santoni, A., Riccardi, C., Barlozzari, T. and Herberman, R.B. (1980). In "Natural Cell-Mediated Immunity Against Tumors" (R.B. Herberman, ed.), p. 753. Academic Press, New York.
Santoni, A., Riccardi, C., Barlozzari, T. and Herberman, R.B. (1980). Int. J. Cancer 26:837.
Savary, C.A. and Lotzova, E. (1978). J. Immunol. 120:239.
Sljivic, V.S. and Watson, S.R. (1977). J. Exp. Med. 145:45.
Wolfe S.A., Tracey, D.E. and Henney, C.S. (1976). Nature 262:584.

NATURAL SUPPRESSOR CELLS FOR MURINE NK ACTIVITY

A. Santoni
C. Riccardi
T. Barlozzari
R.B. Herberman

Biological Research and Therapy Branch
National Cancer Institute
Frederick, Maryland

The levels of natural killer (NK) activity have been found to be affected by multiple factors such as age, genetic background and environment. External and endogenous stimuli may exert their effects either by acting directly on NK cells or their precursors, or on accessory or suppressor cell populations. Soluble mediators as well as regulatory cells have been found to be involved in the regulation of NK activity.

Many of the inhibitory effects on NK activity appear to be due to the presence of suppressor cells. Cells capable of inhibiting the lytic activity of mouse NK cells have been generated after treatment with carrageenan (Hochman et al., 1981), corticosteroids (Hochman and Cudkowicz, 1979), x-irradiation (Hochman et al., 1978), adriamycin (Santoni et al, 1980a), Corynebacterium parvum (Lotzova, 1980), pyran copolymer (Santoni, et al., 1980b) and BCG (Ito et al., 1980). The nature of the suppressor cells has not been adequately defined, but macrophage-like (Santoni et al., 1980b) as well as nonadherent cells lacking easily detectable markers have been usually found to be involved. The mechanism(s) by which these suppressor cells inhibit NK activity has not yet been determined, but pyran-induced suppressor cells appear to act by release of soluble mediators (Santoni et al., 1980b). In addition to induction of suppressor cells by various agents, suppressor cells have been also detected in some natural situations. The low NK activity in newborn or aged mice may be attributable to presence of suppressor cells (Cudkowicz and Hochman, 1979). Furthermore some strains of mice with genetically determined

ISBN 0-12-341360-5

low NK activity have been found to have suppressors for NK activity (Riccardi et al., 1982). Finally a suppressor cell was isolated from the thymus of young adult mice (Nair et al., 1981).

We summarize here our studies dealing with the presence of suppressor cells in some natural situations and their possible role in the physiologic regulation of NK activity.

I. Age

In mice, NK activity is absent at birth, begins to appear around 3 weeks of age, peaks at 6 to 8 weeks and declines to low levels after 12 weeks of age (Herberman et al., 1975; Kiessling et al., 1975; Herberman and Holden, 1978). To analyze whether suppressor cells in the spleen of infant or aged mice might account for the low levels of NK activity, mixture experiments were performed as previously described by Cudkowicz and Hochman, (1979) and by us (Santoni et al., 1980b). Table 1 summarizes the results from a representative experiment with infant spleen cells. Addition of infant splenocytes to those of normal high NK-reactive mice resulted in a substantial inhibition of activity. In agreement with the findings reported by Cudkowicz and Hochman (1979), this inhibitory activity was associated with both adherent and nonadherent cells.

In contrast, the low NK activity of spleen cells of aged mice was not associated with detectable suppressor cells. Unseparated or adherent spleen cells of older mice were not found to be capable of inhibiting NK activity, suggesting that other mechanisms determined the low levels of NK activity in aged mice.

II. Genetics

Normal mice of various inbred strains have been shown to vary considerably in their levels of spontaneous NK activity (Herberman et al., 1975; Kiessling et al., 1975; Herberman and Holden, 1978). The mechanisms responsible for the low NK activity of strains such as SJL/J, A/J and ASW have not been determined. We explored the possibility that the low splenic activity in such strains was due, at least in part, to the presence of naturally occurring suppressor cells, capable of inhibiting the effector phase of NK activity. Adherent spleen cells from low NK-reactive SJL caused a substantial inhibition of NK activity, whereas no inhibitory activity was exerted by adherent spleen cells from high NK reactive CBA/J mice (Riccardi et al., 1982). Similarly, suppressor cells were detected in the spleen of other low

Table 1

In Vitro Inhibition Of NK Activity By Spleen Cells From Infant Mice

	% cytotoxicity			
	NK activity[a)]	Suppressor activity[b)]		
	50:1[c)]	4:1	2:1	1:1[d)]
Exp. 1 11 day-old $B6C3F_1$				
spleen cells	10.5	25.4	26.8	27.9
adherent spleen cells	1.5	17.9	24.5	26.0
Exp. 2 9 day-old $B6C3F_1$				
spleen cells	3.5	18.1	19.2	20.0
nylon nonadherent spleen cells	2.9	19.0	22.9	23.5

a) NK activity was assayed in a 4 hour-^{51}CRA against YAC-1 targets.
b) Mixture of spleen cells of high NK-reactive (E/T 50:1, 37%, exp. 1; 30%, exp. 2) mice with infant splenocytes.
c) E/T ratio.
d) S/E ratio.

NK-reactive strains such as A/J and ASW, but not in that of high NK-reactive strains C57BL/6 and $BD2F_1$ mice. Characterization studies indicated that the SJL suppressor cells were macrophages, since adherence procedures substantially enriched for inhibitory activity, whereas removal of phagocytic cells by treatment with iron and magnet removed most of inhibitory activity (Riccardi et al., 1982). Further more a possible role for suppressor T cells was virtually ruled out by the resistance of suppressor cells to treatment with monoclonal Thy 1.2 plus complement. Moreover neonatal thymectomy didn't lead to augmentation of levels of NK activity in SJL mice nor to an impairment of inhibitory activity of splenic suppressor cells, further ruling out a role for T lymphocytes in the suppression of the lytic phase of NK activity either as effector or regulatory cells (Fig. 1). The inhibition of NK activity by SJL macrophages was found to be mediated by production or release of soluble factors (Riccardi et al., 1982). The nature of the soluble factors is still unknown and experiments are in progress to determine the possible role of prostaglandins.

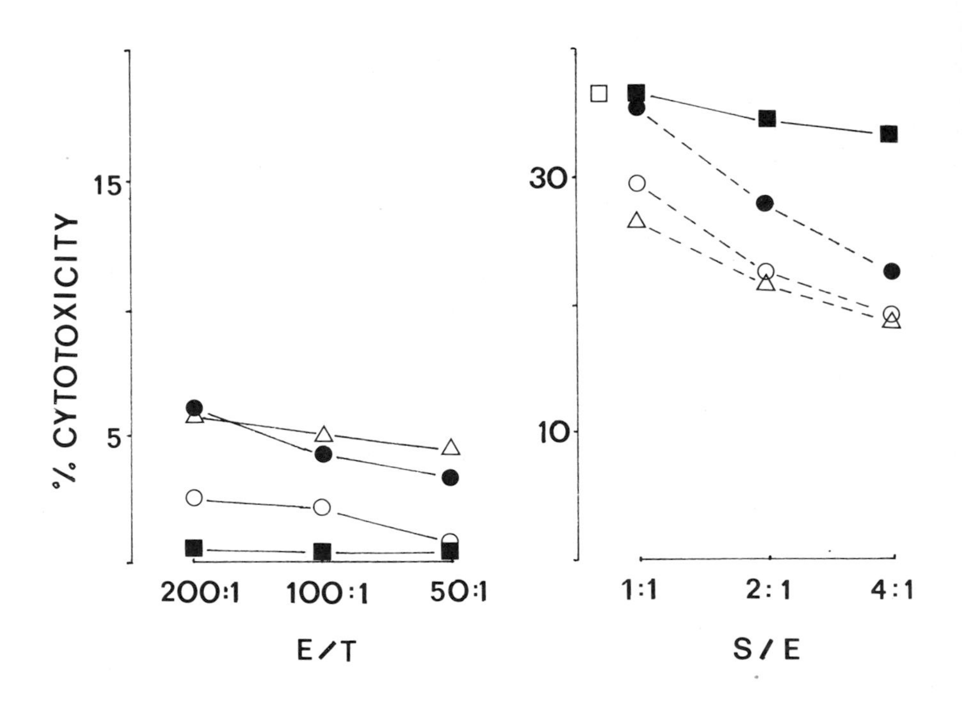

FIGURE 1. Suppression of NK activity by adherent spleen cells from normal, sham operated, and neonatally thymectomized 8 week-old SJL/J mice. NK activity of normal and treated mice (left panel) and test for suppression of NK activity by mixture of normal 8 week-old CBA spleen cells (□) (E/T ratio 50:1) and adherent spleen cells from SJL mice (right panel). Spleen cells of untreated SJL/J (△); spleen cells of sham operated SJL/J (●); spleen cells of thymectomized SJL/J mice (○); normal thymus cells (■); adherent spleen cells from normal SJL mice (△--△); adherent spleen cells from sham-operated SJL/J mice (●---●); adherent spleen cells from thymectomized SJL mice (○---○).

III. Organ Distribution

Murine NK activity has a characteristic tissue distribution, with relatively high NK activity present in the spleen and peripheral blood and little or no activity in the thymus or peritoneal cavity (Herberman *et al.*, 1975; Kiessling *et al.*, 1975; Herberman and Holden, 1978). We were interested in examining whether macrophages, which have been found to mediate the inhibition of the lytic phase of NK cells in various circumstances, could be also responsible in part for the low levels of NK activity exhibited by peritoneal exudate cells (PEC). Adherent cells with the characteristics of macrophages capable to suppress NK activity were found in the peritoneal cavity, but not in the spleen of high NK-reactive mice (C57BL/6, $BD2F_1$, CBA/J) as well as low (A/J and SJL/J) NK-reactive, and also in athymic nude mice, indicating that the thymus may not play a role in the generation of naturally occurring peritoneal suppressor cells (Table 2). Suppressive activity could be modulated by either *in vivo* or *in vitro* treatment with interferon (IFN) or IFN-inducers. Thus, *in vivo* treatment with C. parvum, which resulted in augmentation of NK cell activity in the peritoneal cavity, caused also a reduction of inhibitory activity by adherent PEC. Furthermore adherent PEC from normal mice had a reduced ability to suppress NK activity when treated *in vitro* with IFN or Poly I:C. With regard to the mechanisms of inhibition, as in other situations, peritoneal macrophages seemed to suppress NK activity via production or release of soluble mediators.

CONCLUDING REMARKS

The frequent association between low NK activity and suppressor cells suggests that inhibition by macrophages or other cells at the effector phase may account for, or at least contribute to, the low cytolytic activity. A possible causal relationship between the suppressor cells and low NK activity in SJL mice is further supported by our recent demonstration of *in vivo* suppression of NK activity in SJL recipients of NK cells from high NK strains and *in vivo* transfer of suppressor activity by adherent cells of SJL (Riccardi *et al.*, in preparation). The failure to detect suppressor cells in the spleen of old mice of high NK strains suggests that other factors, such as a reduced pool of NK cells, a lower proportion of target-binding cells, or a decreased efficiency of lytic machinery, may contribute to the decline of NK activity in the aged mice.

Table 2

In Vitro Inhibition Of NK Activity By Adherent PEC From High NK-Reactive CBA/J and Low NK-Reactive SJL Mice From BALB/c Nudes

		% cytotoxicity		
		NK activity	Suppressor activity[a]	
		50:1	2:1	1:1
Exp. 1	8 wk CBA/PEC,	5.7	14.0	18.5
	spleen cells	23.5	21.5	22.7
	thymus cells	0.2	23.0	23.4
Exp. 2	8 wk SJL/PEC,	1.7	12.1	18.3
	thymus cells	0.3	29.5	28.8
Exp. 3	8 wk nu/nu PEC,	5.5	9.9	11.4
	8 wk nu/+ PEC	0.0	9.8	14.5
	thymus cells	0.0	18.1	18.5

a) Mixture of adherent PEC, adherent splenocytes or thymus cells with normal splenocytes (E/T 50:1) of high NK-reactive strains.

The presence of suppressor cells in the peritoneal cavity, a site with low NK activity, but not in the spleen, an organ with high NK activity, suggests a possible regulatory role for these cells. Indeed PEC have been found to be sensitive targets for NK activity, whereas no such targets are found in the spleen (Nunn *et al.*, 1977). Therefore suppressor cells could be needed in the peritoneal cavity to inhibit the expression of cytolytic activity against normal cells. However the inhibition of adherent PEC could be partially overcome by some agents. *In vivo* treatment with C. parvum caused boosting of NK activity in the peritoneal cavity as well as a reduction of suppressor activity mediated by peritoneal suppressor cells. The decrease in inhibitory activity following treatment with IFN and IFN-inducers could be due to effects of IFN directly on suppressor cells or to activation of other cells present in the population, to produce factors capable of balancing the suppressive effects. Taken together, our results indicate that the presence of naturally occurring suppressor cells could be one of the mechanisms by which NK activity is regulated *in vivo*.

REFERENCES

Cudkowicz, G. and Hochman, P.S. (1979). Immunol Rev. 44:13.

Herberman, R.B., Nunn, M.E. and Lavrin, D.H. (1975). Int J Cancer. 16:216.

Herberman, R.B. and Holden, H.T. (1978). Adv Cancer Res. 27:305.

Hochman, P.S., Cudkowicz, G. and Dausset, J. (1978). J Natl Cancer Inst. 61:265.

Hochman, P.S., Cudkowicz, G. (1979). J Immunol. 123:968.

Hochman, P.S., Cudkowicz, G. and Evans, P.D. (1981). Cell Immunol. 61:200.

Ito, M., Ralph, P. and Moore, M.A.S. (1980). Clin Immunol Immunopathol. 16:30.

Kiessling, R., Klein, E. and Wigzell, H. (1975). Eur J Immunol. 5:112.

Lotzova, E. (1980). In "Natural cell-mediated immunity against tumors." Ed. by R.B. Herberman. Acad Press. p. 753.

Nair, M.P.N., Schwartz, S.A., Fernandes, G., Pahwa, R., Ikehara, S. and Good, R.A. (1981). Cell Immunol. 58:9.

Nunn, M.E., Herberman, R.B. and Holden, H.T. (1977). Int J Cancer. 20:381.

Riccardi, C., Santoni, A., Barlozzari, T., Cesarini, C. and Herberman, R.B. (1982). Int J Cancer, in press.

Santoni, A., Riccardi, C., Sorci, V. and Herberman, R.B. (1980a). J Immunol. 126:2329.

Santoni, A., Riccardi, C., Barlozzari, T. and Herberman, R.B. (1980b). Int J Cancer. 26:837.

SUPPRESSION OF MURINE NATURAL KILLER CELL ACTIVITY BY NORMAL PERITONEAL MACROPHAGES

Michael J. Brunda
Donatella Taramelli
Howard T. Holden
Luigi Varesio

Biological Development Branch
Biological Response Modifiers Program
National Cancer Institute - FCRF
Frederick, MD

It has been suggested that natural killer (NK) cells have a wide range of biological activities, including immunosurveillance against neoplastic cells (Herberman, 1981), protection against viral infections (Bancroft et al, 1981), and production of lymphokines, such as interferon (IFN) (Trinchieri et al, 1978). Regulatory systems most probably exist to control the cytolytic potential of these cells, both in normal animals and in animals which have been manipulated by experimental treatment or which have pathological conditions. In analogy with other branches of the immune system, NK activity might be controlled through several types of interactions with immunoregulatory cells.

One cell type that has been shown to influence, either positively or negatively, a variety of immune responses is the macrophage (MØ) (Unanue, 1978; Oehler et al, 1978). There are several lines of evidence suggesting that MØ might be involved in suppression of NK activity: 1) MØ can produce factors such as prostaglandins, which have been shown to inhibit NK activity (Brunda et al, 1980); 2) MØ suppress a wide range of other immunological parameters, such as the generation of cytotoxic T cells (Klimpel and Henney, 1978) and the production of lymphokines (Varesio and Holden, 1980), at relatively low numbers (5 to 10% relative to the number of effector cells); 3) MØ are generally found in high numbers locally at the site of tumor growth, where it has been difficult to demonstrate NK activity (Gerson, 1980);

ISBN 0-12-341360-5

TABLE I. Inhibition of Spontaneous or IFN-Augmented NK Activity by Peritoneal MØ[a]

Percentage of Peritoneal MØ added	% Cytotoxicity	
	-IFN	+IFN
0	21	45
10	0	8
3.3	8	23
1.1	19	28

[a]CBA/J peritoneal MØ were incubated for 18 hr with nylon wool nonadherent CBA/J spleen cells (shown as % relative to the number of spleen cells) in the presence or absence of 1000 U of mouse IFN. The cells were washed and tested against YAC-1 in a 4 hr assay. Shown are data from 100:1 effector cell:target (E:T) cell ratio.

and 4) splenic MØ from mice with low NK activity can, at high suppressor cell to effector cell ratios (1:1 to 4:1), inhibit NK activity during the cytotoxicity assay (Cudkowicz and Hochman, 1979; Santoni et al, 1980).

Since NK activity is present in normal animals, we examined the possibility that normal macrophages regulate NK activity. Using a system where NK activity is either maintained in vitro or augmented by the addition of IFN, we have found that peritoneal MØ from normal animals can markedly suppress NK activity. Nylon wool nonadherent splenic effector cells were incubated for 18 hr either alone or in the presence of 1.1 to 10% syngeneic peritoneal MØ, enriched by adherence, from normal unstimulated mice (Table I). With 10% MØ, a striking decrease in NK activity was observed over a wide range of E:T ratios. As the precentage of MØ added was decreased, less suppression was observed. Similar experiments were performed in the presence of exogenous mouse IFN. In the absence of MØ, IFN induced a significant augmentation of NK activity, as previously described (Djeu et al, 1979). However, when MØ were added to the 18 hr incubation mixture, there was a marked dose-dependent inhibition of cytotoxicity. At lower numbers of MØ, a greater inhibition was seen with IFN augmented cells than with spontaneously maintained cells. In more than 20 experiments, the median inhibition of cytotoxicity by 10% MØ was greater than 65%.

It has been previously reported that suppressor cells of varying cell types, including MØ (Cudkowicz and Hochman, 1979; Santoni et al, 1980; Savary and Lotzova, 1978; Nair et

TABLE II. Failure of Normal Peritoneal MØ to Suppress NK Activity During the Effector Phase

MØ added during pre-incubation	MØ added to assay	Exp.1[a]	Exp.2[b]
-	-	23,25[c]	41
+	-	6	25
-	+	21	41

[a]CBA/J peritoneal MØ (10%) were either incubated for 18 hr with effector cells which were then tested for cytotoxicity or were added directly to a 4 hr cytotoxicity assay. Data shown are % cytotoxicity at 100:1 E:T vs. YAC-1.

[b]Effector cells were incubated for 18 hr in the presence of 1000 U of IFN. MØ were added during the pre-incubation or MØ, incubated alone with 1000 U of IFN for 18 hr, were added to the assay.

[c]Shown are results of direct cytotoxicity assay and following 18 hr pre-incubation.

al, 1981), can inhibit NK activity at the effector phase. The ability of normal peritoneal MØ to inhibit the effector phase of NK activity was examined by mixing effector cells in a 4 hr cytotoxicity assay with 1) freshly harvested peritoneal MØ (Table II, Exp. 1) or with 2) MØ incubated for 18 hr alone (Exp. 1) or with IFN (Expt. 2). In all combinations, addition of MØ to the cytotoxicity assay in these numbers resulted in no significant inhibition of cytotoxicity while incubation of the same MØ for 18 hr with the effector cells caused a marked inhibition of cytotoxicity. These results indicated that the inhibition of cytotoxicity observed in our experiments was not due to suppression during the effector phase.

Since resident peritoneal MØ inhibited NK cell-mediated cytotoxicity, we tested whether the function of other cytolytic cells, i.e. alloimmune T cells, were similarly inhibited by peritoneal MØ. To generate cytotoxic T cells, CBA spleen cells were immunized *in vitro* with P815, a target cell that is not sensitive to murine NK activity in a 4 hr *in vitro* assay. As shown in Table III, pre-incubation for 18 hr with MØ suppressed NK activity but had little inhibitory effect on alloimmune T cells, indicating that this MØ-mediated suppression was not directed against all cytolytic cells.

TABLE III. Comparison of the Ability of MØ to Suppress NK and T Cell-Mediated Cytotoxicity[a]

Effector cells	MØ added	% Cycotoxicity	Target Cell
Normal spleen	-	19	YAC-1
	+	0	
Anti-P815 spleen	-	28	P815
	+	24	

[a]CBA/J spleen cells or CBA/J spleen cells immunized _in vitro_ with P815 were incubated for 18 hr with 1000 U of IFN in the presence (+) or absence (-) of 10% CBA/J peritoneal MØ. Cytotoxicity was assayed against YAC-1 (E:T of 100:1) or P815 (E:T or 12:1). Normal splenic effector cells exhibited no cytotoxicity against P815.

To examine the ability of various MØ populations to suppress NK activity, MØ from the peritoneal cavity and spleen of both normal and _C. parvum_ injected mice were compared. As seen in Table IV, the peritoneal cavity of normal mice had a low level of NK activity while high levels were found in the spleen. However, following injection of _C. parvum_, a shift occurred resulting in high NK activity in the peritoneal cavity and a low level in the spleen. No cytotoxicity was found with MØ- enriched populations from these organs. When these MØ were tested for their ability to suppress NK activity, normal peritoneal MØ from _C. parvum_-injected mice had a reduced ability to suppress. MØ from _C. parvum_-injected mice were functionally activated since they could inhibit mitogen-induced lymphoproliferation and lyse tumor target cells _in vitro_ (data not shown), properties absent in normal MØ. Splenic MØ from either normal or _C. parvum_-injected mice had little or no inhibitory effect. Two interesting inverse correlations were seen in these data: 1) in normal mice, the spleen had high NK and low MØ suppressor activity, while the peritoneal cavity had low NK and high MØ suppressor activity; and 2) in _C. parvum_-injected mice, the peritoneal cavity had high NK and low MØ-mediated suppression. There was no correlation between NK activity and suppression in the spleens of _C. parvum_-injected mice; these results do not rule out the possibility that other cell types or mechanisms are responsible for suppression in the spleens of these animals (see Santoni _et al_, this book).

Our results demonstrate that normal peritoneal MØ have the ability to suppress either spontaneously maintained or IFN-augmented NK activity _in vitro_. Suppression by normal

TABLE IV. Comparison of the NK Activity and MØ Mediated Suppression Present in Normal or _C. Parvum_-Injected Mice

Population	NK activity[a] Whole population	MØ	% Cytotoxicity[b] of effectors with MØ
-	-	-	33
Normal peritoneal	3[c]	0	11
C. parvum peritoneal	24	0	24
Normal spleen	22	0	30
C. parvum spleen	2	0	26

[a]CBA/J peritoneal or spleen cell populations were obtained from normal mice or from mice injected 10 days earlier with 700 μg of _C. parvum_. NK activity of the whole cell population or of MØ was assayed against YAC-1 in 4 hr assay at E:T of 100:1 and is expressed as % cytotoxicity.

[b]MØ (10%) were added to nylon wool nonadherent spleen cells in the presence of 1000 U of IFN and incubated for 18 hr. The cells were washed and assayed for NK activity in a 5 hr assay at E:T of 100:1.

peritoneal MØ was obtained at lower suppressor cell to effector cell ratios than in other studies showing suppression at the cytolytic effector phase. Whether suppression of the maintenance and IFN augmentation of NK activity and suppression of the cytolytic effector phase work through common pathways is not known and will await future studies on the precise mechanism(s) of suppresion in each of these systems.

Although the potential _in vivo_ regulatory role of MØ on NK activity has not been established, it is tempting to speculate that peritoneal MØ may control the level of NK cell-mediated cytotoxicity. Such a possibility is strengthened by the inverse correlation between the levels of cytotoxicity in the peritoneal cavity and spleen and the ability of MØ from these sites to suppress NK activity. Peritoneal MØ may function to inhibit levels of NK activity and prevent the lysis of normal cells in the peritoneal cavity. It has been demonstrated that peritoneal cells, but not spleen cells, from normal mice are sensitive to NK cell-mediated lysis (Welsh _et al_, 1979).

Following stimulation with _C. parvum_, NK activity in the peritoneal cavity is increased while the suppressor activity of MØ is simultaneously decreased. Again, this inverse correlation suggests an interrelationship between MØ and NK cells in the peritoneal cavity. In the experimentally treated animal or in certain pathological states, the altered functional properties of MØ may be of value to the host. For example, in tumor bearing mice injection *C. parvum* could result in 1) a direct activation of MØ leading to enhanced cytolysis of tumor cells by this population, and 2) a change in the suppressor function of MØ causing an increase in NK activity which in turn could contribute to the destruction of neoplastic cells. Such an effect may also occur with other biological response modifiers.

In summary, normal peritoneal MØ have the ability to suppress NK activity *in vitro* and this ability is altered by injection of *C. parvum*. This *in vitro* system may be of value in analyzing both the functional interrelationships of MØ with NK cells and the effect of various biological response modifiers on the regulation of NK activity.

REFERENCES

Bancroft, G.J., Shellam, G.R., and Chalmer, J.E. (1981). J. Immunol. 126:988.

Brunda, M.J., Herberman, R.B., and Holden, H.T. (1980). J. Immunol. 124:2682.

Cudkowicz, G. and Hochman, P.S. (1979). Immunol. Rev. 44:13.

Djeu, J.Y., Heinbaugh, J.A., Holden, H.T., and Herberman, R.B. (1979). J. Immunol. 123:182.

Gerson, J.M. (1980). *In* "Natural Cell-Mediated Immunity Against Tumors" (R.B. Herberman, ed.), p. 1047. Academic Press, New York.

Herberman, R.B. (1981). Clin. Immunol. Rev. 1:1.

Klimpel, G.R. and Henney, C.S. (1978). J. Immunol. 120:563.

Nair, M.P.N., Schwartz, S.A., Fernandes, G., Pahwa, R., Ikehara, S., and Good, R.A. (1981). Cell. Immunol. 58:9.

Oehler, J.R., Herberman, R.B., and Holden, H.T. (1978) Pharmacol. Ther. (A) 2:55.

Santoni, A., Riccardi, C., Barlozzari, T., and Herberman, R.B. (1980). Int. J. Cancer 26:837.

Savary, C.A., and Lotzova, E. (1978). J. Immunol. 120:239.

Trinchieri, G., Santoli, D., Dee, R.R., and Knowles, B.B. (1978). J. Exp. Med. 147:1299.

Unanue, E.R. (1978). Immunol. Rev. 40:227.

Varesio, L. and Holden H.T. (1980). Cell Immunol. 56:16.

Welsh, R.M., Jr., Zinkernagel, R.M., and Hallenbeck, L.A. (1979). J. Immunol. 122:475.

REGULATION OF NK REACTIVITY BY SUPPRESSOR CELLS

M. Zöller[1]
S. Matzku[2]
G. Andrighetto[3]
H. Wigzell[1]

[1]Department of Immunology, University of Uppsala
[2]Institute of Nuclear Medicine, German Cancer Research Center
[3]Istituto Immunopatologia, University of Padua, Verona.

INTRODUCTION INTRODUCTION

In the preceeding volume (1), four mechanisms of "down regulation" of NK activity are discussed: a) elimination of NK cells, b) inhibition of NK cells, c) inhibition of accessory cells, and d) generation of suppressor cells. All four mechanisms seem to be possibly effective under certain circumstances (reviewed in 2,3).

MATERIAL AND METHODS

Animals: CBA/H mice, DA and BD X rats entered the experiments at the age of 6-8 (mice) and 8-12 (rats) weeks, respectively.

Target cells: In NK assays YAC (A/Sn lymphoma) and G-1 (W/Fu lymphoma), for T-killer cells Con-A blasts and in the ADCC system chicken red blood cells (CRBC) coated with rabbit anti-CRBC hyperimmune serum were used.

Effector cells: Thymus, lymph nodes (LN), spleen, bone marrow (BM), peripheral blood lymphocytes (PBL), and peritoneal cells (PC) were prepared like usual and fractionated by discontinuous Percoll gradient centrifugation (4). Fractions of d 1.067, 1.077, 1.090, and 1.121 were harvested.

Cytotoxicity assay: A 4h 51-Cr release assay was used throughout. Cytotoxicity was calculated as % cytotoxicity = 100 x ((%test release - % medium release):(% maximum release - % medium release)). In inhibition experiments putative suppressor

ISBN 0-12-341360-5

cells were added together with the effector cells and inhibition was calculated as % inhibition = 100 x (% "expected cytotoxicity" - % observed cytotoxicity): % "expected cytotoxicity". "Expected cytotoxicity" is identical with the % cytotoxicity of test wells without inhibitory cells, in case the inhibitory cell does not display any cytotoxicity by itself. In case the inhibitory population shows some cytolytic potential the relative cytolytic capacity of the inhibitory population in relation to the effector population was calculated based on the expectation of a linear slope changing effector to target ratios. Linear slopes are normally observed in the range of specific release from 15%-85%. This calculation of comparing cytotoxic values using different numbers of effector cells is somewhat problematic with respect to crowding effects, but this possibility was excluded, as described elsewhere (M. Zöller and H. Wigzell, submitted).

RESULTS

Normally occurring inhibitory cells for NK activity in lymphoid organs: When analysing NK activity in various Percoll density fractions we observed a) that some fractions contained higher NK activity per unit cell number than the original population. This was at first glance not surprising as NK cells have been found to display densities allowing them to be separated from certain other cell types (5); b) significant NK activity could be recovered also from organs which were normally almost or completely devoid of NK activity before fractionation such as the thymus. Again this could have a trivial explanation such as contamination by blood or minor adherent lymph nodes, but this seemed too simple due to a second finding. The peak activity of NK function was not the same in the various lymphoid organs, being skewed to the lighter density fractions in the "low" NK organs. Experiments demonstrating the above observations are shown in Table 1.

Among the explanations that exist for the findings in Table 1, one would be that there exists a relatively high density inhibitory cell for NK lysis in some lymphoid organs which would then automatically cause the skewing, depending on the relative ratio of NK cells versus inhibitory cells. A search for such hypothetical inhibitory cells was carried out by admixing various density fractionated cells to the 1.077 fraction of normal spleen cells which contains the highest NK activity. Results of such experiments using a fixed 100:1 effector: target cell ratio with admixture of various numbers of potential inhibitory cells are shown in Table 2.

Table 1

NK ACTIVITY OF PERCOLL FRACTIONATED LYMPHOID CELLS

Effector cell donor	Percoll fraction	% cytotoxicity with*					
		thymus	LN	spleen	BM	PC	PBL
BD X	unfract.	7	9	59	13	66	80
	d 1.067	31	26	70	70	77	76
	d 1.077	7	8	80	14	81	94
	d 1.090	-2	0	13	5	78	72
	d 1.121	2	4	10	3	19	18
CBA/H	unfract.	5	10	58	34	nt**	nt
	d 1.067	11	17	69	47		
	d 1.077	4	15	72	27		
	d 1.090	1	3	12	8		
	d 1.121	1	4	7	7		

* Target cells for BD X: G-1, for CBA/H: YAC, E:T = 100:1;

** nt: not tested

Table 2

ORGAN AND DENSITY DISTRIBUTION OF NK ACTIVITY INHIBITORY CELLS

Effector cells[*]	"Inhibitory" cells	E: I:T	% cytotoxicity (G-1): "inhibitory" cells: unfract.	1.067	1.077	1.090	1.121
-	Thymus	-:50:1	2	7	-1	-3	0
	LN	-:50:1	3	16	12	2	1
	Spleen	-:50:1	36	37	49	3	3
	BM	-:50:1	8	29	13	1	-1
	PC	-:50:1	49	nt[**]	86	61	44
	PBL	-:50:1	36	76	69	36	19
Spleen d 1.077	-	100: -:1			61		
"	Thymus	100:50:1	53(13)[***]	43(34)	12(80)	8(87)	13(79)
"	LN	100:50:1	57(8)	59(8)	12(81)	10(84)	15(75)
"	Spleen	100:50:1	56(16)	61(9)	69(-2)	13(79)	24(61)
"	BM	100:50:1	51(15)	61(8)	53(17)	12(80)	0(100)
"	PC	100:50:1	69(0)	nt	69(4)	73(-4)	69(-4)
"	PBL	100:50:1	68(-3)	76(7)	70(7)	69(-4)	66(-3)

[*] Effector and "inhibitory" cells from BD X rats; [**] nt: not tested;

[***] in brackets: % inhibition

Strong indications of inhibitory cells being present in density fractions 1.090 and 1.121 were found in cell populations regardless of origin, except PC and PBL, with inhibition extending into the lower density fractions in the case of thymus and LN cells in particular. It is unlikely that the present results were due to crowding since significant inhibition was noted using admixtures as low as 5% of 1.090 thymus, spleen, LN or BM cells (data not included). Likewise, similar experiments at 50:1 effector:target ratios yielded similar inhibition results.

It could be argued that the Percoll fractionation as such induced suppressor cell activity. This is shown not to be the case, as fractionation followed by recombination of all fractions yielded close to identical NK levels as the starting population (Table 3).

Table 3

PERCOLL FRACTIONATION IN ITSELF DOES NOT GENERATE SUPPRESSOR CELLS*

% cytotoxicity (G-1) with DA spleen cells (E:T = 100:1)			
unfract.: 67	d 1.067: 72	d 1.067 +	: 68
	d 1.077: 78	d 1.077 +	
	d 1.090: 32	d 1.090 +	
	d 1.121: 10	d 1.121	

*to be published elsewhere (M.Zöller and H.Wigzell, submitted)

As the density fractions 1.090 and 1.121 comprise around 60-70% and 3-7% respectively of the investigated cell populations obtained from the various organs we must then conclude that the inhibition observed represents a true physiological one from the point of numbers of cells available under in vivo conditions.

Further experiments disclosed the existence of a similar type of inhibitory cell in rat lymphoid organs, with no changes in the inhibitory cell activity in relation to age of the individual and no influence on the inhibitory cell by interferon. The latter study thus excluded that interferon activation of NK activity in part may be due to reduction of inhibition. No MHC or species restriction of the inhibition was noted (M. Zöller and H. Wigzell, to be published).

Mechanism of inhibition: We have failed to any evidence for a soluble mediator released into the supernatant by suppressor cell-enriched populations. Thus, contact between cells is seemingly required. Likewise, we have failed to obtain evidence that the suppressor cells can serve as targets for NK cells using 51-Cr-labelled, suppressor-enriched populations (data not included). Data have been obtained, however, which we interpret to indicate that the present suppressor cell may be endowed with a more general inhibitory capacity for cytolytic cells of mononuclear types. This is shown in Table 4, where the very same density profile of inhibitory cells were found to exist when testing for inhibition against effector cells in IgG-induced ADCC against erythrocytes and against allo-MHC specific cytolytic T cells. Our data as to surface markers and other features of the inhibitory cells in the three systems indicate that they most likely are one and the same type of cells.

Table 4

INHIBITION OF ALLO-MHC T-KILLER CELLS AND ADCC EFFECTOR CELLS

a:

Effector cell	Inhibitory cell	E: I:T	% cytotoxicity BN blasts	BD X blasts	G-1
BD X à BN	-	20: -:1	28	2	1
-	BD X thymus d 1.090	-:10:1	0	0	0
BD X à BN	BD X thymus d 1.090	20:10:1	10	-1	0

b:

Effector cell(CBA/H)	Inhibitory cell(CBA/H)	E: I:T	% cytotoxicity CRBC à CRBC
Spleen	-	100: -:1	33
-	Thymus	-:50:1	0
	d 1.090	-:25:1	0
Spleen	Thymus	100:50:1	7
	d 1.090	100:25:1	22

Surface markers on the inhibitory cell: We have used several procedures to delineate the suppressive cell as to possible lineage using surface markers. Table 5 represents a summary of our results. The inhibitory cell appears to be a high density cell, lacking conventional markers for B cells (surface immunoglobulin) or monocytes (Fc receptor for IgG), whilst having some markers more in line with T cells (Thy-1 positive in the mouse, Helix pomatia A agglutinin positive). However, Lyt-1 and 2 markers have been negative for the mouse inhibitory cell, making the similarity to T lymphocytes less clearcut. In fact, with the presently studied markers there is nothing to exclude that the inhibitory cells are very similar to NK cells as to surface features. More select antisera (anti-asialoGM_1, anti-NK-1) will be applied when available to study this further.

Table 5

CHARACTERIZATION OF NK ACTIVITY INHIBITORY CELLS

	Mouse	Rat
Density	1.090	1.090
Adherence	-	-
Phagocytosis	-	-
Radioresistance	+	+
In vitro survival	<12h	<12h
Cortisone resistance	+	+
Surface immunoglobulin	-	-
Fc-receptor (IgG)	-	-
Helix pomatia agglutinin receptor	+	+
Thy-1*	+	-
Lyt-1	-	nt**
Lyt-2	-	nt
Ly-1*	nt	+
Ly-2*	nt	+

*Thy-1.1 is found on rat pre-T cells (7), Lyt-1 and Lyt-2 are peripheral T cell markers in the rat (8).
**nt; not tested.

DISCUSSION

In the present study we have presented evidence for the existence of a cell endowed with suppressive features for most mononumclear cytolytic effector cell types (here analyzed as NK, CTL, or K cell functions). The mode of inhibition is unknown but does seem to require cell contact. It is reversible as indicated by the fractionation experiments but it is not induced by the density fractionation itself. The fraction of cells in which this suppressor cell is found constitutes the dominating cell group in the lymphoid/hemopoetic cells studied and there is little doubt that this cell can severely influence the outcome of quantitative tests in vitro. The fact that PBL and PC seem very low in this cell type may well explain why the most efficient CTLs can be recovered from PC and PBL after immunization (6). The inhibitory cell is not regulated by the same physiological mechanisms as NK cells (age, interferon), yet it has superficially comparatively similar surface features as to the so far analyzed markers. The suppressor cell is also very short lifed in vitro (almost lost within 12h, unpublished observation). We would thus conclude that the determination of NK activity in various organs/cell populations may be significantly influenced by normally occurring cells within the same population acting in vitro as suppressor cells. We have no evidence that the present inhibitory cell plays any role in vivo as this would require selective means (mutants, reagents) that would enhance/deplete these cells. The fact, however, that they can seemingly serve as general inhibitors for mononuclear killer cells may provide an inlet to a study for a common pathway in cell-mediated cytolytic reactions.

ACKNOWLEDGMENT

Grant support: This investigation was supported by the Deutsche Forschungsgemeinschaft (M.Z.), the Wilhelm-Meyenburg Stiftung (M.Z., S.M.), the Swedish Cancer Society and NIH grant NIH-26752-02.

REFERENCES

1. Herberman, R. B., in Herberman, R. B. (ed.). Natural cell-mediated immunity against tumors. Academic Press, New York, 1980, p.779.
2. Cudkowicz, G. and Hochman, P. S., Imm. Rev. 44, 13 (1979).
3. Herberman, R. B., et al., in Serrou, B., Rosenfeld, C., and Herberman, R. B. (eds.), NK cells: Fundamental aspects and role in cancer. Human cancer immunology. Elsevier/North-Holland, Amsterdam, in press.
4. Zöller, M. and Matzku, S., J. Nat. Can. Inst. 65, 769 (1980).
5. Kurnick, J. T., et al., Scand. J. Imm. 10, 536 (1979).
6. Berke, G., Sullivan, K. A., and Amos, B., J. Exp. Med. 135, 1334 (1972).
7. Williams, A. F., Eur. J. Imm. 6, 526 (1976).
8. Wonigeit, K., Transplant. Proc. 11, 1334 (1979).

REGULATION OF IN VIVO REACTIVITY OF NATURAL KILLER (NK) CELLS*

C. Riccardi,[1] T. Barlozzari,[1,2] A. Santoni[1]
C. Cesarini[1] and R.B. Herberman[2]

[1]University of Perugia, Perugia, Italy
and
[2]Biological Research and Therapy Branch
National Cancer Institute
Frederick, Maryland

In the last few years we have studied the possible *in vivo* role of NK cells, using a rapid *in vivo* assay to measure the spontaneous reactivity of non-immune mice against ^{125}IUdR-labeled tumor cells (Riccardi *et al.*, 1981b). We found a series of correlations between the ability of mice to clear tumor cells *in vivo* and the levels of *in vitro* NK activity (Riccardi *et al.*, 1979; Riccardi *et al.*, 1980a,b). Parallelism between NK activity and *in vivo* destruction of tumor cells was evident for a variety of characteristics such as: 1) Rapidity of killing, 2) Distribution of reactivity among strains, 3) Age dependency, 4) Susceptibility to various treatments known to modulate the *in vitro* NK activity.

More recently we have used the same technique to examine the role of NK cells in the *in vivo* clearance of syngeneic bone marrow cells from normal mice (Riccardi *et al.*, 1981d). The degree of clearance, after inoculation of radiolabeled bone marrow cells, again correlated with the levels of NK activity in the recipients. In this case also, there was a close parallelism between the *in vivo* rapid reactivity and the NK activity for all the parameters analyzed (Riccardi *et al.*, 1981b; Riccardi *et al.*, 1981d). Those involved: 1) Rapidity of killing, 2) Distribution of reactivity among strains, 3) Age dependency, 4) Susceptibility to various

*Supported in part by "Progetto Finalizzato Controllo della Crescita Neoplastica" PFCCN, Contract No. 81.01449.96 CNR, Italy.

ISBN 0-12-341360-5

treatments known to modulate the *in vitro* NK reactivity, 5) Bone marrow dependency of natural reactivity.

The results with normal bone marrow cells are consistent with previous indications that NK cells mediate *in vivo* natural resistance against bone marrow transplants (Kiessling *et al.*, 1977), but differ in that some reactivity has been found against syngeneic cells as well as the stronger reactivity of allogeneic or parental cells, suggesting an autoreactive nature of NK cells (also see Hansson *et al.*, this volume).

Taken together these data point to a heterogeneity in the *in vivo* functions of NK cells, perhaps attributable to different subpopulation of effector cells responsible for each type of activity.

ADOPTIVE TRANSFER OF NK CELLS AND IN VIVO NATURAL REACTIVITY

In view of the accumulating evidence for a possible *in vivo* role of NK cells in defenses against tumor cells, and virus infections, and in homeostatic regulation of normal non-neoplastic tissues, it is particularly important to develop procedures to more definitively document their *in vivo* importance and to determine how to manipulate the levels of *in vivo* natural reactivity. We have recently performed studies to evaluate the possibility of adoptively transferring *in vivo* natural reactivity (Riccardi *et al.*, 1981a). As shown in Table 1, both *in vitro* NK activity and *in vivo* natural reactivity could be reconstituted by *in vivo* transfer of NK-reactive spleen cells into immunodepresssed animals. The transfer of these cells caused a significant reconstitution of both the *in vivo* and *in vitro* reactivity in recipients pretreated with cyclophosphamide (Cy) or γ-irradiation. To determine whether the reconstitution of both *in vivo* natural reactivity and NK activity were mediated by the same cell population, we performed experiments using different donor cell populations (Riccardi *et al.*, 1981a). The characteristics of the donor cells responsible for reconstitution of both *in vivo* and *in vitro* activities are summarized in Table 2. Those data, previously reported (Riccardi *et al.*, 1981), support the hypothesis that both *in vivo* and *in vitro* natural reactivities against YAC-1 labeled cells are mediated by NK cells.

IN VIVO REGULATION OF THE ACTIVITY OF MOUSE NK CELLS

An important aspect of the understanding of NK cell function has been the study of the possible regulatory factors

Table 1. Reconstitution of NK Activity in Depressed Mice

Strain	Treatment d.-4	Treatment d.-1	In vivo reactivity[a)] lungs	%NK activity[b)] lungs
Host				
$BD2F_1$	-	-	0.4	11.0
(8 wks)				
	Cy[c)]	-	15.5	1.5
		Spl.,7.5×10^7	3.0	9.0
Donor				Spleen
$BD2F_1$				45.6
(8 wks)				
Host	d.-2	d.-1		
CBA	-	-	0.9	38.0
(8 wks)				
	850R		15.0	2.0
		Spl.,7.5×10^7	5.0	12.0
Donor				Spleen
CBA				49.5
(8 wks)				

a) % recovery of radioactivity, geometric mean at 3 hours after iv injection of 0.5×10^6 YAC-1 cells.
b) 100:1 E:T vs YAC-1, 4 hr CRA.
c) 250 mg/kg ip, day-4.

involved in determining the levels of reactivity (Herberman *et al.*, 1981). In particular, suppressor cells have been demonstrated in several situations in which NK activity was depressed by drugs and other agents (Cudkowicz and Hochman, 1979) and they have also been detected in untreated animals (Cudkowicz and Hochman, 1979; Riccardi *et al.*, 1981c).

We have recently shown that macrophages, activated by pyran copolymer or from untreated low-NK reactive mice, are

Table 2. Characteristics of Donor Cells Responsible for NK Reconstitution

- -Organ distribution: in spleen, not in thymus.
- -Strain distribution: present in spleens of high NK-reactive strains, but not of low NK-reactive strains.
- -Age dependency: in young NK-reactive, but not old mice.
- -Thymus independency: in spleens of athymic nu/nu mice.
- -Non-phagocytic, nonadherent, Thy 1-positive, asialo GM-1 positive.

capable of inhibiting NK cell activity when mixed *in vitro* with effector cells (Riccardi *et al.*, 1981c; Santoni *et al.*, 1980).

We have now begun a series of experiments to determine the possible *in vivo* relevance of the suppressor effects of those cells. The data in Table 3 indicate that the efficiency of reconstitution of *in vivo* and *in vitro* NK reactivity could be influenced by the recipient mice. Low NK-reactive SJL/J mice were less efficiently reconstituted than high NK reactive CBA/J mice, after depression of reactivity by Cy and transfer of spleen cells. This observation suggests that the host's environment can influence the function of donor NK cells. Such results are in agreement with other observations, showing the presence of adherent suppressor cells in the spleen of untreated SJL/J mice (see Santoni *et al.*, this volume). Suppressor cells could be removed by iron plus magnet treatment, were Thy.1 negative and were enriched by plastic adherence, suggesting that they are macrophages (Riccardi *et al.*, 1981c). This suppression was also seen with soluble factors obtained by a 4-hour incubation of such suppressor cells (Riccardi *et al.*, 1981c). These data are similar to our previous observations that activated splenic macrophages from mice treated with pyran copolymer (Santoni *et al.*, 1980) are capable of inhibiting NK activity upon *in vitro* mixing with effector cells.

The suppressor cells present in the spleen of normal SJL/J mice could account for the lower reconstitution of such mice, as well as for the low NK activity of this strain.

Table 3. Reconstitution of NK Activity in Low-reactive SJL and High-reactive CBA Mice (8-10 Weeks Old)

Strain	Treatment		In vivo reactivity[a]
Host	d.-4	d.-1	lungs
SJL			7.8
	Cy[b]		21.4
		7.5×10^7 spl.	20.9
CBA			0.9
	Cy		28.2
		7.5×10^7 spl.	15.7
Donor			% NK activity[c]
CBA			48.9

a) % recovery of radioactivity (geometric mean) at 3hrs after iv inoculation of YAC-1.
b) 240 mg/kg ip.
c) 4-hr CRA against YAC-1 targets (100:1).

Another experiment, summarized in Table 4, showed that spleen cells from pyran-treated mice (75 mg/kg i.v. 7 days before) could reconstitute the in vivo reactivity if the suppressor macrophages were first removed by nylon adherence, indicating again that the inoculation of a mixture of both effector and suppressor cells can reduce the reconstitution of depressed animals. Taken together, these observations point to the possible in vivo regulatory role of non-NK cells and indicate that these regulatory cells can actually be involved in determining the in vivo levels of reactivity of NK cells.

To further evaluate the in vivo relevance of the suppressor cells detected in in vitro experiments, we performed experiments of in vivo mixing of suppressor and effector cells. The results showed that suppression of NK activity can occur in vivo. As shown in Figure 1, mice depressed by treatment with Cy were significantly reconstituted for their reactivity by adaptive transfer of NK-reactive spleen cells, but such reconstitution was completely abolished by injection

Table 4. Reconstitution of NK Activity with Pyran Copolymer-treated Spleen Cells After Removal of Adherent Cells.

Strain	Treatment d.-4	Treatment d.-1	In vivo reactivity[a)] lungs
Host			
B6	-	-	0.6
(7 wks)			
	Cy[b)]	-	19.7
		Normal Spl. $5x10^7$	11.1
		Spl., Pyran-treated $5x10^7$	21.3
		Spl., Pyran-treated, Nylon-non-adh. $5x10^7$	9.6
Donor			NK activity[c)]
B6		None	23.7
(10 wks)		Pyran 75 mg/Kg day 7	10.8
		Pyran, Nylon-Passed	20.9

a) % recovery of radioactivity (geometric mean) at 4 hrs after iv YAC-1.
b) 240 mg/Kg ip day-4.
c) 4-hr CRA against YAC-1 targets (at 100:1).

of SJL/J nylon-adherent spleen cells 12 hours before assay. Similarly, the same suppressive effect *in vivo* was obtained with nylon-adherent spleen cells from pyran copolymer-treated mice.

FIGURE 1. *In vivo* natural reactivity of CBA mice as % recovery at 4 hours of total radioactive YAC-1 injected (geometric mean). Left panel: [dotted] control; [solid] mice pretreated with Cy (250 mg/Kg ip) 4 days before assay; [lined] mice pretreated with Cy and injected with $5x10^7$ CBA spleen cells 1 day before assay; [squares] mice pretreated with Cy, injected with $5x10^7$ spleen cells 1 day before assay and with $5x10^7$ SJL thymus cells (as filler cells) 12 hrs before assay; [fine dots] mice pretreated with Cy, injected with $5x10^7$ CBA spleen cells 1 day before assay and $5x10^7$ nylon-adherent SJL spleen cells 12 hrs before assay.

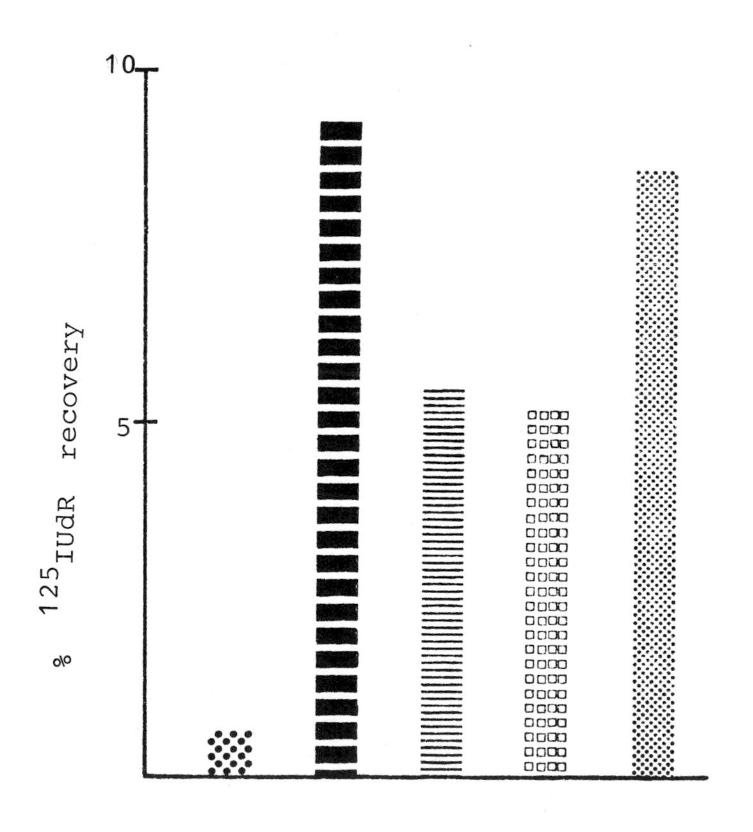

CONCLUSIONS

The data discussed here support the *in vivo* relevance of *in vitro* defined suppressor cells for NK activity. These suppressor cells were able to modulate the levels of *in vivo* natural reactivity as measured by the rapid *in vivo* elimination of YAC-1 tumor cells.

Since the relative contribution of mature NK cells or pre-NK cells for the *in vivo* reconstitution has not been defined, the question remains open as to whether this effect is mediated by actual suppression of NK cells or regulation of their development from pre-NK cell precursors, or both. It has been reported that Ly 5^- pre-NK cells can be boosted by interferon (IFN) to the lytic differentiated state of mature Ly 5^+ NK cells (Minato *et al.*, 1980). In our studies, we have obtained indications that suppressor cells can actively inhibit the lytic function *in vitro* of IFN-boosted spleen cells as well as their *in vivo* reconstitution of NK activity (data not shown). Thus, it seems more likely that *in vivo* regulation by suppressor cells is mediated by inhibition of the lytic function of mature NK cells.

Overall, the present report demonstrates that it is possible to study the *in vivo* regulation of NK activity by suppressor cells, and suggests that these cells can play a physiological role *in vivo* by influencing the levels of NK reactivity in low-reactive strains, or in mice treated with depressive agents.

REFERENCES

Cudkowicz, G. and Hochman, P.S. (1979). Immunol Rev 44:13.

Herberman, R.B., Brunda, M.J., Djeu, J.Y., Domzieg, W., Goldfarb, R.H., Holden, H.T., Ortaldo, J.R., Reynolds, G.W., Riccardi, C., Santoni, A., Stadler, B.M., Timonen, T. (1981). "Human Cancer Immunology", (B. Serrou and R.B. Herberman, Eds.). Elsevier North-Holland, Vol. 6.

Kiessling, R., Hochman, P.S., Haller, O., Shearer, G.M., Wigzell, H. and Cudkowicz, G. (1977). Eur J Immunol 7:655.

Minato, N., Reid, L., Cantor, H., Lengyal, P. and Bloom, B. (1980). J Exp Med 152:126.

Riccardi, C., Barlozzari, T., Santoni, A., Herberman, R.B., and Cesarini, C. (1981a). J Immunol 126:1284.

Riccardi, C., Puccetti, P., Santoni, A. and Herberman, R.B. (1979). J Natl Cancer Inst 63:1041.

Riccardi, C., Santoni, A., Barlozzari, T., Cesarini, C. and Herberman, R.B. (1981b). "Human Cancer Immunology" (B. Serrou and R.B. Herberman, Eds.), Elsevier North-Holland, Vol. 6.

Riccardi, C., Santoni, A., Barlozzari, T., Cesarini, C., Herberman, R.B. (1981c). Int J Cancer 28:775.

Riccardi, C., Santoni, A., Barlozzari, T. and Herberman, R.B. (1980a). In: Natural Cell-Mediated Immunity Against Tumors. (R.B. Herberman, Ed.) Academic Press, New York., p. 1121.

Riccardi, C., Santoni, A., Barlozzari, T. and Herberman, R.B. (1981d). Cell Immunol 60:136.

Riccardi, C., Santoni, A., Barlozzari, T., Puccetti, P. and Herberman, R.B. (1980b). Int J Cancer 25:475.

Santoni, A., Riccardi, C., Barlozzari, T. and Herberman, R.B. (1980). Int J Cancer 26:837.

LACK OF SUPPRESSOR CELL ACTIVITY FOR NATURAL KILLER CELLS IN INFANT, AGED AND A LOW RESPONDER STRAIN OF MICE.[1]

Anthony G. Nasrallah[2]
Michael T. Gallagher[3]
Surjit K. Datta
Elizabeth L. Priest
John J. Trentin

Division of Experimental Biology
Baylor College of Medicine
Houston, Texas

It has been proposed that the low response observed in adult AKR mice and in infant and aged mice of other strains is due to presence of suppressor cells specific for NK cells (Hochman et al., 1978; Savary and Lotzova, 1978). Although some evidence exists for such suppression, most of these studies involve prior experimental manipulation of mice using procedures such as sublethal total body irradiation (Hochman et al. 1978) or injection of carrageenan, hydrocortisone, or C. parvum (Savary and Lotzova, 1978; Cudkowicz and Hochman, 1979; Hochman and Cudkowicz, 1979). Other experiments reporting NK cell suppressors in infant spleens relied on the observed decrease in lytic activity of a normal effector suspension after the addition of infant spleen cells, with no control for possible non-specific decrease in lysis caused by other factors, such as cell "crowding", i.e. by increasing the total cell numbers and cell concentration (Savary and Lotzova, 1978).

In the present study, the inhibition of NK activity obtained by adding spleen cells of low responding mice was compared with the inhibition obtained by adding a variety of "filler" cells. Both produced "suppression"; in no case was the former significantly greater than the latter.

---

[1] Supported by USPHS grants CA 03367, CA 05021 and K6 CA 14219.
[2] Present address: Cook Children's Hospital, Fort Worth, Texas.
[3] Present address: City of Hope National Medical Center, Duarte, California.

ISBN 0-12-341360-5

Mice were from our caesarean-derived, barrier-sustained SPF colony. Eight to ten-week-old (C57Bl/6xA)F_1 mice (CAF_1) were used as controls with moderate NK cell activity. Spleens from 8 to 10-week-old AKR, 1 to 2-week-old (infant) CAF_1 and 12 to 18-month-old (aged) CAF_1 mice were examined for possible suppressor activity since these 3 groups have low or absent levels of natural cytotoxicity.

Effector, putative suppressor and "filler" cells were prepared by passing the tissue through first a 50 mesh then a 200 mesh stainless steel screen, pooling, suspending in Gey's balanced salt solution (BSS), centrifuging and washing once, and resuspending in RPMI 1640 with supplements (10% fetal calf serum, 25 mM Hepes buffer, 50 μg erythromycin/ml, and 100 μg streptomycin/ml). The number of nucleated cells was adjusted to 25 x 10^6 cells per ml.

Effector cells and putative suppressor cells were prepared from pooled spleens of 2 to 3 adult (10 weeks or 1 year old), or approximately 5 infant mice (1 to 2 weeks old) of the CAF_1 responder strain, and from 2 to 3 10-week old low responder AKR mice.

Filler cells were prepared from:

a) 2-3 thymuses of 10-week-old CAF_1 mice, both normal or (1-2 hr) preirradiated with 2000 R (^{137}Cs), since most suppressor cells are reportedly radiosensitive (Nair et al., 1981), and 2000 R is lethal to most suppressor cells (Holan et al., 1978). Because a subpopulation of normal thymocytes are susceptible to NK lysis, and may competitively inhibit NK lysis of YAC-1 cells (Hansson et al., 1980), spleen cells were also used as filler cells after culturing to deplete NK activity.

b) Spleen cells of 10-week old CAF_1 mice, prepared as described, diluted to 1 x 10^6 cells per ml and incubated in 5% CO_2 at 37°C for 24 hours to deplete NK cells, then resuspended to 25 x 10^6 viable cells/ml.

Each of the above suspensions (effector, putative suppressor or filler cells) was mixed 1:1 with either medium or each other as indicated; 200 μl (2.5 x 10^6 cells of 8 to 10-week old CAF_1 spleen effector cells and zero or 2.5 x 10^6 putative suppressor or filler cells) were plated with 50 μl (5 x 10^4) labelled target cells, for an E:T ratio of 50:1.

Target Cells - YAC-1 cells were maintained in RPMI 1640 supplemented with 10% fetal calf serum, 25 mM Hepes buffer, and antibiotics (50 μg erythromycin/ml and 100 μg streptomycin/ml), and labelled with ^{51}Cr by incubating 5 to 10 x 10^6 cells with 100 μCi of sodium chromate (^{51}Cr) at 37° for 30 minutes.

Cytotoxicity Assay - Natural cytotoxicity was measured using the ^{51}Cr release assay. Cell suspensions were prepared and mixed as described above to give a ratio of 50 CAF_1 spleen effector cells to 1 target cell. Four replicates of each sample were plated and were incubated at 37° for 4 hours. Only assays in which spontaneous isotope

TABLE I. LACK OF SUPPRESSOR ACTIVITY IN INFANT SPLEENS

Group	% Cytolysis			Grand Mean ± S.D.[a]	Cytolysis as % of Control
	Exp. 1	Exp. 2	Exp. 3		
CAF_1 Control spleen (C)[b]	6.9	13.2	7.8	9.3 ± 2.0	100
" Infant spleen (I)	0	1.0	0	0.3 ± 0.3	3
" Thymus (T)[c]	nt[e]	nt	nt		
C + I[d]	5.5	7.5	5.0	6.0 ± 0.8	65
C + T[d]	5.9	8.0	4.7	6.2 ± 1.0	67

a S.D. = Standard Deviation

b Spleen effector cells from 8 to 10-week-old CAF_1 hybrid mice were plated at a 50:1 E:T ratio

c Thymocytes were obtained from 8 to 10-week-old CAF_1 hybrid mice

d Thymocytes or infant spleen cells were mixed 1:1 with control effector cells then plated at a 50:1 CAF_1 effector spleen:target cell ratio

e nt = Not tested in this experiment. See Table II.

TABLE II. LACK OF SUPPRESSOR ACTIVITY IN INFANT SPLEENS

Group	% Cytolysis	Cytolysis as % of Control
CAF_1 Control spleen (C)	9.6	100
" Infant spleen (I)	0.9	9
" Unirradiated thymus (T)	0.7	7
" Irradiated thymus (RT)[a]	0.2	2
C + I	4.4	46
C + T	4.1	43
C + RT	4.4	46

a Irradiated thymocytes (RT) were obtained from 8 to 10-week-old CAF_1 mice that had received 2000 R, 1 to 2 hours previously.

TABLE III. LACK OF SUPPRESSOR ACTIVITY IN SPLEENS FROM INFANT AND AGED CAF_1 MICE

Group	% Cytolysis	Cytolysis as % of Control
CAF_1 Control spleen (C)	3.6	100
" Infant spleen (I)	0.4	11
" Aged spleen (A)[a]	0.8	22
" Cultured spleen (CS)[b]	0	0
C + I	2.2	61
C + A	1.8	50
C + CS	2.1	58

[a] Spleen cells from 12 to 18-month-old CAF_1 mice.
[b] Spleen cells from an 8 to 10-week-old CAF_1 mouse were incubated for 24 hours at 37°C in an atmosphere of 5% CO_2. Mean % lysis of YAC-1 by these cell suspensions prior to incubation ranged from 97.8 to 107.7% of control.

TABLE IV. LACK OF SUPPRESSOR ACTIVITY IN AKR SPLEENS

Group	% Cytolysis	Cytolysis as % of Control
CAF_1 Control spleen (C)	8.0	100
AKR spleen[a]	2.6	33
CAF_1 Cultured spleen (CS)	4.7	59
C + AKR	7.6	95
C + CS	5.9	74

[a] AKR mice 8 to 10-weeks of age were used as spleen cell donors.

release was less than 25% were reported. In addition to incubation with effector cells, labelled target cells were incubated with unlabelled target cells at a ratio of 100:1 (unlabelled:labelled). This autologous control (AC) served as the base line for non-specific release of ^{51}Cr. Percent lysis was calculated according to the formula:

$$\% \text{ Lysis} = \frac{\text{Sample cpm} - \text{Autologous control cpm}}{\text{Maximum release cpm} - \text{Autologous control cpm}} \times 100$$

RESULTS: Although mixing of low-NK active infant or aged CAF_1 spleen cells with adult control CAF_1 spleen cells did indeed produce a significant inhibition of the NK activity, filler thymus cells, either irradiated or unirradiated, or cultured spleen cells produced a comparable inhibition (Tables I, II and III).

In the experiment of Table IV, mixing of low-NK active 8 to 10-week old AKR spleen cells with control CAF_1 spleen cells produced only a small decrease in NK activity. In this experiment, unlike that of Table III, culturing CAF_1 spleen cells did not completely eliminate NK activity, but reduced it to 59% of control. However mixture of such cultured spleen cells with control spleen cells reduced the NK lytic activity to 74% of control.

Thus, there was no suggestion of specific suppression of natural cytotoxicity in adult AKR spleens. It is apparent, however, that in all of these experiments, the addition of a variety of "filler cells" non-specifically lowers the percent cytolysis produced by the constant ratio of 50 CAF_1 spleen effector cells per target cell.

Although suppressor cells have been reported to play an important role in the regulation of many branches of the immune response, our data fail to detect specific suppressor cells involved in the regulation of NK cells in the low responding mice studied, and emphasize the need to use filler cells to control for non-specific suppression.

REFERENCES

Cudkowicz, G., and Hochman, P.S. (1979). Immunol. Res. 44:13.

Hansson, M., Kiessling, R., and Welsh, R. (1980). In "Natural Cell-Mediated Immunity against Tumors" (R.B. Herberman, ed.), Academic Press, New York.

Hochman, P.S., and Cudkowicz, G. (1979). J. Immunol. 123:968.

Hochman, P.S., Cudkowicz, G., and Dausset, J. (1978). J. Natl. Cancer Inst. 61:265.

Holan, V., Hašek, M., and Chutna, J. (1978). Transplantation 25:27.

Nair, M.P.N., et al. (1981). Cell. Immunol. 58:9.

Savary, C.A., and Lotzova, E. (1978). J. Immunol. 120:239.

SUPPRESSOR CELLS ACTIVE AGAINST NK-B BUT NOT NK-A CELLS IN MICE TREATED WITH RADIOACTIVE STRONTIUM[1]

Vinay Kumar[2]
Paul F. Mellen
John A. Lust
Michael Bennett[2]

Department of Pathology
Boston University School of Medicine
Boston, Massachusetts

I. INTRODUCTION

Two subsets of natural killer (NK) cells can be distinguished from each other on the basis of their relative functions in ^{89}Sr-treated mice, in beige mice or after treatment of spleen cells with antisera directed against NK 1.2 antigens (1 and R.C. Burton, et al., this volume). Thus NK-A cells, reactive against YAC-1 target cells, are inactive in mice following destruction of the bone marrow by the administration of the bone seeking isotope, leukemia cells, EL-4 lymphoma cells or WEHI-164.1 fibrosarcoma cells remain normal(1,2). We have tested for the presence and/or induction of suppressor cells for NK-A and NK-B cells in mice treated with ^{89}Sr. The data obtained suggest that NK-B suppressor cells may mediate in vivo susceptibility to certain tumors, and that NK-A suppressors are not present in ^{89}Sr-treated mice. The data also confirm the concept of heterogeneity of NK cells.

[1]Supported by NIH grants CA21401, CA31792 and AI18811.
[2]Present address: Department of Pathology, Southwestern Medical School, Dallas, Texas.

ISBN 0-12-341360-5

II. MATERIALS AND METHODS

The sources of mice, tumor cell lines and reagents have been given (1). B6D2F1 mice were injected with 100 μCi ^{89}Sr 3-8 weeks before testing. A ^{51}Cr-release assay was used to measure NK cell function in vitro (1), and the clearance of infused tumor cells labelled with 5-iodo-2'-deoxyuridine-^{125}I (IUdR) from the lungs was used to study NK cell function in vivo (3).

We employed three separate cell mixture protocols in vitro to investigate the presence of suppressor cells. In one, the total number of normal spleen cells was kept constant at 5 x 10^5/well, but the number of spleen cells from ^{89}Sr-treated mice (or filler cells) was varied between 2.5 and 40 x 10^5/well. In this protocol high suppressor:effector (S:E) ratios are associated with high cell densities. The number of experimental spleen cells was kept constant in the second protocol at 5 x 10^5/well, but the number of normal spleen cells was varied between 1.2 and 40 x 10^5/well. In this case the highest S:E ratios were obtained at the lowest cell densities. In the third protocol, S:E ratios of 1:3, 1:1 and 3:1 were achieved at constant cell densities by varying the proportions of experimental and control spleen cells, as previously described (1,2,4). Normal spleen cells treated with anti-NK 1.2 + C' were used as filler cells. Such treated cells have virtually no NK(YAC-1) activity and are depleted of only 5-10% of the spleen cell population.

To detect any suppressor cells in vivo, we transferred 5 x 10^7 normal spleen cells into ^{89}Sr-treated mice 1 day or 2 hrs. prior to challenge with IUdR-labelled YAC-1 cells. This would allow any suppressor cells in the ^{89}Sr-treated mouse to act upon the adoptively transferred normal splenic NK cells. A second protocol involved the infusion of mixtures of spleen cells from normal and ^{89}Sr-treated mice into syngeneic B6D2F1 mice previously injected with 300 mg/kg cyclophosphamide (CY) to depress NK-A cell function or lung clearance (3). To determine if spleen-dependent suppressor cells are "activated" by treatment with ^{89}Sr so as to suppress NK-A function, mice were splenectomised 15 days prior to ^{89}Sr injection. The mice were infused with 5 x 10^6 normal marrow cells 7 days later to provide marrow-derived cells and lung clearance was measured 2 weeks after cell transfer. Alternatively, ^{89}Sr-treated mice were splenectomised and infused with 5 x 10^6 marrow cells 4 times in the next 2 weeks to provide stem cell functions. These mice were challenged 3 weeks after splenectomy.

III. RESULTS AND DISCUSSION

Suppression of NK-B Cells in ^{89}Sr-treated Mice. The functions of NK-B cells against FLD-3, WEHI-164.1 and EL-4 tumor cells are nearly normal in ^{89}Sr-treated mice. However, when ^{89}Sr-treated B6D2F1 mice are challenged with EL-4 cells (10^7 intraperitoneally), the NK(EL-4) activity is quite suppressed (2). The suppression can be seen within 3 days following injection of viable or irradiated cells, or following administration of heat-killed *Corynebacterium parvum* organisms or polyinosinic:polycytidylic acid. Similar pretreatments of control B6D2F1 mice stimulate NK(EL-4) activity. The NK-A activity against YAC-1 cells is low in ^{89}Sr-treated mice and is not affected by such treatments. When spleen cells from ^{89}Sr-treated mice injected with *C. parvum* organisms are mixed with normal B6D2F1 spleen cells, the NK(EL-4) activity is reduced even at a 1:3 S:E ratio. Thus it appears that administration of potential interferon inducers into ^{89}Sr-treated mice induces suppressors of NK(EL-4) cells. This may be caused by activation of inactive "presuppressor" cells present in the spleens of such mice, or may be caused by de novo induction of suppressor cells. It is important to note that similar stimuli fail to induce suppressor cells in control mice. This suggests that an intact bone marrow exerts a regulatory influence on suppressor cell functions, a point further discussed in this volume (Levy, et al.). The in vivo relevance of these suppressor cells was suggested by the observation that B6D2F1 mice, normally resistant to the growth of parental-strain EL-4 cells inoculated intraperitoneally, became quite susceptible following ^{89}Sr treatment (2). Such susceptibility may be due in part to the NK(EL-4) suppressor cells detected after challenge with the tumor.

Many of the above observations have been confirmed more recently by using FLD-3 erythroleukemia cells, which are lysed by NK-B cells (1). The relatively normal NK(FLD-3) activity in ^{89}Sr-treated B10.D2 or B6D2F1 mice is markedly reduced 10 days after infection of such mice with Friend virus complex. Furthermore, cell mixture experiments indicate the spleens of ^{89}Sr-treated and Friend virus-infected B6D2F1 mice have cells capable of suppressing normal NK (FLD-3) cells efficiently (Table I). The infection of control B6D2F1 mice with Friend virus failed to suppress their NK(FLD-3) activity and failed to induce suppressors for NK (FLD-3) cells. We previously observed that genetically resistant ($Fv\text{-}2^{rr}$) C57Bl/6 mice are rendered susceptible to Friend virus after treatment with ^{89}Sr (5).

TABLE I. Differential Suppression of NK-A and NK-B Cells by Spleen Cells from Mice Treated with ^{89}Sr and Infected with Friend Virus[a]

No. spleen cells (x 10^5)			Mean % specific cytotoxicity	
^{89}Sr + FV	Normal	S:E ratio	FLD-3 (NK-B)	YAC-1 (NK-A)
5	0	1:0	4	3
0	5	0:1	18	24
2.5	5	1:2	7(19)[b]	18(19)
5	5	1:1	5(20)	17(20)
10	5	2:1	4(20)	19(18)
20	5	4:1	not done	12(14)

[a]B6D2F1 mice were injected with ^{89}Sr and infected with 10^3 focus-forming units of Friend virus. Assay performed 10 days after infection.

[b]Numbers in parentheses are values when spleen cells treated with anti-NK 1.2 + C' were used as filler controls.

NK-A Cells in ^{89}Sr-treated Mice. In contrast to NK-B cells, the functional activity of NK-A cells responsible for lysis of YAC-1, RL♂-1, C1-18 and MPC-11 lymphoma cells, is very poor in ^{89}Sr-treated mice(1,4). One mechanism considered was suppressor cells for NK-A cells. It has been reported that spleen cells from ^{89}Sr-treated mice inhibit NK(YAC-1) activity at S:E ratios of 2:1 or higher (6). In contrast, we found either no appreciable suppressive effects, or suppression equal to that of thymocyte filler cells at high S:E ratios (4). To resolve this question, we have explored the issue extensively, both by in vitro and in vivo methods. There are two features which distinguish these from previous similar investigations. (i) In the in vitro experiments, we simultaneously employed 3 separate protocols to achieve varying S:E ratios. These protocols take into account the effect of varying cell densities on the NK(YAC-1) activity of cell mixtures. (ii) The second important feature relates to the nature of filler cell controls. We have chosen to employ normal spleen cells treated with anti-NK 1.2 + complement as filler cells. NK-A cell-depleted spleen cells appear to be the most rigorous filler cell control. The results of these experiments can be summarised as follows: When the total number of cells per well in the mixtures (i.e. cell density) is kept constant, no suppression of normal NK(YAC-1) activity is produced, regardless of the suppressor or filler cell population, or the S:E ratio. High cell densities above 25 x 10^5/well, regardless of the protocol or S:E ratio, resulted in inhibition of NK-A cell function. Thus apparent suppression was observed at high S:E ratios (5 x 10^5 normal cells + 20 x 10^5 filler cells, Table I). However, when a constant number of the same spleen cells from ^{89}Sr-treated mice (5 x 10^5) was used, no suppression was detected when mixed with 1.2 x 10^5 normal cells (S:E 4:1), whereas significant reduction occurred if mixed with 40 x 10^5 normal cells (even though the S:E ratio is reduced to 1:8). At the latter condition, spleen cells treated with anti-NK1.2 + C' were equally inhibitory (data not shown). We therefore conclude that the in vitro experiments did not reveal suppressor cells for NK(YAC-1) in spleens of mice treated with ^{89}Sr. Any inhibition observed at high cell densities should be interpreted with caution.

TABLE II. Lung Clearance of IUdR Labelled YAC-1 Cells in B6D2F1 Mice Treated with ^{89}Sr.

Exp.	Pretreatment of mice	Number (x 10^7) spleen cells infused (day-1)	% Retention IUdR per lung[a]
1	^{89}Sr[b]	0	+++
	None	0	+
	^{89}Sr	5 C[c]	++
2	CY	0	+++
	CY	5 C	++
	CY	5 ^{89}Sr	+++
	CY	5 C + 5 ^{89}Sr	++
	CY	5 C + 7.5 ^{89}Sr	++
3	Control	-	+
	Spl-x	-	+
	^{89}Sr	-	+++
	^{89}Sr, then Spl-x	-	+++
	Spl-x, then ^{89}Sr	-	+++

[a] Determined 4 hours after infusion of 10^6 cells (groups of 4-6 mice). +++ > 10%; ++ = 3-10%; + < 1%.
[b] ^{89}Sr, 100 μCi -3 to 8 weeks; CY, cyclophosphamide 300 mg/kg -4 d.; Spl-x, splenectomy (see text for details).
[c] C, normal spleen cells; ^{89}Sr, spleen cells from ^{89}Sr-treated mice.

As compared with normal mice, ^{89}Sr-treated (or cyclophosphamide-treated) mice have very diminished lung clearance of IUdR-labelled YAC-1 tumor cells (Table II). Infusion of normal spleen cells caused a significant improvement in their ability to clear, suggesting that ^{89}Sr-treated mice did not suppress the adoptively transferred normal splenic NK cells. The ability of adoptively transferred normal spleen cells to restore the clearance of YAC-1 cells in CY-treated mice was not suppressed by the addition of spleen cells from ^{89}Sr-treated mice (Table II). Finally, removal of the spleen either prior to or after ^{89}Sr administration had no effect on the clearance of YAC-1 cells from the lungs, indicating that the NK-lowering effect of ^{89}Sr does not require the presence of splenic suppressor cells. We have not been able to detect suppressors of NK-A in vivo (lung clearance) or in vitro from any source, so far.

IV. CONCLUSION

By using ^{89}Sr-treated mice as a model, we provide an analysis of two examples of lowered NK activity. Reduction of NK-B cell activity can be clearly demonstrated to be due to suppressor cells detected in vitro at the effector level. Employing similar in vitro experiments as well as in vivo techniques, no cellular suppression of NK-A cells could be detected in ^{89}Sr-treated mice. Presumably, functionally mature NK-A cells cannot be generated in the absence of an intact bone marrow microenvironment. These experiments strengthen the concept of NK cell heterogeneity.

REFERENCES

Cudkowicz, G. and Hochman, P.S. Immunol. Rev. 44:13 (1979).

Kumar, V., Ben-Ezra, J., Bennett, M. and Sonnefeld, G. J. Immunol. 123:1832 (1979).

Kumar, V., Bennett, M. and Eckner, R.J. J. Exp. Med. 139:1093 (1974).

Luevano, E., Kumar, V. and Bennett, M. Scand. J. Immunol. 13: 563 (1981).

Lust, J.A., Kumar, V., Burton, R.C., Bartlett, S.P. and Bennett, M. J. Exp. Med. 154:306 (1981).

Riccardi, C., Barlozar, T., Santoni, A., Herberman, R.B. and Cesarini, C. J. Immunol. 126:1284 (1981).

DOES A MARROW DEPENDENT CELL REGULATE SUPPRESSOR CELL ACTIVITY?[1,2]

Elinor M. Levy
Vinay Kumar[3]
Michael Bennett[3]

Departments of Microbiology and Pathology
Boston University School of Medicine
Boston, Massachusetts

Neonatal and aged mice, and mice treated with the bone seeking isotope, ^{89}Sr, all have low NK activity against certain targets and elevated suppressor cell activity (e.g., 1-5). One can speculate that the inverse relationship observed between NK and suppressor cells is the result of a direct relationship between the two cell types. We have investigated this thesis and found that the low NK activity is not due to suppressor cells acting at effector cell or precursor cell levels. However, there is some evidence to support the interpretation that suppressor cell activity may be regulated by a marrow dependent cell, and is elevated when this regulatory cell is absent. This evidence will be reviewed here.

^{89}Sr treatment destroys cells of the bone marrow and hemopoiesis is largely shifted to the spleen. Most immune functions *in vivo* remain close to normal.

[1]Supported by NIH grants CA25039, CA21401, CA31792 and AI18811.
[2]Publication No. 96 of the Hubert H. Humphrey Cancer Research Center at Boston University.
[3]Present address: Dept. Pathology, Southwestern Medical School, Dallas, Texas.

ISBN 0-12-341360-5

However, NK activity against YAC-1 cells and the ability to reject bone marrow grafts are markedly impaired in such mice suggesting that these effector cells are marrow-dependent (2,6). At the same time these mice become susceptible to the induction of both T and non-T suppressor cells. Normally resistant mice treated with ^{89}Sr become susceptible to the leukemogenic effects of Friend virus. The change from a resistant to susceptible status is accompanied by the appearance of suppressor T cells (7). Spleen cells from mice treated with ^{89}Sr also fail to produce an *in vitro* antibody response and can suppress the *in vitro* antibody response of normal spleen cells in a Mishell-Dutton system. The impaired response is attributable to a non-T, non-B suppressor cell (5). The suppressor cells present in the spleen of ^{89}Sr treated mice adhere to Sephadex G-10 but not to nylon wool, are relatively large, are not phagocytic, and are sensitive to 1000 R of *in vitro* irradiation (8). Suppressor cells with similar characteristics are also found in normal murine bone marrow (8). Neonatal and aged mice similarly have both T and non-T suppressor cells (3,9). Interestingly, neonatal thymus and bone marrow, both of which contain unusual suppressor cells, are the only normal tissues with detectable NK sensitive target cells (10,11). Thus, NK cells might regulate suppressor cells by lysing them.

To investigate more directly the relationship between NK (YAC-1) activity and suppressor cells we established a model which involved the transfer of spleen cells from ^{89}Sr treated mice to normal lethally irradiated syngeneic recipients (see Table I). Four to ten weeks later these recipient mice (SrSCR) tended to have low NK (YAC-1) activity and elevated suppressor cell activity (as measured by the ability to suppress *in vitro* antibody synthesis) when compared to recipients of normal spleen cells (NSCR). Thus we were able to transfer precursors of the suppressor cells from the spleen of ^{89}Sr treated donors to normal recipients. However, when individual spleens of mice were assayed for NK and suppressor cell activity, it was found that some spleens had good NK (YAC-1) activity but had high suppressor cell activity (12). The observation that elevated suppressor cell activity could coexist with high or low NK activity implies that NK (YAC-1) does not directly regulate the cells which suppress antibody synthesis. All recipient mice tested had the ability to reject bone marrow allografts, indicating that ^{89}Sr had not eliminated precursors of effector cells.

TABLE I. NK (YAC-1) and Suppressor Cell Activity

Source of assayed spleen cells[a]	NK (YAC-1) activity	Suppression of antibody synthesis
^{89}Sr treated mice	-	+++
SrSCR	+	+++
NSCR	++	+
(N+Sr)SCR	++	++
(2N+Sr)SCR	++	-
BMR	+++	-
(BM+Sr)SCR	+++	+

[a]See text for explanation of abbreviations.

To further explore the hypothesis that marrow dependent cells may regulate suppressor cell activity, irradiated recipient mice were reconstituted with mixtures of spleen cells from ^{89}Sr treated mice and normal bone marrow or spleen cells (13). Addition of normal cells had different effects on the generation of NK (YAC-1) cells and suppressor cell activity. Mice who received mixtures of normal spleen or bone marrow cells along with spleen cells from ^{89}Sr treated mice developed normal levels of NK (YAC-1) activity [equal to that of recipients of normal spleen cells (NSCR) or bone marrow cells (BMR)]. This was true at all ratios of normal to Sr-treated tested, i.e., 1:1 [(N+Sr) SCR], and 2:1 [(2N+Sr)SCR] for spleen cell mixtures and 1:10 [(BM+SrSC)R] for bone marrow mixtures. The results suggest that the spleen cells from ^{89}Sr treated mice did not suppress the development of NK cells from normal precursors. On the other hand suppressor cell activity for antibody synthesis persisted in (N+Sr)SCR recipients. At 2:1 spleen cell ratios and in the bone marrow cell containing mixtures [i.e., (2N+Sr)SCR and (BM+SrSC)R mice], there was a significant reversal of suppressor cell generation. This is consistent with the view that there is an interaction between suppressor cell precursors in the spleen cells of ^{89}Sr treated mice and a marrow dependent regulatory cell in the normal cell population. The ratio of these two cell types would determine whether or not significant suppressor cell activity developed in the recipient mice.

Further indirect evidence for a marrow dependent cell which regulates suppressor cell generation comes from the work of Seaman et. al. (14). They showed that autoimmune disease, which is thought to be partly the result of decreased suppressor cell activity, was ameliorated by ^{89}Sr treatment, i.e., the destruction of marrow dependent cells.

What is the candidate for the regulatory cell? Since it is not present in ^{89}Sr treated mice, it is a marrow dependent cell. The fact that NK (YAC-1) cell activity and marrow allograft reactivity can co-exist with elevated suppressor cell activity in the reconstituted mice points to one of two possibilities: the marrow dependent regulatory cell is different from cells responsible for the NK (YAC-1) or marrow allograft reactivity; or the regulatory cell (one of the above) must prevent the generation of suppressor cells from their precursors. Thus if the suppressor cells developed before the regulatory cell, both could co-exist. The observations that bone marrow cells generate NK (YAC-1) activity and other marrow dependent cell function in recipient mice more rapidly than do normal spleen cells and are also more efficient than spleen cells in preventing the generation of suppressor cells would be consistent with this interpretation. While the evidence for a regulatory marrow dependent cell is inferential rather than direct, its existence is an attractive hypothesis.

REFERENCES

1. Herberman, R.B., Nunn, M.E., Lavrin, D.H. and Asofsky, R. Int. J. Cancer 16:216, 1975.
2. Kiessling, R., Hochman, P.S., Haller, O., Shearer, G.M., Wigzell, H. and Cudkowicz, G. Eur. J. Immunol. 7:655, 1977.
3. Mosier, D.E., Mathieson, B.J. and Campbell, P.S. J. Exp. Med. 146:59, 1977.
4. Singhal, S.K., Roder, J.C. and Duwe, A.K. Fed. Proc. 37:1245, 1978.
5. Merluzzi, V.J., Levy, E.M., Kumar, V., Bennett, M. and Cooperband, S.R. J. Immunol. 121:505, 1978.
6. Bennett, M., Baker, E.E., Eastcott, J.W., Kumar, V. and Yonkosky, D. J. Reticuloendothel. Soc. 20:71, 1976.

7. Kumar, V., Caruso, T. and Bennett, M. J. Exp. Med. 143:728, 1976.
8. Levy, E.M., Corvese, J.S. and Bennett, M. J. Immunol. 126:1090, 1981.
9. Rodriguez, C., Andersson, G., Wigzell, H. and Dech, A.B. Eur. J. Immunol. 9:137, 1979.
10. Hansson, M., Karre, K., Kiessling, R., Roder, J., Andersson, B.and Hayry, P. J. Immunol. 123:765, 1979.
11. Riccardi, C., Santoni, A., Barlozzari, T. and Herberman, R. Cell. Immunol. 60:136, 1981.
12. Levy, E.M., Bennett, M., Kumar, V. Fitzgerald, P. and Cooperband, S.R. J. Immunol. 124:611, 1980.
13. Levy, E.M., Kumar, V. and Bennett, M. J. Immunol. 127:1428, 1981.
14. Seaman, W., Blackman, M., Greenspan, J. and Talal, N. J. Immunol. 124:812, 1980.

SUPPRESSOR LYMPHOCYTES OF HUMAN NK CELL ACTIVITY

Jussi Tarkkanen
Eero Saksela[1]

Department of Pathology
University of Helsinki
Helsinki, Finland

I. INTRODUCTION

The natural killer cell activity present in human cord blood is significantly lower than in adult blood. We have suggested that this is related to the action of suppressor cells occasionally found in the small lymphocytic fractions obtained with Percoll fractionation of Ficoll-Isopaque separated, adherent-cell depleted buffy coats from umbilical cord blood (1). When such small lymphocyte fractions are further enriched for IgG Fc receptor carrying cells, which form a subpopulation of T gamma cells as shown below, suppressors of the NK cell activity are obtained in a majority of cases. Both allogeneic and autochlonous combinations of the suppressor cells with highly enriched NK cell populations of the LGL type lead to partial inhibition of the lysis, using either adherent or non-adherent NK susceptible target cells in the cytotoxicity assays.

We have continued the search for small lymphocytic suppressor cells of the human NK activity in adult normal and tumor-bearing individuals and characterized further these cells functionally and with various surface markers.

[1] Supported by grant No. RO 1 CA 23809 from the National Cancer Institute, NIH, Bethesda, Md., USA

ISBN 0-12-341360-5

II. SUPPRESSOR CELLS IN ADULT PERIPHERAL BLOOD

Some suppressor cell activity was found among the small lymphocytic cells (fraction 5) obtained with Percoll fractionation of buffy coat leucocytes in 7 out of 51 consecutively tested healthy blood donors. Reduction of the original cytotoxic activity of the LGL cell fraction by more than 10 percentage units was statistically significant and 3 of these seven donors reduced the NK cell activity more than 20 percentage units.

The small lymphocytic Percoll fraction 5 from 33 out of the 51 donors above were further fractionated into IgG Fc receptor positive and negative cells (Table I). The fractionation was performed with antibody-coated erythrocyte-rosetting (EA rosetting) or in some cases with adsorption on plastic dishes coated with antigen complexed with specific antibody (2). Using these methods, 20 out of the 33 had significant suppressor cell activity in the IgG Fc receptor positive populations, whereas only 7 of the 33 non-receptor carrying populations were positive (Table I). The average reduction of the LGL-mediated cytotoxicity by the added cell fraction tested was between 24-26 percentage units in the positive cases and the specific suppression calculated from these figures varied from 38-49 %, with the strongest suppression found in the EA+ fraction (Table I).

Similar experiments were performed with blood leucocytes obtained from 22 patients with malignant tumors. The patients represented clinical stage I cases with epidermoid carcinomas of the upper respiratory tract, adenocarcinomas of the kidney and ovarian carcinomas, of whom 30 ml venous blood samples were obtained preoperatively. Among these patients, no increase in the proportion of suppressor cell-positive individuals in comparison to healthy donors could be demonstrated (Table I). Percoll fractionated small lymphocytes suppressed the NK cell activity in only one out of 20 cases and the IgG Fc receptor carrying population was significantly suppressing in 2 out of 6 patients whose Percoll fraction 5 cells were subjected to further separation. The proportion of suppressive peripheral blood samples among tumor patients was thus actually somewhat lower than among normal adult donors (Table I) and clearly lower than in the umbilical cord blood samples tested previously (1).

Quantitative estimates of the cell yields in the various fractions obtained from umbilical cord blood samples, from normal adult donors or from tumor patients were not different. The fraction 5 cells in Percoll gradients represented 20 to 30 % of the Ficoll-Isopaque separated adherent cell depleted input populations and 2 to 15 % of the fraction 5 cells were IgG Fc receptor-bearing cells in these experiments.

TABLE I. Suppression of the LGL-mediated NK activity by small lymphocytes from Percoll gradients (Fr 5) either directly or after separation in IgG Fc receptor carrying (EA+) and (EA-) fraction. 33 normal blood donors and 22 tumor patients were tested.

Cell source	Normal adults			Tumor patients		
Fraction	Fr 5	Fr 5 EA-	Fr 5 EA+	Fr 5	Fr 5 EA-	Fr 5 EA+
Proportion of suppressors[1]	16	21	61	5	14	28
Average LGL-cytotoxicity	62±7	60±6	49±4	77	37	39
Average reduction from the above	24±6	26±5	24±3	15	12	19
Average specific suppression[2]	38	44	49	19	32	49

[1] Suppression of more than ten percentage units of the cytotoxicity exerted by LGL alone. Cells tested were added at 1:1 ratio and a 4 hr chromium release assay with K 562 cells was used.
[2] % reduction from the original cytotoxicity of LGL-cells alone

III. EFFECT OF INTERFERON ON THE SUPPRESSOR CELL ACTIVITY

The IgG Fc receptor positive suppressor cell population obtained from adult normal or tumor-bearing patient blood was pretreated with 1000 IU human leukocyte α-IFN and with immune γ-IFN interferon for 1-3 hrs prior to addition on the NK cells in 3 experiments (Table II). In all instances, the interferon pretreatment reduced the suppressive activity by 35-66 %, as shown in Table II in accordance with our previous results with the umbilical cord blood samples. The human gamma interferon at comparable doses was somewhat less effective than the alpha interferon in these experiments.

TABLE II. Inhibition of the suppressor cell activity by human alpha or gamma interferon.

LGL-mediated cytotoxicity, %		Specific suppression by Percoll Fr 5 EA+ cells, %		
		untreated	+IFNα[1]	+IFNγ[1]
Exp 1	52	21	7 (66)	13 (38)
Exp 2	50	34	18 (47)	22 (35)
Exp 3	68	32	16 (50)	19 (40)

[1]Pretreatment of 3 hrs with 1000 IU/ml of IFNα (spec. activity 4×10^6/ml) or 1000 IU/ml of IFNγ (spec. activity 3×10^4/ml). Inhibition of suppression in percent given in brackets.

IV. SURFACE CHARACTERISTICS OF THE SUPPRESSOR CELL POPULATION

The presence of the various surface markers on Percoll fraction 5, IgG Fc receptor positive (EA+) and Fc-receptor negative cells detectable with monoclonal antibodies is shown in Table III. Analyses were performed with Staph A-rosetting techniques directly on cytocentrifuge slides or by absorption of the cells on plastic petri dishes with immobilized antibody on the bottom (3). As seen in Table III, the only monoclonal antibody detected marker consistently shown on this cell fraction was OKT-3. Removal of the OKT-3 carrying cells from the IgG Fc receptor-positive lymphocyte Percoll fraction also removed the suppressive capacity, which was again recovered among the OKT-3 positive cells (Table IV).

As seen in Table III, practically all of the cells in the suppressor cell fraction formed high-affinity rosettes with sheep red blood cells (E-rosettes), and all showed a characteristic small alpha-naftyl-acetic-esterase (ANAE) dot of the T-cell type which was not inhibited by sodium fluoride. The cells also responded to interleukin 2 (TCGF) and proliferated readily as blasts.

TABLE III. Markers detectable on the Percoll fraction 5 cells before and after fractionation into IgG Fc receptor carrying (EA+) and non-carrying (EA-) cells.

Marker	% marker positive cells in		
	Percoll Fr 5	Fr 5 EA-	Fr 5 EA+
EA-rosettes[1]	8.6 (2-16)	-	-
E-rosettes	67	68	72
OKT-3	90	88	63
OKT-4	36	21	26
OKT-8	23	20	23
OKT-6	4	5	3
OKM-1	5	3	5
HLA-ABC	92	78	70
ANAE	90	94	89
Intracytoplasmic Ig	1	1	1

[1]Average of 17 experiments; other markers: average of 3 experiments.

TABLE IV. Suppressor activity of the OKT-3 positive and negative cells obtained from the Percoll fraction 5 IgG Fc receptor carrying cell population.

LGL-mediated cytotoxicity (%)	56
Specific suppression (%) by	
Percoll Fr 5	1
" EA+	34
" EA+, OKT-3+[1]	21
" EA+, OKT-3-[2]	5

[1]Cells attached and recovered from immobilized anti-OKT-3 monolayers.
[2]Cells not attaching on immobilized anti-OKT-3 monolayers.

V. CONCLUDING REMARKS

The results presented in this paper show that it is possible to find small lymphocytic suppressor cells capable of inhibiting the human NK cell activity in the peripheral blood of normal human adults. In 16 % of blood samples from normal blood donors the small lymphocytic fraction 5 from Percoll gradients inhibited the cytotoxic action of allogeneic LGL fractions, and 61 % of the donors were positive when further fractionation into IgG Fc receptor carrying cells was performed. These latter suppressor cell enriched fractions apparently formed a subpopulation of T-gamma cells. They were OKT-3 positive, formed high-affinity rosettes with sheep red blood cells (E-rosettes) and had the characteristic ANAE positive cytoplasmic T-cell dot, which was not inhibited by sodium fluoride. The suppressor activity of these cells could be inhibited by pretreatment with either alpha or gamma interferon.

The incidence of the suppressor cells in normal adult blood was slightly lower than previously found in human umbilical cord blood samples. A series of 22 tumor patients with various clinical stage I carcinomas tested in preoperative blood samples did not show an increased incidence of these cells, actually the incidence was less, i.e. 5 % in Percoll fraction 5 populations and 28 % among the IgG Fc receptor positive cells from these fractions.

REFERENCES

1. Tarkkanen, J., and Saksela, E., Scand. J. Immunol. (1982) In press.
2. Henkart, P., and Alexander, E., J. Immunol. Methods 20: 155 (1978).
3. Biddison, W. E., Sharrow, S. U., and Shearer, G. M., J. Immunol. 127:487 (1981).

MODULATION OF NK ACTIVITY BY HUMAN MONONUCLEAR PHAGOCYTES: SUPPRESSIVE ACTIVITY OF BRONCHO-ALVEOLAR MACROPHAGES[1]

Claudio Bordignon

Clinica Medica Generale, Ospedale "L. Sacco",
Università di Milano, Milan, Italy

Paola Allavena
Martino Introna
Andrea Biondi
Barbara Bottazzi
Alberto Mantovani

Istituto di Ricerche Farmacologiche
"Mario Negri", Milan, Italy

I. INTRODUCTION

NK activity is subject to positive and negative regulatory factors (1). There is evidence that mononuclear phagocytes are one important regulatory mechanism of NK activity (1,2). Cells of the monocyte-macrophage lineage allow full expression of NK cytotoxicity *in vitro* (3,4) and by producing interferon (IFN) they could be a source of a stimulatory factor (2). On the other hand, under pathological conditions, or after administration of pharmacological agents, cells with macrophage characteristics inhibit the expression of NK activity (5-8). It is not yet known whether within the monocyte-macrophage lineage there are cell populations heterogeneous in their interaction with NK cells. Here we report that populations of

[1]Supported by CNR No. 80.01579.96 (Rome, Italy) and by Grant 1R01 CA 26824 from National Cancer Institute, USA.

ISBN 0-12-341360-5

human mononuclear phagocytes from diverse anatomical sites have distinct capacities to modulate NK activity.

II. EXPERIMENTAL PROCEDURES

A. Macrophages

Mononuclear phagocytes were isolated by adherence as previously described (9-11) from peripheral blood, peritoneal exudates (wash outs from patients undergoing surgery for non-malignant, non-infectious gynecological diseases), early lactation milk (3-5 days after delivery) ovarian carcinomatous ascites and bronchoalveolar lavages. After experiments had shown that unfractioned and adherent lung alveolar cells were equally inhibitory for NK, bronchoalveolar lavage cells (already 85-95% macrophages by morphology and non specific esterase staining) were used without further fractionation (Table I).

B. Assay for Suppressive Activity

^{51}Cr-labelled tumor target cells (K562, $1x10^4$) were cultured with peripheral blood lymphocytes (PBL/K562 ratio routinely employed was 25/1) in 0.2 ml of RPMI 1640 medium with 10% FBS in round-bottomed wells of microplates (Sterilin, Teddington, Middlesex, England). The routinely employed incubation time was 4h. Isotope release was calculated as (A/B) x 100, where A is the isotope in the supernatant and B is the total incorporated radioactivity released by incubation with 1% sodium dodecyl sulfate. Specific lysis was calculated by subtracting spontaneous isotope release of tumor cells alone. Spontaneous ^{51}Cr release from K562 cells was 0.5-1.5% per hour of incubation. Graded numbers of mononuclear phagocytes, suspended in 0.05 ml of medium, were placed in round-bottomed wells of microtiter plates. PBL (usually $2.5x10^5$) was added in 0.05 ml to each well, followed by $1x10^4$ ^{51}Cr-prelabelled target cells in 0.1 ml. The effector (PBL)/target (E/T) ratio of 25/1 was kept usually constant, whereas the suppressor:effector (S/E) ratio ranged from 0.125/1 to 1/1 (occasionally 2/1). Mononuclear phagocytes alone caused little ($<$6%) or no lysis of K562 in a 4 h assay. The percentage of inhibition of NK activity was calculated as (1-A/B) x 100, where A is the specific lysis of samples containing PBL and putative inhibitors and B is the specific lysis of PBL alone. In this series of experiments the mean ($\pm$ S.D.) specific lysis of PBL alone (E/T = 25/1) was 37.4 ($\pm$ 12.3) (median 34.1, range 25.0-55.2) at 4 h.

III. RESULTS

Fig. 1 shows a typical experiment in which various concentrations of broncho-alveolar cells or blood monocytes were mixed with PBL from the same donor. Mononuclear phagocytes lacked consistent lytic activity on K562 cells in this short-term assay. Addition of increasing concentrations of blood monocytes (up to a ratio of 1/1 with PBL) did not appreciably modify the cytotoxicity of PBL. In contrast alveolar macrophages caused a dose-dependent reduction of NK cytotoxicity, which was significantly inhibited (50%) at a S/E ratio of 0.125/1 and virtually abolished at a ratio of 0.5/1. Inhibition of NK activity by alveolar macrophages was consistently observed with all 15 preparations tested, with values ranging from 50 to 100% reduction at a S/E ratio of 0.5/1 (Fig.2,Tab.I). Monocytes, milk and peritoneal macrophages (from nonneoplastic exudates or ovarian carcinoma ascites) were consistently devoid of inhibitory capacity up to the highest S/E ratios tested of 2/1 (Fig. 2 A and B). This observation excludes that the inhibition of NK activity by alveolar macrophages is related to trivial "fillers" effects. Using a fixed number of alveolar macrophages ($2x10^5$/well), inhibition of cytotoxicity was evident irrespective of density of effector PBL.

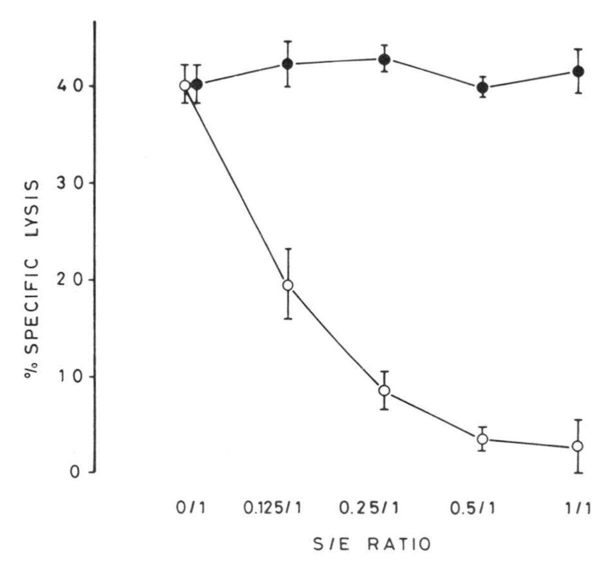

FIGURE 1. Effect of broncho-alveolar macrophages and blood monocytes on NK activity. The E/T ratio was 25/1. Open and closed circles indicate samples with PBL cultured with broncho-alveolar macrophages and blood monocytes respectively.

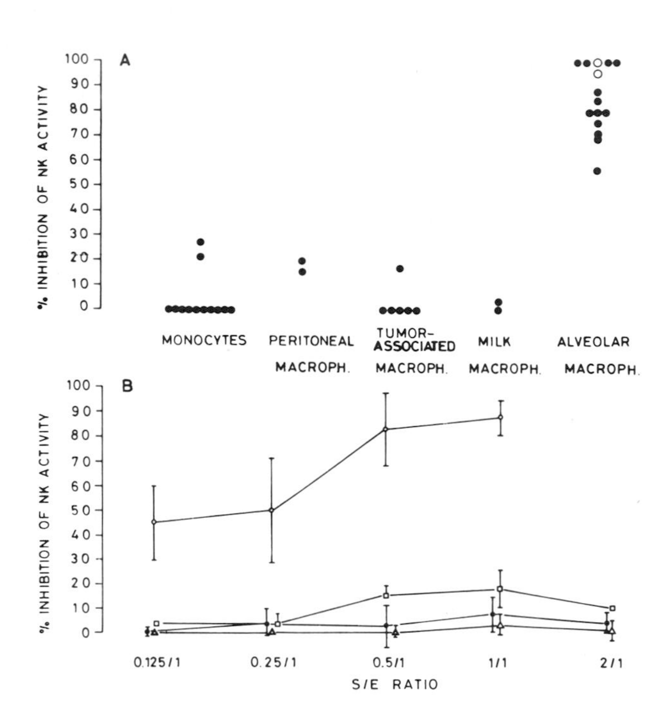

FIGURE 2. Inhibition of NK activity by diverse human mononuclear phagocyte populations. Panel A shows the results of individual experiments, expressed by % inhibition of cytotoxicity at a S/E ratio of 0.5/1. The open circles are data of alveolar macrophages further purified by adherence.
Panel B summarizes the overall inhibitory activity of the various mononuclear phagocytes at different S/E ratios: results are the mean of 2-15 experiments ($\pm$S.D.). Alveolar macrophages (open circles), peritoneal macrophages (squares), milk macrophages (triangles), monocytes (closed circles).

Similarly, using a constant S/E ratio, suppression of NK activity was detected irrespective of the E/T ratio used (from 6/1 to 50/1). The routinely used incubation time was 4 h. When the ^{51}Cr release assay was prolonged to 20 h, inhibition of NK activity by alveolar macrophages was somewhat lower at low S/E ratios.

In vitro or *in vivo* exposure to IFN augments NK activity (12). It was therefore of interest to evaluate whether the suppressive activity of broncho-alveolar cells was expressed on both normal and IFN-stimulated PBL. Effector cells were

TABLE I. Suppressive Activity of Unfractioned Cells from Broncho-Alveolar Lavages and Adherent Alveolar Macrophages on NK Activity of PBL.

Effector cells	% of specific lysis	
	4 h[a]	20 h
PBL alone	34.7 (± 1.7)[b]	44.0 (± 7.2)
PBL + Unfractioned alveolar cells	3.6 (± 1.1)	20.0 (± 3.8)
PBL + Adherent alveolar cells	0.7 (± 1.7)	10.4 (± 1.3)

E/T and S/E ratios were 25/1 and 0.5/1 in this experiment.
[a]Time of culture
[b]Mean (± standard deviation)

precultured for 1 h with 1000 units/ml IFN, a concentration with optimal stimulating activity on PBL under these conditions (unpublished data). IFN-boosted NK activity (as well as the unstimulated one) was virtually abolished at the different A/T ratios (0.2:1 to 1:1).

To obtain indications as to the possible role of cold target competition effects in the inhibition of NK activity by alveolar macrophages, mononuclear phagocytes from blood, milk and broncho-alveolar lavages were labelled with ^{51}Cr and used as targets. As shown in Table 2, broncho-alveolar macrophages were not appreciably lysed by PBL, 4-8% of lysis being only occasionally observed against milk macrophages. Table 3 shows a preliminary experiment of restoration of PBL NK

TABLE II. Susceptibility of Peripheral Blood Monocytes, Alveolar and Milk Macrophages to Lysis by Normal PBL.

^{51}Cr labelled target-cells	Percentage of specific lysis[a]		
	12/1[b]	25/1	50/1
Monocytes	0	0	2.5 (±0.8)
Milk macrophages	4.5 (±0.4)	5.6 (±1.4)	7.8 (±0.8)
Bronchoalveolar macrophages	N.T.[c]	0	0
K562 cells	14.7 (±3.1)	28.5 (±0.3)	34.1 (±0.6)

[a]mean (± standard deviation)
[b]effector to target ratios
[c]not tested

TABLE III. Reversibility of Alveolar Macrophage-Mediated Suppression of NK activity

S/E ratio	% of specific lysis	
	untreated	carbonyl-iron treated
0/1	33.1 (+0.7)	34.5 (+1.1)
0.125/1	19.8 (+0.6)	32.5 (+0.4)
0.5/1	0.7 (+0.9)	0.8 (+0.4)

PBL were cultured for 4 h with bronchoalveolar cells. Then macrophages were removed by carbonyl iron and the cells were further cultured overnight before measuring NK activity.

activity inhibited by alveolar macrophages. After 4 h of co-cultivation of suppressor and effector cells, adherent phagocytes were removed by the carbonyl-iron method. Upon culture overnight, NK activity of non-adherent, non phagocytic cells was restored. However recovery was observed only by the lymphoid populations that had been exposed to the lower concentration of alveolar macrophages.

IV. CONCLUDING REMARKS

The results presented here indicate that human alveolar macrophages are potent inhibitors of the expression of NK cytotoxicity. Inhibition of NK cytolysis was consistently observed at S/E ratios of 0.12-0.2/1. In contrast, blood monocytes and milk or peritoneal macrophages did not affect NK activity appreciably even at S/E ratios of 2/1.

Cells of the monocyte-macrophage lineage can inhibit NK activity in rodents under a variety of experimental conditions (5-8). Moreover, non adherent cells of undefined lineage (5,13) and T cells (14) can interfere with murine NK cytotoxicity. In humans there is preliminary evidence for the occasional occurrence of suppressive activity in lympohid cells associated with various neoplasms (15-17). The high efficacy, with only 10-20% suppressors being required for profound reduction of NK lysis, appears to distinguish the inhibitory activity of human alveolar macrophages from previously described systems of NK suppression in rodents and humans (5-8, 13-17).

The mechanisms by which alveolar macrophages inhibit NK activity are a matter of speculation. Alveolar macrophages were not lysed by blood lymphocytes, whereas low cytotoxicity values were occasionally seen when milk macrophages, devoid of inhibitory activity on NK cells, were used as targets. Therefore it appears unlikely that alveolar macrophages inhibit NK activity by acting as cold competitors for effector cells. In support of this contention, preliminary experiments indicate that suppression by low concentrations of alveolar macrophages is reversible following removal of the adherent suppressors. Alveolar macrophages reporteadly are better producers of prostaglandins than other mononuclear phagocytes (18) and prostaglandins can interfere with NK cytotoxicity (19).

Low levels of NK activity are associated with the human pulmonary tissue (20). Depletion of macrophages by adherence significantly enhanced NK activity and, to an even greater extent, augmented responsiveness of lung cells to INF (20). Therefore, the suppressive capacity of alveolar macrophages could play a role in determining the defective NK cytotoxic capacity in human lungs.

Pulmonary alveolar macrophages share many properties with mononuclear phagocytes from other tissues, but have distinctive structural, metabolic and functional characteristics (21). Alveolar macrophages were unique, among the human mononuclear phagocytes populations tested, in their capacity of inhibiting NK activity effectively at low concentrations. This finding emphasizes the heterogeneity in this respect within the human monocyte-macrophage lineage and suggests the existence of organ related differences in the mechanisms of regulation of NK cytotoxicity.

REFERENCES

1. Herberman, R.B. (ed.) (1980). "Natural Cell-mediated Immunity against Tumors" Academic Press, New York.
2. Herberman, R.B., and Holden, H.T. (1978). Adv. Cancer Res. 27: 305.
3. de Vries, J.E., Mendelsohn, J., and Bont, S.W. (1980). Nature, 283: 574.
4. Reynolds, C.W., Brunda, M.J., Holden, H.T., and Herberman, R.B. (1981). J. Natl. Cancer Inst. 66: 837.
5. Cudkowicz, G., and Hockman, P.S. (1979). Immunol. Rev. 44: 13.
6. Gerson, J.M. (1980). In "Natural Cell-mediated Immunity against Tumors" (R.B. Herberman, ed.), p. 1047. Academic Press, New York.
7. Santoni, A., Riccardi, C., Barlozzari, T., and Herberman, R.B. (1980). In "Natural Cell-mediated Immunity against Tumors" (R.B. Herberman, ed.), p. 753. Academic Press, New York.

8. Tracey, D.E., and Adkinson, N.F. Jr. (1980). J. Immunol. 125: 136.
9. Mantovani, A., Bar Shavit, Z., Peri, G., Polentarutti, N., Bordignon, C., Sessa, C., and Mangioni, C. (1980). Clin. Exp. Immunol. 39: 776.
10. Mantovani, A., Polentarutti, N., Peri, G., Bar Shavit, Z., Vecchi, A., Bolis, G., and Mangioni, C. (1980). J. Natl. Cancer Inst. 64: 1307.
11. Bordignon, C., Avallone, R., Peri, G., Polentarutti, N., Mangioni, C., and Mantovani, A. (1980). Clin. Exp. Immunol. 41: 336.
12. Herberman, R.B., Djeu, J.Y., Kay, H.D., Ortaldo, J.R., Riccardi, C., Bonnard, G.D., Holden, H.T., Fagnani, R., Santoni, A., and Puccetti, P. (1979). Immunol. Rev. 44: 43.
13. Ito, M., Ralph, P., and Moore, M.A.S. (1980). Clin. Immunol. Immunopathol. 16: 30.
14. Lotzova, E. (1980). In "Natural Cell-mediated Immunity against Tumors" (R.B. Herberman, ed.), p. 735. Academic Press, New York.
15. Mantovani, A., Allavena, P., Sessa, C., Bolis, G., and Mangioni, C. (1980). Int. J. Cancer 25: 573.
16. Eremin, O. (1980). In "Natural Cell-mediated Immunity against Tumors" (R.B. Herberman, ed.), p. 1011. Academic Press, New York.
17. Allavena, P., Introna, M., Mangioni, C., and Mantovani, A. (1981). J. Natl. Cancer Inst. 67: 319.
18. Morley, J., Bray, M.A., Jones, R.W., Nugteren, D.H., and van Dorp, D.A. (1979). Prostaglandins 17: 730.
19. Kendall, R.A., and Targan, S. (1980). J. Immunol. 125:2270.
20. Bordignon, C., Villa F., Vecchi, A., Giavazzi, R., Introna, M., Avallone, R., and Mantovani, A. Clin.Exp. Immunol. in press.
21. Cohen, A.B., and Cline, M.J. (1971). J. Clin. Invest. 50: 1390.

SUPPRESSION OF NK CELL ACTIVITY BY ADHERENT CELLS FROM MALIGNANT PLEURAL EFFUSIONS OF CANCER PATIENTS

Atsushi Uchida
Michael Micksche

Institute for Cancer Research
University of Vienna

Accumulated evidence suggests that natural killer (NK) cells play an important role in host resistance against tumors (herberman, 1980, 1981). The most potent antitumor activity may be expected to be present within or around the site of tumor growth. NK cells have been detected in small spontaneous and virus-induced mouse tumors, whereas NK activity has usually been undetectable in large tumors in mice (Gerson, 1980). In humans, lymphocytes isolated from solid tumors (Vose et al, 1977; Moore and Vose,1981) and carcinomatous pleural effusions (Uchida and Micksche, 1981a,b) have been shown to have no NK activity. For a thorough understanding of the interaction of effector cells and tumor cells, it is important to know the manner in which NK cell activity is regulated and the types of cells with which NK cells interact. We report here that the activity of NK cells is regulated by adherent suppressor cells in malignant pleural effusions of cancer patients.

Lymphocyte-rich mononuclear cells isolated from malignant pleural effusions by centrifugation on a discontinuous density gradient showed markedly low or no NK activity against highly sensitive K562 and other target cells in a 4-hour chromium-51 release assay. In contrast, the frequency of large granular lymphocytes in effusion cells was comparable to that in blood lymphocytes from normal donors and cancer patients. Human NK cell activity has been demonstrated to be exerted by a morphological subpopulation, large granular lymphocytes (Timonen et al., 1981). These results suggest that the lack of NK activity in malignant pleural effusions is unlikely to be due to the absence of NK cells but may rather result from the suppression of the activity of NK cells.

One mechanism for the lack of NK activity has been shown

ISBN 0-12-341360-5

TABLE I. De-Suppression of NK Activity of Effusion Cells

Effusion mononuclear cells	% Cytotoxicity*
Fresh unseparated	4
Fresh G10 passed	6
24-h cultured unseparated	2
G10 passed, then 24-h cultured	25**
24-h cultured, then G10 passed	3

* Results represent mean of 6 patients at E:T of 20:1.
** Value is significantly higher compared to other group ($P<0.05$).

to be the presence of suppressor cells in spleens of mice with young age or experimental treatments (Cudkowicz and Hochman, 1979; Hochman and Cudkowicz, 1979; Savary and Lotozova, 1978; Santoni et al., 1980). To detect cells involved in the lack of NK activity in malignant pleural effusions, effusion mononuclear cells were depleted of adherent cells on a Sephadex G-10 column. The mere removal of adherent cells gave no increase in NK activity (Table I). When these effusion cells depleted of adherent cells were cultured for 24 hours, they showed an increase in NK activity. The removal of adherent cells from 24-hour cultured cells resulted in no enhancement of NK activity. On the other hand, blood mononuclear cells from normal donors and cancer patients did not show such increase. These results suggest that carcinomatous pleural effusions contain potentially cytotoxic cells, which are functionally suppressed by adherent cells and can recover their lytic function after 24 hours of culture in the absence of suppressor cells.

To confirm the above suggestion, functional blood NK cells from normal donors and cancer patients were precultured for 24 hours, with or without adherent cells that were isolated from malignant pleural effusions of the same patients by adherence to a serum-coated plastic dish. As previously reported (Uchida and Micksche, 1981b), NK activity and cell composition of 24-hour cultured cells were similar to those of freshly drawn cells. Adherent effusion cells did not kill K562 cells in the assays. NK activity of blood lymphocytes was suppressed by adherent effusion cells (Table II). The suppressor activity of adherent effusion cells was noticeable at a suppressor to effector cell ratio of 0.05:1 and peaked at 0.1:1-0.2:1. In contrast, adherent cells from the peripheral blood of normal controls and cancer patients and from pleural effusions of non-malignant patients did not inhibit NK activity. The presence of adherent suppressor cells in malignant pleural effusions was observed in 31 of 43 cancer patients. In the other five

TABLE II. Suppression of NK Activity by Adherent Effusion Cells

Adherent cells added	% Cytotoxicity at E:T of 10:1	20:1	40:1
None	41	55	68
2.5%	38	51	64
5 %	21*	34*	48*
10 %	10*	20*	28*
20 %	8*	19*	26*

Blood lymphocytes were precultured alone or with adherent effusion cells for 24 hours, then washed and tested.
* Value is significantly lower than that cultured alone ($P < 0.05$).

patients, nylon wool nonadherent T cells from effusions had suppressive activity. Suppressor cells induced by carrageenan or hydrocortisone were found to be macrophage-like cells and suppressor cells from infant or irradiated mice to be null cells (Cudkowicz and Hochman, 1979). The presence of adherent suppressor cells has been observed in the circulation of some cancer patients (Vose, 1980) and nonadherent suppressor cells have been observed in ascites of patients with ascitic ovarian carcinoma (Allavena et al., 1981).

It seems unlikely that tumor cells contaminating in the suppressor cell population are responsible for the suppression of NK activity since adherent effusion cells consisting of 92-99% monocyte/macrophages inhibited NK activity at a suppressor to effector cell ratio of 0.1:1 and tumor cells with 92-98% purity were not able to inhibit NK activity at the same ratio. In addition, 24-hour precultures did not enrich contaminating tumor cells in the effector cell population. A cold inhibition assay based on purified tumor cells from pleural effusions has revealed no inhibition of NK activity against K562 cells in most cases (Uchida and Micksche, 1981a).

To ascertain whether the preculture of suppressor and NK cells is required for the suppression of NK activity, adherent effusion cells were tested for NK suppressor activity with or without preincubation of blood lymphocytes at either 4° C or 37° C. The presence of adherent effusion cells only during a cytotoxicity assay did not suppress NK activity (Table III). Preincubation of adherent effusion cells and blood lymphocytes for 24 hours at 37° C resulted in suppression of NK activity, while the preincubation at 4° C did not. The recovery of viable cells cocultured with adherent suppressor cells was comparable to that cultured alone. Then, blood lymphocytes cultured

TABLE III. Requirement of Preculture for Suppression of NK Activity

	% Cytotoxicity*		Suppression (%)
	PBL	PBL+AEC	
Without preincubation	44	45	-2
24-h preincubation at 4°C	41	38	7
24-h preincubation at 37°C	47	14	70**

Blood lymphocytes (PBL) were added by one-tenth the number of adherent effusion cells (AEC).
* E:T of 20:1.
** Value is significant ($P<0.05$).

for 24 hours with adherent effusion cells were passed through a Sephadex G10 column to remove the adherent suppressor cells. Suppression of NK activity was observed even when the adherent suppressor cells were removed after the preculture. These results suggest that adherent effusion cells suppress NK activity during 24-hour contact with NK cells by affecting but not by killing. It seems likely that events occuring during cytotoxic interactions between NK cells and target cells are not directly inhibited and the maintenance of functional NK cells is suppressed. In mouse models one type of suppressor cells primarily inhibits the lytic events of NK cells (Cudkowicz and Hochman, 1979). For this kind of suppression it is sufficient to mix effector, target and suppressor cells in a cytotoxicity assay. In other mouse models suppression of NK activity occurs in vivo and is reversible (Hochman and Cudkowicz, 1979), as in pleural effusions of cancer patients. In the latter models the mechanism of suppression may be similar to the one observed in the present study. The requirement for prolonged contact of suppressor and NK cells prior to cytotoxicity assays has quite recently been observed in vitro in mice (Brunda et al., 1981).

There is evidence that suppressor substances are involved in suppression of NK activity: Supernatants produced by cultures of suppressor cells from hydrocortisone-treated mouse spleens suppress NK activity (Cudkowicz and Hochman, 1979). Cell-free supernatants of malignant pleural effusions inhibit the generation of culture-induced cytotoxic cells but do not suppress spontaneous NK activity (Uchida and Micksche, 1981a, 1982a). Prostaglandins inhibit NK activity (Droller et al., 1978). However, 24-hour treatment of functional NK cells with effusion supernatants, culture supernatants of adherent effusion cells, or supernatants from 24-hour cultures of adherent effusion cells and blood lymphocytes resulted in no inhibition of NK activity. In addition, the presence of indomethacin, an inhibitor of prostaglandin synthetase, during the preculture and cytotoxicity assay did not abrogate suppression of NK activity by adherent effusion cells. These results suggest that suppression of NK activity by adherent effusion cells requires direct interactions with NK effector cells and is independent of prostaglandin induction. Alternatively, it might be possible that large amounts of prostaglandins are induced and that this process is not inhibited by indomethacin.

It is important to know whether adherent suppressor cells inhibit the recognizing function or killing function of NK cells. The frequency of large granular lymphocytes forming conjugates with K562 target cells was not reduced by adherent effusion cells after 24 hours of preculture, while NK activity was suppressed. The results suggest that NK cells lose their capacity of killing, but not of recognition, during 24 hours of contact with adherent suppressor cells.

Interferon has been shown to play a central role in the augmentation of NK activity (Gidlund et al., 1978). To know whether adherent effusion cells inhibit the enhancement of NK activity by interferon, blood lymphocytes were cultured for 20 hours alone or with interferon either in the presence or absence of adherent effusion cells. In the presence of adherent effusion cells, the baseline level of NK activity was reduced and no augmentation of NK activity by interferon was observed (Table IV). Similarly, effusion mononuclear cells were not able to develop NK activity in response to interferon. These results may indicate that adherent effusion cells inhibit both maintenance and augmentation of NK activity by interferon.

In conclusion, carcinomatous pleural effusions of cancer patients contain adherent cells capable of suppressing the development and maintenance of NK activity, which could be one

TABLE IV. Suppression of Spontaneous and Interferon-Induced NK Activity

	% Cytotoxicity at E:T of 20:1	
	PBL	PBL + AEC
Fresh	40	41
Cultured with medium	43	21*
Cultured with interferon	64	23*

Blood lymphocytes (PBL) and PBL plus one-tenth the nubmer of adherent effusion cells (AEC) were cultured alone or with leukocyte interferon (1,000 U/ml) for 20 hours.
* Value is significantly lower than that of PBL ($P < 0.05$).

cause of the lack of NK activity in the effusions. We have quite recently observed that the circulation of postoperative cancer patients contains adherent suppressor cells similar to the one observed in the present study (Uchida and Micksche, 1982b). Studies in which the modulation of these suppressor cells in vivo will result in a subsequent benefit to cancer patients are described in another chapter.

REFERENCES

Allavena, P., Introna, M., Mangloni, C., and Mantovani, A. (1981). J. Natl. Cancer Inst. 67:319.
Brunda, M. J., Taramelli, D., Holden, H. T., and Varesio, L. (1981). Proc. Am. Ass. Cancer Res. 22:310.
Cudkowicz, G., and Hochman, P. S. (1979). In "Developmental Immunology" (G. W. Siskind, D. D. Litwin, M. E. Weksler, eds.), p. 1. Grune & Stratton, London.
Droller, M. J., Schneider, M. U., and Perlman, P. (1978). Cell. Immunol. 39:615.
Gerson, J. M. (1980). In "Natural Cell-Mediated Immunity Against Tumors" (R. B. Herberman, ed.), p. 1047. Academic Press, New York.
Gidlund, M., Örn, A., Wigzell, H., Senik, A. and Gresser, I. (1978). Nature 273:759.
Herberman, R. B. (ed.) (1980). Natural Cell-Mediated Immunity Against Tumors, Academic Press, New York.
Herberman, R. B. (1981). Human Lym. Differentiation 1:63.
Hochman. P. S., and Cidkowicz, G. (1979). J. Immunol. 120:968.
Moore. M., and Vose, B. M. (1981). Int. J. Cancer 27:265.
Santoni, A., Riccardi, C., Scorci, V., and Herberman, R. B. (1980). J. Immunol. 124:2329.
Savary, C. A., and Lotozova, E. (1978). J. Immunol. 120:239.
Timonen, T., Ortaldo, J. R., and Herberman, R. B. (1981). J. Exp. Med. 153:569.
Uchida, A., and Micksche, M. (1981a). Cancer Immunol. Immuno. ther. 11:131.
Uchida, A., and Micksche, M. (1981b). Cancer Immunol. Immunother. 11:255.
Uchida, A., and Micksche, M. (1982a). Oncology (in press).
Uchida, A., and Micksche, M. (1982b). J. Natl. Cancer Inst. (in press).
Vose, B. M., Vankey, R., Argov, V., and Klein, E. (1977). Eur. J. Immunol. 7:735.
Vose, B. M. (1980). In "Natural Cell-Mediated Immunity Against Tumors" (R. B. Herberman, ed.), p. 1099. Academic Press, New York.

INHIBITION OF NATURAL KILLER CELL CYTOTOXIC REACTIVITY BY TUMOR PROMOTERS AND CHOLERA TOXIN

Ronald H. Goldfarb[1,2]
Ronald B. Herberman

Laboratory of Immunodiagnosis, NCI
NIH, Bethesda, MD

I. INTRODUCTION

Natural killer(NK) cells are a non-phagocytic and non-adherent lymphoid subpopulation that mediates anti-tumor cytolytic activity(1,2). NK cell-mediated anti-tumor immune reactivity is directed against both primary and metastatic cancer(3,4). We have recently reported(5) that NK cells may be isolated to a high state of purity, and that the enriched NK cells have the morphological characteristics of large granular lymphocytes (LGL). Pure populations of NK cells have been utilized to probe both the molecular events involved in the NK lytic mechanism(2,6-8) and the control over NK cell reactivity(9).

A great deal of experimental evidence indicates that NK cells play an important role as a primary immune defense in surveillance against malignant disease(1). NK cells, which bind to, and kill tumor cells, exist in unimmunized individuals and therefore do not require prolonged time periods to be primed or activated(1); in contrast, various other lymphoid effector cells require days or even weeks to be cytolytically activated. If NK cells can indeed play an active role in anti-tumor immune surveillance, than it is possible that either the initiation or promotion of the malignant state is dependent

---

[1] Supported by PHS Fellowship 1F32CA06681 to RGH by NCI. We thank Dr. Takashi Sugimura of the National Cancer Center Research Institute, Tokyo, Japan, for kindly providing us with the teleocidin used in this study.

[2] Present address: Cancer Metastasis Research Group, Department of Immunology and Infectious Disease, Pfizer Central Research, Pfizer Inc., Groton, Connecticut.

ISBN 0-12-341360-5

upon, or associated with, the inhibition or loss of NK cytolytic activity(9). In order to gain insight into the possible role of NK cell function in protection against tumor promotion and in order to further study the regulation of NK cytotoxic action, we have investigated the effects of various agents including the potent tumor promoters phorbol-12-myristate-13-acetate(PMA) and teleocidin, as well as cholera toxin.

II. TUMOR PROMOTERS INHIBIT NK CYTOTOXIC REACTIVITY

Tumor promoting phorbol esters(10,11) and teleocidin(12) are the most effective known tumor promoters ever studied; although teleocidin and PMA appear to have similar mechanisms of action, their chemical structures are unrelated, and distinct(12). Whereas tumor promoters alone are non-carcinogenic they enhance tumorigenesis in animals that have been challenged with a sub-threshhold dose of a carcinogen(12). Tumor promoters have diverse pleiotropic biological effects upon cells treated in culture(14) including lymphoid cell cultures (1). Tumor promoters can function as lymphocyte mitogens or co-mitogens(15-17); induce T lymphocyte activation(18); affect the mixed lymphocyte reaction, and bind to receptors on or within human peripheral blood lymphocytes(19,20). In addition PMA inhibits the _in vitro_ synthesis of antibodies directed against red blood cells(21).

We have recently reported(9) that the tumor promoting phorbol esters, but not non-tumor promoting structural analogs, inhibited both mouse as well as human NK cell activity. PMA inhibited both spontaneous and interferon(IFN) boosted NK cell activity(Table I). The sequence of NK cell treatment with IFN and PMA affected the extent of tumor promoter-mediated inhibition of IFN-boosted NK activity; PMA had greater inhibitory activity when added to NK cells prior to IFN. Pre-treatment of effector cells with PMA for periods as short as 1-30 minutes was adequate for inhibition of NK activity.

NK cells(LGL) showed one of two typical responses to PMA treatment: for some donors both spontaneous and IFN-boosted NK cell activities were diminished, and for others, only the IFN-boosted activity was inhibited. In addition, PMA can also inhibit NK activity augmented by retinoic acid. Our data demonstrates the inhibition of LGL spontaneous, and IFN-boosted, lytic activity by PMA; the results in Table I also demonstrate that a distinct tumor promoter, teleocidin, also inhibits LGL-mediated NK activity.

It has been independently reported by others(22) that tumor promoting phorbol diesters suppress NK activity _in vitro_ by human peripheral mononuclear cells. Whereas we have demonstrated(9) that PMA has a direct effect on isolated LGL populations(90-95% purity), others(22) have suggested that monocytes mediate the PMA suppressive effect. Experiments with more highly purified LGL(99%), prepared by more stringent methods(7) shall be required to further investigate this issue.

TABLE I. Inhibition of NK cell(LGL) Cytolysis by Tumor Promoters and Cholera Toxin

Incubation Conditions	Lytic Units* (-IFN)	Lytic Units* (+IFN 1000μ/ml)
Input (non-purified cells)	56	N.D.
Purified human NK cells(LGL)	160	220
LGL + PMA(100 ng/ml)	20	146
LGL + Teleocidin(100 ng/ml)	16	142
LGL + Cholera Toxin(10 ng/ml)	0	0
LGL + Cholera Toxin A Subunit (100 ng/ml)	158	224
LGL + Cholera Toxin B Subunit (100 ng/ml)	149	216

*1 lytic unit = the number of effector cells required to lyse 30% of K562 cells.

It has also been reported(23) that tumor promoting phorbol esters inhibit rat NK cell activity both in vitro as well as in vivo. Recently, we have extended our murine findings to in vivo studies in which we have also noted profound inhibition of NK reactivity in response to PMA (R.B. Herberman and M. Nunn-Hargrove, unpublished observations).

In contrast to the studies described above, however, it should be noted that one study has suggested that PMA might lead to augmentation of NK activity, rather than inhibition of murine natural cell-mediated cytolysis (J.B. Kolb, A. Sennik and M. Castagna, personal communication); this group has suggested that PMA can alter susceptibility to YAC-1 cells. Interestingly, PMA has been shown to elevate the NK sensitivity of various human lymphoid lines (Daudi, Raji, and Molt-4) when they were cultured in the presence of PMA(24); it was reported that PMA-treated cells showed an increased binding to human but not mouse lymphocytes(24). In recent studies that employed human LGL, we have examined the effect of PMA on binding and lysis of human NK cell-K562 cell conjugates(25); we have observed that PMA can inhibit LGL-tumor cell conjugate formation.

It is clear that the use of enriched NK cells(LGL) as opposed to less pure populations, and the use of different target cells, and the conditions of the time and sequence of PMA incubation conditions, may all have some effect on the PMA-triggered inhibition of NK cytotoxic activity. With the use of pure LGL populations, it is also clear that several classes of potent tumor promoters(PMA and teleocidin) have the capacity to directly inhibit the binding and lytic capacity of human NK cells.

III. INHIBITION OF NK REACTIVITY BY CHOLERA TOXIN

We have also examined the role of cholera toxin in the inhibition of NK cytolytic function(9). Cholera toxin has previously been shown to inhibit T cell-mediated cytolysis(26) and to inhibit the biosynthesis of IFN by lymphocytes(27). The mechanism of cholera toxin inhibition of T lymphocyte lysis has been related to the induction of cyclic AMP(28,29) or to the binding of ganglioside receptors that are apparently common to both IFN and cholera toxin(30,31). IFN inhibits cholera toxin binding to its receptors(30,31) and cholera toxin apparently inhibits IFN activity at the level of a receptor composed of both ganglioside and glycoprotein(32). It is of interest that type I IFN, used in our study, but not type II IFN, can bind to gangliosides, and its anti-growth and its anti-viral properties are inhibited by gangliosides (33). Since NK lytic activity is affected by both cyclic nucleotides(34) as well as by IFN(1) we examined cholera toxin, and its subunits, to probe aspects of the modulation of NK cell regulation. As demonstrated in Table I, cholera toxin profoundly inhibits spontaneous, as well as IFN-boosted, LGL-mediated lysis of K562 tumor cells(9). Cholera toxin also abrogates retinoic acid-boosted NK activity. Recently, we have observed that whereas cholera toxin profoundly inhibited the lytic reaction, it had no effect on the binding of LGL to target cells(25).

The data in Table I also demonstrates that neither cholera toxin A subunit or B subunit, alone, inhibits NK activity; it therefore appears that the intact toxin is required for the suppression of NK activity(9). In contrast to PMA, cholera toxin treatment abrogates IFN activity regardless of the sequence of treatments with boosting stimuli, such as IFN or retinoic acid. It has been independently reported that cholera toxin inhibits spontaneous natural cell-mediated cytolysis(26,27), as well as IFN-augmented activity(26).

It has generally been concluded that the inhibition of T cell cytotoxicity induced by cholera toxin is caused by stimulation of cyclic AMP(28,29). The relationship of cholera toxin and ganglioside receptors for IFN, described above, suggested that an alternative mode of inhibition may be involved for NK cells. Our observations with cholera toxin subunits, however, suggest that the mechanism of cholera toxin inhibition is not merely due to binding to a putative IFN receptor; subunit A, responsible for adenylate cyclase enhancement, or subunit B, which binds to the GM1 ganglioside, each are unable to mimic intact cholera toxin in the inhibition of NK cytotoxic reactivity.

IV. CONCLUSION

Our results demonstrate that the tumor promoters PMA and teleocidin, as well as cholera toxin, lead to potent inhibition of both spontaneous NK activity by human LGL, as well as IFN or retinoic acid-boosted activity. Tumor promoters can be probes for the elucidation of the role of NK cell reactivity relative to tumor promotion. Intervention with tumor-promoter mediated depression of NK cytotoxic reactivity, may be of eventual clinical significance; alternatively, stimulation of NK cytotoxic reactivity may interfere with the onset of tumor promotion.

1. Herberman, R.B., "Natural Cell-Mediated Immunity Against Tumors". Academic Press, New York, 1980.
2. Goldfarb, R.H., and Herberman, R.B., in "Adv. in Inflammation Res." (G. Weissmann, ed.) 4:45 (1982).
3. Gorelik, E., Fogel, M., Feldman, M., and Segal, S., J. Natl. Cancer Inst., 63:1397 (1979).
4. Hanna, N., and Burton, R.C., J. Immunol. 127:1754 (1981).
5. Timonen, T., Ortaldo, J.R., and Herberman, R.B., J. Exp. Med., 153:569 (1981).
6. Goldfarb, R.H., Timonen, T., and Herberman, R.B., Adv. Exper. Biol. and Med. In Press.
7. Goldfarb, R.H., Timonen, T., and Herberman, R.B., Submitted for Publication.
8. Hoffman, T., Hirata, F., Bougnoux, P., Fraser, B.A., Goldfarb, R.H., Herberman, R.B., and Axelrod, J., Proc. Natl. Acad. Sci., USA 78:3839 (1981).
9. Goldfarb, R.H., and Herberman, R.B., J. Immunol., 126: 2129 (1981).
10. Van Duuren, B.L., Prog. Exp. Tumor Res. 11:31 (1969).
11. Hecker, E., Methods Cancer Res. 6:429 (1971).
12. Sugimura, T., Fujiki, H., Mori, M., Nakayasu, M., Terada, M., Umezawa, K., and Moore, R.E., in "Cocarcinogenesis and Biological Effects of Tumor Promoters" (E. Hecker, ed) Raven Press, New York. In Press.
13. Berenblum, I., in "Cancer, A Comprehensive Treatise. Vol. I." (F. Becker, Ed.) Plenum Publishing Co., New York,1975.
14. Goldfarb, R.H., and Quigley, J.P., Cancer Res. 38:4601 (1978).
15. Estensen, R.D., Hadden, J.W., Hadden, E.M., Touraine, F., Touraine, J., Haddox, M.K., and Goldberg, N.G., in "Control of Proliferation of Animal Cells" (B. Clarkson and R. Basenga, Eds.) Cold Spring Harbor Laboratory, N.Y., 1973.
16. Mastro, A.M., and Mueller, G.C., Exp. Cell Res. 88:40 (1974).

17. Skinnider, L., and Giesbrecht, K., Cancer Res. 39:3332 (1979).
18. Rosenstreich, D.L., and Mizel, S.B., J. Immunol. 123:1749 (1979).
19. Mastro, A.M., Krupa, T.A., and Smith, P., Cancer Res., 39:4078 (1979).
20. Estensen, R.D., DeHoogh, D.K., and Cole, C.F., Cancer Res., 40:1119 (1980).
21. Castagna, M., Yagello, M., Rabourdin-Combe, C., and Fridman, W.H., Cancer Letters, 8:365 (1980).
22. Seaman, W.E., Gindhart, T.D., Blackman, M.A., Dalal, B., Talal, N., and Werb, Z., J. Clin. Invest., 67:1324 (1981).
23. Keller, R., Nature, 282:729 (1979).
24. Patarroyo, M., Biberfeld, P., Klein, E., and Klein, G., Cellular Immunol., 63:237 (1981).
25. Ortaldo, U.R., Timonen, T., Goldfarb, R.H., and Herberman, R.B., Submitted for Publication.
26. Fuse, A., Sato, T., and Kuwata, T., Int. J. Cancer, 27:29 (1981).
27. Fuyama, S., Sendo, F., Watabe, S., Seiji, K., and Arai, S., Gann, 72:141 (1981).
28. Lichtenstein, L.M., Henney, C.S., Bourne, H.R., and Greenough, W.D., J. Clin. Invest., 52:691 (1973).
29. Henney, C.S., Gaffney, J., and Bloom, B.R., J. Exp. Med., 140:837 (1974).
30. Grollman, E.F., Lee, G., Ramos, S., Lazo, P.S., Kabak, R., Friedman, R., and Kohn, L., Cancer Res. 38:4172 (1978).
31. Kohn, L.D., Friedman, R.M., Holmes, J.M., and Lee, G., Proc. Natl. Acad. Sci., USA, 73:3695 (1976).
32. Friedman, R.M., and Kohn, L.D., Biochem. Biophys. Res. Comm. 70:1078 (1976).
33. Ankel, H., Krishnamurti, C., Besancon, F., Stefano, S., and Falcoff, E., Proc. Natl. Acad. Sci., USA 77:2528 (1980).
34. Roder, J.C., and Klein, M., J. Immunol. 123:2785 (1979).

TUMOR-PROMOTING DITERPENE ESTERS INDUCE MACROPHAGE DIFFERENTIATION, BUT PREVENT ACTIVATION FOR TUMORICIDAL ACTIVITY OF MACROPHAGE AND NK CELLS[1]

Robert Keller

Immunobiology Research Group
Institute of Immunology and Virology
University of Zurich
Switzerland

INTRODUCTION

Tumor-promoting diterpene esters facilitate the formation of tumors in the skin of mice previously treated ('initiation') with subthreshold doses of chemical carcinogen (1,2). Tumor promoters elicit a considerable array of in vitro effects on cells in culture, often in opposing ways (2,3). Among other attributes, tumor promoters stimulate DNA synthesis, initiate proliferation in various cell types, induce various enzymes and evoke changes in cell membranes and chromosomal damage. It is now increasingly appreciated that tumor promoters also affect mononuclear phagocytes at the differentiation and effector levels, and with opposing consequences. The present paper summarizes recent work in this laboratory, directed primarily towards their efficacy in inducing differentiation in mononuclear phagocyte precursors and the ways in which these diterpene esters affect the spontaneous and induced extracellular cytotoxicity of macrophages and NK cells.

[1]This work was supported by the Swiss National Science Foundation (grants 3.173.77 and 3.609,80) and the Canton of Zurich.

ISBN 0-12-341360-5

Tumor-Promoting Phorbol Esters Induce Macrophage Differentiation in Bone Marrow Precursors

Concurrent work in several laboratories suggests that the tumor-promoting phorbol ester, 12-O-tetradecanoyl-phorbol-13-acetate (TPA), induces in bone marrow cells the proliferation and differentiation of precursors of the mononuclear phagocyte lineage (4,5). Our own studies indicate that TPA, added at an optimal concentration of 10^{-8}M on days 0 and 4, induces in rat and mouse bone marrow cells, cultured in fluid medium, a sequence of events remarkably similar to those initiated by natural growth regulators such as macrophage colony stimulating activity (MCSA). TPA initiates DNA synthesis and cell proliferation in some but by no means all marrow cells. In parallel, the small, round, floating cells increasingly adhere to the substratum and spread, finally attaining a considerable size. DNA synthesis is especially active in the early phase of adherence and spreading. These processes are associated with an increase in lysozyme secretion, increased expression of plasma membrane receptors for the Fc portion of IgG, and enhanced ability to manifest immunologically nonspecific tumoricidal activity upon incubation with macrophage-activating lymphokines (MAL). As TPA and MCSA from different sources are, on the whole, equally effective in eliciting this sequence of events, the findings suggest that TPA and MCSA stimulate precursors of the mononuclear phagocyte lineage preferentially, eliciting their differentiation into mature macrophages. It is noteworthy that these processes proceed in a comparable manner irrespective of whether the same agent, either MCSA or TPA, was present throughout the 5-7 d culture period, or whether the new medium, added on day 3 or 4, was supplemented with the other agent. At the varying stages of differentiation, enhancement of macrophage-type tumoricidal activity by MAL could only be induced when TPA was omitted; in the presence of TPA, such MAL effect was abolished. Moreover, the TPA effect on cell multiplication and differentiation is markedly enhanced by MCSA in concentrations which themselves manifest but weak activity in the absence of the tumor promoter. These findings are most directly explained by postulating a TPA-induced increase in the susceptibility of the cell to MCSA by increasing either the number or the affinity of cell surface receptors for this growth regulator. It is moreover noteworthy that cells taken from 2 day through 8 day rat and mouse bone marrow cultures in no instance manifested NK activity. This finding suggests that differentiation of precursors of the mononuclear phagocyte lineage is not necessarily associated with even a transitory manifestation of NK activity. Furthermore, interaction of adherent bone marrow culture cells with MAL at any time during the differentiation process not only effectively enhanced their tumoricidal activity but in parallel led to a swift and marked suppression of cell proliferation. The major findings are summarized in Table I.

Tumor-Promoting Diterpene Esters Prevent Augmentation of Macrophage and NK Tumoricidal Activity

Interferon. Both NK- and macrophage-mediated spontaneous cytolytic activity is rapidly and profoundly augmented by interferon. In the presence of TPA, this interferon-induced enhancement of NK and macrophage cytotoxic activity is prevented (6). However, the classical antiviral activity and the specific binding of interferon to cell surface receptors remain quite unaffected by TPA, indicating that the abolition of host spontaneous tumoricidal effector function is not exerted via direct interaction with the interferon molecule nor by blocking of interferon binding to its specific receptor (6).

Macrophage-Activating Lymphokines (MAL). Cell-free supernatants from rat spleen cells, cultured for 72 h in serum-free medium supplemented with concanavalin A, consistently enhanced to a marked extent macrophage-type long-term cytotoxicity. The presence of TPA and other tumor promoters during the activation phase abolished this MAL-induced augmentation of macrophage tumoricidal activity (6-8). It should be noted that in no instance did MAL preparations themselves augment NK activity (6).

CONCLUSIONS

The present findings together with earlier work showing that tumor promoters elicit in mononuclear phagocytes diverse in vitro effects such as the secretion of toxic products of oxidative metabolism (e.g. superoxide anion, hydrogen peroxide; 9,10), enhancement of prostaglandin E (11) and enzyme production (e.g. plasminogen activator; 12), attest to the manifold effects of these agents operative at various levels, and with

TABLE I. Kinetics of the Events Induced by TPA and MCSA in Cultures of Rat and Mouse Bone Marrow Cells

culture (days)	total cell number	lysozyme secretion	nonadherent cells				adherent cells			
			number	DNA synthesis	Fc receptors	MAL-ind. cytotoxicity*	number	DNA synthesis	Fc receptors	MAL-ind. cytotoxicity*
0		+	+++				(+)	(+)	(+)	
2	-									
3	-	+	+++	(+)	(+)	+	+	+	+	++
4	±	+	++	(+)	(+)	+	+	++	++	++
5	± → +	++	++	(+)	(+)	+	++	+++	++	+++
6	+	++	+	(+)	(+)	+ → ++	+++	+++	+++	+++
7	+	+++	(+)	(+)	(+)	+ → ++	+++	++	+++	+++
8	+	+++	(+)	(+)	(+)	++	+++	+	+++	+++

± constant; - decrease; + increase; (+)→+ low number or function; ++→+++ high number or function. * MAL are operative only in the absence of TPA.

different consequences. Especially noteworthy is the evidence indicating that tumor promoters selectively stimulate precursors of the mononuclear phagocyte lineage to proliferate and differentiate (4,5). In sharp contrast, induction in these cells of elevated cytolytic capacity by interferon or macrophage-activating lymphokines is prevented by TPA (6). The different and partly opposing ways by which tumor promoters interfere with the capacity of macrophages and NK cells to manifest extracellular cytotoxicity (Table II) provide a compelling example of the complexity of the interactions between host cellular defense and tumor. They show moreover the ease with which this delicate balance can be shifted in favour of, or to the detriment of the host.

TABLE II. Principal Ways in Which Tumor Promoters Interfere with the Acquisition and Manifestation of Extracellular Cytotoxicity by Mononuclear Phagocytes and NK Cells

1. Suppression of the spontaneous cytolytic activity manifested by NK cells and macrophages (7,13).
2. Induction of macrophage-mediated cytostasis (14).
3. Triggering in mononuclear phagocytes of short term killing of target cells susceptible to toxic products of oxidative burst (H_2O_2; 9).
4. Prevention of the interferon-induced augmentation of NK and macrophage cytolytic activity (6).
5. Prevention of lymphokine-induced enhancement of macrophage cytolytic activity (6-8).
6. Inhibition of the cytolytic capacity manifested by induced, activated macrophages (8,13).

ACKNOWLEDGMENTS

I am grateful to Ms. R. Keist, F. Chytil and R. Werschler for skilled technical assistance, and Dr. M. Landy for critical review of the manuscript.

REFERENCES

1. Boutwell, R.K. (1974). CRC Crit. Rev. Toxicol. 2:419.
2. Slaga, T.J., Sivak, A., and Boutwell, K. (1978). "Mechanisms of Tumor Promotion and Carcinogenesis." Academic Press, New York.
3. Emerit, I., and Cerutti, P.A. (1981). Nature (Lond.) 293: 144.
4. Stuart, R.K., and Hamilton, J.A. (1980). Science 208:402.
5. Keller, R., and Keist, R. (submitted).
6. Keller, R., Aguet, M., Tovey, M., and Stitz, L. (1982) Cancer Research (in press).
7. Keller, R., Keist, R., and Hecker, E. (1982). Br. J. Cancer (in press).
8. Keller, R., Keist, R., Adolf, W., Opferkuch, H.J., Schmidt, R., and Hecker, E. (1982). Exp. Cell Biol. (in press).
9. Nathan, C.F., Silverstein, S.C., Brukner, L.H., and Cohn, Z.A. (1979). J. exp. Med. 149:100.
10. Johnston, R.B., Chadwick, D.A., and Pabst, M.J. (1980). In "Mononuclear Phagocytes, Functional Aspects." (R. van Furth, ed.), p. 1143. M. Nijhoff, The Hague.
11. Brune, K., Glatt, M., Kälin, H., and Peskar, B.A. (1978). Nature (Lond.) 274:261.
12. Vassalli, J.-D., Hamilton, J., and Reich, E. (1977). Cell 11:695.
13. Keller, R. (1979). Nature (Lond.) 282:729.
14. Grimm, W., Bärlin, E., Leser, H.-G., Kramer, W., and Gemsa, D. (1980). Clin. Immunol. Immunopathol. 17:617.

INCREASE IN INTRA-CELLULAR LEVELS OF CYCLIC AMP INHIBITS TARGET CELL RECOGNITION BY HUMAN NK CELLS

Måns Ullberg[1]
Mikael Jondal
Gunnar Klein

Department of Tumor Biology
Karolinska Institutet
Stockholm, Sweden

Fred Lanefeldt
Bertil B. Fredholm

Department of Pharmacology
Karolinska Institutet
Stockholm, Sweden

I. INTRODUCTION

Cyclic adenosine 3´, 5´-monophosphate (cAMP), being a "second messenger" to several regulatory agents acting on lymphocytes, has a pronounced effect on several aspects of lymphocyte function (Parker et al., 1974; Goffstein et al., 1980). Thus it has been demonstrated that increased intracellular levels of cAMP inhibits both NK and T cell cytotoxicity (Strom et al., 1973; Wolberg et al., 1978; Targan, 1981; Roder and Klein, 1981). These studies were generally done on the whole effector population level. Few studies have tried to dissect the finer mechanism behind the cAMP induced inhibition, i.e. whether it acts on the level of target cell recognition, on the lytic event or on the effector cell recycling level. Targan has reported that prostaglandin E_2 inhibits

[1] Supported by the Swedish Cancer Society.

ISBN 0-12-341360-5

interferon boosting of NK cytotoxicity at the single cell level and that this inhibition is unrelated to target cell recognition, but did not relate this to cAMP levels (Targan, 1981). Roder and Klein (1981) have studied the influence of cAMP and drug that elevate cAMP levels, on NK cell killing and NK cell recognition. They found that dibutyryl cAMP and the cAMP elevating agents prostaglandin E_1, theophylline and histamine markedly suppressed murine NK activity in a dose- and rate-dependent manner without affecting the frequency of NK-target cell conjugates.

In the present work we have further investigated the role of cAMP in the regulation of human NK cell cytotoxicity. We have increased intra-cellular levels of cAMP by stimulating adenylate cyclase activity with prostaglandin E_1 (PGE_1) or by inhibiting phosphodiesterase break-down of cAMP using the inhibitor ZK62711 (Schwabe et al., 1976), or we have mimicked the effects of cyclic AMP by adding the dibutyryl derivative.

All three procedures produced a statistically significant inhibition of target binding cells (TBC:s). This cAMP dependent effect contrasts to the effect that interferon has on human NK cells. Interferon stimulation does not affect the percentage of TBC:s but effectuates its enhancing influence by increasing the efficiency of the lytic and recycling events as described elsewhere in this publication (Ullberg and Jondal, this publication).

II. METHODS

Nylon wool purified lymphocytes were used in all experiments to minimize nonspecific binding. Chromium release assays were performed with quadruplicate wells in V-shaped microplates. Pre-incubation time was one hour with all drugs. Molt-4 was used as target cell in all experiments. A dilution series of target cells with six different concentrations (diluted 3:5 with 2×10^5 target cells as the highest concentration) was used together with a constant number of effector lymphocytes (10^5). Cytotoxicity assays were run for 3 hours at 37°C. Percentage lysis was determined by using the formula:

$$100 \times \frac{\text{(test release) - (spontaneous release)}}{\text{(80\% of total label) - (spontaneous release)}}$$

The number of killed target cells in each well is obtained by multiplying the percentage lysis by the initial number of target cells. The results from the chromium assays fit to the Michaelis-Menten equation expressed as:

$$V = \frac{V_{max} \times T}{Km + T}$$

In this equation V represents the number of killed target cells at the end of the assay, T the initial number of target cells, Vmax the maximal lytic potential (when the number of target cells approaches infinity) and Km the Michaelis-Menten constant. The Vmax values are determined by using the Lineweaver-Burk plot. In this the reciprocal values of V are ploted as a function of the reciprocal values of T (Ullberg and Jondal, 1981). Linear regression analysis is used to obtain a straight line and the Vmax values are obtained from the reciprocal of the Y intercept.

Cytotoxic assays in agarose were performed as described by Grimm and Bonavida (1979) with slight modifications. Equal numbers of lymphocytes and target cells (2×10^5 in a total volume of 0.2 ml of medium) were spun together and incubated for 10 minutes. The pellet was subsequently resuspended by firm shaking for 4 sec on a whirl-mixer. 0.5 ml of 0.5% agarose solution kept at 48°C was precooled for 20 sec and added to the cells. The suspension of cells in agarose was mixed with a Pasteur pipette and the solution dripped from the height of 30 cm onto plastic Petri dishes (Nunc 60 mm) to form a thin layer. After solidifying, 6 ml of medium with 10% FCS was added. The dishes were incubated for 3 hours at 37°C. During this incubation the active killer cells bound to target cells would kill their targets. The dishes were subsequently stained with trypan blue, washed 3 times in PBS for 10 min and finally fixed with 1% formic aldehyde solution. The dishes were scored using an inverted microscope. The percentage TBC:s (target binding cells) was determined by counting the number of conjugates in 400 lymphocytes. The percentage dead conjugates was determined by counting the number of dead target cells in 100 conjugates. By multiplying these two parameters the fraction of active NK cells of all lymphocytes was calculated.

The maximal recycling capacity (MRC) was estimated by combining data in the chromium and agarose assays. The Vmax values were divided with the number of active NK cells (fraction of active NK cells multiplied with the number of lymphocytes in the chromium assays (10^5 cells). The calculated MRC values reflect the average recycling potential among the active NK cells in three hours given a satur-

TABLE I. Effect of increased intra-cellular levels of cAMP by addition of exogenous dibuturyl cAMP on human NK cell activity[a]

Treatment of cells	Parameters				
	Vmax[b] ($x10^3$)	TBC:s[c] (%)	Cytotoxic TBC:s[d] (%)	Active NK cells[e] (%)	Mean recycling capacity[f] (cells/3h)
Dibutyryl[h] cAMP	5.2	4.4	40	1.7	3.2
Control	11.9	8.1	42	3.3	3.5
Significance[g]	$p<0.025$	$p<0.01$	N.S.	$p<0.025$	N.S.

[a]Values given are means from four separate experiments. [b]Estimated as described in the Methods section. [c]Target binding cells. [d]The percentage of TBC:s that actually kill in a 3 h assay in agarose. [e]The percentage of active NK cells in the whole tested cell population. [f]The mean number of target cells that an average single NK cell can kill in a 3 h ^{51}Cr release assay under optimal conditions. [g]Statistical significance of differences between treated and untreated cell fractions (Student's paired t-test). [h]Concentration of dibutyryl cAMP was 5×10^{-4} M.

ated system.

The levels of cyclic AMP were determined in short term cultures of lymphocytes essentially as described earlier (Hynie *et al.*, 1980; Fredholm *et al.*, 1978).

III. RESULTS

Irrespective if intra-cellular cAMP levels were increased by addition of dibuturyl cAMP (Table I), adenylate cyclase stimulation (Table II) or by inhibition of phosphodiesterases (Table III) it was clear that the over-all lytic capacity (Vmax) decreased as a consequence of impaired NK cell recognition. The remaining TBC fraction comprised a similar percentage of lytic cells, compared to the TBC fraction of untreated control cells, which also recycled normally.

When nylon wool column-passed peripheral lymphocytes

TABLE II. Effect of increased intra-cellular levels of cAMP by adenylate cyclase stimulation with prostaglandin E_1 on human NK activity[a]

Treatment of cells	Parameters				
	Vmax[b] ($x10^3$)	TBC:s[c] (%)	Cytotoxic TBC:s[d] (%)	Active NK cells[e] (%)	Mean recycling capacity[f] (cells/3 h)
PGE_1[h]	4.4	3.8	42	1.6	2.6
Control	13.6	7.7	45	3.4	4.0
Significance[g]	$p < 0.005$	$p < 0.0005$	N. S.	$p < 0.0005$	N. S.

[a]Values given are means from five separate experiments. [b-g]See Table I. [h]Concentration of PGE_1 was 5×10^{-7} M in 0.01% ethanol.

TABLE III. Effect of increased intra-cellular levels of cAMP by inhibition of phosphodiesterase breakdown of cAMP on human NK cell activity[a]

Treatment of cells	Parameters				
	Vmax[b] ($x10^3$)	TBC:s[c] (%)	Cytotoxic TBC:s[d] (%)	Active NK cells[e] (%)	Mean recycling capacity[f] (cells/3 h)
ZK62711[h]	15.6	5.8	74	4.3	3.8
Control	28.5	10.3	71	7.3	3.9
Significance[g]	$p < 0.025$	$p < 0.005$	N. S.	$p < 0.005$	N. S.

[a]Values given are means from four separate experiments. [b-g]See Table I. [h]Concentration of ZK62711 was 5×10^{-6} M in 0.01% ethanol.

were fractionated according to OKT3 expression and stimulated with PGE_1 and isoprenalin, the OKT3-negative fraction was found to be comparatively highly responsive in cAMP production. This fraction is known to contain most of the NK activity in peripheral blood (Platsoucas and Good, 1981;

Zarling *et al.*, 1981).

IV. DISCUSSION

The activity of the human NK cell is quite labile and susceptible to external regulatory signals. The system is not easily standardized, a circumstance which may be related to the fact that intracellular levels of cAMP are known to vary rapidly *in vivo*, possibly as a consequence of regulatory agents such as prostaglandins, adenosine and adrenergic substances (Goffstein *et al.*, 1980; Strom *et al.*, 1973; Wolberg *et al.*, 1978; Targan, 1981). Also, *in vitro*

TABLE IV. Induction of cAMP in lymphocyte subpopulations by stimulation with prostaglandin E_1 and with isoprenalin.

Treatment of cells[a]	Lymphocyte population[b]				
	Ficoll-Isopaque isolated cells	Nylon wool column passed cells	OKT3+ cells	OKT3- cells	OKT3+[c] and OKT3- cells
	(pmol/10^6 cells $\pm$ S.E.)[d]				
PGE_1	33.4$\pm$2.9	6.4$\pm$2.2	1.8$\pm$0.8	14.9$\pm$1.2	6.4$\pm$0.5
Isoprenalin	18.3$\pm$2.3	4.3$\pm$1.2	1.9$\pm$0.7	9.2$\pm$1.1	4.3$\pm$0.8
Control	1.8$\pm$0.3	<1.0[e]	<1.0	<1.0	<1.0

[a]Cells were treated for 15 min at 37°C in standard microplates, PGE_1 was used at a concentration of 10^{-6} M and isoprenaline at 3×10^{-5} M. [b]Nylon wool column passed peripheral lymphocytes were fractionated with commercial OKT3 monoclonal antibodies purchased from Ortho Diagnostics using plastic Petri dishes coated with immunosorbent purified rabbit anti-mouse Ig antibodies as earlier described (Ullberg and Jondal, in press). [c]As control for cAMP induction caused by the fractionation procedure, OKT3- and OKT3+ cells were mixed back in initial proportions. [d]Measured by the competitive protein binding technique of Brown *et al.* (1979). [e]Not detectable levels below 1 pmol/10^6 cells.

manipulations during fractionation procedures are likely to influence the activity (Fredholm et al., 1978). Goffstein et al. have shown that a 2 h incubation period at 27°C is needed to reach stable plateau levels of cAMP (Goffstein et al., 1980). The relative hyperresponsiveness of NK cells to prostaglandin E_1 and isoprenalin , is clearly demonstrated by our data (Table IV) which is in accordance to earlier results (Goodwin et al., 1979a, b).

The target binding paralysis imposed on part of the NK population by increased intra-cellular levels of cAMP is practically instantaneous (data not shown) and probably not caused by impaired membrane mobility as it has been shown that high cAMP levels favor cellular "capping" (Butman et al., 1981). In fact, cAMP has been visualized, by fluorescence technique, underlying the "cap" on the inside of the cell membrane and has been postulated to be involved in a phosphorylation process connecting membrane proteins with cytoskeletal proteins (Earp et al., 1977).

Why increased levels of intracellular cAMP should impair NK recognition remains unclear. However, a similar recognition block has been described for T suppressor cells with Fc receptors for IgG which after treatment with theophylline lose their capacity to form rosettes with sheep red blood cells (Limatibul et al., 1978).

ACKNOWLEDGEMENTS

The excellent technical assistance of Miss Maj-Britt Alter is gratefully acknowledged.

REFERENCES

Brown, B.C., Albano, J.D.M., Ekins, R.P., Sqherzi, A. M., and Tampion, W. (1979). Biochem. J. 121:561.

Butman, B.T., Jacobsen, T., Cabatu, O.G., and Bourguignon, L.Y.W. (1981). Cellular Immunol. 61:397.

Earp, H.S., Utsinger, P.D., Yont, W.J., Louge, M., and Steiner, A.L. (1977). J. Exp. Med. 145:1087.

Fredholm, B.B., Sandberg, G., and Ernström, U. (1978). Biochem. Pharmacol. 27:2675.

Goffstein, B.J., Gordon, L.K., Wedner, H.J., and Atkinson, J.P. (1980). J. Lab. Clin. Med. Dec:1002.

Goodwin, J.S., Wiik, A., Lewis, M., Bankhurst, A.D., and Williams, R.C. (1979a) Cellular Immunol. 43:150.

Goodwin, J. S., Kaszubowski, P. A., and Williams, R. C. (1979b). J. Exp. Med. 150:1260.
Grimm, E., Thoma, J., and Bonavida, B. (1979). J. Immunol. 123:2870.
Hynie, S., Lanefeldt, F., and Fredholm, B. B. (1980). Acta Pharmacol. 47:58.
Limatibul, S., Shore, A., Dosch, H. M., and Gelf, E. W. (1978). Clin. Exp. Immunol. 33:503.
Parker, C. W., Sullivan, T. J., and Wedner, H. J. (1974). Adv. Cyclic Nucleotide Res. 4:1.
Platsoucas, C. D., and Good, R. A. (1981). Proc. Natl. Acad. Sci. USA 78:4500.
Roder, J. C., and Klein, M. (1981). J. Immunol. 127:1424.
Schwabe, U., Miyake, M., Ohga, Y., and Daly, J. W. (1976). Mol. Pharmacol. 12:900.
Strom, T. B., Carpenter, C. B., Garovoy, M. R., Austen, K. F., Merrill, J. P., and Kaliner, M. J. (1973). J. Exp. Med. 138:381.
Targan, S. R. (1981). J. Immunol. 127:1424.
Ullberg, M., and Jondal, M. (1981). J. Exp. Med. 153:615.
Ullberg, M., and Jondal, M. (in press). Clin. Lab. Immunol.
Wolberg, G., Zimmerman, T. P., Duncan, G. S., Singer, K. H., and Elion, G. B. (1978). Biochem. Pharmacol. 27:1487.
Zarling, J. M., Bach, F. H., and Kung, P. C. (1981) J. Immunol. 126:375.

REGULATION OF CYTOTOXIC REACTIVITY OF NK CELLS BY INTERFERON AND PGE_2[1]

Kam H. Leung
Hillel S. Koren[2]

Division of Immunology
Duke University Medical Center
Durham, North Carolina

I. INTRODUCTION

It is becoming increasingly evident that the natural killer (NK) cells may play an important role in immunosurveillance against tumor cells (Roder et al., 1980; Herberman et al., 1981). Recently a subpopulation of lymphocytes described as large granular lymphocytes (LGL) were identified as those cells responsible for mediating NK activity (Herberman et al., 1981). Interferon (IFN) has been shown to be a major biological modifier of NK cytolytic activity (Bloom, 1980). On the other hand, prostaglandin E_2 (PGE_2) produced primarily by macrophages (Kennedy et al., 1980) and some tumor cells (Goodwin et al., 1980) has been shown to suppress endogenous NK activity (Droller et al., 1978). Little is known about the effect of PGE_2 on IFN-activated NK cells. We report that IFN-activated NK cells are partially resistant to the PGE_2-mediated suppression.

II. METHODS

Human peripheral blood lymphocytes depleted of adherent cells by plastic and nylon wool adherence were used routinely. To enrich for the NK population consisting of LGL a 5-step discontinous Percoll density gradient was used (Timonen et al.,

---

[1] Supported by NCI grant CA 23354
[2] Recipient of a Research Career Development Award CA 00581 from the National Cancer Institute.

ISBN 0-12-341360-5

1981) in some experiments.

In experiments, involving murine effector cells, splenocytes were passed on nylon wool columns and erythrocytes eliminated by osmotic shock (Roder and Klein, 1979).

To test the effect of various reagents on NK activity, effector cells ($2x10^6$/ml) were incubated in RPMI-1640 medium supplemented with 10% FCS in 12x75mm polypropylene tubes at 37°C in a 5% CO_2 atomosphere. Poly I:C (Sigma), human fibroblast interferon (HEM), human lymphoblastoid interferon (Burroughs Wellcome), indomethacin (Sigma) and PGE_2 (Upjohn) were dissolved in the culture medium. After 18 hr the cultured cells were collected, washed and resuspended in the medium. Human NK cytotoxicity was measured in a 1-2 hr ^{51}Cr release assay against K562 target cells (Koren et al., 1981). Mouse NK cytotoxicity was measured in a 4 hr assay against ^{51}Cr-labeled YAC-1 or RL♂ 1 target cells (Roder and Klein, 1979). All data presented are from triplicate determinations.

III. RESULTS

A. Susceptibility of IFN-activated NK cells to PGE_2-mediated Suppression.

To examine if PGE_2 can inhibit cytolytic activity of IFN-activated NK cells, nonadherent (NA) lymphocytes cultured with IFN or poly I:C for 18 hr were washed and tested in the presence or absence of PGE_2 in the NK assay. The data presented in Table 1 show that PGE_2 inhibited the endogenous NK activity of fresh (Day 0) or cultured (Day 1) cells. In contrast, poly I:C, IFNα and IFNβ activated cells partially lost their sensitivity to suppression by PGE_2. Similar results were obtained when cells were treated with TCGF (1L-2) under the same experimental conditions (data not shown).

We next determined whether activation of NK cells and loss of susceptibility to PGE_2-mediated suppression can be induced in a population of cells highly enriched for LGL. Table II shows that the endogenous (control) NK activity of all the mononuclear cells (unfractionated, NA and LGL enriched cells) was inhibited by PGE_2. Moreover, NK activity of the 3 populations was augmented after 18 hr incubation in the presence of IFNβ or poly I:C. The LGL fraction as expected exhibited the highest level of cytotoxic activity. The loss of susceptibility to suppression by PGE_2 was also observed with all the cell fractions, suggesting that both activation and acquisition of partial resistance to suppression by PGE_2 are autonomous characteristics of LGL.

TABLE I. The Effect of Activation on NK Sensitivity to Suppression by PGE_2

Day of Culture	Stimulant[a]	PGE_2 in the assay (nM) 0	30	300	3,000
		Percent Cytotoxicity against K562 ± S.E.M.[c]			
Day 0	None	17.0±2.8	8.6±1.0 (49)[b]	6.5±0.4 (62)	4.8±0.5 (72)
Day 1	None	25.9±2.8	15.7±0.9 (3)	12.6±0.6 (51)	8.6±0.6 (67)
Day 1	IFNβ	33.3±1.5	31.0±1.8 (7)	27.4±1.2 (17)	20.8±1.1 (20)
Day 1	IFNα	37.2±2.2	35.2±1.8 (5)	31.2±1.5 (16)	23.6±1.1 (37)
Day 1	Poly I:C	42.2±2.1	43.5±2.2 (0)	40.7±1.9 (4)	38.1±3.4 (10)

[a] IFNα (1,000U/ml), IFNβ (1,000U/ml) and poly I:C (100ug/ml)

[b] The numbers in parentheses represent percent of inhibition by PGE_2

[c] E:T = 10:1 in a 2 hr assay.

TABLE II. The Effect of PGE_2 on Interferon-activated LGL

Cell Preparation	Control –	Control +	IFNβ (10^3U/ml) –	IFNβ (10^3U/ml) +	Poly I:C (100ug/ml) –	Poly I:C (100ug/ml) +
	PGE_2 (300nM) in Assay — Percent Cytotoxicity against K562[a]					
UNF	13.0	6.9 (47)	27.9	24.5 (12)	23.5	24.9 (-6)
NA	15.0	7.8 (48)	24.8	20.3 (18)	26.2	25.9 (1)
LGL-enriched	28.4	18.2 (36)	46.1	43.3 (6)	42.5	43.6 (-3)

[a] E:T = 5:1 in an 1 hr assay. The numbers in parentheses represent percent of inhibition by PGE_2

B. Is Endogenous PGE_2 Responsible for the Loss of Sensitivity of Activated NK Cells to Suppression by PGE_2 ?

The data presented in Table III show that PGE_2 can induce a loss of susceptibility to PGE_2- mediated suppression by NK cells. PGE_2 did not interfere with the activation of NK cells by IFNβ. Although NA cells do not synthesize appreciable amounts of PGE_2 (Koren et al., 1981), indomethacin was added to the NA cell culture to ensure a complete absence of PGE_2 synthesis in the local environment of the cells. Indomethacin had little effect on the IFN-induced resistance to suppression by PGE_2, indicating that *in vitro* acquired resistance is not due to PGE_2.

Table III. The Effect of Indomethacin on Loss of Sensitivity to PGE_2 - Mediated Suppression of NK Activity Induced by Interferon

Pretreatment (18 hr)	None		IFNβ (10^3U/ml)	
	PGE_2 (300nM) in Assay			
	Percent cytotoxicity against K562[a]			
	-	+	-	+
None	28	15 (45)	47	39 (17)
PGE_2 (3,000nM)	15	14 (8)	45	51 (-3)
Indomethacin (3,000nM)	28	15 (45)	51	37 (28)

[a] E:T ratio = 10:1 in a 2 hr assay. The numbers in parentheses represent percent of inhibition by PGE_2

C. The *In Vivo* Effect of Poly I:C on the Susceptibility of Mouse Spleen Cells to Suppression by PGE_2

Since IFN and poly I:C are currently employed in clinical trials of cancer patients, it seemed important to examine whether the protective effect observed *in vitro* is also operative *in vivo*. Therefore, mice from different strains (CBA, C3H and Balb/c nudes) were treated with poly I:C or saline. After 18 hr, their spleen cells were assayed for NK cytotoxicity against YAC-1 or RL♂1 targets in the presence or absence of PGE_2. As shown in Table IV, the NK activity of control spleen cells from the 3 strains of mice was suppressed by PGE_2. The *in vivo* treatment with poly I:C augmented the NK activity and caused a partial resistance to suppression by PGE_2.

TABLE IV. Susceptibility of *In Vivo*-Activated Mouse NK Activity to Suppression by PGE_2

Strain	Control		Poly I:C (100ug/mouse)	
	PGE_2 (300nM) in Assay			
	Cytotoxicity [a]			
	-	+	-	+
CBA	10.7	5.2 (49)	15.9	13.1 (18)
C3H	18.9	9.8 (49)	60.7	43.8 (28)
Nude	22.6	14.2 (37)	80.8	(64.8) (20)

[a] E:T is 50:1 in a 4 hr assay. The target cells for CBA and C3H effector cells were RL♂1 and for nude mice were YAC-1. The numbers in parentheses represent percent of inhibition by PGE_2.

IV. DISCUSSION

In this chapter we have shown that activation of human NK cells with IFN or poly I:C leads to a concomitant loss of sensitivity to suppression by PGE_2. This partial loss of inhibition was not due to endogenous PGE_2 production since the addition of indomethacin to cultures stimulated with IFN or poly I:C did not prevent the partial loss of sensitivity to PGE_2. Using LGL obtained by Percoll gradient sedimentation, we showed that both the activation by IFN and the subsequent loss of sensitivity to suppression by PGE_2 are most probably autonomous functions of LGL independent of monocytes.

The findings reported here may have implications for the regulation of NK reactivity *in vivo*. Within the local environment of a tumor, PGE_2 could be produced in intratumural macrophages (Mantovanni et al., 1980) and/or tumor cells (Pelus and Bockman, 1979). Interferon or other yet unidentified factors could provide the required signals necessary for the activation of NK cells which will be at the same time less susceptibile to the negative effects. Our preliminary results obtained with spleen cells from mice injected with poly I:C provide further evidence for the possible *in vivo* relevance of this phenomenon.

The possible mechanisms for induction of resistance to suppression by PGE_2 may be due to alteration of effector cell surface characteristics (Gresser, 1977) and/or change in the cellular metabolism of cyclic nucleotides (Tovey et al., 1979). These possibilities are currently under investigation in our laboratory.

ACKNOWLEDGMENTS

The authors thank Ms. Connie Hayes for the preparation of the manuscript.

REFERENCES

Bloom, B.R., Nature 284:593 (1980).
Droller, J.J., Schneider, M.U. and Perlmann, P., Cell. Immunol. 39:165 (1978).
Goodwin, J.S., Husby, G. and Williams, R.C., Jr., Cancer Immunol. Immunother. 8:3 (1980).
Gresser, I., Cell. Immunol. 34:406 (1977).
Herberman, R.B. and Ortaldo, J.R., Science 211:24 (1981).
Kennedy, M.S., Stobo, J.D. and Goldyne, M.E., Prostaglandins 20:135 (1980).
Koren, H.S., Anderson, S.J., Fischer, D.G., Copeland, C.S. and Jensen, P.J., Immunol. 127:2007 (1981).
Mantovani, A., Allavena, P., Sessa, C., Bolis, G. and Mantioni, C., Intl. J. Cancer 25:573 (1980).
Pelus, L.M. and Bockman, R.S., J. Immunol. 123:2118 (1979).
Roder, J.C. and Klein, M., J. Immunol. 123:2785 (1979).
Roder, J.C. and Haliotis, T., Immunol. Today 1:96 (1980)
Timonen, T., Ortaldo, J.R. and Herberman, R.B., J. Exp. Med. 153:569 (1981).
Tovey, M.G., Rochette-Egly, C. and Castagna, M., Proc. Natl. Acad. Sci., U.S.A. 76:3890 (1979).

NEGATIVE REGULATION OF HUMAN NK ACTIVITY BY MONOMERIC IgG

Andrei Sulica,[1] Maria Gherman,[1] Moiara Manciulea,[1] Cecilia Galatiuc[1] and Ronald Herberman[2]

[1]Department of Immunology
Victor Babes Institute
Bucharest, Romania
and
[2]Biological Research and Therapy Branch
National Cancer Institute
Frederick, Maryland

Various humoral factors such as interferon and prostaglandin E (Droller *et al.*, 1978; Brunda *et al.*, 1980) have been shown to play a role in the expression of the cytolytic activity of NK cells. Immunoglobulin G has also been suggested to affect NK cells, with natural IgG antibodies possibly mediating the interaction between effector and target cells, either by "arming" the NK cells (Koide *et al.*, 1977) or by coating the targets (Troye *et al.*, 1977). In addition, controversial results have been reporting the effects of immune complexes, containing IgG, with some groups finding inhibition of activity by these Fcγ receptor-bearing (FCR) cells (Pape *et al.*, 1979; West *et al.*, 1979), and others finding no effect (Kay 1980; Barada *et al.*, 1980).

To determine the possible impact of IgG on NK cells, we have studied the relationship of cytotoxic activity to the presence of this molecule in the culture medium used for preincubation of human peripheral blood lymphocytes (PBL) at 37°C or on the surface of these cells, as detected by a rosette assay with ox erythrocytes coated with protein A of *S. aureus* (ES) (Ghetie *et al.*, 1976). PBL were incubated at 37°C, usually for 2 hours at a concentration of 2 x 10^6 cells/ml in medium supplemented with 10% fetal bovine serum (FBS) in Petri dishes, the non-adherent lymphoid cells were harvested, and washed at 4°C, and tested in complete culture medium containing 10% FCS in a 4-hour cytotoxic assay against chromium labeled target cells for NK activity

ISBN 0-12-341360-5

against K 562 or Molt-4, and for antibody-dependent cell-mediated cytotoxicity (ADCC) against IgG-antibody coated targets (Ab/Rl♂1 and Ab/Chang). Data were calculated as lytic units (LU)/10^7 cells required to produce 30% specific lysis.

AUGMENTATION OF CYTOTOXIC ACTIVITY UPON INCUBATION OF EFFECTOR CELLS AT 37°C

Lysis of K562 targets by PBL preincubated at 37°C under the above conditions was 2 to 3 times higher than that obtained with cells kept at 4°C for 2 hours. A similar enhancement of cytotoxic activity also was obtained when Molt-4 cells or antibody-coated targets were employed. Since this augmentation of NK activity was paralleled by a decrease in the proportion of cells bearing surface IgG detectable by rosetting with ES, experiments were done to further examine the possible association between these changes. We tested various conditions which at least partially reduced the dissociation of labile cell-bound IgG. Incubation in test tubes at a higher concentration of 20 x 10^6 PBL/ml, previously shown to maintain the number of ES-rosette-forming cells (RFC), comparable to that obtained with freshly isolated cells (Moraru *et al.*, 1978), prevented the augmentation of NK upon *in vitro* incubation. Significant reduction of 37°C-induced enhancement of cytotoxic activity also was seen when PBL were incubated under the usual conditions in the presence of added IgG (in either 20% human autologuous serum or 10% newborn calf serum). Incubations in IgG-containing medium also led to a significantly smaller decrease in % ES-RFC than that seen with PBL incubated in absence of IgG. Further, when PBL were preincubated in the presence of autologous serum depleted of IgG by passage over a protein A-Sepharose column, we saw augmentation of NK activity above the levels of nonincubated cells, and reduction of ES-RFC, comparable to that seen in FBS-containing medium.

The increase in NK activity was apparent by 30 minutes and peaked at 3 hours after incubation at 37°C, and this was not related to activating factors in FBS, since incubation in serum-free medium supplemented only with 0.2% gelatin also led to about a 2-fold increase in NK activity. Removal of monocytes by adherence on Petri dish also did not account for the augmentation in NK activity since: (i) a similar pattern of enhancement was seen after preincubating PBL at 2 x 10^6 cells/ml in test tubes which provided little surface area for depletion by adherence; (ii) reconstitution of the initial unfractionated cell population, by mixing non-adhering PBL harvested from the Petri dish with monocytes at ratios as high as 2.7:1, did not reduce the augmented lytic activity of preincubated cells; (iii) removal of monocytes by an alternative procedure, in which whole blood was incubated at 37°C with carbonyl iron prior to isolation of non-phagocytic PBL, failed to lead to an increase in reactivity above that of unseparated PBL.

Our results are in good agreement with previous reported data about the enhanced cytotoxicity of cells incubated for a few hours or overnight, as compared to that of unincubated cells or cells incubated in medium containing human serum (Levin *et al.*, 1976; Zielske *et al.*, 1976). In addition the augmentation appeared to be attributable, at least in part, to the dissociation of labile cell-bound IgG from the PBL surface during the preincubation in medium lacking IgG.

ANALYSIS OF IgG AS AN INHIBITOR OF NK ACTIVITY

The culture-induced augmentation of cytolytic reactivity of NK cells was found to be efficiently diminished by human monomeric IgG (mIgG) prepared usually by passage twice of IgG through Sephadex G-200 and collecting finally the descending limb of the 7S peak. The degree of dose-dependent inhibition of NK activity obtained with 0.5 mg mIgG/2 x 10^6 cells/ml was about 30-40% of the value obtained with PBL preincubated for two hours under similar conditions in medium containing 10% FCS only. The same inhibitory activity was observed with ultracentrifuged mIgG or with rabbit IgG, but was not detected when other classes of human immunoglobulins were used. IgG_1 and IgG_3 myeloma proteins isolated in monomeric state provided strong inhibition, while IgG_2 and IgG_4 subclasses were not inhibitory at 0.5 mg/ml (Table 1). The Fc fragment of human mIgG also was capable of inhibiting considerably the activity of preincubated effector cells, whereas the $F(ab')_2$ and Fab fragments inhibited poorly if at all.

The inhibitory capacity of mIgG has properties consistent with those previously described for binding of cytophilic IgG by the FcR on mouse peritoneal macrophage (Sulica *et al.*, 1979a), namely: (i) inhibition was greatly diminished following mild reduction and alkylation of mIgG; (ii) the inhibitory capacity of mIgG was abolished when cytophilic molecules were selectively removed from the IgG preparation by absorptions with mouse macrophages or human granulocytes; (iii) as previously shown for binding of mIgG to the "cytophilic" FcR of mouse lymphocytes and macrophages (Sulica *et al.*, 1979b), rabbit IgG-protein A complexes had inhibitory activity identical to that of non-complexed rabbit IgG.

Table 1. Effect on NK Activity Against K562 by Prior Incubation of PBL in the Presence of Human Myeloma IgG Subclasses or Their Fragments

Treatment[a)]	% inhibition of NK cytotoxicity
mIgG	48
IgG_1	42
IgG_2	6
IgG_3	63
IgG_4	0
Fc	64
Fab	9
$F(ab')_2$	12

a) PBL (2×10^6 cells/ml) preincubated at 37°C for 2 hours in medium with the inhibitors at 0.5 mg/ml.

Our results suggest that regulation of NK activity by IgG is mediated by binding to receptors for cytophilic IgG on NK cells. Additional indirect evidence for the presence of cytophilic FcR on NK cells was provided by experiments in which PBL were fractionated on a SpA-Sepharose 6MB column. The nonadherent population, depleted of IgG-bearing cells, had NK activity 70% lower than that of unfractionated input cells. It should be noted that among the IgG-bearing cells present in PBL are also the L cells, defined by their capacity to interact with mIgG through cytophilic FcR and to lose the labile IgG upon incubation at 37°C (Lobo et al., 1975; Lobo et al., 1976). Attempts are presently being made to demonstrate directly the ability of highly enriched populations of NK cells, isolated by Percoll discontinuous gradient centrifugation (Timonen et al., 1981), to bind cytophilic mIgG by specific FcR.

CHARACTERISTICS OF mIgG-INDUCED INHIBITION

The inhibition by mIgG occurred at 0°C as well as 37°C and this could be correlated with the binding of mIgG to the FcR^+ cells, as detected by means of ES-rosette assay (Sulica et al., 1982). The lowest NK activity and the highest percentage of ES-RFC were seen following the treatment of PBL with mIgG in the presence of 12% polyethylene glycol, a "franking" procedure which enhances the attachment of IgG to the cell surface (Jones et al., 1980).

Inhibition of NK activity by mIgG or autologous serum was not limited to the prevention of augmentation upon incubation at 37°C, since it was seen even after culture-induced augmentation of NK activity. Moreover, treatment of freshly isolated PBL with these reagents at 0°C for 1 hour markedly decreased NK activity and increased the number of IgG-bearing cells, especially when "franked" effectors were employed.

IgG-induced inhibition was also shown to be reversible, upon exposure at 37° for 1 hour, in medium supplemented with FBS only, of PBL previously coated *in vitro* with mIgG. Again the least enhancement of NK activity or decrease in % ES-RFC was observed when more stable binding of mIgG was performed by the "franking" procedure. Since the dissociation of cell-bound IgG occurred progressively during incubation at 37°C, a long-term NK assay (16 hours) performed in medium containing FCS failed to detect any significant inhibition of cytotoxicity by effector cells preincubated in the presence of mIgG.

The presence of mIgG during the cytotoxicity assay was less inhibitory than when PBL were preincubated with the same concentration of mIgG and washed prior to the ^{51}Cr-release assay, suggesting that effector cells had to be exposed to IgG for at least some period prior to the interaction with target cells in order to lose their killing capacity. As shown by a single-cell assay in agarose (Grimm *et al.*, 1979), the inhibition of NK activity by mIgG occurred at some post-recognition or target binding step, since the binding of K562 targets was not reduced following pretreatment of PBL with mIgG.

Preliminary results have been obtained with regard to the biochemical events which may be involved in this NK cell regulation. The inhibition of NK activity was still observed when PBL were simultaneously exposed to mIgG and indomethacin during the preincubation at 37°C (Table 2), suggesting that it is not mediated through the release of prostaglandins. However, results recently obtained suggest that induction of elevated levels of cyclic AMP plays a significant role in the NK inhibition induced by mIgG (A. Sulica, T. Goto and R. Herberman, unpublished observations).

RELATIONSHIP TO OTHER REGULATORY FACTORS

As Table 2 shows, inhibition of PGE release, by preincubation of PBL at 37°C in the presence of indomethacin, enhanced the augmentation of NK activity by more than 50% of the control value obtained with medium-pretreated cells. Similarly, inhibition of NK by PGE secreted by cells adhering on Petri dishes during the incubation of human PBL at 37°C was recently described by others (Koren *et al.*, 1982).

Table 2. Effects of Indomethacin on the Regulation of NK Activity by mIgG

Preincubation in presence of[a)] mIgG	indomethacin	NK activity LU/10^7 cells	% mIgG induced inhibition
-	-	26.9	-
+	-	19.2	28.4
-	+	55.0	-
+	+	38.5	30.0

a) PBL (2 x 10^6 cells/ml) were incubated for 2 hours at 37°C in Petri dishes in complete culture medium containing mIgG (0.2 mg/ml) and/or indomethacin (1.5 x 10^{-6}M).

Treatment (37°C, 1 hour) with human leukocyte interferon (1000 IU/ml) of PBL, previously charged *in vitro* at 0°C with mIgG according to the "franking" technique, reduced by 50% the inhibited cytolytic response seen with cells preincubated at 37°C in medium alone. It is also of note that mIgG was able to inhibit the augmented NK activity that was induced by prior or simultaneous exposure of PBL to interferon during the preincubation procedure to approximately the same extent as it inhibited spontaneous NK activity. It is likely that the negative regulatory signal provided by cell-bound mIgG acts under physiological conditions *in vivo* in balance with positive regulatory signals such as interferon.

DISCUSSION

As described here and previously (Sulica *et al.*, 1982; Sulica *et al.*, 1981) we have obtained evidence for a new mechanism for negative regulation of human NK activity, by cytophilic IgG in monomeric form. Short-term incubation in medium lacking IgG allows dissociation of some of the labile cell-bound IgG from PBL and consequently a release from inhibition takes place. The culture-induced augmentation of NK activity is prevented by preincubation at 37°C in the presence of added IgG or its Fc fragment. The inhibition of NK cytotoxicity by mIgG is in good agreement with previous reports that prior incubation at 37°C of human (Levin *et al.*, 1976; Zielske *et al.*, 1976) or mouse (Nair *et al.*, 1980) effector cells in medium containing homologous serum induced a reduction of NK activity. It also has been suggested that the induced inhibitory effect of normal or cancer sera on murine NK cells was due to a factor with a molecular weight of 150,000 daltons (Nair et al., 1980), quite consistent with our evidence for a role of mIgG.

The inhibition of NK cytotoxicity may be due to either a direct or indirect effect of mIgG on the effector cells. Although our data do not rule out indirect effects, they suggest a direct interaction with NK cells, via FcR for cytophilic mIgG. In fact, expression of FcR for mIgG has been reported recently by others (Saal et al., 1980). A functional role of FcR involved in the binding of particulate immune complexes was recently observed (Merrill et al., 1981) when human PBL were exposed to IgG-containing complexes prior to the conjugation with target cells. Since the substantial NK inhibition occurs at the post binding stage of the killing process, as shown by us and others (Merrill et al., 1981), it can be assumed that attachment of IgG to the specific binding sites on NK cells provide a negative regulatory signal, which seems to be mediated through elevation of cyclic AMP and, which, in some yet to be defined way, turns off the killing. The reversible inhibition of the cytotoxic activity induced by the binding of mIgG to human PBL was not blocked by indomethacin, an inhibitor of prostaglandin synthesis, suggesting therefore that the inhibitory effect of cytophilic IgG is carried out by a novel mechanism for negative regulation of NK activity, distinct from the previously described inhibitory effects of prostaglandins (Droller et al., 1978; Brunda et al., 1980) or suppressor cells (Cudkowicz et al., 1979).

Our findings raise a series of interesting issues that can be approached in future studies: 1) The possibility of an important role of cytophilic IgG in in vivo regulation of NK activity. 2) The possibility for accounting for some differences in the levels of NK activity among individuals by their levels of cytophilic IgG, as well as by their levels of NK cells. In this regard, it is of interest that some patients with agammaglobulinemia tended to have NK activity above that of normal donors (Koren et al., 1978; Pross et al., 1979) and that B lymphocyte-deprived mice possess a heightened in vivo resistance to a carcinogen-induced tumor, which correlated with an augmentation of NK activity (Brodt et al., 1981). 3) The negative regulatory influence of mIgG on NK activity might lead to a novel approach to the in vivo enhancement of NK function, by depletion of cytophilic IgG [e.g., by passage of plasma over insolubilized protein A (Terman et al., 1980; Ray et al., 1981)].

ACKNOWLEDGMENTS

The excellent technical assistance provided by Mr. T. Regalia and Mrs. M. Feher is gratefully acknowledged. We thank Dr. M. Klein for the supply of human IgG myeloma proteins.

REFERENCES

Barada, F.A., Kay, H.D., Emmous, R., Davis, J.S. and Horowitz, D.A. (1980). J Immunol 120:865.
Brodt, P., Kongshavn, P., Vargas, F. and Gordon, J. (1981). J Reticuloendothel Soc 30:283.
Brunda, M.J., Herberman, R.B. and Holden, H.T. (1980). J Immunol 124:2682.
Cudkowicz, G. and Hochman, P.S. (1979). Immunol Rev 44:13.
Droller, M.J., Schneider, M.U. and Perlmann, P. (1978). Cell Immunol 39:165.
Ghetie, V., Moraru, I., Sulica, A., Gherman, M. and Sjoquist J. (1976). Rev Roum Biochem 13:263.
Grimm, E. and Bonavida B. (1979). J Immunol 123:2861.
Kay, H.D. (1980). In "Natural Cell-Mediated Immunity Against Against Tumors." (R.B. Herberman, ed.), p. 329, Academic Press, New York.
Koide, Y. and Takasugi, M. (1977). J Natl Cancer Inst 59:1099.
Koren, H.S., Amos, D.B. and Buckley, R.H. (1978). J Immunol 120:796.
Koren, H.S., Anderson, S.J., Fischer, D.G., Copeland, C.S. and Jensen, P.J. (1982). J Immunol, in press.
Jones, J.F. and Segal, D.M. (1980). J Immunol 125:926.
Levin, A.C., Massey, R.J., Deinhart, F. (1976). Fed Proc 35:472.
Lobo, P.I., Westervelt, F.B. and Horwitz, D.A. (1975). J Immunol 114:116.
Lobo, P.I. and Horwitz, P.A. (1976). J Immunol 117:939.
Merrill, J.E., Ullberg, M. and Jondal, M. (1981). Eur J Immunol 11:536.
Moraru, I., Gherman, M., Sulica, A., Bancu, A.C., Sjoquist, J. and Ghetie, V. (1978). Ann Immunol 129 C:89.
Nair, P.N.M., Fernandes, G., Onoe, K., Day, N.K. and Good, R.A. (1980). Int J Cancer 25:667.
Pape, G.R., Moretta, L., Troye, M. and Perlmann, P. (1979). Scand J Immunol 9:291.
Pross, N.F., Gupta, S., Good, R.A. and Baines, M.G. (1979). Cell Immunol 43:160.
Ray, P.K., McLaughlin, D., Mohammed, J., Idiculla, A., Rhoads, J.E., Mark, R., Bassett, J.G. and Cooper, D.R.

(1981). In "Immune Complexes and plasma exchanges in cancer patients." (B. Serrou and C. Rosenfeld, eds.), p. 197, Elsevier North-Holland Biomedical Press, Amsterdam.
Saal, J.G., Hadam, M., Feucht, H. and Rautenstrauch, H. (1980). Immunobiol 157:272-273.
Sulica, A., Gherman, M., Medesan, C., Sjoquist, J. and Ghetie, V. (1979). Eur J Immunol 9:979.
Sulica, A., Medesan, C., Laky, M., Onica, D., Sjoquist, J. and Ghetie, V. (1979). Immunol 38:173.
Sulica, A., Gherman, M., Manciulea, M., Galatiuc, C. and Herberman, R.B. (1981). Immunolbiol 160:119.
Sulica, A., Gherman, M., Galatiuc, C., Manciulea, M. and Herberman, R.B. (1982). J Immunol, in press.
Terman, D.S., Yamamoto, T., Mattioli, M., Cook, G., Tillquist, R., Henry, J., Poser, R. and Daskal, Y. (1980). J Immunol 124:795.
Timonen, T., Ortaldo, J.R. and Herberman, R.B. (1981). J Exp Med 153:569.
Troye, M., Perlmann, P., Pape, G.R., Spiegelberg, H.C., Nashlund, I. and Gidlof, A. (1977). J Immunol 119:1061.
West, W.H., Cannon, G.B., Kay, H.D., Bonnard, G.D. and Herberman, R.B. (1979). J Immunol 118:355.
Zielske, J.V. and Golub, S.H. (1976). Cancer Res 36:3842.

PRESENCE OF FcR FOR IgG AND IgM ON HUMAN NK CELLS: THE ROLE OF IMMUNE COMPLEXES IN THE REGULATION OF NK CELL CYTOTOXICITY

Jean E. Merrill[1,2]
Sidney Golub[3,4]

[1]Department of Neurology
[3]Departments of Surgery and Microbiology and Immunology
UCLA
Los Angeles, California

Mikael Jondal[5,6]
Fred Lanefeldt[7,8]
Bertil Fredholm[7,8]

[5]Department of Tumorbiology
[7]Department of Pharmacology
Karolinska Institute
Stockholm, Sweden

[2]Supported in part as a postdoctoral fellow of the National Multiple Sclerosis Society and in part by USPHS 2-P50-NS0-8711 10A1 grant from the NIH.

[4]Supported by CA 12582.

[6]Supported by the Swedish Cancer Association and Karolinska Institute.

[8]Supported by King Gustav Vth Research Foundation, the Swedish Society for Medical Science, and the Karolinska Institute.

ISBN 0-12-341360-5

I. INTRODUCTION

It is known that the majority of human natural killer (NK) cells in the peripheral blood have Fc_GR (Möller, 1979; Jondal and Pross, 1975; Timonen et al., 1981). Increasing evidence for the role of Fc_MR^+ cells in cytotoxic responses such as mitogen-induced cellular cytotoxicity (MICC) (Pichler et al., 1979) or antibody-dependent cellular cytotoxicity (ADCC) (Perlmann et al., 1981; Fuson and Lamon, 1977), and the observation that purified, monoclonal IgM could block NK (Timonen and Saksela, 1977), led us in search of an NK cell with the Fc_MR. In addition, we examined the basis of inhibition of NK activity by immune complexes (IC) shown to include particulate (Kay et al., 1979; Pape et al., 1979; Härfast et al., 1980; Merrill et al., 1981) and soluble IC (Merrill et al., 1981; West et al., 1979; Barada et al., 1980; Fink et al., 1977) as well as aggregated IgG (Saksela et al., 1979; Perlmann et al., 1979). By using the single cell cytotoxicity assay in agarose (Grimm and Bonavida, 1979), we could 1) directly identify a small population of NK cells bearing Fc_MR and, 2) demonstrate that NK inhibition by IgG-containing complexes is not related to steric hindrance of target cell contact, but rather to a cell surface signal to which the NK cell is susceptible only _before_ target cell conjugation. In general, IgM-containing complexes did not inhibit NK activity unless in great antibody excess. This inhibition appeared to be due to inhibition of binding of effector to target.

II. DISTRIBUTION OF Fc_GR AND Fc_MR ON HUMAN NK

FcR^+-NK cells were determined by first conjugating effectors and targets for 10' at 37°C (Grimm and Bonavida, 1979) and then rosetting with $OxRBC_G$ or $OxRBC_M$ (Moretta et al., 1978) for 30-60' at 4°C. Resuspension and addition to warm agarose were done in such a way as to prevent disruptions of rosettes (Merrill et al., 1981) and target binding cells (TBC), and those TBC that could kill were scored as rosetted or nonrosetted (Merrill et al., 1981). Figure 1 shows 55% of TBCs are Fc_GR^+ and 24% are Fc_MR^+. Fc_GR^+ TBCs account for 60% of the total killing in fresh nylon wool passed (NWP) peripheral blood lymphocytes (PBL) while Fc_MR^+ TBCs account for 17% of the total killing. This means that 20% of NK cells are FcR^- since there are about 10% TBC in the total NWP population whether the TBC are rosetting or not. Although Moretta et al (1978) have shown that overnight incubation is required for maximal expression of Fc_MR^+-cells, there was no increase in spontaneous Fc_MR^+-NK cells (Merrill et al., 1981).

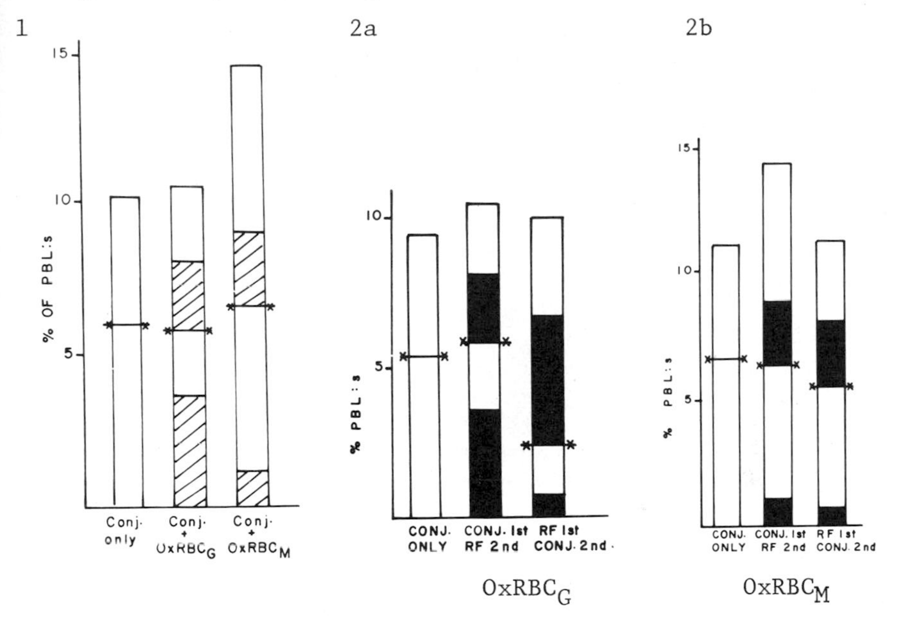

FIGURE 1. FcR^+-NK cells to Molt 4: Conjugates with living targets above x—x, conjugates with dead targets below x—x. ▨ FcR^+-TBC, ▭ FcR^--TBC. Results of 17 exps with $OxRBC_G$, 8 exps with $OxRBC_M$.

FIGURE 2. Effect of order of conjugation or rosette formation (RF) with OxRBC $_G$ or $_M$ on NK activity. Conjugates with living targets above x—x; conjugates with dead targets below x—x. ■ FcR^+-TBC, ▭ FcR^--TBC. a. Results of 17 exps with $OxRBC_G$. b. Results of 8 exps with $OxRBC_M$.

III. THE EFFECT OF IC ON NK ACTIVITY

A. Particulate Complexes (OxRBC-Rabbit Anti-OxRBC)

Figures 2a and 2b show the effect of exposure of NK cells to IgG- and IgM-containing IC respectively before and after target cell conjugation. Fig 2a shows that 10% of the total NWP PBL bind Molt 4 and 57% of these are NK cells. Conjugation before Fc_GR involvement did not inhibit %TBC nor %TBC killers (NK). However, exposure of effectors to $OxRBC_G$ before conjugation resulted in a 57% reduction in total NK activity and an 80% reduction in the killing by Fc_GR^+-NK cells without inhibiting their binding. There was only a 26% reduction in Fc_GR^--NK cells. This demonstrates that $OxRBC_G$ IC specifically inhibit the FcR^+-killer TBCs. Figure 2b shows that exposure of effector cells to $OxRBC_M$ either before or after conjugation to the target does not significantly change the proportion of TBCs, the proportion of NK cells, nor the proportion of Fc_MR^+-TBCs or NK cells (Jondal and Merrill, 1981). Because Fc_MR rosettes, following conjugation, must be resuspended gently and not by vigorous vortexing, there is a nonspecific increase in nonkiller, non Fc_MR^+-TBC (Merrill et al., 1981).

Figure 3 shows the influence of IC modulation of Fc_GR on NK inhibition. Incubating rosettes at 37°C leads to the loss of most Fc_GRs as a consequence of surface capping and shedding. It was thus of interest to see how longstanding the $OxRBC_G$-induced NK inhibition was and if the cells could kill in the absence of detectable Fc_GRs. Figure 3b demonstrates that by 3 hr, 62% of Fc_GR^+-cells incubated with $OxRBC_G$ had modulated their Fc receptors completely, and of the remaining 38%, 34% were capped. Even with immune complex clearing of the Fc_GR from the majority of the Fc_GR^+-cells, those effectors exposed to the immune complexes prior to conjugation were inhibited in their NK cell activity (Figure 3a). However, after overnight incubation when almost 80% of the Fc_GR^+-cells had modulated their receptors, the NK activity had recovered to control levels.

B. Soluble Complexes (KLH-Human or Rabbit Anti-KLH)

KLH-hyperimmune rabbit anti-KLH sera and immune plasma from a female volunteer were separated on a Biogel A 5M column (200-400 mesh) into IgM and IgG fractions whose purity

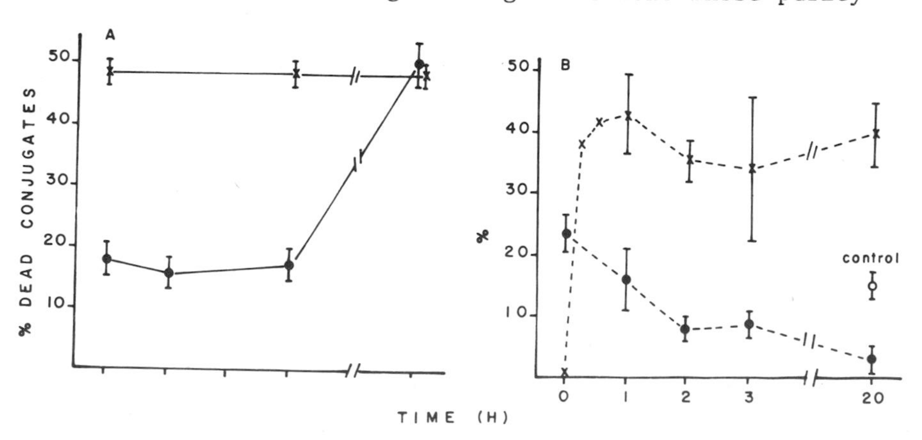

FIGURE 3. Fc_GR modulation effect on NK activity. a. ●—● NK activity after rosetting with $OxRBC_G$. ×—× NK activity of lymphocytes without pretreatment. b. ●····● Fc_GR^+-lymphocytes of total NWP PBL. ×····× Fc_GR-capped cells of total Fc_GR^+-population.

were assessed by IEP and SDS-PAGE. Antibody activity was characterized by hemagglutination titers to KLH-OxRBC and quantitative precipitin analysis. Assessment of IC formation and activity was by inhibition of FcR binding of $OxRBC_G$ or $_M$ and by complement consumption.

Table I shows one representative experiment of 5 on the ability of soluble rabbit and human IC, when effectors are pretreated, to inhibit human NK activity. Ratios of antibody to KLH were chosen within ranges giving optimal FcR inhibition and C3 consumption.

Pretreatment of effectors with IgG-containing IC of either rabbit or human origin inhibited killing but not binding to targets. In great antibody excess, IgM-containing IC inhibited NK activity. Unlike inhibition seen with IgG-IC, these complexes inhibited target cell recognition and binding, not killing.

IV. MECHANISM BEHIND IC INHIBITION OF NK ACTIVITY

The cytolytic process involves three distinct steps: target cell recognition and binding, the lytic event, and finally the effector cell-independent target cell disintegration. Consequently, any inhibitory substance may independently act on one or more of these steps. IgM-containing complexes appear to inhibit the initial binding step, since both rabbit and human IgM complexes reduce the number of TBCs (Table I). Possibly this is merely the consequence of steric hindrance. Alternatively, IgM complexes may induce a state of membrane

TABLE I. Influence of Human and Rabbit IgG and IgM Immune Complexes on Target Binding and Cytotoxicity Capacity of Human NK Cells

Pretreatment	Concentration (μg/ml)	TBCs %	Inhibition of TBCs %	Cytotoxic TBCs %	Inhibition of cytotoxic TBCs %
None	-	8.0	0	64	0
HuIgM/KLH	750/0.25	5.3	34	66	-3
HuIgG/KLH	125/0.25	9.6	-20	35	45
RabIgM/KLH	500/0.25	4.1	49	57	11
RabIgG/KLH	500/0.25	7.7	4	46	28

Pretreatment with human or rabbit IgG or IgM separately in the absence of KLH or with KLH alone did not inhibit TBCs or cytotoxic TBCs. Effectors were NWP lymphocytes; targets were Molt 4.

TABLE II. IgG Immune Complex-Mediated Inhibition of Human NK Cell Activity is not a Consequence of Increased Levels of Intracellular cyclic AMP

Pre-treatment	Induction of cAMP* ($pM/10^6$ cells)			NK Cell Activity at 3 Hr**	Inhibition of NK activity
	5'	15'	35'	% ^{51}Cr release	%
None	1.7	1.9	2.3	33.2	0
PGE_2	11.7	16.8	11.4	17.5	47
$OxRBC_G$ IC	1.8	2.8	3.2	5.2	84

*Measured by the technique of Brown et al. (2).
**Tested against Molt 4 target cells at E:T ratio 20:1

paralysis by a mechanism designated by Edelman as "anchorage modulation" (Edelman, 1976), in which the cell surface mobility is impaired through inhibition of the cell membrane microfilament system. Such reduced mobility would explain the inability of the NK cells to conjugate properly to the target cells.

IgG-containing complexes have a much more profound effect on the human NK system than IgM-containing complexes: the recognition phase is not impaired but the lytic event is inhibited. As increased intracellular levels of cyclic AMP (cAMP) have been shown to turn off NK cells, we investigated the influence of IC in Fc_GR on increased cAMP levels. Table II shows that when particulate immune complexes bind to NK cells and turn off killing, they do not increase intracellular levels of cAMP in the effector population. This is in contrast to the effect of the adenyl cyclase stimulator, prostaglandin E_2. The underlying mechanism behind IgG complex inhibition of NK activity thus remains unclear. It is interesting to note, however, that the Fc_GR expressed on human NK cells has a dual effect on their cytolytic potential. At optimum antibody concentrations, NK cells are induced to mediate ADCC, whereas at superoptimal concentrations, NK cell activity is turned off (Merrill et al., 1981). Recent experiments in this laboratory have shown that optimal antibody concentrations bound to human hematopoetic target cell lines will increase the lytic event and recycling capacity of the NK cells whereas higher concentrations inhibit the cytotoxicity (Ullberg and Jondal, unpublished). Under neither of these conditions is target cell recognition, as visualized as TBCs, altered.

REFERENCES

1. Barada, F.A., Kay, H.D., Emmons, R., Davis, J.S., and Horwitz, D.A., J. Immunol. 1980. 125:865.
2. Brown, B.C., Albano, J.D.M., Ekins, R.P., Sgherzi, A.M., and Tampion, W. Biochem. J. 1971. 121:561.
3. Edelman, G.M., Science. 1976. 192:218.
4. Fink, P.C., Schedel, I., Peter, H.H., and Deicher, H., Scand. J. Immunol. 1977. 173:183.
5. Fuson, E.W., and Lamon, E.W., J. Immunol. 1977. 118:1907.
6. Grimm, E., and Bonavida, B., J. Immunol. 1979. 124:2861.
7. Härfast, B., Andersson, T., Alsheikhly, A., and Perlmann, P., Scand. J. Immunol. 1980. 11:357.
8. Jondal, M. and Merrill, J.E., Eur. J. Immunol. 1981. 11:531.
9. Jondal, M. and Pross, H.F., Int. J. Cancer. 1975. 15:596.
10. Kay, H.D., Fagnani, R. and Bonnard, G.E., Int. J. Cancer. 1979. 24:141.
11. Merrill, J.E., Ullberg, M., and Jondal, M., Eur. J. Immunol. 1981. 11:536.
12. Möller, G. (Ed.) Immunol. Rev. 1979. 44.
13. Moretta, L., Ferrarini, M., and Cooper, M.D., Curr. Top. Immunobiol. 1978. 8:19.
14. Pape, G.R., Moretta, L., Troye, M., and Perlmann, P., Scand. J. Immunol. 1979. 9:291.
15. Perlmann, P., and Cerottini, J.C., Sela, M. (Ed.) in The Antigens, Academic Press, New York. 1979. p. 173.
16. Perlmann, H., Perlmann, P., Moretta, L., and Rönnholm, M. Scand. J. Immunol. 1981. 14:47.
17. Pichler, W.J., Gendelman, F.W., and Nelson, D.L., Cell. Immunol. 1979. 42:410.
18. Saksela, E., Timonen, T., Ranki, A., Hayry, P., Immunol. Rev. 1979. 44:71.
19. Timonen, T., Ortaldo, J.R., and Herberman, R.B., J. Exp. Med. 1981. 153:569.
20. Timonen, T. and Saksela, E., Cell. Immunol. 1977. 33:340.
21. West, W.H., Camron, G.B., Kay, H.D., Bonnard, G.D., and Herberman, R.B., J. Immunol. 1979. 118:335.

NK Activity in Mice is controlled by the Brain Neocortex

Gérard Renoux[1]
Katleen Bizière[2]
Pierre Bardos
Danielle Degenne
Micheline Renoux

Laboratoire d'Immunologie
Faculté de Médecine
Tours, France

I. INTRODUCTION

The observation that the immunopotentiator sodium diethyldithiocarbamate, DTC, has a late anabolic effect and can also increase the synthesis of selective inducers of prothymocytes, suggested a mediation by the Central Nervous System, CNS, (Renoux and Renoux 1979), and prompted us to investigate the role of the CNS on various immune functions.

Several investigators have observed relationships and mutual influences between the thymus and the pituitary gland or the hypothalamus (Pandian and Talwar 1971 ; Pierpaoli and Sorkins 1972 ; Stein et al. 1976), which are obligatory paths for most brain-triggered hormonal signals. In view of the fundamental and practical importance of determining the influence of the brain on the immune system, we thought it of interest to evaluate the role of the brain cortex, prior to examination of the physiologic pathways, since the neocortex controls the sensorial perceptions and comportmental behavior in ape and man, whereas the immune system controls the maintenance of homeostasis and body integrity in response to environment.

[1]Supported by grants from INSERM n° 806010, Institut Mérieux and Clin-Midy.
[2]Present adress : Centre de Recherches Clin-Midy, rue du Pr. J. Blayac, 34082 Montpellier Cedex.

ISBN 0-12-341360-5

Preliminary experiments have shown that a surgical lesion of the left cerebral cortex causes a 50 % reduction in the number of splenic Thy-1$^+$ cells and a severe depression of T-cell-mediated events, without affecting B cell numbers (Renoux et al. 1980 a, b ; Bizière et al. 1980).

Here, we report that an intact left cerebral cortex is essential for the expression of spleen NK activity, as well as for the influence of DTC, whereas these effects are not modified by lesioning the right cortex.

MATERIALS AND METHODS

Female C3H/He mice were used throughout these assays ; treatment with DTC, cytotoxicity assay and evaluation of lytic units (LU), as well as statistical analysis, were as described elsewhere in this volume (Renoux et al. "Sodium diethyldithiocarbamate (DTC)-induced modifications of NK activity in the mouse").

Surgical lesions of the cerebral cortex. Similar lesions were performed on the left or right cerebral cortex, or on both cortex, of 6-week old mice by a technique previously described in rats (Bizière and Coyle 1978). In brief, animals were anaesthetized by an intraperitoneal administration of 0.085 ml/g of a 5 % solution of chloral hydrate. The scalp was reclined and the skull overlying the cerebral cortex was removed ; care was taken not to damage the superior sagittal sinus. Portions of either the right or the left cerebral cortex were removed by shallow knife cuts ; the lesions involved dorsal and lateral aspects of the frontal, parietal and occipital cortex, without penetrating the corpus callosum. The ablated area was gently packed with sterile gel-foam and the scalp was apposed with sutures. After surgery mice received a single i.p. injection of ampicillin (100 mg/kg). Sham operated mice, in which the cortex was not lesionned during surgery, were used as controls, as were normal untreated mice of the same age, sex and strain.

RESULTS

Surgery and anaesthesia can temporarily depress some immune parameters in man (Slade et al. 1975). We, therefore, observed a 8-week interval between surgery and test to minimize the influence of these stresses in mice. As a consequence, the younger, usable animals were of 14 weeks of age.

Fortunately, we have previously evidenced that DTC induced a rise in natural cell-mediated cytotoxicity in normal, 14-week-old C3H/He mice (this volume), a circumstance which permitted the experiments shown in the accompanying Table.

Table I. Influences of lesioning the brain neocortex on NK activity, and on treatment with 25 mg/kg of DTC.

	Treatment[a]			
	Saline		DTC	
Brain cortex lesion	LU/10^7 cells	% of controls	LU/10^7 cells	% of controls
none	50	—	84	168*
left	25	50*	24	48*
right	42	82	92	184*
bilateral	13	27*	18	36*

[a]Fourteen-week-old mice (8 weeks after surgery) ; test : 4 d after treatment.

*$P < 0.01$ in comparison with untreated, sham-operated controls by Student's *t*-test. Four mice per group.

The levels of NK activity in sham operated mice are undistinguishable from those of normal mice of the same age (see, Renoux et al. Table III, this volume), evidencing that the potential impact of the surgical stress has disappeared in a 8-week interval. By means of spleen cells from mice with a left neocortical lesion, the NK activity against YAC-1 cells is strikingly reduced in comparison with control. A bilateral lesion, involving both the left and the right cortex, similarly inhibits the NK activity. In contrast, lesioning the right cortex do not induce a noticeable change in the spontaneous spleen cell cytotoxicity of 14-week-old mice. The differences in NK levels of activity in mice with different brain cortex lesions cannot be attributable to surgical trauma, since all animals underwent similar surgical stress and, as previously reported (Renoux et al. 1980 a), the partial removal of cerebral neocortex do not influence gross behavior and organ weights. Present results confirm previous findings (Bardos et al. 1981), that the NK reactivity of mouse spleen cells is controlled by the left brain neocortex and not by the right symmetrical brain area.

The influence of neocortical lesions on the effects of DTC was examined, in assays where spleen cells were tested for NK activity 4 days after a subcutaneous injection of 25 mg/kg DTC. As shown in Table, a significant rise in NK activity beyond that of untreated controls was observed in sham operated mice, as well as in right decorticated mice. In contrast, DTC was unable to modify the spontaneous spleen cell cytotoxicity impaired in mice by a left or a bilateral cortical lesion. It seems, therefore, that a intact left neocortex could be needed for the DTC-induced enhancement of NK activity.

DISCUSSION

Present findings provide further evidences for a brain localization to control spleen NK activity (Bardos et al. 1981). It is of interest that NK activity and T-cell-mediated events are both controlled by the brain neocortex. An intact left neocortex seems essential to maintain the production of selective inducers, the T-cell responses and the NK activity, whereas the right neocortex apparently is not involved in these immune functions. The influences of the neocortex concern the T-cell lineage, and do not affect B cells (Renoux et al. 1980 a, b ; current studies).

DTC is selectively active on the T-cell lineage through the enhanced production of specific factors (Renoux and Renoux, 1977, 1981 a) : it can recruit cells bearing T-cell markers from bone-marrow cells (Renoux and Renoux 1981 b), and induces T cells to mature. The use of DTC in brain-lesioned mice gives a further insight in the role of the neocortex. In the present study, mice deprived of the right neocortex are stimulated by DTC to increased NK responses, as well as normal, intact mice; animals with either a left or a bilateral lesion become unresponsive. In other studies, an intact right cortex (left-lesioned mice) was a prerequisite to mediate DTC-induced, augmented T-cell events and production of specific factors (Renoux et al. 1980 a, b ; current studies).

These observations suggest that an imbalance between two symmetrical areas in the brain neocortex would control the synthesis of mediators, yet to be identified, triggering the hormonal productions needed for the induction of subsets and functions in the T-cell lineage. This imbalance will also modulate the ability of DTC to induce interrelated modifications in NK-subsets and T-cell subpopulations.

REFERENCES

Bardos, P., Degenne, D., Lebranchu, Y., Bizière, K. and Renoux, G. (1981). Scand. J. Immunol. 13 : 609.
Bizière, K. and Coyle, J.T. (1978). Neurosci. Lett. 8 : 303.
Bizière, K., Renoux, G., Renoux, M., Gyenes, L., Degenne, D., Guillaumin, J.M., Bardos, P., and Lebranchu, Y. (1980). Neurosci. Abst. 6 : 31.
Pandian, M.R., and Talwar, G.P. (1971). J. Exp. Med. 134 : 1095.
Pierpaoli, W., and Sorkin, E. (1972). Nature 215 : 834.
Renoux, G., and Renoux, M. (1977). J. Exp. Med. 145 : 466.
Renoux, G., and Renoux, M. (1979). J. Immunopharmacol. 1 : 247.
Renoux, G., and Renoux, M. (1981 a). In "Augmenting Agents in Cancer Therapy" (E.M. Hersh, M.A. Chirigos, M.J. Mastrangelo, eds.), p. 247. Raven Press, New York.
Renoux, G., and Renoux, M. (1981 b). Abst. 8th Internat. Congress Pharmacol., p. 437.
Renoux, G., Bizière, K., Renoux, M., and Guillaumin, J.M. (1980 a). C.R. Acad. Sci. (Paris) 290 D : 719.
Renoux, G., Bizière, K., Renoux, M., Gyenes, L., Degenne, D., Guillaumin, J.M., Bardos, P., and Lebranchu, Y. (1980 b). Internat. J. Immunopharmacol. 2 : 156.
Slade, M.S., Simmons, R.L., Yunis, E., and Greenberg, L.J. (1975). Surgery 78 : 363.
Stein, M., Schiavi, R.C., and Camerino, M. (1976). Science 191 : 435.

DECLINE OF MURINE NATURAL KILLER ACTIVITY IN RESPONSE TO STARVATION, HYPOPHYSECTOMY, TUMOR GROWTH, AND BEIGE MUTATION: A COMPARATIVE STUDY

Rajiv K. Saxena
Queen B. Saxena
William H. Adler

Gerontology Research Center
National Institute on Aging
National Institutes of Health
Baltimore, Maryland

In view of the suggested role of natural killer (NK) activity in immunesurveillance, we have been interested in determining the factors which regulate NK activity in vivo. Our results regarding the NK status in chronic alcoholics have been discussed elsewhere in this volume (1). In the present chapter we have summarized the results of our studies on NK regulation in starved (2), hypophysectomized (3), tumor bearing (4) and beige mutant mice (5). After a separate brief discussion of each system, we shall present a comparative assessment of the NK changes seen in these situations.

I. DECLINE OF NK ACTIVITY IN STARVED MICE

The effect of starvation was examined on the spleen NK activity in female CBA mice (2). Mice were deprived of food for up to six days but water was given ad libitum. Significant decline of anti-YAC NK activity was observed 3 days after food deprivation and after six days of starvation the NK activity was almost totally absent. Addition of spleen cells from starved mice did not inhibit the NK activity of control spleen

ISBN 0-12-341360-5

cells as assayed in a 4 hr ^{51}Cr release assay using YAC target cells. Generation of suppressor cells which could actively suppress the activity of NK cells, was therefore unlikely to be the mechanism of NK decline in starved mice. Recovery of NK activity in starved mice after the reinstitution of normal diet was also studied. After 3 days of feeding the NK activity of spleen cells from the previously starved mice reached a level even higher than controls. Eight days after the refeeding, the NK activity of spleen cells from recovering mice was not significantly different from normal.

II. DECLINE OF NK ACTIVITY IN HYPOPHYSECTOMIZED MICE

Three week old female C57BL/6 mice were hypophysectomized by the parapharyngeal approach and were used 4 to 8 weeks after the operation. Spleen NK activity of hypophysectomized and sham operated animals was studied in a 4 hr ^{51}Cr release assay using YAC target cells (3). Significant loss of NK activity was detectable four weeks after the operation which was the earliest time point studied, and the NK activity in the hypophysectomized mice further declined to very low levels eight weeks after operation (3).

Low spleen NK activity in hypophysectomized mice was not due to the generation of NK suppressor cells, since the addition of hypophysectomized mouse spleen cells did not suppress the NK activity of normal control mouse spleen cells. Even though hypophysectomy resulted in a marked loss of spleen NK activity, the proportion of T, B and null cells and the YAC target binding capacity of spleen cells from hypophysectomized mice were comparable to normal levels. A normal target binding capacity along with a lack of cytotoxic activity might suggest that the pre-NK cells (6) may be present in hypophysectomized mice. Our results however indicated that NK levels in hypophysectomized mouse spleen cells could not be significantly enhanced by interferon or the interferon inducer Poly I:C. Since pre-NK cells can be induced to differentiate into cytotoxic NK cells by interferon, low NK activity in hypophysectomized mouse spleen was not due to an accumulation of pre-NK cells. Administration of growth hormone (GH, 100 μg/day for 10 days) resulted in a marked recovery of spleen NK activity in hypophysectomized mice which indicate that the loss of NK activity in hypophysectomized mice is reversible. By cell fractionation studies, the cytotoxic activity generated in hypophysectomized mouse spleen following GH treatment could be ascribed to the induction of NK cells.

III. NK STATUS IN P815 TUMOR BEARING MICE

Administration of 10^7 P815 tumor cells intraperitoneally to DBA/2 mice results in the development of P815 ascites tumor which grows rapidly and kills the host between 13 to 17 days after tumor initiation. Anti-YAC spleen NK activity of the tumor bearing (TB) mice declined and was undetectable by the 13th day of tumor bearing. Administration in vitro or in vivo of spleen cells from TB mice did not suppress the NK activity of normal spleen cells. Ascites fluid or serum from TB mice were also devoid of NK suppressive activity in vivo or in vitro. Unlike another tumor system (7), prostaglandins were also not involved in the NK suppression observed in P815 TB DBA/2 mice since indomethacin treatment in vivo or in vitro did not elevate the low NK activity in TB mouse spleen cells. Target (YAC) binding capacity of TB mouse spleen cells was normal but anti-YAC cytotoxic activity could not be boosted by interferon, indicating a lack of pre-NK cells in TB mouse spleen. Though the spleen and bone marrow cells from TB mice lacked any detectable NK activity, their administration (iv) to lethally x-irradiated (1000R) syngeneic mice resulted in the generation of variable but significant levels of NK activity in the spleens of recipient mice. The basic stem cells capable of differentiating into NK cells are therefore present in TB mice but are apparently unable to differentiate normally to cytotoxic NK cells in the altered internal melieu of the tumor bearing host.

IV. BEIGE MUTATION

A recessive beige mutation (bg/bg) of the C57BL/6 strain of mouse has been reported which (a) lacks significant spleen and systemic NK activity and (b) has normal levels of several other immune parameters, especially T-cell responsiveness (8-10). A combination of total lack of NK activity or a normal T cell response backgrund made the beige mutant an ideal model system to study the role of NK cells in immune surveillance. Our results however do not appear to support completely either of the two propositions mentioned above.

Though we could confirm a total deficiency of anti-YAC NK activity of bg/bg spleen cells in a 4 hr ^{51}Cr release assay (5), significant anti-K562 NK activity could be detected in bg/bg spleen cells in longer duration (20 hr) assays. A considerable mouse to mouse variation was found in basal anti-K562 NK activity of bg/bg and bg/+ spleen cell preparations and higher NK activities in some bg/bg mice could reach the lower levels of spleen NK activity in some bg/+ mice (Figure 1). In addition, several agents which are known to enhance the NK activity of mouse spleen cells (11) could markedly enhance the levels of anti-K562 NK activity in bg/+ as well as bg/bg spleen cell preparations (Figure 1). It is interesting to note that the augmented NK activity in each case was strongly correlated to the basal NK activity in bg/bg and bg/+ spleen cell preparations (Figure 1). In view of the fact that augmented NK activity of bg/bg mouse spleen cells could reach the basal spleen NK levels in bg/+ mice, it is possible that a lack of some NK sustaining factor(s) in vivo may contribute to the low levels of NK activity in bg/bg mice. Our recent results have also shown that a cytotoxic T-cell (CTL) response to an in vitro or in vivo challenge with P815 tumor cells were markedly deficient (about 1/3 of normal response) in bg/bg mice (5). Significantly lower than normal CTL response to lymphocytic choriomeningitis virus infection has been demonstrated in bg/bg mice (12) whereas the response to Vesicular Stomatitis Virus was comparable to normal (13). It is therefore possible that a normal or subnormal CTL response in bg/bg mice may depend upon the antigen utilized. In any case, our results stress the need for caution against using bg/bg mice as a model for a selective NK deficiency along with a normal T-cell function.

V. CONCLUSION

Though widely diverse systems have been discussed, the mechanism for depressed NK activity in each case appears to be similar in many respects. Thus an active suppression at the level of NK effector cells has not been demonstrated in any of the systems and in spite of depressed NK activity, target binding capacities of spleen cell preparations were generally normal. It has been our experience that the interferon activated NK levels are proportional to the basal NK levels. In the first three models discussed in this chapter, NK loss was gradual. Interferon could always augment the spleen NK activity in these systems, proportional to the residual NK activity and ultimately when the NK levels reached near zero, no interferon augmentation could be shown. Since significant NK levels persisted in bg/bg mice, interferon could induce a proportional NK enhancement. In none of the systems studied (with a possible exception of bg/bg mice), was the NK depression irreversible. Low spleen NK activity in starved mice could be boosted by refeeding, and GH treatment could induce considerable NK levels in hypophysectomized mice. Similarly bone marrow and spleen cells from NK deficient TB

FIGURE 1. Basal and augmented NK activity in bg/bg and bg/+ mice. Cytotoxic activities of eight spleen cell preparations each from bg/bg and bg/+ mice were studied against K562 target cells at an E:T of 100:1 in a 20 hr ^{51}Cr release assay (11). Activating agents were added to the assay medium at following concentrations. Anti H-2^b antiserum, 1/200 dilution; IF, 1000 U/well; PWM (pokeweed mitogen), 2 μg/well; Con-A (Concanavalin-A), 0.5 μg/well. Correlation of basal and augmented cytotoxic activity for each case is shown and value of correlation coefficient (r) is given.

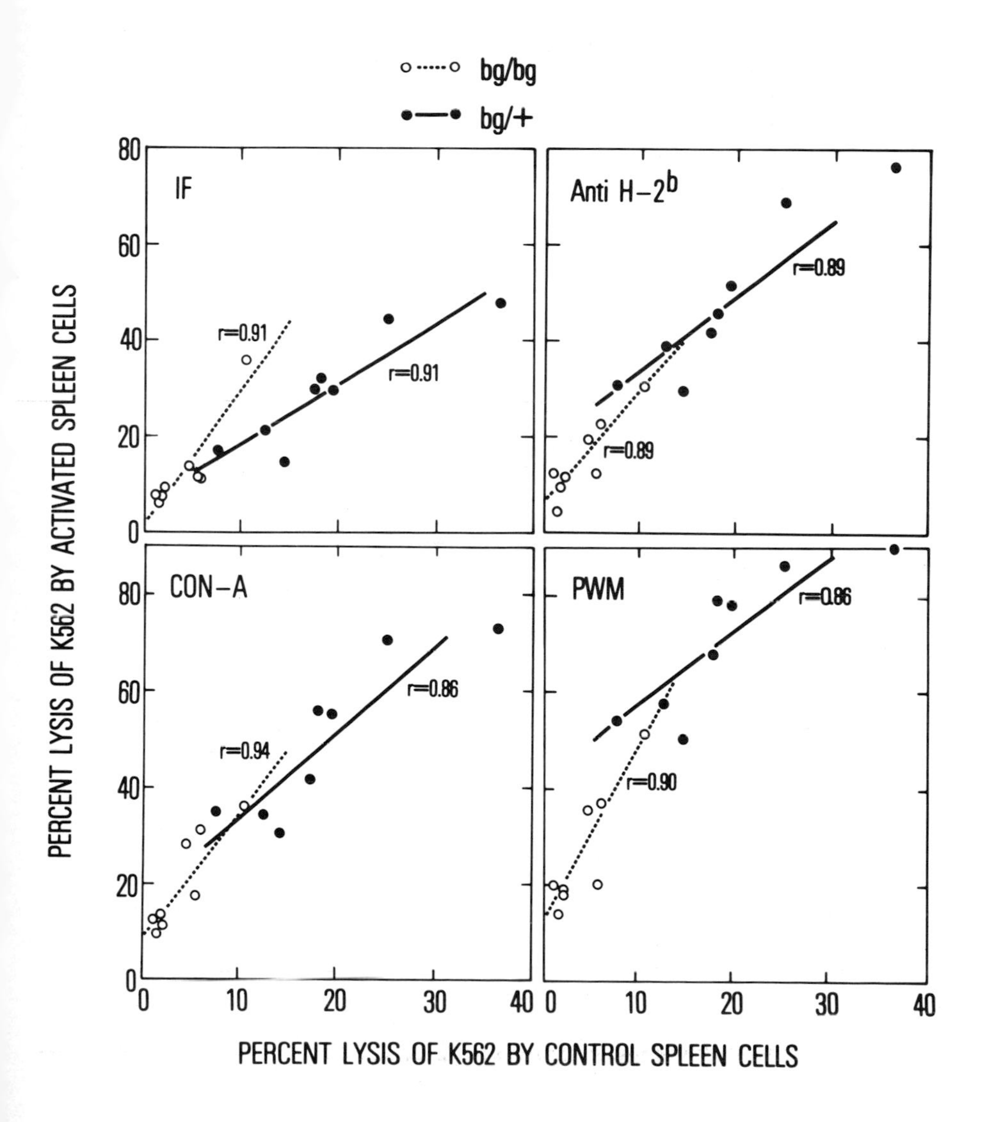

mice could induce significant spleen NK levels in lethally x-irradiated mice. It therefore appears that the basic NK stem cells persisted in each system and given the proper melieu, could readily differentiate into active NK cells. Our results therefore demonstrate that in several experimental model systems associated with low NK activity, an impairment of normal NK cell differentiation in the altered internal melieu may be an important contributory factor responsible for the expression of lower levels of NK activity.

ACKNOWLEDGMENTS

We wish to thank Ms. Eleanor Wielechowski for typing this manuscript and to Ms. Charlotte Adler for preparing the illustration.

REFERENCES

1. Saxena, Q. B., Saxena, R. K., and Adler, W. H., (in this volume).
2. Saxena, R. K., Saxena, Q. B., and Adler, W. H., Indian J. Exp. Biol. 18:1383 (1980).
3. Saxena, Q. B., Saxena, R. K., and Adler, W. H., Int. Arch. Allergy Applied Immunol. Vol. 67 (1982) in press.
4. Saxena, Q. B., Saxena, R. K., and Adler, W. H., (submitted for publication).
5. Saxena, R. K., Saxena, Q. B., and Adler, W. H., Nature (1982), (in press).
6. Roder, J. C., Kiessling, R., Biberfeld, P., and Andersson, B., J. Immunol. 121:2509 (1978).
7. Brunda, M. J., Herberman, R. B., and Holden, H. T., J. Immunol. 124:2682 (1980).
8. Roder, J., and Duwe, A., Nature 278:451 (1979).
9. Roder, J. C., J. Immunol. 123:2168 (1979).
10. Roder, J. C., Lohmann-Matthes, M-L., Domzig, W., and Wigzell, H., J. Immunol. 123:2174 (1979).
11. Saxena, R. K., Adler, W. H., and Nordin, A. A., Cell. Immunol. 63:28 (1981).
12. Welsh, R. M., and Kiessling, R. W., Scand. J. Immunol. 11:363 (1980).
13. McKinnon, K. P., Hale, A. H., and Ruebush, M., Infection and Immunity 32:204 (1981).

ETHANOL AND NATURAL KILLER ACTIVITY

Queen B. Saxena
Rajiv K. Saxena
William H. Adler

Gerontology Research Center
National Institute on Aging
National Institutes of Health
Baltimore, Maryland

A higher than normal incidence of certain types of tumors has been reported in chronic human alcoholics (1,2). In addition, evidence has been accumulating which points to a role of natural killer (NK) cells in the first line of defense against spontaneously arising tumors in vivo (3). In view of these observations, the influence of chronic alcohol ingestion on the NK activity levels was examined. Our results of the in vivo and in vitro effects of alcohol on NK activity in human and mouse systems are summarized in the present chapter.

I. IN VIVO EFFECT OF CHRONIC ALCOHOL INGESTION

A. Human System

Natural killer activity of human peripheral blood lymphocytes (HPBL) was studied in a 4 hr ^{51}Cr release assay using K562 target cells. HPBL preparations derived from a total of 15 normal healthy volunteers (3♀, 12♂, mean age 38.6 years, alcohol intake < 15 g/week) and 32 chronic alcoholics (2♀, 30♂, mean age 39.9 years, mean alcohol intake 218 g/day) were examined for NK activity in 11 experiments. In each experiment at least one control HPBL preparation was tested along with two to six HPBL preparations from alcoholics. Wide variations in the levels of NK activity were found in HPBL

ISBN 0-12-341360-5

preparations from control and alcoholics, but upon statistical analysis of variance, levels of NK activity in PBL derived from alcoholics were significantly higher than the control levels ($p < 0.001$) (4). The difference in the cytotoxic activity of PBL from control and alcoholic subjects was not abolished as a result of depletion of macrophages, T cells and B cells and therefore the cytotoxic activity of PBL from normal as well as alcoholic subjects resided in the NK population (4).

B. Mouse System

Since a human population is subject to numerous variables not able to be controlled or experimentally manipulated, we attempted to develop a mouse model system in which the effect of alcohol feeding on NK activity could be studied (5). Different groups of C57BL/6J mice were given mixtures of alcohol (2%, 4%, 8% or 16% alcohol v/v) in their drinking water and the spleen NK activity of control or alcohol drinking mice was studied 2 weeks later. Spleen NK activity of alcohol drinking mice was significantly higher than control mice. The maximum increase was noted in mice drinking 16% alcohol. In another experiment mice drinking 10% alcohol for 2 or 8 days did not show an enhanced NK activity in their spleen cells whereas, 2 to 3 fold higher NK activity was observed in spleen cells derived from mice drinking 10% ethanol for 4 to 6 weeks. The difference between the levels of spleen NK activity in control and alcohol drinking mice was not abolished when spleen cells were subjected to anti thy-1 + C treatment or passed through nylon wool columns, indicating that the increased spleen NK activity in alcohol drinking mice was not due to changes in T cell, B cell or macrophage populations (5). Even though the same mouse strain was used, considerable variations in the lag phase before NK activation, were observed in subsequent experiments. In some experiments NK activation did not occur in mice drinking alcohol for up to two months. Other undefined factors are therefore involved which regulate the NK activation in response to alcohol and attempts are currently underway to understand these factors.

II. IN VITRO EFFECT OF ETHANOL ON NK ACTIVITY

In view of higher than normal cancer incidence in human alcoholics, a lower level of NK activity was expected in alcholics. Our studies however revealed significantly higher than normal levels of NK activity in human alcoholics as well as in mice drinking alcohol (4,5). In view of these results, it was of interest to determine if ethanol, especially at the concentration present in the blood of human alcoholics, may directly influence the NK activity. Our results indicated that the in vitro addition of ethanol to NK cell assays induced a dose dependent inhibition of anti-K562 NK activity of HPBL (6). In human alcoholics, blood ethanol level may reach up to a level of 500 mg/100 ml blood. Higher than 50 mg/100 ml levels of ethanol were found to induce significant inhibition of NK activity in vitro, and at 400 mg/100 ml concentration, ethanol inhibited the NK activity by approximately 60% (6). The inhibitory effect of ethanol was not due to a direct toxic effect on HPBL since the viability of effector cells was not altered in the presence of ethanol even up to a 2000 mg/100 ml concentration. K562 target binding capacity of HPBL was not altered in the presence of ethanol indicating that ethanol may inhibit a subsequent step(s) in the process leading to the lysis of the target cell after its binding with the effector cell (6). Our results suggest that any putative advantage conferred by the higher levels of NK activity in HPBL from human alcoholics may be offset by the actual presence of alcohol in their blood.

The NK inhibitory effect of ethanol could be confirmed in the mouse system (Table 1). Target (YAC) binding capacity of mouse spleen cells however was not altered in the presence of 2% ethanol. Short term preincubation of mouse spleen cells with ethanol had no effect on their subsequent NK activity. Longer incubations of mouse spleen cells is known to result

TABLE I. In Vitro Effect of Ethanol on the NK Activity of Mouse Spleen Cells

Ethanol Concentration in Assay Medium (% v/v)	% Lysis of YAC[a]
None	42.7 ± 1.3
0.125	37.9 ± 2.6
0.250	36.8 ± 1.8
0.500	29.2 ± 3.0
1.000	14.7 ± 2.0
2.000	4.2 ± 1.3

[a] NK activity of C57BL/6 spleen cells against YAC cells was studied at an E:T = 100:1 in a 4 hr ^{51}Cr release assay. Each value of lysis is a mean of 4 replicate assay wells ± SD.

TABLE 2. Effect of Alcohol on the Activation of Mouse Spleen NK Activity in Response to Interferon and Interleukin-2[a]

Activating Agent Added During Preincubation[b]	Concentration of Ethanol During Pre-incubation	% Lysis of YAC	
		100:1	50:1
A. None	None	14.8	7.9
	0.5% (v/v)	8.5	4.5
	1.0%	6.4	3.1
	2.0%	3.6	1.7
B. Interferon (2500 U/ml)	None	26.9	15.5
	0.5%	11.3	6.2
	1.0%	7.4	3.9
	2.0%	2.1	1.8
C. Interleukin-2 (10 U/ml)	None	88.0	73.3
	0.5%	73.1	56.0
	1.0%	51.3	27.2
	2.0%	2.1	0.7

[a] C57BL/6 spleen cells (10^7/ml in RPMI + 10% FCS) were incubated for 20 hr (37° C, 5% CO_2, 95% air) with interferon or interleukin-2, in presence of different given concentrations of ethanol. Each cell preparation was washed two times with fresh medium and assayed for anti-YAC NK activity in a 4 hr ^{51}Cr release assay. Each value of target lysis is a mean of three replicate assay wells.

[b] Mouse fibroblast interferon (Lee Biomolecular, San Diego, CA) and rat interleukin-2 (Collaborative Research, MA) were used.

in a loss of NK activity (3). In those experiments in which significant residual NK activity could be demonstrated in mouse spleen cells preincubated for 20 hr. preincubation along with ethanol had significant NK inhibitory effect (Table 2A). This is in contrast to the results in the human system in which HPBL preincubated with ethanol for 1 or 20 hrs had NK activity comparable to control levels (6). NK inhibitory effects of prolonged incubation with alcohol on mouse spleen cells were not due to a toxic effect of alcohol since the recovery of viable spleen cells was the same in control and alcohol treated spleen cell cultures. Results in Table 2 also show that the activation of mouse spleen NK activity by interferon and interleukin-2 was also blocked by ethanol. We are

TABLE 3. Effect of Ethanol on the Anti-P815 Cytotoxic T-cell (CTL) Activity

Mode of Ethanol Treatment to Effector CTL	% Lysis of P815 at 100:1 E:T ratio[b]	
	Expt. 1	Expt. 2
No treatment	87.6	80.6
CTL pretreated with 2% ethanol for 1 hr at 37° C and washed	89.7	86.7
2% ethanol added to the cytotoxicity assay medium	16.7	29.5

[a] CTL containing spleen cell preparations were derived from C57BL/6 mice immunized with a single intraperitoneal dose of 10^7 P815 cells, 11 days before sacrifice.

[b] Anti-P815 lytic activity of CTL was studies in a 4 hr ^{51}Cr release assay.

also investigating whether ethanol is a nonspecific inhibitor of cytolytic effector cell function. In this regard we have found that murine immune-T cell activity is inhibited by ethanol (Table 3), whereas bactricidal activity of human monocytes is not influenced by 2% ethanol (our unpublished data).

III. CONCLUSION

Our studies on the effect of ethanol on NK activity have shown that a chronic intake of ethanol results in an elevation of NK activity levels in the cell populations from human blood and mouse spleen. NK activation is not a direct effect of ethanol since the direct addition of ethanol to cytotoxicity assays is markedly inhibitory to the activity of human or mouse NK cells. Moreover, ethanol also blocks the NK activation in response to interferon and interleukin-2. The mechanisism of ethanol induced NK activation is vivo is obscure. Speculation regarding the NK cell being a subpopulation of T-cells or even a type of pre-T cell have been made (3). If this view is correct, ethanol might increase the level of NK cells by preventing their further differentiation to T cells. In this regard, it is interesting to note that a depressed T cell function has been demonstrated in alcoholics (7).

ACKNOWLEDGMENTS

We wish to thank Ms. Eleanor Wielechowski for typing this manuscript.

REFERENCES

1. Lieber, C. S., Seitz, H. K., Garro, A. J., and Woones, T. M., Cancer Res. 39:2863 (1979).
2. Tuyns, A. J., Cancer Res. 29:2840 (1979).
3. Herberman, R. B., and Holden, H. T., Adv. Cancer Res. 27:305 (1978).
4. Saxena, Q. B., Mezey, E., and Adler, W. H., Int. J. Cancer 28:413 (1980).
5. Saxena, Q. B., Saxena, R. K., and Adler, W. H., Indian J. Exp. Biol. 19:1001 (1981).
6. Saxena, Q. B., Saxena, R. K., and Adler, W. H., (submitted for publication).
7. Lundy, J., Raff, J. H., Deakins, J., Wanebo, H. J., Jacobs, R. A., and Lee, T. D., Surg. Gynec. Obstet. 141:212 (1975).

REGULATION OF HUMAN NK ACTIVITY AGAINST ADHERENT TUMOR TARGET CELLS BY MONOCYTE SUBPOPULATIONS, INTERLEUKIN-1, AND INTERFERONS

Jan E. de Vries
Carl G. Figdor
Hergen Spits

Division of Immunology
The Netherlands Cancer Institute
Amsterdam, The Netherlands

I. INTRODUCTION

Human natural killer (NK) cell activity has been shown to be mediated by non-adherent cells which have receptors for the Fc portion of IgG (FcR^+) and which express at least partially, low affinity receptors for sheep erythrocytes (E) (1,2). In addition, recently evidence has been presented that human NK cells belong to the large granular lymphocytes (LGL) population (3,4).

We previously reported that small to medium sized lymphocytes (SML) isolated from peripheral blood lymphocytes (PBL) by 1 x G velocity sedimentation in spite of the presence of FcR^+ cells failed to exhibit NK cell activity against adherent target cells. However, these SML efficiently killed tumor target cells growing in suspension cultures (K562, Molt-4) and antibody coated Chang liver cells (5,6). In addition, we reported that NK activity against adherent target cells could be induced if autologous monocytes or monocyte culture supernatants were added to the SML, indicating that soluble monocyte products are required for the generation of this type of NK activity. In the present communication we demonstrate that SML isolated by centrifugal elutriation (CE) (7,8) lacked NK activity, in spite of the presence of large granular lymphocytes (LGL). Reconstitution experiments with various autologous monocyte fractions that differed in specific density and

ISBN 0-12-341360-5

function (8) revealed that the NK activity inducing capacity is predominantly restricted to the monocyte population with the lowest specific density. In addition it is shown that significant NK activity could be induced if partially-purified Interleukin-1 or interferon was added to the SML.

II. MATERIAL AND METHODS

A. Target Cells

The T_{24} bladder carcinoma cell line and the melanoma cell lines NKI-4 and K which were established in our laboratory, were used as target cell sources (6).

B. Effector Cells

Mononuclear leukocytes were isolated from buffy coats obtained from the Central Laboratory of the Netherlands Red Cross Blood Transfusion Service (Amsterdam, The Netherlands) as described previously (8). The mononuclear leukocytes were fractionated by a modified centrifugal elutriation technique (7,8). Four different lymphocyte fractions (LF) were obtained at speeds of rotation of 3300, 3100, 2900 and 2800, respectively. The monocyte population was subsequently fractionated by increasing the density of the elutriation buffer (8). Three different monocyte fractions (MF) were obtained which differed 0.0028 g/ml in their specific density. These monocyte fractions have been shown to differ in esterase and peroxidase activity, ADCC activity and their capacity to induce T-cell proliferation in mixed leukocyte cultures (8).

C. Characterization of the Fractionated Lymphocyte and Monocyte Populations

The % LGL in the various lymphocyte fractions was determined in cytocentrifuge preparations stained with May-Grünwald Giemsa (MGG) stain. The lymphocyte fractions were further characterized with the monoclonal antibodies OKT-3, 4 and 8 directed against T-cell differentiation antigens (9-13), and a monoclonal antibody (OKM-1) reacting with monocytes and NK cells (14). The binding of the monoclonal antibodies to the various lymphocyte subsets was determined by indirect immunofluorescence with fluoresceinated goat-anti-mouse IgG (Nordic Diagnostics, Tilburg, The Netherlands) utilizing a Fluorescence Activated Cell Sorter (FACS IV, Becton and Dickinson, Rutherford, N.J.). The monocytes were characterized in cytocentrifuge preparations stained for non-specific cytoplasmic esterase and by indirect immunofluorescence with OKM-1 and RUPI-5, a monoclonal antibody produced in our laboratory reacting with a 35,000 D determinant on monocytes and granulocytes. The percentage FcR^+ cells in the various lymphocyte fractions was determined by rosette formation with beads coated with rabbit anti-human IgG (Immunobead, Biorad Laboratories, Richmond, Cal.).

D. Pretreatment of the Lymphocytes with IFN

The various lymphocyte fractions were incubated at concentrations of 5 x 10^6 cells/ml (1½ hr in a 37°C waterbath under shaking) in the absence or presence of 3000 IU interferon. Two different sources of IFN were used. Fiblaferon L. is a human fibroblast IFN with a specific activity of 1 x 10^6 IU/mg protein purchased from Flow Laboratories (Irvine, Scotland). PIF is a human leukocyte IFN with a specific activity of 1 x 10^6 IU/mg protein, obtained from Dr. K. Cantell, Public Health Laboratories, Helsinki, Finland. Both interferons were kindly provided by Dr. H. Schellekens (Radiobiological Institute TNO, Rijswijk, The Netherlands). After two washings the lymphocytes were used as effector cells in the test.

E. Purification of IL-2 from Monocyte Culture Supernatants

Monocytes (92% pure) were isolated from buffy coats by CE. 2 x 10^6 monocytes/ml were cultured in serumfree medium (15) in the presence of 2 x 10^{-5} M indomethacin (Sigma Chemical Company, St. Louis, Mo.) for 24 hrs at 37°C and 5% CO_2. Cell free supernatants were collected, precipitated with ammonium-sulphate to a final concentration of 85% saturation, dialyzed and concentrated to a volume of 4 ml with Aquacide IIA (Calbiochem, San Diego, Cal.). This sample was further fractionated on a Sephacryl S200 column. The various fractions were filter sterilized and tested for their ability to stimulate mouse thymocyte proliferation in the presence and absence of PHA. Thymocyte proliferating stimulating capacity was recovered in 2 major peaks with MW of approximately 70,000 D and 16,000 D respectively. The fractions in the 70,000 D MW range were pooled (40 ml) and bioassayed for its IL-1 activity. At concentrations of 5% (v/v) the thymocyte proliferation in the presence of PHA was enhanced 20 - 40 fold.

F. ^{51}Cr-Release Assay for NK Activity

The NK activity of the various lymphocyte and monocyte fractions was measured in a ^{51}Cr-release assay as described previously (6). The target cells were plated at a concentration of 5×10^3 cells/well and the effector cells were added at concentrations of 5, 2.5 and 1.25×10^5/well. The tests were harvested after 24 h incubation at 37°C and 5% CO_2. The NK activity was expressed in lytic units (LU) where 1 LU was defined as the number of effector cells required to give 25% specific ^{51}Cr release from 5×10^3 target cells.

III. RESULTS

A. Characterization of the Various Lymphocyte and Monocyte Fractions

The various lymphocyte fractions are characterized in Table I. The separation according to size did not result in a significant enrichment or depletion of OKT-3$^+$, OKT-4$^+$ and OKT-8$^+$ lymphocytes (not shown). The slight increase in FcR$^+$ bearing cells in LF-4 was probably caused by the contaminating monocytes. Although OKT-1 only seems to react with a proportion of the LGL, the increase in the % of OKM-1$^+$ cells in the LF-3 and LF-4 correlates with the enrichment of LGL in these fractions and cannot be attributed to OKM-1$^+$ monocytes, since less than 0.5% and 10% RUPI-5$^+$, esterase$^+$ monocytes were present in LF-3 and LF-4, respectively. The monocyte fractions were 90% OKM-1$^+$, RUPI-5$^+$, and esterase$^+$ and contained < 0.5% LGL. The average lymphocyte and granulocyte contaminations in these fractions were 4% and 3%, respectively (the latter not shown). The lymphocytes recovered in LF-1 and LF-2 represented 57% of the lymphocytes initially present in the buffy coat, whereas these recoveries for LF-3 and LF-4 were 13% and 8% respectively.

B. NK Activity of the Various Lymphocyte and Monocyte Fractions

The NK activity of the various lymphocyte and monocyte fractions against 3 different adherent tumor target cells is shown in Table II. T_{24} target cells show the highest susceptibility for NK activity whereas the NKI-4 target cells are the most resistent. Compared to the corresponding unfractionated lymphocytes (UL) the small to medium sized lymphocytes (SML)

TABLE I. Characterization of the Unfractionated Mononuclear Leukocytes and the Various Lymphocyte and Monocyte Fractions Isolated by Centrifugal Elutriation

	Lym.[a]	Mon. (Est)	LGL	OKM-1+	RUPI-5+	FcR+
	mean and (range) of % (positive) cells of 4 expts					
UL[b]	57 (52-62)	25 (24-28)	9 (6-12)	32 (20-37)	22 (20-28)	26 (21-21)
LF-1	95 (90-97)	0	5 (3- 8)	1.5 (1- 3)	0	13 (10-18)
LF-2	90 (82-95)		9 (5-13)	3 (2- 6)	0	11 (10-15)
LF-3	81 (79-84)	< 0.5 (0-0.5)	19 (15-22)	11 (8-16)	0	16 (13-20)
LF-4	70 (65-75)	9 (5-15)	15 (12-20)	22 (17-29)	7 (6-11)	19 (15-23)
MF-1	4 (3- 6)	89 (85-92)	< 0.5	90 (87-92)	89 (85-92)	-
MF-2	4 (3- 5)	90 (84-93)	< 0.5	92 (89-93)	91 (89-92)	-
MF-3	3 (2- 5)	89 (84-92)	< 0.5	92 (90-93)	89 (88-92)	-

[a]% lymphocytes excluding LGL (200 cells counted)
[b]unfractionated mononuclear leukocytes

TABLE II. NK Activity of the Various Lymphocyte and Monocyte Fractions

	T_{24}	K	NKI-4
	% ^{51}Cr release (mean ± S.E.M. of 4 expts)		
UL	41 ± 3	32 ± 3	18 ± 4
LF-1	3 ± 2	1 ± 2	-2 ± 2
LF-2	6 ± 2	2 ± 2	1 ± 2
LF-3	45 ± 4	30 ± 4	20 ± 4
LF-4	68 ± 5	49 ± 4	35 ± 5
MF-1	-3 ± 1	-4 ± 1	-2 ± 1
MF-2	-2 ± 1	-2 ± 1	-3 ± 1
MF-3	-2 ± 2	-1 ± 2	-2 ± 2

Only the results with 5 x 10^5 effector cells/well are shown. The average spontaneous ^{51}Cr release of T_{24}, K and NKI-4 was 19%, 16% and 11%, respectively.

of the LF-1 and LF-2 showed negligible NK activity, indicating that the LGL present in these fractions failed to exert NK activity. The NK activity of LF-3 was approximately of the same level or higher than the NK activity achieved with the UL. Maximum NK activity was consistently observed with LF-4 cells in spite of the fact that LF-4 contained significantly lower concentrations of LGL than LF-3. The three monocyte fractions had no detectable NK activity.

C. NK Activity of LF-1 and LF-2 after Reconstitution with Autologous Monocytes

We previously showed that SML isolated by 1 x G velocity sedimentation lacked NK activity against adherent target cells, but that NK activity could be induced by reconstitution with autologous monocytes (5,6). To investigate the NK inducing capacity of the monocyte subpopulations, graded numbers of the 3 monocyte fractions were added to the combined LF-1 and LF-2. In order to measure induction optimally (and not augmentation of minimal existing NK activity) only NKI-4 melanoma cells were used as target cells in these studies. The NKI-4 cells were rather resistent for NK-activity and LF-1 and LF-2 cells failed to exert any NK activity at effector : target cell ratios up to 400 : 1 (results not shown). In Table III is shown that the NK-inducing capacity is predominantly confined to the monocyte fraction with the lowest specific density (MF-1) representing 25% to 31% of the monocytes initially present in the buffy coat. Considerable variation in the NK-inducing capacity of the various monocyte fractions is observed. In Expt 1 optimal effects were obtained at monocyte concentrations of 25% whereas in the Expts 2, 3 and 4 addition of 12.5% monocytes induced optimal effects. MF-2 in which 18 - 23% of the monocytes were recovered, was less effective in inducing NK activity and in 3/4 experiments a significant reduction in its NK inducing capacity was observed at monocyte concentrations of 25%. MF-3 representing 13 - 16% of the monocytes generally failed to induce significant NK activity.

D. Effect of Interleukin-I on NK Activity

We previously showed that monocyte culture supernatants could partially replace monocytes in the induction of NK activity by SML isolated by velocity sedimentation at 1 x G (6). Supernatants from cultures of monocytes isolated by centrifugal elutriation were found to contain IL-1 activity, but lacked Interleukin-2 and interferon activity (results not

TABLE III. NK Activity of LF-1 + LF-2 after Reconstitution with the Different Monocyte Fractions

			Monocytes added in concentration of		
			6.25%	12.5%	25%
		% specific ^{51}Cr release (mean ± S.D.)			
Expt 1	UL	13 ± 3	-	-	-
	LF-1.2[a]	-1 ± 2			
	LF-1.2 + MF-1		NT	27 ± 4	73 ± 5
	LF-1.2 + MF-2		NT	5 ± 2	11 ± 2
	LF-1.2 + MF-3		NT	1 ± 3	5 ± 2
Expt 2	UL	18 ± 4	-	-	-
	LF-1.2	0.5 ± 2			
	LF-1.2 + MF-1		9 ± 2	12 ± 2	12 ± 4
	LF-1.2 + MF-2		8 ± 2	10 ± 3	5 ± 3
	LF-1.2 + MF-3		2 ± 1	-2 ± 2	-2 ± 3
Expt 3	UL	33 ± 4	-	-	-
	LF-1.2	-2 ± 2			
	LF-1.2 + MF-1		15 ± 3	40 ± 4	24 ± 3
	LF-1.2 + MF-2		6 ± 3	14 ± 4	2 ± 2
	LF-1.2 + MF-3		3 ± 2	2 ± 2	-5 ± 3
Expt 4	UL	22 ± 3	-	-	-
	LF-1.2	-1 ± 2			
	LF-1.2 + MF-1		12 ± 3	17 ± 4	14 ± 3
	LF-1.2 + MF-2		7 ± 2	9 ± 3	5 ± 3
	LF-1.2 + MF-3		2 ± 2	-1 ± 3	-3 ± 2

All tests were carried out at effector cell concentrations of 2.5×10^5 / well. [a]LF-1.2 = combined LF-1 + LF-2.

shown). Partially-purified IL-1 induced significant NK activity in LF-1 and LF-2, suggesting that monocyte help which is required for the expression of NK activity of LF-1 and LF-2 cells is mediated by IL-1 (Table IV). In addition, IL-1 is shown to enhance NK activity of UL, LF-3 cells and less consistently of LF-4 cells.

TABLE IV. Spontaneous and IL-1 or Interferon Boosted NK Activity in the Various Lymphocyte Fractions

fraction	% LGL[a]	LU/10^7 effector cells				total LU / fraction[b]			
		spont	IL-1	PIF	HFIF	spont	IL-1	PIF	HFIF
UL	10 ± 2	7 ± 2	38 ± 5	27 ± 1	21 ± 3	510 ± 199	2649 ± 401	1882 ± 342	1464 ± 272
LF-1	4 ± 1	< 1	14 ± 3	9 ± 2	7 ± 2	< 1	208 ± 28	134 ± 34	104 ± 20
LF-2	8 ± 2	< 1	18 ± 3	13 ± 2	10 ± 3	< 1	191 ± 25	128 ± 16	106 ± 18
LF-3	21 ± 2	10 ± 2	79 ± 8	51 ± 5	42 ± 8	67 ± 20	529 ± 64	342 ± 24	281 ± 38
LF-4	13 ± 2	94 ± 37	135 ± 54	205 ± 41	149 ± 39	391 ± 147	562 ± 211	853 ± 156	620 ± 108

Mean ± S.E.M. of 3 different experiments. IL-1 was tested at a concentration of 10% (v/v).

[a] % of LGL in 200 cells counted

[b] total LU/10^7 cells multiplied by the number of recovered cells

E. Effect of Interferon on NK Activity

Since it has been suggested that interferon can enhance NK activity by induction of NK activity in inactive NK precursor cells (16-18), we investigated whether interferon could induce LF-1 and LF-2 cells to exhibit NK activity. In Table IV it is shown that preincubation of LF-1 and LF-2 cells with interferon resulted in the induction of significant NK activity, whereas interferon also enhanced the existing NK activity of LF-3 and LF-4 cells, in which 13% and 75% respectively of the spontaneous NK activity was recovered. Approximately 20% of the interferon induced NK activity of the unfractionated lymphocytes was recovered in LF-1 and LF-2. PIF was found to be more effective than HFIF in inducing or enhancing NK activity.

IV. DISCUSSION

Recently it has been demonstrated that human NK activity against target cells growing in suspension cultures and anchorage dependent target cells is strongly associated with LGL (3,4). We previously reported that small to medium sized lymphocytes isolated by velocity sedimentation at 1 x G failed to exert NK activity against adherent tumor target cells, but that NK activity could be induced by reconstitution with autologous monocytes or (less effectively) by the addition of monocyte culture supernatants (6). These observations are extended in the present communication in which a different cell separation technique was used. It is demonstrated that small to medium sized lymphocytes (LF-1 + LF-2 cells) depleted of monocytes did not exhibit NK activity, indicating that the LGL in these fractions were inactive. Reconstitution with monocyte fractions that differed in specific density resulted in induction of NK activity. Considerable variation among individual donors was observed, but in general the capacity to induce NK activity was predominantly confined to the monocyte fraction with the lowest specific density. Compared to the other monocyte fractions, MF-1 monocytes have been previously shown to be the most effective in the induction of T-cell proliferation in mixed leukocyte cultures, to contain the lowest esterase and peroxidase activity and to mediate the lowest ADCC activity against antibody coated erythrocytes (8). In addition, we recently found that the IL-1 production by MF-1 was 2 - 3 times higher than that by MF-2 and MF-3 (De Vries & Leemans, unpublished results). Since partially-purified IL-1 also induced NK activity in the SML, it seems fair to conclude that monocyte induced NK activity is mediated by IL-1. Therefore, it is not unlikely that the optimal NK-inducing capacity of MF-1 monocytes can be attributed to the higher IL-1 production of these cells. MF-2 monocytes were less effective in inducing NK activity. The reduction in their NK-inducing capacity at concentrations of 25% indicates that MF-2 monocytes in addition to their NK-inducing capacity also can have suppressive effects on NK activity. Furthermore, these data indicate that the suppressive effects generally are more pronounced with monocytes of higher densities. Although MF-3 monocytes produced IL-1, they failed to induce significant NK activity. These and our previous results (5,6) are in contrast with the data obtained by De Landazuri et al. (4) who reported that monocytes neither induced reactivity in non-cytotoxic lymphocyte fractions nor enhanced or suppressed the NK activity of already cytotoxic lymphocyte fractions separated on Percoll gradients. These differences are most probably due to the different monocyte separation techniques employed and the way in which the monocytes were handled.

Significant NK activity of the SML was also induced by interferon. This interferon-induced NK activity reflected 20% of the activity induced in the unfractionated cells. These data, taken together with the observation of Timonen et al. (3) who demonstrated that interferon enhanced NK activity was exclusively associated with LGL, suggest that interferon triggers the inactive LGL present in the SML to become cytotoxic effectors.

The mechanism by which monocytes or IL-1 induce NK activity in SML remains to be investigated. Since it has been reported that IL-1 can induce T-cell differentiation (19,20) it seemed not unlikely that IL-1 acts on NK precursor cells to become effective killers (LGL). However, the finding that the percentage of LGL present in the SML remained constant after preincubation with IL-1 or interferon and incubation overnight indicates that under influence of these factors generation of new LGL from precursors did not occur (results not shown). Since Farrar et al. (21) reported that IL-1 can induce the production of immune interferon and interferon was found to be produced in long-term cytotoxicity assays (22), it may also be possible that via this pathway the non-reactive LGL present in the LF-1 and LF-2 are triggered to become cytotoxic.

Finally, it is demonstrated that LGL were enriched in LF-3 and LF-4. Although the highest concentrations of LGL were consistently recovered in LF-3 no correlation with NK activity was observed, since maximal spontaneous- and interferon enhanced NK activity was recovered in LF-4 which contained less LGL. These results suggest that "large" LGL present in LF-4 are more efficient killers than the smaller LGL present in

LF-3, whereas the smallest LGL in LF-1 and LF-2 lack NK activity. On the other hand, it cannot be excluded that the contaminating monocytes present in LF-4 play a role in the high Nk activity of this fraction. This possibility is currently under investigation.

ACKNOWLEDGMENTS

We gratefully acknowledge the expert technical assistance of Martin Poelen, Betty Honing, Elisabeth Martens, Jacques Leemans, and Eric Philippus for running the FACS. We thank Dr. Jan Trapman (Dept. of Pathology, Erasmus University, Rotterdam, for screening the monocyte culture supernatants for interferon activity, Dr. W. Bont for his help with the elutriation centrifugation procedure, and Marie Anne van Halem for her excellent secretarial assistance.

REFERENCES

1. West, W.H., Cannon, G.B., Kay, H.D., Bonnard, G.D., and Herberman, R.B., J. Immunol. 118:355 (1977).
2. Kay, H.D., Bonnard, G.D., West, W.H., and Herberman, R.B., J. Immunol. 118:2058 (1977).
3. Timonen, T., Ortaldo, J.F., and Herberman, R.B., J. Exp. Med. 153:569 (1981).
4. De Landazuri, M.O., López-Botet, M., Timonen, T., Ortaldo, J.R., and Herberman, R.B., J. Immunol. 127:1380 (1981).
5. De Vries, J.E., Mendelsohn, J., and Bont, W.S., Nature 283:574 (1980).
6. De Vries, J.E., Mendelsohn, J., and Bont, W.S., J. Immunol. 125:396 (1980).
7. Figdor, C.G., Bont, W.S., De Vries, J.E., and Van Es, W.L., J. Immunol. Meth. 40:275 (1981).
8. Figdor, C.G., Bont, W.S., Touw, I., Roosnek, E., and De Vries, J.E., Blood, in press (1982).
9. Kung, P.C., Goldstein, G., Reinherz, E.L., Schlossman, S.F., Science 206:349 (1979).
10. Reinherz, E.L., Kung, P.C., Goldstein, G., and Schlossman, S.F., J. Immunol. 123:1312 (1979).
11. Reinherz, E.L., Kung, P.C., Goldstein, G., and Schlossman, S.F., Proc. Natl. Acad. Sci. USA 76:4061 (1979).

12. Reinherz, E.L., Kung, P.C., Goldstein, G., and Schlossman, S.F., J. Immunol. 123:2894 (1979).
13. Reinherz, E.L., Kung, P.C., Goldstein, G., and Schlossman, S.F., J. Immunol. 124:1301 (1980).
14. Breard, J.M., Reinherz, E.L., Kung, P.C., Goldstein, G., and Schlossman, S.F., J. Immunol. Immunopathol. 18:145 (1980).
15. Spits, H., IJssel, H., Terhorst, C., and De Vries, J., J. Immunol. 128:95 (1982).
16. Saksela, E., Timonen, T., and Cantell, K., Scand. J. Immunol. 10:257 (1979).
17. Targan, S., and Dorey, F., J. Immunol. 124:2117 (1980).
18. Bloom, B., Minato, N., Neighbour, A., Reid, L., and Marcus, D., in "Natural Cell Mediated Immunity against Tumors" (R.B. Herberman, ed.), p. 505. Academic Press, New York, 1980.
19. Beller, D.I., and Unanue, E.R., J. Immunol. 118:1780 (1977).
20. De Vries, J.E., Vyth-Dreese, F., Van der Hulst, R., Sminia, P., Figdor, C.G., Bont, W.S., and Spits, H., Immunobiology, in press (1982).
21. Farrar, W.L., Johnson, H.M., and Farrar, J.J., J. Immunol. 126:1120 (1981).
22. Trinchieri, G., and Santoli, D., J. Exp. Med. 147:1314 (1978).

PRODUCTION OF INTERFERON BY HUMAN NATURAL KILLER CELLS IN RESPONSE TO MITOGENS, VIRUSES AND BACTERIA

Julie Y. Djeu[1], Tuomo Timonen[2] and Ronald B. Herberman[2]

[1]Division of Virology, Bureau of Biologics, Bethesda, Maryland. [2]Biological Research and Therapy Branch, National Cancer Institute-FCRF, Frederick, Maryland

The ability of interferon (IFN) to augment natural killer (NK) cell activity has been well established and substantial evidence to date suggests that most biological agents that enhance NK cytotoxicity do so via their common ability to induce IFN (1). The recent demonstration that virtually all human NK cells have the morphology of large granular lymphocytes (LGL) (2), accompanied by the ability to highly purify these cells using density gradient centrifugation (3), have provided the opportunity to examine whether mitogens, viruses, bacteria and tumor cells which can augment NK activity may do so by stimulating IFN production in the NK cell population.

METHODS

Human peripheral blood lymphocytes were obtained by Ficoll-Hypaque density gradient centrifugation of lymphoid cell concentrates from the Plateletpheresis Center or Blood Bank of the National Institutes of Health, Bethesda, MD. Adherent cells were rigorously removed by incubation on Petri dishes for 2 hr at 37°C and subsequent incubation on nylon wool columns for 30 min at 37°C. The remaining nonadherent cells, enriched in NK and T cells, were then placed on a 7 step discontinuous Percoll density gradient (2). After centrifugation at 550 g for 30 min, the fractions were collected and checked for LGL morphology on cytocentrifuged slides. Cells in Fraction 3, which invariably contained 70-90% LGL, were then incubated overnight with various agents at 37°C. The following day, the supernatants were collected for IFN testing and the cells were tested for NK activity against K562 tumor cells.

RESULTS

Table 1 shows a representative experiment on IFN induction in LGL. Within 18 hr of exposure to a variety of agents,

ISBN 0-12-341360-5

LGL exhibited significantly enhanced levels of NK activity and IFN was present in their culture supernatants. Mitogens such as SEA, Con A, and PHA induced 50-250 units of IFN/ml, as did the bacteria, BCG and C. parvum. Higher levels, often over 3000 units/ml, were induced by viruses, particularly influenza A/PC. Several human tumor cell lines, K562, Molt-4, T-24, and ALAB also induced 50-125 units of IFN/ml. However, these lines were found to be mycoplasma-contaminated, and mycoplasma-free lines induced low or no detectable levels of IFN. There was no quantitative correlation between the level of augmentation of NK activity and the level of IFN induced by each agent. Higher production of IFN did not induce a correspondingly higher level of augmentation of cytotoxicity, suggesting that under the conditions of the preincubation, small amounts of IFN may be sufficient to produce maximum cytotoxicity in LGL.

TABLE 1. Augmentation of NK activity and induction of IFN in LGL

Stimulus		Conc.	% NK	IFN (U/ml)
Medium			25.5	0
Mitogens-SEA		0.1 ug/ml	38.7	125
Con A		1 ug/ml	40.1	50
PHA		1 ug ml	38.8	250
Viruses-Influenza A/PC		1:1000 dil	36.5	3125
HSV-1		1:100 dil	39.9	625
Bacteria-BCG		10^5/ml	39.4	50
C. parvum		50 ug/ml	33.8	200
Tumor cells-	K562	1%	34.1	125
(Mycoplasma	Molt-4	1%	35.4	125
contaminated)	T-24	1%	38.6	125
	ALAB	1%	38.4	50
Tumor cells-	K562	1%	28.1	0
(Mycoplasma	MOLT-3	1%	24.2	0
-free)	T-24	1%	23.9	0
	ALAB	1%	24.8	0

The purity of the LGL population is critical for definitive identification of these cells as the source of IFN. To leave little doubt about the capability of LGL to produce IFN, the LGL fractions were subjected to further purification by removal of contaminating T cells, macrophages and B cells before stimulation with the various agents. A summary of these data is represented in Table 2. Removal of T cells with monoclonal OKT3 plus complement from the LGL had no effect on the IFN production or enhancement of NK activity. Similarly, incubation of LGL with silica or carrageenan, which inactivates macrophages, did not interfere with the two functions. Indeed, simultaneous treatment of LGL with OKT3 plus complement and silica provided further proof of the ability of LGL to function alone in response to all the stimulating agents. Passage of lymphocytes through nylon wool is an efficient means of removing B cells, which is usually performed before LGL are isolated on Percoll gradients. To further eliminate any contaminating B cells from the LGL, they were treated with anti-IgM plus complement and again, this had no effect on LGL responses to the stimulating agents.

TABLE 2. Effect of removal of contaminating cells on LGL response to stimulating agents.

Treatment of LGL with:	Cell type removed	NK boost	IFN produced
OKT3 + complement	T cells	NI[a]	NI
silica	macrophages	NI	NI
carrageenan	macrophages	NI	NI
OKT3 + complement/ silica	T cells + macrophages	NI	NI
anti-IgM + complement	B cells	NI	NI

[a]No Inhibition

Characterization of the IFN by neutralizing antibodies (4) to IFN-α, β and γ is summarized in Table 3. In most experiments, the IFN induced by SEA, Con A or PHA was resistant to anti-α, although occasionally slight sensitivity to this antibody was seen. The SEA and Con-A induced IFN was consistently neutralized 30-35% by anti-β . The remainder of IFN activity was resistant to anti-α , but susceptible to anti-γ. Viruses such as influenza A/PC and HSV-1 clearly induced IFN-α as shown by their complete sensitivity to anti-α and resistance to anti-β γ. In a few individuals, their LGL responded to influenza A/PC by producing IFN that was resistant to both anti-α and anti-β , suggesting that IFN-γ was produced, perhaps due to prior exposure and sensitization. Bacteria-induced IFN was similar to tumor-induced IFN, having the characteristics of IFN-α .

TABLE 3. Characterization of IFN produced by LGL

Stimulus		% inhibition of IFN		
		anti-α	anti-β	anti-γ
Controls -	IFN-	100	0	0
	IFN-	0	100	0
	IFN-	0	0	100
Mitogens -	SEA	0	35	0
	Con A	0	30	ND[b]
	PHA	0	0	100
Viruses -	Influenza A/PC	100[a]	0	0
	HSV-1	100	0	0
Bacteria -	BCG	100	0	0
	C. parvum	100	0	0
Tumor cells -	K562-mycopl	100	0	0

[a] In a few individuals, the IFN induced in LGL by influenza A/PC was resistant to both anti-α and anti-β .

[b] Not Done.

DISCUSSION

The property of LGL to respond to a variety of stimuli by IFN production is an important finding because this suggests that NK cells are capable of positive self regulation of their function, with a variety of foreign materials first inducing production of IFN which in turn increases the cytotoxic reactivity of these cells. Furthermore, these findings broaden the range of natural host defense capabilities of these effector cells beyond their ability to spontaneously lyse certain target cells. The production of IFN from NK cells in response to a wide range of foreign materials could result in a direct increase in antiviral resistance of cells and could also have a number of indirect effects on host resistance, via the ability of IFN to activate other cells, such as macrophages (5) and T cells (6). Another aspect of considerable interest in the present findings was the ability of LGL to produce a variety of types of IFN. It would be of interest to determine whether the different types of IFN are made by separate subpopulations of LGL or whether, depending on the nature of the stimulus, the same cells can produce different IFN species. In any event, these data strengthen the suggestion that NK cells may be important in natural resistance to various microorganisms as well as to tumor cells (1,7,8).

Other cell types also possess the capability to produce IFN, notably T cells upon stimulation by mitogens or specific antigens (9,10). This response requires macrophage cooperation and a longer period of culture, 3-5 days, for IFN production, and consists mainly of IFN-γ. It is of interest that a recent study on Con A stimulation of unseparated human peripheral blood lymphocytes obtained similar findings to ours, in that the IFN that appeared within 24 hr was a mixture of IFN-β and γ (11). Macrophages may also produce IFN in response to viruses such as influenza (12). Moreover, B cells or Fc receptor positive, surface immunoglobulin negative cells have been implicated as a source of IFN in cultures containing Sendai virus (13) or tumor cells (14,15) while null cells have been observed to produce IFN in response to HSV-1, C. parvum, and tumor cells (16). The possibility exists, however, that the IFN production by many of these non-T cell preparations may in fact be attributable to LGL. Another report on the induction of IFN-γ by Staphylococcus aureus Protein A in Fc receptorpositive, non-T cells (17) would be consistent with the production by LGL. Recently, K562 cells have been shown to induce IFN production by LGL (18) but our data suggest that mycoplasma contamination may be a major factor in the stimulation.

Clearly, various classes of lymphoid cells can produce IFN but in most cases this capacity may be restricted to certain stimuli, e.g., T cells to mitogens or recall antigens. Our data suggest that NK cells are free from such restriction, ready for rapid IFN response to a wide array of materials and subsequent augmentation of NK activity.

REFERENCES

1. Herberman, R.B., Ed. Natural Cell-Mediated Immunity against Tumors. (Academic, New York, 1980).
2. Timonen, T., Ortaldo, J.R., and Herberman, R.B. J. Exp. Med. 153, 569 (1981).
3. Timonen, T. and Saksela, E. J. Immunol. Methods 36, 285 (1980).
4. Djeu, J.Y., Stocks, N., Zoon, K., Stanton, G.J., Timonen, T., and Herberman, R.B. Submitted for publication.
5. Mantovani, A., Peri, G., Polentarutti, N., Allavena, P., Bordignon, C., Sessa, C., and Mangioni, C. in Natural Cell-Mediated Immunity against Tumors (ed. Herberman, R.B.) (Academic, New York, 1980).
6. Zarling, J.M., Sosman, J., Eskra, L., Borden, E.C., Horoszewica, J.S., and Carter, W.A. J. Immunol. 121, 2002 (1978).
7. Cudkowicz, G. and Hochman, P.S. Immunol. Rev. 44, 13 (1979).
8. Welsh, R.M. Curr. Top. Microbiol. Immunol. (in the press).
9. Epstein, L.B., Kreth, H.W., and Herzenberg, L.A. Cell. Immunol. 12, 407 (1974).
10. Rassmussen, L.E., Jordan, G.W., Stevens, D.A., and Merigan, T.C. J. Immunol. 112, 728 (1974).
11. DeLey, M., Van Damme, J., Claeys, H., Weening, H., Heine, J.W., Billiau, A., Vermylen, C., and De Somer, P. Eur. J. Immunol. 10, 877 (1980).
12. Roberts, N.J., Douglas, R.G., Simons, R.M., and Diamond, M.E. J. Immunol. 123, 365 (1979).
13. Yamaguchi, T., Handa, K., Shimizu, Y., Abo, T., and Kumagai, K. J. Immunol. 118, 1931 (1977).
14. Weigent, D.A., Langford, M.P., Smith, E.M., Blalock, J.E., and Stanton, G.J. Inf. & Immunity 32, 508 (1981).
15. Trinchieri, G., Santoli, D., Dee, R.R., and Knowles, R.B. J. Exp. Med. 147, 1299 (1978).
16. Peter, H.H., Dalluger, H., Zawatzky, R., Euler, S., Leibold, W., and Kirchner, H. Eur. J. Immunol. 10, 547 (1980).
17. Catalona, W.J., Ratcliff, T.L., and McCool, R.E. Nature 291, 77 (1981).
18. Timonen, T., Saksela, E., Virtanen, I., and Cantell, K. Eur. J. Immunol. 10, 422 (1980).

INTERFERON PRODUCTION AND NATURAL CYTOTOXICITY BY HUMAN LYMPHOCYTES FRACTIONATED BY PERCOLL DENSITY GRADIENT CENTRIFUGATION[1]

Eda T. Bloom

Geriatric Research, Education and Clinical Center
VA Wadsworth Medical Center
and Department of Medicine
University of California
Los Angeles, California

Mitsuo Takasugi

Department of Surgery
University of California
Los Angeles, California

I. INTRODUCTION

Interferons (IFN) are potent immunoregulatory molecules which are produced by most nucleated cells. Their stimulatory effect for natural cell-mediated cytotoxicity (NCMC) has been well documented (Santoli and Koprowski, 1979; Herberman and Ortaldo, 1981). Human IFN have been classified into fibroblast (α), leukocyte (β), and immune (γ) types, depending upon the cells producing them and according to their physical, biological, and antigenic characteristics (Pestka et al, 1981).

Koide and Takasugi have previously shown that human peripheral blood lymphocytes produce a factor, N-cell-activating factor or NAF, which augments NCMC (Koide and Takasugi, 1978). Its physico-chemical properties are consistent with its identity as interferon (Koide and Takasugi, 1980). NAF or IFN

[1]Supported by NIH grant CA30187, NIH contract NO1-CP43211, and contract EY-76-S-03-0034 from the Department of Energy.

ISBN 0-12-341360-5

production by fresh lymphocytes could be stimulated with viruses, polynucleotides, mitogens, cultured lymphoblastoid cells, or mixed leukocyte culture. Trinchieri first reported that lymphocytes could be induced to synthesize IFN by co-culture with tumor cells and suggested that natural killer (NK) cells were producing the IFN (Trinchieri et al., 1978). Subsequently, Timonen et al. (1980) reported IFN production in response to K562 cells. The authors supported the conclusion that interferon was produced by NK cells, the mediators of NCMC. They reported that large granular lymphocytes (LGL), believed to be NK cells, stained with fluoresceinated antibody to IFN.

Recently NCMC effector cells in humans have been characterized as large granular lymphocytes (LGL) (Saksela et al., 1979). Subpopulations of lymphocytes enriched for NCMC activity have been obtained by density gradient centrifugation techniques using Percoll (Pharmacia, Piscataway, NJ) by Timonen et al. (1980) and Bloom (1981). In the present study, populations of lymphocytes isolated by Percoll centrifugation were co-cultured with Raji cells to stimulate interferon production. The same populations were tested against K562 and CEM for NCMC, to investigate correlations between NCMC and IFN production.

II. METHODS

A. Effector Cells

Lymphocytes from healthy donors were isolated from approximately 100 ml of blood using Ficoll-Hypaque centrifugation and carbonyl iron treatment to remove phagocytic cells (Bloom, 1981). Lymphocyte preparations were $\geq$ 99% pure when examined on slides prepared with the cytocentrifuge and stained with Giemsa and for α-naphthyl acetate esterase activity. Approximately 10^8 lymphocytes were layered over a Percoll continuous density gradient and separated by centrifugation as described previously in greater detail (Bloom, 1981). Briefly, a 35 ml gradient was used with Percoll concentrations ranging from 46% to 77%. Following centrifugation for 30 minutes at 400 g, 3 ml aliquots were collected. Fraction II (divided into IIa and IIb) was composed of the two 3 ml aliquots with the greatest cell yield, and had densities corresponding to refractive indices of 1.3440 to 1.3450. The peak cell yield was a consistent feature of the gradient profile. Aliquots from the gradient above IIa contained lighter and larger cells than those in IIa and IIb and were pooled to form fraction I. Any monocytes still present in the original cell suspension were found only within this fraction. The small, denser cells sedimenting below IIb were divided into IIIa,

IIb, . . . , with the number of divisions depending on yield. NCMC was consistently increased in the lightest 5%-15% of the cells which were contained in fraction I, with some extension into fraction IIa. We also observed a slight but significant increase in activity with cells sedimenting toward the bottom of the gradient.

B. Cytotoxicity Test

Natural cytotoxicity was measured in the ^{51}chromium-release assay as described previously (Bloom, 1981). The tests were performed with effector:target cell ratios of 20:1 and 10:1 against two target cells, K562 (Lozzio and Lozzio, 1975) and CCRF-CEM [abbreviated here as CEM (Smith et al., 1974)]. Percent cytotoxicity was calculated as follows:

$$\% \text{ cytotoxicity} = 100 \times \frac{\text{Exp} - \text{SR}}{\text{TR} - \text{SR}}$$

where Exp is the mean ^{51}Cr release from three experimental wells; SR is the spontaneous release tested in wells with medium and target cells; and TR is the total release from target cells in wells with 1 N HCl. The data obtained with an effector:target ratio of 10:1 are reported in the results.

C. Interferon Induction and Assay

Lymphocytes were stimulated to produce interferon by cocultivation with Raji cells at a ratio of 8 to 1 (Koide and Takasugi, 1978,1980). Two million lymphocytes from peripheral blood, or Percoll fractions were incubated with 2.5×10^5 Raji cells in 0.5 ml RPMI 1640 medium containing 10% fetal calf serum (Sterile Systems, Logan, UT). Control cultures contained lymphocytes alone, Raji cells alone, or neither cell type. Supernatants were harvested after 64 hours of incubation at 37°C and stored at -70°C until assays for IFN activity were performed.

The antiviral activity of IFN was assayed using vesicular stomatitis virus (VSV) and G-11 cells in Microtest I plates (Falcon, Oxnard, CA) (M. Takasugi, manuscript in preparation). Supernatants were titrated for putative IFN antiviral activity in quadrupling dilutions beginning with 1:4. Interferon titers were designated as the dilution of supernatant protecting 25% of the G-11 cells from destruction by VSV. The titers were related to a simultaneous titration of 1,000 units of standard fibroblast interferon obtained from the Research Resources Branch, NIAID.

TABLE 1. NCMC and Interferon Production by Lymphocytes obtained by Percoll Fractionation

Fraction	% of Recovered cells	NCMC at an effector:target ratio of 10:1 K562	CEM	Interferon production Titer[a]	Units[b]
Unfract.	-	6.5	1.3	11.3	706
I	3	36.7	12.9	325.0	20,312
IIa	26	17.5	6.7	13.0	812
IIb	45	4.7	1.9	0	0
IIIa	11	0.6	1.1	0	0
IIIb	6	2.2	1.5	0	0
IIIc	9	4.8	1.9	0	0

[a]Dilution of co-culture supernatant protecting 25% of G-11 cells from VSV.
[b]Based on the standard of 1,000 units of IFN which had a titer of 16.

III. RESULTS

Peripheral blood lymphocytes were obtained from two different donors, and subpopulations were obtained by Percoll density gradient centrifugation. The donors were selected for their weak and strong cytotoxicity against K562 target cells. Each subset and the original suspension were tested for NCMC against K562 and CEM and for interferon production. The results obtained with lymphocytes from the donor reacting weakly against K562 is shown in Table 1. The data show that the cytotoxic activity was strongest in fraction I, with declining activity toward the bottom of the gradient. The most dense fraction exhibited a slight increase in activity against K562 compared to the fractions immediately above it. This was observed in an additional 16 of 21 gradients and is consistent with previous findings (Bloom, 1981) and with those shown in Table 2. IFN production was greatest among the lighter cells harvested from the top of the gradient. The correlation coefficient for NCMC against the target cell K562 and IFN production by each subset was 0.990; for NCMC against CEM and IFN production it was 0.910; and for NCMC against the two target cells it was 0.910. These correlation coefficients

TABLE 2. Interferon Production by Highly Cytotoxic Lymphoctyes

Fraction	% of Recovered cells	NCMC at an effector:target ratio of 10:1		Interferon production	
		K562	CEM	Titer[a]	Units[b]
Unfract.	-	20.7	5.4	11.3	706
I	6	43.6	10.8	333.1	20,820
IIa	43	30.3	10.2	35.8	2,240
IIb	33	19.5	8.2	8.9	556
IIIa	8	21.5	8.4	9.8	612
IIIb	10	26.6	9.5	24.3	1,519

[a,b]See footnotes, Table 1.

were significant at the $p < 0.001$, $p < 0.01$, and $p < 0.01$ levels, respectively, by the student's t statistic. No protective effect against VSV was detected in control supernatants from all fractions of lymphocytes or Raji cells, each cultured alone.

The data shown in Table 2 were obtained with a more active effector cell suspension from a different donor. Although the cytotoxic activity was dispersed throughout the gradient, the IFN-producing cells separated according to a similar pattern to those in Table 1.

IV. DISCUSSION

Purified lymphocyte subsets were separated by Percoll density gradient centrifugation, and each population was tested for NCMC against K562 and CEM. Each subset was also stimulated with Raji cells to assess IFN production. The data obtained from a weakly cytotoxic donor (Table 1) showed that IFN production was closely correlated with NCMC. This suggests that IFN is produced by the same cells which mediate the cytotoxicity or that the cells which produce IFN sediment together with the NCMC effectors or NK cells. The results agree with those of Timonen et al. (1980) who have reported that LGL produce IFN in response to K562 cells which can be neutralized with anti-IFN antibody. Djeu et al. (in press) have also concluded that LGL produce IFN in response to a number of stimuli.

The cytotoxicity by the second donor was more evenly distributed over the Percoll gradient (Table 2). The pattern of interferon production was similar to that obtained with the first donor, i.e., dramatically enriched over only the lightest portion of the gradient. Thus, the cells producing IFN again separated with the lighter densities of Percoll. NCMC was also stronger in the lighter cells but was produced by cells sedimenting throughout the gradient. The existence of cells producing IFN generally paralleled the occurrence of cells experessing NCMC over the gradient, except that cells in the first fraction produced very great amounts of IFN.

NCMC activity and at times IFN-producing activities were slightly increased among cells in the heaviest fraction. We have observed increased NCMC activity against K562 in the bottom fraction of 16 of 21 gradients. By using a higher percentage of Percoll than others, a finer separation of activities among the denser lymphocyte populations may have been achieved.

In summary, effector cells from two donors with varying NCMC activities were separated by Percoll density gradient centrifugation. In agreement with findings by others, we detected enriched NCMC activity and IFN production in the cells of the lightest fractions. However, a weaker but significant increase in NCMC and possibly interferon production was also observed in the heaviest fraction.

REFERENCES

Bloom, E. T., Cell. Immunol. 61:231-244 (1981).
Djeu, J. Y., Timonen, T., and Herberman, R. B., in "Role of Natural Killer Cells, Macrophages, and Antibody Dependent Cellular Cytotoxicity in Tumor Rejection and as Mediators of Biological Response Modifier Activity" (M. Chirigos, ed.), Raven Press, New York (in press).
Herberman, R. B., and Ortaldo, J. R., Science 214:24 (1981).
Koide, Y., and Takasugi, M., J. Immunol. 121:872 (1978).
Koide, Y., and Takasugi, M., in "Natural Cell-mediated Immunity against Tumors," (R. Herberman, ed.), p. 537. Academic Press, New York, 1980.
Lozzio, C. B., and Lozzio, B. B., Blood 45:321 (1975).
Pestka, S., Maeda, S., and Staehelin, T., in "Annual Reports in Medicinal Chemistry - 16" (E. H. Cordes, ed.), p. 229. Academic Press, New York, 1981.
Saksela, A., Timonen, T., Ranki, A., and Häyry, R., Immunol. Rev. 44:71 (1979).
Santoli, D., and Koprowski, H., Immunological Rev. 44:125

(1979).
Smith, L. W., Smith, R. M., Hathcock, et al., in "Proc. 8th Leukocyte Culture Conference" (K. Lindahl-Kiessling and D. Osaba, eds.), p. 127. Academic Press, New York, 1974.
Timonen, T., and Saksela, E., J. Immunol. Methods 36:285 (1980).
Timonen, T., Saksela, E., Virtanen, I., and Cantell, K., Eur. J. Immunol. 10:422 (1980).
Trinchieri, G., Santoli, D., Dee, R. R., Knowles, B., J. Exp. Med. 147:1299 (1978).

RECOGNITION STRUCTURES FOR NATURAL KILLER CELLS ON HUMAN LYMPHOCYTES: A PANEL STUDY

Miklós Benczur
Tamás Laskay
Győző Petrányi

National Institute of Haematology and Blood Transfusion
Budapest

I. INTRODUCTION

Natural killer cells may lyse a limited assortment of normal cells such as subpopulations of thymocytes, bone marrow cells, cultured fibroblasts, etc. /Nunn et al.,1977; Petrányi and Hollán, 1980; Hansson et al., 1980/. Some authors have found mitogen-stimulated target cells insensitive for NK-killing and subsequently poor competitors in cross competition assays / Seeley and Karre, 1980; Herberman and Ortaldo, 1980/. With the use of in vitro cultivated tumor cell lines of human origin as target cells, and peripheral blood lymphocytes as effector cells, a limited specificity of NK cells has been demonstrated /Koide and Takasugi, 1977; 1978; Herberman and Ortaldo, 1980/. The existence of similar structures on non-malignant cells follows from the foregoing results, although this assumption has not been verified yet. The authors of this paper studied the role of the supposed surface structures in cytotoxic reactions on peripheral lymphocytes, uesed either as effector cells or as PHA-stimulated target cells.

II. METHODS

1. Effector Cells. Peripheral lymphocytes of healthy volunteers were separated on Ficoll-Uromiro gradient. Monocytes were depleted on a plastic surface and the cells were separated on nylon wool column /Greawes and Brown, 1974/. The nylon column passed /Ncp/ T+O lymphocytes were used as effector cells.

2. Target Cells. Ncp cells were stimulated in 10% FCS-RPMI-1640 medium by PHA /3 µl/ml for 48 hours. In vitro cultivated target cell lines /kindly provided by G.Klein, Karolinska Institutet/that we used were NPG-39 and IM-266/EBV transformed

ISBN 0-12-341360-5

lymphoblastoid lines/. After washing, the target cells were labelled with 200 μCi 51-chromium for 60 min washed four times and kept at 37 °C until use.

<u>3. Cytotoxicity assay.</u> Our method has been published elsewhere /Benczur <u>et al.,</u> 1980/. The cytotoxic value for 20:1 ratio was computed from the dose-response line and used as a stabilized comparative value.

<u>4. Anti Beta2 Microglobulin /anti-BMi/ treatment.</u> In some experiments effector cells were preincubated with reduced-alkylated monospecific antibodies/goat IgG Dacopatts 50 μg per ml 2 h, 37 °C/and washed 3 times.

<u>5. HLA Typing.</u> The standard NIH lymphocyte cytotoxicity method and the long incubation time lymphocytotoxicity methods were used /Terasaki, 1972; vanRood <u>et</u> <u>al</u>., 1975/

III. RESULTS

<u>1. Anti-beta 2 Microglobulin /anti-BMi/ treatment.</u> Using lymphoblastoid cell lines as target pre-treatment of effector cells inhibited or decreased the NK-activity. Similarly ADCC, against the same target cells sensitized by specific HLA antibodies, was also inhibited. The cytotoxic effect on autologous PHA blasts of CTL,and autologous and allogeneic effect of NK-cells generated in a 6 day old mixed lymphocyte culture was not affected by anti-BMi, although an inhibitory effect was observed on Daudi target cells /Table I./

<u>2. Autologous and allogeneic cytotoxicity on PHA blast cells.</u> Using a panel of 29 autologous and allogeneic PHA target cells and measuring the cytotoxic activity of Ncp effector cells of the same donors, we recorded distinct cytotoxic responses in the short term assay: the highest cytotoxicity was 56% and spontaneous release was under 17%. Target cell sensitivity and effector cell reactivity varied four times and seven times within the respective maximum values. To get rid of these variations the interaction analysis was employed /Koide and Takasugi, 1977/. The selective cytotoxycity scores in autologous and allogeneic combinations can be seen in Table II. Autologous reactivity was nil or very low /up to+4/ in 21 donors who are thus regarded as non-reactive persons. However in 8 donors, distinct cytotoxicity, up to a score of +38 was seen against autologous target cells. We also compared average cytotoxicity against the target panel and

the selective autologous cytotoxicity scores. Obviously no sign of correlation was observed; effector cells with high /and generalized "strong"/ reactivity exerted no or low selective cytotoxicity and vice versa.

3. Computer analysis of selective cytotoxicity scores. Correlation was calculated according to the individual reaction patterns of Ncp cells. As the summary of this analysis,a best fitting picture, is seen in Fig 1. which shows that a minimum of 7 different groups of target cells may be distinguished.

4. Analysis of HLA-antigens. All donors involved in this study were HLA-typed. The computer analysis revealed no correlation between any of the particular HLA antigens and group of target cells.

5. Autologous rosette formation /ARFC/. Rosette formation of T-cells with autologous erythrocytes is a measure of autologous reactivity in MLC/Palacios et al., 1980/. The average ARFC number of 20 donors was 19 per cent /8-20 %/. No correlation was observed between low or high autologous cytotoxicity scores /Table II/ and number of RFCs.

TABLE I. Anti-beta 2 Microglobulin and Interferon treatment of NK, K and CTL Effectors Cells

Effector cells	Anti beta 2 Mi	IFN	Daudi[a]	Daudi +anti DR-6	NPG-39[b]	NPG-39 +anti 2,27	LY C PHA	LY D PHA
	no	no	14	61	-	-	-	-
LY A	yes	no	2	19	-	-	-	-
	no	yes	2o	64	-	-	-	-
	no	no	19	67	26	44	-	-
LY B	yes	no	o	17	2	o	-	-
	no	yes	2o	75	-	-	-	-
LY C +	no	no	25	-	-	-	13	16
	yes	no	9	-	-	-	13	17
LY D[MMC]	no	yes	5o	-	-	-	13	19
LY D +	no	yes	32	-	-	-	25	7
	yes	no	1o	-	-	-	2o	4
LY C[MMC]	no	yes	4o	-	-	-	15	o

[a]Daudi: HLA DR6, [b]NPG-39: HLA 2.31-15,27,CW3

TABLE II. Specific cytotoxicity scores of 29 nylon column passed effector cells reacted against autologous and allogeneic PHA blast cells.

Target cells/48 PHA blast/

	1.	2.	3.	4.	5.	6.	7.	8.	9.	10.	11.	12.	13.	14.	15.	16.	17.	18.	19.	20.	21.	22.	23.	24.	25.	26.	27.	28.	29.	react.[a]
1.	-34[b]	-35	9	-6	49	20	-4	-5	-7	24	-29	-10	-32	-7	125	-30	-9	-55	-8	-23	13	8	-1	53	1	-4	5	-19	12	10.8
2.	1	20	48	23	56	40	41	38	50	28	-33	11	-3	10	16	-53	-20	-9	-27	-43	-53	11	-19	0	-25	-31	11	-16	-35	8.6
3.	-58	-55	38	-9	-59	-5	-38	-35	14	45	63	132	5	23	-74	76	-53	-44	13	35	88	-30	-39	-43	-55	46	-46	13	53	10.1
4.	42	-18	-32	-26	9	-26	23	21	-37	-28	20	-6	-6	-52	20	-12	28	17	5	47	-3	9	34	39	10	-30	-23	-19	-3	5.1
5.	-14	-21	53	-31	-51	20	-29	-56	7	0	38	52	-2	-12	-21	35	-13	-10	3	-2	19	-3	-9	-22	-15	39	0	31	15	4.4
6.	-22	-6	-12	36	44	22	-10	-12	1	22	-61	0	3	3	-14	39	-60	-44	-23	-18	73	-25	-24	-20	-27	53	-11	12	84	8.4
7.	-67	82	39	56	22	-64	-98	99	-18	41	74	32	26	114	-53	-31	23	-16	54	-24	115	-58	-48	-31	-49	-97	-41	-22	-61	14.2
8.	10	41	15	56	4	-8	-13	18	40	7	14	-57	-35	31	-22	2	19	9	25	-17	14	-17	-15	-11	-15	-13	11	-54	-40	8.6
9.	24	-5	1	-26	-2	0	-31	0	-10	-9	-5	-40	-79	0	-46	-4	15	30	31	0	-21	27	14	6	24	17	28	65	-7	4.9
10.	88	-11	-10	-39	-23	32	94	-12	24	-49	-6	-62	-66	71	-14	24	44	53	-18	20	-74	9	-17	-5	-20	11	-27	-12	0	6.0
11.	-6	1	-1	21	37	82	32	-37	-1	28	-49	-10	7	1	16	-38	-10	6	-21	-40	-47	41	-7	-3	5	-15	20	11	-22	6.7
12.	13	-19	-6	-23	0	18	20	-30	10	-8	-52	-26	-29	-19	60	-15	-9	24	10	-21	-18	61	-3	23	3	6	31	14	-14	6.1
13.	36	4	14	-32	-21	0	-23	-3	-10	-11	-11	-52	-60	8	1	-11	26	23	36	8	-11	-39	23	0	17	0	48	44	-7	5.6
14.	58	3	-21	-33	-6	0	32	-2	-1	-20	7	-46	-11	-32	6	-21	17	37	-1	57	-34	15	25	3	11	-21	1	0	-22	3.0
15.	-1	28	-58	-4	56	-5	-24	-8	-26	6	-3	-27	68	-14	3	-34	19	-5	22	40	-52	-4	31	9	45	-34	-31	-11	17	5.6
16.	5	6	-1	16	51	28	8	-11	7	31	-38	11	-24	12	26	-24	-16	14	-23	-27	-52	7	-14	3	2	-25	27	15	-19	6.0
17.	42	2	-2	0	4	0	-45	-9	3	2	-25	-35	-29	26	1	8	8	15	-16	-24	-44	14	13	3	19	-11	37	48	-7	5.5
18.	-66	14	28	-47	-62	-20	-15	70	-22	39	17	60	-6	-7	-20	24	-34	-24	-28	-17	52	-6	30	-23	-17	46	0	-2	36	5.8
19.	-12	49	-47	33	56	-50	-2	61	28	5	35	-12	105	-15	-87	-11	0	-27	4	53	-11	-26	13	9	-10	-49	-17	-24	-51	9.1
20.	-25	43	-117	55	54	-70	23	44	15	21	-1	-1	275	-21	-7	1	2	-44	10	-9	-47	-41	52	-40	12	-40	93	-65	15	20.6
21.	-31	-26	20	-7	-51	-14	-43	-60	-1	-24	22	8	-8	28	-31	24	-8	26	-26	-22	10	5	35	21	47	30	4	20	50	5.4
22.	36	23	14	-3	0	0	-23	13	17	-20	-4	-30	-37	18	-5	-46	35	5	15	-34	17	7	5	6	-3	-23	13	25	-22	7.1
23.	-27	-49	96	34	-97	-2	52	-36	0	-8	40	114	16	-113	49	-6	-11	30	-30	-10	-16	45	-43	13	-5	-5	37	-49	-18	11.8
24.	-12	-24	-8	-24	-51	10	-12	-42	-37	-38	-9	3	-32	23	0	51	0	4	-24	-7	34	2	16	4	13	71	5	40	44	4.0
25.	-4	-12	103	17	-98	-3	89	4	34	-18	29	118	-30	-32	-46	101	-88	-77	-54	-10	90	-27	-62	-82	-51	89	-40	-27	87	12.6
26.	47	15	-1	-3	30	8	-12	-28	-2	-12	-9	-30	-13	21	4	-31	16	18	28	-8	-5	-44	-43	-1	4	3	29	29	-10	6.0
27.	0	-9	-69	-24	-8	-27	-35	3	-46	-35	-6	-52	6	-39	50	10	46	38	23	71	-12	10	52	17	28	-1	14	-4	1	4.4
28.	-19	11	-68	-16	43	-12	15	31	-37	-9	32	-41	13	-24	22	-6	34	-8	39	42	-34	2	26	14	36	-12	-27	-29	-20	6.5
29.	-1	-13	-27	10	13	24	29	-11	1	-13	-46	-3	-18	-1	40	-19	-1	12	-19	-15	8	45	-24	52	9	3	28	-10	-51	4.1

[a] Mean cytotoxic reactivity of effector cells against the panel target cells, in per cent.

[b] Autologous cytotoxicity, underline.

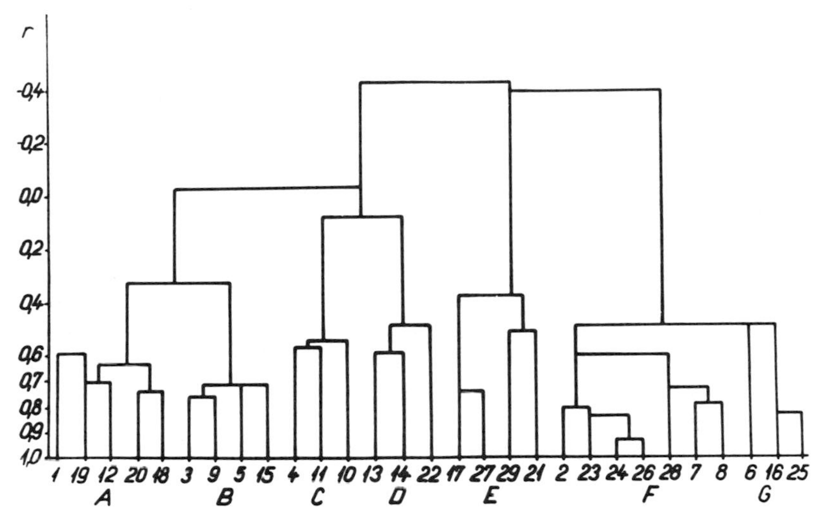

Figure 1. Correlations of the reaction patterns.
r = correlation coefficient

III. DISCUSSION

After anti-BMi treatment of human peripheral lymphocytes a co-shedding of FcR and beta 2 Mi structures was observed /Sármay *et al.*, 1980/. It is coupled with a dose dependent inhibitory effect on ADCC if sensitized Rh positive RBC were used as target. Depending on the target cell in use a variation of anti-BMi treatment was registered; with the same dose applied, no inhibitory effect on K-562 was measured in our previous experiments /Sármay *et al.*, 1982/. In this study NK-cells generated during MLC were not inhibited if reacted allogeneic PHA blasts but inhibition on Daudi target was observed /Table I/. Some experiments support the assumption that NK effect is partially a consequence of antibodies either arming the effector cells or secreting during incubation in the assay. These observations, though not confirmed by other laboratories /Koide and Takasugi, 1977; Troye *et al.*, 1977; Herberman *et al.*, 1979/, suggest that the effect of anti-BMi may be mediated through Fc-receptors. On the other hand capping of Fc-IgG completly inhibited K-cell activity but only partially decreased NK-cytotoxicity /Bolhuis *et al.*, 1978/. Other laboratories also concluded that FcR does not participate in the NK killing mechanism /Kalden *et al.*,1977/. These results support another possibility, viz. that anti-BMi treatment exerted its effect on other structures via HLA-antigens or similar molecules. This is also substantiated by our observation that anti-BMi treatment of lymphocytes abolished or decreased the detectablity of HLA-antigens /Benczur et al.,/

Recent studies of autologous lymphocyte mediated killing on separated tumor biopsy cells revealed 28% autologous killing and 5% allogeneic effect, suggesting that histocompatibility antigens are involved in natural killing /Vánky et al., 1980/. Since there is not a reliable way to verify this hypothesis on a battery of freshly separated tumor biopsy cells, a model system for this purpose would be necessary.

It was found that only a proportion of killer cells recognize autologous structures: more than half of autologous Ncp cells exerted no cytotoxicity. The fact that IFN activation increased selective cytotoxicity but no change in reaction patterns was observed /data not shown here/ indicates that effector cells belong to the NK-system.

Recently de novo generated NK-cells have been shown in MLC/Bolhuis and Schellekens, 1981/. Correlation between activity of these cells and number of ARFC-s has also been observed /Palacios and Alacron Segovia, 1981/. The lack of association between autologous NK-reactivity and number of ARFC may reflect the difference in recognition during MLC and in the short term cytotoxicity assay.

Earlier data in mouse and man demonstrated a broad specificity of NK cells against tumor cells of T and B origin /Koide and Takasugi, 1977; 1978/. Thus it seems likely that other and possibly more complicated structures for NK recognition exist. With such structures, working on a battery of target and effector cells of the same origin is clearly an advantage. Our data indicated the existence of at least 7 different structures. Cross competition analysis inside group B disclosed competition inside though not outside the group, providing evidence of the existence of the structure/s. We will, however, be able to better understand this matter if preliminary competition assay is extended.

REFERENCES

Benczur,M.,Petrányi,G.Gy.,Pálffy,Gy.,Varga,M.,Tálas,M., Kotsy,B.,Földes,I.,Hollán,S.R. /198o/ Clin.Exp.Immunol. 39:657

Benczur,M.,Sármay,G.,Laskay,T.,GyódiE.,Petrányi,G.Gy., Gergely,J. submitted.

Bolhuis,R.L.H.,Schuit,H.R.E.,Nooyen,A.M., and Ronteltap,C.P.M. /1978/ Eur.J.Immunol. 8:731

Bolhuis,R.L.H.,and Schellekens,H., /1981/ Scand,J.Immunol. 13:401/

Greaves,M.F., and Brown,G., /1974/ J.Immunol. 112:420

Hansson,M.,Kiessling,R.,and Welsh,R., /198O/ In "Natural Cell-Mediated Immunity against Tumors"/R.B.Herberman,ed./, p.855.Academic Press,New.York
Herberman,R.B.,Djeu,J.Y.,Kay,H.D.,Ortaldo,J.R.,Riccardi,J., Bonnard,G.D.,Holden,H.T.,Fagnani,R.,Santoni,A., and Puccetti,P., /1979/ Immunol.Rev. 44:43
Herberman,R.B.,Ortaldo,J.R. /198o/ In "Natural Cell-Mediated Immunity against Tumors" /R.B.Herberman,ed./, p. 873. Academic Press, New York
Kalden,J.R.,Routin,R.,and Cesarin,I.P. /1977/ Eur.I.Immunol. 7:537
Koide,Y.,and Takasugi,M. /1977/ J.Natl.Cancer Inst. 59:1o99
Koide,Y.,and Takasugi,M. /1978/ Europ.J.Immunol. 8:818
Nunn,M.E.,Herberman,R.B.,and Holden,H.T., /1977/ Int.J.Cancer 2o:381
Palacios,R.L.,Alacron-Segovia,D.,Ruiz-Arguelles,A.,and Diaz-Jouane E. /198o/ J.clin.Invest. 65:1527
Palacios,R.,Alacron-Sagovia,D., /1981/ Scand.J.Immunol.13:499
Petrányi,G.Gy.,and Hollán,S.R. /198o/ Tissue Antigens 16:1
Sármay,G.,Iványi,J.,and Gergely,J. /198o/ Cell.Immunol.56:452
Sármay,G.,Benczur,M.,Petrányi,G.,Klein,E.,Gergely,J.,submitted
Seeley,J.K.,and Karre,K., /198o/ In "Natural Cell-Mediated Immunity against Tumors"/R.B.Herberman,ed./, p.477. Academic Press,New York
Terasaki,P. /1972/ In "Manual of Tissue Typing" /Terasaki,P. ed./, p.50. NIH, Bethesda
Troye,M.,Perlman,P.,Pape,G.R.,Näslund,I.,Gidlöf,A., /1977/ J.Immunol. 119:1O61
Vánky,F.T.,Argov,S.A.,Einhorn,S.A.,and Klein,E., /198o/ J.Exp.Med. 151:1151
vanRood,J.J.,vanLeeuwen,A.,Keuning,J.J.,vanOud,B., and Alblas,A., /1975/ Tissue Antigens 5:73

SPECIFICITY OF FRESH AND ACTIVATED HUMAN CYTOTOXIC LYMPHOCYTES

Farkas Vánky
Eva Klein

Department of Tumor Biology
Karolinska Institute
and
Radiumhemmet
Karolinska Hospital
Stockholm, Sweden

I. INTRODUCTION

Our experiments were directed for the demonstration of the immune recognition of autologous tumor cells in patients with sarcomas and carcinomas. The occurrence of such recognition was suggested by the induction of DNA synthesis of blood lymphocytes after in vitro exposure to autologous tumor cells (ATS) (Vánky et al., 1978). In fewer cases a direct cytotoxic effect on autologous tumor cells has also been observed (Vose et al., 1977).

When evaluating a direct lymphocyte-mediated cytotoxic effect, it has to be clarified whether it is due to the recognition of antigens on the target surface - in this case lymphocytes with receptors for the antigen exert the effect i.e. it is performed by clones of lymphocytes or whether the lysis is indiscriminant - in this case the effector population is polyclonal and the clonally distributed antigen receptors may not be involved in the lytic interaction with the target (Klein, 1980).

Supported by Federal Funds from the DHEW, Contract Number NO1-CM-74144, by NIH grant 1 RO1 CA-25184-01A1, and by the Swedish Cancer Society.

ISBN 0-12-341360-5

Lymphocytes cultivated in MLC and in mixed cultures with autologous tumor cells (ATS) also lysed the patients' tumor cells. The highest frequency of positive experiments were obtained after ATS activation (Vánky et al., 1981).

Allogeneic tumor cells were lysed by lymphocytes treated with interferon IFN-α and as "third party" targets by MLC lymphocytes (Vánky et al., 1980, 1981). In the ATS cultures no allo-tumor lysis was generated. K562, the prototype NK target was lysed by lymphocytes of MLC and ATS.

When different targets are killed by the same lymphocyte population, the effects can still be antigen specific if they are exerted by different sets of effectors in each case.

On the population level, the following lytic effects might be generated in antigen-containing lymphocyte cultures: 1) Enlargement of the specific clone results in cytotoxicity against a) cells which carry the stimulating antigen; b) cells which carry cross-reactive antigens. 2) Transactivation, induced by soluble factors produced in the mixed cultures, recruits lymphocytes with other specificities. Thus targets unrelated to the stimulus may be killed if the relevant clones are sufficiently enlarged. 3) Activated lymphocytes kill certain type of targets independently of antigen recognition. The targets which are sensitive for the "activated killers, AK" are usually cultured lines.

In order to analyse the cytotoxicities exerted on freshly separated tumor cells we performed experiments with the cold target competition assay (Vánky and Klein, 1982a, b; Vánky et al., 1982).

II. MATERIAL AND METHODS

The tumor cells were separated from surgical specimens of patients with solid tumors (7 osteo-, 6 fibrosarcomas, 3 oat cell-, 8 adeno-, 11 squamous cell carcinomas of the lung, and 5 hypernephromas). They were used either directly or after frozen storage. The cell separation procedure, the conditions of the mixed cultures and the cold target competition assays are described in our previous publications (Klein et al., 1980; Vánky et al., 1978, 1981, 1982a, b).

The 6 days MLC and ATS cultures contained 95-100% OKT3, 23-30% OKIal and 0-2% OKM1 positive cells.

Cytotoxicity was measured in 4 h ^{51}Cr release assay. In addition to tumor cells lymphocytes cultured either with PHA (0.5 μg/ml - Purified Phytohaemagglutinin, Wellcome Reagents Ltd.) or with 50 μg/ml Con-A (Sigma Chemical Co.) for 3-4 days were used as competitor targets. They were treated with chilled formaldehyde (0.4% in PBS, pH 7.2) for 30 sec at 0°C. Con-A was removed from the blasts by washing with 0.1 M α-Methyl-d-Mannoside (Sigma).

III. RESULTS

A. Lysis of Autologous Tumor Cells

Table I summarizes our results on the lysis of autologous tumor cells exerted by fresh, ATS- or MLC-activated lymphocytes. This was inhibited by the identical cells, while autologous lymphocytes (formaldehyde trated Con-A or PHA blasts) and allogeneic tumor biopsy cells or lymphocytes had no or weak effect.

The auto-tumor lysis exerted by the fresh and ATS-activated lymphocytes was strongly inhibited by K562 at all ratios (Table I).

TABLE I. Cold Target Competition Experiments with Freshly Harvested or Activated Auto-Tumor Cytotoxic Systems[a]

Mode of activation	Unlabeled targets[b]	Number of tests	% Inhibition of ^{51}Cr-release ratio labeled / cold targets 5:1	1:1	1:5
-	auto T	10	30±3[c]	67±11	93±7
-	allo T	15	0	6±5	26±7
-	auto L	5	0	9±6	13±8
-	allo L	5	7±7	10±5	20±3
ATS	auto T	15	28±7	77±6	96±1
ATS	allo T	21	4±2	15±8	19±17
ATS	auto L	10	0	11±10	18±12
ATS	allo L	17	0	16±6	23±7
ATS	K562	3	89±11	91±9	100±0
MLC	auto T	13	21±7	58±15	92±8
MLC	allo T	17	3±1	15±6	18±6
MLC	auto L	7	3±2	5±2	16±3
MLC	stimulator L	6	0	8±5	16±7
MLC	third party L	8	0	0	10±5

[a]Specific ^{51}Cr release measured at effector:target ratio 50:1 in 4h cytotoxicity test without cold targets varied between 20% and 47%.

[b]T = tumor biopsy cells; L = PHA or Con-A blasts.

[c]Mean ± standard deviation.

B. Lysis of Allogeneic Tumor Cells

Table II summarizes our results on the lysis of allogeneic tumor cells by lymphocytes from healthy donors or tumor patients. Lymphocytes activated by short IFN treatment and in MLC (using unrelated lymphocytes as stimulators) were used as effectors.

In contrast to the auto-tumor cytotoxicity, allo-tumor lysis by IFN-treated freshly separated lymphocytes was inhibited by the identical tumor cells and by the Con-A blasts from the same donor. A weak inhibition with allogeneic third party tumor cells occurred but only with high dose of unlabeled targets.

TABLE II. Cold Target Competition Experiments with Freshly Harvested or Cultured Activated Lymphocytes Acting on Allogeneic Tumor Cells[a]

Activation	Unlabeled targets[b]	Number of tests	% Inhibition of ^{51}Cr-release ratio labeled / cold targets 5:1	1:1	1:5
IFN prior to the test[d]	identical T	24	35±6[c]	66±13	92±5
	responder T	8	8±2	11±3	14±3
	other T	27	6±1	14±4	34±1
	target donor L	7	16±5	72±11	87±18
	responder L	5	0	4±1	14±2
	other L	21	0	5±2	17±4
	K562	20	61±9	82±5	99±1
	Daudi	6	33±2	62±3	91±2
MLC	identical T	13	46±12	66±19	90±17
	responder T	7	0	6±2	16±4
	other T	24	4±1	6±2	21±4
	target donor L	10	51±11	74±16	88±16
	responder L	6	5±2	5±3	16±2
	stimulator L	7	0	8±3	10±4
	other L	9	5±2	8±2	9±2
	K562	4	57±6	79±5	98±2

[abc]See Table I. [a]Specific ^{51}Cr release without cold targets varied between 23-43%.

[b]The relationship of the various cells were: Responder, stimulator and target from different donors. Thus the target represented a "third party". "Other" tumor cells and lymphocytes are derived from a fourth donor.

[d]Fresh blood lymphocytes were treated with 1000 IU-IFN-α/ml for 3 hs.

The "third party" allo-tumor lysis exerted by the MLC population was inhibited by the identical tumor and also by lymphocytes from the patient whose tumor was the labeled target. The following cells had only slight inhibitory effect: tumor cells from the responder, Con-A blasts derived from the stimulator or from the responder and other donors.

In the allo-tumor killing systems also, K562 and Daudi cells were general inhibitors at all ratios. Daudi cells inhibited the IFN-activated lysis as strongly as the identical targets and K562 was even more effective.

C. Lysis of NK Sensitive Cell Lines

Table III summarizes our results on the lysis of Daudi and K562 exerted by IFN-activated lymphocytes. K562 was a potent inhibitor for the Daudi lytic system. Tumor biopsy cells were also able to inhibit by with lower efficiency and allogeneic Con-A blasts inhibited the K562 lysis only slightly.

IV. CONCLUSIONS

The competition experiments suggest that when freshly isolated tumor cells were the targets the cytotoxicities were specific on the level of the functioning subsets, because the different targets were acted on by different sets being present simultaneously in the effector population.

TABLE III. Cold Target Competition Experiments with The Fresh IFN Activated Lymphocytes Acting on Daudi and K562[a,b]

Target	Unlabeled targets[b]	Number of tests	% Inhibition of ^{51}Cr-release ratio labeled / cold targets 5:1	1:1	1:5
Daudi	Daudi	11	32±8[c]	54±12	96±4
Daudi	K562	2	45±11	71±7	99±1
Daudi	allo. T	24	7±3	23±15	50±16
K562	K562	4	32±21	52±9	99±1
K562	allo. L	3	2±1	7±3	24±7
K562	allo. T	12	5±	19±6	43±3

[abc]See Table I. [a]Specific ^{51}Cr release without cold target varied between 33-79%.

The lysis of autologous and allogeneic tumor cells seems to occur on a different basis. Tumor (and/or organ) specific antigens may be recognized on the autologous tumor cells. The competition by lymphocytes in the allo tumor lysis indicates that in this event alloantigens are recognized.

Some of the findings need further clarification, however. The competition exerted by allogeneic tumor cells in the anti Daudi and anti K562 system is noteworthy. Among the lymphocytes which act on these cell lines, those which express receptors for alloantigens of the tumor cells may react with these and therefore we see a quite efficient competition, even if the basis of the recognition of the two targets by the same cells is on different basis (Klein, 1981). However, in this case the competition by allogeneic lymphocytes is also expected but this occurred at considerably lower level. A similar dilemma appears with the IFN-activated effectors (Table II). Here the competitive capacity of allogeneic, unrelated tumor cells other than the labeled target (other T) is higher than that of allogeneic lymphocytes (other L) and of autologous tumor cells. The question to solve is thus to what extent the competition is due to crossreactive alloantigens on the two cells and if so, why are the allo-lymphocytes less effective.

The lack of crossreactivity between the autologous and allogeneic tumors (even with similar histological types) may be due to a real antigenic difference. Alternatively, antigenic crossreactivities may exist but in the short term T cell cytotoxicity assays they may not be detectable due to the MHC restriction.

The emergence of auto-tumor cytotoxicity in the MLC is probably due to production of IL-1 and IL-2 as a corollary of antigen recognition which leads to activation and proliferation of cells which do not bear receptors for the stimulating antigens.

In the ATS cultures, alloreactivity was not generated. This may be due to the relatively lower proportion of auto-tumor reactive lymphocytes compared to those that react with the MHC antigens. In accordance the blastogeneic response is considerably lower in the ATS cultures compared to the MLC-s (Vánky et al., 1978). The proportion of antigen reactive cells in the culture probably determines the quantity of the soluble amplifying factors produced.

We consider the lysis of K562 cells (the prototype of the human NK-sensitive line) to be a sign of the activated state of the lymphocytes (Klein and Vánky, 1981). The main difference between this activity and the lysis of freshly harvested tumor cells is that interaction with K562 is polyclonal as shown by the general competing potential of the K562.

In several series of experiments, the natural killing of blood lymphocytes were monitored in various categories of patients using K562 as targets. The impetus was given by the proposition that the natural killer cells contribute to tumor surveillance. Since we postulate that the anti-K562 effect is a measure of the activation profile of the lymphocyte population with regard to lytic function, it provides indeed an information about the potentialities of these cells. The lytic potential is then manifested against targets which are damaged on the basis of antigen recognition and are insensitive to the indiscriminant cytotoxicity. For the in vivo antitumor response, the relevant information is whether the tumor cells express antigens that are recognized by members of the T cell repertoire. If such lymphocytes with specific receptors exist, they are likely to be represented in the subsets with varying functional characteristics. The blood lymphocyte population of individuals with a high proportion of lytic cells may thus exert antigen specific cytotoxicity.

REFERENCES

Klein, E. (1980). Immunol. Today 1, IV.

Klein, E., and Vãnky, F. (1981). Cancer Immun. Immunotherapy 11:183.

Klein, E., Vãnky, F., Galili, U., Vose, B.M., and Fopp, M. (1980). Contemp. Top. Immunbiol. 10:79.

Vãnky, F., and Klein, E. (1982a). Immunogenetics, in press.

Vãnky, F., and Klein, E. (1982b). Int. J. Cancer, in press.

Vãnky, F., Klein, E., Stjernswärd, J., Rodrigez, L., Pẽteffy, A., Steiner, L., and Nilssone, U. (1978). Int. J. Cancer 22:679.

Vãnky, F., Argov, S., Einhorn, S., and Klein, E. (1980). J. Exp. Med. 151:1151.

Vãnky, F., Argov, S., and Klein, E. (1981). Int. J. Cancer 27:273.

Vãnky, F., Gorsky, T., Gorsky, Y., Masucci, M-G., and Klein, E. (1982). J. Exp. Med., in press.

Vose, B.M., Vãnky, F., and Klein, E. (1977). Int. J. Cancer 20:512.

FRACTIONATION OF NATURAL KILLER CELLS ON TARGET CELL MONOLAYERS[1]

Pamela J. Jensen[2]
Patricia A. Weston[3]
Hillel S. Koren[4]

Division of Immunology
Duke University Medical Center
Durham, North Carolina

I. INTRODUCTION

Natural killer (NK) cells can lyse a variety of target cells, including virally infected and uninfected tumor lines, but the mechanisms by which NK cells recognize these different target cells are not understood. In addition, the precise relationship between NK and K cells, which are responsible for antibody-dependent cellular cytotoxicity (ADCC), is not clear.

To investigate the question of NK and K cell heterogeneity, we fractionated human peripheral blood lymphocytes (PBL) on NK-sensitive target cell monolayers. The adherent and nonadherent fractions obtained in this manner are, respectively, enriched for or depleted of both ADCC and NK activities against several target cells.

---

[1]Supported by Public Health Service Grants CA23354 and CA09058-05.
[2]Present address: NCI-Frederick Cancer Research Facility, P.O. Box B, Frederick, Md. 21701, U.S.A.
[3]Present address: Department of Pharmacology, Duke University Medical Center, Durham, N.C. 27710, U.S.A.
[4]Recipient of Research Career Development Award CA00581.

ISBN 0-12-341360-5

II. METHODS

A. Effector Cells

Human PBL, depleted of cells adherent to plastic (Koren and Williams, 1978), were prepared for all experiments. Large granular lymphocytes (LGL), which were greatly enriched for NK cells (Timonen et al, 1981), were also used for comparison in some cases.

B. Cell Lines

The following human cell lines were employed both as NK target cells and for monolayer fractionation: HSB, a T cell line (Royston et al, 1974), and K562, a myeloid line (Lozzio and Lozzio, 1975), both grown in suspension culture; and SV-Hep2, a monolayer cell line that is persistently infected with Sendai virus (Lavappa, 1978). All of these lines are readily lysed by NK cells, but Hep2, the uninfected parent of SV-Hep2, is only very slightly sensitive to lysis by NK cells (Weston et al, 1980). For ADCC assays, the target cells were either CEM-NKR, an NK-cell-resistant variant of CEM, a human T cell line (Foley et al, 1977), or SB, a human B cell line (Royston et al, 1974).

C. Monolayer Fractionations

As described previously (Weston et al, 1980; Jensen et al, 1979), nonadherent and adherent fractions were recovered from lymphocytes incubated on confluent SV-Hep2 monolayers or on K562 or HSB monolayers that had been prepared on poly-L-lysine-coated plates. Cells adherent to HSB or K562 monolayers were recovered by further incubation of the monolayers at 4°C in 5mM ethylenediaminetetraacetic acid for 2 hours. Fractions were tested for NK and ADCC activities immediately and, in some cases, also after overnight incubation at 37°C.

D. Cytotoxicity Assays

NK and ADCC activities were measured at 3 or 4 effector cell to target cell ratios (EC:TC) in 2-4 hour ^{51}Cr release assays. To determine ADCC activity, SB or CEM-NKR cells were labelled with trinitrobenzene sulfonic acid (SB-TNP or CEM-NKR-TNP), and a rabbit hyperimmune anti-TNP serum was added to the assay (Koren and Williams, 1979). Chromium-labelled SV-Hep2 target cells were used in monolayer form, but the other target cells were added in suspension.

III. RESULTS

A. NK Activity of PBL Adherent and Nonadherent to NK-sensitive Monolayers

Fractionation on SV-Hep2 monolayers effectively separated PBL into populations depleted of or enriched for NK activity (Table I, Weston et al, 1980). When tested either on the day of fractionation (day 0) or after overnight incubation (day 1), PBL nonadherent to SV-Hep2 cells showed depleted NK activity against SV-Hep2, K562, and HSB targets. The adherent fractions were enriched relative to control cells for NK activity against all three target cells when tested on day 1 and showed intermediate activity on day 0. Incubation of PBL on uninfected Hep2 monolayers did not generate fractions enriched in or depleted of NK activity.

Results similar to those with SV-Hep2 monolayers were found when PBL were fractionated on HSB monolayers (Table I). Specifically, nonadherent cells were depleted of NK activity, and adherent cells were enriched for NK activity relative to control populations when they were tested on day 1.

When PBL were incubated on K562 monolayers and the fractions were tested for NK activity on day 0, results similar to those with HSB monolayers were found. However, consistent enrichment of NK activity relative to controls was not found in K562-adherent populations after overnight incubation, in contrast to the results with HSB- and SV-Hep2-adherent populations.

B. ADCC Activity of PBL Adherent and Nonadherent to NK-sensitive Monolayers

In contrast to NK activity, ADCC of PBL populations adherent to SV-Hep2, HSB, or K562 monolayers was enriched relative to control cells even when tested on day 1 (Table II). Nonadherent cells had lower ADCC.

C. NK Activity of Large Granular Lymphocytes (LGL) Adherent and Nonadherent to NK-sensitive Monolayers

LGL fractions nonadherent and adherent to HSB monolayers, respectively, were depleted of or enriched for NK activity compared to unfractionated LGL (Table III). Fractionation of LGL on K562 monolayers also yielded results qualitatively similar to those found with whole lymphocyte populations (data not shown).

TABLE I. NK Activity of PBL Adherent and Nonadherent to Target Cell Monolayers

Effector cells	Day 0 % cytotoxicity vs.		Day 1 % cytotoxicity vs.		
	SV-Hep2[a]	HSB[a]	SV-Hep2[a]	HSB[a]	K562
Expt. 1					
Control	26.2	9.7	21.1	4.8	24.3[a]
Nonadherent to SV-Hep2	0.7	1.9	6.5	0.8	3.8
Adherent to SV-Hep2	6.6	10.6	31.7	19.6	40.1
Expt. 2					
Control	NT[c]	10.9	NT	10.2	11.1[b]
Nonadherent to HSB	NT	1.2	NT	0.6	5.9
Adherent to HSB	NT	5.4	NT	20.3	21.0

[a]EC:TC was 10:1 [b]EC:TC was 5:1 [c]Not tested

TABLE II. NK and ADCC Activities of PBL Adherent and Nonadherent to Target Cell Monolayers

Effector cells[a]	NK Activity % cytotoxicity vs.	ADCC Activity % cytotoxicity vs.
Expt. 1	HSB[b]	CEM-NKR-TNP[c]
Control	10.4	9.3
Nonadherent to HSB	4.1	7.4
Adherent to HSB	5.0	15.8
Expt. 2	SV-Hep2[b]	SB-TNP[b]
Control	23.1	23.9
Nonadherent to SV-Hep2	3.7	6.6
Adherent to SV-Hep2	4.0	33.4
Expt. 3	K562[d]	SB-TNP[d]
Control	15.1	21.2
Nonadherent to K562	2.1	10.3
Adherent to K562	7.5	36.2

[a]Assays done on day 0. [c]EC:TC was 1.2:1.
[b]EC:TC was 10:1. [d]EC:TC was 5:1.

TABLE III. NK Activity of LGL Adherent and Nonadherent to HSB Cell Monolayers

Effector cells[a]	% cytotoxicity vs. HSB target cells[b]
LGL control	15.2
LGL nonadherent to HSB monolayers	7.0
LGL adherent to HSB monolayers	19.8

[a]Assays done on day 1
[b]EC:TC was 10:1. Assay time was two hours.

D. Possible Involvement of Interferon in Augmented NK Activity

We considered the possibility that interferon plays a role in the augmented cytotoxic activity observed in the adherent fractions on day 1. Antiviral activity, therefore, was determined in the supernatants of all lymphocyte fractions after 18 hours incubation. HSB-monolayer-adherent lymphocytes produced 4-40 units/ml in each of four experiments; the non-adherent fractions produced equivalent amounts of interferon in two of the experiments but did not show detectable quantities in the other two. Fractions adherent to SV-Hep2 monolayers consistently produced 20-200 times more interferon than the nonadherent cells. The addition of 1-1,000 units/ml of exogenous α-interferon to SV-Hep2-monolayer-nonadherent cells during the 18-hour incubation did not enhance NK activity in this fraction (Weston et al, 1980).

IV. DISCUSSION

Fractionation of human PBL on NK-sensitive target cell monolayers yielded qualitatively similar results whether virally infected (SV-Hep2) or uninfected (HSB) tumor cell lines were used. Lymphocytes were generally found to be enriched for or depleted of NK activity against a variety of sensitive target lines, not just the one used for monolayer fractionation. This suggests that overlapping populations of NK effector cells recognize and lyse the infected (SV-Hep2) and noninfected (HSB, K562) target lines investigated. However, a more quantitative analysis (Jensen et al, 1979) of depletion of cytotoxic activity in PBL nonadherent to K562 or HSB monolayers revealed donor-dependent heterogeneity within the population of NK effector cells and suggested that different target cells each bear their own complement of one or more NK recognition sites.

In light of recent data (Perussia and Trinchieri, 1981), it is likely that NK cells are inactivated on the monolayers used for fractionation. The enhancement of NK activity in adherent fractions on day 1 may be induced by interferon generated during overnight incubation. However, interferon clearly acts preferentially on the adherent and not the nonadherent fractions.

Finally, since qualitatively similar data were generated when either LGL populations or whole PBL were fractionated on target cell monolayers, it is unlikely that cell types other than NK cells have a major role in the cellular interactions responsible for the depletion and enrichment of NK activity shown here.

ACKNOWLEDGMENTS

The authors thank Dr. Marc Golightly and Mr. Philip Brandt for assistance with the LGL preparations and Mr. David Howell (Immunology Division, Duke University) for donating the CEM-NKR cell line.

REFERENCES

Foley, G.E., Lazarus, H., Sidney, F., Uzman, B.G., Boone, B.A., and McCarthy, R.E., Cancer 18:522 (1977).

Jensen, P.J., Amos, D.B., and Koren, H.S., J. Immunol. 123:1127 (1979).

Koren, H., and Williams, M., J. Immunol. 121:1956 (1978).

Lavappa, K.S., In Vitro 14:469 (1978).

Lozzio, C.B., and Lozzio, B.B., Blood 45:321 (1975).

Perussia, B., and Trinchieri, G., J. Immunol. 126:754 (1981).

Royston, J., Smith, R.W., Buell, D.N., Huant, E.S., and Pagano, J.S., Nature 251:745 (1974).

Timonen, T., Ortaldo, J.R., and Herberman, R.B., J. Exp. Med. 153:569 (1981).

Weston, P.A., Levy, N.L., and Koren, H.S., J. Immunol. 125:1387 (1980).

Weston, P.A., Jensen, P.J., Levy, N.L., and Koren, H.S., J. Immunol. 126:1220 (1981).

NON-NK LEUKOCYTES DEMONSTRATE NK-PATTERNED BINDING

Gerald E. Piontek
Rolf Kiessling
Alvar Grönberg
Lars Ährlund-Richter

Department of Tumor Biology
Karolinska Institute
Stockholm, Sweden

I. INTRODUCTION

The mechanism of the binding and lytic events between NK cells and tumor targets is poorly understood. It could be asked to what extent the binding of an NK cell to a target involves a specific recognition event of an "antigen-antibody" type or whether it involves some alternative recognition event (1). Several non-specific properties of tumor targets influence their lytic sensitivity, including cell surface hydrophobicity (2), sialic acid levels (3), and active membrane repair mechanisms (4, 5). Target cell selectivity in the NK system is inferred from cytotoxicity assays, competition assays, and single cell target binding assays. In this report we have studied the selective binding between some leukocyte populations and a series of mouse and human tumor targets. Our results suggest that the ability of leukocytes to bind tumors in an "NK patterened way" is an ubiquitous property of several different types of leukocytes, irrespective of their NK activity.

II. MATERIALS AND METHODS

The target binding cell assay was a modification of the technique described by Roder and Kiessling (6). Leukocyte effector cells were labeled with carboxyfluorescein diacetate,

ISBN 0-12-341360-5

C-FDA, 15 minutes at 37°C in balanced salt solution (7). The labeled effectors were washed and admixed with tumor cells at a 1 to 4 effector target ratio, spun 1 minute at 300 g, incubated 10 minutes at 37°C, then resuspended by vortexing 5 seconds and placed onto microscope slides. The percentage of fluorescent effector cells bound to nonfluorescent target cells was scored under a UV microscope. The binding assays were scored blind and at least 150 cells were counted. The C-FDA was a generous gift of Dr. Peter Perlmann, University of Stockholm. The NK cytotoxicity assay using nylon passed CBA spleen cells and the competition assays have been described (8).

The Hy-3 cytotoxic cell line and a non-cytotoxic subline was produced by Dr. Hans Hengartner, University of Zurich, and a description of this line is given elsewhere in this volume. Briefly, Hy-3 is an Lyt-2 positive, IL-2 dependent cell line derived from a C57Bl/6 female anti-C57Bl/6 male mixed lymphocyte reaction. During *in vitro* culture, this line lost its specific H-2 restricted cytolytic activity and acquired cytolytic activity against NK sensitive targets.

Peritoneal exudate cells were taken from normal adult CBA mice and were adhered onto plastic petri dishes 2 h at 37°C. Following three washes to remove nonadherent peritoneal cells, the adherent cells were removed by use of a rubber policeman and employed in the target binding assays.

III. RESULTS AND DISCUSSION

A panel of mouse and human tumor targets with different levels of sensitivity to mouse NK activity were used. A positive but weak correlation (r = 0.5) was observed between spleen NK lysis and the percentage of target binding cells (% TBC) among these targets by nylon passed CBA spleen cells (Table I), in agreement with previous results (6). Cold target competition assays were performed with isotope labeled YAC-1 targets and the other tumor lines as competitors of CBA nylon passed spleen cells. The competing capacity of the cell lines against YAC-1 targets consistently correlated better with % TBC than did the lytic sensitivity (r = 0.7). In our hands, this occurs because some of the lytically insensitive targets were bound effectively (*i.e.* MPC-11, GM-86, and L1210, Table I), in agreement with previous reports (5, 9, 10).

We also performed binding studies with a cloned cytotoxic Lyt-2 positive cell line Hy-3 (described by Hengartner *et al.*, herein). The degree of binding between Hy-3 and several

TABLE I. Binding Pattern of Nylon Wool Passed Spleen Cells and Thymocytes to Various Tumor Targets[a]

Target	NK lysis	Number of experiments	% TBC	
			Nylon-passed spleen	Thymocytes
RL♂1	38	4	16	16
YAC-1	30	4	17	22
MOLT-4	22	4	7	13
YAC-Ascites	5	3	9	5
RBL-5	4	4	7	7
P52	2	2	5	10
MPC-11	4	2	13	15
GM-86	4	2	-	11
L1210	1	4	14	9
YWA	∅	1	10	3
Daudi	∅	1	-	5

[a]NK-lysis (4 h assay) is evaluated at a 33 to 1 effector to target ratio with CBA nylon wool passed spleen cells. The correlation coefficient (r) of NK-lysis versus % TBC was 0.5 and 0.8; r of % competition (3 to 1 competitor to target ratio) versus % TBC as 0.7 and 0.8 for spleen and thymus, respectively.

tumor targets correlated (r = 0.9) strongly with the ability of these tumor targets to bind spleen NK cells, as defined by a cold target competition assay (Fig. 1b). A lytically inactive variant of Hy-3 had a similar binding pattern (Fig. 1a), showing that NK patterned binding can occur in the absence of lysis and suggesting that binding and lysis are independently regulated events (11).

As a new and surprising finding, the NK-patterned binding occurred with mouse thymocytes from several genotypes and from newborn thymocytes, as exemplified by CBA adult thymocytes (Table I). The ability of leukocytes to display NK-patterned binding was not unique to cells of the T-lineage, since adherent peritoneal cells also bound accordingly (Fig. 2). These cells had low NK lytic activity (< 5% lysis of YAC-1) and high % TBC activity (>50% binding to YAC-1). A positive correlation between the competing ability of these tumors and the % TBC with adherent peritoneal cells was observed (r = 0.8).

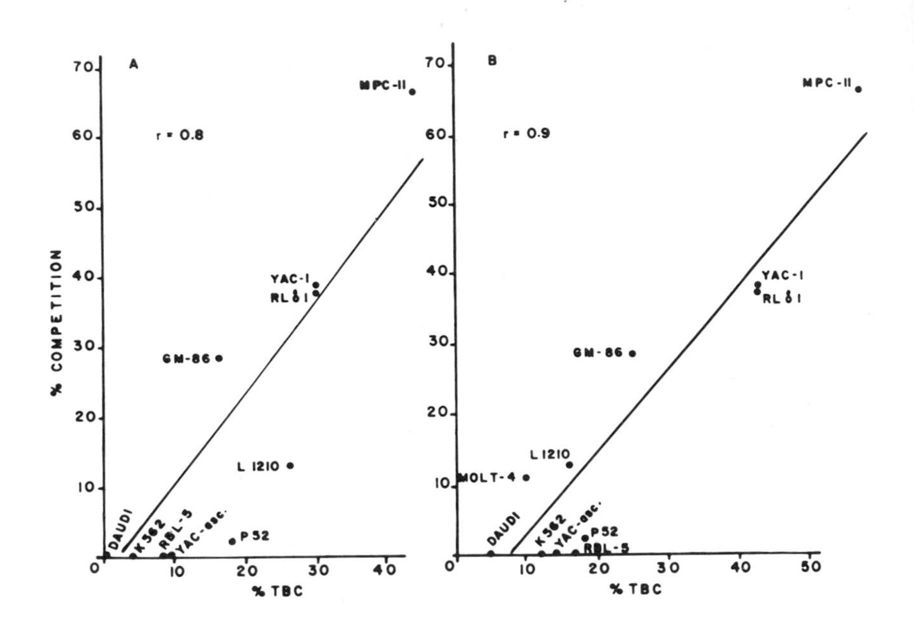

FIGURE 1. Binding pattern of a non-cytotoxic variant (A) of a cytotoxic cell line, Hy-3 (B) to various mouse and human tumors correlated with the competing ability of these tumors against Cr-labeled YAC-1 targets with CBA spleen cells as effectors at a ratio of 30:1 (8). Hy-3 lysed 45% of YAC-1 cells in a 3 h cytotoxicity assay at a 3 to 1 effector target ratio.

The levels of NK sensitivity on an individual tumor were also varied by selecting NK resistant variants from the NK sensitive YAC-1 lymphoma (12). These variants had a decreased NK sensitivity as well as lower % TBC by nylon passed spleen cells, adherent peritoneal cells, and adult and newborn thymocytes (Table II).

IV. CONCLUSIONS

The fact that % TBC correlates weakly with lysis but strongly with the competing ability of these tumor cells suggests that the target binding assay is an adequate measurement of selectivity in the NK system. The findings also

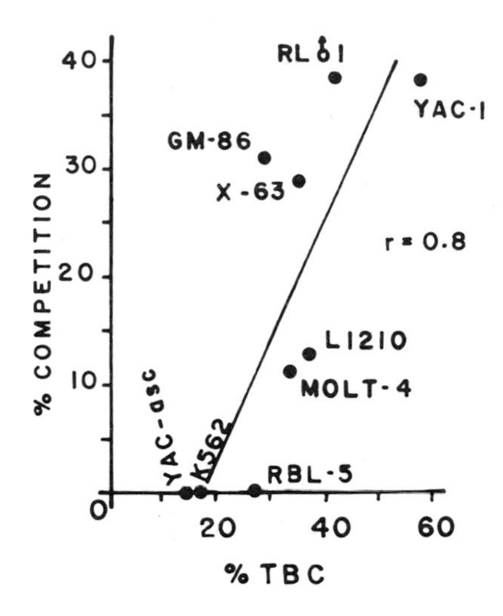

FIGURE 2. Binding pattern of CBA adherent peritoneal cells to several tumor targets versus the competing ability of these tumors (3 to 1 competitor target ratio) against Cr-labeled YAC-1 targets (8). The correlation coefficient (r) is 0.8.

TABLE II. Binding pattern of nylon wool passed spleen cells, adherent peritoneal cells, and adult and newborn thymocytes to NK-resistant variants of YAC-1[a]

Target	NK lysis	% TBC			
		Nylon-passed spleen	PEC	Thymocytes Adult	Thymocytes Newborn
YAC-1	31	17	56	20	15
ACC	12	9	46	8	10
ACK	11	9	26	8	7
ACG	10	10	37	5	9

[a]Data are means from 2-4 experiments. The NK-lytic assay is described in Table I. Newborn CBA mice were less than 3 days old.

illustrate that an interpretation of NK cell frequency based upon % TBC is misleading, unless purified NK populations are used. The finding that NK-patterned binding occurs with leukocytes devoid of NK lytic activity shows that binding and lysis are separately controlled events. Since there is no evidence that thymocytes have "pre-NK" cells (i.e. thymocytes cannot be lytically activated by IFN), the NK patterned binding is not restricted to NK cells. We consider it probable that qualitative differences exist among the binding of lytically active and inactive effector cells (13).

It has been suggested (19) that cytotoxic T-cells express specific T-cell receptors and NK-like receptors. Our results would support this hypothesis (see Fig. 1), and also suggest an explanation for an NK-patterned selectivity occurring with monocytic-lineage cells (15). We are presently investigating the nature of the NK-patterned binding.

ACKNOWLEDGMENTS

This work was supported by grants: 1 RO1 CA 267 82-02, and the Swedish Cancer Society. GEP was supported by a Cancer Research Campaign International Fellowship awarded by the International Union Against Cancer and a Fellowship from the Leukemia Society of America.

REFERENCES

1. Ahrlund-Richter, L., Klein, G., and Masucci, G. (1980). Som. Cell Genet. 6:89.
2. Becker, S., Stendahl, O., and Magnusson, K. (1979). Immunol. Commun. 8:73.
3. Yogeeswaran, G., Grönberg, A., Hansson, M., Dalianis, P., Kiessling, R., and Welsh, R.M. (1981). Int. J. Cancer 28:517.
4. Hudig, D., Djebadge, M., Redelman, D., and Mendelsohn, J. (1981). Cancer Res. 41:2803.
5. Collins, J.L., Patek, P.Q., and Cohn, N. (1981). J. Exp. Med. 153:89.
6. Roder, J.C., and Kiessling, R. (1978). Scand. J. Immunol. 8:135.
7. Bruning, J.W., Kardol, M.J., and Arentgen, R. (1980). J. Immunol. Methods 33:33.
8. Hansson, M., Kärre, K., Bakacs, T., Kiessling, R., and Klein, G. (1978). J. Immunol. 121:6.

9. Kunkel, L.A., and Welsh, R.M. (1981). Int. J. Cancer 27:73.
10. Kiessling, R., Klein, E., and Wigzell, H. (1975). Eur. J. Immunol. 5:112.
11. Roder, J.C., Kiessling, R., Biberfeld, P., and Andersson, B. (1978). J. Immunol. 121:2509.
12. Grönberg, A., Kiessling, R., Eriksson, E., and Hansson, M. (1981). J. Immunol. 127:1734.
13. Shortman, K., and Golstein, P. (1979). J. Immunol. 123:833.
14. Klein, E. (1980). Immunol. Today Dec:1.
15. Lohmann-Matthes, M.L., Domzig, W., and Roder, J. C. (1979). J. Immun. 123:1883.

TARGET CELL RECOGNITION BY NATURAL KILLER AND NATURAL CYTOTOXIC CELLS[1]

Edmund C. Lattime
Gene A. Pecoraro
Osias Stutman

Cellular Immunology Section
Memorial Sloan-Kettering Cancer Center
New York, New York

I. INTRODUCTION

Natural cell-mediated cytotoxicity (NCMC) is the function of at least two distinct effector cell populations (1,2), natural killer (NK) and natural cytotoxic (NC) cells. Studies from a number of laboratories have resulted in significant characterization of the effector populations; however, little is known regarding the nature and specificity of their respective target cell recognition structures. Cross cold target inhibition studies (3) have been valuable in comparing a number of tumor cell lines for the presence or absence of such "antigens"; however, with few exceptions, little progress has been made in characterizing such molecules. The inhibition of NK and NC cell activities by monosaccharides has led us to suggest that lectin-like receptors may be involved in the effector-target interaction (4,5). On the other hand, the isolation of a resistant variant of the NK-susceptible L-5178 lymphoma has resulted in the association of the neutral glycolipid asialo-GM2 (ASGM-2) with NK-susceptibility (6,7).

This report describes two series of experiments which were undertaken: 1) to further examine the target cell re-

[1]Supported by National Institute of Health Grants CA-08748, CA-17818, CA-25932, and American Cancer Society Grant IM-188.

ISBN 0-12-341360-5

quirements for NCMC lysis, and 2) to compare the recognition requirements involved in NK and NC cell-mediated cytotoxicity. The role of H-2 antigen expression on target cell susceptibility was examined by use of the R1.1 thymoma and its H-2$^-$ variant. We examined the relationship between NK and NC recognition structures through the use of cross cold target inhibition studies and NK and NC-susceptible target cell lines.

II. TARGET CELL H-2 EXPRESSION MAY AFFECT NK SUSCEPTIBILITY?

Antigens encoded by the major histocompatibility complex are involved in cell-cell interactions leading to the generation and function of cytotoxic T lymphocytes (8). *In vivo* positive selection of the C58 thymoma R 1.1 resulted in the isolation of a variant (R1E/TL8x.1.Oua) (R1.1E) which failed to express either the H-2 or Tl antigens characteristic of the parental line (9). We have used the two cell lines in studies aimed at examining the role of H-2 expression in NK tumor susceptibility. Significant NK cell-mediated lysis [asialo-GM1$^+$ (ASGM-1) effector cell] of the R1.1 tumor was measured in splenic populations from poly IC boosted (100 μg i.p. + 24 hrs.) syngeneic C58, C57BL/6J, BALB/c and CBA/H mice, while no lysis of the R1.1E variant was observed (Table I). No lysis of either target was measured in non-boosted

TABLE I.
The H-2$^-$R1.1E Tumor Cell Line is Resistant to NK Cell Lysis[a]

Effector	Treatment	% Cytotoxicity	
		R1.1	R1.1E
C58	C	24	0
	ASGM-1 + C	0	0
C57BL/6J	C	24	0
	ASGM-1 + C	0	0
BALB/c	C	44	0
	ASGM-1 + C	5	0
CBA/H	C	53	0
	ASGM-1 + C	18	0

animals (data not shown). Although these findings do not rule out other target-related explanations for the resistance of the R1.1E line to NK lysis, and in fact other investigators have shown that teratocarcinoma cell lines are variably susceptible to NK lysis (10), they do suggest a role for H-2 expression. Studies have shown that xenogeneic cytotoxic T cells kill the R1.1E variant, as do certain lectin-activated populations (9,11), demonstrating that the target is susceptible to certain cell-mediated cytotoxicity mechanisms and supporting a role for H-2 in NK cell recognition.

Since an association between the neutral glycolipid ASGM-2 and NK lysis has been described (7), we examined the R1.1 and R1.1E cell lines for expression of ASGM-2. Both lines manifested comparable ASGM-2 expression when assayed either by complement mediated lysis using anti-ASGM-2 antisera (our unpublished results) or by surface labelling followed by thin layer chromatography (D. Urdal, Fred Hutchinson Cancer Research Center, personal communication). These findings show that the NK resistance of R1.1E is not a function of ASGM-2 expression and also that ASGM-2 expression does not, in itself, bestow NK cell susceptibility. The above studies show that variant tumors with differences in susceptibility to NCMC represent possible models for the definition of NCMC target "antigens".

III. NK AND NC TARGET CELLS SHARE RECOGNITION DETERMINANTS

We have previously reported results showing that the NC susceptible Meth-A and Meth-113 target cell lines cross compete with the NK cell lysis of the YAC-1 lymphoma suggesting that NK and NC target cells share certain recognition structures (4,12). The geometry of the adherent cell assay required for the NC cell precluded the reverse experiments, i.e. the inhibition of NC cell activity by NK susceptible

Footnote for Table I.
a. All experiments were done using 18 hr. ^{51}Cr release assays and either the H-2$^+$ R1.1 or the H-2$^-$ variant R1E/TL8 x 1.0ua (R1.1E) target cell lines. 5x10^3 labelled targets were assayed with 2.5x10^5 C or Ab + C treated whole spleen populations from the indicated mouse strains (8-10 wks. of age). Aliquots of the same effector population were used with each target. Pretreatment cell numbers were not restored. The R1.1 and R1.1E cell lines were provided by the Salk Tissue Bank.

targets (4). The non-adherent WEHI-164.1 sarcoma cell line has been shown to be sensitive to NC cell lysis (13) and has thus enabled us to further examine the relationship between the two effector populations.

NC cell cytotoxicity, as measured in 18 hr. ^{51}Cr release assays using the WEHI-164.1 target, is significantly inhibited by the addition of the non-labelled NK-susceptible YAC-1 lymphoma (Table II), demonstrating shared NK-NC determinants and supporting our previous studies (4,12).

TABLE II. NK and NC Cells Share Target Cell Determinants[a]

Effector	CT	CT:LT	% Cytotoxicity	% Inhibition
C57BL/6J	--	--	44	--
	WEHI-164	10:1	23	48
		3:1	23	48
		1:1	25	44
	YAC-1	10:1	15	66
		3:1	23	48
		1:1	24	45

a. All experiments were done using 18 hr. ^{51}Cr release assays and the NC susceptible WEHI-164.1 [labelled target (LT)] tumor cell line. $5x10^3$ labelled target cells were assayed with $2.5x10^5$ (50:1) whole spleen cells and the indicated ratios of non-labelled [cold target (CT)] tumor cells (NK susceptible YAC-1 or NC susceptible WEHI-164.1).

As described above, a resistant variant of the NK susceptible L-5178 cl. 27v lymphoma has been isolated (6). In order to further compare NK and NC target recognition, the NK susceptible cl. 27v and resistant cl. 27av lines were compared for their ability to cross compete with the NC lysis of the WEHI-164.1 target. Addition of the NK-susceptible cl. 27v cells to the NC assay resulted in significant inhibition (Table III). The NK resistant cl. 27av variant also inhibited the lysis of the NC target, but to a significantly lesser degree. The preferential inhibition of NC cell activity by the cl. 27v, seen in each of three experiments, is similar to that seen when NK cell activity is examined (6). Clone 27 av has been shown to lack the neutral glycolipid asialo-GM2, and it has been suggested that this molecule may play a role in

NK target recognition (7). Whether the presence or lack of asialo-GM2 is involved in the differential inhibition of NC cell activity is currently under study.

TABLE III. NC Cell Activity is Preferentially Inhibited by the NK Susceptible L-5178 cl. 27v [a]

Effector	CT	CT:LT	%Cytotoxicity	% Inhibition
C57BL/6J	--	--	26	--
	WEHI-164.1	10:1	17	35
		3:1	16	39
	L-5178 cl. 27v	10:1	5	81
		3:1	9	66
	L-5178 cl. 27av	10:1	17	35
		3:1	21	20

a. See Table II legend for methodology. At equal CT:LT ratios, cl. 27v. inhibited target lysis significantly ($p<0.05$) greater than did the cl. 27av.

The above findings support our previous studies in demonstrating significant cross competition between NK and NC target cells, and in doing so give further evidence of shared recognition structures (4, 12). We do not interpret these findings as precluding NK or NC-restricted recognition. On the contrary, previous studies have shown that although both NK and NC cell activities are inhibited by the addition of monosaccharides to the assay cultures (4,5), different "inhibition profiles" were observed in these studies with a number of sugars inhibiting NK activity while preferential inhibition of NC activity was obtained with D-mannose. This would suggest the presence of distinct, although perhaps overlapping, recognition structures involved in NK and NC cell lysis.

IV. CONCLUSIONS

The studies described in this report demonstrate that 1) the expression of H-2 molecules on tumor cell lines may play a role in lysis by NK cells, and 2) on the basis of cross cold target inhibition studies, NK and NC cells share common target recognition structures.

ACKNOWLEDGMENTS

We wish to thank Dr. C. Henney of the Fred Hutchinson Cancer Research Center for the L-5178 cell lines as well as specific antisera; Dr. R. Burton of the Massachusetts General Hospital for the WEHI-164.1 cell line; Ms. Mary Lou Devitt for expert technical assistance; and Ms. Linda Stevenson for preparing this manuscript.

REFERENCES

1. Herberman, R.B., and Holden, H.T., Adv. Cancer Res. 27: 305 (1978).
2. Stutman, O., Paige, C.J., and Feo Figarella, E., J. Immunol. 121: 1819 (1978).
3. Ortiz de Landazuri, M.D., and Herberman, R.B., Nature (New Biol.) 238: 18 (1972).
4. Stutman, O., Dien, P., Wisun, R., Pecoraro, G.A., and Lattime, E.C., in "Natural Cell-Mediated Immunity Against Tumors" (R.B. Herberman, ed.), p. 949. Academic Press, New York, 1980.
5. Stutman, O., Dien, P., Wisun, R.E., and Lattime, E.C., Proc. Natl. Acad. Sci. 77: 2895 (1980).
6. Durdik, J.M., Beck, B. N., Clark, E.A., and Henney, C. S., J. Immunol. 125: 683 (1980).
7. Young, W.W., Durdik, J.M., Urdal, D., Hakomori, S.-I., and Henney, C.S., J. Immunol. 126: 1 (1981).
8. Miller, J.F.A.P., Immunol. Rev. 42: 76 (1978).
9. Hyman, R., and Stallings, V., Immunogenetics 3: 75 (1976).
10. Stern, P., Gidlund, M., Orn, A., and Wigzell, H., Nature 285: 341 (1980).
11. Dennert, G., and Hyman, R., Eur. J. Immunol. 7: 251 (1977).
12. Lattime, E.C., Pecoraro, G.A., and Stutman, O., J. Immunol. 126: 2011 (1981).
13. Burton, R.C., in "Natural Cell-Mediated Immunity Against Tumors" (R.B. Herberman, ed.), p. 19. Academic Press, New York, 1980.

THE RELATIONSHIP BETWEEN NK AND NATURAL ANTIBODY TARGET STRUCTURES[1]

A. H. Greenberg
B. Pohajdak

Manitoba Institute of Cell Biology
University of Manitoba
Winnipeg, Manitoba, Canada

It has recently become apparent that a relationship exists between the tumor target structures (TS) recognized by natural antibody and NK cells (Grönberg et al, 1980; Chow et al, 1981). Grönberg et al (1980) found that rabbit natural antibody (RNAb) preferentially lysed NK sensitive tumors in the presence of complement, and that RNAb anti-YAC-1 was absorbed by tumors in a manner corresponding to their NK sensitivity. We have confirmed these observations, and in earlier work described a similar relationship between murine CBA/J strain NK cells and NAb using NK^s and NK^r tumors, as well as randomly selected clones of the YAC-1 tumor (Chow et al, 1981) (Fig 1). In these experiments absorption of NAb and inhibition of NK lysis by YAC-1 clones showed quantitatively similar patterns.

The relationship of murine NAb (MNAb) and RNAb to NK lytic sensitivity leads one to speculate that the target structures of these effectors might be the same, or coexpressed on the membrane. However, our further studies on a second NK sensitive (SL2) lymphoma, which cross inhibits the NK lysis of the YAC-1, do not support this interpretation. In these experiments, MNAb reactivity and NK lytic sensitivity of tumor clones were inversely expressed (Fig 1). The inverse relation was also noted with the RNAb, and a comparison of the RNAb to the MNAb reactivity of the clones from both tumor lines revealed a highly significant positive correlation (Fig 2). That the NAb TS may be structurally related was suggested by cross absorption studies in which the SL2-5 tumor could readily deplete the RNAb and MNAb reactivity to the YAC-1, and vice versa (Fig 3). The RNAb and MNAb target structures, then, appear to be similar on the tumors linked with the expression of the NK TS, however, in the case of the SL2-5 this was a reciprocal relationship.

---

[1]Supported by the NCI of Canada.

ISBN 0-12-341360-5

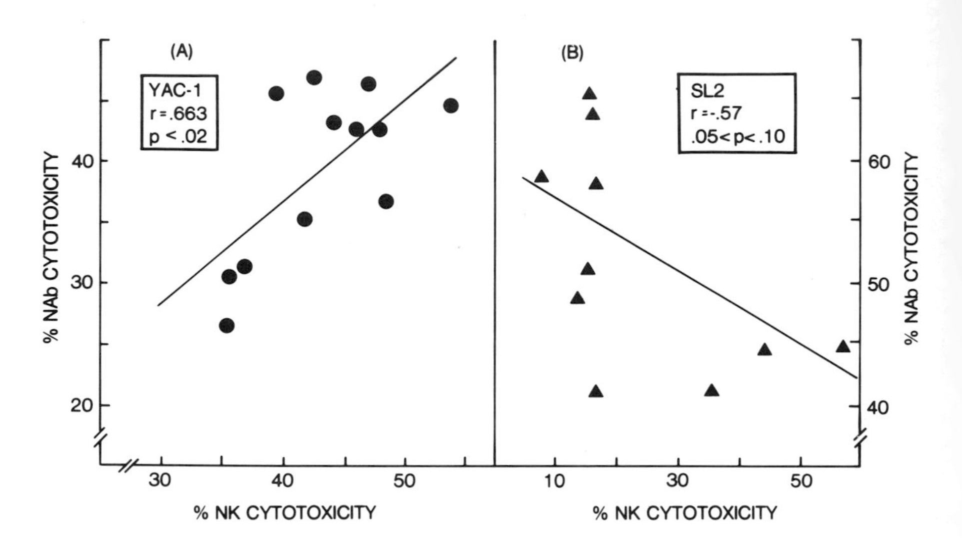

FIGURE 1: Correlation of CBA/J NK cell and NAb cytotoxicity using clones of the YAC-1 and SL2 lymphomas. (from Chow et al, 1981).

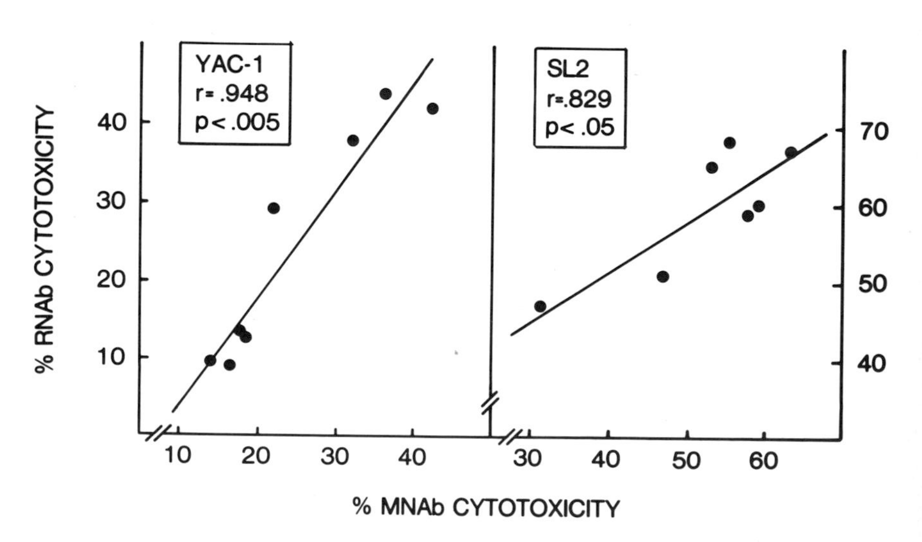

FIGURE 2: Correlation of murine (CBA/J) and rabbit NAb cytotoxicity using YAC-1 and SL2 clones.

A dissociation of the expression of NAb and NK TS was noted when examining the membrane modulation induced by treatment with interferon (IFN). NK sensitivity of the SL2-5 was profoundly inhibited by small doses of IFN, while both RNAb and MNAb mediated lysis were enhanced (Fig 4). This inverse relationship of NK and NAb was similar to that identified earlier when comparing the reactivity of SL2 clones. The YAC-1 tumor, on the other hand, exhibited a slight decrease in NK sensitivity and RNAb lysis after IFN treatment, while MNAb lysis rose significantly (Fig 4). This result was surprising considering that the NAb TS appeared to be antigenically very similar in absorption studies, and a corresponding IFN modulation would have been predicted. An explanation of this differential response could be a defect in the IFN regulation of one of the tumors, however, the SL2 and YAC-1 responded well to IFN in other respects such as NK-TS suppression and growth inhibition. Another possibility is that the MNAb and RNAb antigenic determinants are on different membrane moieties that were independently regulated by IFN. For example, if the TS are small cross-reacting molecules found on a number of different membrane structures, they may have appeared to be antigenically similar in absorption studies, yet the moieties on which they are predominantly located could respond to IFN in a distinct manner.

One can find considerable support in the literature for the notion that membrane perterbations associated with malignant transformation promotes reactivity to both NK cells and NAb. Since carbohydrate (CHO) antigens are frequently cited as NAb target structures (Rogentine and Plocinik, 1974; Springer et al, 1976; Sela and Edelman, 1977; Arend and Nijssen, 1977), and there is also some evidence that NK and NC cells interact with these molecules (Stutman et al, 1980; MacDermott et al, 1981; Forbes et al, 1981) we will consider, in the following discussion, the possibility that membrane CHO may be a common TS. The idea that malignant cells bear incomplete or deleted carbohydrate chains on glycolipids or glycoproteins, as a result of a defect in enzymes participating in the metabolism of membrane sugars, was suggested by Hakamori and Murakami (1968) many years ago and may be the modification on which the NK cells and NAb are mutually dependent.

NAb are more widely reactive than NK cells, and have been demonstrated for cell types which are not lysed by NK cells, for example, murine and sheep erythrocytes. In addition, NAb activity to soluble antigens such as immunoglobulin, DNA and oxazolone have been identified (reviewed in Elson et al, 1979). The relationship between NAb and NK TS must therefore be limited to only that portion of the NAb population which binds to

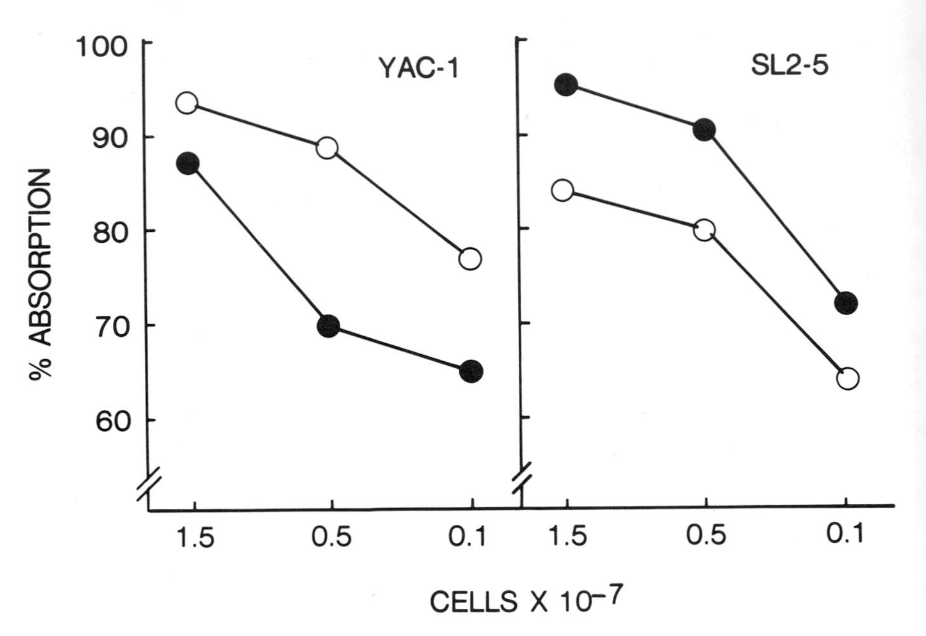

FIGURE 3: Cross-absorption of RNAb activity by YAC-1 (○) and SL2-5 (●).

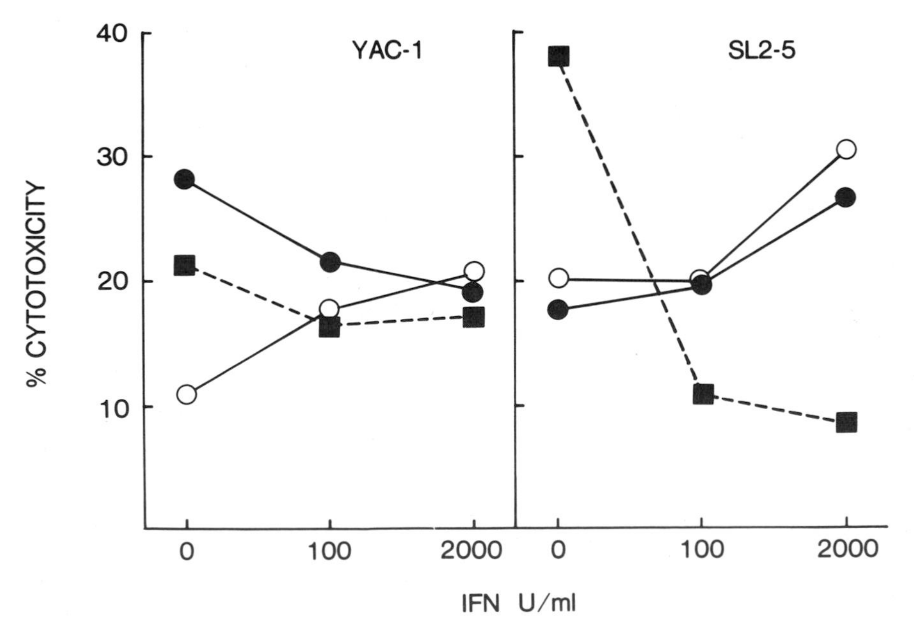

FIGURE 4: Interferon-induced modulation of sensitivity to lysis by CBA/J NK cells (--■--), RNAb (——●——) and CBA/J MNAb (——○——). Tumor cells were incubated with IFN for 20 hours.

tumor membrane. The tumor membrane reactive NAb we will consider fall into three groups: autoantibodies, antibody to virally or chemically modified cells, and interspecies natural antibody. NAb to MHC-linked antigens have also been described (Wolosin and Greenberg, 1981; Longenecker and Mosmann, 1980) but space limitations do not allow us to discuss the possible relation to NK cells which may also be directed at MHC antigens on some human tumors (Vanky et al, 1980).

One of the most prevalent autoreactive NAb we have been able to detect on the SL2-5 and YAC-1 lymphomas used to study NK cytotoxicity are thymocytotoxic (Greenberg, A. H. and Wolosin, L. B., manuscript in preparation). Thymocytes are themselves good NK targets although it is not clear whether the NK sensitive subpopulation (Hansson et al, 1979) is the same as that lysed by thymocytotoxic autoantibodies. Schlesinger and Bekesi (1977) noted that lysis of thymocytes by normal AKR sera was directed at membrane carbohydrates and could be inhibited by simple sugars, and by neuraminidase treated tumor cells. One of the several sugar inhibitors they identified was D-mannose, which Stutman et al (1980) has also shown is an inhibitor of NK and NC cells. In the NZB two forms of thymocyte autoantibodies have been identified, one of which is reactive only with desialized lymphocytes, and another that is cytotoxic for both intact thymocytes and asialated lymphocytes (Imai et al, 1980). A relationship between sialation and NK sensitivity has also been proposed by Yogeeswaran et al (1981), after finding that neuraminidase released less sialic acid on NK sensitive than resistant targets and increased tumor NK sensitivity somewhat. Neuraminidase-treated cells are also excellent targets for NAb (Rogentine and Plocinik, 1974), and lysis could be inhibited by oligosaccharides containing beta-D-galactosyl, although both the antigen and antibodies appeared to be heterogeneous in these studies.

Another cellular autoantigen which has been described as a target for both NAb and NK cells is present on normal fetal fibroblasts. Both anti-tumor NAb (Ménard et al, 1974), and NAb reactive with primary human skin cells (Thorpe et al, 1977) can be removed by fetal tissue while NK cells can lyse and are inhibited by fetal cells (Shellam and Hogg, 1977; Nunn et al, 1977) and preferentially lyse less differentiated cell lines (Stern et al, 1980). In that regard, there are several oncofetal developmental antigens that have been described with carbohydrate specificity (Willison and Stern, 1978; Solter and Knowles, 1978; Gooi et al, 1981). In one case, the stage-specific embryonic antigen (SSEA-1) is formed by simple glycosylation, and it has been suggested that, in general, glycosylation may form the basis of stage specific expression of embryonic antigens.

It was the observation that tumors bearing endogenous type C viral particles are good NK targets that led to the interest in these antigens as NK target structures (Herberman et al, 1975). Attempts to define specific viral antigens as NK targets have generally not been successful, however, there are noteworthy exceptions in particular the observation of Kende et al (1979) who have implicated gp 70. It is also true that virally transformed tumors are NAb reactive and antigens have been characterized which are both viral (Aoki et al, 1966; Martin and Martin, 1975) and non-viral (Barbacid et al, 1980). Murine (Aoki et al, 1966) and human natural antibodies (Kurth et al, 1977) have been described which react with gp 70. However, one interesting study suggested that the antigenic specificities may not be of viral origin. Using tumors expressing the SSV-SSAV complex, Snyder and Fleissner (1980) found that the NAb precipitating gp 70 molecules reacted non-specifically with the carbohydrate moiety of the viral tumor antigen. They also suggested that since the addition of oligosaccharide to viral glycoprotein is a host-dependent process, glycosylation and CHO determined antigenicity will vary qualitatively among the cells in which the virus is found. It is interesting to speculate that the inconsistent expression of NK TS on viral glycoprotein antigens could be, in part, the result of such qualitative variation in host-derived oligosaccharides.

Reactivity of tumors with heterologous NK cells (Roder et al, 1977) and NAb (Schlesinger et al, 1966; Sela and Edelman, 1977; Grönberg et al, 1980) have been described for a wide variety of tumors. Analysis of membrane antigen profiles by NK rosette inhibition (Roder et al, 1979) suggested that murine and human NK cells recognize different sets of glycoprotein surface molecules. Antigens reacting with heterologous NAb have been partially characterized on thymocytes (Schlesinger et al, 1966) and some tumors (Sela and Edelman, 1977). The most striking feature of this work is that these antibodies seem to be predominantly directed against carbohydrate antigens (D-galactose, N-acetylneuraminic acid, N-acetylglucosamine). Human NAb which bind to blood group substance, another membrane saccharide, have also been detected on a number of homologous tumors (Springer et al, 19676; Arend and Nijssen, 1977).

In summary, there appears to be a relationship between the expression of NK, murine and rabbit NAb reactivity on tumor targets. The observation that NK and NC cells are inhibited by simple carbohydrates and that the NK target structures are also associated with glycoprotein and glycolipid moieties, as well as the degree of sialation, has led to the suggestion that NK cells have a lectin-like recognition unit. It is also true that one of the NAb target structures most commonly identified is carbohydrate in nature. A reasonable hypothesis to account for the target structure relationship is that both of these effectors can detect certain membrane glycosylation defects. This idea has, in addition, an appealing evolutionary perspective since NK and NAb are thought to be very early immune phenomena and both intercellular (Siu et al, 1978) and humoral recognition factors in invertebrates (Yeaton, 1981), as well as antibody in early vertebrates (Gold and Balding, 1975), utilize carbohydrate determinants for cellular recognition.

REFERENCES

Aoki, T., Boyse, E. A., and Old, L. J. (1966). Cancer Res. 26:1415.

Arend, P., and Nijssen, J. (1977). Nature 269:255.

Barabacid, M., Bolognesi, D., and Aaranson, S. A. (1980). Proc. Natl. Acad. Sci. USA 77:1617.

Chow, D., Wolosin, L., and Greenberg, A. H. (1981). JNCI 67:445.

Elson, C. J., Naysmith, J. D., and Taylor R. B. (1979). Int. Rev. Exp. Path. 19:137.

Forbes, J. T., Bretthauer, R. K., and Oeltmann, T. N. (1981). Proc. Natl. Acad. Sci. USA 78:5797.

Gold, E. R. and Balding, P. (1975). "Receptor specific proteins, plant and animal lectins." Excerpta Medica, American Alsevier Publishing Co., New York.

Grönberg, A., Hansson, M., Kiessling, R., Anderson, B., Kärre, K., and Roder, J. (1980). JNCI 64:113.

Gooi, H. C., Feizi, T., Kapadia, A., Knowles, B. B., Solter, D., and Evans M. J. (1981). Nature 292:156.

Hakomori, S., and Murakami, W. T. (1968). Proc. Nat. Acad. Sci. USA 59:254.

Hansson, M., Kiessling, R., Andersson, B., Kärre, K., and Roder, J. (1979). Nature 278:174.

Herberman, R. B., Nunn, M. E., and Lavrin D. H. (1975). Int. J. Cancer 16:216.

Imai, Y., Nakano, T., Sawada, J., and Osawa, T. (1980). J. Immunol. 124:1556.

Kende, M., Hill, R., Dinowitz, M., Stephenson, J. R., and Kelloff, G. J., (1979). J. Exp. Med. 149:358.

Kurth, R., Teich, N. M. Weiss, R., and Oliver, R. T. (1977). Proc. Natl. Acad. Sci. USA 74:1237.

Longenecker, B. M., and Mosmann, T. R. (1980). Immunogenet. 1.
MacDermott, R. P., Kienker, L. J., Bertovich, M. J., and Muchmore, A. V. (1981). Immunol. 44:143.
Martin, S. E., and Martin, S. J. (1975). Nature 256:489.
Ménard, S., Colnaghi, M. I., and Della Porta, G. (1974). Br. J. Cancer 30:524.
Nunn, M. E., Herberman, R. B., and Holden, H. T. (1977). Int. J. Cancer 20:381.
Roder, J., Ahrlund-Richter, L., and Jondal, M. (1979). J. Exp. Med. 150:471.
Rogentine, G. N., and Plocinik, B. A. (1974). J. Immunol 113:848.
Schlesinger, M., and Bekesi, J. G. (1977). JNCI 59:945.
Schlesinger, M., Cohen, A., and Hurvitz, D. (1966). Israel J. Med. Sci. 2:616.
Sela, B., and Edelman, G. M. (1977). J. Exp. Med. 145:443.
Shellam, G. R., and Hogg, N. (1977). Int. J. Cancer 19:212.
Siu, C. H., Loomis, W. M., Lerner, R. A. (1978). In "The Molecular Basis of Cell-Cell Interaction", (R. A. Lerner and D. Bergsma ed.), p. 459, Alan R. Liss. Inc., New York.
Snyder, H. W., and Fleissner, E. (1980). Proc. Natl. Acad. Sci. USA 77:1622.
Solter, C., and Knowles, B. B. (1978). Proc. Natl. Acad. Sci. 75:5565.
Springer, G. F., Desai, P. R., and Scanlon, E. F. (1976). Cancer 37:169.
Stern, P., Gidlund, M., Orn, A., and Wigzell, H. (1980). Nature 285:341.
Stutman, O., Dien, P., Wisun, R. E., and Lattimer, E. C. (1980). Proc. Natl. Acad. Sci. USA 77:2895.
Thorpe, W. P., Parker, G. A., and Rosenberg, S. A. (1977). Immunol. 119:818.
Vanky, F. T., Argov, S., Einhorn, S., and Klein, E. (1980). J. Exp. Med. 151:115.
Willison, K. R., and Stern, P. L. (1978). Cell 14:785.
Wolosin, L. B., and Greenberg, A. H. (1981). J. Immunol. 126:1456.
Yeaton, R. W. (1981). Develop. Comp. Immunol. 5:535.
Yogeeswaran, G., Grönberg, A., Hansson, M., Dalianis, T., Kiessling, R., and Welsh, R. M. (1981). Int. J. Cancer 28:517.
Young, W. Y., Durdik, J. M., Urdal, D., Hakomori, S.-I., and Henney, C. S. (1981). J. Immunol. 126:1.

EFFECT OF SODIUM BUTYRATE AND HEMIN ON NK SENSIVITY OF K562 CELLS

Marie-Christine Dokhélar
Thomas Tursz

Laboratoire d'Immunologie Clinique
Institut Gustave Roussy
Villejuif, France

I. INTRODUCTION

In human NK cell assays, the leukemic K562 cells are commonly used as targets for their high and reproducible sensitivity to NK lysis. The K562 cell line was originally considered as a highly undifferenciated cell line of the granulocytic lineage (Lozzio and Lozzio, 1975). However, several erythroid markers, such as glycophorin A are also expressed in K562 cells and hemoglobin (Hb) synthesis has been demonstrated either spontaneously in various culture conditions, and after sodium butyrate (SB) (Anderson et al., 1979) or hemin induction (Rutherford et al., 1979). Exposure to hemin increases Hb synthesis four to fivefold but does not permit K562 cells to reach the final maturation steps of the erythroid lineage (Guerrasio et al., in press; Benz et al., 1980). On the other hand, the presence of a marker of the megakaryocytic lineage, i.e. a peroxidase activity (PA) identical to the platelet peroxidase has also been demonstrated (Vainchenker et al., in press). This PA is present in rare cells of the K562 cell line. SB induction greatly increases the number of cells exhibiting this PA, but does not induce a further maturation towards the megakaryocytic lineage.

All together, these data suggest that K562 cells are leukemic stem cells with at least bipotential capacities of differentiation, expressing few embryonic and fetal markers (Benz et al., 1980; Vainchenker et al., in press). In this regard, the K562 cell line could provide a unique model to investigate the expression of the target structure(s) recognized by NK cells.

ISBN 0-12-341360-5

We compared the sentivity to NK cells lysis of non-induced, hemin or SB-induced K562 cells. It was possible to demonstrate that only SB was able to modify NK lysis of K562. The sensitivity to NK lysis of different K562 subclones was slightly heterogenous but could not be correlated to their Hb synthesis or the presence of a PA, suggesting that the NK target structure(s) are not directly dependent upon the cell lineage commitment.

II. MATERIALS AND METHODS

A. NK Assays

Effector cells were peripheral blood lymphocytes isolated by centrifugation on Ficoll-Paque, and tested for spontaneous cytotoxicity against ^{51}Cr-labeled K562 cells. Effector to target ratio was 50:1. Incubation was at 37°C for 4 hrs.

B. Incubation with Inducers

SB was added to the culture mediumat a final concentration of 1 mM, during 5 days. Control cells consisted of cells coming from the same flask, and cultivated without inducer in the same conditions.

Hemin was added at 0.1 mM final concentration, during 5 days.

C. Target Binding Cell Assay

The method used was described by Roder et al.(1978). Nylon-wool passed lymphocytes were pelleted onto K562 cells at 25:1 ratio. After 30 min incubation, the percentage of K562 cells binding to lymphocytes were counted.

D. Monolayers

For the preparation of the monolayers, the method used was that of Phillips et al. (1980). Peripheral blood lymphocytes ($5x10^6$) were added to each monolayer of K562 cells. After 2 hrs of incubation at 37°C, non adherent lymphocytes were recovered and used as effector cells in NK assays.

III. RESULTS AND DISCUSSION

The results have been reported in detail elsewhere (Dokhélar et al., 1982).

A. Effects of Hemin and SB on Susceptibility of K562 Cells to Lysis

A clear decrease of K562 cell lysis was observed after treatment of the target cells with SB, whereas hemin had no or slight effect (Table I).

The effect of SB on K562 sensitivity to NK lysis was dose-dependent (maximal effect at 2mM concentration). Decreased NK sensitivity occurred after 2 days-culture of K562 cells with SB and was maximal after 4 days. When SB was removed from the culture medium, normal NK activity was progressively restored, and completely recovered on day 4.

These kinetics completely parallelled the appearance or disappearance of the PA in the nuclear envelope and endoplasmic reticulum, which are early differentiation markers of the megakaryocytic lineage. It thus could be assumed that during the differentiation process, some NK target structures are lost by K562 cells. The absence of effect of hemin could be due to a more limited effect of hemin on mammalian cells when compared to that of SB. These differences were already seen in Friend erythroleukemia cells in which SB induces a complete erythroid commitment, whereas hemin only modifies Hb synthesis (Mager and Bernstein, 1979). On K562 cells, i and I antigen expression is not modified after hemin induction whereas SB clearly decreases i expression.

TABLE I. Effect of Diffentiation Inducers on K562 Cells

	K562 cells cultivated 5 days with		
	Normal medium	Sodium butyrate	Hemin
Mean % specific lysis	41.0	25.1	41.4
% hemoglobin producing cells	2-3	2-3	15-20
% cells exhibiting platelet peroxidase activity	1-2	10-40	1-2

A complete abrogation of K562 sensitivity to NK lysis after SB treatment could never be achieved. Whether the observed NK resistance was due to decreased expression of NK target structures on the cell menbrane of every cell, or whether only some of the cells were affected remains to be established. With regard to various hematopoietic markers, the K562 line appeared quite heterogeneous, some cells exhibiting erythroid features either spontaneously or after induction and some other cells carrying megakarycytic markers. Cloning of K562 cells resulted in different subclones appearing more homogenous in their cellular commitment (Vainchenker et al., in press). Roughly, an inverse relationship was found between the capacity of Hb synthesis and the presence of the PA. The susceptibility to NK lysis of the different clones was slightly different but could not be correlated to their Hb synthesis or the presence of a PA. For instance, Table II reports the results obtained with the clone 48 with predominent erythroid features and with the clone 209-23 which is the more "megakaryocytic" one.

It can be seen that both clones appeared equally sensitive to NK lysis and that SB treatment induced a similar NK protection in the two clones. These results suggest that the NK target structure expression is not directly dependent upon the cell lineage commitment predominant in each clone.

B. Mechanisms of SB Effect

Cold target competition assays were performed (Table III). Untreated K562 cells acte as better competitors whether radiolabeled targets were induced or not. However, SB-induced cold targets were able to compete with K562 cells. Both types of cold targets could completely abrogate the ^{51}Cr release, but twice as many SB treated cells were needed for competition when compared to K562 control cells. These data suggest that some target structures are shared by induced and non-induced cells,

TABLE II. NK Sensitivity of K562 Subclones. Effect of SB

Subclone	% lysis cloned cells / % lysis uncloned cells	% lysis inhibition[a]
48	1.28	41.9
209-23	1.31	53.5

[a]After culture in SB-supplemented medium

TABLE III. Cold Target Competition Assay

Radiolabeled targets:	untreated		treated[a]	
Competitor K562 cells[b]:	untreated	treated	untreated	treated
% inhibition	60.1	44.4	66.1	53.9

[a] K562 cells treated with SB
[b] Competitor to target ratio was 5:1

but present in lower amount on SB-induced K562 cells.

In addition, SB induced K562 cells were less efficient in binding nylon wool-passed lymphocytes than control K562 cells.

Finally, NK effectors were completely retained on monolayers of control K562 cells, whereas absorption of lymphocytes on monolayers of SB induced cells only led to a slight decrease of NK activity (data not shown).

Taken together these results indicate that SB affects the expression of some binding structure(s) during the differentiation process.

As interferon has been shown to reduce the sensitivity of target cell to NK lysis, we investigated whether SB effect was mediated through interferon (Table IV). No interferon could be detected in the culture medium after 5 days of culture with SB. Addition of interferon in the culture medium after 4 days of culture with SB led to a greater decrease of K562 sensitivity to NK lysis than with SB alone. Finally, culture of K562 cells in the presence of both SB and anti-interferon serum (kindly provided by M. Tovey, Villejuif, France) gave the same results as with SB alone. Thus, it can be concluded that SB effect was not dependent upon the production of interferon.

TABLE IV. Effects of SB and Interferon on NK sensitivity of K562 Cells

K562 cultivated with	% specific lysis
normal medium	60.9
SB (1 mM, 5 days)	22.8
interferon (7500 U/ml, 24 hrs)	33.8
SB + interferon	15.9
SB + anti-interferon	21.6

Whatever the mechanism whereby SB affects NK sensitivity, the present results underline that NK susceptibility can be modulated by a differentiation inducer. The NK sensitivity of various malignant cells could mainly be a reflection of their immaturity.

ACKNOWLEDGMENTS

We wish to thank U. Testa and W. Vainchenker for their help in this study, and Y. Finale and C. Tétaud for expert technical assistance.

REFERENCES

Anderson, L. C., Jokinen, M., and Gahmberg, C. G. (1979). Nature 278:364.

Benz, E. J., Murane, M. J., Tonkonow, B.L., Berman, B. W., Mazar, E. M., Cavallesco, C., Jenko, T., Snyder, E. L., Forget, B. J., and Hoffman, R. (1980). Proc. Natl. Acad. Sci. USA. 77:3509.

Dokhélar, M. C., Testa, U., Vainchenker, W., Finale, Y., Tétaud, C., Salem, P., and Tursz, T. (1982). J. Immunol. in press.

Guerrasio, A., Vainchenker, W., Breton-Gorius, J., Testa, U., Rosa, R., Thomopoulos, P., Titeux, M., Guichard, J., and Bezard, Y., Blood Cells, in press.

Lozzio, C. B., and Lozzio, B. B. (1975). Blood 45:321.

Mager, D., and Bernstein, A. (1979). J. Cell. Phys. 100:467.

Phillips, W. H., Ortaldo, J. R., and Herberman, R. B.(1980). J. Immunol. 125:2322.

Roder, J. C., Kiessling, R., Biberfield, P., and Andersson, B. (1978). J. Immunol. 124:2491.

Rutherford, T. R., Clegg, J. B., and Weatherhall, B. J. (1979). Nature 280:164.

Vainchenker, W., Testa, U., Guichard, J., Titeux, M., and Breton-Gorius, J. Blood Cells, in press.

ANALYSIS OF DIFFERENTIATION EVENTS CAUSING CHANGES IN NK CELL TUMOR-TARGET SENSITIVITY

Magnus Gidlund
Masato Nose
Inger Axberg
Hans Wigzell

Department of Immunology, Uppsala University, Box 582, S-751 23 Uppsala, Sweden

Thomas Tötterman
Kenneth Nilsson

Department of Pathology, Wallenberg Laboratory, Uppsala, Sweden

INTRODUCTION

The specificity shown by NK cells, and thus, features expressed on the potential target allowing it to be lysed, remains unclear. In part this can be attributed to the complexity of the system itself: 1) The target repertoire for NK cells include certain normal, transformed and virusinfected cells (1-4) 2) there is a considerable fluctuation, in the expression of NK cell activity, due to genetic and systemic regulations (5-7) and in the target cells maintained in vitro 3) NK cells can lyse tumor targets across the species barrier (8). 4) The non-apparent (9) or apparent (10) clonality of the NK cell pool. 5) The bifunction of the cells to act as K-cells or as a direct lytic cell (4). 6) The lytic system is poorly understood and genetically distinguished from cytotoxic T-cells and activated macrophages (11).

In a previous report we analysed the relation between the stage of differentiation of the target cells and sensitivity for

ISBN 0-12-341360-5

NK lysis (12). We concluded that particular features governed by the stage of differentiation were important in determining the NK susceptibility of the tumor. In the present communication we present data obtained in analysing freshly isolated human chronic lymphocytic leukemia (CCL) cells and a murine myeloid leukemia cell line undergoing induced controlled differentiation in vitro (13,14). We have analysed how the induction to differentiation altered the NK sensitivity of the tumor cell population, its ability to be lysed by homologous or xenogeneic effector cells and whether the alteration in sensitivity affected the competing capacity of such cells when known NK target cells were used.

CHRONIC LYMPHOCYTIC LEUKEMIA (CLL) CELLS INDUCED TO DIFFERENTIATION SHOW AN INCREASE IN NK CELL SENSITIVITY

Freshly obtained tumor cells from B-CLL patients can, in certain cases, be induced to differentiate in vitro, if treated with the phorbol ester 12-O-Tetradecanoyl-13-Phorbolacetate (TPA) (13,15,16). The phenotypical properties expressed by such cells are given in table 1.

Table 1. Phenotypic properties of CLL cells induced to differentiation

Morphological (lymphoblastoid)
Increase in cytoplasmic Ig (heavy and light chain)
Increase in cytoplasmic HLA-Dr
Ig secretion (heavy and light chain)
Increase in B_2-microglobulin secretion
Decrease in surface Ig expression
Decrease in Fc receptor expression
Decrease in sialic acid
C3b receptor unaltered
Disappearance of mouse erythrocyte receptors
CALLA not expressed

For experimental details and assays for individual markers see (13,15,16). For determination of sialic acid see Yogeeswaaran et al. in present volume.

Freshly obtained CLL tumor cells can be lysed by allogeneic NK cells (17). We therefore analysed the sensitivity expressed by such cells after treatment with TPA. In table 2, one representative experiment is shown where two different tumor populations were examined, having been cultured in the presence of TPA for three days, for sensitivity to interferon-activated human effector cells.

Table 2. Sensitivity of CLL cells after treatment with TPA

Type of Response[1]	NK sensitivity (% Lysis) Control 100:1	25:1	TPA-treated 100:1	25:1	No of/[2] Total Tests	Active[3] Disease
Non-Responder	10.4	5.2	10.5	2.7	16/35	1 (6%)
Responder	15.2	4.8	34.8	15.2	19/35	16 (84%)[4]

[1] *Tumor cells were cultured in the presence or absence of $1.6x10^{-7}$M TPA for three days and tested in a 4 Hr ^{51}Cr release assay against interferon-activated normal peripheral blood as in (17), at indicated ratios between effector and target cells. The total material was subdivided in tumor populations showing no increase (non-responders) and in responder populations where a significant increase, comparable to the given example, was noted.*

[2] *Frequence of the responses* [3] *Clinical signs monitored as in (18)* [4] *One patient showing stable activity in this group progressed within one month after being tested.*

The total material included 18 B-CLL patients tested at 35 different occasions during a time period of 18 months (19). A substantial increase in NK cell susceptibility was noted in 19 cases (responders). This increase peaked at the third day of culture (data not shown). No change in background sensitivity was shown by the cells cultured in the absence of TPA. Furthermore, the increase when induced was not correlated to the background sensitivity of the CLL cells (P=0.3, paired t-statistics). The tumor material could thus, on basis of lytic increase after TPA treatment be divided into two tumor populations: one, where no changes could be detected and a second where a significant increase in NK sensitivity was found. This latter increase was positively correlated to the induced amount of cytoplasmic Ig (P >0.005 (19)) and thus "linked" to a known parameter of the B-cell differentiation. A similar correlation was seen when the patients were clinically evaluated. In 31 out of 34 cases (or maybe 32, see legend table 2, footnote 3) did the enhanced sensitivity coincide with clinical stages in growth characteristics (i.e. stable-progression). Furthermore, if the patients were retested at different time points, the close correlation between NK sensitivity, disease activity, and amount of induced cytoplasmic Ig was maintained (6 patients (19)). That indeed the effector cells mediating the lytic effects on the TPA-induced target were of NK type was verified as summarized in table 3.

Table 3. NK characteristics of the effector cell mediating lysis against CLL cells responding to TPA (20)

Nonadherent
Responsive to interferon
Efficiency of lysis of TPA-induced CLL (CLL-TPA) positively correlated to that of known NK targets (K-562, U-266)
CLL-TPA competes better than untreated CLL when K-562 and U-266 are used as targets
Lysis of CLL-TPA can be blocked by "cold" U-266 and K-562
Lysis is undisturbed by OKT-3 monoclonal antibodies which remove T-killer cell activity (21,22)

Furthermore, human CLL-TPA displayed a corresponding increase in lytic susceptibility using interferon-activated mouse spleen cells as effector cells. We thus, conclude, that induction to differentiation will lead to an increase in sensitivity in NK cell-mediated lysis. Furthermore, this "permissivness" to display change in susceptibility, is correlated with a clinically aggressive leukemia.

ANALYSIS OF A MURINE LEUKEMIA UNDERGOING TUNICAMYCIN-INDUCED DIFFERENTIATION: EFFECT ON SENSITIVITY AGAINST MURINE AND HUMAN NK CELLS

The mouse myeloid leukemia cell line, M1, was originally isolated from the SL strain of mice (23). It has since then been an accepted model to study myeloid differentiation, using a variety of different inducing agents (14,24). Tunicamycin specifically inhibits synthesis of N-acetylglucosaminyl pyrophosphoryl polyisoprenol, leading to inhibition of protein glycosylation (25). Tunicamycin induces the M1 cell line to express phagocytic activity, increased expression of Fc-receptors, and to show macrophage morphology (14). In table 4, we studied the effect of tunicamycin on the M1 cell line, using human and mouse effector cells.

In experiment B and C we in addition analysed how inhibition of cellular division affected the background sensitivity and the tunicamycin-induced alteration in NK cell sensitivity.

Table 4. Sensitivity of M1 cell line before and after treatment with tunicamycin: Effect of cell division inhibitors

Treatment[1]	Mouse effectors[2] 100:1	Mouse effectors[2] 50:1	Human effectors[3] 100:1	Human effectors[3] 50:1
Exp A				
Control	50.4[4]	37.1	12.9	4.7
Tunicamycin	23.6	17.3	36.5	27.7
Exp B				
Control	51.0	31.9		
Tunicamycin	31.6	18.6		
Irradiation+ tunicamycin	49.6(53.2)	33.2(35.4)[5]		
Exp C				
Control	37.3	26.9		
Tunicamycin	14.2	10.0		
Mitomycin C+ tunicamycin	30.3(33.7)	18.6(22.1)[6]		

1) M1 tumor cells were put in culture with or without tunicamycin (0.33 µg/ml) as described in (14) for two days. In exp. B and C cells were irradiated (2500r) or Mitomycin C treated (60 µg/ml) prior to further culturing. After culturing, the cells were used as target cells as in table 2.
2) Mouse spleen effectors were obtained and prepared from mice treated with the interferon inducer Tilorone as in (6).
3) Human effectors were prepared and interferonactivated as in table 2.
4) Percent lysis
5) 6) In parenthesis; lysis value with only irradiation or Mitomycin C treated target cells

Tunicamycin treatment resulted in a substantial decrease in sensitivity to mouse NK cells, but paradoxically also in an increase in sensitivity when using human NK cells. Irradiation and Mitomycin C treatment to prevent cell division inhibited the tunicamycin mediated effect on NK sensitivity. Accordingly we analysed if the irradiation per se eliminated the primary effect of tunicamycin, that is, inhibition of protein glycosylation. Results are shown in table 5.

Table 5. Effect of irradiation on cell division and sugar incorporation in M1 cell line before and after tunicamycin treatment

Parameter[1)]	Control		Tunicamycin	
	0	2500r	0	2500r
Cell number	3.1	1.2	2.1	0.6
^{3}H-Gal($x10^{-3}$ cpm)	2.7	2.0	0.8	0.9
^{14}C-GlcN(-"-)	0.7	0.6	0.1	0.2

[1)] *Number of cells were determined after 48 hrs in culture and given as the increase. Radiolabelled Galactose (Gal) and Glucoseamin (GlcN) were added at 1 µCi/well, in triplicate, incubated for 6 hours, and harvested on Skatron harvesting machine at time of cell number determination. Cpm values are expressed per 10^5 cells. Tunicamycin treatment as in table 4.*

Tunicamycin treatment resulted in a partial reduction in the cell number, and this can in higher concentrations lead to a complete elimination (14). 2500r irradiation caused a close to complete inhibition of cell division, leaving additional tunicamycin treatment without effect as to cell numbers. Irradiation, however, failed to have any impact on the inhibition of glycosylation by tunicamycin.

DISCUSSION

NK cells can mediate efficient lysis against undifferentiated, MHC-lacking embryonic carcinoma cells (9). Furthermore, K-562, a widely used and highly susceptible NK target (26), is a cell line considered to represent an erythroid leukemia at an early differentiation stage, as indicated by, for instance, studies on the hemoglobin biochemistry. Normal stem cells of the bone marrow and immature thymocytes can serve as quite sensitive targets for NK cells (1,27). When several of these NK susceptible cell types are induced to differentiate by a variety of distinctly different inducers, there will be a parallel change in NK sensitivity towards resistance (12 and present study). It was thus considered that NK cells for unclear reasons do display a tendency to kill cells in more early stages of differentiation (9).

Subsequently, we have come across a few exceptions where the opposite may occur, that is, induction of differentiation is accompanied by an increase in sensitivity to NK cells (19, 20 and present study). This included a neuroblastoma cell line (Gidlund to be published), CLL cells, and a murine myeloid leukemia cell line, when this was used as a xeno target cell. CLL cells, for instance, could be made quite sensitive to NK lysis, if obtained from patients with progressive disease, and subsequently induced to differentiate *in vitro*. Why progressive disease was required is unclear but this may merely represent the existance of physiological stages of responsiveness of the tumor cells to the differentiation inducing agent. Interestingly, the same inducing agent (TPA) that could induce differentiation and *increase* in NK susceptibility in CLL cells would do the opposite, that is, *decrease* NK sensitivity, when the histiocytic U-937 line was studied. This finally excluded that the inducing agent *per se* directly changed the target cell with regard to sensitivity towards lysis. Studies with the murine M1 line demonstrated that tunicamycin could cause differentiation parallelled by a decrease in NK sensitivity.

However, this was only true for murine NK cells, whereas human NK cells displayed an increased effector capacity for the differentiated cells. This merely confirm and further pinpoint the issue that, although NK cells can kill across species barriers, there is a species preference under most normal conditions (8). Interpretations of the variation in the species preference have lead to the hypothesis that this is being conveyed upon the cells by surface glycoproteins (28).

Specificity of NK cells remains a complicated issue and there is no doubt that this represent a multifactorial issue at the level of interaction between effector and target cells as to binding, lytic susceptibility, repair mechanisms etc. The controlled differentiation studies do allow a parallel analysis of biochemical changes and changes in susceptibility to NK lysis, which may finally solve the issue of relevant binding structures for NK cells on target cells. It is here noteworthy that analysis of glycolipid composition of the presently studied cell types indicate opposite changes of certain glycolipids (in particular with regard to sialic acid) in cells becoming more susceptible versus more resistant towards NK killing (29,30). Thus, it is already clear that the findings of different changes in NK lytic susceptibility can be found to be parallelled at the molecular level in a most promising manner.

ACKNOWLEDGEMENTS

This work was supported by NIH grants R01-CA-26752-02 and the Swedish Cancer Society. M.N. has a fellowship from UICC. We also wish to thank Prof. Y. Ichikawa for providing of M-1 cell line and Profs. G. Tamura and A. Takatsuki for the donation of Tunicamycin. We also thank Ms Birgitta Ehrsson-Lagerkvist for typing the manuscript.

REFERENCES

1. Hansson, M., Kiessling, R., Andersson, B., Kärre, K., and Roder, J., Nature 278, 174 (1979).
2. Kiessling, R., Klein, E., and Wigzell, H., Eur.J.Immunol. 5, 112 (1975).
3. Herberman, R.B., Nunn, M.E., and Lavrin, D.H., Int.J.Cancer 16, 216 (1975).
4. Natural cell mediated immunity against tumors. Ed. R.B. Herberman, Academic Press (1980).
5. Petranyi, G.R., Kiessling, R., and Klein, G., Immunogenetics 2, 53 (1975).
6. Gidlund, M., Örn, A., Wigzell, H., Senik, A., and Gresser, I., Nature 273, 759 (1978).
7. Herberman, R.B., Djeu, J.Y., Jay, H.F., Ortaldo, J.R., Riccardi, C., Bonnard, G., Holden, H.T., Fagnani, R., Santoni, A., and Pucetti, P., Immunol.Rev. 44, 43 (1979).
8. Hansson, M., Bakacs, K.K.T., Kiessling, R., and Klein, G., J.Immunol. 121, 6 (1978).
9. Stern, P., Gidlund, M., Örn, A., and Wigzell, H., Nature 285, 341 (1980).
10. Herberman, R.B., and Ortaldo, J.R., In Natural cell mediated immunity against tumors. Ed. R.B. Herberman, Academic Press (1980), p. 873.
11. Roder, J.C., and Duwe, A.K., Nature 278, 452 (1979).
12. Gidlund, M., Örn, A., Pattengale, P., Jansson, M., Wigzell, H., and Nilsson, K., Nature 292, 848 (1981).
13. Tötterman, T.H., Nilsson, K., and Sundström, C., Nature 288, 176 (1980).
14. Nakayasu, M., Terada, M., Tamura, G., and Sugimura, T., Proc.Natl.Acad.Sci. 77, 409 (1980).
15. Tötterman, T.H., Nilsson, K., Claesson, L., Simonsson, B., and Åman, P., Human Lymphocyte Differentiation 1, 13 (1981).
16. Tötterman, T.T., Nilsson, K., Sundström, C., and Sällström, J., Human Lymphocyte differentiation, in press.
17. Pattengale, P.K., Gidlund, M., Nilsson, K., Sundström, C., Sällström, J., Simonsson, B., and Wigzell, H., Int.J.Cancer (1982) in press.
18. Levin, W.C., Gehan, E., Griffith, K., Hughley, C.M., Silcer, R.T., Steinfeldt, J., Winer, L., and Cahn, E.L., Cancer Chemother.Reports 3, 1 (1968).
19. Tötterman, T.H., Gidlund, M., Nilsson, K., and Wigzell, H., Submitted.
20. Gidlund, M., Tötterman, T.H., Nilsson, K., Alm, G., Kabelitz D., and Wigzell, H., Submitted.
21. Chang, T.W., Kung, P.C., Gingras, S.P., and Goldstein, G., Proc.Natl.Acad.Sci. (Wash.) 78, 1805 (1981).
22. Landegren, U., Ramstedt, U., Axberg, I., Örn, A., and Wigzel H., Int.J.Cancer 28, 725 (1981).

23. Ichikawa, Y., J.Cell Physiol. 74, 223 (1969).
24. Krystosek, A., and Sachs, L., Cell 9, 675 (1976).
25. Waechter, C.J., and Lennarz, W.J., Ann.Rev.Biochem. 46, 95 (1976).
26 West, W.H., Cannon, G.B., Kay, H.D., Bonnard, G.D., and Herberman, R.B., J.Immunol. 118, 355 (1977).
27. Hansson, M., Kiessling, R., In "NK cell: Fundamental aspects and role in cancer", Human Cancer Immunology vol. 6. North-Holland Publisher, Amsterdam (1981).
28. Axberg, I., Gidlund, M., Örn, A., Pattengale, P.K., Riesenfeldt, I., Stern, P., and Wigzell, H., In "Thymus hormones and T-lymphocytes." Ed. F. Aiuti and H. Wigzell. Academic Press (1980), p. 155.
29. Yogeeswaaran, G., Grönberg, A., Hansson, M., Dalianis, T., Kiessling, R., and Welsh, R.M., Int.J.Cancer 28, 517 (1981).
30. Yogeeswaaran, G., Welsh, R.M., Grönberg, A., Kiessling, R., Patarroyo, E., Klein, G., Gidlund, M., Wigzell, H., and Nilsson, K., In the present volume.

SPECIFICITY OF NATURAL KILLER (NK) CELLS: NATURE OF TARGET CELL STRUCTURES

Jerome A. Werkmeister[1]
Stephen A. Helfand[2]
Tina Haliotis
Hugh Pross
John C. Roder

Departments of Microbiology & Immunology
and Radiation Oncology
Queen's University
Kingston, Ontario, Canada K7L 3N6

I. INTRODUCTION

Natural killer cells have now been demonstrated to kill a wide variety of malignant cells as well as some normal cells to a lesser extent without any apparent need for prior stimulation (reviewed in 1).

While it is true that NK cells have a broad degree of reactivity, it is equally apparent that there exists a fine degree of selectivity, since some target cells are highly sensitive to cytolysis whereas others are not (1). Most studies have suggested that NK activity is directed against multiple antigenic determinants, some of which may be shared by more than one target cell. The precise nature of the target antigens recognized by NK cells remains poorly understood, however. In this chapter we will briefly summarize and update the early work on specificity of NK cells and then discuss recent evidence as to the nature of

---

[1]Supported by the Ontario Cancer Treatment & Research Foundation, the National Cancer Institute of Canada and the Medical Research Council of Canada.

[2]Present address: Department of Neurology, Harvard Medical School, Boston, Mass.

ISBN 0-12-341360-5

the target cell structure. In particular, we will concentrate to a large extent on the relationship between target cell differentiation and susceptibility to lysis by NK cells.

II. COMPETITIVE INHIBITION OF NK-TARGET INTERACTION USING UNLABELLED TARGET CELLS AND PUTATIVE NK TARGET STRUCTURES

Early work on the specificity of NK activity indicated that these cells from individual donors exhibited distinct differing patterns of reactivity against a variety of different target cells. Studies using unlabelled targets in cross competition assays with a panel of ^{51}Cr labelled target cells have shown that cytotoxicity could be inhibited by some but not all target cells (reviewed in 1). Those cells most sensitive to NK lysis proved to be the best competitors and a good correlation was found between susceptibility to direct lysis and competition among a variety of targets. The ability of different cell lines to compete inferred but did not prove the existence of cross reacting target antigens and a complementary receptor on the effector cell.

Early work by Roder _et al_ (2,3) has shown that soluble extracts from NK-sensitive target cells could inhibit binding but not lysis of susceptible targets. These glycosylated target structures (130,160 and 240,000 d) were isolated from susceptible target cells and found to have both cross-reactive and unique NK specificities (240 K molecule).

More recent studies on human target cell lines (melanoma MM200 and Chang cell cultures) have demonstrated the presence of neutral 140,000 d glycoproteins which inhibited NK but not antibody-dependent cellular cytotoxicity (ADCC) Similar structures were not found in supernatants of NK resistant cell lines. While a detailed analysis of the specificity of this inhibition was not performed, it was evident that these glycosylated (D-mannose and N-acetyl-D-glucosamine specific) putative membrane antigens were not extensively cross-reactive (4).

The above results from both murine and human systems demonstrate that glycoproteins are effective in inhibiting target cell-effector cell (NK) interactions and subsequent cytolysis, suggesting carbohydrate moieties as possible target structures. This conclusion is further supported by several findings of inhibition of NK activity with a variety of different sugars (reviewed in 1).

An alternative approach to the problem of the molecular nature of the NK target structure has been the biochemical analysis of selectively derived NK-sensitive (NK^S) or NK-resistant (NK^R) variants. Durdik _et al_ (5) have isolated two NK^S variants of the L5178Y lymphoma which, unlike the NK^R lines, expressed the neutral glycosphingolipid asialo GM2 (6). Gronberg _et al_ (7), using repeated cycles of immunoselection _in vitro_ and _in vivo_, have described NK^R variants of the YAC-1 lymphoma which were impaired in their NK-target cell binding ability. A positive association was found between NK sensitivity and asialo GM2 and certain gangliosides (8), which suggests but does not prove that these glycolipid structures may be directly involved in NK-target cell interactions. We have recently isolated partially but selectively NK^R-YAC-1 variants by treatment with the mutagen nitrosoguanidine (9). The basis for this resistance differs from NK^R variants described above since the binding efficiency was found to be normal (9). It is suggested that the alteration lies in an acceptor site (distinct from the NK target structure) for a putative lytic molecule released from the NK cell after contact with the target. Preliminary studies of the biochemical nature of this difference by two-dimensional gel electrophoresis of whole cell lysates revealed only a few protein differences (unpublished data).

III. THE NATURE OF THE TARGET STRUCTURES

Although the exact nature of the NK target structure remains obscure, it appears unlikely that they are associated entirely with endogenous type C viral determinants or MHC antigen products (reviewed in 1). The remaining part of this review will focus on current findings from our own and other laboratories showing the association of differentiation of the target cell and NK sensitivity.

The intriguing possibility that the NK-target cell interaction may be dependent on the stage of differentiation of the malignant target cell has evolved from a series of investigations on murine and human normal tissues (thymus and bone marrow) showing increased susceptibility during embryonic compared with adult life (reviewed in 1). Additional support for this hypothesis was extended to malignant target cells where it was recently shown in the mouse that embryonal carcinoma cells are more susceptible to NK lysis compared with more differentiated endodermal cell lines arising spontaneously from the parental teratoma (10).

It has become increasingly apparent in recent years that a variety of cultured malignant cells can be induced to differentiate along distinct cell lineages using a variety of different agents. The association of human target cell differentiation and susceptibility to NK cytolysis has been subsequently analysed in detail under more controlled conditions using well known target cells and chemical inducers.

Studies on the K562 (erythroleukemia) and HL-60 (promyelocytic leukemia) cell lines by ourselves (11,12) and on the GM-86 (mouse Friend erythroleukemia), U-937 (histiocytic lymphoma) and K562 cell lines by others (13) have provided evidence to support the idea that target cell susceptibility to NK lysis may in part be dependent on the stage of differentiation of the tumor cell.

The kinetics of cellular differentiation was paralleled by a concomitant decrease in susceptibility to NK lysis. K562 cells were found to differentiate within 4 days of culture in optimal concentrations of 1 mM sodium butyrate or 0.1 mM haemin and could be serially maintained in culture for 2-3 months in the presence of the inducer. Differentiated K562 cells (Benzidine +ve and increase in glycophorin A) were partially 'cloned' by limiting dilution techniques, seeding 50 cells per well in microtitre plates in the presence of 1 mM butyrate. All differentiated 'clones' had lowered susceptibility to NK-mediated cytolysis (11). K562 'clones' isolated and maintained in the absence of butyrate produced variable sensitivities ranging from 2- to 3-fold increases to 2- to 3-fold decreases. These latter 'clones' were unstable. Differentiated K562 (NK-resistant) cells grown in the absence of butyrate reverted to the NK sensitive phenotype within 1 week following removal of the inducing agent (11). This is consistent with data showing that embryonal and fetal hemoglobin accumulation in induced K562 cells is completely reversible and is not accompanied by terminal differentiation (14). In contrast to the K562 system, DMSO- and TPA-induced HL-60 cultures underwent terminal differentiation (mature granulocytes and macrophages, respectively), peaking at 5-7 days and 2-3 days, respectively. DMSO, one of the most widely used chemical inducers, did not induce differentiation of K562 (unlike murine erythroleukemic cell lines) and did not alter the sensitivity of the cells to lysis. Similarly we found that NK susceptibility of non-induceable Molt-4 cells was unaffected by butyrate, haemin or DMSO (11). Our results suggest that the decrease in NK susceptibility is affected by the nature of the differentiated target cells and not the

TABLE I. DIFFERENTIATION-INDUCED NK TARGET CELL RESISTANCE: SPECIFICITY AND RELATION TO NK BINDING

Target cell	LU[a]/10^7			%CFC[b]	I_{50}[c] x 10^3
	NK	ADCC	Mono		
K562 untreated	160	317	40	42	22
K562-butyrate	29	295	40	8	110
HL-60 untreated	198	392	0	6	14
HL-60-DMSO	74	432	0	1	58
HL-60-TPA	66	400	0	1	-

[a]A lytic unit (LU) was defined as the number of effector cells required for 20% lysis (NK and monocyte lysis) or 40% lysis (ADCC).

[b]Number of Percoll-fractionated large granular lymphocytes (LGL's enriched 10 fold in NK cytolytic activity) binding to target cells.

[c]Number of unlabelled competitor cells required for 50% inhibition of ^{51}Cr release by untreated target cells.

inducer itself. Furthermore, we found comparable differences in NK sensitivity in the untreated and differentiated K562 cells in assays containing 1 mM sodium butyrate (11).

The reduction in NK sensitivities of the differentiated target cells appeared to be specific and was not due to a generalized change in the plasma membrane. As shown in Table I, no significant differences were observed in ADCC activity or monocyte-mediated cytolysis against the untreated and induced cultures.

Certain target cells induce interferon which is known to augment NK cytolysis (reviewed in 1). It was thus possible that the differentiated (NK-resistant) target cells were less capable of inducing interferon production in culture. As shown in Table II, our results demonstrate that the differentiated tumor cell lines, K562 and HL-60, are less susceptible to both untreated as well as interferon (α or β)-activated NK cells. Furthermore, there were no differences in the amount of interferon produced in culture using highly enriched NK cells from Percoll gradients with resistant or sensitive K562 or HL-60 cells.

TABLE II. DIFFERENTIATION-INDUCED NK TARGET CELL RESISTANCE IN RELATION TO INTERFERON (IF) INDUCTION. EFFECT OF EXOGENOUS INTERFERON.

Target cell	$LU^{a}/10^{7}$		IF titre[b]
	Untreated NK	IF-activated NK	
K562 untreated	168	317	1/500
K562-butyrate	40	88	1/500
HL-60 untreated	118	332	n.d.
HL-60-DMSO	40	118	n.d.
HL-60-TPA	35	131	n.d.

n.d. not determined

[a]30% lysis.

[b]highest dilution of supernatant causing 50% inhibition of Vesicular Stomatitis Virus infection of fibroblast monolayers. Supernatants from Percoll-fractionated LGL's co-cultured for 18 hr with differentiated-induced or untreated target cells.

It appeared that the predominant reason for the decrease in NK susceptibility of the differentiation-induced target cells (K562 and HL-60) was a direct decrease in the NK binding ability of these cells as assessed by both competitive cold target inhibition and conjugate forming cell assays (Table I). In addition we have evidence to suggest that the NK-resistant differentiated K562 'clones' may be defective in an early post-recognition step of the NK lytic pathway. We have recently shown that macrophage-depleted highly enriched human NK cells generate a rapid burst of superoxide radicals within seconds of contact with susceptible target cells (15). The generation of these superoxide radicals (detected by luminol-dependent chemiluminescence) was directly related to NK mediated cytolysis since free radical scavengers that reduced chemiluminescence (superoxide dismutase, mannitol, DMSO, benzoate) also blocked NK killing (16). The differentiated K562 cell 'clones' were less capable of inducing oxygen intermediates compared with the untreated K562 cells (11).

There is now enough evidence to indicate that upon differentiation (either natural or induced) a particular cell may acquire distinct phenotypic changes which may be associated with a concomitant resistance to NK mediated cytolysis.

These findings reported here pose a potential dilemma in the use of such inducing agents in the therapy of certain leukemias, since these differentiated cells would have a selective survival advantage in relation to the postulated NK surveillance mechanism.

Differences in analysis of total cellular proteins, particularly membrane glycoproteins with respect to molecular size, affinity for selected sugars and isoelectric point may lead to a better understanding of the nature of these antigens. It should be emphasized that this phenomenon of lowered NK sensitivity may not be universal for all differentiated tumor cells. We have preliminary data showing that differentiation of some human melanoma lines with 1 mM theophylline can cause an increase in NK susceptibility (17). Theophylline-differentiated melanoma cells have a 4-fold increase in melanoma-associated antigens and a parallel decrease in the expression of HLA-Dr (18) and nerve growth factor (NGF) receptors (Riopelle, personal communication). This differentiated-related increase in NK susceptibility and association with certain membrane antigens is consistent with recent findings in melanoma showing that expression of certain melanoma antigens (perhaps differentiation antigens) on the cell surface are significantly associated with a favourable prognosis (19).

REFERENCES

1. Roder, J.C., Karre, K., and Kiessling, R., Prog. Allergy 28:66 (1981).
2. Roder, J.C., Rosen, A., Fenyo, E.M., and Troy, F.A., Proc. Natl. Acad. Sci. 76:1405 (1979).
3. Roder, J.C., Ahrlund-Richter, L., and Jondal, M., J. exp. Med. 150:471 (1979).
4. Zaunders, J., Werkmeister, J., McCarthy, W.H., and Hersey, P., Br. J. Cancer 43:5 (1981).
5. Durdik, J.M., Beck, B.N., Clark, E.A., and Henney, C.S., J. Immunol. 125:683 (1980).
6. Young, W.H., Durdik, J.M., Urdal, D., Hakomori, S., and Henney, C.S., J. Immunol. 126:1 (1981).

7. Gronberg, A., Kiessling, R., Eriksson, E., and Hansson, M., J. Immunol. 127:1734 (1981).
8. Yogeeswaran, G., Gronberg, A., Hansson, M., Dalianis, T., Kiessling, R., and Welsh, R.M. Int. J. Cancer 28:517 (1981).
9. Roder, J.C., Beaumont, T., Kerbel, R.S., Haliotis, T., and Kozbor, D., Proc. Natl. Acad. Sci. 78:6396 (1981).
10. Stern, P., Gidlund, M., Orn, A., and Wigzell, H., Nature 285:341 (1980).
11. Werkmeister, J., Helfand, S., Haliotis, T., Pross, H., and Roder, J. (submitted).
12. Werkmeister, J., Pross, H., Roder, J., Rubin, P., and Eaton, B., J. Cell. Biol. 91:24a, Abstract 1041 (1981).
13. Gidlund, M., Orn, A., Pattengale, P.K., Jansson, M., Wigzell, H., and Nilsson, K., Nature 292:848 (1981).
14. Dean, A., Erard, F., Schneider, A.B., and Schnecter, A.N., Science 212:459 (1981).
15. Roder, J.C., Helfand, S., Beaumont, T., and Duwe, A., (submitted).
16. Helfand, S., Werkmeister, J., and Roder, J., (submitted).
17. Werkmeister, J., Helfand, S.L., Haliotis, T., Rubin, P., Pross, H., and Roder, J., (submitted).
18. Liao, S.K., Kwong, P.C., and Dent, P.B., Proc. Can. Fed. Biol. Soc. 24: 280, Abstract 920 (1981).
19. Werkmeister, J., Edwards, A., McCarthy, W., and Hersey, P., Cancer Immunol. Immunother. 9:233 (1980).

EFFECTS OF INTERFERON AND TUMOR PROMOTER, 12-O-TETRADECANOYLPHORBOL-13-ACETATE, ON THE SENSITIVITY OF TROPHOBLAST CELLS TO NATURAL KILLER CELL ACTIVITY

Kenneth S.S. Chang and Kenichi Tanaka

Laboratory of Cell Biology, National Cancer Institute,
Bethesda, Maryland 20205

I. INTRODUCTION

We have established trophoblast cell culture lines from murine placenta and demonstrated their lack of expression of H-2 antigen, and their ability to grow as carcinomas in different strains of mouse across histocompatibility barriers (1). The differentiation status of trophoblast cells are somewhat similar to that of embryonal carcinoma cells in that they are H-2 nonexpressive, resistant to type C retrovirus infection and SV40 T antigen formation, and insensitive to interferon in terms of protection against vesicular stomatitis virus (VSV) replication. It has been postulated that NK cells function as a surveillance system against the emergence of certain tumors and hemopoietic cells with aberrent, nonself antigen (2,3). Because our preliminary test indicated that, as in the case of embryonal carcinoma cells (4), alloimmune T cells failed to kill trophoblast cells but NK cells were able to kill, a possible role of NK cells in regulating the emergence and progression of trophoblastic diseases as well as other fetomaternal relationships may be considered.

Since IFN has a regulatory role in NK-sensitivity of some target cells (5-7) and since IFN is found to be present in murine placentas (8), we studied the role of IFN in the interaction of NK and trophoblast cells. The effect of 12-O-tetradecanoylphorbol-13-acetate (TPA), one of the tumor promotors, on trophoblast cells has also been studied, because it has pleiotropic, and regulatory effects on differentiation of a variety of cells *in vitro* (9-11).

ISBN 0-12-341360-5

II. EFFECTS OF IFN ON NK CELLS AND NK-SENSITIVITY OF TROPHOBLAST CELLS

Trophoblast cell lines designated PL/CB and PL/B6 etc., indicating their (BALB/c x C57BL/6)F_1 and C57BL/6 origins respectively, were labeled with [^{125}I]-iododeoxyuridine and used as targets for cell-mediated cytotoxicity tests (12). The N:NIH(s)-nu (nude, with NIH Swiss background) mouse spleen cells were used as NK effector cells. For IFN-treatment of effector cells, 6 x 10^7 cells in 0.6 ml were treated with 180 units of IFN for 1 hr at 37 °C and diluted to desired concentrations of cells (13). IFN-treatment of target cells was done by incubating for 4 hr at 37° the mixture of 5 x 10^5 cells and 60 units of IFN in 0.2 ml in the presence of 2 µCi [I^{125}]-iododeoxyuridine, and by washing 3 times before mixing with effector cells.

As shown in Table I, treatment of effector cells with IFN (partially purified from culture supernate of L cells infected with UV-inactivated Sendai virus) markedly enchanced the cytotoxic effects of macrophage-depleted spleen cells. By contrast, treatment of both trophoblast and YAC target cells with IFN decreased markedly their NK-sensitivity. These effects were observed also when purified L cell IFN containing IFN α and β (obtained from NIH) was used, and IFN-antiserum could be shown to neutralize these effects.

TABLE I. NK-Sensitivity of Trophoblast Cells treated with IFN

Treatment of Effector Cells[a]	Effector/ target ratio	PL/CB		PL/B6		YAC	
		none	IFN	none	IFN	none	IFN
none	200	16.3[b]	3.1	20.6	14.1	28.4	12.3
	100	12.3	1.0	14.5	6.8	22.3	11.8
	50	12.1	0.2	13.4	2.0	18.6	12.6
IFN	200	33.4	25.7	45.6	32.6	30.4	17.6
	100	27.5	17.3	32.4	21.1	27.3	11.3
	50	9.6	9.7	22.7	10.2	21.7	12.0

[a] N:NIH(s)-nu mouse spleen cells cultured on plastic plates at 37 °C for 1 hr to remove macrophages

[b] net % cytotoxcity

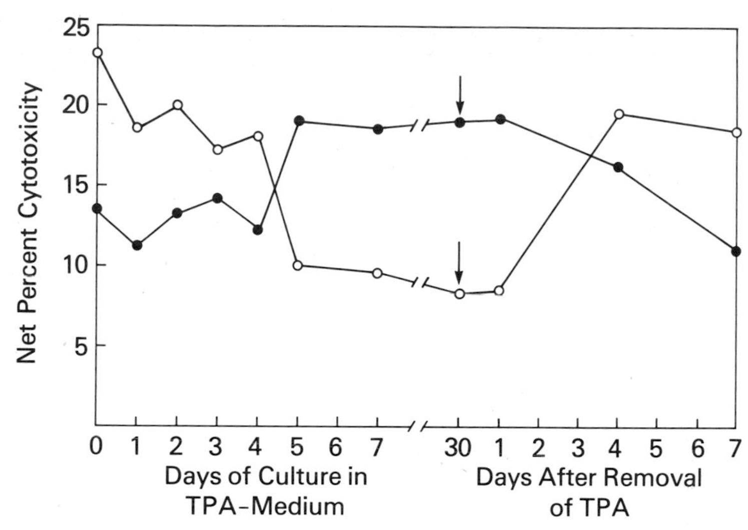

Figure 1. Reversibility of TPA-effects on NK-sensitivity of trophoblast cells (PL/CB) and antagonistic effects of IFN. Arrows indicate the time when TPA was removed from the culture. Effector/target ratio was 200. (o—o), cells treated with TPA; and (●—●), those treated with TPA followed by IFN.

The sensitivity of trohoblast cells to the protective activity of IFN against VSV replication was very low, if any, even when treated with a high dose (300 units). By contrast, YAC cells were as sensitive as L cells to the protective action of IFN against VSV infection (14).

III. INTERACTION OF TPA AND IFN ON THE NK-SENSITIVITY OF TROPHOBLAST CELLS

The trophoblast cells cultured in the presence of TPA (1.6×10^{-7} M) for 3-4 weeks also showed reduced NK-sensitivity; the reduction was often more marked than that caused by the 4 hr treatment with IFN. However, no such effect was observed for other targets including RBL-3, $S^{+}L^{-}$ mink cells superinfected with a dual-tropic virus (Chang, unpublished), ($S^{+}L^{-}$ mink-V), and NIH Swiss mouse embryo cell cultures (data not shown). As illustrated in Fig. 1, there was a gradual decrease in NK-sensitivity of PL/CB cells during the first 4 days of culture in TPA-medium, and this decrease was further augmented by the 4-hr IFN treatment immediatly before NK test. During the following weeks of culture in TPA-medium, these cells showed a marked reduction in NK-sensitivity. However, this effect was counteracted by the 4-hr IFN-treatment before each NK test, resulting in near-normal NK-sensitivity. After removal of TPA from the culture medium at day 30, the NK-sensitivity of TPA-treated cells became normal again, and the NK-sensitivity-reducing effect of IFN became apparent again. The mechanism for these effects is not clear. In a competitive inhibition assay, the unlabeled PL/CB cells treated with IFN, TPA, or both effectively competed against the labeled target cells resulting in marked inhibition of NK activity. These results indicate the presence of NK-recognition structure on the membrane of TPA- or IFN-treated trophoblast cells.

IV. INCREASED TUMORIGENICITY OF TPA-TREATED TROPHOBLAST CELLS

Since TPA-treatment appeared to decrease the NK-sensitivity of PL/CB cells but not S^+L^- mink-V cells, these cells were tested for their tumorigenicity by s.c. injection in N:NIH(s)-nu mice. As shown in Table II, the S^+L^- mink-V cells as well as (the spontaneously transformed) L cells with or without TPA-treatment readily produced tumors in these

TABLE II. Increased Tumorigenicity of TPA-Treated Trophoblast Cells in N:NIH(s)-nu Mice

Cells inoculated	Treatment	Dose of s.c. inoculation	Days of observation	Mean tumor size (cm^3)[a]
PL/CB	TPA	$1x10^7$	40	41.0[b]
	none	$1x10^7$	40	0.7
	TPA	$3x10^6$	40	27.7[b]
	none	$3x10^6$	40	<0.1
L	TPA	$1x10^7$	28	58.8[c]
	none	$1x10^7$	28	42.4
S^+L^--Mink-V	TPA	$1x10^7$	14	41.9[c]
	none	$1x10^7$	14	42.8

[a] 0.4 (ab^2) where a and b represent the larger and smaller diameter respectively disecting the tumor at right angle
[b] P<0.01 as compared with the control
[c] P>0.05 as compared with the control

mice. By contrast, the untreated PL/CB cells which scarcely induced tumors became more tumorigenic after TPA-treatment. Furthermore, when these PL/CB cells were injected i.p. into (BALB/c x C57BL/6)F_1 mice, TPA-treated cells produced ascitic tumors and killed mice more rapidly than the untreated cells. Thus, the NK-sensitivity-decreasing effect of TPA-treatment appeared to correlate with the augmented tumorigenicity of trophoblast cells in N:NIH(s)-nu and (BALB/c x C57BL/6)F_1 mice.

V. DISCUSSION

The presence of H-2 nonexpressive trophoblast cells in the placenta may serve to provide a barrier against afferent and efferent mechanisms of maternal immunologic response against the fetus which bears allogeneic paternal antigen. Since the alloimmune T cells were not effective, the presence of functionally efficient macrophages and an NK cell system would be an important defense mechanism for the maternal host. The presence of high titers of IFN in murine placenta (8) was confirmed by us, and would contribute to maternal defense mechanism mediated by IFN-enhanced NK system. However, the IFN-treated trophoblast cells acquire some resistance to the NK cell cytotoxicity, thereby counteracting the excessive augmentation of maternal NK cells. Obviously, there must be a balance between these two forces so that a homeostasis is maintained.

Because the reduced NK-sensitivity of TPA-treated trophoblast cells was restored to normal sensitivity by treatment with cycloheximide (10 μg/ml) for 1 hr at 37°C prior to NK test, it was suggested that the TPA-induced protein(s) played some role in the phenotypic manifestation of NK-resistance of trophoblast cells. TPA was reported to reduce the NK-sensitivity of a lymphoma cell line (15) and also inhibit spontaneous or IFN-boosted NK effector activities (16). The facts that N:NIH(s)-nu mice did not easily allow the untreated trophoblast cells to form a tumor, despite the absence of H-2 antigen expression on the cell membrane and deficiency of T cells in the host, and that TPA-treated cells became more NK-resistant and tumorigenic may suggest the important role of NK cells to reject tumors. By contrast, S^+L^- mink-V cells with or without TPA treatment were able to grow as tumors although both were as NK-sensitive as untreated trophoblast cells. Therefore, the decreased NK-sensitivity of trophoblast cells as a result of TPA-treatment may be a necessary but not sufficient condition for these cells to produce tumors in N:NIH(s)-nu mice.

VI. SUMMARY

The NK-sensitivity of murine trophoblast cells was reduced either by a 4 hr treatment with murine IFN or after culturing for more than 4 days in the presence of a tumor promotor, TPA. This effect of TPA was reversible. However, the 4 hr IFN-treatment before NK test of those cells cultured previously in TPA-medium resulted in restoration of their normal NK-sensitivity. The decrease in NK-sensitivity of IFN- or TPA-treated trophoblast cells was not due to the loss of NK-recognition membrane structures. The possible contribution of de novo synthesized protein(s) which could confer NK-resistance was suggested. The TPA-treated trophoblast cells became more tumorigenic than the untreated controls when inoculated into mice. The possible roles of NK cells in trophoblast diseases and fetomaternal relationships are discussed.

REFERENCES

1. Log, T., Chang, K.S.S., and Hsu, Y.C. Int. J. Cancer 27: 365, (1981).
2. Kiessling, R. and Wigzell, H. Immunol. Rev. 44:165 (1979).
3. Cudkowicz, G. and Hochman, P.S. Immunol. Rev. 44:13 (1979).
4. Stern, P., Gidlund, M., Orn, A., and Wigzell, H. Nature 285:341 (1980).
5. Hansson, M.R., Kiessling, R., Andersson, B., Karre, K., and Roder, J. Nature 278:174 (1979).
6. Trinchieri, G., Santoli, D., Granato, D., and Perussia, B. Fed. Proc. 40:2705 (1981).
7. Welsh, R.M., Karre, K., Hansson, M., Kunkel, L.A. and Kiessling, R.W. J. Immunol. 126:219 (1981).
8. Fowler, A.K., Reed, C.D., and Giron, D.J. Nature 286: 266 (1980).
9. Blumberg, P.M. CRC Crit. Rev. Toxicol. 8:153 (1980).
10. Blumberg, P.M. CRC Crit. Rev. Toxicol. 8:199 (1981).
11. Diamond, L., O'Brien, T.G., and Baird, W.M. Adv. Cancer Res. 32:1 (1980).
12. Chang, K.S.S. and Log, T. Int. J. Cancer 25:405 (1980).
13. Djeu, J.Y., Heinbaugh, J.A., Holden, H.T., and Herberman, R.B. J. Immunol. 122:175 (1979).
14. Tanaka, K., Log, T. and Chang, K.S.S. In: Adv. Comparative Leukemia Research, 1981, Yohn, D.S. and Blakeslee, J.R. (Eds.), Elsevier/North-Holland (in press).
15. Gidlund, M., Orn, A., Pattengale, P.K., Jansson, M., Wigzell, H. and Nilsson, K. Nature 292:848 (1981).
16. Goldfarb, R.H. and Herberman, R.B. J. Immunol. 126:2129 (1981).

SEROLOGICAL APPROACHES TO THE ELUCIDATION OF NK TARGET STRUCTURES

David L. Urdal,
Ichiro Kawase,
Christopher S. Henney[1]

Program in Basic Immunology
Fred Hutchinson Cancer Research Center
Seattle, Washington

I. INTRODUCTION

In the previous edition of this book, we reported the isolation of a pair of lymphoma cell variants that differed markedly in their susceptibility to lysis by NK cells (1). Clone 27v (27v) of the L5178Y lymphoma was susceptible to NK cell-mediated lysis, whereas clone 27av (27av) was resistant. Despite the large difference in the abilities of the two cell lines to serve as targets for NK cell mediated lysis, 27v and 27av were equivalently susceptible to alloimmune cytotoxic T cells, cytotoxic macrophages and K cells (2).

Not only were the 27av cells insusceptible to NK cell-mediated lysis, but 27av did not inhibit the lysis of 27v cells by NK effector populations. These findings suggested that the susceptible variant binds to NK cells, but that the insusceptible line does not.

Having obtained two variant cell lines which functionally can be considered as NK susceptibility variants, we have been interested in pursuing a systematic biochemical analysis of these cells, with the intention of identifying possible cell surface "hallmarks" of NK susceptibility.

[1] Supported by grants CA 24537 and AI 15384 from the National Institutes of Health and IM-247 from the American Cancer Society. David L. Urdal is supported by a human cancer-directed fellowship DRG-528 of the Damon Runyon-Walter Winchell Cancer Fund.

ISBN 0-12-341360-5

Two serological approaches that we have taken towards the elucidation of NK target structure have been 1) the characterization of the antibodies in normal rabbit sera responsible for the lysis of NK sensitive target cells, and 2) the preparation of monoclonal antibodies directed against the NK-susceptible cell, that could block NK mediated lysis. The rationale behind both of these approaches was the impression that the antigens recognized by these antibodies might correlate with the target structures recognized by murine NK cells.

A. Characterization of the Antibodies in Normal Rabbit Sera

Recently, Gronberg et al. (3) have described an interesting serologic similarity amongst a panel of murine lymphoid cells that are susceptible to NK cell mediated lysis. NK susceptible cells were found to be readily lysed by normal rabbit serum, whereas NK resistant cells were not affected by such treatment. We made the same observation using the L5178Y variants, 27v and 27av. That is, 27v was susceptible to normal rabbit serum and 27av was completely resistant (2).

With the hope that the antigens detected by normal rabbit sera might correlate with target structures recognized by murine NK cells, we sought to characterize the specificity of the naturally occuring antibodies in normal rabbit sera that were responsible for the lytic activity of those sera towards NK sensitive target cells. Five different normal rabbit sera were examined, with essentially the same results. The results from one serum are summarized in Tables I and II.

Table I. Lysis of NK Target Cells by Rabbit Serum 2941

Target cell treatment with NAN-ase[b]	% Specific lysis[a]		
	27v-1C2	27av	YAC-1
-	69.7	0.3	0.9
+	70.8	2.3	73.8

(a) The serum was used at final concentrations of 1:32, 1:128, and 1:512. The value for the specific ^{51}Cr release seen at a dilution of 1:32 is recorded in the table. Guinea pig serum at a final dilution of 1:64 was used as a source of complement.

(b) 10^6 cells were labeled with ^{51}Cr and then washed. The cell pellet was suspended in 0.1 ml BME (without serum) PH 6.7 and 0.01 ml of Vibrio cholera neuraminidase (NAN'ase; 500 U/ml, Calbiochem-Behring) was added. After 45 minutes at 37°C the cells were washed and used in the assay.

Table II. Carbohydrate and Glycolipid Inhibition of Cell Lysis by Serum 2941

Inhibitor	Concentration of inhibitor resulting in a 50% decrease of target cell lysis[a] Target: 27v-1C2	NAN'ase YAC-1
GalNAc[b]	3.2	>17.5
Lactose	>17.5	3.2
Gal	>17.5	>17.5
Glc	>17.5	>17.5
Man	>17.5	>17.5
Gg_3cer	0.06	>0.5
Gg_4cer	>0.5	0.03
Gb_4cer	>0.5	>0.5
Lcn_4cer	>0.5	>0.5
GM_1	>0.5	>0.5

(a) The numbers in the table represent the sugar concentration (mM) or the glycolipid concentration (µg/well) that resulted in 50% inhibition of target cell lysis. Sugar preparations (dilutions of iso-osmotic solutions) or liposomes containing glycolipid (prepared as described in (4) were incubated with serum 2941 (diluted 1:4) for 45 min before the addition of ^{51}Cr labeled target cells and guinea pig complement.

(b) GalNAc, N-acetylgalactosamine; Gal, D-galactose; Glc, D-glucose; GlcNAc, N-acetylglucosamine; Man, D-Mannose; Gg_3cer, gangliotriosylceramide; Gg_4cer, gangliotetraosylceramide; Gb_4cer, Globotetraosylceramide; Lcn_4cer, lactoneotetraosylceramide.

We found that the susceptibility of 27v (subclone 1C2) cells to lysis by normal rabbit serum (Table I) was due to antibodies directed to gangliotriosylceramide (Gg_3cer). This conclusion was reached based on the following observations: 1) the anti-27v activity in normal rabbit serum was removed by absorption on guinea pig red blood cells (4), which express Gg_3cer as their major glycolipid (5); 2) the lysis of 27v-1C2 cells was specifically inhibited by N-acetylgalactosamine, the immunodominant sugar on Gg_3cer (Table II), and 3) of the glycolipids tested, pure Gg_3cer was the most effective glycolipid inhibitor of cell lysis (Table II), see also (4).

To our surprise, (given the previously published report of Gronberg et al. (3)), when we extended these results to another NK sensitive target, YAC-1, we found that these cells were quite insensitive to normal rabbit serum (Table I). Gronberg et al. (3) had found that YAC-1 cells grown in their laboratory were highly susceptible to rabbit serum. Whether this discrepancy was due to the source of normal rabbit sera, or to the fact that the YAC-1 cells were propagated under different conditions remains unclear.

We pursued this line of inquiry further and found that YAC-1 cells could be lysed by the same panel of normal rabbit sera that lysed 27v-1C2 cells if the YAC-1 cells were first exposed to neuraminidase (NAN'ase) (Table I). The NK insusceptible cell line, 27av, remained insusceptible to normal rabbit sera even after the cells had been treated with neuraminidase . Hence the difference observed between YAC-1 and 27v-1C2 may have been the trivial one of differences in sialylation. However, we found that the antibodies responsible for the lysis of neuraminidase treated YAC-1 had a different specificity from those responsible for the lysis of 27v-1C2. Several observations supported this conclusion: 1) YAC-1 cells were insensitive to monoclonal antibodies directed to Gg_3cer before or after NANase treatment (4), 2) the anti-neuraminidase treated YAC-1 activity in normal rabbit sera was not absorbed by guinea pig red blood cells (4) and 3) Gg_3cer was not detected on the surface of neuraminidase treated YAC-1 cells by the galactose-oxidase-tritiated sodium borohydride procedure (4).

Several observations suggested however that the activity towards neuraminidase treated YAC-1 cells was due to anti-carbohydrate antibodies: 1) Lactose and, to a lesser extent, galactose, were effective inhibitors of YAC-1 cell lysis (GalNAc was not effective) (Table II); 2) of the glycolipids tested, purified gangliotetraosylceramide (Gg_4cer) which has galactose as the terminal sugar, was the most effective glycolipid inhibitor of neuraminidase treated YAC-1 cell lysis (Table II); and 3) Gg_4cer was detected on the surface of neuraminidase treated YAC-1 cells by the galactose oxidase-tritiated sodium borohydride procedure (4).In spite of the dramatic increase in the sensitivity of YAC-1 to normal rabbit serum after neuraminidase treatment, this enzyme had no effect on the susceptibility of the three cell lines studied to NK cell mediated lysis, nor did it affect the ability of these cells to inhibit NK-cell mediated lysis (4).

Although it has proven possible to identify the fine specificities of the antibodies responsible for lysis of several murine lymphomas, these studies have revealed a previously unencountered complexity in the relationship between susceptibility to serum and to NK cells. Firstly, as is detailed above, the antibodies responsible for lysing NK susceptible target cells are not of a single specificity. Secondly, some cells, notably the YAC-1 lymphoma cell line grown in this laboratory, was highly susceptible to NK cells, but was unaffected by normal rabbit serum.

Hence the carbohydrate coat "worn" by a cell is extremely important in determining the sensitivity of the cell to the antibodies in normal rabbit serum. The significance of that coat in determining the sensitivity of a cell to NK cell mediated lysis is much less clear. Certainly the reported correlation between sensitivity to normal rabbit serum and to NK cells is a tenuous one and does not hold for all cells, or indeed apparently for the same cell line propagated in different laboratories.

B. Blocking of NK mediated lysis by monoclonal antibodies

An alternative approach that we have taken to probe the nature of the NK target structure has been to prepare monoclonal antibodies

Table III. Inhibition of cell mediated lysis by monoclonal antibody E

	% Specific ^{51}Cr release[c]		
Effector cell	Target cell	No antibody	Antibody 1:16
NK[a]	27v-1C2	42.9	12.3(71)[d]
NK	YAC-1	60.7	64.6
CTL[b]	27v-1C2	60.4	6.4(89)
CTL	P815	69.1	62.7

(a) Spleen cells from CBA mice injected 18 hours previously with 100 µg poly I-C were used as NK effector cells. The E/T ratio used was 40:1.

(b) CTL specific for H-2^d were generated by in vitro mixed lymphocyte culture with C57Bl/6 as responders and DBA/2 spleen cells as stimulators at a ratio of 1:1. The effector to target cell ratio used was 10:1.

(c) A 4 hour ^{51}Cr release assay was used and antibody was present for the whole assay. Monoclonal antibody E was the ascites fluid from nude mice that had been inoculated with the E hybridoma. This fluid was heat inactivated (56°, 30 min) before use in the assay. 50% inhibition of NK-mediated lysis was seen at an antibody dilution of 1:1600.

(d) The numbers in parentheses indicate the percent inhibition of lysis by antibody.

directed against 27v and to screen for those that block the NK mediated lysis of the target cell. Here too, the 27v and 27av lines have proven useful, for antibodies can be screened for their ability to bind to 27v and not to 27av. Antibodies that demonstrated this specificity were subsequently examined for their ability to block the NK mediated lysis of 27v.

One hybridoma supernate was found that inhibited the lysis of 27v by NK spleen cells (Table III). Cloning of this hybridoma culture produced several hybridomas that secreted an antibody, termed E, that blocked the NK-mediated lysis of 27v. The monoclonal antibody secreted by these clones was characterized as an IgM antibody by serology and by its exclusion with the void volume on a Sephacryl S-300 column. Blocking was demonstrated to be dependant on the pentameric nature of the antibody, since reduction and alkylation of the molecule eliminated the ability of the antibody to block the NK cell mediated lysis of 27v cells.

Our excitement at the discovery of this antibody was lessened however when we observed, as shown in Table III, that, 1) it only blocked the NK mediated lysis of 27v; and not the NK cell mediated lysis of other cells and, 2) it also blocked the CTL mediated lysis of this cell line. This pattern of specificity, binding only to 27v and not to 27av or YAC-1, suggested that the antibody might be directed to Gg_3cer, a molecule that we had already shown to be expressed on 27v but not on the other two cells.

This possibility was tested by absorption of hybridoma supernates onto guinea pig red blood cells. Absorption onto sheep red blood cells, which do not express Gg_3cer was used as a control. All blocking activity was eliminated after absorption of culture supernatants onto guinea pig red blood cells, but not by absorption onto sheep red blood cells. The results of this initial experiment, were confirmed by demonstrating that the monoclonal E reacted only with Gg_3cer when tested by gel diffusion, complement fixation and the inhibition of guinea pig red blood cell agglutination against a battery of purified glycolipids.

These results were intriguing to us, because we had previously reported (6) that another monoclonal antibody directed against Gg_3cer, termed 2C2, failed to block the NK mediated lysis of 27v. Comparison of the fine specificities of these two monoclonals suggested that the major difference between the two IgM antibodies was one of avidity: 1) using a series of Gg_3cer derivatives previously used by Young _et al_. in the characterization of 2C2 (7) we found that, E fixed complement with a much broader range of Gg_3cer analogues (8). Furthermore, the hemagglutination of guinea pig red blood cells by 2C2 could be inhibited by 12 mM N-acetyl galactosamine, the immunodominant sugar on Gg_3cer, whereas hemagglutination by E was not inhibited by concentrations of N-acetyl galactosamine approaching 100 mM. Finally, the hemagglutination of guinea pig red blood cells by 2C2 was more temperature sensitive than was agglutination caused by E. The hemagglutinin titer of 2C2 decreased more than two-fold when the assay was performed at room temperature compared to when the assay was performed at 4°C. On the other hand the titer of E was not affected by such a temperature shift.

Interestingly, it seems likely that the inhibition of NK cell-mediated lysis by E was due to interference with some stage after the binding fo NK cells to its target. This conclusion was suggested by the observation that cells precoated with E antibody were still effective competitive inhibitors of the lysis of ^{51}Cr-labeled 27v cells. These results suggested that antibody to Gg_3cer did not prevent the binding of target cell to effector cell, but still prevented the lysis of the bound cell. We are therefore hopeful that prospectively, E will prove a useful tool in further dissecting the mechanism of NK cell-mediated lysis.

The results from both of the serological approaches outlined above have intimated a central role for gangliotriosylceramide in explaining the phenomena under study. Interestingly, one of the first biochemical differences that we detected between 27v and 27av was the presence of this glycolipid on the NK susceptible 27v variant and its complete absence from 27av (6). However, all of our attempts to conclusively demonstrate a function for Gg_3cer in target cell recognition by NK cells have proven unsuccessful: 1) subclones of 27v that varied tenfold in the expression of Gg_3cer were all lysed to an equivalent extent by NK cells (6); 2) direct implantation of Gg_3cer into 27av cells, demonstrated by uptake of tritium labeled Gg_3cer and acquisition of reactivity to anti-Gg_3cer, did not result in susceptibility to NK cells (8), and 3) we have recently found other tumor cells that express this glycolipid (R1.1 and RIE/TL8X.1) but that are insusceptible to NK cell mediated lysis and fail to inhibit the lysis of NK-susceptible target cells. We are thus forced to conclude that Gg_3cer is a molecule important to understanding the biology of the 27v cell, but it is not a universal NK target structure. Our finding however that a monoclonal antibody to this molecule can block CTL and NK mediated lysis may prove useful for the elucidation of the mechanism of cell mediated lysis.

REFERENCES

1. Durdik, J., Beck, B.N., Henney, C.S., in: Natural cell-mediated Immunity Against Tumors. Herberman, R.B. (ed), New York, Academic Press, pp 805-817, 1980.
2. Durdik, J., Beck, B.N., Clark, E.A., Henney, C.S., J. Immunol. 125:683 (1980).
3. Gronberg, A., Hansson, M., Kiessling, R., Andersson, B., Karre, K., Roder, J., J. Nat'l Cancer Inst. 64:1113 (1980).
4. Urdal D.L., Henney C.S., Molecular Immunol. In press, 1981.

5. Seyama, Y. and Yamakawa, T., J. Biochem. (Tokyo) 75:837 (1974).
6. Young, W., Durdik, J., Urdal, D.L., Hakomori, S-I, Henney, C.S., J. Immunol. 126:1 (1981).
7. Young, W., MacDonald, E., Nowinski, R., Hakomori, S-I., J. Exp. Med. 150:1008 (1979).
8. Urdal, D.L., Kawase, I., and Henney, C.S., Cancer Metastasis Rev. vol. 1, in press, 1982.

SURFACE SIALIC ACID OF TUMOR CELLS INVERSELY CORRELATES WITH SUSCEPTIBILITY TO NATURAL KILLER CELL MEDIATED LYSIS[1]

Ganesa Yogeeswaran

Department of Microbiology, Boston University
School of Medicine, Boston, Massachusetts

Raymond M. Welsh

Department of Pathology, University of Massachusetts
Medical School, Worcester, Massachusetts

Alvar Gronberg, Rolf Kiessling, Manuel Patarroyo,
George Klein

Department of Tumor Biology, Karolinska Institute
Stockholm, Sweden

Magnus Gidlund, Hans Wigzell, Kenneth Nilsson

Departments of Immunology and Pathology,
Uppsala, Sweden

Sialic acid is bound to terminal galactose and/or N-acetyl galactosamine residues in the oligosaccharide chains of glycoproteins (GP) and certain glycosphingolipids (GSL). Over 60% of sialic acid bound to cells is localized at the cell surface and contributes to the net negative charge (Warren, 1976). Several studies in the past, comparing transformed cells with non-transformed cells, have suggested that the elevated levels of sialic acid in the transformants (Yogeeswaran, 1980) might play a role in decreasing the antigenicity (Ray, 1977), increasing intravenous survival (Weiss *et al.*, 1974), invasiveness (Yarnell and Ambrose, 1969) and metastatic potential (Yogeeswaran and Salk, 1981). Since natural killer (NK) cells are thought to inhibit metastases (Talmadge *et al.*, 1980), we have conducted three corre-

[1]Supported by NIH grants CA19312, AI12438, CA26782, CA-26752, and a grant from Swedish Cancer Society.

ISBN 0-12-341360-5

lative studies of cell surface sialic acid and NK sensitivity of tumor target cells. In this chapter we will present data on 1) the level of sialic acid in related NK sensitive and insensitive YAC-1 lymphoma variant lines, 2) the changes in cell surface sialic acid accompanying interferon (IFN)-mediated protection of mouse target cells against NK lysis and 3) the changes in surface sialic acid that accompany phorbol ester mediated suppression or augmentation of NK-sensitivity of human tumor cell lines. The significance of the results of these series of experiments is discussed in the end.

The cell surface sialic acid was analyzed by metabolic labeling of cells with ^{3}H-N-acetyl mannosamine (ManNAc) which specifically labels sialic acid, followed by neuraminidase (NANase) hydrolysis (Yogeeswaran _et al._, 1981), or analyzed by periodate oxidation at 0-4°C (in the absence of anion transport, periodate ion oxidizes surface sialic acid)

Table I. Comparison of NK-susceptibility and NANase Releasable Sialic Acid of YAC Lymphoma Variant Lines

Cell Lines	NK-Susceptibility % Lysis[a]	% NANase Releasable Sialic Acid[b]
YAC-1	31.7	2.0
YAC-IR	33.8	4.2
YAC-ACA8	15.8	9.8
YAC-ACC8	12.6	10.5
YAC-8	0.4	19.0
A9HT	1.0	12.6
A9HT/YAC-IR	3.8	13.9
YAC anti-A7	4.3	16.5
YAC anti-A11	4.6	14.6

[a]A standard 4 hr NK assay was performed with C3H spleen cells as effector(endogenous NK cells). Values represent an average of triplicates obtained at 100:1 E/T ratio, but 3 ratios were used during the experiment.

[b]Percent NANase releasable sialic acid of total cell surface was calculated as a percent of radioactivity released by _Vibrio cholerae_ NANase from ManNAc labeled whole cells: [(C-N)/C] X 100, where C=cpm present in control cells and N=cpm in NANase treated cells. The results are an average of duplicates.

followed by tritiation by NaB^3H_4 (Gahmberg and Andersson, 1977).

In the first set of studies, we compared the cell surface sialic acid in a series of variants derived from the highly NK-sensitive YAC-1 lymphoma. The variants, which had widely diverging sensitivities of NK cells (Table I), were obtained by a number of methods, including selection in the presence of NK cells, antibody to H-2 or antibody to MuLV induced antigen and by fusion of sensitive cells to NK-resistant cells. The sensitivity of these cells to NK lysis did not correlate with their sensitivities to anti-H-2^a cytotoxic T cells. While no correlation could be made between the NK-sensitivity of these variants and total sialic acid (Yogeeswaran et al., 1981), a significant inverse correlation ($r=-0.88$, $p<0.001$) was observed between levels of percentage NANase releasable sialic acid of total labeled sialyl components (Table I) and NK sensitivity. Both cell surface sialyl-GP and sialyl-GSL showed similar inverse correlations to NK sensitivity ($r=-0.87$ and $r=-0.74$).

Table II. Comparison of NK susceptibility and NANase Releasable Sialic acid of Control and IFN-Treated Mouse Tumor Cells

Cell Lines	NK-Susceptibility % Lysis		% NANase Releasable Sialic Acid	
	Control	IFN[a]	Control	IFN[a]
L929 (sarcoma)	30.0	5.6	12.1	19.2
YAC-1 (Lymphoma)	73.4	45.8	17.8	48.5
P52 (Lymphoma)	9.7	6.1	15.8	56.7
L1210 (Leukemia)	6.1	3.6	8.0	55.0

[a]Cells were grown in the presence of 10^3U/ml Type 1 IFN (Calbiochem, CA) or in the absence (control) for 24 hr, and labeled in the same medium with 2 μCi per ml for the next 24 hr. A 4 hr NK assay was performed using LCMV induced effectors for L929 and Pichindei virus-induced C3H spleen (i.e., NK) cells for YAC-1, P52 and L1210 cells. %lysis at 150:1 E/T.

In a second series of experiments we have correlated the changes in NANase releasable sialic acid of mouse sarcoma, lymphoma and leukemic target cells which were rendered NK-resistant following treatment with IFN. As seen in Table II L929 target cells decreased in their NK sensitivity several-fold accompanied by a significant increase in sialic acid. YAC-1 and P52 lymphomas and L1210 leukemic target cells showed similar inverse correlations between NK sensitivity and NANase releasable sialic acid. In all the systems examined, we found an IFN-dose dependent increase in NANase releasable sialic acid following the protection of target cells from NK lysis by IFN treatment. However, we have not studied the time kinetics of changes in NK sensitivity and sialic acid during IFN treatment.

In a third series of experiments we have examined the changes in cell surface sialic acid of several human tumor cell lines following the treatment with 12-0-tetradecanoyl phorbol-13-acetate (TPA) which mediated an induction of certain differentiated functions and changed their NK sensitivity. Again we found a striking inverse correlation between surface sialic acid and NK sensitivity (Table III). In B-cell lymphoma (Daudi) and T-cell lymphoma (Molt-4) and a human neuroblastoma cell line, TPA treatment resulted in marked increase in NK sensitivity accompanied by a significant decrease in surface sialic acid. In a human histiocytic lymphoma cell line (U937), TPA treatment resulted in a decrease in NK sensitivity accompanied by an increase in surface sialic acid.

Thus we found a consistent inverse correlation between surface sialic acid and NK sensitivity in all the three series of experiments using several cell lines. In order to test the functional significance of elevated surface sialic acid in increased resistance of target cells to NK lysis, we treated highly sialylated target cells with NANase and tested their NK sensitivity as compared to untreated control cells. NANase treatment of insensitive YAC-variants or IFN protected targets or phorbol treated target cells caused a moderate increase in their sensitivity (Yogeeswaran _et al._, 1981). This approach has a potential problem in the regeneration of sialic acid following its removal and certain sialyl groups are resistant to cleavage by bacterial enzyme due to structural reasons (e.g., G_{M2} and G_{M1}). The results therefore suggest that cell surface sialic acid plays a partial role in reducing NK sensitivity and other factors, such as a lack of target structure also may contribute to resistance. Alternatively, other phenomena such as the ability of target cells to repair their membranes or to activate NK cells _in vitro_ may play a role in determining their sensitivity

Table III. Comparison of NK-susceptibility and Mild Periodate Oxidation-Tritiation (sialic acid labeling) of Control and TPA Treated Human Cell Lines

Cell Lines	NK-Susceptibility % Lysis[a] Control	TPA[b]	Periodate Labeling[c] (T-C) $1x10^{-6}$cpm/mg Protein Control	TPA[b]
Daudi (B-Lymphoma)	18.7	41.2	8.2	4.6
Molt-4 (T-Lymphoma)	21.7	59.4	11.6	5.7
U937-parent (Hist. Lymph.)	42.5	0.2	5.1	9.1
SH-SY-SY (Neuroblastoma)	28.5	55.0	9.1	5.3

[a]A 4 hr NK assay using human peripheral blood mononuclear cells were used as effectors for U937 and nueroblastoma at 100:1 E/T. Nylon wool non-adherent cells were used as effectors for Daudi and Molt cells at 100:1 E/T in a 5 hr assay.

[b]Daudi and Molt cells were grown in control or $8x10^{-8}$M TPA for 48 hours and taken for the different assays. U937 cells were grown in $1.6x10^{-7}$M TPA for 4 days and SH-SY-SY cells were grown for 12 days using the same concentration of TPA.

[c]$2x10^{7}$ cells were reduced in the presence of 1 mg/ml fresh $NaBH_4$ in medium for 10 min. and divided into two aliquots. One portion of cells was oxidized at 0-4°C with 0.5 mM $NaIO_4$ in balanced salt solution for 30 min. Another aliquot was kept as control without periodate. Cells were washed in ice cold medium and reduced with 50 µCi NaB^3H_4 in 100 µl for 15 min. Cells were washed 3 times and TCA precipitated 3 times and solubilized in 5% SDS/0.1N NaOH and taken for protein and radioactivity determinations. Specific periodate labeling = Treated minus control cpm.

(Kunkel and Walsh, 1981). Therefore, we conclude that elevated cell surface sialic acid is a common phenotype of target cells that are resistant to NK cells. The functional significance of this carbohydrate moiety in NK-target interactions still needs to be resolved.

REFERENCES

Gahmberg, C.G., Andersson, L.C. (1977) J. Biol. Chem. 252:5888.
Kunkel, L.A., Welsh, R.M. (1981) Int. J. Cancer 27:73.
Ray, P.K. (1977) Adv. Appl. Microbiol. 21:227.
Talmadge, J.E., Meyers, K.M., Prieur, D.J., Starkey, J.R. (1980) J. Nat. Cancer Inst. 65:929.
Warren, L. (1976) in "Biological Roles of Sialic Acid" (Rosenberg, A. and Schengrund, C.-L., eds.) p. 103, Plenum Press, New York.
Weiss, L., Glaves, D., Waite, D.A. (1974) Int. J. Cancer 13:850.
Yarnell, M.M. and Ambrose, E.J. (1969) Eur. J. Cancer 5:265.
Yogeeswaran, G. (1980) in "Cancer Markers: Diagnostic and Developmental Significance" (Sell, S., ed.), p. 371, Humana Press, Clifton, New Jersey.
Yogeeswaran, G. and Salk, P.L. (1981) Science 212:1514.
Yogeeswaran, G., Gronberg, A., Hansson, M., Dalianis, T., Kiessling, R., Welsh, R.M. (1981) Int. J. Cancer 28:517.

INHIBITION OF SPONTANEOUS AND ANTIBODY-DEPENDENT CELLULAR CYTOTOXICITY USING MONO- AND OLIGOSACCHARIDES

Peter Kaudewitz[1]

Hans Werner (Löms) Ziegler[1]

Gerd R. Pape[1,2]

Gert Riethmüller[1]

[1]Institute for Immunology
University of Munich

[2]II. Med. Clinic Grosshadern
University of Munich

INTRODUCTION

In order to elucidate the as yet unsolved question of the target recognition system of natural killer (NK) cells, several investigators have employed various mono- and oligosaccharides for inhibition of murine and human NK cells (Stutman et al., 1980; MacDermott et al., 1981). Mouse cytotoxic T-lymphocytes (CTL) were found not to be affected under conditions that exerted strong inhibitory effects on NK cells (Stutman et al., 1980). Furthermore antibody-dependent cell-mediated cytotoxicity (ADCC) against K562 and Chang tumor cells was not inhibited under these conditions (MacDermott et al., 1981). The reduction of NK cell activity in the presence of a variety of sugars was interpreted as competitive blockade of a lectin-like recognition unit on the effector cell (Stutman et al., 1980; MacDermott et al., 1981; Ades et al., 1981). In view of the rather large number of various saccharides inhibiting lysis of a single target, the question of specificity remained open.

ISBN 0-12-341360-5

In order to detect nonspecific inhibitory mechanisms, we monitored the osmolalities of the single saccharides. Only saccharides of comparable osmotic effects were chosen for the final experiments. Variations on the effector cell level were controlled by using batches of cryopreserved lymphocytes of single donors.

Since all previous investigators stressed the point of NK-related recognition, we tested in parallel the effects of saccharides on ADCC as a cytotoxic system of which the primary recognition is deemed to be mediated by antibodies. CTLs, sensitized against allogeneic lymphocytes, were used as a further control.

METHODS

Peripheral blood mononuclear cells were isolated by Ficoll-Urovison centrifugation according to the method of Böyum (1968) and frozen under controlled conditions in the presence of dimethylsulfoxide using a PTC 200 machine (Planer Ltd., Sunberry-on-Thames, Great Britain). Aliquots of cells were kept in liquid nitrogen until use.

K562 leukemia cells, P815 mouse mastocytoma cells and WEHI 164 mouse fibrosarcoma cells (Röllinghoff, Warner, 1973) were labeled with ^{51}Chromium for 1 h and used at 5×10^3 (K562) or 10^4 (WEHI164, P815) cells per well as targets in a 5 h release assay. As antibody in the ADCC system, we used rabbit anti-P815 immunoglobulin fraction (kindly provided by R. Bolhuis, Rijswijk, The Netherlands).

Cytotoxic T-lymphocytes were generated by in vitro stimulation of 5×10^6 mononuclear cells (A) with 5×10^6 irradiated allogeneic mononuclear cells (B) for six days. B-targets were cultured for the same time without stimulation.

Osmolality was measured with a Knaur Halbmicroosmometer.

Stock solutions of sugars were used at final concentration of 35 and 70 mM.

Regression analysis was employed to define straight lines representing dose response curves of six effector cell titrations. In all experiments presented correlation coefficients were significant at $p < 0.05$. Lytic units (LU) representing the number of effector cells required to mediate 20 % chromium release in the NK and ADCC experiments and 40 % in the CML experiments were calculated from these lines.

EFFECTS OF SACCHARIDES ON SPONTANEOUS AND ANTIBODY DEPENDENT CELL MEDIATED CYTOTOXICITY

Table I demonstrated that the osmolalities of the single saccharides used in our experiments are similar, but all were used under hyperosmotic conditions.

TABLE I Osmolality of different sugar solutions (m-osmol/kg)

Sugar	35 mM	70 mM
None	295	295
Mannose	324	350
Methyl-Mannoside	325	352
N-Acetyl-D-Glucosamine	324	362
Raffinose	328	356

Inhibition studies using liquid nitrogen stored lymphocytes from one individual (donor W.) as effectors versus K562 are summarized in table II.

TABLE II Effects of sugars on natural cytotoxicity against K562 target cells

	LU per 10^6 cells at different sugar concentrations	
Sugar	35 mM	70 mM
None	10.6±2.3* (6)	10.6±2.3 (6)
Mannose	9.2±3.3 (4)	5.6±1.1 (6)
Methyl-Mannoside	4.3±1.1 (6)	3.3±1.2 (6)
N-Acetyl-D-Glucosamine	3.3±0.4 (4)	1.8±0.3 (4)
Raffinose	3.0±1.5 (4)	2.0±0.7 (4)

*Mean of the number of experiments (in parentheses)

± standard deviation

The concentrations of 35 and 70 mM were chosen since higher concentrations resulted in osmolalities higher than 400 mOsm. As can be seen from table II, mannose at a concentration of 70 mM inhibited the natural cytotoxicity against K562 targets by 47 % of the control values, whereas methyl-mannoside and raffinose at 70 mM were consistently more effective in all experiments. N-acetyl-D-glucosamine was even more inhibitory (83 % inhibition). At 35 mM the degree of inhibition was smaller but the pattern remained the same.

As a rather insensitive target for NK cells, the WEHI fibrosarcoma cells were used after pretreatment with Actinomycin D (Ziegler et al., in prep.). In this system the effector cells were activated by Interferon-ß for 2 h. As shown in table III for two saccharides, a similar degree of inhibition was found. Mannose inhibited by 42 % and raffinose by 41 %.

TABLE III Effects of sugars on spontaneous cytotoxicity against WEHI164 target cells

Sugar	LU per 10^6 effector cells at 70 mM sugar concentration
None	18.1
Mannose	10.5
Raffinose	10.6

To assess the inhibitory effect of saccharides on another cytotoxic system, using a presumed different recognition mechanism than NK cell-mediated lysis, antibody-dependent cell-mediated cytotoxicity was assayed in the presence of various sugars. Representative data of experiments performed in parallel with those described in table II are given in table IV. Effector cells of donor W. and of another donor (A.) were used. In contrast to the findings of MacDermott in a different system (Mac Dermott et al., 1981), ADCC against P815 target cells using a rabbit anti-mouse antiserum was inhibited like the NK activity. In the experiments presented, however, the inhibition by mannose was less than the inhibition seen in the NK system. In a further series of experiments using effector cells of several different donors, ADCC was likewise inhibited by the applied saccharides.

In order to study possible effects of saccharides on the Fcγ-receptor, we examined rosette formation of T-lymphocytes and K562 tumor cells with IgG-coated bovine erythrocytes. Up to a concentration of 100 mM mannose, rosette formation by both types of Fcγ -receptor bearing cells was not impaired.

TABLE IV Effects of sugars on ADCC against P815 target cells

	LU per 10^6 effector cells at different sugar concentrations			
	(A)[1]	(W)[2]	(A)	(W)
Sugars	35 mM		70 mM	
None	10.8	10.4	10.8	10.4
Mannose	10.8	-	9.5	7.1
Methyl-Mannoside	5.4	7.4	4.8	5.9
Raffinose	6.4	-	2.1	-
N-Acetyl-D-Glucosamine	4.5	-	3.5	-

[1] Donor A. [2] Donor W.

In contrast to NK and ADCC, allostimulated CTL of two different donors were not inhibited even at 100 mM sugar concentration (mannose):

TABLE V Effect of 100 mM Mannose on CML

	LU per 10^6 effector cells	
	Donor S.	Donor M.
No sugar	37.5	32.3
Mannose	35.9	35.7

CONCLUDING REMARKS

From a panel of 10 inhibitory saccharides four (mannose,methyl-mannoside, raffinose, and N-acetyl-D-glucosamine) were studied in more detail. Using liquid nitrogen stored effector cells, we found that natural cytotoxicity against K562 target cells and Actinomycin D-pretreated WEHI tumor cells was inhibited by these saccharides. When a single donor of effector cells was used a reproducible pattern of inhibition was found, e. g. raffinose had a greater inhibitory effect than mannose for the donor tested. In addition to natural cytotoxicity, antibody-dependent cell-mediated cytotoxicity was inhibited to a similar degree. In contrast allospecific CTL were resistant to inhibition.

The conclusions as to the NK recognition mechanism reached by others on the basis of similar experiments are in agreement with our results. However, ADCC in the P815 system is inhibited to a similar degree under comparable conditions. This fact needs to be further analysed because the rabbit antiserum used may by chance be directed against similar carbohydrate determinants as the saccharides inhibiting the cytotoxic reaction. The parallel inhibition of NK and ADCC of a single donor may however be taken as a reflection of the close relationship of NK and K cells belonging to the same subpopulation of lymphocytes.

ACKNOWLEDGEMENTS

The authors acknowledge the expert technical assistance of B. Seidel and K. Haynes, the secretarial help of S. Förster and the advice on statistical procedures by A. König.

REFERENCES

Ades, E.W. Hinson, A., and Decker, J.M., Immunobiol. 160:248 (1981)

Böyum, A., Scand. J. Clin. Lab. Invest. 21:77 (1968)

MacDermott, R.P., Kienker, L.J., Bertowich, M.J. and Muchmore, A.V., Immunology 44:143 (1981)

Röllinghoff, M., Warner, N.L., Proc. Soc. Exp. Biol. Med. 144: 813 (1973)

Stutman, O., Dien, P., Wisun, R.E., Lattime, C., Proc. Natl. Acad. Sci. USA 77:2895 (1980).

THE USE OF A MONOCLONAL ANTIBODY TO ANALYSE HUMAN NK CELL FUNCTION

Walter Newman

Program in Basic Immunology
Fred Hutchinson Cancer Research Center
and Department of Microbiology and Immunology
University of Washington
Seattle, Washington

INTRODUCTION

A major unresolved isssue in the study of natural killer (NK) cells is the specificity of the effector target cell interaction. Thus far, neither the nature of the NK cell receptor nor the corresponding target cell antigen has been elucidated. Given the mounting evidence from murine studies that NK cells may inhibit establishment of tumors (Haller, et al., 1977; Collins, et al., 1981; Kasai, et al., 1979) and that cells uniquely resistant to NK lysis are more tumorigenic than their NK susceptible counterparts (Collins, et al., 1981; Kawase et al., 1981; Minato, et al., 1979), the specificity and nature of the NK tumor target interaction becomes of special interest. We have approached the issue of human NK cell recognition capability by development and analysis of murine monoclonal antibodies which block human NK cell cytolysis. As a first order approximation, such reagents may be instructive as to the nature of the recognition and/or lytic process performed by these cells.

Antibody 13.1 has been described previously, and is of interest because it inhibits NK cell lysis solely at the effector cell level and only for some NK sensitive targets and not others (Newman, 1981). Those targets whose lysis is inhibitable are primarily although not exclusively of the myeloid and erythroid lineage. Moreover, blocking

[1] This work was supported by NIH grant AI 16496 and NSF grant PCM 8008749.

ISBN 0-12-341360-5

is accomplished with nanogram quantities of antibody. In this report we detail the results of experiments designed to analyze the mechanism by which inhibition is accomplished. Our results show that antibody 13.1 must be added before or soon after effector-target cell interaction to be inhibitory, that effector-target cell conjugation is not affected, and that the 13.1 determinant on the 13.1 molecule is probably not an active recognition site, but may be linked to a recognition-like structure.

METHODS

A. Monoclonal Antibodies

Antibody 13.1 was derived from mice immunized with an enriched population of human NK cells. The nature of the 13.1 antigen has not yet been established, but it is trypsin sensitive and is present on all peripheral blood mononuclear cells. Antibody 13.1 is an IgG1 with a single light chain. Monoclonal antibody 34/28 is an IgG1 reactive with the 45,000 dalton chain of HLA ABC antigens (Truco, et al., 1979).

B. ^{51}Cr release assay

Effector PBL from normal individuals were isolated by centrifugation over Ficoll-Hypaque. Effectors and targets were mixed at different ratios and cultured for four hours, and the percent lysis was calculated according to the standard techniques as previously described (Fast, et al., 1981). Spontaneous release values for K562 targets were always 10% or less of total releasable counts.

C. Enrichment of NK Cells

Density gradient centrifugation of nylon wool non-adherent peripheral blood mononuclear cells was performed using 2.5% stepwise gradients of Percoll from 40-57%. Discrete bands of mononuclear cells from the two lightest density fractions contained the great majority of NK activity (Timonen, et al., 1981). These cells constituted 5% to 10% of all mononuclear cells applied to the Percoll gradient and were used in these experiments as the NK enriched fraction. Prior to use, they were washed three times in BME containing 10% calf serum.

D. Calculation of V_{max} and K_m Values

Ten X 10^4 effector PBL were incubated with decreasing twofold dilutions of target K562 cells, beginning at 10^5 targets per well (E:T = 1:1). Experiments were performed in the presence of either medium alone, 7 micrograms/ml or 7 nanograms/ml of antibody 13.1, or 6 X 10^4 or 1 X 10^4 unlabelled K562 targets. All experiments were performed in triplicate. Preliminary experiments established that the release of ^{51}Cr was linear with time until three hours at several E:T ratios. Therefore, all wells were harvested and counted at the three-hour time point. Percent specific lysis was converted to velocity (V) values by multiplication times the initial number of targets (T) added per well. Values for T/V versus T (Hanes-Woolf plot) were fitted to a straight line by linear regression analysis. The values for V_{max} were calculated from the slope ($\frac{1}{V_{max}}$) and the values for $-K_m$ were calculated from the x intercept. The correlation coefficient was always 0.98 or greater.

RESULTS

Table I shows the effects of antibody 13.1 on the lysis of K562, MOLT 4 and Chang plus antibody (ADCC) targets. Substantial inhibition is achieved by nanogram quantities of antibody. Previous experiments have shown that inhibition is accomplished by pre-treatment of effector NK cells but not by pre-treatment of targets (Newman, 1981). Also shown in Table I is the inability of 13.1, at 10 µg/ml, to block lysis of the NK sensitive T-leukemic cell line MOLT 4. To date, antibody 13.1 has failed to block the lysis of five other human NK sensitive T leukemia cell lines (Newman, 1981) over a range of antibody concentration and effector to target ratios. Table I also shows that no blocking of NK lysis of either K562 or MOLT 4 cells is accomplished by 10 micrograms/ml of anti-HLA antibody 34/28. This control establishes that mere binding of murine IgG1 antibody to NK cells is insufficient to block NK lysis. Also, no blocking of ADCC versus antibody coated Chang targets was observed.

TABLE I

Selective Blockade of Lysis by Monoclonal Antibody 13.1

	Percent lysis of Targets		
	K562	MOLT 4	Chang +Ab (ADCC)
medium alone	60	42	45
13.1, 10µg/ml	10	43	41
13.1, 10 ng/ml	15	--	42
34/28, 10µg/ml	59	54	--

Antibody or medium was added to PBL effector cells for one half hour at room temperature prior to addition of target cells. E:T ratio equals 100:1.

We first examined the effect of delayed addition of antibody to the ^{51}Cr release assay. Figure 1 shows that for maximum inhibition of NK lysis by 13.1 antibody to occur, the antibody must be added within the first 15 minutes after the admixture of effectors and targets. Some inhibition is achieved if the antibody is added within the first hour and one half of a four-hour chromium release assay, but by two hours, no further inhibition is achieved by addition of monoclonal reagent.

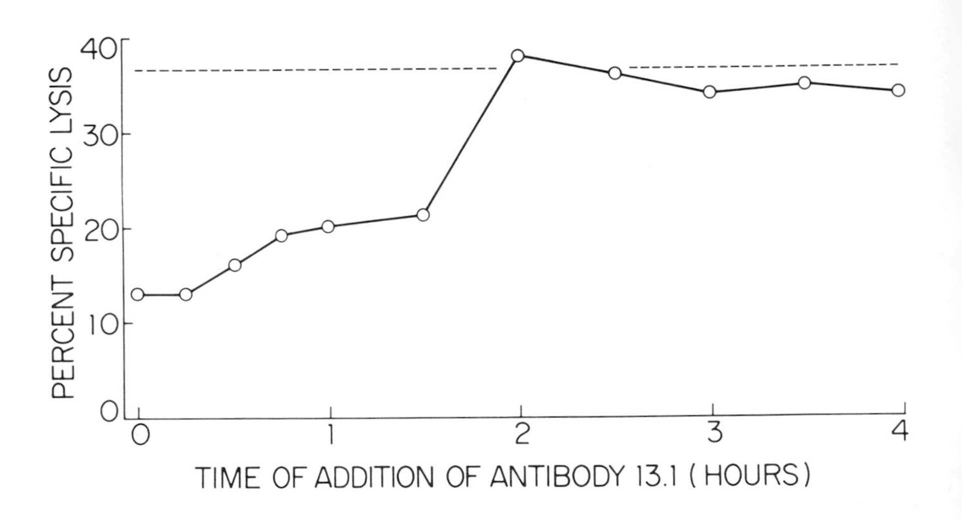

FIGURE 1. Effect of delayed addition of antibody 13.1 on NK lysis of K562 targets. Two µg antibody ( ) or medium ( ) was added at the indicated times either before (time 0) or after gentle centrifugation of effectors plus targets to initiate the lytic assay. All wells were harvested at 4 hours.

Hence, the dominant effect of antibody mediated inhibition occurs within the first few minutes of the chromium release assay. The antibody does not require 3-4 hours to exert its effects, as chromium release assays of short duration, sufficient to observe significant lysis (approximately 1/2 hour) also showed dramatic inhibition by antibody (data not shown). Hence, the antibody appears to inhibit at an early phase in the lytic cycle and is analogous to the results obtained by treatment of murine CTL with anti-Lyt-2 serum (Shinohara, et al., 1981).

To examine the blocking effect in more detail, populations of NK cells were enriched by density gradient centrifugation on Percoll and subjected to analysis of conjugate forming ability in the presence or absence of antibody 13.1. Results show that in medium alone, approximately 30% of the Percoll enriched cells formed conjugates with K562 targets, and that of these, 53% contained a dead target cell. When the assay is done in the presence of antibody 13.1, however, there is no increase or decrease in the percentage of conjugate forming cells. However, within these conjugates, only 2% dead targets are observed. No effect on conjugate formation *per se* was seen in this and five other experiments with the same results. Parallel experiments with Percoll enriched effector cells in a 4 hr assay showed dramatic inhibition of lysis by antibody 13.1, from 94% lysis to 34% lysis at an E:T ratio of 10:1.

In the final series of experiments, we took advantage of the analogy between NK cell lysis of targets and the enzyme substrate interaction to probe the mechanism whereby antibody inhibits NK lysis. Unlabelled targets are a control in these experiments, as one would predict that their mode of inhibition is competitive. The percent inhibition caused either by antibody or unlabeled targets was approximately the same. Hence any differences in the mode of inhibition by these two agents is not a result of substantial differences in the percent inhibition of lysis.

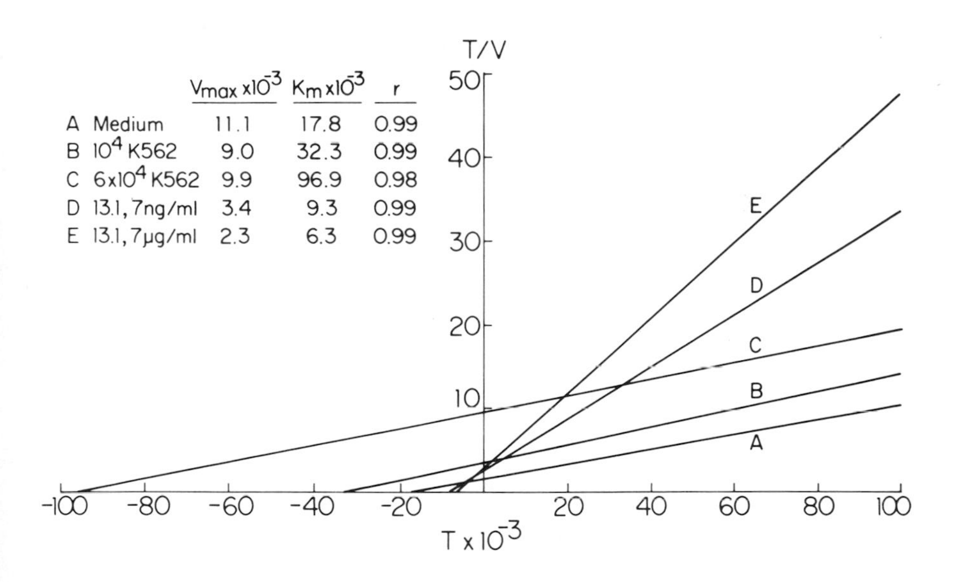

FIGURE 2. Linear regression analysis of NK cell assay in presence of medium (A), 10^4 K562(B), 6 X 10^4 K562(C), 7 ng/ml 13.1 (D) or 7 µg/ml antibody 13.1; n = 8.

The results presented in Figure 2 indeed confirm that for the NK-K562 system, unlabelled K562 targets behave as competitive inhibitors whether using 1 X 10^4 or 6 X 10^4 competitor cells. The lack of effect on V_{max} (I_{slope}) and the apparent increase in K_m (-x intercept) values is characteristic of this form of inhibition. However, on examination of the effects of 13.1 antibody at 7 micrograms/ml or 7 nanograms/ml, competitive inhibition was not seen.

Figure 2 presents a representative experiment which shows the antibody clearly reduced V_{max}, suggesting that, unlike the effect of unlabelled competitors, no additional numbers of labelled targets could compensate for antibody induced inhibition. In addition, there was usually a decrease in the K_m value, but this was not uniformly observed. It is more difficult to summarize the effects of K_m and to comment on its significance and hence, it is not possible to state unequivocally that the mode of inhibition of NK lysis by antibody 13.1 is either non- or un-competitive. It may in fact be neither of these. Additional analyses will be necessary to clarify this point. However, in all experiments we noticed a dramatic decrease in the V_{max} with no increase in K_m, and feel confident in concluding that inhibition is therefore not competitive and is achieved by some means distinct from that of unlabelled targets.

DISCUSSION

This report concerns the mechanism whereby nanogram quantities of monoclonal antibody 13.1 are capable of causing dramatic decreases in the ability of human NK cells to kill K562 tumor targets. The control performed with antibody 34/28 excludes that mere binding of murine IgG1 antibody to NK cells inhibits lysis. It has previously been shown that antibody 13.1 inhibits lysis by interaction with effector cells, not targets (Newman, 1981). To examine whether antibody 13.1 inhibited lysis at a discreet stage of the lytic cycle, we tested the effect of delayed addition of antibody to the ^{51}Cr release assay. This analysis is complicated by the ability of NK cells to recycle, and hence these cultures are not synchronized. However, it was possible to establish that late addition of antibody was without effect. For maximum inhibition to be achieved, antibody 13.1 must either be added to effectors before addition of targets, or within fifteen minutes after target cell addition. Addition of antibody at later times resulted in a progressive diminution in the inhibitory effect such that by 2 hours further addition of antibody was without effect.

Since it appeared, therefore, that antibody was acting at an early stage in the lytic cycle, we examined its effects upon conjugate formation. Since the frequency of conjugate forming cells in nylon-wool non-adherent populations exceeds the frequency of NK cells, that is lymphoid cells bound to dead K562 target cells, we used Percoll enriched populations of killer cells. This gave us greater assurance that effects caused by antibody were on NK cells and not on cells with mere conjugation ability. Our results showed that antibody 13.1 had no effect upon the frequency of conjugate forming cells. However, within those conjugates, only 4% dead targets were observed after one hour of incubation. In the absence of antibody, 53% of conjugates possessed dead K562 targets. These results are of interest because they demonstrate that conjugate formation per se is insufficient for lysis of target cells. Additional step(s) must occur, at least one of which is inhibited by antibody 13.1, for lysis to proceed. It is not yet possible to state whether this involves additional recognitive events which require target-effector proximity, or lytic events. It may be that mere conjugation lacks the "effective recognition" which can trigger the lytic event. Alternatively, NK cells may conjugate to targets via a recognition process distinct from the conjugation to K562 achieved by non-NK cells, and that it is this recognition which leads to effective lysis of targets. In this analysis conjugation assumes recognition, and implies that antibody 13.1 is interfering with some step in the programming for lysis or lytic stage. However, the requirement for early addition of antibody, the inability of antibody to block lysis of T lymhphoma targets, the inability to block antibody dependent lysis by ADCC effector cells, and experiments presented elsewhere which suggest antibody 13.1 can diminish the strength of the K562-NK bond (Newman, 1981), lead us to favor the hypothesis that antibody 13.1 interferes with a critical recognition process which can occur subsequent to conjugate formation, and not with the lytic machinery itself. Whether 13.1 reacts with such a recognition molecule at its combining site, or binds in a less direct manner was addressed in the kinetics analysis. Through use of the Michaelis-Menten enzyme-substrate-inhibitor equations, we have examined the mode of 13.1 mediated inhibition. It was evident that antibody 13.1 inhibited lysis in a manner distinct from unlabelled K562. Antibody 13.1 caused highly reproducible decreases in V_{max} values in a dose responsive manner, suggesting no excess numbers of labelled K562 targets could overcome the inhibition observed. The effects of antibody on K_m were somewhat variable and harder to interpret but never led to increases. Hence it is possible to conclude that the mode of inhibition is not strictly competitive, and it is unlikely therefore that antibody is binding to an active site of an NK recognition structure. However, should such a recognition structure exist for NK cells as is suggested by the selectivity of blocking, our results do not exclude that antibody is bound to another site on such a molecule. Our results would also be consistent with antibody binding to another cell surface molecule either covalently or non-covalently bound to such a recognition structure. Biochemical analysis of the 13.1 molecule should help elucidate the basis for the results reported here.

ACKNOWLEDGMENTS

I thank Geraldine Shu and Kevin Draves for excellent technical assistance.

REFERENCES

Collins, J.L., P.Q. Patek and and M. Cohn, J. Exp. Med. 153; 89-106 (1981).

Fast, L.D., J.A. Hansen and W. Newman, J. Immunol. 127; 448-452 (1981).

Haller, O., M. Hansson, R. Kiessling and H. Wigzell, Nature 270; 609-611 (1977).

Kasai, M., J.C. LeClure, L. McKay-Bourdreau, F.W. Shen and H. Cantor, J. Exp. Med. 149, 1260-1264 (1979).

Kawase, I., D.L. Urdal, C.G. Brooks and C.S. Henney, submitted for publication (1981).

Minato, N., B.R. Bloom, C. Jones, J. Holland and L. Reid, J. Exp. Med. 149; 1117-1133 (1979).

Newman, W., submitted for publication (1981).

Shinohara, N., M. Taniguchi and M. Kojima, J. Immunol. 127; 1575-1578 (1981).

Timonen, T., J.R. Ortaldo and R.B. Herberman (1981) J. Exp. Med. 153; 569-582.

Truco, M.M., G. Garotta, J.W. Stocker and R. Ceppellini, Immunol. Rev. 47; 219-252 (1979).

HETEROGENEITY OF MLC-GENERATED NK-LIKE CELLS[1]

Paula J. D'Amore[2,3]
Marc G. Golightly[3,4]
Sidney H. Golub[3,4]

Departments of Pathology[2], Surgery (Division of Oncology)[3]
and Microbiology and Immunology[4]
UCLA School of Medicine
Los Angeles, California

I. INTRODUCTION

Allogeneic mixed lymphocyte culture (MLC) has been shown to generate NK-like cells in addition to CTL. This *in vitro* generation of NK-like activity allows us to characterize NK-like effectors, trace their development and maturation and examine the cytotoxic specificity of these cells. In addition, a MLC provides an excellent model to compare the development and killing behavior of CTL and NK-like cells.

The present report is a brief summary of our work using the MLC system to investigate the following: a) the phenotype characterization of the effector cells responsible for MLC augmented NK-like cytotoxicity, b) whether natural killing is a function of the developing CTL and c) the cytotoxic specificity of NK-like cells.

II. KINETICS OF NK AND ALLOSPECIFIC KILLING

MLC were maintained for a period of one to seven days. The effector cells generated were tested in the single cell cytotoxicity assay to determine the kinetics of natural and allospecific killing. As shown in Table I, the maximum response to the NK-sensitive target, K562, was found on Day 3, with an increase in both the number of cells binding K562 and the num-

[1]This work was supported by USPHS grant CA17013 and fellowship support from training grant CA-9120 and CA 09120.

ISBN 0-12-341360-5

TABLE I. Binding and Killing Activity of MLC-generated NK-like and Alloreactive Cells

Days in MLC	% Binders[a] with: K562	Allospecific	Doubles (K562 & allospecific[b])	% Killers[a] with: K562	Allospecific
0	6.0	1.0	0	2.0	0
1	9.0	2.5	0	3.5	0.5
2	10.5	4.5	0	5.5	1.0
3	13.5	6.5	0	6.5	2.5
4	10.5	8.0	0.5	5.0	3.0
5	11.0	10.0	0	4.5	6.0
6	9.5	10.5	0	4.0	7.0
7	5.5	10.5	0	2.5	5.5

[a]Results are the average of 2 different slides scored and 300 PBL counted on each slide. Results are expressed as the % of total PBL. To determine the frequency of doubles, at least 200 binding cells were scored per slide.

[b]Allospecific cells are PBL of the original stimulator type. In each determination, allospecific cells were trypsinized and fluoresceinated with FDA.

ber of cells killing this NK-sensitive target.

To measure the allospecific response, PBL of the original stimulator type were used as targets in the single cell assay. Stimulator cells were first trypsinized to eliminate the ability of these target cells to bind to either responder cells or K562. The maximum allospecific response was found on Days 6 and 7. There was a progressive increase in the number of binders and killers of the alloantigen target (Table I).

III. NATURAL KILLING FUNCTION OF CTL?

MLC generates two cytotoxic events--natural and allospecific Killing. The question is whether MLC-induced natural killing is a separate cytotoxic activity executed by a subpopulation of cells distinct from the alloreactive T cells or whether NK and allospecific killing are mediated by the same effector cell population. Our approach to this problem was to determine whether MLC-generated effector cells could simultaneously bind both allospecific and NK-sensitive targets or whether separate populations were binding each target. MLC-generated cells were tested daily for 7 days in the single cell assay in which both targets (K562 and alloantigen) were used together. The alloantigen target cells (PBL of the stimulator type) were trypsinized to prevent their binding to other cells and also fluoresceinated to distinguish them from the MLC-generated effectors. During the 7 days of MLC, no simultaneous binding of both targets was seen (Table I). We have shown that this assay can detect cells capable of simultaneously binding two targets (e.g. CTL binding two targets bearing the same HLA determinants or NK cells binding two K562 cells) (D'Amore and Golub, submitted for publ.). If NK is a function of the developing alloreactive lymphocyte, there should be a point in time when a cell changes from NK reactivity to alloreactivity. For this reason, the MLC-generated cells were tested daily to assure us that there was no time when the lymphocytes would bind both targets concurrently. Our findings indicate that there are two separate cell populations mediating allospecific and NK-like cytotoxicity.

IV. CHARACTERIZATION OF CELLS MEDIATING MLC-INDUCED NK-LIKE CYTOTOXICITY

It is clear from the kinetic studies that MLC results in augmented anti-K562 cytotoxicity. We then investigated which cells were responsible for the enhanced NK-like cytotoxicity. While it is generally accepted that standard (not _in vitro_ stimulated) NK activity is mediated by lymphocytes with Fcγ receptors, other reports have indicated that cells with Fcμ receptors may contribute to NK activity (Merrill, Ullberg and Jondal, 1981; Jondal and Merrill, 1981). Therefore, we wanted to test the hypothesis that Fcγ and/or Fcμ receptor cells were involved in MLC-generated NK-like activity. As reported by Golightly _et al_. (Golightly, D'Amore and Golub, in press), a significant amount of the anti-K562 activity (as measured by ^{51}Cr-release) of MLC-generated cells was mediated by the Fcμ^{+} population. To confirm our finding that Fcμ^{+} cells can kill the NK-sensitive target, K562, we utilized a combined rosetting and single cell cytotoxicity assay. Fresh PBL, prior to MLC stimulation, contained less than 1% K562 binding cells that were also EAμ rosette forming (Table II). With time in MLC, the number of EAμ rosetting binders and EAμ rosetting killers increased with peak binding on Day 3 (7.3% of all PBL). In agreement with our ^{51}Cr-release data (Seeley and Golub, 1978), maximum numbers of killer cells were found on Days 3-5 (3.6%). The number of EAγ rosette forming K562 binders showed a slight increase on Day 3 (10.3%) while the number of EAγ rosette forming killers remained around 3% at all times. Thus our results at the single cell level confirm our findings

TABLE II. Single Cell Analysis of MLC-induced NK-like Cytotoxicity

Days in MLC	% Binders[a]			% Killers[a]		
	Total[b]	EAγ	EAμ	Total[b]	EAγ	EAμ
0	13	7.3 $\pm$0.3	0.67 $\pm$0.3	5	3.4 $\pm$0.3	0 $\pm$0.0
1	14	8.3 $\pm$0.3	1.7 $\pm$0.3	5	3.3 $\pm$0.3	0.67 $\pm$0.3
3	19	10.3 $\pm$0.3	5.7 $\pm$0.3	8	3.6 $\pm$0.3	3.6 $\pm$0.3
5	16	6.7 $\pm$0.3	7.3 $\pm$0.3		2.3 $\pm$0.3	3.6 $\pm$0.3

[a]Results are the average of 3 different slides scored with 300 lymphocytes counted on each slide. Results are expressed as % of total PBL $\pm$ standard error.

[b]Total % of K562 binding, or cytolytic, cells as determined in the single cell assay without added rosetting reagents. 300 PBL were scored as results are expressed as % of total PBL.

that during MLC the number of $Fc\gamma^+$ killer cells appear to be stable or declining, while the number of $Fc\mu^+$ killers increase.

V. HETEROGENEITY OF NK-LIKE CELLS

Does heterogeneity in cytotoxic specificity exist within the MLC-generated NK-like cell population whereby subpopulations recognize different target specificities or do all NK-like cells have the capability to react with more than one target type? By using the single cell assay in conjunction with two different NK-sensitive targets, we could distinguish between an effector cell binding to one target and an effector cell simultaneously binding two different NK targets. To compare the specificity of target binding prior to MLC stimulation and at the peak of NK-like activity, fresh PBL and Day 3 MLC cells were chosen as the effector cells. Targets of varying NK- sensitivity, including MOLT-4 (highly sensitive) and two moderately sensitive EBV-transformed B cell lines (NLA and L14) were used in conjunction with the highly sensitive

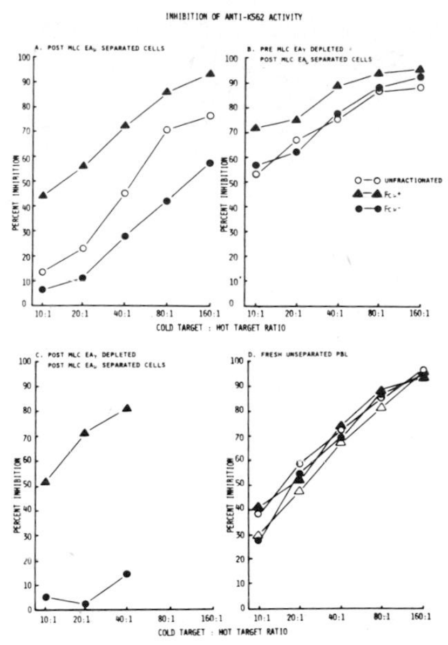

FIGURE 1. Inhibition of anti-K562 cytotoxicity of various cell populations by cold K562 target cells. EAμ separations were performed on MLC-generated cells in figures a-c; open circles indicate unfractionated PBL, closed triangles--Fcμ$^+$, and closed circles--Fcμ$^-$. (1a) EAμ separation post MLC. (1b) EAγ depletion prior to MLC and EAμ separation post MLC. (1c) EAγ depletion post MLC followed by EAμ separation. (1d) Results of four donors' fresh unseparated PBL.

K562. K562 was fluoresceinated to distinguish it from the other targets. We found that there was little or no simultaneous binding of K562 and one of the other targets by fresh PBL or Day 3 MLC cells (D'Amore and Golub, submitted for publ.). The failure to find double binders of two NK-sensitive targets argues for a clonal distribution of NK-like effectors with regard to various NK-sensitive targets. Double binding would be expected if there was a recognition structure common to many NK-sensitive targets or if NK cells possessed multiple receptors, each recognizing a unique specificity on the surface of the target cell.

In addition to the heterogeneity found among NK-like cells in regards to cytotoxic specificity, heterogeneity was found in the interaction of NK-like cells with one target type. When unfractionated, Fcμ$^+$, and Fcμ$^-$ MLC-generated cells were tested in cold target competition analyses, the anti-K562 activity of the Fcμ$^+$ population (Figure 1a) was readily inhibited by cold K562 at competitor doses similar to those seen with fresh PBL (Figure 1d). In contrast, the Fcμ$^-$ population was difficult to inhibit with homologous K562 targets and the inhibition of the unfractionated MLC cells by cold K562 was intermediate to that of the $Fc\mu^+$ and $Fc\mu^-$ cells (Figure 1a). It appears that there are at least two populations of MLC-generated NK-like cells ($Fc\mu^+$ and $Fc\mu^-$) with different patterns of inhibition by homologous targets. To further characterize the $Fc\mu^-$ population, MLC-generated cells were depleted of $Fc\gamma^+$ cells on an $EA\gamma$ monolayer and then separated into $Fc\mu^+$ and $Fc\mu^-$ cells. The $Fc\mu^-$ anti-K562 activity remained difficult to inhibit with cold K562 (Figure 1c), allowing us to classify these cells as $Fc\mu^-\gamma^-$. The lack of a difficult to inhibit $Fc\mu^-$ population, when PBL were depleted of $Fc\gamma^+$ cells on $EA\gamma$ monolayers prior to MLC stimulation (Figure 1b), identifies the precursor of these cells as $Fc\gamma^+$. In Figure 1b, the presence of the easily inhibited $Fc\mu^+$ population identifies one precursor of these cells to be $Fc\mu^-\gamma^-$. Therefore, MLC-generated anti-K562 effectors can be characterized as one of three types: $Fc\mu^+\gamma^-$, $Fc\mu^-\gamma^-$ and $Fc\mu^-\gamma^+$ derived from $Fc\mu^-\gamma^-$ and $Fc\mu^-\gamma^+$ precursors. The $Fc\mu^-\gamma^-$ effectors differ in that their interaction with a target leads to minimal inhibition with a homologous target.

VI. CONCLUSIONS

Our data suggests that NK-like cells generated in MLC represent a heterogeneous population. Among the cells responsible for the enhanced cytotoxic activity of NK-like effectors are $Fc\mu^+$ cells as well as $Fc\gamma^+$ cells. Our studies also indicate heterogeneity in cytotoxic specificity among NK cells, perhaps by clonally-distributed subpopulations, each recognizing a different target specificity. Even within one specificity, heterogeneity can be detected in terms of sensitivity to cold target competition with homologous targets. It has been proposed that NK is a function of the developing CTL (Klein, 1980). Our studies indicate, to the contrary, that MLC-generated NK is a separate cytotoxic activity executed by a subpopulation of cells distinct from alloreactive T cells.

REFERENCES

D'Amore, P.J. and Golub, S.H., submitted for publication.
Golightly, M.G., D'Amore, P.J., and Golub, S.H., Cell. Immunol., in press.
Jondal, M. and Merrill, J., Eur. J. Immunol., 11:531, 1981.
Klein, E., Immunol. Today, 1(6), iv, 1980.
Merrill, J., Ullberg, M., and Jondal, M., Eur. J. Immunol., 11:536, 1981.
Seeley, J.K. and Golub, S.H., J. Immunol., 120:1415, 1978.

STIMULATION OF LYMPHOCYTES WITH ALLOGENEIC NORMAL CELLS AND AUTOLOGOUS LYMPHOBLASTOID CELL LINES: DISTINCTION BETWEEN NK-LIKE CELLS AND CYTOTOXIC T LYMPHOCYTES BY MONOCLONAL ANTIBODIES[1]

Joyce M. Zarling[2]
Fritz H. Bach

Immunobiology Research Center
Departments of Laboratory Medicine/Pathology and Surgery
University of Minnesota
Minneapolis, Minnesota

Patrick C. Kung

Centocor, Inc.
Malvern, Pennsylvania

I. INTRODUCTION

Epstein-Barr virus (EBV), the causative agent of infectious mononucleosis and also probably of Burkitt's lymphoma (Klein, 1972) transforms B cells which can grow as lymphoblastoid cell lines (LCLs). Effector cells, lytic for autologous EB transformed LCLs, can be generated by stimulating lymphocytes with pooled allogeneic normal cells in mixed leukocyte cultures (MLC) (Zarling and Bach, 1978). Since effector cells lytic for NK-sensitive targets, including the

[1]This research was supported by NIH grants CA 26738 and CA 27826. This is paper #285 from the Immunobiology Research Center, University of Minnesota, Minneapolis, MN.
[2]J.M. Zarling is a Scholar of the Leukemia Society of America.

ISBN 0-12-341360-5

HLA- K562 cell line, are also generated in MLC (Seeley and Golub, 1977, 1978; Ortaldo and Bonnard, 1977; Zarling and Kung, 1980); it was not known whether lysis of the autologous LCLs was mediated by NK-like cells or cytotoxic T lymphocytes (CTLs). Effectors lytic for LCLs can also be generated by stimulating lymphocytes with autologous LCLs. Effectors from such cultures can, in addition, lyse certain cell lines lacking the EBV genome (Svedmyr, et al., 1974; Tanaka et al., 1980; Sugamura and Hinuma, 1980; Biron et al., 1981) suggesting that EB+ and EB- cell lines may share common recognition structures for the same effector cells or that different effector cells may be directed against the EB+ LCLs and the EB- NK-sensitive targets.

This chapter summarizes our recent findings that monoclonal antibodies, OKT8 or OKT3, are useful for distinguishing NK-like cells from CTLs generated following stimulation with allogeneic lymphocytes or autologous LCLs (Zarling and Kung, 1980; Zarling et al., 1981a,b).

II. METHODS

Effector cells were generated by culturing 8 x 10^6 responding lymphocytes with 8 x 10^6 x-irradiated (2500R) allogeneic normal cells pooled from 20 individuals (poolx), with 2 x 10^6 x-irradiated (4000R) LCLs (LCLx) or with 2 x 10^6 glutaraldehyde (0.001%) treated LCLs (LCLg). Effector cells, harvested on day 6 or 7, were treated with control ascites or monoclonal antibodies at a final dilution of 1:100 followed by treatment with non-toxic rabbit complement (C) as detailed (Zarling et al., 1981a,b). Dead cells were removed by Ficoll-hypaque gradient centrifugation and the recovered viable cells were tested at several effector:target (E:T) ratios in 6-7 hr ^{51}Cr release assays. Lytic units (LU)/10^6 cells and total LU recovered were calculated as described (Zarling and Kung, 1980). The antibodies used included OKT3 which reacts with 90-95% of peripheral sheep red blood cells (E) rosetting cells (Kung et al., 1979), OKT11A which reacts with 100% peripheral E rosetting cells (Verbi et al., in press) and OKT8 which reacts with the cytotoxic/suppressor cell population (Kung et al., 1980).

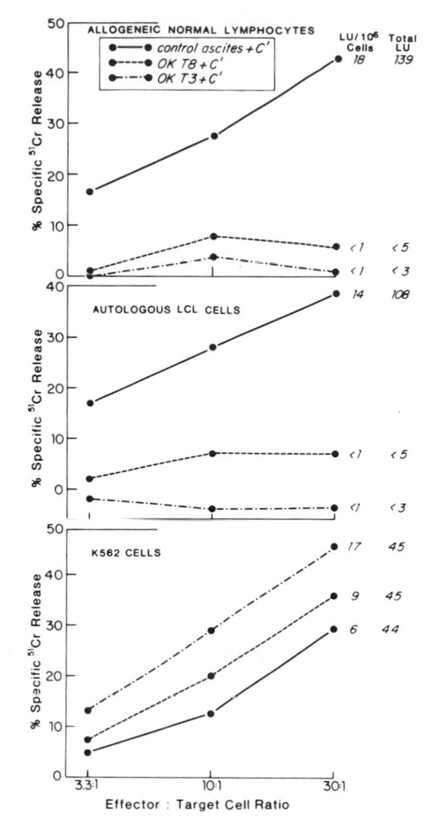

FIGURE 1. Effects of Treatment of Pool-stimulated Cells with OKT3 or OKT8 plus C on Lysis of Allogeneic Normal Cells, Autologous LCLs and K562.

Pool-stimulated cells were treated with monoclonal antibodies and C as detailed in the Methods section and lytic units (LU) were calculated as previously detailed (Zarling et al., 1981a).

III. DISTINCTION BETWEEN NK-LIKE CELLS AND CTLs GENERATED IN MLC

Results in Figure 1 demonstrate that treatment with OKT3 or OKT8 plus C reduces lysis of pool-sensitized cells against allogeneic normal lymphocytes as well as against autologous LCLs but that depletion of OKT3+ or OKT8+ cells results in increased lysis of K562 cells at equivalent E:T ratios, presumably due to enrichment for the OKT3- OKT8- NK-like cells. In this experiment and in several others (Zarling and Kung, 1980; Zarling et al., 1981a) the total LU of cytotoxic activity against allogeneic normal targets and autologous LCLs recovered after treatment of 1 x 10^7 cells with these antibodies were minimal whereas the LU of cytotoxic activity against K562 recovered following elimination of the OKT3+ or OKT8+ cells were usually not different from the LU recovered after treatment with control ascites and C. These results thus indicate that CTLs can be distinguished from NK-like cells by treatment with monoclonal antibodies OKT3 or OKT8 and C and that lysis of autologous EB transformed LCLs is mediated by effectors which are phenotypically CTLs.

IV. STIMULATION WITH AUTOLOGOUS LCLs LEADS TO EFFECTORS LYTIC FOR EB+ AND EB- TARGETS

Results in Table I represent our observations (Zarling et al., 1981b) that stimulation of lymphocytes from EBV seropositive individuals with autologous LCLs leads to the generation of effectors lytic for autologous and allogeneic LCLs as well as for EB- NK-sensitive targets including 2 T cell leukemia lines (HSB-2, MOLT-4) and K562. This apparent non-EB restricted killing has also been reported by others (Svedmyr et al., 1977; Viallat et al., 1978; Sugamura and Hinuma, 1980). Stimulation of lymphocytes from EB seronegative individuals with autologous LCLs gave rise to minimal, if any, cytotoxicity against autologous or allogeneic LCLs while the effectors were lytic for the EB- NK-sensitive

TABLE I. Lysability of EB+ and EB- cell lines by interferon-treated lymphocytes and LCL-stimulated lymphocytes[a]

Effector cells from EBV seropositive donor	% Specific ^{51}Cr release					
	EB+ LCLs			EB- targets		
	M	S	SB	HSB-2	MOLT-4	K562
M+IF	4.2	4.3	7.8	25.2	25.3	57.7
M+M LCLx	81.2	59.5	57.6	78.8	82.7	78.7

[a]Lymphocytes were cultured for 16 hr with 150 units/ml highly purified human fibroblast interferon (IF) (kindly provided by Dr. William Carter), or were stimulated for 6 days with or without x-irradiated autologous LCL. E:T ratio = 30:1. (Modified from Zarling et al., 1981b).

targets (Zarling et al., 1981b). Others have likewise found that when xenogeneic serum is excluded from culture, LCL-stimulated cells from seronegative donors do not lyse LCLs whereas EB- K562 cells are lysed (Tanaka et al., 1980; Misko et al., 1981). Thus, lysis of EB+ targets by LCL-stimulated cells may reflect a secondary CTL response in seropositive donors. Interferon (IF) is produced as a result of stimulation of lymphocytes with LCL (Biron et al., 1981) and the IF could activate NK-like cells to lyse the NK-sensitive EB- targets. Evidence to support this contention derives from findings that lymphocytes treated with IF are cytotoxic for the EB- targets HSB-2, MOLT 4 and K562 but the EB+ LCLs are highly refractory to lysis by these effectors (Table I).

V. DIFFERENTIAL EFFECTS OF TREATMENT WITH OKT8 AND C ON LYSIS OF EB+ AND EB- TARGETS BY LCL STIMULATED CELLS

More direct evidence that lysis of EB+ LCLs and EB- targets is mediated by CTLs and NK cells, respectively, derives from results of experiments where LCL-stimulated effectors were treated with OKT8 and C. Lysis of EB+ LCLs was markedly reduced whereas lysis of the EB- targets was not (Zarling et al., 1981b). Results of two experiments are shown in Table II and indicate that cytotoxic activity, in

TABLE II. Lytic activity against various EB+ and EB- cell lines recovered from LCL-stimulated effectors after depletion of OKT8+ cells

Lytic units recovered after treating 1 x 10^7 effectors[a]						
Effector cell treatment	EB+ LCLs			EB- targets		
	S	J	SB	HSB-2	K562	MOLT-4
A. C alone	250	250	182	111	125	188
OKT8 + C	<21	<21	25	80	133	179
	S			HSB-2	K562	MOLT-4
B. C alone	217			388	139	136
OKT8 + C	28			312	132	142
OKT11A + C	2			37	16	23

[a]LU recovered after treating 1 x 10^7 S's LCL-stimulated effectors was calculated as detailed (Zarling et al., 1981b).

terms of LU recovered against S's autologous LCLs following treatment with OKT8 and C was reduced 88 to >93% and lysis of the allogeneic LCLs (J's LCLs and SB) was reduced by more than 86%. In contrast, treatment with OKT8 and C either failed to reduce, or reduced minimally, the total number of LU recovered against HSB-2, MOLT-4 or K562. Treatment with OKT11A, which reacts with all peripheral E rosetting cells, in the presence of C diminished cytotoxicity against both EB+ and EB- targets demonstrating that both CTL and NK-like cells are, for the vast majority, OKT11A+.

VI. CONCLUDING REMARKS

The studies summarized here show first that monoclonal antibodies OKT3 or OKT8 are useful for discriminating CTLs from NK-like cells generated in MLC. The vast majority of the lytic activity against allogeneic normal cells is mediated by OKT3+ OKT8+ effectors while that against K562 is mediated by cells resistant to lysis by OKT3 or OKT8 antibodies and C. Lysis of autologous EB transformed LCLs by lymphocytes stimulated with pooled allogeneic normal cells is mediated by effectors which are phenotypically CTLs. Second, findings that stimulation with autologous LCLs gives rise to effectors lytic for both EB+ and EB- cell lines can be explained on the basis of a mixture of CTLs and NK-like cells. Lysis of both autologous and allogeneic EB+ LCLs was reduced by more than 85% whereas lysis of the EB- NK-sensitive targets was decreased minimally, if at all, following treatment with OKT8 and C.

It would be premature to conclude from these findings that all CTLs are OKT3+ or OKT8+ and that all NK-like cells are OKT3- or OKT8-, since, for one reason, clones which lyse both LCLs and K562 cells have recently been derived which appear to be OKT3+ OKT4+ OKT8- (Bach et al., in press; Bunzendahl et al., in preparation). Nonetheless, treatment of effector cells, generated following stimulation with alloantigens, tumor cells or viruses, with OKT3 or OKT8 and C should be helpful in evaluating the relative contribution to lysis of a variety of target cells by CTLs and NK-like cells.

ACKNOWLEDGMENTS

We thank M.S. Dierkins and E. Sevenich for technical assistance and J. Ritter and J. Gfrerer for help in preparation of this manuscript.

REFERENCES

Biron, C.A., Hutt-Fletcher, L.M., Wertz, G.T., and Pagano, J.S. (1981) Int. J. Cancer 27:185.

Bach, F.H., Leshem, B., Kupperman, O., Bunzendahl, H., Alter, B.J., and Zarling, J.M. In: The Potential Role of T Cell Subpopulations in Cancer Therapy (Fefer, A. and A.L. Goldstein, Eds.) Raven Press, New York. In press.

Klein, G. (1972) Proc. Natl. Acad. Sci. 69:1056.

Kung, P.C., Goldstein, G., Reinherz, E.L., and Schlossman, S.F. (1979) Science 206:347.

Kung, P.C., Talle, M.A., DeMaria, M.E., Butler, M.S., Lifter, J., and Goldstein, G. (1980) Transplant. Proc. 12:141.

Misko, I.S., Kane, R.G., and Pope, J.H. (1981) Int. J. Cancer 27:513.

Ortaldo, J.R., and Bonnard, G.D. (1977) Fed. Proc. 35:1325. Seeley, J.K., and Golub, S.H. (1977) Am. Assoc. Cancer Res. 18:174.

Seeley, J.K., and Golub, S.H. (1978) J. Immunol. 120:1415.

Sugamura, K., and Hinuma, Y. (1980) J. Immunol. 124:1045.

Svedmyr, E.A., Deinhardt, F., and Klein, G. (1974) Int. J. Cancer 13:891.

Tanaka, Y., Sugamura, K., Hinuma, Y., Sato, H., and Okochi, K. (1980) J. Immunol. 125:1426.

Verbi, W., Greaves, M.S., Schneider, C., Koubek, K., Janossy, G., Stein, H., Kung, P.C., and Goldstein, G. Europ. J. Immunol. In press.

Viallat, J., Svedymr, E., Steinitz, M., and Klein, G. (1978) Cell. Immunol. 38:68.

Zarling, J.M., and Bach, F.H. (1978) J. Exp. Med. 147:1334.

Zarling, J.M., Bach, F.H., and Kung, P.C. (1981a) J. Immunol. 126:375.

Zarling, J.M., Dierckins, M.S., Sevenich, E.A., and Clouse, K.A. (1981b) J. Immunol. 127:2118.

Zarling, J.M., and Kung, P.C. (1980) Nature 288:394.

GENERATION OF LYTIC POTENTIAL IN MIXED CULTURES AND ITS MODIFICATION BY IFN-α

Shmuel Argov
Eva Klein

Department of Tumor Biology
Karolinska Institute
Stockholm, Sweden

I. INTRODUCTION

Lymphocyte populations cultured without antigenic stimulus lose their natural killer (NK) activity (Masucci et al., 1980). However, if stimulated in the culture, the cytotoxic activity against NK-sensitive cell lines is maintained and even such lines which are insensitive in the short term NK assays can be affected. The lytic effect in such cultures is newly generated and is not performed by cells which keep NK activity (Bolhuis and Ronteltap, 1980; Seeley et al., 1979). This assumption is substantiated by the difference of fresh and cultured killer lymphocyte populations with regard of phenotypic characteristics (Poros and Klein, 1978).

Presentation of antigen to lymphocyte populations in vitro triggers the receptor carrying T cells to divide. After a few days such mixed cultures develop cytotoxic potential. On the population level the following lytic effects can be generated in antigen-containing lymphocyte cultures: 1) Enlargement of the specific clone, with resultant cytotoxicity against a) cells which carry the stimulating antigen; b) cells which carry cross-reactive antigens. 2) Transactivation induced by soluble

---

Supported by Federal Funds from the DHEW, Contract Number NO1-CM-74144, by NIH grant 1 RO1 CA-25184-01A1, and by the Swedish Cancer Society.

ISBN 0-12-341360-5

factors, produced in the mixed cultures, will recruit lymphocytes with other specificities. Thus targets not cross reactive with the stimulus may be killed if the proportion of lymphocytes with receptors recognizing their antigens is high. In addition activated lymphocytes can kill certain types of targets independently of antigen recognition. This effect is polyclonal.

II. MATERIAL AND METHODS

A. Medium

The medium used in the cultures and cytotoxic assays was RPMI-1640 supplemented with 10% heat inactivated autologous serum.

B. Cell Separation

Mononuclear cells were separated from heparinized blood of healthy donors by the Ficoll-Isopaque method. They were twice washed, transferred into 250 ml Falcon plastic flasks (10^8 cells/15ml medium) and kept in 5% CO_2 at 37°C for one hour. Non-adherent and adherent cells were separately collected, the latter after addition of PBS-2 mM EDTA. This population consisted mainly of monocytes and was used as stimulators in the mixed cultures. The non-adherent cells were placed on nylon wool column (10^8 cells/0.85 gr nylon wool), and kept in 5% CO_2 at 37°C for one hour. Thereafter the cells were allowed to pass and these were used as responders in the mixed cultures. The nylon wool-adherent fraction, known to be enriched in B cells, was recovered from the fibers by washing with PBS-2mM EDTA. This B cell fraction was also used as stimulator in AMC.

C. Mixed Cultures - AMC and MLC

15×10^6 cells of the T-enriched subpopulation were mixed with $7.5 - 15 \times 10^6$ mitomycin C-treated (25 µg/ml, at 37°C for 30 minutes) or irradiated (4000 rad) stimulator cells, autologous B cells and monocytes in AMC or allogeneic total population of lymphocytes in MLC, in 15 ml medium.

D. Assessment of Level of DNA Synthesis

10^5 cells from 7 days-cultures were exposed to 1 μCi of ^{3}H-thymidine. The samples were harvested after 8 hours. The reactivity index (R.I.) was expressed as the ratio between the isotope uptake in the mixed culture and the control culture.

E. Interferon Preparation

The partially purified Sendai virus induced human leukocyte interferon (IFN-α) preparation was the gift of Dr. Kari Cantell (Central Public Health Laboratory, Helsinki, Finland).

F. Target Cells in The Cytotoxic Experiments

Cell lines - K562 and Daudi were maintained in RPMI-1640 medium supplemented with 10% FBS. PHA blasts - T enriched lymphocyte fractions were cultured for 3 days with 1 μg/ml PHA (Wellcome Research Laboratories, Beckenham, England). STA blasts - 10^6 cells of the B enriched subset was cultured with 10^6 formalinized Staphylococcus aureus (Cowan strain I) for 3 days.

III. RESULTS

A. Autologous Mixed Culture, AMC

Lymphocytes activated in AMC lysed K562 and Daudi, allogeneic and autologous blasts (Tomonari, 1980). The two latter targets were similarly affected and there was no indication for selective effects (Table I). Blasts induced by Staphylococcus aureus (STA) were not lysed (not shown).

2-3 hours IFN treatment of the effectors prior to the lytic assay enhanced the anti-Daudi and anti-K562 activity but it did not affect activity against the PHA blasts. The results with Daudi are shown in Table II.

Interferon-α added to the cultures inhibited cell proliferation. The generation of lytic activity was unaltered.

TABLE I. The Effect of The Presence of IFN-α in The Cultures. Cell Multiplication, DNA Synthesis and Lytic Potential Estimated on The 7th Day of The Cultures

Effectors	Cytotoxicity against				Reactivity index	Cell yield[a]
	Cell lines		PHA blasts			
	K562	Daudi	Autologous	Allogeneic		
Control[b]	22 ± 11[c] (10)[d]	6 ± 6 (18)	4 ± 3 (10)	5 ± 2 (10)		75 ± 17 (10)
With IFN[e]	27 ± 13 (10)	9 ± 8 (18)	6 ± 2 (10)	5 ± 4 (10)		66 ± 19 (10)
AMC	39 ± 19 (15)	28 ± 26 (26)	16 ± 17 (20)	14 ± 17 (23)	39 (22)	98 ± 32 (32)
With IFN	36 ± 19 (15)	25 ± 20 (26)	14 ± 12 (20)	13 ± 12 (23)	18 (22)	75 ± 25 (32)
MLC	21 ± 5[f] (3)	37 ± 18 (14)	19 ± 13 (14)	27 ± 12 (14)	79 (11)	280 ± 164 (11)
With IFN	39 ± 28 (3)	48 ± 20 (14)	29 ± 19 (14)	53 ± 18 (14)	84 (11)	147 ± 47 (11)

[a]Cell recovery after 7 days expressed as percentage of input.
[b]T lymphocytes cultured alone for 7 days.
[c]Percent specific ^{51}Cr release ± standard error. 4 hours assay. The effector/target ratios were either 50:1 or 20:1, they were identical in the parallel experiments.
[d]Number of experiments.
[e]1000 i.u/ml IFN-α present during the culture period.
[f]The mean specific ^{51}Cr release exerted by the control cultures in these 3 experiments was 13.

TABLE II. The Effect of Interferon on The Anti-Daudi Lytic Potential of Cultured Cells

Culture	Nr. of expts.	% specific ^{51}Cr release	
		-	IFN pretreated effectors[b]
Control	10	9±8	23±10
with IFN[a]		11±8	25±10
AMC	12	25±18	41±19
with IFN		28±17	30±18
MLC	3	24±8	34±10
with IFN		32±7	33±10

[a]1000 i.u/ml IFN-α added at the initiation of the culture.
[b]The effector cells were incubated for 2h with 1000 i.u/ml IFN-α prior the lytic assay.

B. Allogeneic Mixed Cultures, MLC

Lymphocytes activated in MLC also lysed K562, Daudi and PHA blasts. The allospecific component could be seen by the stronger effect against the stimulator alloblasts (Table I). When Staphylococcus aureus (STA) blasts were used as targets lysis was only seen against the stimulator cells (Table III).

With these cells IFN-α treatment prior to the lytic assay enhanced some of the activities. Comparison of the extent of increase with PHA blast targets - autologous and allogeneic - and the lack of IFN-induced lysis of STA-blasts suggests that the non-specific component of the cytotoxicity was enhanced.

The proliferation of the lymphocytes decreased when IFN was added at the initiation of the culture because the cell yields were lower. However the DNA synthesis of the cells measured on day 7 was higher in the IFN-containing MLC-s, which suggests a change in the kinetics of the cell multiplication in these cultures (Table I).

The cultures with IFN had a higher lytic potential on a per cell basis. There was an elevation of the specific component of the cytotoxicity which was more substantial than the increase of the nonspecific component. The increase of the specific effect was detectable with PHA and STA blasts and it was quantitatively similar when evaluated on both targets (Tables I, III).

The cytotoxic potential of cells cultured in presence of IFN could not be enhanced by renewed IFN exposure prior the assay (Table II).

TABLE III. Lysis of Staphylococcus Aureus Blasts by Lymphocytes Activated in MLC[a]

	STA blasts	
	autologous	stimulator
MLC	5[c] ± 4	28 ± 10
With IFN[b]	4 ± 3	54 ± 9

[a]6 experiments.
[b,c]See Table I.

IV. CONCLUSIONS

The results with the AMC and MLC differed. There was no enhancing effect on the lysis of PHA blasts when the AMC lymphocytes were pretreated with IFN while this occurred with the cells of the MLC. Furthermore, when IFN was added to the culture it did not influence on the generation of lytic potential in the AMC while it was regularly enhancing in MLC. Since the activation, as judged by the cell proliferation, in the AMC is considerably weaker compared to MLC, the population is likely to differ in the differentiation profile. It is possible that in the AMC cultures lysis against the PHA blast were not performed by cells which can respond to IFN. On the other hand the anti-Daudi and anti-K562 effects were increased after IFN treatment of the lymphocytes. It can be assumed, therefore, that lysis of the cell lines and the blasts were not performed by the same effector subsets. That this may be the explanation is suggested by the experiments of Santoli et al. (1982). They have analyzed the cytotoxicities of MLC cultures and found, in tests against autologous PHA blasts and K562 - which are both "anomalous" targets in the system - were not performed by completely overlapping subsets, when these were characterized by the expression of E receptors.

The lower cell proliferation coinciding with higher specific cytotoxicity in IFN containing MLC-s, is in line with the results of Spits et al. (1981). In their experiments the extent of the generation of nonspecific cytotoxicity was related to the initial DNA synthesis in the culture. They obtained differences in the DNA synthesis by varying the number of stimulator cells.

It is possible that the suppression of cell multiplication by IFN in the MLC-s is selective and the specific clones were not or less severly affected. This selectivity may be determined by the change in the differentiation state of the alloreactive lymphocytes when they encounter the antigen. Another factor may be the alterations of the growth kinetics in the culture which is known to be influenced by the cell concentration. An initial growth inhibition may change the culture conditions and when growth is resumed the proportion of "irrelevant" cells is lower.

It was a general finding that lymphocytes which have been cultivated in the presence of IFN were not responsive to reexposure to IFN in the assay, i.e. no change in their function was imposed by the renewed IFN treatment. At the present stage we do not know the reason for this. By the continuous presence of IFN, the cells may have been optimally stimulated for cytotoxic function. However even populations with low activity were refractory for the IFN-induced enhancement. Further experiments are needed to clarify the mechanism of this phenomenon. It has to be noted that IFN-γ is produced during the MLC and still the lytic potential of cells can be enhanced by IFN-α exposure. The question arises therefore whether the refractoriness occurs only when the same type of IFN is used on both occasions or the differences in the IFN quantities and the kinetics determine the differences found.

REFERENCES

Bolhuis, R.L.H., and Ronteltap, C.P.M. (1980). Immunol. Letters I:191.
Masucci, M.G., Klein, E., and Argov, S. (1980). J. Immunol. 124:2458.
Poros, A., and Klein, E. (1978). Cell. Immunol. 41:240.
Santoli, D., Francis, M.K., and Trucco, M. (1982). Cell. Immunol., in press.
Seeley, J.K., Masucci, G., Poros, A., Klein, E., and Golub, S.H. (1979). J. Immunol. 123:303.
Spits, H., De Vries, J.E., and Terhorst, C. (1981). Cell. Immunol. 59:435.
Tomonari, K. (1980). J. Immunol. 124:1111.

CULTURED NATURAL KILLER CELLS

Scott P. Bartlett
Robert C. Burton[1]

Transplantation Unit
Massachusetts General Hospital
Boston, Massachusetts

I. INTRODUCTION

In a separate chapter of this volume we have given evidence for the presence of two natural killer (NK) cell subsets that exist in fresh murine spleen cell preparations: NK_A cells and NK_B (or NC) cells (Burton, Bartlett, Lust, and Kumar). These effectors appear to account for all of the NK activity of these cell suspensions. Culture of murine spleen cells, both "stimulated" and "unstimulated" produces a variety of effects. We reported in 1977 that cytotoxic cells which are both Thy-1^+ and Thy-1^- arise spontaneously in unstimulated cultures (1). It was shown at that time that the phenomenon was fetal calf serum (FCS) batch dependent and required the presence of Thy-1^+ cells, as no cytotoxic activity arose in cultures of nude spleen cells (It should be noted, however, that NK_B activity at that time had not been fully characterized, and NK_B sensitive targets were unrecognized). Subsequently, phenomena of this type have been related to natural killing

[1]This study was supported by Grants CA-17800 and CA-20044 awarded by the National Cancer Institute, Bethesda, Maryland and by Grants AM-07055 and HL-18646 from the National Institutes of Health, Bethesda, Maryland. RCB was supported by the John Mitchell Crouch Fellowship of the Royal Australasian College of Surgeons and a National Health and Medical Research Council of Australia Fellowship in Applied Health Sciences.

ISBN 0-12-341360-5

(2) and it has been suggested that culture might be a way of enriching for NK cells, especially if stimuli such as those provided by a concomittant MLC or tumor necrosis serum are included (2-7). The purpose of this review is to summarize investigations performed in this laboratory on the phenotype of cytotoxic cells arising in "unstimulated" FCS supplemented cultures and to relate these activities to those generated under "stimulated" culture conditions. The mice used, reagents employed, tumor cell lines, assay conditions and *in vitro* culture conditions have been described in detail previously (1,8-16).

II. RESULTS

A. Generation of Cytotoxic Effector Cells in "Unstimulated" Cultures

To determine the effect of culture on cytotoxic activity of murine spleen cells, cultures were prepared under the described conditions and assayed on various days of culture against a panel of tumor cell targets which are sensitive to NK_A-mediated lysis (YAC) and NK_B-mediated lysis (WEHI-164.1) and resistant to the NK activity of fresh murine spleen (P-815.1, EMT-6) (17). The cells were harvested and subjected to treatment with anti-NK-1.2 plus complement (C) or monoclonal (MC) anti-Thy-1.2 + C before overnight assay against these targets. A representative experiment is shown in Figure 1. Several conclusions become apparent: (i) the "unstimulated" culture of spleen cells results in an increased killing of tumor cell targets up to a maximum on day 6 (data not shown), (ii) tumor cell targets not sensitive to lysis by fresh murine spleen (P-815.1, EMT-6) become sensitive to lysis by 6 day cultured effectors, (iii) although the killing of the NK_A sensitive target YAC continues throughout the culture period, the cytotoxic effector changes phenotype - killing effected by the NK-1.2$^+$ population of fresh murine spleen cells is replaced by an NK-1.2$^-$, Thy-1.2$^+$ cell, (iv) the NK insensitive target P-815.1 becomes susceptible to a Thy-1.2$^+$ effector, (v) although NK_B activity (as measured against WEHI-164.1 and by resistance to anti-NK-1.2 and MC anti-Thy-1.2 + C) continues throughout the culture period, its activity increases some fourfold (see below) (vi) EMT-6, a target insensitive to lysis by fresh murine spleen becomes susceptible to lysis by a cell of similar phenotype to that responsible for the lysis of WEHI-164.1 (Thy-1.2$^-$, NK-1.2$^-$), making it possible therefore that

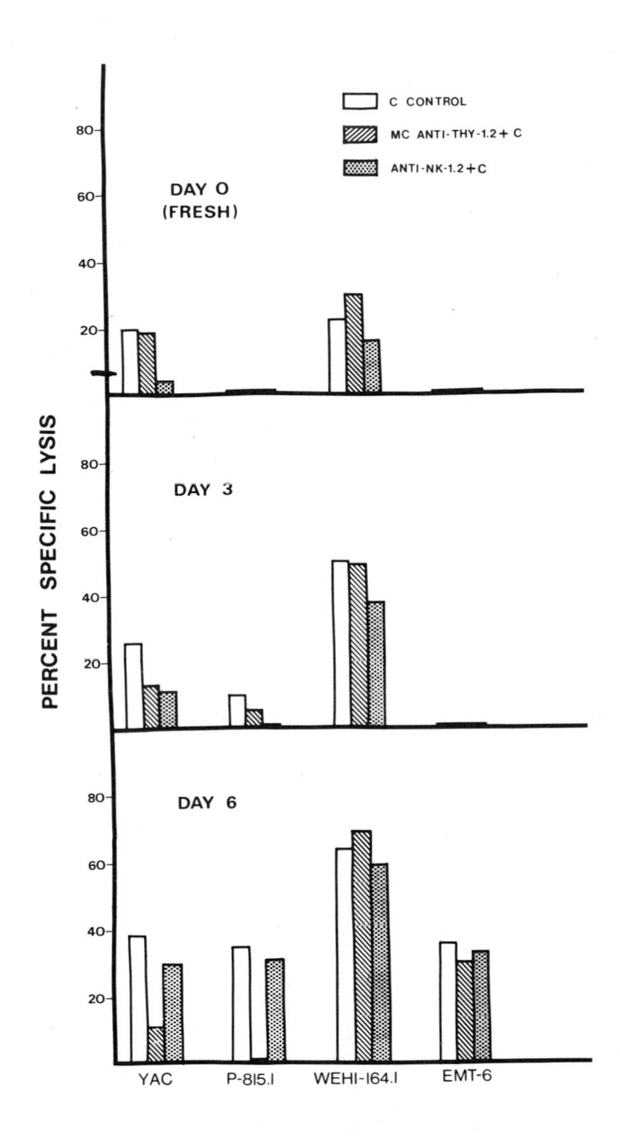

FIGURE 1. Activity of cytotoxic effector cells against tumor targets on days 0,3, and 6 of culture following treatment with C, MC anti-Thy-1.2 plus C or anti-NK-1.2 plus C.

the same effector lyses both WEHI-164.1 and EMT-6 or, alternatively, that two different cells with this common phenotype may be present in the 6 day cultures.

B. Phenotype of "Unstimulated" Culture-Induced Killer Cells

The cell curface antigens present on culture induced effectors were further defined by treatment with specific alloantisera or antibodies and C prior to assay (Table I). A fourfold reduction in cytotoxicity or thymidine incorporation compared to C controls established the presence of these antigens, and concomittant positive controls (data not shown) established these as active reagents. Lysis of YAC and P-815.1 is mediated largely by Thy-1.2$^+$ effectors, and that of WEHI-164.1 and EMT-6 by Thy-1.2$^-$ effectors. All cultured killer cells expressed H-2 and xenoantigens whereas none of the effectors expressed NK-1.2 or the B cell/macrophage antigen LyM-1.2. The Thy-1.2$^+$ culture induced killer is also Ly-1.2$^+$ and Ly-2.2$^+$ as is the allospecific Tc induced *in vitro* and tested with these reagents (data not shown). The antigenic profiles of fresh murine spleen cells mediating the lysis of these same targets are shown for comparison, and as mentioned previously, P-815.1 and EMT-6 are resistant to lysis by this preparation.

C. Survival of NK_A and NK_B Cells in Culture

To determine the survival of NK_A cells in culture, murine spleen cells were assayed on various days of culture against YAC following treatment with specific anti-NK-1.2 antiserum and C. The reduction in lytic activity against this NK_A sensitive target on day 0 of culture (fresh) following such treatment was assigned a value of 100%; on subsequent days of culture NK-1.2$^+$ killing was compared to this standard. It was observed that NK-1.2$^+$ activity is rapidly lost in culture and by day 6 only 8% of this activity remains (data not shown). That this does not merely represent a change in phenotype of these cells will be commented upon subsequently.

The fate of NK_B cells upon culture is more diffcult to study as these cells do not as yet possess a specific cell marker. However, these effectors are partially sensitive to high titre anti-H-2 serum and C treatment (17,18) and upon culture their sensitivity is significantly increased (up to fourfold) suggesting more expression of this surface antigen(s) (data not shown). Additionally,

TABLE I. Cell Surface Markers on Fresh Spleen Cells and Unstimulated Culture-Induced Effectors

Cell Surface Antigen	Phenotype[a] of Effectors Mediating Lysis of: YAC	P-815.1	WEHI-164.1	EMT-6
Unstimulated 6 day cultured spleen				
Thy-1.2	+	+	–	–
NK-1.2	–	–	–	–
H-2	+	+	+	
LyM-1.2	–	–	–	
Xenoantigen(s)[b]	+	+	+	
Ly-1.2	+	+	–	
Ly-2.2	+	+	–	
Fresh spleen				
Thy-1.2	–(+)[c]	0	–	0
NK-1.2	+	0	–	0
H-2	+	0	–	0
Xenoantigen(s)	+	0	+	0
LyM-1.2	–	0	–	0

[a]Based on at least a fourfold reduction in cytotoxicity as compared to C control

[b]As defined by treatment with RAMS, Rabbit anti-mouse serum, plus complement

[c]Two fold reduction

when one examines the lytic activity against WEHI-164.1 of fresh and 6 day cultures, based on a comparison of the corresponding points of the effector to target curves, and taking into account cell loss consequent upon culture, a 200-400% increase in activity is regularly observed (data not shown). These results suggest that the NK_B cell undergoes significant change in culture and that the activity observed is not merely due to persistence of these cells in culture.

D. Antiserum Pretreatment and the Generation of Cytotoxic Effector Cells in Culture

Fresh murine spleen cell preparations were treated with specific antisera or antibodies and C prior to culture and their 6 day activity assessed (data not shown). It was observed that the 6 day cultured Thy-1.2$^+$ effector requires the presence of, or is derived from Thy 1.2$^+$, H-2$^+$, NK-1.2$^-$ cells found in fresh spleen preparations, whereas the cultured Thy-1.2$^-$ effector(s) requires only the presence of the H-2$^+$ precursor(s).

III. DISCUSSION

From the data presented in this review and elsewhere (1,16), we have shown that the "unstimulated" culture of murine spleen cells has diverse and complex effects. In summary: (i) NK-1.2$^+$ NK_A cells do not survive under these culture conditions (ii) NK-1.2$^+$ cells are not necessary for, nor do they give rise to any of the cytotoxic activities found in these cultures (iii) NK-1.2$^-$, Thy-1.2$^-$ cytotoxic NK_B activity is increased by culture some fourfold and the range of target cells lysed increases. This increase in the range of target cells killed suggests that perhaps two different cells, both NK-1.2$^-$, Thy-1.2$^-$ are involved (iv) Thy-1.2$^+$, NK-1.2$^-$, Ly-1.2$^+$, Ly-2.2$^+$ "Tc" spontaneously arise in culture from, or in the presence of Thy-1.2$^+$, NK-1.2$^-$ precursors, and kill a variety of NK sensitive and NK-insensitive targets, both of the same and different H-2 type (v) None of these *in vitro* derived cells expresses the B cell/macrophage alloantigen LyM-1.2. In this "unstimulated" culture system, no role for NK-1.2$^+$ NK_A cells has been demonstrated. Surprisingly, this has not turned out to be the case when "stimulated" culture conditions are employed. Seely and Karre (2,6,) in their studies of conventional murine MLC have demonstrated the generation of "anomalous killer" (AK) cell activity in these cultures which is non-H-2 restricted and peaks 2 days before maximum allospecific Tc activity. Further collaborative studies with these investigators have shown that these AK cells are Thy-1.2$^+$ and NK-1.2$^-$ like the spontaneous Tc referred to above, but they are very clearly derived from NK-1.2$^+$, Thy-1.2$^-$ precursors (19). This is the first deomonstration that NK-1.2$^+$ cells can "differentiate" in culture. Clearly, the presence of an MLC significantly alters the *in vitro* environment, and this phenomenon is quite different to that observed in "unstimulated" cultures. Minato et al (20) in a culture system similar to ours have demonstrated the presence of two distinct Thy-1.2$^+$ cytotoxic activities. Only one of these killer cell types develops when nude (*nu*/*nu*) spleen cells are cultured, suggesting a degree of clonal restriction or a different maturation lineage. Competitive inhibition assays may help elucidate this issue and determine whether subpopulations exist within the Thy-1.2$^+$ population.

These findings all emphasize the complex effects that *in vitro* culture may have on normal lymphohemopoietic cells. Only by following these effects in kinetic studies with a range of tumor targets and a variety of reagents which detect cell surface markers can an appreciation of the diversity of effector cells which may be generated be obtained. In particular, when weak Tc responses induced *in vitro* are being studied e.g. to tumor associated antigens, it is necessary to take full cognizance of the presence of spontaneous culture-induced effector cell activity and to define their relationship to the specific Tc under study (21,22).

REFERENCES

1. Burton, R.C., Chism, S.E. and Warner, N.L. (1977) J. Immunol. 119:1329.
2. Karre, K. and Seeley, J.K. (1979) J. Immunol. 123: 1511.
3. Chun, M., Pasanen, V., Hammerling, U., Hammerling, G.F. and Hoffman, M.K. (1979) J. Exp. Med. 150:426.
4. Paciucci, P.A., MacPhail, S., Zarling, J.M. and Bach, F.H. (1980) J. Immunol. 124:370.
5. Bach, F.H., Paciucci, P.A., MacPhail, S., Sondel, P.M., Alter, B.J. and Zarling, J.M. (1980) Transplant. Proc. 12:2.
6. Seeley, J.K. and Karre, K. (1980) In "Natural Cell-Mediated Immunity Against Tumors" (R. Herberman, Ed) p. 477 Academic Press, New York.
7. Chun, M., Fernandes, G. and Hoffman, M.K. (1981) J. Immunol. 126:331.

8. Burton, R.C., Thompson, J. and Warner, N.L. (1975) J. Immunol. Methods 8:133.
9. Kiessling, R., Klein, E. and Wigzell, H. (1975) Eur. J. Immunol. 5:112.
10. Chism, S.E., Burton, R.C. and Warner, N.L. (1976) J. Natl. Cancer Inst. 57:377.
11. Tonkonogy, S.L. and Winn, H.J. (1976) J. Immunol. 116:835.
12. Jeekel, J.J., McKenzie, I.F.C. and Winn, H.J. (1972) J. Immunol. 108:1017.
13. Sarmeinto, M., Glasebrook, A.L. and Fitch, F.W. (1980) J. Immunol. 125:2665.
14. McKenzie, I.F.C. and Potter, T. (1981) Adv. Immunol. 27:179.
15. Burton, R.C. and Winn H.J. (1981) J. Immunol. 126:1985.
16. Bartlett, S.P. and Burton, R.C. J. Immunol. (In Press).
17. Burton, R.C., Bartlett, S.P., Kumar, V. and Winn, H.J. (1981) Trans. Proc. 13:783.
18. Burton, R.C., Bartlett, S.P. and Winn, H.J. (1980) In "Genetic Control of Natural Resistance to Infection and Malignancy" (E. Skamene and P.A.L. Kongshavn, Eds) p. 413 Academic Press, New York.
19. Karre, K. and Burton, R.C. (1981) Submitted for publication.
20. Minato, N., Reid, L. and Bloom, B.B. (1981) J. Exp. Med. 154: 750.
21. Burton, R.C., Chism, S.E. and Warner, N.L. (1978) Cont. Topics Immunobiol. 8:69.
22. Burton, R.C. and Plate, J.M.D. (1981) Cell Immunol. 58:225.

SPONTANEOUS MONOCYTE MEDIATED CYTOTOXICITY IN MAN: EVIDENCE FOR T HELPER ACTIVITY

Andrew V. Muchmore

Metabolism Branch
National Institutes of Health
Bethesda, Maryland

Eugenie S. Kleinerman

Biological Response Modifiers Program
National Institutes of Health
Frederick, Maryland

I. INTRODUCTION

The primary purpose of the immune system is to protect the host against invasion by pathogenic organisms. A variety of host defense mechanisms have evolved to combat such pathogens and are generally classified as specific or non-specific. Specific host defenses require previous exposure to a particular antigen, usually take time to generate, and exhibit a form of memory. Antibody responses and sensitized T cell responses are examples of specific immune responses. Non-specific immune responses conversely require no previous exposure to antigen, have a broad responsiveness, and do not involve a memory system. Recently a great deal of interest has been accorded non specific models of cellular cytotoxicity because in murine systems assays of non specific killing such as (NK) have seemed to correlate well with in vivo tumor resistance as summarized in (Herberman 1980). Furthermore, non-specific forms of cell mediated cytotoxicity appear to phylogenetically antedate the appearance of more specific forms of host defense (Decker 1981).

A variety of in vitro models to study non-specific immunity have been characterized, including NK or natural killer cells, NC or natural cytotoxicity, spontaneous monocyte mediated cytotoxicity and synergistic cytotoxicity. (Herberman 1975, Stutman 1980, and Muchmore 1979). We have extensively characterized a human model of monocyte mediated cytotoxicity which has many unique and useful characteristics. Develop-

ISBN 0-12-341360-5

ment of cytotoxocity does not require the addition of antigen, antibody, or lectin to the culture system. Through a variety of cellular depletion studies the killer cell has been shown to be a monocyte (Muchmore 1979). Since no exogenous stiumli are used to activate the killer cell the system serves as a good model to examine not only the interaction that takes place between the effector cell and the target cell, but also the mechanism by which human monocytes develop into cytotoxic effector cells. We have previously shown that cytotoxic monocytes are under the influence of suppressor cells. We have now further characterized these regulatory cells and find evidence that cytotoxic monocytes require helper influences as well as being under the control of suppressor cells.

RESULTS

Our previous data demonstrated that although fresh mononuclear cells fail to spontaneously lyse xenogeneic red blood targets, after 3-7 days of _in vitro_ culture the monocytes in these mononuclear cell preparations become spontaneously cytotoxic; furthermore we have shown that such cytotoxic monocytes, once generated, can be dramatically suppressed by the addition of fresh human lymphocytes. The nature of the suppressor cell has remained obscure since both SRBC rosette positive (i.e. T cells) and rosette negative cells can suppress cytotoxic monocytes. We have attempted to further characterize the suppressor cell responsible for inhibition of monocyte killing, using monoclonal antisera but so far we have been unsuccessful.

We now have evidence which suggests that monocytes, in addition to being controlled by a suppressor cell, require a helper cell to develop their cytotoxic function. This hypothesis stemmed from observations that purified enriched monocyte preperations (presumable depleted of suppressor lymphocytes) failed to develop earlier and enhanced cytotoxicity as would have been predicted if control of monocyte function was only due to suppressor influences. An example of this is shown in Table I. Since it is technically difficult to obtain monocytes completely devoid of contaminating lymphocytes two methods of monocyte preparation were utilized (plastic adherence and Percoll gradients). Note that cytotoxicity decreased as lymphocyte contamination decreased. Since we have previously shown that the killer cell is a monocyte, these data suggested that lymphocytes might play an important helper role in the generation of spontaneous monocyte-mediated cytotoxicity.

Table I

Lack of enhanced killing after suppressor lymphocyte depletion

	%^{51}Cr Release			
	Day 2	Day 3	Day 4	Day 5
Monocytes[1]	5	5	11	16
Monocytes[2]	5	5	16	30
Whole mononuclear cells[3]	5	7	23	70

Legend: [1] monocytes seperated on Percoll gradients, 78% esterase positive [2] monocytes seperated by plastic adhenence, 60% esterase positive [3] whole mononuclear cells 20% esterase positive. Results expressed as %^{51}Cr release using chicken red blood cell targets at a 1:5 target to effector cell ratio.

To directly address this hypothesis lymphocytes were depleted of monocytes by plastic adherence (less than 1% esterase positive cells). These purified lymphocytes were cultured for three days, harvested, counted for viability and then added to fresh percol enriched autologous monocytes (80% esterase positive) for 48 hours. Monocyte cytotoxicity was then assayed and as can be seen in Table II, the precultured lymphocytes were able to induce earlier killing mediated by fresh autologous monocytes while neither fresh monocytes nor precultured lymphocytes killed by themselves.

TABLE II

Precultured lymphocytes lead to early monocyte mediated cytotoxicity

Cells added	%^{51}Cr Release
Fresh monocytes + fresh lymphocytes	11
Fresh monocytes + precultured lymphocytes	47
Fresh mopnocytes	5
Precultured lymphocytes	7

Legend: Lymphocytes depleted of monocytes were precultured in NCTC 109 with added penicillin, streptomycin, L-glutamine, and 2% fetal calf serum for 72 hours. After this period of time the cells were harvested, counted for viability, and added in a one to one ratio to freshly obtained autologous monocytes. These cultures were incubated for 48 hours and subsequently assayed for cytotoxicity using ^{51}Cr labelled chicken red blood cell targets in an 18 hour assay at a 1:5 target to effector ratio, with results presented as percent of total releasable ^{51}Cr.

Using monoclonal antisera we have begun to characterize the cell responsible for this helper activity. The cell responsible for help appears to have the phenotype of a T cell since both the monoclonal antisera OKT 3 and Leu 1 both completely abrogate the appearance of cytotoxicty as seen in Table IV. Neither mouse ascites nor anti-IgG have a suppressive effect (data not shown).

TABLE III
Spontaneous monocyte mediated cytotoxicity requires T cells

Treatment to cells	Percent Inhibition
none	0
OK T3	100
Leu 1	100

Legend: 10 microliters of Leu 1 or OKT 3 were added to 7 X 10^6 mononuclear cells and allowed to incubate for 6 days, washed and assayed for cytotoxicity against chicken red blood cells as previously described. Results are presented as percent inhibition compared to control cultures.

DISCUSSION

Our current data broadens our understanding of the potential cellular defects which may result in depressed monocyte-mediated cytotoxicity. A variety of human disorders have been reported to have defective monocyte mediated cytotoxicity (Kleinerman 1980a, Blaese 1980). One important group of patients with markedly deficient killing includes adults with diseminated solid tumor malignancies. Interestingly we have shown that some chemotherapeutic agents which act to enhance cytotoxicity in vitro also enhance cytotoxicity in vivo (Kleinerman 1980b, Kleinerman 1981). Some of these agents act to inhibit suppressor lymphocyte influences while others act to directly stimulate monocyte activity. Still others act through mechanisms which are not understood. It would be of interest and potentially of therapeutic importance if drugs could be found which directly activate helper activity resulting in enhanced spontaneous monocyte mediated cytotoxicity.

Patients with various immunodeficiencies also exhibit defective monocyte killing. For example individuals with the Wiskott Aldrich syndrome routinely fail to exhibit in vitro cytotoxic activity. We interpret these data as further evidence that these patients have defective monocyte function. Another example of immunodeficient patients which fail to exhibit spontaneous monocyte mediated cytotoxicity are those with the autosomal recessive disorder ataxia telangiectasia. In a series of unpublished experiments we have been unable to show other monocyte killing defects nor have we been able to demonstrate hyperactive suppressor influences on monocyte killing functions. Since these patients are known to have a variety of T cell defects it is tempting to speculate that they have a helper T cell defect which results in ineffective monocyte activation and deficient moncyte killing. Experiments addressing these possibilities are in progress.

In summary we have presented data suggesting that spontaneous monocyte-mediated cytotoxicity is under both helper and suppressor influences. Thus defective in vitro killing may be the result of defective monocyte function, hyperactive suppressor function, or inadequate helper function. We believe that this system offers many opportunities to further our understanding of regulatory influences on human monocyte activity.

REFERENCES

Blaese, R.M., Muchmore, A.V., Lawrence, E.C., and Poplack, D.G., in Primary Immunodeficiencies (M. Seligman and W.H. Hitzig, ed.) p. 391 Elsevier/North Holland Biomedical Press 1980

Decker, J.M., Elmholt, A., and Muchmore, A.V., Cellular Immunology 122:161 (1981).

Herberman, R.B., Nunn, M.E., and Lavrin, D.H., Int. J. Cancer 15: 216 (1975).

Herberman, R.B. editor Natural Cell-Mediated Immunity Against Tumors, Academic Press, New York (1980).

Kleinerman, E.S., Zwelling, L.A., Howser, D., Barlock, A., Young, R.C., Decker, J.M., Bull, J. and Muchmore, A.V., Lancet 22:1102 (1980a).

Kleinerman, E.S., Zwelling, L.A., and Muchmore, A.V., Cancer Research 40:3099 (1980b).

Kleinerman, E.S., and Muchmore, A.V., Proc. Amer. Assoc. for Cancer Res. 22:1101 abstract (1981)

Muchmore, A.V., Decker, J.M., and Blaese, R.M., Jo. Immunol. 119: 1680 (1977a

Muchmore, A.V., Decker, J.M., and Blaese, R.M., Jo. Immunol. 122: 1146 (1979b).

Stutman, O., Dien, P., Wison, R., and Lattime, E.C., Proc. Natl. Acad. Sci. 77: 1680 (1980).

CULTURES OF PURIFIED HUMAN NATURAL KILLER CELLS

Tuomo Timonen,[1] John Ortaldo and Ronald Herberman

Biological Research and Therapy Branch
National Cancer Institute, Frederick, MD

The discovery of T-cell growth factor (interleukin 2, IL-2), along with the development of methods for continuous cultures and cloning of T-cells, have shown that cytotoxic (Lutz *et al.*, 1981), helper (Kurnick *et al.*, 1981) and suppressor T-cells (Fresno *et al.*, 1981) can be induced to proliferate *in vitro*. Although NK cells do not appear to be conventional T-cells, with regard to their surface antigenic phenotype, function and morphology, analyses of the IL-2-propagated cultured lymphoid cells (CLC) from mice have revealed both functionally and phenotypically NK-like cells among them (Kuribayashi *et al.*, 1981; Dennert *et al.*, 1981). Also, human CLC have been shown to exert NK-like cytotoxicity (Ortaldo *et al.*, 1980). It has, however, remained unclear whether NK cells themselves are capable of proliferating in the presence of IL-2, or whether conventional T-cells gain NK-cell like properties when cultured. We have studied this question by culturing highly purified large granular lymphocytes (LGL), which mediate NK activity, and small T-cells, both obtained by discontinuous density gradient centrifugation, in the presence of supernatants containing IL-2. The results strongly suggest that LGL can proliferate and maintain their cytotoxic characteristics in the cultures.

Initiation of the Cultures

LGL and T-cells were enriched by discontinuous density gradient centrifugation on Percoll as described in more detail elsewhere in this volume (Timonen *et al.*). LGLs were further purified from the low density fractions by depletion of high affinity sheep erythrocyte (E)-rosette-forming cells.

[1]Present address: Department of Pathology, University of Helsinki, Finland

ISBN 0-12-341360-5

The final purity of LGL was $\geq 90\%$, the contaminating cells being mostly monocytes. The contamination of LGL by T-cells was monitored by measuring the frequency of cells reacting with the monoclonal antibody OKT3, using the Staphylococcus aureus-binding assay (Ranki et al., 1976). If T-cells were detected, an additional cycle of depletion of high affinity E-rosette forming cells was performed, until undetectable frequencies of OKT3-positive cells were reached. The contamination of T-cells by LGL was monitored by morphologic analysis. LGL frequencies among the high density cells were always less than 1%. The cultures were initiated using 0.5 x 10^6 cells/ml medium (RPMI-1640 supplemented with 10% heat-inactivated human AB serum and antibiotics), 19% autologous peripheral blood mononuclear cells (4000R-irradiated) as feeder cells, 0.5 ug/ml PHA and 10% IL-2 (lectin free, Associated Biomedic , Buffalo, NY). When the cell density reached or approached 1 x 10^6 cells/ml, the cultures were washed and resuspended to 0.25 x 10^6 cells/ml in fresh medium with IL-2 (Figure 1). No feeder cells or PHA were added after the initiation of the cultures. By this method, it has been possible to culture both LGL and T-cells for about one month, after which the rapid proliferation regularly ceased. Within this time, LGL had multiplied 20-40, and T-cells 50-80 fold.

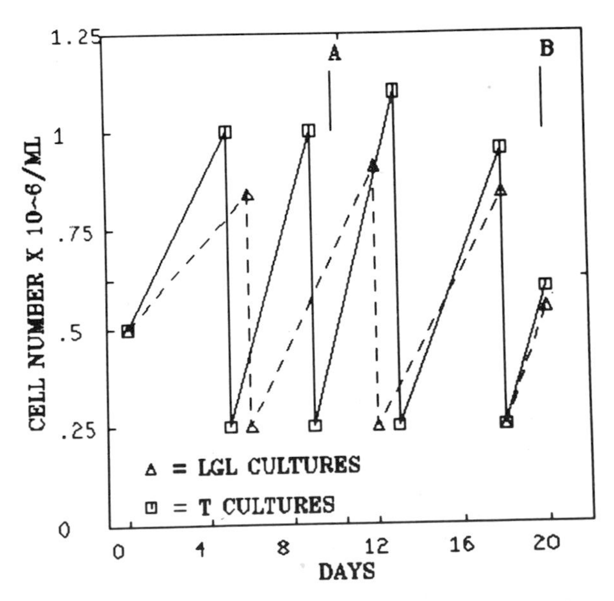

FIGURE 1. Growth curves of LGL and T-cells cultures. A and B stand for time points when samples were taken from the culture for morphological and cytotoxicity analyses.

Morphology

The morphology of both LGL and T-cell cultures was analyzed from Giemsa-stained cytocentrifuged cell smears. The uncultured cells were ≥90% pure LGL and T-cells. At both day 10 and 20 after the initiation of the cultures, almost all cells in LGL cultures showed a morphology which was distinguishable from cultured T-cells (Figure 2). Cultured LGL were blasts (15-20 μ diameter in contrast to 10-15 μ in fresh LGL) with a relatively pale cytoplasm and kidney-shaped nuclei, and with azurophilic cytoplasmic granules in an average of 78% of the cells. In contrast, cultured T-cells had a strongly basophilic cytoplasm and cytoplasmic granules in less than 5% of the cells. Thus, although the size of the cultured cells and the frequency of granular cells in the cultures had changed among LGL, there was a clear morphological similarity between fresh and cultured LGL, distinguishing them from T-cells. The results therefore support the contention that LGLs are capable of proliferating in the presence of IL-2.

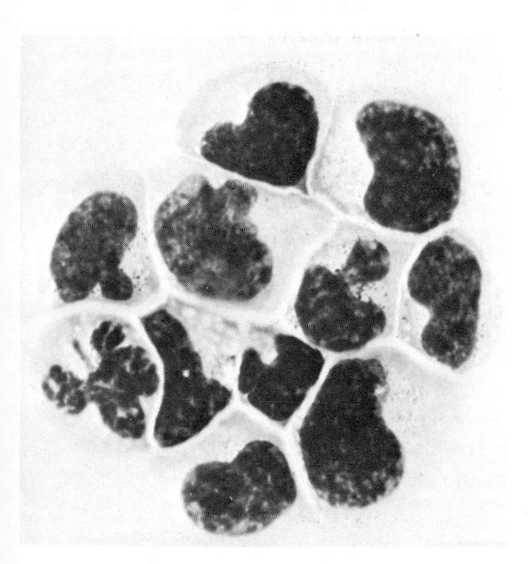

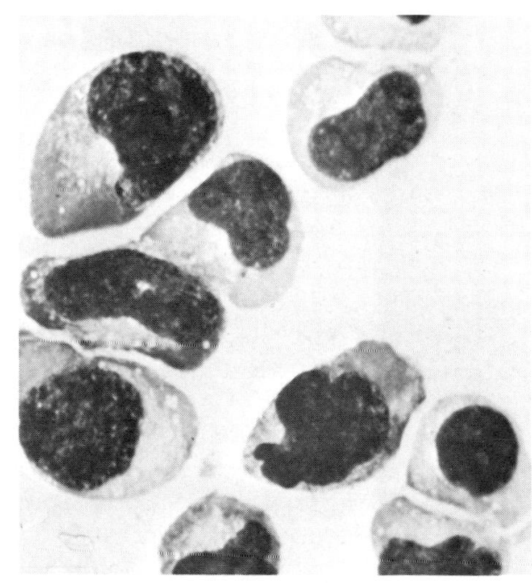

FIGURE 2. Morphology of cultured LGL (left panel) and T-cell (right panel), at day 20.

Cytotoxic Activity

Earlier experiments have demonstrated that fresh uncultured NK cells are capable, in addition to NK activity, to perform antibody-dependent cellular cytotoxicity (ADCC) and lectin-induced cellular cytotoxicity (LICC), whereas they do not significantly lyse allogeneic blasts or xenogeneic mouse lymphoma cells (for example, RLδ1). Fresh small T-cells exert only LICC. As shown in Figure 3, the pattern of cytotoxic activity by LGL was maintained in the cultures, further supporting the proliferative capacity of LGL. The small T-cells gained reactivity against anchorage-dependent target cells in the cultures, but they did not lyse the suspension-grown target cells. Cultured T-cells exerted weak cytotoxicity also against allogeneic blasts. As a whole, the cytotoxic profile of LGL was completely different from that of T-cells.

Surface Markers

As described elsewhere (Ortaldo et al., 1981), most freshly isolated LGL express antigens detected by monoclonal antibodies OKM1 and OKT10 and partially reactive with anti-Ia antibodies. Furthermore, almost all LGL are known to carry receptors for the Fc part of IgG molecule (Timonen et al., 1981). LGL do not react with the monoclonal antibody OKT3, which detects almost all of the peripheral blood T-cells. Small T-cells, on the other hand, are negative for OKM1 and OKT10 specificities, negative for Ia, and positive for OKT3. The analysis of the surface markers from the cultured cells has so far given somewhat discrepant results. Using the Staphylococcus aureus binding assay (Ranki et al., 1976), cultured LGL gradually lost the antigens detected by OKM1 and OKT10, began to express Ia and partially the antigens detected by OKT3 (approximately 30-50% of total cultured cells). With regard to the receptors for the Fc part of IgG, the ADCC activity of the cultured LGL clearly demonstrates the presence of this receptor on them. Although rosetting of cultured LGL with antibody-coated ox erythrocytes has given negative results, the expression of this receptor on cultured LGL is detectable by using fluorescein-labeled antigen-antibody complexes and flow cytometry analysis (J. Titus and J.R. Ortaldo, unpublished observation). According to both flow cytometry and Staphylococcus aureus binding assays, cultured T-cells remain positive for OKT3, become positive for Ia, and remain negative for OKM1 and Fc receptors for IgG.

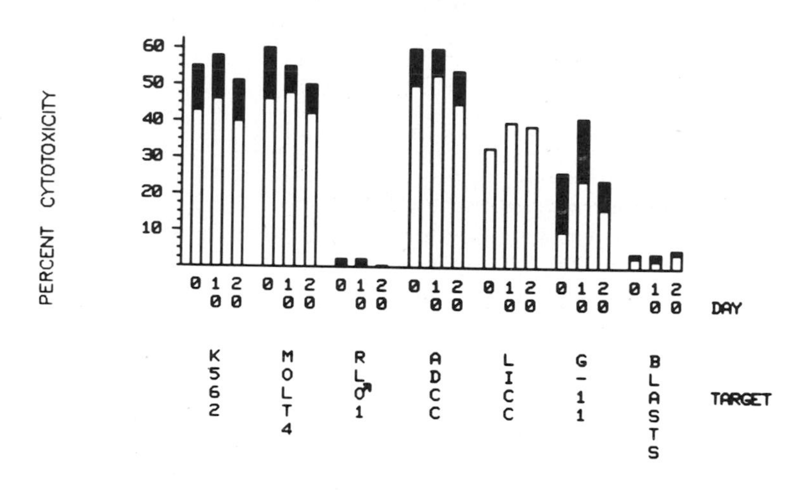

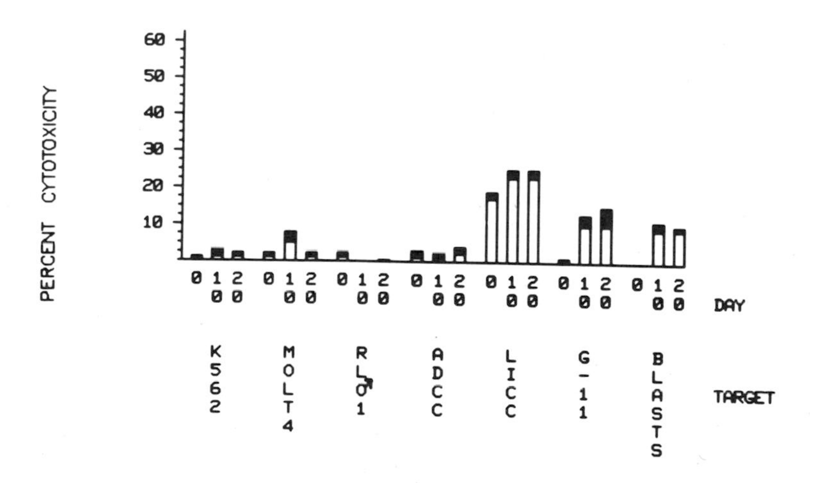

FIGURE 3. Cytotoxic activity of fresh (day 0) and cultured LGL and T-cells (days 10 and 20). Percent cytotoxicity figures from 4 hour chromium^{51}-release experiments (effector:target cell ratio 20:1). Mean from three experiments, black bars indicate activity induced by the pretreatment of the effector cells for 3 hours with 800 IU/ml of human β (fibroblast) IFN (Hem Research, Rockville, MD).

CONCLUSIONS

Both morphological and functional analyses of cultured LGL and T-cells strongly suggest that LGLs are capable of proliferating *in vitro* when stimulated initially with PHA and maintained in culture in the presence of IL-2. The ability to expand LGL will obviously be helpful in the attempts to clone NK cells and in their biochemical analyses. The results also associate LGL with T-cells; this association is even strengthened by the fact that the attempts to culture LGL in the presence of colony stimulating factor (CSF) have been unsuccessful (W. Domzig, unpublished observation). The present results raise two major questions: is the growth factor responsible for the proliferation of LGL really IL-2, and are the cells proliferating in the LGL fractions really LGL? To address the first problem, we have successfully used highly purified IL-2 for culturing LGL (Timonen *et al*., 1982). As regards the cells initially proliferating in the LGL fractions, the evidence for the precursor cells being LGL is not conclusive, although strongly suggestive. Both morphologically and functionally the proliferating cells closely resembled LGL. The surface marker analysis, on the other hand, has not particularly clarified the LGL identity in the growing cells. The limiting dilution analysis on the precursor frequencies among LGL fractions have indicated that 1 in 70 cells proliferates, whereas the frequency of precursors among T-cells was 1 in 10 (Timonen *et al*., 1982). When cultures were initiated in the absence of lectins, the frequency of LGL remained unchanged (1/20-1/100), whereas the T-cells demonstrated a significant shift in proliferation (1/3500-1/4000) (Timonen *et al*., 1982). The cytotoxic patterns, morphology and phenotype of LGLs were similar to lectin-initiated cultures, with the major difference being the LGL growth rate. The T-cells did not grow without lectin.

Current studies, examining the growth of antigenic subpopulations of fresh LGL and cloning of LGL, will elucidate the apparent heterogeneity of NK cells.

ACKNOWLEDGMENT

This work has been supported by Grant No. 1 RO1 CA 23809-01 from the National Cancer Institute, National Institutes of Health, Bethesda, MD.

REFERENCES

Dennert, G., Yogeeswaren, G. and Yamagata, S. (1981). J Exp Med 153:545.

Fresno, M., Nabel, G., McVan-Boudreau, L., Furthmayer, H. and Cantor, H. (1981). J Exp Med 153:1246.
Kuribayashi, K., Gillis, S., Kern, D.E. and Henney, C.S. (1981). J Immunol 126:2321.
Kurnick, J.T., Hayward, A.R. and Altevogt, P. (1981). J Immunol 126:1307.
Lutz, C.T., Glasebrook, A.L. and Fitch, F.W. (1981). J Immunol 126:1404.
Ortaldo, J.R., Timonen, T. and Bonnard, G.D. (1980). Behring Inst Mitt 67:258.
Ortaldo, J.R., Sharrow, S.O., Timonen, T. and Herberman, R.B. (1981). J Immunol 127:2401.
Ranki, A., Totterman, T.H. and Hayry, P. (1976). Scand J Immunol 5:1129.
Timonen, T., Ortaldo, J.R. and Herberman, R.B. (1981). J Exp Med 153:659.
Timonen, T., Ortaldo, J.R., Vose, B.M., Henkart, M., Alvarez, J. and Herberman, R.B. (1982). Submitted.
Timonen, T., Ortaldo, J.R., Stadler, B.M., Bonnard, G.D. and Herberman, R.B. (1982). Submitted for publication.

LYTIC EFFECT OF T-CELL CULTURES AGAINST TUMOR BIOPSY CELLS AND K562

Farkas Vánky
Eva Klein

Department of Tumor Biology
Karolinska Institute
and
Radiumhemmet
Karolinska Hospital
Stockholm, Sweden

I. INTRODUCTION

Our goal was the generation of auto-tumor killer T-cell cultures from patients with solid tumors. Different strategies were used for establishment of the cultures: 1) The lymphocytes were first exposed to autologous tumor biopsy cells. These cultures were designated ATS. 1a) IL-2 was added at the initiation of the cultures or 1b) 3-10 days later. 2) The lymphocytes were activated first in MLC, and exposed to IL-2 2-3 days later. The details of part of these experiments are in press (Vánky et al., 1982)

II. MATERIAL AND METHODS

The tumor cells were separated from surgical specimens of patients with solid tumors (3 osteo-, 1 fibrosarcomas, 1 oat cell-, 3 adeno-, 4 squamous cell carcinomas of the lung, and

---

Supported by Federal Funds from the DHEW, Contract Number NO1-CM-74144, by NIH grant 1 RO1 CA-25184-01A1, and by the Swedish Cancer Society.

ISBN 0-12-341360-5

3 hypernephromas). They were used either directly or after frozen storage. The separation procedure is described in our previous publications (Vánky et al., 1979).

Lymphocytes were separated from heparinized blood on Ficoll-Isopaque (F.I.) followed by removal of plastic- and nylon-adherent cells.

Aliquots of nylon column passed lymphocytes (10-15 x 10^6/10-15ml flasks) were mixed either with 0.5-1.5 x 10^6 autologous tumor biopsy cells - for the ATS- or with 5-7 x 10^6 mitomycin C treated, allogeneic, F.I. separated lymphocytes - for the MLC.

The 6 days MLC and ATS cultures contained 95-100% OKT3, 23-30% OKIa1 and 0-2% OKM1 positive cells. After the first 7-10 days, and up to 6 weeks, the doubling time of the cells was between 2 and 3 days. Thereafter it became longer and then the cells ceased to divide.

The source of IL-2 was the supernatant of cultured F.I. separated lymphocytes of healthy blood donors. The lymphocytes were incubated with Wellcome PHA-P (100 μg/10^8 cells/ml) for 30 min at 37°C followed by 6 washes and thereafter kept (10^6 cells/ml) in serum free RPMI-1640 for 48 h.

The IL-2 containing medium was made up of this supernatant mixed with equal volume of RPMI-1640 (Flow Laboratories Ltd., Arvine, Ayrshire, Scotland) containing 10% heat inactivated serum from healthy male donors. Half of the culture medium was exchanged on day 7 and 10, thereafter two third of the volume twice weekly, when the cultures also were divided.

III. RESULTS

Table I presents the lytic characteristics of 5 ATS-IL-2 cultures propagated for 4-6 weeks. They were selected in order to show the different patterns of reactivity obtained. We consider specific ^{51}Cr values over 20% as positive results. Culture no. 2285 was cytotoxic for the autologous but not for 2 allogeneic tumors or K562. Culture no. 2293 killed the autologous tumor cells and K562, no. 1138 killed auto- and allogeneic tumor biopsy cells and K562. Two cultures initiated from the same patient (1047a and 1047b) did not affect either autologous or allogeneic tumors. However, 1047a, to which IL-2 was added on day 5, lysed K562, while 1047b, with IL-2 added on day 0, lacked reactivity.

In all the material - 13 ATS-IL-2 cultures - seven had the first type of reactivity. One of these received IL-2 at the start of ATS and six 3-10 days later. Auto-tumor and K562 lysis was obtained with 2 cultures. Auto-, allo-tumor and

TABLE I. Cytotoxic Potential of ATS-IL-2 Cultures[a]

Patient	IL-2 added on day	Tumor biopsy targets								K562
		2285	2293	1138	1047	1045	2207	2294	2295	
2285	10	56	-	-	-	2	10	-	-	2
2293	0	-	22	-	0	0	-	12	-	28
1138	0	31	-	27	-	-	-	10	25	58
1047a	5	-	-	0	0	-	-	-	0	48
1047b	0	-	-	0	0	-	-	-	0	0

[a]% specific ^{51}Cr release in 4 h assay. Autologous combinations underlined. All targets were sensitive for lysis by activated allogeneic lymphocytes.

K562 lysis was seen with 2 cultures. There was a regularity, in that the cultures which reacted with allogeneic tumors were intiated in the presence of IL-2.

Auto-tumor reactive T cell cultures could also be generated when the lymphocytes were first activated in the conventional MLC (Table II). These cultures had regularly broader cytotoxicity, 4 of the 8 cultures lysed at least one allogeneic tumor biopsy target and all were highly active against the K562 cells.

TABLE II. Cytotoxic Potential of MLC-IL-2 Cultures[a]

Patient	Target cells		K562
	Autologous tumor	Allogeneic tumor	
2293	9	9, 4, 4	56
2294	12	17, 27, 6	56
2296	26	3	81
2297	29	4	59
2285	47	19, 25	33
1045	24	13, 24	45
513	24	30, 39	41
518	17	0, 2	-

[a]% specific ^{51}Cr release in 4 h assay. In the group of allogeneic tumors each value represents different targets, all unrelated to the stimulator in MLC, i.e. they represent "third parties".

A summary of the results with the IL-2 lines is presented in Table III. The reactivity of the cultures prior to the IL-2 initiated growth period is also given. In 2/14 cases auto-tumor lysis was detectable with the freshly harvested lymphocytes. None of the 14 unmanipulated effector population lysed allogeneic tumor biopsy cells (all susceptible to the lysis of activated allogeneic and some lysed by unmanipulated autologous lymphocytes) but they lysed K562 cells in 8/9 cases.

Lymphocytes of ATS cultures tested on the 7-day lysed the autologous tumor cells in 7/7 cases. None of these cultures damaged allogeneic tumor cells but lysed K562. The majority of the propagated ATS cultures tested 4-6 weeks later lysed the autologous tumor cells and only 2 acted on allogeneic tumors. Similarly to the ATS, the MLC maintained their characteristics after propagation with TCGF. All lysed K562, some lysed autologous and third party tumor cells.

On day 39, ATS-IL-2 culture no. 518 was first cultured for 24 h in absence of IL-2 and then reexposed to autologous tumor biopsy cells. Cytotoxicity was measured 3 days later (Table IV). After restimulation, the auto-tumor lysis was enhanced. Autologous Con-A blasts, allogeneic tumor biopsy, K562 cells were not damaged. There was a weak effect, 10%, which is under the value which we usually consider to be positive against cultured kidney cells.

TABLE III. Summary of The Cytotoxic Pattern of ATS and MLC Cultures Prior to and Following Propagation with IL-2[a]

Mode of activation	Targets		
	Autologous tumor	Allogeneic tumor	K562
-	2/14	0/14	8/9
ATS	7/7	0/7	6/6
ATS-IL-2	11/13	2/13	4/13
MLC	3/8	4/8	8/8
MLC-IL-2	5/8	4/8	7/7

[a]The values represent no. of positive populations/total numbers tested. Positivity: % specific ^{51}Cr release $\leq$20.

TABLE IV. Cytotoxicity of The ATS-IL-2 Culture of Hypernephroma Patient 518 on Day 42

Restimulation	Targets					
	T518	L518	N518	T513	T2291	K562
-	21[a]	0	0	0	0	4
+	44	0	10	0	0	1

[a]% specific ^{51}Cr release in 4 h assay.

T = tumor cells; L = Con-A blasts; N = kidney cells cultured for 42 days. T513 is another hypernephroma and T2291 squamous cell carcinoma of the lung.

IV. CONCLUSION

The cultures differred in their lytic potential against autologous and allogeneic tumor biopsy cells and K562. The reactivity of the growing T-cell cultures reflected the characteristics of the population from which they were initiated.

Some of the cultures lysed exclusively the autologous tumor cells. Others were seemingly non-specific because they acted also on several allogeneic tumors. However, on the level of the functioning set of lymphocytes the cytotoxicities were probably selective. Analysis of such cultures prior to the IL-2 driven growth indicated that the effect on the different biopsy cells was exerted by different set of cells in the culture (Vánky and Klein, 1982; Vánky et al., 1982). Thus several antigen recognizing clones may be present also in the growing cultures which exhibit broader reactivity.

The activation of "anomalous" clones, i.e. those not directed against the stimulator cells, occurred when the stimulation was strong such as in MLC or when IL-2 was added at the initiation of the ATS cultures. This is in accordance with the study of Spits et al. (1981), in which the development of the non-specific killer potential in MLC showed a positive correlation with the quantity of the stimulator cells. Furthermore, in parallel experiments with "single" and "pool" stimulated MLC, efficient generation of auto-tumor killing occurred under the latter condition (Strausser et al., 1981).

We assume that the anti K562 effect is polyclonal, nonselective which reflects the differentiation state of the lymphocytes (Klein, 1980).

REFERENCES

Klein, E. (1980). Immunol. Today 1, IV.
Spits, H., de Vries, J.E., and Terhorst, C. (1981). Cell. Immunol. 59:435.
Strausser, J.L., Mazumder, A., Grimm, E.A., Lotze, M.T., and Rosenberg, S.A. (1981). J. Immunol. 127:2666.
Vãnky, F., and Klein, E. (1982). Int. J. Cancer, in press.
Vãnky, F., Vose, B.M., Fopp, M., Klein, E., and Stjernswärd, J. (1979). In "Immunodiagnosis and Immunotherapy of Malignant Tumors - Relevance to Surgery?" (H.D. Flad, Ch, Erfarth, M. Betzler, eds), p. 143. Springer Verlag, Berlin.
Vãnky, F., Argov, S., Einhorn, S., and Klein, E. (1980). J. Exp. Med. 151:1151.
Vãnky, F., Gorsky, R., Gorsky, Y., Masucci, M-G., and Klein, E. (1982). J. Exp. Med., in press.

NATURAL KILLER CELL-LIKE CYTOTOXICITY MEDIATED BY HERPESVIRUS TRANSFORMED MARMOSET T CELL LINES

Donald R. Johnson[1]

Department of Pathology and Laboratory Medicine
University of Nebraska
Omaha, Nebraska

Mikael Jondal[2]

Department of Tumor Biology
Karolinska Institute
Stockholm, Sweden

Peter Biberfeldt[3]

Department of Pathology
Karolinska Institute
Stockholm, Sweden

I. INTRODUCTION

In contrast to human T lymphocytes, cotton-topped marmoset T cells can easily be transformed in vitro with Herpesvirus ateles (HVA) (Falk et al., 1978). Also Herpesvirus saimiri (HVS) may transform marmoset T cells although such transformation only occurs in vivo, ie. cell

[1]Fellowship provided by the Cancer Research Institute Inc.
[2]Supported by grants from The Swedish Cancer Association and the Karolinska Institute.
[3]Supported by the Swedish Cancer Society.

ISBN 0-12-341360-5

lines are established from explanted thymomas (Falk et al., 1972). Our recent studies in continuous HVA/HVS transformed cell lines have indicated that they retain some functions typical for T cell immunity, in particular the capacity to mediate cytotoxicity reactions against certain types of target cells, in the absence of added antibodies or lectins (Johnson et al., 1981; Johnson et al., 1981). The characteristics and specificity of this reaction has led us to suggest that HVA/HVS cell line-mediated cytotoxicity may represent a variant of NK killing that can be expressed in the T cell lineage, possibly in parallel with specific reactivity.

As the HVA/HVS cell lines easily can be grown to the large numbers needed to perform meaningful biochemistry on the levels of the cell surface receptors, and the lytic machinery, they may prove useful for characterization of cell-mediated cytotoxicity. For these reasons they are included in the present publication.

II. METHODS

Peripheral blood lymphocytes from adult cotton-topped and white-lipped marmosets (Saguinus oedipus and S. nigricollis) (generously supplied by L. Wolfe, Rush Medical Center, Chicago) and from normal human donors were separated by centrifugation on Ficoll/Isopaque and nylon wool as described earlier (Julius et al., 1973).

Permanent lymphoid cell lines were established by transformation of marmoset lymphocytes with cell-free HVA obtained from supernatants of HVA-infected marmoset kidney fibroblasts or cocultivation with HVA-producing tumor or lymphoid cells (Falk et al., 1974). All cell lines were propagated in RPMI 1640 medium supplemented with 10% fetal calf serum, 100 units of penicillin per ml. and 100 μg of streptomycin per ml.

The cytotoxicity assays were done in standard V-shaped microplates (Linbro) using 10^4 target cells per well in a total volume of 150 μl of medium with 10% newborn calf serum. Target cells were labelled with sodium (^{51}Cr)chromate (100 μCi per 5×10^6 cells for 1 hour at 37°C). Effector cells were added at different ratios and incubated together with the target cells for 6 hours before the test was ended by harvesting 50 μl of supernatant for calculation of released radioactivity. Cytotoxicity was calculated by the formula:

$$100 \times \frac{\text{(test release)} - \text{(spontaneous release)}}{\text{(80\% of total label)} - \text{(spontaneous release)}}$$

III. RESULTS

Seven different HVA-HVS transformed cell lines were tested for surface markers, killing of different target cells and other functional characteristics (Table I and II) as reported in more detail elsewhere (Johnson et al., 1981; Johnson et al., 1981). The cytotoxic specificity was primarily directed against human T cell lines derived from ALL. The activity of the individual lines was found to be different in intensity and also varied in time when serially tested on a monthly basis (data not shown here). The lines had surface markers typical for the T cell phenotype and lacked the ability to mediate LDCC and ADCC. In cold target inhibition tests (against hot target Molt-4), K-562 was most inhibitory followed by Molt-4 itself. No inhibition was seen with cold targets Daudi and YAC-1.

Electron microscopy gave the following results. The general ultrastructure of the cells was lymphoblastoid (Figs. 1, 2) with relatively abundant cytoplasm, rich in polyribosomes, mitochondria and a moderately well developed Golgi zone including some "dense bodies/granules." Short profiles of endoplasmic reticulum were present in varying numbers. The nuclear contour was irregular to convoluted or occasionally "ceribriform" (Fig. 2) and displayed usually one rounded nucleolus. The surface of the cells showed varying degrees of villousness. Relatively many cells showed cytoplasmic-nuclear polarization with uropod for-

TABLE I. Characteristics of HVA/HVS transformed marmoset T cell lines.

Surface marker / Function	+/-
SRBC receptor	+
T antigens	+-
IgG receptors	-
IgM receptors	-
Complement (C3) receptors	-
Lectin-dependent killing (LDCC)	-
IgG dependent killing (ADCC)	-
Inhibited by cold targets:	
K-562	+
Molt-4	+
Daudi	-
YAC-1	-

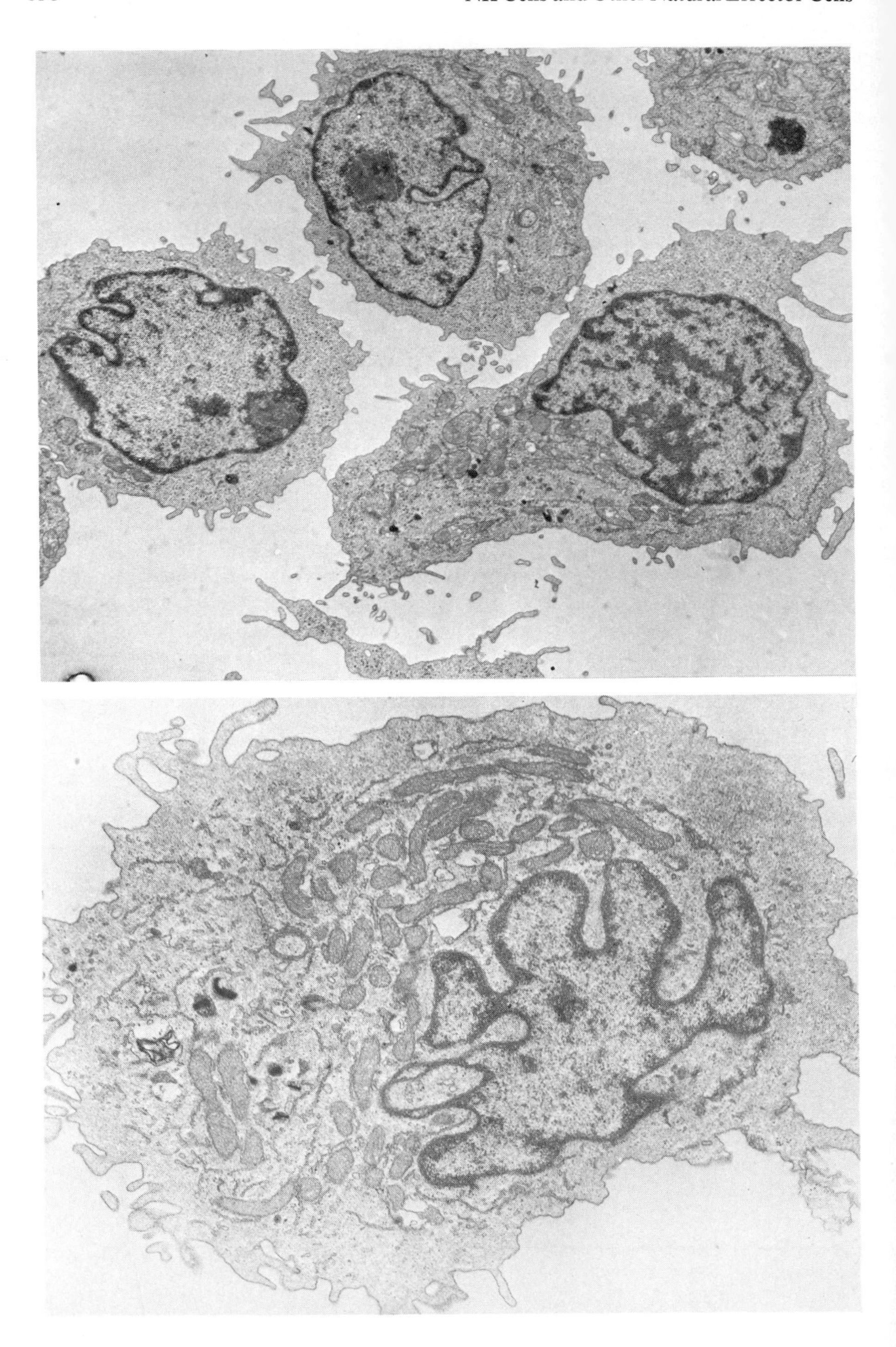

mation and "hand-mirror" configuration, suggestive of well developed surface motility and preambulatory activity.

Cytocentrifuged cells showed relatively strong "spotty" positivity for acid phosphatase, but weak reactivity for alphanaphtyl acid esterase (ANAE).

Killing by the 77-DI-2 effector line resembled killing by cotton-topped marmoset NK cells more closely than killing by white-lipped marmoset NK cells or human NK cells (Table III).

Using a panel of inhibitors of cell-mediated cytotoxicity reactions it became clear that HVA/HVS cell line mediated

TABLE II. Target cell spectrum of HVA/HVS transformed killer cell lines.

Origin and characteristics tested target cells	EBV[a] +/-	Susceptibility[b] to killing
T cell lines from acute lymphocytic leucemia[c]	-	+++
K-562, derived from erythroleukemia	-	+
B cell lymphomas	-	+
B cell lymphomas	+	-
Human B type lymphoblastoid cell lines	+	-
Marmoset B type lymphoblastoid cell lines	+	-
Human T lymphoblasts	-	-
YAC-1, mouse prototype NK target	-	-

[a]Positive or negative for the Epstein-Barr virus.
[b]+++ indicates between 25-75%, + indicate between 10-25% and - below 10%. [c]Including cell lines Molt-4, 1301, CCRF-CEM and HSB-2.

killing was much more difficult to inhibit than human NK cells (Table IV). Only cytochalasin B and low temperature gave absolute inhibition. Also a combination of papain treatment followed by inhibition of protein synthesis with puromycin gave an absolute inhibition in this system (data not shown).

IV. DISCUSSION

The over-all impression from our results is that the HVA/HVS cell lines represent highly active marmoset T cells in a differentiation stage in which they can mediate a variant of NK-like cytotoxicity. To our disappointment we have so far not been able to fully exploit the access to

TABLE III. Cytotoxicity by the HVA transformed effector cell line 77-DI-2 in comparison with NK killing by human and marmoset lymphocytes.

Effector[a] cell	Target cell[b]								
	1301			K-562			Raji		
	20	10	5	20	10	5	20	10	5
	(specific 51-Cr release)								
77-D1-2 cell line	62	53	41	15	11	6	3	1	0
Marmoset NK (cotton-top)	49	38	29	21	10	4	6	4	1
Marmoset NK (white-lip)	98	81	72	96	79	63	62	51	36
Human NK	71	56	42	73	41	30	11	8	2

[a]Marmoset and human NK cells were isolated from peripheral blood.
[b]Origin and characteristics of target cells is given in the Methods section. Numbers 20, 10 and 5 indicate different effector: target ratios 20:1, 10:1 and 5:1.

large numbers of virally transformed killer cells as they have a tendency to vary, in an uncontrollable way, in their cytotoxic potential during tissue culture. This seem to be a problem with all in vitro maintained killer cell cultures, also those dependent in interleukin-2. Our efforts to control HVA/HVS cell line-mediated killing, using published methods including lectins, phorbol eaters, calcium ionophores, exogenous cylcic nucleotides etc., have only marginally increased killing. Possibly, the variations we

TABLE IV. Influence of inhibitors of cell-mediated cytotoxicity on human NK activity and on killing mediated by HVA/HVS transformed cell lines.

Inhibitor	Site of action	Concentration[a]	HVA/HVS[b] cell line	Human[b] NK
Trypsin	Membrane	10 mg/ml	0.8	0.2
Papain	Membrane	2.5 mg/ml	0.3	0.1
DMSO	Membrane	0.1 M	0.9	0.2
Cytochalasin A	Cytoskeleton	1 ug/ml	0.9	0.0
Cytochalasin B	Cytoskeleton	10 ug/ml	0.0	0.0
Colchicine	Cytoskeleton	5 ug/ml	0.6	0.1
Puromycin	Protein synth	50 ug/ml	1.0	0.4
Cyclohexamide	Protein synth.	10 ug/ml	0.9	0.2
Dibuturyl-cAMP	Cyclic nuc.	10 mM	0.5	0.0
Theophylline	Cyclic nuc.	10 mM	0.3	0.0
Low temp.	Glycolysis	18°C	0.0	0.0
NaF	Glycolysis	5 mM	0.8	0.3
NaN	Respiration	5 mM	0.5	0.2
DNP^{3}	Respiration	1 mM	0.9	0.1
EDTA	Respiration	10 mM	0.3	0.0

[a]Highest tested concentration that would not be toxic for target cells in the absence of effector cells. [b]Influence on killing given as an index obtained by dividing specific ^{51}Cr release in the presence of inhibitor by control release in the absence of inhibitor.

see in cytotoxic potential may be explained by on-going "auto-killing" in the cultures as we know that these effector cell lines are themselves susceptible to human NK cell lysis. A shift in the target-effector balance may then explain the variation as we see them.

A final point of interest may be the relatively low reactivity the HVA/HVS transformed cell lines show against Epstein-Barr virus positive B type lymphoma cells (Table III). This may reflect a specific immunological defect in these animals which can partly explain their general susceptibility to viral oncogenesis and to Epstein-Barr virus induced lethal lymphoproliferation in particular (Miller et al., 1979).

AKNOWLEGEMENTS

The excellent technical assistance of Miss Maj-Britt Alter is gratefully acknowledged. Also, we thank Dr. L. Wolfe, Rush Medical Center Chicago, for a continuous supply of marmoset blood.

REFERENCES

Falk, L., Johnson, D.R., and Deinhardt, F., Int. J. Cancer, 21, 652, (1978).

Falk, L., Wolfe, L.G., and Deinhardt, F., Bacteriol. Proc. 38, 191, (1972).

Falk, L., Wright, J., Wolfe, L., and Deinhardt, F., Int. J. Cancer, 14, 244, (1974).

Johnson, D.R., and Jondal, M., Nature, 291, 81, (1981).

Johnson, D.R., and Jondal, M., Proc. Nat. Acad. Sci. 78, 6391 (1981).

Julius, M.H., Simpson, E., and Herzenberg, L.A., Eur. J. Immunol. 3, 645, (1973).

Miller, G., in "The Epstein-Barr virus" (Editor, Epstein, M.A., and Achong, B.G.), 351, Springer Verlag, New York, (1979).

CLONED CELL LINES WITH NATURAL KILLER ACTIVITY[1]

Gunther Dennert

Department of Cancer Biology
The Salk Institute for Biological Studies
San Diego, California

I. INTRODUCTION

The characterization of natural killer (NK) cells and the respective factors which modulate their cytolytic activity has been seriously hampered in the past by inadequate methodology to separate NK cells from other classes of lymphocytes. Recent advances in cell culture technique have made possible the establishment *in vitro* of permanent cell lines with natural killer activity (Dennert,1980; Dennert et al.,1981). This progress can now be exploited in order to characterize the function, specificity, and cell surface markers of NK cells. In this article, a summary of some of the properties of cloned NK cell lines will be reported.

II. RESULTS

A. Establishment of Cell Lines and Clones with Natural Killer Activity

Previous studies had indicated that NK activity may increase during mixed lymphocyte reactions which give rise to a substantial release of T cell growth factor (TCGF). It was

[1]This work was supported by U. S. Public Health Service Grants CA 15581 and CA 19334, and The American Cancer Society Grant IM-284.

ISBN 0-12-341360-5

therefore reasoned that NK cells might be able to proliferate in TCGF containing cell supernatants. In order to avoid the proliferation of thymus derived lymphocytes under the influence of TCGF, initially spleen from nude mice which lack T cell function but are high in NK activity (Herberman et al. 1978) were employed. Results showed that such cultures indeed express high levels of NK activity, as shown by lysis of NK sensitive targets such as YAC-1. After several months in culture, NK activity had increased by about a factor of twenty on YAC-1 targets and cloning of the effector cells was initiated (Dennert,1980). The procedure of choice was cloning by limiting dilution in microtiter wells making use of TCGF containing media prepared by stimulating spleen cells with Con A. NK cell lines were isolated and cloned from a number of different mouse strains including BALB/c nu/nu, BALB/c, C57Bl/6, C3H, (B6xDBA/2)F_1 and (B6xC3H)F_1. Preliminary tests of these cell lines revealed very similar, if not identical, specificity patterns in the cytolytic reaction on various NK sensitive and resistant target cells (Dennert,1980; Dennert et al.,1981).

B. Effects of Lymphokines on Cell Proliferation and Cytolytic Activity of NK Cells

While NK cell lines grow very rapidly in tissue culture media containing TCGF, they stop synthesizing DNA in normal medium and eventually die. Lymphokines contained in conditioned media therefore appear to be essential for the continuous proliferation and well being of these cells. In an attempt to delineate which lymphokines may be responsible for NK cell proliferation and expression of cytolytic activity, a number of lymphokine preparations of various purity were tested. Results showed that preparations containing TCGF, like Con A conditioned media (Con A CM), supernatant (SN) of the cell line LBRM 33, or TCGF purified from LBRM 33 SN, stimulate NK cell proliferation. In contrast, supernatants containing lymphocyte activating factor (LAF) or interferon did not stimulate (Table 1). Furthermore, cell supernatants depleted of TCGF by absorption with cells failed to stimulate DNA synthesis (data not shown). These results provide compelling but not conclusive evidence that TCGF is indeed necessary and sufficient for NK cell proliferation.

Another question of interest is whether lymphokines may have modulatory effects on the cytolytic activity of NK cell lines. To that end it is known already (Gidlund et al., 1978; Trinchieri and Santoli,1978) that interferon (IF) may

TABLE I. Effect of Lymphokines on Proliferation of NK 11 6D6[a]

Supernate	Concentration	Proliferation ^{3}H-TdR Incorporation (CPM)
None	None	780 ± 108
Con A CM	10%	20,424 ± 347
LBRM-33	10%	21,710 ± 1,097
TCGF (LBRM-33)[b]	500 U/ml	27,761 ± 608
LAF ($P388_{D1}$)[c]	10%	363 ± 131
IF[d]	1000 U/ml	425 ± 139

[a]NK 11 6D6 is a cloned NK cell line derived from BALB/c mice (Dennert et al., 1981)

[b]TCGF partially purified from LBRM-33 (supplied by J. Watson, University of California, Irvine)

[c]Supernate from $P388_{D1}$ cell line (supplied by G. W. Wetzel, University of California, Irvine)

[d]Mouse fibroblast interferon obtained from L929 cells (supplied by Lee Biomolecular, San Diego)

stimulate NK cytolytic activity *in vitro* or *in vivo*. NK cell lines cultured for 24 hr in normal medium show decreased cytolytic activity as compared to cultures left in Con A CM. This showed that lymphokines contained in Con A CM are responsible for maintenance of optimal cytolytic activity since the viability of the cultures in normal medium was not affected. NK cells were incubated for 1-24 hr in various lymphokine preparations and subsequently assayed for cytolytic activity. Results showed that TCGF containing cell supernatants but not supernatants containing LAF or IF boosted cytolytic activity. This suggests that TCGF is not only necessary and sufficient for NK cell proliferation but also for the maintenance of cytolytic activity. This is also supported by the finding that LBRM 33 SN, which contains TCGF (Gillis et al., 1980) but not LAF or IF, maintains the growth and cytolytic activity of various cloned NK cell lines like NK 11 6D6 and NKB61A2 for several weeks.

Further support for the conclusion that TCGF is an important lymphokine for proliferation of NK cell lines and their activity comes from the observation that NK cells appear to express TCGF receptors. Thus incubation of cloned NK cell lines with TCGF-containing medium at 4°C for 2-4 hr results in disappearance of TCGF activity from the SN, as shown by its failure to support the proliferation of TCGF-dependent cell lines such as CTLL-2 (Baker et al., 1979). These data taken together suggest that cell proliferation and cytolytic activity of NK cell lines depend on TCGF for which these NK cells express cell surface receptors and, furthermore, that LAF and IF have little effect.

C. Specificity and Target Receptors of NK Cells

The availability of NK cell clones provided the unique possibility to study the specificity of NK cells on a clonal level. Since NK cells show lytic preference for some tumor cells as opposed to others, they appear to express target specificity which may point to specific target receptors. One question which could be examined is whether target specificity of NK cells is clonally distributed or not. A number of NK cell clones from various mouse strains were therefore assayed on a battery of target cells. Results showed that the patterns of lysis were indistinguishable between NK clones and normal spleen cells containing NK activity (Dennert 1980; Dennert et al., 1981). This suggested that NK cells do not have clonally distributed target receptors like conventional T killer cells.

The receptor (or receptors) NK cells display could have either of two functions. Either it binds to target cells only or it binds to the target and transmits the lethal hit as well. By employing various tumor cells as cold target inhibitors, it was attempted to distinguish between these two possibilities (Dennert, 1981). In cold-target inhibition assays it is seen that there are targets which inhibit lysis of YAC-1 as efficiently as YAC-1, i.e., Chang and B/C-N. Others are intermediate inhibitors like S49 and Raji, and still others inhibit very poorly, such as EL4, P815 and R1G1 (Table 2). Testing of these targets for lysis reveals that YAC-1, Chang, S49 and Raji are lysable but all others are not. Hence one of the targets, B/C-N, which inhibits in a cold-target inhibition assay is not lysable. This clearly shows that the binding of the effector to the target is not sufficient for lysis to occur. Several explanations for this result are possible. Thus, the lethal hit could either not be received by the target or its action could be inactivated and the damage subsequently repaired regardless of whether the lethal hit is transmitted by the target-specific receptor or a site different from the receptor. Interestingly, among the targets which are not lysable (R1G1, B/C-N, EL4 and P815), there are some (EL4 and P815) which are lysable by NK cells

TABLE II. Effectivity of Targets in Cold-Target Inhibition Assays and Lysability of Targets in the Absence or Presence of Con A

Inhibitor/Target	Cold Target Inhibition: % Inhibition	Target Lysability: % Cytotoxicity	
		w/o Con A	w/Con A
YAC-1	55	55	100
Chang	45	63	65
B/CN	59	1	23
S49	24	52	63
Raji	23	22	48
P815	5	8	100
EL4	5	11	83
R1G1	7	11	21

[a]Data shown are taken from titration curves performed at various cold target inhibitor ratios. The effector cell was NK 11 6D6 and target was YAC-1. Cold target to labeled target ratio was 10:1 as was the effector to labeled target ratio. Incubation time was 4 hr.

[b]Attacker to target ratio was 10:1 and a 4 hr incubation was done. Con A concentration was 5 μg/ml and NK 11 6D6 was used as effector cell.

in the presence of Con A (Table 2). This clearly shows that there are targets which can receive, or are susceptible to, the lytic signal while others are not. There are, therefore, several possible reasons why targets are not lysed by NK cells: 1) The target does not express a structure required for NK cell binding but can receive the lytic signal (P815, EL4). 2) The target expresses the structure to which NK cells bind, but does not receive, or is not susceptible to, the lytic signal (B/C-N). 3) The target neither allows effector cell binding nor lysis (R1G1). While these results exemplify how NK targets can escape NK lysis, by mutations affecting either the receptor binding site or the acceptor site for the lytic signal, they do not distinguish between the possibilities that the target receptor site transmits the lethal signal or whether this function is executed by a different site. But it is clear from these experiments that target binding and lysis are two independent events (Dennert, 1981). This conclusion is also supported by the finding that hexose phosphates like mannose 6-(P) inhibit target lysis by NK cells but not their binding to the targets (data not shown).

One as yet unknown question regarding the NK cell receptor is what its nature might be. To that end we know already that it is not antibody attached to NK effectors via Fc receptors. This is because these NK cell lines do not appear to function in an antibody-dependent cell mediated cytolytic reaction (Dennert et al., 1981). It rather seems that NK cells synthesize their target receptors because NK cell lines grown in LBRM 33 SN, which is presumably free of specific antibody, are specifically cytolytic.

D. Cell Surface Markers of NK Cells

Since the cell surface antigenic phenotype of NK cells is a somewhat controversial issue, NK cell lines could provide a useful material to characterize the cell surface antigens of NK cells. So far only a limited number of cell surface antigens have been examined. It was shown that our NK cell lines are Thy 1^+, $T200^+$, Lyt 1^-2^-, asialo GM 1^+, and asialo GM 2^+ (Dennert, 1980; Dennert et al., 1981). It is interesting that NK cell lines from nude mice are Thy 1^+ which suggests that these Thy 1^+ cells do not require thymic differentiation for their function. The absence of Lyt 1 and Lyt 2 antigens clearly distinguishes these cells from conventional T cells. Similarly the presence of considerable amounts of asialo GM 1 and asialo GM 2 distinguishes NK cells from specific T killer cells, which occasionally express only very small amounts of asialo GM 1.

III. SUMMARY AND PERSPECTIVES

Cell lines with natural killer activity can be established and cloned in TCGF containing cell supernatants starting from spleen cells of normal or nude mice. These cells, because of their expression of Thy 1 antigen, may belong to the T cell compartment, although processing in the thymus is apparently not required for their function. These cell lines are considered to be of NK type because of their lytic specificity on a number of target cells as well as their cell surface phenotype. But it has to be kept in mind that the parameters currently available for the definition of NK cells are at best rather superficial. Hence it is possible that several distinct cell types with NK activity exist of which only one is being studied here (Minato et al., 1981).

NK cell lines require TCGF for their proliferation and TCGF is able to boost their cytolytic activity. The importance of TCGF for NK cells is reflected in the finding that NK cells absorb this lymphokine from supernatants which is compatible with the assumption that NK cells possess cell surface receptors for TCGF. A search for other lymphokines able to stimulate NK cell proliferation or able to boost NK cytolytic activity was not successful. Both IF and LAF had no effects on NK cells. The failure of IF to boost NK activity was unexpected since IF has been reported to have stimulatory effects on NK cytolytic activity. Hence it is possible that our NK cell lines are either of a type of NK cells or in a state of differentiation in which they are refractory to IF action. Assay of NK cell clones for their cytolytic specificity on various targets revealed that their target receptors are not clonally distributed. Binding of the NK cell to the target cell surface antigen does not necessarily result in target lysis because there are tumor targets which bind to NK effectors but are not lysed. Other targets are lysable if coupled with lectin to the effector cells and still others exist which neither can be lysed by, nor do they bind to, NK cells. This, of course, shows that binding and killing are two events independent from each other and that there are several means by which tumor cells may escape NK cell lysis.

While cloning of NK cells has already contributed significantly to the knowledge of NK cells, it is now possible to assess the *in vivo* functional role of NK cells in respect to tumor rejection and regulation of hematopoiesis. Furthermore, the putative defects in certain clinical situations involving NK cells can now be studied by establishing NK cell lines from respective patients.

ACKNOWLEDGMENTS

I thank Cheryl Bry for technical assistance.

REFERENCES

Baker, P. E., Gillis, S., and Smith, K. A. (1979). *J. Exp.Med. 149:*273.

Dennert, G. (1980). *Nature 287:*47.

Dennert, G. (1981). In " Mediation of Cellular Immunity in Cancer by Immune Modifiers" (Chirigos et al., ed.), p. 153. Raven Press, New York.

Dennert, G., Yogeeswaran, G., and Yamagata, S. (1981). *J. Exp. Med. 153:*545.
Gidlund, M., Orn, A., Wigzell, H., Senik, A., and Gresser, I. (1978). *Nature (Lond.) 273:*759.
Gillis, S., Scheid, M., and Watson, J. (1980). *J. Immunol. 125:*2570.
Herberman, R. B., Nunn, M. E., and Holden, H. T. (1978). *J. Immunol. 121:*304.
Minato, N., Reid, L., and Bloom, B. R. (1981). *J. Exp. Med. 154:*750.
Trinchieri, G., and Santoli, D. (1978). *J. Exp. Med. 147:*1314.

CHARACTERIZATION OF CLONED MURINE CELL LINES HAVING HIGH CYTOLYTIC ACTIVITY AGAINST YAC-1 TARGETS

Colin G. Brooks
Kagemasa Kuribayashi
Susana Olabuenaga
Mei-fu Feng
Christopher S. Henney[1]

Program in Basic Immunology
Fred Hutchinson Cancer Research Center
Seattle, Washington

Despite intensive study of natural killer (NK) cells, many of the major questions concerning these cells, such as the lineage to which they belong, the nature of the target structure(s) recognized, the nature of the receptors they bear, and the biochemistry of target cell lysis, still remain unanswered. A major advance towards answering these questions could be made if it were possible to grow cloned populations of NK cells in long-term culture. Such an approach has already been successful in studies of various T cell subsets (1), and preliminary studies in this and other laboratories have indicated that long-term growth of NK cells may indeed be possible (2-4). Here we describe the production and characterization of a series of murine cell lines which have extremely high cytotoxicity against YAC-1 targets and which share other key characteristics with splenic NK cells.

A. Materials and Methods

Growth medium (GM) for killer cell lines was composed of 6 parts of Click's medium with 10% fetal calf serum (FCS) and 4 parts of supernatant (SN) from Concanavalin A (Con A) stimulated BALB/c

[1] Supported by grants CA 24537 and AI 15384 from the National Institutes of Health and IM-247 from the American Cancer Society.

ISBN 0-12-341360-5

spleen cells (5).

For production of cloned killer cell lines, spleen cells (or occasionally BCG-stimulated peritoneal exudate cells (PEC)) from various strains of mice were first treated with monoclonal anti-Thy-1.2 and guinea-pig complement under conditions which eliminated >90% of their Con A responsiveness but which minimally affected NK activity. The Thy-1.2-depleted cells were resuspended at 5 million/ml in RPMI-1640/10% FCS containing 2-ME and 50 µg/ml poly I:C and 25 ml aliquots cultured overnight. The poly I:C stimulated cells were then transferred to GM, and cloned usually 2 days later by limiting dilution in microplate wells containing 50 thousand syngeneic thioglycollate-induced PEC. Wells were re-fed with GM every 3-4 days. After 10-14 days, colonies were visible, developing at a cloning efficiency of about 1%.

For cytotoxicity assay, Cr-labelled targets were mixed with effectors in V-bottomed microtest plates and after 4 hr at 37° aliquots of supernatant were taken. Specificity was determined by titrating effector cells on a panel of target cells. The dose of effector cells giving closest to 30% lysis of YAC-1 was used to determine the relative specificity on the test target cells: relative specificity = 100 X (%lysis of test cells)/(% lysis of YAC-1).

Blocking of cytotoxicity with antisera was performed in flat-bottomed microplate wells. The anti-Ly-5 serum (used at 1/12) was generously donated by Dr. M.R. Tam, and has been described elsewhere (6). The anti-Lyt-2 antibody (used at 1/50) was prepared from culture supernatants of hybridoma 53.6.72 (7) and concentrated by ammonium sulfate precipitation. Antibody dependent cytotoxicity was measured according to Pollack et al. (8), using SL-2 target cells treated with anti-Thy-1.1 monoclonal antibody.

Cell surface antigen display was evaluated by fluorescence staining using monoclonal antibodies wherever available. Reagents were spun at 100,000g for 20-60 min to remove aggregates. The anti-NK alloantigen reagent used (generously donated by Dr. R.C. Burton) was an (NZB X CE) F1 anti-CBA serum which specifically reacts with NK cells and defines the NK-1.2 antigen (9, 10). Anti-Qa-5 antibody was kindly donated by Dr. G.C. Koo.

B. Results

Following treatment of spleen cells with anti-Thy-1 antibody and complement, stimulation with poly I:C, and a brief period of culture in medium containing supernatant from Con A stimulated spleen cells, approximately 1% of cells (irrespective of the strain of mouse used) were able to form colonies in the presence of appropriate feeder cells (such as syngeneic PEC). Without feeder cells the colony forming frequency was about 10-fold lower. To investigate the possibility that at least some of these clones were of NK lineage, colonies were expanded by culture in GM and tested for cytolytic activity towards the NK sensitive target cell, YAC-1. Cloned cell lines were considered to have high cytolytic activity against YAC-1 if they gave greater than 20% lysis at an E:T ratio of 1:1. A large proportion of the cloned cell lines prepared from CBA, C57BL/6 and C57BL/6-bg/+ mice were of this type (Table 1). The lytic activity of these clones was often 100-1000 fold greater than that of normal or poly I:C activated spleen cells. By contrast, with spleen cells from DBA/2, BALB/c and C57BL/6-bg/bg mice, highly cytolytic clones were obtained only rarely.

Table 1. Cytotoxicity Against YAC-1 of Clones Obtained From Different Strains of Mice

Strain	Spleen cell cytotoxicity[a]	No. clones studied	% clones cytotoxic[b]
Experiment 1			
CBA	30	18	28
C57BL/6	15	14	80
DBA/2	19	40	5
BALB/c	12	23	0
SJL	2	40	20
Experiment 2			
C57BL/6 bg/+	11	20	85
C57BL/6 bg/bg	2	19	10

[a] Mean % cytotoxicity of spleen cells from 6-8 week old animals against YAC-1 at E:T ratio 25:1 (4 animals/group in Experiment 1, 2 animals/group in Experiment 2).

[b] % of clones having significant cytotoxic activity (greater than 20% cytotoxicity against YAC-1 at E:T ratio 1:1).

Growth of the cell lines was absolutely dependent on the presence of supernatant from Con A stimulated spleen cells. Con A itself was not required, as growth was unimpeded by an excess (25 mM) of α-methyl-D-mannoside (α-MM) which was shown to inhibit the binding of Con A to these cells. The supernatants used to promote growth contained interleukin-2 (IL-2), and when purified IL-2 (kindly provided by Dr. S. Gillis) was tested, it was indeed found capable of promoting thymidine uptake in these cells (Table II). By contrast, mouse interferons (both β and γ) had no growth stimulating activity, either alone or in admixture with IL-2 (data not shown).

Over a period of months, cytotoxic activity of the clones declined. This was probably due to a gradual overgrowth of cultures by variants with reduced cytotoxicity as (a) maximal cytolytic activity could be maintained by periodical re-cloning and selection of the daughter clone having highest cytotoxicity, and (b) at each reclon-

Table II. Stimulation of thymidine uptake by IL-2[a]

Units/ml of IL-2	Source of IL-2 Whole SN[b]	Purified[c]
0	1234	
0.1	2553	2762
0.5	7554	6911
2.0	11206	13341

[a] Five thousand cloned killer cells (B10G7) were incubated in flat-bottomed microplate wells for 16 hr with various concentrations of IL-2, then pulsed for 4 hr with ^{3}H-thymidine. Figures show cpm uptake.
[b] SN from Con A stimulated BALB/c spleen cells.
[c] IL-2 from mouse tumor cells purified by successive gel filtration, ion exchange chromatography, and isoelectric focussing.

ing daughter clones having both higher and lower cytotoxic activity than the parent clone (with a mean approximating that of the parent clone) were obtained. Typical results for B10G7 and its daughter clones are shown in Table III.

Table III. Maintenance of Cytolytic Activity by Recloning

Cell line	Day tested	Percent cytotoxicity (E:T ratio 1:20 on YAC-1)
B10G7 (parent)	0	42
B10G7 (parent)	217	2
C2F10 (1st recloning)	217	13
J10E6 (2nd re-cloning)	217	25
L5B3 (3rd re-cloning)	217	38
L5B7 (3rd re-cloning)	217	21
L5B9 (3rd re-cloning)	217	40

A number of trivial explanations for the cytotoxic activity of these cloned lines were ruled out. 1. Cytotoxicity was not an artefact associated with mycoplasma infection (11), as the target cells were mycoplasma-free; 2. Cytotoxicity was not directed against FCS components, as identical cytotoxicity was found against _in vivo_ grown target cells assayed in the presence of syngeneic mouse serum or FCS; 3. Lectin bound to the killer cells was not required for cytotoxicity, as killer cells grown or assayed in the presence of 25 mM αMM had similar cytotoxicity and specificity to control cells; 4. Cytotoxicity required intimate contact between effector and target cells because (a) much higher killing occurred when assays were set up in V-bottomed rather than flat-bottomed plates and (b) cytotoxicity was inhibited by low numbers of unlabelled (competitor) target cells.

The surface markers detected on five randomly selected cloned killer cell lines is shown in Table IV. All the lines bore Thy-1, Ly-5 and asialo-GM1, and lacked Ig, Lyt-1, FcR and Mac-1. Qa-5 was present on both C57BL/6 lines, but absent from the CBA lines as expected (CBA mice are a Qa-5 negative strain). The NK-1.2 antigen, defined by (NZB X CE) F1 anti-CBA serum, was found on four of the five cell lines. The presence of Lyt-2 was studied by flow cytofluorimetry on a FACS II. Two of the lines expressed high quantities of this antigen, while the other three lines were either completely negative or had only a trace amount.

Table IV. Surface Membrane Antigens on Cloned Killer Lines

Characteristic	A3F4 (C57BL/6)	A3B3 (C57BL/6)	B10G7 (CBA)	B12C8 (CBA)	B13D7 (CBA)
Ig	-	-	-	-	-
Thy-1	+	+	+	+	+
Lyt-1	-	-	-	ND[a]	-
Lyt-2	+	+	-	-	±[b]
Ly-5	+	+	+	+	+
Asialo-GM1	+	+	+	+	+
Qa-5	+	+	-[c]	ND	ND
FcR	-	-	-	-	-
Mac-1	-	-	-	-	-
NK-1.2	+	+	-	+	+
Inhibition by anti-Lyt-2[d]	25	0	0	ND	15
Inhibition by anti-Ly-5[d]	57	95	23	ND	49

[a] ND = not determined
[b] Possible trace amounts of Lyt-2 detected by FACS analysis
[c] Cells of CBA genotype do not express Qa-5
[d] % inhibition of cytotoxicity measured on YAC-1 targets

Studies in this laboratory and others have shown that anti-Lyt-2 antibodies inhibit CTL but not NK cytolysis, whereas anti-Ly-5 antibodies inhibit NK but not CTL cytolysis (12, 13). It was therefore of interest to find that cytotoxicity by the cloned killer cells was inhibited by anti-Ly-5 serum (Table IV). By contrast anti-Lyt-2 antibody generally did not inhibit, although some inhibition of cytotoxicity of one of the Lyt-2-bearing clones was observed.

Initially the specificity of the cloned killer cell lines was studied on a panel of 6 murine lymphoid tumor cells. The 5 cloned lines described above all had similar specificity, with strongest reactivity against YAC-1 (Table V). However, in contrast to normal or activated splenic NK cells, the cloned lines could not distinguish between clone 27v and clone 27av variants of the L5178Y tumor (14), and generally exhibited higher lysis of P815 and EL4 than was observed with spleen cells.

The specificity of the B10G7 line was studied more extensively using a panel of more than 40 target cells. Like spleen cells, B10G7 cells were generally not cytotoxic to normal cells (fibroblasts, PEC), although some low lysis of LPS and Con A blast cells was seen. In addition, of 9 xenogeneic leukemia target cells, none were lysed by B10G7 cells, even those which were susceptible to normal and/or activated murine spleen cells. Most interestingly, with 6/6 solid tumor-derived target cells tested, all of which were moderately susceptible to normal or activated spleen cells, no lysis was

Table V. Specificity of Killer Cell Lines

Effectors	YAC-1	L5178Y cl27v	L5178Y cl27av	R1.1	P815	EL4
N CBA[a]	100[c]	24	5	10	5	6
Poly I:C CBA[b]	100	59	20	53	14	10
B10G7	100	10	12	20	30	37
B12C8	100	6	9	14	40	35
B13D7	100	5	7	10	23	18
A3F4	100	34	19	46	90	78
A3B3	100	12	9	10	67	77

[a] Normal CBA spleen cells
[b] Mice injected ip with 200 µg poly I:C 1 day earlier
[c] Relative specificity (YAC-1 = 100). Figures are the mean values obtained in at least 3 separate experiments with each effector cell.

obtained with B10G7 cells even when assays were run for 24 hr. In additional experiments, none of the 5 cloned killer cell lines described here showed antibody dependent lytic activity (data not shown).

C. Discussion

When murine spleen cells were depleted of T cells by treatment with anti-Thy-1 antibody and complement, then cultured for 1 day with poly I:C and subsequently cloned in medium containing supernatant from Con A stimulated murine spleen cells, many of the cloned cell lines had high cytolytic activity against YAC-1. It was interesting to note that the clones with highest cytolytic activity were generally obtained with spleen cells from CBA and C57BL/6 mice. By contrast, clones prepared with spleen cells from C57BL/6 bg/bg mice possessed poor lytic ability. Interestingly, spleen cells from another NK deficient strain, SJL, produced many highly cytotoxic clones, suggesting the possibility that there may be no defect in the NK compartment in these animals, but rather a suppressor cell or inhibitor which was lost during cloning.

Growth of these cell lines required continual presence of SN from Con A stimulated spleen cells. Such SN contained high concentrations of both IL-2 and interferon, both of which are potent NK cell stimulators. Experiments using a short-term thymidine uptake assay demonstrated that IL-2, but not interferon was able to induce proliferation of these cell lines. Like CTL cell lines, they therefore appear to be dependent on IL-2 for continuous growth. In addition, in order to maintain maximal cytolytic activity over long periods, re-cloning was necessary at about 3 month intervals to avoid overgrowth of cultures with variants of lower activity.

To characterize these killer cell lines, 5 clones were selected for detailed study. A key marker for distinguishing NK cells and CTL is Lyt-2, which is absent from NK cells but present, with but a single recorded exception (15), on all CTL and CTL lines. Three of the cloned killer cell lines totally lacked, or had only a minute amount, of this antigen as determined by flow cytofluorimetry. The other two lines bore high quantities of Lyt-2. Both these lines were of C57BL/6 origin, but many other C57BL/6 lines produced in a similar manner were $Lyt\text{-}2^-$. The principal factor determining whether $Lyt\text{-}2^-$ or $Lyt\text{-}2^+$ lines were produced was the timing of cloning. Thus, if cultures were cloned immediately after poly I:C stimulation, all the clones were $Lyt\text{-}2^-$ whereas if cloning was delayed for more than a week many $Lyt\text{-}2^+$ clones were obtained.

It might be argued that the $Lyt\text{-}2^+$ clones were CTL and the $Lyt\text{-}2^-$ clones were of NK lineage. However, when a large series of other markers were studied, no other differences between these two types of clone was found. For example, both bore very large amounts of asialo-GM1, a selective marker for NK cells rather than CTL (16). The lines also generally reacted with (NZB X CE) F1 anti-CBA serum which defines the NK cell specific marker, NK-1.2 (9, 10). Most persuasively, lysis by all the cell lines was inhibited by anti-Ly-5 serum, which in our hands and those of others (12, 13), specifically inhibits lysis by NK cells but not CTL.

The observation that $Lyt\text{-}2^+$ colony forming cells (CFC) developed later in the pre-cloning culture period than the $Lyt\text{-}2^-$ CFC suggests either that they are of separate lineage, or that the $Lyt\text{-}2^-$ CFC could differentiate into $Lyt\text{-}2^+$ CFC in the presence of other cell types. In either case, the finding of cytotoxic cell lines bearing both a CTL-specific marker (Lyt-2) and several NK cell markers (asialo-GM1, NK-1.2, blocking by anti-Ly-5) is a provocative observation which may point to there being a close relationship between NK cells and CTL.

The specificity of all 5 clones studied in detail was similar. Although highest activity was seen against YAC-1 targets, the specificity of the cloned lines differed in a number of ways from that of whole spleen NK cells. Most strikingly, when one of the lines was tested on a large panel of targets, its specificity was restricted to murine lymphoid tumors with little or no lysis of xenogeneic lymphomas or solid tumor derived target cells, including those which were lysed moderately well with whole spleen NK cells. These findings are interesting in view of the apparent heterogeneity of murine NK cells (17-19), and are consistent with the view that the cloned killer cells described here belong to a subpopulation of NK cells primarily reactive against certain murine lymphomas.

REFERENCES

1. Moller, G. (ed.), Immunol. Rev. 54 (1981).
2. Kuribayashi, K., S. Gillis, D.E. Kern, and C.S. Henney, J. Immunol. 126:2321 (1981).
3. Dennert, G., G. Yogeeswaran, and S. Yamagata, J. Exp. Med. 153:545 (1981).
4. Nabel, G., L.R. Bucalo, J. Allard, H. Wigzell, and H. Cantor. J. Exp. Med. 153:1582 (1981).
5. Rosenberg, S.A., P.J. Spiers, and S. Schwarz. J. Immunol. 121:1951 (1978).
6. Pollack, S.B., M.R. Tam, R.C. Nowinski, and S.L. Emmons. J. Immunol. 123:1818 (1979).
7. Ledbetter, J.A., and L.A. Herzenberg, Immunol. Rev. 47:63 (1979).
8. Pollack, S.B., S.L. Emmons, L.A. Mallenbeck, and M.R. Tam. In: Herberman, R.B. (ed.) Natural cell-mediated immunity against tumors. P. 153. Academic Press. New York (1980).
9. Burton, R.C. In: Herberman, R.B. (ed.) Natural cell-mediated immunity against tumors. P. 19. Academic Press, New York (1980).
10. Burton, R.C., and H.J. Winn. J. Immunol. 126:1985 (1981).
11. Brooks, C.G., R.C. Rees, and R.H. Leach. Eur. J. Immunol. 9:159 (1979).
12. Minato, N., L. Reid, H. Cantor, P. Lengyel, and B.R. Bloom. J. Exp. Med. 152:124 (1980).
13. Seaman, W.E., N. Talal, L.A. Herzenberg, L.A. Herzenberg and J.A. Ledbetter, J. Immunol. 127:982 (1981).
14. Durdik, J.M., B.N. Beck, E.A. Clark, and C.S. Henney, J. Immunol. 125:683 (1980).
15. Swain, S.L., G. Dennert, S. Wormsley, and R.W. Dutton, Eur. J. Immunol. 11:175 (1981).
16. Young, W.W., S-I. Hakomori, J.M. Durdik, and C.S. Henney, J. Immunol. 124:199 (1980).
17. Minato, N., L. Reid, and B.R. Bloom, J. Exp. Med. 154:750, (1981).
18. Paige, C., E.F. Figarella, M.J. Cuttito, A. Cahan, and O. Stutman, J. Immunol. 121:1827 (1978).
19. Lust, J.A., V. Kumar, R.C. Burton, S.P. Bartlett, and M. Bennett, J. Exp. Med. 154:306 (1981).

INDUCTION OF NK-LIKE ANTI-TUMOR REACTIVITY IN VITRO AND IN VIVO BY IL-2[1]

Eli Kedar[2] and Ronald B. Herberman[3]

Laboratory of Immunodiagnosis
National Cancer Institute
Bethesda, Maryland

There is considerable evidence that natural killer (NK) and cytotoxic (NC) effector cells are involved, to some extent, in the primary line of defense against both the local outgrowth as well as the metastatic spread of transplanted syngeneic experimental tumors and possibly also in surveillance (Herberman, 1980; Herberman and Ortaldo, 1981; Herberman, 1982).

Recent studies have pointed to the wide heterogeneity of the NK/NC cells with respect to cell surface makrers, morphology, range of susceptible target cells, and modulation by various agnets (Herberman, 1980; Stutman _et al._, 1981). Data from various laboratories strongly suggest that at least a proportion of these cells belong to the T-cell lineage (Herberman and Ortaldo, 1981; Stutman _et al._, 1981; Minato _et al._, 1981) and it has been shown that mouse NK reactivity can be augmented markedly by brief exposure _in vitro_ to T-cell growth factor (interleukin-2, IL-2) (Minato _et al._, 1981; Kuribayashi _et al._, 1981). Furthermore, similar to activated T-cells, NK cells with potent anti-tumor reactivity of both mouse and human origin, can be maintained and expanded in the presence of IL-2 (Dennert _et al._, 1981; Nabel _et al._, 1981; Timonen _et al._, 1982).

In view of these findings, there exists the possibility that administration _in vivo_ of IL-2 and/or of large number of IL-2-propagated, highly activated NK/NC cells may be

[1] Supported in part by grants to the Lautenberg Center from the Dr. "I" Fund Foundation, New York, and the late Mr. Harold B. Abramson, New Jersey.

[2] Present address: The Lautenberg Center for General and Tumor Immunology, Hebrew University-Hadassah Medical School, Jerusalem 91 010, Israel.

[3] Present address: Biological Research and Therapy Branch, NCI-FCRF, Frederick, Maryland.

ISBN 0-12-341360-5

beneficial in therapy of experimental and human tumors. The data presented herein show that: (a) Lymphoid cells from various lymphoid tissues of both normal and tumor-bearing hosts (man, mouse) that had been maintained in long-term cultures with IL-2 showed strong cytotoxic reactivity *in vitro* against a wide array of tumor cells, including targets known to be resistant to fresh NK/NC effector cells, and expressed surface marker characteristics of both T and NK cells; (b) clones obtained from such cultured lymphoid cell (CLC) lines had a more restricted target cell reactivity; (c) IL-2-grown CLC were effective in preventing local tumor growth in neutralization (Winn) assays and less effective in mice with disseminated neoplasms; (d) lymphoid cells treated briefly (24 h) *in vitro* with IL-2, or taken from mice injected with IL-2, exhibited increased anti-tumor cytotoxic capacity *in vitro*; and (e) IL-2 by itself rendered nude mice resistant to a transplant of human tumor (Daudi) cells; in addition, IL-2 had some therapeutic action in mice with established tumors when administered together with chemotherapy. A portion of the data presented here, and a detailed description of the methodology, have been reported elsewhere (Gorelik *et al.*, 1981; Kedar *et al.*, 1982a; Kedar *et al.*, 1982b).

MATERIALS AND METHODS

Preparation of IL-2-Containing Medium (IL-2-M)

Crude rat IL-2 (C-IL-2-M) was prepared from splenocytes of W/Fu or Lewis rats by culturing ($4\text{-}7 \times 10^6$/ml) with 5 µg/ml Con A for 36-48 h. In some preparations, Con A was depleted and these preparations were designated mitogen-depleted IL-2-M (MD-IL-2-M). Human C-IL-2-M and MD-IL-2-M were prepared by cultivating pooled human PBL with 1% PHA-M for 2-3 days.

Establishment of CLC Lines and Clones

Cultures of mouse cells were initiated in 10-30% rat MD-IL-2-M or C-IL-2-M. Cultures of human PBL or their subsets were initiated in 15% human serum (autologous or AB) and 10-15% human C-IL-2-M or MD-IL-2-M. Mouse cultures were examined repeatedly for cytotoxic activity and other functions for up to 12 months and human cultures up to 6 weeks after initiation. Feeder cells were not added at any time.

Monoclonal murine cultures were established from 3 mo. old spleen CLC by the limiting dilution or soft agar techniques (Kedar *et al.*, 1982a). Human clones were derived from

PBL of a normal donor stimulated with PHA (clone 1) and from normal PBL stimulated with irradiated allogeneic leukocytes (clones 5 and 9).

RESULTS

In Vitro Amplification of Cytotoxic Capacity by MD-IL-2-M

Splenocytes obtained from normal mice of several strains were incubated with mouse fibroblast IFN or rat MD-IL-2-M and then tested for cytotoxic activity against an NK-sensitive target (RLσ1 lymphoma) and an NK-resistant target (M109 lung carcinoma). Whereas a 1 h exposure to either agent did not affect the activity significantly, a marked elevation was observed after a 24 h exposure, against both targets (Table 1). The stimulatory effect of IL-2 was consistently more pronounced than that of IFN.

In another series of experiments, human and murine lymphocyte cultures, and mixed lymphocyte-tumor cultures, were carried out for 6-7 days in the presence or absence of IL-2 and cytotoxicity was assayed against the stimulating tumor cells and against a panel of unrelated targets. It was noted that cultures containing both autologous human or syngeneic mouse tumor cells and IL-2 had the highest reactivity against the corresponding tumor, but an appreciable cytotoxicity also developed against unrelated targets.

In Vivo Stimulation of Cytotoxic Activity by MD-IL-2-M

Having demonstrated a profound stimulatory effect of IL-2 *in vitro*, it was of interest to evaluate its influence *in vivo*. As shown in Table 2, mice injected with IL-2 showed

Table 1. Potentiation by MD-IL-2-M and IFN of Cytotoxic Activity in Murine Spleen Cell Cultures

Effector cells	Pretreatment of effectors	%Lysis (E/T=20/1)[a] RLσ1	M109
BALB/c	None, fresh	16	7
	Medium, 24 hr	12	15
	MD-IL-2-M, 24 hr	53	59
	IFN, 24 hr	27	32
C3H	None, fresh	24	10
	Medium, 24 hr	21	18
	MD-IL-2-M, 24 hr	69	62
	IFN, 24 hr	48	37

a) Means of 3 experiments. Rat MD-IL-2-M used at 10 or 20% v/v, and IFN at 5,000 or 10,000 U/ml. Assays were 4 hr with RLσ1 and 18 hr with M109.

Table 2. Amplification of Cytotoxic Activity by In Vivo Administration of IFN or MD-IL-2-M

Mice	Treatment	Day[a)]	Mean % Lysis[b)]			
			YAC-1(4 hr)		M109(18 hr)	
			Spl	P	Spl	P
C3H	Medium	1		12		
	IFN(25,000U,ip,d.0)	1		17		
	IL-2(1ml,ip,d.0)	1		39		
	Medium	3			19	8
	IFN	3			26	19
	IL-2	3			85	36
			RL♂1 (4 hr)		M109(18 hr)	
			Spl	P	Spl	P
BALB/c	Medium	3	18	5	3	
	IL-2(0.5ml ip,d.0)	3	26	28	12	
	(" " d.-2,-1,0)	3	32	57	14	
	(0.5 ml iv, d.0)	3	39	19	36	

a) Effector cells assayed for cytotoxic activity.
b) Splenocytes (Spl) and peritoneal exudate cells (P).

within one day a marked elevation in reactivity against lymphoma and carcinoma cells. As in the *in vitro* experiments, IL-2 was more effective than IFN. In general, multiple inoculations of IL-2 was more effective than a single dose; i.v. administration activated splenocytes to a greater extent, whereas i.p. inoculation produced the opposite pattern. The IL-2-induced augmentation was more evident during the first 6 days after the last inoculation, with only a modest increase on days 10-12.

In addition to the potentiation of cytotoxic activity, IL-2 also induced a marked increase in spleen size per spleen (150-275x10^6/spleen, compared to 90-110x10^6 in control mice). Thus, the actual increase in cytotoxic activity was even greater in IL-2-treated mice. In contrast, the number of peritoneal cells was similar in both groups. Furthermore, a large proportion (up to 30%) of spleen and peritoneal cells in IL-2-treated mice were large lymphoblasts.

Long-Term Murine Cultures in IL-2-Containing Media

Over 50 long-term cultures lines from various mouse strains (carried for 1-12 months) were established in media containing IL-2. The following observations were made: (a) normal spleen cultures in C-IL-2-M expanded 10^5-10^6-fold per month with a generation time of 15-24 h; (b) cultures of splenocytes from mice with progressive disseminated tumors and from nude mice, and spleen cultures maintained in MD-IL-2-M showed a rather slow growth rate during the first 7-10 days of culture, but thereafter growth was usually similar to that of cultures originated in normal splenocytes and maintained in C-IL-2-M; (c) cells of spleen, peripheral blood and peritoneal exudate had similar growth rate, thymocytes showed a slightly lower proliferative capacity, and bone marrow cells had poor growth capacity, with most of the cultures dying after 2-3 months.

All of the cultures of spleens, peritoneal, and peripheral blood cells from either tumor-bearing mice, normal mice, or nude mice, maintained in either C-IL-2-M or MD-IL-2-M, showed a similar pattern of strong cytotoxic reactivity (even at 1/1 E/T ratio) against most tumor targets tested (Table 3). Freshly explanted tumors were slightly less sensitive than the corresponding cultured cells. CLC originated from thymus and bone marrow cells had slightly lower activity (Kedar *et al.*, 1982a).

Low but significant cytotoxicity by CLC was demonstrated against fresh syngeneic normal murine lymphoid target cells, especially thymus and adherent peritoneal exudate cells, and when the ^{51}Cr assay was extended to 18 hr. Con A-induced spleen lymphoblasts were markedly more susceptible targets to lysis than fresh splenocytes, and allogeneic blasts were more sensitive than syngeneic blasts. In addition, xenogeneic rat and human cultured tumor cells, which are rather resistant to fresh murine NK effector cells, were also lysed by the murine CLC to a considerable extent.

Table 3. Cytotoxic Activity of BALB/c CLC and CLC Clones

Target Cells		%Lysis (E/T = 10/1) by:			
	CLC	Cl.1	Cl.22	Cl.11	Cl.39
Mouse cells					
YAC-1 lymphoma[a]	60[c]/88[d]	86[d]	92[d]	77[d]	94[d]
RL♂1 lymphoma[a]	49/81		93	90	95
M109 lung carcinoma[b]	57/92	53	76	33	22
MT mammary carcinoma[b]	17/41	31	39	25	12
B16 melanoma[a]	23/65				
3LL lung carcinoma[b]	15/31				
Meth 27A sarcoma[a]	14/24				
EL4 lymphoma[b]	41/80	91	90		28
P815 mastocytoma[b]	33/67	72	92		16

a) Cultured cells.
b) Freshly explanted cells. Percent cytotoxicity measured at 4-6 hr[c] and 16-18 hr[d].

With most of the CLC, maximum cytotoxicity for various targets was observed after 2-3 weeks in culture and this activity was quite stable during the first 3-4 months. Later on, however, some lines showed a gradual increase in activity, particularly against the NK-resistant solid tumor cells.

To characterize the nature of the effector cells having such a broad spectrum of cytotoxic reactivity, a series of over 60 clones were established from tumor-bearing BALB/c and C3H spleen CLC. As shown for some of the BALB/c clones in Table 3, the patterns of cytotoxicity could be categorized into two main classes. Whereas some clones (e.g., 11, 39) had strong activity against the highly NK-sensitive lymphoma cells (RLð1 and YAC-1) and relatively low activity against NK-resistant lymphomas (EL4) and solid tumors (MT, M109), other clones (e.g., 1, 22) were strongly cytotoxic against both types of targets. The two classes of clones were therefore designated CNK-L (cultured NK cells, lymphoma targets) and CNK-SL (cultured NK cells, solid and lymphoma targets). In general, both types of clones had similar reactivity against normal lymphoid target cells, when tested in an 18 hr assay. Clones maintained in MD-IL-2-M showed the same pattern and magnitude of cytotoxicity.

In light of the relatively limited degree of heterogeneity among the clones in their patterns of cytotoxicity, it was of interest to determine whether the reactivity of a given clone against various targets was due to discrete or to common target recognition sites. For this purpose a cold target inhibition assay was employed. The results (Kedar _et al._, 1982b) strongly suggested that most of the clones, of both CNK-L and CNK-SL types, had heterogeneity of receptors for various targets; regardless of the level of susceptibility of each target to lysis, the homologous target inhibited cytotoxicity substantially more than the other targets.

Human CLC and CLC Clones

Using human C-IL-2-M, bulk cultures were established from PBL of 38 health donors and 27 cancer patients. Over a period of 3 weeks, about 80% of normal cultures expanded 200-2500-fold (mean, 638), whereas only 37% of patient cultures showed similar growth (mean, 322-fold). Large granular lymphocytes (LGL) (Timonen _et al._, 1981), isolated by the Percoll gradient fractionation technique, had a low growth rate (generation time 50-60 hr) particularly in MD-IL-2-M.

There was no direct correlation between the growth capacity of the individual cultures and their cytotoxic activity. When CLC were tested repeatedly, maximum reactivity was reached by 1-2 weeks after culture initiation, and thereafter the pattern and magnitude of cytotoxicity was quite stable. Similar patterns of results were obtained with all CLC, derived from either tumor patients or normal donors (Table 4 shows results with patient CLC). Most sensitive to lysis were K562 and the adherent lines, MCF-7 and G-11. Fresh lung tumors were intermediately susceptible to both autologous and allogeneic CLC and less sensitive than the adherent tumor lines. Normal lymphoid target cells (fresh autologous or allogeneic PBL and Con A-induced blasts) were considerably less sensitive than the tumor cells when tested in a 6 hr ^{51}Cr assay. The cytotoxic activity of the CLC against all target cells was invariably higher than that demonstrated by the corresponding fresh PBL and by PBL cultivated for 1 week in FCS or AB serum-containing medium, without IL-2 (data not shown).

In an attempt to gain more information on the characteristics of the killer CLC, nylon nonadherent PBl were separated on a Percoll gradient. When tested fresh, the LGL population exhibited higher reactivity than the unseparated cells against all the targets, whereas the small T-cell fraction expressed very low or no cytotoxicity. Cultured LGL maintained strong reactivity toward all cultured tumor targets and had increased cytotoxicity against the fresh tumors. In contrast, IL-2-cultured T-cells acquired a remarkably elevated activity (compared with day 0) against the adherent solid tumors (fresh and cultured), yet only a low level of cytotoxicity was detected against K562.

Table 4. Cytotoxic Activity of Human CLC and Clones

Target Cells	% Lysis (E/T = 20-30/1) by:					
	CLC	CLC[c]	CLC[d]	Cl.1	Cl.5	Cl.9
Human cells						
K562	45[a]/61[b]	15[b]	74[b]	79[b]	9[b]	26[b]
SK-MES-1	29/45	53	64	93	55	48
9812	28/62					
HT-29	16/43	44	65	82	6	16
MCF-7	25/52	58	66	64		
G-11	43/66	59	72	80	62	85
Autologous lung ca.	14/32					
Allogeneic lung ca.	15/28	23	32	20		
Autologous PBL	3/14					
Autologous Con A lymphoblast	5/20					

Cytotoxicity measured at [a] 6 h or [b] 18 h.
[c] NK-depleted T cells; [d] NK-enriched (LGL) cells

Cultured T-cells also expressed higher cytotoxicity against normal lymphoblasts than the cultured LGL. These findings point to the possibility of the existence in the T-cell fraction of nonspecific cytotoxic cell precursors that can be IL-2-activated to strongly react with adherent solid neoplasms (and lymphoblasts) preferentially. These cells were quite different from LGL in their cytotoxicity patterns.

Similar to the murine clones, human clones also could react against several tumor target cells, although some clones (e.g., 5 and 9, Table 4) showed a degree of heterogeneity, compared to uncloned cultures, in their pattern of cytotoxicity.

Surface Markers and Other Characteristics of Murine and Human CLC

To further characterize our cultured cells, we assessed their surface markers and other properties. In the mouse, virtually all cells were nonadherent, nonphagocytic, Thy 1.2^+, Ig^-, asialo Gm 1^+ and expressed the receptor for peanut agglutinin. Killing of CLC by anti-Thy 1.2 and C' required higher amounts of these reagents than those required for lysing fresh splenic T-cells and thymocytes. A considerable proportion of CLC and clones expressed Lyt 2, a few were Lyt 1^+, and some cells possessed neither of these markers. In addition, it appeared that a considerable proportion possessed NK2, T200 and Qa 2,3 markers, and, surprisingly, a receptor for SRBC. Following treatment with various antibodies and C' there was a good correlation between the number of killed CLC and the reduction in cytotoxic activity.

In addition to the expression of nonspecific cytotoxicity and the NK2 antigen, our CLC also resembled classic NK/NC cells in other properties, such as morphology (dense azurophilic granules), ADCC activity (although Fc receptor could not be detected by rosetting), production of IFN (50-500 U/ml), boosting of cytotoxicity by IFN, and inhibition of cytotoxicity by simple sugars like D-mannose and D-galactose. Some of the cultured cells have been shown to contain proteoglycans (and possibly heparin) but not histamine. Furthermore, the murine CLC seemed to produce a lymphotoxin-like material, both spontaneously and in higher quantities following stimulation with mitogens (Kedar _et al_., 1982b). The binding capacity to various tumor target cells was 2-5-fold greater by CLC than by unsensitized fresh splenocytes, with 10-30% of the former cells able to form conjugates.

Recently we have established a clone of adherent cytotoxic CLC of BALB/c origin, that is not phagocytic and which expresses most of the properties described above.

Similar to murine CLC, human CLC exhibited markers and properties characteristic of both NK and T-cells. Thus cells had the morphology of LGL, were active in ADCC (again with no directly detectable Fc receptor), were boosted by IFN, expressed a receptor for SRBC, and lacked surface Ig. Furthermore, like mouse CLC, human CLC had a strong binding affinity for various tumor target cells: 15-38% of the CLC could form conjugates, compared to only 3-10% of fresh PBL.

Anti-Tumor Activity In Vitro of IL-2 and CLC

In light of the ability of IL-2 to elicit strong anti-tumor reactivity _in vitro_ and _in vivo_, and the potent cytotoxic reactivity of CLC, it was of interest to determine whether they could also affect tumor growth _in vivo_.

In a preliminary experiment, we have found that nude mice pretreated with rat MD-IL-2-M became refractory to challenge with a supralethal dose of Daudi tumor cells (a lymphoblastoid line derived from a patient with Burkitt lymphoma) (Table 5). It should be emphasized that splenocytes freshly obtained from untreated mice were devoid of appreciable cytotoxicity _in vitro_ against Daudi target cells (<5% lysis in a 4 hr assay and <12% in a 16 hr assay at E/T = 50/1).

In the local Winn neutralization assay, human and mouse CLC afforded significant protection against syngeneic and allogeneic lymphomas and carcinomas. When sufficient numbers of CLC were given (i.e., ⩾ 10/1 human CLC in nude mice, or ⩾ 30/1 mouse CLC), tumors appeared later and ⩾ 50% of the experimental mice remained tumor-free, whereas ⩾ 80% of the control mice died of tumor. These findings indicated that tumor cell killing by CLC can also occur _in vivo_.

The above observations encouraged us to initiate a study on the therapeutic effects of IL-2 and CLC in hosts with advanced malignant disease. Since our recent studies (Gorelik _et al._, 1981) showed that transfused CLC can only survive transiently _in vivo_ and their cytotoxic effects against tumor cells last only 1-2 days, attempts were made to administer IL-2 and/or CLC repeatedly over a period of 1-3 weeks. The experiments summarized in Table 6 indicated that: (a) multiple systemic administrations of either IL-2 alone or CLC alone had significant therapeutic effects in mice with established tumors; the survival time and cure rate of mice treated with chemotherapy plus immunotherapy were greater than those of mice treated with chemotherapy only; (b) combined treatment with CLC and IL-2 was more effective than either treatment alone, probably by allowing the infused CLC to survive longer in the recipient mice. Under the conditions employed, no signs of graft-versus-host disease were noticed in any of the groups. Studies are

Table 5. Prevention of Growth of Daudi Tumor in Nude Mice by Rat MD-IL-2-M

Group[a)]	Incidence of tumors (and mean tumor size, mm^2) on:[a)]		
	Day 15	Day 30	Day 60
Control	5/10 (36)	8/10 (355)	8/10 (1040)
IL-2-treated	0/10	2/10 (112)	3/10 (265)

a) IL-2, 1 ml ip on d.0, 3 and 5; 10^6 Daudi cells sc on d.6. On day 90 4/10 control mice and 9/10 of treated mice were alive.

Table 6. Chemoimmunotherapy with IL-2 and/or CLC

Tumor and treatment[a)]	Median survival time (days)	Tumor-free survivors/total (day 120)
M109 $5x10^5$ ip, day 0	37	0/8
M109 + Cy 200 mg/kg, day 3	48	2/10
M109 + Cy + IL-2 1 ml, ip, days 5,7,9	67	4/10
Thymoma 136.5 $1x10^5$ ip, day 0	14	0/10
136.5 + Cy day 3	22	0/10
136.5 + Cy + IL-2, days 4,6,8	27	2/10
136.5 + Cy + $15x10^6$ CLC ip, days 4,8,14	28	3/10
M109 $1x10^5$ ip, day 0	35	0/8
M109 ip + Cy 250 mg/kg, day 7	57	0/8
M109 ip + Cy + $15x10^6$ CLC ip days 8,15,22	72	2/8
M109 ip + Cy + IL-2 1 ml ip, days 8,15,22	69	2/8
M109 ip + Cy + CLC + IL-2, as above	102	3/8
M109 $1x10^5$ sc, day 0	61	0/8
M109 sc + CCNU 35 mg/kg, day 7	78	1/8
M109 sc + CCNU + CLC, as above	>120	2/8
M109 sc + CCNU + IL-2 i.t. and ip, as above	89	2/8
M109 sc + CCNU + CLC + IL-2 i.t. and ip, as above	>120	5/8

a) In last expt., mice inoculated with tumor sc were inoculated with CLC and IL-2 into the tumor site (i.t.) (1/4 dose) and ip (3/4 dose).

underway to define the optimal dose, timing, and sequence of CLC and IL-2 administrations into tumor-bearing mice in combination with other therapeutic measures.

DISCUSSION AND CONCLUSIONS

Although the mechanism by which IL-2 activates lymphoid cells to such a high level of anti-tumor reactivity is not known at present, this approach may have practical clinical implications. Thus, lymphoid cells from cancer patients, with or without prior specific tumor stimulation *in vitro*, could be expanded with IL-2 and subsequently infused in large numbers to the autologous donor. Several limitations exist, however, to the immediate employment of IL-2-grown cells in the clinical setting. These include: (a) limited survival *in vivo* and complete dependence on a continuous supply of IL-2 (Gorelik *et al*., 1981); (b) abnormal or impaired circulation *in vivo* - the majority of the cells administered systemically have been reported to be trapped rapidly in the lungs and liver (Lotze *et al*., 1980); (c) the existence within the propagated population of clones with anti-normal tissue reactivity (Lotze *et al*., 1981; this work), with a potential of provoking autoimmune pathologic reactions; and (d) the posibility of introducing to the patient large numbers of cells with suppressor and/or tumor enhancing capacity. It is therefore suggested that extensive studies should now be initiated in experimental models aimed at: (i) improving the survival and circulation *in vivo* of IL-2-grown CLC; (ii) evaluating the *in vivo* behavior and anti-tumor action of more selected populations and of cloned cells with either helper or cytotoxic reactivity; and (iii) determining the various effects of IL-2 administered *in vivo*.

The observation made in this study that IL-2 can elicit a much stronger anti-tumor reactivity than IFN, both *in vitro* and *in vivo*, may suggest that IL-2 can be applied *in vivo* as a potent biological response modifier. Preliminary studies conducted in normal and in immunocompromised mice substantiate this possibility (Kedar, to be published).

Apart from the practical importance of IL-2 and IL-2-grown CLC, of major interest to us was the nature of the cytotoxic effector cells. In other studies (Ortaldo *et al*., 1980), several types of cytotoxicities could be detected in human CLC: NK-like, ADCC, lectin-induced, and polyclonally activated T-cell cytotoxicity. The observations in this work that murine and human CLC express classic T-cell markers, and the relatively strong cytotoxicity also against allogeneic lymphoblasts, is suggestive of polyclonally activated cytotoxic T-cells (CTL). On the other hand, the strong reactivity also against syngeneic normal lymphoid and tumor cells and against xenogeneic targets including cells which lack MHC antigens (e.g., K562), suggest that other types of cytotoxic cells are also present in our long-term cultures.

Several observations strongly suggest that a major contribution to the cytotoxic activity seen here was by cells more characteristic of NK/NC cells: (a) The broad cytotoxic capability against a wide array of syngeneic, allogeneic, and xenogeneic target cells; the particularly high sensitivity of the NK-sensitive target cells (Yac-1, RLo1 in the mouse and K562 in man); and the reactivity toward normal thymus, peritoneal, and bone marrow cells are all reminiscent of the classical NK cells. (b) The finding of ADCC activity by the CLC; the augmentation of cytotoxicity by pretreatment with IFN; and the inhibition of cytotoxicity of murine CLC by sugars are properties associated more with NK cells than with CTL (Herberman, 1980). (c) The presence in the murine CLC of NK2 antigen and high amounts of asialo GM1. (d) The presence of Lyt 2 marker in a large proportion of murine CLC, but not in fresh NK effector cells, is not surprising since this antigen has recently been found on a subpopulation of NK cells which are responsive to IL-2 (Minato *et al.*, 1981); and Thy 1.2 antigen has been shown also on murine NK cells (Herberman, 1980). (e) The morphology of the human and mouse CLC and CLC clones with the distinct azurophilic granules is quite similar to fresh LGL with strong non-specific cytotoxic activity (Timonen *et al.*, 1981).

Finally, although we ascribed here the cytotoxicity-stimulating action to IL-2, it is highly likely that other lymphokines and monokines also contributed to this effect. Substances such as immune IFN, lymphocyte-activating factor (LAF or IL-1), macrophage-activating factor, residual mitogen, and other immunomodulating products are plentiful in the C-IL-2-M and MD-IL-2-M preparations employed in our work. Whether or not similar effects *in vivo* and *in vitro* can be elicited with purified IL-2 is a matter for further investigation.

ACKNOWLEDGEMENTS

We wish to thank Drs. G. Bonnard, E. Gorelik, T. Timonen, B. Sredni, B. Bonavida, B. Mathieson, G. Koo, J. Reid and N. Navarro for invaluable contribution to this work, and to Ms. B.L. Ikejiri, E. Chriqui, A. Gross and H. Bercovitz for skillfull technical assistance.

REFERENCES

Dennert, G., Yogeeswaran, G. and Yamagata, S. (1981). J Exp Med 153:545

Gorelik, E., Kedar E., Sredni, B. and Herberman R.B. (1981). Int J Cancer 28:157.
Herberman, R.B. (ed.) (1980). "Natural Cell-Mediated Immunity Against Tumors." Academic Press, New York.
Herberman, R.B. and Ortaldo, J.R. (1981). Science 214:24.
Herberman, R.B. (1982). In: "Clinical Immunology Reviews," Vol. I, (R. Rocklin, ed.) Marcel Dekker Inc., New York.
Kedar, E., Herberman, R.B., Gorelik, E., Sredni, B., Bonnard, G.D. and Navarro, N. (1982a). In: The Potential Role of T-cell Subpopulations in Cancer Therapy, (A. Fefer, ed.) Raven Press, in press.
Kedar, E., Ikejiri, B.L., Sredni, B., Bonavida, B. and Herberman, R.B. (1982b). Cell Immunol, in press.
Kuribayashi, K., Gillis, S., Kern, D.E. and Henney, C.S. (1981). J Immunol 126:2321.
Lotze, M., Line, B., Mathieson, D. and Rosenberg S.A. (1980). J Immunol 125:1487.
Lotze, M.T., Grimm, E.A., Mazumder A., Strausser, J.L. and Rosenberg, S.A. (1981). Cancer Res 41:4420.
Minato, M., Reid, L. and Bloom B. (1981). J Exp Med 154:750.
Nabel G., Bucalo L.R., Allard, J., Wigzell, H. and Cantor, H. (1981). J Exp Med 152:1582.
Ortaldo, J.R., Timonen, T.T. and Bonnard, G.D. (1980). Behring Inst Mitt 67:258.
Stutman, O., Lattime, E.C. and Figarella, E.F. (1981). Fed Proc 40:2699.
Timonen, T., Ortaldo, J.R. and Herberman, R.B. (1981). J Exp Med 153:569.
Timonen, T., Ortaldo, J.R., Stadler, B.M., Bonnard, G.D. and Herberman, R.B. (1982). J Immunol, in press.

CLONED LINES OF MOUSE NATURAL KILLER CELLS

Carlo Riccardi[1,2], Paola Allavena[1], John R. Ortaldo[1], and Ronald B. Herberman[1]

[1]Biological Research and Therapy Branch
National Cancer Institute
Frederick, Maryland

[2]University of Perugia
Perugia, Italy

I. INTRODUCTION

Recent studies in a number of laboratories have identified soluble T cell growth-promoting factors (IL-2). Using IL-2, successful cloning of functional T cells (both cytolytic and helper subpopulations) has recently been achieved (Nabbolz *et al*, 1978; Schreier *et al*, 1980).

Recently cultured lines and clones of K cells have been established with IL-2 from the spleen of non-immunized mice (Dennert, 1980; Nabel *et al*, 1981; Dennert *et al*, 1981; Kedar and Herberman, this volume). Their cytotoxic activity had similar patterns of reactivity as fresh spleen cells against a panel of NK-sensitive and NK-resistant tumor lines. More detailed studies of such NK clones could be very useful to understand the specificity of NK reactivity, and particularly to determine whether NK clones demonstrate heterogenous patterns of reactivity.

As part of a study of the factors regulating the expression of NK activity, we established clones of NK cells in vitro, by culturing spleen cells from SJL/J mice which have been shown to be low NK-reactive, as measured against YAC-1 and other NK-sensitive tumor lines. Analysis of these clones has indicated considerable heterogeneity in both phenotype and patterns of cytotoxic activity.

ISBN 0-12-341360-5

II. GROWTH OF SJL/J SPLEEN CELLS FROM NONIMMUNE MICE IN IL-2 CONDITIONED MEDIUM

Cell lines were initiated by culturing spleen cells at the concentration of 10^6 cells/ml in a 30 ml Costar flask, with RPMI 1640 medium containing 5% fetal bovine serum, 10^{-4} M 2-mercapatoethanol, 1% glutaminine, 1% sodium pyruvate, 0.5% non-essential amino acids, antibiotics and 5 1/2 units/ml of rat IL-2 (Collaborative Research). The IL-2 was purified by ammonium sulfate precipitation and ion exchange chromatography. Cultures were assayed by resuspending the appropriate number of cells in fresh medium containing IL-2 in the presence of irradiated (3000 R) syngeneic spleen cells as feeders.

III. FREQUENCY OF PROLIFERATION OF SJL/J SPLEEN CELLS AND SJL/J CULTURED CELLS

To evaluate the frequency of cells proliferating in the presence of IL-2, we used the limiting dilution assay. Graded numbers of cells were cultured in a microtiter plate in conditioned medium containing 5 1/2 u IL-2/ml and 5 x 10^6 irradiated (3000 R) syngeneic spleen cells. Replicate wells (12-24) were seeded for each cell concentration and the number of positive wells was evaluated by measuring the incorporation of ^{3}H-thymidine. Initially the frequency of proliferating cells from the SJL/J spleen cells was 1/900. After 20 days of culture, the frequency was 1/1100 and after 45 days the frequency reached 1/8.

IV. CLONING OF CULTURE CELLS

After 2 months, when the proliferative frequency was high and the bulk culture showed cytotoxic activity against YAC-1 (35% cytotoxicity at 100:1 in 18 hr assay cf. 2.9% by fresh SJL spleen cells), the culture was cloned by limiting dilution at a concentration of 0.5 cells/well in a microtiter plate with conditioned medium containing feeder cells and IL-2. At the time of cloning, the frequency of proliferating cells was 1/6, resulting in a total of 36 clones.

V. CYTOTOXIC ACTIVITY OF CLONES

The cytotoxic activity of the clones was tested against 5 different targets in an 18 hour ^{51}Cr-release assay. Table 1 shows the reactivity of those clones against NK-sensitive and NK-resistant tumor cells. When cultured cells from bulk culture were tested, they showed relatively high levels of cytotoxicity against 2 of the NK-sensitive tumor lines but no reactivity against K562. However, cytotoxicity was measurable against the NK-resistant line M109 but not against EL4⁻, another NK-insensitive tumor line. Of the 7 clones shown in Table 1, all had some cytotoxicity against M109 cells, but only 4 of them (Clones 1, 2, 3 and 5) were able to kill NK-sensitive YAC-1 and RL♂1 tumor cells. Among the targets tested, the other clones (6, 8 and 9) showed appreciable reactivity only against M109 targets.

Table 1. Cytotoxic Activity of Cultured Spleen Cells From SJL/J Mice

	E/T Ratios	NK-sensitive targets			NK-insensitive targets	
		YAC-1	RL♂1	K562	EL4-	M109
Bulk culture	100	31.0	30.1	1.5	0.8	28.1
	50	28.0	14.5	1.3	1.0	22.0
	10	22.0	3.5	1.8	0.3	2.5
Clone 1	10	30.0	7.0	1.2	1.0	30.1
	1	11.0	6.9	2.2	0.9	17.0
Clone 2	10	25.0	11.0	0.7	0.5	18.0
	1	21.0	6.0	0.8	1.0	10.0
Clone 3	10	35.0	9.3	6.2	0.2	32.0
	1	29.0	5.0	3.4	1.1	15.0
Clone 5	10	55.8	7.8	4.4		
	1	8.4	4.1	2.2		
Clone 6	10	0.0	0.1	1.9	0.8	20.0
	1	0.7	0.0	0.5	0.0	13.0
Clone 8	10	-2.0	2.7	8.5	1.3	20.1
	1	0.0	0.9	0.3	0.1	12.5
Clone 9	10	0.0	0.0	5.0	1.0	16.0
	1	-1.1	-0.5	3.6	1.2	12.0

VI. PHENOTYPE OF THE NK CLONES

Five of the clones have been studied in regard to their morphology on Giemsa-stained cytocentrifuge slides and their expression of T cell associated antigens, as detectable by monoclonal antibodies in fluorescence flow cytometry (Table 2). Three of the clones studied had morphology similar to that of large granular lymphocytes (LGL), with azurophilic granules in their cytoplasm, whereas the other clones lacked detectable granules and had the appearance of lymphoblasts. There was also considerable heterogeneity in expression of Thy 1.1, Lyt 1 and Lyt 2 antigens, with only up to 30-50% of the cells of some clones expressing one or more of these antigens.

Table 2. Phenotype of Some SJL Clones with NK Activity

	Expression of:			
Clone	Lyt 1	Lyt 2	Thy 1.1	Morphology
1	+(30%)[a]	+(30%)	±(10%)	LGL
2	-	-	-	non-LGL
3	±(12%)	-	+(50%)	LGL
5	±(10%)	+(30%)	-	non-LGL
9	-	+(20%)	±(18%)	LGL

[a]Expression assessed with monoclonal antibodies and fluorescent flow cytometry. In parenthesis is the percentage of cells positive, i.e., with fluorescence above the background controls. - = < 10% of cells +.

VII. DISCUSSION

The initial frequencies of proliferating cytotoxic cells from the spleens of SJL mice were quite low, as might be expected for a strain with low NK activity. However, after two months of culture, a high frequency of proliferating cells was obtained and of these, at least 20% had cytotoxic activity against one or more target cells. The levels of cytotoxic activity by some of the clones were quite high, comparable to those seen with cultures from some high or intermediate NK strains (e.g., Kedar and Herberman, this volume). The results indicate that a low NK strain like SJL has some cells with the inherent ability to be highly cytotoxic, and the initially low reactivity may be due to either a low frequency of cytotoxic cells and their progenitors, or to the presence of suppressor cells, that may interfere with the proliferation of cytotoxic cells as well as with the effector phase (Santoni _et al_, this volume).

Although the cytotoxic clones from SJL mice have not yet been fully characterized, it seems likely that they represent NK cells or related natural effector cells. They were derived from normal mice and were grown from the outset with an IL-2 preparation depleted of mitogen. Such growth of IL-2-dependent cytotoxic progenitors, in the absence of mitogenic or antigenic stimulation, is analogous to observations that human LGL can be propagated in a similar way (Timonen et al and Vose et al, this volume). In contrast, typical small T cells have only been grown on IL-2 after stimulation. The NK nature of at least some of the clones was also supported by their LGL-like morphology.

The patterns of cytotoxicity associated with the SJL clones are similar to those of Kedar and Herberman (this volume) in that heterogeneity among the clones was seen and some were reactive against both NK-sensitive lymphoma cells and an NK-resistant solid tumor target cell. It was of interest that a few of the clones (clones 6, 8 and 9), initially considered to be non-cytotoxic because of lack of reactivity against YAC-1 or other NK-sensitive targets, were selectively cytotoxic against M109 target cells. It might be suggested that such clones are more analogous to NC cells than NK cells; however, one of the clones (clone 9) had LGL morphology and M109 has been shown to have some susceptibility to mouse NK cells (Riccardi et al, 1980). In addition to their heterogeneity in cytotoxic activities, the SJL clones also developed considerable hterogeneity in cell surface antigens. In contrast to most previously reported NK clones, only one was clearly Thy 1 positive. Also, as seen by Kedar and Herberman (this volume) with some BALB/c clones, some of the SJL clones expressed Lyt 2 antigens and at least one was positive for Lyt 1 as well. These findings provide further evidence for the derivation of NK cells from the T cell lineage.

REFERENCES

Dennert, G. (1981). Nature 287:47.

Dennert, G., Yogeeswaren, G., and Yamagata, S. (1981). J. Exp. Med. 153:565.

Nabbolz, M., Engers, H.D., Collavo, D., and North, M. (1978). Cur. Top. Microbiol. Immunol. 81:176.

Nabel, G., Bucalo, L.R., Allard, J., Wigzell, H., and Cantor, H. (1981). J.Exp. Med. 152:1582.

Riccardi, C., Santoni, A., Barlozzari, T., Puccetti, P., and Herberman, R.B. (1980). Int. J. Cancer 25:475.

Schrier, M.H., Iscove, N.N., Tees, R., Aarden, L., and von Boehmer, H. (1980). Immunol. Rev. 51:315.

CLONAL ANALYSIS OF HUMAN NATURAL KILLER CELLS

B.M. Vose[1,3], C. Riccardi[2,3], R.J. Marchmont[1], G.D. Bonnard[3].

[1] Department of Immunology, Paterson Laboratories, Christie Hospital and Holt Radium Institute, Manchester, England.

[2] Institute of Pharmacology, University of Perugia, Perugia, Italy.

[3] Laboratory of Immunodiagnosis, National Cancer Institute, Bethesda, Maryland, USA.

I. INTRODUCTION

Cells cultured in interleukin-2 (IL-2) show heterogeneous effector functions in cytotoxicity assays. Specific T cell lysis in appropriately sensitized cultures, natural killing (NK) and antibody-dependent cellular cytotoxicity (ADCC) are all readily detectable (Alvarez et al, 1978; Bonnard et al, 1978; Schendel et al, 1980). It has recently been possible to establish cultures of large granular lymphocytes (LGL) in IL-2, following their separation on discontinuous Percoll gradients. These cultured lymphoid cells (CLC) maintain the morphology and function of native NK cells (Ortaldo et al). CLC specific for human tumours have also been derived after in vitro sensitization and separation of T cell blasts from LGL on similar density gradients (Vose and Bonnard, 1982). In the present report the optimal conditions for initiating IL-2 dependent cultures from LGL and small T cells were examined by limiting dilution analysis. The role of interferon in modulating IL-2-driven population expansion and the mechanism of that effect were also studied, using the limiting dilution analysis.

II. LIMITING DILUTION ANALYSIS

Limiting dilution experiments enabled us to analyse the frequency of the cells proliferating in conditioned media (CM) containing IL-2 (prepared as in Bonnard et al, 1980, by PHA pulsing of lymphocytes and passage of CM through an anti-

NK CELLS AND OTHER NATURAL EFFECTOR CELLS

ISBN 0-12-341360-5

PHA immunoabsorbent column. The frequency of resulting microcultures which showed cytotoxic activity against the prototype NK target K562 was also monitored. Small T cells and LGL, freed of cells forming rosettes with sheep erythrocytes (E) at $29^{o}C$, were isolated on discontinuous Percoll gradients, as detailed previously (Timonen and Saksela, 1980). Cells were dispensed into 0.2ml round bottomed microtest plates (12-24 wells/responder concentration), together with irradiated (3000R) feeder cells and conditioned medium (at double the concentration giving maximal proliferation of IL-2-dependent CLC).

Proliferation was measured by uptake of ^{3}H-thymidine over the last 18 hours of a 7 day culture period, and cytotoxicity against ^{51}Cr labelled K562, RLσ1 mouse lymphoma and PHA-induced alloblasts (5000 cells/well) was measured in an 18 hour assay. The frequency of the limiting cell in each test was determined as that concentration at which 36% of wells were negative (Taswell, 1981).

A. Large Granular Lymphocytes

Low density Percoll fractions from peripheral blood leucocytes further depleted of $29^{o}C$ E-rosette-forming cells (Timonen and Saksela, 1980), and showing greater than 95% cells with LGL morphology and <1% OKT3+ cells, proliferated in lectin free condition medium. Approximately 1/200 cells produced microcultures under these conditions (Table 1). Addition of small amounts of PHA (0.10 μg/ml) was sufficient to induce maximal frequency of proliferating LGL (1/70) and thereafter addition of increasing quantities of lectin was without effect on the proliferative frequency. A similar frequency of proliferating LGL was also seen in the total absence of added feeder cells, i.e. the precursors of proliferating LGL appeared to express the IL-2 receptor spontaneously.

A high proportion of the proliferating LGL microcultures were lytic for K562 (but not for NK-resistant RLσ1 or PHA alloblasts Table 1. Lysis of K562 was at least 29 fold more frequent than lysis of the other targets.

B. Small T Cells

Proliferation of the small T cells under limiting dilution conditions was critically dependent upon the presence of lectin and less than 1/4000 small T cells expanded in conditioned medium in the presence of feeder cells alone (Table 1), possibly reflecting the frequency of *in vivo* activated T cells. When PHA was added, T cell proliferation increased to reach a maximum of 1/6. This proliferation of T cells with PHA occurred only in the presence of feeder cells. These data suggest that PHA alone does not activate T cells or induce

TABLE 1. Limiting Dilution Analysis of Proliferative and Cytotoxic Cells in LGL and Small T Cells.

	Frequency of Cells Responding			
PHA Conc. (μg/ml)	Proliferation	Cytotoxicity		
		K562	RL♂1	Alloblast
LGL				
-	1/ 300	1/560	-	-
0.05	1/ 105	1/170	$1/5.10^{3}$	$<10^{-4}$
0.10	1/ 72	-	-	-
0.25	1/ 70	-	-	-
0.5	1/ 74	1/102	$<1/7.10^{3}$	$<10^{-4}$
Small T Cells				
-	1/4216	$<1/10^{-4}$		
0.05	1/ 905	$<1/10^{-4}$	$<1/10^{-4}$	$<10^{-4}$
0.10	1/ 22	-	-	-
0.25	1/ 11	-	-	-
0.5	1/ 6	$<1/10^{-4}$	-	$<10^{-4}$

All data represent the mean of at least 3 determinations.

TABLE 2. Effect of Interferon on Proliferative and Cytotoxic Cell Frequencies in LGL and Small T Cells

IFN (500 U IFNα/ml)	PHA Conc./μg	Feeder Cells	Proliferative Frequency of: LGL	Small T Cells
None	0.05	PBMC[3]	1/ 50	1/1280
Pretreat[2]			1/ 25	1/ 640
In culture			1/ 237	1/4825
None	0.1	PBMC	1/ 46	1/ 150
Pretreat			1/ 18	1/ 100
In culture			1/ 174	1/ 950
None	0.05	Monocyte	1/ 200	1/ 720
In culture		Monocyte	1/ 275	1/ 950
In culture		Monocyte + T Cell	1/1100	1/3600
None		Monocyte	1/ 125	1/ 5
In culture		Monocyte	1/ 120	1/ 5
In culture		Monocyte + T	1/6000	1/ 82
		Monocyte + OKT4+	1/ 800	1/ 22
		Monocyte + OKT8+	1/ 800	1/ 20

1) 500 units IFN/ml; 2) For 18 hours; 3) Peripheral blood mononuclear cells.

the IL-2 receptor but that T cell-accessory cell interactions are required (Grönvik and Andersson, 1980). The cultured T cells usually lacked cytotoxic activity against K562, RL♂1 or alloblasts even when large numbers of cells proliferated. This emphasises the clear differences in lytic activity between LGL and T cell cultures. To avoid cytotoxicity induced by possible low levels of PHA, the effector cells had been routinely washed and supplemented with fresh medium 24 hours prior to the tests (Ortaldo et al, 1982).

This series of experiments established conditions under which selective cultures of LGL-like activity could be initiated, i.e. in lectin-depleted IL-2, in the absence of irradiated feeder cells, where proliferative frequency of LGL exceeded that of small T cells by >20-fold. To facilitate T cell growth it was necessary to activate (by lectins, or by autologous tumour; Vose and Bonnard, 1982). These data indicate that, when PBL are expanded with lectin-free conditioned media, NK-like cells are likely to predominate. Such cultures have been shown to mediate widespread lysis of several tumour target cells (Vose and Moore, 1981).

III. EFFECT OF INTERFERON

Interferon (IFN) has a major immunomodulatory role (Trinchieri et al, 1978; Vánky and Argov, 1980), it has been recently suggested that IFN and IL-2 have synergistic boosting effects on NK, with IFN inducing IL-2 receptors on mouse spleen cells (Kuribayashi et al, 1981). We observed that overnight pretreatment of LGL with IFN-α or β, followed by extensive washing, increased the frequency of cells proliferating in conditioned medium (Table 2). This effect was dose-dependent, with maximal augmentation (approximately 2-fold) at 500 units IFN/ml. Increased proliferative frequency was paralleled by increased frequency of anti-K562, but not anti-alloblast cultures. IFN also increased the frequency of T cells responding to a suboptimal dose of PHA but did not influence maximal proliferative frequency. No cytotoxicity against K562 was induced by IFN pretreatment of T cell microcultures.

Growth stimulation by IFN was not seen when it was present throughout the 7 day culture. Rather, proliferative and K562 cytotoxic frequencies were reduced in a dose dependent manner by both IFN α and β. Maximum inhibition (at least 4-fold) of proliferation with both LGL and T cells was seen with 500 u/ml IFN. The inhibitory activity of IFN was detectable only when T cells were present in the cultures (Table 2, Expt.3 and 4). Thus with LGL as responding cells, and irradiated adherent monocytes as feeders, no IFN-induced growth suppression was recorded. Addition of irradiated T cells to feeder monocytes induced suppressor activity. Separation of T cells into $OKT4^+$ and $OKT8^+$ populations (by attachment to anti-mouse immunoglobulin-coated petri dishes) suggested that both were essential for maximal suppression.

These data indicate that IFN (α and β) can induce more IL-2 receptors on LGL or synergize with a lectin for optimal induction of IL-2 receptor on small T cells. When present throughout the assay period, IFN can induce T cell-dependent suppression of proliferation.

IV. CLONED POPULATIONS OF NK CELLS

We have recently examined the activity of a number of cloned cell populations with cytotoxic activity against the K562 cell line. Peripheral blood lymphocytes from a healthy donor were stimulated in mixed leukocyte reaction (MLR) for 5 days and surviving cells plated at one cell/3 wells in the presence of conditioned medium and 2.5×10^4 irradiated stimulator cells as feeders. After 14 days, 48 wells from 4 plates showed macroscopic growth (cloning efficiency 50%) and cells were transferred to Costar 24 well cluster plates and maintained by twice weekly feeding with conditioned medium and restimulation with irradiated stimulator cells at weekly intervals. Of the 48 cloned cell populations, 6 consistently showed killing against K562 (Table 3).

TABLE 3. Phenotype, Cytotoxic and Proliferative Activity of NK-Like and Other Cloned CLC Derived from MLC

Clone Number	% Cytotoxicity[1] Against:		% Cells Staining With:		PLT Response[2]
	K562	Raji	OKT4	OKT8	
LH 1*	8.1	-1.3	45	0	-
LH 9*	25.3	-5.4	50	0	-
LH12*	61.4	NT	55	0	-
GB18*	52.1	-0.4	50	1	-
LH18	0.1	-2.3	95	2	-
LH13	-4.4	NT	100	0	+
GB 2	7.5	0	95	2	+
LH 4	0	-1.7	0	98	-
GB12	0	1.1	2	98	-

[1] Effector:target ratio: 20:1; 18/hr assay. Only 4 of the 6 NK-reactive clones are shown*.

[2] Positive when proliferation in the cultures stimulated with the original stimulating cells was significantly higher than in the controls cultured with 3rd party cells (Students t test).

Cloned cells with NK-like activity did not show primed lymphocyte test (PLT) responses to the specific, original, stimulating cells, nor any proliferation to third party lymphocytes, and they did not lyse the relatively NK resistant Raji cells. Surprisingly, they differed markedly in the phenotype by monoclonal antibody from NK-negative clones. All NK clones were weakly $OKT4^+$ (approximately 50% cells) and were negative for OKT8 (Table 3). In other non-NK clone 95-100% cells were strongly stained by either OKT4 or OKT8. The number of clones examined is currently being expanded to assess the generality of these findings, with recloning to document the clonality of the CLC.

V. CONCLUSIONS

These studies suggest that, starting from fresh PBL, NK cells with distinct cytotoxic potential are susceptible to growth in lectin free conditioned medium, and that very few, if any, of the postulated *in vivo* activated T cells grow. Furthermore, sensitization in MLR does not decrease, but rather increases the frequency of NK cell precursors, as well as providing newly activated T cells. NK-cells grown from MLC has a distinct OKT4 phenotype that has not been seen so far in NK cells grown from fresh LGL. The effect of purified human IFN on cultures started from fresh PBL suggests that the elevated frequency of both NK and T cell precursors in MLR could be due in part to the production of IFN-γ, in addition to the effects of alloantigens and macrophage products.

ACKNOWLEDGEMENTS

This study was supported by grants from the Medical Research Council and the Cancer Research Campaign of Great Britain. The authors are grateful to Sheldon Grove and Wendy White for technical assistance and Drs R.B. Herberman and M. Moore for encouragement and critical reading of the manuscript.

REFERENCES

Alvarez, J.M., Landazuri, M.O., Bonnard, G.D., and Herberman, R.B., J. Immunol. 121:1270-1275 (1978).

Bonnard, G.D., Schendel, D.J., West, W.H., Alvarez, J.M., Maca, R.D., Yasaka, K., Fine, R.L., Herberman, R.B., de Landazuri, M.O., and Morgan, D.A., in "Human Lymphocyte Differentiation: Its Application to Cancer" (Serrou, B., Rosenfeld, C. eds), pp.319-326. Elsevier/North Holland Biomedical Press, Amsterdam, (1978).

Bonnard, G.D., Yasaka, K., and Maca, R.D., Cell Immunol. 51: 390-401 (1980).

Grönvik, K-O., and Andersson, J., Immunol. Rev. 51:35-59 (1980).
Kuribayashi, K., Gillis, S., Kern, D.E., and Henney, C.S., J. Immunol., 126:2321-2327 (1981).
Minato, N., Reid, L., and Bloom, B.R., J. Exp. Med. 154:750-762 (1981).
Ortaldo, J.R., Timonen, T.T., Vose, B.M., and Alvarez, J.M., in "The Potential Role of T Cell Subpopulations in Cancer Therapy" (Fefer, A. ed). Raven Press, New York, in press.
Schendel, D.J., Wank, R., and Bonnard, G.D., Scand. J. Immunol. 11:99-107 (1980).
Taswell, C., J. Immunol. 126:1614-1619 (1981).
Timonen, T., and Saksela, F., J. Imm. Meth. 36:285-291 (1980).
Trinchieri, G., Santoli, D., and Koprowski, H., J. Immunol. 120:1849 (1978).
Vánky, F., and Argov, S., Int. J. Cancer 26:405-413 (1980).
Vose, B.M., and Moore, M., Imm. Lett. 3:237-241 (1981).
Vose, B.M., and Bonnard, G.D., Int J. Cancer, in press (1982).

CONTINUOUS CULTURE OF HUMAN NK-T CELL CLONES

Daniel Zagury[1]
Doris Morgan

Université Paris VI
Paris, France

I. INTRODUCTION

NK lymphocytes, which spontaneously lyse *in vitro* a variety of tumor and normal target cells have been observed among normal and pathological human PBL and in T cell cultures (Alvarez *et al.*, 1979). These NK cells have not yet been well characterized. Many questions have yet to be answered: 1) What are the molecules involved in the recognition process of the target cells? 2) Do the NK T cells (NK-TC) represent a separate lineage with a specific *in vivo* function or are they an intermediate maturation stage of a different cell type such as cytolytic T lyphocyte (CTL)?

The lack of conclusive data can be explained in part by experimental limitations. Pure populations of KN lymphocytes have not been available, although cell separation techniques have been used to concentrate them (Ortaldo *et al.*, 1981). Until recently assays to identify killer cells directly in effector cell suspensions were not available either (Zagury *et al.*, 1975).

In this chapter we shall discuss first several of these new techniques, one of which yields homogeneous populations of NK-TC *in vitro*. As reported here, the availability of these pure NK-TC populations allowed us to define their antigenic characteristics and analyse their cytotoxic function.

[1]The work was supported by grants from DRET, INSERM, ADRC (Ville-juif) and Ligue Nationale Francaise contre le Cancer.

ISBN 0-12-341360-5

II. MATERIAL AND METHODS

Reagents

TCGF: Semi-purified preparations deprived of PHA by ammonium sulfate precipitations were used.

Monoclonal antibodies: OKT4 (helper-induced), OKT4 (suppressor-cytotoxic), and OKT10 were purchased (or generously supplied by Dr. G. Goldstein, ORTHO Pharmaceutical Corporation, Raritan, New Jersey).

Cells

T cells were obtained either from continuous cultures (CTC) derived from PBL with TCGF or from a mixed lymphocyte culture (MLC) in which PBL were repeatedly stimulated by irradiated allogenic lymphocytes. Normal human spleen cells (Ra) were used as stimulator cells. T cell clones were produced by isolating a single cell from either CTC or MLC (Zagury *et al.*, 1980) and culturing it subsequently with a filter layer of irradiated lymphoic cells and TCGF. Single cloned cells were successfully recloned by using the same conditions as for primary cloning.

Cell-mediated cytotoxicity (CMC)

CMC was assayed by the chromium release test using allogenic targets for detection of CTL, PHA-treated cells for lectin-induced cell cytotoxicity (LICC), and K562 cells for natural killer (NK) activity. In addition, the killer cell numeration test (Zagury *et al.*, 1980a) was used to determine the percentage of killer cells within a cell suspension. Briefly, an effector cell (E) and a target (T) were microassociated to form an E-T doublet. During incubation at 37°C killer cells were identified by the lysis of the associated target. Target cell lysis was visualized in phase microscopy by loss of cell refringency. In order to evaluate whether a same effector cell can lyse two different targets, the test was performed in some instances by association of two targets (T1 and T2) to single effector cells (E) to form T1-E-T2 doublets.

T cell antigens

Antigenic markers on T cell suspensions as well as single T cells were identified by the murine monoclonal antibodies OKT4, OKT8, and OKT10, using two assay systems: 1) complement dependent cytotoxicity, and 2) rosetting with ox-red blood cells coated with antimouse Ig antibody (Bernard *et al.*, 1982). These assay techniques were also used to select for cloning cells with defined antigens. Elimination of cells carrying unwanted antigens was accomplished by using corresponding monoclonal antibodies with complement. In the remaining subpopulation, cells carrying the desired antigen were identified by rosetting and individually isolated for cloning.

III. RESULTS

Production of T lymphocyte clones with NK activity

When clones were first produced by random isolation of T cells from CTC or MLC, NK activity was detected in some, where 5-20% of the cells lysed K562 targets (Zagury et al., 1981). Moreover 65% of cells in clone 45B9 expressed NK activity. This result prompted us to preselect cultured T cells prior to cloning to see whether the frequency of NK clones could be increased from selected subsets. Selected T cell clones were produced by isolation of single cells with a well defined surface pattern relative to T4, T8, T10 antigens. These cells were identified in T cell cultures, after complement dependent cytotoxicity and specific rosette testing (Section II). NK-TC clones with high lytic activity against K562 targets were produced from parents expressing a variety of antigenic phenotypes (Table I).

Cell surface antigens in NK clones

Table I indicates that no specific T4, T8, T10 antigenic pattern could be associated to NK clones; in addition, rosette tests performed directly on single killer cells demonstrated that neither T4, not T8, nor T10 antigens were consistently associated with killing of K562 targets. T10 was expressed on a percentage of cells; however this percentage was variable from clone to clone and often inferior to the percentage of NK killer cells. Indeed in clone 203C11 when T10 antigens were detected directly on isolated NK-TC just after they killed their associated target, only 4 out of 11 tested cells carried T10 antigens. Table I indicates also that cells with a clone often did not exhibit the parental antigenic pattern and can express new antigens.

CMC in NK-TC clones

Table I indicates that NK clones express a high cytolytic activity against K562 targets detected by chromium release test. This activity was performed by 55-65% of the NK-TC. Repeated testing of some clones show that a high level of lytic activity persisted for at least 2 weeks. However cytolytic activity against K562 was not expressed in all secondary clones obtained

TABLE I. NK-TC CLONE ANTIGENS

Clones	Origin	Parent cell phenotype	Chromium release test (%)[1] Target panel				Antigens[2] (% cells)		
			K 562	Ra EBV+	N4 EBV+	PHA-coated cells	T4	T8	T10
45 B9	CTC	ND[4]	95[3]	ND	ND	40	ND	ND	ND
			70			46			
194 F2	MLC	T10+ T8- T4-	60[3]	0	0	28	8	0	82
			40	0	ND	39			
201 E3	MLC	T4+ T8- T10-	86	26	0	30	1	48	11
			75	30	0	49	0	59	13
			60	32	0	44			
203 D9	MLC	T4- T8- T10-	67	10	0	47	15	32	52
							18	37	38
203 C11	MLC	T4- T8- T10-	48	0	ND	23	6	10	46
							4	16	42
204 F2	MLC	T10+ T8- T4-	56	0	0	36	10	21	12

[1] The chromium release test was performed with an effector target ratio of 5 (except for 45 B9 where the ratio was 30) and a 3 hour incubation at 37°C. Results are expressed according to the formula (Specific lysis - spontaneous lysis) : (Maximal lysis - spontaneous lysis).
[2] T4, T8 and T10 antigens were identified by monoclonal antibodies detected by a rosette test (sec.2)
[3] Multiple results for a clone represent sequential tests performed at 4-7 days intervals.
[4] ND = not done.

by recloning 201E3 (only 3 out of 5). CMC against other targets was also investigated.

1) All NK clones expressed a LICC activity directed against PHA coated L1210 or P815 cells. This non-specific activity ranged however lower than that reported in continuous T cell culture (Zagury _et al._, 1980b). Double target-effector association technique demonstrated that a substantial number of NK cells lysed also PHA treated target (LICC activity).

2) CTL activity was observed only in clone 201E3, in which 15% of the cells lysed Ra cells carrying sensitizing antigens. In addition, double target-effector association technique showed that some CTL—even though scarce (6%)—were able to lyse K562 and Ra cells.

3) NK-TC clones which lysed K562 target did not express cytotoxic activity against allogenic B^{ebv+} cell (N4cell). This result eliminates the possibility that killing of EBV transformed cells from Ra individuals by NK-TC clones was due to an NK activity.

IV. DISCUSSION

In this chapter we have reported three sets of information concerning NK activity in continuous NK-TC clone cultures. These data were obtained using recent methodological advances which permitted isolation, long-term culture, and direct analysis of killer cells.

1) An experimental system to produce pure NK-TC populations from human fresh PBL: T cell clones with high NK activity—NK-Tc clones—were produced by isolation of either random or preselected cells from CTC or MLC. These clones proliferated for several weeks in presence of TCGF and expressed high NK activity for at least two weeks. It should be emphasized that not all cells expressed killing function and, in addition, recloning of 201E3 clone cells did not always generate clones with NK activity. Thus lytic function in NK-TC is not fixed as it occurs when lymphoid cells are immortalized into cell lines, such as plasmocytomas. In T cell clones which are not cell lines, it would appear that cells are not exactly all at an identical level of maturation. Thus they do not all express the same function at the same time (Zagur _et al._, 1981). Immortalization of human NK-TC by either hybridization to a human T lymphoma (Irigoyen _et al._, 1981) or viral transformation would be required in order to fix NK function in culture.

NK-TC clones today represent a unique _in vitro_ system to study NK-TC, as demonstrated here, by the direct investigations of their antigenci characteristics and cytotoxic functions. Also if NK-TC were to play some role in tumor cell destruction, NK-TC clones from cancer-patients would be available. Such clones would represent an innocuous source of NK cells for adoptive cytotherapy.

2) Experimental evidence that NK-TC clones can originate from parents with different antigenic pattern. In addition none of tested antigen (T4, T8, or T10) could be directly correlated with NK activity in NK-TC clones. Direct analysis of isolated NK cells demonstrated that T10 antigens were not necessary for NK-TC activity. Differences in antigenic expression among cells within a clone can well be explained by differences in maturation of these cells which all carry the same parental genetic programme.

3) Experimental evidence to consider that NK-TC could represent a step in CTL differentiation. Functional differences among cells in the NK-TC clone 201E3 was also observed. 15% of the cells exhibited specific CTL lysis against targets carrying sensitizing antigens. Single cell experiments using two targets associated to one effector cell demonstrated that a small number of killer cells did express both NK and CTL activity. This result is consistent with the concept that NK cells represent an early stage in CTL differentiation. This interpretation would also explain why NK activity is found at an early stage of MLC before specific CTL.

REFERENCES

Alvarez, J. M., de Landazuri, M. O., Bonnard, G. D., and Herberman, R. B. (1979). J. Immunol. 121, 1270-1275.

Bernard, J., Ternynck, T., and Zagury, D. (1981). Immunology Letters, in press.

Irigoyen, O. H., Rizzolo, P. V., Thomas, Y., Rogozinski, L., and Chess, L. (1981). J. Expl. Med. 154, 1827-1837.

Kung, P. and Goldstein, G. (1981). Vox Sang. 39, 121.

Ortaldo, J. R., Sharrow, S. O., Timonen, T., and Herberman, R. B. (1981). J. Immunol., in press.

Zagury, D., Bernard, J., Thiernesse, N., Feldman, M., and Berke, G. (1975). Europ. J. Immunol. 5, 818-822.

Zagury, D., Fouchard, M., Morgan, D. A., and Cerottini, J. C. (1980a). Immunology Letters 1, 335-339.

Zagury, D., Morgan, D. A., and Fouchard, M. (1980b). Biomed. 33, 272-276.

Zagury, D., Morgan, D. A., and Fouchard, M. (1981). J. Immunol. Methods 43, 43-78.

PERMANENTLY GROWING MURINE CELL CLONES WITH NK LIKE ACTIVITIES

Hans Hengartner
Hans Acha-Orbea
Rosmarie Lang
Lothar Stitz
Kendall L. Rosenthal

Dept. of Exp. Pathology
University Hospital Zürich
Zürich, Switzerland

Peter Groscurth

Institute of Anatomy
University of Zürich
Zürich, Switzerland

Robert Keller

Immunobiology Research Group
University of Zürich
Zürich, Switzerland

I. INTRODUCTION

Function, surface markers and additional characteristics of specific effector cells are often very difficult to determine in complex mixtures of cells. The development of improved tissue culture systems and the discovery of specific growth factors allowed the establishment of various permanently growing delicate cell lines (deWeck, 1979). Functional T lymphocytes, cytotoxic and helper T cells, are one type amongst them.

We established several murine T cell clones after specific stimulation in secondary mixed lymphocyte cultures and subsequent growth in the presence of T cell growth factor (TCGF or IL 2). Thereafter we discovered a drifting of the initial specific cytotoxicity towards a cytolytic activity against NK-sen-

ISBN 0-12-341360-5

sitive targets. In this chapter we describe the derivation of and some observations on such cloned cell lines.

II. DERIVATION OF ANTIGEN SPEICIFIC CLONED TCGF DEPENDENT CELLS

We established several permanently growing cloned cell lines. Antigen-specific cells were initially enriched by stimulating primed or unprimed spleen cells with irradiated stimulator cells in a mixed lymphocyte culture. Seven to ten days later the cells were restimulated with X- irradiated stimulator cells. After 24 to 48 hrs proliferating cells were cloned, either on soft agar or by the limiting dilution technique in 0.2 ml cultures in the presence of IL-2. Growing cells were expanded in 0.2 ml cultures in the presence of peritoneal exudate cells as feeder cells. After several initial growth crises, some of these cloned and recloned cells could be adapted and maintained in culture for more than 3 years. By this procedure we selected cells which were strictly IL-2 dependent but grew without further restimulation with the original antigen. The cloning efficiency of the first cloning was always far below 1%, whereas the efficiencies of subsequent reclonings (cloned at 0.2 to 0.3 cells per well) increased to 10 - 50%. At this stage we dealt with real specific cytotoxic T cell clones. Most of these cells displayed abnormal chromosome numbers (F. Wiener, personal communication).

III. ANALYSIS OF SPECIFICITY

A. Observations on Male Antigen (HY) Specific T Cell Clones

Five independent HY-specific T cell clones were selected by distributing cells from mixed lymphocyte cultures C57 BL/6 female anti C57 BL/6 male under limiting dilution conditions (von Boehmer et al., 1979). The HY-specific cytotoxicity of all the clones was H-2D^b restricted. In addition they killed H-2D^d antigen expressing cells from both sexes. Within weeks, cultures of such cells could lose completely their cytotoxic activity or they killed male and female LPS blasts equally well. The maintenance of specific cytotoxic T cell clones was only possible by the regular subcloning of the cells. The specific killer activity was better preserved by growing the cells in a completely serum-free medium (Schreier et al, 1980). Three clones which derived from independent mixed lymphocyte cultures were maintained over one year in Iscove's modified DMEM (Iscove and Melchers, 1978) supplemented with 10% fetal bovine serum and 10% Con A stimulated rat spleen cell supernatant as IL-2 source. After this period, some of the cell lines exhibited neither of the two original cytolytic activities ($H-2D^b$ restricted HY or $H-2D^d$). But by screening a collection of tumor cell lines as target cells we discovered a newly acquired NK-like cytotoxic activity of the three cloned T cell lines. Table 1 represents examples of the cytotoxic activities of the three lines against target cells in a 3 hr ^{51}Cr-release assay.

These results clearly demonstrated the loss of the initial $H-2\ D^d$ specific cytotoxicity and an acquired strong activity against YAC-1. The loss of the HY-specificity is not shown here. NK sensitive target cells like P52 ($H-2^b$), RBL-5 ($H-2^b$) were equally good targets. WEHI-164 ($H-2^d$) was lysed after incubation periods longer than 10 hr, which is in agreement with observations on enriched NK cells in bulk cultures. The common feature of the three independent cell lines HY-1, HY-2 and HY-3 is that the same target cells are killed or not killed to the same degree.

Furthermore we selected several subclones of HY-3 which either lost the ability to kill YAC-1 or in contrast exhibited an extremely high specific cytotoxicity towards YAC-1. The question remains still open whether the former cells lack the specific receptor structures or whether they lost the cytotoxic elements such as enzymes or other cytocidal products.

TABLE I. Cytotoxic Activities of Cloned Cell Lines

Target	H-2	HY - 1			HY - 2			HY - 3			B10 anti B10.D2MLC		
YAC-1	a	72	61	43	77	64	40	62	45	33	32	15	0
EL-4	b	7	5	3	14	6	6	22	15	8		———	
P815	d		———		3	2	3	15	10	7	42	18	8
B10.D2 Con A Blast	d	11	10	4	6	5	2	15	9	4	63	48	35

Results tabulated are % specific lysis at effector to target cell ratios of 10:1, 3:1, 1:1; 10^4 target cells per well. The results represent the mean of 3 values of at least 3 independent experiments.

B. Observations on Cloned Cytotoxic Cells from other Mixed Lymphocyte Cultures

During our attempt to grow clones of allospecific or virus-specific cytotoxic T cells, we selected many permanently growing cytotoxic cell lines which killed NK-sensitive targets. It seems that our tissue culture conditions or cloning procedures favour the growth of a special cytotoxic cell type. This is documented in the following typical cloning experiment. Fourteen days after the initiation of the mixed lymphocyte culture, B10 anti-B10.D2 ($H-2^b$ anti $H-2^d$), we selected 59 clones. Only 2 clones killed the NK insensitive target P815 ($H-2^d$) and a surprisingly high number of 20 clones killed the NK-sensitive target YAC-1. Another example is the growth of VSV-specific cytotoxic killer T cells. In vivo primed spleen cells were stimulated with VSV infected BALB/c macrophages in a secondary mixed lymphocyte culture. The $H-2^d$-restricted VSV cytotoxicity of this bulk culture could be maintained over months by biweekly restimulations of diluted proliferating T cell cultures with X-irradiated VSV infected macrophages. We selected six permanently growing VSV specific cytotoxic T cell clones by the described cloning procedure under conditions of limiting dilution.

Three clones were grown as permanent cell lines which were routinely diluted every 2 to 3 days into fresh medium. As soon as the effectors cells were not restimulated with infected stimulator cells any more, we monitored a continous decline of the virus-specific cytotoxicicity over the following 2-3 months until all three cell lines did not exhibit any virus specific cytotoxicity at all. At the same time we noticed that one of the cell lines showed a steadily increasing cytotoxicity against the NK sensitive target cell line YAC-1. This activity has remained stable for over 5 months.

This observation is parallel to the one on the male antigen specific T cell clones, where an initial H-2 restricted antigen specificity was lost and a less specific cytotoxicity appeared. The newly acquired activity stayed at a constantly high level over many months in culture. This is in contrast to the specific cytotoxicity generated in the secondary mixed lymphocyte cultures. Allospecific and hapten-specific cytotoxic T cell clones retain their specificity if they are restimulated with the original stimulator cells regularly every 2 to 3 weeks (H. Acha-Orbea and H.P. Pircher, unpublished results) in the presence of IL-2.

The target specificity of the clones with NK-like activities seems to differ. Whereas YAC-1 as targets were killed by all such clones there was some heterogeneity in respect of killing of the rat cell line P-12 (derived from DA rat), RAJI (huma Burkitt Lymphoma), WEHI-164 ($H-2^d$), L 1210, P52 ($H-2^b$) and RBL5 ($H-2^b$). But the structure of the recognized antigen on the target cell has still to be elucidated.

IV. MORPHOLOGY AND SURFACE MARKERS

Electron microscopically, both HY-2 and HY-3 clones with NK cell-like activity did generally not differ from each other. At their surface the round shaped cells displayed short pseudopode-like cellular processes (Fig. 1). They were charc- terized by an irregularly formed nucleus rich in euchromatin with one or two excentric nucleoli. The cytoplasm contained few mono- and polyribosomes and a large number of elongated profiles of rough endoplasmic reticulum (rER) filled with an electron dense substance. The numerous mitochondria often dis- played both transversely and longitudinally orientated cristae. A well defined Golgi complex and small bundles of microfila- ments were regularly observed in close vicinity to the nucleus. In addition, all HY-2 and HY-3 cells studied contained a large number of membrane-bound lysosomal granules in their cyt- plasm (Fig. 2). The size of these granules varied between 500 and 1200 nm. At higher magnification, each lysosome exhibited in the centre a fine granular electron-dense material while the periphery was formed by small vesicles. Light microscopi- cal and enzyme histochemical studies revealed high activity of chloroaecetate esterase but no acid phosphatase in most of these granules.

Cells of HY-3 clones without NK-like activity differed from other clones particularly by ultrastructural morphology of their lysosmes (Fig. 3). Thus they contained less but larger granules in their cytoplasm. Most of the granules were filled by numerous electron dense or translucent vesicles. Histochemically, the lysosomes of these cells showed moderate activity of both chloroacetate esterase and acid phosphatase.

Surface markers of the cloned cytotoxic cells were tested with monoclonal antibodies. All expressed the H-2K, H-2D Lyt-2 and Thy 1 antigen. Whereas all were Lyt-1 negative. The presence of the monoclonal antibody against Lyt-2 in the ^{51}Cr release assay was inhibitory for alloreactive or hapten- specific cytotoxic T lymphocytes but not inhibitory for the described cytotoxic cells against YAC-1.

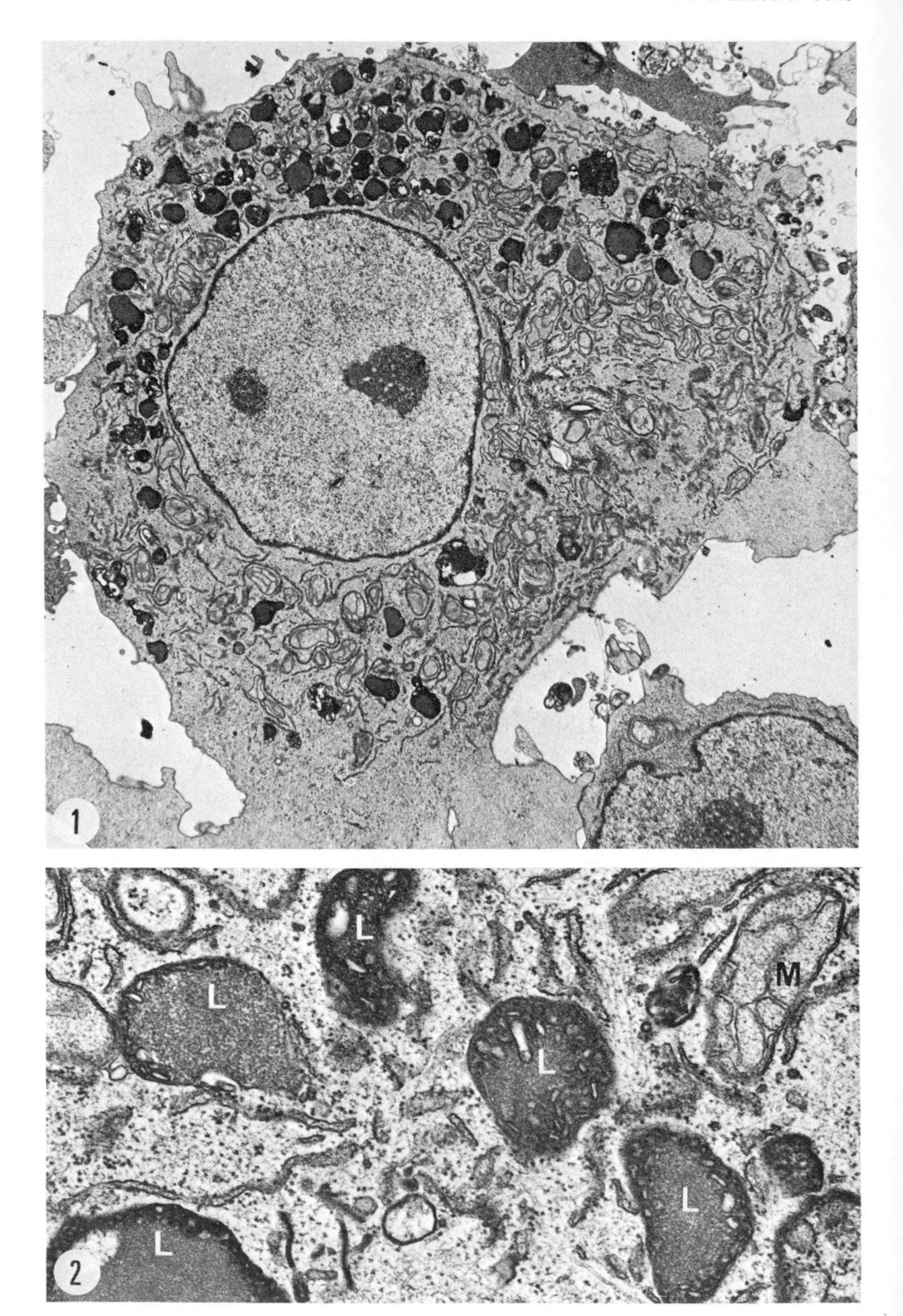
1
L
L
M
L
L
L
2

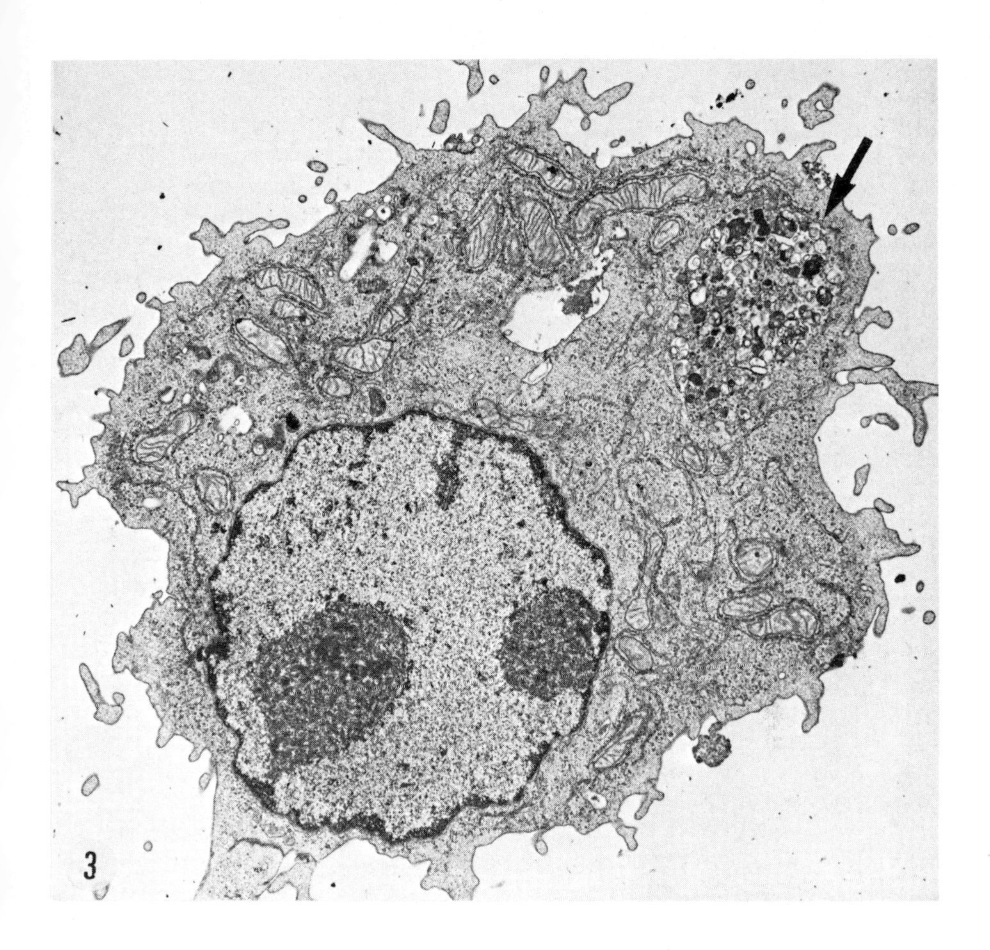

Fig. 1: HY-3 cell with NK like activitiy, x 10'000

Fig. 2: Part of an HY-3 cell (enlargement of Fig. 1) The lysosomes (L) display numerous small vesicles, x 35'000

Fig. 3: HY-3 cell without NK like activity. The cytoplasm contains one large lysosomal granule (arrow), x 9'500

V. CONCLUDING REMARKS

Several antigen-specific cytotoxic T cells were cloned. The low cloning efficiency provided evidence that we selected an extremely rare cell type which changed the target cell specificity regularly towards NK-sensitive targets within relatively short growth periods. In addition we observed a change of the growth behavior, characterized by increased adherance of the cells and by the appearance of abnormal chromosome patterns. The specific cytotoxicity against YAC-1 lymphoma cells was extremely high and there was no lag phase detectable in a cytotoxicity assay. For example 26 % target cells were killed at an effector to target ratio of 10 : 1 within 30 min. In summary it seems that our tissue culture conditions without the repetitive restimulation of effector cells with the original antigen favours a change of the initial antigen specificity of the cytotoxic cell. We observed a very limited variability of the NK-like specificity which is compatible with the idea that NK cells do not recognize rare antigen structures but rather common determinants expressed on a wide range of tumor cell lines. Our cloned cytotoxic cells are Thy 1^+, Lyt-2^+ and Lyt-1^-. They are strictly IL-2 dependent and their cytotoxicity could not be modulated by various periods of interferon treatment

Morphologically, NK-like clones differed from those without NK-like activity predominantly by the ultrastructure and enzyme pattern of lysosomal granules. The classification of NK cells is a very difficult task (Burton, 1980; Minato et al., 1981). Certainly the target specificity of the effector cell is not sufficient. Surface markers, growth requirements and the ability to be activated by interferon and IL-2 are presumably also important criteria.

IV. ACKNOWLEDGMENTS

We wish to thank Dr. Rolf Zinkernagel for the support and the critical reading, Mrs. Margrit Bucher for the preparation of the manuscript, Mrs. M. Balzer for excellent technical assistance and Dr. Rolf Kiessling for supplying us with the NK-sensitive target cell lines. This work was supported by the Swiss National Science Foundation grant 3.485-0.79.

REFERENCES

von Boehmer, H., Hengartner, H., Lernhardt, W., Schreier, M. H., and Haas, W. (1979) Eur. J. Immunol. 9, 592

Burton, R. C. (1980) In Natural Cell-Mediated Immunity against Tumors. (R. B. Herberman, ed.), p.105, Academic Press, New York

Iscove, N. N. and Melchers, F. (1978) J. exp. Med. 147, 923

Minato, N., Reid, L. and Bloom, B. R. (1981) J. Exp. Med. 154, 750

Schreier, M. H., Iscove, N. N., Tees, R., Aarden, L., and von Boehmer, H. (1980) Immunological Rev. 51, 315

DeWeck (1979) Proc. Sec. Lymphokine Conf. Academic Press, New York

NATURAL KILLER ACTIVITY IN CLONED IL-2 DEPENDENT ALLOSPECIFIC LYMPHOID POPULATIONS

John R. Neefe[1] and Robin Carpenter

Divisions of Immunologic Oncology
and Medical Oncology
Lombardi Cancer Center
Georgetown University
Washington, D.C.

I. INTRODUCTION

Natural killer (NK) cells have been defined by their capacity to lyse certain allogeneic tumor targets without priming or prior exposure. A recognition and binding event is involved, and it appears that at least several widely distributed specificities may be involved in the recognition (Herberman and Ortaldo, 1981). Although there has been no evidence for memory, it has been observed that antigen-specific cytolytic populations expanded with interleukin 2 (IL-2) commonly express high levels of NK activity (Kornbluth et al., 1981). We have made the same observation in murine systems. This observation may be explained in two ways: (1) an antigen specific clone, perhaps broadly cross-reactive, is capable of killing NK-susceptible targets through recognition of shared or "public" determinants; (2) NK cells grow well, or even preferentially, in the presence of the IL-2 and feeders used to expand the cytolytic population. We have investigated murine populations resulting from allosensitization to an H-$2K^b$ mutant and subsequently cloned in an effort to distinguish these possibilities. The data suggest that NK-like activity is a property of some allosensitized clones.

---

[1]Supported by NIH grants 5R01 CA25041 and 2P30 CA14626.

ISBN 0-12-341360-5

II. MATERIALS AND METHODS

Mice of the strains C57BL/6J (B6) and B6.C-H-2^{bm1}/ByJ (bm1) were obtained from Jackson Laboratories and housed in our colony. The H-2^{b} mutants bm8 and bm11, congenic with B6, were kindly provided by Dr. Roger Melvold.

Lymphocytes were obtained as a single cell suspension from the spleens of sacrificed mice and cleared of erythrocytes with red cell lysing buffer.

Allogeneic sensitization was performed by coculturing $100x10^6$ responder lymphocytes with $200x10^6$ irradiated stimulator lymphocytes in a tissue culture flask with 100 ml Eagle's Minimal Essential Medium containing supportive fetal calf serum, nutrients and antibiotics.

Cell sorting or analysis was performed on the Cytofluorograf (Ortho 50H). Cells to be processed were cleared of debris by flotation over Ficoll-Hypaque and stained with fluoresceinated monoclonal anti-Lyt 2 (Becton Dickinson) for 45 min at 4^oC in buffered saline containing fetal calf serum and azide. Cells were washed and analyzed or sorted at 1000-2000 cells per second after gating out debris.

Interleukin-2 was generated in cultures of rat spleen cells stimulated with Concanavalin A. The supernatants were harvested at 24-48 hr, filtered and frozen. Aliquots were concentrated by ultrafiltration, assayed for activity against a standard IL-2 dependent cell line, and used at the defined optimally supportive strength.

Cloning was performed according to described procedures (Braciale et al., 1981; Lamb et al., 1982). Positively selected allosensitized lymphocytes were plated in 20ul wells at an average of 0.5 cells per well with irradiated allogeneic feeder lymphocytes at $4x10^6$/ml and IL-2 containing supernatant. Wells were examined microscopically for growth at 5-10 days and wells with growth were transferred serially to larger wells at 5 day intervals or as needed. Fresh feeders and IL-2 were added every 5 days.

Cytotoxicity was assayed in a standard 4 hr ^{51}Cr-release system. Targets were the murine lymphomas EL-4 (H-2^b) and YAC-1 (H-2^a) which were maintained in tissue culture or in frozen aliquots; the IL-2 dependent cell line D76, of bm1 origin; and normal or Concanavalin A stimulated spleen cells of H-2^b mutant origin. D76 is highly susceptible to lysis by H-2^{bm1}-specific allogeneic killers but is not lysed by natural killers. Targets were labelled for 1-2 hr with ^{51}Cr, washed and plated at $5-50x10^3$ per well in 200ul wells with attackers at ratios of 0.1 to 50:1. Plates were centrifuged lightly, incubated for 4 hr, harvested with the Titertek Supernatant Collection System (Flow Laboratories), and counted in a gamma counter. Data are expressed in Table I as

$$\% \text{ Specific Release} = (E - SR)/(MAX - BG)$$

where E is the mean counts of triplicate samples, SR is the counts released in wells with medium only, MAX is the counts released by inbcubation of targets with 0.1N HCl, and BG is machine background. Data in parentheses are calculated standard errors.

III. RESULTS

B6.C-H-2^{bm1} were primed to C57BL/6J for 10 days _in vitro_. Three days following boosting, Lyt 2^+ large cells were collected on the Cytofluorograf in an effort to purify dividing cells of the cytotoxic phenotype. The sorted cells were homogeneous with regard to size and with regard to high levels of expression of Lyt 2. The sorted cells were plated at 0.5 cells/well in the presence of feeders and IL-2. 900 cells were plated from a suspension which consisted of single cells and no visible clumps. By the Poisson distribution, the chance of a well containing two or more cells was 9%. The number of wells with growth was 50. This indicates a precursor frequency of 11%. The likelihood of two precursors in a well was 0.002. After serial passage to larger wells, fourteen clones grew to sufficient numbers for assaying cytolytic activity and half had significant levels of cytolytic activity at attacker-to-target ratios of 5:1 or 1:1. Seven of the most active were selected for detailed specificity testing shown in Table I. These clones appeared to recognize the "strong" CML determinants previously described (Melief et al., 1980). B6 possesses one such determinant not possessed by bm1 and this determinant is expressed on bm11 but not bm8. Thus the allogeneic specificity displayed by these clones is very precise and must detect the strong specificity previously deduced by other methods. Relevant is the observation that five of seven clones also lysed the NK-susceptible target YAC-1. The feeders or IL-2 were not responsible for the NK activity, since all clones were maintained in an identical fashion: H44 and the negative clones showed no activity against YAC-1. The NK activity was not related to total activity of the clones, since several highly active clones (H28, H32) were devoid of NK activity. Selection of "active" clones for study did not introduce a bias toward selection of populations with contaminating NK clones capable of boosting activity in the screening procedure, since all clones were screened against the appropriate allogeneic target and also against YAC-1. In addition, the allogeneic specificity was very strong. A single allogeneic specificity expressed on mutant targets and the NK-susceptible target is not sufficient to explain the data since H12 lyses b, bm8, bm11 and YAC-1, but H32, which also lyses b and YAC-1, fails to lyse bm8 and bm11.

TABLE I.

Clone	Expt.	Targets				
#	#	EL-4	bm1	bm8	bm11	YAC-1
H2	2	5 (2)	6 (5)			22 (3)
H12	1	23 (1)	6 (4)			23 (6)
	3	66 (3)	-3 (1)	19 (1)	54 (4)	24 (1)
H14	1	21 (2)	5 (4)			14 (1)
	3	24 (1)	-1 (10	-4 (1)	29 (4)	2 (2)
H28	2	30 (3)	7 (6)			4 (3)
	3	70 (8)	0 (2)	0 (3)	49 (6)	0 (0)
H32	1	75 (4)	11 (4)			4 (1)
	3	76 (9)	-1 (1)	-1 (1)	-2 (1)	9 (1)
H40	1	40 (2)	2 (4)			21 (1)
	3	74 (3)	-4 (1)	-3 (1)	42 (2)	1 (1)
H44	1	77 (2)	11 (4)			0 (3)
	3	63 (2)	-3 (0)	2 (1)	27 (1)	0 (1)

[a] % specific release (standard error)

IV. DISCUSSION

The cytotoxic populations described here were cloned by limiting dilution under conditions that make it unlikely that very many of them arose from more than one cytotoxic precursor. Formal proof of clonality by subcloning has not yet been obtained. It is possible that the unit of limiting dilution was not a single cell but a pair of cells and that growth occurred only in wells with two interacting populations. We infer the possibility of such a phenomenon from studies of a feeder-independent population of allosensitized bm1 anti-B6 derived totally independently from the clones described here. This cell was obtained from a well originally seeded at 30 cells per well. For over two years, it has maintained significant allogeneic specificity for H-2^b but also kills the NK-susceptible target. Various measures to separate the activities, such as cold target cross inhibition or inhibition with oligosaccharides have failed to separate them. Despite vigorous growth of the bulk population, subcloning has only been successful when wells were seeded at more than 10 cells per well. The subclones have always retained the activities of the parent line and the levels of the two activities have maintained their proportion in the parental line (unpublished data). Although it is difficult to rule out absolutely the possibility that cooperating populations may be necessary for the growth of the "clones" it seems quite unlikely that the "helper" population should have natural killer properties. Additional evidence against the possibility of the natural killer activity arising from contaminating natural killer clones is the very high frequency of NK activity in the allospecific clones (present on at least one occasion in 5 of 7 cytotoxic clones). Thus, we cannot totally eliminate the possibility that the NK activity we observed was a property of NK clones contaminating our allospecific clones. We consider this unlikely as the sole explanation of our results.

If the NK activity is then a property of the allosensitized clone, it may be asked: Is the activity identical to that found in unprimed populations? Is it a manifestation of an allogeneic cross-reactivity between the appropriate allogeneic target and the allogeneic NK-susceptible target? Is it a property of a subset along the maturational line of the clone? On the first question, the most relevant data come from the cell sorter analysis. The population cloned was highly positive for Lyt 2 and bulk populations grown from that as well as those clones tested to date have all been Lyt 2^+ and homogeneous in their fluorescence profile. Since NK cells are not Lyt 2^+ (Herberman and Ortaldo, 1981) it is unlikely that the NK activity we are observing is identical to that seen in unprimed populations. The CML specificities definable with the H-$2K^b$ mutants are numerous and in several cases are known to result from one or a few amino acid changes in the H-2K molecule (Brown and Nathenson, 1977). The data in Table I and other unpublished data with similar sets of B6 anti bm1 clones suggest that the great majority of our clones recognize the "strong" specificity definable in this system. Assuming that this specificity truly constitutes a single epitope, then all clones should have the same reactivity with YAC-1. The observation is different; thus, the clones react with multiple epitopes in the bm1 anti B6 combination, some of which are shared by YAC-1, or the combining site for the mutant targets is different from the combining site for the NK-susceptible target. One possibility is that growing cytolytic populations may have NK-like activity during certain stages of the cell cycle or at times during their maturation from growing to cytolytic cells. Variations in growth properties could then explain differences in NK activity.

V. CONCLUSION

We have studied murine splenocytes sensitized to a discrete allogeneic stimulus resulting from one or two amino acid differences between the responding and stimulating strain. Cytolytic clones generated from this sensitization possessed exquisite allogeneic specificity. Many of them also possessed lytic activity against the NK-susceptible target YAC-1. Our data could result from shared allogeneic specificities between YAC-1 and the appropriate allogeneic targets, but we also raise the possibility that NK-like activity may be a general property of allospecific cytolytic clones or perhaps of cells in the clonal line at some point during the cell cycle or the maturational cycle. The cells mediating the NK-like activity are probably different from NK cells in unprimed populations.

REFERENCES

Braciale, T.J., Andrew, M.E. and Braciale, V.L. (1981). J. Exp. Med. 153:910.

Brown, J.L. and Nathenson, S.G. (1977). J. Immunol. 118:98.

Herberman, R.H. and Ortaldo, J.R. (1981). Science 214:24.

Kornbluth, J., Silver, D.M. and Dupont, B. (1981). Immunological Rev. 54:111.

Lamb, J.R., Eckels, D.D., Lake, P., Johnson, A.H., Hartzman, R.J. and Woody, J.N. (1982). J. Immunol. 128:23.

Melief, C.J.M., DeWaal, L.P., van der Muelen, M.Y., Melvold, R.W. and Kohn, H.I. (1980). J. Exp. Med. 151:993.

REGULATION BY INTERFERON AND T CELLS OF IL-2-DEPENDENT GROWTH OF NK PROGENITOR CELLS: A LIMITING DILUTION ANALYSIS

Carlo Riccardi[1,2], B.M. Vose[1,3], R.B. Herberman[1]

[1]Biological Research and Therapy Branch
National Cancer Institute
Frederick, Maryland

[2]University of Perugia
Perugia, Italy

[3]Christie Hospital
Manchester, England

I. INTRODUCTION

Several mechanisms have been reported to be involved in regulation of NK cells (Herberman *et al*, 1981 , 1982). These include inhibition of NK activity by various kinds of suppressor cells, normally present in the spleen of infant or low NK-reactive mice (Cudkowicz and Hochman, 1979; Riccardi *et al*, 1981), or induced by various agents in high-reactive animals (Cudkowicz and Hochman, 1979; Riccardi *et al*, 1981; Santoni *et al*, 1980). Conversely, accessory macrophages have been shown to be important in maintaining the normal levels of natural reactivity or in mediating the boosting of reactivity by production and release of interferon (IFN) (Puccetti *et al*, 1979; Djeu *et al*, 1979). IFN has been reported to be a potent stimulator of NK activity in various species (Trinchieri *et al*, 1978; Herberman *et al*, 1981). More recently a similar effect on mouse NK activity has been reported to be mediated by interleukin 2 (IL-2) (Kuribayashi *et al*, 1981).

The above studies dealt with the regulation of NK activity of spontaneously active NK cells or of IFN-boostable cells, called pre-NK cells (Minato *et al*, 1980). It has been recently reported that cells showing the same characteristics as mouse NK cells can be grown in vitro in medium containing mouse or rat 1L-2 (Dennert, 1980; Nabel *et al*,

ISBN 0-12-341360-5

1980). This observation opens the possibility of analyzing the regulatory role of cells and factors on the growth and differentiation of NK cells.

We report here data showing the possible role of IFN and T cells in the regulation of the growth of murine NK cells in vitro, by a limiting dilution analysis, measuring the frequency of NK cell progenitors.

II. LIMITING DILUTION ANALYSIS

Limiting dilution analysis allows the enumeration of cells growing in medium containing IL-2 and of the frequency of cells showing cytotoxic activity. We applied this approach to evaluate the frequency of progenitors of cytotoxic cells against YAC-1 tumor cells, upon culturing of murine spleen cells. Mouse spleen cells were cultured in round-bottomed microwells, together with irradiated (3000 R) feeder cells, and partially purified (lectin-depleted) rat IL-2.

Proliferation was measured by uptake of ^{3}H-thymidine over the last 18 hours of a 7-day culture period and cytotoxicity against ^{51}Cr-labelled YAC-1 or other target cells in a 4-hour assay. Limiting frequency was determined as the concentration of cells at which 36% of wells were negative.

III. FREQUENCY OF NK PRECURSORS IN SPLEENS OF DIFFERENT MOUSE STRAINS

In Table 1 are reported the frequency of proliferating and NK progenitor cells in the spleens of 6-week-old mice of various high and low NK strains (medians of 3 experiments). Whereas the frequency of proliferating cells from CBA, BALB/c and SJL mice was similar, the frequency of progenitors of cytotoxic cells against YAC-1 was much lower in SJL. CBA/J mice had only a very low frequency of cells cytotoxic for the NK-insensitive P815 tumor line.

Those data show that in high reactive strains, the ratio betweean cytotoxic and proliferating cells for YAC-1 was approximately 1/20, whereas in a low NK strain, a much smaller proportion of proliferating cells have detectable cytotoxic activity.

Surprisingly we detected only a low number ($1/10^{6}$) of NK cells developing from BALB/c nu/nu mice (a high NK-reactive strain). The frequency of proliferating cells was also lower than that of the other strains, with a ratio of cytotoxic to proliferating cells of approximately 1/64.

These data confirm that spleen cells from athymic mice have a lower proliferation capability, when cultured with IL-2, exposed to conventional or nu/+ mice (Lipsick and

Kaplan, 1981) and also indicate a lower capability of NK cell progenitors to grow under these conditions.

TABLE 1. Limiting Dilution Analysis of Proliferative and Pre-NK Cells in Spleens of Various Mouse Strains

Strain	Frequency of responding cells		
	Proliferation	Cytotoxicity vs. YAC-1	Cytotoxicity vs. P-815
CBA/J	1×10^{-3}	5×10^{-5}	1×10^{-6}
SJL/J	1×10^{-3}	1×10^{-6}	-
BALB/c	1×10^{-3}	5×10^{-5}	-
BALB/c nu/nu	7×10^{-5}	1×10^{-6}	-

Feeder cells: 5×10^4 irradiated spleen cells of same strain.

IV. EFFECT OF IFN PRETREATMENT ON FREQUENCY OF NK CELL GROWTH

It has been recently reported that mouse NK activity can be boosted by IL-2 (Kuribayashi et al, 1981). This boosting appeared to be independent of IFN, but IFN appeared to augment the expression of IL-2 receptors of NK cells (Kuribayashi et al, 1981) resulting in an augmented boosting by IL-2. To further analyze this phenomenon, we studied the influence of IFN pretreatment on the growth of NK cells. As shown in Table 2, pretreatment of spleen cells with IFN (600 u/ml) for 6 hours caused an augmentation of the frequencies of both proliferation and anti-YAC-1 cytotoxic cells.

TABLE 2. Effect of IFN Pretreatment on Frequency of Proliferating and Cytotoxic Cells

Strain	IFN	Frequency of responding cells: Proliferation	Frequency of responding cells: Cytotoxicity
CBA/J	-	1×10^{-3}	5×10^{-5}
CBA/J	+	3×10^{-3}	2×10^{-4}
BALB/c	-	8×10^{-4}	5×10^{-5}
BALB/c	+	5×10^{-3}	2×10^{-4}
nu/nu	-	4×10^{-5}	6×10^{-7}
nu/nu	+	2×10^{-4}	4×10^{-6}

[a]Cells pretreated for 6 hrs before culture with 600 u/ml IFN.

V. EFFECT OF IFN PRESENT THROUGHOUT THE CULTURE PERIOD

To further analyze the regulatory role of IFN, we cultured spleen cells in the presence of IL-2, IFN and syngeneic feeder cells for 7 days. As shown in Table 3, IFN caused significant diminution in the frequency of proliferating and cytotoxic cells using CBA/J and BALB/c nu/+ mice. Thus, under these conditions, IFN played a negative, inhibiting role on the growth of NK cell progenitors, and this effect was dose dependent (data not shown). To better define the mechanism responsible for this effect, we performed the same experiment using BALB/c nu/nu spleen cells. As shown in Table 3, the presence of IFN in the culture caused a significant augmentation, rather than inhibition, of both proliferating and cytotoxic cells. These results suggest that the inhibition of NK cell growth in the presence of IFN is dependent on the presence of a thymic-dependent population of cells.

TABLE 3. Effect of IFN Present Throughout the Culture Period

Strain	Interferon	Frequency of responding cells Proliferation	 Cytotoxicity
CBA/J	-	1×10^{-3}	5×10^{-5}
CBA/J	+	7×10^{-5}	5×10^{-7}
nu/+	-	1×10^{-3}	5×10^{-5}
nu/+	+	1×10^{-4}	1×10^{-6}
nu/nu	-	7×10^{-5}	7×10^{-7}
nu/nu	+	3×10^{-4}	5×10^{-6}

[a]600u/ml
[b]4 hr ^{51}Cr-release assay against YAC-1

VI. ROLE OF T CELLS IN THE DEPRESSIVE EFFECT OF IFN ON THE FREQUENCY OF CYTOTOXIC NK CELLS

To further study the possible involvement of T cells in IFN-mediated inhibition of NK cell growth, we cultured BALB/c nu/nu spleen cells in the presence of different feeders. As shown in Table 4, when nu/nu spleen cells were cultured in the presence of histocompatible feeder cells from euthymic mice, NK cell frequency was substantially increased, indicating that non-proliferating T cells can help the IL-2-dependent growth of NK cell progenitors. However, when IFN was

TABLE 4. Role of T-Cells in Mediating the Depressive Effect of IFN on Frequency of Cytotoxic NK Cells

Feeder cells	IFN	Frequency of NK cells
BALB/c nu/nu	-	1×10^{-6}
BALB/c nu/nu	+	5×10^{-6}
BALB/c nu/+	-	4×10^{-6}
BALB/c nu/+	+	1×10^{-6}
BALB/c thymocytes	-	4×10^{-6}
BALB/c thymocytes	+	5×10^{-7}

BALB/c nu/nu responder cells cultured with 5×10^4 feeder cells/well
[a]Interferon 600 u/ml
[b]4 hours ^{51}Cr-release assay against YAC-1

added to such cultures, in contrast to the augmenting effect obtained with cultures of BALB/c nu/nu responder cells on nu/nu feeders, there was an inhibition of NK cell growth. Both the helper effects obtained with nu/+ feeders and the inhibiting effects induced by IFN in such cultures were abrogated by pretreatment with Anti-Thy 1 plus complement (data not shown).

VII. CONCLUSIONS

The present results point to an important role of T-cells and IFN in regulation of the growth and maturation of progenitors of NK cells and indicate greater complexity for the overall regulation of NK activity than had been previously realized. Figure 1 represents a model which attempts to incorporate the present and previous information. IFN can act as a positive regulator of NK activity, not only by augmenting the reactivity of pre-NK cells or NK cells with low spontaneous activity (Timonen *et al*, this volume), but to boost cytotoxic reactivity of mature NK cells, also by increasing the potential of progenitor cells to grow in the presence of IL-2, apparently by inducing increased receptors for this lymphokine (Kuribayashi *et al*, 1981). On the other hand, when IFN is present in a culture containing T cells, it can appreciably inhibit NK cell proliferation by acting in some way on the T cells. As shown in Table 4 this IFN-induced inhibition is not dependent on proliferative T cells, since irradiated T cells can mediate this effect. Whereas pretreatment of T cells with anti-Thy 1 + C eliminated this

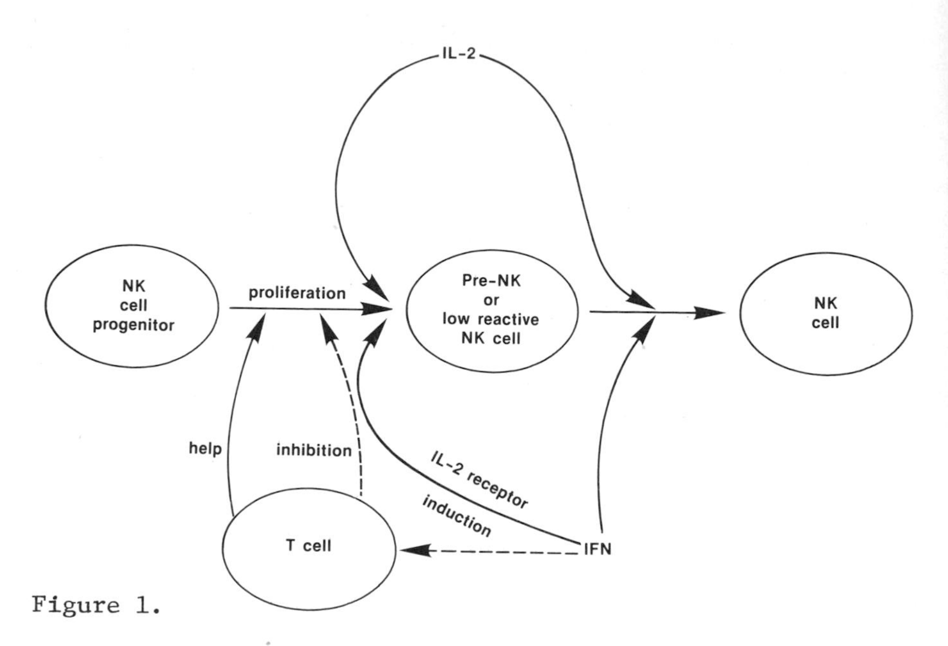

Figure 1.

IFN-inducible inhibition, similar pretreatment with anti-Lyt 1 or anti-Lyt 2 + C had no effect (data not shown) suggesting either Lyt 1^+ or Lyt 2^+ could mediate inhibition, or that the regulatory cell has a Thy 1^+Lyt 1^-Lyt 2^- phenotype.

An unexpected result was the lower frequency obtained by culturing of nu/nu spleen cells. This appeared to be due, at least in part, to the lack of T cells that can cooperate with NK cell progenitors and augment their proliferation. These results are clearly in contrast with the well known high NK-reactivity in athymic mice and rats, and recent evidence for increased numbers of NK cells in athymic rats (Reynolds et al, this volume). In light of the recent indications for the heterogeneity of NK cell subpopulations (Minato et al, 1980), one could hypothesize that the IL-2-dependent NK cells that proliferate and differentiate in our system only represent a small portion of the NK cells in the spleen of nu/nu mice, and that another, IL-2-dependent NK cell subpopulation predominates in those mice. This hypothesis would also imply that there are other, yet-to-be-defined, growth factors for some NK cells.

A further intriguing element to consider in this discussion is that some NK cells themselves may produce IFN (Djeu et al, this volume) and possibly also IL-2 (Domzig and Stadler, this volume). This suggests that certain stimuli could induce positive autoregulation of NK cell development.

In addition to the need for further studies to dissect out the complexities of NK cell regulation in vitro, it will also be important to begin to perform studies to determine the in vivo relevance of these phenomena.

REFERENCES

Cudkowicz, G. and Hochman, P.S. (1979). Immunol. Rev. 44:13.

Dennert, G. (1980). Nature 287:47.

Djeu, J.Y., Heinbaugh, J.A., Holden, H.T., and Herberman, R.B. (1979). J. Immunol. 122:175.

Herberman, R.B., Brunda, M.J., Cannon, G.B., Djeu, J.Y., Nunn-Hargrove, M.E., Jett, J.R., Ortaldo, J.R., Reynolds, C, Riccardi, C., and Santoni, A. (1981). In "Augmenting Agents in Cancer Therapy" (E.M. Hersh _et al_, eds.), p. 253, Raven Press, New York.

Herberman, R.B., Brunda, M.J., Djeu, J.Y., Domzig, W., Goldfarb, R.H., Holden, H.T., Ortaldo, J.R., Reynolds, C.W., Riccardi, C., Santoni, A., Stadler, B.M., and Timonen, T. (1982). In "Human Cancer Immunology" Vol. 6 (R. Herberman, B. Serrou and C. Rosenfeld, eds.), North-Holland, in press.

Kuribayashi, K, Gillis, S., Kern, D.E., and Henney, C.S. (1981). J. Immunol. 126:2321.

Lipsick, J.S. and Kaplan, N.O. (1981). Proc. Natl. Acad. Sci. USA 78:2398.

Minato, N., Reid, L., Cantor, H., Lengyel, P., and Bloom, B. (1980). J. Exp. Med. 152:126.

Nabel, G, Fresno, M., McVey-Boudreau, L., Chessmen, A., Buscalo, R., and Cantor, H. (1980). In "Progress in Immunology IV" (M. Fourgerau and J. Dausset, eds.), p. 315. Academic Press, New York.

Puccetti, P., Santoni, A., Riccardi, C., Holden, H.T., and Herberman, R.B. (1979). Int. J. Cancer 24:819.

Riccardi, C., Santoni, A., Barlozzari, T., Cesarini, C., and Herberman, R.B. (1981). Int. J. Cancer 28:775.

Santoni, A., Riccardi, C., Barlozzari, T., and Herberman, R.B. (1980). In "Natural Cell-Mediated Immunity Against Tumors" (R.B. Herberman, ed.), p. 753. Academic Press, New York.

Trinchieri, G., Santoli, D., Dee, R.R., and Knowles, B.B. (1978). J. Exp. Med. 147:1299.

NATURAL CYTOTOXIC ACTIVITY OF MOUSE SPLEEN CELL CULTURES MAINTAINED WITH INTERLEUKIN-3

Julie Y. Djeu
Emanuela Lanza

Division of Virology
Bureau of Biologics
Bethesda, Maryland

Andrew J. Hapel
James N. Ihle

Biological Carcinogenesis Program
Frederick Cancer Research Center
Frederick, Maryland

I. INTRODUCTION

A newly discovered lymphokine, Interleukin-3 (IL-3), isolated from Con A-conditioned medium, induces the T cell-associated enzyme, 20α hydroxysteroid dehydrogenase (20α SDH), in cultures of nu/nu splenic lymphocytes, and has the ability to promote the proliferation of certain classes of lymphocytes from the spleen but not the thymus of the mouse (Ihle et al., 1981a,b and 1982). Culture conditions for its production are identical to those required for Interleukin-2 (IL-2) production and both lymphokines appear to be released from T cells with the phenotype: Thy 1.2^+, $Lyt1^+$, $Lyt2^-$. The responder cells that are induced to proliferate by the two factors, however, are different. A variety of cell types, which differ in their surface phenotype, can be maintained in IL-3, but they uniformly acquire 20α SDH enzyme activity. These cells can be derived from spleen (Hapel et al., unpublished) or bone marrow (Greenberger et al., 1982), and can be separated by fluorescence flow cytometry (Hapel et al., unpublished) into Thy 1^+, $Lyt1^+$ and Thy 1^-, $Lyt1^-$ populations. Analysis of the function of Thy 1^-, $Lyt1^-$, $Lyt5^+$ and $H\text{-}11^+$ cells cultured in IL-3 has now indicated that they mediate natural cytotoxic (NC) activity but lack natural killer (NK) activity. Ability of cells with NC activity to proliferate in IL-3 thus provides a means of separating this effector cell type from a heterogenous population of leukocytes that possess spontaneous cytotoxicity to a variety

ISBN 0-12-341360-5

of tumor cells. This may also enable us to expand NC cells in culture for the purpose of investigating their cell lineage as well as the role of these cells in host defense.

II. METHODS

Spleen cells from several strains of mice were passed through nylon wool and the nonadherent cells were seeded at $2-4 \times 10^6$ cells/ml in RPMI 1640 medium containing 10% fetal bovine serum and supplemented with purified IL-3 at a level that yields maximum 20 SDH induction (Hapel *et al.*, 1981). The medium was changed every 3 days. The usual pattern in the initiation of these cultures was cell proliferation during the first 2 days followed by a 10 day span of drastic cell loss. The cells subsequently recovered and began to proliferate slowly until about day 20 when they became readily transferable by serial passage in medium containing IL-3.

III. RESULTS AND CONCLUSIONS

Several cell lines established from BALB/c, NIH Swiss, and CBA/N mice were tested for NK activity against YAC-1, RLo1, and RBL-5 (Djeu *et al.*, 1979; Lanza and Djeu, 1982). At the same time they were tested for NC activity against WEHI-164 (Lust *et al.*, 1981). Table 1 shows the lack of ability of these cell lines to lyse any of the NK susceptible targets, either after 4 hr of incubation or a longer 18 hr assay.

TABLE 1. Cytotoxicity of IL-3 Dependent Cell Lines Against NK and NC Target Cells

Strain	Time of incubation	% Cytotoxicity[a]			
		NK targets			NC target
		YAC-1	RLo1	RBL-5	WEHI-164
BALB/c	4 hr	0	0	0	1.8
NIH/Swiss		0	0	0	3.4
CBA/N		0	0	0	2.8
BALB/c	18 hr	0	0	0	25.6
NIH/Swiss		0	0	0	27.9
CBA/N		0	0	0	35.2

[a]Effector/target ratios were all 200/1.

In contrast, all the cell lines exhibited significant lysis of WEHI-164 by 18 hr. Pretreatment of these cell lines with interferon (IFN) for 1 hr did not change their pattern of nonreactivity against YAC-1 cells, indicating that there were no IFN-activatable pre-NK cells in these cultures (Table 2). It is also interesting to note that IFN did not enhance the activity against WEHI-164.

TABLE 2. Effect of Interferon on Cytotoxicity of IL-3 Dependent Cell Lines

Strain of cell line	Pre-incubation with IFN	% Cytotoxicity[a] NK target YAC-1	NC target WEHI-164
NIH/Swiss-1	-	0	20.2
	+	0	21.0
NIH/Swiss-2	-	0	32.4
	+	0	31.9

[a] 4 hr assay with YAC-1 target cells and 18 hr assay with WEHI-164.

Treatment of the cell lines with anti-asialo-GM1 plus complement, which effectively removes NK cells (Kasai *et al.*, 1980), did not inhibit their cytotoxicity against WEHI-164. The IL-3-dependent cell lines thus differ from NK cells in their lack of detectable surface asialo GM1. These cells also lack detectable Thy 1.2, which can be found in low density on NK cells (Mattes *et al.*, 1979). In other aspects, however, they are similar to NK cells in that they are $Lyt5^+$, $H\text{-}11^+$ (A.J. Hapel, H.C. Morse, and J.N. Ihle, unpublished observations), but $Lyt1^-$, $Lyt2^-$, and Ig (Koo *et al.*, 1980).

Increased awareness of the heterogeneity of leukocyte subclasses that can lyse tumor cells without prior sensitization has produced more questions than answers as to the real reason for their existence in nature. Of these effector mechanisms, it has been proposed that NK cells detected on YAC-1, RLo1 or RBL-5 tumor cells might serve in surveillance against leukemic cells while NC cells detected on anchorage-dependent Meth A or WEHI-164 target cells might be active against solid non-lymphoid tumors (Stutman and Cuttito, 1981). However, their relationship to each other and their interactions with other host defense mechanisms are still unclear. Many of the properties of NK and NC effector functions are the same, e.g. pre-existence in high levels in the host without priming, lack of H-2 restriction, ability to lyse more than one tumor cell type, no classical immunological memory, and no well-defined markers of T cells, B cells, or macrophages (Stutman *et al.*, 1980). Other properties distinguish NK cells from NC cells, e.g. lack of strain and age dependency of NC cells, although it now appears that NK activity, absent from the adult spleen, is maintained throughout life in peripheral blood (Lanza and Djeu, 1982). NC cells also show resistence to inactivation by *in vivo* treatment with ^{89}Sr, cyclophosphamide, estrogen, and irradiation (Stutman *et al.*, 1980; Lust *et al.*, 1981), all of which can provoke suppression of NK cells (Herberman, 1980).

NK and NC cells are closely related in terms of function but enough differences exist to pose the question of whether they are from separate cell lineages or merely two variants of an effector cell type that mediate most or all spontaneous cytotoxicity. The difficulty in solving this question stems from the inability to physically separate these two cell types due to lack of known specific markers. Our observation of the selectivity of IL-3 to maintain proliferation of cells with NC activity, but not NK activity now provides an important feature that may be helpful to separate the two activities. Conversely, NK cells have been shown by others to proliferate in IL-2 (Kedar *et al.*, 1982). The different lymphokine requirements for NK and NC cells, although suggestive of separate lineage, can still be taken as two maturation pathways or stages of differentiation of a common stem cell. NK and NC cells may belong to a unique class of lymphocytes that exhibits little or no Thy 1, Lyt1, Lyt2, but express Lyt5 and H-11, that, under the influence of IL-2, differentiate into NK cells and, in the presence of IL-3, become NC cells. After differentiation, they may then acquire further markers or properties that distinguish them from each other, e.g. NK 1.1 antigens on NK cells and cortisone or other drug resistance in NC cells.

REFERENCES

Djeu, J.Y., Heinbaugh, J.A., Holden, H.T., and Herberman, R.B. *J. Immunol.* 122, 141 (1979).

Greenberger, J.S., Hapel. A., Nabel, G., Eckner, R.J., Newburger, P.E., Ihle, J., Denburg, J., Moloney, W.C.,

Sakakeeny, M., and Humphries, K. Exp. Hematol. Today 1982 Yearbook (1982).
Hapel, A.J., Lee, J.C., Greenberger, J., and Ihle, J.N. In Mechanisms of Lymphocyte Activation (eds., Resch, K., and Kirchner, H.) p. 353. (Elsevier/North-Holland Biomedical Press, Amsterdam, 1981).
Herberman, R.B. In Natural Cell-Mediated Immunity Against Tumors (ed., Herberman, R.B.) (Academic Press, New York, 1980).
Ihle, J.N., Pepersack, L., and Rebar, L. J. Immunol. 126, 2184 (1981a).
Ihle, J.N., Lee, J.C., and Rebar, L. J. Immunol. 127, 2565 (1981b).
Ihle, J.N., Rebar, L., Keller, J., Lee, J.C., and Hapel, A.J. Immunol. Rev. Vol. 63 (1982).
Kasai, M., Iwarori, M., Nagai, T., Okumura, K., and Tada, T. Eur. J. Immunol. 10, 175 (1980).
Kedar, E., Ikejirim, B.L., Gorelik, E., and Herberman, R.B. (1982) Cancer Immunol. Immunother. In press.
Koo, G.C., Jacobson, J.B., Hammerling, G.J., and Hammerling, U. J. Immunol. 125, 1003 (1980).
Lanza, E., and Djeu, J.Y. (1982) this volume.
Lust, J.A., Kumar, V., Burton, R.C., Bartlett, S.P., and Bennett, M. J. Exp. Med. 154, 306 (1981).
Stutman, O., Figarella, E.F., Paige, C.J., and Lattime, E.C. In Natural Cell-Mediated Immunity Against Tumors (ed., Herberman, R.B.) p. 187. (Academic Press, New York, 1980).
Stutman, O., and Cuttito, M.J. Nature 290, 254 (1981).

EVIDENCE FOR PROTEASES WITH SPECIFICITY OF CLEAVAGE AT AROMATIC AMINO ACIDS IN HUMAN NATURAL CELL-MEDIATED CYTOTOXICITY

Dorothy Hudig[1]
Doug Redelman[2]
Lory Minning[1]

Department of Medicine
University of California, San Diego
La Jolla, California

I. INTRODUCTION

The biochemical mechanism of killing in lymphocyte-mediated cytotoxicity is not known. We considered the possibility that an amphipathic lytic structure is formed by the lymphocytes during cell-mediated killing, and that formation of this structure might be initiated by one or more proteases, whether or not the final lytic mechanism is complement. To test this hypothesis, we commenced an extensive series of experiments to implicate proteases in cell-mediated killing (1-4). "Antiprotease" reagents were included in the lytic assays. The most useful reagents we tested were human plasma antiproteases and alternate substrates. The plasma antiproteinases, such as alpha-1-antitrypsin (a-1-AT) and alpha-1-antichymotrypsin (a-1-X) are macromolecules with m. wts. greater than 50,000 which couple covalently to the active site of serine-dependent proteases, thus inactivating the enzyme. Alpha-1-antitrypsin has a broad specificity and can inactivate many proteases, but alpha-1-antichymotrypsin inactivates only those serine-dependent proteinases which

[1]Supported in part by NIH CA 28196.
[2]Supported in part by NIH CA 24450.

ISBN 0-12-341360-5

cleave proteins at amino acid aromatic sites. Ester and amide derivatives of amino acids may serve as alternate substrates for proteases with corresponding specificities. When present with the natural substrate, these derivatives serve as competitive inhibitors. Derivatives of D-amino acids also serve as competitive inhibitors, but are not hydrolyzed.

In this paper we present some of our data, obtained with these natural and synthetic reagents, which implicate a crucial role for at least one serine-dependent protease in human NK. Furthermore, these data suggest that this protease has a restricted specificity of cleavage for the aromatic amino acid residues of its natural substrate.

II. MATERIALS AND METHODS

A. Assay of NK

Effector lymphocytes were human peripheral blood mononuclear cells isolated by the method of Boyum (5). Either "slow" (T24 human bladder carcinoma) or "fast" (K562 cells derived from a patient with myelogenous leukemia) tumor cells were used as target cells. Cell death was assayed by release of ^{51}Cr from the target cells as previously described in detail (2,6). The lytic activities of control and "antiprotease" treated assays were converted into lytic units (7) to facilitate comparison. The 50% inhibitory dose (ID_{50}) of a reagent is that concentration which reduces lytic activity by an amount equivalent to a 50% reduction in the effector cell number.

B. Reagents

1. Plasma Antiproteinases. An a-1-AT preparation containing several other proteins (Wa-1-AT) was obtained from Worthington Biochemical Corp. (Freehold, NJ). Electrophoretically pure a-1-AT was the generous gift of Dr. Carol Fulcher. One unit of trypsin inhibitory capacity (TIC) is the amount of antiproteinase needed to inactivate 1 mg of trypsin. Alpha-1-antichymotrypsin was a generous gift of Dr. James A. Travis. This preparation did not inactivate trypsin.

2. Synthetic Substrates and Inhibitors. Acetyl-L-tyrosine ethyl ester (Ac-L-TyrOEt) was obtained from Sigma Chemical Co. (St. Louis, MO). Ac-D-TyrOEt, AcTyr, acetyl-L-phenylalamine ethyl ester (Ac-L-PheOEt), Acetyl-L-tyrosine

amide (Ac-L-TyrNH_2) and Ac-D-TyrNH_2 were obtained from Vega Biochemicals (Tucson, AR). The esters were dissolved in 0.1 ml 100% ethanol prior to the addition of medium and diluted in medium containing an identical final concentration (0.15% or less) of alcohol. Control assays also contained identical concentrations of ethanol.

C. Conjugate Formation

Lymphocyte-K562 conjugates were formed by centrifuging 2:1 mixtures of peripheral blood mononuclear cells and K562 cells for 3 minutes at 40g at room temperature. The resultant conjugates were kept on ice prior to gentle suspension and analysis by direct counts on a hemocytometer. The K562 cells could be readily distinguished by their larger size from the lymphocytes. All assays were coded by random numbers and read blind.

III. RESULTS

The plasma antiproteinases a-1-AT and a-1-X both inhibited "slow" NK. Although pure a-1-AT did inhibit NK, a crude preparation of a-1-AT was more inhibitory when the two preparations were used at similar trypsin inhibitory capacities. See reference 9 for further details.

Alpha-1-antichymotrypsin would not have been detected in the trypsin inhibitory assay used to normalize a-1-AT and the Wa-1-AT. Antichymotrypsin containing no antitrypsin activity was a potent inhibitor of NK (Figure 2). Furthermore, unlike a-1-AT, a-1-X consistently inhibited both "slow" and "fast" NK (data not shown). Thus it is possible that a-1-X was one of the contaminants of the Wa-1-AT preparation that made this reagent more inhibitory than pure a-1-AT.

Ester and amide derivatives of aromatic acids inhibited NK (Table I), whereas similar derivatives of basic and aliphatic amino acids did not (data not shown). Furthermore, the relative ability of these aromatic compounds to inhibit NK correlated with their K_m or K_i for chymotrypsin (Table I).

Several lines of evidence support our contention that these compounds are affecting the actual lytic mechanism and not other auxiliary cell functions necessary for cytotoxicity. These reagents are nontoxic to lymphocytes as monitored by trypan blue exclusion. Pretreatment of lymphocytes followed by washing had no effect on subsequent NK activity (data not shown). These compounds did not increase the spontaneous ^{51}Cr release of the targets. Although it is possible that hydrophobic molecules may affect membrane functions, this seems unlikely in this situation as there is stereospecificity to inhibition of NK. The L forms of AcTyrOEt (Table I and Figure 3) and AcTyrTyrNH_2 (Table I) were twofold better competitive inhibitors for NK than the D forms.

TABLE I. Comparison of Inhibition of Human NK and Reactivity with Chymotrypsin[a]

Reagent	ID_{50}[b] (mM)	K_m or K_i (mM)
L-TyrOEt	0.04	N.A.[c]
Ac -PheOEt	0.1	1.8
Ac-L-TyrOEt	0.3-0.4	0.7
Ac-D-TyrOEt	0.8	N.A.
Ac-L-TyrNH_2	5.0	32
Ac-D-TyrNH_2	10.0	N.A.
AcTyr	>20.0	110

[a]Pancreatic chymotrypsin; [b]ID_{50} based on comparison of lytic units of NK activity; [c]Not available.

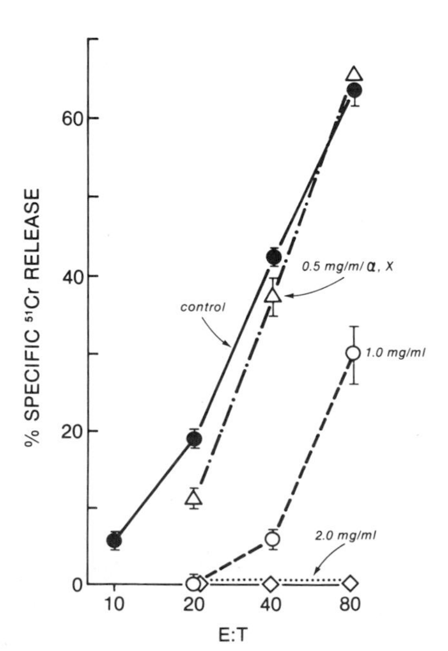

Figure 1. Inhibition of NK to T24 cells by alpha-1-antichymotrypsin. Assays were performed as in Reference 9.

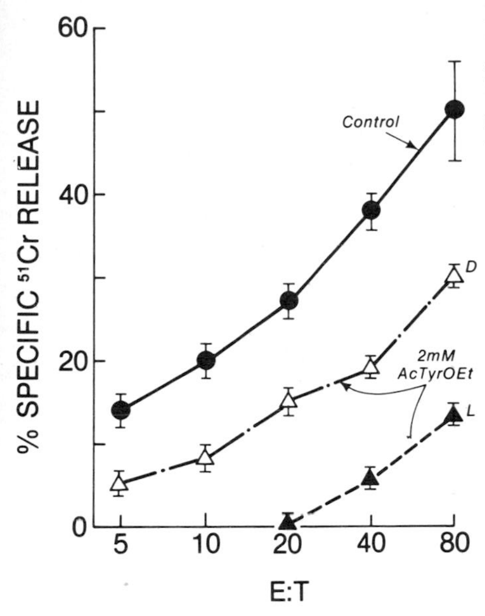

Figure 2. Stereospecificty of AcTyrOEt inhibition of NK to K562 cells. Assays were terminated after 4 hours of incubation.

AcTyrOEt at 0.8 mM did not affect conjugate formation (Figure 3B), although this concentration depressed NK fourfold (Figure 3A). Higher concentrations (e.g., 3.2 mM) depressed conjugate formation slightly in 1 of 3 experiments (data not shown). Thus it seems likely that these aromatic amino acid derivatives inhibit the lytic state of killing, and not the binding events which may occur prior to activation of the lytic mechanism.

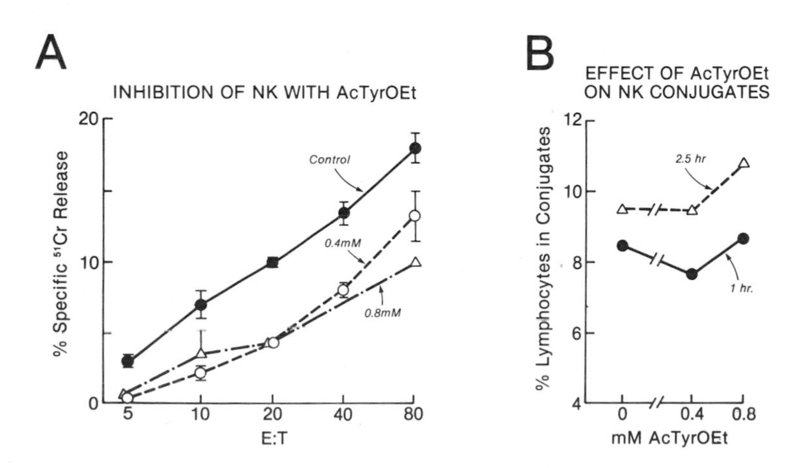

FIGURE 3. Inclusion of AcTyrOEt during NK conjugate formation does not affect the number of conjugates formed.

V. DISCUSSION

The inhibition of NK by plasma antiproteinases implies that a protease of the serine-dependent type is required for killing since these antiproteinases inactivate <u>only</u> this class of protease. Furthermore, the marked inhibitory activity of a-1-X to both "slow" and "fast" NK implies that at least one serine-dependent protease with chymotrypsin-like activity (i.e., cleavage at aromatic amino acids) is crucial to NK. In addition, it is likely that the protease activity is either membrane-associated or released upon activation of the lytic mechanism, because plasma antiproteinases are large acidically charged molecules which are unlikely to cross lymphocyte membranes. Since pretreatment of lymphocytes with plasma antiproteases followed by washing has no subsequent effect, this indicates that the "NK" protease is either not activated and/or sequestered prior to killing.

This conclusion that a protease with aromatic amino acid specificity is required for NK is further supported by the data from the experiments using ester and amide derivatives of aromatic amino acids to inhibit NK to K562 cells. The stereospecificity of this effect resembles the stereospecificity of interactions of these compounds with pancreatic a-chymotrypsin (8). That these synthetic substrates do not inhibit conjugate formation adds additional support to our hypothesis that we are inhibiting the actual lytic mechanism. However, a technical caveat should be considered in these conjugate experiments, as only ~20% of the conjugate forming cells are actually killer cells. Thus if formation of only the killer conjugates were inhibited, this effect could be obscured by the higher frequency of formation of nonkiller conjugates.

An additional series of experiments using AcTyrOEt as a competitive inhibitor indicate that activated NK cells required more of the reagent to achieve 50% inhibition of killing than did control freshly isolated cells or control cells cultured for 24 hours in the absence of activating agents (4). Activation was stimulated by preincubation with either 0.32% ethanol, 10 μg/ml lipopolysaccharide or a mixed lymphocyte response. The shift in the amount of AcTyrOEt to achieve an ID_{50} after activation was directly proportional to the degree of <u>in vitro</u> activation. These observations also make it much less likely that the AcTyrOEt is interfering with cell functions that are not directly relevant to killing.

The approach of the experiments described in this paper provides indirect evidence for one or more proteases in human NK. Experiments are in progress to identify and isolate "NK" proteases. However, ultimate success will reside in identifying not only the protease(s), but also identification of the natural substrate of the protease(s) and identification of the final lytic substance.

ACKNOWLEDGMENTS

The authors gratefully acknowedge the helpful discussions of Drs. Walter Brocklehurst and Russell F. Doolittle, the illustration of Phyllis Stookey and the manuscript preparation of Ms. Paige Gilman.

REFERENCES

1. Redelman, D., and Hudig, D., J. Immunol. 124:870 (1980).
2. Hudig, D., Haverty, T., Fulcher, C., Redelman, D., and Mendelsohn, J., J. Immunol. 126:1569 (1981).
3. Hudig, D., Redelman, D., and Mendelsohn, J., Fed. Proc. 39:359 (1980).
4. Minning, L., Hudig, D., and Redelman, D., Fed. Proc. (1982).
5. Boyum, A., Scand. J. Clin. Lab. Invest. 21, Suppl. 97:77 (1968).
6. Hudig, D., Djobadze, M. D., Redelman, D., and Mendelsohn, J., Cancer Res. 41:2803 (1981).
7. Brunner, K. T., Engers, H. D., and Cerottini, J. C., in "In Vitro Methods in Cell-Mediated and Tumor Immunity" (B. R. Bloom and J. R. David, eds.), p. 423. Academic Press, New York, 1976.
8. Niemann, C., Science 143:1287 (1964).

THE ROLE OF NEUTRAL SERINE PROTEASES IN THE MECHANISM OF TUMOR CELL LYSIS BY HUMAN NATURAL KILLER CELLS

Ronald H. Goldfarb[1,2]
Tuomo T. Timonen[3]
Ronald B. Herberman[4]

Laboratory of Immunodiagnosis
National Cancer Institute
NIH, Bethesda, MD

I. INTRODUCTION

Natural killer(NK) cells comprise the key lymphoid effector cell subpopulation involved in natural cell-mediated immunity against primary and metastatic tumors(1-5). Nevertheless, only a modest understanding of the molecular and biochemical events of NK cell-tumor cell interaction(6) and subsequent tumor cell cytolysis has been achieved. A number of studies have probed the biochemistry of the NK lytic mechanism, and studied the role of lysosomal enzymes(7), cellular hydrolases(8), and phospholipases and phospholipid methylation(9). Recently, we have also examined the role of oxidative metabolism(10) and a role for neutral serine proteolytic enzymes(11) in the mechanism of NK cell-mediated cytolysis; the former pathway does not appear to play an important role in NK lysis, whereas proteases appear to contribute to NK immune cytolytic activity(3,11,12).

In this chapter, we review the experimental evidence that documents the production of neutral serine proteolytic enzymes by human large granular lymphocytes(LGL), which account for NK activity(13), and the role of neutral serine proteases in the NK lytic mechanism.

---

[1]Supported by PHS Fellowship 1F32CA06681 to RHG by NCI, DHHS.

[2]Present address: Cancer Metastasis Research Group, Dept. of Immunology & Infectious Disease, Pfizer Central Research, Pfizer Inc., Groton, CT 06340.

[3]Present address: Dept. of Pathology, University of Helsinki, Finland.

[4]Present address: Biological Research and Therapy Branch, Biological Response Modifiers Program, NCI-FCRF, Frederick, Maryland 21701.

ISBN 0-12-341360-5

II. PROTEASES IN CELL-MEDIATED IMMUNE LYSIS

Proteases are often associated with cellular migration, invasiveness, and tissue remodeling of intracellular proteins, cytoskeletal elements, and cell-surface components(14); for example, neutral serine proteases contribute to the degradation of cell matrix glycoproteins(15).

Neutral serine proteolytic enzymes have therefore been examined in the lytic mechanisms of various lymphoid effector cells, particularly cytolytic T lymphocytes and activated macrophages(3,12).

Diisopropyl fluorophosphate(DFP), the specific active site inhibitor of neutral serine proteases, has been shown to inactivate target cell lysis mediated by cytolytic T lymphocytes(16). Moreover, it has been demonstrated that cell lysis by cytotoxic T lymphocytes requires cell-surface associated proteolytic enzymes(17). Indeed, a cytotoxic protease, which is capable of killing radiolabeled tumor cells has been isolated from human peripheral leukocytes(18). Earlier, it had been noted that lymphocyte plasma membranes may be cytocoxic to tumor cells, which led to the idea that a membrane-associated protease may be responsible for the cytolytic activity(19). Interestingly, low levels of exogenously added proteolytic enzymes have been reported to enhance the lysis of tumor cells by cytotoxic T lymphocytes(20).

The role of neutral serine proteases in tumor cell lysis has also been examined relative to the mechanism of cytolytic, activated macrophages(21,22,23); it appears that tumor cell lysis mediated by macrophages is dependent, to some extent, on the secretion of neutral serine proteases(21,22). It appears that the ability of macrophages to bind to tumor cells and to secrete a cytolytic protease are independent functions, each of which are required for macrophage mediated lysis(24); it has also been reported that the macrophage cytolytic protease interacts with hydrogen peroxide in a synergistic fashion in the cytolytic mechanism of this effector cell(25).

It has indeed been demonstrated that activated macrophages produce several neutral serine proteases including plasminogen activator(26). A regulatory cascade of DFP-sensitive neutral serine proteases have in fact been reported to be involved in the functional regulation of activated macrophages (27).

A role for neutral serine proteases have also been suggested for the mechanism of K cell killing in antibody-dependent cell-mediated cytotoxicity(ADCC)(28). Protease inhibitors suppressed ADCC by human peripheral blood mononuclear cells against a lymphoblastoid target cell(28); furthermore, the addition of exogenous proteases enhanced ADCC(20).

It therefore appears that proteolytic enzymes play a role in tumor cell lysis by various effector cells including cytolytic T cells, activated macrophages, and K cells. Considerable interest has therefore centered upon a possible role for proteolytic activity as a potential mechanism in NK mediated cell lysis, as described below.

III. NEUTRAL SERINE PROTEASES AND THE NK LYTIC MECHANISM

Several reports have suggested that neutral serine proteases are also involved in the NK lytic mechanism(29,30,31). Whereas an inhibitor of serine proteases was reported to have no effect on the binding of NK cells to target cells, subsequent tumor cell lysis was reported to be inhibited(29); it was suggested that following NK cell-tumor cell conjugate formation, a proteolytic lytic activity which functions as a lytic unit is expressed(29). It was suggested that energy was required to either expose the enzyme active site, or to create a degradative interaction with the tumor cell(29).

Support for the idea that proteases are involved in the NK lytic mechanism has also come from observations that human NK activity was inhibited by protease inhibitors of low or high molecular weight, or by proteinase substrates(30). It was found(30) that NK activity was inhibitable by substrates for chymotrypsin and plasmin but not by substrates for elastase or collagenase. A preliminary report(31) has suggested that cell-surface associated tryptic and chymotryptic proteases play a crucial role in natural cell-mediated tumor cell lysis(31); furthermore, it was suggested that an enzyme of elastase-like specificity may also play some role in NK lytic activity.

These interesting reports suggested, but did not provide, direct evidence for a role of proteolytic activity in the mechanism of NK lysis. Direct evidence would require, as a minimum criterion, the demonstration that a pure population of NK cells is capable of producing proteolytic activity. Studies to date have not employed pure NK populations, and the results can not be attributed to only NK cells. Therefore, the inhibitors used could have indirectly affected NK lytic activity by modulation of an additional lymphoid sub-population. In addition, a protease produced by contaminating lymphoid cells could be degradative for enzymes produced by NK cells, and could therefore mask the detection of, or interfere with the isolation of, putative NK proteolytic enzymes.

A. Isolation of Pure NK Cells

In order to further evaluate the exact role of proteases in the NK lytic mechanism, we therefore utilized pure populations of NK cells through the use of methodology recently developed in our laboratory(13); we have found that the lymphoid sub-population that mediates NK activity can be morphologically defined as large granular lymphocytes(LGL)(13); it appears that K cells, which mediate ADCC, are also LGL(13). A number of purification steps, including Percoll gradient discontinuous gradient centrifugation, and elimination of effector cells that form rosetts with sheep erythrocytes at 29°C, lead to the isolation of human NK cells with a purity of at least 95-99%(3,11,12). We have studied highly purified LGL to determine whether NK cells produce either cell-associated or extracellular proteolytic activity, as described below.

B. Production of Proteolytic Activity by LGL

We examined LGL for their ability to produce proteolytic activity through the use of iodinated fibrin, a well characterized substrate for the assay of neutral serine proteases (32,33). We examined both culture supernatants and cellular homogenates, derived from LGL cultured overnight under serum-free conditions, for proteolytic activity by previously descrived methods(34,35). In addition, we examined freshly isolated LGL, or long-term cultures of LGL grown in IL-2(T cell growth factor), for their ability to produce enzymatic activity; the intact cells were grown on iodinated fibrin under serum-free conditions.

We have found that human NK cells(LGL) isolated to a high state of purity, produce plasminogen-dependent fibrinolytic activity, i.e. the neutral serine protease plasminogen activator(PA)(11,12). PA is produced by either intact or cultured LGL, and is found in both an extracellular, soluble form as well as in a cell-associated form(11,12). Therefore, NK cells (LGL), as well as macrophages(26), B cells(37) or thymocytes (38), produce PA.

The cell-associated form of the enzyme appears to be in a cell fraction enriched for cell-surface membrane, when fractionated by well-established methodology(36). This fraction contained less than 10% of the total celluar protein and was also enriched for plasma membrane markers (e.g. 5' nucleotidase and $Na^{+}K^{+}$ ATPase).

As for other PAs(14,35), the LGL extracellular enzyme appears to exist in several molecular weight forms: 78,000, 73,000, 52,000, 45,000, 28,000, and 26,000 daltons(11).

We have studied the expression of LGL PA under conditions where NK activity is either enhanced or diminished. Upon interferon(IFN) treatment, only the cell-associated form of the LGL PA was augmented(11). We have also examined LGL isolated from patients with Chediak-Higashi syndrome for proteolytic activity, since patients with this syndrome are known to have low NK activity(39). We have observed that LGL from Chediak-Higashi syndrome patients are atypical in morphology, with only large azurophilic granule(s); these morphologically unusual LGL display defective NK lytic activity and produce only very low PA levels(11).

In order to gain insights into the specificity of the LGL protease, and in order to determine whether other proteases were involved in the regulation of PA activity, we utilized a variety of non-toxic protease inhibitors. We have observed that inhibitors of neutral serine proteases(DFP and NPGB), and inhibitors of tryptic proteases, including PA, (e.g. leupeptin), inhibit LGL PA; aprotinin, a plasmin inhibitor, has some inhibitory effect of LGL PA, whereas inhibitors of chymotrypsin (e.g., chymostatin and TPCK) or elastase (e.g. elastatinal) had no ability to inhibit PA from LGL.

C. Effect of Exogenously Added Protease Inhibitors and Proteases on NK Cell Lysis of Tumor Cells

In order to probe whether the association of a specific protease with LGL was of physiological significance in the lytic mechanism, selective protease inhibitors. and proteases, were employed.

In order to examine this issue in a critical fashion, we established both a ^{51}chromium release assay, as well as a single cell conjugate assay, in serum-free conditions in order to eliminate serum-containing protease inhibitors, We therefore utilized serumless medium supplemented with highly purified bovine serum albumin(11) which was tested for, and found to be free of, contaminating preteases or inhibitors.

We have found that exogenously added enzymes (e.g. trypsin and chymotrypsin) in ng. quantities, substantially increased LGL-mediated lysis of K562 cells(11, and Table I). Whereas plasmin induced a lsight enhancement of reactivity, urokinase(PA), elastase, or alpha thrombin had no effect.

Conversely, specific inhibitors of serine proteases led to profound inhibition of NK-mediated cytolysis. In addition to the active site titrant/inhibitors DFP and NPGB, the tryptic inhibitors(benzamidine, p-aminobenzamidine, and leupeptin) and the chymotryptic inhibitors (chymostatin and TPCK) profoundly inhibited LGL-mediated lytic activity. In contrast, inhibitors of plasmin (aprotinin), elastase (elastátinal) and thrombin (hirudin) had little or no effect in inhibiting NK lysis. Our data also demonstrates that protease inhibitors can also inhibit conjugate formation between LGL and K562 cells.

TABLE I. Effect of Proteases and Inhibitors on NK Cells(LGL).

Incubation Conditions	LYTIC UNITS* (-IFN)	LYTIC UNITS* (+IFN) 1000 units/ml	K562 CONJUGATES (-IFN)	K562 CONJUGATES (+IFN)
Input(non-purified cells)	62	N.D.	N.D.	N.D.
Purified NK cells(LGL)	148	202	56%	57%
LGL + chymotrypsin(.1μg)	190	N.D.	N.D.	N.D.
LGL + trypsin (.1μgm)	205	N.D.	N.D.	N.D.
LGL + chymostatin (10μg)	6	14	48%	45%
LGL + TPCK (10μg)	10	38	25%	7%
LGL + Leupeptin (10μg)	2	14	N.D.	N.D.
LGL + DFP (100mM)	1	18	N.D.	N.D.
LGL + TLCK (10μg)	2	34	25%	22%
LGL + NPGB (10μg)	8	16	N.D.	N.D.
LGL + benzamidine (10μg)	4	30	N.D.	N.D.
LGL + p-amino" (10μg)	1	26	N.D.	N.D.
(μg amounts refer to μg/ml)				

* 1 lytic unit = #Effector cells lysing 30% of K562 cells.

IV. CONCLUSION

It therefore appears that human NK cells(LGL) produce PA activity. Furthermore, tryptic and chymotryptic proteases can enhance NK cell killing, and inhibitors of these enzyme classes can inhibit NK reactivity. The tryptic inhibitors leupeptin, benzamidine, and p-aminobenzamidine, as well as DPF and NPGB inhibited LGL PA in agreement with inhibitory profiles reported for other PAs(35,40). It is of interest that chymotryptic membrane-associated proteases are known to regulate the release of PA(41); it is therefore possible that a chymotryptic enzyme regulates LGL PA function.

Proteases are degradative enzymes, and it appears that LGL PA may play a significant role in the NK lytic mechanism. NK cell proteolytic activity might act in concert with other lytic pathways including phospholipases and lymphotoxins, hydrolases, or lysosomal enzymes, in the destruction of the tumor target cell(3,11,12). It is possible that neutral proteases, through limited proteolysis at physiological pH, at both the cell surface and within cellular domains, might be the major regulators in the control of a number of lytic processes.

It is quite clear that further work is yet to be done in order to elucidate the role of LGL PA and other proteases, that mediate direct degradative actions and indirect regulatory effects, in elucidating the NK-mediated lytic mechanism.

1. Herberman, R.B., "Natural Cell-Mediated Immunity Against Tumors". Academic Press, New York, 1980.
2. Herberman, R.B., and Ortaldo, J.R., Science. 214:24 (1981)
3. Goldfarb, R.H., and Herberman, R.B., in "Adv. in Inflammation Res." (G. Weissmann, ed.) 4:45 (1982).
4. Gorelik, E., Fogel, M., Feldman, M., and Segal, S., J. Natl. Cancer Inst. 63:1397 (1979).
5. Hanna, N., and Burton, R.C., J. Immunol. 127:1754 (1981).
6. Roder, J.C., Kiessling, R., Biberfeld, P., and Andersson, B., Proc. Natl. Acad. Sci., USA. 76:1405 (1979).
7. Roder J. D., Argov, S., Klein, M., Petersson, C., Kiessling, R., Anderson, K., and Hansson, M., Immunol. 40:107 (1980).
8. Zagury, D., Maziere, J.C., Morgan, M., Fouchard, M., and Hosli, P. Biomedicine. 34:82 (1981)
9. Hoffman, T. Hirata, F., Bougnoux, P., Fraser, B.S., Goldfarb, R.H., Herberman, R.B., and Axelrod, J. Proc. Natl. Acad. Sci., USA. 78:3839 (1981).
10. Goldfarb, R.H., Pick, E., Timonen, T., and Herberman, R.B. In Press, This Volume.
11. Goldfarb, R. H., Timonen, T., and Herberman, R.B., Submitted for Publication.
12. Goldfarb, R.H., Timonen, T., and Herberman, R.B., Adv. Exper. Biol. and Med. In Press.
13. Timonen, T., Ortaldo, J.R., and Herberman, R. B., J. Exp. Med. 153:569 (1981).
14. Goldfarb, R.H., in "The Biology of Metastasis" (L.A. Liotta and I. Hart, Eds.). Martinus Nijhoff Publishers, Boston. In Press.
15. Liotta, L.A., Goldfarb, R.H., Brundage, R., Siegal, G.P., Terranova, and Garbisa, S., Cancer Res. 41:4629 (1981).
16. Ferluga, J., Asherson, G.L, and Becker, E.L., Immunol. 23:577 (1972).
17. Redelman, D., and Hudig, D., J. Immunol. 124:870 (1980).
18. Hatcher, V.B., Oberman, M.S., Lazarus, G.S., and Grayzel, A.I., J. Immunol. 120:665 (1978).
19. Ferluga, J., and Allison, A.C., Nature. 250:673 (1974).
20. Kedar, E., Ortiz de Landazuri, M., and Fahey, J.L., J. Immunol., 112:26 (1974).
21. Adams, D.O., J. Immunol. 124:286 (1980).
22. Adams, D.O., Kuo-Jang, J., Farb, R., and Pizzo, S.B., J. Immunol. 124:293 (1980).
23. Piessens, W.F., and Sharman, S.D. Cellalar Immunol. 56: 286 (1980).
24. Adams, D.O., and Marino, P.A., J. Immunol. 126:981 (1981).
25. Adams, D.O., Johnson, W.J., Fiorito, E., and Nathan, C.F., J. Immunol. 127:1973 (1981).
26. Unkeless, J.C., Gordon, S., and Reich, E., J. Exp. Med. 139:834 (1974).

27. Chapman, H.A., Vavrin, Z., and Hibbs, J., Proc. Natl. Acad. Sci., USA. 76:3899 (1979).
28. Trinchieri, G., and DeMarchi, M., J. Immunol. 116:885 (1976).
29. Roder, J.C., Kiessling, R., Bibberfeld, P., and Andersson, B., J. Immunol. 121:2509 (1978).
30. Hudig, D., Haverty, T., Fucher, C., Redelman, D., and Mendelsohn, J., J. Immunol. 126:1569 (1981).
31. Lavie, G., Weiss, H., Pick, A.I., and Franklin, E.C., Fourth Cong. Immunol. Abstracts. 11:4.30 (1980).
32. Shulman, N.R., and Tangon, H.J., J. Biol. Chem. 186:69 (1950).
33. Unkeless, J.C., Dano, K., Kellerman, G.M., and Reich, E., J. Biol. Chem. 249:4295 (1974).
34. Goldfarb, R.H., and Quigley, J.P., Cancer Res. 38:4601 (1978).
35. Goldfarb, R.H., and Quigley, J.P., Biochem. 19:5463 (1980).
36. Quigley, J.P., Goldfarb, R.H., Scheiner, C.J., O'Donnel-Tormey, J., and Yeo, T., Prog. in Clinic. and Biol. Res. 41:773 (1980).
37. Maillard, J.L., and Favreau, C., J. Immunol. 126:1126 (1981).
38. Fulton, R.J., and Hart, D.A., Biochem. Biophys, Acta., 642:345 (1981).
39. Haliotis, R., Roder, J., Klein, M., Ortaldo, J., Fauci, A., and Herberman, R.B., J. Exp. Med. 151:1039 (1980).
40. Zimmerman, M., Quigley, J.P., Ashe, B., Dorn, C., Goldfarb, R.H., and Troll, W., Proc. Natl, Acad. Sci., USA, 75:750 (1978).
41. O'Donnel-Tormey, J., and Quigley, J.P., Cell. 27:85 (1981).

THE ROLE OF SURFACE ASSOCIATED PROTEASES IN HUMAN NK CELL MEDIATED CYTOTOXICITY. EVIDENCE SUGGESTING A MECHANISM BY WHICH CONCEALED SURFACE ENZYMES BECOME EXPOSED DURING CYTOLYSIS.

Gad Lavie[1]

The Division of Hematology & Center of Transfusion,
The Beilinson Medical Center,
Petah Tiqva, Israel

I. INTRODUCTION

Three stages distinguishable in the cytotoxicity reaction are a binding stage in which the effector lymphocytes attach to target cells, presumably via specific receptors, a lytic stage in which damage is inflicted upon the target cells and detachment of the effector cells from the damaged targets (1,2).

Among the variety of factors involved in this reaction cell surface associated proteolytic enzymes are believed to play a major role in the lethal hit stage (3). The cell surface membrane of peripheral blood lymphocytes indeed contains numerous types of serine proteases with trypsin-like (4), chymotrypsin-like and elastase-like (5) properties which can account for this activity.

From these different groups of enzymes only those sensitive to N-α-p-tosyl-L-lysine chloromethyl-ketone (TLCK) and to 1-1-tosylamide-2-phenylethyl chloromethyl-ketone (TPCK) participate in the natural cell mediated cytotoxicity reaction

[1]Fellow in Cancer Research supported by George and Rose Blumenthal Cancer Research fellowship of "The Israel Cancer Research Fund".

Supported by a grant from the Israeli Ministry of Health.

ISBN 0-12-341360-5

against target tumor cells in vitro. The administration of each of these two inhibitors separately leads to an almost complete inhibition of cytotoxicity suggesting that the two enzyme types act on different substrates upon the target cell and perhaps also in a two stage cascade-like manner where the inhibition of each stage halts the entire process. The elastase type enzymes remain uninvolved in this reaction.

While cytotoxicity is completely inhibited when protease inhibitors are applied during the course of the reaction, the inhibitors exert almost no effect when added as a pretreatment of effector cells prior to the addition of target cells. This diminished inhibitory activity suggests that surface membrane components which carry those enzymes which are engaged in cytotoxicity are concealed within the surface membrane of the cells and become exposed to the external environment only after membrane receptors are attached to the target cells. This hypothesis is further supported by the selective nature of the cytotoxicity reaction in terms of the target cells which are chosen for attack. Natural killer cells which are brought in contact with target tumor cells from different sources may elicit a marked cytotoxic response against one target culture and leave others intact. This pattern is also maintained when tight contact between effector lymphocytes and target tumor cells is established by centrifugation, further indicating that in the absence of specific binding of target cells the cytotoxic enzymes remain inactive. Since other outer cell membrane associated proteases such as elastases as well as other non proteolytic enzymes are continuously active (6), it seems that enzymes engaged in cytotoxicity are localized on specific structures on the surface membrane.

The exposure of these structures following attachment to appropriate target cells may be mediated by tubulin as colchicine is a potent inhibitor of the cytolytic phase of the reaction (7).

II. CHARACTERIZATION OF PROTEOLYTIC ENZYMES ENGAGED IN CYTOTOXICITY

The profile of serine enzymes localized on the surface of lymphocytes was obtained from an autoradiograph of a preparation of human peripheral blood lymphocytes which were treated with ^{3}H-diisopropyl-fluorophosphate (DFP). The labeling was for a short period of three seconds at 2°C, conditions in which predominantly surface associated enzymes become labeled (8), followed by detergent solubilization, separation on

SDS-PAGE and development for autoradiography (method described in 5).

The resulting profile of such an experiment is shown in Fig. 1. Two major components with molecular weights of

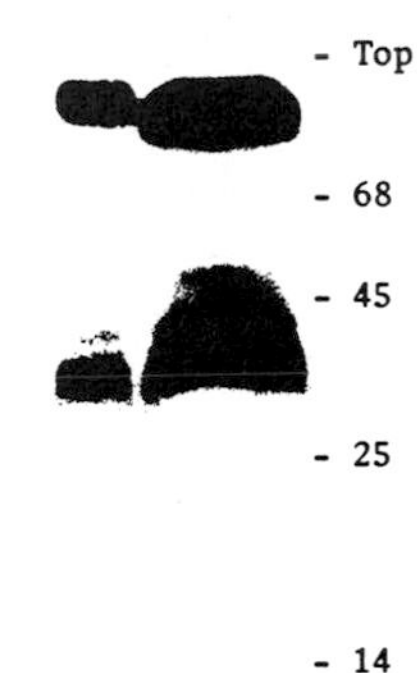

FIGURE 1. An autoradigram of a 10% SDS-polyacrylamide slab gel on which a ^{3}H-DFP labeled NP-40 lysed preparation of peripheral blood lymphocytes was separated. The left panel is a preparation of 10^6 cells and the right panel 2×10^7 cells.

200×10^3 and 35×10^3, and two minor components with molecular weights of 45×10^3 and 58×10^3 can be detected on the film. The 35×10^3 band is a TLCK sensitive trypsin-like proteinase likely to be the enzyme first isolated and characterized by Hatcher et al (4) and shown to be engaged in the cytotoxic activity of lymphocytes against a transitional cell carcinoma target cell line. The 45×10^3 and 58×10^3 bands have been found to bind ^{3}H-acetyl.ala.ala.pro.val.chloromethyl-ketone (5), a synthetic inhibitor which binds elastase-type enzymes with high affinity (9).

In order to test the role of each enzyme type in the cytotoxicity reaction of NK lymphocytes, the effects of group specific protease inhibitors were studied. TLCK which inhibits trypsin-like enzymes, TPCK which inhibits chymotrypsin-like enzymes and acetyl.ala.ala.pro.val.chloromethyl-ketone (AcAPVCK) which binds elastases were applied to four hour cytotoxicity assays directed against chromium-51 labeled K-562 targets. The kinetics of inhibition of the reaction are shown in Fig. 2.

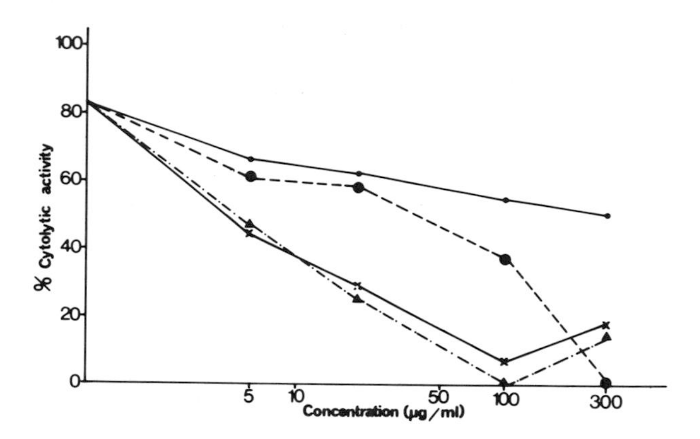

FIGURE 2. The effect of inhibitors of trypsin, chymotrypsin and elastase-like enzymes on the cytotoxicity of NK cells against K-562 myeloid leukemia target cells.
10^6 human peripheral blood lymphocytes were incubated with 10^4 Cr^{51} labelled K-562 cells in a 4 hour chromium release assay in the presence of different concentrations of:
Tosyl lysine chloromethyl-ketone (TLCK) ●----●.
Tosyl phenyl ethyl chloromethyl-ketone (TPCK) ▲-.-.-.-.▲
Acetyl.ala.ala.pro.val.chloromethyl-ketone (Ac.APVCK) •———•
Equal conc. of both TPCK and Ac.APVCK X———X

Both TLCK and TPCK were found to be potent inhibitors of the reaction when each was applied separately. While the inhibition by TPCK assumed a linear correlation with drug concentration, reaching a maximal effect at approximately 100μg/ml, the curve of inhibition in response to TLCK was repeatedly observed to acquire a sigmoid shape. Levels of inhibition by this drug were found to be negligible at concentrations of up to 100μg/ml and rose sharply thereafter reaching a maximum at approximately 300μg/ml.

Fig. 2 also indicates that the elastase enzyme inhibitor AcAPVCK had almost no effect on the cytotoxic activity of the cells and did not alter the dose response inhibition curve of TPCK when the two were applied concomitantly. This observation is of particular interest since elastase type enzymes have broad substrate specificities, and the inhibition of the mononuclear cell enzymes by AcAPVCK is notably effective.

III. EVIDENCE IN SUPPORT OF THE HYPOTHESIS THAT CONCEALED CYTOLYTIC PROTEASES BECOME EXPOSED FOLLOWING TARGET CELL BINDING

The protease inhibitor trasylol (Sigma aprotinin) has been added to a cytotoxicity reaction between human peripheral blood NK cells and K-562 targets as a 10% supplement of the medium. It was applied to part of the cultures for the entire length of the reaction, and to others as a 30 minute pretreatment of the effector lymphocytes prior to the addition of the target cells. Samples of the supernatant were collected for the determination of chromium-51 release after 2, 3 and 4 hours of incubation and the effect of the inhibition on the cytotoxicity reaction is shown in Fig. 3. While the presence

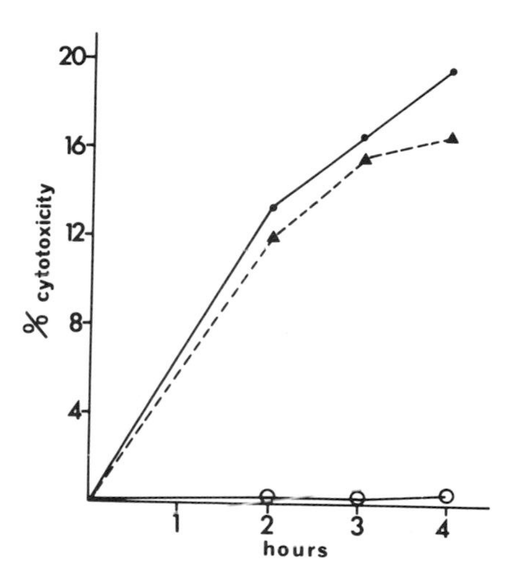

FIGURE 3. The effect of Trasylol on the cytotoxicity reaction when applied to effector cells prior to or during contact with the target cells. Human peripheral blood lymphocytes were pretreated with Trasylol (Sigma Aprotinin) for 30 minutes. The cells were then washed x2 with PBS and tested for their ability to lyse Cr^{51}-labelled K-562 target cells in a 4 hour chromium release assay ▲------▲. The cytotoxic capacity of the pretreated cells is compared with that of cells incubated with Trasylol for the entire assay o———o and with untreated cell •———• .

of trasylol during the entire course of the experiment resulted in the complete suppression of cytotoxicity, a 30 minute pretreatment of the effector cells with this preparation had almost no effect. In other similar trasylol pretreatment experiments in which samples of the supernatant were collected 10, 30 and 60 minutes after the initiation of the reaction by the addition of target cells, the level of chromium-51 release induced by trasylol pretreated lymphocytes was of a higher order of magnitude in comparison with the level of isotope release induced by untreated cells (data not shown).

This lack of inhibition of cytotoxicity by trasylol pretreated lymphocytes in short assays seems to speak against the possibility that redistribution of de novo synthesized enzymes were the cause for this phenomenon.

A similar decline in the inhibition of cytotoxicity has also been observed when effector lymphocytes were preincubated with small molecular weight protease inhibitors such as TLCK, however, while the pretreatment of effector cells with trasylol induced no inhibitory effect on the reaction, residual inhibitory activity of an order of magnitude of 30-40% did remain after preincubation of the lymphocytes with TLCK.

The results of these experiments suggest that proteases engaged in cytotoxicity are at least partly enclaved within the surface of resting NK cells in a manner which provides effective barrier from small molecular weight synthetic inhibitors. The attachment to target cells seems to trigger marked conformational changes on the surface of the killer lymphocytes leading to the exposure and activation of the cytotoxic enzymes to form contact and inflict damage upon the bound target cells. A schematic diagram which illustrates this concept is shown in Fig. 4.

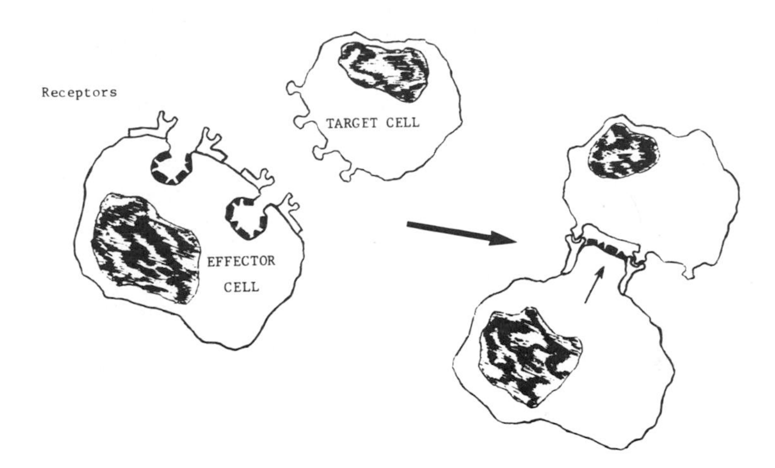

FIGURE 4. A suggested model for the exposure of surface associated proteases on effector lymphocytes during cytolysis.

Additional support to this hypothesis comes from experiments which show that cytotoxic cells exhibit selectivity in their reaction against different target cultures. Table I

TABLE I. A comparison of the cytotoxic activity of lymphocytes from three age matched healthy donors against a long term TCC target cell culture N.S., a primary TCC target K.H. and against K-562 myeloid leukemia cells under conditions of tight contact between effector and target cells.

	% Cytolysis		
	N.S.(TCC)	K.H.(TCC)	K-562
Donor #1	12.7	0	60.7
Donor #2	2.3	10.5	38.4
Donor #3	0.7	1.3	46.8

summarizes the results of an experiment in which peripheral blood lymphocytes from healthy individuals were tested against a long term human transitional cell carcinoma of the bladder (TCC) target culture N.S., a primary TCC target culture K.H.

and against K-562 myeloid leukemia cells. The reaction tubes were centrifuges at 1000 rpm immediately after the addition of the labelled targets to assure that tight contact with the effector cells has been formed. The mixed cultures were then incubated for 4 hours at 37°C in a 5% CO_2 atmosphere. Although this centrifugation may be a routine procedure in many laboratories, the lack of cytotoxic activity against some TCC targets by tightly attached cells shown to be reactive against K-562 confirms and emphasizes the idea that the enzymes are not exposed on the surface of resting NK cells.

It is not unlikely that the process of exvagination of cytotoxic protease-containing structures may require the action of the contractile apparatus of the cell. The active process of exposing defined structures may explain the energy requirements of the lytic process (10,11) and its sensitivity to cytochalasin B and to colchicine (7).

The lack of involvement of elastase-type enzymes in the reaction is somewhat unexpected. A possible speculation to account for the neutrality of these enzymes is to assume that they are distributed in sites distant from the specialized cell surface regions which encompass the hypothesized cytotoxicity structures which carry those enzymes engaged in cytolysis.

The concept of enzyme clustered surface membrane invaginations linked to the contractile apparatus of the cell and triggered to undergo marked changes by the binding of an extracellular ligand is not new. It has been described in detail to occur during the induction of receptor mediated endocytosis of the cholesterol-carrying-low density lipoprotein in coated pits (12).

REFERENCES

1. Wagner, H., and Rollinghoff, M. Eur. J. Immunol. 4:745 (1974).
2. Ferluga, J., and Allison, A.C. Nature. 250:673 (1974).
3. Ferluga, J, Asherson, G.L., and Becker, E.L. Immunology 23:577 (1972).
4. Hatcher, V.D., Oberman, M.S., Lazarus, G.S., and Grayzel, A.I. J. Immunol. 120:665 (1978).
5. Lavie, G., Zucker-Franklin, D., and Franklin, E.C. Immunology 125:175 (1980).
6. Wachsmuth, E.D., and Stoye, J.P. J. Reticuloendothel. Soc. 22:469 (1977).
7. Roder, J.C., Kiessling, R., Biberfeld, P., and Anderson, B. J. Immunol. 121:2509 (1978).

8. Heck, L.W., Remold-O'Donnell, E., and Remold, H.G. Biochem. Biophys. Res. Commun. 83:1576 (1978).
9. Powers, J.C., Gupton, B.F., Harley, A.D., Nishino N., and Whitley, R.J. Biochem. Biophys. Acta 485:156 (1977).
10. Berke, G., and Gabison, D. Eur. J. Immunol. 5:671 (1975).
11. Roder, J.C., Argov, S., Klein, M., Pettersson, C., Kiessling, R., Anderson, K., and Hansson, M. Immunology 40:107 (1980).
12. Goldstein, J.L., Anderson, R.G.W., and Brown, M.S. Nature 279:676 (1979).

THE RELATIONSHIP BETWEEN SECRETION OF A NOVEL CYTOLYTIC PROTEASE AND MACROPHAGE-MEDIATED TUMOR CYTOTOXICITY.

William J. Johnson
James E. Weiel
Dolph O. Adams

Departments of Pathology and Microbiology-Immunology
Duke University Medical Center
Durham, North Carolina

INTRODUCTION

Cytolysis of neoplastic cells *in vitro* and *in vivo* can be efficiently accomplished by cells of the mononuclear phagocyte system, when these cells are activated (Adams and Snyderman, 1979; Hibbs, Remington and Stewart, 1980). The lytic mechanisms operative in macrophage-mediated tumor cytotoxicity remain the subject of considerable research and controversy (Evans and Alexander, 1976; Keller, 1977). Most workers agree that tumor cytolysis by activated macrophages is contact dependent, and lytic substances have not generally been recovered from cultures of activated macrophages interacting with tumor cells (Keller, 1977). Macrophages are known, however, to secrete a variety of substances capable of injuring tumor cells (Nathan, Murray and Cohn, 1980). We have recently discovered a novel serine protease secreted by activated macrophages which lyses neoplastic but not normal target cells. We here review the properties of this enzyme and the evidence implicating it as a major effector of macrophage-mediated tumor cytotoxicity.

The Nature of Cytolytic Factor

Macrophages activated for tumor cytolysis by BCG secrete a potent lytic substance, which we have termed cytolytic factor (CF) (Adams, *et al.*, 1980). Supernatants from BCG-activated macrophages, when obtained in the absence of serum, contain considerable lytic activity against neoplastic but not normal target cells (i.e. lysis of over 50% of MCA-I sarcoma targets by a 1/200 dilution of CF)

ISBN 0-12-341360-5

(Adams, *et al.*, 1980). Lytic activity is quantified by culturing macrophage supernatants with radiolabeled target cells and then quantifying release of radioactivity from the killed targets. CF is quite potent as the concentration of CF which can lyse one half of a population of tumor target cells is estimated to be $\sim 1.0 \times 10^{-9}$ M. CF, when compared with equally proteolytic amounts of trypsin is $\sim 10^{5}$ times more cytolytic than trypsin (Adams, *et al.*, 1980). Secretion of CF is readily inhibited by heat, sodium azide and cycloheximide treatment of the macrophages (Adams, *et al.*, 1980). The lytic effects of the CF itself are readily inhibited by fetal calf serum, by heating the CF to 56°C for 30 minutes, and by the serine protease inhibitors bovine pancreatic trypsin inhibitor (BPTI), diisopropylfluorophosphate (DFP) and alpha-2-macroglobulin (Adams, *et al.*, 1980; Johnson, Whisnant and Adams, 1981). CF is not, however, inhibited by arginine, catalase, supernatants of inflammatory macrophages, glucose, protein hydrolysate, or extensive dialysis against fresh medium (Adams, *et al.*, 1980).

Table 1

Characteristics of Cytolytic Factor (CF)

1. Co-chromatographs with a novel protease of $M_r \sim 40{,}000$
2. Inhibited by three inhibitors of serine proteases
3. Secretion correlates with activation of macrophages
4. Kills tumor cells selectively and efficiently
5. Inhibited by fetal calf serum or by 56°C for 30 minutes

Upon separation of supernatants of activated macrophages by gel filtration, the lytic activity is contained in a fraction of molecular weight of $\sim$ 40,000 daltons. The lytic activity co-chromatographs with a novel neutral protease which is secreted by BCG-activated macrophages but not by resident or inflammatory macrophages (Adams, *et al.*, 1980). The lytic activity of the separated protease is inhibited by DFP and BPTI (Adams, *et al.*, 1980). The lytic substance present in supernatants of activated macrophages (i.e., CF) thus appears to be a neutral serine protease of molecular weight $\sim$ 40,000 daltons.

Secretion of Cytolytic Factor

CF is spontaneously secreted by cultures of macrophages which are activated for tumor cytolysis by BCG (Adams, et al., 1980). Secretion of CF is not observed in supernatants of cultures of resident macrophages under any circumstance, or of cultures of macrophages elicited by various inflammatory stimuli, unless the latter are cultured with relatively large (microgram) amounts of endotoxin (Adams, et al., 1980; Adams, Johnson, and Marino, 1982). Macrophages primed for activation by lymphokines in vitro or by pyran or BCG in vivo do not spontaneously secrete CF but do so when pulsed with but traces (nanograms) of endotoxin (Adams, Johnson, and Marino, 1982). The secretion of CF, like that of many other secretory products of macrophages, appears to be regulated in a two step fashion: an initial preparatory or priming signal prepares the macrophages so that a second triggering signal can elicit actual release of CF.

Activated macrophages selectively bind tumor targets, and this binding is an initial and necessary step in cytolysis (Marino and Adams, 1980a and 1980b). Such binding induces increased secretion of CF, over basal levels, from activated macrophages (Johnson, Whisnant and Adams, 1981). This induction of secretion is selective, in that only the binding of tumor cells to activated macrophages initiates enhanced secretion. If normal targets are cultured with activated macrophages or if resident or inflammatory macrophages are cultured with tumor targets, augmented secretion of CF does not occur. Dead or metabolically inactive tumor cells and plasma membrane preparations (Marino, Whisnant and Adams, 1981) of these cells also induce augmented secretion (Johnson, Whisnant and Adams, 1981). The augmented secretion itself has some degree of specificity, in that release of plasminogen activator, another serine protease, is not increased. The enhanced secretion of CF induced by target binding is quite rapid, occuring in less than 30 minutes after addition of tumor targets to activated macrophages (Johnson, Whisnant and Adams, 1981).

Effect of CF on Targets

CF exhibits some degree of target selectivity (Adams, et al., 1980). A wide variety of tumor targets including leukemias, lymphomas, sarcomas, carcinomas and a melanoma have been demonstrated to be susceptible to the lytic effects of CF, although the degree of target lysis does vary from one tumor to another (Adams, et al., 1980). Normal target cells, on the other hand, are not generally susceptible to CF (Adams, et al., 1980). The cytolytic activity of CF for tumor targets, when monitored by release of either ^{51}Cr or ^{3}H-thymidine, is first observed after five to six hours of interaction with the targets (Adams, et al., 1980; Weiel and Adams, 1981). Target lysis then proceeds progressively in a sigmoidal fashion until maximal lysis is observed at 16 to 20 hours. The final degree of target lysis observed is dependent on the concentration of CF used and can reach nearly 100%, if sufficient CF is added to the tumor targets (Weiel and Adams, 1981).

The injurious effects of CF, which appear to depend on direct interaction with the targets and not upon interaction with medium constituents (Adams, et al., 1980), have some of the characteristics of an enzyme-mediated reaction (Weiel and Adams, 1981). CF does not appear to induce irreversible injury (programming of tumor targets for lysis), as incubation of tumor targets with CF for less than six hours followed by removal of CF does not result in appreciable lysis of targets (Weiel and Adams, 1981). After six hours, lysis begins but its progression depends on the continued presence of CF. The ultimate degree of lysis is dependent on the concentration of CF and is relatively independent of the number of targets. CF is not adsorbed by exposure to either neoplastic or normal target cells (Weiel and Adams, 1981). Studies with trypan blue further indicate target cell injury and death occur before adherent targets are released from the culture vessels (Weiel and Adams, 1981). In addition, the action of CF on tumor targets does not appear to be dependent on cofactor(s) secreted by the tumor cells themselves (Weiel and Adams, 1981). Taken together, the data suggest the interaction of CF with tumor cells has several characteristics of an enzyme-substrate interaction.

Involvement of CF in Tumor Cytotoxicity

The secretion of CF correlates strongly with activation of macrophages for tumor cytolysis. Macrophages activated in vivo for tumor cytolysis by agents such as C. parvum or BCG secrete CF spontaneously (Adams and Marino, 1982). Macrophages primed for cytolysis by lymphokine treatment in vitro, or by Pyran or BCG treatment in vivo are not spontaneously cytolytic and do not secrete CF spontaneously (Adams and Marino, 1982). Upon pulsing with nanogram amounts of endotoxin, the primed macrophages become cytolytic and secrete CF. More importantly, dose curves for the amount of endotoxin required to induce macrophage-mediated cytolysis and secretion of CF are essentially identical (Adams and Marino, 1982). Macrophages which have been activated for cytolysis by BCG and then held in culture for 24 hours lose the capacity to kill tumor cells and to secrete CF; pulsing these macrophages with nanograms of endotoxin restores both functions (Adams and Marino, 1981). Macrophages elicited by sterile inflammatory agents are not cytolytic for tumor cells and do not secrete CF; pulsing these macrophages with micrograms of endotoxin activates both secretion of CF and tumor cytolysis (Adams, et al., 1980). As before, dose curves of the amount of endotoxin required to induce the two functions are virtually identical (Adams and Marino, 1982). Finally, macrophages from C3H/HeJ mice, which can bind tumor cells but which are not tumoricidal, are unable to secrete CF (Adams, Marino and Meltzer, 1981).

The protease inhibitors BPTI and DFP, which inhibit CF, likewise inhibit macrophage-mediated cytotoxicity and do so at concentrations which inhibit CF (Adams, 1980). These inhibitors are not toxic to macrophages and are effective at inhibiting cytotoxicity only if added to the macrophages during the cocultivation with tumor cells. Pretreatment of the macrophages with these inhibitors is not effective. In fact, the inhibitors act not during binding but during the target injury stage of macrophage-mediated cytotoxicity (Adams and Marino, 1981).

Two other lines of evidence implicate a role for CF in tumor cytolysis. First, the time course of target injury produced by CF is very similar to the time course of target injury produced by cocultivation of activated macrophages with tumor targets (Adams, et al., 1980). Second, the binding of neoplastic targets to activated macrophages triggers the release of CF (Johnson, Whisnant and Adams, 1981).

SUMMARY

We have reviewed the properties of a novel cytolytic serine protease and discussed the evidence for its participation in the cytolysis of tumor cells effected by activated macrophages. Taken together, the evidence strongly suggests secretion of CF is necessary for completion of macrophage-mediated tumor cytotoxicity. Many other macrophage-derived, lytic molecules have been described (for reviews, see Nathan, Murray and Cohn, 1980 or Adams and Marino, 1982). The possibility that two or more such molecules can act in concert to effect lysis of tumor cells bound to macrophages should not be ignored. For example, the ability of CF to synergize with H_2O_2 in effecting tumor cytolysis has recently been described (Adams, et al., 1981). Cooperation between various effector molecules thus may well operate in macrophage-mediated tumor cytolysis and, indeed, the molecules producing target injury may vary from target to target. Demonstration of the effector molecules actually involved in the lysis of various types of tumor cells bound to macrophages now appears to be an important next step in unraveling the molecular basis of macrophage-mediated cytotoxicity.

ACKNOWLEDGEMENTS

The studies from this laboratory described here are supported by USPHS Grants CA-16784, CA-29589, CA-14236, and ES-07031.

REFERENCES

1. Adams, D.O. 1980. J. Immunol. 124:286-292.
2. Adams, D.O., W.J. Johnson, E. Fiorito and C.F. Nathan. 1981. J. Immunol. 127:1973-1977.
3. Adams, D.O., W.J. Johnson and P.A. Marino. 1982. (Manuscript in preparation).
4. Adams, D.O., K.J. Kao, R. Farb and S.V. Pizzo. 1980. J. Immunol. 124:293-300.
5. Adams, D.O. and P.A. Marino. 1981. J. Immunol. 126:981-987.
6. Adams, D.O., P.A. Marino and M.S. Meltzer. 1981. J. Immunol. 126:1843-1847.
7. Adams, D.O. and R. Snyderman. 1979. J. Natl. Cancer Inst. 16:1341-1345.
8. Adams, D.O. and P. Marino. 1982. (Manuscript in preparation for Cont. Topics Hematol/Oncology, Volume III).
9. Evans, R. and P. Alexander. 1976. In, Immunobiology of the Macrophage. D.S. Nelson (ed.). Academic Press, p. 536-576.
10. Hibbs, J.B., J.S. Remington and C.C. Stewart. 1980. Pharmacol. Ther. 8:37-69.
11. Johnson, W.J., C.C. Whisnant and D.O. Adams. 1981. J. Immunol. 127:1787-1792.
12. Keller, R. 1977. In, The Macrophage and Cancer. K. James, B. McBride and A. Stewart, (eds). p. 31-49.
13. Marino, P.A. and D.O. Adams. 1980a. Cell. Immunol. 54:11-25.
14. Marino, P.A. and D.O. Adams. 1980b. Cell. Immunol. 54:26-35.
15. Marino, P.A., C.C. Whisnant and D.O. Adams. 1981. J. Exp. Med. 154:77-87.
16. Nathan, C.F., H.W. Murray and Z.A. Cohn. 1980. New Engl. J. Med. 303:622-626.
17. Weiel, J.E. and D.O. Adams. 1981. Abstracts of the 18th National Meeting of the Reticuloendothelial Society. p. 1a.

PHOSPHOLIPID METABOLISM DURING NK CELL ACTIVITY: POSSIBLE ROLE FOR TRANSMETHYLATION AND PHOSPHOLIPASE A_2 ACTIVATION IN RECOGNITION AND LYSIS

Thomas Hoffman
Philippe Bougnoux
Toshio Hattori
Zong-liang Chang
Ronald B. Herberman

Biological Research and Therapy Branch
National Cancer Institute
Frederick, Md.

I. Introduction

The presumption that NK function might be mediated in part by certain pathways of phospholipid metabolism is derived largely by analogy from other biolgical systems. Historically, phospholipase A_2 activity, i.e. generation of lysophosphatidylcholine (and unsaturated fatty acids) from phosphatidylcholine, has been suggested as a possible mechanism of lysis of targets by cytotoxic T lymphocytes (Berke, 1977). More recently, phospholipid transmethylation, i.e. the step-wise addition of methyl groups to phosphatidylethanolamine via s-adenosylmethionine (SAM), resulting in phosphatidylcholine synthesis, has been implicated in a wide range of receptor-mediated biological functions (Hirata, 1980). Amongst these, in addition to cytotoxic T lymphocyte lysis (Zimmerman, 1978), is release of histamine from mast cell granules after interaction with IgE immune complexes (Ishizaka, 1980).

II. Methods

Cytotoxicity was measured in a four hour ^{51}Cr release assay. Phospholipid methylation was measured using the incorporation of L-[methyl-^3H]methionine into cellular phospholipids as described (Hoffman, 1980). Lipids were quantitated by thin layer chromatography using the method of Skipski (1969). Phospholipase A_2 activity was measured by

ISBN 0-12-341360-5

the release of [1-^{14}C] arachidonic acid or selective decrease in the level of phosphatidylcholine in prelabeled cells (Hoffman, 1980).

III. Results

When peripheral blood mononuclear cells were treated with 3-deazadenosine (DZA), an inhibitor of transmethylation reactions, in the presence of homocysteine thiolactone, a dose-dependent inhibition of NK activity was observed. In the dose range of effective inhibition of NK activity, phospholipid methylation was inhibited by >50%, without significant effects on protein or nucleic acid methylation.

The incorporation of the radioactive methyl group of methionine into peripheral blood mononuclear effector cell phospholipids increased substantially, when NK-susceptible target cells were added. Stimulation of phospholipid methylation was not observed when human NK-resistant mouse lymphoma cells were used.

Inhibitors of phospholipase A_2, including mepacrine, tetracaine, corticosteroids, and Rosenthal's inhibitor uniformly inhibited NK activity by human peripheral blood cells. Inhibition of NK activity by Rosenthal's inhibitor was also observed when isolated large granular lymphocytes were used as effector cells.

Additional evidence for a role of phospholipase A_2 activation in NK activity was obtained using Lipomodulin, a naturally-occurring 40,000 Dalton molecular weight protein inhibitor of the enzyme (Hirata, _et. al._, 1980). Treatment of effector cells with lipomodulin throughout the period of incubation with K562 target cells resulted in a dose-dependent inhibition of lysis (Table 1).

TABLE 1. Effect of Lipomodulin (LM) on NK activity

Cell	Treatment	Cytotoxicity (% ^{51}Cr release)
PBL[a]	medium	36.9
	LM (2.5 x 10^{-7}M)	11.1 (70)[b]
	LM (5.0 x 10^{-8}M)	26.3 (29)
	LM (5.0 x 10^{-9}M)	35.1 (5)
K562	LM (5.0 x 10^{-7}M)	-0.1

[a]E/T ratio 100:1

[b]percent decrease from medium control

Direct measurement of arachidonic acid release as an indicator of phospholipase A_2 activity proved to be technically unfeasible due to avid metabolism of the fatty acid by tumor target cells. When incorporation of arachidonate into phosphatidylcholine of peripheral blood cells was measured, less label was incorporated by effector cells exposed to susceptible target cells, than when they were incubated with resistant cells.

IV. DISCUSSION

The observations of increased phospholipid methylation, when peripheral blood cells were incubated with NK-susceptible targets, and the inhibition of NK activity by agents which inhibit phospholipid transmethylation, indicate that phospholipid methylation is a concomittant of NK effector-target interaction. Possibly, as has been shown for other receptor-mediated biological phenomena, this increase is a consequence of binding of receptors on NK cells to target structures. The end-product of this reaction, phosphatidylcholine, is a principal substrate for phospholipase A_2. As shown in our previous studies (Hoffman, 1980) and extended by the observations reported here using a naturally-occurring inhibitor, agents which inhibited this enzyme also inhibited NK activity. Either lysophosphatidylcholine *per se* or fatty acids or their metabolic products (i.e., prostaglandins, lipid peroxides, leukotrienes) directly injure the target cell, or lead to other reactions within the effector or target which ultimately lead to lysis of the target. At this time, data to clearly discriminate between these possibilities are unavailable.

Although the experiments described, when taken together, establish a role for pathways of phospholipid metabolism in NK activity, they should also be viewed as circumstantial, in the absence of direct measurement of the enzymatic activities induced by target molecules isolated from NK targets. The involvement of other pathways is not necessarily excluded by these observations. The results obtained, using a limited number of target cells in a human NK system, should only with great caution be applied to other cytotoxic mechanisms (antibody-dependent, complement-mediated, or other cell-mediated lysis) and should not be extrapolated to other species.

The pitfalls inherent in the interpretation of data as are summarized here characterize those prevalent in this complicated field. Despite extensive studies of the mode of action of any of the inhibitors tested, controversy often exists regarding the specificity of the inhibitors. For example, DZA has been claimed to alter the metabolism of cyclic-AMP (Zimmermann, 1980), in addition to its effects on transmethylation reactions. Tetracaine is well-known to affect Ca^{++} metabolism and lysosomal "stability" (Dawson, 1979), and steroids have a wide-range of effects on multiple enzyme systems within the cell. Furthermore, even specific inhibitors may have unanticipated effects. For example, Rosenthal's inhibitor has been claimed to inhibit binding of cytotoxic T cells to their targets, yet enhance lysis after conjugates have formed (Berke, G., personal communication).

When the system studied entails the interaction between multiple actively metabolizing cells, unequivocal establishment of the site of action of any agent or treatment becomes hazardous. Even under "pretreatment" conditions, it is often difficult to rule out extravasation of the agent into the medium, or passage into other cell populations in a given system.

Additional difficulties of interpreatation are presented by the generally widespread use of heterogenous populations of effectors or targets. Until recently, purified populations of effectors have been unavailable. With the advent of methods for large scale purification and cultivation (Ortaldo, 1981) of populations of effector cells, new technologies for obtaining clones, and the application of single cell assays, these problems may ultimately be circumvented.

We are presently evaluating cell-free products of effector cells with cytotoxic capacity for their phospholipase activity and are attempting to mimic specific cytotoxicity with agents which induce these enzymatic activities. Hopefully, this approach will provide us with new, direct evidence on the biological role of phospholipid methylation and phospholipase A_2 activity, and either establish or refute the hypothesis that lysophosphatidylcholine is the lethal molecule involved in lysis by human NK cells.

REFERENCES

Berke, G. (1977). In "Regulatory Mechanisms in Lymphocyte Activation" (E. Lucas, ed.), P. 812. Academic Press, New York.

Dawson, A.P., Selwyn, M.J., and Fulton, D.V. (1979). Nature 277: 484.

Hirata, F., and Axelrod, J. (1980). Science, 209:1082.

Hirata, F., Schiffmann, E., Venkatsubramanian, K., et. al. (1980). Proc. Natl. Acad. Sci. USA 77:2533.

Hoffman, T., Hirata, F., Bougnoux, P., et. al. (1981). Proc. Natl. Acad. Sci. USA 78:3839.

Ishizaka, T., Hirata, F., Ishizaka, K., and Axelrod, J. (1980). Proc. Natl. Acad. Sci. USA 75:1718.

Ortaldo, J.R., Timonen, T.T., and Vose, B.M. (1981). In "The Potential Role of T-Cell Subpopulations in Cancer Therapy" (A. Fefer, ed.), p. 869. Raven Press, New York.

Skipski, V.P., and Berclay, M. (1969). In "Methods in Enzymology" (E. Lowenstein, ed.). Vol. XIV. p. 530. Academic Press, New York.

Zimmermann, T.P., Wolberg, G., and Duncan, G.S. (1978). Proc. Natl. Acad. Sci. USA 75:6220.

Zimmermann, T.P., Schmitges, C.J., Wolberg, G., et. al. (1980). Proc. Natl. Acad. Sci. USA 77:5639.

ROLE OF NATURAL KILLER CYTOTOXIC FACTORS (NKCF) IN THE MECHANISM OF NK CELL MEDIATED CYTOTOXICITY[1]

Susan C. Wright and Benjamin Bonavida

Department of Microbiology and Immunology
UCLA School of Medicine
Los Angeles, California

I. INTRODUCTION

We have previously reported that murine spleen cells or human peripheral blood lymphocytes stimulated with lectin will release soluble cytotoxic factors which selectively lyse NK-sensitive target cells in a 48-hour assay (1). It was also found that RAT*, a xenoantiserum directed against cytotoxic cells, could inhibit both NKCMC reactions as well as neutralize the lytic activity of the soluble factors (2). These data suggested that the cytotoxic factors might be involved in the lytic mechanism of NKCMC and subsequent studies were designed to examine this possibility. The approach taken was to examine the functional characteristics of the natural killer cytotoxic factors (NKCF) to determine if they conform with the known characteristics of the NKCMC system. Answers were sought to the following questions in order to establish if there was any correlation between lysis of targets by NKCF and lysis in NKCMC reactions: (1) Are NKCF released by NK effector cells? (2) Are NKCF released during coculture with NK target cells? (3) What target cells are susceptible to lysis by NKCF? (4) What effect does interferon (IFN) have on the production of NKCF? (5) Is the lytic activity of NKCF inhibited by the same monosaccharides shown previously to inhibit NKCMC? (6) Are YAC-1 clones selected for their resistance to NKCF

[1] Supported in part by Grant CA-12800 awarded by the National Cancer Institute and by National Cancer Institute Training Grant CA-09120 to S.C. Wright.

ISBN 0-12-341360-5

also resistant to NKCMC? The answers to these questions derived from experiments summarized in this report indicate that there is a strong correlation between NK target cell lysis by NKCF and in NKCMC reactions.

II. MATERIALS AND METHODS

Marbrook cultures were performed as described previously (1). Briefly, effector cells were cultured in the lower chambers along with a stimulatory agent which could be PHA, Con A, or NK-sensitive cells. The viability of target cells cultured in the top chamber was determined after 48 hours of culture by trypan blue exclusion. NKCF could also be generated in large-scale cultures of murine spleen cells and YAC-1 stimulator cells and a 50:1 spleen to stimulator cell ratio for 24 to 48 hours. Cytotoxic activity in the cell-free supernatants from these cultures was assessed either by trypan blue exclusion following a 48-hour incubation with target (micro supernatant assay) or by a 16-hour ^{51}Cr-release assay. All calculations, the CMC assay, the use of anti-Thy-1.2 and anti-asialo GM1 have been previously described (3). Mouse fibroblast interferon (10^7 U/mg protein) was obtained from Calbiochem. Spleen cells (20×10^6/ml) were pretreated with interferon (IFN) for 1.5 hours and washed prior to use in an assay. NK conjugate formation was performed according to the method of Roder *et al.* (4). Adsorption experiments were performed by incubating 1×10^8 cells/ml of cell-free supernatants containing NKCF for 30 minutes on ice. Residual NKCF activity in the supernatant was assessed against YAC-1 targets in a 16 hour ^{51}Cr-release assay. Percent adsorption was calculated as follows:

$$\% \text{ adsorption} = \frac{\% \text{ cytotoxicity unadsorbed NKCF} - \% \text{ cytotoxicity adsorbed NKCF}}{\% \text{ cytotoxicity unadsorbed NKCF}}$$

III. RESULTS

A. Release of NKCF by NK-Like Effector Cells

The characteristics of the effector cell which releases NKCF during coculture with YAC-1 stimulator cells are compared to those of the NK effector cell in Table I. Both cell types are plastic nonadherent spleen cells and are not present in thymocytes. Mutant beige (bg/bg) mouse spleen cells show a decrease in their ability to release NKCF as well as in their

TABLE I. Characteristics of the Eeffector Cell Which Releases NKCF

Effector Cell Population	% Cytotoxicity YAC-1 Targets	
	NKCF[1]	NKCMC[2]
Exp.1 Unfractionated CBA spleen	35 ± 3.6	23 ± 1.8
Plastic non-adherent spleen	33 ± 5.8	26 ± 2.0
CBA thymocytes	0 ± 0	2 ± 1.0
Exp.2 Bg/ + spleen cells	26 ± 2.0	25 ± 2.2
Bg/Bg spleen cells	14 ± 1.7	2 ± 0.6
Exp.3 nu/nu spleen + NRS + C'	38 ± 2.6	52 ± 3.8
nu/nu spleen + αasGM_1(1:1000)+C'	19 ± 2.3	22 ± 1.7
Exp.4 CBA spleen cells	25 ± 3.8	27 ± 3.6
CBA spleen cells pretreated with Emetine (10^{-5}M)	32 ± 3.5	25 ± 3.2
Exp.5 CBA spleen cells	25 ± 3.8	27 ± 3.6
CBA spleen cells pretreated with IFN ($2x10^4$ U/ml)	41 ± 3.5	68 ± 2.9
CBA spleen cells pretreared with IFN and cycloheximide	21 ± 2.6	29 ± 2.1

[1] Cytotoxicity was measured in the 48-hour Marbrook culture. Effector cells were stimulated with YAC-1 cells.

[2] NKCMC was determined using a four-hour ^{51}Cr-release assay with an effector to target cell ratio of 50:1.

ability to lyse NK targets in a four-hour CMC assay. The effector cell in both assays is equally sensitive to treatment with anti-asialo GM1 and complement and is not a mature T cell since nude mouse spleen cells show high activity. Emetine treatment of effector cells did not inhibit NKCF release nor did it inhibit NKCMC reactions (Table I, Exp.4). Though interferon-pretreated spleen cells release increased levels of NKCF, this increase required protein synthesis during the pretreatment period (Table I, Exp.5). This is in agreement with previous results which indicated that protein synthesis was required during the pretreatment period to observe IFN-induced augmentation of NK activity (5,6). These data indicate that the cell type which releases NKCF shows many characteristics in common with the NK effector cell and that release of NKCF is independent of protein synthesis. Furthermore, interferon enhances the production of NKCF, suggesting a mechanism of NKCMC augmentation by IFN.

B. NK Target Cells Stimulate Release of NKCF

If NKCF are involved in the lytic mechanism of NKCMC, it would be predicted that they would be released from the NK cell upon interaction with any NK target cell. Thus, coculture of murine spleen cells with NK_A target cells (YAC-1) or NK_B target cells (WEHI-164 or FLD-3), resulted in the release of NKCF into the supernatant (data not shown). Although two NK-resistant tumor cells (P815 and EL4) were found to be poor stimulators, the NK-resistant AKSL2 tumor cell proved to be a very effective stimulator cell. These results showed that NK-sensitive targets and some NK-resistant target cells stimulate NKCF production. Therefore, the ability to stimulate NKCF release cannot be the sole criteria to determine whether a tumor cell is NK-sensitive.

C. NKCF are Selectively Cytotoxic to NK-Sensitive Target Cells

The sensitivity to lysis by either murine or human NKCF of all target cells tested thus far is listed in Table II. All target cells which were found to be sensitive to NKCF were also sensitive to NK activity in either a four-hour assay (YAC-1, RL♂1, Molt-4 or K562), or an 18-hour assay (WEHI-164 and FLD-3). All those targets found resistant to NKCF were also resistant to NKCMC in this laboratory. Although Molt-4 was not lysed by murine NKCF in 48 hours, there was a significant cytotoxic effect if the incubation was extended to 72 hours. K562, however, remained resistant to NKCF in a 72-hour assay. Although human NK targets are slightly sensitive to murine NK cells, only Molt-4 cells are lysed by NKCF in an extended assay. The NK specificity of NKCF is further emphasized by the observation that YAC or RL♂1 cells passaged *in vivo* become resistant to both NKCF and NKCMC.

D. Characteristics of NKCF

Several characteristics of NKCF are consistent with their possible involvement in NKCMC reactions. NKCF is adsorbed by NK-sensitive target cells, and its lytic activity is inhibited by several monosaccharides at the same concentrations as that

TABLE II. Target Cell Specificity of NKCF Released from Murine Spleen Cells or Human Peripheral Blood Lymphocytes

Targets Sensitive to Murine NKCF (% Cytotoxicity)*		Targets Resistant to Murine NKCF (% Cytotoxicity)	
YAC-1	(30)	YAC-ascites	(3)
RL♂1	(28)	RL♂1-ascites	(0)
WEHI-164	(24)	P815	(4)
FLD-3	(14)	EL4	(2)
		AKSL2	(0)
		R1.1	(0)
		Molt-4	(5)
		K562	(4)
		RAJI	(1)

Targets Sensitive to Human NKCF (% Cytotoxicity)		Targets Resistant to Human NKCF (% Cytotoxicity)	
Molt-4	(50)	RAJI	(1)
K562	(15)	IM-9	(0)

* Cytotoxicity was determined using the 48-hour miniaturized Marbrook culture. Murine effector cells were stimulated with either PHA or YAC-1 cells. Human lymphocytes were stimulated with PHA-coated L929 fibroblasts.

previously reported to inhibit NKCMC reactions (7). NKCF is also sensitive to proteolytic digestion, stable at -20°C, and of MW > 10,000.

E. Development of YAC-1 Clones Resistant in NKCMC are Also Found Resistant to NKCF

After 32 days of culture in cell-free supernatants containing NKCF diluted 1:2 with fresh media, YAC-1 cells become relatively resistant to the toxic effects of NKCF. These cells were cloned by limiting dilution in the presence of NKCF and two out of the nineteen clones were subcloned. The three resulting subclones (YAC-R1.1, YAC-R1.2, and YAC-R15.1) were resistant to lysis by NKCF as well as by NK cells in a four-hour ^{51}Cr-release assay (Table III). However, the resistant subclones were not totally resistant to any type of CMC since they were lysed by alloimmune T cells contained in peri-

TABLE III. YAC-1 Clones Resistant to Both NKCF and NK CMC

Target	% Cytotoxicity NKCF[1]	NKCMC	Alloimmune T Cell CMC	% Conjugate Formation
YAC-1	20 ± 2.0	38 ± 4.0	29 ± 3.6	26 ± 3.5
YAC-R1.1	0 ± 0	4 ± 1.2	26 ± 2.9	23 ± 4.5
YAC-R1.2	4 ± 2.1	4 ± 0.6	31 ± 4.5	31 ± 2.1
YAC-R15.1	0 ± 0	5 ± 1.7	27 ± 3.6	33 ± 3.2

	% Adsorption of NKCF Activity
YAC-1	45 ± 5.7
YAC-R1.1	0 ± 0
YAC-R15.1	0 ± 0
YAC-R15.1	0 ± 0

[1] Cytotoxic activity in cell-free supernatants containing NKCF was determined in a 16-hour ^{51}Cr-release assay.

toneal exudates from C57Bl/6 mice injected ten days previously with P815 cells to the same extent as the parental YAC-1. The subclones were recognized and bound by spleen cells which indicates that their NK-resistance is probably not due to a lack of NK target structures, unlike other NK-resistant clones recently described (8,9). The mechanism by which the subclones are NK-resistant is suggested by the adsorption experiments which indicate that the resistant subclones lack a binding site for NKCF (Table III). These results provide compelling evidence that the mechanism of lysis in NKCMC is similar to lysis of NK targets by NKCF.

IV. DISCUSSION

Several lines of evidence have suggested that there exists a good correlation between lysis by NKCF and in NKCMC. Taking this into consideration, we would like to propose a model for the mechanism of NKCMC in which NKCF participates as the lytic mediator. According to this model, target cell lysis is the result of a series of discrete events, some of which may be modulated to obtain an increase or decrease in the levels of cytotoxicity. The first event is the recognition and binding of effector to target cell. At present, it is unclear what the NK recognition antigen is. The target cell must then activate the lytic mechanism which results in the release of NKCF. NKCF binds to the target cell membrane and causes cell death independent of the continued presence of the effector cell. Freshly isolated spleen cells contain a sufficient quantity of NKCF to lyse target cells in the absence of protein synthesis. However, it appears that IFN can induce effector cells to synthesize increased amounts of NKCF which could account for the IFN-induced enhancement of NKCMC. An alternative explanation is that IFN-pretreated cells contain the same amount of NKCF but are more efficient at releasing it than normal cells. After NKCF is released, it must bind to the target cell to cause lysis. The observation that certain monosaccharides inhibit NKCF activity suggests that these factors may be binding to simple sugars on the target cell membrane. The mechanism by which NKCF lyse the target cell is not understood; however, it would be predicted that these factors are more lytically active when effector is in contact with target cells since lysis in CMC occurs in four hours, whereas lysis by cell-free supernatants containing NKCF requires 16 hours.

This model allows us to establish certain criteria which will determine whether a tumor cell will be NK-sensitive. An NK-sensitive target cell must be recognized and bound by the effector cell, it must signal the effector to release NKCF, and it must bind NKCF and be sensitive to its cytotoxic effect. If any characteristic is lacking, the cell will be NK-resistant. Thus, YAC-R1.1, YAC-R1.2, and YAC-R15.1 subclones are all bound by NK effector cells, stimulate release of NKCF, but lack a binding site for NKCF and are therefore NK-resistant.

Although the available evidence is consistent with this model, further work is necessary to prove that NKCF is the lytic mediator of NKCMC.

REFERENCES

1. Wright, S.C. and B. Bonavida, J. Immunol. 126:1516, 1981.
2. Wright, S.C. and B. Bonavida, Transpl. Proceed. 13:770, 1980.
3. Wright, S.C. and B. Bonavida, J. Immunol., Submitted.
4. Roder, J.C., R. Kiessling, P. Biberfeld and B. Anderson, J. Immunol. 121:2509, 1978.
5. Senik, A., J.P. Kolb, A. Orm and M. Gidlund, Scan. J. Immunol. 12:51, 1980.

6. Ortaldo, J.R., W. Phillips, K. Wasserman and R.B. Herberman, J. Immunol. 125:1839, 1980.
7. Stutman, O., E. Lattime, P. Dien, M. Cuttito and R. Wisun, Fed. Proc. 39:1151, Abstract 4654, 1980.
8. Durdik, J.M., B.N. Beck, E.A. Clark and C.S. Henney, J. Immunol. 125:683, 1980.
9. Gronberg, A., R. Kiessling, E. Eriksson and M. Hanson, J. Immunol. 127:1734, 1981.

ROLE OF LYMPHOTOXINS IN NATURAL CYTOTOXICITY

Robert S. Yamamoto
Monica L. Weitzen
Karen M. Miner
James J. Devlin
Gale A. Granger

Department of Molecular Biology and Biochemistry
University of California, Irvine
Irvine, California

I. INTRODUCTION

Lymphotoxins (LT) are cytostatic or cytotoxic glycoproteins produced in vitro by activated lymphocytes (Jeffes and Granger 1976; Klostergaard et al 1979). Lymphotoxins within each animal species examined are heterogeneous with respect to both size and charge (Granger et al 1978; Hiserodt et al 1979; Ross et al 1979). Immunological and biochemical studies in both the human and murine LT systems have demonstrated that the various molecular weight (MW) LT classes are an interrelated family of molecules (Hiserodt et al 1979; Yamamoto et al 1978). Functional studies with the various MW human LT classes have shown that these molecules have different cell lytic capacity when tested on a variety of target cell types in vitro (Yamamoto et al 1979). The cell lytic capacity of some of these LT forms parallel those observed for natural killer cells (NK), where normal lymphocytes and PHA-blasts are resistant to lysis and other target cells vary in their susceptibility to NK (Bonnard and West 1978). Previous work with two of the high MW LT classes alpha heavy (α_H, 120,000 - 150,000 daltons [d]) and Complex (Cx >200,000 d) obtained from lectin-activated nonimmune human lymphocytes indicated that these two LT forms are lytic for the NK-sensitive target cell K-562 (Yamamoto et al 1978; Harris et al, manuscript in preparation). Because these lymphocytes were polyclonally stimulated with lectins to release these cell-lytic molecules, the possibility exists that these LT forms were produced by the subpopulation(s) of lymphocytes which mediates NK. The present report examines the possibility that populations of human lymphocytes which have been treated in vitro to contain activated NK-like effectors produce LT forms when stimulated

ISBN 0-12-341360-5

with lectins that cause lysis of NK sensitive cells. Human peripheral blood lymphocytes (PBL) are preactivated by coculturing in the presence of fetal calf serum (FCS) (Jondal and Targan 1978; Zielske and Golub 1976). After five days the cells are stimulated with Concanavalin A (Con A) for a short interval. The supernatants from these lectin-stimulated preactivated lymphocytes were then used in cytolytic assays on NK sensitive K-562 or MOLT-4F cells (Hansson et al 1978; Tsutsui and Everett 1974; West et al 1977).

A. Generation of LT Forms

Human PBL were obtained from defibrinated peripheral blood by the Ficoll-Hypaque technique (Boyum 1976). The PBL were then cultured at a cell density of 2 X 10^6 cells/ml in RPMI-1640 medium supplemented with 10% FCS (RPMI-10%) at 37°C for 5 to 7 days. The PBL cultured by this method will be referred to as cPBL. At the end of the 5 to 7 days of culture the cPBL were then used in either a direct cell killing assay or restimulated with 10 μg of Con A/ml at a cell density of 2 X 10^6 cells/ml in fresh RPMI-10% for 5 hrs at 37°C. In addition effector cells generated by this method were also tested for their ability to cause lysis of ^{51}Cr-labeled NK sensitive target cells K-562 and MOLT-4F or NK-insensitive target cells Jy and RPMI-1788 in a 4 hr direct cell killing ^{51}Cr release assay. The effector cells generated by this method caused significant lysis of the NK-sensitive target cells K-562 and MOLT-4F and little or no significant lysis of the continuous B cell lines Jy or RPMI-1788 (see Table I).

TABLE I. Percent Specific Lysis of ^{51}Cr-Labeled Allogeneic Target Cells by cPBL Effectors *In Vitro*

Exp. no.	% Specific ^{51}Cr release of allogeneic at a 25:1 effector:target cell ratio[a]			
	K-562	MOLT-4F	Jy	RPMI-1788
1	41 ± 2	48 ± 5	6 ± 2	8 ± 5
2	35 ± 1	37 ± 3	4 ± 1	13 ± 3

[a]Percent specific ^{51}Cr release (± SE) of allogeneic target

cells in a 4 hr ^{51}Cr release assay as calculated by the following formula: % Lysis = A-B/C-B X 100 where:

A is equal to experimental release
B is equal to spontaneous release
C is equal to total release

The cPBL not used in the direct cell killing assay were stimulated with Con A for 5 hrs. At the end of the 5 hr stimulation period the cultures were cleared of cells and debris by centrifugation. These supernatants were then immediately tested for LT activity on L-929 cells (Spofford et al 1974) and lysis of ^{51}Cr labeled NK sensitive target cells K-562 and MOLT-4F in an 18 hr ^{51}Cr release assay. While supernatants that expressed lytic activity for L-929 cells were also lytic for the ^{51}Cr-labeled NK sensitive K-562 and MOLT-4F target cells, they did not always correlate. However, supernatants which expressed very little lytic activity for L-929 cells contained no significant amount of lytic activity for the K-562 or MOLT-4F cells (see Table II). While not shown, these supernatants were not lytic for the other NK-resistant targets.

TABLE II. Lysis of Various Target Cells by Supernatants from Con A Activated cPBL Cultures

Exp. no.	LT units on L-929 cells[a]	% Specific ^{51}Cr release from[b]: K-562	MOLT-4F
1	60	10 ± 2	45 ± 3
2	48	28 ± 4	62 ± 5
3	14	3 ± 4	3 ± 2

[a]Units of LT activity are expressed as the reciprocal of the dilution of the LT containing supernatant necessary to lyse 50% of the target L-929 cells *in vitro*.

[b]Cells that were ^{51}Cr-labeled (10^4 in 0.02 ml) were added to microtiter wells containing 0.05 ml of supernatant or control media plus 0.05 ml of RPMI-10% for a total volume of 0.120 ml/well and incubated at 37°C for 18 hrs. Percent specific lysis was calculated by the formula in Table I a.

B. Inhibition of Natural Killing of K-562 and MOLT-4F Cells by Saccharides *In Vitro*

It has been previously demonstrated that various saccharides are inhibitory in murine NK systems *in vitro* (MacDermott et al 1980; Stutman et al 1980). A number of saccharides were tested were extensively tested for their ability to inhibit lysis of ^{51}Cr-labeled K-562 and MOLT-4F cells mediated by freshly isolated PBL in a 4 hr direct cell killing assay. Two saccharides, N-acetylglucosamine and α-methyl-mannoside, were chosen because they gave the most reproducible inhibitory effects in these assays. An experiment which is representative of six total experiments demonstrate that 50 mM concentrations of the two saccharides cause significant inhibition of lysis of MOLT-4F and K-562 cells mediated by fresh PBL (see Table III).

Table III. The Ability of N-acetylglucosamine and α-methyl-mannoside to Inhibit Natural Killing of K-562 and MOLT-4F Cells *In Vitro*

Target cell	% Lysis at effector:target cell ratio 50:1[a]	% Inhibition of lysis in the presence of 50 mM concentrations of [b]: N-acetylglucosamine	α-methyl-mannoside
K-562	38 ± 3	44	25
MOLT-4F	44 ± 5	51	54

[a]Same as Table I a.

[b]Percent inhibition of lysis was calculated by the following formula: % Inhibition = (1 - x/y) X 100 where:

X is equal to percent specific ^{51}Cr release in the presence of sugar
Y is equal to percent specific ^{51}Cr release

The PBL not used in the direct cell killing assay were then used to generate supernatants. Cells were cultured for 5 days and then stimulated with Con A for 5 hrs. The supernatants from these cultures were assayed on ^{51}Cr-labeled MOLT-4F cells in an 18 hr microplate assay. The two sugars which reproducibly inhibited NK activity of MOLT-4F cells were tested for their ability to inhibit supernatant lysis of MOLT-4F cells. An experiment which is representative of three experiments shows that the two sugars, N-acetylglucosamine and α-methyl-mannoside, at 50 mM concentrations significantly inhibited supernatant lysis of the ^{51}Cr-labeled MOLT-4F cells (see Table IV).

Table IV. N-acetylglucosamine and α-methyl-mannoside Inhibit Lymphocyte Supernatant Lysis of MOLT-4F Cells *In Vitro*

% Lysis[a]	% Inhibition of lysis in the presence of 50 mM concentrations of[b]:	
	N-acetylglucosamine	α-methyl-mannoside
45 ± 3	61	68

[a]Same as Table II b.

[b]Same as Table III b.

C. Inhibition of Natural Killing by Polyspecific Rabbit Anti-human Complex and Monospecific Rabbit Anti-human α_2 LT Antisera

Two antisera capable of neutralizing two different cell lytic forms of LT activity from human lymphocytes were tested for their ability to inhibit natural killing of MOLT-4F and K-562 target cells. Antisera were developed in rabbits immunized with Cx obtained from molecular sieving columns or with a purified α_2 subclass of human LT as previously described (Yamamoto et al 1978). Freshly isolated human PBL were tested for their ability to cause lysis of ^{51}Cr-labeled MOLT-4F or K-562 cells at a ratio of 50:1 effector to target cell in the presence of either normal rabbit serum (NRS), anti-Cx, or anti-α_2 LT sera in a 4 hr assay. The serum was added to a final concentration of 10% in these cultures. The results of 6 separate studies were evaluated, three are shown in Table V. In all studies anti-Cx serum caused a significant inhibition of ^{51}Cr release, in some cases inhibition reached 100%. In contrast, inhibition of ^{51}Cr release with anti-serum was variable; both in amount of inhibition observed and the particular aggressor-target cell combination that was inhibited.

Table V. Inhibition of Natural Killing by Heterogeneous Anti-LT Sera *In Vitro*[a]

Exp. no.	Target cell	% Lysis in NRS[b]	Percent Inhibition of Lysis in the presence of [c]: Anti-Cx	Anti-α_2 LT
1	K-562	40 ± 2	100	43
	MOLT-4F	64 ± 5	45	37
2	K-562	43 ± 2	88	0
	MOLT-4F	50 ± 4	57	35
3	K-562	30 ± 4	98	85
	MOLT-4F	44 ± 4	84	0

[a]Freshly isolated PBL were assayed against ^{51}Cr-labeled MOLT-4F and K-562 cells in the presence of a 1:10 dilution of NRS or antisera in a 4 hr ^{51}Cr release assay.

[b]Same as Table I a.

[c]Percent inhibition of lysis was calculated by the following formula: % Inhibition = (1 - x/y) X 100 where:

X is equal to percent specific ^{51}Cr release in the presence of antisera
Y is equal to percent specific ^{51}Cr release in NRS.

The present studies were designed to examine the role of cytotoxic mediators produced by human lymphocytes in natural killing *in vitro*. The PBL used in these studies were preactivated by culturing with fetal calf serum (cPBL) to generate effector cells with enhanced natural killing activity. We found these preactivated PBL would rapidly (5 hrs) release cell lytic forms into the supernatant when stimulated with Con A *in vitro*. These supernatants contained cell lytic forms for K-562, MOLT-4F, and L-929 cells. But supernatant cell lytic forms caused a more protracted lysis (10-16 hrs) of MOLT-4F and K-562 cells than intact NK effector cells. Data to be reported elsewhere indicates that these cell-lytic forms are heterogeneous. It is clear certain of these LT forms are lytic for the L-929 cell (an NK resistant target) and not the K-562 or MOLT-4F, however, it is not yet clear if those lytic for the latter targets are still lytic for the former. This question is currently being investigated. In addition, collaborative studies are underway with Dr. Wright and Dr. Bonavida to determine the relationship between cell lytic forms that have described (Wright and Bonavida 1980) and those detected in our studies. A possible explanation for the protracted lysis is the toxins are in a more lytically active form when delivered directly to the target cell by the NK effector and may be unstable or incomplete when released into the supernatant upon lectin stimulation. Alternatively the effector cell may deliver a high local concentration of these materials to the target cell in the microenvironment between cells when contact is made between the NK effector and target cell membranes.

Several probes were used to determine the relationship between lysis by soluble mediators and lysis by the intact lymphocytes. The first probe was the use of saccharides because they have been demonstrated to inhibit NK mediated lysis in the murine natural killing system. The two saccharides used in these studies were N-acetylglucosamine and α-methylmannoside. These two sugars significantly inhibited lysis of MOLT-4F and K-562 cells by fresh PBL and also inhibited supernatant lysis of MOLT-4F cells. It is still not clear whether the sugars which are inhibitory affect the binding or lytic stages of each reaction. Studies are currently in progress to examine this question. A second probe employed in these studies was antisera which neutralize human Cx and α LT forms (Yamamoto et al 1978). These results indicate that antibodies generated against LT forms do inhibit the NK reaction *in vitro*. Because the antiserum reactive with the Cx LT forms is polyspecific, it cannot be equivocally stated that its ability to block NK lysis is due solely to its ability to neutralize the components in the Cx form. This will require more specific serum. However, a serum has been developed, at the time of writing this manuscript, against a purified 140,000 MW α_H LT form which is a subunit of the Cx and its ability to block NK is currently being tested. While anti-α_2 serum is inhibitory in these reactions, its effects are quite variable. This may indicate these determinants have limited expression during lysis. Alternatively, this serum is not strong and may be easily overwhelmed. Stronger serum and/or monoclonal reagents will be necessary to adequately answer these questions.

These findings are provocative and provide support for the role of LT cell-lytic effectors in the lysis of K-562 and MOLT-4F targets by human NK effectors.

REFERENCES

Bonnard, G.D., and West, W.H., in "Immunodiagnosis in Cancer" (R.B. Herberman and K.R. McIntire, eds.) p. 1032. Marcel Dekker, New York, 1978.

Boyum, A., Scand. J. Immunol. 5:9 (1976).

Granger, G.A., Yamamoto, R.S., Fair, D.S., and Hiserodt, J.C., Cell. Immunol. 38:388 (1978).

Hansson, M., Bakacs, K.K.T., Kiessling, R., and Klein, G., J. Immunol. 121:6 (1978).

Hiserodt, J.C., Tiangco, G.J., and Granger, G.A. J. Immunol. 123:311 (1979).

Hiserodt, J.C., Tiangco, G.J., and Granger, G.A., J. Immunol. 123:317 (1979).

Jeffes, E.W.B. III, and Granger, G.A., J. Immunol. 117:774 (1976).

Jondal, M., and Targan, S., J. Exp. Med. 147:1621 (1978).

Klostergaard, J., Yamamoto, R.S., and Granger, G.A., Mole. Immunol. 17:613 (1979).

MacDermott, R.P., Kienker, L.J., and Machmore, A.V., Fed. Proc. 39:1198 (1980).

Ross, M.W., Tiangco, G.J., Horn, P., Hiserodt, J.C., and Granger, G.A., J. Immunol. 123:325 (1979).

Spofford, B., Daynes, R.A., and Granger, G.A., J. Immunol. 112:2111 (1974).

Stutman, O., Dien, P., Wisun, R.E., and Latteme, E.C., Proc. Natl. Acad. USA 77:2895 (1980).

Tsutsui, I., and Everett, N.B., Cell. Immunol. 10: 359 (1974).

West, W.H., Cannon, G.B., Kay, H.D., Bonnard, G.D., and Herberman, R.B., J. Immunol. 118:355 (1977).

Wright, S.C., and Bonavida, B., Fed. Proc. 39:359 (1980).

Yamamoto, R.S., Hiserodt, J.C., Lewis, J.E., Carmack, C.E., and Granger, G.A., Cell. Immunol. 38:403 (1978).

Yamamoto, R.S., Hiserodt, J.C., and Granger, G.A., Cell. Immunol. 45:261 (1979).

Zielske, J.V., and Golub, S.H., Cancer Res. 36: 3842 (1976).

CARBOHYDRATE RECEPTORS IN NATURAL CELL-MEDIATED CYTOTOXICITY[1]

James T. Forbes
Thomas N. Oeltmann

Division of Oncology
Department of Medicine
Vanderbilt University
Nashville, Tennessee

Cytolysis of susceptible target cells mediated by natural killer cells can be divided into four discrete stages (Roder and Haliotis, 1980): 1) recognition of the target cell by the effector cell and the subsequent binding. This stage can be visualized microscopically in the form of tumor cell-effector cell conjugates. 2) The second stage involves as yet unknown mechanisms of triggering and activation of the effector cell to elicit the third stage: 3) The lethal cytolytic event. 4) The fourth stage is terminal and results in target cell death. Stutman *et al* (1980) demonstrated that murine natural cell-mediated cytotoxicity (NCMC) can be inhibited by several sugars including D-mannose, D-galactose, and L-fucose and suggested that the recognition structure may be lectin like in nature. A receptor, specific for D-mannose 6-phosphate, on the surface of human fibroblasts mediates the pinocytosis of β-glucuronidase and alpha-l-iduronidase (Kaplan *et al*., 1977; Sando and Neufeld, 1977). Man-6-P has been implicated as the common recognition marker for intracellular transport of many glycosylated lysosomal hydrolases (Fischer *et al*., 1980). This receptor could also participate in the recognition between cells, as in the case of effector and target cell or could be involved in the lytic phase of NCMC.

We have demonstrated (Forbes *et al*., 1981) that mannose 6-phosphate, (Man-6-P); fructose 1-phosphate, (Fru-1-P); and fructose 6-phosphate, (Fru-6-P); can inhibit human NCMC to K-562 when added directly to the incubation mixture of effector cells and target cells. When tested in the same manner glucose 6-phosphate, galactose 1-phosphate, galactose 6-phosphate, mannose 1-phosphate, galactose, glucose, mannose,

[1]Supported by NIH grant CA 23477

ISBN 0-12-341360-5

and fucose had no effect. Inorganic phosphate was not inhibitory in the same concentrations (1-50 mM). Inhibition by Fru-1-P and Man-6-P may occur by the same mechanism as these two sugars are stereochemically similar (Kaplan et al., 1977). Fru-6-P may be converted to Man-6-P by enzymes in the culture media but the evidence for this is not compelling (Forbes et al., 1981).

Addition of Man-6-P to the effector cells prior to the addition of labeled target cells enhances slightly the inhibitory activity. Pretreatment of effector cells for one or two hours at 4° or 37° with either Man-6-P, Fru-1-P, or Fru-6-P followed by washing prior to the addition of target cells had no effect on the level of subsequent NCMC. These results demonstrate that Man-6-P is not directly toxic to effector cells. The addition of Man-6-P after the initiation of culture greatly decreases the efficacy and the level of this decrease is proportional to the length of time between the initiation of culture and the addition of Man-6-P.

In order to establish the presence of a Man-6-P receptor on the target cell, a covalent conjugate composed of gelonin, a toxic seed protein, and monophosphopentamannose (Bretthauer et al., 1973) was prepared. Gelonin is a single chain glycoprotein with biologic properties similar to those of the A-chains of abrin, ricin, and modeccin. Gelonin is non-toxic towards whole cells presumbably because it lacks the B-chain necessary for toxin binding. Gelonin was coupled to monophosphopentammanose by the reduction of the Schiff base between the C-1 reducing terminal sugar residue and a free amino group of the gelonin protein (Schwartz and Gray, 1977). This conjugate should be toxic towards any cell type which has a functional mannose 6-phosphate receptor. K-562 and F-265 target cell lines were all shown to be susceptible to this hybrid. This susceptibility could be abrogated by incubation with Man-6-P or either of the lysosomal inhibitory amines, NH_4Cl (10 mM) and chloroquine (25 μM).

Table I demonstrates that the formation of effector cell-target cell conjugates was not significantly inhibited in the presence of Man-6-P, Fru-6-P, or Fru-1-P at a concentration of 50 mM suggesting that the sugar phosphates do not affect the recognition phase of this reaction. When either effector cells or target cells were incubated in the presence of 50 μM chloroquine for 18 hours at room temperature or 37°C and tested both for conjugate formation and levels of NCMC, only the level of NCMC was significantly decreased while the number of conjugates formed was unaffected.

High uptake acid hydrolase (purified spleen β-glucoronidase provided by Dr. William Sly), was also effective in blocking human NCMC against K-562 (Table II). This molecule

TABLE I

Effect of Sugars on Effector-Target Conjugate Formation*

	Target Cell	
	K-562	F-265
Control	4.0+	12.0+
Glc	3.5	12.5
Man-6-P	3.8	12.0
Man	3.8	11.0
Glc-6-P	3.6	11.5
Fru-6-P	4.0	12.5
Man-1-P	4.1	13.0

+Conjugate x 10^{-4}/ml

*10^6 PBL + 10^5 K-562 or F-265 in 0.2 ml centrifuged at 200 x g for 5 min., incubated 30 min. at 4°C, resuspended and counted.

is known to be internalized by recognition of a terminal phosphorylated mannose. Furthermore, polyphosphomonoester fragments of phosphomannans secreted by the yeast, H. holstii, have also been shown to inhibit the cytotoxicity of K-562 by human effector cells (Table III). The inhibition of NCMC by Man-6-P probably involves local inhibition of the binding and uptake into the target cell of lytic molecules which contain Man-6-P like residues. This molecule may then be processed in the target cell by the lysosome. The data presented here suggest that a receptor for Man-6-P is involved in the cytolytic phase of NCMC. This is an especially interesting hypothesis since receptors for Man-6-P have been shown to be a mechanism by which certain lysosomal hydrolases can be reintroduced into some cells(Kaplan et al., 1977; Sando and Neufeld, 1977; Fischer et al., 1980). In that case, just as here, Man-6-P and Fru-1-P are effective in blocking the process. This strengthens the argument for a target cell lysosomal pathway in the lytic event of NCMC.

We have used macromolecular protease inhibitors to inhibit human NCMC as seen in Table IV. Chymostatin and alpha-1-antichymotrypsin, which inhibit chymotrypsin (Adyagi and Umejava, 1975), also inhibit NCMC while leupeptin, which inhibits trypsin, does not. These data suggest that the lytic molecule is chymotrypsin-like in nature.

TABLE II. INHIBITION OF NCMC BY HIGH UPTAKE FORM OF β-GLUCURONIDASE

Additions to Culture	% NCMC*	% Inhibition
None	56	-
Control+	53	5
β-glucuronidase (25,000 units)	35	38

*10^6 human PBL incubated at 37°C for 4 hrs with 10^4 ^{51}Cr-K-562

+Equal volume of same buffer used for β-glucuronidase

TABLE III INHIBITION OF NCMC* BY PPME+

Additions to Culture	% NCMC*	% Inhibition
None	50	-
10^{-5}M PPME	50	0
5 x 10^{-5}M	33	34
10^{-4}M	16	68

*10^6 PBL incubated 4 hrs at 37°C with 10^4 ^{51}Cr-K-562

+PPME = polyphosphomonoester fragment of phosphomannan secreted by the yeast Hansenula holstii (MW ∿ 10^6) [1 mg/ml RPMI = 1 mMP: assuming 1000 P: per molecule PPME]

Further studies have shown that the increased levels of cytotoxicity produced by pretreatment of the effector cells with interferon are also inhibited in the presence of Man-6-P.

Man-6-P inhibition of murine NCMC was also tested. The results in Figure 1 demonstrate that NCMC by C57BL/6 spleen cells to YAC target cells can be blocked by Man-6-P.

This provides a mechanism with which to test the C57BL/6 (bg/bg) mutation which is characterized by markedly reduced NCMC (Roder and Duwe, 1979) and by defects in lysosomal transport (Brandt et al., 1975). The human counterpart of this mouse model is Chediak-Higashi syndrome (Windhorst and Padgett, 1973) which also demonstrates impairment of NCMC along with demonstrable lysosomal defects.

Our data suggest involvement of lysosomal-like enzymes and pathways in the lytic mechanism of NCMC. This provides a method for the study of genetic defects in which the lytic event of NCMC is defective and the mechanism of the correction of these defects by biological response modifiers such as interferon. The identification of the lytic pathway in this reaction is of interest for several reasons: 1) it provides

TABLE IV. Effect of Protease Inhibitors on NCMC

Inhibitor	% Cytotoxicity	% Inhibition
None	38	-
0.5 μg leupeptin	37	3
0.5 μg chymostatin	19	50
0.5 μg alpha-1-antichymotrypsin	23	39

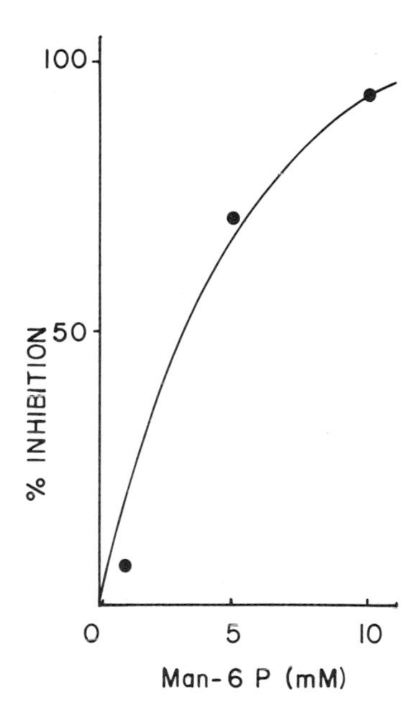

Figure 1. Inhibition of NCMC to Yac 1 by C57BL/6 spleen cells by Man-6-P. 1×10^4 ^{51}Cr-Yac 1 incubated 4 hrs. at 37°C with 1×10^6 spleen cells. NCMC in the absence of Man-6-P = 28%.

a model in which a lytic molecule may be isolated and identified and 2) it provides a model for the study of a mechanism for other lymphocyte mediated killing. This pathway may or may not be similar to other types of lymphocyte mediated target cell destruction. Since a number of agents which augment natural cytotoxicity both *in vitro* and *in vivo* appear to do so by increasing the effectiveness of the effector cell and not be increasing the number of effector cells capable of reacting with target cells, then knowledge of the physiology of such lytic events is important for predicting ways to manipulate NCMC or other lymphocyte mediated cytotoxicity by extrinsic agents.

ACKNOWLEDGEMENTS

The authors wish to thank Dr. W. Sly for his gift of β-glucuronidase, Dr. R. Bretthauer for his gift of polyphosphomonoester fragments of phosphomannans secreted by *H. holstii*, and Dr. J. Braatz for his gift of alpha-1-antichymotrypsin.

REFERENCES

Aoyagi, T., and Umejava, H. in "Proteases and Biological Control" (Reich, E., Rifkin, D.B., and Show, E., eds.) p. 429. Cold Spring Harbor Laboratory, Cold Spring Harbor, 1975.

Bretthauer, R.K., Kaczorowski, G.J., and Weise, M.J. Biochemistry 12:1251 (1973)

Brandt, E.J., Elliot, R.W., and Swank, R.T. J. Cell Biol. 67:774 (1975)

Fischer, H.D., Gonzalez-Noreiga, A., Sly, W.S., and Morre, D.J. J. Biol. Chem. 255:9608 (1980)

Forbes, J.T., Bretthauer, R.K., and Oeltmann, T.N. Proc. Natl. Acad. Sci. (USA) 78:5797 (1981)

Kaplan, A., Achord, D.T., and Sly, W.S. Proc. Natl. Acad. Sci. (USA) 74:2026 (1977)

Roder, J.C. and Duwe, A. Nature 278:451 (1979)

Roder, J.C., and Haliotis, T., in "Natural Immunity against Tumors" (R. Herberman, ed). p. 379. Academic Press, New York, 1980.

Sando, G.N. and Neufeld, E.F. Cell 12:619 (1977)

Schwartz, B.A. and Gray, G.R. Arch. Biochem. Biophys. 181: 542 (1977)

Stutman, O., Sien, P., Wisner, R.F., and Lattime, E.C. Proc. Natl. Acad. Sci. (USA) 77:2895 (1980)

Windhorst, D.B. and Padgett, G. J. Invest. Derm. 60:259 (1973)

CELLULAR SECRETION ASSOCIATED WITH HUMAN NATURAL KILLER CELL ACTIVITY[1]

Eero Saksela
Olli Carpén
Ismo Virtanen

Department of Pathology
University of Helsinki
Helsinki, Finland

I. INTRODUCTION

Active cellular motility and effector-target conjugate formation are necessary prerequisites for the human natural killer (NK) cell mediated lysis (1). These functions can be considered important in bringing the effector cells into a close proximity with the target cells and are similar in many respects with the requirements on mouse allospecific T-cell killing and mouse NK activities as well as with the antibody-dependent cellular cytotoxicity system (2, 3). The actual lytic step is energy-dependent but the specific mechanism has remained controversial in all cell-mediated cytotoxic systems studied (3, 4).

In the present experiments we investigated the lytic step of human NK cells using an enriched effector cell fraction which consisted of 75-85 % large granular lymphocytes (LGL) known to be the mediators of human NK activity (5). About 30-50 % of these cells form conjugates with K-562 cells. Of the conjugate-forming cells, 85-90 % have been shown to be capable of mediating interferon-activated lysis during an 18 h incubation in agarose, indicating that a great majority of K-562 binding LGL are endowed with NK lytic potential (6).

[1]Supported by grant No. RO 1 CA 23809 from the National Cancer Institute, NIH, Bethesda, Md., USA, by the Finnish Medical Research Council and by a research contract with Finnish Life Insurance companies.

ISBN 0-12-341360-5

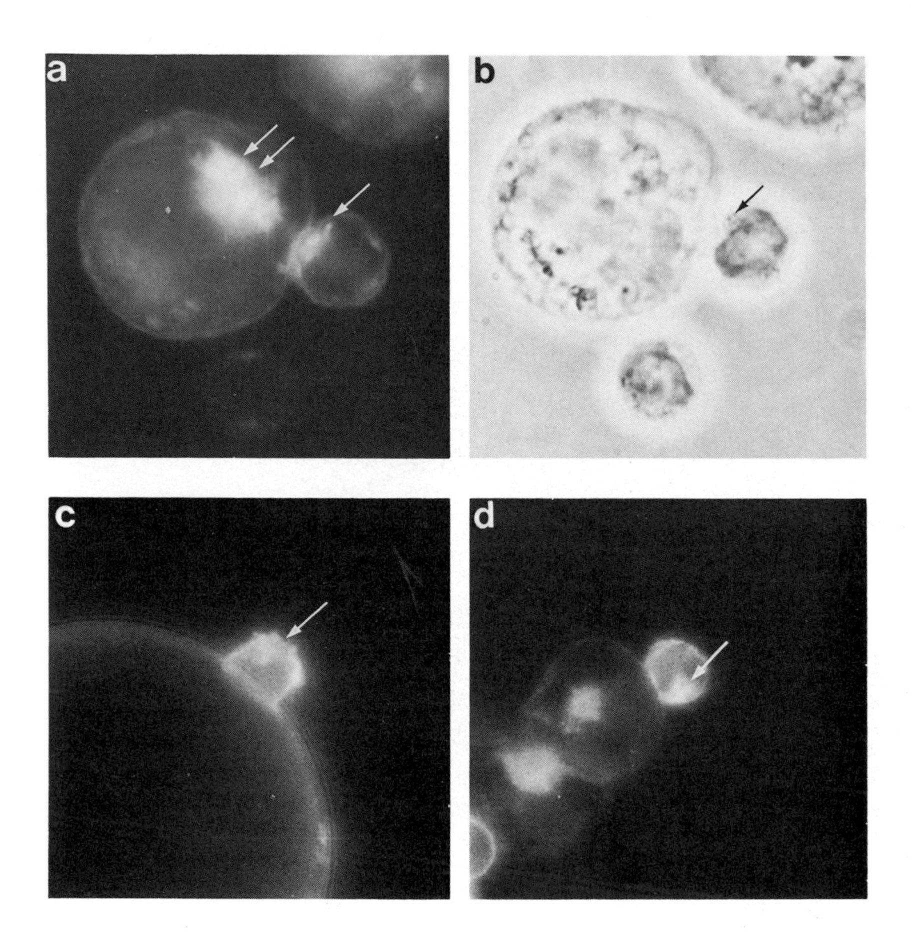

FIGURE 1. a-b. NK effector cell/K 562 target cells conjugates labeled with TRITC-WGA. The bright juxtanuclear granular fluorescence (a) represents the region of Golgi apparatus in both the attached NK cells (arrows) and in the K 562 cell (double arrows). The same cells are shown in phase contrast (b), where the characteristic NK cell granules are also visible (arrows). c-d. NK cell/poly-l-lysine bead (c) and NK cell/Raji (d) conjugates labeled with TRITC-WGA. Note the lack of orientation of the Golgi apparatus (arrows) towards the contact area. x 800.

II. MORPHOLOGY OF THE GOLGI APPARATUS

In transmission electron microscopy of conjugates, formed by highly enriched human NK cells and the sensitive K-562 erythroleukemia cell targets, a distinct polarisation of the Golgi apparatus (GA) was seen in the effector cells towards the target-contact area in each case where the orientation of the GA could be visualized. In order to substantiate this finding in a larger number of NK effector cell/target cell conjugates, we took advantage of the known preferential binding of certain fluorochrome-coupled lectins, such as Ricinus communis agglutinin (RCA) and wheat germ agglutinin (WGA) to the Golgi apparatus when suitably fixed and detergent treated cells are used (7, 8). In all conjugates where the orientation of the effector cell GA could thus be visualised in relation to the target cells, the GA was located immediately adjacent to the contact area (Fig. 1). Such orientation of the Golgi did not occur when the LGL were allowed to bind on poly-l-lysine coated Sepharose particles or on NK insensitive target cells.

III. BLOCKING OF THE SECRETORY PATHWAY

The carboxylic ionophore, monensin, (Eli Lilly) is known to specifically block cellular secretion by causing a partial Na/K equilibrium within cells leading to an interruption of the vesicular traffic of Golgi derived vacuoles to the cell

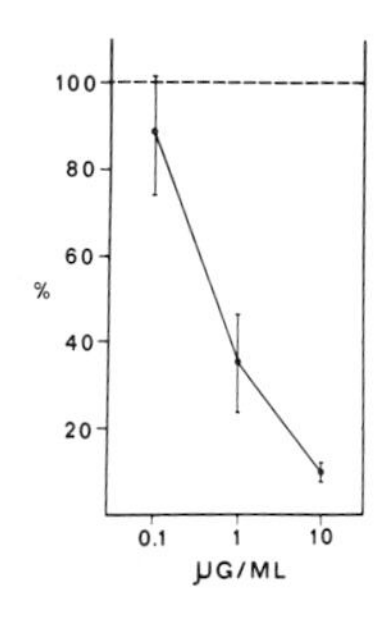

FIGURE 2. The effect of monensin on human NK cell mediated cytolysis. The LGL effector cells were treated with various concentrations of monensin for 30 to 90 min and washed prior to addition of ^{51}Cr-labelled K 562 cells. 7 experiments ± SD.

surface (9). Upon monensin treatment, secretion ceases and large vacuoles accumulate in the cytoplasm in a fashion similar in all cell types studied so far, including lymphocytes (10, 11, 12).

As seen in Fig. 2, pretreatment of the effector cells with monensin blocked the human NK cell-mediated lysis in a dose-dependent fashion. At higher doses, the effect was irreversible for at least two days after washing the cells and incubation in medium. At lower doses, when the monensin containing medium was removed after 30 min and the incubation continued in normal medium, some recovery of the cytotoxic activity occurred. The recovery could be enhanced and the cytotoxicity returned close to the original levels by addition of 100-1000 IU of alpha-interferon in the incubation medium (Fig. 3). All concentrations of monensin shown in the figures were non-toxic to the cells, as measured by trypan-blue exclusion and labeled methionine uptake in the cellular protein. In agreement with the findings of Tartakoff and Vassalli (10, 12, 13), large vesicles accumulated in the cytoplasm of the monensin-treated effector cells. The vacuoles were detectably enlarged already after 9 hrs and were apparently filled with glycoprotein rich material as revealed with fluorochrome coupled lectins. Monensin treatment did not seem to affect the target binding stage of the lytic cascade of LGL, as no differences were noticed in the amount of effector/target cell conjugates nor in the ultrastructure of the contact area of treated cells when compared to untreated effectors.

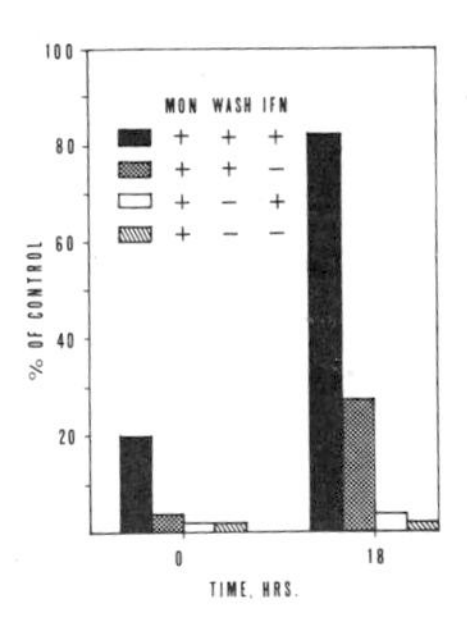

FIGURE 3. The effect of monensin and interferon of human NK activity. Purified LGL effector cells were pretreated for 60 min with 2 μg/ml monensin. The cells were then washed twice and 1000 IU of alpha-interferon was added as shown in the figure. A 4 h ^{Cr}K562 assay with E:T ratio 50:1 was performed immediately or after 18 h incubation in +37°C. The values are percentages of the cytotoxicity of untreated effector cells. A representative of 4 experiments.

IV. ROLE OF CELLULAR SECRETION IN NK LYSIS

These studies are compatible with a suggested role for an active cellular secretion in the lytic step mediated by human NK cells. In the effector-target cell conjugates, an orientation of the effector cell Golgi apparatus occurred towards the contact surface and inhibition of the fusion of Golgi derived vacuoles with the effector cell membrane could prevent lysis. The orientation was apparently specific for LGL to cellular targets since binding to Sepharose particles or to NK insensitive target cells was not associated with this phenomenon. The nature of the putative secretory substances released have not been characterized so far. There is, however, some recent evidence suggesting that cytotoxic material is released in the supernatants of dense co-incubation cultures of human lymphocytes and NK-sensitive targets (17). The cytotoxic action of the supernatant factor could be inhibited with mannose-6-phosphate. This substance has been also shown to inhibit the NK activity of human effector cells preincubated in its presence (18). Fuctose-1- and 6-phosphates were also inhibitory and impairment of the lysosomal functions with NH_4Cl or chloroquine likewise inhibited the NK cell mediated lysis. In the mouse NK system, several coat-color mutants have been identified with defective NK cell functions and these mutants have been regularly linked with deficient lysosomal enzyme action (14). All these findings are compatible with the interpretation that a secretory product, apparently associated with lysosomal functions, is involved in the lytic attack of human NK cells. The putative chemical mediators of the lysis are apparently very precisely delivered since third-party target cells are not killed and even an acceptor site on the target cell surface may be required (19).

ACKNOWLEDGMENTS

The technical assistance of Mrs. Maija-Liisa Mäntylä and Mrs. Raili Taavela is gratefully acknowledged.

REFERENCES

1. Carpén, O., Virtanen, I., and Saksela, E., Cell Immunol. 58:97 (1981).
2. Cerottini, J. C., and Brunner, K. T., in "B and T Cells in Immune Recognition" (F. Loor, and G. E. Roelants, eds.), p. 319. Wiley & Sons, 1977.
3. Allison, A. C., and Ferluga, J., New Engl. J. Med. 295: 165 (1976).
4. Henney, C. S., Immunol. Today 1:36 (1980).
5. Timonen, T., Ortaldo, J. R., and Herberman, R. B., J. Exp. Med. (in press 1980).
6. Timonen, T., Ortaldo, J. R., and Herberman, R. B., J. Immunol. (in press 1982)
7. Virtanen, I., Ekblom, P., and Laurila, P., J. Cell Biol. 85:429 (1980).
8. Kääriäinen, L., Hashimoto, K., Saraste, J., Virtanen, I., and Penttinen, K., J. Cell Biol. 87:783 (1980).
9. Tartakoff, A., Int. Rev. Exp. Pathol. 22 (in press 1981).
10. Tartakoff, A., and Vassalli, P., J. Exp. Med. 146:1332 (1977).
11. Smilowitz, M., Cell 19:237 (1980).
12. Tartakoff, A., Vassalli, P., J. Cell Biol. 79:694 (1978).
13. Tartakoff, A., Vassalli, P., J. Cell Biol. 83:284 (1979).
14. Örn, A., Håkansson, E. M., Gidlund, M., Ramstedt, R., Axberg, I., Wigzell, H., and Lundin, L. G., J. Exp. Med. (in press).
15. Timonen, T., and Saksela, E., J. Immunol. Methods 36: 285 (1980).
16. Timonen, T., Saksela, E., Ranki, A., and Häyry, P., Cell Immunol. 48:133 (1978).
17. Wright, S. C., and Bonavida, B., J. Immunol. 126:1516 (1981).
18. Forkes, J. T., Butthauer, R. K., and Oeltmann, T. N., Proc. Natl. Acad. Sci. (USA) 78:5797 (1981).
19. Roder, J. C., Beanmont, T. J., Kerbel, R. S., Haliotis, T., and Kozbor, D., Proc. Natl. Acad. Sci. (USA) 78: 6396 (1981).

MECHANISM OF NK CELL LYSIS

Phuc-Canh Quan, Teruko Ishizaka and Barry R. Bloom
Department of Microbiology and Immunology
Albert Einstein College of Medicine,Bronx, N.Y.
and Subdepartment of Immunology,
Johns Hopkins Medical School, Baltimore, Md.

Cell-mediated cytotoxicity can be effected by a variety of cells of the immune system, including cytotoxic T lymphocytes (CTL), natural killer (NK) cells and mononuclear phagocytes. The precise molecular mechanism by which target cell lysis is achieved by any of these effector cells remains unknown, and it is not even clear whether the different effector cells induce lysis by distinct or related mechanisms. A great deal of information is available on the biochemical process required for lysis by CTL that has permitted the lytic process to be analyzed in terms of definable phases: i) recognition - adhesion; ii) the lethal hit (also termed programming for lysis); and iii) lysis, which is independent of the presence of CTL. One important approach to elucidating the biochemical mechanisms required for lysis by CTL, has been the use of metabolic inhibitors(Chang and Eisen, 1980; Golstein and Smith, 1977; Henney, 1973; Martz, 1977) There are severe limitations on the ability to define precise steps in the lytic mechanism acted upon by individual metabolic inhibitors, since: i) it is difficult to be certain drugs act only on the known step for which they are chosen; ii) the effector cell populations are not homogeneous; and iii) the lytic process is asynchronous and the assay is of long duration. Nevertheless, using this approach it has been possible to delineate a number of important requirements for cytolysis by cytotoxic cells.

The purpose of the present work was to analyze the lytic process mediated by murine NK cells by methods, primarily the use of metabolic inhibitors, that have proven informative in studies of CTL and of histamine release by mast cells, in order to elucidate the similarities and differences in the effector mechanisms of these cells

Effects of protease inhibitors and substrates on NK cell-mediated lysis: Numerous studies on CTL have indicated that target cell lysis can be inhibited by protease inhibitors, particularly with specificity for trypsin-like serine esterases (Chang and Eisen, 1980; Martz, 1977). Consequently we examined the effect of a variety of protease inhibitors of known specificity on NK cell-mediated lysis of YAC cells (Quan et al, 1982). It emerged that NK cell lysis of YAC-1 cells was inhibited by 3 chymotrypsin inhibitors (TPCK, indole and

ISBN 0-12-341360-5

NCDC) and by 2 competitive substrates for chymotryptic-like enzymes (ATEE and APNE), TPCK being the most effective (Table 1). In contrast, 5 inhibitors of trypsin-like proteases (TLCK, BAME, AMCA, EACA and lima bean trypsin inhibitor) as well as inhibitors of leucine amino-peptidase or pepsin, or plasmin or plasminogen activator (leupeptin, pepstatin, NPGB, EACA, AMCA) were ineffective, except at doses which produced toxic effects on the target cells, and presumably on the effector cells as well. From the dose inhibition studies it was clear that even 10-fold greater amount of TLCK relative to TPCK, for example, produced only a marginal inhibition of NK cell lysis. These data indicate that, as in the case of CTL, a protease is required at some stage in the lytic process for NK cells,but the specificity of NK cell protease would appear different, namely chymotryptic-like, rather than trypsin-like in CTL. Curiously, chymostatin, a peptide inhibitor of chymotrypsin, was ineffective at blocking lysis, just as natural trypsin inhibitors are ineffective at blocking lysis by CTL suggesting that they may not have access to the critical active site after target cell binding.

Role of calcium and effects of other metabolic inhibitors. Calcium ions are known to be critical for programming the "lethal hit" by CTL, and the effects of cations or other drugs which can affect calcium activity was explored (Table 1). Strontium ions proved to be

TABLE I

Effects of Various Reagents on NK Cell Mediated Cytolysis

	% Inhibition of NK lysis (range)	Minimum toxic (t) or highest (h) concentration tested (mols/liter)
Chymotrypsin Inhibitors and Substrates		
Tos-phe-CH_2Cl (TPCK)	75-98	2×10^{-5} (t)
Indole	52-60	5×10^{-4} (t)
2-Nitro-4-carboxyphenyl-N-N-diphenylcarbamate (NCDC)	38-45	5×10^{-4} (t)
N-Ac-Tryp-OEt (ATEE)	55-62	5×10^{-4} (t)
N-Ac-phe-ONaph (APNE)	65-95	1×10^{-4} (t)
chymostatin	0-10	200 μg/ml (h)
Trypsin Inhibitors and Substrates		
Tos-lys-CH_2Cl (TLCK)	20-50	2×10^{-4} (t)
	0	2×10^{-5}
Nitrophenylguanidinobenzoate (NPGB)	30-45	1×10^{-4} (t)
N-Benzoyl-Arg-OMe (BAME)	35-50	5×10^{-3} (h)
transaminomethylcyclohexane carboxylic acid (AMCA)	0	1×10^{-2} (h)
ε-NH_2-caproic acid (EACA)	0-5	1×10^{-2} (h)
Lima bean trypsin inhibitor	0-8	2 mg/ml (h)
Other Protease Inhibitors		
leupeptin	0-12	200 μg/ml (h)
pepstatin	0-20	200 μg/ml (h)
Membrane-and Cytoskeletal-Active Agents		
deaza SIBA	50-70	1×10^{-4} (h)
chloroquine	60-80	1×10^{-4} (h)
diamide	40-50	1×10^{-5} (t)
glutathione (GSSG)	50-65	3×10^{-2} (t)
cytochalasin B	60-90	1×10^{-5} (h)
colchicine	60-92	1×10^{-3} (h)
PMA	55-70	1×10^{-5} (h)
Ca^{2+} and cAMP-Related Agents		
$SrCl_2$	90-99	1×10^{-2} (h)
$ZnCl_2$	0	1×10^{-4} (h)
Cholera toxin	70-90	10 μg/ml (h)
8 Br-cAMP (+ MIBX)	30-50	1×10^{-3} (h)
trifluoperazine (stelazine)	65-85	1×10^{-5} (t)

Data present the range of specific cytotoxicity at lymphocyte/target ratios of 50:1 in at least 4 experiments for each agent.

a very potent inhibitor of NK cells in this system, whereas they were found to be without effect on CTL or substituted for Ca^{2+} in supporting CTL-mediated lysis (Gately and Martz, 1979). Zinc cations, which can affect the conformation of microtubules or inhibit membrane proteases were without effect. Perhaps of greatest interest trifluoperazine (stelazine), which is known to block the binding of Ca^{2+} to calmodulin (Teo and Wang, 1973) exerted a profound inhibitory effect on NK cell-mediated cytolysis. The effect of this drug on CTL activity has not yet been described.

5'-deoxyisobutylthio-3-deaza-adenosine (deaza SIBA) an analog of S-adenosyl homocysteine which is an inhibitor of membrane methyl transferases (Hirata et al, 1979) exerted an inhibitory activity on NK cell mediated lysis, suggesting that methylation of phospholipids in the membrane may be a critical activation step in the lytic process.

It has been well established that agents which disrupt cytoskeletal functions, such as colchicine and cytochalasin B, as well as cyclic AMP-inducing agents have a profound inhibitory effect on cytolysis by CTL. As shown in Table 1, cytochalasin B and colchicine, as well as cholera toxin and 8 Br-cAMP similarly inhibited NK cell mediated lysis. Chloroquine, which has been reported to stabilize lysosomes in PMN, inhibited NK cell lysis although at relatively high doses. Thiol-reactive compounds, such as diamide and oxidized glutathione, had some inhibitory activity, but only at relatively high concentrations that produced some toxicity in the target cells alone.

Finally, phorbol myristate acetate, known to activate T cells for mitogenesis and macrophages for oxidative metabolism inhibited rather than augmented NK activity, although it is difficult to ascribe its effect to any single metabolic step.

Analysis of the site(s) of action of inhibitors on the sequence of NK cell-mediated cytolysis. While it is clear that there are severe limitations on the ability to pinpoint the precise step in the lytic mechanism acted upon by inhibitors in a 4 hr cytotoxicity assay, attempts were made to ascertain some sequential relationships of inhibitor action in the lytic pathway. Most of the agents at subtoxic concentrations had no effect on conjugate formation and target cell binding, and results for six agents are shown in Figure 1. Target cell binding, as reported for CTL cells, was found to be dependent on Mg^{2+} but did not require Ca^{2+}. Addition of calcium ions triggerred the program for lysis. The protease inhibitors, cholera toxin, stelazine, 3-deaza-SIBA, 8 Br-cAMP + MIBX and chloroquine were without effect on target cell binding, yet significantly inhibited NK cell-mediated lysis (Fig.1). Of the drugs examined, only cytochalasin B inhibited the target cell binding (data not shown).

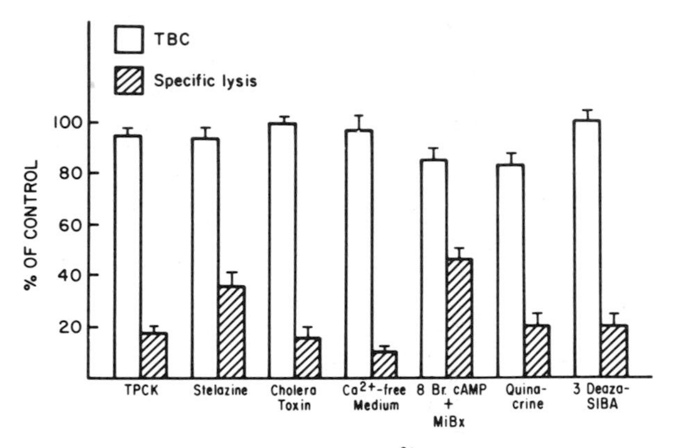

Fig. 1 - Effects of inhibitors and Ca^{2+}-free medium on target cell binding and cytolysis.

It thus became possible to analyze whether the various inhibitory agents could act prior to concordant with, or subsequent to the initiation of the lytic event by Ca^{2+}. The experimental design was similar to that developed by Martz, Henney and Chang and Eisen in which target cell binding is allowed to occur in the presence or absence of drugs prior to calcium addition or at various times after addition of calcium. As shown in Figure 2, the effect of

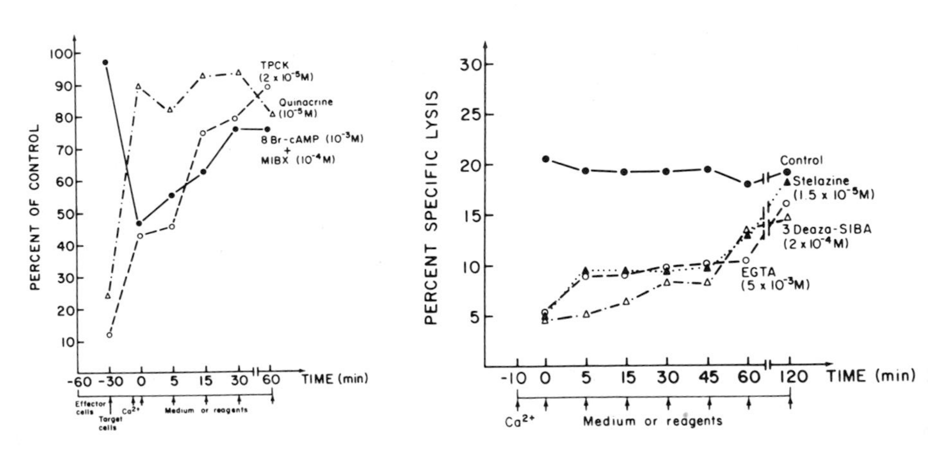

Fig. 2 - Effects of addition of TPCK, 8 Br-cAMP+MIBX(an inhibitor of phosphodiesterase)and quinacrine at different times during the NK assay .

Fig. 3 - Effects of addition of EGTA, stelazine and 3-Deaza-SIBA added at various times during NK assay.

TPCK (and other chymotryptic inhibitors) was greatest prior to addition of Ca^{2+}, whereas the effects of cAMP appeared to be closely associated with the calcium dependent step. This result indicates that the role of protease is likely to be related to membrane activation of the effector cell rather than as a mediator of target cell lysis.The results shown in Figure 3, while not definitive, suggest that the time course of inhibition of NK activity by stelazine are similar and may affect the same phase of the sequence, namely the calcium-dependent program for lysis, while deaza SIBA appears to inhibit NK activity to an extent prior to, as well as during the calcium-dependent step. Obviously, these stages could be more precisely delineated if homogeneous effector populations were available.

Similarities and differences between NK,CTL and mast cells. A model for the mechanism of lymphocyte-mediated cytolysis favored by many is the "stimulus-secretion" model in which appropriate binding of an effector cell to a target cell stimulates membrane changes which lead to a vesicular secretory event resulting in lysis of the target cell. Perhaps the best understood paradigm of vesicular secretion is that of histamine release induced by antigen or anti-IgE antibodies from mast cells or basophils, an event which is complete within two minutes after appropriate stimulation (Ishizaka et al, 1980). Such a model seems particularly relevant to NK cell lysis, since human and rat NK cells have been found to be associated with a large granular lymphocyte (LGL) fraction. The effects of some of the agents studied here on NK cells, CTL and on histamine release by mast cells are summarized in Table 2.

TABLE 2

Patterns of Inhibition of NK, CTL and Mast Cell Secretion of Histamine

	NK	CTL	Histamine release
Protease inhibitors			
of trypsin-like enzymes	-	+	+
of chymotrypsin-like enzymes	+	-	+
of other enzymes	-	-	±
Cytoskeletal and membrane active agents			
deaza SIBA	+	+	+
quinacrine	+	?	+
cytocholasin B	+	+	+
colchicine	+	+	+
chloroquine	+	+	+
Calcium related agents			
Sr^{+2}	+	-	+
trifluoperazine	+	?	±
cAMP	+	+	+

In studies on histamine release, the earliest definable event following appropriate stimulation is a protease-requiring step, that can be blocked by inhibitors of both trypsin and chymotrypsin-like enzymes. Within 15 seconds after triggering there is methylation of the membrane phospholipids, the step inhibited by deaza-SIBA, followed in 30 seconds by a calcium flux, then steps inhibited by microtubule-active drugs leading to secretion of histamine. Sr^{2+} ions spontaneously trigger histamine release from mast cells and basophils (Foreman et al, 1979) and, in the case of large granular lymphocytes with which human NK activity is associated, Sr^{2+} appears similarly to cause degranulation which parallels inhibition of NK activity(Neighbour and Huberman, 1982). This is schematized for NK cells in Fig. 4.

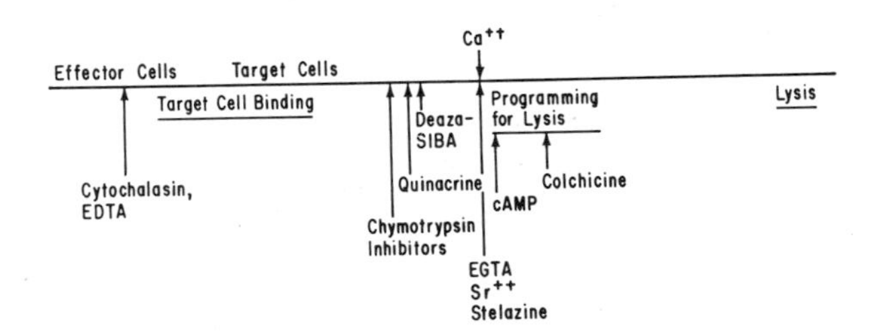

The only discrepancy at present appears to be the ability of trifluoperazine, which blocks the binding of Ca^{2+} to calmodulin, to block NK cell lysis more effectively than release of histamine. This would suggest that the effects of Ca^{2+} in the histamine release pathway are probably ionic, whereas they may require calmodulin-dependent regulatory activity in the case of NK lysis. While the available data are consistent with a stimulus-secretion model for T cell and NK cell mediated cytolysis, definitive proof must await the use of homogeneous effector cell populations and identification of the products of membrane activation or secretion involved in the lytic event.

1. Chang, T.W. and Eisen, H.N. J. Immunol. 124:1028, 1980.
2. Foreman, J.C., Sobotka, A.K. and Lichtenstein, L.M. J. Immunol. 123:153, 1979.
3. Gately, M., and Martz, E. J. Immunol. 122:482, 1979.
4. Golstein, P. and Smith E.T. Contemp Top. Immunobiol. 7:273, 1977.
5. Henney, C.S. Transpl. Rev. 17:37, 1973.
6. Hirata, F., Axelrod, J. and Crew, F.T.P.N.A.S. 76:4813, 1979.
7. Ishizaka, T., Hirata, F.,Ishizaka, K. and Axelrod, J. P.N.A.S. 77:1903, 1980.
8. Martz, E. Contemp. Top. Immunobiol. 7:301, 1977.
9. Neighbour, P.A. and Huberman, H.S.J. Immunol. 1982, in press.
10. Quan,P-C., Ishizaka, T. and Bloom, B.R. J. Immunol. 1982, in press.
11. Teo, T.S. and Wang, J.H. J. Biol. Chem. 248:5950, 1973.

MECHANISMS OF NK ACTIVATION: MODELS TO STUDY INDUCTION OF LYSIS AND ENHANCEMENT OF LYTIC EFFICIENCY[1]

Stephan R. Targan[2]

Geriatric Research and Education Center
Medical and Research Services
Wadsworth VA Medical Center
and
Department of Medicine
UCLA Center for the Health Sciences
Los Angeles, California

I. INTRODUCTION

The cytolytic activity of a population of natural killer cells can be enhanced by several modalities. These include *in vitro* activation by interferon (IFN), prostaglandin E_2 (PGE_2), ethanol (Kendall and Targan, 1980), and *in vivo* activation by short bursts of exercise (Targan, et al., 1981). Even though recent studies show differences in inhibition of target binding and lysis, no specific or defineable mechanism has been implicated in any phase of the actual lytic process of these cells. Therefore, we asked what changes occur in binding and postbinding NK lytic events following these pharmacologic activations of total NK lysis.

[1]This work was supported in part by a Clinical Investigator Award from the Wadsworth VA Hospital and in part by USPHS Grant CA-12800-08 and NIH Grant "Inflammatory Bowel Disease: NK Activity in Gut Mucosa."

[2]Present Address: Department of Medicine, UCLA Center for the Health Sciences, Los Angeles, California

ISBN 0-12-341360-5

II. RECRUITMENT OF PRE-NK CELLS TO ACTIVE NK CELLS

Based on our research studies, one mechanism of NK activation occurs by induction of cytolytically inactive pre-NK cells to lytically active NK cells. We have used this model to ascertain what types of manipulation would inhibit and/or enhance expression of this lysis. In this manner, there will be a better understanding of what facets of the post recognition and binding stages of the NK lytic mechanism are present on pre-NK cells and what alterations are required for the completion and/or linkage of the components of this system so that lysis occurs. Nonetheless, mechanisms which invoke either neoexpression of molecules important in the lytic unit or rearrange and linkage of existing membranes are all consistent with three recent observations that give insight into the requirements for this lytic expression.

NK cells and pre-NK cells can be exposed to interferon for brief periods (5 minutes) without being activated. If free interferon is removed by washing and the cells allowed to further incubate at 37°C for 30 minutes to 2 hours, NK cells are activated. Therefore, it is during this time that the cellular processing required for pre-NK cells to express their lytic potential occurs. Various perturbations (pharmacologic, enzymatic, immunologic, or physical) of NK and pre-NK cells can be performed during these 2 hours and their effect on the activation of the lytic potential of pre-NK cells can be ascertained.

Pre-NK activation is inhibited if these cells are bound to target cells prior to expression of their lytic potential (Silva, et al., 1980). It seems that after binding, the effector cells become committed to either kill or not kill and are no longer inducible by interferon. The mechanisms of this block is not clear. It may be that the binding of the effector cell to targets neutralizes adenylcyclase activity or perturbs the membrane in such a fashion so that the exogenous signals are not transmitted into this cell.

We have recently demonstrated that the activation of NK lysis can be blocked by the interaction of two distinct activators prostaglandin E_2 and interferon. By first priming NK cells with interferon or prostaglandin and allowing expression of enhanced lytic potential to occur over a three hour time span, this activation can be prevented by subsequent exposure of these cells to the other modulator prior to completed activation. It appears to occur by the actual blockage of recruitment of pre-NK cells into the lytic pool (Targan, 1981). There is a temporal relationship to the occurrence of this inhibition, as inhibition only occurs during pre-activation cellular processing. This suggests that each agent may produce activation through two separate pathways and that the lack of activation may be due to competition of each of these for some essential substrate. Thus, with simultaneous activation neither pathway would have enough of this molecule available to complete the processes needed for full expression of activation. One other explanation for these above modulations could involve cellular membrane changes. If membrane alterations produced by prostaglandin and interferon separately, occurs simultaneously before final expression, it could cancel each others effects.

We have produced a heterologous antisera and monoclonal antibody to lymphocyte populations containing alloimmune periodate activated NK and T-cell cytotoxic cells which we term RH_2 (Hiserodt, et al., unpublished results). The antisera RH_2 and monoclonal antibody 17.2 block 50% to 100% of NK lysis if present during the cytotoxic reaction. These antisera have been absorbed on target cells and thus most likely do not directly affect the blockage of target cell molecules (Hiserodt, et al., 1981). RH_2 blocks NK killing both at the level of target recognition and post recognition events (Fig. 1). The monoclonal antibody 17.2 blocks at the post binding events (Table 1) (Hiserodt, et al., 1981). If NK cells are pulsed with these reagents directly following interferon priming, the usual activation of NK lysis is inhibited (Fig. 2 and Table 2). This suggests that whatever these agents react with are important in the production and/or linkage of post binding pre-NK lytic components. Furthermore, neoexpression of these molecules at the end of an *in vitro* activation is unlikely since RH_2 or 17.2 is washed out of the system and not present for the entire two hour incubation. This does not preclude, however, the possibility that more of these molecules are produced intracellularly and that the antibody blocks their linkage into the proper lytic sequence in the membrane.

Initial target cell recognition by cells of NK lineage does not necessarily lead to lysis. It is clear from these data that cellular processing requiring some membrane modulation is necessary for completing the cellular machinery that leads to lysis. How much of this machinery is lacking from pre-NK cells and how many alterations must occur before lysis proceeds is unknown. Since this differentiation is not terminal, however, in that 6 hours after activation under proper conditions the pre-NK cells return to their baseline recognition-non-lytic state, this suggests that lytic activation of these pre-NK cells most likely results from rearrangement and linkage of existing structures into a functional unit (Silva, et al., 1980).

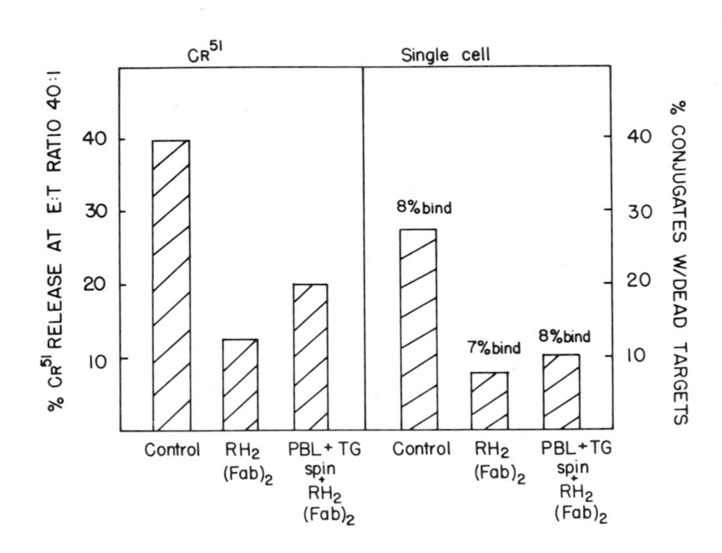

FIGURE 1. $(Fab)_2$ produced from RH_2 antisera was added to the cytolytic assay either before or following conjugation formation. Its effect on overall NK lysis was measured in a 3 hour ^{51}Cr release assay and its effect on target binding and/or single hit lysis was measured in a 3 hour single cell cytotoxic assay. TG = K562.

TABLE I

	Control		17.2	
Donor	% ^{51}Cr Release	Bind	% ^{51}Cr Release	Bind
1	59	24/400	12	$24/400^{+}$ $21/400^{o}$
2	50	N.D.	21	N.D.
3	40	38/400	20	$37/400^{+}$ $37/400^{o}$

To analyze at what phase of NK lysis monoclonal 17.2 inhibited, we used a ^{51}Cr release assay and a single cell assay with K562 as target cells. By adding 17.2 either prior to forming conjugates (+) or following conjugation (o) in a single cell assay, we could establish that there was no inhibition of the initial effector-target interaction. Bind = No. lymphocytes binding targets/No. lymphocytes counted.

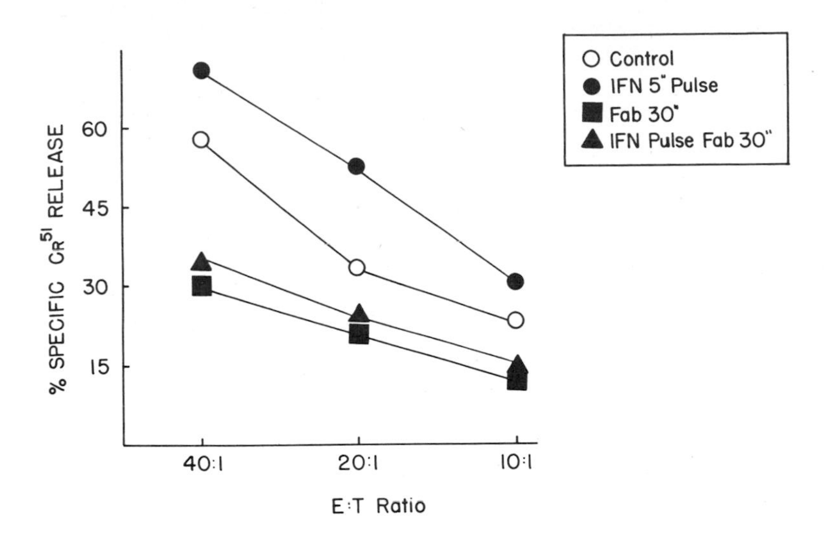

FIGURE 2. PBL were treated with 20 units of interferon for 5 minutes. The cells were then washed and followed by addition of RH_2 and then allowed to incubate at 37°C for two hours prior to testing in a ^{51}Cr release assay. These manipulations were done to demonstrate that inhibition of interferon activation was not due to the prevention of interferon binding to these cells.

TABLE II
Inhibition of IFN Activation by Pretreatment with 17.2

Group	E/T Ratio*		
	40	20	10
Control	31	25	18.
IFN	63	52	35
17.2**	37	27	11
17.2 + IFN+	41	40	16

*^{51}Cr E/T Ratios

**PBL pretreated with monoclonal antibody 17.2 for 30 min at 37°C and washed x2 with PBS

+17.2 treated cells exposed to 50 units of IFN for 1 hour at 37°C

III. ENHANCEMENT OF NK CELL'S LYTIC CAPACITY

A second mechanism by which interferon activates NK cells is to enhance each individual cell's lytic capacity. A population of NK cells kill in a random fashion over three hours with all lytic events being completed by three to four hours (Targan and Dorey, 1980; Targan, et al., 1980). After interferon treatment, these cells now kill by ten to fifteen minutes of the assay as if all lytic machinery had been enhanced so that they kill in a synchronous fashion, i.e., more rapidly. This observation at the single cell level probably correlates well with interferon enhancement of the NK cells ability to recycle and kill multiple targets (Kendall and Targan, 1980; Targan, et al., 1981).

These particular phenomena have been demonstrated in two systems using a combination of a chromium release and a single cell cytotoxic assay. By using two separate *in vitro* modulators, interferon and prostaglandin E_2, or using a combination of *in vivo* activations using moderate exercise and subsequent *in vitro* activation with interferon, the mechanism of enhanced lytic efficiency can be studied. By measuring the frequency of NK cells by using a single cell cytotoxic assay, we have demonstrated that the amount of NK cytotoxic enhancement measured by a chromium release assay produced by IFN alone or a combination of activators is not explained completely by the recruitment of new cytotoxic cells into the lytic population. If one calculates the number of lytic events that would occur if only one effector cell could kill one target it does not explain the data adequately. Thus, with interferon alone there is an enhancement of the number of target cells that are killed per active killer cells, i.e., there is enhancement of each cells lytic capacity. Furthermore, by using EDTA to inhibit subsequent lysis and comparing this to the normal kinetics of chromium release, it is clear that more lytic events occur in the interferon activated population than in the endogenous NK population. Moreover, when two different modulators sequentially activate an NK population, the further enhanced lysis seen is not due to recruitment of different subsets of pre-NK cells but due to the further enhancement of the lytic capabilities of the same population of NK cells.

There is also a temporal relationship to whether this additive activation occurs. If prostaglandin is added after completion of full expression of the interferon induced activated lytic process, there is enhancement of the number of lytic target interactions that each NK cell can perform. Since there is a different temporal relationship between this enhancement of recycling and the recruitment of pre-NK cells to become lytically active, this suggests that the sequential interferon-prostaglandin enhancement of recycling may be generated by different cellular mechanisms than the initiation of the lytic potential produced by each agent alone.

In summary, the enhancement of the ability of NK cells to kill more than one target may represent a "loaded" NK cell. Once the NK cell is killed it is capable of immediately binding and killing another target. In contrast, it appears that in the endogenous state there is a relative refractory period before the second event could occur. It is not clear what the fate of the lytic unit of the NK cell is following initiation of lysis, i.e. are they cleaved and transferred from the membrane or do they remain intact. If by addition of interferon or other modulators to culture of NK cells more lytic units are made available or their stability to be replaced in the NK membrane are enhanced, then the already active cells become more efficient killer cells. Thus, more lytic machinery would be available after an individual lytic event for the cell to rapidly proceed on to a second lytic event.

IV. CONCLUSION

In conclusion, there are multiple facets of the NK cell lytic mechanism that can be studied by using these activation models. What is clear is that it is important not only to investigate the components of single hit, nonaugmented NK lysis but also what is involved in enhancing an individual effector cells lytic efficiency, i.e. the rate of the single target lysis and the number of targets lysed per NK effector cell. Therefore, equal amounts of chromium release in two populations of NK cells, as measured by the total lytic units per lymphocyte population, may very well be measuring activity of individual effector cells with vastly different lytic capabilities. The specific cellular and molecular components responsible for these alterations in the cytotoxic functioning of the NK cells should be actively pursued in future studies.

ACKNOWLEDGMENTS

The author wishes to thank Ms. Linda Rodman for the typing of this manuscript.

REFERENCES

Hiserodt, J., Britvan, L., and Targan, S. (1981). Submitted for publication.
Kendall, R., and Targan, S. (1980). J. Immunol. 125:2770.
Silva, A., Bonavida, B., and Targan, S. (1980). J. Immunol. 125:479.
Targan, S., and Dorey, F. (1980). J. Immunol. 124:2157.
Targan, S., Grimm, E., and Bonavida, B. (1980). J. Clin. Lab. Immunol. 4:165.
Targan, S., Britvan, L., and Dorey, F. (1981). J. Clin. Exp. Immunol. 45:352.
Targan, S. (1981). J. Immunol. 127:1424.

AUGMENTED BINDING OF TUMOR CELLS BY ACTIVATED MURINE MACROPHAGES AND ITS RELEVANCE TO TUMOR CYTOTOXICITY

Scott D. Somers and Dolph O. Adams

Department of Pathology and Microbiology-Immunology
Duke University Medical Center
Durham, N.C.

Macrophages (MØ) activated by a variety of infectious agents are able to destroy tumor cells efficiently in vitro (for reviews, see Adams and Marino, 1982 or Evans and Alexander, 1976). Macrophage-mediated cytotoxicity is target selective, as neoplastically transformed cells are lysed in preference to their non-transformed counterparts (Adams and Marino, 1982; Evans and Alexander, 1976). Most workers have found the cytolysis by activated MØ to be contact-dependent (Adams and Marino, 1982; Evans and Alexander, 1976). Both microscopic and cine-microscopic observations of activated MØ-tumor cell cultures have further revealed clustering of activated MØ about tumor cells that are lysed (Meltzer, Tucker and Breuer, 1975; Steward, Adles and Hibbs, 1975). After vigorous washing of such cultures, the tumor cells remain attached to the MØ, so the two can be considered physically bound to one another (Piessens, 1978; Marino and Adams, 1980a). The binding between activated MØ and tumor cells currently appears to be an important part of the recognition system which MØ employ to discriminate neoplastic and non-neoplastic cells for lysis (Marino, Whisnant and Adams, 1981). This chapter reviews work in this laboratory over the past several years directed at elucidating the biology of the interaction between activated murine MØ and neoplastic target cells.

Augmented Versus Low-Level Binding

The extent of MØ-target binding can be quantified. Piessens initially measured this interaction by adding nonadherent tumor cells labelled with tritiated thymidine to monolayers of guinea pig MØ activated in vitro (Piessens, 1978). He found increased binding of tumor cells to activated as opposed to control MØ (Piessens, 1978). We subsequently confirmed and extended these observations by examining the interaction between murine MØ activated in vivo and various types of nonadherent targets labelled with $Na^{51}CrO_4$ (Marino and Adams, 1980a).

ISBN 0-12-341360-5

Macrophages activated for cytolysis by bacillus Calmette Guerin (BCG) demonstrate a high level of binding: 30-60% of tumor cells added are bound to the activated MØ (Marino and Adams, 1980a). This augmented binding, which is saturable by addition of excess numbers of targets, is highly selective for the cell combination of activated MØ and neoplastic targets (Marino and Adams, 1980a). It is to be contrasted with a low level of binding, observed between numerous cell pairs, such as activated MØ plus lymphocytes; inflammatory MØ plus tumor cells or lymphocytes; and adherent tumor cells or fibroblasts plus lymphocytes or nonadherent tumor cells (Marino and Adams, 1980a). The low level binding, where 4-12 percent of added targets are bound, is not saturable (Marino and Adams, 1980a). The high level or augmented binding, though not the low level binding, can be competitively inhibited by addition of excess unlabelled targets (Marino and Adams, 1980a). Once established, the augmented binding can be disrupted by addition of large numbers of unlabelled targets (Somers and Adams, 1982).

Recent evidence indicates other qualitative differences between augmented and low-level binding. First, augmented binding is dependent upon divalent cations such as Ca^{++} or Mg^{++}, while the low level binding is not (Somers and Adams, 1982). Second, augmented binding is dependent upon trypsin-sensitive surface structures on the activated MØ, while the low level binding is not (Somers and Adams, 1982). Third, the two differ morphologically. Tumor cells bind in clusters of 3-5 to the central portion of activated MØ, while single tumor cells bind to the periphery of inflammatory MØ (Somers and Adams, 1982). Fourth, the augmented binding has a functional consequence, while the low level binding does not. The binding of tumor cells or of plasma membranes from tumor cells to activated MØ initiates secretion of a cytolytic protease (Johnson, Whisnant and Adams, 1981). Thus, the augmented binding of neoplastic cells to activated MØ has several characteristics of specific (i.e., high-affinity, low capacity) binding of ligands to receptors, while the binding of tumor cells to inflammatory MØ has many characteristics of nonspecific (i.e., low-affinity, high capacity) binding observed between ligands and cells or plastic (Hollenberg and Cuatracasas, 1979).

Specific binding of soluble ligands is classically distinguished by quantifying the binding of radioactively labelled ligand in the presence of a large (50:1) excess of unlabelled ligand (Hollenberg and Cuatracasas, 1979). In this circumstance, the large excess of unlabelled ligand competitively inhibits almost all binding of labelled ligand to receptors, so that any residual binding of labelled ligand is not to receptors and, hence, nonspecific. Employing ^{111}In-oxine labelled tumor cells which have a high specific activity (2-4 cpm/cell), we have been able to perform similar studies (Somers and Adams, 1982). Binding of ^{111}In-oxine labelled tumor targets is measured in the presence of a large excess of unlabelled targets (250 to 1) and is taken to be an estimate of nonspecific binding. Given that the binding agent in these studies does not follow the laws of mass action, we have found the great predominance of augmented binding to be specific. By contrast, the low level binding to inflammatory MØs is almost totally nonspecific (Somers and Adams, 1982).

Cell Biology of Augmented Binding

Methyl-transfer reactions appear to be necessary for completion of augmented binding (Adams, Pike and Snyderman, 1981). Transmethylation is vital for numerous cellular functions, including membrane fluidity and capping of membrane structures (for review, see Adams, Pike and Snyderman, 1981). The observation that inhibitors of methylation inhibits augmented binding raises the possibility that the augmented binding of tumor cells by activated MØ requires membrane fluidity. This possibility is now being tested more directly. Preliminary evidence indicates a series of aliphatic alcohols, which we have established to enhance the fluidity of MØ-cell membranes, can increase the binding of tumor cells by activated macrophages (Adams, Somers and Snyderman, 1982). This is of interest, because we have previously observed that augmented binding results in a close and sinuous apposition of MØ and tumor cell plasma membranes (Marino and Adams, 1980).

The structure(s) responsible for binding of neoplastic targets to activated MØ appear to be contained within plasma membranes of tumor cells (Marino, Whisnant and Adams, 1981). Plasma membrane preparations from three types of tumor cells (a lymphoma, a leukemia, and a mastocytoma) bind to BCG-activated MØ and effectively inhibit subsequent binding of any of these three neoplastic cells (Marino, Whisnant and Adams, 1981). Equal amounts of membranes from lymphocytes, however, do not inhibit augmented binding. Of particular interest, membrane preparations from tumor cells do not inhibit the low level binding to inflammatory MØ (Marino, Whisnant and Adams, 1981). These observations are important, because they suggest structures contained within and common to the plasma membranes of 3 different tumor cells are responsible for binding of these cells to activated MØ.

Once established, the binding of tumor cells to activated MØ is physically quite strong. Employing a novel system devised by McClay et al. (McClay, Wessel and Marchase, 1981), we have measured the forces required to disrupt MØ-tumor cell interactions (Adams, Somers and Whisnant, 1982). An initial weak interaction, requiring less than 10^{-5} dynes to disrupt, occurs at 4°C between all cell pairs studied to date (Adams, Somers and Whisnant, 1982). If such cultures are raised to 37°C, a firmer binding is established, requiring greater than 2×10^{-4} dynes to disrupt, but only between the combination of activated

MØ and tumor cells (Adams, Somers and Whisnant, 1982). This implies that MØ metabolism must be active to convert weak binding to strong. We have found microtubules and microfilaments necessary for augmented binding (Somers and Adams, 1982).

Augmented Binding and Cytolysis

Several lines of evidence implicate augmented binding as an integral part of MØ-mediated tumor cytotoxicity. First, the selectivity of binding very closely resembles the selectivity of cytolysis, in regard to both type of target and type of MØ (Adams and Marino, 1980a). Specifically, maximum binding is observed between tumor cells and activated MØ. Only low level binding is seen between activated MØ and lymphocytes or between inflammatory MØ and tumor cells or non-neoplastic cells (Marino and Adams, 1980; Somers and Adams, 1982). Second, various experimental manipulations which increase or decrease the extent of binding respectively increase or decrease the subsequent extent of cytolysis and do so proportionately (Marino and Adams, 1980b). Third, competitive inhibition of binding with either unlabelled tumor cells or with membrane preparations from tumor cells leads to proportionate decreases in subsequent cytolysis of the targets (Marino and Adams, 1980b; Marino, Whisnant and Adams, 1981). Fourth, BCG-elicited MØ from A/J mice are deficient in their ability to complete augmented binding (Adams, Marino and Meltzer, 1981). These MØ cannot effect MØ-mediated tumor cytotoxicity, though they secrete copious amounts of a lytic proteinase (Adams, Marino and Meltzer, 1981). Fifth, interposition of a porous filter between activated MØ and tumor cells completely inhibits binding and completely inhibits cytolysis as well (Adams and Marino, 1981). Taken together, these data strongly suggests that binding is necessary for completion of cytolysis.

Binding, however, is not sufficient for cytolysis. BCG-elicited MØ from C3H/HeJ mice are fully competent to complete augmented binding but are incapable of mediating cytolysis (Adams, Marino and Meltzer, 1981). In addition, MØ which have been primed in a variety of ways can bind neoplastic targets extensively but cannot lyse them (Adams and Marino, 1981). Binding thus appears to be an initial necessary but insufficient step toward completion of MØ-mediated cytolysis (for reviews of the stages of lysis, see Adams and Marino, 1982 or Marino and Adams, 1981).

Induction of Capacity for Augmented Binding

The capacity to complete augmented target-cell binding is not observed in either resident peritoneal MØ or in inflammatory MØ elicited by a wide variety of phlogistic stimuli (Marino and Adams, 1980; Somers and Adams, 1982). Development of this capacity occurs in MØ, which are primed in a variety of ways in vivo (see chapter by Adams and Dean, this volume). We have recently found that MØ responsive to lymphokines complete only low-level binding of tumor cells (Marino and Adams, 1982). When these MØ are exposed to lymphokines for 6-8 hours, they develop the capacity for completing augmented binding (Marino and Adams, 1982). The interaction of lymphokines in inducing binding capacity have recently been studied in some detail (Marino and Adams, 1982). To summarize these studies briefly, the kinetics and characteristics for induction of binding closely resemble those for induction of priming for cytolysis by macrophage activating factor (MAF).

CONCLUSIONS

Macrophages activated in a variety of ways in vivo and in vitro selectively bind neoplastic targets to an augmented degree. The augmented binding, which is inhibited by plasma membranes of tumor cells, is saturable, competitively inhibitable, dependent upon trypsin-sensitive surface components and has two functional consequences: cytolysis of the bound targets and secretion of a lytic proteinase. The augmented binding thus has many characteristics of the specific interaction between ligand and receptor. By contrast, the low-level binding seen between numerous cell pairs in culture has many characteristics of nonspecific binding. Current evidence thus suggests that the capacity for augmented binding is mediated by receptors on the macrophages for structures contained within the plasma membranes of neoplastic cells.

The cell biology of augmented binding is beginning to emerge. Morphologic studies have shown that the two cells become closely apposed to one another and that the MØ become molded to the surface of the target. This development appears to coincide with development of a vigorous physical attachment between the two cell types. Development of the augmented binding appears to require membrane fluidity on the part of the MØ, a hypothesis consistent with the cell-to-cell molding described above. Activated MØ thus appear to be able to interact with tumor cells to develop vigorous bonding between the two cells. Other MØ do not appear to be able to complete this transition and thus do not appear to be able to retain the tumor cells, which are initially attached weakly to their surface. The specific MØ properties that permit this firm attachment are currently under investigation.

The capacity for completion of augmented binding develops as MØ become activated. Several lines of evidence indicate that this occurs when MØ become primed. Recent evidence further indicates that lymphokines, and lymphokines alone, can induce in vitro the capacity for augmented binding.

The augmented binding of neoplastic tumor cells to activated MØ appears to be a necessary step for MØ to lyse tumor cells. Augmented binding thus appears to be an important part of the recognition system which activated MØ employ to discriminate between neoplastic and non-neoplastic cells for subsequent cytolysis.

ACKNOWLEDGEMENTS

The studies from this laboratory described here are supported by USPHS Grants CA14236, CA16184, CA29589, and ES07031.

REFERENCES

1. Adams, D.O., Johnson, W.J. and Marino, P.A. 1982. Fed. Proc. (In Press).
2. Adams, D.O. and Marino, P.A. 1981. J. Immunol. 126:981.
3. Adams, D.O. and Marino, P.A. 1982. Cont. Topics Hematology/Oncology - III (In preparation).
4. Adams, D.O., Marino, P.A. and Meltzer, M.S. 1981. J. Immunol. 126:1843.
5. Adams, D.O., Pike, M.C. and Snyderman, R. 1981. J. Immunol. 127:8225.
6. Adams, D.O., Somers, S.D. and Snyderman, R. 1982. (Manuscript in preparation).
7. Adams, D.O., Somers, S.D. and Whisnant, C. 1982. (Manuscript in preparation).
8. Evans, R. and Alexander, P. 1976. p. 536. In Immunobiology of the Macrophage. D.S. Nelson, ed. Academic Press, New York.
9. Hollenberg, M.D. and Cuatrecasas, P. 1979. In The Receptors, Vol. 1, General Principles and Procedures, R. O'Brien, ed. New York, Plenum Publishing Corp.
10. Johnson, W.J., Whisnant, C. and Adams, D.O. 1981. J. Immunol. 127:1787.
11. Marino, P.A. and Adams, D.O. 1980a. Cell. Immunol. 54:11.
12. Marino, P.A. and Adams, D.O. 1980b. Cell. Immunol. 54:26.
13. Marino, P.A. and Adams, D.O. 1982. (Submitted for publication).
14. Marino, P.A., Whisnant, C. and Adams, D.O. 1981. J. Exp. Med. 154:77.
15. McClay, D.R., Wessel, G.M. and Marchase, M.R. 1981. PNAS, USA 78:4975.
16. Meltzer, M.S., Tucker, R.W. and Breuer, A.C. 1975. Cell. Immunol. 17:30.
17. Piessens, W.F. 1978. Cell. Immunol. 35:303.

18. Stewart, C., Adles, C. and Hibbs, J.B. 1975. In, The RES in Health and Disease. H. Friedman, M. Escobar and S. Reichard, eds. Plenum Press, New York.
19. Somers, S.D. and Adams, D.O. 1982. (Manuscript in preparation).

THE ROLE OF FREE OXYGEN RADICALS IN THE ACTIVATION OF THE NK CYTOLYTIC PATHWAY[1]

Stephen L. Helfand*+
Jerome Werkmeister*
John C. Roder*

*The Dept. of Microbiology and Immunology
Queen's University
Kingston
Canada

+The Dept. of Neurology
Harvard Medical School
Massachusetts General Hospital
Boston, Massachusetts

INTRODUCTION

The mechanism by which the NK cell kills its target cell is poorly understood. Previous work has shown that the cytolytic pathway can be divided into 4 discrete stages: (i) recognition and binding to the target; (ii) triggering and activation of the lytic mechanisms; (iii) delivery of the lethal hit; and (iv) target cell death. The tumor cell surface glycoproteins involved in stage 1 have been isolated (1,2) and cyclic nucleotides have been implicated in stage 2 (3) as well as interferon (4) and perhaps phospholipase A (5). NK-resistant tumor cell variants have been made which express NK target structures but lack putative receptors for an NK toxic moiety delivered in stage 3 (6). MHC-linked genes may regulate stage 1 (7) and non-MHC genes (beige) have been

[1]This work was supported by the MRC and NCI of Canada and a Queen's University development grant.

ISBN 0-12-341360-5

described which block stage 3 but not stage 1 (8). Recent work has elucidated the role of oxygen radicals in early activation events, stage 2, in the NK cytolytic pathway.

Oxygen Radicals in Other Systems

Granulocytes and monocytes are known to generate oxygen radicals ($O_2^{\cdot -}$, H_2O_2, $OH^{\cdot}$) when encountering and phagocytosing appropriate targets such as bacteria or parasites (9,10). These oxygen radicals are known to be an early step in activation of the killing mechanism and may be employed as toxic substances in the lethal hit itself. The nonphagocytic killing of tumor cells by macrophages also involves oxygen radicals derived from H_2O_2 (11). Since the killing of target cells by the NK cell is also nonphagocytic in nature and thus reminiscent of macrophage killing, we investigated the possible role of oxygen radicals in NK cytolysis and have found they are an essential element in the activation of the cytolytic pathway.

NK Cells Produce Oxygen Radicals

Large granular, HNK-1^+, lymphocytes were isolated from peripheral blood and mixed with K562 tumor cells. Within seconds of the NK-target cell interaction, there was a burst of oxygen radicals generated (Fig. 1). These highly reactive molecular species were easily detected in the presence of luminol or other readily oxidized substrates by the emission of light. Oxygen radicals interact with luminol which transduces their energy into photons that are readily measured in an ordinary scintillation counter in the out-of-coincidence mode.

The fact that oxygen radicals were induced by glutaraldehyde-fixed target cells and cell-free, plasma membrane vesicles made from NK-sensitive but not NK-insensitive targets, shows that it is the effector cells rather than the targets that are responsible for the generation of these oxygen radicals (12). Extensive target panel studies have shown a direct correlation ($r > 0.93$) between target-effector conjugate formation, induction of oxygen radical production in NK cells and susceptibility to NK-mediated cytolysis. Highly sensitive NK target cells (human fetal fibroblasts, K562) induce a large amount of oxygen radicals whereas insensitive NK targets (P815, L5178Y) induce very little. In addition, glutaraldehyde-fixed K562 cells that differ in the degree of preservation of the NK target structure, as determined by ability to compete with unfixed K562 cells in a cold target

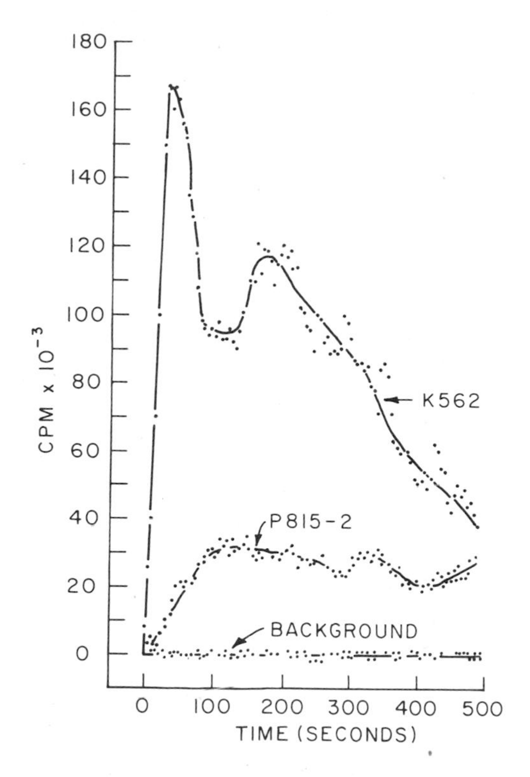

FIGURE 1. Chemiluminescence follows the NK-target cell interaction. Monocyte-depleted lymphocytes were separated on Percoll density gradients as described previously (37). Fraction 2 cells were enriched 10-fold in NK cytolytic activity and consisted of 60% large granular lymphocytes, 55% $HNK-1^+$ cells and <0.1% monocytes or granulocytes as assessed by immunofluorescent staining with Mo2 monoclonal antibody (monocyte specific) and Geimsa cytocentrifuge smears (1000 cells counted). 10^6 Percoll fr.2 cells were dark-adapted in 10% luminol saturated FCS and 10^7 tumor cells or medium alone was added and mixed. Chemiluminescence was measured at ambient temperature (20°) in a liquid scintillation counter in the out-of-coincidence mode. Values represent CPM above background measured at 5 sec intervals. Lymphocytes depleted of $HNK-1^+$ cells did not generate chemiluminescence.

inhibition assay, show a strong correlation ($r > 0.98$) between the ability to compete for the NK receptor and the ability to trigger oxygen radical production. Finally, several differentiated K562 clones which were selectively resistant to NK-mediated cytolysis also induced less oxygen radicals (13). These data suggest that it is the target structure binding to the NK receptor itself that may be the trigger for oxygen radical production.

Oxygen Radicals Are a Necessary Step in the NK Cytolytic Pathway

We then asked whether oxygen radicals were merely an epiphenomena or if they were involved directly in the NK cytolytic pathway. Specific enzymes and scavengers were used to selectively remove free radicals from the system.

Superoxide dismutase (SOD) is an enzyme whose apparent sole function is to scavenge free superoxide radicals, $O_2^{\cdot -}$, by the dismuation reaction $O_2^{\cdot -} + O_2^{\cdot -} + 2H^+ \longrightarrow O_2 + H_2O_2$ (14,15) whereas mannitol (16), benzoate (16,17) or DMSO (18) are $OH^{\cdot -}$ scavengers. All of these agents cause a marked decrease in the chemiluminescence resulting from the NK-target cell interaction. Catalase (19), an enzyme which specifically degrades H_2O_2 to $OH^- + H^+$, has a lesser effect on the chemiluminescence. These results indicate that all three molecular species ($O_2^{\cdot -}$, H_2O_2, $OH^{\cdot -}$) are generated during NK activation.

The addition of SOD, mannitol, benzoate or DMSO to the NK cytolytic assay markedly inhibited cytolysis (80%) in a dose-dependent reversible reaction. Catalase, on the other hand, had no effect on NK-mediated cytolysis. Further analysis has shown that none of these agents, with the possible exception of DMSO, interfered with the ability of the NK cell to recognize and bind to its target cell. This data suggests that the highly reactive molecular species $O_2^{\cdot -}$ and $OH^{\cdot -}$ are necessary for cytolysis to proceed. It is also interesting to note that this is the converse of what is known to occur in monocyte-mediated cytolysis where H_2O_2 has been shown to be important (11). In our hands, catalase but not SOD, mannitol, benzoate nor DMSO inhibited cytolysis of tumor cells by human monocytes.

Oxygen Radical Production is Dependent on the State of NK Activation

Since oxygen radicals are essential in NK cytolysis, it was not surprising to find that the state of NK activation and the generation of oxygen radicals are closely related. Interferon pre-treatment of NK cells caused a concomitant increase in oxygen radical production and cytolysis. In addition, the degree of augmentation in both oxygen radical production and cytolysis by interferon-boosted effectors versus non-boosted effectors correlated directly (12).

An examination of low-NK responsiveness in humans has shown that individuals with this stable, selective defect have normal frequencies of HNK-1$^+$ NK cells which bind normally to the tumor targets and are boosted normally by interferon, but fail to become activated as judged by a 60% decrease in the burst of oxygen radicals that are normally generated following target-effector interaction (20). Preliminary data suggest that these low NK responders are also more sensitive to SOD-induced inhibition of their NK-mediated cytolysis than normals.

Possible Mechanisms of Oxygen Radical Generation

Since NK recognition and binding to target determinants induces the NK cell to generate oxygen radicals, we also investigated the possible mechanisms of the trigger event. Agents that effect the NK cell's ability to generate oxygen radicals and kill tumor cells include azide and chloroquine. Sodium azide inhibits oxidative metabolism completely and consequently blocks the generation of oxygen radicals and NK-mediated cytolysis. Chloroquine, on the other hand, is a selective inhibitor of lysosomes and also leads to a parallel decrease in oxygen radical production and NK-mediated cytolysis as summarized in Table I. Neither of these agents inhibits target-effector binding. Therefore, both oxidative metabolism and normal lysosome function are necessary for stage 2 and 3 in the cytolytic pathway as postulated previously (21).

In leukocytes, the oxidase enzyme which generates these oxygen radicals appears to be associated with the plasma membrane (22,23). In these cells, an appropriate perturbation of their plasma membrane leads to activation of the oxygen-radical producing system. The depolarization of the membrane and resultant Ca^{++} influx caused by the perturbation are felt to be responsible for this activation (24,25,26). Therefore, we studied the effect of inducing a calcium ion influx on the NK cell. It was found that when a calcium ionophore, A 23178, was added to the NK cells in the presence of Ca^{++}, there was a production of oxygen radicals (Fig. 2). This presumed triggering of the oxidase enzyme system in the NK cell is accomplished without the need for the NK receptor-target structure interaction and suggests that with the NK cell, as with other leukocytes, the appropriate perturbation of their plasma membrane, (as would occur following NK-target cell binding or interferon-interferon receptor binding) leads to (i) depolari-

TABLE I. The Effect of Oxygen Radical Modifiers on Chemiluminescence and NK-mediated cytolysis.

Agents	Chemiluminescence	NK Cytolysis
IFN (200 U/ml)	↑↑↑	↑↑↑
SOD (500 U/ml)	↓↓↓	↓↓↓
Catalase (500 U/ml)	↓	-
Mannitol (50 mM)	↓↓↓	↓↓↓
Benzoate (100 mM)	↓↓↓	↓↓↓
DMSO (100 mM)	↓↓↓	↓↓↓
Azide (1 mM)	↓↓↓↓	↓↓↓↓
Chloroquine (0.1 mM)	↓↓	↓↓
A 23178 (1 μg/ml; 1.5 μM)	↑↑	-*
Low NK donors	↓↓	↓↓↓

The agents shown were added to Percoll fractionated, NK-enriched lymphocytes in a 10 min chemiluminescence assay or a 4-hr ^{51}Cr release assay against K562 targets. Each arrow represents approximately 25% enhancement or suppression from control values. Dashes (-) indicate no effect. IFN, interferon; SOD, superoxide dismutase; DMSO, dimethylsulfoxide; A 23178, calcium ionophone; 4 low NK donors with stable activity were selected from 600 normal donors screened.

*The inophore had no effect on normal NK cells but could reverse the depression caused by fluphenazine.

zation of the membrane; (ii) calcium ion influx; (iii) activation of the oxidase enzyme system; and (iv) the subsequent generation of oxygen radicals.

The data suggest that the generation of oxygen radicals is an early event in the triggering and activation of the lytic mechanism, stage 2 (12,20). A further piece of evidence to support this hypothesis comes from studies of patients with Chediak-Higashi disease who have normal frequencies of $HNK\text{-}1^+$ cells (27) which do not function (28). The block in NK cytolysis appears to be in stage 3 since binding to the target (stage 1) is normal and is followed by normal triggering and production of oxygen radicals (stage 2) thereby indicating that the oxygen radicals normally appear prior to the step in the cytolytic pathway which is blocked in the Chediak-Higashi syndrome (unpublished observation).

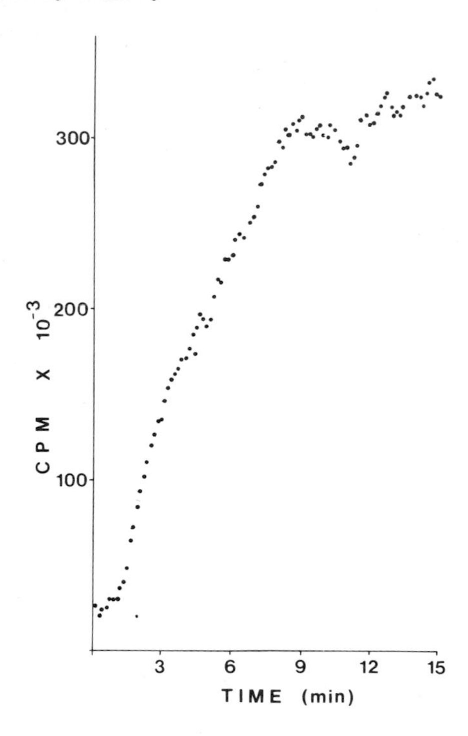

FIGURE 2. A calcium ionophore induces oxygen radical production in NK-enriched lymphocytes. NK cells were prepared as described in Fig. 1 and 10^{-6} M calcium ionophore A 23178 was added at time 0. Values represent mean CPM above background.

Oxygen radicals are known to be intimately involved in the biosynthesis of several important compounds by triggering and participating in both the lipooxygenase and cyclooxygenase pathway of arachidonic acid oxidation (29,30). These two independent pathways which generate such potent biological compounds as 5 HETE, leukotrienes, hydroxy fatty acids, prostaglandins, prostacyclins and thromboxanes have already been implicated in the NK cytolytic pathway. Steroids (31) and phospholipase inhibitors (5) inhibit NK activity and also block the production of arachidonic acid by the phospholipases. ASA (aspirin) and NSAID's block the cyclooxygenase step and are known to inhibit NK (32). cAMP, theophylline, and PGE also inhibit NK function (3) and are known to inhibit phospholipase A (33) and may inhibit prostaglandin synthesis later in the pathway (34).

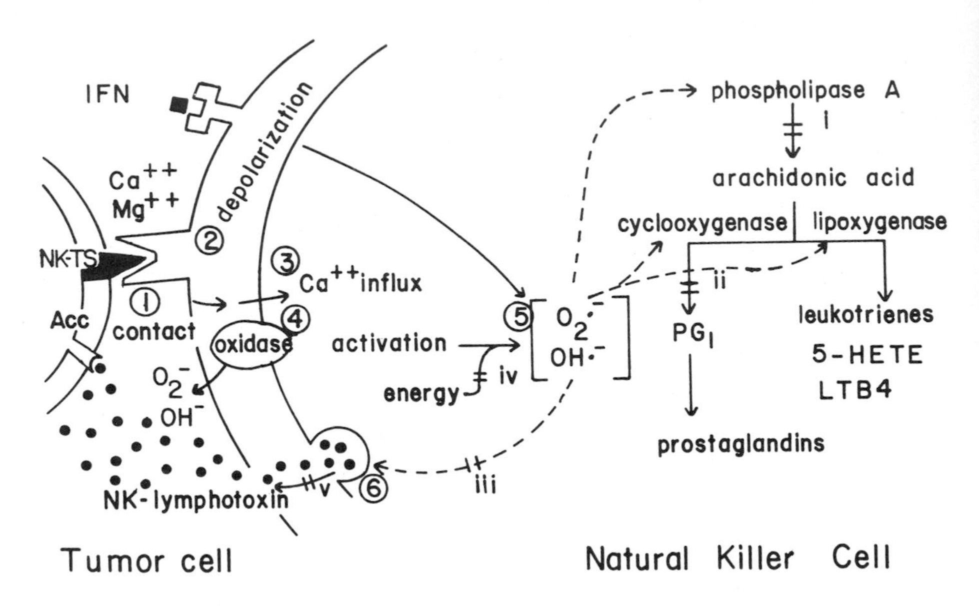

FIGURE 3. A hypothetical scheme of events during NK-target cell interaction. Evidence suggests that a recognition structure on the NK cell recognizes and binds to a glycoprotein target structure (NK-TS) or the tumor cell in a divalent cation-dependent process (1,2,21). Based on analogy with other leukocyte systems studied, we then postulate that membrane depolarization ensues, followed by a calcium influx and activation of the oxidase enzyme system which generates the oxygen radicals we detect by chemiluminescence (Fig. 1). These oxygen radicals activate a cascade of other biochemical pathways and lead to the release of a hypothetical NK lymphotoxin which binds to acceptor sites (ACC) on the tumor cell membrane which were inferred from selectively NK-resistant tumor variants with normal NK-TS expression (6). Both NK-mediated cytolysis and biochemical pathways were blocked at the indicated points by (i) steroids, cAMP, theophylline and phospholipase inhibition (5); (ii) ASA (aspirin), NSAID; and (iii) superoxide dismutase, mannitol, benzoate but not catalase; (iv) sodium azide and (v) chloroquine.

SUMMARY

The earliest event in the NK cytolytic pathway (stage 1) involves recognition and binding to appropriate tumor cell targets. If NK cells are like many other cell types studied, then this target-effector interaction would be expected to lead to a perturbation of the cell membrane, depolarization of the membrane, a calcium ion influx and subsequent activation of the oxidase enzyme system to generate oxygen radicals which we can measure within seconds of NK-target cell contact. Some of these free radicals ($O_2^{\cdot -}$, $OH^{\cdot -}$) are necessary since their removal with specific enzymes (superoxide dismutase) or hydroxyl scavengers inhibits NK-mediated cytolysis. We feel that the NK-derived oxygen radicals activate other events in the NK cytolytic pathway and are not themselves involved in the lethal hit. We hypothesize that these oxygen radicals may activate phospholipases to produce arachidonic acid as suggested by others (35,36). Alternatively, arachidonic acid may be oxidized by the NK-derived oxygen radicals in the lipo-oxygenase or cyclooxygenase pathways. The result of this cascade of events is the arming of the NK cell and delivery of the lethal hit to the target cell, stage 3. Hence the oxygen radicals by themselves or by virtue of their oxidative progeny, may be important as bioregulators and as participants in the NK cytolytic pathway as schematized in Fig. 3.

REFERENCES

1. Roder, J.C., Rosen, A., Fenyo, E.M. and Troy, F.A. PNAS USA 76:1405 (1979).
2. Roder, J.C., Ahrlund-Richter, L. and Jondal, M. J. Exp. Med. 150:471 (1979).
3. Roder, J.C. and Klein, M. J. Immun. 123:2785 (1979).
4. Herberman, R.B., Ortaldo, J.R. and Bonnard, G. Nature 277:221 (1979).
5. Hoffman, T., Kirata, F., Bougnouk, P., Fraser, B.A., Goldfarb, R.H., Herberman, R.B. and Axelrod, J. PNAS USA 78:3839 (1981).
6. Roder, J.C., Beaumont, T.J., Kerbet, R.S., Haliotis, T., and Kozbor, D. PNAS USA 78:6396 (1981).
7. Roder, J.C. and Kiessling, R. Scand. J. Immunol. 8:135 (1978).
8. Roder, J.C. and Duwe, A.K. Nature 278:451 (1979).
9. Babior, B.B. N. Engl. J. Med. 298:659 (1978).
10. Weissmann, G., Smolen, J.E. and Kordiak, H.M. N. Engl. J. Med. 303:27 (1980).
11. Nathan, C.F., Silverstein, S.C., Brukner, L.H. and Cohn, Z.A. J. Exp. Med. 149:100 (1979).
12. Helfand, S., Werkmeister, J., Beaumont, T. and Roder, J. Submitted.
13. Werkmeister, J., Helfand, S., Haliotis, T., Pross, H. and Roder, J. Submitted.
14. McCord, J.M. and Fridovich, I. J. Biol. Chem. 244:6049 (1969).
15. McCord, J.M., Keele, Jr., B.B. and Fridovich, I. PNAS USA 68:1024 (1971).

16. Kellogg, E.W. and Fridovich, I. J. Biol. Chem. 250:8812 (1975).
17. Neta, P. and Dorfuran, L.M. Adv. Chem. Ser. 81:1:222 (1968).
18. Anabor, M. and Neta, P. Int. J. Appl. Radiat. Isot. 18:493 (1967).
19. Beers, R.F. and Sizer, I.W. J. Biol. Chem. 195:133 (1952).
20. Helfand, S., Werkmeister, J., Roder, J., Abo, T., Balch, C. and Pross, H. Submitted.
21. Roder, J.C., Argov, S., Klein, M., Petersson, C., Kiessling, R., Andersson, K. and Hanssen, M. Immunology 40:107 (1980).
22. Dewald, B., Baggiolini, M., Curnutte, J.T., Babior, B.B. J. Clin. Invest. 63:21 (1979).
23. Cohen, H.J., Chovaniec, M.E. and Davies, W.E. Blood 55: 355 (1980).
24. Lew, P.D., Stossel, T.P. J. Clin. Invest. 67:1 (1981).
25. Whitin, J.C., Chapman, C.E., Simons, E.R., Chovaniec, M.E. and Cohen, H.J. J. Biol. Chem. 255:1874 (1980).
26. Seligman, B.E., Grallin, J.I. J. Clin. Invest. 66:493 (1980).
27. Roder, J.C., Haliotis, T., Klein, M., Korec, S., Jett, J., Ortaldo, J., Herberman, R., Katz, P. and Fauci, A.S. Nature 284:553 (1980).
28. Abo, T., Roder, J.C., Abo, W., Cooper, M. and Balch, C. Submitted.
29. Lewis, R.A. and Austen, K.F. Nature 293:103 (1981).
30. Kuehl, Jr., F.A. and Egan, R.W. Science 210:978 (1980).
31. Hochman, P. and Cudkowicz, G. J. Immunol. 119:2013 (1977).
32. Grohman, P.H., Porzsolt, F., Quirt, I., Miller, R.G. and Phillips, R.A. Clin. Exp. Immunol. 44:611 (1981).
33. Gerrard, J.M., Peller, J.D., Krich, T.P. and White, J.G. Prostaglandin 14:39 (1977).
34. Lepetina, A.G., Schmitges, C.J., Chandrabose, K. and Cuatrecasus, P. Biochem. Biophys. Res. Commun. 76:828 (1977).
35. Del Maestro, R. Ph.D. thesis, Uppsala University, Sweden (1979).
36. Seligman, M.L. and Demopoulos, H.B. Annals N.Y. Acad. Sci. 222:640 (1973).
37. Timonen, T. and Saksela, E. J. Immunol. Methods 36:285 (1980).

CELL SURFACE THIOLS IN HUMAN NATURAL CELL-MEDIATED CYTOTOXICITY

Dorothy Hudig[1]
Doug Redelman[2]
Lory Minning[1]

Department of Medicine
University of California, San Diego
La Jolla, California

I. INTRODUCTION

The biochemical events of lymphocyte-mediated killing of nucleated cells are poorly understood. In this paper, we present data that indicate that a lymphocyte cell surface thiol is required for target binding and killing. Several types of reagents were used as inhibitors of cellular thiols. Although chloromethyl (CH_2Cl) ketone derivatives of amino acids specifically inactivate serine-dependent proteases with affinity for the amino acid component of the compound (1,2) by alkylating the histidine the active site of the enzyme, these chloromethyl ketones can also nonspecifically alkylate thiol residues (2). Substrate specificity is expected among the chloromethyl ketone derivatives of different amino acids when a specific protease is inactivated, but not when -SH groups of most proteins are alkylated. There is little information concerning the ability of these compounds to penetrate cells. Iodoacetamide (IA) does penetrate cells, and alkylates the -SH of many enzymes and also intracellular, reduced glutathione. In particular, IA can alkylate glyceraldehyde-3-phosphate dehydrogenase (3), thereby blocking oxidative phosphorylation as a cellular source of energy. The diazine compound diamide (4) penetrates cells and reversibly oxidizes sulfhydryl groups,

[1]Supported in part by NIH CA-28196.
[2]Supported in part by NIH CA-24450

ISBN 0-12-341360-5

including glutathione (5). The bromobimane reagent with a quartenary NH_2 group (qBBr, or Thiolyte Monoquat^R) couples covalently to -SH groups and fluoresces only in the coupled state (6,7). In addition, it does not penetrate either red cell (7) or lymphocyte membranes (8). Some of the NK experiments performed with these compounds are described in the following sections.

II. MATERIALS AND METHODS

A. NK Assay

Effector cells were normal human peripheral blood mononuclear cells isolated by the methods of Boyum (9). Cell death was assayed by release of ^{51}Cr from the K562 tumor cell line as described previously (10).

B. Conjugate Formation

Lymphocyte-tumor cell conjugates were formed by centrifuging 2:1 ratios of peripheral blood mononuclear cells and K562 cells for 3 minutes at 40g at room temperature. The resultant conjugates were kept on ice prior to gentle resuspension. Analysis was by direct counts with a hemocytometer. The large K562 cells could be readily distinguished from the lymphocytes.

C. Reagents

Tosyl lysine chloromethyl ketone ($TosLysCH_2Cl$) was obtained from Sigma Chemical Co. (St. Louis MO) and tosyl leucine chloromethyl ketone ($TosLeuCH_2Cl$) from Vega Biochemicals (Tucson AZ). The leucine derivative was first dissolved in 0.1 ml 100% ethanol prior to the addition of medium and diluted in medium containing an identical (0.15% or less) final concentration of alcohol. Control assays also contained identical concentrations of ethanol. Iodoacetamide (IA) was obtained from Aldrich Chemical Co. (Milwaukee WI). Diazenedicarboxylic acid bis(N,N-dimethylamide) or diamide was obtained from Calbiochem-Behring (La Jolla CA), as was the quartenary ammonium salt of bromobimane (qBBr), Thiolyte Monoquat^R. The qBBr solutions were carefully protected from light during use. All solutions were freshly made for each experiment and kept on ice prior to use.

III. RESULTS

Both $TosLysCH_2Cl$ and $TosLeuCH_2Cl$ were potent inhibitors of NK. Lymphocytes were pretreated for 15 minutes with these reagents, then washed several times with medium prior to assay (Fig. 1). Contrary to what would be expected if either chloromethyl ketone were reacting with a specific protease, both reagents were equally effective (Fig. 1) and inhibited NK by two-fold at $\sim 5 \times 10^{-5}$M (data not shown).

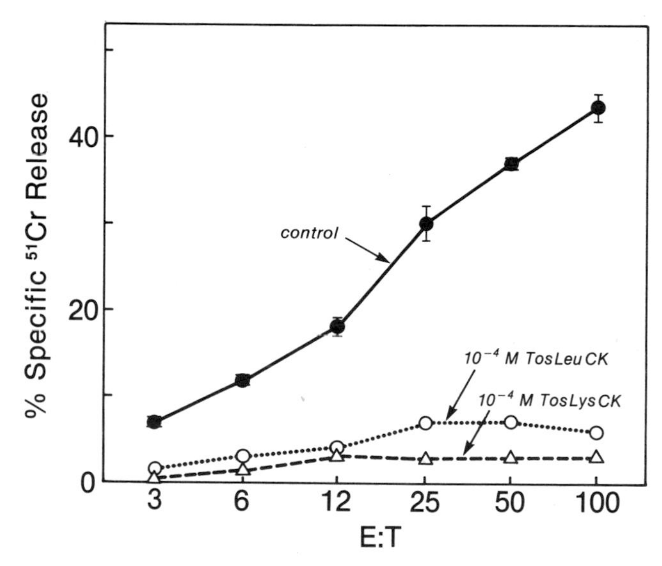

FIGURE 1. Both $TosLysCH_2Cl$ and $TosLeuCH_2Cl$ inhibit NK. Cells were pretreated for 15 minutes at 37°C and washed twice prior to assay.

An additional series of experiments also suggests that these chloromethyl ketones were inhibiting killing by alkylating -SH groups as opposed to inactivating proteases. It is known that $TosLysCH_2Cl$ inhibition of isolated proteases can be prevented by the inclusion of excess substrate or inhibitor for the protease (2). However, when a similar experimental design was employed using a combination of 4mM of the reversible inhibitor p-aminobenzamidine and 5×10^{-5}M $TosLysCH_2Cl$ for pretreatment of lymphocytes, the inhibition of NK was identical to that achieved by 5×10^{-5}M $TosLysCH_2Cl$ alone (data not shown). The 4mM p-aminobenzamidine can inhibit NK when in the assay, but not by pretreatment. Thus since it does not protect NK activity from pretreatment with $TosLysCH_2Cl$, it is unlikely that the $TosLysCH_2Cl$ is inactivating a protease with specificity for basic amino acid residues.

Our next experiments examined the effects of the -SH reactive agents IA and diamide. IA, when included in the assay, completely inhibited NK at $1x10^{-3}M$ (data not shown). Diamide also inhibited NK (Fig. 2). However, whereas IA-pretreated and washed lymphocytes were inhibited in subsequent assays, cells similarly pretreated with diamide were not irreversibly inactivated (data not shown). Furthermore, in experiments in the mouse CTL system, the effects of diamide varied directly with the final cell concentrations in the assays. This effect can be observed by the greater inhibition of NK at lower E:T ratios than at higher E:T ratios in which more cells were present (Fig. 2).

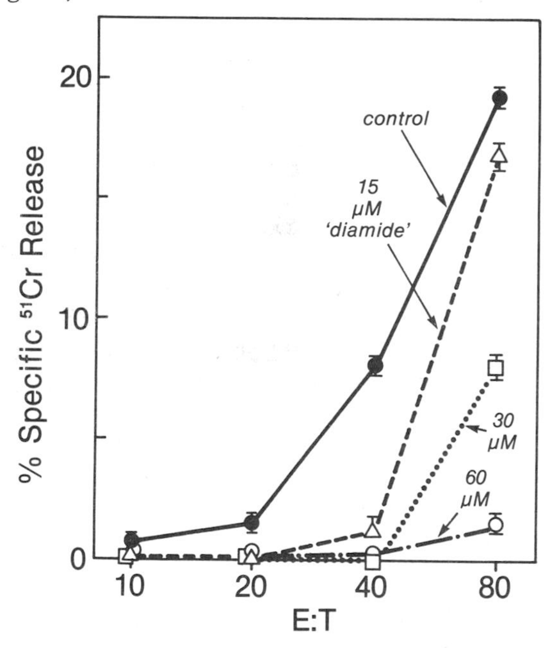

FIGURE 2. Diamide inhibition of NK. The concentrations indicated were included in a standard 4-hour assay.

The charged, nonpenetrating reagent qBBr, also inhibited NK (Fig. 3). The effects were irreversible; furthermore, the inclusion of qBBr throughout the assay was only marginally more effective than pretreatment (data not shown). This lack of difference may be due to the light sensitivity of qBBr. In addition, qBBr affects conjugate formation, although the effect is not as pronounced on conjugate formation as it is on NK activity; e.g., in Fig. 3, 1mM qBBr inhibited NK 8-fold and inhibited conjugation by only 2-fold. Iodoacetamide and

TosLysCH$_2$Cl also inhibited conjugate formation (data not shown). Conjugate experiments have not yet been performed with diamide. Thus it appears that reagents affecting thiols, unlike those amides and esters which can inhibit proteases, inhibit NK conjugation.

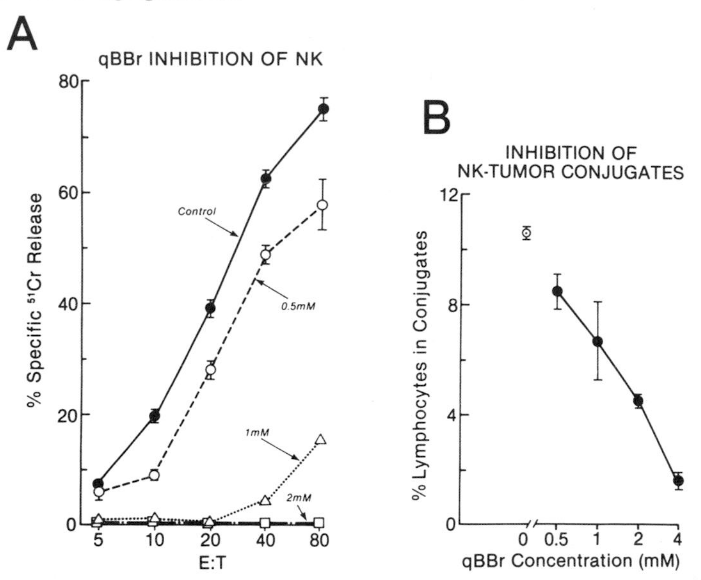

FIGURE 3. qBBr inhibition of NK. Lymphocytes were treated with qBBr for 15 minutes at 37°C in tubes protected from light, and then washed twice prior to assays for NK and conjugate formation. Duplicate conjugate determinations were made, and 3600 lymphocytes were scored per determination.

IV. DISCUSSION

The experiments presented suggest that a lymphocyte cell-surface thiol is required for NK lytic activity. If this sulfhydryl is oxidized or alkylated with a nonpenetrating reagent such as qBBr, both human natural killing and mouse T cell killing (11) are inhibited. Such plasma membrane thiols are extremely rare under oxidative extracellular conditions (5). Thus it is likely that a relatively small number of surface molecules could be involved. These molecules may not be uniquely associated with NK function. The only tenable conclusion is that at least one biochemical event required for NK is dependent upon a free cell-surface thiol. This -SH group could be required for a crucial enzymatic activity, or for essential disulfide cross linkages. Such cross linkages could contribute to either stable conjugate formation and/or release of the lytic substance. As NK proceeds in a matter of minutes at 37°C, does not require de novo protein synthesis (10), and is only partially reduced when oxidative phosphorylation is poisoned with NaN_3 (Hudig, personal observations), it would appear that only a narrow repertoire of events and cell functions are crucial to NK. The nature and role of lymphocyte plasma membrane sulfhydryl containing molecules is under further investigation by one of us (D.R.) who has generated antisera specific for qBBr.

V. ACKNOWLEDGMENTS

We gratefully acknowledge the helpful discussions with Dr. R. Fahey of the U.C. San Diego Chemistry Department, the illustrative services of Ms. Phyllis Stookey, and the manuscript preparation of Ms. Leslye Rucker.

V. REFERENCES

1. Schuellmann, C., and Shaw, E., Biochem. 2:252 (1963).
2. Shaw, E., Mares-Guia, M., and Cohen, W., Biochem. 4:2219 (1965).
3. Segal, H.L., and Boyer, P.D., J. Biol. Chem. 204:265 (1953).
4. Kosower, N.S., Kosower, E.M., and Wertheim, B., Biochem. Biophys. Res. Comm. 37:593 (1969).
5. Kosower, E.M., and Kosower, N.S., in "Glutathione: Metabolism and Function" (I.M. Arias and W.B. Jakoby, eds.), p. 139. Raven Press, New York, 1976.
6. Kosower, E.M., Pazhenchevsky, B., and Hershpowitz, E., J. Amer. Chem. Soc. 100:6516 (1978).
7. Kosower, N.S., Kosower, E.M., Newton, G.L., and Ranney, H.M., Proc. Natl. Acad. Sci.(USA) 76:3382 (1979).
8. Noelle, R.J., and Lawrence, D.A., Fed. Proc. 38:914 (1979).
9. Boyum, A., Scand. J. Clin. Lab. Invest. 21 suppl. 97:77 (1968).
10. Hudig, D., Djobadze, M.D., Redelman, D., and Mendelsohn, J., Cancer Res. 41:2803 (1981).
11. Redelman, D., and Hudig, D., J. Immunol. 124:870 (1980).

ACTIVATED MACROPHAGE MEDIATED CYTOTOXICITY FOR TRANSFORMED TARGET CELLS[1]

John B. Hibbs, Jr.
Donald L. Granger[2]

Veterans Administration Medical Center and
Department of Medicine, Division of Infectious Diseases
University of Utah School of Medicine
Salt Lake City, Utah

I. INTRODUCTION

Recent work from this laboratory has focused on metabolic changes that develop in transformed target cells after contact with activated macrophages. These studies demonstrate that activated macrophages induce major bioenergetic perturbations in transformed target cells that includes inhibition of mitochondrial respiration (1). There are two clearly defined transformed cell phenotypic responses to activated macrophage induced cytoxicity (2). The first, the lytic phenotype, progresses to lysis after activated macrophage induced inhibition of mitochondrial respiration (Granger, D.L. and Hibbs, J.B., Jr., unpublished). The second, the nonlytic phenotype, adapts metabolically to activated macrophage induced inhibition of mitochondrial respiration and eventually recovers normal mitochondrial respiratory capacity as well as normal proliferative capability (3, Granger, D.L. and Hibbs, J.B., Jr., unpublished).

---

[1] Supported by Veterans Administration, Washington, D.C. and American Cancer Society Grant CH-139.

[2] Present Address: Department of Physiological Chemistry, Johns Hopkins University, School of Medicine, Baltimore, MD.

ISBN 0-12-341360-5

II. EVIDENCE FOR TWO TRANSFORMED CELL PHENOTYPES TO ACTIVATED MACROPHAGE-INDUCED INHIBITION OF MITOCHONDRIAL RESPIRATION

Cytotoxic activated macrophage-induced inhibition of O_2 consumption has occurred in all transformed target cells we have tested (1, Granger, D.L. and Hibbs, J.B., Jr., unpublished). Inhibition of oxygen consumption occurred regardless of species, tissue of origin, or whether the transformation event was spontaneous, induced by radiation, by a chemical carcinogen, or by an oncogenic virus. There are at least two target cell responses to activated macrophage induced inhibition of mitochondrial respiration: stasis without progression to lysis (L1210 cell is an example of this phenotype) and stasis followed by lysis (murine P815 mastocytoma cell is an example of the second phenotype). When compared to control cells, endogenous respiration of both L1210 cells and P815

TABLE I. Endogenous respiration of L1210 and P815 cells that have been co-cultivated with cytotoxic activated macrophages.

Cell line	Time of co-cultivation with activated macrophages prior to respiration measurement (hours)[a]	Oxygen consumption (µl O_2 per hour per 10^6 cells)[a] endogenous
L1210	0	7.2 ± 0.5
	10	3.8 ± 0.3
	20	1.0 ± 0.2
P815	0	5.9 ± 0.3
	10	2.0 ± 0.5

[a] See reference 1 for experimental details.

(D.L. Granger and J.B. Hibbs, Jr., unpublished data.)

TABLE II. Aerobic and anaerobic glycolysis of L1210 and P815 cells that have been co-cultivated with cytotoxic activated macrophages.

Cell line	Culture environment prior to glycolysis measurement[a]	Lactate produced (μmoles per hour per 10^5 cells)[b]	
		aerobic	anaerobic
L1210	Medium alone	33 ± 5	91 ± 8
	Medium + activated macrophages	65 ± 6	65 ± 4
P815	Medium alone	17 ± 2	42 ± 4
	Medium + activated macrophages	54 ± 10	47 ± 10

[a] L1210 cells were cultivated alone or with activated macrophages for 20 hours and P815 cells with activated macrophages for 10 hours prior to removal. Removed target cells were washed three times with medium without serum prior to glycolysis incubation.

[b] See reference 1 for experimental details.

(D.L. Granger and J.B. Hibbs, Jr., unpublished data.)

cells was decreased after removal from monolayers of cytotoxic activated macrophages (Table I). Decreased O_2 consumption of L1210 and P815 cells was not due to cell death because there was always > 90% viability as determined by trypan blue exclusion when the measurements were completed. There are other similarities in the response of L1210 and P815 cells to cocultivation with cytotoxic activated macrophages. Proliferation of both L1210 cells and P815 cells was unaffected by normal or stimulated mouse peritoneal macrophages. In addition, cytotoxic activated macrophage-induced cytotoxicity for both L1210 cells and P815 cells required macrophage target cell contact. Cytotoxicity could not be reproduced by incubating L1210 cells or P815 cells in medium from activated macrophage cultures.

In spite of these similarities, the ultimate cytotoxic fate of L1210 cells and P815 cells is different. The difference between these two transformed cell phenotypes become apparent after an initial period of cytotoxic activated macrophage induced cytostasis and inhibition of mitochondrial respiration which occurs in both phenotypes. L1210 cells did not lyse during 72-96 hours of cocultivation with activated macrophages in culture medium with a non-limiting supply of glucose. The number of viable cells did not decrease and there was not an increase in trypan blue staining cells. In contast, by 30 hours of coculture with cytotoxic activated macrophages, 80-99% of P815 cells underwent lysis in culture medium supplemented with a non-limiting supply of glucose (Granger, D.L. and Hibbs, J.B., Jr., unpublished).

Glycolysis rates were determined for the nonlytic transformed cell phenotype (L1210 cells) and the lytic transformed cell phenotype (P815 cells) (Table II). Both L1210 and P815 cells had increased aerobic glycolysis following a period of co-cultivation with cytotoxic activated macrophages when compared to control cells, and the Pasteur effect was absent. Although L1210 cells achieve a greater glycolytic rate compared to P815 cells following co-cultivation with cytotoxic activated macrophages, the difference is small and the percent increase of glycolytic rate in P815 cells (~70%) is grater than the percent increase of glycolytic rate in L1210 cells (~50%). These results do not explain why P815 cells die and L1210 cells do not die following a period of co-cultivation with cytotoxic activated macrophages. It is possible that P815 cells utilize their glycolytically-produced ATP less efficiently than L1210 cells and hence, are unable to survive once inhibition of mitochondrial respiration has occurred. It is also possible that cytotoxic activated macrophage-mediated cytolysis of P815 cells occurs as a result of other metabolic effects not measured in these studies.

III. EVIDENCE THAT CYTOTOXIC ACTIVATED MACROPHAGE-INDUCED L1210 CYTOSTASIS AND INHIBITION OF L1210 CELL MITOCHONDRIAL RESPIRATION IS REVERSIBLE

The cytotoxic activated macrophage-induced metabolic changes that develop in cells with the nonlytic phenotype such as L1210 cells are not lethal (3, Granger, D.L. and Hibbs, J.B., Jr., unpublished). This conclusion is based on the findings that L1210 cells do not release significant amounts of ^{3}H-thymidine label during the period of cytotoxic activated macrophage-induced cytostasis, that periodic examination of these cytostatic L1210 cells for trypan blue exclusion shows no evidence of cell death, and that individually cultured L1210 cells that had been rendered cytostatic by cytotoxic activated macrophages formed colonies with 75% the efficiency of normal L1210 cells. As a result, using L1210 cells that had been co-cultivated with cytotoxic activated macrophages, it was possible to evaluate how development of altered energy metabolism and its eventual return to a baseline level correlate with the resumption of L1210 cell proliferation (Granger, D.L. and Hibbs, J.B., Jr., unpublished).

Figure 1 shows the correlation between recovery of L1210 cell division and resumption of mitochondrial respiration following a 21 hour period of co-cultivation with cytotoxic activated macrophages. After removal from the activated macrophage monolayer, the cytostatic L1210 cells were cultured in fresh medium with daily replenishment. At the intervals shown in Figure 3, aliquots of the population were removed from culture, washed by centrifugation, and respiration and aerobic glycolysis were measured. Viable cell counts were also made during this experiment. During the initial 21 hour co-culture period, respiration fell to a low level and during the same time, the rate of aerobic glycolysis rose by more than threefold. During subsequent culture, while the L1210 cells remained nondividing (hours 21-68), respiration was inhibited and glycolysis was elevated. However, once cell division resumed (at about 90 hours), respiration had risen to the control rate and aerobic glycolysis had fallen to the control level. The recovery of L1210 cell division was associated with a return of L1210 cell energy metabolism to a normal pattern. It was not possible from these experiments to determine whether mitosis proceeded or followed resumption of mitochondrial respiration. Krahenbuhl _et al_. showed previously that EMT-6 adenosarcoma cells rendered cytostatic by cytotoxic activated macrophages eventually recover the ability to proliferate (4).

IV. THE MOLECULAR EFFECTORS OF CYTOTOXIC ACTIVATED MACROPHAGES MUST BE CAPABLE OF CAUSING INHIBITION OF METABOLIC PATHWAYS IN MEMBRANE BOUND ORGANELLES OF TRANSFORMED TARGET CELLS

Cytotoxic activated macrophages inhibit proliferation and can cause eventual destruction of a wide variety of target cells (1-11). In parallel with inhibition of proliferation,

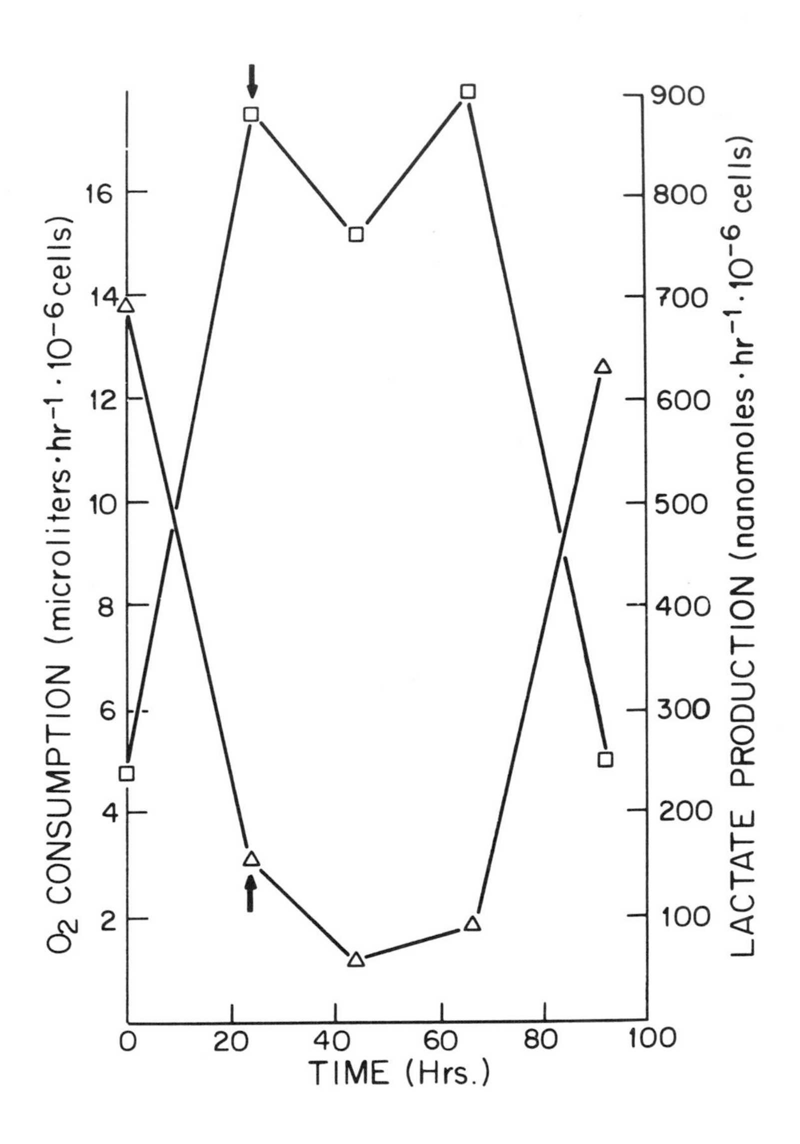

Figure 1.-- Respiration and glycolysis of L1210 cells injured by CM. At the time shown, L1210 cells were removed from culture and the rates of dinitrophenol-uncoupled respiration (Δ) and aerobic glycolysis ([]) were measured. The arrows show the time at which L1210 cells were separated from CM and reincubated in fresh culture medium (D. L. Granger and J. B. Hibbs, Jr., unpublished data).

there is, in a population of unsynchronized target cells, inhibition of DNA synthesis (4,8-11). Krahenbuhl showed that EMT-6 adenosarcoma target cells synchronized in discrete phases of the cell cycle, continued forward through the cell cycle for 2-6 hours after contact with cytotoxic activated macrophages and then further progression ceased (10). Kaplan et al. have provided further evidence of dysfunction within the nucleus of transformed cells co-cultivated with cytotoxic activated macrophages (11). Using the technique of microfluorometry, these investigators showed that Lewis lung carcinoma cells, in contact with cytotoxic activated macrophages undergo one round of cytokinesis in the absence of DNA replication ("reductive" cell division) before progressing to eventual lysis.

Therefore, it is clear that cytotoxic activated macrophages influence metabolic pathways within the nucleus as well as within mitochondria of transformed target cells. The biochemical effectors of the contact mediated and nonphagocytic activated macrophage cytotoxic reaction must be able to modify transformed cell metabolism in a way that explains the following observations: 1) The highly reproducible pattern of metabolic inhibition that is induced in transformed target cell membrane bound organelles--the nucleus and mitochondria; 2) the two transformed cell phenotypic responses (lytic and nonlytic) to this pattern of metabolic pathway inhibition, and; 3) the reversibility of the prolonged metabolic inhibitions induced in transformed cells with the nonlytic phenotype, i.e., L1210 cells.

ACKNOWLEDGEMENTS

We thank Gwenevere Shaw for typing the manuscript.

REFERENCES

1. Granger, D. L., Taintor, R. R., Cook, J. L., and Hibbs, J. B., Jr., J. Clin. Invest. 65:357 (1980).
2. Cook, J. L., Hibbs, J. B., Jr., and Lewis, A. M., Jr., Proc. Natl. Acad. Sci. USA, 77:6773 (1980).
3. Granger, D. L., and Hibbs, J. B., Jr., Fed. Proc. 40:761 (1981).
4. Krahenbuhl, J. L., Lambert, L. H., Jr., and Remington, J. S., Cell. Immunol. 25:279 (1976).

5. Hibbs, J. B., Jr., Lambert, L. H., Jr., and Remington, J. S., Nature New Biology 235:48 (1972).
6. Hibbs, J. B., Jr., Lambert, L. H., Jr., and Remington, J. S., Science 177:998 (1972).
7. Meltzer, M. S., Tucker, R. W., Sanford, K. K., and Leonard, E. J., J. Natl. Cancer Inst. 54:1175 (1975).
8. Keller, R., J. Exp. Med. 138:625 (1973).
9. Krahenbuhl, J. L., and Remington, J. S., J. Immunol. 113:507 (1974).
10. Krahenbuhl, J. L., Cancer Res., 40:4622 (1980).
11. Kaplan, A. M., Collins, J. M., Morahan, P. S., and Snodgrass, M. J., J. Immunol. 121:1781 (1978).

THE UROPOD AS AN INTEGRAL AND SPECIALIZED STRUCTURE OF LARGE GRANULAR LYMPHOCYTES[1]

Kenneth E. Muse
Hillel S. Koren

Division of Immunology
Duke University Medical Center
Durham, North Carolina

I. INTRODUCTION

Large granular lymphocytes (LGL), a minor subpopulation (3-5%) of peripheral blood leukocytes, may be the main nonadherent effector cells implicated in mediating natural killer (NK) cell activity against tumor target cells.

Numerous attempts have been directed toward the identification of markers restricted to or highly selective for NK cells (Kay et al., 1979; Targan and Dorey, 1980). The best such markers to date has been the morphological associations of NK cells with LGL which have a high cytoplasmic/nuclear ratio and azurophilic cytoplasmic granules (Timonen and Sakesela, 1980; Timonen et al., 1981; Reynolds et al., 1981).

Since the morphology of LGL may serve as a characterizing feature for NK cells, we have initiated studies to further characterize the functional morphology of LGL. Morphological evidence is presented that a large proportion of LGL, highly enriched by Percoll density gradient centrifugation, are characterized by an elongated, posterior appendage, similar in description to the "uropod" previously described for the hand-mirror or motile lymphocyte (Lewis, 1931; MacFarland, 1969). Furthermore, we show that the uropod is a prominent morphological feature of the LGL during the interaction with tumor cells. The possible relevance of this characteristic morphological feature of the LGL and its possible role in effector-target cell interactions are discussed.

---

[1]Supported by NCI grant CA 23354.
[2]Recipient of a Research Career Development Award CA 00581 from the National Cancer Institute.

ISBN 0-12-341360-5

II. RESULTS

A. Identification of LGL by Independent Criteria

Table I summarizes the distribution of LGL obtained in bands from nonadherent mononuclear cells of human peripheral blood (PBL) separated by discontinuous gradient centrifugation on Percoll (Timonen and Saksela, 1980) according to various criteria. The ^{51}Cr-release data are in accordance with previous reports (Timonen et al., 1981; Reynolds et al., 1981) and clearly indicate a significant increase in NK activity in the low density fractions against K562 tumor cells, primarily in bands 2-3 and 3-4 which represent approximately 35% of the total recovered cells. These findings are in good agreement with our single-cell conjugate data obtained according to the method of Bradley and Bonavida (1981). Specifically, bands 2-3 and 3-4 had the highest number of conjugates with K562 target cells (27.4 and 19.8% respectively) and the highest proportion of dead conjugated target cells (35 and 21% respectively).

Two additional criteria, azurophilic granules and the presence of a marker detected by the OKT10 monclonal antibody, previously shown to be selective for LGL in human PBL (Ortaldo et al., 1981), correlated well with the above mentioned functional criteria. In fact, the estimates of the proportion of LGL by Giesma staining and by OKT10 were indistinguishable. The typical uropod-associated morphology of unstained, unfixed cells which we observed to be associated with LGL (described in detail below, B), also correlated very well with the other morphological and functional criteria described above. From this point on our work focused on the innate morphological features peculiar to the LGL and their association with tumor cells during the LGL-target cell interaction.

B. Morphology of Uropod-bearing LGL

LGL enriched by Percoll density gradients were examined by light (LM), scanning (SEM) and transmission electron microscopy (TEM).

When suspensions of LGL, incubated at 37°C, were introduced into hemocytometer chambers, the uropod-bearing LGL were easily recognized even at low magnification with the LM. The fact that the LGL expressed the uropod as an integral part of its morphology, with no apparent stimulatory effect, had not been previously described. Figure 1 depicts a phase contrast photomicrograph of the uropod-bearing LGL. These cells are characterized by irregular surface contours and an elongated appendage which corresponds to the uropod. In contrast, the pseudopods at the opposite pole of the cell present a thin,

TABLE I. Distribution of Cells, LGL, Conjugates, and NK Activity Among Fractions Obtained by Discontinuous Density Gradient Centrifugation

Fraction[a]	% Distribution of recovered cells	% LGL[b] (Giemsa stain)	% Uropod-bearing cells[b]	% OKT10+ cells[b]	% Bound LGL[c]	% Dead conjugated targets[c]	% Specific lysis[d]
UNF	-	14.5	10.5	13.5	8.5±2.6	16.5±3.6	20.5±4.1
1-2	1.66	4.5	5.5	6.8	5.0±2.2	7.0±2.3	10.4±2.3
2-3	7.75	40.5	32.5	43.0	27.4±4.0	35.0±4.2	39.4±3.1
3-4	27.66	35.5	26.5	32.0	19.8±3.8	21.0±3.9	34.2±4.4
4-5	39.37	18.5	12.5	19.5	9.8±3.2	10.5±2.5	22.1±2.5
5-6	19.58	6.0	6.0	7.4	6.0±1.9	5.0±1.9	13.6±2.4
6-7	3.98	2.0	1.5	3.5	2.5±1.1	1.5±0.5	5.7±1.2

[a]7 discontinuous density Percoll bands of nonadherent mononuclear cells were obtained in 9 experiments. The numbers represent the mean of these experiments.
[b]Average number of LGL per 200 cells counted per fraction. The % uropod-bearing LGLs was assessed by bright field microscopy and the % OKT10 + cells by fluorescence microscopy.
[c]Conjugate data was determined by the conjugate/agarose assay with K562 tumor cells. E:T ratio = 1:1 and the data expressed as mean ±SD.
[d]Activity was determined in a 1 hr ^{51}Cr-release assay at an effector to K562 target ratio of 5:1 and expressed as percent specific lysis±SD.

more veil-like appearance. As evidenced in Figure 2, the typical LGL morphology can be preserved in Giemsa-stained, cytocentrifuge smears if prepared under optimal conditions (700 rpm, 30 sec.).

LGL were viewed in the SEM to determine the 3-dimensional surface topography of the cell. As shown in Figure 3, the uropod appears as a smooth-surfaced appendage, distally marked by short villi. The opposite pole of the cell is characterized by a smooth surface and anteriorly extending, surface folds. The preparation also included rounded, smooth-surfaced cells (Figure 3); such cells may represent the non LGL-like cells. present in the Percoll-enriched preparations.

A thin section electron micrograph (TEM) of a typical uropod-bearing LGL is shown in Figure 4. The uropod is most easily recognized and clearly identified by its ultrastructural features. The uropod is distally characterized by numerous terminal villi or "microspikes" (MacFarland, 1969) which have a diameter of 150-250 nm and a length not exceeding 0.85 um. Cytoplasmic organelles such as mitochondria, golgi complex and associated vesicles, and much of the rough endoplasmic reticulum are concentrated in the uropod. Cytoplasmic inclusions, characterized by parallel tubular arrays (Payne et al., 1981), were observed within the uropod region and may serve as a cytoplasmic marker for the LGL. Further studies on the nature of these granules and their possible significance in tumor cell cytolysis are in progress. The uropod also contained microfilaments arranged in irregular arrays and microtubules, radiating from centrioles, which extended parallel to the longitudenal axis of the uropod. The opposite pole of the LGL exhibited surface folds void of villi and limited in cytoplasmic organelles except for ribosomes and 4-6 nm microfilaments.

C. Interaction Between LGL and K562 Tumor Cells

Percoll gradient enriched LGL were cultured with K562 tumor cells at an E:T ratio of 2:1 and their interaction was monotored by video phase-contrast, time-lapse microscopy. Events that could be identified precisely at the LM level were: (1) the initial contacting of the target cell via the anterior terminus of the LGL, (2) the repositioning of the LGL which brought the uropod into repeated contacts with the surface of the tumor cell (Figure 5), (3) the ensuing lysis of the tumor cell which occured within 30-60 minutes following the initial contact, and (4) the release of the LGL from the lysed target and the ability of the detached LGL to recycle.

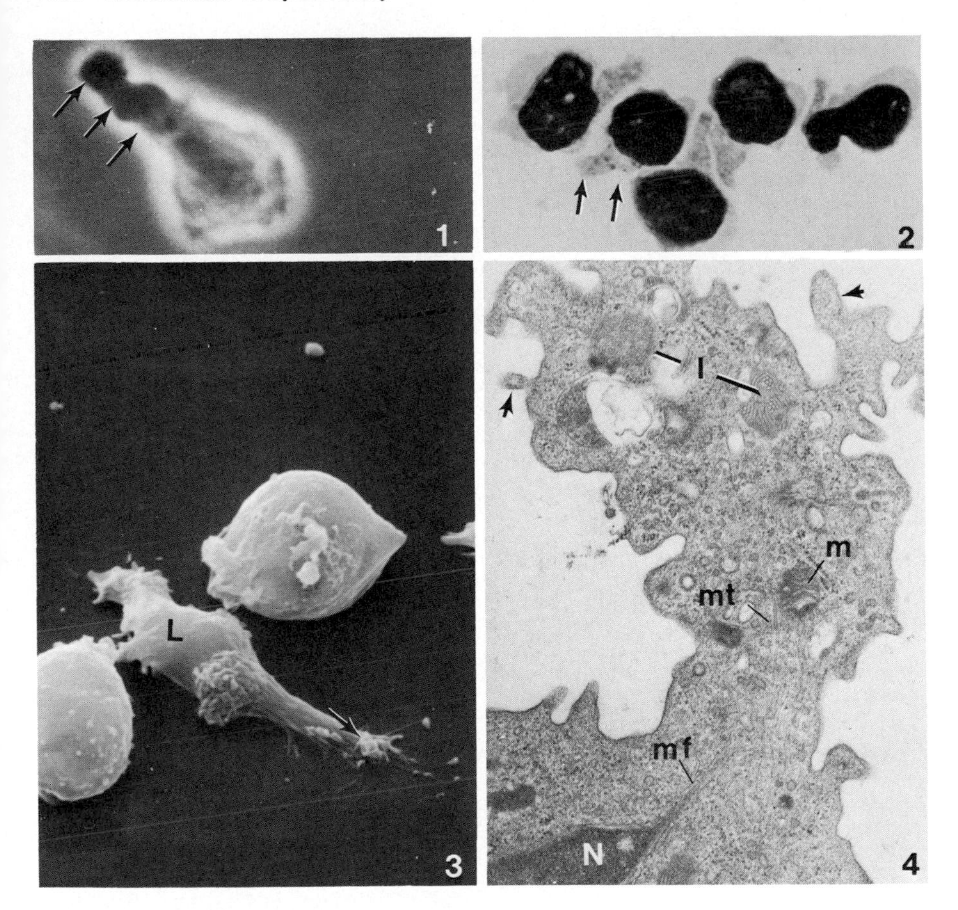

Figure 1. Phase-contrast micrograph of a uropod-bearing LGL prepared from human peripheral blood. Uropod (arrows). x1700. Figure 2. Oil emmersion LM of a Giemsa-stained LGL cytocentrifuge smear. Note the presence of the characterizing azurophilic granules within the uropod region (arrow). x1350. Figure 3. SEM showing the surface topography of the LGL (L). The uropod can be easily recognized from the concentrating microvilli (arrow) at the terminal position. x4000. Figure 4. TEM longitudenal profile of a LGL uropod. Note the terminal microvilli (arrows) and the characteristic inclusions (I) defined by parallel tubular arrays. [mitochondria (m), microtubules (mt), microfilaments (mf), nucleus, (N)]. x21,000.

III. DISCUSSION

In the present study we have demonstrated that the uropod is an integral and specialized part of the LGL and that it can serve as an easily identifable and reliable marker. The distribution of this uropod-bearing LGL in a 7 phase Percoll den-

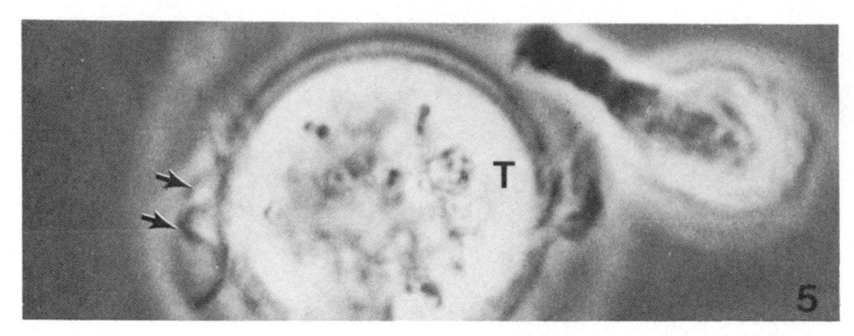

Figure 5. A selected frame from a video, time-lapse sequence showing an LGL, via its uropod, contacting a K562 tumor cell (T) demonstrating early signs of lysis (arrows). x1450.

sity gradient corellates well with the distribution of other known morphological and functional activities for LGL. Furthermore, our morphological observations suggest that the uropod of the LGL may serve as the anatomic component responsible for mediating cytotoxicity.

The present work raises another important point central to several major issues in cell biology--what are the mechanisms governing the modulation of cell morphology? One possibility is that the asymmetric morphology of the LGL is mediated by extracellular factors; such as cell-cell contact, extracellular matrices, or diffusable signals which modulate cell shape. An alternative is that the LGL, arising from a distinct cell lineage, contains endogenous information which specifies its morphology. The LGL may provide an excellent *in vitro* cell model for such investigations.

ACKNOWLEDGMENTS

We would like to thank Dr. J. R. Ortaldo (NIH) and Dr. G. Goldstein (Ortho Diagnostic Corp.) for their gift of the OKT10 monoclonal antibody and Mr. Philip Brandt for his technical assistance.

REFERENCES

Bradley, L.P., and Bonavida, B., J. Immunol. 126:208 (1981)
Kay, H., R., and Bonnard, G.D., Int. J. Cancer 24:141 (1979).
Lewis, W.H., Bull. Johns Hopkins Hosp. 49:29 (1931).
MacFarland, W., Science 163:818 (1969)
Ortaldo, J.R., Sharrow, S.O., Timonen, T., and Herberman, R.B. J. Immunol. 127:2401 (1981).
Payne, C.M., and Glasser, M., Blood 57:567, (1981).
Reynolds, C.W., Timonen, T., and Herberman, R.B., J. Immunol. 127:282 (1981).
Targan, S., and Dorey, F., J. Immunol. 24:2157 (1980)
Timonen, T., and Sakesela, E., J. Immunol. Methods 36:285(1980).
Timonen, T., Ortaldo, J.R., and Herberman, R.B., J. Exp. Med. 153:569 (1981).

MORPHOLOGICAL CHARACTERISTICS OF LYMPHOCYTE-TARGET CELL INTERACTIONS AND THEIR RELATIONS TO CYTOLYTIC ACTIVITY IN THE HUMAN NK SYSTEM

Győző G. Petrányi
Miklós Benczur
Miklós Varga

National Institute of Haematology and
Blood Transfusion, Budapest, Hungary

I. INTRODUCTION

Cell to cell contact between target cells and all kind of cytotoxic lymphocytes (CTL,NK,K) is a fundamental prerequisite for the killer effect (Koren and Ax, 1973; Herberman and Ortaldo, 1981; Rothstein et al., 1978). In the last few years however, an increasing body of evidence substantiate the possibility that often there is no direct correlation between the target binding function and cytolytic potential of various killer lymphocytes (Fiskelson and Berke, 1978; Grimm and Bonavida, 1977; Roder and Kiessling, 1978; Saksela et al., 1979). Since it was observed in our previous microcinematographic study that the attachement of motile lymphocytes with a large surface area to the target cells is the shape of binding which associates mostly with the destruction of targets (Petranyi et al., 1978), we analyzed further the correlation between the shapes of binding and the lytic function of lymphocyte populations to various NK tumor targets.

II. METHODS

Peripheral blood lymphocytes of healthy individuals were separated by Ficoll/Uromiro gradient, and depleted from monocytes by iron-magnet treatment. After nylon-wool column passage, the nonadherent cells were fractionated by E-rosetting to T and non-T non-B populations, and the B cells eluted from the

ISBN 0-12-341360-5

nylon wool were further depleted, by E-rosetting of T cells. This fractionation resulted in 85-90% pure T and B cells populations, detected by E-rosette and surface Ig determination, respectively, and in the non-T non-B population, fewer than 3-5% T cells and 1-2% B cell contamination were demonstrable.

Lymphocyte-target cell conjugate formation was studied as follows: a/$5x10^6$ lymphocytes with target cells in a ratio of 2:1, in 0.2 ml volume, were incubated at $37C^o$ for 90 min., in conventional tissue culture conditions. Thereafter the pellet was gently resuspended and counted for the percentage of target cell bound lymphocytes; b/morphokinetic characterization of the binding of lymphocytes to target cells was performed in a similar way as above, in a migration chamber (volume 1,5 ml, Sterilin, Ltd. Middlesex). The percentage of motile lymphocytes as well as the different shapes of target cell binding lymphocytes were registered in an inverted phase contrast microscope (Wild, M-40 Heerbrugg, Switzerland) on $37C^o$ hot stage.

The cytotoxic assay using ^{51}Cr isotope labelled target cells, and serial lymphocyte target cell ratio has been detailed elsewhere (Benczur et al., 1979). Based on the dose response cytotoxic release curve, killing activity for 20:1 ratio was computed and used as a comparative value either in the short (4 hours for K-562 and Molt-4 target cells) or long term (16 hours for Raji and Daudi target cells) assays.

III. RESULTS

Parallel time kinetics studies on lytic and binding function of monocyte-depleted peripheral blood lymphocytes were performed by "sensitive" (K-562 and Molt-4) and "resistant" (Raji and Daudi) target cells. A high and early lytic activity (4-6 hours) on K-562 and Molt-4 target cells was found in association with a low number of target cell bound lymphocytes, in contrast to the Raji target cells, where the initially low activity (for 4-10 hour period) was associated with a high number of target cell bound lymphocytes (Figure 1).

The highest lytic activity was observed in the case of non-T non-B populations as found by others. However both T and B enriched fractions displayed significant cytotoxic activity, which could be due to contamination by various NK effector cell populations. Comparing the binding and lytic function of the three lymphocyte populations on each target cells indirectly, a discrepancy could be observed in the case of non-T non-B cells; thus relatively lower numbers of binding cells paraleled with higher lytic activity. This was most pronounced in the comparison of both activity in the case of B, and non-T non-B cells on Raji and Daudi target cells (Table 1).

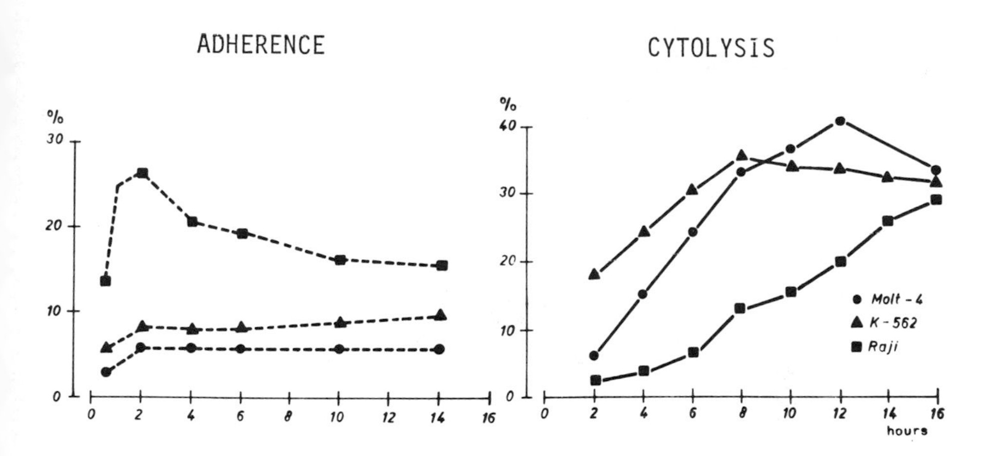

FIGURE 5. Time kinetics of target binding function and natural cytotoxicity on various target cells. (Each point represents the mean of 4-6 experiments in wich the same cell population was measured for both parameters parallel).

TABLE 1. Comparison between binding and lytic function of human lymphocytes on various target cells.

Lymphocyte populations	K-562	Target cells Molt-4	Raji	Daudi
T cells				
binding % [1]	7.8±1.8	7.2±2.1	14.2±1.9	8.9±2.0
cytolysis % [2]	17.9±4.3	16.3±4.6	6.2±0.8	13.7±3.2
B cells				
binding %	8.3±1.4	6.3±0.4	17.9±1.4	12.9±1.5
cytolysis %	20.9±3.3	16.8±3.0	9.7±2.4	12.5±3.4
non T non B cells				
binding %	10.2±0.8	8.9±1.5	16.8±1.8	8.9±1.4
cytolysis %	40.6±5.5	33.9±4.6	13.1±2.0	26.5±4.5

[1] The same lymphocyte fractions were investigated for binding and lytic function. Results of visual counting of more than 400 lymphocytes. Data are the mean of 6-8 experiments ± SE.

[2] Cytolysis %, expressed for 20:1 lymphocyte to targetcell ratio.

Concerning the three most characteristic shapes of binding (Figure 2) the majority of "cap-like" contact of non-T non-B lymphocytes on K-562 target and the "rosette-like" attachment of B cells on Raji target was the most remarkable feature in this study. It was also conspiciuous that almost 50% of non-T non-B cells were motile in contrast to the 11% of T and B cells respectively (Table 2).

Comparison between the characteristic shape of binding and thecytolytic activity of lymphocyte populations (Tables 1 and 2) suggests a possible association of NK activity with the "cap-like" shape of attachement. Namely the number of the "cap-like" adherence is the only one that consequently paralelled with the cytolytic potential of the lymphocyte populations. Lymphocyte binding to the Raji cells exceeded two to three times that of K-562. This was most conspicuous in the case of "rosette-like" attachement of B and non-T non-B lymphocytes. Since the higher binding rate did not run parallel with the lysis, it was concluded that the "rosette-like" binding shape may not be a representative form of NK function.

IV. DISCUSSION

Our present work gives evidence that the dissociation of the target cell binding and lytic function depends either on the functional property and population characteristics of lymphocytes or on the behaviour of target cells. Raji and Daudi tumor cells bind more T and B cells than K-562 and Molt-4, without the reflection of a parallel killing effect. This pattern became more clear in the study comparing the binding and lytic function of lymphocytes of 16 healthy individuals. Non-T non-B cells showed significant association between these two functions on K-562 target cells (r=0.76) while negative correlation appeared in the case of B cells investigated on Raji target (r=-0.31, unpublished results). Most recent work of Landazuri et al. (1981) reports the correlation between conjugate formation and cytotoxicity in the case of highly purified NK cells (Large Granular Lymphocyte) separated by Percoll discontinuoud gradient. In this experiment however, conjugate formation was observed by non-cytotoxic fractions of lymphocytes as well. In addition, fewer adherent target cells (G-11) were killed in comparison to Molt-4, despite a similar percentage of cnjugate formation.

The shape of binding appears to reflect the function and characteristics of both lymphocytes and target cells. We found no association between "uropodapsis" attachement (termed by McFarland and Heilman (1965)). and killing in the NK system. The most frequent shape of target cell binding lymphocytes with a parallel killing function observed, is the binding with a large surface area of motile lymphocytes, without the parti-

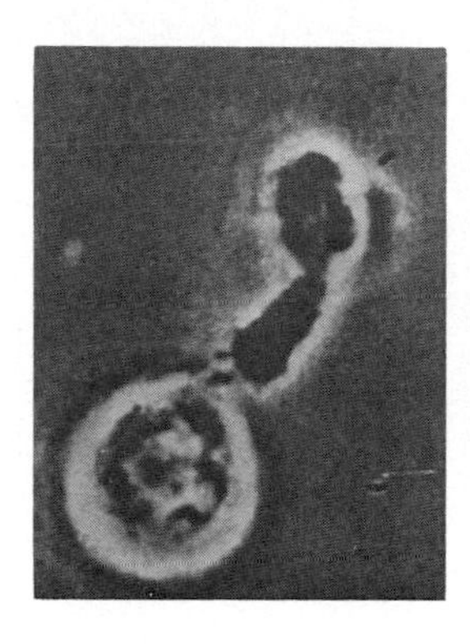

"UROPODAPSIS"

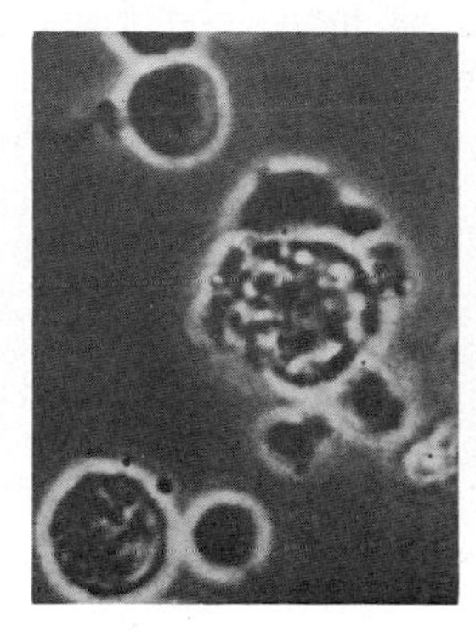

"CAP LIKE"

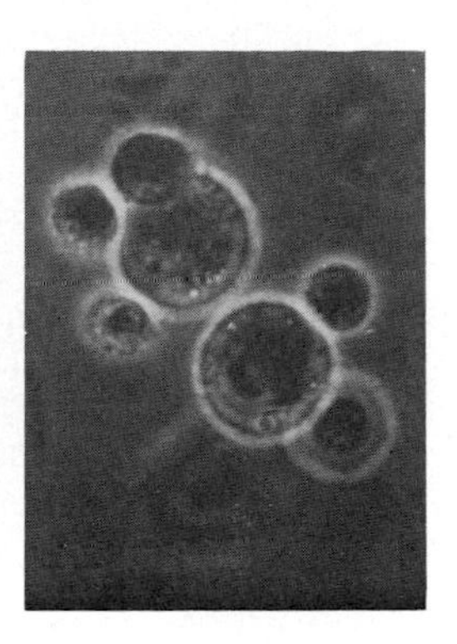

"ROSETTE LIKE"

FIGURE 2. Characteristic shapes of target cell binding lymphocytes. "Uropodapsis":adherence with the uropode to K-562 target cell (1000 x magnification);"Cap-like"adherence: killed K-562 target cell (400 x Magnification); "Rosette-like" binding: rosette formation on living K-562 target cell (200 x magnification).

TABLE 2. Shape of binding of human lymphocytes to K-562 and Raji target cells.

lymphocytes population	characteristic shape of binding motile ly.	uropodapsis	cap-like	rosette-like
T cells				
K-562	10.6±3.5[1]	1.0±0.5	1.0±0.3	1.9±0.8
Raji	11.6±2.7	2.9±0.6	3.3±1.7	4.5±1.2
B cells				
K-562	13.9±0.4	1.5±0.9	3.0±1.3	3.4±0.6
Raji	17.0±1.0	3.1±0.6	3.9±1.4	23.0±3.2
non T non B cells				
K-562	44.3±4.2	1.5±0.3	4.0±0.6	0.4±0.2
Raji	46.3±8.8	3.5±0.5	7.3±0.7	6.0±3.0

[1] Percent of binding lymphocytes with various features (shape) of total lymphocytes fraction. Data are the mean of 4-6 experiments, two samples in each case, more than 400 cells counted / sample.

cipation of the bottom of the uropode ("cap-like" adherence).

The adherence of round lymphocytes to the target cells ("rosette-like") seems to have the less relevance to NK killing. Since the characteristics of target cells (i.e. Raji) could influence significantly the number of lymphocytes showing this type of binding, receptors (EBV,C3b, C3d, Fc) and other additive factors (C', Ig, lymphokines) may be involved in this phenomenon. Taken together these and our microcinematographic observations it is supposed that "rosette-like" adherence reflects binding by non effector populations, while "uropodapsis" is mostly performed by effector cells in a relative inactive stage of the activation processes.

REFERENCES

Benczur, M., Györffy, Gy., Garam, T., Varga, M., Medgyessi, Gy., Sándor, M., and Petrányi, Gy. (1979). Immunobiol. 156:320.

Fiskelson, Z., and Berke, G. (1978). J. Immunol. 120:1121.

Grim, E., and Bonavida, B. (1977). J. Immunol. 119:1041.

Herberman, B. R., and Ortaldo, J. R. (1981). Science. 214:24.

Koren, S. H., and Ax, W. (1973). Eur. J. Immunol. 3:32.

Landazuri, M. E., Lopez-Botel, M., Timonen, T., Ortaldo, R.J., and Herberman, R. B. (1981). J. Immunol. 127:1380.

McFarland, W., and Heilman, D. H. (1965). Nature. 205:887.

Petrányi, Gy., Benczur, M., and Varga, M. (1978). In "Manipulation of Immune Response in Cancer" (A. Mitchison and R. Landy, ed.), p. 165. Academic Press, New York.

Roder, J., and Kiessling, R. (1978). Scand J. Immunol. 8:135.

Rothstein, T., Mage, M., Jones, G., and McHugh, L. (1978). J. Immunol. 121:1562.

Saksela, E., Timonen, T., Ranki, A., and Häyry, P. (1979). Immunol. Rev. 44:71.

CYTOTOXIC AND CYTOSTATIC ACTIVITY OF HUMAN LARGE GRANULAR LYMPHOCYTES AGAINST ALLOGENEIC TUMOR BIOPSY CELLS AND AUTOLOGOUS EBV INFECTED B LYMPHOCYTES

Maria G. Masucci
Maria T. Bejarano
Farkas Vánky
Eva Klein

Department of Tumor Biology
Karolinska Institute
Stockholm, Sweden

I. INTRODUCTION

Human lymphocytes responsible for the cytotoxic activity exerted against NK-sensitive cultured tumor cell lines have been assigned the morphology of large granular lymphocytes (LGL) (Timonen et al., 1979; Timonen et al., 1980). We have investigated the lytic potential of LGL-enriched subsets obtained by centrifugation on discontinuous Percoll density gradients against low NK-sensitive targets such as freshly isolated sarcoma and carcinoma cells and lymphoblastoid cell lines (LCL). The latter were established by in vitro infection of B lymphocytes with Epstein-Barr virus (EBV). Our data confirm

This investigation was in part supported by Grant no. 5R01 CA-25250-02 awarded by the National Cancer Institute DHEW and also by the Swedish Cancer Society.
Maria Grazia Masucci is recipient of a fellowship from the Foundation Blancefor-Boncompagni-Ludovisi född Bildt, Stockholm Sweden.

ISBN 0-12-341360-5

that both spontaneous and IFN-induced cytotoxicity are exerted by lymphocytes with low cell density. The same cells were also active against targets which were not lysed by the total population, such as the fresh biopsy cells.

II. MATERIAL AND METHODS

A. Fractionation of Lymphocytes on Percoll Density Gradients

The procedure described by Timonen et al. (1979) was used with minor modifications.

B. Target Cells

Lymphoblastoid cell lines established by infecting B enriched lymphocyte populations with the B95-8 strain of EBV (Miller et al., 1973). The transformed cells were used as targets between the fourth and the eight week of in vitro culture.

The methods used to separate the viable tumor cell fractions from human tumor biopsies have been described in detail elsewhere (Vose et al., 1977; Klein et al., 1980). Human erythrocytes sensitized with 1:100 final dilution of anti-D antibodies (Human, Budapest, Hungary) were used as targets for the ADCC tests.

C. Interferon Treatment of The Effector Lymphocytes

Parallel aliquots of 2×10^6 cells from unfractionated and Percoll gradient separated lymphocyte populations were incubated for 3 h at 37°C in a 5% CO_2 incubator to one aliquot 1000 IU/IFNα (kindly provided by Dr. Wolfgang Berthold, Biberach, Germany) were added. The untreated and IFN-treated cells were washed once in medium before use as effectors.

D. Cytotoxicity Tests

The details of the ^{51}Cr release assay performed with in vitro growing tumor cell lines and with freshly separated tumor biopsy cells have been previously described (Klein et al., 1980; Masucci et al., 1980).

E. Inhibition of The Outgrowth of EBV Infected B Cells

The outgrowth inhibition was assayed according to a modification of the method described by Moss et al. (1978). B enriched lymphocytes were infected with EBV and reconstituted at 1:9 ratio with autologous lymphocyte fractions separated on discontinuous density gradients. 0.2 ml of the reconstituted mixtures were placed in flat-bottomed microwells (10 wells/dilution). The cultures were evaluated weekly for transformation using the following score: single B blasts +, small clamps ++, big clumps +++, overgrowth ++++. An average score of ++++ in 10 wells corresponded to 100% growth (0% inhibition).

III. RESULTS

A. Characterization of Lymphocyte Subpopulations

The cellular composition of the different fractions varied considerably (Table I). Fraction 1 was depleted of small lym-

TABLE I. Characterization of Lymphocyte Fractions Obtained by Discontinuous Percoll Gradient Centrifugation[a]

Fraction density	Distribution of recovered cells[b]		Morphological characterization			
	total	LGL	small ly	LGL	monocytes	granulocytes
Unfractionated	-		77.6	14.2	3.3	4.7
1	1.0	4.3	21.8	28.7	41.2	8.7
2	8.7	44.5	43.8	50.0	8.0	1.7
3	29.4	28.4	82.9	13.0	3.0	1.7
4	32.1	15.5	86.4	7.7	1.8	3.8
5	14.4	5.4	91.2	4.7	0.6	9.6
6	13.3	1.7	88.3	2.2	1.5	17.0

[a]Mean of 8 experiments.
[b]Total recovery 94.9 ∓ 6.0.

phocytes, contained virtually all the contaminating monocytes which remained after nylon passage, and was twofold enriched in LGL. About half of the total LGL-s were recovered in fraction 2 (interface of the 40% Percoll). This fraction was depleted of small lymphocytes and contained 8% of monocytes. The fractions with high cell density were enriched in small lymphocytes and progressively depleted in LGLs. When present, contaminating granulocytes appeared in the dense fractions.

B. Cytotoxic Activity Against Cultured Tumor Cell Lines and Antibody-Coated Erythrocytes

We confirmed that low density fractions which were enriched in LGL showed significant enrichment in cytotoxic activity against tumor cell lines and antibody-coated erythrocytes (Table II). However, in spite of the relative enrichment in LGL, the cytotoxic activity of fraction 1 was low (in some of the experiments lower than in the unfractionated lymphocytes). This was probably due to the inhibitory activity

TABLE II. Cytotoxic Activity Against Long Term Cultured Tumor Cell Lines and Antibody-Coated Erythrocytes[a]

Fraction number	% LGL	Cytotoxic activity				
		K562[b]		Daudi[b]		ADCC[c]
		-IFN	+IFN	-IFN	+IFN	
Unfractionated	10	27	36	23	38	7
1	19	23	28	20	35	13
2	55	44	62	33	56	42
3	30	23	37	26	46	17
4	5	15	17	10	12	7
5	1	7	8	3	5	4
6	0	5	4	2	4	2

[a]One representative experiment
[b]25:1 effector/target ratio in a 5h ^{51}Cr release assay.
[c]Human ORh^{+} red blood cells were labeled with ^{51}Cr and treated with anti-D antibodies (1:100 dilution) for 30' targets. Untreated red blood cells are not lysed by blood lymphocytes.

of macrophages present in this fraction. This phenomenon did not occur in ADCC. In accordance with previously reported results, only the natural cytotoxic fractions were consistently increased in activity by treatment with 1000 U/ml of IFN.

C. Cytotoxic Activity Against Tumor Biopsy Cells

Seven tumor biopsy preparations were used as targets in parallel with K562 in thee different experiments (Table III). As previously shown, cytotoxic activity of unfractionated effectors was low but IFN treatment induced such effect (^{51}Cr release 20%) is some but not all experiments. Cytotoxic cells were enriched in fraction 2. Significant cytotoxicity was obtained in 3 of 7 cases with untreated lymphocytes and in all cases after IFN treatment. In parallel with the anti-K562 cytotoxicity, only low levels of spontaneous and IFN-inducible lysis was exerted by the more dense fractions against this type of target.

The overlapping phenotypic characteristics of the lytic cells against the NK-sensitive K562 and the seemingly specific effectors are in line with results of cold target competition experiments (Vãnky, F., and Klein, E., this volume).

TABLE III. Cytotoxic Activity of Lymphocytes from Healthy Donors Against Freshly Isolated Tumor Biopsy Cells[a]

Fraction number	IFN treatment	Expt. 1 K562	Expt. 1 A	Expt. 2 K562	Expt. 2 B	Expt. 2 C	Expt. 2 D	Expt. 3 K562	Expt. 3 E	Expt. 3 F	Expt. 3 G
Unfractionated	-	37[b]	9	22	0	10	1	57	5	3	7
	+	53	20	49	12	25	5	58	23	13	19
2	-	71	29	59	7	33	5	72	13	8	19
	+	75	36	63	21	52	22	75	32	27	31
3	-	51	9	46	7	33	3	67	13	8	17
	+	59	24	61	19	49	12	75	32	33	31
4	-	27	9	24	2	15	3	48	9	0	17
	+	38	18	47	14	36	8	55	25	17	22
5	-	25	8	14	0	6	3	20	0	6	10
	+	37	17	22	2	16	8	38	9	14	21
6	-	27	6	2	0	0	4	15	0	5	2
	+	34	18	12	0	4	4	18	0	11	5

[a]4h assay at 50:1 effector/target ratio. A-G different tumors used as targets. [b]% specific ^{51}Cr release.

D. Cytotoxic Activity Against Autologous Lymphoblastoid Cell Lines

Freshly established EBV genome positive lymphoblastoid cell lines are usually insensitive to the cytotoxic activity of unmanipulated blood lymphocytes. Low levels of cytotoxicity can however be demonstrated after IFN treatment of the effector cells. Cytotoxic activity against these cells was also detected in the subsets with low cell density (Table IV). The effect was more pronounced when the effectors were pretreated with IFN.

E. Capacity to Inhibit The Outgrowth of EBV Infected B Cells

The addition of autologous T cell enriched subsets fractionated on density gradients resulted in variable degree of inhibition of the EBV induced B cell proliferation. The results of one representative experiment are shown in Table V. The extent of inhibition observed after 2 weeks in cultures seeded at 2 x 10^6 cells/ml correlated with the level of anti K562 cytotoxicity and the enrichment of LGL observed on day 0.

TABLE IV. Cytotoxic Activity Against Freshly Established Autologous Lymphoblastoid Cell Lines[a]

Donor	Fraction density	Cytotoxic activity[b]		
		-IFN	+IFN	% LGL
	Unfractionated	8	20	17
IE	2	18	47	47
	4	6	16	15
	Unfractionated	0	5	22
LS	2	1	17	57
	4	0	1	6

[a]The autologous LGL was produced by in vitro infection of purified B cells with B95-8 strain of EBV. The cell lines were in cultures for less than 2 months.

[b]25:1 effector/target ratio, 5h ^{51}Cr release assay.

TABLE V. Inhibition of EBV Infected B Lymphocyte Proliferation[a]

Fraction number	Percentage of		
	anti K562 cytotoxicity	LGL	inhibition of transformation
Unfractionated	40	18	60
2	65	72	100
3	52	17	75
4	20	9	25
5	10	8	5
6	4	1	5

[a]One representative experiment.

IV. CONCLUSIONS

Our results confirm that cells with spontaneous and IFN inducible cytotoxic potential are in lymphocyte subpopulations with low cell density. Their lytic capacity was not limited to in vitro growing tumor cell lines since allogeneic freshly isolated tumor biopsy cells from sarcomas and carcinomas were also killed with high efficiency. EBV transformed LCL which are usually insensitive to the cytotoxic activity of unmanipulated lymphocytes were killed by autologous LGL enriched lymphocyte subpopulations suggesting the possible role of these cells in the in vivo control of EBV infection. This may be realized via several mechanisms. Our results demonstrate that LGL enriched lymphocyte fractions are involved in the regression of the EBV induced transformation system.

The EBV induced growth of B cells in vitro may be considered to be a good model for the studies of cellular immune responses toward transformed cells. Though EBV infection alone does not render the cells malignant,their proliferative capacity with fatal consequence can be manifested under special conditions such as inherited (Sullivan et al., 1980) or acquired (Hanto et al., 1981) immunosuppressed patients. In immunocompetent individuals the proliferative potential is suppressed. Using this model we have shown for the first time that NK cells have the capacity to limit the growth of cells being in a very early stage of transformation before they have undergone phenotypical and chromosomal changes occurring in long term cultures.

REFERENCES

Hanto, D.W., Frizzera, G., Purtillo, D.T., Sakamoto, K., Sullivan, J.L., Saemundsen, A.K., Klein, G., Simmons, R.L., and Najarian, J.S. (1981). Cancer Res. 41:4253.

Klein, E., Vânky, F., Galili, U., Vose, B.M., and Fopp, M. (1980). In "Contemporary Topics in Immunobiology" (I.P. Witz and M.G. Hanna, Jr., eds.), Vol. 10. Plenum Publ. Corp.

Masucci, M.G., Masucci, G., Klein, E., and Berthold, W. (1980). Proc. Nat. Acad. Sci. USA. 77:3620.

Miller, G., and Lipman, M. (1973). Proc. Nat. Acad. Sci. USA. 70:190.

Moss, D.J., Rickinson, A.B., and Pope, J.H. (1978). Int. J. Cancer 22:662.

Sullivan, J.L., Byron, K., Brewster, F., and Purtillo, D.T. (1980). Science 210:543.

Timonen, T., Saksela, E., Ránki, A., and Häyry, P. (1979). Cell. Immunol. 48:133.

Timonen, T., and Saksela, E. (1980). J. Immunol. Methods 36:285.

Vose, B.M., Vânky, F., and Klein, E. (1977). Int. J. Cancer 20:512.

ASSOCIATION OF HUMAN NATURAL KILLER CELL ACTIVITY AGAINST HUMAN PRIMARY TUMORS WITH LARGE GRANULAR LYMPHOCYTES

Susana A. Serrate, Brent M. Vose, Tuomo Timonen,
John R. Ortaldo and Ronald B. Herberman

Biological Research and Therapy Branch
National Cancer Institute
Frederick, Maryland

Human natural cell mediated cytotoxicity has been found against allogeneic leukemia-lymphoma targets as well as against tumor cells freshly isolated from solid tumors. Normal peripheral blood leukocytes (PBL) lysed tumor cells isolated from the blood of leukemia patients (Zarling *et al.*, 1979). In that study, the effector cell population had characteristics compatible with natural killer (NK) cells. Conflicting results have been reported regarding solid tumors. Klein *et al.* (1980) have found reactivity against freshly isolated targets from carcinomas and sarcomas, only after the *in vitro* treatment with interferon (IFN) of the effector cells, and suggested that polyclonally activated T-cells were the effector population for the natural cytotoxicity. Other studies (Vose *et al.*, 1980) have also shown reactivity of normal PBL, after boosting with IFN, against primary tumors, but the effector cells were not characterized. Recently, human NK cells have been identified as a subpopulation of lymphoid cells, with high cytoplasmic-nuclear ratio and azurophilic granules in their cytoplasm, termed large granular lymphocytes (LGL) (Timonen *et al.*, 1979; Timonen *et al.*, 1981). These cells have been shown to be responsible for the lysis of adherent as well as non-adherent tumor cell lines (Landazuri *et al.*, 1981). We have now analyzed the reactivity of different subpopulations of lymphoid cells from peripheral blood of normal donors, against tumor cells freshly isolated from human solid tumors.

REACTIVITY OF NORMAL ALLOGENEIC PBL AGAINST PRIMARY TUMORS

In this study, carcinomas of the lung, colon, breast or neck, and melanomas were examined. Tumor cell suspensions

ISBN 0-12-341360-5

were prepared from surgical specimens by enzymatic digestion with collagenase and purification on Ficoll and Ficoll-Triosil gradients, as previously described (Moore et al., 1981). ^{51}Cr-labelled tumor cells were used as targets in 18 h microcytotoxicity assays. Unseparated PBL, plastic-adherent cells (AC) and cells isolated from low and density fractions of Percoll gradients from allogeneic normal donors were tested as effector cells.

The reactivity of allogeneic PBL against several solid tumors is shown in Table 1. Five out of 9 tumors were significantly lysed by PBL from 4 of 9 normal donors. As expected, the % cytotoxicity against many of them was not as high as it was against K562 or some adherent cell lines. The reactivity appeared to be restricted to some donors who were able to kill several targets. In order to know whether the reactivity against these tumors was mediated by NK or other effector cells, we performed Percoll-gradient fractionation of nonadherent PBL and used cells from the different fractions as effectors in the microcytotoxicity assays against spontaneous tumors. Effector cells from the low density fractions were contained ≥60% LGL, whereas those from the high density fractions were usually 80% small lymphocytes. We found that the cytotoxicity against 7/12 primary tumors was mediated by LGL-containing cell populations (Table 2). Furthermore, using LGL-enriched fractions, cytotoxic reactions were observed with cells from some donors whose unseparated PBL were not reactive (9/11 donors were

Table 1. Reactivity of Normal Allogeneic PBL Against Human Primary Tumors

TARGETS	% LYSIS (100:1)
TUMOR 3	6[a]; 10*
TUMOR 5	9; 13
TUMOR 8	-2; 7
TUMOR 11	-4; -10; 2
TUMOR 16	6; 4; 23*
TUMOR 12	14*; -3
TUMOR 14	11*; 9
TUMOR 22	16*; 2; 15*
TUMOR 25	4

Tumors positive/Total number of tests: 5/19

* Significant above baseline, $p < 0.05$.

a) Values indicate results obtained with peripheral blood leukocytes from 1 or more normal donors. The mean % lysis against K562 was 50.6% (38-61%).

Table 2. Reactivity of Allogeneic LGL Against Primary Solid Tumors

TARGETS	% LYSIS (25:1)
TUMOR 3	10[a]; 12*
TUMOR 5	10; 12*
TUMOR 8	7; 4
TUMOR 11	20*; 20*; 16*; 16*; 23*
TUMOR 12	27*; 6
TUMOR 14	16*; 5
TUMOR 16	19*; 26*; 28*
TUMOR 22	34*; 6; 26*
TUMOR 25	5

Number of positive tumors/Total number of tests: 15/21

* Significant above baseline, $p < 0.05$.

a) Values indicate results obtained with LGL from 1 or more normal donors. The mean % lysis against K562 in the same assays was 56% (72-44%).

positive). Our results (Table 3) also showed that tumors that were significantly killed by PBL or LGL were not lysed by small T-cell-containing fractions from the same donors, even after treatment with IFN (not shown). These results indicate that the reactivity found against spontaneous human tumors is strongly associated with cells with the typical morphology of NK cells/LGL, and is restricted to the cell population that also lyse K562. They also suggest that the heterogeneity of NK cells among donors against cell lines might also exist against primary tumors. Further, our data indicate that the distinction between resistant and sensitive tumors cannot be made on the basis of the reactivity of PBL since some tumors were only lysed by LGL-enriched effector cells. We do not know whether the lack of reactivity of some PBL is due to suppressor cells/factors, as suggested for mouse NK activity (Serrate, submitted for publication), or to some other mechanism.

It has been reported (DeVries *et al.*, 1980) that purified lymphocyte populations do not exert cytotoxic activity against non-lymphoid targets, but upon addition of monocytes the reactivity can be restored. Using purified LGL against adherent cell lines (Landazuri *et al.*, 1981) it was shown that monocytes were not required for the cytotoxicity. Besides, some reactivity of activated monocytes against fresh tumor cells was reported (Mantovani *et al.*, 1980). In order to know the effects of adherent cells in our system, we studied the ability of AC to lyse primary human tumor cells

Table 3. Reactivity of Cells from High and Low Density Fractions of Percoll Gradients Against Human Primary Tumors

TARGETS	% Cytotoxicity mediated by	
	Low Density (LGL)	High Density (small T-cells)
TUMOR 11	19[a)]*	7
TUMOR 12	27*	1
TUMOR 14	16*	0
TUMOR 16	26*	5
K562	39*	4

* Significant above baseline $p < 0.05$
a) Results are expressed as percent cytotoxicity in 18 h assay.

and their effect on the activity of LGL. As shown in Table 4, normal AC did not exhibit cytotoxic activity against tumors sensitive to lysis by LGL from the same donors. We did not consistently find significant increase or decrease in the cytotoxic ability of LGL, after the addition of autologous monocytes, although concentrations of monocytes higher than 15% have not been tested.

We have also examined the reactivity of PBL and the different Percoll fractions from some tumor patients against the autologous tumor cells and the cell line K562. Some representative results are shown in Table 5. Tumors 11 and 22 were lysed by allogeneic LGL, and the latter also by autologous effectors. Tumor 25 was not significantly lysed by either allogeneic nor autologous effectors. Further studies will be required to establish the frequency of tumors sensitive to lysis by autologous LGL.

Table 4. Reactivity of Normal Adherent Cells and LGL Against Human Primary Tumors

TARGETS	% Cytotoxicity mediated by		
	LGL	Adherent cells	LGL/adherent cells[a)]
TUMOR 3	10*	5	11*
TUMOR 16	26*	6	17*
TUMOR 14	16*	3	7
TUMOR 22	35*	-6	30*
K562	61*	5	51*

a) 10% adherent cells.
* Significant above baseline at $p < 0.05$.

TABLE 5. Reactivity of Allogeneic and Autologous Effector Cells Against Primary Tumor Targets

	EFFECTORS					
	Allogeneic			Autologous		
FRACTION:	PBL	LGL	T-Cells	PBL	LGL	T-Cells
	E/T 50:1	25:1	25:1	50:1	25:1	25:1
TARGETS						
K562	27[a)]*	73*	--	13	53*	10
T11	18*	25*	7	2	8	5
K562	38*	58*	1	11	16*	0
T22	14*	35*	3	1	15*	7
K562	--	43*	7	38*	28*	0
T25	--	3	0	3	4	5

* Significant above baseline, $p < 0.05$.
a) Results expresssed as % cytotoxicity in 18 h assays.

CONCLUDING REMARKS

In this study, we have found that most human primary tumors can be lysed by normal lymphoid cells. This activity is strongly associated with LGL, known to be the effector cells for human NK activity. From our data it appears that tumors can be divided in 2 main categories:

a. Tumors resistant to lysis by all donors tested, possibly because of intrinsic properties of the tumor cells.
b. Tumors that are sensitive to lysis by some or all donors.

The lack of reactivity of the effector cells from some donors, might be related to some suppressor mechanism, operating in PBL populations, or to the absence of the appropriate recognition structures on the surface of LGL.

REFERENCES

DeVries, J.E., Mendelson, J., and Bont, W.S. (1980). Nature 283:574.

Klein, E., Masucci, M.C., Masucci, G. and Vanky, F. (1980). In "Natural Cell Mediated Cytotoxicity Against Tumors" (R. Herberman, ed.), p. 909, Academic Press, New York.

Landazuri, M.O., Lopez-Botet, M., Timonen, T., Ortaldo, J.R. and Herberman, R.B. (1981). J Immunol 127:1380.

Mantovani, A., Peri, G., Polentarutti, M., Allavena, P., Bordignan, C., Sessa, C. and Mangioni, C. (1980). In "Natural Cell Mediated Cytotoxicity Against Tumors" (R. B. Herberman, ed.), p. 1271, Academic Press, New York.
Moore, M. and Vose, B.M. (1981). Int J Cancer 27:265.
Timonen, T., Saksela, E., Ranky, A. and Hayry, P. (1979). Cell Immunol 48:133.
Timonen, T., Ortaldo, J.R. and Herberman, R.B. (1981). J Exp Med 153:569.
Vanky, F.T., Argov, S.A., Einhorn, S.A. and Klein, E. (1980). J Exp Med 151:1151.
Vose, B.M. and Moore, M. (1980). J Natl Cancer Institute 65:257.
Zarling, J.M., Eskra, L., Borden, E.C., Horozewicz, J. and Carter, W.A. (1979). J Immunol 123:63.

AUTO-TUMOR LYTIC POTENTIAL OF LYMPHOCYTES SEPARATED FROM HUMAN SOLID TUMORS

Farkas Vánky
Eva Klein

Department of Tumor Biology
Karolinska Institute
Stockholm, Sweden

I. INTRODUCTION

Lymphocytes isolated from solid tumors (TIL) were shown to have signs of activation such as stable E rosette formation, capacity for "natural attachement" (Galili et al., 1979) and high ^{3}H-Th incorporation compared to blood lymphocytes (Klein et al., 1980). On the other hand their lytic potential against the autologous tumor biopsy cells and NK sensitive cell lines was weak (Vose et al., 1977).

We describe here additional experiments in which the cytotoxic potential of the TIL was examined.

II. MATERIAL AND METHODS

The separation of TIL from surgical specimens of solid tumors is described in detail (Klein et al., 1980) where we have provided details of several experiments in order to emphasize that the various subsequent steps in the procedure were applied on the basis of the qualities of the fractions.

Supported by NIH Grant 1-R01-CA-25184-01A1 and by the Swedish Cancer Society.

ISBN 0-12-341360-5

III. RESULTS

A. Characterization of Nylon Passed Subsets of TIL with Monoclonal Antibodies

Five TIL populations were tested. Compared to the patient's blood lymphocytes these contained higher proportion of OKIa1 and OKT8 reactive cells (Table I).

B. Cytotoxic Potential of TIL

2/17 TIL samples caused higher than 10% specific ^{51}Cr release from autologous tumor cells (Tables II and III). Pretreatment of the effectors with Hu-IFN-α (1000 U/ml, for 2-3h) did not change this effect.

Higher proportion, 4/8, of allogeneic tumors were lysed, and in this case IFN pretreatment enhanced the efficiency. The mean ^{51}Cr release value increased from 7 to 20.

Ten of 13 TIL samples lysed K562. Also this activity increased with IFN pretreated effectors as seen in the elevated ^{51}Cr release; from 17 to 30.

For comparison, the results obtained with blood lymphocytes of the same patients are given. These were similar to the results with TIL and confirm our previous results with regard to the IFN induced elevation of allo-tumor reactivity (Vánky and Klein, 1982).

C. Auto-tumor Cytotoxicity of TIL Subsets Separated on The Basis of Adherence to Nylon Fibre

Auto-tumor lysis was exerted by nylon fibre column passed (NCp) subsets of TIL populations in 3 of 4 tests. In 2 cases the effect increased after IFN treatment (Table IV). The adherent cells (NCa) were not lytic and IFN did not induce cytotoxicity. Two of the 4 simultaneously tested PBL were positive.

TABLE I. Reactivity of TIL with Monoclonal Antibodies

	% of positive cells				
	OKT3	OKT4	OKT8	OKIa1	OKM1
TIL	76±3[a]	48±8	32±5	30±7	14±4
PBL	69±10	45±5	19±4	6±2	12±5

[a]Assayed by indirect immunofluorescence; mean ± S.E.

TABLE II. Cytotoxicity of The Tumor Derived Lymphocytes

Patient no.	IFN[a]	Auto-tumor		Allo-tumor		K-562	
		TIL	PBL	TIL	PBL	TIL	PBL
510	–	0	2[b]	17	0	19	0
	+	3	4	27	12	37	49
508	–	0	0	–	–	32	17
	+	5	4	–	–	40	36
512	–	4	20	–	–	29	28
	+	7	35	–	–	49	78
801	–	0	–	–	–	40	–
	+	0	–	–	–	58	–
2280	–	18	1	0	–	–	–
	+	19	0	0	–	–	–
1137	–	6	1	14	2	–	–
	+	7	6	25	16	–	–
2272	–	0	–	0	–	20	–
	+	0	–	19	–	27	–
2265	–	0	0	–	18	13	13
	+	0	4	–	22	32	40
1132	–	0	18	10	–	26	15
	+	0	22	19	–	30	31
1034	–	11	2	–	–	24	74
	+	11	5	–	–	27	92
1035	–	2	8	–	–	6	39
	+	0	0	–	–	58	44
2261	–	5	19	–	–	15	31
	+	2	23	–	–	13	71
2298	–	0	0	13	2	0	30
	+	0	0	29	40	0	47
1038	–	0	31	–	–	10	15
	+	0	28	–	–	24	35
792	–	0	3	–	–	12	23
	+	0	2	–	–	20	29
518	–	0	7	0	8	0	1
	+	0	8	0	9	–	25

[a] Aliquots of lymphocytes were incubated with Hu-IFN- (1000/ml) for 2-3h prior to the test.
[b] 4h assay, effector:target ratio 50:1. % specific ^{51}Cr release.

TABLE III. Cytotoxicity of TIL. Summary

	Effectors		Targets K-562	Tumor biopsy autologous	Tumor biopsy allogeneic
No of[a] tests			11	14	5
Mean % ^{51}Cr release	TIL	-	17[b]	3	7
		IFN	30	5	20
	PBL	-	26	8	6
		IFN	50	13	23
Tests[c] with ^{51}Cr release >10%	TIL	-	10/13	2/17	4/8
		IFN	12/13	2/17	5/8
	PBL	-	10/12	4/14	1/6
		IFN	12/12	5/14	4/6

[a]TIL and PBL were tested in parallel.
[b]See Table I.
[c]All tests.

TABLE IV. Auto-tumor Lysis of TIL Subsets Separated by Passage through Nylon Fibre Column

Patient no.	IFN[a]	Effectors TIL Total	TIL NCp	TIL NCa	PBL Total	PBL NCp	PBL NCa
510	-	0[b]	0	3	2	0	3
	+	3	3	-	4	5	5
512	-	4	10	2	20	-	-
	+	7	22	6	35	-	-
2261	-	5	21	5	19	23	16
	+	2	37	0	23	29	30
792	-	0	12	2	3	3	0
	+	0	14	2	2	4	0

[a,b]See Table I.

D. Suppression of The Auto-tumor Lysis by Admixture of Nylon Adherent Fraction

Table V presents two experiments in which auto-tumor cytotoxicity of the NCp TIL was inhibited when these were recombined with the autologous NCa fraction. Addition of NCa cells from the blood had no inhibitory effect.

E. Suppression of The Generation of Auto-tumor Cytotoxicity by NCa TIL

In 2 experiments in vitro generation of auto-tumor lysis in the blood lymphocyte population was inhibited by admixture of the NCa TIL (Table VI). In this system the NCp blood lymphocytes were cultured with the autologous tumor biopsy cells for 6 days.

TABLE V. Effect of The NCa Fraction on The Auto-tumor Cytotoxicity of The NCp TIL

Patient no.	Effector	populations[a]	% specific ^{51}Cr release
2261	NCp TIL	-	21
	NCp TIL	NCa TIL	3
	-	NCa TIL	5
	NCp TIL	NCa PBL	25
	-	NCa PBL	16
512	NCp TIL	-	10
	NCp TIL	NCa TIL	0
	-	NCa TIL	2
	NCp TIL	NCa PBL	28

[a]Ratio NCp TIL:NCA TIL or PBL 2:1.

TABLE VI. Effect of The NCa TIL Fraction on The In Vitro Generation of Auto-tumor Cytotoxicity

Patient no.	Lymphocytes cultured with autologous tumor biopsy cells[a]		% specific ^{51}Cr release
1032	NCp PBL	-	52
	NCp PBL	NCa TIL	5
	-	NCa TIL	0
	NCp PBL	NCp TIL	55
1034	NCp PBL	-	54
	NCp PBL	NCa TIL	15
	NCp PBL	NCp TIL	47

[a]The NCp fraction of PBL was cultured for 6 days with the patient's tumor cells. NCa or NCp fractions of TIL were added at 2:1 PBL:TIL fraction ratio.

IV. CONCLUSIONS

The outcome of various functional assays point to an impairment of the immunocompetence of TIL. At least in part the presence of suppressor cells and the suppressive effect of residual tumor cells seems to be responsible for this (Niitsuma et al., 1981; Vose and More, 1979). The separation procedure which involves several steps is likely to influence the composition and consequently the functional properties of the resulting TIL populations.

The monoclonal profile of the TIL suggested that its T cell compartment differs in composition from that of the blood lymphocytes. The relatively high proportion of OKIa1 and OKT8 positive cells in the TIL substantiates our previous results which indicated T cell activation on the basis of functional properties because these antigens on T cells appear in activated populations (Reinherz et al., 1979).

In several series of experiments TIL exerted no or weak NK (anti K562) potential (Gerson, 1980; Mantovani et al., 1980; Niitsuma et al., 1981; Totterman et al., 1978; Vose et al., 1977; Vose and Moore, 1979). In the present assays also the mean ^{51}Cr release value of K562 cells exposed to TIL was lower than that obtained with the patient's PBL, though the performance of TIL seemed to be better than in the previous reports.

IFN pretreatment of the TIL effectors enhanced the lysis of K562.

Allogeneic tumor cells were affected by some of the TIL populations and their function was stronger after IFN pretreatment. We proposed earlier that the lytic potential against allogeneic tumor cells is based on recognition of alloantigens (Vânky and Klein, 1981). The presence of alloreactive cells in the TIL population was also shown by their response in the conventional MLC (Klein et al., 1980).

In the tests performed with the total TIL there was no evidence for enrichment of lymphocytes which lyse the autologous tumor cells. The few experiments in which we tested the nylon column passed TIL suggests that in some the auto-tumor killer cells are suppressed. Once the coexistent nylon adherent suppressor cells were removed the lytic effect could be enhanced by IFN.

Blood lymphocytes are stimulated to divide when exposed to autologous tumor cells in vitro and auto-tumor lytic potential develops in the majority of such cultures (Vânky and Stjernswärd, 1976; Vose et al., 1978).

In addition to the suppression of the direct cytotoxicity, the nylon adherent TIL compartment affected the generation of auto-tumor lytic potential in blood lymphocyte populations, when added to lymphocyte tumor cultures.

In our previous experiments the total TIL population did not respond with blastogenesis when cultivated with autologous tumor cells. However, when the nylon wool adherent cells were removed such reactivity occurred in 6 of 10 cases (Klein et al., 1980). The present experiments reveal thus that in addition to its effect on the proliferative response the adherent subset suppresses also the cytotoxic response of auto-tumor recognizing cells present in the TIL.

The effect of TIL on the function of PBL was studied previously. Vose and Moore (1979) demonstrated that admixture of TIL to blood lymphocytes abrogated the blastogenesis induced in autologous lymphocyte-tumor mixed cultures. The present experiments indicate that the adherent subset in the TIL is responsible for the inhibitory effect imposed on the blood lymphocytes.

REFERENCES

Galili, U., Vânky, F., Rodrigez, L., and Klein, E. (1979). Cancer Immunol. Immunother. 6:129.

Gerson, J.M. (1980). In "Natural Cell Mediated Cytotoxicity against Tumors" (R. Herberman ed.) p 1047. Academic Press, New York.

Klein, E., Vánky, F., Galili, U., Vose, B.M., and Fopp, M. (1980). In "Contemporary Topics in Immunobiology" (M.G.Jr. Hanna and I.P. Witz eds.) p 79. Plenum Publ. Co. New York.

Mantovani, A., Allavena, P., Sessa, C., Bolis, G., and Margioni, C. (1980). Int. J. Cancer 25:573.

Niitsuma, M., Golub, S.H., Edelstein, R., and Holmes, C.E. (1981). J. Nat. Cancer Inst. 67:997.

Reinherz, E.L., Kung, P.C., Pesando, J.M., Ritz, J., Goldstein, G., and Schlossman, S.F. (1979). J. Exp. Med. 150:1472.

Tötterman, T.H., Häyry, P., Saksela, E., Timonen, T., and Eklund, B. (1978). Eur. J. Immunol. 8:872.

Vánky, F., and Klein, E. (1982). Immunogenetics 15:31.

Vánky, F., and Stjernswärd, J. (1976). In "In Vitro Methods in Cell Mediated and Tumor Immunity" (B. Bloom and J.R. David eds.) p 597.

Vose, B.M., and Moore, M. (1979). Int. J. Cancer 24:579.

Vose, B.M., Vánky, F., and Klein, E. (1977). Int. J. Cancer 20:895.

Vose, B.M., Vánky, F., Fopp, M., and Klein, E. (1978). Int. J. Cancer 21:588.

NATURAL CELL-MEDIATED CYTOTOXICITY AGAINST SPONTANEOUS MOUSE MAMMARY TUMORS

Susana A. Serrate and Ronald B. Herberman

Biological Research and Therapy Branch
National Cancer Institute
Frederick, Maryland

The possible role of natural killer (NK) cells in the immunosurveillance of spontaneous tumors has been a subject of considerable discussion (Herberman *et al.*, 1980). However, although the ability of NK cells to regulate the local growth (Kiessling *et al.* 1976; Sendo, *et al.*, 1975; Kiessling *et al.*, 1975) and the number of metastases (Petranyi *et al.*, 1976), of *in vivo*-injected tumor cell lines has been extensively documented, little information is available regarding the activity of these effector cells on spontaneously arising tumors in mice. In addition, conflicting results were obtained when human effector cells were tested *in vitro* against human primary tumor cells (Zarling *et al.*, 1979, Klein *et al.*, 1980, Mantovani *et al.*, 1980; Vose *et al.*, 1980). In order to determine whether natural effector cells may play any role in the defense against autochthonous tumors, we studied the ability of normal lymphoid cells from allogeneic and syngeneic normal donors to react *in vitro* against tumor cells freshly isolated from spontaneous primary mammary tumors.

REACTIVITY OF NORMAL LYMPHOID CELLS AGAINST PRIMARY MAMMARY TUMORS

The study of spontaneous mammary tumors in mice provided an attractive system, not only because of their many similarities with human breast cancer, but also because it was possible to examine large numbers of tumors and study the reactivity of effector cells from syngeneic normal donors. The spontaneous mammary tumors arose in C3H/HeN ex-breeder female mice at approximately 40 weeks of age. Cell suspensions, containing $\geq 75\%$ tumor cells, were prepared by mild enzymatic digestion with collagenase and separation from host cells by discontinuous Percoll density gradient centri-

ISBN 0-12-341360-5

fugation as previously described (Serrate et al., submitted for publication). Effector cells were nylon-wool nonadherent spleen cells, from 8-10 week old C3H/HeN mice. Long term (18h) ^{51}Cr-release microcytotoxicity assays were performed against 25 mammary tumors, and some representative results are shown in Table 1.

Overall, syngeneic effector cells had significant cytotoxicity against 60% of the tumors tested. This reactivity differed from that observed against the commonly used highly sensitive cell lines, like YAC-1, in that:
(1) There was no dose-response curve. Cytotoxicity was not seen at high E/T ratios (i.e., 200:1, 100:1 or 50:1) against most of the tumors. In contrast, low but significant lysis was induced by low numbers of effector cells (i.e., E/T of 25:1, 12:1, etc.). The lack of reactivity of high concentrations of effector cells might be caused by inhibitory cells or factors present in the tumor or the effector cell population. Experiments using lower numbers of tumor target cells/ well showed the same dose-response pattern, indicating that the biphasic results were not due to crowding. Although macrophages have been found to suppress NK activity, it is unlikely that enough macrophages remained on the nonadherent effector cell population to account for these results. PBL from the same donors also showed a similar pattern. Interestingly, when the effector cells were fractionated in

Table 1. Reactivity of Syngeneic Normal Lymphoid Cells Against Primary Mammary Tumors

	% Cytotoxicity against[a]				
E/T[b]	FMTC1[c]	FMTC2	FMTC3	FMTC4	FMTC5
200:1	2.0	0.8	-1.1	-4.1	4.9
50:1	5.1	15.0[d]	2.7	-1.9	5.8
12:1	17.3[d]	20.6[d]	1.3	1.8	6.1
3:1	--	4.2	14.0[d]	12.6[d]	-1.8

a) Measured in 18 hr. assays.
b) Effector to Target Ratio. Effector cells were syngeneicnormal nylon-wool non-adherent spleen cells.
c) $FMTC_1$, $FMTC_2$, etc. = Freshly isolated tumor cells from tumor 1, 2, 3, 4 and 5.
d) Significant above baseline at $p < 0.05$. $FMTC_1$, $FMTC_2$, and $FMTC_3$ were considered sensitive since significant cytotoxicity against them was found at 2 or more E/T ratios.

Percoll density gradients (see below), low density fractions were able to mediate lysis in a dose-response fashion. These results suggest that inhibitory cells might be present within the unseparated effector cell population.
(2) The sensitivity to lysis was considerably lower. The range of percent cytotoxicity was 7 to 17, as evaluated by the means of the significant values, at each E/T ratio. This was not due to a general lack in sensitivity to lysis, since these tumor cells were almost as sensitive as YAC-1 or normal lymphoid cells to cytotoxicity by antibody (Ab) plus effector cells or Ab plus complement. The tumor cells acquired increased sensitivity to lysis after several days of *in vitro* culture, as shown in Table 2, and tumor cell lines derived from primary tumors were also more sensitive. In contrast, the frequency of tumors sensitive to endogenous or IFN-activated effector cells, decreased after serial transplantation in normal syngeneic hosts (Table 3).

In conclusion, the sensitivity of primary tumors was lower than that of cell lines, and appeared to be modulated by environmental conditions.

EFFECTS OF INTERFERON AND INTERFERON INDUCERS ON THE REACTIVITY AGAINST AUTOCHTHONOUS TUMORS

NK-cell activity can be increased by *in vitro* pretreatment with IFN, or after *in vivo* induction with viruses or polyribonucleotides (Djeu *et al.*, 1981). Using spleen cells from syngeneic mice injected with Newcastle disease virus (NDV) or lymphochoriomeningitis virus (LCMV), we found: a) significant cytotoxicity ($p < 0.05$) against 90% of the tumors, as compared to 60% lysed by unstimulated natural effector cells, and b) a

Table 2. Effect of Culture of Primary Tumor Cells on Sensitivity to Natural Cell-Mediated Cytotoxicity

	% Cytotoxicity[a] against FMTC after culture for				
E/T	0 days	1 day	10 days	30 days	Cell line[c]
25:1	4.9	-8.0	0.1	9.2[b]	25.2[b]
3:1	2.6	-3.1	5.4	10.1[b]	13.6[b]

a) 18 hr assay.
b) Significantly above baseline control, $p < 0.05$.
c) From a primary tumor cultured for more than 2 months.

Table 3. Natural Cell-mediated Cytotoxicity Against Primary Tumors After Serial Transplantation in Normal Hosts

	Number of tumors sensitive/ total number of tumors tested					
Number of *in vivo* passages[a]:	0	1	2	3	4	5
Effector cells from:						
Normal C3H/HeN	14/22	0/4	1/8	0/4	0/4	0/5
NDV-injected C3H/HeN	9/10	1/4	1/8	1/4	0/4	1/5

a) Tumor fragments of primary tumors were subcutaneously implanted in 8-week-old syngeneic mice. After 30 days, tumors were excised and tested in cytotoxicity assays, and also reimplanted in syngeneic 8-week-old mice. This procedure was repeated four times.

significant increase in the percent cytotoxicity against 5 out of 10 tumors studied (representative data shown in Table 4).

Short *in vitro* treatment of nonadherent spleen cells, with several doses of IFN (crude mouse fibroblast IFN), showed very slight or no effects on the reactivity, whereas the presence of IFN during the 18 hr assay caused significant reduction in the reactivity against 2/3 tumors.

Thus, these results indicate that *in vivo* IFN induction, possibly through the recruitment of more effector cells, was able to augment the reactivity against primary tumors. They also suggest that, as described for some tumor cell lines (Trinchieri *et al.*, 1978), IFN might "protect" tumor cells from lysis by natural effector cells. Interestingly, we also found that cell lines derived from primary tumors lost their sensitivity to natural cell-mediated cytotoxicity, after *in vitro* culture with IFN.

ASSOCIATION OF THE EFFECTOR CELLS ACTIVE AGAINST PRIMARY TUMORS WITH LARGE GRANULAR LYMPHOCYTES

Human, rat and mouse NK activity have been associated with a morphologically defined subpopulation of lymphocytes, termed large granular lymphocytes (LGL). Human and rat LGL

Table 4. Cytototoxic Reactivity of Normal and IFN-Induced NA-spleen Cells Against Syngeneic Primary Tumors

	% Cytotoxicity[a] mediated by NA-spleen cells from					
	Normal C3H/HeN			NDV-injected C3H/HeN		
E/T:	100:1	25:1	3:1	100:1	25:1	3:1
TARGETS						
$FMTC_1$	-2.3	-3.9	2.7	13.4[b]	23.4[b]	32.1[b]
$FMTC_2$	3.5	20.1[b]	17.7[b]	12.9[b]	31.0[b]	18.4[b]
$FMTC_3$	-4.8	-1.7	0.5	1.7	3.8	11.0[b]

a) 18 hr assay
b) Significantly above baseline control, $p < 0.05$

are also responsible for the lysis of adherent cell lines (Landazuri *et al.*, 1981; Reynolds, Rees and Herberman, this volume) but this point has not been established for mouse NK activity. We have examined the ability of Percoll fractionated non-adherent peripheral blood leukocytes to lyse syngeneic primary mammary tumors (Table 5). Enrichment for LGL to 36% was achieved in the low density fractions, and it was accompanied by an increase in the reactivity, in a dose response fashion, against 2 out of 3 tumors. No cytotoxicity was observed with cells from the high density fractions, composed mainly by small lymphocytes. Thus, these data strongly suggest a relationship of LGL (and NK cells) to the effectors for lysis of primary tumors.

COLD TARGET INHIBITION OF THE LYSIS OF YAC-1 BY PRIMARY TUMORS

Cold target competition assays were done in order to know if YAC-1 and primary mammary tumor shared common target structures.

Freshly isolated mammary tumor cells from 3/6 tumors were able to compete for the lysis of YAC-1, although less than cold YAC-1 cells (Fig. 1). With 2 of the 6 tumors, an increase in the cytotoxicity against YAC-1 was observed. Preliminary results indicate that the lysis of radiolabeled fresh tumor cells was also inhibited by cold YAC-1 cells. These results suggest that primary tumors might share with YAC-1 cells the structures recognized by NK effectors.

Table 5. Cytototoxic Reactivity of Cells Separated on Percoll Gradients Against Primary Tumors

	% Cytotoxicity[a] (±SE) mediated by:		
TARGET	PBL (input)	Fraction 2-3[b]	Fraction 5[b]
YAC	16.1(4)	23.3(6)[b]	--
$FMTC_1$	6.8(3)	24.1(4)[b]	2.5(3)
$FMTC_2$	4.9(1)	38.0(3)[b]	1.8(2)
$FMTC_3$	2.2(2)	3.1(2)	0.9(2)

a) 18 hr assay.
b) Fraction 2-3 = low density (36% LGL); Fraction 5 = high density (mostly small lymphocytes).

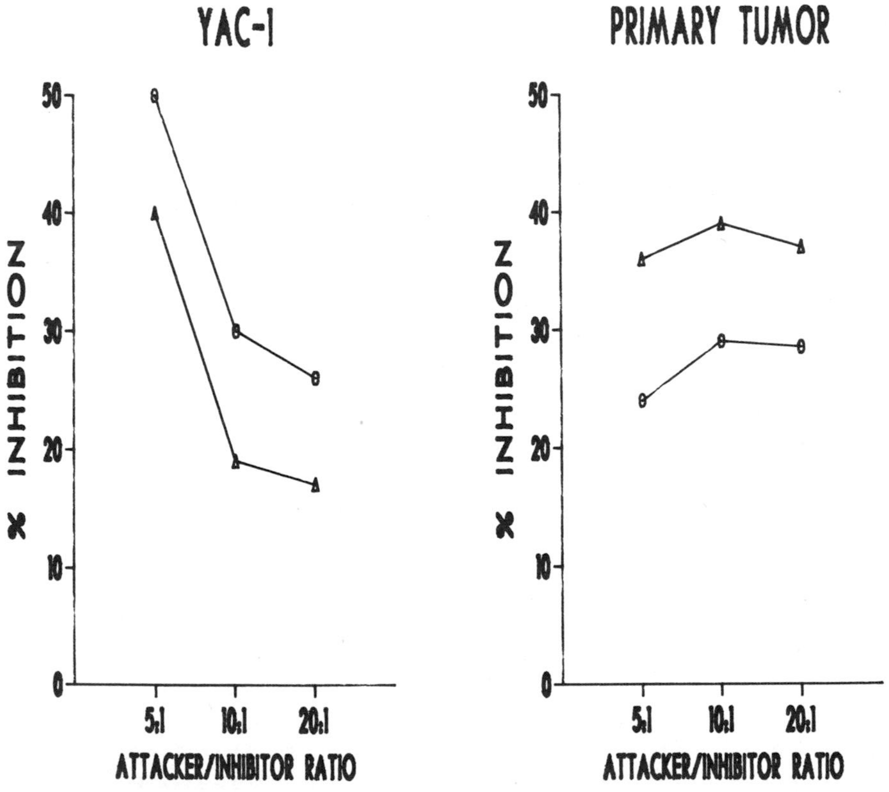

FIGURE 1. Cold target inhibition by YAC-1 (o) or fresh primary tumor cells (Δ) of the cytotoxocity (18 hr assay) of syngeneic normal lymphoid cells against YAC-1 or primary tumor. The inhibited cytotoxicity (at 25:1) against YAC-1 was 23.1% and against primary tumor, 14.8%.

STRAIN AND AGE DISTRIBUTION OF THE REACTIVITY AGAINST PRIMARY TUMORS

Lysis of freshly isolated mammary tumor cells was induced by NA-spleen cells from both high (C3H/HeN, BALB/c nude, C3H/HeN nude) and intermediate (C57BL/6), or low (SJL) NK strains. Thus, the reactivity does not appear to be strain distributed, as it is for some NK targets, like YAC-1 (Herberman *et al.*, 1975). We also found that it is not age-related, since similar levels of cytotoxicity were obtained with NA-spleen cells from syngeneic mice at 8, 16 or 32 weeks of age, not only against YAC-1 targets but also against primary tumors. Furthermore, NA-spleen cells from 3 tumor-bearing animals were active against primary tumors, and 2 of them were also reactive against the autologous tumor cells.

CONCLUDING REMARKS

We studied the reactivity of normal syngeneic NA-spleen cells against mouse spontaneous mammary tumors, and found a considerable proportion of them sensitive to NCMC. This reactivity was possibly mediated by NK cells since it was associated with LGL, it was increased after IFN induction, and the primary tumor cells could compete with the NK-target cell line, YAC-1. It remains to be determined whether these effector cells play an important role *in vivo*, in the defense against primary tumors.

REFERENCES

Djeu, J.V., Timonen, T. and Herberman, R.B. (1981). In "Role of Natural Killer Cells, Macrophages and Antibody Dependent Cellular Cytotoxicity in Tumor Rejection and/as Mediators of Biological Response Modifiers Activity" (M.A. Chirigos, ed.), Raven Press, New York, in press.

Herberman, R.B. (1980). In "Natural-Cell Mediated Cytotoxicity Against Tumors" (R.B. Herberman, ed.) pg. 1141, Academic Press, New York.

Herberman, R.B., Nunn, M.E. and Lavrin, D.H. (1975). Int J Cancer 16:216-229.

Kiessling, R., Petranyi, G., Klein, G. and Wigzell, H. (1975). Int J Cancer 15:933-940.

Kiessling, R., Petranyi, G., Klein, G. and Wigzell, H. (1976). Int J Cancer 17:275-281.

Klein, E., Masucci, M.C., Masucci, G. and Vanky, F. (1980). In "Natural Cell Mediated-Cytotoxicity Against Tumors" (R.B. Herberman, ed.) Academic Press, p. 909.

Landazuri, M.O., Lopez-Botet, M., Timonen, T., Ortaldo, J.R. and Herberman, R.B. (1981). J Immunol 127:1380-1383.

Mantovani, A., Allavena, P., Sessa, C., Bolis, G. and Mangioni, C. (1980). Int J Cancer 25:573-582.

Petranyi, G., Kiessling, R., Povey, S., Klein, G., Herzemberg, E. and Wigzell, H. (1976). Immunogenetics, 3:15-28.

Sendo, F., Aoki, T., Boyse, E.A. and Buofo, O.K. (1975). J Natl Cancer Inst 55:603-609.

Talmadge, J.E., Meyers, K.M., Prieur, D.J. and Starkey, J.R. (1980). Nature 284:622-624.

Trinchieri, G. and Santoli, D. (1978). J Exp Med 147:1314-1333.

Vose, B. M. and Moore, M. (1980). J Natl Cancer Inst 65:257-263.

Zarling, J.M., Eskra, L., Borden, E.C., Horoszewicz, J. and Carter, W.A. (1979). J Immunol 123:63-70.

NATURAL KILLING OF HEMATOPOIETIC CELLS[1]

Mona Hansson
Rolf Kiessling

Department of Tumorbiology
Karolinska Institutet
Stockholm, Sweden

Miroslav Beran

Department of Radiobiology
Karolinska Institutet
Stockholm, Sweden

I. INTRODUCTION

The major emphasis on NK cells has been their role as an immunosurveillance mechanism against tumor growth. There is however also suggestions that NK cells may have a more physiological role as regulators of normal hematopoietic cells. This hypothesis is based on the findings that mouse NK cells can lyse BM cells and a subpopulation of immature thymocytes from young mice (1,2), together with the findings that the *in vivo* mechanism of bone marrow graft resistance shares several characteristics with the NK system (3,4). In favour of this hypothesis is also the recent demonstration that NK cells are responsible for the rapid *in vivo* clearence of syngeneic and allogeneic BM grafts (5). In this chapter we will demonstrate further evidence in line with these findings, dealing with the ability of human NK cells to lyse normal human thymocytes and BM cells, and to inhibit *in vitro* differentiation and maturation of granulocytic progenitor cells.

[1]Supported by National Cancer Institute,DHEW CA 26782 and the Swedish Cancer Society.

ISBN 0-12-341360-5

II. RESULTS

We have recently described that also human NK cells can lyse certain normal cells, preferably of immature type (6). When comparing the degree of lysis between fetal thymocytes and fetal BM[1] with that of adult BM, we found the fetal cells more sensitive than the adult BM cells. This difference in sensitivity to macrophage depleted PBL in 12 hrs cytotoxicity assays is consistent and seen at all effector to target cell ratios. The low level of cytotoxicity is increased when IFN-pretreated effector cells are used, but the 3-fold difference in sensitivity between fetal and adult BM remains (fig. 1).

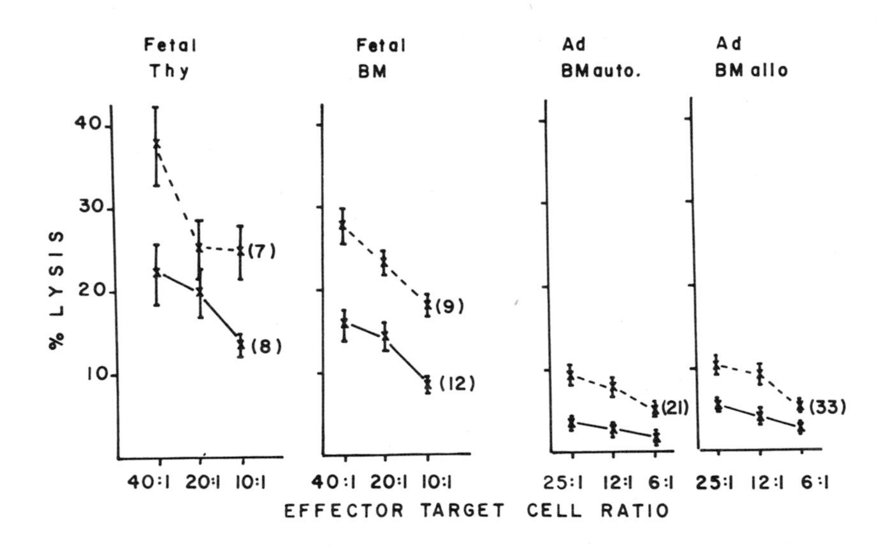

FIGURE 1. Mean % lysis of fetal thymocytes, BM and adult BM in 12 hrs ^{51}Cr-release assays, by macrophage depleted PBL without (x——x) or after IFN-pretreatment (x---x). Both autologous and allogeneic PBL:BM combinations are shown with adult BM targets. Number of experiments is indicated in parenthesis.

---

[1]Abbreviations: BM, bone marrow; PBL, peripheral blood leukocytes; SIg, surface immunoglobulin; IFN, interferon; GM-CFC, granulocytic-macrophage colony forming cell; E/T, effector to target.

Fetal BM contains mainly immature hematopoietic cells, no SIgG positive B-lymphocytes, < 20% SIgM positive cells and they do not respond to T-cell mitogens (i.e. no mature T-cells). In contrast, adult BM contains also $SIgG^+$ B-lymphocytes, respond to B- and T-cell mitogens and only a minor population are immature progenitor cells (table I). Data from both mouse and human tumor target systems, have shown that when certain tumors (erythroleukemias, teratocarcinomas and a histiocytic cell line) are induced to differentiate into more mature stages, they become less sensitive or resistant to NK lysis (7,8). Our interpretation of the consistent difference in sensitivity between fetal and adult BM is that the stage of differentiation is of importance also in NK lysis of normal targets.

The monoclonal antibody OKT 10 is presently believed to define an early stage of pre-thymocytes in the BM, and is a common marker for all thymocytes (9). When we compared fetal and adult BM for reactivity with the OKT 10 antibody, we found approximatly three times more positive cells in fetal BM than in adult BM as seen in table I (M.Jondal and M.Hansson unpublished observations).

TABLE I. Comparison between Fetal and Adult BM

	OKT 10[a]	PHA[b]	PWM[b]	NK-lysis[c]
Fetal BM	33	0.8	1.0	24
Adult BM	10	2.0	3.1	10

[a]Mean % flourescence positive cells with the use of a FITC conjugated goat anti mouse Ig as the second antibody (6 exp.)

[b]Mitogen reactivity index: ^{3}H-thymidine incorporation of mitogen stimulated cultures/ ^{3}H-thymidine incorporation of control cultures after 7 days stimulation.

[c]Mean % lysis at 20:1 (fetal BM) or 25:1 (adult BM) E/T ratio with IFN activated lymphocytes (complete data presented in fig. 1).

We believe that the higher frequency of OKT 10^+ cells in fetal BM indicate a higher number of cells in a certain stage of differentiation, where they are susceptible to NK lysis. In a recent paper, Crawford and coworkers have shown (10), that they recovered most of the colony forming precursor cells (70-80%) among the OKT 10^+ cells in adult BM, while only 15-20 % of the CFC were among the OKT 10^- cells when sorted in a FACS.

Experiments are in progress to characterize the NK-targets in BM with regard to different surface markers including the OKT 10 antigen.

In a series of experiments, we have taken a more direct approach to see whether the NK sensitive targets in adult BM are of stem cell nature. BM from healthy adults have been cultured for in vitro granulopoiesis in a soft agar colony assay with leukocyte conditioned medium as a source of colony stimulating factors (11). We have compared in parallel different effector cell populations for their degree of cytotoxicity against BM targets with their capacity to inhibit GM-CFC growth of the same BM samples. This "GM-CFC inhibition assay" is schematically described in fig. 2.

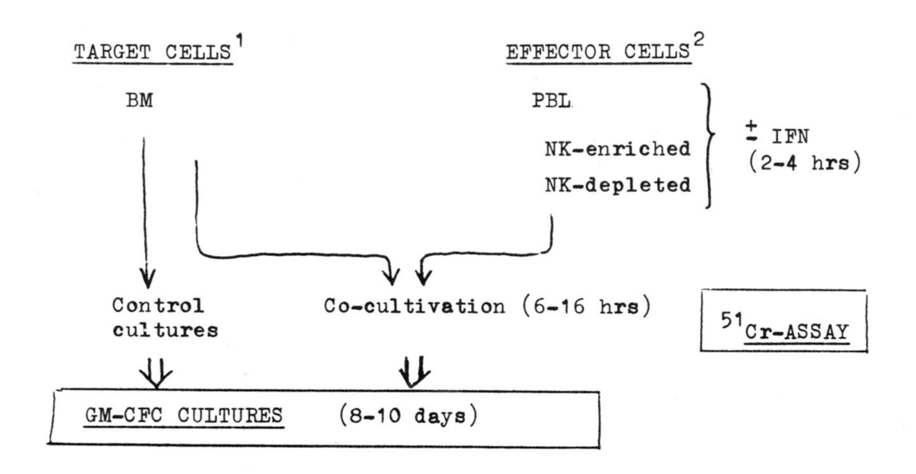

FIGURE 2. [1]Ficoll-Isopaque reparated BM cells obtained by posterior iliac crest aspiration.
[2]Autologous or allogeneic PBL, only macrophage depleted or subjected to various fractionation procedures, with or without IFN pretreatment.

When BM cells were subjected to autologous or allogeneic macrophage depleted PBL this way, and then cultured for GM-CFC growth, the number of colonies/ 10^5 BM cells were significantly reduced as compared to control cultures (i.e. inhibition). The inhibition was even more pronounced if PBL were preactivated with IFN, which also enhanced lysis of adult BM, as can be seen in table II.

TABLE II. GM-CFC Inhibition by PBL

AUTOLOGOUS PBL[a]		ALLOGENEIC PBL[a]	
-	IFN	-	IFN
37.5 ± 5.9[b]	56.1 ± 4.7	54.2 ± 8.9	62.9 ± 8.1
n=17		n=8	

[a]Macrophage depleted PBL with or without IFN treatment at an E/T ratio of 5:1, with 12 hrs co-incubation were used in all assays.

[b]Mean % inhibition ± SE

Since the GM-CFC inhibitory cell is present in freshly harvested PBL, shows no autologous or allogeneic preference and is more efficient after IFN activation, the lytically active NK cell is the most likely candidate. However, _in vitro_ granulopoiesis can be inhibited by both soluble factors and other cell mediated inhibitors like macrophages (12) and null cells (13). One must point out the complexity in regulation of granulopoiesis, in the sense that different subsets of leukocytes can via different mechanisms act on differentiation and proliferation of precursor cells. Therefore it is important to distinguish the NK-mediated inhibition from that effectuated by other subsets of leukocytes.

Human NK cells have been characterized morphologically as large granular lymphocytes (LGL) and can be enriched for among non adherent lymphocytes on a Percoll density gradient (14). We used the Percoll gradient to obtain one NK-enriched population, the low density fraction (LD), and one NK-depleted high density fraction (HD). As can be seen in table III, the LD fraction contained the lytically active NK cells, whereas the HD cells were devoid of NK activity and no significant lysis was seen even with the IFN treated HD cells. With these two fractions differing in NK activity, we were able to better evaluate the role of NK cells in the GM-CFC assay.

The NK active LD fraction was consistently more inhibitory than the HD fraction, containing mainly small and mediumsized T-lymphocytes, although some inhibition was often seen also with the lytically inactive HD cells at the highest E/T ratio. It is not likely that the GM-CFC inhibition by HD cells is due to "contaminating" NK cells since IFN pretreatment did not increase this inhibition. This is in contrast to the NK-mediated inhibition where a two-fold increase in inhibition was seen at all E/T ratio after IFN treatment (fig. 3).

TABLE III. Lysis of BM by Different Percoll Fractions

Effector cells[a]	% Lysis of Adult BM Cells			
	no treatment		IFN pretreatment	
	12:1	6:1	12:1	6:1
Non adherent PBL	3.1	2.3	12.8	7.6
Low density	19.2	16.3	29.8	27.5
High density	0.2	0.7	2.7	1.4

[a]Allogeneic lymphocytes were used as effector cells in a 12 hrs ^{51}Cr-release assay, with or without IFN pretreatment.

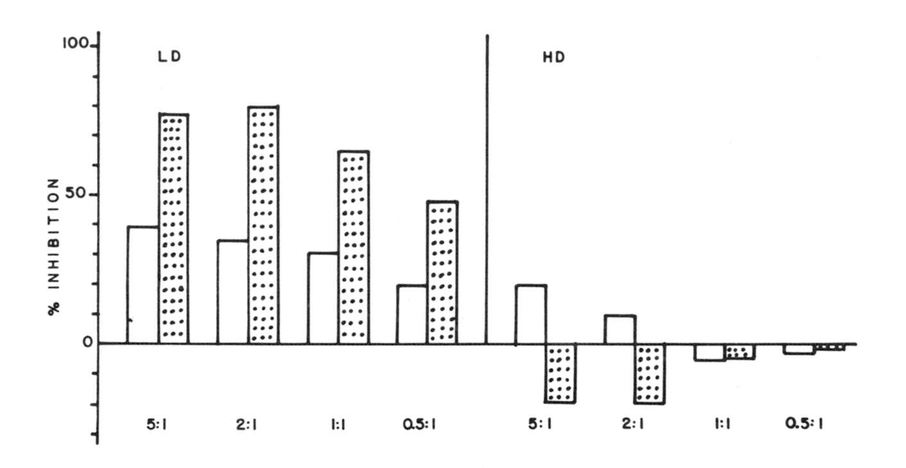

FIGURE 4. Allogeneic LD and HD lymphocytes without ( ) or after ( ) IFN pretreatment, tested for GM-CFC inhibition at different E/T ratio in a 6 hrs incubation assay.

Another difference between the LD- and HD-mediated GM-CFC inhibition was seen when we studied preincubation time requirements of the two fractions. The LD cells were only inhibitory when cocultured with BM for at least 4 hrs with an increased inhibition if longer times were used. On the other hand, HD cells showed a constant but low level of inhibition (20-30%) whether seeded directly with the BM in the agar assay or cocultivated for 2, 4 or 16 hrs before (data not shown).

III. CONCLUSIONS

In summary, we have described that human NK cells can lyse an immature cell present in normal BM. We have compared the NK lysis of BM cells with inhibition of *in vitro* granulopoiesis as a result of co-incubation with different NK-active lymphocyte populations. The cell fractions containing NK cells are the best inhibitors if *in vitro* granulopoiesis, in an IFN-time- and cell-contact dependent manner that resembles NK lysis. There are at least two equally possible mechanisms by which NK-mediated GM-CFC inhibition occurs. From these experiments it can not be concluded whether actual lysis of GM-CFC progenitors, or some growth inhibitory lymphokine production by NK cells is the reason for GM-CFC inhibition. NK cells do not only respond to exogenously IFN, but they can also be triggered by the target cell to produce IFN themselves (15), and we are presently investigating whether BM cells under these conditions can trigger IFN production by NK cells. It is quite clear that NK cells are responsible to a large extent for the GM-CFC inhibition, and this must be taken in to consideration both when studying patients with hematopoietic disorders, and when performing BM transplantations.

REFERENCES

1. Nunn,M.E.,Herberman,R.B. and Holden,H.T. Int.J. Cancer 20:381, (1977).
2. Hansson.M.,Kärre,K.,Kiessling,R.,Roder,J.,Andersson,B. and Häyry,P. J.Immunol. 123:765, (1979).
3. Kiessling,R.,Hochman,P.S.,Haller,O.,Shearer,G.M., Wigzell H. and Cudcowicz,G. Eur.J.Immunol. 7:655, (1977).
4. Datta,S.K.,Gallagher,M.T.,Trentin,J.,Kiessling,R. and Wigzell,H. Biomedicine 31:62, (1979).
5. Riccardi,C.,Santoni,A.,Barlozzari,T. and Herberman,R.B. Cell. Immunol. 60:136, (1981).
6. Hansson,M.,Kiessling,R. and Andersson,B. Eur.J.Immunol. 11:8, (1981).
7. Stern,P.,Gidlund,M.,Örn,A. and Wigzell,H. Nature 285:341, (1980).
8. Gidlund,M.,Örn,A.,Pattengale,P.K.,Jansson,M.,Wigzell,H. and Nilsson,K. Nature 292:848, (1981).
9. Janossy,G.,Tidman,N.,Papageorgiou,E.S.,Kung,P.C. and Goldstein,G. J.Immunol. 126:1608, (1980).
10. Crawford,D.H.,Francis,G.E.,Wing,M.A.,Edwards,A.J.,Janossy G.,Hoffbrand,A.V.,Prentice,H.G.,Secher,D.,McConell,I., Kung,P.C. and Goldstein,G. Brit.J.Haemat.49:209, (1981).

11. Beran,M.,Reizenstein,P. and Uden,A.M. Brit.J.Haematol. 44:39, (1980).
12. Spitzer,G.,Verma,D.,Beran,M.,Zander,A.,McCredie,K.O. and Dickie,K.A. J.N.C.I. in press. (1982)
13. Morris,T.C.M.,Vincent,P.C.,Sutherland,R. and Hersey,P. Brit.J.Haematol. 45:541, (1980).
14. Timonen,T. and Saksela,E. J.Immunol.Meth.36:285, (1980).
15. Timonen,T.,Saksela,E.,Virtanen,T. and Cantell,K. Eur.J.Immunol. 10:422, (1980).

HUMAN BONE MARROW MONONUCLEAR CELLS AS EFFECTORS AND TARGETS OF NATURAL KILLING[1]

Sudhir Gupta[2]
Gabriel Fernandes[3]

Memorial Sloan-Kettering Cancer Center
New York, New York

I. INTRODUCTION

Natural killer (NK) cells have been implicated in immune survillence against tumors, defense against certain virus and hematopoietic allograft rejection (1,2). Lopez et al (3) have reported that patients with acute leukemias or aplastic anemia who underwent stem cell allograft transplantation, if had normal or increased NK activity during pretransplant period, invariable developed graft-versus-host disease. Therefore we examine bone marrow mononuclear cells from healthy subjects for their effector functions in NK and ADCC against cells of K-562 cell line and autologous or allogeneic bone marrow mononuclear cells as targets. Studies were also performed to investigate if bone marrow mononuclear cells could also be target for NK or ADCC by autologous or allogeneic peripheral blood T or non T cells.

II. MATERIAL AND METHODS

A. Separation of Mononuclear and T and non T Cells

Heparinized peripheral blood and bone marrow aspirates were obtained from 6 young healthy volunteers. A written

---

[1]Supported by NIH grant CA-17404, AG-00541, AG-03417
[2]Present address: Div. of Basic and Clinical Immunology U. of California, Irvine, CA 92717
[3]Present address: Div. of Clinical Immunology, U. of Texas Health Science Center San Antonio, TX

ISBN 0-12-341360-5

consent was obtained. Mononuclear cells were obtained on Ficoll-Hypaque density gradient. Cells were washed x3 in Hanks' balanced salt solution and resuspended in medium RPMI-1640. From the peripheral blood mononuclear cells, monocytes were depleted by incubating with carbonyl iron solution (Lymphocyte Separator Reagent Technicon, Terrytown, N.Y.) and separating on Ficoll-Hypaque gradient. Lymphocytes were further enriched into T and non T cells by rosette-formation with neuraminidase-treated SRBC (4). Cells were washed and resuspended in medium RPMI-1640.

B. Natural Killer Cell Assay

Bone marrow mononuclear cells and cells of K-562 cell lines were used as targets for NK. Target cells were labeled with ^{51}sodium chromate, washed and resuspended in medium RPMI-1640 supplimented with 10% heat-inactivated fetal calf serum at 1×10^5/ml. Effector bone marrow mononuclear cells or peripheral blood T or non T lymphoid cells were also washed and resuspended in the same medium and were added to the target cells at effector to target ratios of 50:1 and 25:1. Plates were incubated for 4 hours at 37°C and then centrifuged and ^{51}Cr release was measured and percent lysis was calculated as described (5).

C. Antibody-dependent Cytotoxicity Assay

ADCC was measured by a method described earlier (5). Appropriate dilutions of bone marrow mononuclear cells or peripheral blood T or non T cells in RPMI-1640 were mixed with 1×10^5/ml of ^{51}Cr labelled chicken RBC (CRBC) or cells of SB cell line and anti CRBC or anti-SB IgG antibodies. In controls, antibody was replaced by medium. The plates were centrifuged and incubated at 37°C for 3 days. Following incubation, plates were centrifuged and supernatants were examined for ^{51}Cr release and percent lysis was calculated (5).

III. RESULTS

D. Bone Marrow Mononuclear Cells as Effectors in NK and ADCC Assay

Results from 4 such experiments are shown in Fig. 1. Bone marrow mononuclear cells produced 13.4 ± 8.6% (mean ± SD)

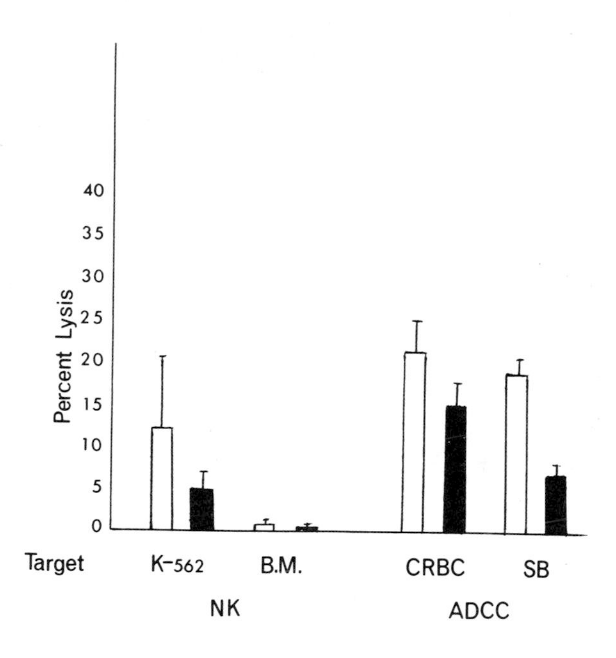

FIGURE 1. Bone marrow mononuclears as effectors in NK and ADCC.

lysis of cells of K-562 at 50:1 (□) effector:target ratio and 5 ± 2% lysis at 25:1 (■) effector:target ratio. However, no lysis was observed against autologous bone marrow mononuclear cells as targets. Bone marrow mononuclear cells mediated ADCC against Chicken RBC as well as cells of SB cell line with a mean lysis of 23% (50:1) and 15% (25:1) against CRBC and mean lysis of 18% (50:1) and 6% (25:1) against cells of SB cell lines as targets.

E. Bone Marrow Cells as Targets for NK by Autologous or Allogeneic Peripheral Blood T and non T Cells

Data pooled from 6 experiments of bone marrow mononuclear cells as targets against autologous T and non T lymphocytes are shown in Fig. 2. Cells of K-562 cell line were used as positive controlled. T cells had no NK activity against autologous bone marrow mononuclear cells target. A mean lysis of

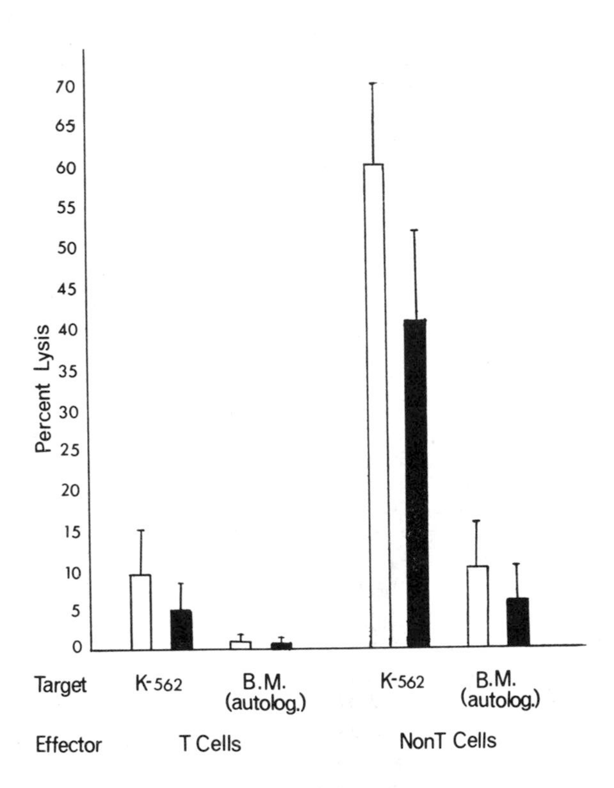

FIGURE 2. Bone marrow cells as targets of NK in autologous system.

10% was observed against K-562 cells. Non T cells produced minimal lysis (mean 10% and 6% at 50:1 and 25:1 effector: target ratio) of autologous bone marrow cells as compared to mean lysis of 61% and 41% (50:1 and 25:1 effector:target ratio respectively) of K-562 cell line cells.

Results of data pooled from 6 experiments demonstrating T and non T cell-mediated NK activity against allogeneic bone marrow mononuclear cell targets are shown in Fig. 3. T cells failed to demonstrate any NK activity against allogeneic bone marrow mononuclear cell targets. However, non T cells produced significant lysis (mean lysis of 31% and 24% at 50:1 and 25:1 effector:target ratio respectively) of allogeneic bone marrow mononuclear cells. Both T and non T cells demonstrated good NK activity against K-562 (positive control).

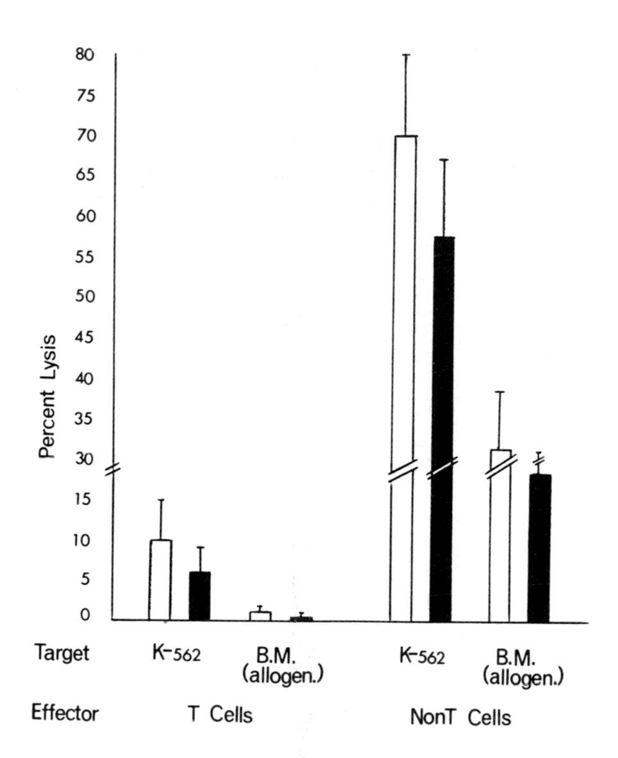

FIGURE 3. Bone marrow cells as target of NK in allogeneic system.

IV. DISCUSSION

In this study we have demonstrated that human bone marrow mononuclear cells were able to kill cells of K-562 cell line as well as allogeneic bone marrow mononuclear cells. However no spontaneous killing (NK) was observed against autologous bone marrow mononuclear cells by autologous T or non T cells. This observation could explain the role of NK cells in hematopoietic allograft reject in mice (2) and possibly in man (3). We have also demonstrated that bone marrow mononuclear cells could also be targets for NK by peripheral blood non T lymphoid cells but not by T cells. We have observed a marked inhibition of colony forming units of granulocyte-monocyte lineage, in co-cultures of bone marrow mononuclear cells with allogeneic non T cells (unpublished observations). It is likely that bone marrow precursor cells are target of NK by peripheral blood non T cells. It is very likely that this could be one of mechanism of granulocytopenia in a subgroup of patients with persistant neutropenia.

Attempts have been made to characterize NK cells using a variety of surface markers (1,4,6). It appears that phenotypically NK cells are heterogeneous and share antigens/markers of cells of T cell lineage as well as of monocyte-macrophage lineage.

In the present study we have also demonstrated that bone marrow mononuclear cells are effectors in ADCC as well. It is not surprising because these mononuclear cells contains cells of different lineages, lymphoid and non lymphoid, possess IgG Fc receptors. Furthermore, third population lymphoid cells, T cells, polymorphonuclear neutrophils, monocytes and eosinophils have been shown to mediate ADCC (4,7-9). Attempts are now been made to separate bone marrow mononuclear cells into multiple fractions and examined them morphologically, phenotypically with classical markers and monoclonal antibodies and correlate them with NK or ADCC functions.

REFERENCES

1. Herberman, R.B., Timonen, T., Ortaldo, J.R., Bonnard, G.D., and Gorelik, E. Prog. Immuno. 4:691 (1980).
2. Kiessling, R., and Haller, O. Contemp. Top. Immunobiol. 8:171 (1978).
3. Lopez, C., Kirkpatrick, D., Sorell, M., O'Reilly, R.J., and Chang, C. Lancet 11:1103 (1979).
4. Gupta, S., Fernandes, G., Nair, M., and Good, R.A. Proc. Natl. Acad. Sci. (USA) 75:6437 (1978).
5. Fernandes, G., and Gupta, S. J. Clin. Immunol. 1:141 (1981).
6. Kraft, D., Rumpold, H., Steiner, R., Radaskiewicz, P.S., and Wiedermann, G. In: Mechanisms of Lymphocyte Activation (K. Resch and H. Kirschner, eds.), p. 279. Elsivier/North-Holland, Biomed. Press, Amsterdam, 1977.
7. Pross, H.F., and Jondal, M. Clin. Exp. Immunol. 21:226 (1975).
8. Nelson, D.L., Bundy, B.M., Picheon, M.E., Blaese, R.M., and Strober, W. J. Immunol. 117:1472 (1976).
9. Parillo, J.E., and Fauci, A.S. Blood 53:457 (1978).

NATURAL KILLER (NK) CELL ACTIVITY AGAINST EXTRACELLULAR FORMS OF TRYPANOSOMA CRUZI[1]

Frank M. Hatcher[2]
Raymond E. Kuhn

Department of Biology
Wake Forest University
Winston-Salem, North Carolina

I. INTRODUCTION

Humans infected with the protozoan parasite, *Trypanosoma cruzi*, present a number of symptomatic and pathologic consequences known as Chagas' disease (American trypanosomiasis). In infected mammals (human or experimental animal systems) the parasites multiply within host cells (primarily muscle and mononuclear phagocytes) as amastigotes with extracellular trypomastigotes found prinicipally in the blood.

The mechanisms by which immune resistance to *T. cruzi* is manifested are not understood and sterile immunity apparently does not develop (Brener and Chiari, 1971; Krettli, 1977). It is known, however, that resistance to *T. cruzi* is dependent on the presence of an intact thymus (Kierszenbaum and Pienkowski, 1979; Trischmann et al., 1978) and mature peripheral T lymphocytes (Roberson et al., 1973). However, antibodies produced during infection and specific for *T. cruzi* are also important in resistance as shown by transfer of immune serum to newly infected mice (Krettli and Brener, 1976) and indirectly by studies using B-cell-deficient mice. (Rodriguez et al., 1981).

---

[1]Supported by NIH grant A1-14841
[2]Present address: Department of Microbiology, Meharry Medical College, Nashville, TN.

ISBN 0-12-341360-5

The activation of nonspecific effector mechanisms during infection may also be important in resistance to *T. cruzi*. Nogueira, Cohn and their coworkers have demonstrated that macrophages from infected mice become activated and that the activation requires the presence of specifically sensitized T lymphocytes (Nogueira et al., 1977; Nogueira and Cohn, 1979). Such activated macrophages have increased ability to destroy the parasite (Nathan et al., 1979). Results from our recent studies suggest that infection with T. cruzi also augments cytotoxic activity of lymphocytes against allogeneic tumor cells (Hatcher and Kuhn, 1981). The host cell population mediating this activity was found to be natural killer (NK) cells (Hatcher et al., 1981), and Thy 1^+ NK-like cells (see our other paper in this volume for review of *T. cruzi*-induced lytic activity). The results summarized here are from our investigations on host effector activity against *T. cruzi*, and indicate that augmented NK cell activity develops against extracellular forms of *T. cruzi in vitro*. We also report that the addition of specific antibodies to the assay medium further increases the anti-*T. cruzi* activity of augmented NK cell populations.

II. MATERIALS AND METHODS

The experiments which provided the basis for this summary of NK cell activity against *T. cruzi* were done using either C57B1/6 (B6; highly resistant to *T. cruzi*) or C3H(He) Dub (highly susceptible) female mice (8-12 weeks of age). Spleen or peritoneal exudate cells (SC or PEC, respectively) were collected from normal mice, or mice infected for 2 days with *T. cruzi* (i.p.; Brazil strain) or stimulated 18-24 hours previously with poly I:C (100 μg in PBS, i.p.) to augment NK cell activity as described (Hatcher et al., 1981; Djeu et al., 1979). Blood-form trypomastigotes (BFT) were obtained from the blood of infected mice as described (Hatcher and Kuhn, 1981); culture-form epimastigotes were obtained from axenic LIT culture medium inoculated 2-3 weeks previously with BFT (Powell and Kuhn, 1980); fibroblast-derived trypomastigotes (FDT) were obtained from C3H embryo fibroblast cultures infected *in vitro* with BFT 5-7 days after infection was initiated (Kuhn and Murnane, 1977). Parasites to be used as targets in cytotoxicity assays were washed extensively in RPMI-1640 (10% FBS) and viable, motile parasites enumerated with a hemocytometer.

Cytotoxicity assays were done using the method described by others (Kierszenbaum and Hayes, 1980) except the final volume was 0.2 ml and included 2 x 10^4 viable parasites. Cytotoxic activity of cells from mice was determined as the reduction in numbers of parasites in cultures containing host cells plus parasites from that observed in cultures of parasites and medium only. The data are presented as the mean percent cytotoxic activity from duplicate cultures. Each individual experiment reported here has been repeated at least two times with similar results.

III. RESULTS

The results of several experiments are presented in Table I where the activity of splenic NK cells of B6 mice against epimastigotes of *T. cruzi* was examined. Marginal cytotoxic activity was detected in cultures containing cells from normal mice, whereas cells from poly I:C stimulated or infected mice had significant activity against parasites. Similar results were obtained using PEC from these mice or using cells from similarly treated C3H mice. In other experiments it was observed that the cytotoxic activity of cells from stimulated or infected mice was labile in short-term culture conditions and attained maximum activity against the parasite by 6 hours (data not shown). No reduction of parasite viability was observed in cultures containing medium alone, or culture fluid from spleen cells (normal or stimulated) incubated for 5 hours without the parasite. Also, the effector cells capable of killing *T. cruzi* were in the population non-adherent to plastic.

After treatment with antiserum plus complement the cells responsible for NK activity were shown to have a surface antigen phenotype of NK 1.2^{+}, Thy $1.2^{\pm}$, Ig^{-}(Table I; results

TABLE I. Cytotoxic activity of NK cells from B6 mice against culture-derived epimastigote forms of *T. cruzi*.

Source of cells	Pretreatment	Mean % cytotoxic activity 5:1E/T	50:1E/T
Normal	None	6.0	16.5
Poly I:C stim.	None	7.5	57.5
2-day infection	None	-	53.4
Normal	Plastic NA	6.0	15.0
Poly I:C stim.	Plastic NA	11.0	69.5
Poly I:C stim.	C' alone	-	53.1
" "	anti-NK 1 + C'	-	8.5
" "	anti-Thy 1 + C'	-	29.6

from assays using anti-mouse immunoglobulin plus complement are not shown). These results are similar to that previously observed for NK cells from normal or T. cruzi-infected mice tested against YAC-1 tumor target cells (Hatcher et al., 1981 and this volume). The evidence to date from these and other experiments indicates that murine NK cells from stimulated or recently infected mice have increased cytotoxic activity against T. cruzi.

It is important to note here, that the epimastigote forms of T. cruzi are not observed in the blood of infected mice. Thus, the susceptibility of various infective trypomastigote preparations in cultures containing augmented NK activity was examined. The results (Table II) indicate that trypomastigotes from the blood of infected mice (BFT) are susceptible to the cytotoxic activity of cell populations containing augmented NK activity. However, FDT are only minimally affected by these effector cells. In other experiments we (Gwinn, Hatcher, and Kuhn, unpublished) as well as others (Krettli et al., 1979) have demonstrated immunoglobulin on the surface of trypomastigotes from mice infected with T. cruzi. Thus the activity exhibited against BFT in these studies by cells from poly I:C stimulated mice could reflect antibody-dependent, cellular cytotoxicity. Results in Table II show that when serum from immunized (anti-T. cruzi) rabbits was included in the culture medium host cells from

TABLE II. Cytotoxic activity of murine spleen cells (B6) from normal or poly I:C stimulated mice against epimastigotes, BFT, or FDT with or without antibodies specific for T. cruzi.

Source of cells	Serum supplement	Mean % cytotoxic activity against[b]		
		epimastigotes	BFT	FDT
Normal	NRS	8.3	9.1	10.2
Poly I:C	NRS	57.3	34.3	18.4
Normal	IRS	42.1	7.5	30.6
Poly I:C	IRS	84.2	37.5	40.8

[a]A dilution of rabbit serum (1/1000) from normal animals (NRS) or rabbits immunized with sonicated T. cruzi (IRS) was used in these assays (titer of these sera against epimastigotes by indirect immunofluorescence was 1/16 and 1/2048, respectively).

[b]The mean percent cytotoxic activity in duplicate cultures containing murine cells and parasites at a E/P ratio of 50:1 after a 5 hr incubation period.

normal or stimulated mice had increased cytotoxic activity against epimastigotes or FDT compared to that observed in cultures containing normal rabbit serum. The addition of specific antibodies did not, however, increase the effector activity of cells against BFT targets.

IV. Discussion

Our results show that NK cells from poly I:C stimulated mice can kill epimastigote forms and BFT of T. cruzi, but have little activity against FDT. In the presence of specific antibodies poly I:C stimulated cells had significant levels of cytotoxic activity against the trypomastigote. Thus, it might be proposed that the stimulation of NK cell activity immediately following murine infection with this parasite (Hatcher et al., 1981) would be beneficial to the infected host. It is important to note however, that both the stimulation of NK activity by T. cruzi and in vitro NK cell activity against this parasite has been observed in resistant and susceptible strains of mice. No direct correlation between NK activity with resistance, therefore, is seen in an experimental situation. It may be that the experiments studying ADCC against T. cruzi are of more importance in this regard. Other workers have observed the susceptibility of epimastigotes or trypomastigotes to mammalian host cells in in vitro assays for ADCC (Sanderson et al., 1977; Okabe et al., 1980; Kierszenbaum and Hayes, 1980; Kipnis et al., 1981). The present studies extend previous studies and indicate that murine NK cells may be involved in ADCC against this parasite. Thus the level and activity of NK cells during infection could be important to the infected host in the elimination of circulating trypomastigotes following the production of antibodies. Furthermore, such cells may contribute to the pathology of the disease by acting on host cells which have passively acquired T. cruzi antigens on their surface. Such a mechanism has been proposed by Ribeiro dos Santos and Hudson (1980), who concluded that cytotoxic cells from infected mice specific for T. cruzi-antigen coated host cells were cytotoxic T lymphocytes.

The apparent differences in susceptibility of FDT and epimastigotes to augmented NK cell activity raises several questions which may be important to the immunobiology of mammalian infection with T. cruzi. If BFT, which contain surface immunoglobulin, can eliminate or shed bound antibody after injection into experimental mice as has been reported to occur in vitro (Kloetzel and Deane, 1977), then this might explain why infection-induced NK activity has no apparent relationship with resistance to infection with BFT of T. cruzi. Second, the lack of parallel susceptibility of FDT and epimastigotes may reflect a parasite adaptation to a potentially hostile host environment and as such represent an immune evasion mechanism by the infective form of T. cruzi. An important aspect of an investigation of these questions would be to examine epimastigotes and trypomastigotes for the presence or absence of components which might serve as NK cell recognition sites. Finally, if a change in the ability of naturally cytotoxic effector cells to interact with circulating forms of T. cruzi as infection proceeds, such changes might be associated with the resistance phenotype of inbred strains of mice to this parasite.

ACKNOWLEDGMENTS

The authors wish to express their appreciation to Dr. Robert C. Burton for the anti-NK 1.2 alloantiserum, and to Michael C. Cerrone for excellent technical assistance.

REFERENCES

Brener, Z., and Chiari, E. (1971) Trans. Roy. Soc. Trop. Med. Hyg. 65: 629.
Djeu, J.Y., Heinbaugh, J.A., Holden, H.T., and Herberman, R.B. (1979) J. Immunol. 122: 175.
Hatcher, F.M., and Kuhn, R.E. (1981) J. Immunol. 126: 2436.
Hatcher, F.M., Kuhn, R.E., Cerrone, M.C., and Burton, R.C. (1981) J. Immunol. 127: 1126.
Kierszenbaum, F., and Hayes, M.M. (1980) Immunol. 40: 61.
Kierszenbaum, F., and Pienkowski, M.M. (1979) Infect. Immun. 24: 110.
Kipnis, T.L., James, S.L., Sher, A., and David, J.R. (1981) Amer. J. Trop. Med. Hyg. 30: 47.
Kloetzel, J., and Deane, M.P. (1977) Rev. Inst. Med. Trop. Sao Paulo 19: 397.
Krettli, A.W. (1977) J. Protozool. 24: 514.
Krettli, A.W., and Brener, Z. (1976) J. Immunol. 116: 755.
Krettli, A.W., Weisz-Carrington, P., and Nussenzweig, R.S. (1979) Clin. Exp. Immunol. 37: 416.
Kuhn, R.E., and Murnane, J.E. (1977) Exp. Parasitol. 41: 66.
Nathan, C., Nogueira, N., Juanbnanich, C., Ellis, J., and Cohn, Z.A. (1979) J. Exp. Med. 149: 1056.
Nogueira, N., and Cohn, Z.A. (1979) J. Exp. Med. 148: 288.
Nogueira, N., Gordon, S., and Cohn, Z. (1977) J. Exp. Med. 146: 172.
Okabe, K., Kipnis, T.L., Calich, V. L. G., and Dias da Silva, W. (1980) Clin. Immunol. Immunopath. 16: 344.

Powell, M.R., and Kuhn, R.E. (1980) J. Parasitol. 66: 399.
Ribeiro dos Santos, R., and Hudson, L. (1980) Clin. Exp. Immunol. 40: 36.
Roberson, E.L., Hanson, W.L., and Chapman, W.L. (1973) Exp. Parasitol. 34: 168.
Rodriquez, A.M., Santoro, F., Afchain, D., Bazin, H., and Capron, A. (1981) Infect. Immun. 31: 524.
Sanderson, C.J., Lopez, A.F., and Bunn-Moreno, M.M. (1977) Nature 268: 340.
Trischmann, T., Tanoqitz, H., Wittner, M., and Bloom, B. (1978) Exp. Parasitol. 45: 160.

MECHANISMS OF NATURAL MACROPHAGE CYTOTOXICITY AGAINST PROTOZOA*

Santo Landolfo
Giovanna Martinotti
Pancrazio Martinetto

Institute of Microbiology
University of Torino
10126 - Torino, Italy

I. INTRODUCTION

A large body of evidence stresses the importance of humoral and cellular natural immunity against tumors and microorganisms (1). As far as the cellular component is concerned, the cell populations involved in this type of resistance appear quite heterogeneous. Natural killer cells (NK), natural cytotoxic cells (NC) and macrophage-like cells (Mø) are a few examples of effector cells capable of mediating natural cytotoxicity depending upon the target used (1). Despite several data accumulated so far, the exact mechanisms of target destructions are however still unknown.

Previous studies at our laboratory have demonstrated the occurrence of natural cytotoxicity in vitro against an extracellular protozoon, _Trichomonas vaginalis_ (_T. vaginalis_), both in murine and human system (2,3). The effector cells in these

*Supported by CNR - PFCCN No 80.01548.96

ISBN 0-12-341360-5

systems appear to belong to the 'mononuclear phagocyte system' (4) inasmuch as they are phagocyting, adherent and positive to esterase staining. In the present study we have extended these observations by analyzing some aspects of the molecular mechanisms involved in T.vaginalis lysis in vitro.

II. MATERIALS AND METHODS

The procedures employed to isolate, culture and label T. vaginalis protozoon in vitro with ^{3}H-Thymidine (^{3}H-Thy) have been previously described (2). MØ enriched populations were obtained by plating resident peritoneal cells from normal (C57 BL/6 CR1 xDBA/2 CR1) F1 mice. The 24 hr cytotoxicity assay, measuring the release of (^{3}H-Thy) from labelled protozoa was performed as described by Landolfo et al. (2).

The following reagents were used in the present experiments: a) Colchicine (10^{-5}M), Cyclohexamide (40 ug), Puromycin (40 ug) (Sigma, St. Louis, Mo). b) Vinblastin (10^{-5}M) (Lilly, Indianapolis, Ind.). c) Cytochalasin B (5 ug) (Aldrich Co., N.Y.). d) D_2O = deuterium oxide (37%) (Carlo Erba Milano, It.) e) BLTI = Aprotinin = Bovine lung trypsin inhibitor (750 KIU/ml) (Sigma)

III. RESULTS AND DISCUSSION

In order to find out if MØ need direct contact or release a soluble factor for protozoon lysis, supernatants from MØ, cultured for 24 hr with T. vaginalis were tested for their ability to lyse in vitro labelled protozoa. As controls supernatants from MØ T. vaginalis, cultured alone were included. As shown in Table 1 supernatants derived from MØ - T. vaginalis mixtures displayed the highest cytotoxicity levels. Supernatants from MØ cultured alone

Table 1. Cytolysis of *T.vaginalis* by supernatants from Mø cultures.

Supernatants from:	% Cytolysis		
	1:4*	1:8	1:16
Mø + *T.vaginalis* ‡	++ §	+	- §
Mø	+	-	-
T.vaginalis	-	-	-

* Supernatants dilutions

‡ Cultured for 24 hr at 37°C in 5% CO_2

§ ++, 20%-40%; +, 10%-20%; -, 0-5%

were still effective but at much lower level, whereas supernatants from *T.vaginalis* alone did not display any cytotoxicity. From this data it seems that normal Mø produce a toxic factor continuously *in vitro* but they increase the release when in the presence of the protozoa.

To analyze some features of the cytotoxic factor (s), the supernatants from Mø-*T.vaginalis* cultures were subjected to different treatments. As shown in

Table 2. Effects of various treatments on CF.

Treatments	% Cytolysis
--	++ *
Dialysis	++
Heat 56°Cx30min	++
90°Cx30min	-
BLTI(Aprotinin)	++

* See Table 1.

Table 2, cytotoxic activity was retained after extensive dialysis against the culture medium, it was resistant to heating for 30 min at 56°C, but sensitive to 90°C for 30 min.

The last set of experiments was deviced to establish if the Mø cytotoxic factor was actively synthetized and secreted. Table 3 demonstrates that culturing Mø and T.vaginalis together in the presence of two protein synthesis inhibitors such as Cycloeximide and Puromycin, completely prevented the appearance of cytotoxicity in the culture supernatants.

Table 3. Effects of Mø treatments on CF production.

Mø pretreatment	% Cytolysis
—	++ *
Heat (56°Cx30min)	—
Cyclohexamide	—
Puromycin	—
Colchicin	+++
Vinblastin	+++
D_2O	++
Cytochalasin B	--

* See Table 1.

These results indicate that cytotoxic factor released by Mø requires active protein synthesis. In addition it appears that Mø microtubules and microfilaments are involved in T.vaginalis lysis. This conclusion comes from the findings that microtubule and microfilament binding drugs strongly affect protozoon lysis, either by stimulating or inhibiting it. When Mø and T.vaginalis were cultured in presence of

Colchicine and Vinblastin and D_2O supernatant cytotoxicity was significantly increased. These results are in agreement with reports showing that the above reagents can stimulate secretion of different Mø products by induced fusion of cytoplasmic lysosomes with Mø membrane (5). Therefore in our system, the increased cytotoxicity after Colchicine or Vinblastin tretment may be a consequence to enhanced release of lysosome content in the extracellular compartment by Mø. As far as Cytochalasin B is concerned, it is known that this drug disrupts microfilaments which play a crucial röle in the movement of cells and may thus prevent contact formation with target cells. So, it might be that Citochalasin B blocks cytotoxicity preventing contact between macrophages and T. vaginalis.

In conclusion in the present study we have demonstrated that natural Mø cytotoxicity against an extracellular protozoon, T.vaginalis, is mediated by one or more soluble factors actively synthetized and secreted by normal resident Mø.

ACKNOWLEDGMENTS

We thank Dr. G. Forni for critical review of the manuscript and Mrs Rachele Nobile for typing the manuscript.

REFERENCES

1. Herberman, R.B. (ed.) Natural cell-mediated immunity against tumors. Academic Press, New York 1980.
2. Landolfo, S., Martinotti, G., Martinetto P., and Forni,G., J. Immunol. 124: 508 (1980).
3. Mantovani,A., Peri, G., Polentarutti, N., Martinotti, G., and Landolfo, S., Clin. Exp. Immunol., (1981), in press.

4. Van Furth, R. (ed.) Mononuclear Phagocytes. Martinus Nijhoff Publishers, The Hague, 1980.
5. Gordon, J., and Werb, Z. Proc. Natl. Acad. Sci. 73: 872 (1976).

IN VITRO EFFECTS OF NATURAL KILLER (NK) CELLS ON CRYPTOCOCCUS NEOFORMANS[1]

Juneann W. Murphy
Department of Botany-Microbiology
University of Oklahoma
Norman, Oklahoma

D. Olga McDaniel
Department of Microbiology & Public Health
University of Alabama
Birmingham, Alabama

I. INTRODUCTION

Natural killer (NK) cells may have a much broader target range than has been previously considered. It is well established that NK cells can effectively eliminate tissue cell targets, such as certain neoplastic cells (Herberman and Holden, 1978; Herberman and Holden, 1979; Herberman et al., 1979; Riccardi et al., 1979; Riccardi et al., 1980a; Riccardi et al., 1980b; Riccardi et al., 1981; Talmage et al., 1980), viral infected cells (Bloom et al., 1980; Santoli et al., 1980; Welsh et al., 1979) and even normal bone marrow (Nunn et al., 1977; Riccardi et al., 1981), adherent peritoneal (Welsh et al., 1979), and thymus (Hansson et al., 1979; Nunn et al., 1977) cells. Very little has been done concerning the effects of NK or NK-like cells on nontissue targets (Murphy and McDaniel, 1982). There are some published data (Cauley and Murphy, 1979; Cozad and Lindsey, 1974; Cutler, 1976; Graybill and Mitchell, 1978; Tewari et al., 1981) that suggested to us that NK cells should be considered as possible effectors against certain fungi; therefore, we have been studying the abilities of cells with characteristics of NK cells to inhibit the *in vitro* growth of a yeast-like organism, *Cryptococcus neoformans*. Using unstimulated murine or rat splenic cells, we have simultaneously determined (1) the *in vitro* NK reactivity and (2) the *C. neoformans* growth inhibitory ability *in vitro*, to demonstrate a correlation between the two activities. In addition, we have partially characterized the effector cells in the *C. neoformans* growth inhibition assay and have begun to study the kinetics of cryptococci growth inhibition.

[1]This study was supported by an NIH Biomedical Research Support grant No. 2 507 RR07078, a Brown-Hazen Research Corporation Grant and Public Health Service Grants AI-15716 from the National Institutes of Allergy and Infectious Diseases.

ISBN 0-12-341360-5

II. CORRELATION OF NK CELL REACTIVITY WITH ABILITY TO INHIBIT GROWTH OF C. NEOFORMANS IN VITRO

In a series of experiments, NK cell reactivities of splenic cell populations were varied by nylon wool passage, varying donor mouse strains, using different aged mice, treating splenic cell donors with poly I:C or Corynebacterium parvum, and discontinuous Percoll gradient fractionation of the spleen cells. In each case, the NK reactivity of the effector cell pool was determined by using a 4 hr ^{51}Cr release assay with YAC-1 targets with an effector:target (E:T) ratio of 50:1, and the ability of the effector cells to inhibit the growth of cryptococci was measured by incubating C. neoformans isolate 184 with the effector cells at an E:T ratio of 500:1 for 18 hr prior to enumerating the viable cryptococci by a plating technique (Murphy and McDaniel, 1982). The results of several typical experiments using murine effector cells are shown in Table 1.

TABLE 1. Correlation of NK reactivity with C. neoformans growth inhibitory ability

Effector Cell Source (mouse-age, strain)	% Specific ^{51}Cr Release ± SEM	% Cryptococci Growth Inhibition ± SEM
Normal Splenic Cells (7 wk old CBA/N)	53.0 ± 2	56.0 ± 4
Nylon Wool Nonadherent (NWN) Splenic Cells	67.0 ± 3	79.0 ± 3
NWN Splenic Cells (7 wk old CBA/N)	58.6 ± 1.5	98.0 ± 1.6
NWN Splenic Cells (7 wk old A.TH)	14.0 ± 3.2	23.3 ± 3.2
NWN Splenic Cells (7 wk old CBA/N)	56.4 ± 2.8	91.6 ± 0.2
NWN Splenic Cells (20 wk old CBA/N)	37.9 ± 0.8	6.2 ± 0.2
NWN Splenic Cells (20 wk old CBA/N)	39.4 ± 0.4	64.7 ± 2.1
NWN Splenic Cell (20 wk old poly I:C treated CBA/N)	78.5 ± 2.0	81.9 ± 0.5
NWN Splenic Cells (20 wk old CBA/N)	36.6 ± 0.3	9.0 ± 2.6
NWN Splenic Cells (20 wk old CBA/N C. parvum treated:)		
3 days, previously	45.0 ± 1.5	85.8 ± 3.4
7 days, previously	22.6 ± 0.7	27.9 ± 0.4

We found that removal of nylon wool adherent cells (macrophages and B cells) enhanced the NK and cryptococci growth inhibitory activities of the cell pools. These results not only showed a correlation between NK and growth inhibitory activities but also indicated that the effector cells were not macrophages or B cells. In all situations examined, a correlation was observed between the NK and growth inhibitory activities. For example, nylon wool nonadherent (NWN) splenic cells from seven week old CBA/N mice had high levels of NK activity and inhibited C. neoformans growth better than cells from age-matched A.TH animals that had low NK activity. Effector cells from young (6-8 week old) mice with high NK reactivity were more efficient in inhibiting the growth of C. neoformans than NWN splenic cells from old (16-20 week old) animals. Poly I:C, an NK augmenting agent, simultaneously boosted the NK and cryptococci growth inhibitory activities in spleens of mice treated 24 hr previously. The growth inhibitory ability of splenic cells paralleled the rise and fall of NK cell reactivity after treating CBA/N mice iv with 0.7 mg of formalin-killed C. parvum. Taken together these results represented a substantial amount of indirect evidence supporting the contention that NK cells were responsible for the reduction of C. neoformans in the in vitro assays using murine effector cells.

Two other facets were considered in comparing NK and cryptococci growth inhibitory activities. One was to eliminate the possibility that the cryptococci growth inhibitory capability was a unique activity of mouse cells, and the other was to be certain that C. neoformans isolate 184 was not an unusually sensitive target. To check the former, splenic cells from young outbred rats were separated into subpopulations by discontinuous Percoll gradient fractionation (Reynolds et al., 1981), prior to determining the % large granular lymphocytes (LGL), % effector cell: C. neoformans conjugates, % ^{51}Cr release with YAC-1 targets, and the % C. neoformans growth inhibition per cell fraction. Table 2 shows the results. NK cells which are typically large granular lymphocytes

TABLE 2. Characteristics of various Percoll gradient fractions of rat splenic cells

Fraction	% LGL[a]	% Effector:*C. neoformans* Conjugates	% ^{51}Cr Release	% *C. neoformans* Growth Inhibition
Unfractionated	16	8	12.6	57.8
3	72	23	28.1	64.5
4	24	10	17.6	INS[b]
5	61	14	21.1	59.3
6	11	4	8.0	36.9
7	1	4	1.9	4.5

[a]LGL - large granular lymphocytes

[b]INS - insufficient sample

(Reynolds et al., 1981) were found predominately in fraction 3 and that fraction of cells also exhibited the highest % effector cell: *C. neoformans* conjugates, % specific ^{51}Cr release and % cryptococci growth inhibition. In all other fractions these parameters varied directly with the variance of % LGL. These data provided additional support for the concept that NK cells were the cells responsible for the growth inhibition of cryptococci, and they also indicated that the cryptococci inhibitory activity was not an isolated phenomenon occurring only with murine effector cells. To ascertain whether or not isolate 184 was an unusually sensitive target, 6 other *C. neoformans* isolates were used in the growth inhibition assay with effector cells from 7 week old CBA/N or A.TH mice. The CBA/N splenic cell pools with high NK activity inhibited the growth of all isolates of *C. neoformans* significantly better than A.TH splenic cells which had low NK reactivity (Murphy and McDaniel, 1982), demonstrating that isolate 184 was not a unique target.

III. PROPERTIES OF THE EFFECTOR CELLS RESPONSIBLE FOR IN VITRO CRYPTOCOCCI GROWTH INHIBITION

To characterize the effector cells in the growth inhibition assay, CBA/N splenic cells were treated with various antisera and complement prior to being tested for their cytolytic ability against YAC-1 targets and growth inhibitory ability against C. neoformans. Table 3 summarizes the results. These data plus those showing enrichment of NK and growth inhibitory activities by nylon wool passage defined the effector cells in the ^{51}Cr release and growth inhibition assays as: nonadherent, Thy 1^-, surface Ig^-, Ia^-, and asialo GM_1 positive, characteristics which are typical of murine NK cells.

TABLE 3. Effects of various treatments on NK and cryptococci growth inhibition activities

	Effect on:	
Pretreatment of Effector Cells	NK Reactivity	Cryptococci Growth Inhibitory Ability
anti-Thy 1 + C'	none	none
anti-MIg + C'	none	none
anti-Ia^k + C'	none	none
anti-asialo GM_1 + C'	significantly reduced	significantly reduced

Competitive inhibition experiments have been done using a procedure similar to those described by Herberman et al. (1976), in an attempt to determine whether or not the same subpopoulation of effector cells was responsible for the cytolytic activity on YAC-1 targets and inhibition of cryptococci growth; however, at this stage we have obtained conflicting data. C. neoformans did not competitively inhibit in the ^{51}Cr release assay with YAC-1 targets; whereas, YAC-1 cells did significantly reverse the cryptococci growth inhibition. Before any conclusions can be drawn, considerably more work must be done in this area.

IV. PRELIMINARY STUDIES ON IN VITRO GROWTH INHIBITION KINETICS

Our initial work on NK cells against C. neoformans was done by incubating the effector cells with C. neoformans at an E:T ratio of 500:1 for 18 hr before determining the number of cryptococci CFU. Other E:T ratios ranging from 100:1 to 500:1 have been used in the 18 hr assay with NWN splenic cells from 7 week old CBA/N mice as effector cells. As the E:T ratios were increased, the % growth inhibition also increased (Figure 1); therefore in studies varying other parameters, an E:T ratio of 500:1 was consistently used. Another parameter studied was the incubation time required before cryptococci growth inhibitory effects could be observed. Five hours has been the shortest time evaluated, and we found that cryptococci growth was inhibited 72% by 7 week old CBA/N splenic NWN cells and 34% by similar populations of effector cells from A.TH mice.

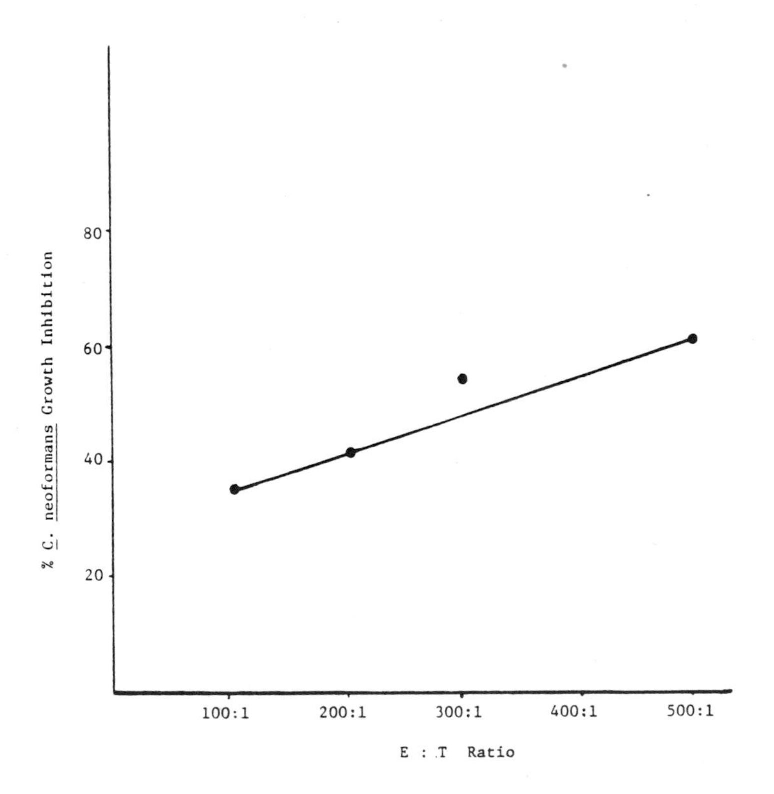

Figure 1. Effect of effector cell concentration on inhibition of cryptococci growth.

Although the numbers of C. neoformans cells increased in the assay vessels (control and test) over the next 20 hr, the % inhibition of cryptococci growth did not change significantly. It appears that the event which effects C. neoformans growth occurs relatively early in the incubation period (within 5 hr) and is detectable up to 24 hr after mixing of effector and target cells. This event is most likely fungicidal rather than fungistatic, since with acridine orange and fluorescent microscopy, we have been able to observe dead (red) cryptococci cells adjacent to viable lymphocytes (green) after incubating the two cell populations together for 5 hr.

V. CONCLUSIONS

The results presented here strongly suggest that NK cells are cytotoxic, in vitro, against a target of nontissue origin, namely a yeast-like organism C. neoformans. This discovery may be a first step in changing our previous concepts concerning the spectrum of NK activity and first line cellular defense mechanisms against mycotic and possibly bacterial and parasitic infectious agents.

ACKNOWLEDGEMENTS

We thank Mr. Lee Mosley and Ms. Mitra Rahimi for technical assistance and Ms. Susan Nelson and Ms. Beverly Richey for typing of the manuscript.

REFERENCES

Bloom, B., Minato, N., Neighbour, A., Reid, L., and Marcus, D. (1980). In "Natural Cell-Mediated Immunity Against Tumors" (R. B. Herberman, ed.), p. 505. Academic Press, New York.

Cauley, L. K. and Murphy, J. W. (1979). Infect. Immun. 23:644.

Cozad, G. C. and Lindsey, T. J. (1974). Infect. Immun. 9:261.

Cutler, J. E. (1976). J. Reticuloendothel. Soc. 19:121.

Graybill, J. R. and Mitchell, L. (1978). Infect. Immun. 21:674.

Hansson, M., Karre, K., Kiessling, R., Roder, J., Anderson, B., and Hayry, P. (1979). J. Immunol. 123:765.

Herberman, R.B. and Holden, H.T. (1978). In "Advances in Cancer Research" (G. Klein and S. Weinhouse, ed.) p. 305. Academic Press, New York.

Herberman, R. B. and Holden, H. T. (1979). J. Natl. Cancer Inst. 62:441.

Herberman, R. B., Nunn, M. E., and Holden, H. T. (1976). In "In Vitro Methods In Cell Mediated and Tumor Immunity" (B. R. Bloom and J. R. David, ed.) p. 489. Academic Press, New York.

Herberman, R. B., Djeu, J. Y., Kay, H. D., Ortaldo, J. R., Riccardi, C., Bonnard, G. D., Holden, H. T., Fagnani, R., Santoni, A., and Puccetti, P. (1979). Immunol. Rev. 44:43.

Murphy, J. W. and McDaniel, D. O. (1982). J. Immunol. In press.

Nunn, M. E., Herberman, R. B., and Holden, H. T. (1977). Int. J. Cancer 20:381.

Reynolds, C. W., Timonen, T., and Herberman, R. B. (1981). J. Immunol. 127:282.

Riccardi, C., Puccetti, P., Santoni, A., and Herberman, R. B. (1979). J. Natl. Cancer Inst. 63:1041.

Riccardi, C., Santoni, A., Barlozzari, T., Puccetti, A., and Herberman, R. B. (1980a). Int. J. Cancer 25:475.

Riccardi, C., Santoni, A., Barlozzari, T., and Herberman, R. B. (1980b). In "Natural Cell-Mediated Immunity Against Tumors" (R. B. Herberman, ed.), p. 1121. Academic Press, New York.

Riccardi, C., Barlozzari, T., Santoni, A., Herberman, R. B., and Cesarini, C. (1981). J. Immunol. 126:1284.

Riccardi, C., Santoni, A., Barlozzari, T., and Herberman, R. B. (1981). Cell. Immunol. 60:136.

Santoli, D., Perussia, B., and Trinchieri, G. (1980). In "Natural Cell-Mediated Immunity Against Tumors" (R. B. Herberman, ed.), p. 1171. Academic Press, New York.

Talmage, J. E., Meyers, K. M., Prieur, D. J., and Starkey, J. L. (1980). Nature 284:622.

Tewari, R. P., Mkwananzi, J. B., McConnachie, P., Von Behren, L. A., Eagleton, L., Kulkarni, P., and Bartlett, P. C. (1981). ASM Abstracts. p. 324.

Welsh, R. M., Zinkernagel, R. M., and Hallenbeck, L. A. (1979). J. Immunol. 122:475.

NK ACTIVITY OF TUMOR INFILTRATING AND LYMPH NODE LYMPHOCYTES IN HUMAN PULMONARY TUMORS[1]

Sidney H. Golub
Masayuki Niitsuma
Norihiko Kawate
Alistair J. Cochran
E. Carmack Holmes

Department of Surgery, Division of Oncology,
UCLA School of Medicine, Los Angeles, California

I. INTRODUCTION

One of the general assumptions among those studying NK cells is that these cells are important in the host defense against tumors. However, little evidence is available to indicate that NK cells actively restrict tumor growth of established tumors or can lead to tumor regression. One approach to this question has been to study the cells actually at the tumor site. Such studies have generally shown that the human lymphoid cells infiltrating tumor sites have little NK activity (Vose *et al.*, 1977; Totterman *et al.*, 1978; Moore and Vose, 1981). Our purpose was to determine if NK cells are even present at tumor sites and the influence of the tumor microenvironment on NK function. To do this, we studied both tumor infiltrating lymphocytes (TIL) from human pulmonary tumors and TIL from human tumors that had received an intralesional injection of BCG. As BCG can augment NK activity in humans (Thatcher and Crowther, 1978) and since this agent can provoke intense granulomatous reactions at injected tumor sites, we

[1]These studies were supported in part by USPGS grants CA12582 and CA29938.

ISBN 0-12-341360-5

felt that this would be a useful way to study the regulation of NK activity in the presence of a tumor mass.

II. NK ACTIVITY OF TIL

We have previously reported (Niitsuma et al., 1981) that TIL from human pulmonary tumors have decreased proportions of T cells and decreased activity in assays of T cell function including alloantigen induced proliferation and generation of cytolytic T cells. However, we did find considerable proportions of cells with Fcγ receptors among these TIL (average of 18.8% EAγ rosette-forming cells) and therefore it is possible that NK activity would be present among TIL. As shown in Table I, ^{51}Cr-release assays against K562 targets showed low activity among TIL from pulmonary tumors of a variety of histologic types. The TIL were obtained by mechanical disaggregation, Ficoll-Hypaque density gradient centrifugation, and further fractionated by filtration through nylon monofilament mesh to remove tumor cells. We have shown that removal of residual tumor cells did not result in improvement of NK activity (Niitsuma et al., 1981), indicating that the low NK activity is not due to cold target competition by residual tumor cells. However, TIL obtained from tumors injected two weeks prior to surgery with intralesional BCG (Holmes et al., 1979) did exhibit higher levels of NK activity than TIL from uninjected tumors. Similarly, hilar lymph node lymphocytes (LNL) and peripheral blood lymphocytes (PBL) from the same patients showed modestly higher NK activity (Table I). It is interesting to note that in three cases we were able to obtain BCG injected TIL and uninjected TIL from the same patients. These patients had multiple pulmonary metastases, only one of which was BCG injected. In these cases, the NK activity of uninjected TIL remained low while the NK activity of TIL from the adjacent BCG injected tumors was high (Table I). These preliminary results suggest that the induction of augmented NK activity at one tumor site, even in combination with augmented systemic NK activity among PBL and LNL, does not necessarily ensure the delivery of active NK cells at other tumor deposits. In this sense, our results are compatible with the suggestion of Hanna and Fidler (Hanna and Fidler, 1980) that NK activity is most effective against circulating tumor cells and of less relevance in defenses against established tumor deposits.

Since we had shown the presence of relatively high proportions of Fcγ receptor cells among TIL, we then asked whether there were target binding and killing cells among the TIL. For this assessment we utilized the single cell assay of Grimm and Bonavida (Grimm and Bonavida, 1979). Preliminary results with this analysis are shown in Table II. Thus far we have

TABLE I. NK Cytotoxicity of TIL and BCG-TIL

Cell Source		Number of Samples	Mean % Cytotoxicity ± S.E. (50:1)
No BCG:	TIL	20	7.7 ± 1.9
	LNL	7	15.4 ± 4.8
	PBL	19	28.6 ± 4.0
BCG:	BCG-TIL	5	20.5 ± 7.7
	TIL (not injected)	3	5.7 ± 4.7
	LNL	4	20.7 ± 4.5
	PBL	5	37.0 ± 6.1

seen equivalent numbers of target binding cells among TIL as we see among PBL from the same patients. The proportion of those binding cells actually mediating cytolysis in a three hour assay is also similar among TIL as compared to PBL (Table II). Thus, the low activity in ^{51}Cr-release assays is probably not due to a lack of NK cells among the TIL but instead appears to be due to a functional impairment of these cells. Since the proportion of K562 killing cells among the binders is comparable to PBL, the functional impairment is most likely in the recycling ability of the NK cells among the TIL. In line with this hypothesis are the results of Moore and Vose (Moore and Vose, 1981) and our unpublished observations that TIL cannot be activated to further NK activity with interferon. Interferon, which has been shown to accelerate the kinetics of NK-induced lysis (Silva et al., 1980), would presumably promote recycling and might have little effect on cells unable to mediate repeated lytic events.

Freeing the TIL from the tumor environment did not restore their NK activity. Incubation of TIL for 24 hours at 37° failed to restore NK activity to levels comparable to those in the PBL. For example, four TIL specimens we tested had NK activity of 5%, 4%, 3%, and 2% vs K562 at a 50:1 effector:target ratio. Incubation for 24 hours at 37° followed by washing resulted in very slight increases in NK cytotoxicity (6%, 10%, 4%, and 4% for the same four samples). The addition of 1×10^{-5} M indomethacin in the incubation medium did not improve cytotoxic activity of the TIL against K562 targets. Thus, the TIL appear to be profoundly suppressed in NK function and cannot be restored with activators of NK activity or by allowing them a period of time to shed tumor cell products or other suppressive substances.

TABLE II. Target Binding and Killing Cells

Cell Source	Number of Samples	% Binding Cells[a] ± S.E.	% Killers[b] ± S.E.	% Total NK[c] ± S.E.
TIL	6	10.7 ± 1.9	33.4 ± 6.6	3.6 ± 0.8
LNL	4	15.3 ± 4.2	48.6 ± 12.8	7.6 ± 2.6
PBL	6	9.3 ± 2.1	39.8 ± 6.3	4.2 ± 1.6

[a] % cells binding K562. 300 lymphocytes scored.
[b] % binding cells that kill. 100 binders scored.
[c] % of all cells that kill. Product of a x b.

III. SUPPRESSOR CELL ACTIVITY OF TIL AND LNL

One possible mechanism for the suppressed activity in NK assays of TIL would be the presence of suppressor cells of NK function. We have examined this possibility by the addition of TIL to autologous PBL and have not found the mixtures to show significantly diminished cytotoxic activity relative to the PBL. Although TIL have been reported to contain suppressor cells for proliferative responses (Vose and Moore, 1979), our results (Table III) show very little suppressive influence of the TIL on NK activity. Furthermore, removal of adherent cells on nylon fiber columns also failed to restore NK activity (Niitsuma _et al._, 1981) indicating that an adherent suppressor of NK activity appears to be unlikely in this situation. Of course, it is possible that suppressor cells for the development of NK cells from their precursors exist or that the conditions we chose to test for suppressor cells are not able to detect their presence. However, it is also possible that suppressor functions, like proliferative and cytotoxic functions, are not fully expressed by TIL. The microenvironment of the tumor may not be conducive to any differentiated cell functions including suppression. If that is the case, one might expect that cells at a site more distant from the tumor, but still under the influence of tumor cell products, would be more likely to exert suppressive functions. We have begun preliminary experiments to test this hypothesis. We have begun to examine LNL for suppressive activity as a function of anatomical location in relationship to the tumor. One such experiment is shown in Figure 1. In this experiment inguinal LNL were obtained from a patient with malignant melanoma and divided into nodes that were in close proximity to a node that had been completely replaced by melanoma cells ("superficial" nodes) and another group of nodes that were more distant from the same tumor-bearing node ("deep" nodes). The LNL tested were from nodes that did not contain tumor cells. The LNL from both node groups showed equivalent levels of NK activity.

TABLE III. NK Activity of TIL and PBL Mixtures

Sample Number	% Cytotoxicity vs K562[a] of:		
	TIL[b]	PBL[b]	TIL + PBL[c]
1	5	56	49
2	2	30	24
3	0	9	10
4	2	43	44
5	6	31	25

[a] 5×10^3 K562/well.
[b] 2.5×10^5 cells/well.
[c] 2.5×10^5 TIL + 2.5×10^5 PBL/well.

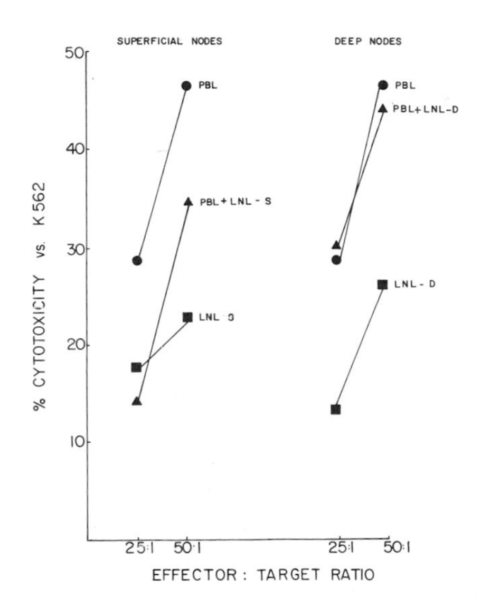

Figure 1. NK cytotoxicity of PBL (circles), LNL (squares) and mixtures of PBL and LNL (triangles). Effector:target ratio refers to PBL or LNL when tested alone and to PBL when tested in a 1:1 mixture with LNL.

However, the "deep" node LNL had little effect upon the NK activity of PBL while the "superficial" node LNL, which were obtained from nodes in close proximity to the tumor, suppressed the PBL activity so that the mixture of cells had less activity than the PBL alone. These results, while quite preliminary, do suggest that suppressors of NK effector function can exist and may be influenced in their development by proximity to the tumor site. Such cells may be similar to the NK suppressors described by Mantovani et al.(Manovani *et al.*, 1980) in carcinomatous effusions. It is our working hypothesis that the cells actually infiltrating solid tumors are too profoundly influenced by the microenvironment of the tumor to express a differentiated function such as suppressor cell activity while lymphoid tissues draining the tumor site or effusions would be more likely locations to find such suppressor cells.

ACKNOWLEDGMENTS

The technical assistance of Ms. Veronica Routt is gratefully acknowledged.

REFERENCES

Grimm, E. and Bonavida, B., J. Immunol., 123:2861-2869, 1979.
Hanna, N. and Fidler, I.J., J. Natl. Cancer Inst., 65:801-809, 1980.
Holmes, E.C., Ramming, K.P., Bein M.E., Coulson, W.F. and Callery, C.D., J. Thor. Cardiovasc. Surg., 77:362-368, 1979.
Mantovani, A., Allavena, P., Sessa, C., Bolis, G. and Mangioni, C., Int. J. Cancer, 25:573-582, 1980.
Moore, M. and Vose, B.M., Int. J. Cancer, 27:265-272, 1981.
Niitsuma, M., Golub, S.H., Edelstein, R. and Holmes, E.C., J. Natl. Cancer Inst., 67:997-1003, 1981.
Silva, A., Bonavida, B. and Targan, S., J. Immunol., 125:479-484, 1980.
Thatcher, N. and Crowther, D., Cancer Immunol. Immunother., 5:105-107, 1978.
Totterman, T.H., Hayry, P., Saksela, P., Timonen, T. and Eklund, B., Eur. J. Immunol., 8:872-875, 1978.
Vose, B.M. and Moore, M., Int. J. Cancer, 24:579-585, 1979.
Vose, B.M., Vanky, F., Argov, S. and Klein, E., Eur. J. Immunol., 7:753-757, 1977.

NATURAL KILLER ACTIVITY IN HUMAN OVARIAN TUMORS[1]

Martino Introna
Paola Allavena
Raffaella Acero
Nicoletta Colombo
Pierangela Molina
Alberto Mantovani

Istituto di Ricerche Farmacologiche
"Mario Negri"
Milan, Italy

I. INTRODUCTION

The analysis of NK activity within tumors should provide information relevant to understanding its possible role in the control of established malignancy and it should provide a conceptual framework for therapeutic attempts aimed at activating this host defence mechanism at the tumor site. Lymphoid cells isolated from murine sarcomas have low but significant NK activity (1,2,3). Human neoplasms are to some extent heterogeneous in the NK effector capacity of tumor-associated lymphoid cells (TAL) but, in general, TAL have defective NK cytotoxicity, or none at all (4,5,6). Our attention is focused on human ovarian carcinoma, which, in addition to solid tumor masses, frequently causes ascites, an easily accessible source of TAL, tumor associated macrophages (TAM) and carcinoma cells in suspension (6,7,8). Here we will discuss our current understanding of the regulation of NK activity in these human tumors.

---

[1] Supported by Italy-Great Britain Cooperation program (CNR, Rome, Italy), by CNR (No. 80.01579.96) and by Grant 1R01 CA 26824 from the National Cancer Institute, USA.

ISBN 0-12-341360-5

II. EXPERIMENTAL PROCEDURES

Methods for the isolation of TAL, TAM and carcinoma cells from human ovarian tumors have been detailed elsewhere (6,7, 8). Unless otherwise specified NK activity was measured in a 4 h ^{51}Cr release assay against K562 cells at an attacker: target cell (A:T) ratio of 25:1. Large granular lymphocytes (LGL) were identified in Giemsa stained preparations (9) and they were enriched in the less dense fractions of discontinuous (32.5-45%) Percoll gradients as described by Timonen and Saksela (9).

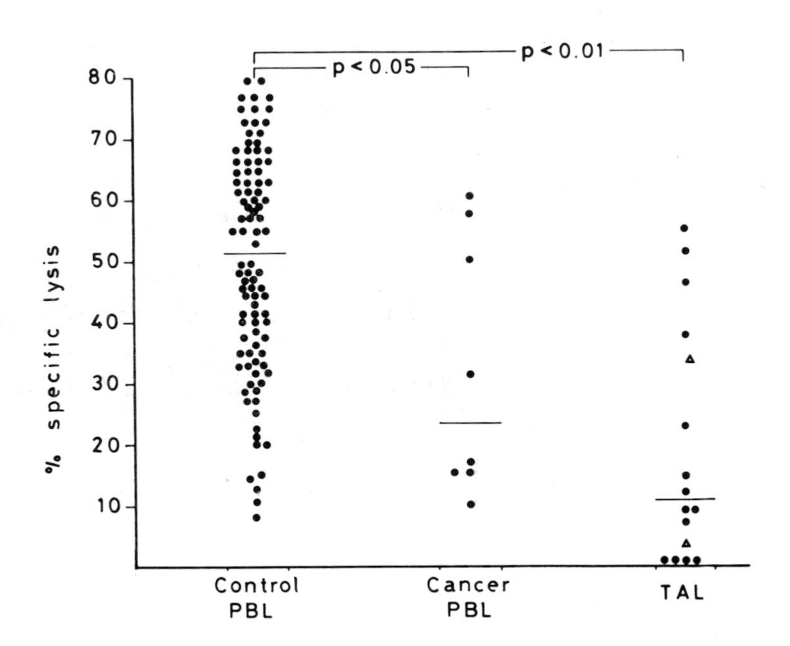

FIGURE 1. NK activity against K562 cells of TAL from ovarian cancer patients. The open triangles indicate TAL from solid ovarian carcinomas. Results are presented as percentage of specific lysis after 4h at an A:T ratio of 25:1.

III. NK ACTIVITY OF TAL FROM ASCITES OR SOLID OVARIAN TUMORS

In previous investigations, we repeatedly found that TAL from ascitic ovarian tumors have usually very low, but significant, NK activity against K562 (6,7,10). This reactivity was lower than that of peripheral blood lymphocytes (PBL) from the same patients and these in turn were less reactive than normal PBL. These previous observations on ascitic TAL are confirmed by results shown in Fig. 1. Moreover, we have separated sufficient TAL from 2 solid ovarian carcinomas (triangles in Fig. 1) to permit testing NK activity: one of the two solid tumor TAL was completely devoid of NK cytotoxicity, whereas the other caused low, but appreciable, killing of K562 (34% at an A:T ratio of 25:1). PBL and TAL from ovarian cancer patients exhibited susceptibility to augmentation of NK activity against K562 by in vitro interferon (IFN) comparable to control PBL (6).

IV. SEARCH FOR SUPPRESSOR CELLS FOR NK CELLS WITHIN OVARIAN CARCINOMAS

The defective NK activity of TAL could in principle be accounted for by at least two mechanisms, one being the presence of suppressor cells or factors and the other being a low number of the relevant effectors, recently identified in blood as LGL (11), which enter the tumor. Preliminary evidence for the presence in situ of cells inhibiting NK activity was obtained in 3 peritoneal effusions from ovarian cancer(6),in 2 breast carcinomas(12)and in pleural carcinomatous effusions(13). We have examined a relatively large number of ovarian cancer patients for the presence of suppressor TAL or TAM. In part of the experiments TAL or TAM were mixed in varying ratios with PBL and lysis of K562 cells was measured immediately (7), whereas in others, suppressors were allowed to interact with PBL for 20 h before adding indicator PBL (Table 1 and 2). Results obtained with the first protocol are summarized in Table 1 and have been detailed recently (7). TAM were devoid of suppressive activity on NK cells when the assay was performed immediately after mixing, whereas in a minority of TAL preparations (7/27) we confirmed our preliminary indications (6) for the presence of inhibitory cells at the tumor site. Filler controls devoid of NK activity used in these experiments were murine thymocytes, human polymorphs and, in some experiments autologous TAM (ref. 7 and see above). Preincubation of TAL with PBL overnight did not increase the number of

TABLE I. Inhibition of NK Activity by TAL and TAM from Ovarian Carcinomatous Ascites

	Immediately tested		Tested after 20 h preincubation with PBL	
	Suppressor/ total	Inhibition (%)	Suppressor/ total	Inhibition (%)
TAL	7/27	36 (14-60)	1/5	27
TAM	1/6	17	5/8	44 (13-80)

Results summarized here refer to a ratio of putative suppressor to PBL of 1:1. Inhibition of NK activity is presented as median % with range in parenthesis. TAL or TAM were mixed with PBL and either tested immediately for NK activity or cultured overnight before the assay. A typical experiment of the latter protocol is shown in Table II.

TABLE II. Effect of Cocultivation of TAM and PBL on NK Activity

PBL (A:T ratio)	TAM (A:T ratio)	Specific lysis (%)	Inhibition (%)
12:1 (a)	-	14.0 $\pm$ 2.1	-
12:1 (a)	12:1	6.9 $\pm$ 2.0	50
12:1 (a)	6:1	12.2 $\pm$ 2.1	13
12:1 (a)	3:1	14.1 $\pm$ 2.9	-
12:1 (b)	-	23.6 $\pm$ 5.5	-
12:1 (b)	12:1	11.4 $\pm$ 2.1	52
12:1 (b)	6:1	13.6 $\pm$ 3.9	42
12:1 (c)	-	49.3 $\pm$ 7.6	-
12:1 (c)	12:1	10.1 $\pm$ 0.5	80

TAM from one ovarian cancer patients were tested for inhibition of NK activity of autologous PBL (a), allogeneic ovarian cancer PBL (b) and normal PBL (c). TAM were cultured 20 h with PBL before testing cytotoxicity. TAM alone caused less than 5% specific lysis. Specific lysis is mean % $\pm$ S.D.

TAL showing inhibitory capacity on NK activity or the levels of inhibition (Table I). In contrast, when TAM were cultured overnight with PBL, appreciable inhibition of NK activity was observed in part of the patients (5/8, Table I). Table II shows one of these experiments: TAM caused a significant dose-dependent inhibition of NK lysis at a ratio to PBL of 0.5:1-1:1, and suppression was detected with both autologous and allogeneic PBL.

V. LGL IN ASCITIC AND SOLID OVARIAN CARCINOMAS

Blood effector cells of human NK activity have been positively identified as LGL (9,11). Therefore we evaluted the concentration of LGL in TAL morphologically (Fig. 2). Normal PBL had a median percentage of LGL of 10% (range 5-22 mean ± S.D. 10.9 ± 4.7) and the limited number of ovarian cancer PBL preparations examined showed values in the same range (Fig.2). In contrast the median LGL concentration in TAL was 2% (range 0.5-10%, mean ± S.D. 2.8±2.7, $p < 0.01$ versus normal or cancer PBL). Only one TAL preparation showed a relatively high number of LGL (10%) and it also had relatively high NK activity (46.5% specific lysis). Comparison of Fig. 1 and 2 reveals that some TAL preparations with very low LGL concen-

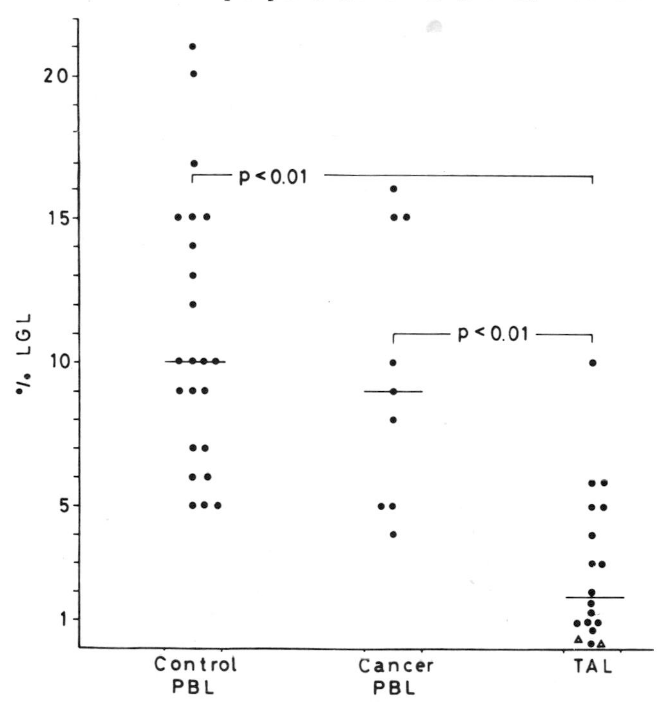

FIGURE 2. Percentage of LGL in TAL from ovarian cancer patients. The open triangles indicate TAL from solid ovarian carcinomas.

trations had appreciable levels of NK cytotoxicity: for instance, the two lymphoid cell preparations from solid tumors had less than 0.5% LGL (Fig. 2), but one of the two had appreciable cytotoxic activity (34% specific lysis, Fig. 1).

In humans, blood cells with LGL morphology and NK activity can be enriched by use of discontinuous gradients of Percoll (9,11). Inasmuch as it is not known whether at sites other than blood effectors of NK activity in humans retain the LGL morphology, it was of interest to evaluate whether the low NK activity of TAL was related to LGL, as expected on the basis of studies with PBL (9,11,14-16). In agreement with previous reports (9,11), blood LGL sedimented in low density Percoll fractions, with a percentage of 75% in fraction 2 compared to 17% of the original preparation (4.4 times enrichment) (Fig.3).

When the same procedure was applied to TAL, the distribution of LGL and NK cytotoxicity was the same as for blood, NK peaking in fractions 2-3, where the highest LGL concentration was found without gross contamination with tumor cells or macrophages, and no activity in fraction 5 and 6. It is of interest that enrichment for LGL on Percoll permitted detection

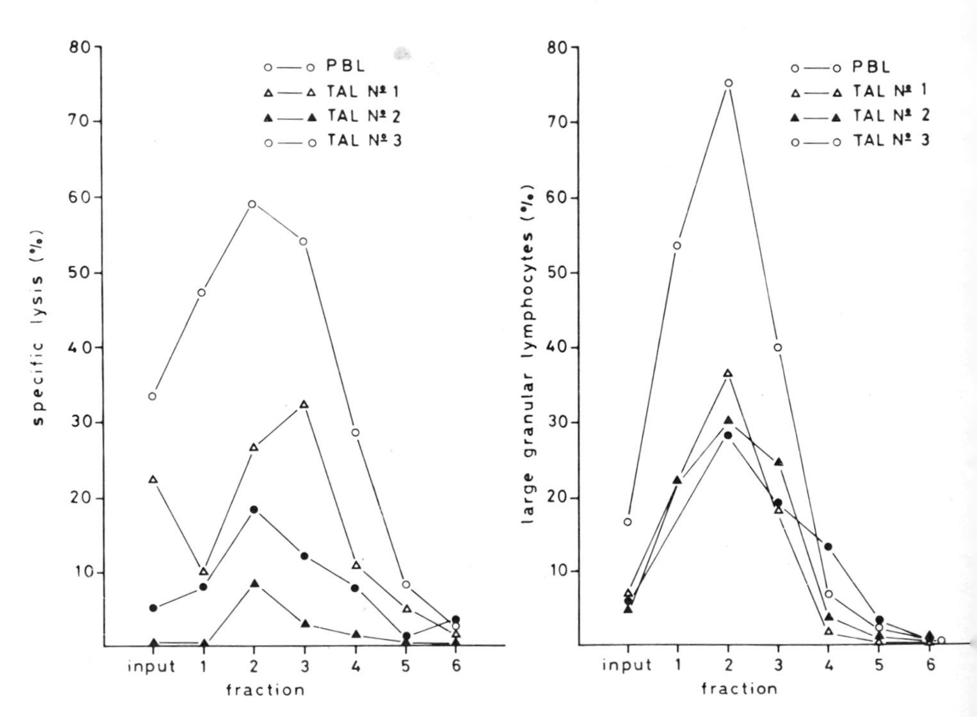

FIGURE 3. Sedimentation on Percoll gradients of LGL and NK activity from TAL. NK activity was measured at an A:T ratio of 25:1 after 20 h of incubation.

of appreciable NK cytotoxicity in 2 TAL preparations (o and Δ symbols, Fig. 3) virtually devoid of activity before separation. Although relative to unseparated cells enrichment for LGL in fraction 2 was the same for lymphocytes from the tumors as for cells from blood (four to sixfold), because of the low representation of LGL at the tumor site, the final LGL concentration attained in fraction 2 from TAL was relatively low (28-37%).

VI. CONCLUDING REMARKS

Effector cells involved in NK activity in humans have been positively identified as LGL (9,11,14-16). Upon separation on Percoll gradients previously used for blood NK effectors (9,11), NK activity of TAL and cells with LGL morphology sedimented in the same fractions and these were the same as for blood NK cells. The codistribution of LGL and NK activity of TAL sedimented on Percoll is compatible with the hypothesis that also in these tumors, as observed with peripheral blood (9,11,14-16), LGL mediate NK activity. NK activity is usually low or absent within human and murine tumors (1-6,17-19) and various factors could play a role in this defective reactivity. In ovarian carcinoma we found evidence for cell-mediated inhibition of NK activity only in a minority of the patients (7). Moreover TAL were found to inhibit NK activity in 2 breast carcinomas (12) and suppressors for NK activity were found in the blood of occasional melanoma patients (20). Vose and Moore found no evidence for suppressive activity on NK in lymphoid cells from lung tumors (18). In the present investigation, purified TAL from ascitic or solid ovarian noeplasms had a very low representation of cells with LGL morphology. Similarly, Saksela et al. (16,18) reported a low frequency of LGL in lymphoid cells infiltrating 3 seminomas and 1 ovarian carcinoma. Hence, it is likely, that a relative lack of relevant effector cells is the major determinant of defective NK cytotoxicity within these human tumors. This does by no means exclude a contributory role, in individual patients, of other pathogenetic mechanisms.

Occasional TAL preparations (e.g. one of the two from solid tumors, Fig. 1 and 2) had appreciable NK activity in spite of having low LGL levels. Cells other than LGL could contribute to killing of K562 in these subjects. However, macrophages, a likely candidate for such cytotoxicity, were a minor contaminant of TAL and K562 proved relatively resistant to monocytes and macrophages (21). Moreover, tumor-associated macrophages from ovarian carcinomas showed no evidence of activation in terms of enhanced cytolytic capacity (8). One possibility is that the few LGL present were very active. Separation of LGL from these particular neoplasms should help to elucidate this issue.

REFERENCES

1. Becker, S., and Klein, E. (1976). Eur. J. Immunol. 6:892.
2. Flannery, G.R., Robins, R.A., and Baldwin, R.W. (1981). Cell. Immunol. 61: 1.
3. Moore, K., and Moore, M. (1979). Br. J. Cancer 39: 636.
4. Tötterman, T.H., Häyry, P., Saksela, E., Timonen, T., and Eklund, B. (1978). Eur. J. Immunol. 8: 872.
5. Vose, B.M., Vánky, F., Argov, S., and Klein, E. (1977). Eur. J. Immunol. 7: 753.
6. Mantovani, A., Allavena, P., Sessa, C., Bolis, G., and Mangioni, C. (1980). Int. J. Cancer 25: 573.
7. Allavena, P., Introna, M., Mangioni, C., and Mantovani, A. (1981). J. Natl. Cancer Inst. 67: 319.
8. Peri, G., Polentarutti, N., Sessa, C., Mangioni, C., and Mantovani, A. (1981). Int. J. Cancer 28: 143.
9. Timonen, T., and Saksela, E. (1980). J. Immunol. Methods 36: 285.
10. Allavena, P., Introna, M., Sessa, C., Mangioni, C., and Mantovani, A..J. Natl. Cancer Inst., in press.
11. Timonen, T., Ortaldo, J.R., and Herberman, R.B. (1981). J. Exp. Med. 153: 569.
12. Eremin, O. (1980). In "Natural Cell-Mediated Immunity Against Tumors" (R.B. Herberman, ed.), p.1011. Academic Press, New York.
13. Uchida, A., and Micksche, M. (1981). Proc. Am. Ass. Cancer Res. 22: 272.
14. Timonen, T., Ranki, A. Saksela, E., and Häyry, P. (1979). Cell. Immunol. 48: 121.
15. Timonen, T., Saksela, E., Ranki, A., and Häyry, P. (1979). Cell. Immunol. 48: 133.
16. Saksela, E., Timonen, T., Ranki, A., and Häyry, P. (1979). Immunol. Rev. 44: 71.
17. Moore, M., and Vose, B.M. (1981). Int. J. Cancer 27: 265.
18. Tötterman, T.H., Parthenais, E., Häyry, P., Timonen, T., and Saksela, E. (1980). Cell. Immunol. 55: 219.
19. Gerson, J.M. (1980). In "Natural Cell-Mediated Immunity Against Tumors" (R.B. Herberman, ed.), p.1047. Academic Press, New York.
20. Masserini, C., Villa, M.L., Rovini, D., and Clerici, E. (1981). Boll. Ist. Sieroter. Mil. 60: 307.
21. Mantovani, A., Peri, G., Polentarutti, N., Allavena, P., Bordignon, C., Sessa, C., and Mangioni, C. (1980). In "Natural Cell-Mediated Immunity Against Tumors" (R.B. Herberman, ed.), p.1271. Academic Press, New York.

NATURAL CYTOTOXIC EFFECTORS IN HUMAN TUMOURS AND TUMOUR DRAINING NODES

B.M. Vose
M. Moore

Department of Immunology
Paterson Laboratories
Christie Hospital and Holt Radium Institute
Manchester
England

I. INTRODUCTION

A subpopulation of blood and spleen lymphocytes of distinct morphology (LGL) from cancer patients frequently has the ability to lyse certain human tumour cell lines. Attempts to demonstrate similar reactivities in extravascular sites such as tumour-draining lymph nodes and the tumour itself have been less successful (Vose, 1980; Vose et al, 1977). This deficit of natural effector function may indicate the presence of suppressor cells at these sites, compromise of activity by tumour derived factors or a failure of LGL to localise at the tumour site. In this report the boosting of NK cell activity at extravascular sites by interferon (IFN) and interleukin-2 (IL-2) is examined. These studies were initiated to investigate the possibility that low levels of NK might be rendered detectable once augmented by these agents.

II. NATURAL EFFECTOR FUNCTION IN COLON CARCINOMA

In a series of colon carcinoma patients, significant natural effector function in a 4 hour ^{51}Cr release assay at effector:target ratios of 40:1 was recorded in 32/38 (84%)

ISBN 0-12-341360-5

blood preparations, 11/72 (15%) mesenteric lymph nodes from 40 patients and 2/13 (15%) tumour infiltrating lymphocyte preparations against the K562 cell line. No fall in blood reactivity was found with advancing disease. Positive node reactivity ranged from 11% - 26% lysis. It was not related to disease stage (Dukes stage A - 1 patient reactive, B - 5 reactive, C - 3 reactive and C + metastases - 1 reactive with the percentage of positive patients in each group approximately equal). Lymph node preparations reacting with K562 did not show any characteristic histology (2 sinus histiocytosis, 4 tumour involved and 4 normal) and were at variable distances from the tumour (6 within 2 cm, 4 distant mesenteric nodes). Subpopulation analysis on 7 positive nodes showed that greatest reactivity resided in the SRBC rosetting population. Similar data have been reported previously (Eremin et al, 1978). Two TIL populations with natural effector function came from patients with stage C disease. Even in preparations with K562 killer activity, levels of lysis did not reach those of blood lymphocytes from the same individual.

III. EFFECT OF INTERFERON ON EXTRAVASCULAR NATURAL EFFECTORS

Pretreatment of LNC or TIL from colon cancer patients with α interferon under conditions giving optimal augmentation of blood and spleen NK (250 IU/ml for 18 hour) failed to induce cytotoxicity in initially inactive preparations even at high effector:target ratios (Table 1). Similarly in LNC which showed low level reactivity no augmentation of lytic activity against K562 was evident. Admixture of TIL or LNC with autologous blood lymphocytes at ratios of up to 4:1 did not, in our hands inhibit K562 killing (Vose, 1980; Moore and Vose, 1981).

IV. EFFECT OF CONDITIONED MEDIA CONTAINING IL-2 ON NATURAL EFFECTORS

Blood lymphocytes TIL and LNC from cancer patients can be maintained in lectin-free conditioned media containing IL-2. Cultured lymphoid cells undergo rapid expansion of cell numbers with doubling times between 12 and 48 hours. After culture for 10-21 days in conditioned media cells were tested for cytotoxicity against a range of tumour targets. Cultured lymphoid cells killed autologous tumour, allogeneic tumour and in some cases K562 (Table 2). Studies reported in other sections of this book established that both activated T cells and LGL proliferate in mitogen-depleted conditioned media. Murine NK is boosted by purified IL-2 (Kuribayashi et al 1981). Studies are in progress with CLC initiated from isolated T cells or LGL to determine the nature of the effector cell in these reactions.

TABLE 1. Effect of Interferon[1] on Killing of K562 by Lymphocytes from Different Sites

Site	Effector:Target Ratio	Diagnosis	% Cytotoxicity Against K562 -IFN	+IFN
Blood	100:1	Normal	33.5*	60.9*
	20:1	Normal	55.2*	69.4*
	20:1	Normal	37.1*	68.3*
Lymph Node	100:1	Colon Ca	12.1*	11.5*
	20:1	Colon Ca	5.1	5.4
	20:1	Colon Ca	2.2	6.8
TIL	20:1	Colon Ca	0	1.7
	20:1	Lung Ca	0	0
	20:1	Lung Ca	4.1	5.4
Spleen	20:1	Gastric Ca	59.2*	80.2*

1 IF pretreatment 250 IU/ml for 18 hrs.
* $P<0.05$.

TABLE 2. Cytotoxicity of Cultured Lymphoid Cells Against K562,

Patient	Site	Effector:Target Ratio	% Cytotoxicity Against: Autologous Tumour	K562
Head & Neck Carcinoma	Blood	10:1	6	32*
	TIL	10:1	22*	88*
Colon Carcinoma	Blood	4:1	9*	24*
	Lymph Node	4:1	13*	21*
Lung Cancer	Blood	4:1	15*	4
	TIL	4:1	13*	3
Gastric Carcinoma	Blood	4:1	11*	3
	Node	4:1	14*	5
	Spleen	4:1	11*	4
	TIL	4:1	11*	3

* $P<0.05$

V. QUANTITATION OF LYTIC PRECURSORS IN TIL

Recent effort has concentrated upon the enumeration of cytotoxic precursors in blood and TIL by limiting frequency analysis. Blood lymphocytes and TIL were isolated as previously described (Vose, 1980). They showed differences in T cell subsets identified by the OKT series of monoclonals (Table 3). In 5/6 TIL tested to date, there is an increase in the percentage of cells staining with the OKT8 (suppressor/cytotoxic) monoclonal compared with peripheral blood from the same individual. The Pan T reagent T28 (a gift from Dr P C L Beverley, University College, London) stained approximately the same proportion of cells in both PBL and TIL preparations (62% $\pm$ 11% S.D). Lymphocytes from 3 of the 6 patients underwent a proliferative response upon cocultivation with autologous tumour under limiting dilution conditions in the presence of conditioned media containing IL-2 and irradiated autologous blood mononuclear cells as feeders (Table 3).

TABLE 3. Composition and Proliferative Frequency of Lymphocytes from Blood and Tumour to Autologous Tumour

Tumour	Site	% T Cells Staining With[1] OKT4	OKT8	Proliferative Frequency in MLTI
Lung 101	PBL	58	28	1/1150
	TIL	60	24	1/ 200
Lung 104	PBL	71	28	1/ 630
	TIL	41	51	1/ 260
Lung 105	PBL	63	31	1/ 600
	TIL	33	66	1/ 180
Lung 102	PBL	80	16	
	TIL	71	29	

[1] Indirect immunofluorescence tests. Total T cells was determined by staining with the monoclonal T28 and by rosetting with SRBC.

In each case the frequency of tumour reactive lymphocytes was greater at the tumour site (1/180, 1/200, 1/260) than in the blood (1/600, 1/630, 1/1150). Analysis of frequency of cytotoxic precursors was made against autologous and allogeneic tumour and K562. Against each target, except K562, the greater frequency of response was found in the TIL but the absolute number varied widely from 1/250 against autologous tumour to 1/2350 against allogeneic tumour. Tumour derived lymphocytes also showed the higher production of IL-2 upon stimulation with autologous tumour. These data suggest the accumulation of tumour reactive cells at the antigenic site.

VI. DISCUSSION

Extravascular lymphocytes show a marked deficit in natural effectors capable of lysing the K562 cell line. In extensive studies of lymphocytes from tumour-draining lymph nodes or from the tumour site; natural effectors (NE) have been found in only a minority of cases (approximately 15%) and reactivity was always less than that determined in the blood lymphocytes of the same patient. We have not been able to augment low level killing or induce killing in initially non-reactive samples by treatment with α interferon under conditions which uniformly boost PBL activity. In addition, TIL and LNC did not inhibit PBL activity upon admixture (Moore and Vose, 1981) nor did freshly isolated tumour cells (Vose and Moore, 1980).

However, in at least some cases, lysis of K562 was induced by expansion of TIL/LNC in conditioned media containing IL-2. It is not clear if this represents the maturation of NK precursors, a manifestation of polyclonal T cell activation or whether the active principle is IL-2 or immune IFN. Under the conditions used, lysis of autologous and allogeneic tumour was also apparent. Limiting dilution analysis of the frequency of cytotoxic precursors in populations stimulated by autologous tumour showed that a relatively high number of precursors of K562 killers was present at the tumour site, but that this was exceeded by the number reactive against the autologous tumour (Table 4).

TABLE 4. Frequency of Cytotoxic Precursors Against Different Targets in Blood and Tumour

Lymphocytes From	Frequency of Response Against: Autologous Tumour	Allogeneic Tumour	K562
Blood	1/1500	1/4650	1/2100
Tumour	1/ 250	1/2350	1/1250

In the only other study of this type, an <u>in situ</u> concentration of specific cytotoxic effectors in Murine Sarcoma Virus-induced tumours was noted (Brunner <u>et al</u>, 1981) (1/83 in tumour 1/430 in blood). It remains to be established if the killer population of K562 at extravascular sites is the same as that found in the blood. It has not been possible to identify cells of large granular lymphocyte morphology in LNC/TIL. Preliminary cell separation studies suggest that extravascular NK resides in the T cell population (SRBC rosetting). Studies are in progress to examine the characteristics of these cells more fully.

ACKNOWLEDGEMENTS

This study was supported by grants from the Medical Research Council and the Cancer Research Campaign of Great Britain. The authors are grateful to Dr G D Bonnard (NIIH) for conditioned media, to surgical colleagues for supply of material and Dr K H Fantes (Wellcome Laboratories) for Interferon. Wendy White and Roger Ferguson contributed technical expertise.

REFERENCES

Brunner, K.T., McDonald, H.R., and Cerottini, J-C., Exp. Med. 154:362-373 (1981).
Eremin, O., Coombs, R.R.A., Plumb, D., and Ashby, J., Int. J. Cancer 21:42-50 (1978).
Kuribayashi, K., Gillis, S., Kern, D.E., and Henney, C.S., J. Immunol. 126:2321-2327 (1981).
Moore, M., and Vose, B.M., Int. J. Cancer 27:265-272 (1981).
Vose, B.M., in "Natural Killer Cells" (Herberman, R.B., ed.), pp.1081, Academic Press, New York (1980).
Vose, B.M., and Moore, M., J. Nat. Cancer Inst., 65:257-263 (1980).
Vose, B.M., Vanky, F.T., Argov, S., and Klein, E., Europ. J. Immunol., 7:753-757 (1977).

CONTROL OF NATURAL CYTOTOXICITY IN THE REGIONAL LYMPH NODE IN BREAST CANCER

Susanna Cunningham-Rundles

Clinical Immunology Laboratory
Sloan-Kettering Institute for Cancer Research
New York, New York

I. INTRODUCTION

The potential importance of natural cell-mediated cytotoxicity or natural killer, NK, activity in host defense against spontaneous neoplastic growth has been clearly indicated in studies of natural resistance to tumor growth in experimental animals (1,2). Similarly, susceptibility to transplanted tumors has been found in association with low NK activity. Although analysis of the role of natural cytotoxicity in human tumor resistance has been intrinsically more difficult to carry out, there are indications that presence of the NK system may be protective in some primary immunodeficiency diseases and the primary absence of natural cytotoxicity may be associated with malignancy.

Depression of NK activity in advanced cancer has been demonstrated (3,4), but lack of correlation between natural cytotoxicity and clinical status has also been reported (5). Resolution for these disparate findings is suggested by the observations of Forbes *et al*. (6) that patients with extensive small cell lung cancer in remission had lower natural cytotoxicity than did patients with equivalent disease in relapse, although all of the patients had generally low NK activity. The apparent direct correlation between clinically evident tumor burden and natural cytotoxicity observed in this study was held to be a reflection of host reaction against tumor. Reduced natural cytotoxicity in advancing malignancy might therefore occur as a result of exhaustion of NK function. We have recently reported depressed natural killer activity in primary, untreated breast cancer (7), compared to healthy controls ($p < .001$), a finding which could be interpreted as

ISBN 0-12-341360-5

reflecting pre-existing immune deficiency underlying cancer development or as depletion of NK function during a pre-cancerous or pre-clinical phase.

Examination of NK activity at the tumor site is essential both for assessment of the potential antitumor capacity of this mechanism and for the development of practical immunotherapy. Animal experiments have shown that NK cells do exist within the tumor, at least in virally or chemically induced tumors, and can be functionally active against the NK-sensitive K562 target cell. In contrast, low or absent natural cytotoxicity has been found in lymphocytes isolated from human tumors (8,9), which may indicate that the NK system is not an effective host defense mechanism for spontaneously arising tumors. Alternatively, it may be that NK cells within the tumor domain are subject to rapid tumor mediated inactivation so that NK cells isolated from this source might not be able to retain activity long enough to permit assessment of their _in vivo_ potential _in vitro_. Instead, the tumor draining lymph node would appear to be the crucial lymphoid compartmental interface for lymphocyte-mediated host defense. However, in contrast to the numerous reports of significant NK activity in the regional lymph node cell (RLNC) in animals, there have been conflicting reports in investigations of natural cytotoxicity of RLNC in cancer patients. Moore and Vose (9) have reported low or absent NK activity in RLNC and Eremin _et al_. (8) have found significant if variable natural cytotoxicity. We have recently reported (7) that 25% of breast cancer patient RLNC have significant anti-K562 cytotoxicity. In these studies we examined RLNC from several axillary levels in the same patient and found that NK activity could be strongly present at one level and totally absent at another. These observations suggested that functional expression of NK activity at the tumor site might reflect the interaction of host defense mechanisms resulting in activation and concentration of NK cells in the region of the tumor, with tumor anti-lymphocyte suppressive mechanisms and tumor growth promoting factors acting to limit, impede, and exhaust the NK system. This view would imply that activation and inhibition or repression of NK activity would occur sequentially in different axillary levels during tumor development. One would not, however, be able to predict whether at the specific time of single sampling the RLNC of a particular level would contain 1) no NK cells; 2) pre-NK cells; 3) active NK cells with or without pre-NK cells; or 4) operationally suppressed NK cells. In our initial studies we found evidence for type 1 and type 3 RLNC. Type 3 RLNC generally, but not always, contained some pre-NK cells as evidenced by interferon inducibility. We found no type 2 RLNC and could not distinguish between type 1 and type 4 RLNC.

In the experiments described here, we identify type 2 RLNC and define conditions in which type 4 RLNC may be derepressed, thus establishing the existence of type 4 RLNC as distinct from type 1 RLNC.

II. MATERIALS AND METHODS

RLNC were prepared by teasing lymph nodes from defined axillary levels (level I most proximal, level III most distal to tumor) from patients with primary, untreated breast cancer at the time of mastectomy. Adherent cell depletion was carried out by incubation in plastic tissue culture flasks at 37°C for 45 min. Indomethacin pretreatment was carried out for 2 hr followed by washing before use in the NK assay. Cell lines used as targets were K562 and AlAb and BT-20, adherent breast cancer cell lines obtained from Dr. J. Føgh (MSKCC). Orthoclone OKT-3 was obtained from Dr. G. Goldstein (Ortho Pharmaceuticals). Cloned interferon (IFC-rA) was obtained from Hoffman-LaRoche in collaboration with Dr. H.F. Oettgen (MSKCC). Overnight 18 hr incubations were carried out using RLNC at 10 X 10^6 lymphocytes in 10 ml RPMI 1640 with 10% pooled normal human serum in a humid 37°C incubator with 5% CO_2. Addition of αIFN was made at 800 U/10^6 lymphocytes, OKT-3 at 500 ng/10^6 lymphocytes. In interferon inductions carried out during the 4 hr chromium (Cr) release assay, IFN was added at 1,000 U/well (1,000 U/10^6 lymphocytes at an E:T ratio of 100:1). Six day cultures were similarly prepared. Cell lines were irradiated and used as inducers of NK activity during 6 day culture at a concentration of 1 X 10^6 stimulators to 10 X 10^6 responders. At the end of the 6 day culture period, cells were removed, washed, counted using trypan blue dye exclusion to assess viability and plated for assay against ^{51}Cr labeled target cells. The natural cytotoxicity assay was used as described previously (7) at effector-target ratios (E:T) as given.

III. RESULTS AND DISCUSSION

In the earlier studies, we reported that interferon did not induce anti-K562 activity in RLNC which were initially inactive. We interpreted these results as suggesting that precursor NK cells were not present in RLNC without active NK cells, or that pre-NK cells were repressed and therefore could not respond to interferon. Whereas in these studies we used human leukocyte interferon (αIFN) derived from buffy coat leukocytes, we have more recently used the purified, cloned αIFN. In Table I are shown the results of an 18 hr incubation of RLNC from three axillary levels with 2,000 U of αIFN/5 X 10^6 RLNC assayed against ^{51}Cr labeled K562 target cells. In this experiment, inducibility appeared to increase with decreased initial activity and initial activity was greater at level III, most distant from the tumor. In other RLNC (data not shown), no induction whatsoever could be obtained, thus confirming our original observation that some RLNC cannot be induced.

If tumor proximity exerts suppressive effects upon expression of cytotoxicity, preculture alone might be expected to weaken this inhibition. As shown in Table II for three RLNC preparations from 3 different patients, this does appear to occur in some cases. A typical negative RLNC (#3) is shown for comparison. In all cases shown, 18 hr viability was greater than 98%, thus indicating that direct cytotoxic effects of the tumor on RLNC do not explain NK negative RLNC.

Tumor invasion of the regional lymph node would be expected to exert a regulatory influence on RLNC functional expression that might be fundamentally different from RLNC

TABLE I. Cytotoxicity at Different Axillary Levels

RLNC Level	Condition	% ^{51}Cr release: E-T ratio 100:1	50:1	25:1
I	Without IFN	11.4	8.2	N.D.
	With IFN	23.6	N.D.	N.D.
II	Without IFN	10.2	7.6	4.5
	With IFN	18.8	10.8	5.9
III	Without IFN	14.6	N.D.	N.D.
	With IFN	17.8	9.9	N.D.

TABLE II. Effect of Preculture on RLNC Cytotoxicity

Cell source	No preculture % ^{51}Cr release E-T ratio			18 hour preculture % ^{51}Cr release E-T ratio		
	100:1	50:1	25:1	100:1	50:1	25:1
RLNC #1	1.8	-1.6	-2.5	9.3	13.2	11.2
RLNC #2	2.1	0.4	0.0	15.3	5.9	3.0
RLNC #3	4.1	2.8	0.6	3.1	4.3	0.9

TABLE III. Effect of Tumor Involvement on NK Activity in RLNC

Cell Source		Preculture	4 hr Assay	% ^{51}Cr release E-T ratio 100:1	50:1	25:1
RLNC I	uninvolved	none	- IFN	2.8	2.1	2.5
RLNC I	uninvolved	none	+ IFN	3.5	1.7	0.0
RLNC I	uninvolved	18 hr	- IFN	25.3	N.D.	N.D.
RLNC I	uninvolved	18 hr	+ IFN	38.9	N.D.	N.D.
RLNC I	involved	none	- IFN	1.0	1.2	0.0
RLNC I	involved	none	+ IFN	0.4	0.0	0.0
RLNC I	involved	18 hr	- IFN	10.8	N.D.	N.D.
RLNC I	involved	18 hr	+ IFN	20.6	N.D.	N.D.
RLNC II	uninvolved	none	- IFN	6.8	2.6	3.4
RLNC II	uninvolved	none	+ IFN	7.6	3.4	3.8
RLNC II	uninvolved	18 hr	- IFN	28.6	35.4	N.D.
RLNC II	uninvolved	18 hr	+ IFN	43.9	48.6	N.D.

that were not in direct contact with tumor. In the experiment shown in Table III, tumor involved and tumor uninvolved lymph nodes from the same patient were studied. As shown here, neither involved nor uninvolved RLNC from level I, nor RLNC from level II, had anti-K562 activity initially. Furthermore, NK activity could not be induced with αIFN during the 4 hr assay in any of these preparations. However, parallel samples incubated for 18 hr, washed, and tested against the K562 target were quite different. Both level I and level II uninvolved RLNC were now observed to have substantial anti-K562 activity (level I, 2.8% to 25.3% post culture; level II, 6.8% to 35.4% maximum increase post culture). Level I RLNC with tumor involvement also showed increased anti-K562 activity (1.2% to 10.8%) but the observed post culture cytotoxicity was still below the level of positive cytotoxicity. In contrast to studies without preculture, both tumor involved and tumor uninvolved RLNC from level I and RLNC from level II could now be induced by IFN against K562. These results suggest that tumor may exert a crticial regulatory effect at the level of both pre-NK and active NK function in the regional lymph node.

Evidence that negative NK activity which might be related to direct tumor mediated suppression can be distinguished from adherent cell suppression is shown in Table IV. In the experiment with RLNC #1, preculture did not increase endogenous anti-K562 activity but monocyte depletion carried out after the 18 hr preculture phase just before assay against K562 did result in augmented activity. Possible evidence for

TABLE IV. Effect of Adherent Cell Depletion on RLNC NK

Cell source	Cell treatment	Pre-culture	% ^{51}Cr release: E-T ratio 100:1	50:1	25:1
RLNC #1	none	none	4.1	2.8	0.6
RLNC #1	none	18 hr	3.1	4.3	0.9
RLNC #1	monocyte depleted	18 hr	13.2	9.7	4.6
RLNC #2	none	none	7.3	0.2	0.0
RLNC #2	10^{-6} Indo-methacin	2 hr	12.0	N.D.	N.D.

prostaglandin-associated suppression is suggested from a related experiment shown in this table in which brief pre-culture with 10^{-6} M Indomethacin led to augmented NK activity.

Further experiments have been carried out to ascertain inducibility of the NK system in the regional node. In the experiments summarized in Table V, both αIFN and OKT-3 which induces γ interferon (10) were used. RLNC from patients 1 and 2 had significant endogeneous NK activity following 18 hr pre-culture. Addition of αIFN during preculture had an incremental effect on the anti-K562 activity of RLNC #2 but not RLNC #1. The simultaneous addition of αIFN and OKT-3 during preculture was additionally augmenting. RLNC #3 did not have NK activity but was susceptible to some augmentation by αIFN that was slightly greater when OKT-3 was simultaneously added. OKT-3 alone during 18 hr preculture did not augment the cytotoxicity of RLNC #3. The anti-K562 activity of RLNC #4 was significantly increased by OKT-3 and OKT-3 with IFN but not by IFN alone. The augmentation observed here is important since an initially NK negative population of cells was converted to NK positive functional expression.

Finally, RLNC were studied for NK activity following 6 day culture with and without activators against both K562 and two disease-related targets. *E. coli* was used as a reference augmenting agent which has been found to induce IFN and to be a powerful inducer of NK activity (Cunningham-Rundles and Rubin, in preparation).

As shown in Table VI, unstimulated cytotoxic activity toward all targets was negative. *E. coli* activation led to expression of cytotoxicity toward the A1Ab target and to lysis of BT-20. Other RLNC (not shown) could not be induced to lyse BT-20. K562 induced cytotoxicity toward itself and A1Ab but not BT-20. Neither A1Ab nor BT-20 were NK activators to any target studied. Lack of induction of cytotoxicity by A1Ab and BT-20 may be ascribed to lack of activation of IFN

TABLE V. Augmentation of NK Activity by αIFN and OKT-3

Cell source	Preculture condition	% ^{51}Cr release E:T ratio 100:1	50:1	25:1
RLNC #1	18 hr	38.3	N.D.	N.D.
RLNC #1	18 hr + IFN	35.4	12.9	11.5
RLNC #1	18 hr + IFN + OKT-3	46.6	N.D.	N.D.
RLNC #2	18 hr	23.4	14.2	5.9
RLNC #2	18 hr + IFN	31.5	17.4	6.9
RLNC #2	18 hr + OKT-3	38.0	24.5	11.6
RLNC #3	18 hr	0.6	0.9	0.0
RLNC #3	18 hr + IFN	6.3	8.9	4.4
RLNC #3	18 hr + OKT-3	1.1	0.0	0.0
RLNC #3	18 hr + IFN + OKT-3	12.1	6.4	3.4
RLNC #4	18 hr	5.9	N.D.	N.D.
RLNC #4	18 hr + IFN	6.6	N.D.	N.D.
RLNC #4	18 hr + OKT-3	14.0	6.4	2.0
RLNC #4	18 hr + IFN + OKT-3	17.7	10.1	6.8
Normal control PBL	18 hr	23.8	13.4	7.2
	18 hr + IFN	46.9	34.7	24.7
	18 hr + OKT-3	54.4	48.1	30.9
	18 hr + IFN + OKT-3	79.8	67.3	54.9

TABLE VI. Induction of Cytotoxicity in RLNC

Cell source	Inducer	% ^{51}Cr release at E-T ratio 100:1 for targets K562	A1Ab	BT-20
RLNC	none[a]	3.4	1.4	0.0
RLNC	*E. coli*	40.7	27.0	37.2
RLNC	K562	51.7	50.1	3.0
RLNC	A1Ab	2.3	3.7	2.0
RLNC	BT-20	2.1	2.5	0.0
PBL normal control	none[b]	53.8	9.9	3.2

[a]Lymph node cells were cultured for 6 days with or without inducer as designated.
[b]Peripheral blood mononuclear cells (PBM) were tested fresh without culture.

production of RLNC or lack of secretion of IFN by these lines. The results obtained with K562 suggest that there may be shared recognition structures between K562 and A1Ab. The fact that *E. coli* activation led to lysis of BT-20 by some RLNC but not others may indicate that some NK augmenting agents enable RLNC lysis of targets having recognition structures related to *in vivo* sensitizing antigens but for which endogenous cytotoxicity is either too weak or has been suppressed.

These studies indicate that inhibition of cytotoxic activity does occur in the regional lymph node of breast cancer patients and that both culture and augmenting agents may be used to release expression of natural cytotoxicity. Furthermore, these studies indicate that biological response modifiers which induce anti-K562 activity may also augment lysis of disease-related targets.

ACKNOWLEDGMENTS

The excellent technical assistance of Mrs. K.M. Smith is gratefully acknowledged. These studies were supported in part by USPHS grants CB 84228-34 and CA 08748-17, and the Richard Molin Memorial Foundation for Cancer Research.

REFERENCES

1. Herberman, R., and Holden, H.T., Adv. Cancer Res. 27: 308 (1978).
2. Habu, S., Fukui, H., Shimanara, K., Kasai, M., Nagai, Y., Okumara, K., and Tamaoki, N., J. Immunol. 127:34 (1981).
3. McCoy, J., Herberman, R.B., Perlin, E., Levine, P., and Alford, C., Proc. Amer. Assoc. Cancer Res. 14:107 (1973).
4. Pross, H.F., and Baines, M.G., Int. J. Cancer 18:593 (1976).
5. Vose, B.M., Vanky, E., Ropp, M., and Klein, E., Br. J. Cancer 38:375 (1978).
6. Forbes, J.T., Greco, F.A., and Oldham, R.K., in "Natural Cell Mediated Immunity Against Tumors" (R.B. Herberman, ed.), p. 1031. Academic Press, New York, 1980.
7. Cunningham-Rundles, S., Filippa, D.A., Braun, D.W., Jr., Antonelli, P., and Ashikari, H., J. Nat. Cancer Inst. 67:585 (1981).
8. Eremin, O., in "Natural Cell Mediated Immunity Against Tumors" (R.B. Herberman, ed.), p. 1011. Academic Press, New York, 1980.
9. Moore, M., and Vose, B.M., Int. J. Cancer 27:265 (1981).
10. Cunningham-Rundles, S., Rubin, B., and Goldstein, G. (in preparation).

LEUKEMIC BLASTS AS EFFECTORS IN NATURAL KILLING AND ANTIBODY DEPENDENT CYTOTOXICITY ASSAY[1]

Sudhir Gupta[2]
Gabriel Fernandes[3]

Memorial Sloan-Kettering Cancer Center
New York, New York

I. INTRODUCTION

Since the descovery of natural killer (NK) function about 8 years ago, attempts have been made to characterize NK cells. Cells with lymphoid characteristics of T cell lineage as well cells of monocyte-macrophage lineage have shown to mediate NK activity (1-3). The best marker appears to be morphologic, that is, large granular lymphocytes (LGL), which comprise of approximately 5% peripheral blood mononuclear cells (4). These cells account for a high proportion of TG cells and react with OKT10 monoclonal antibody (5,6). However, none of these markers are confined to cell of a single cell lineage. There is controversy regarding the distinction between NK and K cells that mediate antibody-dependent cytotoxicity. Because of these conflicts we chose to examine leukemic blasts of monocytic, meyloid and T cell lineage from patient's with acute leukemias, for effector functions in NK and ADCC.

II. MATERIALS AND METHODS

A. Separation of Mononuclear Cells

Bone marrow or peripheral blood leukemic blasts from 19 patients with acute leukemias that were separated on Ficoll-Hypaque density gradient and cryopreserved in an automated

---

[1]Supported by NIH grants CA17404, AG-00541, AG-03417.
[2]Present address: Div of Basic and Clinical Immunology U. of California, Irvine, CA 92717
[3]Present address: Division of Clinical Immunology, U. of Texas Health Science Center, San Antonio, TX

ISBN 0-12-341360-5

liquid-freezing apparatus,were thawed rapidly in a 37°C water bath. Cells were washed x3 in medium containing barbital buffer with 5% bovine serum albumin. Cell viability was >90% as determined with trypan blue dye exclusion test and each preparation contained >92% blasts.

Diagnosis of acute leukemias was made by immunologic, morphologic, histochemical and clinical criteria. Patients were classified according to FAB classification (7). Included were 3 patients with undifferentiated acute myeloblastic leukemia (M1), 4 with differentiated myeloblastic leukemia (M2), 3 with undifferentiated acute monocytic leukemia (M5a), 5 with differentiated acute monoblastic leukemia (M5b), and 4 with acute lymphoblastic leukemia of T cell origin (L2).

B. Natural Killing Activity

Cells of K-562 cell lines were used as targets after labeling with ^{51}chromium. A fixed number of viable blast cells in RPMI-1640 with 10% fetal calf serum supplimented with antibiotics were prepared in varying dilution to give a effector to target ratios of 50:1, 25:1 and 12.5:1. The details of technique and calculation of cytotoxicity are described elsewhere (8).

C. Antibody-dependent Cellular Cytotoxicity (ADCC)

ADCC was determined by a method described previously (8). In brief, varying concentrations of leukemic blasts were added to 50 µl of RPMI-1640 containing 10^4 ^{51}chromium labeled chicken RBC (CRBC) or cells of SB cell line and 100 µl of appropriate dilution of anti-CRBC or anti-SB IgG antibody. Cytotoxicity was calculatd in 3 hour ^{51}Cr release assay (8).

Receptors for IgG Fc were assayed by a previously described method (9).

III. RESULTS

Data regarding NK in acute myeloblastic (AML) and acute monoblastic leukemia (AMOL) are shown in Table 1. Leukemic blasts from all patients with AML or AMOL demonstrated very low NK activity even at high effector: target ratio. No correlation was observed between NK activity and the proportion of leukemic blasts with IgG Fc receptors.

TABLE 1. Natural Killing by Myeloblastic and Monoblastic Leukemic Cells from Acute Leukemias

Patients	Diagnosis	Percent Lysis Effector/Target Ratio 50:1	25:1	12.5:1	% Blasts with Fc Receptors
1	AML (M1)	1	0	ND	19.0
2	AML (M2)	2	2	2	20.0
3	AML (M2)	5	4	3	22.0
4	AML (M2)	8	6	4	15.0
5	AMOL (M5a)	1	1	0	9.0
6	AMOL (M5a)	3	2	2	97.0
7	AMOL (M5b)	1	1	1	52.0
8	AMOL (M5b)	6	6	4	17.0

Results of ADCC by leukemic blasts from patients with AML and AMOL are given in Table 2. Five of 7 patients with

TABLE 2. Antibody-dependent Cytotoxicity by Myeloblastic and Monoblastic Leukemic Cells

Patients	Diagnosis	Percent Lysis Effector/Target Ratio 50:1	25:1	12.5:1	% Blasts with Fc Receptors
1	AML (M1)	92	91	84	28
2	AML (M1)	15	12	5	19
3	AML (M1)	16	5	7	4
4	AML (M2)	9	7	4	15
5	AML (M2)	10	6	4	20
6	AML (M2)	32	28	12	10
7	AML (M2)	12	8	7	22
8	AMOL (M5a)	45	34	19	9
9	AMOL (M5a)	84	70	67	97
10	AMOL (M5a)	54	50	51	80
11	AMOL (M5b)	47	42	39	52
12	AMOL (M5b)	84	81	79	76
13	AMOL (M5b)	47	40	26	17
14	AMOL (M5b)	84	74	68	76
15	AMOL (M5b)	44	34	32	44

AML had low ADCC activity (< 20% lysis). Only 4-28% leukemic blasts expressed IgG Fc receptors. Leukemic blasts from all 8 patients with AMOL demonstrated high ADCC activity (45-84% lysis) and 6 of 8 patients had 44-80% leukemic blasts expressing IgG Fc receptors. No difference in ADCC was observed between undifferentiated (M5a) and differentiated (M5b) leukemic blasts from patients with AMOL.

Data regarding NK and ADCC in T cell ALL are shown for 25:1 effector/target ratios in Fig. 3. Leukemic blasts

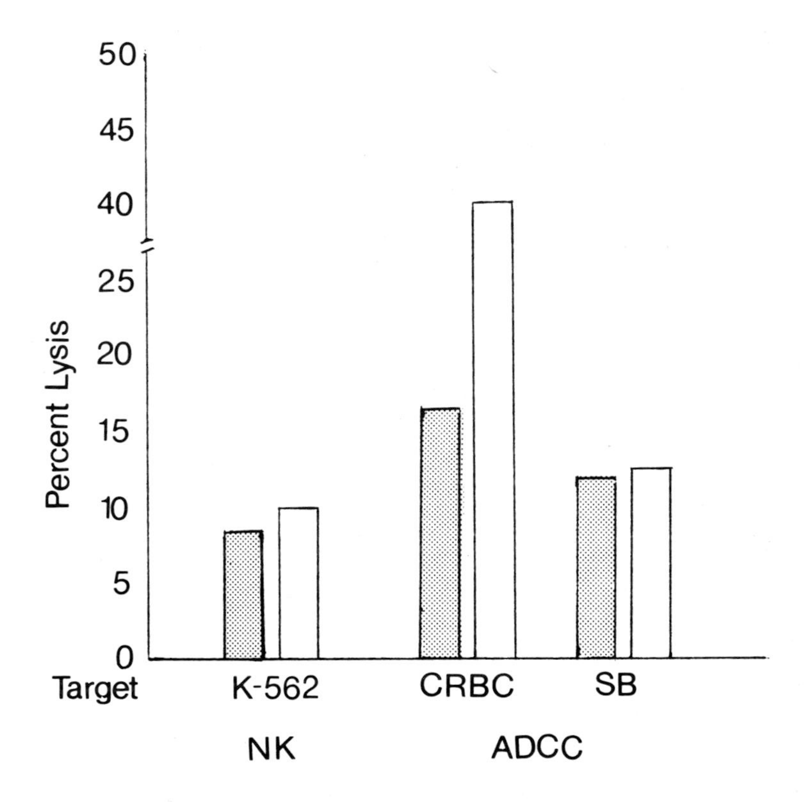

FIGURE 3. NK and ADCC by leukemic blasts from T ALL.

demonstrated a mean lysis of 9.6% (range 4-14%) against K562 target. This was comparable to mean lysis by purified T cells from peripheral blood of healthy controls. ADCC against CRBC target was 16.2% (range 6-35%), that was significantly ($p < 0.001$) less when compared to ADCC by peripheral blood T cells (49%) from healthy controls. However, ADCC against SB cell line target was comparable (mean lysis of 12%) to T cells from peripheral blood of healthy controls. Twenty to 42% of leukemic blasts expressed receptors for IgG Fc (data not shown).

IV. DISCUSSION

Numerous attempts have been made to characterize NK cells with regard to cell lineage, using a variety of surface phenotypic marker. These have recently been reviewed (10). Some investigators have claimed that NK cells belong to T cell lineage as demonstrated by receptors for SRBC or reactivity with monoclonal antibodies defining lymphocytes of T cell lineage (1,2,4,5). Others have demonstrated them to be cells of monocytic-myeloid lineage based on cytochemistry and reactivity with monoclonal antibodies against myeloid-monocytic lineage (3,5). It has, however, became evident that some of the monoclonal antibodies (OKM1) said to define cells of monocyte-lineage, also react with lymphocytes reacting with monoclonal antibodies (OK10, 9.6, Leu 2a) that define T cells (5,6). It appears that NK cells have surface antigens that are shared by monocytes. Recently Lohmeyer et al have reported two monoclonal antibodies that were raised following immunization with Tγ cells from patients with T cell chronic lymphocytic leukemia (11). One of these antibody reacted with adherent cells and the other with a subpopulation of T cells. NK activity was found in population of cells that reacted with monoclonal antibody which was found to be staining adherent cells. A subpopulation of these cells reacted with monoclonal antibody defining T cells. In the present study we have demonstrated that T cell leukemic blasts were efficient natural killers as well as were effectors in ADCC. This supports the T cell lineage of NK cells. We did not find any significant NK activity in leukemic blasts from patients with acute myeloblastic or monocytic leukemia. However, Hokland et al reported NK activity in monoblasts from patients with acute leukemias (12). The difference in the results between two studies could be due a possibility that monoblasts from a subpopulation of acute monoblastic leukemia are effectors in NK or cryopreservation, that we used in our study, resulted in a loss of NK activity, though the ADCC activity was preserved. Cytophilic attached IgG does not inhibit ADCC activity of leukemic blasts, suggesting that these antibodies might not mask Fc receptors completely (13). We have also demonstrated that leukemic blasts were not the targets for NK by peripheral blood lymphocytes from healthy subjects (13).

Lohmann-Matthes reported NK activity in human monocytes from peripheral blood that is increased after activation with macrophage activating factor (3). There are no definite data to support that mature unactivated human monocytes mediate NK. In summary we have demonstrated that T leukemic blasts are effectors in NK and ADCC, whereas leukemic myeloblasts or monoblasts are effectors in ADCC but lack NK activity.

REFERENCES

1. Gupta S., Fernandes G., Nair M. and Good R.A. (1978). Proc. Natl. Acad. Sci. (USA) 75:6437.
2. West W.H., Boozer R.B. and Herberman R.P. (1978). J. Immunol. 120:90
3. Lohmann-Matthes M. In: Natural and Induced Cell-Mediated Cytotoxicity (G. Riethmuller, P. Wernet, G. Cudkowicz, Eds.), Academic Press, New York, p. 221, 1979.
4. Timonen T., Ortaldo J.R. and Herkerman R.B. (1981). J. Exp. Med. 153:569.
5. Platsoucas C.D. In: Mechanisms of Lymphocyte Activation (K. Resch and H. Kirschner, Eds.) Elsevier/North-Holland Biomed. Press Amsterdam, p. 290, 1981.
6. Kraft D., Rumpold H., Steiner R., Radaskiewicz T., Swetly P. and Wiedermann. In: Mechanisms of Lymphocyte Activation (K. Resch and H. Kirschner, Eds.) Elsevier/North-Holland Biomed. Press, Amsterdam, p. 279, 1981.
7. Bennett J.M., Catovsky D., Daniel M.T., Flandrin G., Galton D.A.G., Gralnick H.R. and Sultan C. (1976). Brit. J. Haematol. 33:451.
8. Fernandes G. and Gupta S. (1981). J. Clin. Immunol. 1:141.
9. Gupta S. and Gupta S. (1977). Cell. Immunol. 34:10.
10. Herberman R.B. and Ortaldo J.R. (1981). Science 214:24.
11. Lohmeyer J., Rieber P., Feucht H., Hadam M., Riethmuller G. In: Mechanisms of Lymphocyte Activation (K. Resch and H. Kirschner, Eds.), Elsevier/North-Holland Biomed. Press, Amsterdam, p. 282, 1981.
12. Hokland P., Hokland M. and Ellegaard (1981). Blood 57:972.
13. Fernandes G., Garrett T., Nair M., Straus D., Good R.A. and Gupta S. (1979). Blood. 54:573.

NATURAL KILLER CELL ACTIVITY ASSOCIATED WITH SPONTANEOUS AND TRANSPLANTED RETICULUM CELL NEOPLASMS

Kenneth S. S. Chang and Richard Kubota

Laboratory of Cell Biology, National Cancer Institute, Bethesda, Maryland 20205

I. INTRODUCTION

The SJL/J strain of mouse is known to develop in later life a high incidence of spontaneous Hodgkin's-like reticulum cell neoplasm (RCN), type B. These tumors are pleomorphic, and contain histiocytes, giant cells, lymphocytes, granulocytes and plasmacytes as well as reticulum cells (1,2). It was discovered that the nonadherent lymphoid cells derived from the spleen and lymph nodes (LN), in which the transplantable RCN such as D4 (3), RCS5 or RCS19 (4) was growing, showed a marked increase in natural killer (NK) cell activities (3,5,6).

This article summarizes these results and compares with NK activities exhibited by lymphoid cells derived from spleens and LN of mice bearing spontaneous RCN, either primary or at the early passages *in vivo*. The origin and heterogeneity of NK cells associated with RCN (RCN-NK) will be considered.

II. DETECTION OF RCN-NK ACTIVITIES

Although most of our work was done with [^{125}I]iododeoxyuridine (IdU)-labeled target cells, our initial experiments that led to discovery of RCN-NK activity were done with a short term (4 hr) ^{51}-Cr-release assay. Normal SJL/J mice exhibit rather low NK activity (with a peak at 6-9 weeks of age), and it is only after the growth of RCN has reached certain extent in spleens and LN that these organs show a marked increase in NK activity. *In vitro* cultured lines of YAC, RBL-3, EL4G$^-$ and D2 cells were commonly used as targets. The former three lines showed various degress of sensitivity to killer

ISBN 0-12-341360-5

cells derived from mice beearing D4, RCS5, and RCS19 tumors, but D2 cells never showed any significant lysis by these effector cells (Table I). The killer cells were non-adherent on culture plates or nylon wool columns, non-phagocytic, relatively resistant to X-ray and only partially inactivated by treatment with anti-Thy 1.2 serum plus complement; the killer activity which covers a wide range of target tumor cells, was selective, and was not restricted by a histocompatibility barrier. They were IgG^- and Fc-receptor$^+$, but the presence of Fc-receptor$^-$ subpopulation is not excluded.

TABLE I. NK Activity of Spleens and Lymph Nodes from RCN-bearing SJL/J Mice

Source of NK cells[a]		Target/ Effector Ratio	Net % cytotoxicity[b] against			
			YAC	RBL-3	EL-4G$^-$	D2
Transplant. RCN lines	D4	200	10-20	10-45	10-30	0-4
	RCS5	200	10-20	10-45	10-30	0-4
	RCS19	200	10-20	10-45	10-30	0-4
Spontan. RCN	8 RCNs	200	0-7	2-4	2-8	0-8
	D20[c]	200	n.d.	0	n.d.	38
		100	n.d.	0	n.d.	31
	D21	200	n.d.	n.d.	31	21
		100	n.d.	n.d.	24	20
	D22	200	n.d.	0	1	10
		100	n.d.	0	0	5
	D28	200	1	1	12	3
		100	0	0	4	0
	D33	200	16	2	2	21
		100	19	0	0	18
	D27	200	16	7	5	48
		100	25	5	2	45
Spontan. RCN at early passages	D27 (P-1)	200	0	6	14	4
		100	1	4	15	4
	D27 (P-7)	200	3	4	2	16
		100	2	2	0	6

[a] Spleens and/or lymph node cells, depleted of macrophages.
[b] % cytotoxicity subtracted with that shown by normal SJL/J mice. n.d.: not done.
[c] Individual mouse.

No such augmented NK activity was associated with other types of tumors such as lymphocytic lymphoma and mammary adenocarcinomas that grew in SJL/J mice.

The RCN-NK cells were also shown to be effective killers by *in vivo* experiments where a modified Winn test (7) was used. As shown in Table II, the effector cells (depleted of macrophages) present in RCS5 were mixed at various ratios with EL4G⁻ cells. The mixtures, after 4 hr incubation, were injected s.c. into groups of C57BL/6 mice. The tumor incidence, mortality due to progressive EL4G⁻ growth, and the mean tumor sizes were significantly decreased in these groups as compared with the control groups. Similar results were obtained when effector cells from D4-bearing mice were used. These results suggest that the target cells' tumorigenicity was seriously compromised by a short term *in vitro* incubation

TABLE II. Modified Winn Test For RCN-NK Activities

Effector cells[a]	Effector/target[b] ratio	Tumor Incidence[c] day 12	Mortality day 21	Mean tumor volume[d] mm^3	Mean tumor volume[d] % control
SJL/J	200	3/8	0/8	158	9.4
mice	100	3/8	2/8	122	7.3
bearing	50	4/8	0/8	114	6.8
RCS5	25	8/8	5/8	549	32.7
Normal	200	7/8	3/8	630	37.5
SJL/J	100	7/8	4/8	691	41.2
mice	50	7/8	8/8	805	48.0
	25	8/8	8/8	1678	77.0
None (medium)		8/8	8/8	1678	100.0

[a] Effector cell suspensions prepared from pooled spleens and lymph nodes of mice were mixed with 10^6 target cells at the ratios indicated. After incubation in RPMI1640 medium for 4 hr at 37°C in CO_2-incubator, the mixtures were injected s.c. into C57BL/6 mice.

[b] EL4G⁻ culture cells were used as the target.

[c] Since RCS5 cells alone do not make tumor in C57BL/6 mice, these tumors are considered to have been formed by EL4G⁻ cells.

[d] Mean tumor volume at day 12.

with effector cells, although the possibility of subsequent host cell participation was not ruled out.

III. HETEROGENEITY OF NK CELLS ASSOCIATED WITH SPONTANEOUS RCN

In contrast to NK cells associated with transplantable RCN, more variable and heterogenous NK activities were observed with those associated with spontaneous RCN in old SJL/J mice. As summarized in Table I, some primary RCNs (from 8 mice) showed no NK activity against any of the four target cell lines, whereas some others showed killer effect on D2 only, $EL4G^-$ only, or both D2 and $EL4G^-$, or both D2 and YAC, but scarcely against RBL-3. The target cell spectrum for D27 tumor changed also after 1 passage and 7 passages in syngeneic SJL/J mice. Because the NK activities associated with transplantable RCN lines tested at the same time showed a different target spectrum, these differences could not be explained by day-to-day variations. The possibility of involvement of immune T cells cytotoxic for the primary RCN and cross-reacting with D2 cells was considered remote because treatment of nonadherent effector cells with anti-Thy 1.2 serum plus complement resulted in only a partial (40%) loss of cytotoxic activity. Removal of macrophage and nylon-wool adherent cells did not diminish the cytotoxicity.

These observations may suggest that different subpopulations of NK (or pre-NK) cells are being activated during primary development and subsequent transplant passages, each subpopulation representing a clone of NK cells with different repertoire of target antigen-recognition sites (or receptors) on cell surface. This is consistent with reports on the heterogeneity of NK cells (8,9). Since the primary RCN is pleomorphic, and as the subsequent transplant passages tend to change the histology to monomorphism with reticulum cells as the dominant feature, it is possible that there are changes in populations of putative suppressor cells or other regulatory cells in the RCN site, which may be reflected to changes in populations of activated NK cells. Further investigation is necessary to elucidate the significance of these observations in terms of immunologic surveillance and evolution of RCN.

IV. ORIGIN OF NK CELLS ASSOCIATED WITH RCN

In our first report in was postulated that the RCN itself was derived from NK cells neoplastically transformed (5). This view has been pursued further by Ponzio _et al_. (9).

In order to test the alternate hypothesis that RCN-associated NK cells were derived from the host, the following experiment was conducted. (C57BL/6 x SJL/J)F_1 mice pretreated with silica to abrogate Hh (Hemopietic-histocompatibility) restriction (10,11) and subsequently injected with RCN of SJL/J ($H-2^s$) origin supported the growth of transplanted RCN. The high NK activity associated with this RCN was markedly reduced by *in vitro* treatment with anti $H-2^b$ serum (pre-absorbed with SJL/J RCN) plus complement, indicating that the NK cells were of the host origin (3). The remaining NK activity after such treatments (9) may be explained by the carry-over of SJL/J NK cells in the RCN preparation. This can be avoided by inoculating smaller doses of RCN and/or by passaging at least twice in the F1 host. The close association of RCN growth with elevated NK activity may indicate a special function of RCN in promoting NK activity by an unknown mechanism(s) of cellular interactions, because repeated immunization with RCN cells in mice could not induce such a high NK activity.

Recnt investigations revealed the presence of immune interferon (IFN γ)(acid-labile) in the RCN-homogenate (12, Chang, unpublished). Intraperitoneal injection of RCN into SJL/J mice induces a transient production of interferons detectable in the serum of mice 1 day after injection, but the IFN γ appears in the lymphoid organs only when the RCN has grown to a certain extent. The increase in NK activity in these RCN preparations may be accounted for by the action of IFN γ triggering activation of pre-NK cells to become mature NK cells as well as enhancing the killing efficiency of pre-existing NK cells. The mechanism by which IFN γ is induced in the RCN-carrying organs is not clear.

We observed that if transplantable RCN cells were cloned by *in vivo* passage through limiting dilution procedures, one can get a line of RCN with little NK activity associated. If the transplantable RCN lines are of monoclonal origin, as most other tumors are, the above finding is not consistent with the idea that the NK cells were of RCN origin. The facts that the NK activities exhibited by primary RCN are heterogeneous, and that the IFN-stimulation result in polyclonal activation of pre-NK cells are consistent with the hypothesis that the RCN-associated NK cells were of host origin.

The role of host NK cells in coexistence with the growing cells of established RCN transplants *in vivo* is not clear since it was found that the ^{51}Cr- or ^{125}IdU-labeled tumor cells were not sensitive as targets to the RCN-NK cells. However, it is possible that the IFN γ or other agents induced at the tumor site may have exerted a protective effect on the tumor cells against NK cells. On the other hand, the difficulties experienced in establishing RCN lines by in vivo passage may be explained by the combined cytotoxic activities of NK cells carried over from the donor and those derived from the recipient mice. Only those tumor cell clones that are insensitive to NK cells or those highly sensitive to the protective action of IFN may survive and become established. These possibilities are currently being evaluated.

SUMMARY

The spleens and LN of SJL/J mice bearing transplantable RCN showed a markedly high NK activity which could be demonstrated by both in vitro and in vivo tests. Available evidence suggests these NK cells were of the host origin. The primary RCN and its early in vivo passages exhibited either no marked NK activity or NK activities against a range of target cells either narrower than or different from that for NK cells associated with transplantable RCN. In vivo passages resulted in a change in target cell range. The possible roles of RCN-induced IFN and heterogenous NK cells in RCN development were discussed.

REFERENCES

1. Murphy, E. D. Proc. Amer. Assoc. Cancer Res. 4:46 (1963).
2. Dunn, T. B. and Deringer, M. K. J. Natl. Cancer Inst. 40:771 (1968).
3. Chang, K. S. S. and Log, T. Int. J. Cancer 25:405 (1980).
4. Carswell, E. A., Wanebro, H. J., Old, L. J., and Boyse, E. A. J. Natl. Cancer. Inst. 44:1281 (1970).
5. Chang, K. S. S. Advances in Comparative Leukemia Research, 1977, p. 327, ed. Bentvelzen, P., Hilgers, J., and Yohn, D. S., Elsevier/North Holland, Amsterdam (1978).
6. Fitzgerald, K. L. and Ponzio, N. M. Cell Immunol. 43:185 (1979).
7. Winn, H. J. J. Immunol. 84:530 (1960).
8. Lust, J. A., Kumar, V., Burton, R. C., Bartlett, S. P., and Bennett, M. J. Exp. Med. 154:306 (1981).
9. Ponzio, N. M., Fitzgerald, K., and McMaster, J. Fed. Proc. 40:1092 (1981).
10. Cudkowicz, G. and Hochman, P. S. Immunol. Rev. 44:13 (1979).
11. Lotzova, E. and Cudkowicz, G. J. Immunol. 113:798 (1974).
12. Ponzio, N. M., Fitzgerald, K. L., Vilcek, J., and Thorbecke, G. J. Ann. N.Y. Acad. Sci. 350:157 (1980).

RETICULUM CELL SARCOMAS OF SJL/J MICE: PRE-B CELL LYMPHOMA WITH APPARENT NATURAL KILLER CELL FUNCTION

Nicholas M. Ponzio

Department of Pathology
CMDNJ-New Jersey Medical School
Newark, New Jersey

I. INTRODUCTION

SJL/J mice exhibit a high incidence of spontaneous reticulum cell sarcomas (RCS), with 90% of animals developing tumors at an average age of 13 months (Murphy, 1963; Siegler and Rich, 1968). The majority of primary RCS are classified as Dunn type B neoplasms, being composed of diverse cell types (Murphy, 1969). Upon successful serial transplantation into syngeneic recipients, however, there is a progressive loss of cellular heterogeneity and the predominant cell type is, morphologically, lymphoblastic. With respect to the transplantable RCS that we have studied, they do not express the characteristic properties of mature T cells, B cells or macrophages (Lerman, et al., 1979; Ponzio, et al., 1977a; Scheid, et al., 1981; unpublished observations). However, data which support a B cell lineage for transplantable RCS of SJL mice have been reported. Thus, following systemic injection, they exhibit localization properties into lymphoid follicles that are characteristic for B cells (Carswell, et al., 1976).

These studies were supported in part by N.I.H. Grant CA22544, funds from the American Cancer Society, New Jersey Division, CMDNJ Foundation Grant 28-82 and the Lawrence Wilkins Cancer Research Fund. The author is the recipient of Research Career Development Award No. CA00833, from the National Cancer Institute, DHHS.

ISBN 0-12-341360-5

Using purified tumor cells, we confirmed our original description of Ia^s determinants on the majority of RCS cells (Scheid, et al., 1981). Lyb-2 and Fc receptors were also detectable, but on a minority (5 - 15%) of RCS cells. However, following incubation of purified RCS with lipopolysaccharide (LPS) or Interleukin (IL)-1, a 5-fold increase in cells expressing Lyb-2 and a 3-fold increase in cells expressing Fc receptors was noted. Moreover, significant and reproducible increases in surface Ig expression by RCS cells were obtained after incubation with LPS or IL-1 (Scheid et al., 1981). These data, together with evidence that SJL mice, chronically suppressed from birth with anti-μ antibody, fail to develop spontaneous tumors (Katz, et al., 1980a), strongly indicate a B cell lineage for RCS in this strain.

The host-tumor relationship of SJL mice to syngeneic RCS also exhibits several unique features. RCS cells fail to grow in irradiated or cytoxan-treated, as compared to normal, SJL mice (Lerman, et al., 1976). Partial reconstitution of tumor growth in such mice can be obtained, with prior injection of normal lymphoid cells, indicating the requirement of normal cells to support RCS growth. Another characteristic quality of RCS cells is their ability to induce marked proliferation of Lyt-1^+, 2^- T cells in vitro (Ponzio, et al., 1977a; Lerman, et al., 1979). A similar proliferative response occurs in the popliteal lymph node (LN), following footpad injection of irradiated RCS cells (Ponzio, et al., 1977b). This proliferative response is unaccompanied by the development of RCS-specific cytotoxic cells (Ponzio, et al., 1977b), appears to be under strict IR gene control (Katz, et al., 1980b) and is intimately related to the capacity of RCS cells to grow progressively in certain SJL F_1 hybrids.

Another recent finding is the high degree of NK activity present in lymphoid cell suspensions prepared from tumor-bearing SJL mice (Chang, 1978; Fitzgerald and Ponzio, 1979). Since many agents, including tumor cells, have been shown to stimulate increased NK activity, it was possible that RCS-induced host NK cells were responsible for the NK activity seen SJL mice injected with viable RCS cells. It must be noted, however, that SJL mice are classified as a low NK strain, with NK lytic ability that is notoriously difficult to augment, even with known NK inducers and extended assay intervals (Fitzgerald and Ponzio, 1981; G. Cudkowicz, personal communication). Thus, this raised a second possibility, namely, that the tumor cells themselves contributed to the observed NK activity. Given the B cell lineage of RCS tumors, it was important to clarify the nature of the responsible NK effector cells in this system.

II. RESULTS AND DISCUSSION

A. NK Activity in RCS Cell Suspensions

The levels of NK activity against YAC targets observed in cells obtained from RCS bearing donors are presented in Table I. When compared to effector cells taken from normal SJL donors, the NK activity present in RCS cell suspensions was markedly increased. This increased NK activity was exhibited by RCS tumors that were well established transplantable cell lines (RCS-5 and RCS-X), by transplantable RCS lines of more recent vintage (CH-2 and CH-5), as well as by primary RCS tumors, tested prior to initial transplantation (CH-11 and CH-12). The target cell range of RCS effectors was similar to that seen with CBA/J effector cells. Thus, YAC-1, RLδ1 and RBL-5 were all lysed by RCS effectors, however, several other lymphoma targets (BW5147, C1498 and L1210) were resistant to lysis by both RCS and CBA/J effector cells (Fitzgerald and Ponzio, 1981).

TABLE I

PRESENCE OF NK ACTIVITY AGAINST YAC TARGETS IN LYMPHOID TISSUES OF NORMAL AND RCS-BEARING SJL/J MICE.

Strain	RCS Tumor	No. of Passages	Tissue	% ^{51}Cr Release[a] 4hr.	16hr.
SJL/J	5	> 200	LN	20.9±3.9	57.1±6.3
			SPL	ND	53.5
	X	> 200	LN	24.6±4.1	52.3±7.6
			SPL	ND	77.1
	CH-2	< 50	LN	14.0	47.0
	CH-5	< 50	LN	ND	27.0
	CH-11	None	LN	ND	56.0
	CH-12	None	LN	ND	31.0
	-	-	Normal LN	2.0±0.8	5.5±1.4
	-	-	Normal SPL	3.2±1.3	5.7±3.3
CBA/J	-	-	Normal SPL	28.5±4.0	47.9±3.0

[a]E/T = 50. Values with S.E. were obtained from 7 - 10 individual experiments and those without S.E. represent the mean of 2 individual experiments. ND = Not determined.

B. Capacity of RCS to Induce Host NK Activity

It became readily apparent that RCS cells could stimulate NK activity in SJL host cells. As shown in Table II, following i.v. injection of 10^7 irradiated tumor cells (X-RCS), there was a transient rise in splenic NK activity in SJL recipients, when compared to uninjected controls. The levels obtained on day 3 after injection of 10^7 X-RCS was 4-fold higher than levels obtained in high NK strain CBA/J effector cells. However, X-RCS injected SJL mice never demonstrated levels of NK activity that approached those seen in SJL recipients injected with viable RCS cells. A number of other agents, including *C.parvum*, BCG and poly I:C, which stimulated increased NK activity in other low NK strains, failed to augment NK activity in SJL mice following injection over a broad dose range (Fitzgerald, et al., 1981). X-RCS, on the other hand, reproducibly augmented NK activity in SJL mice, as well as in other low NK strains. X-RCS was also capable of inducing NK activity in SJL lymphoid cells *in vitro*. Thus, SJL LN, spleen and, surprisingly, even thymus cells, when co-cultured for 3 days with X-RCS, showed significant NK activity against YAC targets (Ponzio, et al., 1980).

TABLE II

INDUCTION OF NK ACTIVITY IN SJL MICE BY IRRADIATED RCS CELLS

Strain	No. of X-RCS injected[a]	$LU/10^8$ Cells on Day[b] 1	3	4
SJL/J	None	1.89	1.89	1.89
	10^6	1.93	1.93	1.86
	$5x10^6$	1.97	2.13	3.16
	10^7	3.72	19.47	6.40
CBA/J	None	4.92	4.92	4.92

[a] Normal SJL mice were injected i.v. with irradiated RCS (10,000R) cells and splenic NK activity was determined on the days indicated, using YAC-1 targets in a 16hr. assay (mean of 3 mice per interval). NK activity of uninjected SJL/J (low NK) and CBA/J (high NK) mice are shown for comparison (cumulative mean of 9 mice; 3 mice per interval).

[b] One Lytic unit (LU) calculated for the number of effectors required to obtain 20% lysis.

C. RCS Tumor Cells Possess NK Function

Strong evidence that RCS cells themselves are responsible for the majority of NK activity seen in tumor cell suspensions was obtained, following purification of tumor cells. RCS cells will grow in only certain SJL F_1 mice, depending on the I-region haplotype of the non-SJL parent (Katz, et al., 1980b). Therefore, purified RCS cells were obtained after growth of T cell depleted tumor cells in permissive (SJL x A.TH)F_1 recipients and removal of F_1 host cells with anti-H-$2D^d$ serum + C. Such purified RCS cells retained 80% of NK activity, when compared to normal serum + C treated controls. Identical results were obtained when RCS were sequentially passaged in (SJL x A.TH)F_1 mice to avoid cytotoxic effects contributed by carry-over of parental SJL non-tumor cells that would not be killed by antiserum pretreatment (Fitzgerald, et al., 1981).

D. Distinguishing Characteristics of RCS-associated NK Cells

Since cell suspensions prepared from RCS-bearing donors would likely contain RCS-induced, as well as RCS-mediated NK activity, experiments were designed to distinguish between these activities. In these experiments, advantage was taken of the potent NK-inducing capacity of X-RCS. Thus, the NK activity of three effectors populations was compared: Conventional NK activity, present in normal SJL mice; RCS-induced NK activity, present in SJL mice injected with X-RCS; RCS-mediated NK activity, present in cells taken from tumor bearing SJL mice. The distinguishing properties of these three effector populations are summarized in Table III.

As indicated earlier, the target cell preference of the 3 effector populations was similar . It is unlikely that natural cytotoxic (NC) cells (Stutman, et al., 1978) contributed to overall lysis, since cytotoxicity against the NC-susceptible fibrosarcoma target cell WEHI-164.1 was minimal with all 3 effector cells (Ponzio, unpublished observations). The effects of various antisera distinguished between RCS-mediated and the other two forms of NK activities. Anti-Ia^s serum + C pretreatment of effector cells significantly reduced NK activity in the RCS-mediated population, but had absolutely no effect on conventional or induced populations (Fitgerald, et al., 1981). In contrast, 3 antisera (NK-1, Ly-11.2 and Asialo GM_1) that effectively removed NK activity in the conventional and induced populations, had little, or no effect on RCS-mediated NK activity

Table III

CHARACTERISTICS OF RCS-ASSOCIATED NK CELLS

Property	Type of NK Cell Activity[a]		
	Coventional	RCS-Induced	RCS-mediated
Magnitude of NK Lysis:	Low	Moderate	High
Target Cell Preference:			
Lymphoid (YAC, RLδ1, RBL5)	++	++	++
Solid (WEHI-164.1 Fibrosarcoma)	±	±	±
Optimal Assay Time			
4 hr	+	+	+
16 hr	++	++	++
Optimal E/T Ratio	50-100	50-100	5-25
Susceptibility to Antisera + C			
$H-2^s$	+	+	+
Thy 1.2	±	±	−
Ia^s	−	−	+
NK-1	+	+	−
Ly 11.2	+	+	−
Asialo GM_1	+	+	−
Radiosensitivity	Moderate	Moderate	Low
Activity in ^{89}Sr-injected mice	−	−	++
Effect of removing machrophages	None	None	None

[a]Except where indicated, assays were performed against ^{51}Cr-labeled YAC targets for 16 hr.

Conventional NK activity refers to that present in normal, 6-8 week old SJL mice. RCS-induced NK activity refers to that present in SJL mice injected with 10^7 irradiated (10,000R) RCS cells i.v. and tested 3 days later. RCS mediated NK activity refers to that obtained when effector cells are taken from the grossly enlarged lymphoid tissues of SJL mice, injected with 10^7 viable RCS cells 7 days previously.

(Fitzgerald, et al., 1981). In vitro irradiation of cell populations prior to use as effectors demonstrated a significantly greater radioresistance of RCS-mediated NK activity. Finally, the NK activity of the 3 effector populations was assessed, using ^{89}Sr-injected (NK deficient) SJL recipients. Whereas ^{89}Sr-injected SJL mice demonstrated little, if any, NK activity and were unable to show increased NK activity after injection of X-RCS, ^{89}Sr-treated SJL mice injected with F_1 purified, viable RCS cells showed both levels of tumor growth and levels of NK activity in LN and spleen that were indistinguishable from those seen in non-^{89}Sr-injected controls (Fitzgerald, et al., 1981).

Overall, our results indicate that RCS tumor cells of SJL/J origin possess NK cytotoxic function. However, their B cell lineage is at variance with the data of others, which supports a T cell origin for NK cells. Experiments designed to specifically isolate NK effector cells from RCS tumor cell suspensions are currently underway and may help resolve this question. Preliminary results lend further support for NK function in transplantable RCS lines. Thus, YAC-binding cells within tumor cell suspensions, when separated on density gradients, show high NK activity in vitro and, after adoptive transfer to SJL recipients, grow as typical RCS tumors (Ponzio, unpublished observations). In addition, analysis of the NK effectors within RCS tumors is being performed, using a number of anti-B cell and anti-pre-B cell reagents. In this way it will be possible to definitively demonstrate the nature of RCS-associated NK effector cells.

It is possible that the differences that exist between RCS-mediated NK activity and both conventional and induced NK activities reflect neoplastic expansion of a unique population of NK cells. Alternatively, as a result of neoplastic transformation, RCS cells may have acquired NK function via a derepression - type mechanism. It has been shown that pre-T cells (Herberman, et al., 1978) and normal pro-monocytes (Lohmann-Matthes, et al., 1979), as well as a malignant pro-monocyte cell line (Kerbel, et al., 1981), can exert NK cell function. Since available evidence strongly indicates a B cell lineage for RCS of SJL mice, most probably at a pre-B-cell stage of differentiation, our results suggest that NK activity may be expressed by pre-B lymphocytes as well, and may be a characteristic quality of immature lymphoid cells in general.

REFERENCES

Carswell, E.A., S.P. Lerman and G.J. Thorbecke: Cell Immunol., 23:39, 1976.

Chang, K.S.S. In: Advances in comparative leukemia research (S. Bentvelzen, ed.) p. 327, Elsevier/North Holland Biomedical Press, Amsterdam, 1978.

Fitzgerald, K.L. and N.M. Ponzio: Cell. Immunol., 43:185, 1979.

Fitzgerald, K.L. and N.M. Ponzio: Int. J. Cancer 28:627, 1981.

Fitzgerald, K.L., J. McMaster and N.M. Ponzio: Int. J. Cancer 28:635, 1981.

Herberman, R.B., M.E. Nunn and H.T. Holden: J. Immunol. 121:304, 1978.

Katz, I.R., R. Asofsky and G.J. Thorbecke: J. Immunol., 125:1355, 1980a.

Katz, I.R., S.P. Lerman, N.M. Ponzio, D.C. Shreffler and G.J. Thorbecke: J. Exp. Med., 151:347, 1980b.

Kerbel, R.S., J.C. Roder and H.F. Pross: Int. J. Cancer 27:87, 1981.

Lerman, S.P., E.A. Carswell, J. Chapman and G.J. Thorbecke: Cell Immunol., 23:53, 1976.

Lerman S.P., J. Chapman-Alexander, D. Umetsu and G.J. Thorbecke.: Cell. Immunol., 43:209, 1979.

Lohmann-Matthes, M-L., W. Domzig and J. Roder: J. Immunol., 123:1883, 1979.

Murphy, E.D.: Proc. Amer Assoc. Cancer Res., 4:46, 1963.

Murphy, E.D.: J. Natl. Cancer Inst. 42:797, 1969.

Ponzio, N.M., C.S. David, D.C. Shreffler and G.J. Thorbecke: J. Exp. Med., 146:132, 1977a.

Ponzio, N.M., K.L. Fitzgerald, G.J. Thorbecke and J. Vilcek: Ann. N.Y. Acad. Sci., 350:157-167, 1980.

Ponzio, N.M., S.P. Lerman, J.M. Chapman and G.J. Thorbecke.: Cell. Immunol., 32:10, 1977b.

Scheid, M.P., J.M. Chapman-Alexander, T. Hayama, I.R. Katz, S.P. Lerman, C. Nagler, N.M. Ponzio and G.J. Thorbecke. In: B lymphocytes in the immune response: Functional, Developmental and Interactive Properties. (ed. by N. Klinman, D. Mosier, I. Scher and E. Vitetta) p. 475 Elsevier North Holland, Amsterdam, 1981.

Siegler R. and M.A. Rich: J. Natl Cancer Inst., 41:125, 1968.

Stutman, O., C.J. Paige and E. FeoFigarella: J. Immunol., 121:1819, 1978.

IDENTIFICATION AND CHARACTERIZATION OF LARGE GRANULAR LYMPHOCYTE (LGL) LEUKEMIAS IN F344 RATS

Craig W. Reynolds,[1] Jerrold M. Ward,[2]
Alfred C. Denn III[1] and E. William Bere, Jr.[1]

[1]Biological Research and Therapy Branch
and
[2]Tumor Pathology and Pathogenesis Section
Laboratory of Comparative Carcinogenesis
National Cancer Institute - FCRF
Frederick, Maryland

INTRODUCTION

Recently, a number of reports have shown that a distinct subpopulation of large granular lymphocytes (LGL) are capable of mediating natural killer (NK) activity in both rats and humans (Reynolds, et al., 1981a; Reynolds et al., 1981b; Timonen et al., 1981). Although some general characteristics of these cells have been examined, it has been very difficult to obtain sufficiently large numbers of highly enriched cells to define, in detail, such important characteristics as granule composition, cell surface receptors and lytic machinery. One means of overcoming this difficulty would be to find naturally occurring LGL leukemias which maintain their functional characteristics. This manuscript describes a series of spontaneously occurring LGL leukemias in rats and briefly examines their functional and cell surface characteristics. The results suggest that these tumors might be highly useful models for those studies which require a large number of cytotoxic cells.

MATERIALS AND METHODS

Animals

All experiments were performed with aged (24-34 months old) F344 rats. The histological and pathological characteristics of the tumors which appear in these animals have previously been described (Ward, 1982).

ISBN 0-12-341360-5

Target Cells

The rat G1-TC and (C58NT)D-TC and mouse YAC-1 lymphomas were maintained *in vitro* as suspension cultures (Oehler *et al.*, 1978).

Preparation of Normal Spleen Cells and Splenic Tumor Cells

Spleens from normal or tumor-bearing animals were prepared by washing in Hanks' balanced salt solution (HBSS) as previously described (Ortiz de Landazuri and Herberman, 1972; Oehler *et al.*, 1977). For the morphological evaluation of lymphocyte preparations, air-dryed cytocentrafuge preparations were fixed in methanol and stained with 10% Giemsa (pH 7.4).

^{51}Cr-Release Cytotoxicity Assay

Two-fold serial dilutions of effector cells were mixed with ^{51}Cr-labeled target cells and incubated at 37°C for 4 hours, as previously described (Reynolds and Herberman, 1981).

Flow Microfluorometry (FMF)

FMF analysis was performed as previously described (Reynolds *et al.*, 1981b; Miller *et al.* 1978; Ortaldo *et al.* 1981) using cell sorters (FACS II or III) (Becton-Dickinson FACS Systems).

RESULTS AND DISCUSSION

Previous studies have demonstrated that there are spontaneous large granular lymphocyte (mononuclear cell leukemia) tumors in a high percentage of aged (>24 months old) F344 rats (Ward, 1982). However, the functional characteristics or origin of these tumors have not been known. Since the morphology of these leukemias has a general similarity to rat LGL and because previous studies in the mouse (Chang *et al.* 1980; Fitzgerald *et al.* 1979) and the human (Hokland *et al.* 1981) have demonstrated high tumor-associated NK activity, we examined a series of rat LGL leukemias for the ability to kill NK-sensitive and NK-resistant target cells. Three examples of these leukemias are shown in Table 1. The results suggest that a wide range of cytotoxic activity can be seen in rat LGL tumors. Of the 15 leukemias examined, 3 were found to have high activity, 8 were classified as intermediate and 4 had very low cytotoxic potential. Although

Table 1. Cytotoxic Activity of Rat LGL Tumors

		% Cytotoxicity[a]		
Effector Cells	E/T	YAC-1	G1-TC	(C58NT)D
RNK-16	50:1	56.8	22.5	0.0
(High)	6:1	45.7	16.3	0.7
RNK-13	50:1	37.9	5.9	2.2
(Intermediate)	6:1	11.2	2.5	3.7
RNK-9	50:1	12.2	NT	NT
(Low)	6:1	4.5	-	-
Normal	50:1	41.8	7.5	NT
spleen	6:1	2.3	0.0	-

a) 18 hour ^{51}Cr-release cytotoxicity assay.

the pattern of this cytotoxicity correlates well with the known specificity of rat NK cells (Reynolds and Herberman, 1981), the activity of these cells differs from normal LGL in that maximal lysis has been seen only in an 18 hour ^{51}Cr-release assay. The high percentage of tumor cells in the spleen (60-90%), the unusually high activity seen with some leukemias and the relatively low activity in normal age-matched controls suggested that the cytotoxic activity was related to the tumor cells themselves. To further investigate the characteristics of these tumor cells we separated spleen cells from tumor-bearing animals on discontinuous density gradients of Percoll (Reynolds et al., 1981a). The results demonstrated that the leukemia cells could be enriched in the fraction normally containing LGL, and cells in this fraction were highly active in cytotoxicity assays (data not shown).

Because most of the LGL leukemias shared some characteristics with both macrophages (adherent and erythrophagocytic) and NK cells (granular and cytotoxic), we examined the tumor cells for cell surface antigens associated with various cell lineages. We have recently shown that normal rat LGL are W3/13 OX-1, OX-8 and BC-84 positive (Reynolds et al. 1981b). The results in Table 2 indicate that the leukemias are quite heterogenous with respect to most cell surface antigens but that many of these tumors share cell surface antigens with normal LGL and macrophages (W3/13, OX-1, BC-84, M1/70). At present, however, no clear phenotypic distinction can be drawn between those tumors with high or low cytotoxic activity. Interestingly, one-third of the tumors examined did not have the leukocyte common (L-C) antigen found on essentially all normal leukocyte populations (Reynolds et al., 1981b). In addition, six of eight tumors had easily

Table 2. Cell Surface Antigenic Characteristics of Rat LGL Leukemias

Antibody (Antigen)	No. Positive / Total	Antibody (Antigen)	No. Positive / Total
W3/13	7/7	OX-8 (suppressor/ cytotoxic T-cells)	5/9
W3/25 (helper T cells)	0/9	BC-84 (ART-1[a])	6/9
OX-1 (leukocyte common)	6/9	M1/70 (Mac-1)	5/9
OX-7 (Thy 1.1)	6/8	Ia	2/6

detectable Thy 1.1, an antigen found in the rat only in early leukocyte precursors (Goldschneider *et al.*, 1980).

The results from these studies indicate that spontaneous LGL leukemias occur in aged rats, and that some of these tumors have very high cytotoxic activity. However, these leukemias are clearly heterogeneous with respect to cytolytic potential and cell surface antigens.

Present studies are now underway to culture these cell lines *in vitro* and to passage these tumors *in vivo*. Preliminary results suggest that these cells will not grow *in vitro* either alone or with exogenous IL-2. However, we have been able to successfully passage some of these tumors *in vivo* and to maintain their cytolytic potential through the first passage. It seems likely that these leukemias will provide an excellent source for a large quantity of highly enriched cytotoxic cells for detailed analysis.

REFERENCES

Chang, K.S.S. and Log, T. (1980). Int J Cancer 25:405.
Fitzgerald, K.L. and Ponzio, N.M. (1979). Cell Immunol 43:185.
Goldschneider, I. *et al.* (1980). J Exp Med 152:419.
Hokland, P. *et al.* (1981). Blood 57:972.
Miller, M.H., *et al.* (1978). Rev Sci Instrum 49:1137.
Oehler, J.R. *et al.* (1977). Cell Immunol 29:238.
Oehler, J.R. *et al.* (1978). Int J Cancer 21:204.

Ortaldo, J.R. et al. (1981) J Immunol 127:2401.
Ortiz de Landazuri, M. and Herberman, R.B. (1972). J Natl Cancer Inst 49:147.
Reynolds, C.W. and Herberman, R.B. (1981). J Immunol 126:1581.
Reynolds, C.W. et al. (1981a). J Immunol 127:282.
Reynolds, C.W. et al. (1981b). J Immunol 127:2204.
Timonen, T. et al. (1981). J Exper Med 153:569.
Ward, J.M. (1982). Prog Exp Tumor Res, in press.

TUMOR RELATED CHANGES AND PROGNOSTIC SIGNIFICANCE OF NATURAL KILLER CELL ACTIVITY IN MELANOMA PATIENTS[1]

Peter Hersey
Anne Edwards
William McCarthy
Gerald Milton

Medical Research Department
Kanematsu Memorial Institute and
Melanoma Unit, Department of Surgery
Sydney Hospital
Sydney, N.S.W., Australia

I. INTRODUCTION

Although there is now considerable evidence that natural killer (NK) cells may be important in control of tumor growth in experimental animals, (1,2) there is as yet very little evidence that they have a similar role in man. This is largely because experiments analogous to those carried out in inbred animal strains are not feasible in human subjects but also because epidemiological approaches to the question have been limited by a paucity of information about genetic and environmental factors which have major influences on NK activity in man (3).

In this respect it appeared of particular relevance to understand the effects that tumor-host interactions may have on NK activity and on prognosis in tumor bearing patients. These considerations led to the work summarized below in which the influence of small localized tumors on NK activity was

[1]Supported in part by grants from the NSW State Cancer Council and in part under NCI Contract NO1-CB-74120.

ISBN 0-12-341360-5

assessed by studies before and after surgical removal of the tumor. An indirect assessment of the importance of NK activity in prognosis was then made by relating this information to histological features of the tumor known to be important in prognosis and to the recurrence free period of the patients under study. Although the latter studies are as yet incomplete the results suggest the importance of NK activity in control of tumor growth in this human model is far from proven.

II. SEQUENTIAL STUDIES ON NK ACTIVITY BEFORE AND AFTER SURGICAL REMOVAL OF MELANOMA

The patients selected for these studies were patients with stage I (primary) or stage II melanoma. Measurement of NK activity was based on ^{51}Cr release from a variety of cultured melanoma and non-melanoma target cells. Full details of the study are given elsewhere (4). The results, summarized in Table I, indicated that in patients with primary melanoma there was a marked increase in NK activity against the cultured melanoma cells after surgical removal of the tumor that was not seen against the non-melanoma cells.

TABLE I. Mean NK Activity Against Different Cell Lines Before and After Surgical Removal of Tumor

	Stage I primary			Stage II		
Cell Line	A	B	C	A	B	C
MM200	14.1 ±10.4 n=23	21.5[a] ±12.7 n=19	15.3 ±8.6 n=17	14.5 ± 8.8 n=63	11.3[a] ± 8.5 n=51	12.6 ± 9.3 n=40
Chang	12.5 ± 7.0 n=23	11.2 ± 6.3 n=19	10.7 ± 6.3 n=17	11.3 ± 7.1 n=60	8.7[a] ± 6.5 n=50	9.3 ± 5.4 n=39
MCF-7	21.8 ±12.3 n=16	18.0 ±11.9 n=12	15.6 ±8.4 n= 9	20.8 ±11.5 n=37	15.5 ±10.9 n=26	15.5 ±11.9 n=19

Time A, B and C = before, 2-4 and 6-8 weeks after surgery. Values indicated are means ± I S.D. a, A cf B, $P < .05$

These changes were not seen in patients with stage II melanoma who instead had a significant decrease in NK activity against certain of the target cells after removal of the involved lymph nodes. NK activity in the latter groups tended to remain lower in subsequent studies than in patients with primary melanoma. No significant changes were seen in melanoma patients admitted for WEG or in the control patients with non-melanoma carcinomas. (These findings are discussed below).

III. CORRELATION BETWEEN HISTOLOGICAL FEATURES OF PRIMARY MELANOMA AND NK ACTIVITY OF BLOOD LYMPHOCYTES

Several histological features of primary melanomas are known to have importance in prognosis from melanoma (5). The single most important factor was that of tumor thickness which was associated with a high incidence of local recurrences and poor overall survival (6). The influence of lymphoid infiltration on prognosis is more controversial and some reports suggest that when tumors were matched for thickness lymphoid infiltration below primary melanomas may be associated with a poorer prognosis (7). The relation between NK activity and tumor thickness or lymphoid infiltration below primary melanomas was therefore examined to determine if these prognostic indicators may operate by association with NK activity.

When NK activity of blood lymphocytes from patients was measured against melanoma target cells it was found there was a direct correlation with tumor thickness (4). This was particularly marked shortly after surgical removal of the melanoma and less so at times before removal of the tumor. There was no correlation between tumor thickness and NK activity against non-melanoma target cells. These observations suggested the correlation with tumor thickness was due to cytotoxicity induced against the melanoma cells by the growing tumor.

Analysis of the relation between lymphoid infiltration below the primary tumor and NK activity of blood lymphocytes in 189 patients revealed a significant inverse correlation i.e. when there were "moderate" or "prominent" degrees of lymphoid infiltration, NK activity to melanoma and Chang target cells was lower than would be expected if there were no association between NK activity and lymphoid infiltration. Conversely, "absent" or "few" lymphocytes below the tumor was associated with high NK activity of blood lymphocytes (8).

These results are summarized in the contingency table

TABLE II. Association of NK activity with Lymphocytic Infiltration at the Base of the Tumor.

Patient Groups	Lymphocytic Infiltration: Tumors > 1.5 mm (a)		Tumors 1.5 mm or less		Total
	Absent/Few	Mod./Prom.	Absent/Few	Mod./Prom.	
0-9	6(10.25)	9(9.44)	7(9.44)	29(21.86)	51
10-21	15(19.3)	22(17.78)	20(17.78)	39(41.14)	96
22 and over	17(8.44)	4(7.78)	8(7.78)	13(18.00)	42
TOTALS	38	35	35	81	189

$X^2 = 19$ 6 df $.001 < P < .005$ (a) = 9.87 2 df $P < 0.01$

analysis in Table II. Patients were grouped according to the level of their NK activity against the melanoma target cells on the basis of analysis of the mean and standard error of the values for the 189 patients (15.07±2.96). Patients with low NK activity (0-9% ^{51}Cr release) had values more than 2 standard errors below the mean and patients with high NK activity (22% ^{51}Cr release and over) had values more than 2 standard errors above the mean. The expected numbers in the table were calculated on the basis of no association between NK activity and lymphoid infiltration. The strongest inverse correlation was apparent in patients with tumors greater than 1.5 mm in thickness but the same trend was evident in those with tumors less than 1.5 mm in thickness.

IV. RELATION BETWEEN NK ACTIVITY AT THE TIME OF SURGICAL REMOVAL OF MELANOMA AND DURATION TO RECURRENCE

A group of over 227 patients were followed after surgical removal of their primary tumor to determine whether their NK activity against melanoma cells measured shortly after removal of their tumor could be related to the recurrence free interval from melanoma. Life table analysis of the cumulative proportion of patients free of recurrent melanoma at a minimum and maximum follow-up period of 6 and 42 months respectively, revealed that the trend appeared opposite to that expected in that there was a lower proportion of recurrences in patients with "low" compared to those with "high" initial NK activity. (Proportion of male patients free of recurrences at 2 years was 0.91, 0.81 and 0.74 in those with "low", "normal" and "high" NK activity respectively. These differences were not significant by log rank analysis. There were insufficient recurrences in the female patients at this time for analysis).

Because of the low number of recurrences in the patients studied at the time of their primary treatment, the influence of NK activity in patients at the time of treatment of the first local recurrence on the duration to development of distant metastases was determined. As shown by Figure 1, there was a lower incidence of distant recurrences over a 30 month follow-up period in patients who had "high" compared to those with "low" or "normal" NK activity at the time of removal of their first recurrence.

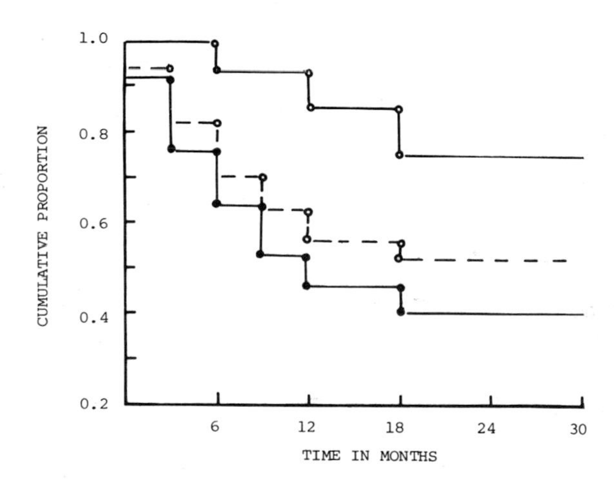

FIGURE 1. Cumulative proportion of patients free of distant metastases after removal of local melanoma recurrences. 0—0 patients with high, 0--0 normal, ●—● or low NK activity. Numbers in each group were 15,33 and 28 respectively. (By log rank analysis, X^2 values for comparison of high and low or high and normal NK activity groups were 4.32 and 4.5 respectively. ($.025 < P < .05$).

V. DISCUSSION

The studies outlined above indicate that the level of NK activity in melanoma patients was influenced, often to a major degree, by the presence of their tumor. In patients with primary melanomas the increase in NK activity following surgical removal of the tumor suggested that one effect of tumor host interactions was suppression of NK activity. Shortly after tumor removal the level of NK activity appeared to be related to tumor thickness and to be directed against melanoma cells rather than to non-melanoma cells used in the study. The latter changes were not seen in patients with melanoma involving regional lymph nodes which suggested that a different set of tumor host interactions occurred in these patients.

Several possible mechanisms could account for the suppression of NK activity in patients with primary melanomas. One is that suppressor cells may have been induced by tumor host interactions which inhibited NK activity. Previous studies have shown that tumor growth in melanoma patients was associated with induction of suppressor T cells against immunoglobulin synthesis (9) and these or similar suppressor cells may have been responsible for inhibition of NK activity. The latter findings would be consistent with the demonstration that lymphoid infiltration below primary tumors was inversely correlated with NK activity in blood lymphocytes if it were assumed that the lymphoid cells below the tumors were suppressor cells. There is some evidence for the latter in that most of the lymphocytes below melanomas were reported to express antigens identified by OKT8 monoclonal antibodies which are considered to identify human suppressor and cytotoxic T cells. (R. Mackie meeting report). These results suggest that even with small primary melanomas the predominant effect of tumor host interactions is suppression of NK activity in the circulation presumably from recirculation of suppressor lymphocytes from the site of the tumor. Whether this mechanism is also responsible for depression of NK activity in patients with advanced tumor growth is uknown (10,11).

These findings appear to have important implications as to the prognostic significance of NK activity in melanoma patients. Hence the correlation of NK activity measured shortly after removal of the primary tumor with thickness of the tumor implies that high NK activity at this time is also a bad prognostic feature. (This may not apply when NK activity is measured at other times in relation to tumor removal). This is supported by studies relating the level of

NK activity after removal of the primary tumor to the interval to development of local recurrences. The results in male patients show a trend for this interval to be shorter in those with high compared to those with low NK activity.

The relation of lymphoid infiltration below primary melanomas to prognosis is less well defined because there is also an association of the latter with tumor thickness (7). When this is taken into account some studies suggest lymphoid infiltration at this site may be a bad prognostic feature. This would be consistent with the association between the latter and low NK activity in the blood shown in the present study which may predispose to systemic as compared to regional spread of melanoma. The higher incidence of distant metastases in patients with low compared to high NK activity at the time of treatment of their first local recurrence supports this view. Alternatively, these results may indicate that the patients with low NK activity had more tumor present at the time of detection of the recurrence and hence developed distant metastases more quickly because of the latter. We are unable to differentiate between these possibilities at present.

The question of the prognostic significance of NK activity in melanoma patients is hence largely unresolved. The evidence that is available must however raise some doubts about its significance in preventing spread of melanoma to regional lymphatic areas. They do not exclude that NK activity may have an important role in prevention of blood born spread of tumor cells or that the apparent unfavourable prognosis of lymphoid infiltration below primary melanomas may be due to the association of this histology with low NK activity in blood lymphocytes. Further study of the natural history of melanoma in relation to natural killer cell activity of the host may further clarify these questions.

REFERENCES

1. Herberman, R.B., Djeu, J.Y., Kay, H.D., Ontaldo, J.R., Riccardi, C., Bonnard, G.D., Holden, H.T., Fagnani, R., Santoni, A., and Puccetti, P., Immunological Rev. 44:43 (1979).
2. Kiessling, R., and Haller, O., in "Contemporary topics in Immunobiology" (N.L. Warner, ed.) p 171. Plenum Press, 1978.
3. Hersey, P., Aust. N.Z. J. Med. 9:464 (1979).
4. Hersey, P., Edwards, A., and McCarthy, W.H., Int. J. Cancer 25:187 (1980).

5. McGovern, V.J., Shaw, H.M., Milton, G.W., and Farago, G.A., Histopathology 3:385 (1979).
6. Milton, G.W., Shaw, H.M., Farago, G.A., and McCarthy, W.H., Brit. J. Surgery 67:543 (1980).
7. McGovern, V.J., Shaw, H.M., Milton, G.W., and Farago, G.A., in "Pathology of Malignant Melanoma" (A.B. Ackerman, ed.) Masson Publishing, New York, 1981 (in press).
8. Hersey, P., Hobbs, A., Edwards, A., McCarthy, W.H., and McGovern, V.J., Cancer Res., 1981 (in press).
9. Werkmeister, J., McCarthy, W.H., and Hersey, P., Int. J. Cancer 28:1 (1981).
10. Pross, H.F., and Baines, M.G., Int. J. Cancer 18:593 (1976).
11. Takasugi, M., Ramseyer, A., Takasugi, J., Cancer Res., 37:413 (1977).

THE ASSESSMENT OF NATURAL KILLER CELL ACTIVITY IN CANCER PATIENTS

Hugh F. Pross
Peter Rubin

Departments of
Radiation Oncology and Microbiology & Immunology
Queen's University
Kingston, Ontario, Canada

Malcolm G. Baines

Department of Microbiology & Immunology
McGill University
Montreal, Quebec, Canada

I. INTRODUCTION

The study of human natural killer (NK) cell activity has, by necessity, been characterized by a dependence on indirect evidence to support hypotheses based on *in vitro* observations. Because the assays used to detect human NK cells involve the lysis of tumor target cells, it is not unreasonable to postulate that this is the role of NK cells *in vivo* (1,2). The question of whether or not this postulate is indeed valid has been the subject of our research program for many years. In 1976 we reported that the NK activity of patients with advanced cancer was significantly lower than normal (3) and this was confirmed concurrently by Takasugi *et al* (4). More recent studies

---

Supported by the Ontario Cancer Treatment and Research Foundation and the Medical Research Council of Canada.

ISBN 0-12-341360-5

have also confirmed these results, the degree of NK "deficiency" observed depending on the malignancy under study, and not necessarily showing a correlation with disease progression. All of these papers could be criticized for several reasons. The reporting of data generated from small numbers of patients, who have been followed for short periods of time, has been a major defect. Other serious criticisms have been directed at the use of chromium release values to express relative cytotoxicity, failure to use well-defined controls, failure to keep the study "blind" with respect to patient diagnosis until after completion of the study, failure to adequately review the charts, especially in so far as other variables such as drugs and infections are concerned, and finally, failure to use an NK target that is as unrelated as possible to the type of cancer that the patient has. In this chapter, we would like to discuss how we have dealt with some of these problems.

II. FACTORS AFFECTING THE ASSESSMENT OF NK ACTIVITY

In our early work (3) we showed that there is a relationship between the presence of malignant disease and alterations in NK activity. It has always been unclear, however, whether these alterations are the result of the malignant disease or whether NK cells actually influence the course of the disease. In an attempt to answer this question, we have studied approximately 1650 cancer patients and 550 normal adults with respect to their NK activity. The patients have been tested from 1-30 times over the course of their disease and, at present, we are assembling the data for computer analysis while waiting for the majority of the patients to either expire or pass the 5 year survival mark. The long term and sequential nature of these studies has introduced a number of factors which had to be considered before the data could be properly evaluated. These factors could be termed "mathematical", "physiological" and "clinical".

A. Mathematical Aspects of NK Data Handling

It is customary to plot NK data as per cent cell-mediated lysis *vs* log. LT ratio, which results in a sigmoid curve. Because the curve is not linear, the proportional difference between two donors varies markedly, depending on which LT ratio is used as the comparison point. Similarly, the use of lytic units calculated from these curves will also give different relative cytotoxicity values, depending on the per cent lysis value chosen. It is absolutely essential, therefore, that cytotoxicity data be rendered linear, and that comparison between two sets of values yield data that reflect the proportional difference between the two donors. The equation $y = A(1-e^{-kx})$ can be used to explain cytotoxicity data (y) generated from a series of LT ratios (x) (5). The "A" value is the plateau that the conventional sigmoid curve attains at high cell ratios, and can be set at 100% or, as is more often the case in practice, at 75-80%. The equation can be used to solve for cytotoxicity at a particular LT ratio, for x at a defined y (i.e. lytic units) or for k, the negative of the slope of ln(A-y) vs x. The ratio of k values to each other, or to the mean k of several lines, is an accurate expression of the proportional NK activity of the donors (5) and gives similar data to that obtained using kinetic analysis (6). In order for comparisons to accurately reflect differences in effector function, the A value must be the same for each series of lines under comparison, i.e. within any one experiment. Since the plateaus for different donors may not be precisely the same, setting the A at a fixed value involves a compromise to some extent. In human NK assays, using a sensitive target such as K562, the lines very rarely fail to converge at plateaus which are not significantly different from each other. Artifactually low plateau values will occur when very high LT ratios are required to demonstrate cytotoxicity, while suppressor cells would cause the curve to dip rather than to plateau.

In practice, routine NK assays using patient's lymphocytes are handled as follows (Table I): at least two fresh or cryopreserved normal control lymphocyte preparations are tested with every group of patients. These controls have been previously well characterized so that their NK activity is known relative to the mean of many normals over several hundred assays. This information is used to arrive at a correction factor for the particular control donor. Within each assay, the k values for the normal controls are corrected by their appropriate factors, and the mean of these normal values is used as the denominator in calculating the relative cytotoxicity of each individual patient (Table I). An alternative to this method is to use a larger number of normal controls so that the correction factor is not necessary. By handling the data in this way, it is possible to arrive at a reliable approximation of a patient's NK activity, relative to normal, which is independent of day-to-day variations in the sensitivity of the assay system. The method is dependent on the observation that individual

TABLE I. CONVERSION OF CHROMIUM RELEASE DATA[a] TO RELATIVE NK

Donor	% NK @ 5/1	LU/10^6 (20%)	k $\times 10^{-3}$	Corr. factor	Corr. k $\times 10^{-3}$	Relative NK
N1[b]	19	38	62	1.11	68	0.95
N2	34	78	127	0.59	75	1.95
N2 F/T	33	74	120	0.59	71	1.85
N3	32.0	72	118	0.45	54	1.82
N3 F/T	35	81	132	0.45	60	2.03
				Geo. Mean:	65	
P1	43	111	181	-	-	2.78[c]
P2	11	20	33	-	-	0.51

[a]based on analysis of four LT ratios with 5000 K562 targets per well. The correction factors were calculated from the mean k value for each normal donor derived from at least 10 previous assays, divided by the overall mean of all normal assays done to date. Donors N1-N3 had been tested 39, 213 and 98 times prior to calculation of their respective correction factors.

[b]N1-N3, normal donors; N2 F/T and N3 F/T, lymphocytes from donors N2 and N3 cryopreserved 13 and 9 weeks previously, respectively; P1,P2 - patients. Relative NK (RNK) is calculated as follows:

$$RNK = \frac{k \text{ (test)}}{\text{Geo. Mean } k \text{ (Controls)}}$$

[c]Standard deviations are $\pm$ .02 and $\pm$ .04 respectively.

normal donors are very consistent in their level of NK activity relative to others (7).

B. Physiological Variables Affecting NK Activity

Many investigators have studied the relationship between age, sex, blood group and Rh type on NK activity (reviewed in (7)) and for the most part there is little difference between these subgroups when one considers the range of activity between different individuals. On analysis of our own data from about 550 normal donors (7), we were surprised to find that there is a slight but significant increase in NK activity with age ($p < 0.01$), ranging from about 50% of normal adult levels in cord blood to maximum levels after age 30. The overlap between the groups is extensive, however. No significant differences were found between donors having different blood groups, Rh type or, in a small subgroup of 26 well characterized donors, HLA type. In contrast with our earlier impression, we also found a significant difference between the NK levels of males and females in most of the subgroups, female NK activity being approximately 0.8 that of males. Again there was a large degree of overlap in the groups. On the basis of these observations, we have concluded that it is not necessary to stringently control for age in NK studies of cancer patients. It is necessary, however, to ensure that the proportion of males and females are similar if different patient subgroups are being compared.

C. Clinical Variables Affecting NK Activity

The accurate assessment of NK function in cancer patients is dependent to a great extent on being able to correct for, or eliminate, the NK modifying effects of a myriad of clinical conditions which may affect cytotoxicity. Some of these, such as alcoholism, smoking, salicylate ingestion, immunization and viral infection have been more or less documented by a number of investigators (referenced in (5,7)). The effects of other factors are unknown. It is apparent that many patients have concurrent problems unrelated to their malignancy, and for which they are on medication. Also, patients with newly diagnosed, untreated cancer are frequently taking analgesics, antiemetics or sedatives, and the likelihood of this increases with the stage of disease. The use of these "symptomatic" drugs also confounds the analysis of data from patients who are being followed sequentially to assess the effect of cancer or various types of cancer therapy on NK function.

In the last two years, we have also been interested in the mechanism by which overall NK activity may be altered in cancer patients, particularly the question of whether decreased NK levels are due to a reduced frequency of active cells or to an overall inhibition of cytotoxic activity per cell. We have found (8) that normal donors with high or low NK activity differ in the frequency of lytic NK cells in a fairly predictable way (as determined by the single cell cytotoxicity assay (9)). As yet we do not know to what extent this is also true in patients, and these studies are ongoing. In the course of these studies, two interesting

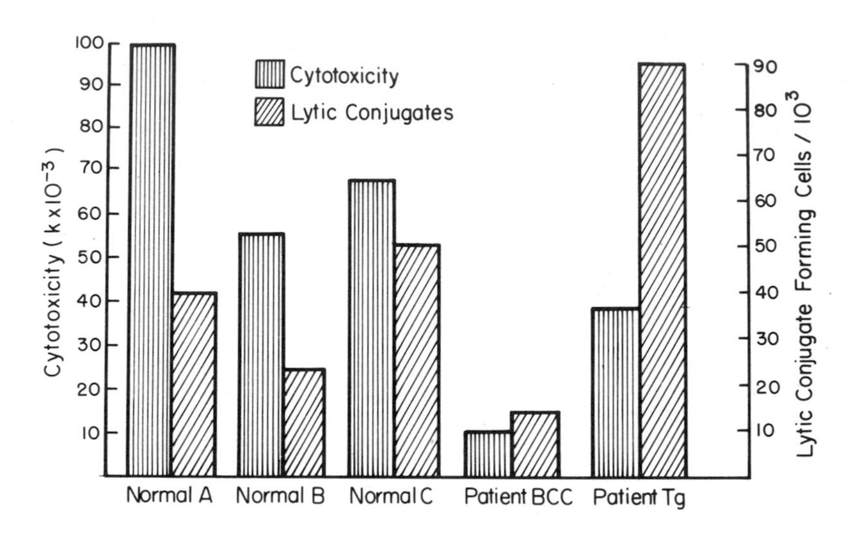

FIGURE 1. Cytotoxicity (vertical histograms) and NK frequency (diagonal histograms) of 3 normal donors (A-C) and 2 unique patients with cancer: BCC - multiple basal cell çarcinomas with immunosuppression; Tg - T gamma lymphoproliferation with neutropenia.

patients have been observed, both of whom are extreme examples of abnormalities in NK frequency and/or function (8,10). Fig 1 illustrates the data obtained from a patient (BCC) who has been on immunosuppressive therapy (prednisone/azathiaprine) to prevent renal transplant rejection. Although his renal function is normal, he suffers from multiple recurrent basal cell carcinomas. It can be seen in this figure that the patient's overall NK activity is extremely low, and that this is attributable to a defect in lytic activity per cell (recycling ability) in combination with a low frequency of NK cells. This low frequency of NK cells is still within normal limits however (9). The second patient (Tg) was diagnosed as having a lymphoproliferative disorder characterized by elevated Tgamma levels (45%), recurrent infections and neutropenia (10). In contrast to the enhanced NK activity obtained with E rosette depleted normal lymphocytes, all of this patient's NK activity was removed by E-rosette depletion. Furthermore, the "leukemic" cells were capable of antibody-dependent cell mediated cytotoxicity against both tumour target cells and normal granulocytes. In spite of a high frequency of NK cells (Fig 1), the recycling ability of these cells was abnormally low, giving rise to an overall level of NK activity within normal limits. These patients illustrate two of the numerous combinations of NK frequency and recycling ability which can occur - low NK frequency with low recycling due to chronic immunosuppression, and high NK frequency with low recycling due to the functional immaturity of abnormal NK cells.

REFERENCES

1. Jondal, M., and Pross, H.F., Int. J. Cancer 15:596 (1975).
2. Pross, H.F., and Jondal, M., Clin. exp. Immunol. 21:226 (1975).
3. Pross, H.F., and Baines, M.G., Int. J. Cancer 18:593 (1975).
4. Takasugi, M., Ramseyer, A., and Takasugi, J., Cancer Res. 37:413 (1977).
5. Pross, H.F., Baines, M.G., Rubin, P., Shragge, P., and Patterson, M., J. Clin. Immunol. 1:51 (1981).
6. Callewaert, D., Smeekens, S.P., and Mahle, N.H., J. Immunol. Methods (in press).
7. Pross, H.F., and Baines, M.G., Int. J. Cancer (in press).
8. Rubin, P., Pross, H.F., and Roder, J., Submitted for publication.
9. Grimm, E., and Bonavida, B., J. Immunol. 123:2861 (1979).
10. Pross, H.F., Pater, J., Dwosh, I., Giles, A., Gallinger, L.A., Rubin, P., Corbett, W.E.N., Galbraith, P., and Baines, M.G., J. Clin. Immunol. (in press).

NK CELL ACTIVITY IN PATIENTS WITH HIGH RISK FOR TUMORS AND IN PATIENTS WITH CANCER[1]

Marc Lipinski
Marie-Christine Dokhélar
Thomas Tursz

Laboratoire d'Immunologie Clinique
Institut Gustave Roussy
Villejuif, France

I. INTRODUCTION

Numerous studies done in the mouse suggest that various groups of spontaneously cytotoxic cells might play a role as anti-tumor agents. Similarly, human natural killer (NK) cells might possibly be involved in a primary system of defense against neoplastic growth. To explore this hypothesis, we found it interesting to select well-defined groups of patients with high risk for tumors and to compare their NK cell activity with that of appropriate controls. We thus investigated patients with congenital immunodeficiency (ID) and kidney allograft recipients treated with immunosuppressive drugs, two categories of subjects with strikingly high incidence of tumors (Kersey et al., 1973; Penn, 1978). In these groups, it is well known, however, that the overall elevation of cancer frequency does not reflect a higher incidence of every type of tumors, but rather a drasticaly increased risk for certain types, especially malignant lymphomas and to a lesser extent, leukemias. This observation prompted us to also measure NK cell activity in patients with malignant lymphomas, and to compare the results to those obtained in control populations of normal subjects and of patients affected with other types of non-lymphoid tumors.

[1] Supported by DGRST, IGR and ADRC grants.

ISBN 0-12-341360-5

II. METHODS

Effector cells were peripheral blood lymphocytes isolated by centrifugation on Ficoll-Paque and tested for spontaneous cytotoxicity against ^{51}Cr-labeled K562 cells. Effector to target cell ratio was 50:1. Incubation was at 37°C for 4 hrs. Supernatants were harvested and radioactivity was counted. Percent specific lysis (% SL) was calculated as follows:

$$\% \, SL = \frac{\text{cpm assay} - \text{cpm min}}{\text{cpm max} - \text{cpm min}} \times 100$$

where min and max indicate spontaneous (without effectors) and total (with HCl 6N) releases of ^{51}Cr by targets, respectively. Results were compared by statistical methods with mean specific lysis calculated in control populations tested in the same conditions.

III. RESULTS

A. NK Cell Activity in Patients with High Risk for Tumors

1. Patients with Congenital ID. Peripheral blood lymphocytes of a series of patients with various ID were tested for NK cell activity. Results have been partially reported elsewhere (Lipinski et al., 1980). Individual patients with NK cell activity in the normal range are presented in Table I. They were affected with X-linked agammaglobulinemia, hypogammaglobulinemia or some common variable ID mainly involving humoral immunity. From these data, it can be postulated that ID involving B cell immunity do not affect NK cell activity.

TABLE I. Congenital ID Patients with Normal NK Activity

Diagnosis	%SL
X-linked agammaglobulinemia	46.7
X-linked agammaglobulinemia	33.2
Hypogammaglobulinemia	45.5
Common variable ID	40.5
Common variable ID	50.7
Normal population (112 subjects)	49.4±14.0

TABLE II. Congenital ID Patients with Low NK Activity

Diagnosis	% SL
Severe Combined ID	4.7
Severe Combined ID	0.6
Severe Combined ID	10.5
Ataxia-telangectasia	0
Ataxia-Telangectasia	14.0
Wiskott-Aldrich syndrome	6.3
Common variable ID	9.8
Common variable ID	2.7

ID patients with low NK cell activity are listed in Table II. These included patients with severe combined ID, ataxia-telangectasia, common variable ID and Wiskott-Aldrich syndrome. All these patients shared severe defects in various T cell-mediated functions, in agreement with experimental data suggesting that NK cells could be of T lineage.

2. Patients with Drug-Induced ID. NK cell activity was sequentially measured in 15 kidney allograft recipients treated with decreasing doses of immunosuppressive therapy including azathioprine and steroids. NK cell activity was consistently decreased in all patients, even when drug doses were minimal months after the graft (Lipinski et al., 1980). It is noteworhty that in patients studied months after the transplant, NK cell function still appeared drastically impaired, whereas other classical immunological parameters such as in vitro responses to mitogens or allogeneic cells were roughly normal or minimally affected. The mean NK cell activity calculated in the kidney recipient group (7.8 ± 6.7%) differed significantly ($p<10^{-8}$) from that calculated in a control population of uremic routinely hemodialyzed patients (45.2 ± 6.2%) as well as in a control group of healthy subjects (49.4 ± 14.0%).

B. NK Cell Activity in Patients with Malignancies

NK cell activity was investigated in a total of 59 patients affected with malignant lymphomas and 75 patients affected with various types of non-lymphoid tumors, either localized or with overt metastases. Contrasting with the results obtained in the latter group which did not significantly differ from controls in healthy subjects (47.3 ± 17.3% vs 46.3 ± 16.9%), the mean NK cell activity in the group of patients with lymphomas was strikingly decreased 27.5 $\pm$ 19.1%, $p < 10^{-3}$), with a significantly larger distribution ($p<10^{-2}$) of NK activity values in the lymphoma group than in the control group. No correlation was observed with the extent or histological type of the lymphoma (Tursz et al., in press).

IV. DISCUSSION

Data summarized here and reported elsewhere in more detail (Lipinski et al., 1980a; Lipinski et al., 1980b; Tursz et al., in press) show that groups of patients with high risk for malignancies present with a strikingly low NK cell activity against K562 target cells. These patients include kidney allograft recipients treated with azathioprine and steroids, and patients with primary ID. There is an increased incidence of malignancy in patients treated with immunosuppressive agents, about 100 times greater than in a control population in the same age range, solid lymphoma being by far the most frequent tumor arising in these patients (Penn, 1978). Statistical data are more scarce for patients with congenital ID, since early deaths from severe infections are frequent. In the surviving ID patients, there is nonetheless a major risk for malignancy when all ID are taken together; when every type of ID is considered separately, it appears that tumors arise principally in patients with severe combined ID, Wiskott-Aldrich syndrome and ataxia-telangectasia (Kersey et al., 1973). Among the tumors observed in these patients, lymphomas predominate, with leukemias and epithelial tumors as second risks. It is striking to observe that this population, heterogeneous from the point of view of ID classification, has in common a strongly decreased NK cell activity.

In contrast, hemodialyzed patients, who exhibit various abnormalities in T cell-mediated immunity but do not seem to be at high risk for malignancy, were found to have normal NK cell activity in this study. All together, these data suggest that NK cells play a role in the surveillance against lymphomas in the humans. After the concept of immunosurveillance had been presented more than 20 years ago (Thomas, 1959), T cells appeared to be best suited for such a hypothetical role, but reports in the last ten years did not bring strong evidence in this favor. NK cells would appear now as better candidates for such an action.

In view of our measurements of NK cell activity, the effectiveness of NK cells as anti-tumor agents could well be restricted to the surveillance against the apparition of lymphoid tumors, primarily lymphomas, as the groups we have studied presented concomitantly with a high risk for lymphomas and a very decreased NK cell activity. With this in mind, it is noteworthy that most cancer patients reported here did not show any deficiency in their NK cell activity. Numerous types of spontaneously cytotoxic cells are likely to exist in man, as already shown in the mouse (Minato et al., 1981). Other systems of measurement of spontaneous cytotoxicity have to be designed in order to answer the question whether spontaneously cytotoxic cells have a restricted or more general action against tumors.

ACNOWLEDGMENTS

We wish to thank J.-L. Amiel, J.-L. Virelizier, C. Griscelli and H. Kreis for allowing us to study their patients. The expert assistance of Y. Finale is gratefully acnowledged.

REFERENCES

Kersey, J.H., Spector, B.D., and Good, R.A. (1973). Int. J. Cancer 12:333.

Lipinski, M., Tursz, T., Kreis, H., Finale, Y., and Amiel, J.L. (1980a). Transplantation 29:214.

Lipinski, M., Virelizier, J.-L., Tursz, T., and Griscelli, C. (1980b). Eur. J. Immunol. 10:246.

Minato, N., Reid, L., and Bloom, B.R. (1981). J. Exp. Med. 154: 750.

Penn, I. (1978). Surgery 83:492.

Thomas, L. (1959). In "Cellular and Humoral Aspects of the Hypersensitive State" (Harper, ed.), p. 529. New York.

Tursz, T., Dokhélar, M.-C., Lipinski, M., and Amiel, J.-L. (in press). Cancer.

NK AND K CELL ACTIVITY IN MAMMARY AND CERVIX CARCINOMA PATIENTS IN RELATION TO RADIATION THERAPY AND THE COURSE OF DISEASE

Tamás Garam

National Institute of Haematology and Blood Transfusion
Budapest

Tamás Pulay

Semmelweis University, Clinic of Gynecologic and Obstetrics
Budapest

Tibor Bakács
Egon Svastits
Gábor Ringwald

National Institute of Oncology
Budapest

Klára Tótpál
Győző Petrányi

National Institute of Haematology and Blood Transfusion
Budapest

I. INTRODUCTION

Increasing evidence both in the exprerimental and clinical approach suggests that the natural

ISBN 0-12-341360-5

cellular system is involved in the defence against tumor cells. Many attempts have been made to clarify the role of NK cells in human tumors.However, only a few contradictory data are avaible to substantiate the conclusion that NK cells inhibit the growth or spreading of human tumors /Herberman and Holden, 1978; 1979/. To gather data in this field we investigated the NK and K cell activity in mammary carcinoma and cervical carcinoma patients with respect to radiotherapy and the course of the disease. In addition, we attempted to clarify the effect of patients, sera on NK and K cell function.

II. PATIENTS AND METHODS

79 patients with mammary tumors and 40 cervical cancer patients were tested before and after surgical and X-ray therapy.
The patients were categorized in stage I-IV according to the definition of UICC, and FIGO. The patients were not given anti-tumor therapy before the surgical intervention.

The control group consisted of 70 healthy women.
We used two methods in examining to cytotoxicity with the lymphocytes separated from the peripheral blood. One of them was the conventional cytitoxic assay, /Benczur *et al.*, 1979/ and the other method was the other method was the cytotoxic capacity test developed by us /Garam *et al.*, 1981; Bakács *et al.*, 1981/. Briefly this test consists of the following: we dilute the target cells and determine the maximal target cell number killed by a constant number of lymphocytes. The calculations were made using the following formula:
Cytotoxic capacity:

$$\frac{\text{\%release at the saturating dose of target cells} \times \text{number of saturating target cells}}{100}$$

The saturating target cell number is defined as the smallest number of target cells which enables the determination of the maximum target killing capacity of lymphocytes. We used K-562 leukaemia cells, human red blood cells /HRBC/ and chicken red blood cells /CRBC/ as target cells.
We added the decomplemented serum of the patient in

a 1:4 final dillution to the cultures containing his own lymphocytes and measured the inhibiting and stimulating effects of the serum on the cytotoxic activity.

III. RESULTS

The data in Table I. and Table II. show a general decrease of the NK and K cell activity of mammary carcinoma patients, and a decrease of the NK cell activity of cervix carcinoma patients, in comparison with the healthy controls. Irradiation did not influence the NK and K cell function in comparison to the values obtained before radiotherapy. NK activity is increased by average of 72.6% by the serum of mammary tumor patients and 41.6% by the serum of cervix carcinoma patients. The serum of mammary tumor patients increased the K-562 ADCC reaction by an average of 1.5%, however they decreased the HRBC ADCC reaction by an average of 34.7%.

Table III. and IV. show the connection between the different stages of the disease and the cytotoxic activity.

TABLE I. Cytotoxic capacity of mammary tumor patients

	Controls	Mammary tumor patients		
		before irrad.	after irrad.	effect of the sera
Cytotoxic capacity NCMC/Mean/	7.3[x]	4.3[++]/100%/	4.2	7.4 /172.6%/
± SE	o.6	o.5	o.4	o.6
Cytotoxic capacity K-562 ADCC /mean/	21.o	13.5[++]/100%/	13.2	13.7 /101.5%/
± SE	1.5	1.1	1.o	o.8
Cytotoxic capacity HRBC ADCC /mean/	83.5	6o.1[++]/100%/	54.2	39.2 / 65.3%/
± SE	6.1	5.4	7.5	4.o

x killed target cells by $2x10^6$ ly.$x10^4$;
++ significantly below controls $p < 0.001$

TABLE II. Cytotoxic activity of cervical cancer patients

	Controls	Cervix carcinoma patients		
		before irra- diation	after irra- diation	effect of the sera
NK /mean/	3o.8[x]	2o.9[+]/100%/	19.6[++]	29.2/141.6%/
± SE	1.6	3.o	1.9	2.2
K-562 ADCC /mean/	4o.2	38.3	36.8	n.d.
±SE	1.6	5.3	4.5	
CRBC ADCC /mean/	48.8	47.6	55.7	n.d.
±SE	2.3	5.9	6.6	

x specific release in % at 20:1, effector to target cell ratio

\+ significantly below controls, $p < 0.01$

++ significantly below controls, $p < 0.001$

The NK activity was the only parameter revealing a significant association with the progress of the tumor growth.

TABLE III. Connection between the in vitro lymphocyte activity and the extent of the disease in the mammary tumor patients

	N	Cytotoxic capacity NCMC ±SE	Cytotoxic capacity K-562 ADCC ±SE	Cytotoxic capacity HRBC ADCC ±SE
Contr.	70	7.3±0.6	21.0±1.5	83.5±6.1
St.I.	42	5.6±1.2	20.1±2.7	65 ±10.8
St.II.	33	3.9±0.8	12.2±1.7	56.5±7.1
St.III.	3	1.7±0.5	6.7±0.2	31.6±24.6
St.IV.	1	-	-	18.4±1.8

TABLE IV. Connection between the in vitro lymphocyte activity and the extent of the disease in cervical cancer patients

	N	NK specific	K-562 ADCC release in % at 20:1 effector to target cell ratio	CRBC ADCC
Control	70	30.8±1.6	40.2±1.6	48.8±2.3
St.I.	15	23.4±3.4	41.2±3.3	53.3±5.2
St.II.	11	++18.2±2.4	38.5±7.7	47.0±9.1
St.III.	6	++14.6±3.1	38.8±6.9	45.6±11.0
St.IV.	8	12.0±3.0	34.0±6.7	40.7±10.0

++ $p < 0.001$

IV. DISSCUSSION

In the course of our experiments we found in both groups of patients a lower cytotoxic activity than in the healthy controls. We must emphysize that patients had not been treated previously. With the mammary tumor patients, the cytotoxic capacity test was employed. In earlier studies we have shown that this test is capable of mearusing cytotoxic activity more sensitivily than the conventional citotoxicity assay. /Garam <u>et al</u>.,1981; Bakács <u>et al</u>.,1981/. Although the number of patients in each stage is not great, investigation has indicated that the aggravation of the desease leads to a decrease in cytotoxic activity. It is not known whether this decrease is a primary or secondary phenomenon, though the latter is supported by the fact that in the first stage the reaction did not decreased substantially as compared to that of healthy controls. The investigated parameters of cytotoxic activity of lymphocytes was modified by several factors in vivo. Autologous sera obtined from mammary and cervix carcinoma patients contain varions factors, like immunoglobulins, soluble antigens, immune complexses and complement componens, that may influence the cytotoxic effect of lymphocytes. It is worth noting that the irradiation treatment does not modify the cytotoxic activity in any of the tests.

Our data suggest that besides the follow up of other parameters, a systematic control of the NK and K cell activity in mammary and cervical cancer patients could help in the immunological monitoring of the progression of the disease.

REFERENCES

Herberman,R.B.,and Holden,H.T. /1978/ Adv.Cancer Res. 27:3o5

Herberman,R.B.,and Holden,H.T. /1979/ J.Natl.Cancer Inst. 62:441

Benczur,M.,Győrffy,Gy.,Garam,T.,Varga,Gy., Medgyesi,Gy.,Sándor,M.,and Petrányi,Gy./1979/ Immunobiol. 156:32o

Garam,T.,Bak,M.,Bakács,T.,Döbrentei,E.,and Petrányi,Gy., /1981/ Cancer, 46:285o

Bakács,T.,Garam,T.,Ringwald,G.,Tótpál,K.,and Svastits,E., /1981/ Archiv für Geschwulstsforschung 51:327

DEFICIENT NK AND ADCC MEDIATED BY PURIFIED E-ROSETTE POSITIVE AND E-ROSETTE NEGATIVE CELLS FROM PATIENTS WITH B-CELL CHRONIC LYMPHOCYTIC LEUKEMIA. AUGMENTATION BY *IN VITRO* TREATMENT WITH HUMAN LEUKOCYTE INTERFERON.[1]

Chris D. Platsoucas
Sudhir Gupta
Robert A. Good
Gabriel Fernandes

Memorial Sloan-Kettering Cancer Center
New York, New York

I. INTRODUCTION

Several investigators have demonstrated that both purified E-rosette positive and E-rosette negative cells from normal donors mediate natural killer (NK) and antibody-dependent cell-mediate cytotoxicity (ADCC) (1-4). Cells with Fc receptors for IgG (T_γ cells) are those among the E-rosette forming cells responsible for NK and ADCC. Immunoglobulin negative cells, forming high affinity rosettes with Ripley serum-coated human erythrocytes, are those among the E-rosette negative population that mediate NK and ADCC (1-5). Human leukocyte interferon augments both the NK and ADCC of human peripheral blood mononuclear leukocytes (6-8).

We have examined the NK and ADCC mediated by purified E-rosette positive and E-rosette negative cells from untreated patients with CLL, and we found these functions defective in the majority of the patients. *In vitro* treatment with human leukocyte interferon restored or significantly augmented the NK and ADCC mediated by E-rosette forming cells in the majority of the patients examined. However, NK and ADCC by E-rosette negative cells was not affected by this treatment.

---

[1]Supported by grant CH-151 from the American Cancer Society and NIH grants AI11843, NS11457 and AG03417.

ISBN 0-12-341360-5

II. METHODS

E-rosette positive and E-rosette negative peripheral blood lymphocytes from patients with CLL and normal donors were prepared by the method described elsewhere (9). E-rosette positive cells contained more than 95% E-rosetting cells, without nonspecific esterase-positive cells and less than 2% Ig-positive cells. E-rosette negative cells contained more than 92% leukemic cells, were devoid of nonspecific esterase-positive cells and contained less than 2% E-rosetting cells. E-rosette negative, Ripley rosette-forming cells were prepared from E-negative populations by rosetting with Ripley serum-coated Rh+ human erythrocytes. Human leukocyte interferon (sp. act. 1×10^6/mg of protein) was provided by the Interferon Laboratories of this Center. NK cytotoxicity was determined against the K562 targets (9). ADCC was determined against anti-chicken antibody-coated chicken erythrocytes and anti-SB antibody-coated cells of the SB line (9). $T\mu$ and $T\gamma$ cells were determined by rosetting techniques (10).

III. RESULTS AND DISCUSSION

A. NK and ADCC Mediated by Purified E-Rosette Positive and E-Rosette Negative Cells from Patients with CLL.

NK and ADCC mediated by purified E-rosette positive and E-rosette negative cells from untreated patients with CLL are

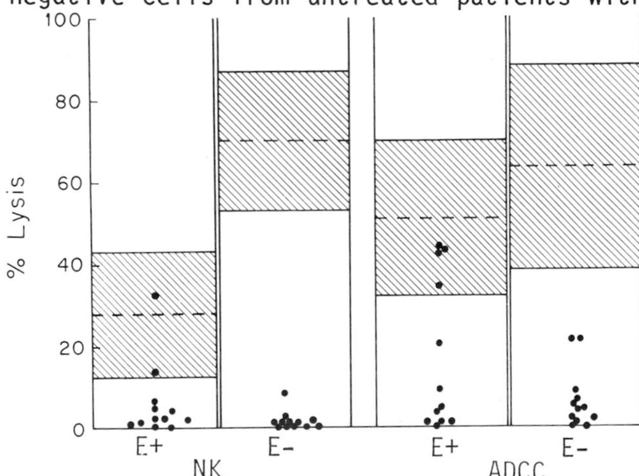

Figure 1. NK and ADCC mediated by E-rosette positive and E-rosette negative cells from patients with CLL. Normal Controls: (%lysis, mean±s.d.): E^+ NK (n=12) 28±14.5; E^+ ADCC (n=15) 51±19; E^- NK (n=15) 70±17; E^- ADCC (n=15) 64±25. [Reproduced from ref. 9 with permission of the publisher.]

TABLE I

Natural killer cytotoxicity mediated by purified E-rosette negative, Ripley rosette-forming cells from patients with CLL

Patients	E-Rosette Negative Ripley Rosette-Forming Cells	Natural Killer Cytotoxicity
	%	%
1	5	0.6[a]
2	2	0.0[a]
3	8	0.9[a]
Normal Controls	41[b]	83.0[b,c]

[a]Effector to target ratio, 100:1. [b]CLL patients vs normal controls, $p<0.001$. [c]Effector to target ratio, 10:1. [Reproduced from ref. 9 by permission of the publisher.]

shown in figure 1. Only two of twelve patients exhibited normal levels of NK mediated by purified E-rosette positive cells. Four of twelve patients exhibited normal levels of ADCC mediated by E-rosette positive cells (figure 1). All patients exhibited defective NK and ADCC mediated by E-rosette negative cells. To investigate the possibility that defective NK and ADCC by the largely leukemic E-rosette negative population was due to dilution of effector cells by the leukemic cells, we examined: (1) the NK and ADCC of these cells at very high effector to target ratios (500:1, 250:1, 125:1, etc), and observed that these cells were not able to lyse targets even at these high effector to target ratios (Data not shown) (9); (2) the NK mediated by cells isolated from the E-rosette negative cell fraction, by rosetting with Ripley serum-coated Rh+ human RBC. Cells from patients with CLL did not exhibit any natural killer cytotoxicity, in contrast to those isolated from normal donors (Table 1).

B. Effect of *In Vitro* Treatment with Interferon on the NK and ADCC Mediated by Purified E-Rosette Positive and E-Rosette Negative Cells from Patients with CLL.

Natural killer cytotoxicity mediated by purified E-rosette forming cells from patients with CLL was restored to normal, or significantly augmented in eight of twelve patients (66%) (fig. 2). Augmentation of NK mediated by E-rosetting peripheral blood lymphocytes was observed in approximately 85% of the normal donors examined. These donors exhibited persistently normal but low levels of E-rosette NK (fig. 2). NK mediated by E-rosetting cells in the

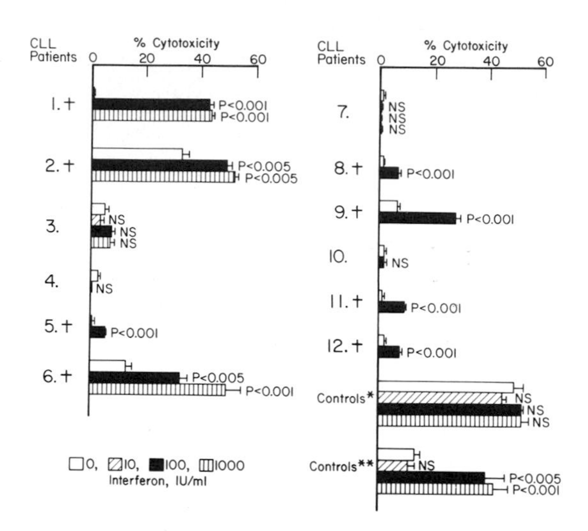

Figure 2. NK mediated by purified E-rosette positive cells from patients with CLL. Effect of in vitro treatment with interferon. *Normal donors with high NK levels. **Normal donors with low NK levels. +CLL patients with E-rosette NK significantly augmented by in vitro treatment with interferon. Effector to target ratio 25:1. [Reproduced from ref. 9 by permission of the publisher.]

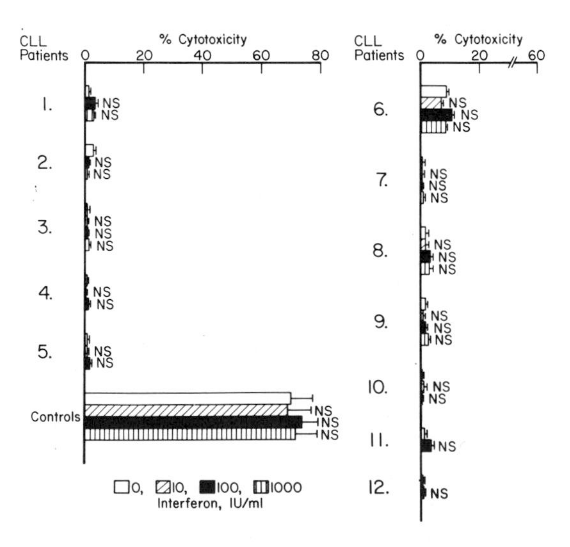

Figure 3. NK mediated by purified E-rosette negative cells from patients with CLL. Effect of in vitro treatment with interferon. Effector to target ratio 25:1. [Reproduced from ref. 9 by permission of the publisher.]

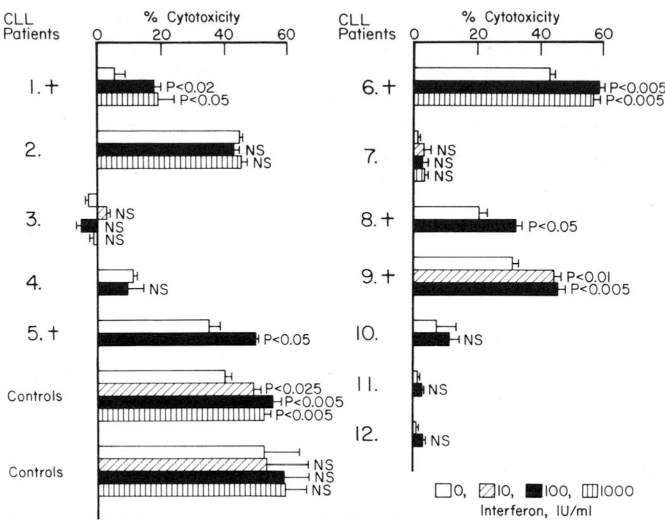

Figure 4. ADCC against antibody-coated CRBC mediated by purified E-rosette positive cells from patients with CLL. +CLL patients with ADCC significantly augmented by *in vitro* treatment with interferon. Augmentation of ADCC by E-rosetting cells from normal donors was observed in 30% of those examined (n=12). E/T ratio, 25:1. [Reproduced from ref. 9 by permission of the publisher.]

remaining donors (15%), who exhibited high levels of cytotoxicity, was not augmented by interferon treatment *in vitro* (11,12). NK mediated by E-rosette negative cells from patients with CLL was not affected by this treatment in any of the patients (fig. 3). Similar observations were made with Ripley rosette-forming cells isolated from E-rosette negative populations from patients with CLL (data not shown). Natural killer cytotoxicity mediated by purified E-rosette negative cells from normal donors was not augmented significantly by *in vitro* treatment with interferon in approximately 80% of the donors examined, even at very low effector to target ratios (1.25:1; 6.75:1; etc.). However, significant augmentation was observed in the remaining 20% of the normal donors (11,12).

ADCC by purified E-rosette positive cells from patients with CLL, against antibody-coated CRBC was significantly augmented by *in vitro* treatment with interferon in five of twelve patients (fig.4). Similar results were obtained using anti-SB antibody-coated cells of the SB line as targets. All these patients also had their NK mediated by E-rosette forming cells augmented by interferon. Furthermore, *in vitro* treatment with interferon resulted in significant augmentation of ADCC mediated by E-rosette forming cells in approximately 30% of the normal donors examined (11,12). ADCC by purified E-rosette negative cells from patients with CLL or normal donors was not augmented by in vitro treatment with interferon (data not shown) (9).

Simultaneous determination of the proportions of the Tμ and Tγ cells in the E-rosetting population of patients with CLL demonstrated significantly increased proportions of Tγ cells (9). These cells are effectors in NK and ADCC mediated by the E-rosetting population (1-4). In vitro treatment of E-rosetting cells with human leukocyte interferon for 14 hrs at 37°C did not alter the proportions of Tμ and Tγ cells in the normal controls or the patients with CLL, although it resulted in significant augmentation of the NK and ADCC in most of the cases. Significant but transient (within six hours) increase in the proportions of Tγ cells and decrease in the proportions of Tμ cells has been reported following in vitro treatment with interferon (12,13). The proportions of these cells return to normal levels after 12 hrs. However, ADCC remains augmented, suggesting that enhancement of Fc receptor expression is not necessarily associated with augmentation of ADCC by interferon.

REFERENCES

1. Kay M.D., Bonnard G.D., West W.H. and Herberman R.B. J. Immunol. 118:2058 (1977).
2. Kall M.A. and Koren H. Cell. Immunol. 40:58 (1978).
3. Gupta S., Fernandes G., Nair M. and Good R.A. PNAS 75:5137 (1978).
4. Pickler W.J., Gendelman F.W. and Nelson D.L. Cell Immunol. 42:410 (1979).
5. Natvig J.B. and Froland S.S. Scand. J. Immunol. 5 (Suppl. 5):83 (1976).
6. Einhorn S., Blomgren H. and Strander H. Int. J. Cancer 22:405 (1978).
7. Trinchieri G. and Santoli D. J. Exp. Med. 147:1314 (1978).
8. Herberman R.R., Ortaldo J.R. and Bonnard G.D. Nature 227:221 (1979).
9. Platsoucas C.D., Fernandes G., Gupta S.L., Kempin S., Clarkson B., Good R.A. and Gupta S. J. Immunol. 125:1216 (1980).
10. Gupta S. and Good R.A. Cell. Immunol. 34:10 (1977).
11. Platsoucas C.D., Gupta S., Good R.A. and Fernandes G. Fed. Proc. 39:932 (1980).
12. Platsoucas C.D., Fernandes G., Good R.A. and Gupta S. Submitted.
13. Itoh K., Inoue M., Kataoka S. and Kumagai K. J. Immunol. 124:2589 (1980).

NATURAL KILLING IN PATIENTS WITH HODGKIN'S DISEASE[1]

Sudhir Gupta[2]
Gabriel Fernandes[3]

Memorial Sloan-Kettering Cancer Center
New York, New York

INTRODUCTION

Immunological studies in Hodgkin's disease have revealed a wide variety of cell-mediated immune defects (reviewed in 1,2). Patients with Hodgkin's disease have increased risk of developing second malignancy and are susceptible to increase frequency of viral infections. Natural killer cells have been implicated in defence against tumors and viral infections (3,4). T cells with IgG Fc receptors (Tγ) have shown to be effectors in NK and antibody-dependent cytotoxicity (ADCC) (5,6). In Hodgkin's disease there is maldistribution of T cell subsets between peripheral blood and spleen (7,8). Therefore, we examined peripheral blood and splenic T and non T lymphocytes in untreated patients with Hodgkin's disease for NK activity.

MATERIAL AND METHODS

Twenty-eight untreated patients with Hodgkin's disease were the subjects of the present study. Histopathologic staging was done according to the Rye system (9). The majority of patients underwent staging laparotomy and were pathologically staged according to Ann Arbor classification (10). They included 4 patients with stage I, 13 with stage

[1]Supported by grants from NIH-CA17404, AG-00541, AG-03417.
[2]Present address: Div. of Basic and Clinical Immunology, University of California, Irvine, CA 92717.
[3]Present address: Div. of Clinical Immunology, U. Texas Health Science Center, San Antonio, TX.

ISBN 0-12-341360-5

II, 9 with stage III and 2 with stage IV Hodgkin's disease. Twenty-six had nodular sclerosis and 2 had mixed cellular histologic type of Hodgkin's disease. Thirty age and sex matched healthy donors served as controls.

Spleens were cut into small pieces and passed through wire mesh to obtain a single cell suspension. Mononuclear cells from spleen as well as from peripheral blood were separated on Ficoll-Hypaque (FH) density gradient. Phagocytic cells were depleted by carbonyl iron particle ingestion and separation on FH gradient. T and non T cells were separated by rosette-formation with neuraminidase-treated sheep RBC. Purified subpopulations were incubated in medium containing 10% fetal calf serum at 37°C for 10-15 hours prior to NK assay.'

Natural killing was assayed against cells of K-562 cell line in 4 hour ^{51}Cr release assay and ADCC was measured against chicken RBC in the presence of anti-CRBC-IgG antibody in a 3 hour ^{51}Cr release assay as described (11).

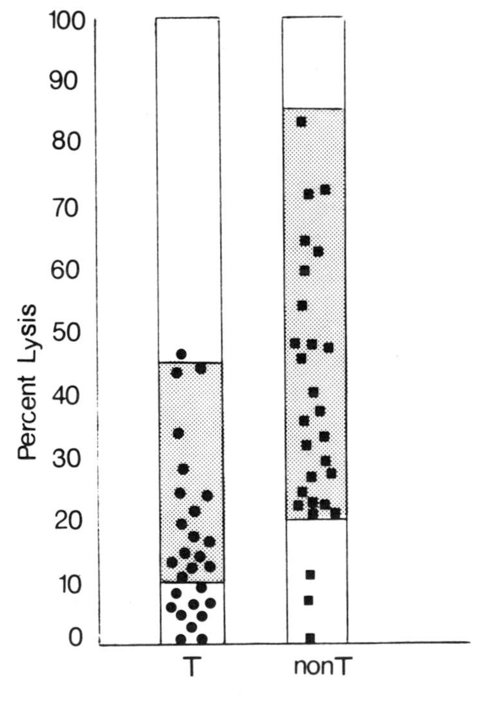

FIGURE 1. NK activity in peripheral blood of Hodgkin's disease.

RESULTS

Data of natural killing in the peripheral blood T and non T cells are shown in Fig. 1. Shaded areas show the range (mean±2SD) for healthy controls. 10 of 26 patients had decreased T cell mediated NK activity when compared to range for controls, whereas only 3 of those 10 patients had decreased non T cell-mediated NK. No patient demonstrated deficiency of non T cell NK without a deficiency of T cell NK.

Results of T and non T cell NK with regard to stage of disease are shown in Table 1. No significant difference was observed between T-or non T cell-mediated NK in patients with stage I and II, and stage III and IV Hodgkin's disease.

TABLE 1. Natural Killing by Peripheral Blood T and Non T Cells in Hodgkin's Disease: Relationship with the Stage of Disease

Stage of disease	Effector cells	
	T 50:1	Non T 50:1[a]
I and II (17)	15.3 ± 6.8[b]	40.1 ± 10.8
III and IV (11)	17.4 ± 6.2	43.4 ± 6.9
P volume	NS	NS

[a]Effector:target ratio
[b]Mean±SD

Data of T and non T cell NK in the peripheral blood and spleen from 17 patients are summarized in Table 2. T cell-mediated NK was comparable between peripheral blood and spleens of same patients. However, significantly ($P<0.025$) decreased non T cell-mediated NK was observed in the spleens from patients with Hodgkin's disease when compared to NK of their peripheral blood non T cells. This difference was even more striking ($P<0.001$) when results for effector:target ratio of 25:1 were compared.

TABLE 2. NK in Peripheral Blood and Splenic T and non T Cells from Patients with Hodgkin's Disease

Source of effector cells	T 50:1	T 25:1	Non T 50:1	Non T 25:1[a]
Peripheral blood (17)	15.1±6.2	8.1±3.2	31.4±9.4	30.5±6.8
Spleen (17)	13.5±4.8	5.5±2.1	21.6±6.2	17.7±4.8
P value	NS	NS	<0.025	<0.001

[a]Effector:Target ratio

Results of the influence of splenic involvement on NK activity are summarized in Table 3. Neither in the peripheral blood nor in the splenic T or non T cells any difference in NK activity was observed between the group with involved spleens and those without splenic involvement.

TABLE 3. Effect of Splenic Involvement by Tumor on T and Non T Cell-Mediated NK

Source of effector cells	T 50:1	T 25:1	Non-T 50:1	Non-T 25:1[a]	Spleen Involvement
Peripheral blood (6)	14.1±4.2	8.9±2.4	31.9±9.4	30.2±10.2	+
Peripheral blood (11)	15.6±3.8	7.5±2.2	31.1±8.8	30.7± 9.6	-
P value	NS	NS	NS	NS	
Spleen (6)	12.6±3.4	5.9±1.9	20.5±8.4	14.1± 6.8	+
Spleen (11)	13.8±4.5	5.4±2.0	25.8±7.6	20.8± 4.5	-
P value	NS	NS	NS	NS	

[a]Effector:target ratio

DISCUSSION

In this study we have demonstrated that approximately 35% patients with untreated Hodgkin's disease had deficiency of T cell-mediated NK activity in their peripheral blood. Only 3 of 10 patients, however, demonstrated deficient non T cell NK. Comparable T cell mediated NK was observed between peripheral blood and spleens from same patients. Splenic non T cells had significantly lower NK activity when compared to that of peripheral blood non T cells. Stage of the disease or splenic involvement had no influence on the splenic or the peripheral T or non T cell-mediated NK.

Immunologic studies of patients with Hodgkin's disease have demonstrated a variety of T cell defects, B cell defects, defects of monocytes and abnormal circulating humoral factors (1,2). Circulating immune complexes are well documented in Hodgkin's disease (12,13) and they are known to inhibit NK function (14). Increased suppressor monocytes, and increased products PGE_2 have been reported in Hodgkin's disease (15). They have been shown to modulate NK activity (16). Therefore, a possibility should be entertained that either circulating immune complexes or/and regulatory influence of monocytes and PGE_2 could be responsible for decreased NK activity in Hodgkin's disease. Although there is a maldistribution of Tγ cells (effectors in NK) in Hodgkin's disease but no direct correlation was observed between the proportion of Tγ cells and NK activity in Hodgkin's disease (17). Impaired interferon production in Hodgkin's disease could also be responsible for decrease NK (18). The decrease in NK function in patients with Hodgkin's disease might be responsible for increase susceptibility to viral infection and to the risk of developing second malignancy. The defect of NK is observed as early as in stage IA disease. It would be of interest to investigate the NK in healthy family members of patients with Hodgkin's disease to investigate the possibility of horizontal transmission of this immune defect.

REFERENCES

1. Gupta, S., Clin. Bull. 11:58 (1981)
2. Kaplan, H.S., Hodgkin's disease (2nd Ed.), Oxford University Press, Cambridge, MA, 1980.
3. Keissling, R. and Haller, O., Contemp. Top. Immunobiol. 8:171 (1978).
4. Herberman, R.B., Human Lymph. Diff., 1:63 (1981).

5. Gupta, S., Fernandes, G., Nair, M. and Good, R.A., Proc. Natl. Acad. Sci. (USA), 75:5137 (1978).
6. Kall, M.A. and Koren, H.S., Cell. Immunol., 40:58 (1978).
7. Gupta, S. and Tan, C., Clin. Immunol. Immunopath., 15:133 (1980).
8. Gupta, S., Clin. Exp. Immunol., 42:186 (1980).
9. Lukes, R.J., Craver, L.F., Mall, T.C., Rappaport, H. and Ruben, P., Clin. Res., 26:1311 (1966).
10. Carbone, P.P., Kaplan, H.S., Musshoff, K., Smithers, D.W. and Tubiana, M., Cancer Res., 31:1860 (1971).
11. Fernandes, G. and Gupta, S., J. Clin. Immunol., 1:141, (1981).
12. Amlot, P.L., Slaney, J.M. and Williams, W.D., Lancet, 1:449 (1976).
13. Brandeis, W.E., Tan, C., Wang, Y., Good, R.A. and Day, N.K., Clin. Exp. Immunol., 39:551 (1980).
14. Nair, P.M.N., Fernandes, G., Onoe, K., Day, N.K. and Good, R.A., Int. J. Cancer, 25:667 (1980).
15. Goodwin, J.S., Messner, R.P., Bankhurst, A.D., Peake, G.T., Saiki, J.M. and Williams, R.C., N. Eng. J. Med., 297:963 (1977).
16. Santoni, A., Riccardi, C., Barlozzari, T. and Herberman, R.B., in "Natural Cell-Mediated Immunity" (R.B. Herberman, ed.), p. 753, Academic Press, New York, 1978.
17. Gupta, S. and Fernandes, G., Clin. Exp. Immunol., 45: 205 (1981).
18. Stevens, D. and Merigan, T., J. Clin. Invest., 51:1170 (1972).

NATURAL KILLER CELLS IN HAMSTERS AND THEIR EARLY AUGMENTATION AND LATE SUPPRESSION DURING TUMOR GROWTH[1]

Surjit K. Datta
John J. Trentin
Takanobu Kurashige[2]

Division of Experimental Biology
Baylor College of Medicine
Houston, Texas

CHARACTERISTICS OF HAMSTER NK CELLS

We have reported the existence, in both randombred and LSH inbred hamsters, of natural spleen-cell-mediated cytolytic activity against an SA7 virus-induced lymphoma of the LSH inbred hamster (Datta et al., 1979). Lytic activity was higher using spleen cells of randombred (38% lysis) than of inbred hamsters (15%). The effector cells shared the major characteristics of mouse NK cells (Trentin and Datta, 1981). They differed from mouse NK cells in that a) they were found in spleens of the youngest hamsters tested (4 to 7 days old) rather than appearing abruptly after 21 days of age; b) their activity rose progressively with age to high levels at $1\frac{1}{2}$ years (in some 2 year old hamsters tested the activity was at the same level as in 3-4 month old hamsters), and c) their activity was as high in bone marrow as in spleen (Datta et al., 1979; Trentin and Datta, 1981). Transplantation of 2×10^5 SA7 tumor cells into groups of 10 hamsters, $1\frac{1}{2}$ and $3\frac{1}{2}$ months old, of both randombred and inbred strains, resulted in a higher rate of growth and % of takes after 90 days in inbred hamsters than in randombred hamsters, and in younger than in older hamsters of each strain (Trentin and Datta, 1981).

---

[1] Portions of this manuscript have been published under the title "Natural killer (NK) cells in hamsters and their modulation in tumorigenesis" (Trentin and Datta, 1981). Supported by USPHS grants CA 12093, CA 03367, and K6 CA 14219 from the NCI.

[2] Current address: Department of Pediatrics, Kochi Medical School, Kohas Nangoku, Kochi, JAPAN.

ISBN 0-12-341360-5

AUGMENTATION THEN SUPPRESSION OF HAMSTER NK ACTIVITY BY TUMOR GROWTH

Adult randombred SPF hamsters in a closed isolation colony were given a transplant of 10^7 SA7 cultured tumor cells subcutaneously (SC) and their spleen cell-mediated cytolytic activity was measured at various days post-transplant, along with normal age-matched controls. In addition to the SA7 target, against which specific T-cell immunity may also have been induced, cell-mediated cytotoxicity was also tested against an unrelated (human) target, K562, to assess NK-cell activity more directly, rather than a combination of NK and immune T-cell cytotoxic activity. Five and seven days post-transplant of SA7 cells, when there was little or no visible tumor, spleen NK activity was greatly augmented. But on day eleven post-transplant, when the tumor was medium in size, activity was 35 - 39% of control level, and at 21 and 28 days when the SC tumor was large, cytolytic activity was even more severely depressed (Fig. 1).

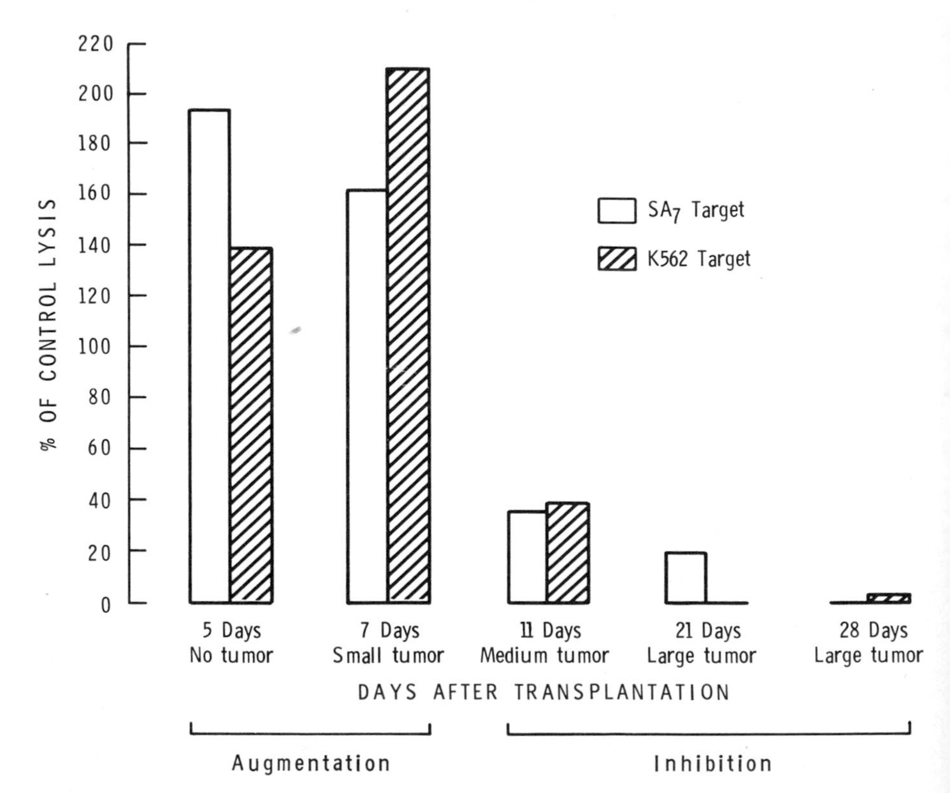

Figure 1. Spleen NK cell activity of randombred hamsters following SC inoculation of 10^7 cultured SA7 tumor cells. E:T = 100:1

The late decrease of NK-cell activity in hamsters bearing large tumors prompted us to investigate whether suppressor cells or serum blocking factors might be involved. A cell mixture protocol was used to test the ability of spleen cells from tumor-bearers to suppress NK cell-mediated lysis by normal spleen cells. The E:T ratio of normal spleen cells was kept constant at 50:1, and to this were added increasing numbers of spleen cells from hamsters bearing large SA7 tumors so that ratios of spleen cells from tumor-bearer to normal spleen cells were 0.25:1, 0.5:1, 1:1 and 2:1. As a control for cell "crowding" effects, normal thymus cells, which have insignificant or no NK-cell activity (Datta et al., 1979) were added to the constant number of normal spleen cells at similar ratios. NK-cell activity of these mixtures was tested against both SA7 and the xenogeneic target K562. Addition of either tumor-bearer spleen cells or normal thymus cells depressed cytolytic activity, but spleen cells from tumor-bearing hamsters at ratios of 1:1 and 2:1 reduced NK-cell lysis of SA7 and K562 target cells beyond the inhibiting effect of normal thymus cells (Fig. 2).

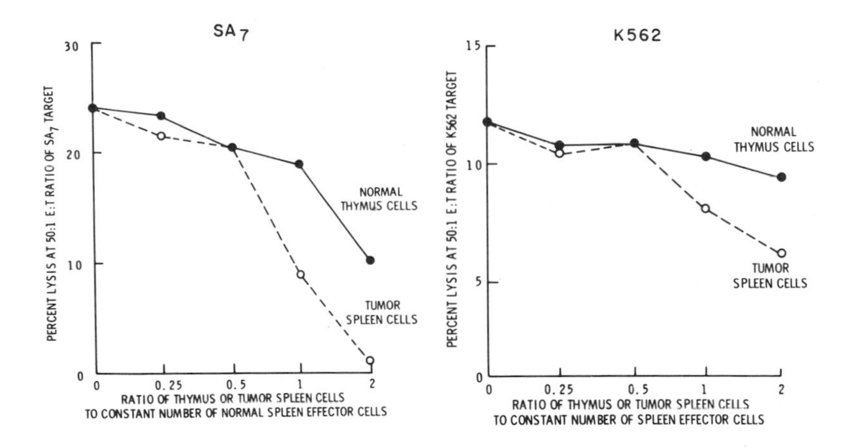

Figure 2. NK activity of mixtures of spleen cells from normal hamsters and from hamsters bearing large SA7 tumors, versus either SA7 or K562 target cells. Normal spleen cells and target cells were kept at a constant ratio of 50:1. In addition were added increasing numbers of either spleen cells from tumor-bearing hamsters, or normal thymus cells, in ratios to the normal spleen cells of 0.25:1, 0.5:1, 1:1 or 2:1. Data combined from 3 experiments, each involving a pool of 3 normal spleens and of 1 or 2 SA7 tumor-bearer spleens per experiment. At the ratio of 2 thymus or tumor-bearer spleen cells to one normal spleen effector cell, the P value by Student's t test for both the SA7 and the K562 targets is $< .05$.

Removal of adherent cells from tumor-bearer spleen cells partially reversed suppression of NK cell lysis, whereas exposure of the spleen cells to 950R had little or no effect. In a preliminary study we found that serum from large SA7 tumor-bearers suppressed NK-cell activity of normal spleen cells against SA7 target cells. These findings are suggestive of the late suppression of NK activity by pyran copolymer-activated macrophages, which is mediated by soluble factors (Santoni et al., 1980). Both suppressor cells and serum factors are known to regulate tumor-immune T-cell responses in many tumor systems (Hellstrom and Hellstrom, 1969; Broder et al., 1978). The relative degree of participation of suppressor cells or serum factors or other factors in the observed late inhibition of NK activity in large tumor bearing hamsters remains to be determined.

Herberman et al. (1977) and Djeu et al. (1980) have demonstrated that a variety of syngeneic and allogeneic tumor cells, injected into either conventional or nude mice, increased both interferon and NK activity. Only if the particular tumor cells induced interferon was there a corresponding increase in NK activity. Both activities peaked after 1 or 2 days, and declined to normal levels in most cases by day 3 (Djeu et al., 1980). We have not yet examined SA7 tumor-inoculated hamsters for interferon production. The 'early' augmentation of NK in tumor bearing hamsters at days 5 and 7 is much later than the interferon-related tumor cell stimulation of NK activity reported for mice. However, the kinetics of tumor-cell-induced interferon response may differ from mouse to hamster. In any case, augmentation or suppression of NK activity in tumor-bearers undoubtedly has implications for the balance between tumor control and tumor progression.

REFERENCES

Broder, S., Muul, L., and Waldmann, T.A. (1978). J. Natl. Cancer Inst. 61:5.

Datta, S.K., Gallagher, M.T., and Trentin, J. J. (1979). Int. J. Cancer 23:728.

Djeu, J.Y., Huang, K.Y., and Herberman, R.B. (1980). J. Exper. Med. 151:781.

Hellstrom, K.E., and Hellstrom, I. (1969). Adv. Cancer Res. 12:167.

Herberman, R.B., Nunn, M.E., Holden, R.T., Staal, S., and Djeu, J.Y. (1977). Int. J. Cancer 19:555.

Santoni, A., Riccardi, C., Barlozzari, T., and Herberman, R.B. (1980). Int. J. Cancer 26:837.

Trentin, J.J. and Datta, S.K. (1981). In "Advances in Experimental Medicine and Biology: Hamster immune responses in infectious and oncologic diseases." (J.S. Streilein, D.A. Hart, J.S. Streilein, W.R. Duncan and R.E. Billingham, eds.), p. 153, v. 134. Plenum Press, New York.

NK DEFICIENCY IN X-LINKED LYMPHOPROLIFERATIVE SYNDROME

Janet K. Seeley[1]

Department of Pediatrics
University of Connecticut Health Center
Farmington, Connecticut

Thomas Bechtold[2]
David T. Purtilo[2]

Department of Pathology and Laboratory Medicine
University of Nebraska Medical Center
Omaha, Nebraska

Tullia Lindsten

Department of Tumor Biology
Karolinska Institute
Stockholm, Sweden

I. INTRODUCTION

Evidence is accumulating for a potential *in vivo* role for natural killer (NK) activity in tumor rejection and in controlling viral infection. For example, Riccardi and his colleagues (Riccardi et al, 1980) have demonstrated a close correlation between NK and the rapid *in vivo* clearance of radiolabeled tumor cells. *In vitro*, viral infection confers considerable NK sensitivity to many human and murine target cells normally resistant to NK lysis (Santoli et al., 1978; Welsh et al., 1980). Similar viral-induced NK sensitivity has also been reported to occur *in vivo* (Minato et al., 1979).

---

[1]Supported by NIH BSRG RR 05712-10.
[2]Supported by NIH CA 30196.

ISBN 0-12-341360-5

Although this is true for a number of RNA and DNA viruses, normal human lymphocytes transformed by the Epstein-Barr virus (EBV) are poor NK targets in short and long term NK assays (Seeley et al., 1981). However, when treated with agents which induce viral production, these cells become highly and selectively NK sensitive (Blazer, 1980).

II. NK DEFICIENCY IN X-LINKED LYMPHOPROLIFERATIVE SYDROME (XLP)

It is important to determine if NK activity is diminished in malignancy and viral disease and to evaluate if this deficiency is integral to the pathogenesis of the disease. In 1980, Sullivan et al. (Sullivan et al., 1980) described deficient NK activity in 80% of patients with X-linked lymphoproliferative syndrome (XLP). Males with XLP are susceptible to life-threatening EBV infection resulting in fatal or chronic infectious mononucleosis, aplastic anemia, acquired agammaglobulinemia and/or malignant lymphoma (Purtilo et al., 1977). Of the 108 males with XLP thus far identified, only 23 survive (Purtilo et al., 1981).

Recently, we have sought to determine: 1) Is the deficient NK function in males with XLP a primary defect, or is it a result of the disease process; and 2) Considering the requisit steps leading to NK target lysis, where does the defect occur? Nine individuals with XLP were available for study to try to answer these questions.

A. Primary or Secondary NK Defect in XLP

None of the surviving males with XLP were tested for NK activity prior to the expression of their disease. Thus, it was not possible to draw conclusions whether their NK deficiency was primary or secondary. However, since all sons of XLP carriers are at 50% risk of developing XLP phenotypes after EBV infections, we would expect that approximately half of the asymptomatic males at risk would have abnormally low NK levels, if the defect were primary.

Figure 1 shows that the mean NK activity for 12 males at risk was 32.5%, ie., very close to that of normal controls and significantly higher than that of the known XLP patients.

This finding suggests that the NK deficiency consistently detected in seven of the nine XLP patients had occurred during the progression of the disease.

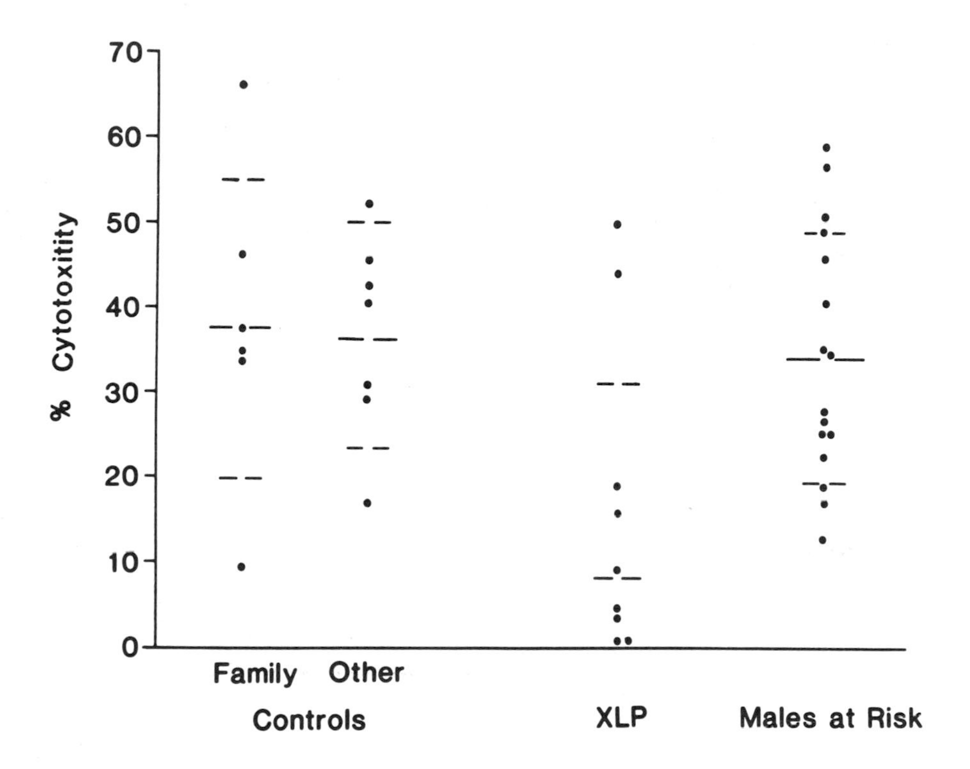

FIGURE 1. NK data on XLP patients and family members at 20:1 E:T ratio in a 4 hour CRA as described in Ref. 5. Horizontal lines indicate the mean $\pm$ one S.D. Donor cells were depleted of plastic-adherent cells prior to the assay. Family controls: three fathers (ages 34-43) and three paternal cousins (ages 8-14); other controls: 7 male university personnel (ages 22-38); XLP: nine affected males (ages 8-23; mean age 13); males at risk: 12 sons and grandsons of XLP carriers (ages 2-28; mean age 8). Four of the younger males at risk tested twice.

Evidence presented elsewhere has demonstrated other immune abnormalities in five of the asymptomatic males at risk, ie. failure to produce antibodies to the EBV nuclear antigen, failure to switch from IgM to IgG during ØX174 immunization, and abnormal ratios of T cell subsets (Seeley et al., 1981; Seeley et al., 1982). These are defects shared by the affected males, suggesting that they reflect primary or congenital abnormalities.

Only one XLP patient thus far has been evaluated for NK during a fulminant and subsequently fatal course of infectious mononucleosis (Sullivan et al., 1979). A vigorous cytotoxic activity against NK sensitive and resistant targets was reported, which was resistant to prednisone therapy. We speculate that this reflected an intensely activated NK function. It is interesting to note that normal individuals express normal to slightly depressed NK levels during mononucleosis (Seeley et al., 1981).

B. Site of the NK Deficiency in XLP

Cytotoxicity by NK is a multi-step process which is subject to augmentation and suppression. We asked three questions in order to determine the mechanisms of the NK defect in males with XLP: 1) Do the patients lack circulating NK effector cells; 2) Are the effectors defective in the target binding or lysing step; or 3) Are defects in NK regulation responsible for their immune purturbations? The following studies were done to evaluate these questions.

1. Surface Markers. XLP patients have normal numbers of lymphocytes with Fc receptors (detected with Ig-coated OX RBC): For XLP patients the mean was 10.6 percent with a range of 5.7 to 13.2 compared to 11.9 percent and a range of 6.8 to 14.6 percent in normal donors. Preliminary experiments indicate that XLP donors also have normal numbers of OKM1 (Ortho Diagnostic Systems) positive cells. Although these markers are not exclusive for NK cells, they do suggest presence of cells with NK phenotype in XLP patients. It is not yet clear if these patients have normal numbers of large granular lymphocytes, shown by Timonen to be the morphologic phenotype of human NK cells (Timonen et al., 1979).

2. Target Binding and Lysis. Table I demonstrates that males with XLP also have cells which can bind K562 (TBC) and that half of these cells were active killer cells in the single cell agarose assay. Pretreatment of the effectors to

deplete Fc receptor positive cells sharply reduced the percent of TBC and eliminated cytotoxic TBC as well as the chromium-release activity. The XLP TBC values were similar to those obtained for the control groups, in sharp contrast to the significantly lower percent cytotoxicity detected in the chromium release assay (CRA).

In the single cell agarose assay, cells are held immobile in a semisolid gel. While in the CRA the effectors are free to recycle, ie. bind, detach, and bind another target. Considering the difference detected in the two assays, it seemed likely that the XLP target-binding cells were unable to recycle within the 4 hour CRA.

TABLE I. Decreased NK Activity in Males with X-Linked Lymphoproliferative Syndrome in the Chromium Release Assay vs. Normal Activity in the Single Cell Assay

Donors		CRA[a]	SCA[b]		
			TBC	Cytx TBC	NK Cells
		%		%	
Normal Controls					
Family	(6)[c]	37.8 ± 8.4	7.4 ± 0.1	41.6 ± 5.8	3.1 ± 0.8
Other	(7)	36.6 ± 7.2	5.3 ± 1.7	43.7 ± 4.6	2.3 ± 0.7
XLP	(9)	12.1 ± 5.5*	5.2 ± 1.3	49.3 ± 8.8	2.6 ± 0.7

[a]CRA (See Figure 1 legend) Values = mean ± SEM.

[b]SCA = three hour single cell assay with 1:1 effector to target ratio as described in reference 14. TBC=Target Binding cells ie, # lymphocyte-target conjugates/total lymphocytes x 100. Cytotoxic TBC's=# conjugates with dead targets/total conjugates x 100. NK cells=TBC value x % cytotoxic TBC's.

[c]Numbers in parenthesis represent the number of individuals tested.

*$p<0.01$ between Family or Other Controls and XLP groups in the CRA, as determined by the students t test. All other differences between groups are not significant.

3. <u>Extended Incubation Times</u>. Therefore, we compared the effects of incubating the CRA cultures 4 or 16 hours (Figure 2). Five of the seven XLP patients who expressed low NK levels as demonstrated in Figure 1 were available for this

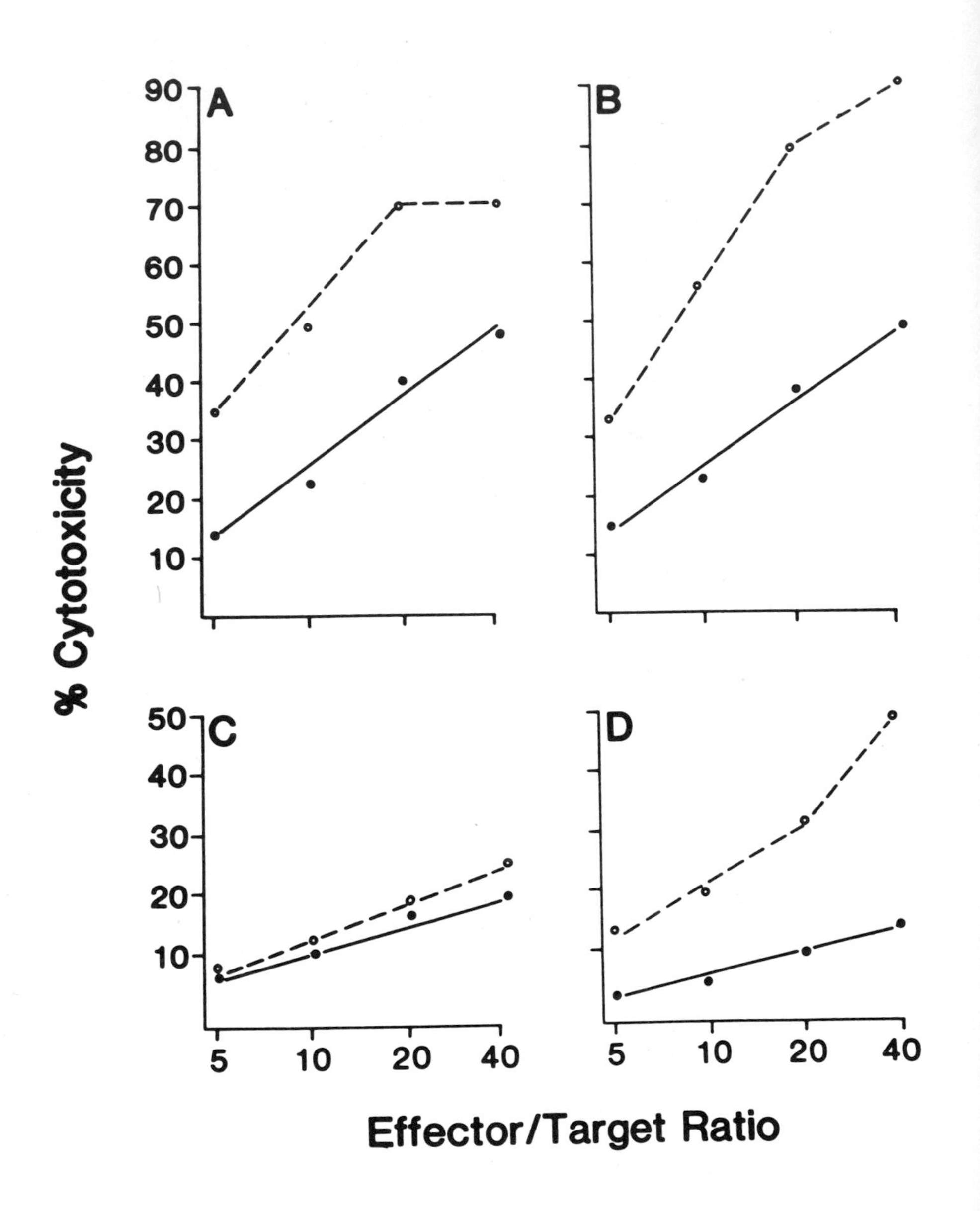

FIGURE 2. Low NK mediated by XLP lymphocytes compared to controls at all E/T ratios in 4 hour (______) and 16 hr (-----) cultures. (A) represents normal control; (B) an age-matched male family control (paternal cousin of patient represented in [C]); (C) and (D) represent two XLP patients.

study. There was a substantial increase in chromium release by effectors from normal and family controls and also by one of the XLP patients (2 A, B and D, respectively). However, the extended incubation resulted in very little augmented chromium release by the other four patients (representative experiment Figure 2C). These results suggested that NK cells from the majority of the low NK XLP donors could not recycle, even during extended incubation periods.

A similar conclusion was suggested by Merrill and her colleagues (Merrill et al., 1981) in their study on multiple sclerosis (MS) patients. Chronic MS patients had normal NK in the single cell assay but significantly low chromium release activity. As in our study, adherent cells were depleted prior to the assay, reducing the possibility of macrophage-mediated NK suppression. They further demonstrated that interferon treatment of the effectors resulted in a two-fold increase in chromium release. In contrast in the XLP study, Sullivan, et al. (Sullivan et al., 1980) described a maximum increase of 33 percent by interferon-treated XLP effectors. Other agents which have been partially successful in correcting the NK defect in Chediak Higashi syndrome (Katz et al., 1980) have not yet been employed with XLP lymphocytes.

In summary, we have presented evidence that the NK defiency in XLP is acquired after EBV infection. The patients had normal numbers of lytic target binding cells but very low chromium release values, suggesting a recycling defect. These studies support the rationale for the original generic name for XLP, X-linked recessive progressive combined variable immunodeficiency disease (Purtilo et al., 1975) and that the underlying immunoregulatory defect(s) become accentuated following infection by the virus.

REFERENCES

Blazer, B., Patarroyo, M., Klein, E. and Klein, G., J. Exp. Med. 151:614 (1980).

Katz, P., Roder, J.C., Herberman, R.B., and Facui, A.S., Clinical Research (1980).

Merrill, J., Jondal, M., Seeley, J.K., Ullberg, M. and Siden, A., Clin. Exp. Immunol., in press (1981).

Minato, N., Blood, B.R., Jones, C., Holland, J., and Reid, L.M., J. Exp. Med. 149:1117 (1979).

Purtilo, D.T., Cassel, C., Yang, J.P.S., Stephenson, S.R., Harper, R., Landing, G.H. and Vawter, G.F., Lancet ii:882-885, (1975).

Purtilo, D.T., DeFlorio, D., Young, J.P.S., Otto, R. and Edwards, W., N. Eng. J. Med 297:1077 (1977).

Purtilo, D.T., Sakamoto, K., Maurer, H.S., Barnabei, V., Seeley, J.K., Kaplan, E., Muralidnaran, K., Sexton, J., Buchanan, G.R., Saemundsen, A.K., Finkelstein, G. and Baker, J.A., Clin. Immunol. Immunopath., in press (1981).

Riccardi, C., Santoni, , Barlozzari, T. and Herberman, R.B., in "Natural Cell-Mediated Immunity Against Tumors" (R.B. Herberman, ed.), p. 1121. Academic Press, New York, 1980.

Santoli, D., Trinchieri, G. and Lief, F.S., J. Immunol. 121:526 (1978).

Seeley, J.K., Harada, S., Bechtold, T., Ochs, H., Ballow, M. and Purtilo, D.T., Fed. Proc., in press (1982).

Seeley, J.K., Sakamoto, K., Ip, S.H., Hansen, P.W. and Purtilo, D.T., J. Immunol, in press (1981).

Seeley, J.K., Svedmyr, E., Weiland, O., Klein, G., Moller, E., Eriksson, E., Anderson, K. and Van der Uval, L., J. Immunol 127:293 (1981).

Sullivan, J.L., Byron, K.S., Brewster, F.E. and Purtilo, D.T., Science 210:543 (1980).

Sullivan, J.L., Sakamoto, K., Byron, K.S., Brewster, F.E. and Purtilo, D.T., Clinical Research 28:641A (1979).

Timonen, T., Ranki, A., Saksela, E. and Hayry, P., Cellular Immunology 48:121-132 (1979).

Welsh, R.M. and Kiessling, R.W., in "Natural Cell-Mediated Immunity Against Tumor" (R.B. Herberman, ed.), p. 963. Academic Press, New York, 1980.

MECHANISMS OF NATURAL KILLER CELL DEPRESSION IN MULTIPLE SCLEROSIS

Ronald C. McGarry
J.C. Roder

Department of Microbiology and Immunology,
Queen's University,
Kingston, Ontario,
Canada.

D. Brunet

Department of Neurology,
Queen's University,
Kingston, Ontario,
Canada.

INTRODUCTION

Multiple sclerosis (MS) is a chronic, degenerative human disease of the CNS whose etiology and pathogenesis is unknown. The disease is characterized by inflammatory demyelination appearing as distinct plaques in the white matter of the brain and spinal cord and by a frequently relapsing clinical course (Raine, and Schaumburg, 1977). The inflammatory nature of demyelination and the similarity of this disease to two immunologically mediated animal models, experimental allergic encephalomyelitis (EAE) and neuritis (EAN) have led to the suggestion that the pathogenesis of MS is autoimmune. However, the search for the identity of the putative autoantigen in extracts of CNS has failed to provide compelling evidence for this theory (Knight, 1977; Caspary, 1977). The second line of in-

Supported by the Multiple Sclerosis Society of Canada.

ISBN 0-12-341360-5

vestigation has involved a search for an infectious etiologic agent based on the discovery of a viral etiology of other chronic neurological diseases and by both epidemiological and serological studies of MS patients (Johnson, 1975). Several chronic diseases of the central nervous system (CNS) are known to be caused by either slow, latent or persistent viral infections of nervous tissue. Examples of these diseases include subacute sclerosing panencephalitis (SSPE), canine distemper demyelinating encephalomyelitis and Visna encephalomyelitis. Many studies have suggested a connection between viral agents (particularly measles) and MS but a definite link remains to be proven. However, the evidence for a viral pathogen coupled with the marked genetic dysequilibrium in the distribution among MS patients of certain HLA haplotypes, namely HLA-A3, B7 and DW2 (McFarlin and McFarland, 1976) suggests that there may be a genetic predisposition among individuals to developing MS and a virus may serve as a precipitating agent.

Recent evidence suggests that spontaneous cytotoxic mechanisms (NK) could not only serve as a defense against tumours but might also be involved in natural resistance to virus infection (Welsh and Zinkernagel, 1977; Minato et al., 1979; Santoli et al., 1979). The possible linkage of NK levels to HLA type plus the possible role of NK in resistance to persistant virus infection has led us and others to consider the possibility that NK may be involved in the MS disease process.

Natural Killer Cells and Multiple Sclerosis

Much disagreement exists in the recent literature with respect to the status of NK activity in MS patients. Generally depressed NK function has been noted in MS patients by Neighbour et al. (1980) and Benczur et al. (1980) against the K562 erythroid tumour cell line. This depression could not be attributed to immunosuppressive therapy or the clinical status of the patients. Interferon production in these patients was found to be depressed by both groups (Neighbour and Bloom, 1979; Benczur et al., 1980). In contrast,Santoli et al. (1981) state that NK activity against K562 and RDMC or LNSV (SV40-transformed human skin fibroblast) cell lines was normal in MS patients, as were virus-induced interferon production and boosting of NK activity by interferon and poly I:C. However, Hauser et al. (1981) also found normal killing of K562 in the presence of a significantly depressed NK response to measles virus persistently-infected Hela cells, suggesting a selective depression of NK activity. They also found normal numbers of NK cells in the peripheral blood of MS patients, as judged by the numbers of non-B lymphocytes bearing avid Fc receptors for IgG.

Antibody-dependent cellular cytotoxicity (ADCC) against nucleated targets, a functional property ascribed to NK cells, is generally felt to be normal in MS patients. However Mar (1980) found significantly higher ADCC activity to rat glial tumour cells in MS patients than in normal subjects or neurological controls.

Recent results from our laboratory suggest that at least a proportion of MS patients examined have reduced levels of NK activity to K562 in 4 hour assays when compared to normal or neurological controls (Figure 1,). This difference was magnified when the less NK-sensitive target MeWo human melanoma cell line, was used as the target in an 18 hour assay (Data not shown). ADCC against antibody-coated P815 cells has been found to be normal in these patients. Analysis of the frequency of markers in the lymphocyte populations of the peripheral blood of these NK-depressed

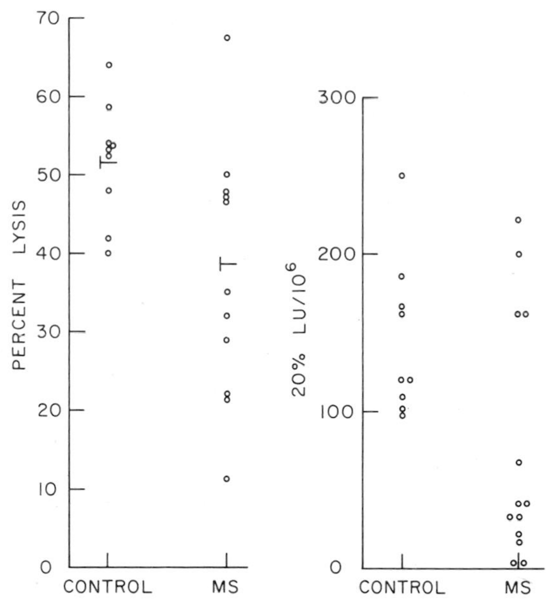

Figure 1. Natural killer cell activity was measured in standard 4 h. ^{51}Cr release assays against K562 tumour cells using effector cells from both control (lab personnel) and definite MS patients. Results are expressed as both percent lysis at a 25:1 effector:target ratio (left) and 20% lytic units per 10^6 effector cells (right).

patients with monoclonal OKT antisera and immunofluorescence suggested that the total numbers of T cells ($OKT3^+$) cells were relatively normal but, as described by Bach et al.(1980), there appeared to be an upward shift in the ratio of OKT4 helper/OKT8 suppressor cells (Figure 2a). We further have found normal to high numbers of $HNK1^+$ cells in these NK depressed individuals (Figure 2b).This monoclonal antibody has been shown to selectively bind to a differentiation antigen present on human NK cells (Abo et al. 1981).

Preliminary results from one patient further suggest that conjugate formation between targets and effector cells was normal in NK-enriched Percoll fractions (data not shown).

Other preliminary results suggested that the NK activity of NK-depressed patients is refractory to interferon boosting (Table 1).We are currently expanding these observations.

Since only a minority of MS patients studied exhibited a decrease in NK activity, it therefore remains to be shown that the underlying NK dysfunction is related to the disease process or is merely a reflection of the genetic makeup of the individuals studied (Pross and Baines, 1976).

It is apparent that shifts in the levels of subpopulations of immunocompetant cells occur during acute MS attacks. Huddlestone and Oldstone (1979) found a decline of T cells bearing Fc receptors for IgG during acute attacks with a return to normal levels during remission. Similarly Reinherz et al.(1979) have shown that OKT5 suppressor cells decline from frequencies of 20% of the peripheral blood mononuclear cells to undetectable levels in most (6/9) MS patients studied. Again, a return to normal occurred during remission. While it appears that the numbers of NK cells may be normal in those patients studied, we are currently performing a more detailed analysis of their frequency in the peripheral blood of MS victims.

It has been shown that NK cells in both the mouse and the human can lyse small subpopulations of immature cells in the neonatal thymus or adult bone marrow (Hansson et al., 1979). In addition, NK cells have been shown to kill neuroblastoma cells (Fiorani et al, 1978) as well as cells from a peripheral neurinoma line (Cornain et al., 1975). The possibility thus exists that NK cells, in addition to immune T cells might participate in the destruction of virally infected brain cells. Indeed, Adams (1977) observed that in 63% of active MS cases studied, the perivascular cuffs within active brain plaques consisted of lymphoid as

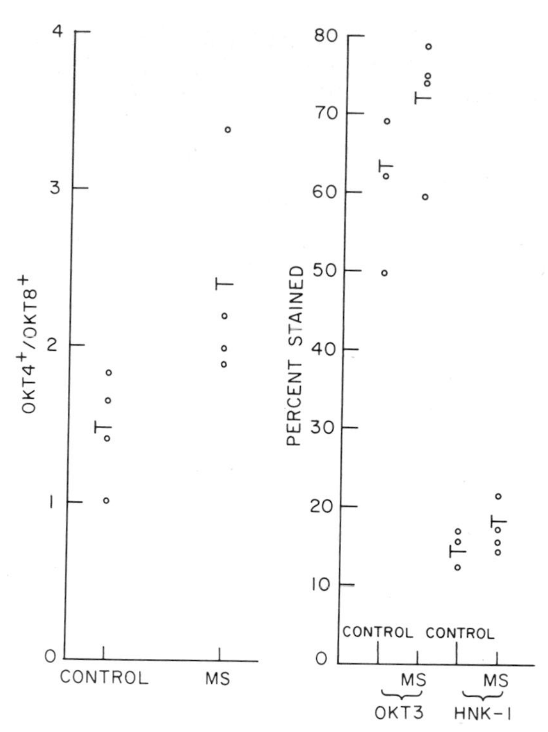

Figure 2-PBL were obtained from either control individuals or patients with definite MS and stained with either OKT-3, OKT-4, OKT-8 (2a,left)(Otho Pharmaceuticals) or HNK-1 monoclonal antibodies (2b,right). Bars represent the mean values.

Table 1.

Subject	Interferon boosted	40% Lytic Units per 10^6 cells
J.R.	-	20.0
J.R.	+	67.7
J.H.	-	14.3
J.H.	+	16.7
G.L.	-	4.0
G.L.	+	4.0

PBL were obtained from either normal (J.R.) or two definite MS patients (J.H. or G.L.). For interferon (IF) boosting, cells were cultured with 500 units of IF/ml for 1h at 37°. NK activity was tested against MeWo targets in 18h assay.

well as phagocytic cells, and formation of these cuffs is considered an early event in the pathogenesis of MS. The possible sequestering of NK cells in the brain and their direct participation in plaque formation must therefore be examined to rule out their possible role in MS pathogenesis by some form of NK autoimmune mechanism.

Alternatively, the observed NK depression may reflect a primary functional defect resulting in the suppression of the NK cytolytic mechanism subsequent to target-effector binding, interferon production or both. While there is a marked HLA association with MS, the inability of MS patients' lymphocytes to produce interferon upon appropriate stimulation (Neighbour and Bloom, 1979) would imply a generalized inability to mount a non-specific antiviral effect. Indeed, mouse strain-specific resistance to murine cytomegalovirus (MCMV) infection has been correlated with levels of NK activity (Bancroft et al., 1981). The NK-deficient beige mouse has been shown to be more susceptible to fatal MCMV infection than their NK normal littermates. Cytotoxic T cells appear late in infection with this virus (Ho, 1980), and NK cells may represent the initial line of defense against fatal infection (Shellam et al., 1981). Since it has also been shown that NK cells can lyse host cells infected with persistent virus (Welsh and Zinkernagel, 1977; Minato et al., 1979; Santoli et al., 1979), a defect in any of the mechanisms which serve to limit the spread of such viruses could place the host in an immunologically compromised situation. The studies of Hauser et al. (1981) suggest such a possibility. Since it has been difficult to link MS with any single viral agent, the disease may result from neonatal infection with any of a series of viruses (eg. measles, canine distemper). In the normal individual such infections are limited by mechanisms such as NK cells and interferonmodulated immunity. If an MS-susceptible individual is deficient in these respnses, this could contribute to the latency of the viral infection with the subsequent disease state appearing sometime later.

By approaching these topics and others, we hope to help clarify some of the questions regarding NK cell function and multiple sclerosis.

REFERENCES

Abo, T. and C.M. Balch. (1981). J. Immunol. 127:1024.
Adams, C.(1977). Brit. Med. Bull. 33:15.
Bach, M.A., F. Phan-Dinh-Tuy, E. Tournier, L. Chatenoud and J.F. Bach.(1980). Lancet. ii:1221.

Bancroft, G.J., G. R. Shellam and J.E. Chalmer. (1981). J. Immunol. 126:988.

Benczur, G., G. Petranyi, G. Palaffy, M. Varga, M.Talas, B. Cotsy, I. Foldes and S.R. Hollan. (1980). Clin. exp. Immunol. 39:657.

Caspary, E.A. (1977). Br. Med. Bull. 33:50.

Cornain, S., C. Carmaud, D. Silverman, E. Klein, and M. Rajewsky. (1975). Int. J. Cancer. 16:301.

Fiorani, M., R. Butler, L. Bertolini, and R. Revottella. (1978). Eur. J. Cancer. 14:217.

Hansson, M., R. Kiessling, B. Andersson, K. Karre, and J.C. Roder.(1979). Nature. 278:174.

Hauser, S.L., K.A. Ault, M.J. Levin, M.R. Garovoy and H.L. Weiner.(1981). J. Immunol. 127:1114.

Ho,M. (1980). Infect.Immun. 27:767.

Huddlestone, J.R. and M.B.A. Oldstone. (1978). Neurology, Minneap. 28:392.

Johnson, R.T. (1975). Adv. Neurol. 13:1.

Knight, S.C., (1977). Br. Med. Bull. 33:45.

Mar, P. (1980). J. Neurolog. Sci. 47:285.

McFarlin, D.E. and McFarland, H.F. (1976). Arch. Neurol. 33:395.

Minato, N., B.R. Bloom, C. Jones, J. Holland, and L.M. Reid. (1979). J. Exp. Med. 149:1117

Neighbour, P.A. and B.R. Bloom. (1979). Proc. Nat. Acad. Sci. (USA) 76:476.

Neighbour, P.A., A.I. Grayzel and B.R. Bloom. (1980). Prog. Immunol. (Proc. Wld. Cong. Immunol., Paris).

Pross,H.F. and M.G. Baines. (1976). Int.J. Cancer. 18:593.

Raine, C.S. and H.H. Schaumburg. (1977). In: P. Morell, ed., Myelin p. 271. Plenum Press, New York.

Reinherz, E., P.C. Kung, G. Goldstein, and S. Schlossman.(1979). Proc. Nat. Acad. Sci. (USA). 76:4061.

Santoli, D. and H. Koprowski. (1979). Immunol. Rev. 44:125.

Santoli, D., W. Hall, L. Kastrukoff, R.P. Lisak, B. Perussia, G. Trinchieri, and H. Koprowski. (1981). J. Immunol. 126:1274.

Shellam, G.R., J.E. Allen, J.M. Papadimitriou, and G.J. Bancroft.(1981). Proc. Nat. Acad. Sci. (USA). 78:5104.

Welsh, R.M. and R.M. Zinkernagel. (1977). Nature. 268:646.

IMPAIRED NATURAL KILLER CELL FUNCTION IN MULTIPLE SCLEROSIS AND ASSOCIATION WITH THE HLA SYSTEM

Miklós Benczur
Éva Gyódi
Győző Petrányi
Susan R.Hollán

National Institute of Haematology and Blood Transfusion
Budapest

György Pálffy

Department of Neurology and Psychiatry Medical School
Pécs

Margarita Tálas
Ivana Stőger
István Földes

Microbiological Research Group National Institute of Health and Hygiene
Budapest

I. INTRODUCTION

Multiple schlerosis /MS/ is a progressive demyelinating process of the central nervous system of unknown origin. Besides geographic, ethnic, nutritional, and other factors, an autoimmune mechanism and /or persistent virus infection were also proposed as a possible pathogenetic basis for the disease /Ross, 1962; Ross and Lenman, 1965; Adams and Imagawa, 1966; Ammitzboll and Clausen, 1972/. The association with certain HLA antigens /A3,B7,DR2,,DW2/ and the familiar occurrence may reflect the genetic basis of the disease susceptibility gene/s/Jersild, et al.,1972; Bertrams et al., 1973; Sander et al., 1976/. Alterations in the cellular immune parameters /Knowlen et al.,1970; Weiner et al.,1978/ and in T cell subpopulations were also reported /Reinher et al.,1980/. The earlier work in our laboratory and in others revealed impaired NK-cell function in this disease /Benczur

ISBN 0-12-341360-5

et al., 1980 a,b; Neighbour and Bloom,1979; Bloom 1980; Neighbour et al., 1981; Gyódi et al; Stőger et al/

II.MATERIALS AND METHODS

1. Multiple Shlerosis patients: 133 patients were studied /52 male and 81 female/. At the time of examination 33 had exacerbation, 57 were in remission and 43 were in a slowly progressive state. Blood samples were taken within the first month of remission. No immunosuppressive drugs were administered, except for 3 patients who were treated by corticosteroids.

2. Cytotoxicity Assay. Our method was described previously /Benczur et al., 1979/. Briefly, Ficoll-Uromiro separated and carbonyl-iron treated peripheral blood mononuclear cells were mixed with 51-chromium labelled target cells /K-562/ in double stepwise dilution from 50:1 to 3:1 effector to target cell ratios /in four parallels/. Supernatants of the plate cultures were taken after 4 hours ot incubation and the level of radioactivity was determined. Based on the dose-response line, the cytotoxicity value for 20:1 effector to target cell ratio was computed and used as a stabilized comparative value.

3. Interferon Stimulation. Human leukocyte interferon /kindly provided by dr.V.P.Kuznecov, Gamaleya Institute Moscow/ was used for preincubating effector cells /3500 IU/ml, in 1:20 dilution/ for 30 min. and the cells were washed three times.

4. Interferon Assay. The cytopathic effect-inhibition method was used for determining the amount of interferon /Finter, 1973/.

5. Tissue Typing. The standard NIH lymphocyte cytotoxicity method for HLA- A,-B,C typing /Tarasaki, 1972/ and the long incubation time lymphocytotoxicity method /VanRood et al,1975/ were used for typing the HLA-DR antigens.

III. RESULTS

In Hungary, 133 MS patients were HLA-typed and compared with a healthy population. The comparison revealed significant association, with B7 /32.3% versus 15.1%/ and DR2 antigen /42.7% versus 19.1%/. Other deviations /A28,B12/ did not reach the level of significance.

Group of patients with well-established diagnosis /"definite"/ showed a considerably decreased NK-cell function compared with healthy controls. Interferon activation of NK-cells did not restore the level of cytotoxic activity to that of the age and sex matched control group; however an increase in cytotoxicity was observed /Table I/. Similarly K-cell activity of MS patients was also depressed.

The in vitro production of IFN after vesicular stomatitis virus induction was investigated in 76 MS patients and in 34 control individuals /Table II/. Peripheral lymphocytes of MS patients produced a markedly decreased amount of IFN.Analysing the individual titre values in the control group, 24 supernatant samples showed positivity at dilution ratios of 1:320 or higher 8 samples at 1:160 and only two samples gave positive results at 1:80. The majority of the supernatants in MS showed a titre of 1:40 /22/; in 20 ceases, 1:80; 14 supernatants gave a positive result at 1:160 and only 5 supernatants had a titre of 1:320.

TABLE I. NK and K Cell Acitivity of MS Patients and the Effect of Interferon

Group	N° of cases	NK cytotoxicity %	NK +IFN cytotoxicity %	ADCC cytotoxicity %
Control	46	18.6±1.6	25.6±1.8	37.7±2.1
MS total	59	11.3±1.1[b]	15.1±1.3[c]	3o.o±1.5[b]
Definite	4o	1o.o±1.1[c]	13.2±1.2[c]	29.1±1.9[b]
Probable	12	16.8±3.o	22.8±4.o	32.2±3.8
Possible	7	1o.1±3.3[a]	11.3±4.6[b]	31.2±3.1

Difference from controls:

a 5 % > P > 1%; b 1% > P > 0.1%; c P < 0.1%

TABLE II. Interferon Production of Lymphocytes of MS Patients and Healthy Individuals after in Vitro Virus Induction

	N^{o}	Interferon titre IU/50 ul	p
MS Patients	76	84.9±9.1 /10-320/	p < 0.001
Controls	34	268.2±14.4 /80-320/	

The decreased IFN production of MS patients was not found to be related to B7 or DR2 antigens characteristic for MS.

Studying the association between B7 /DR2 HLA antigens and low in vitro cytotoxicity, significant associations for ADCC but not NK were found in the MS group. No similar associations were observed in healthy individuals /Table III/.

TABLE III. Association of low in Vitro Cytotoxic Response of Lymphocytes of MS Patients and HLA B7/DR2 Antigens

		NK	P_{corr}	ADCC	P_{corr}
MS Patients					
B7+	n=15	6.8±1.2		22.5±2.1	
B7-	n=3o	8.9±1.2	n.s.	29.1±2.1	<o.o5
DR2+	n=19	6.5±1.1		22.9±2.2	
DR2-	n=26	9.5±1.4	n.s.	29.8±2.2	<o.o5
Healthy control					
B7+	n=7	18.8±3.o		32.3±2.6	
B7-	n=36	17.8±1.9	n.s.	38.4±2.6	n.s.
DR2+	n=7	24.9±4.5		46.2±6.o	
DR2-	n=36	16.7±1.7	n.s.	35.7±2.3	n.s.

IV. DISCUSSION

Various observations verified the dysfunction of the immune system in MS /Batchelor et al., 1978; Weiner et al., 1978; Knowles and Saunders,197o/. Recently imbalances concerning T-lymphocyte subsets have been observed /Santoli et al., 1978; Reinherz et al.,1980/. Publications on decreased ConA and virus-induced suppressor cell activity have also appeared /Antel, 1978/. Earlier investigation in our and other laboratories demonstrated a decreased production of IFN by peripheral lymphocytes, in vitro /Benczur et al.,1980 a,b; Bloom 1980; Neighbour et al.,1981; Gyódi et al/.

Interferon is recognized as a central regulator for the NK system /Herberman et al.,1979; Santoli and Koprowski 1979/. On the other hand, IFN and K-cells play an important role in the defense against virus infections /Quinnan, and Menschewitz, 1979/. Thus the dysfunction of the NK-system reported here further substantiates the possible role of viruses in the pathogenesis of MS.

To detect possible correlations within the disease group, between NK cell acitivity on the one hand and B7 or DR2 HLA antigens on the other is rendered highly difficult by technical limitations in a group with an average NK cytotoxicity of under 10 per cent. Therefore it is not surprising that the difference in NK activity was detected as a tedency while K-cell activity was found significant. The HLA B7-related genetic control of low cytotoxic functions on embryonic fibroblast target cells was reported earlier /Petrányi et al.,1974/ and in ADCC, using HLA antibody-coated tumor target cells, the same was documented for K-cell activity /Santoli et al.,1976/. These data with the correlations reported in this paper may suggest how the HLA-associated disease susceptibility gene/s operates.

There was no correlation between HLA antigens and low or high IFN production in MS. Obviously this is because interferon genes have not been located in the histocompatibility region /Allen and Fantes, 1980/; Nagota et al.,198o/.

Based on the family studies it has been postulated that more than one disease susceptibility genes are operative in MS /Tiwari et al.,1980/. Highly significant defective IFN production may lead one to assume that there is a gene responsible for susceptibility in the interferon controling area. However it is not clear whether low IFN levels are cause or the consequence of multiple sclerosis.

REFERENCES

Adams,J.M.,and Imagawa,D.T./1966/Proc.Soc.Exp.Biol.Med.111:562

Allen,G.,and Fantes,A.K.,/1980/ Nature 287:4o8

Ammitzboll,T.,and Clausen,J.,/1978/Acta Neurol.scand.48:47

Antel,J.P., /1978/ Ann.Neurol. 5:338
Batchelor,J.R.,Compston,A.,and McDonald,W.I.,/1978/ Brit.Med.Bull. 34:279
Benczur,M.,Győrffy,Gy.,Garam,T.,Varga,M.,Medgyesi,Gy., Sándor,M.,and Petrányi,G.Gy.,/1979/Immunobiol. 156:32o
Benczur,M.,Petrányi,G.Gy.,Pálffy,Gy.,Varga,M.,Tálas,M., Kotsy,B.,Földes,I.,and Hollán,S.R. /198o/a/ Clin.exp. Immunol. 39:657
Benczur,M.,Varga.,Gyódi,E.,Petrányi,G.Gy.,Pálffy,Gy.,Kotsy,B, Ónody,C.and Hollán,S.R. Tissue Antigens 16:16
Bertrams,J.,vonFisenne,E.,Hoher,P.G.,and Kuwert,E., /1973/ Lancet 2:441
Bloom,B., /198O/ Nature 287:275
Finter,N.B./1973/ In "Interferon and Interferon inducers" /N.B.Finter ed./p.135. North Holland, Amsterdam
Gyódi,E.,Benczur,M.,Pálffy,Gy.,Tálas,M.,Földes,I.,and Hollán, S.R., Human Immunol /in press/
Herberman,R.B.,Ortaldo,J.R.,and Bonnard,G.D./1979/ Nature 277:221
Jersild,C.,Svejgaard,A.,and Fog,T., /1972/ Lancet 1:124o
Knowles,M.,and Saunders,M., /197o/ Neurology, 2o:7oo
Nagota,S.,Mantes,N.,and Weissmann,C.,/198o/Nature 287:4ol
Neighbour,P.A.,and Bloom,B.,/1979/Proc.Natl.Acad.Sci.76:476
Neighbour,P.A.,Miller,A.E.,and Bloom,B.R.,/1981/ Neurology 31:561
Petrányi,G.Gy.,Benczur,M.,Ónody,C.,Hollán,S.R.,and Iványi,P., /1974/ Lancet 1:736
Quinnan,G.V., and Menschewitz,J.E., /1979/ J.exp.Med.15o:1594
Reinherz,E.L.,Weiner,H.L.,Hauser,S.L.,Cohen,O.J.,Distaso,J.A., and Schlossman,S.F., /198o/ New Eng.Med. 3o3:125
Ross,C.AC.,/1962/ Brit.Med.J. 1:1523
Ross,C.AC.,and Lenman,J.A.R.,/1965/ Brit.Med.J. 1:226
Sander,H.,Kuntz,B.,Scholz,S.,and Albert E.D./1976/ In "HLA and Disease" p.91.Inserm,Paris
Santoli,D.,Trinchieri,G.,Zmijewsky,M.Ch.,and Koprowski,H., /1976/ J.Immunol. 117:765
Santoli,D.,Moretta,L.,Lisok,L.P.,Gilden,D.W.,and Koprowski,H. H., /1978/ J.Immunol. 12o:1369
Santoli,D.,Koprowski,H.,/1979/Immunol.Rev. 44:125
Stőger,I.,Tálas,M.,Benczur,M.,Gyódi,E.,Petrányi,G.Gy., and Kotsy,B., Arch.Virol. /in press/
Terasaki,P./1972/In"Manual of Tissue Typing"/J.G.Ray,ed./ p.5O.NIH, Bethesda
Tiwari,J.L.,Morton,N.E.,Latonel,J.M.,Terasaki,P.I.,Zander,H., Cho,Y.W./198o/In"Histocompatibility Testing"/P.I.Terasaki ed./p.687. Munsgaard,Coppenhagen
vanRood,J.J.,vanLeeuwen,A.,Alblas,A.,/1975/Tissue Antigens 5:73
Weiner,H.L.,Cherry,J.,McIntosch,K.,/1978/Neurology, 28:415

INTERFERON (IFN) PRODUCTION AND NATURAL KILLER (NK) CELL ACTIVITY IN PATIENTS WITH MULTIPLE SCLEROSIS: INFLUENCE OF GENETIC FACTORS ASSESSED BY STUDIES OF MONOZYGOTIC TWINS

Jochen Abb[1]
Peter Kaudewitz[2]
Helmut Zander[3]
Hans-Werner Loems Ziegler[2]
Friedrich Deinhardt[1]
Gert Riethmüller[2]

[1] Max von Pettenkofer-Institut
Munich, Germany
[2] Institut für Immunologie
Munich, Germany
[3] Labor für Immungenetik, Kinderpoliklinik
Munich, Germany

I. INTRODUCTION

Three major factors appear to be involved in the etiology of multiple sclerosis (MS) - environmental, genetic, and immunological. Epidemiological similarities between MS and poliomyelitis point to a possible role for viral infection in the etiology of MS (1,2,3). The genetic influence appears to be related to the HLA system. A weak association of MS with HLA-A3,B7 and a strong association with HLA-Dw2/DR2 is well established in the Caucasian population (4). Recent workshop analysis of multiple-case families revealed a major determinant for susceptibility to MS within the Ir(HLA-DR) region (5). In addition there is at least one major factor not linked to the HLA system (5). Previous studies have provided accumulating evidence to suggest that alterations in cellular immune parameters may have a role in the pathogenesis of the disease. Thus it has been shown that both the absolute number and the functional activity of peripheral blood suppressor T cells is reduced in patients with acute phase MS (6,7). Some investigators,

ISBN 0-12-341360-5

but not others, have found diminished capability of lymphocytes from MS patients to secrete antiviral factors and to exert NK activity (8-14). Since antigens of the major histocompatibility complex (MHC) are known to regulate IFN production (15) and NK cell activity (16), the observed variability of results might possibly be explained by the genetic heterogeneity of patient and control groups. In the present study leukocytes from MS patients were compared with those from their healthy monozygotic twins for the ability to produce IFN in response to viral and nonviral stimuli and to mediate NK activity.

II. MATERIAL AND METHODS

A. Patients and Control Donors

Nine pairs of monozygotic twins discordant for MS were included in the study. Monozygosity was verified by typing for ABO, Rhesus, Kell, MNSs, Duffy, Lewis, Kidd, P1 and HLA-A, -B, -C, and -DR antigens. The recommendations of the Schuhmacher Committee (17) were adopted for the diagnosis and grading of MS patients. None of the patients had actual disease activity at the time of study. Age, sex, HLA phenotype and clinical data of MS patients are shown in Table I. No patients or controls received steroids or other immunosuppressive drugs during the month before testing, and none had a second major illness.

B. IFN Production, IFN Assay, and IFN Characterisation

Mononuclear cells were separated from heparinised peripheral blood of MS patients and control donors by sedimentation on Ficoll-Isopaque gradients. Depletion of adherent cells was achieved by a 1-h incubation in plastic tissue culture flasks. Nonadherent cells (2.5×10^6/ml) were induced to IFN production by the addition of 2.5 ug/ml purified phytohemagglutinin (PHA-P) (Wellcome), 100 hemagglutininating units/ml influenza AX31 virus (kindly provided by Dr.J.Skehel, Mill Hill, England), 100 ug/ml polyinosinic-polycytidilic acid (Poly I:C) (Boehringer, Mannheim, Germany), or 2.5×10^5 heat (56°C, 45 min.)-treated Molt 4 leukemia cells. Supernatants of 24-h cultures were tested for antiviral activity by measuring the inhibition of the cytopathic effect (CPE) of vesicular stomatitis virus on human diploid fibroblast cells as described

TABLE I. Clinical Profiles of Patients with MS

Patient No.	Sex	Age (yrs)	HLA phenotype	Grading of diagnosis	Course of disease	Duration of illness(yrs)	Disability score[a]
1	F	49	A3,w24,B13,w44,Cw6, DR2,4	Definite	Remittant, progressive	11	7
2	M	45	A1,2,Bw51,w62,Cw3, DR2,4	Definite	Remittant, progressive	19	6
3	M	50	A1,3,B7,8,CwX, DR2,3	Definite	Primarily progressive	15	9
4	F	37	A3,28,B7,w62,Cw3, DR2,5	Definite	Remittant, progressive	11	7
5	M	63	A3,w24,Bw35,w51,Cw4, DR1,w9	Definite	Remittant, progressive	44	5
6	F	72	A2,3,B7,CwX, DR2	Definite	Remittant	41	3
7	M	54	A1,3,B7,w62,CwX, DR2,5	Definite	Remittant, progressive	31	8
8	F	59	A2,w32,Bw44,w50,Cw5,w6, DR4,7	Possible	Primarily progressive	26	6
9	F	55	A3,B13,14,CwX, DR2,7	Possible	Primarily progressive	34	6

[a] Disability scores were determined according to Kurtzke (19).

by Campbell et al. (18). Antiviral units are expressed as the reciprocal of the highest dilution inhibiting 50% of the CPE and are equivalent to approximately 1 reference unit of the WHO Human Leukocyte Reference Interferon B 69/19.

Antiviral activity induced by stimulation with PHA-P was characterised as IFN γ : Antiviral activity was sensitive to heat and low pH, inhibition of the CPE was only demonstrated on homologous human cells, and the antiviral activity was not affected by incubation with antibodies to human IFN α or human IFN β (kindly provided by Drs. K.Cantell, Helsinki, Finland, and A. Billiau, Leuwen, Belgium, respectively). Antiviral activity induced by stimulation with influenza AX31 virus, poly I:C, or Molt 4 cells shared several of the characteristics of IFN α : Antiviral activity was insensitive to treatment at 56°C or pH2, inhibition of the CPE was demonstrated on both human and bovine cells, and the antiviral activity was neutralised by antibody to human IFN α but not by antibody to human IFN β .

C. NK Cell Assay

NK cell activity of nonadherent cells was tested in a standard 4-h ^{51}Cr-release assay with K562 target cells. Dose-response curves of specific cytotoxicity were determined by plotting specific ^{51}Cr release vs the number of effector cells. The number of effector cells necessary to lyse 25% of the target cells was determined from the curves and referred to as 1 lytic unit (LU).

III. RESULTS

Production of IFN by nonadherent cells from MS patients or control donors is shown in Table II. The amount of IFN α or IFN γ secreted in response to viral or nonviral stimuli was comparable for MS patients and their healthy monozygotic twins. No correlation could be established between stage of clinical disease or HLA phenotype and the ability to produce antiviral activity, since none of the MS patients studied showed a significant impairment of IFN production or failed to respond to any particular inducer. Table II also shows the comparison of the NK cell activity of nonadherent cells from MS patients and healthy twin donors. Lytic activity of noninduced NK cells was decreased more than 2-fold in 3 out of 6 patients with definite disease. Levels of cytotoxicity of the other patients were identical or even higher than those of MS patients and twin controls did not reach the level of sig-

TABLE II IFN Production and NK Activity in Patients with MS and Their Healthy Monozygotic Twins

Donor	PHA-P	AX31	Poly I:C	Molt4	NK activity
Pat. 1	40[a]	800	320	240	4.5[b]
Twin 1	160	600	320	160	81.5
Pat. 2	40	1600	1280	160	3.6
Twin 2	10	3200	640	160	3.0
Pat. 3	160	1600	640	80	5.5
Twin 3	160	3200	640	80	19.5
Pat. 4	120	3200	640	320	9.5
Twin 4	80	3200	640	480	20.3
Pat. 5	80	6400	640	640	10.5
Twin 5	160	1600	320	640	9.3
Pat. 6	160	800	160	160	8.2
Twin 6	10	800	320	640	10.0
Pat. 7	20	400	640	40	N.D.[c]
Twin 7	30	400	320	40	N.D.
Pat. 8	40	3200	640	640	21.0
Twin 8	80	1600	640	320	7.8
Pat. 9	10	800	320	60	9.4
Twin 9	10	1600	320	80	11.1

[a] IFN induced (IU/ml)
[b] LU/10^6 cells
[c] Not done

nificance when analysed by Wilcoxon matched pairs signed rank test. Similar patterns of lytic activity were obtained after induction of NK cells by the addition of exogenous IFN; non-adherent cells from MS patients and control donors could be stimulated to the same extent (data not shown). The observed impairment of NK cell function in some patients could not be correlated with the clinical status or any particular HLA haplotype.

IV. DISCUSSION

Studies on cell-mediated immune function in vitro of MS patients have been performed by several groups. It has been reported that measles-virus treated lymphocytes of MS patients exhibited impaired or absent suppressor activity on ConA-induced lymphocyte proliferation (6,8). Consistent with their inability to suppress ConA responses, lymphocytes from MS patients also failed to produce significant amounts of IFN in response to challenge with measles virus in vitro (8). More recently, Benczur et al. (9) reported that spontaneous or poly I:C-induced NK activity against K562 target cells is decreased in patients with MS. Santoli et al. (10), however, did not find defects of unstimulated or IFN-induced NK cells in MS patients. In addition, patients lymphocytes responded normally to virus challenge in vitro in terms of both activation of NK cells and IFN production (10). Similarly conflicting results have been reported for the NK-mediated killing of target cell lines persistently infected with measles virus. While 2 groups found decreased NK activity in patients with MS (11,12), others were unable to demonstrate significant changes of NK activity against measles-virus carrying target cells (13,14). The results of Hauser et al. (12) indicated that low NK activity in MS patients was associated with HLA antigens A3,B7, whereas data of Benczur et al. (9) and Santoli et al. (10) suggested an association of NK hyporeactivity with the DR2 antigen. Conflicting results of the cited studies on NK cell function in patients with MS thus may be due to the lack of HLA-matched control groups. To secure optimal genetic homogeneity we have compared NK activity and IFN production in monozygotic twins discordant for MS. Our results clearly demonstrate that patients with MS do not show a constitutive deficiency in the production of IFN α or IFN γ in vitro. We also did not observe consistent alterations in the activity of unstimulated or IFN-induced NK cells in MS patients. In light of these results, the suggested dysfunction of NK cells and IFN-producing cells in the pathogenesis of MS should be reconsidered. This might best be achieved by longitudinal studies of individual patients during the course of their illness.

V. ACKNOWLEDGEMENT

We are grateful to Mrs. H. Abb, Mrs. B. Seidel, Mrs. I. Petersmann and Ms. C. Hilf for expert technical assistance. We thank Mrs. H. Bernd, Bavarian Red Cross Blood Center, Munich, for performing the red blood cell typing. Financial support was provided by DFG Za 59/2/5.

VI. REFERENCES

1. Leibovitz, U., and Alter, M., in "Multiple sclerosis, clues to its cause" (U. Leibovitz and M. Alter, eds.), p. 300. North Holland, Amsterdam, 1973.
2. Poskanzer, D.C., Schapira, K., and Miller, H., Lancet ii: 917 (1963).
3. Kurtzke, J.F., and Hyllested, K, Ann. Neurol. 5:6 (1979).
4. Jersild, C., Dupont, B., Fog, T., Platz, P.J., and Svejgaard, A., Transplant. Rev. 22: 148 (1975).
5. Tiwari, J.L., Morton, N.E., Lalouel, J.M., Terasaki,P.I., Zander, H., Hawkins, B.R., and Cho, Y.W., in "Histocompatibility Testing 1980" (P. Terasaki, ed.), p. 687. University of California, Los Angeles, 1980.
6. Arnason, B.G.W., and Antel, J., Ann. Immunol. (Inst. Pasteur) 129C: 159 (1978).
7. Reinherz, E.L., Weiner, H.L., Hauser, S.L., Cohen, J.A., Distaso, J.A., and Schlossman, S.F., New Engl. J. Med. 303: 125 (1980).
8. Neighbour, P.A., and Bloom, B.R., Proc. Nat. Acad. Sci. U.S.A. 76: 476 (1979).
9. Benczur, M., Petranyi, G., Palffy, G., Varga, M., Talas, M., Kotsy, B., Földes, I., and Hollan, S.R., Clin. exp. Immunol. 39: 657 (1980).
10. Santoli, D., Hall, W., Kastrukoff, L., Lisak, R.P., Perussia, B., Trinchieri, G., and Koprowski, H., J. Immunol. 126: 1274 (1981).
11. Ewan, P.W., and Lachmann, P.J., Clin. exp. Immunol. 30: 22 (1977).
12. Hauser, S.L., Ault, K.A., Levin, M.J., Garovoy, M.R., and Weiner, H.L., J. Immunol. 127: 1114 (1981).

13. Huddlestone, J.R., and Oldstone, M.B.A., Clin. Immunol. Immunopathol. 13: 444 (1979).
14. Rola-Pleszcinski, M., Abernathy, M., Vincent, M.M., Hensen, S.A., and Bellanti, J.M., Clin. Immunol. Immunopathol. 5: 165 (1976).
15. De Maeyer, E., and De Maeyer-Guignard, J., in "Interferon 1" (I. Gresser, ed.), p. 75. Academic Press, London, 1979.
16. Kiessling, R., and Wigzell, H., Immunol. Rev. 44: 165 (1979).
17. Rose, A.S., Ellison, G.W., Myers, L.W., and Tourtellotte, W.W., Neurology 26: 20 (1976).
18. Campbell, J.B., Grunberger, T., Kochman, M.A., and White, S.L., Canad. J. Microbiol. 21: 1247 (1975).
19. Kurtzke, J.F., Acta Neurol. Scand. 46: 493 (1970).

STUDIES OF HUMAN NK CELL FUNCTION IN CHRONIC DISEASES

P. Andrew Neighbour
Elizabeth Reinitz
Arthur I. Grayzel
Aaron E. Miller
Barry R. Bloom

Departments of Pathology, Microbiology & Immunology, Medicine and Neurology
Albert Einstein College of Medicine
and Montefiore Hospital and Medical Center
Bronx, New York

A physiologic role for natural killer (NK) cells in resistance to microbial infections has been proposed. NK cells show a selective preference for many virus-infected target cells in tissue culture (Santoli et al, 1978a; Minato et al, 1979). However, because of the close functional relationship between interferons (IFN) and NK, it has been difficult to dissociate the contribution of IFN from that of NK cells in mediating resistance to viruses *in vivo*. For example, a rise in NK is often accompanied by the appearance of serum IFN during the early stages of acute virus infections (Welsh, 1978). Thus, it is not always clear whether recovery is mediated by IFN-regulated NK of infected cells or by the direct antiviral action of IFN.

One experimental approach to this question has been to use genetically defined animal models of immunodeficiency. Minato et al (1979) have shown that IFN-regulated NK mediates the rejection of virus persistently infected tumor cells in nude mice. More recently, the beige mouse, which has a selective NK impairment (Roder, 1979), is being used to test its inherent susceptibility to virus infection.

Supported by NIH grants NS 15541, NS 11920 and AI 09807, grant RC 1006 from the National Multiple Sclerosis Society and a Clinical Research Center grant from the Arthritis Foundation.

ISBN 0-12-341360-5

We have studied the IFN-NK system of patients with chronic diseases of putative viral etiology in an attempt to identify the possible role of this system against viruses in man, and to learn more of the mechanisms of this immune response. We summarize here our recent studies showing defects in the IFN-NK system in patients with chronic diseases; consider explanations for the observed defects; and evaluate their significance to the pathogenesis of these diseases.

A. Reduced NK activity in patients with chronic diseases

We have tested NK of peripheral blood mononuclear cells (PBMC) against K562 targets using a short term (4 or 5 hr) conventional ^{51}Cr-release cytotoxicity assay (Bloom et al, 1980; Neighbour et al, in press). We observed that approximately 30% of patients with multiple sclerosis (MS) and 40% with systemic lupus erythematosus (SLE) exhibit NK significantly below that of normal donors (i.e. outside the 95% confidence limits for the normal group mean). In contrast, patients with rheumatoid arthritis (RA) do not differ significantly from the normal population. Similar results have been reported by other laboratories for MS and SLE patients (Benczur et al, 1980; Merrill et al, in press; Silverman & Cathcart, 1980; Hoffman, 1980; Oshimi et al, 1980; Goto et al, 1980; Santoli & Koprowski, 1979). Hauser et al (1981) found MS patients' NK to be significantly reduced against HeLa.measles cells, but not against K562 cells. Santoli et al (1981) found normal NK among MS patients when tested on K562 and other NK-susceptible targets.

In addition to "endogenous" NK, we have determined the response of patients' and normal donors' NK to augmentation by IFN and IFN inducers. While the majority of MS patients were augmented with IFN _in vitro_, their levels of augmented NK rarely reached those of normal donors, and the group mean NK was significantly reduced (Neighbour et al, in press). This was observed for both IFN-α and for the IFN inducer Newcastle Disease virus (NDV). Furthermore, many SLE patients did not respond at all to IFN or NDV, and again those whose NK could be augmented rarely killed as well as normal donors. These findings have also been seen by others (Benczur et al, 1980; Merrill et al, in press).

The amount of ^{51}Cr released in the conventional cytotoxicity assay is the accumulation of multiple lethal hits, each consisting of target recognition and binding, followed by target cell lysis (Ullberg & Jondal, 1981). Thus, the reduced cytotoxic activity seen as diminished ^{51}Cr-release could be due to: i) fewer target binding cells; ii) fewer functional cytotoxic cells; iii) reduced rate of lysis; or iv) reduced recycling time. The development of alternative cytotoxicity assays has enabled dissection of the NK cytotoxic process, and these can be used to analyze the basis of the reduced NK in patients. For example, Merrill et al (in press) have used the single cell agarose assay originally described by Grimm and Bonavida (1979) to measure the proportion of functional NK cells in MS patients. They found that acute relapsing MS patients have decreased numbers of effectors in peripheral blood capable of binding and lysing NK-susceptible target cells.

We have also begun to apply this method to PBMC from SLE donors together with the measurement of recycling index (RI) using the method of Ullberg and Jondal (1981). Preliminary data indicate that, as in MS, there are fewer functional NK cells in untreated PBMC from SLE donors (Table 1; manuscript in preparation). However, 60% of the patients had elevated RI

Table 1. Percentage of NK cells and recycling indices of PBMC from patients with systemic lupus erythematosus.

Donor group	Effector treatment	% TBC [a]	% Dead conjugates [b]	% NK [c]	Vmax ($x10^{-4}$) [d]	RI [e]
Normal	Control	19.5±2 [f]	19.2±3	3.02±2.1	1.31±0.3	4.6±2
SLE	Control	22.2±3	13.5±2	2.04±1.0	1.23±0.3	8.5±3
Normal	IFN [g]	18.7±3	22.4±4	3.40±0.7	2.24±0.4	7.1±2
SLE	IFN	23.9±9	19.9±3	3.39±0.6	1.81±0.5	4.5±1

a) Cells bound to K562 after 15 mins incubation expressed as % of total nylon-nonadherent PBMC.
b) Determined at 3 hr by trypan blue staining of conjugates suspended in 1% agarose on microslides.
c) (% TBC x % dead conjugates)/100.
d) Vmax determined from ^{51}Cr-release data using the "enzyme kinetics" cytotoxicity assay (Callewaert et al, 1978).
e) Recycling index calculated as Vmax/(No. of effectors x % NK)(Ullberg & Jondal, 1981).
f) Mean ± s.e.m.
g) 16 hr incubation with $1x10^3$ units of IFN-α.

despite low % NK, indicative of in vivo activation of NK perhaps by IFN. Consistent with these data are the findings of serum IFN in many SLE patients during active disease (Hooks et al, 1979). IFN treatment of normal PBMC in vitro markedly increased recycling with only a slight increase in the % NK. In the SLE group, however, IFN reversed the apparent deficit of functional effectors by raising the % NK to normal levels, and these newly recruited cells had a lower RI. Therefore, the primary NK defect in SLE appears to be an absence of functional cytotoxic cells due to an in vivo blockage of pre-NK to NK differentiation. It also appears that serum IFN in these patients, while readily capable of activating mature NK, is prevented from recruiting NK precursors in vivo. Because this situation can be reversed in vitro, the defect does not appear to be due to an insufficiency of precursor cells, but rather to an as yet unidentified blocking mechanism.

B. Possible explanations for defective NK function

There are at least two explanations for the observed IFN-NK defects of these patients: i) effectors may have been depleted from the circulation; or ii) effectors may be present but their functions might be impaired. Serum immunoglobulins in the form of lymphocytotoxic antibodies with NK specificity, or changes in the traffic of lymphocyte subsets might contribute to the apparent "loss" of NK cells from peripheral blood.

NK cells are included within a morphologically defined population of mononuclear cells termed large granular lymphocytes (LGL) (Saksela et al, 1979). However, not all LGL are capable of spontaneous cytotoxicity and, therefore, determination of the proportion of LGL in NK-defective donors would inadequately address the question of whether reduced NK results from depletion of effectors. Furthermore, most of the cell surface antigens defined by monoclonal antibodies present on NK cells, such as OKT8, OKT10, OKT11a, OKM1, 9.6, and HNK-1, are also expressed by other mononuclear subsets. Therefore, these markers cannot be used to determine the proportion of NK cells per se. There are significant shifts in the balance of PBMC subpopulations in patients with chronic diseases. For example, in MS and SLE there are fluctuations in the proportions of T cells, and of the helper and cytotoxic/suppressor subsets (Santoli et al, 1978b; Moretta et al, 1979; Reinherz et al, 1980). Because of the overlap between some of these populations and NK, we can conclude only that fluctuations in these subsets might also reflect changes in the proportions of NK cells. Clearly, NK-specific identifying markers are needed to resolve this question.

Because of the relationship between IFN and NK, defective NK might be due to an inability to produce necessary IFN. We have shown that PBMC from many patients with MS or SLE, but not RA, produce significantly decreased levels of IFN in response to measles, NDV, poly I.C, and concanavalin A (Neighbour et al, 1981; Neighbour & Grayzel, 1981). Therefore, we attempted to correlate defects in NK with the ability both to produce and to respond to IFN in vitro (Neighbour et al, in press). The lack of correlation between IFN production and endogenous NK by normal donors showed that these two functions can operate independently. Moreover, PBMC from some MS and SLE donors had normal NK but were unable to secrete IFN. Responsiveness of their NK to IFN was also unrelated to their ability to produce IFN. Therefore, defective NK does not appear to be directly attributable to abnormal IFN production.

Circulating immune complexes binding to the Fc receptors of NK cells may impair NK (Silverman & Cathcart, 1980), and such complexes are a feature of many chronic diseases. However, NK is largely resistant to Fc receptor blockage, and the levels of immune complexes in patient sera are below those required to inhibit NK in vitro. Prostaglandins have been shown to inhibit NK (Droller et al, 1978), and prostaglandin-mediated immunoregulation may be altered in many patients with chronic diseases. Finally, spontaneous cytotoxicity and antibody-dependent cellular cytotoxicity (ADCC) appear to be mediated by the same cells under different conditions. Therefore, a loss in NK function might parallel, and be due to, a concommitant increase in K cell activity. Indeed, recent studies indicate that some MS patients have elevated ADCC (Frick & Stickl, 1980).

C. Issues relating to the significance of defective IFN-NK function in chronic disease

A fundamental question that arises from these studies concerns the significance of these in vitro observations to the etiology and progression of chronic disease. A major problem encountered when attempting to answer this question is whether PBMC reflect systemic immunologic function. As mentioned earlier, the observed anomalies may result from altered lymphocyte traffic. The relevant effector cells may have selectively left the periphery to perform their function at a specific site. For example, the presumed target in MS might be the CNS. Studies of NK by lymphocytes recovered from CSF indicate, however, that these cells show little, if any, spontaneous cytotoxicity even after IFN treatment (Merrill et al, in press). In SLE where the lesions are more disseminated, it would be difficult to determine where the NK cells might have migrated. In patients with RA, however, it is clear that the primary lesion is the synovium. We have recently found (manuscript in preparation) that the cells which infiltrate the inflamed synovium contained more functional NK cells than simultaneously tested autologous PBMC (Table 2). In addition, more LGL were seen among these cells and IFN was detected in the joint fluids of more than 80% of these patients.

Another concern is whether or not the targets used in these studies are physiologically appropriate. Clearly, different results are found for different targets. This may be due to differential involvement and contribution of other cytolytic mechanisms. For example, it has been proposed that a proportion of K562 cytotoxicity is K cell-mediated (Pape et al, 1979). More importantly, however, if target selective NK clones exist, then the defects themselves may be restricted to only certain clones. If the primary etiology of a particular disease is viral, perhaps a virus-infected cell line would be the most appropriate target.

In considering whether defective IFN-NK is a primary or secondary consequence, one must not lose sight of the fact that patients with chronic diseases such as MS or SLE exhibit a multitude of apparently related and unrelated immunological anomalies. Major determining factors in these defects are the severity and stage of disease. Our studies found this also true for NK (Neighbour et al, in press). Another factor is the influence of steroid therapy on immune function tested *in vitro*. We have found some correlation between the extent of NK defect in SLE and the level of steroids administered therapeutically (Neighbour et al, in press). However, instances of

Table 2. NK function in active RA patient synovia.

Source of effectors	No. of samples	% TBC	% dead conjugates	% NK	P value[a]
Peripheral blood	13	20.8±4[b]	19.0±3	3.9±0.9	-
Synovial fluid	13	21.2±2	24.9±5	5.3±1.1	<0.01
Synovial membrane	6	25.7±5	23.7±4	6.0±1.2	<0.01

a) % NK cells in synovial samples compared with PBMC (Mann-Whitney test).
b) Mean ± s.d.

individuals on high steroids with normal NK, and those with no NK prior to steroid therapy suggested that while steroids may have some inhibitory effect on NK, they are not the sole explanation for the observed defects. For these reasons, and the fact that not all patients with these diseases are immunodeficient, we contend that most of the anomalies are probably due to generalized disturbance of the peripheral immune system and are not primary to the disease etiology. However, abnormal immune function following the onset of disease is likely to contribute to its subsequent progression.

In conclusion, these studies have not yet provided any firm support for the role of NK in resistance to virus infections and chronic disease. Until we can determine the basic underlying mechanisms of the observed defects, it will remain unclear as to whether or not abnormal IFN-NK function is involved in the development or progression of these diseases. However, the investigation of IFN-NK in patients with apparent deficits in this system has shed some light on the basic interaction between IFN and NK. Clearly the defects are highly selective, some individuals being deficient in only one component of the IFN-NK system. These data suggest that endogenous NK and NDV-induced IFN production are either (a) mediated by distinct subpopulations of cells; or (b) independently regulated functions of the same cell. Lastly, there are presently several attempts in progress to administer IFN to patients with non-tumor-associated chronic diseases. While IFN may prove to be effective therapeutically in these diseases, the studies described here show that, as well as having defective NK, the response of these patients' cells to IFN is also highly abnormal. Therefore, careful monitoring of the IFN-NK system is warranted before, during and after treatment.

REFERENCES

Benczur, M., Petranyi, G.G., Palffy, G., Varga, M., Talas, M., Kotsy, B., Folder, I., Hollan, S.R., Clin. Exp. Immunol. 39: 657 (1980).

Bloom, B.R., Minato, N., Neighbour, P.A., Reid, L.M., and Marcus, D. in "Natural Cell-mediated Immunity Against Tumors" (R.B. Herberman, ed.) Academic. Press, NY, p. 505 (1980).

Callewaert, D.M., Johnson, D.F., and Kearney, J., J. Immunol. 121: 710 (1978).

Droller, M.J., Schneider, M.U., and Perlmann, P., Cell. Immunol. 39: 165 (1978).

Frick, E., and Stickl, H., J. Neurol. Sci. 46: 187 (1980).
Goto, M., Tanimoto, K., and Horiuchi, Y., Arthitis Rheum. 23: 1274 (1980).
Grimm, E., and Bonavida, B., J. Immunol. 123: 2861 (1979).
Hauser, S.L., Ault, K.A., Levin, M.J., Garovoy, M.R. and H.L. Weiner., J. Immunol. 127: 1114 (1981).
Hoffman, T., Arthritis Rheum. 23: 30 (1980).
Hooks, J.J., Moutsopoulos, H.M., Geis, S.A., Stahl, N.H., Decker, J.L., and Notkins, A.L., New Engl. J. Med. 301: 5 (1979).
Merrill, J., Jondal, M., Seeley, J., Ullberg, M., and Siden, A., Clin. Exp. Immunol. In press (1982).
Minato, N., Bloom, B.R., Jones, C., Holland, J., and Reid, L.M., J. Exp. Med. 149: 1118 (1979).
Moretta, A., Mingari, M.C., Santoli, D., Perlmann, P., and Moretta, L., Scand. J. Immunol. 10: 223 (1979).
Neighbour, P.A., and Grayzel, A.I., Clin. Exp. Immunol. 45: 576 (1981).
Neighbour, P.A., Grayzel, A.I., and Miller, A.E., Clin. Exp. Immunol. In press. (1982)
Neighbour, P.A., Miller, A.E., and Bloom, B.R., Neurology 31: 561 (1981).
Oshimi, K., Gonda, N., Sumiya, M., and Kano, S., Clin. Exp. Immunol. 40: 83 (1980).
Pape, G.R., Troye, M., Axelsson, B., and Perlmann, P., J. Immunol. 122: 2251 (1979).
Reinherz, E.L., Weiner, H.L., Hauser, S.L., Cohen, J.A., Distaso, J.A., and Schlossman, S.F., New Engl. J. Med. 303: 125 (1980).
Roder, J.C., J. Immunol. 123: 2168 (1979).
Saksela, E., Timonen, T., Ranki, A., and Havry, P., Immunological Rev. 44: 71 (1979).
Santoli, D., and Koprowski, H., Immunological Rev. 44: 125 (1979).
Santoli, D., Trinchieri G., and Lief, F., J. Immunol. 121: 526 (1978a).
Santoli, D., Moretta, L., Lisak, R.P., Gilden, D., and Koprowski, H., J. Immunol. 120: 1369 (1978b).
Santoli, D., Hall, W., Kastrukoff, L., Lisak, R., Perussia, B., Trinchieri, G., and Koprowski, H., J. Immunol. 126: 1274 (1981).
Silverman, S.L. and Cathcart, E.S., Clin. Immunol. Immunopathol. 17: 219 (1980).
Ullberg, M., and Jondal, M., J. Exp. Med. 153: 615 (1981).
Welsh, R.M., J. Exp. Med. 148: 163 (1978).

PERIPHERAL BLOOD NATURAL CELL-MEDIATED CYTOTOXICITY IN PATIENTS WITH ATOPIC DERMATITIS[1]

Nuha T. Kusaimi[2]
John J. Trentin

Division of Experimental Biology
Baylor College of Medicine
Houston, Texas

I. INTRODUCTION

Immune abnormalities in atopic dermatitis (AD) are well documented at the clinical, humoral and cellular levels. AD patients are prone to viral and bacterial infections of the skin. There is usually elevated serum IgE, decreased or absent delayed-type hypersensitivity to a number of antigens, decreased lymphocytic phytohemagglutinin-induced blastogenesis, defective function of the T cells and of the blood lymphocytes, monocytes and polymorphonuclear leucocytes and reduction in circulating T γ cells.

Therefore natural cell-mediated cytotoxic activity of the peripheral blood lymphocytes was studied in 10 patients with atopic dermatitis, and simultaneously in a healthy control matching each patient for sex, skin color and age ± 6 years. Eight patients gave a family history of allergic disease (as atopic dermatitis, asthma, hives or allergic rhinitis), and four patients had a personal history of asthma. None of the patients had systemic corticosteroids for at least a year prior to the investigation, and, none had anti-histamines for at least 48 hours before blood samples were obtained.

[1] Supported by USPHS Grants CA-03367, CA-12093 and K6 CA-14219 from the NCI.

[2] Present address: Department of Medicine, College of Medicine, University of Mosul, IRAQ.

ISBN 0-12-341360-5

II. METHODS: Mononuclear leucocytes were separated using LSM solution (Bionetics, Kensington, Md.). K562 target cells were maintained in RPMI containing 20% inactivated FCS, were radio-labeled by incubation with ^{51}Cr as sodium chromate (Amersham Corp., Arlington, Ill.) at 37°C for one hour, centrifuged, washed and adjusted to a count of 0.2×10^6 viable cells/ml. Effector and target cells at ratios of 100:1, 50:1, 25:1 and sometimes 12.5:1, in round-bottomed multiwell plates (Linbro Scientific, Inc., Hamden, Conn.) were incubated at 37°C in 5% CO_2 atmosphere for 4 hours. Other details of methodology are in press (Kusaimi and Trentin, Arch. Dermatol.).

III. RESULTS: There was a marked reduction in the natural cytotoxicity of most of the atopic dermatitis patients studied. The three patients with the least or no reduction in natural cytotoxicity (less than 51% reduction) were the 3 with only moderate or mild degrees of atopic dermatitis. When expressed as % of paired control lytic activity, the mean value for all E:T ratios of all 10 AD patients was 41.9 ± 28% of control value, P = < 0.005 (Table I).

TABLE I. Natural Cell-Mediated Cytotoxicity in Ten Patients with Atopic Dermatitis and Their Matched Healthy Controls.

Age, Sex, Color, Severity	Effector: Target Cell Ratio	% Lysis of K562 Cells by			Average % of Control
		Control Cells	Patient Cells	Patient as % of Control	
30 F	100:1	62.2	23.8	38.3	
White	50:1	55.8	15.9	28.5	
Severe	25:1	35	9.4	26.9	
	12.5:1	25.4	4.4	17.3	27.8
20 F	100:1	72	15.8	21.9	
Black	50:1	32	8	25	
Severe	25:1	28	8	28.6	
	12.5:1	16.7	3.6	21.6	24.3
32 F	100:1	73.3	28.7	39.2	
White	50:1	69	28.3	41	
Severe	25:1	50	20.9	41.8	
	12.5:1	40	13.1	32.8	38.7
28 M	100:1	84	77.8	92.6	
Brown	50:1	76	73	96.1	
Mild	25:1	76	42.7	56.1	81.6

TABLE I. (Continued)

Age, Sex, Color, Severity	Effector: Target Cell Ratio	% Lysis of K562 Cells by			Average % of Control
		Control Cells	Patient Cells	Patient as % of Control	
30 F	100:1	59	23	39	
Black	50:1	62	8.6	13.9	
Severe	25:1	35.3	2.9	8.2	20.4
30 M	100:1	56	20	35.7	
Brown	50:1	42	20	47.6	
Severe	25:1	30	14	46.7	43.3
44 F	100:1	70.4	35	49.7	
Black	50:1	45	19	42.2	
Severe	25:1	43	7.2	16.7	36.2
20 F	100:1	66	44	66.7	
Black	50:1	57	26	45.6	
Moderate	25:1	40	18	45	
	12.5:1	24.3	9.7	39.9	49.3
26 F	100:1	66	7.4	11.2	
Black	50:1	57	6.4	11.2	
Severe	25:1	40	4.3	10.8	
	12.5:1	24.3	6.7	27.6	15.2
42 F	100:1	24.7	25.3	102.4	
Black	50:1	17	10.6	62.4	
Mild	25:1	10.9	14.6	134.9	99.9
TOTAL		1665.5	696.1	1465.2	
MEAN		47.6	20	41.9	
± S.D.		20	15.7	28.3	

The mean natural cytolytic activity of the 10 matching healthy controls was 47.6 ± 20%. This value compares very closely with that obtained by Tursz, Dokhelar and Gluckman (1980) for the peripheral NK cell activity in 120 healthy controls versus the same (K562) target cells used in this study (FIG. 1).

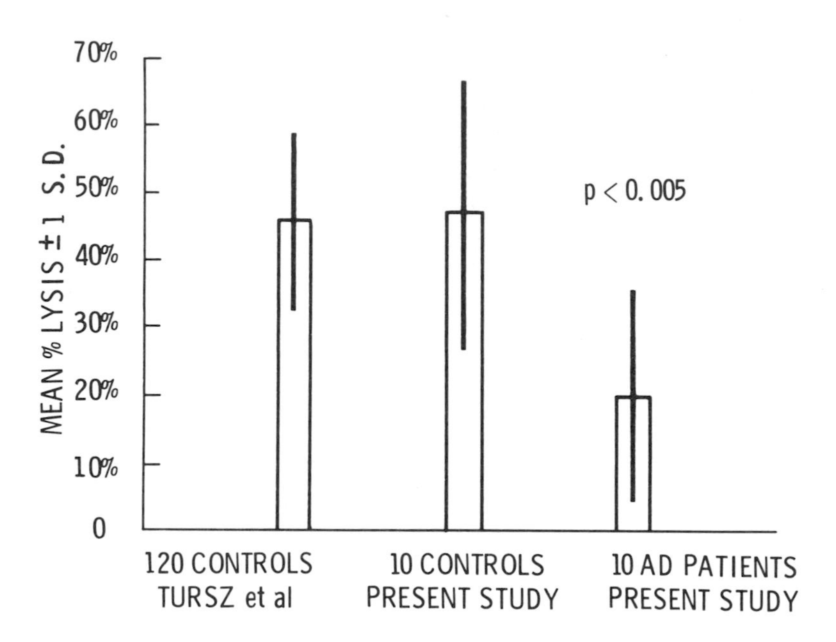

The present findings indicate clearly a decrease in the natural cytotoxic capacity of the AD patients studied. This is in accord with other reported immunologic aberrations in atopic dermatitis patients. (See Kusaimi and Trentin, in press, for detailed references). Natural killer cells, which are the main contributors to natural cytotoxicity, are directed not only against tumor cells, and hemopoietic grafts, but also against microbial infections, including virus infections (Herberman and Ortaldo, 1981). The present results are therefore in accordance with the observed susceptibility of AD patients to viral infections, such as herpes simplex and vaccinia infections (Baer, 1959).

REFERENCES

Baer, R.L. (1959). Med. Clin. North Am. 43:765.
Herberman, R.B., and Ortaldo, J.R. (1981). Science 214:24.
Kusaimi, N.T., and Trentin, J.J. (1982). Arch. Dermatol. 118:568.
Tursz, T., Dokhelar, M.C., and Gluckman, E. (1980). Lancet i: 375.

Portions of this manuscript have been published in the August 1982 issue of Archives of Dermatology, volume 118, pages 568-571.

RECOVERY OF NK CELL ACTIVITY AFTER BONE-MARROW TRANSPLANTATION

Marie-Christine Dokhélar
Marc Lipinski
Thomas Tursz

Laboratoire d'Immunologie Clinique
Institut Gustave Roussy
Villejuif, France

I. INTRODUCTION

Natural killer (NK) cells are responsible for the spontaneous lysis in vitro of a variety of target cells, mainly of tumoral origin (Herberman and Holden, 1978). Several reports have suggested that this cytolysis is mediated through poorly matured lymphocytes (Herberman et al., 1978). However, whether these functionally-defined lymphocytes are in the process of further differentiation or have actually reached their own final stage of maturation remains controversial. NK cells are believed to originate in the bone-marrow because irradiated mice injected with bone-marrow from high or low NK activity donors acquire the level of activity of the donor (Haller and Wigzell, 1977). Also, NK activity is abrogated in previously high responder mice treated with the bone-marrow toxic isotope ^{51}Strontium (Haller et al., 1977). In man, bone-marrow transplantation provides a unique oppotunity for studying the evolution of NK cell activity in recipients whose classical B and T cell functions are deeply affected for several months after the graft. We report here that NK cell activity is recovered by the second month following the graft, much earlier than B or T cell functions. In addition, we present data indicating that, in recipients affected with graft-versus-host disease (GVHD) during the first month following the graft, there is

---

[1] Supported by DGRST, IGR and ADRC grants.

ISBN 0-12-341360-5

an associated reappearance of NK cell activity occurring consistently earlier than in recipients with a normal clinical evolution.

II. MATERIALS AND METHODS

A. Patients

Twenty-four patients undergoing bone-marrow transplantation were studied. They were all grafted at the Hôpital St Louis, Paris, bone-marrow transplant unit (E. Gluckman). Thirteen patients had idiopathic severe aplastic anemia, 4 post-hepatitic aplastic anemia, 1 Fanconi anemia and 6 various leukemias. In 17 cases, the donor was an HLA-identical sibling and an identical twin in 4 cases. The conditioning regimen included in most cases total body irradiation (700 to 1,000 rads) and cyclophosphamide (120 mg/kg). After the transplant, 14 patients received methotrexate and 6 patients cyclosporin A for GVHD prophylaxis. Patients who had identical twins received no treatment after grafting. During the first 30 days, 9 patients had evidence of GVHD. No rejection was observed, and complete chimerism was demonstrated in 17 patients.

B. Methods

Effector cells were peripheral blood lymphocytes isolated by centrifugation on Ficoll-Paque, and tested for spontaneous cytotoxicity against ^{51}Cr-labeled K562 cells. Effector to target cell ratio was 50:1. Incubation was at 37°C for 4 hrs. Supernatants were harvested and radioactivity was counted. Percent specific lysis (% SL) was calculated as follows:

$$\% \text{ SL} = \frac{\text{cpm assay} - \text{cpm min}}{\text{cpm max} - \text{cpm min}} \times 100$$

where min and max indicate spontaneous (without effectors) and total (with HCl 6N) releases of ^{51}Cr by targets, respectively.

III. RESULTS AND DISCUSSION

The results have been reported in detail elsewhere (Dokhélar et al., 1981).

A. Pre-transplant NK Values

NK cell activity was depressed in most cases studied before the conditioning regimen, especially in patients with aplastic anemia (mean value: 7.3%). Such low NK values had already been observed (Livnat et al., 1980) and could reflect some stem cell defect involving NK precursors.

B. Reappearance of NK Cell Activity

Bone-marrow recipients were sequentially tested for NK cell activity. In all patients studied, peripheral NK activity could be detected after bone-marrow transplantation. A host origin of NK cells seems quite unlikely because of the very low pre-transplant NK activity. NK activity could not be attributed to a direct transfusion effect of already matured NK cells, since we have constantly found low NK values in the bone-marrow samples studied. Therefore, our data suggest that the functional NK cells demonstrated in bone-marrow recipients originate from bone-marrow progenitors.

In patients whose clinical evolution was uneventful, virtually no NK activity could be detected in the first month following the graft. By 5 to 6 weeks following the transplant, a sudden recovery of a normal NK activity was observed in all these patients. More than 6 weeks after the graft, ther was no distinguishable difference when NK activities were compared between bone-marrow recipients and a normal control population. This observation is in agreement with previously reported results indicating that NK activity was normal in bone-marrow recipients one month after the graft (Livnat et al., 1980). It is well extablished that other immunological functions such as T proliferation and cell-mediated cytotoxicity are not recovered by this time (Ringden et al., 1979, Noel et al., 1978). Thus, NK cell activity appeared to be restored earlier than most other classical lymphocytic functions. This supports the assumption that NK activity is mediated through poorly matured lymphocytes, or alternatively, it might reflect a more rapid maturation of NK cells than T cells.

C. NK Activity in Acute GVHD

In certain cases, high NK cell activity was measured in bone-marrow recipients less than 4 weeks after the graft. Several patients exhibited a normal level of NK activity as early as 10 to 15 days after the transplant. Strikingly, all of them were affected with an early acute GVHD. Similarly, all the patients affected with a GVHD during the first month following the transplant had recovered a normal or high NK activity during this period. When both groups of bone-marrow recipients were compared less than 4 weeks after the graft, a striking difference between the means of NK activities was observed (10.1% in the non-GVHD group vs 52.5% in the GVHD group, $p<10^{-9}$). Serial interferon levels were measured by virus growth inhibition in several recipients (G. Périès, Hôpital Saint Louis, Paris). No correlation was found between interferon production levels and NK cell activities during GVHD.

It has been suggested that NK cells are immature cells of T lineage (Herberman et al., 1978). The high NK values observed in our study could reflect the activation during GVHD of subsets of T lymphocytes, some of which able to exert NK activity at some stage of their maturation. However, a direct role for NK cells in the pathogenesis of GVHD cannot be ruled out. Lopez et al. have reported normal pre-transplant NK values associated with GVHD occurrence (Lopez et al., 1979). It must be emphasized that they were dealing with another NK detection assay (using Herpes virus-infected fibroblasts as targets). The heterogeneity of spontaneously cytotoxic cells is well documented. In particular, Fitzgerald et al. reported distinct surface marker phenotypes for cells responsible for killing of K562 cells and virus-infected fibroblasts (Fitzgerald et al., 1981).

In bone-marrow recipients, we believe that the measurement of NK cell activity could provide an attractive routine procedure for monitoring the prophylactic treatment for GVHD during the first several weeks after the transplant.

D. NK Activity during Viral Infections

The assessment of NK activity after bone-marrow transplantation could have another clinical interest. In patients with restored NK cell activity, we observed a drastic fall of NK values in 7 cases. These 7 patients were affected with severe viral infections, mainly with cytomegalovirus. NK cells are involved in the control of cytomegalovirus infection in mice (Quinnan and Manischewitz, 1979). Our data suggest that in humans, NK cells could play a role in the defense towards viral agents in such deeply immunodepressed patients.

ACNOWLEDGMENTS

We thank E. Gluckman for allowing us to study her patients and Y. Finale for expert technical assistance.

REFERENCES

Dokhélar, M.-C., Wiels, J., Lipinski, M., Tétaud, C., Devergie, A., Gluckman, E, and Tursz, T. (1981). Transplantation 31: 61.

Fitzgerald, P., Evans, R., and Lopez, C. (1981). In "Mechanisms of Lymphocyte Activation"(K. Resch and H. Kirchner, eds.), p. 595. Elsevier/North Holland, Amsterdam.

Haller, O., Kiessling, R., and Wigzell, H. (1977). J. Exp. Med. 145:1411.

Haller, O., and Wigzell, H., (1977). J. Immunol. 118:1503.

Herberman, R.B., and Holden, H.T. (1978). Adv. Cancer Res. 27: 305.

Herberman, R.B., Nunn, M.E., and Holden, H.T. (1978). J. Immunol. 121:304.

Livnat, S., Seigneuret, M., Storb, R., and Prentice, R.L. (1980). J. Immunol. 124:481.

Lopez, C., Sorele, M., Kirkpatrick, D., O'Reilly, R.J., and Ching, C. (1979). Lancet 2:1103.

Noel, D.R., Witherspoon, R.P., Storb, R., Atkinson, K., Doney-K., Michelson, E.M., Ochs, H.D., Warren, R.P., Weiden, P.L., and Thomas, E.D. (1978). Blood 51:1087.

Quinnan, G.V., and Manischewitz, J.E. (1979). J. Exp. Med. 150:1549.

Ringden, O., Witherspoon, R.P., Storb, R., Ecklund, E., and Thomas, E.D. (1979). J; Immunol. 123:2729.

ENHANCED NK ACTIVITY IN PATIENTS TREATED BY INTERFERON-α. RELATION TO CLINICAL RESPONSE.

Stefan Einhorn, Anders Ahre, Henric Blomgren, Bo Johansson, Håkan Mellstedt and Hans Strander.

Radiumhemmet
Karolinska Hospital
Stockholm, Sweden

INTRODUCTION

Clinical studies using interferon (IFN) in the treatment of human tumors have been going on at Radiumhemmet, Karolinska Hospital since 1969 (Strander and Einhorn, 1982). During the last years IFN trials have also been started at several other hospitals throughout the world. In some of the diseases studied, IFN has been found to induce regressions of the tumors. This has, for instance, been found in multiple myeloma (Mellstedt et al., 1979; Gutterman et al., 1980), lymphoma (Gutterman et al., 1980; Merigan et al., 1978), acute leucemia (Hill et al., 1979), mammary carcinoma (Gutterman et al., 1980) and ovarian carcinoma (Einhorn et al., 1982). Although IFN has been found to excert antitumor effects in man, it is important to stress that IFN has not been shown to be superior to conventional therapy in any malignant tumor studied to date (Strander and Einhorn, 1982). What is then the mechanism behind the antitumor effects? One possibility is a direct cell multiplication inhibitory effect by IFN on the tumor cells (Paucker et al., 1962; Einhorn and Strander, 1978), another is that IFN acts indirectly, i.e. by augmenting possible antitumor activities of the immune system (Epstein, 1977). Since IFN has been shown to enhance natural killer (NK) activity (Trinchieri and Santoli, 1978; Einhorn et al., 1978a), it has been suggested that the enhanced NK activity is a mechanism by which IFN excerts its antitumor effects. To evaluate this possibility, we have performed a series of studies on NK activity in tumor patients undergoing IFN therapy. We have posed two major questions:

ISBN 0-12-341360-5

1) Does *in vivo* therapy with IFN affect human NK activity?
2) Are changes in NK activity of importance for IFN's anti-tumor effects?

Some data from these studies have been published previously (Einhorn et al., 1978b; Einhorn et al., 1980; Einhorn et al., 1981b).

DOES IN VIVO THERAPY WITH IFN AFFECT HUMAN NK ACTIVITY?

NK activity was measured by a 4 hr ^{51}Cr release assay using Chang cells as targets. In 83 patients, NK activity was measured prior to and during IFN-therapy, which in most cases consisted of 3 x 10^6 IFN units daily. In 74 of these patients (90%), NK activity increased following the first injection of IFN, as measured 24 hr after the injection (Table 1). The increase was statistically significant ($p<0.001$). Some patients received 1,5 or 6 x 10^6 IFN units daily instead of 3 x 10^6 units. There was an inverse relationship between the IFN dose

TABLE 1. NK activity before IFN therapy and change after initiation of therapy. Mean ± S.E. of cytotoxic indexes are presented. Lymphocyte:target cell ratio 50:1.

	NK activity		Total number of pat.	No. of pat. in which NK increased	Significance limits
Before IFN	0.19	±0.01	83	-	-
Change after 1 day	+0.18	±0.02	83	75	p<0.001
Change after 1 week	+0.17	±0.03	53	44	p<0.001
Change after 3 months	+0.19	±0.03	31	23	p<0.01
Change after 6 months	+0.15	±0.06	14	11	p<0.05
Change after 9 months	+0.21	±0.06	12	10	p<0.01
Change after 12 months	+0.12	±0.12	6	5	N.S.

and the relative enhancement of NK activity, i.e. 1,5 x 10^6 units gave the highest increase in NK activity, 3 x 10^6 units an intermediate enhancement of NK activity, whereas 6 x 10^6 units gave the lowest increase in NK activity. This difference was statistically significant ($p<0.05$) (Einhorn et al., 1981a).

NK activity was also measured after 1 week, 3 months, 6 months, 9 months and 12 months of IFN therapy. As can be seen in Table 1, NK activity was increased in a majority of the patients at all these time points. The increase was in all cases but one statistically significant.

Addition of IFN to the assay *in vitro* enhanced NK activity in all patients prior to therapy. During IFN therapy, however, the ability of IFN to enhance NK activity *in vitro* was reduced, in some cases to nondetectable levels (Einhorn et al., 1980; Einhorn et al., 1981b).

ARE CHANGES IN NK ACTIVITY OF IMPORTANCE FOR IFN's ANTITUMOR EFFECTS?

To analyse this question, we selected a group of tumor patients undergoing IFN therapy that filled the following criteria: 1) A relatively large group of patients included in the clinical trial. 2) A tumor type in which it is possible to follow the response to therapy in an objective and relatively easy manner. 3) A tumor in which IFN is known to excert antitumor

TABLE 2. Relation between NK activity and response of the tumor to IFN therapy. Response has been defined as > 50% reduction of M-component. Mean ± S.E. are presented.

NK activity	Non-responders (28 patients)		Responders (11 patients)	
Before IFN	0.19	±0.11	0.20	±0.13
IFN in vitro	+0.12	±0.07	+0.11	±0.11
IFN in vivo, after 1 day	+0.20	±0.21	+0.19	±0.20
IFN in vivo, after 1 week	+0.17	±0.23	+0.17	±0.20
IFN in vivo, after 3 months	+0.2.	±0.22	+0.19	±0.19

effects in a proportion of the cases. In Stockholm, a trial using IFN as therapy for multiple myeloma was started in 1977. In multiple myeloma, the product of the tumor clone, the monoclonal immunoglobulin, can be followed serially and used as a measure of the tumor volume. The criteria for response have been described elsewhere (>50% reduction of the pretreatment immunoglobuline value) (Mellstedtet al., 1980). Approximately 15% of multiple myeloma patients have been found to respond to IFN (Mellstedt et al., 1979; Gutterman et al., 1980; Mellstedt et al., 1980).

In 39 myeloma patients NK activity was measured prior to and during IFN therapy, after which changes in NK activity were correlated to clinical response (Einhorn et al., 1981a). No correlations between preinjection levels of NK activity and clinical response (defined as more than 50% reduction of M-component) could be found (Table 2). Neither IFN-induced enhancement of NK activity *in vitro* nor IFN-induced enhancement of NK *in vivo*, as measured after 1 day, 1 week and 3 months of IFN therapy could be found to correlate with clinical response to IFN therapy (Table 2). Using 25% reduction of M-component as a criteria for response gave similar results, i.e. no correlations between changes in NK activity and clinical response to the IFN therapy (Einhorn et al., 1981a). Also by the use of a third method, linear regression analysis, correlating changes in NK activity to changes in M-component 6 weeks after initiation of treatment, no relation between effects on NK activity and clinical effects could be observed (Einhorn et al., 1981a).

CONCLUSIONS

Human NK activity increase following exposure to IFN *in vivo* (Einhorn et al., 1978b; Einhorn et al., 1980; Einhorn et al., 1981a; Einhorn et al., 1981b, Table 1). Whether or not this bears relevans for the antitumor effect excerted by IFN is not known. In multiple myeloma we have found no indication for this being the case (Einhorn et al., 1981a, Table 2).

ACKNOWLEDGEMENTS

The excellent technical assistance of Mrs E, Anderbring and Miss L. Ödin is greatefully acknowledged. We thank professor Kari Cantell for providing some of the IFN used.

These studies were supported by grants from the Swedish Cancer Society and the Cancer Society of Stockholm.

REFERENCES

Einhorn, N., Cantell, K., Einhorn, S., and Strander, H. (1982) Cancer Treat. Rep., In press.
Einhorn, S., Blomgren, H., and Strander, H. (1978a) Int. J. Cancer, 22:405.
Einhorn, S., Blomgren, H., and Strander, H. (1978b) Acta Med. Scand. 204:477.
Einhorn, S., and Strander, H. (1978) Adv. Exp. Med. Biol., 110:159.
Einhorn, S., Blomgren, H., and Strander, H. (1980) Int. J. Cancer 26:419.
Einhorn, S., Ahre, A., Blomgren, H., Johansson, B., Mellstedt, H. and Strander, H. (1981a) Manuscript in preparation.
Einhorn, S., Blomgren, H., Strander, H., and Troye, M. (1981b) Progress in Cancer Research and Therapy, 19:193.
Epstein, L.(1977) In: "Interferons and their actions" (W.E. Stewart, ed), p. 91. CRC Press, Cleveland.
Gutterman, J.U., Blumenshein, G.R., Alexanian, R., Yap, H.Y., Buzdar, A.B., Cabanillas, F., Hortobayi, G.N., Distefano, A., Hersh, E.M., Rasmussen, S.L., Harmon, M., Kramer, M., and Pestka, S. (1980) Ann. Intern. Med. 931:399.
Hill, N.O., Loeb, E., Pardue, A.S., Dorn, G.L., Khan, S., and Hill, J.M. (1979), J. Clin. Hemat. Oncol. 9:137.
Mellstedt, H., Ahre, A., Björnholm, M., Holm, G., Johansson, B., and Strander, H. (1979) Lancet, 3:245.
Strander, H., Brenning, G., Engstedt, L., Gahrton, B., Holm, G., Lerner, R., Lönnqvist, B., Nordenskiöld, B., Killan, der, A., Stalfeldt, A-M., Simonsson, B., Ternstedt, B., and Wadman, B. (1980) 2nd Int. Conf. Immunotherapy, April 28-30, NIH, Bethesda, Md, USA.
Merigan, T.C., Sikora, K., Breeden, H.J., Levy, K., and Rosenberg, S.A., (1978) N. Engl. J. Med. 299:1449.
Paucker, K., Cantell, K., and Henle, W., (1962) Virology 17:324.
Strander, H., and Einhorn, S., (1982) Cancer Clin. Trials. In press.
Trinchieri, G., and Santoli, D. (1978) J. Exp. Med. 147:1314.

NK CYTOTOXICITY IN INTERFERON TREATED MELANOMA PATIENTS[1]

Sidney H. Golub

Departments of Surgery/Oncology and
Microbiology and Immunology
UCLA School of Medicine
Los Angeles, California

I. INTRODUCTION

Interferon (IFN) clearly represents a group of molecules that can exert considerable regulatory activity on the natural killer (NK) system. Much of the work on the modulation of NK activity by IFN has focused on in vitro effects of IFN on NK cells. There have been several reports which indicate that in vivo administration of IFN-α to humans results in augmentation of NK activity (Einhorn _et al_., 1978; Huddlestone _et al_., 1979; Kariniemi _et al_., 1980; Einhorn _et al_., 1980) and similarly, other biologically active agents such as BCG, also appear to augment NK activity (Thatcher and Crowther, 1978). This report will summarize our investigations on the regulation of NK activity in patients treated with IFN-α, and our preliminary results in patients treated with BCG.

II. KINEMATICS OF ALTERATIONS IN NK ACTIVITY IN IFN TREATED PATIENTS

Our IFN study has utilized sixteen patients with metastatic malignant melanoma who had metastases to subcutaneous tissues and/or lungs. The patients were treated with daily intramuscular doses of $1x10^6$, $3x10^6$, or $9x10^6$ units of human leukocyte IFN for 42 consecutive days. The IFN was provided by the American Cancer Society as part of a multi-institutional trial. Peripheral blood lymphoid cells (PBL) were obtained daily for

[1]These studies were supported by USPHS grants CA17013 and CA12582.

ISBN 0-12-341360-5

two days prior to treatment, daily for three days following initiation of treatment, and weekly thereafter. NK cytotoxicity was measured against the K562 myeloid leukemia cell line and in some experiments additional targets were included. We have used both the standard ^{51}Cr-release assay and the single cell binding and cytolysis assay (Grimm and Bonavida, 1979) for measurement of cytotoxicity.

Changes in NK activity from our IFN trial are summarized in Table I. The data are summarized in two forms. Each patient was tested at four effector:target ratios (100:1, 50:1, 25:1, 12.5:1) and the average of all four ratios was taken as an indication of total cytotoxic activity per PBL sample. A mean of these average NK values was then taken for all patients studied at each time point. This mean NK value is reported in Table I along with the change from the pre-treatment values (an average of the two pre-treatment samples). Also included in Table I is the proportion of patients who showed an increase in NK activity compared to that individual patient's pre-treatment values. Several features of the changes in NK activity deserve mention. First of all, we found a rather consistent decline in NK activity 24 hours after the first dose of IFN. Subsequently, cytotoxic activity increased, with the most consistent and largest increase in activity after seven consecutive days of IFN treatment. Cytotoxic activity then gradually declined to pre-treatment levels or slightly above pre-treatment levels. We have run similar sequential assays with healthy volunteers who did not receive IFN and they showed no such increased activity. Furthermore, every cytotoxicity assay was accompanied by a simultaneous assay with cryopreserved PBL from a healthy donor. These repetitive assays with a single sample also showed no time related change. Thus, the pattern observed appears to be a very real and fairly consistent one. We have also seen a similar pattern in a number of patients treated with intralesional BCG. Once again, peak NK activity was found within several days following the initiation of BCG treatment. The increase NK activity was most noticeable against the NK-sensitive target K562 while no increase was detected against the NK-insensitive M14 melanoma cell line. We have observed similar changes with respect to these same targets with the IFN-treated patients, namely an increased activity against NK-sensitive targets such as K562 or MOLT-4, and against moderately NK-sensitive targets such as EBV-transformed lymphoblasts, but no increased activity against NK-resistant targets such as M14. However, we did see increased activity against antibody-coated M14 in assays of antibody dependent cellular cytotoxicity (ADCC) with PBL from IFN treated patients.

TABLE I. NK Cytotoxicity Against K562 of PBL from IFN Treated Melanoma Patients

Day of Treatment	Mean % Cytotoxicity ± S.E.[a]	Mean Change ± S.E. in % Cytotoxicity[b]	Number increased[c]/ Number tested
-1	25.8 ± 4.3	---	---
0	30.3 ± 4.5	---	---
1	21.0 ± 2.4	-7.0 ± 3.7	6/15
2	32.5 ± 5.8	+3.8 ± 5.0	8/14
3	35.3 ± 4.0	+7.7 ± 5.0	9/15
7	43.1 ± 3.5	+15.3 ± 3.7[d]	13/16[d]
14	35.0 ± 2.9	+6.6 ± 4.2	10/15
21	31.8 ± 2.7	+3.9 ± 3.7	9/16
28	32.4 ± 3.4	+3.5 ± 4.3	9/15
35	30.0 ± 3.9	+1.0 ± 3.8	7/14
42	36.7 ± 4.9	+5.7 ± 5.4	9/14

[a]Mean average NK (4 effector:target doses) ± standard error.
[b]Mean (day of test) - Mean $\left(\frac{\text{Day } -1 + \text{Day } 0}{2}\right)$ ± standard error.
[c]Compared to pre-treatment (average of Day -1 + Day 0).
[d]Significantly increased by both paired t-test and signed rank test.

Several other features of the change in NK cytotoxicity should be mentioned, although these features are discussed in greater detail in Golub _et al_. (Golub _et al_., in press). Increases in NK activity were noted at all three dose levels employed in this study, although the lowest dose used ($1x10^6$ units/day) actually gave the most prolonged increase in the cytotoxicity and the least decline at Day 1. We could find no obvious correlation between changes in NK cytotoxicity and clinical response to IFN treatment. Both clinical non-responders and patients who showed some clinical benefit from IFN therapy responded with equivalent increases in NK activity. No correlation could be detected between changes in NK activity and changes in the proportions of T cells (as detected by E-rosettes), complement receptors (as detected by EAC rosettes), Fcγ receptors (as detected by EAγ rosettes) or total lymphocyte counts.

III. REGULATION OF NK ACTIVITY IN IFN-TREATED PATIENTS

We attempted to determine the cellular events involved in the early rise in NK activity during the first week of treatment and the reasons for the subsequent decline in NK activity

TABLE II. K562 Binding and Killing Cells in IFN Treated Patients

Patient Number	Cell Preparation	Day of IFN Treatment: Day 0 %B[a]	Day 0 %NK[b]	Day 2 %B	Day 2 %NK	Day 7 %B	Day 7 %NK	Day 35 %B	Day 35 %NK
8	PBL	9	4	17.7	11.5	12.5	5	11.5	10
12	PBL	9	2.8	27	7.5	31	10.2	12	8
	Nylon fiber column (NFC) passed PBL	15	0	22	4	25	10	17	4
13	PBL	8	1.9	10	3.6	9	3.6	3	0.3
	NFC passed PBL	13	3.6	14	5.6	19	7.6	12	1.6
	EAγ monolayer depleted PBL	4	0.2	7	0.5	8	0.6	1	0.1
14	PBL	12	3	23	8	33	7	15	10
15	PBL	12	9	21	7	29	13	18	11
16	PBL	9	3	18	12	27	13	17.5	11

[a]% PBL that bind K562. 300 PBL scored.

[b]% PBL that kill K562. 100 binders scored (% killers) and multiplied by (a).

to pre-treatment levels despite continued IFN treatment. We have not thoroughly examined the reasons for the decline in activity at 24 hours after treatment, although others have noted this change and they have attributed this deline to either redistribution of NK cells (Kariniemi _et al._, 1980) or the febrile episodes associated with IFN treatment (Einhorn _et al._, 1980). These seem to be reasonable explanations although we think there may also be an artifactual component (perhaps caused by repeated venipuncture) since we see a similar decline in normal controls who had not received IFN (Golub _et al._, in press). Concerning the more important issue of the cellular events involved in the increasing NK activity during the first week of treatment, we attempted to determine whether this was due to an increase confined to mature NK cells (i.e. an increased cytotoxic activity per NK cell) or at an increase in the total number of cells actually mediating cytolytic events. To address this question, we have turned to the single cell assay developed by Grimm and Bonavida (Grimm and Bonavida, 1979). These results are summarized in Table II where selected time points were examined for the number of K562 binding cells and K562 lytic cells. We have previously reported that the number of K562 binders increases during the period of increased NK activity in the first week of IFN treatment (Golub _et al._, in press). The increase in K562 target binders was found among nylon fiber non-adherent cells while the passage of the PBL over immune complex monolayers (EAγ monolayers) removed the majority of binding cells and nearly all the cytolytic cells. Thus, we see an increase in target binders which is concurrent with the increased cytotoxic activity against K562 targets and these target-binding and target-killing cells bear the usual characteristics of NK cells in that they are nylon fiber non-adherent and Fcγ receptor positive. We have also tested for the presence of target-binding cells to the NK-resistant M14 melanoma cell line (data not shown). In general, there are many M14 binding cells (about 10% of the PBL), but few M14 killing cells (less than 1%) and no increase in either binders or killers during IFN treatment. Thus, the increase in target binding seems to be confined to NK-sensitive targets.

We next considered the reasons for the decline in NK activity after peak cytotoxicity was achieved, usually during the first week of IFN treatment. Removal of adherent cells by passage of the PBL over Sephadex G10 columns failed to restore NK activity as did the addition of 1×10^{-5}M indomethacin. Furthermore, the addition of PBL obtained after several weeks of IFN treatment failed to suppress the NK activity of fresh normal PBL or cryopreserved autologous pre-treatment PBL. Thus, it is unlikely that a suppressor macrophage or other prostaglandin secreting cell is suppressing the response in the later time period. Although it is possible that the anatomical distribution of NK cells changes with time during IFN treatment, the fact that a number of Fcγ receptor cells remained relatively constant throughout the treatment argues against any major redistribution of cells after the first week of IFN treatment (Golub _et al_., in press).

The available evidence suggests that a major factor in the failure to show augmented NK activity after the first week of IFN treatment is an increasing resistance of the NK cells themselves to further IFN activation. We have found that the in vitro addition of 100 units of IFN to PBL from melanoma patients obtained prior to the initiation of IFN treatment resulted in a consistent increase in NK activity with an average net increment of 8.4% against K562. However, if the PBL were obtained from IFN treated patients, the degree of activation by IFN in vitro was related to the period of in vivo IFN treatment (Golub _et al_., in press). PBL obtained during the first several days of IFN therapy remained responsive to IFN in vitro (average net increment of 7.6%). In contrast, PBL obtained from the time of peak in vivo activity (Day 7) to the end of treatment (Day 42) were poorly responsive to IFN in vitro with an average net increase of only 1.4%. In contrast, PBL obtained after IFN treatment had been completed regained complete responsiveness to IFN in vitro (average increase of

13.7%). It should be pointed out that during most of this refractory period the NK activity is at approximately pre-treatment levels so the inability to augment this activity is not due to NK activity already being at maximum levels. Subsequent experiments have shown this resistance to IFN activation to extend to higher doses of IFN (up to 1000 units), longer treatment periods in vitro (up to 24 hours in vitro) and other sources of IFN (fibroblast IFN, lymphoblast IFN, and IFN inducers such as poly IC).

IV. CONCLUSIONS

Our results on changes in NK activity in cancer patients receiving IFN-α are similar, but not identical, to other groups who have studied such patients. The reasons for these differences are not immediately apparent although differences in IFN dose, purity of IFN preparations, and the types of patients being treated may well contribute to the variations among laboratories. However, a certain common pattern is apparent in that the patients frequently show a decline in activity immediately after the first dose of treatment, a subsequent rise in activity, and an increasing resistance to subsequent activation with further treatment. This pattern was also noted by Pape *et al*. in hepatitis patients treated with IFN-β. (Pape *et al*., 1981).

It has been clearly established that IFN can activate mature NK cells to increase the proportion of target-binding cells that can actually mediate the cytolysis (Zarling *et al*., 1979; Targan and Dorey, 1980). Our results also suggest that some of the increase in NK activity is due to the recruitment of additional cells from a precursor pool into active NK function. The strongest evidence for recruitment of additional NK cells is the increase in binding activity. The subsequent decline in NK activity appears to be primarily due to increasing cellular resistance to IFN activation. This phenomenon has been noted by others (Einhorn *et al*., 1980) and suggests a limitation on the usefulness of protocols such as the one employed in this study for maintaining elevated levels of NK activity for extended periods of time. Whether elevated NK activity contributes to therapeutic effects of IFN has yet to be established, but even if NK activity is irrelevant to the therapeutic benefits of IFN, the phenomenon of acquired resistance to IFN induced biological changes may be of relevance to other systems.

ACKNOWLEDGMENTS

The IFN used to treat the patients was provided by the American Cancer Society and the IFN used in vitro was generously provided by Dr. M. Krim of the Sloan-Kettering Institute. The author greatly appreciates the cooperation of Dr. Martyn W. Burk and Dr. Donald L. Morton in providing samples from their patients and the technical assistance of Mr. Dean Hara.

REFERENCES

Einhorn S., Blomgren, H., and Strander, H., Acta. Med. Scand., 204:477-483, 1978.

Einhorn S., Blomgran, H., and Strander, H., Int. J. Cancer, 26:419-428, 1980.

Golub, S.H., D'Amore, P.J., and Rainey, M., J. Natl. Cancer Inst., (in press).

Golub, S.H., Dorey, F., Hara, D., Morton, D.L., and Burk, M.W., J. Natl. Cancer Inst., (in press).

Grimm, E., and Bonavida, B., J. Immunol., 123:2861-2869, 1979.

Huddlestone, J.R., Merigan Jr., T.O., and Oldstone, M.B., Nature, 282:417-420, 1979.

Kariniemi, A.L., Timonen, T., and Kousa, M., Scand. J. Immunol., 12:371-374, 1980.

Pape, G.R., Hadem, M.R., Eisenburg, J., Riethmuller, G., Cancer Immunol. Immunother., 11:1-6, 1981.

Targan, S., and Dorey, F., J. Immunol., 124:2157-2161, 1980.

Thatcher, N., and Crowther, D., Cancer Immunol. Immunother., 5:105-107, 1978.

Zarling, J.M., Eskra, L., Borden, E.C., Horoszewicz, J., and Carter, J.M., J. Immunol., 123:63-70, 1979.

PHASE I TRIAL OF IMMUNOMODULATORY ACTIVITIES OF HUMAN LEUKOCYTE INTERFERON IN ADVANCED CANCER PATIENTS

Jerry A. Bash,[1] James N. Woody and John R. Neefe[1]

Divisions of Immunologic Oncology
and Medical Oncology
Lombardi Cancer Center
Georgetown University
Washington, D.C.

I. INTRODUCTION

The present study was initiated as one of several phase I and phase II clinical trials of human interferon (IFN) sponsored by the National Cancer Institute. These were prompted by the reported immunomodulatory effects of IFN in animals and man (Gresser, et al., 1972; Strander et al., 1973) and the reported clinical success in preliminary trials with cancer patients (Gutterman, et al., 1979). It was made feasible by improved industrial production of IFN to allow sufficient quantities for large well-controlled trials.

In our study twenty-two patients with advanced malignancy were treated in a phase I trial of (Meloy) human leukocyte interferon (IFN-α). Patients in groups of 3-5 received a single intramuscular injection of $0.5\text{-}60 \times 10^6$ IU/M^2. Immunomodulatory effects on natural killing (NK), antibody-dependent cellular cytotoxicity (ADCC) and monocyte-mediated cytotoxicity (MMC) of peripheral blood mononuclear cells were evaluated.

---

[1]Supported by NIH Contract NCI-CM-07437-1 and Grant 2P30 CA14626

ISBN 0-12-341360-5

II. MATERIALS AND METHODS

A. Effector Cell Separation

1. Mononuclear Cell Separation. Forty ml of venous blood were collected in heparinized vacutainers, washed, diluted and layered on Ficoll-Hypaque gradients to obtain the mononuclear fraction.

2. Nonadherent Cell Separation. Mononuclear cells suspended in RPMI+20% fetal calf serum ($1-2x10^6$/ml) were layered onto plastic tissue culture flasks previously coated with murine sarcoma line TU5 microexudate as described elsewhere (Ackerman and Douglas, 1978). After 45 minutes of incubation at 37°C nonadherent cells were decanted and further eluted with 5x washing and resuspension in RPMI+10% FCS. Nonadherent cells were counted and used as effector cells in NK and ADCC assays.

3. Adherent Cell Separation. Adherent cells were eluted from microexudate coated flasks by 10 minute incubation in 10mM EDTA in PBS as previously described (Ackerman and Douglas, 1978; Mantovani, et al., 1979). These were washed, counted and used as effector cells in the MMC assay. Collected cells were 90% viable, 90% latex positive and 90% nonspecific esterase positive.

B. Natural Cytolysis Assays

1. NK. NK activity was determined in a standard 4hr assay against ^{51}Cr-labelled K-562 targets at 40:1, 20:1, 10:1, 5:1 effector to target ratios.

2. ADCC. ADCC activity was determined against a murine lymphoma line EL-4 sensitized with rabbit anti-mouse lymphocyte serum. ^{51}Cr-release from 4 hour assays was determined for ADCC and NK simultaneously.

3. MMC. MMC was determined by measuring 48hr ^{3}H-Thymidine release by prelabelled SV40 transformed murine fibroblasts, TU-5 cultured with monocytes according to a method described in detail elsewhere. (Mantovani, et al., 1979).

C. Analysis of Data

1. Percent Release (% R) was determined according to the formula:

$$\% R = \frac{E-S}{T-B} \times 100 \text{ where}$$

E = experimental
S = spontaneous
T = total
B = background

Mean $\pm$ SD were determined from triplicate cultures.

2. Lytic Units (L.U.) were determined by a program which compared regression lines plotted from counts vs. E:T ratio of patient sample compared to frozen controls. After determination that the lines being compared were parallel the horizontal distance between their midpoints was computed and expressed as lytic units $\pm$ 95% confidence interval.

III. RESULTS

A. NK and ADCC

In order to determine the effect of IFN administration on day 0, the pretreatment levels were determined on 2-3 samples on days -6 to 0. Upon comparison of pretreatment and post-treatment levels a number of patients showed significant changes. There was no absolute dose response relationship. However, 6 of 12 patients receiving 10-40 $\times 10^6$/IU/m^2 showed a significant increase in NK or ADCC during the 1st post treatment week.[2] Four responders are shown in Table I.

As shown in Figure 1, the most dramatic increase in NK and ADCC activity was seen with patient MW on day +3 after receiving 25×10^6 IU/M^2 on day 0. Patient RS who received 40×10^6 IU/M^2 exhibited a more modest increase on day +3. Both returned to baseline at day +7. Patient RC who received 60×10^6 IU/M^2 showed a somewhat dramatic loss of NK activity at day 1 which was manifested in both NK and ADCC activity through day 7.

---

[2]Anti-viral activity as determined by manufacturer. Activity as determined in our independent assays involving a different system was 3-8 fold lower.

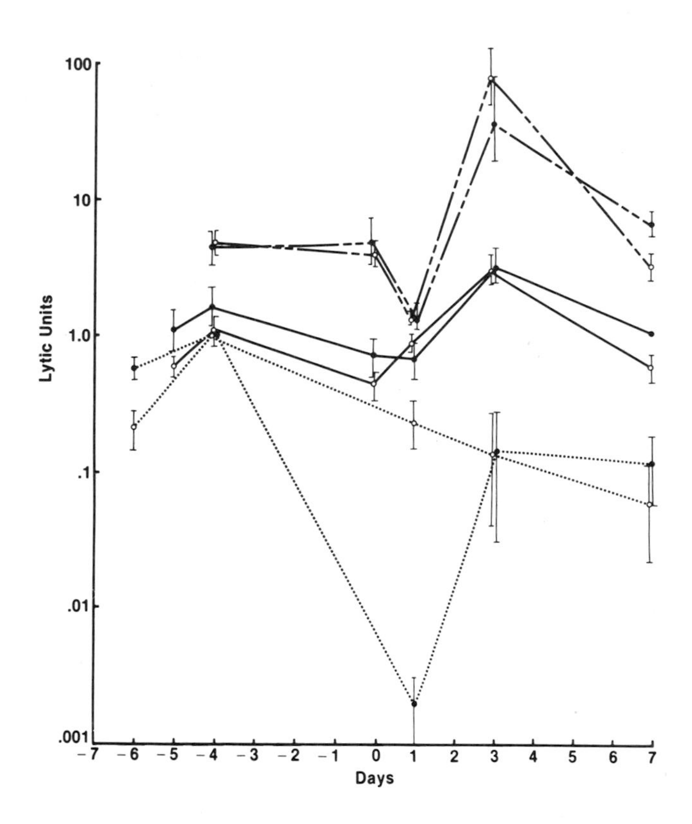

FIGURE 1. Comparison of responses of three patients treated with IFN- at different doses on day 0. NK (●) and ADCC (○) activity in lytic units (with 95% confidence intervals) is shown for patients MW (— — —), RS (——) and RC (······) treated with 25, 40 and 60 x10^6 IU/M^2 respectively.

TABLE I

SUMMARY OF IFN-α-AUGMENTED NK AND ADCC RESPONSES[a]

PT	IFN-α DOSE ($x10^6$ IU/M^2)	PRETREATMENT DAY	PRETREATMENT L.U.	POST-TREATMENT DAY	POST-TREATMENT L.U.
MM	10	0	1.54(1.38-1.74)[b] 2.07(1.07-8.43)[c]	3	2.61(2.16-3.22)[b] 2.77(2.07-3.92)[c]
RA	15	-4	2.09(1.86-2.39) 2.61(1.88-3.77)	7	12.08(7.32-24.12) 15.0(8.27-38.69)
MW	25	0	4.8 (3.4-7.4) 3.9 (3.2-4.9)	3	35.3(19-79) 78(49-126)
RS	40	-4	1.60(1.18-2.24) 1.06(0.829-1.37)	3	3.10(2.43-4.31) 2.96(2.36-3.85)

[a] Maximum Lytic Units (95% Confidence Interval)
[b] NK (top level numbers)
[c] ADCC (bottom level numbers)

B. MMC

It has been more difficult to establish a pretreatment baseline for the MMC assay because of variability inherent in the assay. As a consequence, only one patient showed a trend toward increased activity post-treatment. MMC was not suppressed by IFN. No clear correlation with changes in ADCC or NK was seen.

IV. CONCLUSIONS

The results of this study suggest that a single dose of IFN between 10-40x10^6 IU/M^2 results in enhancement of natural cytolysis in as many as half of patients. The quantitative and qualitative nature of the augmentation seen thus far appears to exhibit individual variation and may reflect different boosting dose-responses as seen *in vitro* (Kadish et al., 1981). Since most of our patients had adequate levels of NK, ADCC and MMC pretreatment, our results could be interpreted to suggest that IFN may be more useful in patients preselected for low activity. Alternatively, it may be necessary to develop an assay predictive of killer augmentation or directly of tumor response. Finally, it may be that alternate schedules of administration or sources or types of IFN may be more efficacious. We are currently pursuing these studies with IFN-α at higher doses administered weekly.

ACKNOWLEDGEMENTS

The authors would like to acknowledge the technical assistance of Daniel Teitelbaum and Mary Harris and thank E. Phillips for writing the program for linear regression analysis.

REFERENCES

Ackerman, S.K. and Douglas, S.D. (1978). J. Immunol. 120:1372.
Gresser, I., Maury, C. and Brouty-Boye, D. (1972). Nature 239:167.
Gutterman, J.U., Blumenschein, G.R., Alexarian, R., Yap, H-Y., Buzdar, A.V., Cabanillas, F., Hortobagyi, G.N., Distefano, A., Hersh, E.M., Rasmussen, S.L., Harmon, M., Kramer, M. and Pestka, S. (1979). In "Report of the IInd International Workshop on Interferons" (Krim, M. et al., ed.) Rockefeller University Press, New York.
Kadish, A.S., Doyle, A.T., Steinhauer, E.H. and Ghossein, N.A. (1981). J. Immunol. 127:1817.
Mantovani, A., Jerrells, T.R., Dean, J.H. and Herberman, R.B. (1979). Int. J. Cancer 23:18.
Strander, H., Cantell, K., Carlstrom, G., and Jakobsson, P.A. (1973). J. Natl. Cancer Inst. 51:733.

MODULATION OF NK ACTIVITY BY RECOMBINANT LEUKOCYTE INTERFERON IN ADVANCED CANCER PATIENTS

Annette E. Maluish
John R. Ortaldo
Ronald B. Herberman

Biological Research and Therapy Branch
Biological Response Modifiers Program
National Cancer Institutes
Frederick Cancer Research Facility
Frederick, Maryland 21701

I. INTRODUCTION

Interferons (IFN) are glycoproteins produced by certain cells after exposure to mitogens, viruses or other agents. They have been shown to possess a wide range of biological activities, in particular the ability to augment natural killer (NK) cell activity *in vitro* (Herberman *et al*, 1979). *In vivo* administration of IFN has also been reported to boost NK activity (Einhorn *et al*, 1978). Clinical studies using IFN in viral diseases and cancer have been carried out with relatively crude material, and some therapeutic benefits have been reported (Merigan *et al*, 1978; Gutterman *et al*, 1980). Little information is available, however, on the effects of the IFN on the patient's immune function.

Recently it has been possible to clone the gene for human leukocyte A IFN, insert it into E. coli and thus produce a recombinant leukocyte A IFN. This material has subsequently been purified to at least 95% homogeneity and is free of detectable endotoxin. This recombinant IFN has been used in the studies reported here.

ISBN 0-12-341360-5

II. MATERIALS AND METHODS

A. Patients

A total of 108 patients were treated in two Phase I studies designed to evaluate the human tolerance, pharmacokinetic profiles and immunomodulatory effects of recombinant leukocyte A IFN. The patients entered into the study had disseminated cancer of a variety of types and were treated by intramuscular injection of interferon either twice daily or three times weekly for 28 days, with each group of five patients receiving a fixed dose of IFN, ranging from 0.5 - 100 million units per injection.

B. Natural Killer Cell Assay

Blood specimens were obtained from the patients at least twice and usually three times prior to the start of IFN treatment (day 0). Subsequent samples were obtained on approximately days 1, 4, 7, 14, 21, 28, and 36. Mononuclear cells were separated on a Ficol-Hypaque gradient and used in a standard 4 hour ^{51}Cr-release NK cytotoxicity assay. The target cell was K-562 and the cells were used at effector: target ratios of 60:1, 20:1 and 7:1 in triplicate.

All assays were standardized for daily assay variation by the use of a set of three standards. The standards were cryopreserved cells obtained by leukophoresis of three normal donors. Aproximately 250 vials of each sample were cryopreserved and in each assay all three standards were assayed along with the patients' cells. Repeated testing of these same standards over several months enabled the calculation of a "set standard mean" ie. the mean cytotoxicity for all three standards together over the whole period of testing . The standardization of the patients NK cytotoxicity was done using the following formula:

$$\text{Standardized NK} = \frac{\text{Observed NK X daily standard mean}}{\text{Set Standard mean}}$$

The daily standard mean was the mean of the three standards on the day of the test.

C. Analysis of change in NK Activity

Earlier studies in our laboratory have shown that normal untreated individuals show considerable variation in NK cytotoxicity when tested repeatedly over a 28 day period. To be confident that any observed changes were due to the IFN being administered, this individual variation needed to be taken into account. In this study the NK cytotoxicity of the samples prior to the start of treatment were used to establish a range for each patient, which would define the limits of that patient's own variability. This range was determined as the mean ± 1.6 S.D. of the pre-treatment values. Any blood samples with cytotoxicity above or below that range were considered to show an increase or decrease in NK activity.

III Results

The results are presented in Table I for all patients together, whether they received IFN twice daily or three times weekly and for all doses of IFN. NK activity did not show any significant or sustained elevation at any dose of either protocol with no more than 20% of the values being increased at any time after the start of IFN. In fact there was a depression of NK activity in some patients (30% of the values). The remaining 50% showed no change. The values for 10 normal donors, tested at the same frequency and for the same length of time as the patients were quite different. Over 80% of the values showed no change, the remaining values were usually increased, only 8% of values were decreased and those points were all in the period 10-28 days after Day 0. There was no difference in results for patients on either the twice daily or three times weekly protocol.

TABLE 1. NK Activity: Change after Start of IFN Therapy.

Time after start of IFN	Percentage of samples showing					
	Increase		Decrease		No Change	
	Pts	Norm	Pts	Norm	Pts	Norm
Days 1-3	21	14	32	0	47	86
Days 4-7	16	20	31	0	53	80
Days 10-28	15	22	29	8	56	76

Whilst the changes were not clearly dose related (Table 2), there was a trend towards a greater and more sustained depression at the higher doses (data not shown). The pretreatment NK values of the patients were within the range observed in our laboratory for normal donors and thus the changes were not due to abnormally high or low pretreatment values.

Table 2. NK Activity: Change related to dose of IFN

Dose (x10 units)	Percentage of post-therapy samples showing		
	Increase	Decrease	No Change
1	20	37	43
3	11	28	60
9	22	8	70
18	5	34	61
36	21	29	50
50	23	0	68
68	20	16	64
86	24	28	48
100	9	26	65
118	0	53	47

A portion of the patients showed some clinical response to IFN (Sherwin et al., 1982) but there was no correlation with the NK activity or change in NK activity. All patients were tested for the ability of the recombinant IFN to boost their NK activity *in vitro* prior to the start of IFN treatment. The NK activity was augmented in almost all patients by the addition of 300 units of IFN to the cytotoxicity assay. There was no correlation, however, between this augmentation and the change in NK following *in vivo* administration of IFN.

IV. Discussion

The results of this study indicate that *in vivo* administration of IFN twice daily or three times weekly did not lead to an augmented NK response but instead to a diminished response in about 30% of the patients. There are a number of possible explanations for this unexpected result. The determination of a range of activity for each patient based on the pretreatment values allows a better discrimination between changes due to treatment and just normal disease-related or technical spontaneous variations in activity. For individuals with large spontaneous variation in activity, the change must be great to be observed and thus the majority of points showed no change. We feel, however, that this is a valid method of analysis and that erroneous conclusions may be drawn if only one pretreatment value is obtained. Another possible reason for the results observed, is that this particular clone of IFN is not able to augment NK activity when administered _in vivo_; it may require another IFN molecule or group of molecules. We have some preliminary evidence that inoculation of this IFN by the intra-muscular route, may induce circulating inhibitory factors which may interfere with the NK-boosting effects of IFN. In some other studies where an augmentation of NK activity has been observed, the IFN was given as a single injection and not repeatedly as in this study. Further studies are being undertaken to determine whether the lack of augmentation of NK observed here was due to the type of IFN, or the route or timing of IFN administration.

REFERENCES

Einhorn, S., Blomgren, H., and Strander, H. (1980). Int. J. Cancer 26:419.

Gutterman, J.U., Blumenschein, G.R., Alexanian, R., Yap, N.Y., Buzdar, A.U., Cabanillas, F., Hortobagyi, G.N., Hersh, E.M., Rasmussen, S.L., Harmon, N., Kramer, M., Petska, S. (1980). Ann. Intern. Med. 93:399.

Herberman, R.B., Ortaldo, J.R., and Bonnard, G.D. (1979). Nature 277:221.

Merigan, T.C., Sikora, K., Breeden, J.H., Levy, R., Rosenberg, S.A. (1978). N. Engl. J. Med. 299:1449.

Sherwin, S.A., Knost, J.A., Fein, S., Abrams, P., Foon, K., Ochs, J., Schoenberger, C., Maluish, A., and Oldham, R. N. Engl. J. Med., Submitted.

IN VIVO EFFECTS OF CORYNEBACTERIUM PARVUM ON NATURAL CELL-MEDIATED IMMUNITY IN ACUTE MYELOID LEUKEMIA PATIENTS[1]

Peter Hokland[2]
Jørgen Ellegaard

University Department of Medicine and Haematology
Aarhus Amtssygehus
Aarhus, Denmark

I. INTRODUCTION

Since it was discovered that Corynebacterium parvum (Cp) could activate the reticuloendothelial system (Halpern et al., 1964), several approaches have been used to investigate the immunological effects of Cp, especially in experimental oncological models, but lately also in humans (Thatcher et al., 1979; Hokland et al., 1980). It is now realized that treatment with such a complex (and structurally poorly defined) compound as Cp results in a number of nonspecific as well as specific effects on cell-mediated immunity. These different effects are furthermore greatly influenced by factors such as the dose of Cp given and its route of injection. Given these limitations, which also apply to the use of BCG and other bacterial immunostimulators, it is not surprising that the literature on Cp is often confusing. With respect to the effect of Cp on NK and ADCC in man, however, a fairly constant finding has been the enhancement of NK activity (Thatcher et al., 1979;Hokland et al., 1980), probably mediated by an <u>in vivo</u> interferon induction by Cp (Hokland et al., 1980). In this chapter, we will present some of our recent results on NK and ADCC activity in Cp-vaccinated AML patients in complete remission with special emphasis on the validity and the advantages of longitudinal testing in therapeutic situations.

[1]Supported by the Danish MRC and the Danish Cancer Society
[2]Present address: Division of Tumor Immunology, Sidney Farber Cancer Institute, 44 Binney Street, Boston, MA 02115 USA

ISBN 0-12-341360-5

II. THE RELEVANCE OF LONGITUDINAL NK AND ADCC MEASUREMENTS

In a previous study, we performed longitudinal measurements of NK and ADCC by testing the effector cells immediately against different suspensions of target cells. However, this approach is very time consuming. An alternative approach would be to use effector cells cryopreserved on each day of blood sampling and test these cells against the same target cell suspension. We decided to compare these protocols,and to this end, blood was drawn from 3 unimmunized volunteers on several occasions, and mononuclear cells were isolated by the Isopaque-Ficoll method. An aliquot of the cells was tested immediately for NK activity against *in vitro* cultured K562 and ADCC activity was determined against P815 cells, propagated in DBA mice, coated with a rabbit anti-P815 IgG antibody. The rest of the cells (usually around 2 x 10^7/sample) were frozen in liquid nitrogen and thawed when the series of investigations from the given donor had been completed. From Table 1 it will be seen that the cryopreserved cells gave somewhat lower but more consistent results, maybe because they were tested against the same target cell suspension and not, as was the case with the fresh cells, against different target cell preparations.

TABLE I. Fresh or Frozen NK and ADCC Effector Cells: A Comparison

		Day						
Donor	Cells	1	2	3	4	5	7	14
A	Fresh	33.5[1]	24.8	32.4	30.8	24.6		
		41.4[2]	42.7	48.5	34.0	41.7		
	Frozen	ND	24.5	23.5	23.5	24.0		
		23.5	27.4	24.6	19.4	24.9		
B	Fresh	27.7				31.0	21.0	23.5
		56.1				53.8	44.5	50.1
	Frozen	21.7				25.5	18.6	20.4
		45.7				44.6	43.7	46.1
C	Fresh	24.9	23.7	33.8	31.8	30.8		
		43.8	40.2	48.8	41.8	46.7		
	Frozen	20.9	23.9	22.8	25.9	25.7		
		33.9	29.5	33.7	29.6	33.4		

[1]NK against K562 target cells, E/T ratio, 25:1

[2]ADCC against P815 target cells coated with anti-P815 IgG antibodies (final dilution 1:2500), E/T ratio, 25:1

III. KINETICS OF NK AND ADCC IN AML PATIENTS

NK and ADCC were tested on several occasions during 4 week periods in 2 unvaccinated and 6 Cp-vaccinated (total doses of 1mg by intravenous and subcutaneous administration) AML patients in complete remission. As seen in Fig. 1, no consistent changes were observed in the unvaccinated patients while most of the vaccinated patients showed a considerable increase in NK with peaks of activity ranging from day 1 to 28. A somewhat more homogeneous NK pattern was seen in 4 donors,which were tested simultaneously against the Raji Burkitt lymphoma line, with NK peaks in 3 of the patients 24 hours after the Cp injections (Fig. 2).

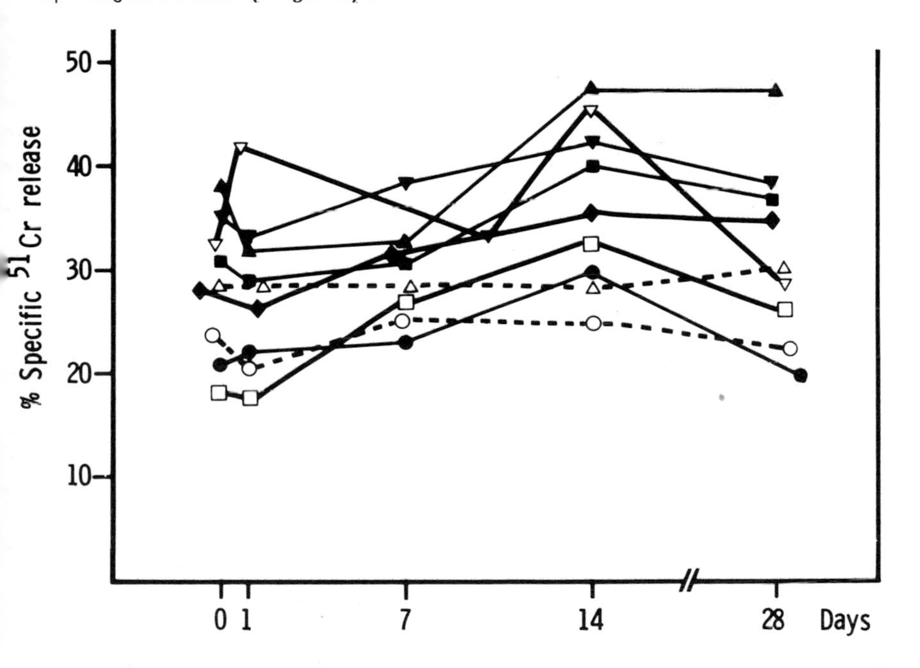

FIGURE 1. NK activities against K562 cells (E/T ratio, 50:1) in 4h cultures of frozen PBL in two unvaccinated (broken lines) and seven Cp-vaccinated AML patients.

The effect of different doses of Cp on NK was tested in one patient in whom the initial Cp dose of 2 mg had to be reduced because of severe local side effects,first to 1mg and then to 0.2mg. From Fig. 3 it is clear that while the two highest doses of Cp resulted in NK peaks at day 1 and 14, respectively, the lowest did not change NK at all.

ADCC against IgG-coated P815 cells was tested in 5 vaccinated and 1 unvaccinated patient. From Fig. 4 it appears that

Cp did not induce consistent changes, though 2 of the patients showed dramatic decreases in ADCC 2 weeks after the injections.

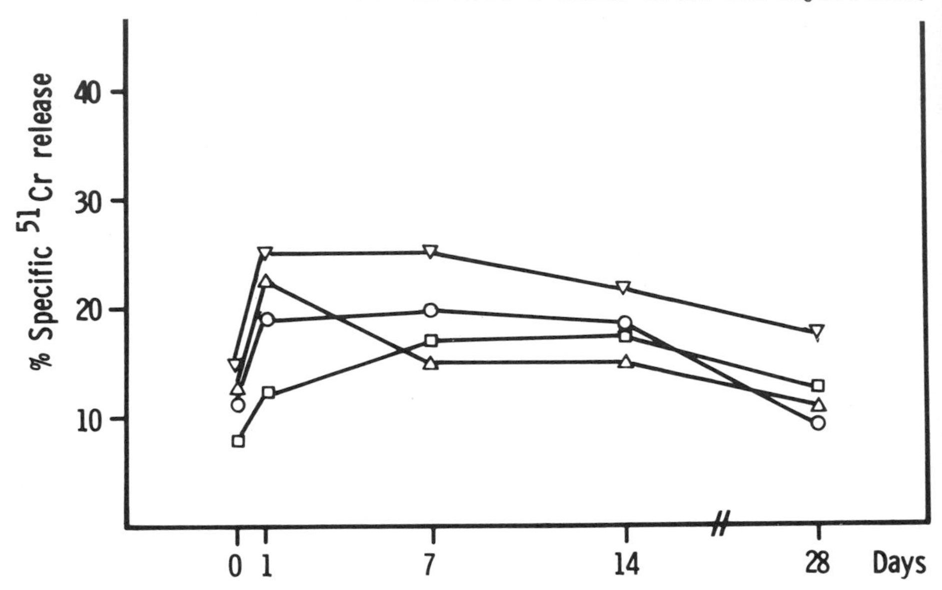

FIGURE 2. NK activities against Raji cells(E/T ratio, 50:1) in 4h cultures of frozen PBL from 4 vaccinated AML patients.

IV. CONCLUSIONS

In a previous study with normal volunteers, we observed a marked enhancement of NK *in vivo* by Cp (Hokland et al., 1980). The reason why the changes in the AML patients are not as clearcut (Fig. 1) may be that the inductive chemotherapy regimen given prior to the Cp injections had depressed the NK function and that this decrease varied from patient to patient. However, Cp did result in increased NK to the more resistant Raji line, and this might suggest that the chemotherapy regimen used could be selectively cytotoxic to the K562 inducible effector cells.

We have not investigated the nature of the NK cells induced by Cp in detail, but preliminary data (not shown here) suggest that they are to be found both in the T and non-T lymphocyte fractions. Whether these Cp-boostable NK cells function via interferon induction *in vivo* (Hokland et al., 1980), as was the case in normal donors, could not be verified, since interferon titrations of serum samples from the patients seldomly showed marked increases (data not shown). With the route of injection used here (i.c. and s.c.) this does not however, preclude a local stimulation of interferon production in the regional lymph nodes and subsequently, an increase in NK. Intravenous infusions of Cp could possibly have resulted in significant titers, but since this way of administration is also known to activate suppressor macrophages, we avoided it. Different routes of administration may probably also explain why we, in contrast to Thatcher et al. (1979), were unable to demonstrate consistent increases in ADCC.

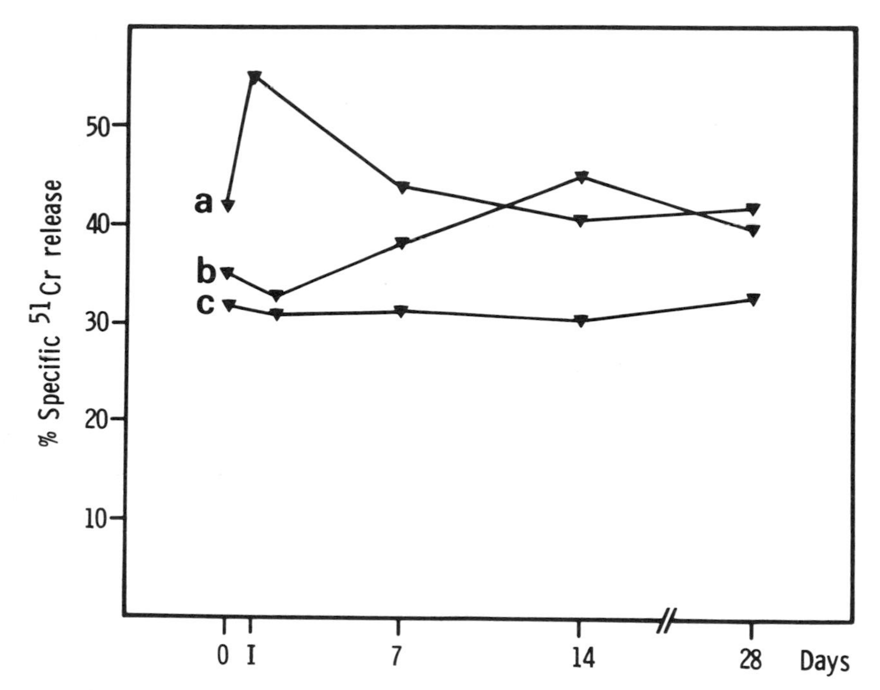

FIGURE 3. NK activity against K562 cells (E/T ratio 50:1) in an AML patient vaccinated with 3 different doses of Cp (a,2mg; b, 1mg; c, 0.1mg).

The implications of these findings can be summarized as follows:

a) The use of cryopreserved effector cells from different days in a vaccinations schedule, tested on the same target cell suspension, is preferable to testing NK on freshly isolated lymphocytes on each day of blood sampling, since day-to-day differences in target cell sensitivity are avoided.

b) An immunization schedule with i.c. and s.c. injections in doses of 1mg seems to result in consistently elevated levels of NK (Figs. 1, 3 and 4). Lower doses should be investigated for NK activation in each patient.

c) Corynebacterium parvum therapy can, its side effects not

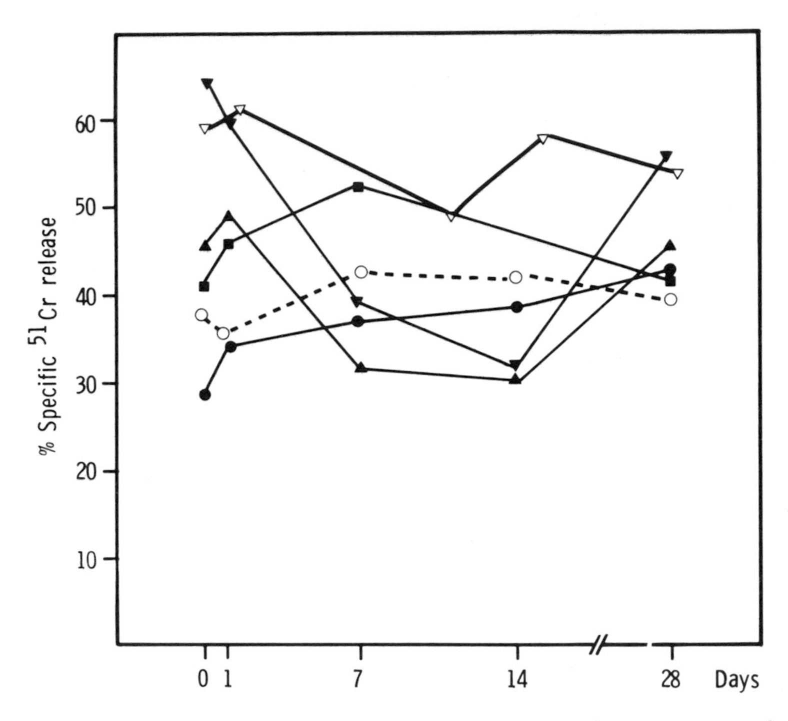

FIGURE 4. ADCC of PBL from 1 unvaccinated (broken line) and 5 Cp-vaccinated AML patients in 4h cultures against IgG-coated P815 cells (E/T ratio, 50:1). Specific lysis in cultures without anti-P815 antibodies never exceeded 10%.

withstanding, represent an easy way of interferon induction *in vivo*, and the subsequent enhancement of NK function is now well established. Our data on AML patients in complete remission do, however, indicate that for these effects to be clearly demonstrable, the immune system of the patients has to be relatively intact (Fig. 1).

REFERENCES

Halpern, B.N. (1975) Corynebacterium Parvum. Plenum Press, New York.

Hokland, P., Ellegaard, J. and Heron, I. (1980) J. Immunol. 124, 2180.

Thatcher, N., Swindell, R. and Crowther, D. (1979) Clin. Exp. Immunol. 35, 171.

MODIFICATION OF HUMAN NATURAL CELL-MEDIATED CYTOTOXICITY BY MVE-2[1]

James T. Forbes
Anne Luck
F. Anthony Greco

Division of Oncology
Department of Medicine
Vanderbilt University
Nashville, Tennessee

Biological response modifiers (BRM) have been defined as those agents which can influence the host response to tumor growth and dissemination. One such agent is a copolymer of maleic anhydride and divinyl ether (MVE-2) with an average molecular weight of 15,500 (Breslow et al., 1973). Bilogical response modification by MVE-2 has been demonstrated in animal tumor models. The object of this study was to determine the level of biological response modification in patients entered into a Phase I study of MVE-2.

A total of 21 evaluable patients with advanced refractory neoplastic disease were entered into this Phase I study of MVE-2. MVE-2 at the appropriate dose level was infused intravenously over two hours at weekly intervals for a total of six weeks. Three patients were entered at 150 mg/m^2, four at 300 mg/m^2, four at 350 mg/m^2, one at 400 mg/m^2, four at 450 mg/m^2, and five at 600 mg/m^2. Patients were monitored for biological response modification by measuring 29°C E rosettes, 4°C E rosettes, PHA induced blast transformation, PWM induced blast transformation, natural cytotoxicty to K-562, plasma interferon levels, and plasma lysozyme levels in the peripheral blood. All tests on cells were performed on Ficoll-Hypaque separated cells using procedures previously reported (Dean _et al_., 1977; Djeu _et al_., 1977; Forbes _et al_., 1981a; Forbes _et al_., 1981b). All assays were performed twice prior to administration of MVE-2 and weekly thereafter for as long as the patient remained in the study. All values except interferon levels and lysozyme levels were determined in conjunction with cryopreserved normal cells from a single leukophoresis run on a single donor. If values for any test

[1]Supported by NIH grants CA 23477, CA 27333, and NOI CM 07438.

ISBN 0-12-341360-5

failed to fall within the usual range for that cryopreserved sample then all associated test values determined on that day were considered invalidated. For convenience the data are expressed as an index of the cryopreserved cell control equal to the value for the test cells divided by the value for the cryopreserved cell standard. This allows comparison of data from the same patient determined at different times and takes into account some of the day to day experimental variation inherent in these biological assays. The reproducibility of these cryopreserved lymphocytes and their use in the standardization of *in vitro* assays has been well documented (Oldham *et al.*, 1976; Ortaldo *et al.*, 1976; Forbes *et al.*, 1981b). The data derived from these studies are presented as follows: The average pretreatment index value is compared with an average of the post treatment index values to determine change during the time of treatment. A change is considered significant if the average values are changed and not if only single post treatment values are changed.

Table I demonstrates the average effect of MVE-2 administration on 4°C E rosettes in refractory cancer patients. Changes in the levels of 4°C rosettes were only noted at the higher dose levels. The levels of 29°E rosettes were not significantly altered.

TABLE I. 4°C E Rosettes

Dose Level	Average Total Dose	# Increased Post Rx Treated	Average Change*
150 mg/m^2	1517 mg	1/3	-0.01
300	2633	1/4	-0.14
350	2754	2/3	-0.12
450	4668	2/4	0.15
600	3691	5/5	0.26

*An Index 1.00 = 65% rosette forming cells.

No significant changes in the levels of response to the T-cell mitogen phytohemagglutinin (PHA) were noted during MVE-2 administration. However, as illustrated in Table II nine of eighteen patients tested showed increased responses to pokeweed mitogen (PWM).

TABLE II. PWM Transformation

Dose Level	Average Total Dose	#Increased Post Rx / # Treated	Average Change*
150 mg/m^2	1517 mg	0/3	-0.26
300	2633	2/4	0.11
350	2754	0/3	-1.31
450	4668	4/4	0.64
600	3691	3/4	0.40

*An Index of 1.00 = 37,000 cpm [^{3}H]-thymidine incorporation

When natural cell mediated immunity (NCMC) to the cell line K-562 was tested in a four hour cytotoxicity assay at an effector:target = 100:1 the data in Table III were derived. There is a post treatment increase in the number of patients with increased NCMC and the level of increase in NCMC appears to increase proportionately with the level of MVE-2 administered. An example of the serial NK values in one of the patients so affected is shown in figure 1.

TABLE III. NCMC to K-562

Dose Level	Average Total Dose	#Increased/#Treated Post Rx	Average* Change
150 mg/m^2	1517 mg	1/3	-0.08
300	2633	2/4	0.35
350	2754	3/4	0.71
450	4668	4/4	0.62
600	3691	4/4	2.19

*An index of 1.00 = 16% cytotoxicity

The results of assays of interferon levels in the plasma from these patients are shown in Table IV. These data reflect the ability of plasma to protect human foreskin fibroblast cultures from infection with vesicular stomatitis virus and were compared to a standard interferon sample assayed concurrently. The values are in units per milliliter. It should be noted that many of these patients had elevated levels (> 100 IU) of interferon before being treated with interferon.

Plasma samples from normal individuals from the general population and from laboratory workers contained typically less than 100 IU of interferon in this assay. At least one-half of the patients tested showed some elevation of plasma interferon levels during the course of this study. Some patients experienced only brief periods of elevated interferon levels while others have had higher average values following MVE-2 administration than before the drug was given.

TABLE IV. Interferon Levels

Dose Level	Average Total Dose	#Increased / #Treated Post Rx	Average Change*
150 mg/m^2	1517	2/3	139
300	2633	3/4	-8
350	2754	2/4	-34
450	4668	2/4	330
600	3691	2/5	10

*International units

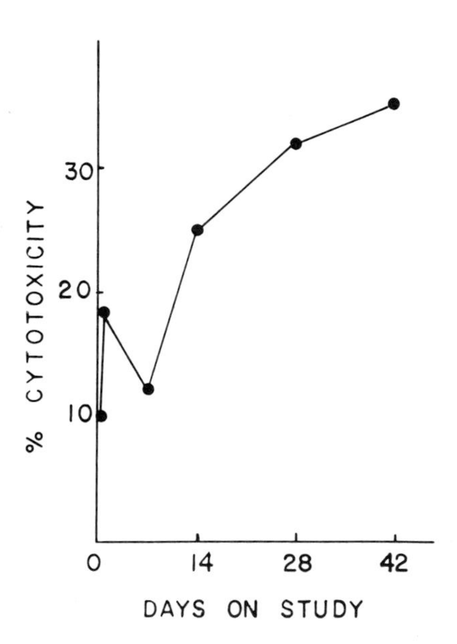

Figure 1. % NCMC of patient treated on days 0,7,14,21,28, and 35 with 450 mg/m^2 MVE-2. The value plotted for day 0 is an average of the two pretreatment values.

Circulating lysozyme levels were measured in heparinized plasma by the turbidometric assay using micrococsis lysodeikticus lysozyme levels correlate with monocyte activity in several diseases and are indicative of active inflammatory responses. None of the levels have been significantly affected in association with MVE-2 administration.

These data suggest that changes in the levels of E rosettes, PWM transformation, NCMC to K-562 and interferon levels are associated with administration of MVE-2. The association of MVE-2 effects on NCMC and plasma interferon levels are most provocative as interferon in known to augment NCMC (Giresser et al., 1976; Djeu et al., 1977a; Djeu et al., 1977b; Herberman et al., 1979). However, close examination of these two parameters on a case by case basis reveals that while NCMC is generally higher in those instances where interferon levels are high increases in circulating interferon levels are not always associated with increased NCMC.

Studies on these same patients also demonstrated that the maximum tolerated dose was a cumulative dose of 2800 mg of MVE-2. Cumulative doses higher than this lead to substantial proteinuria and a reversible nephrotic syndrome. The majority of biological response modification was seen at doses exceeding this 2800 mg level and the degree of modification was not striking suggesting that MVE-2 used at these levels and with this administration schedule does not hold great promise as a biological response modifier.

REFERENCES

Breslow, B.S., Edwards, E.K., and Newberg, N.R. Nature 246:160 (1973)

Dean, J.H., Connoi, RD., Herberman, R.B., Silva, J., McCoy, J.L., and Oldham, R.K. Int. J. Cancer 20:359 (1977)

Djeu, J., Payne, S., Alford, C., Heim, W., Pomeroy, T., Cohen, M., Oldham, R.K., and Herberman, R.B. Clin. Immunol. Immunopath. 8:405 (1977)

Djeu, J.Y., Heinbaugh, J.A., Holden, H.T., and Herberman, R.B. J. Immunol. 122:175 (1979a)

Djeu, J.Y., Heinbaugh, J.A., Holden, H.T., and Herberman, R.B. J. Immunol. 122:182 (1979b)

Forbes, J.T., Greco, F.A., and Oldham, R.K. Cancer Immunol. Immunother. 11:147 (1981a)

Forbes, J.T., Niblack, G.D., Fuchs, R., Richie, R.E., Johnson, H.K., and Oldham, R.K. Cancer Immunol. Immunother. 11:139 (1981)

Gresser, I., Tovey, M.G. Bendu, M.T., Maury, G., and Brontz-Boye, D. J. Exp. Med. 114:1305 (1976)

Herberman, R.B., Ortaldo, J.T., and Bonanrd, G.D. Nature 227: 221 (1979)
Oldham, R.K., Dean, J., Cannon, G.B., Graw, R., Dunston, G., Applebaum, F., McCoy, J., and Herberman, R.B. Int. J. Cancer 18:145 (1976)
Ortaldo, J.R., Oldham, R.K., Holden, H.T., and Herberman, R.B. Cell. Immunol. 25:60 (1976)

INTERACTION BETWEEN INTERFERON AND NATURAL KILLER CELLS IN HUMANS AFTER ADMINISTRATION OF IMMUNOMODULATING AGENTS.

Tadao Aoki
Hideo Miyakoshi
Yoh Horikawa

Research Division, Shinrakuen Hospital
Nishiariakecho 1-27, Niigata 950-21, Japan

Akira Shibata
Yoshitaka Aoyagi

1st Department of Internal Medicine
Niigata University School of Medicine
Asahimachi-1, Niigata 951, Japan

Mikio Mizukoshi

Immunology Division, Research Center
Fujizoki Pharmaceutical Co., Ltd.
Komiyacho 51, Hachiohji 192, Japan

Although there are several reports on the relationship between kinetics of interferon (IFN) production and natural killer (NK) activity after administration of immunomodulating agents (Djeu, Heinbaugh, Holden and Herberman, 1979; Örn, Gidlund, Ojo, Grönvik, Andersson, Wigzell, Murgita, Senik and Gresser, 1980), no descriptions have yet appeared about differences in diverse immunomodulating agents and/or in types of IFN.

Recently, we have extensively studied the acting mechanisms of Lentinan, a glucan extracted from Lentinus edodes, and Staphage Lysate (SPL), mitogenic components of Staphylococcus aureus lysed by a polyvalent staphylococcus bacteriophage (Aoki, Urushizaki, Miyakoshi and Ishitani, 1980; Aoki, Miyakoshi, Horikawa and Usuda, 1981; Miyakoshi, Aoki and Mizukoshi, 1981;

ISBN 0-12-341360-5

Miyakoshi, Aoki and Usuda, 1981). Both of these immunomodulating agents are IFN-inducers as well. Since some interesting findings concerning the subject of this chapter were noticed during these investigations, we like to describe these results and to propose some new problems about the influence of different types of IFN on NK activity in the human system. In other words, NK activity was readily enhanced, following an increase in the IFN titers by Lentinan administration, whereas no significant activation of NK cells was observed even when IFN titers were elevated by SPL administration.

For NK activity assay, an established human myeloid cell line K562 was used as target cells to measure the ^{51}Cr-release from these cells after incubation at 37°C for 6 hrs at the mixture ratio of target cells and attacker cells of 1:20 and/or 1:40. The attacker cells were collected either from peripheral blood or from regional tissues, including malignant tissues, by the density gradient Ficoll-Conray method. Everytime, the attacker cells were smeared on slide glass and stained with Giemsa to examine their morphological features.

International units of IFN in the body fluid were determined by measuring its inhibitory activity against cytopathic effects of vesicular stomatitis virus (VSV) using a human cell line FL. The standard purified α and β IFNs were obtained from the Research Department, Laboratory of Chemistry, Japanese Red Cross Central Blood Center (Tokyo) and the Toray Industries, Inc. (Tokyo), respectively.

By sequential measurements of NK activity and IFN titers after Lentinan administration, the peaks of IFN appeared at 12 to 24 hrs and NK activity gradually increased in inverse proportion to a decrease in IFN titers, peaking at 24 to 48 hrs (Fig. 1). The time difference between the peaks of IFN titers and NK activity was almost always 12 to 24 hrs. Based on the findings of previous studies that IFN and IFN-inducers cause a rapid augmentation of the cytotoxic activity of NK cells both _in vivo_ and _in vitro_ (Djeu, Heinbaugh, Holden and Herberman, 1979; Herberman, Ortaldo and Bonnard, 1979; Heron, Hokland, Möller-Larsen and Berg, 1979; Minato, Reid, Cantor, Lengyel and Bloom, 1980; Örn, Gidlund, Ojo, Grönvik, Andersson, Wigzell, Murgita, Senik and Gresser, 1980), the elevation of NK activity in the present investigation could be induced by IFN production due to Lentinan administration. Because of heat-labile and acid-labile characteristics, this human IFN may be IFN-γ. To interpret a dissociation between the peaks of IFN titers and NK activity, it would be one possible explanation that IFN-activated NK cells gathered and infiltrated into the malignant tissues just after the IFN production by Lentinan administration, resulting in no detection of effective NK cells in the peripheral blood.

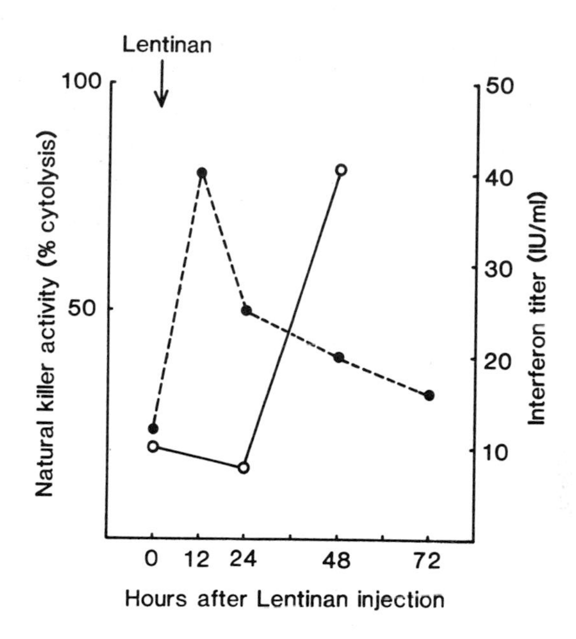

Fig. 1. Alterations in natural killer activity (o——o) of peripheral blood lymphocytes and interferon titers (•- - - -•) in plasma from a patient with gastric cancer after Lentinan administration.

Afterwards, when the number of IFN-activated NK cells reached a surplus, these NK cells appeared in the peripheral blood. Thus, the peak of NK activity may indicate a phenomenon just in the peripheral blood circulation. The finding e.g. that the mononuclear cells from the solid malignant tissues in some cases, metastatic paraganglioma, (Fig. 2), were morphologically quite similar to NK cells, called large granular lymphocytes (LGL) (Timonen, Ortaldo and Herberman, 1981), substantiates this possibility. It is of great interest, however, that these cells from regional foci showed cytotoxic activity against fresh malignant cells of this patient but no NK activity to K562 cells. The same findings were obtained by Dr. Uchida's group (personal communications). This fact may be attributed to the presence of at least two different subpopulations of LGL, being cells with the regionally gathered LGL with no NK activity to K562 cells, or with these LGL losing NK activity to K562 cells after contact with malignant cells (Perussia and Trinchieri, 1981).

Although SPL administration induced IFN production, this treatment did not augment NK activity of mononuclear cells in the peripheral blood (Fig. 3). IFN or IFN-like substance produced

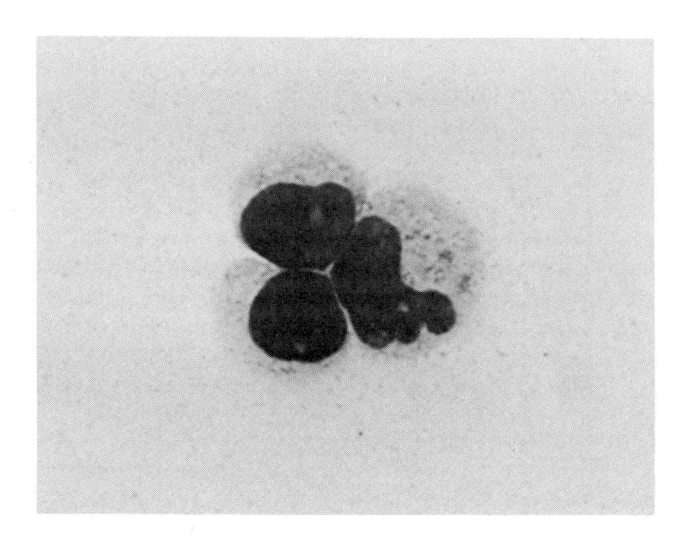

Fig. 2. Large granular lymphocytes (LGL) isolated from the malignant tissues of a patient with metastatic paraganglioma. Giemsa's stain. Oil immersion microscopy (x 1,500).

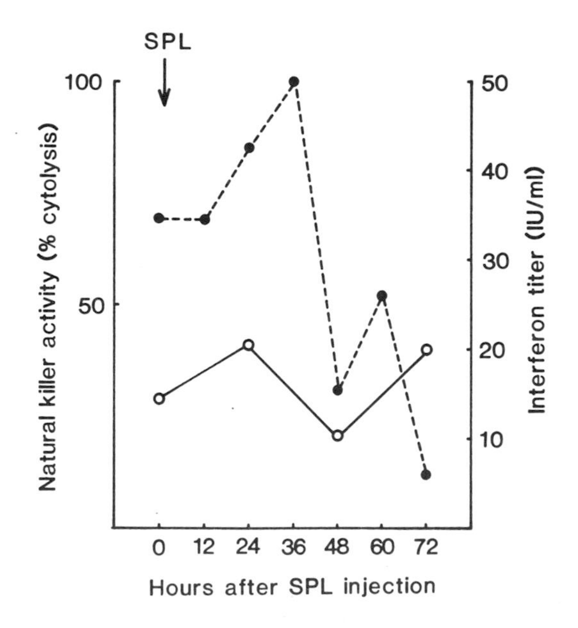

Fig. 3. Alterations in natural killer activity (o——o) of peripheral blood lymphocytes and interferon titers (•- - - -•) in plasma from a patient with multiple sclerosis after SPL injection.

both in vitro and in vivo with SPL may be human IFN-γ(type II), because of its physicochemical characteristics — heat-labile and acid-labile. Now, this finding proposes a new problem in terms of the relationship between the type of IFN or IFN-like substance and NK activation. All known types of human IFN can so far activate NK cells (Djeu, Heinbaugh, Holden and Herberman, 1979; Herberman, Ortaldo and Bonnard, 1979; Heron, Hokland, Möller-Larsen and Berg, 1979; Minato, Reid, Cantor, Lengyel and Bloom, 1980; Örn, Gidlund, Ojo, Grönvik, Andersson, Wigzell, Murgita, Senik and Gresser, 1980) in contrast to the in vivo inability of a new type of IFN or IFN-like substance described here. It is also possible, however, that SPL may stimulate suppressor cells to deppress NK cells. Though this possibility seems unlikely from our previous studies, any conclusions cannot be defined without absolutely ruling out this possibility. Nevertheless, this new type of IFN clearly inhibited the VSV replication. Simultaneously, SPL administration induced significant regression of warts (verruca plana) (Aoki, Miyakoshi, Horikawa and Usuda, 1981), suggesting also the effect of this IFN on wart virus. Moreover, if multiple sclerosis (MS) is caused by viral infection as strongly suggested by epidemiological studies (Kurtzke, 1976; Kurtzke and Hyllested, 1979), SPL may affect MS patients through IFN production. In this context, SPL was administered to 10 MS patients, resulting in the following clinical improvements (Aoki, Miyakoshi, Horikawa and Usuda, 1981): (a) the frequency of exacerbations decreased to about 1/3, and (b) the total amount of prednisolone administered was able to be reduced to 1/3. Since IFN titers were heightened in most of these improved patients, SPL appears effective on MS through the IFN production, supporting the possible viral infection theory of MS. So, it seems most likely that this IFN or IFN-like substance produced by SPL affects virus replication but does not activate NK cells. Because of the existence of IFN-independent pathways for NK activation in the murine system (Brunda, Herberman and Holden, 1980), the search for possible alternative pathways for NK activation also would be a very important project in the clinical study.

REFERENCES

Aoki, T., Miyakoshi, H., Horikawa, Y., and Usuda, Y., in "Augmenting Agents in Cancer Therapy" (E.M. Hersh, M.A. Chirigos, and M.J. Mastrangelo, eds.), p. 101. Raven Press, New York, 1981.

Aoki, T., Urushizaki, I., Miyakoshi, H., and Ishitani, K., in "Cancer Immunology and Parasite Immunology" (L. Israël, P. Lagrange, and J.C. Salomon, eds.), p. 85. Les Colloques de l'INSERM, Paris, 1980.

Brunda, M.J., Herberman, R.B., and Holden, H.T., in "Natural

Cell-Mediated Immunity against Tumors" (R.B. Herberman, ed.), p. 525. Academic Press, New York, 1980.
Djeu, J.Y., Heinbaugh, J.A., Holden, H.T., and Herberman, R.B., J. Immunol. 122:175 (1979).
Herberman, R.B., Ortaldo, J.R., and Bonnard, G.D., Nature 277:221 (1979).
Heron, I., Hokland, M., Möller-Larsen, A., and Berg, K., Cell. Immunol. 42:183 (1979).
Kurtzke, J.F., in "International Symposium on the Aetiology and Pathogenesis of the Demyelinating Deseases" (H. Shiraki, T. Yonezawa, and Y. Kuroiwa, eds.), p. 59. Japan Science Press, Tokyo, 1976.
Kurtzke, J.F., and Hyllested, K., Ann, Neurol. 5:6 (1979).
Minato, N., Reid, L., Cantor, H., Lengyel, P., and Bloom, B.R., J. Exp. Med. 152:124 (1980)
Miyakoshi, H., Aoki, T., and Mizukoshi, M., Biomed. Res. 2:629 (1981).
Miyakoshi, H., Aoki, T., and Usuda, Y., in "Manipulation of Host Defence Mechanisms" (T. Aoki, I. Urushizaki, and E. Tsubura, eds.), p. 118. Excerpta Medica, Amsterdam, 1981.
Örn, A., Gidlund, M., Ojo, E., Grönvik, K.O., Andersson, J., Wigzell, H., Murgita, R.A., Senik, A., and Gresser, I., in "Natural Cell-Mediated Immunity against Tumors" (R.B. Herberman, ed.), p. 581. Academic Press, New York, 1980.
Perussia, B., and Trinchieri, G., J. Immunol. 126:754 (1981).
Timonen, T., Ortaldo, J.R., and Herberman, R.B., J. Exp. Med. 153:569 (1981).

AUGMENTATION OF NK CELL ACTIVITY IN CANCER PATIENTS BY OK432: ACTIVATION OF NK CELLS AND REDUCTION OF SUPPRESSOR CELLS

Atsushi Uchida
Michael Micksche

Institute for Cancer Research
University of Vienna

There is increasing evidence that natural killer (NK) cells play an important role in host resistance against tumors (Herberman and Holden, 1978). The activity of NK cells seems to be regulated in both positive and negative ways, and it has been shown that interferon plays a major role in the augmentation of NK activity (Gidlund et al., 1978), whereas suppressor cells can suppress NK activity (Cudkowicz and Hochman, 1979; Uchida and Micksche, 1981a). The enhancement of NK activity in vivo could result in a subsequent benefit to the host. OK432, a heat- and penicillin-treated lyophilized powder of the low virulent Su strain of streptococcus pyogenes A3, has received much attention as a useful immunotherapeutic agent in cancer patients (Uchida and Hoshino, 1980a,b). In rodents, OK432 are found not only to inhibit directly tumor growth (Okamoto et al., 1967) but also to stimulate host immunity by activating macrophages (Ishii et al., 1976), inducing immune interferon (Matsubara et al., 1979), and activating NK cells (Oshimi et al., 1980). In humans, OK432 immunotherapy has caused an enhancement of lymphoproliferative response to T cell mitogens along with a reduction of mitogenic suppressor cells (Uchida and Hoshino, 1980b). We report here that treatment of cancer patients with OK432 results in the augmentation of NK activity as well as the reduction of NK suppressor activity.

For a clinical application of OK432, it seems important to know which route of administration of OK432 is most effective for boosting of NK activity in cancer patients. Patients with advanced malignant melanoma were each treated with an intravenous, intradermal, or intramuscular single injection of OK432 (1 KE = 0.1 mg dried streptococci) on Day 0. Control patients

NK CELLS AND OTHER NATURAL EFFECTOR CELLS

ISBN 0-12-341360-5

TABLE I. Effects of Systemic Administration of OK432 on Blood NK Activity

OK432*	Cytotoxicity					
	LU/10^7	% of normal control				
	Day 0	0	1	2	3	7
None	31	35	27	31	38	24
Intravenous	14	16	67	127	179	38
Intradermal	12	13	45	87	119	21
Intramuscular	28	31	15	38	45	34

* Patients received 1 KE of OK 432 on Day 0.

were not given any anticancer agents. NK activity was determined in a 4-hour chromium-51 release assay using K562 target cells, and the activity was expressed in lytic units (LU) per 10^7 cells, 1 LU being defined as the number required to produce 33% specific cytotoxicity (Uchida and Micksche, 1982). To reduce the daily variance of the assays, results were also expressed as the percentage of normal donors tested at the same time. During the observation period of 8 days, NK activity of untreated patients was relatively stable (Table I). A single injection of OK432 resulted in a rapid increase in NK activity of blood lymphocytes. Significant augemntation of NK activity was detectable with i.v. and i.d. routes on Day 1, peaked on Day 3, and then returned to pretreatment levels on Day 7. In contrast, an i.m. injection od OK432 gave no enhancement of NK activity. These results indicate that i.v. and i.d. administration of OK432 rapidly increase blood NK activity but this augmentation disappears within 7 days. As recent evidence has indicated that human NK activity is exerted by large granular lymphocytes (Timonen and Saksela, 1980), the number of large granular lymphocytes was determined on Day 0 and 3. There was no difference in the frequency of large granular lymphocytes in blood lymphocytes on Day 0 and Day 3. These results suggest that the augmentation of NK activity in OK432-treated patients is unlikely to be due to an increase in number of NK cells but may rather result from the enhancement of lytic function of NK cells. OK432 has been reported to induce immune interferon in mice (Matsubara et al., 1979). However, no interferon was detected in the peripheral blood of cancer patients treated with OK432, suggesting that the enhancement of NK activity by OK432 is not mediated through interferon.

As lymphocytes with the most potent antitumor activity may

TABLE II. Augmentation of Effusion NK Activity by Intrapleural Administration of OK432

OK432 treatment*	Cytotoxicity (LU/10^7)	
	Day 0	Day 7
None	1	1
Non-responder**	1	1
Responder***	1	183

* I.pl. injection of 10 KE of OK432 on Day 0.
** No therapeutic benefit from OK 432.
*** Reduction of effusion tumor cells.

be expected to be present within or around the site of tumor growth, it seems of practical importance to know whether OK432 can augment NK activity in carcinomatous pleural effusions and reduce the number of tumor cells in effusions. Twelve patients with carcinomatous pleural effusions were treated with intrapleural injections of OK432 (10 KE per week). Two patients showed no clinical evidence of the therapeutic effects of OK-432. Three patients had a complete disappearance of pleural effusions within 7 days. The other seven patients showed a marked reduction of pleural effusions along with a decrease or disappearance of tumor cells in their effusions as early as 7 days after the first treatment. To know the mechanism responsible for the antitumor effects of OK432, mononuclear cells were isolated from pleural effusions by discontinuous density gradient centrifugation. Effusion lymphocytes from untreated patients had markedly impaired or no NK activity (Table II), as previously reported (Uchida and Micksche, 1981a,b). Seven days after the first injection of OK432, two patients who showed no clinical improvement still had no NK activity in their effusions. The other seven patients showed a marked increase in NK activity of effusion lymphocytes on Day 7, which was even higher than that of blood lymphocytes of normal controls. The other three patients were not monitored for effusion NK activity because of the complete disappearance of effusions. In contrast, effusion NK activity of untreated control patients remained negative. It seems unlikely that this augmentation of effusion NK activity is due to an increase in the number of NK cells in effusions, since the frequency of large granular lymphocytes was not increased by the intrapleural administration of OK432. On the other hand, NK activity of blood lymphocytes was not changed in a consistent pattern by the therapy. These results indicate that the intrapleural administration of OK432 may result in significant augmentation of effusion NK activity

TABLE III. Reduction of NK Suppressor Cells by Intrapleural Injection of OK432

Adherent cells added*	Cytotoxicity (LU)		% Suppression
None	93	117	-
Untreated	37	-	60
OK432-treated	-	131	-11

* Blood lymphocytes were cultured for 24 h alone or with 10 % number of adherent effusion cells from patients before and 7 days after the first i.pl. injection of OK432.

along with a reduction or disappearance of pleural effusions and effusion tumor cells. It should be noted that the enhanced effusion NK activity is well correlated with the reduction of effusion tumor cells.

As described in another chapter and in a previous report (Uchida and Micksche, 1981a), malignant pleural effusions of cancer patients contained adherent cells capable of inhibiting the maintenance and development of NK activity, which could be one cause of the lack of NK activity in the effusions. To know the effects of intrapleural administration of OK432 on adherent effusion NK suppressor cells, adherent effusion cells that suppressed NK activity of normal lymphocytes before therapy were again tested for suppressor activity 7 days after the initiation of OK432 therapy. Adherent effusion cells from OK-432-treated patients were no longer able to suppress NK activity of blood lymphocytes (Table III). The recovery of adherent cells from pleural effusions was higher in OK432-treated patients than in untreated patients. These results indicate that adherent effusion cells lose their suppressor activity during the intrapleural OK432 therapy.

To confirm the in vivo effects of OK432 on suppressor cell activity, adherent effusion cells from untreated patients were incubated for 24 hours alone or with OK432 (0.5 KE/ml), washed and further cultured with normal lymphocytes for 24 hours. OK-432-treated adherent effusion cells were no longer able to inhibit NK activity, while 24-hour cultured adherent cells maintained the same level of suppressor function as fresh adherent cells (Table IV). The recovery of viable cells cultured for 24 hours with OK432 was comparable to that cultured alone, suggesting that OK432 is not toxic to adherent cells. These results indicate that OK432 is able to inhibit or abolish the suppressive activity of adherent effusion cells.

TABLE IV. In Vitro Reduction of Suppressor Activity of Adherent Effusion Cells by OK432

Adherent cells added*	Cytotoxicity (LU)	% Suppression
None	81	-
Fresh	27	67
24-h cultured	31	62
OK432-treated**	92	-14

* Blood lymphocytes were cultured alone or with 10% number of adherent effusion cells for 24 hours.
** 24-h treatment with OK432 (0.5 KE/ml).

OK432 is known to have a direct cytostatic or cytotoxic effect on certain tumor cells (Okamoto et al., 1967). It might be possible that the reduction or disappearance of tumor cells in pleural effusions of patients treated with intrapleural administration of OK432 is due to the direct cytotoxic effects of OK432 on effusion tumor cells. However, the dose of OK432 used in the present study (10 KE per effusion, less than 0.01 KE/ml) was not cytotoxic nor cytostatic against effusion tumor cells in vitro.

OK432 has been demonstrated to induce immune interferon in mice in vivo (Matsubara et al., 1979). For the following reasons, however, it seems unlikely that interferon induced by OK432 was responsible for the augmentation of NK activity in OK432-injected effusions: Interferon did not increase NK cell activity of effusion lymphocytes in vitro (another chapter). Supernatants produced by OK432-stimulated lymphocyte cultures did not contain any detectable amounts of interferon (Uchida and Micksche, 1981c). Cell-free supernatants of pleural effusions of OK432-treated pateints failed to enhance NK activity. It might be possible that OK432 blocks suppressive activity of adherent effusion cells and de-suppressed NK cells are activated by interferon produced by OK432-activated macrophages or mesothelial cells.

In conclusion, systemic and local administration of OK432 can augment NK activity in cancer patients independently of interferon induction. Blood NK activity was augmented by an intravenous and intradermal single injection of OK432, but not by an intramuscular route, indicating that the route of administration is an important variable. The long term effects of repeated injections of OK432 on blood NK activity are under investigation. On the other hand, the enhanced NK activity in pleural effusions was maintained by repeated intrapleural injections of OK432 at weekly intervals. This could result from both direct activation of NK cells and inactivation of adherent suppressor cells. Although the mechanism by which OK432 abrogate suppressor function of adherent effusion cells is not understood yet, intrapleural OK432 treatment will result in a subsequent benefit to cancer patients, since interferon cannot increase NK activity of effusion lymphocytes. Indeed, OK432-augmented effusion NK activity was well correlated to the reduction or disappearance of tumor cells in pleural effusions. We have also observed that effusion lymphocytes from cancer patients treated with intrapleural administration of OK432 can show cytotoxicity against autologous effusion tumor cells. Although the characteristics of these effector cells are not yet determined, our results might support a concept that NK cells may interact with autologous tumor cells in vivo.

REFERENCES

Cudkowicz, G., and Hochman, P. S. (1979). Immunol. Rev. 44:13.
Gidlund, M., Örn, A., Wigzell, H., Senik, A., and Gresser, I. (1978). Nature 273:759.
Herberman, R. B., and Holden, H. T. (1978). Adv. Cancer Res. 27:305.
Ishii, Y., Yamaoka, H., Toh, K., and Kiuchi, K. (1976). Gann 67:115.
Matsubara, S., Suzuki, F., and Ishida, H. (1979). Cancer Immunol. Immunother. 6:41.
Okamoto, H., Shoin, S., Koshimura, S., and Shimizu, R. (1967). Jpn. J. Microbiol. 11:323.
Oshimi, K., Kano, S., Takaku, R., and Okumura, K. (1980). J. Natl. Cancer Inst. 65:1265.
Timonen, T., and Saksela, E. (1980). J. Immunol. Methods 36: 285.
Uchida, A., and Hoshino, T. (1980a). Cancer 45:476.
Uchida, A., and Hoshino, T. (1980b). Int. J. Cancer 26:401.
Uchida, A., and Micksche, M. (1982). Immunobiol. (in press).
Uchida, A., and Micksche, M. (1981a). Cancer Immunol. Immunother. 11:255.
Uchida, A., and Micksche, M. (1981b). Cancer Immunol. Immunother. 11:131.
Uchida, A., and Micksche, M. (1981c). Int. J. Immunopharmac. 3:365.

IN VIVO EFFECTS OF BIOLOGICAL RESPONSE MODIFIERS AND CHEMOTHERAPEUTIC AGENTS ON NK ACTIVITY IN CANCER PATIENTS[1]

Didier Cupissol
François Favier
Augustin Rey
Bernard Longhi
Carine Favier
Bernard SERROU

Department of Chemo-Immunotherapy
and
Laboratoire d'Immunopharmacologie des Tumeurs
INSERM U-236 and ERA-CNRS n° 844
Centre Paul Lamarque
Hôpital St-Eloi
Montpellier, France

Cancer immunotherapy remains a controversal approach to this disease (12,13,14,15,16). At least, in man, this arises the fact that tumor antigens are still poorly on completely undefined, and lymphocyte subpopulations as well as the mechanisms governing immunological equilibrium are poorly understood, in spite of recent advances to better characterize lymphocyte subpopulations. In large part, this explains the difficulty of interpreting results following immunostimulation or active immunotherapy (2). A fuller knowledge of lymphocyte subpopulations, along with an ever increasing number of immunomodulating drugs acting on known functions or subpopulations, should lead to an improvement in results. The biological significance of NK cells remains obscure. Nevertheless, they are becoming better characterized and we now know they may play a major role in regulation of the immune response (10). In vivo, there is a clear increase in the survival of mice injected with NK-sensitive tumor cells. Although NK cell activity would seem to be

1. Work supported by funds from INSERM and Ligue Nationale Française contre le Cancer.

ISBN 0-12-341360-5

able to be used as a screening test in cancer patients, there are no reports describing its *in vitro* use to predict *in vivo* results.

Based on these facts, we investigated the possibility that NK cell activity may be of value as an effective screening test for new drugs, as well as for a precise evaluation of the immune status of the cancer patient. We present here the results of new drugs demonstrating remarkable immunorestorative capability and NK activity modulation in the immunodepressed cancer patient.

MATERIAL AND METHODS

1. Patients

In vivo testing was evaluated in immunodeficient, advanced solid tumor patients (25,26). Blood samples were drawn on the morning of the first day and on the day following cessation of treatment. Patients were tested for NK activity.

2. NK cell assay

NK activity was measured by a chromium 51 assay. K 562 target cells were marked with chromium 51 and placed in the wells of a linbro round-bottom microtest plate (Ref. 76-013-05 Flow) with different concentrations of human lymphocytes which have been separated from peripheral blood by the Boyum technique using Ficol-Metrizoate (Pharmacia) at 1.077 g/cm3. The cells were incubated for 4 hours at 37°C in a humid CO_2 incubator (6-26-27).

3. Drugs tested

a Human Fibroblastic Interferon. Interferon was furnished by Dr. Horoszewicz (RPMI Buffalo) and contained 10^7 IU/mg protein. It was administered at 3.3×10^6 IU twice a week (IV) (28).

b. Bestatin. Bestatin is an extract of Streptomyces olivoreticuli which was purified by Umezawa et al. (30). This drug returns the number of circulating T lymphocytes to normal (3), retards tumor growth in animals, and reduces the number of spontaneous tumors in old mice (4).

The drug is not toxic when administered per os 40mg/m2, three times/week for 15 days to immunodeficient solid tumor patients (26-27).

c. RO 10 9359. This substance is a retinoic acid derivative obtained from Roche Laboratory. The dose chosen was based on the known toxicity in mice, rats and dogs (7,19) and concentrations used clinically to treat dermatoses (20). Each patient received 5 mg/kg/day per os for 21 days.

d. Cyclomunine. This drug is a cyclohexadepsipeptide of Fusarium equiseti which has been shown to exert varied in vitro effects on lymphocytes depending on the dose (8,9). At low concentration cyclomunine potentiates Con. A induced suppressor cell activity and is considered a Facteur Thymique Serique (FTS) inducer (8).

e. NPT 15 392. NPT 15 392 is a biologically active isoprinosine analog which exerts its effect on the immune system (11-12).

g. Cis-Platinum. Cis diamino-dichloro-platinum is the first in a serie of platinum derivatives to show in vivo inhibition of various murine tumors. Undesirable side effects include digestive, renal, hématopoïetic and auditive complications. The first clinical trials (phases I and II) demonstrated these same toxicities but also demonstrated a remarkable effectiveness in the treatment of testicular carcinomas, hematosarcomas, head and neck cancers and ovarian cancers. Cis-Platinum is presently employed in assocation with other chemotherapeutic agents. In the present study, we used a single injection of 100 mg/m2 I.V. (1,23).

4. Blood samples

Blood samples were drawn on the morning of the first day and on the day following cessation of treatment.

5. Statistical analysis

Statistical analysis was performed using the paired sample T test.

RESULTS (Table 1)

There was a strong increase in NK activity in patients treated by human fibroblastic interferon, bestatin or cis-platinum. Only a slight increase was noted for immunodeficient patients treated by retinoic acid RO 10 9359.

We found a dose dependent increase in NK activity for cyclomunine but it is still difficult to acertain if there is a correlation with serum level of circulating Facteur thymique Sérique (Table 2). There was no significant change in NK activity in cancer patients treated by NPT 15 392.

TABLE 1

IN VIVO MODULATION OF NK CELL ACTIVITY
(% Cytotoxicity)

	Human Fibroblastic Interferon	Bestatin	RO 10 9359	Cyclomunine (mg/kg) 0.7	Cyclomunine (mg/kg) 0.38	NPT 15 392	CDDP Cis-Platinum
Number of Patients	3	40	40	6	8	10	15
Before treatment	20 ± 6	12 ± 4	43 ± 6.8	10 ± 2.8	45 ± 18	24 ± 2.1	4.8 ± 3.13
After treatment	80 ± 9	32 ± 6	44 ± 10.5	38 ± 9.4	48 ± 17	24 ± 2.4	32 ± 5.5
p value	< 0.001	< 0.05	NS	< 0.05	NS	NS	< 0.001

- Target cell/lymphocyte ratio 25 : 1
- NS = Not Significant

TABLE 2

FTS INCREASED IN CYCLOMUNINE-TREATED PATIENTS
(0.7 mg/kg)

	Patient Number					
	1	2	3	4	5	6
D0	1/4	1/4	1/8	1/64	1/4	1/32 - 1/64
D8	1/64	1/32 - 1/64	1/8	1/64	1/4	1/256 - 1/512
D21	1/128	1/32	1/32	-	-	-
D50	1/128	-	1/256	1/64	1/128	1/128

D = Days after treatment

DISCUSSION

Some authors have suggested a role for NK cells in the defense against primary tumor growth-bearing mice (29). It has also been proposed that increased tumorigenesis occurs when NK activity is selectively depressed, and decreased tumor formation when such deficiencies are selectively reconstitued to normal levels (22). The beige mouse tumor model has been used to test for oncogenesis in NK deficient animals (22). However, beige mice demonstrate residual NK activity which can be augmented by interferon. In addition to increasing NK cell activity, interferon can also alter other immune functions in man (28). A possible clinical approach would attempt to restore depressed NK function in the tumor patient. For this reason we monitored changes in NK activity in response to new drugs which exert an action similar to that of interferon but which are easier to use and less expensive to produce.

The results reported here show that Bestatin, human fibroblastic interferon and cis-platinum can increase NK activity as well as other immune functions, which are particularly modified in advanced solid tumor patients (21,24,25). The return to normal of these functions is desirable since subnormal levels are associated with poor prognosis (11).

The effect of interferon on NK activity has been particularly well documented (10). Moreover immunostimulants such as BCG and Corynebacterium Parvum have been reported to increase NK activity by inducing the production of interferon. But these immunostimulants may also affect other immune functions, acting either directly or indirectly on interferon or other monokine inductions (3). The results for human fibroblastic interferon and bestatin are very similar. This was previously noted by Blomgren et al. for leucocyte interferon and bestatin (3). Nevertheless, the analogy is limited by the fact that interferon will stimulate antibody dependent cell-mediated cytotoxicity whereas bestatin will not (3).

Although Retinoic Acid RO 10 9359 effects on NK activity were minimal, suggesting this drug has no interferon-like activity, it was previous shown that RO 10 9359 can induce reappearence of positive skin reactivity as well as increase autorosette forming T cells (5) in immunodepressed cancer patients (10).

Cyclomunine significantly increases NK activity and induces FTS production. Previous animal studies support this finding which is an additional argument for further investigation, since this substance would be the first to induce FTS which plays a role in T lymphocyte differentiation (7). The increase in FTS is not immediate, as shown in Table 2, being observed 30 to 40 days after treatment in animals (8).

Although there are hypotheses for the mechanism of action of bestatin and cyclomunine on NK cells, the mode of action of Cis-Platinum on NK cells is not known at the present. Within 48 to 72 hours after Cis-Platinum treatment, there was a highly significant increase in NK activity. Our present study is in complete agreement with previous studies with monocyte mediated cytotoxicity(17,18)and offers a promising foundation for planning a broader study to confirm the correlation between changes in immune response and circulating Cis Platinum levels. This could lead to a new protocol in which Cis-Platinum would be used as an anti-tumor cytolytic agent during standard treatment cycles followed by a one month period during which smaller doses would be employed to manipulate immunological activity, especially NK activity, in an attempt to maintain a strong immune response.

Subsequent studies (phases III) should confirm not only the usefulness of monitoring NK activity per se but its value as a means of evaluating the effects of biological response modifiers used in the treatment of cancer and of adapting treatments (chemotherapy, immunomodulating agents).

REFERENCES

1. Alberts, D.S., Hilgers,R.D., Moon, T.E., O'Toole, R., Mantz, F., Martim-Beau, P.W., Stephens, R.L., Rivkin, S. and Mason, N., in "Cisplatin Current Status and New Developments" (A.W. Prestayko, S.T. Crooke and Carter S.K. Eds.), p. 393. Academic press New-York, (1980)
2. Belpomme, D., Marty, M., Gisselbrecht, C., Mignot, L. and Pujade-Lauraine, E., Sem. Hop. Paris, 57, 1547 (1981).
3. Blomgren, H., Strander, H. and Edsmyr, F., in "Small molecular immunomodulators" (H. Umezawa Ed.), p. 71. Pergamon Press, Oxford-New-York, (1981).
4. Bruley-Rosset, M., Florentin, I., Kiger, N., Schultz, G., Mathe, G., Immunology, 38,75 (1979).
5. Caraux, J., Thierry, C., Serrou, B., J. Natl. Cancer Inst, 69, 593, (1979).
6. Cupissol, D., Longhi, B., Rey, A., Serrou, B., Cancer Treatment Report (1982) In press.
7. Cuttner, J., Glidewell, O., Holland, J., in "Immunotherapy of cancer : present status of trials in man" (W.D. Terry Ed.) Elsevier/North-Holland, Amsterdam (1982) In press.

8. Dardenne, M., Niaudet, P., Simon-Lavoine, N., Bach, J.F., Int. J. Immunopharm. 2, 154, (1980).
9. Florentin, I., Simon-Lavoine, N., Bruley-Rosset, M., Mathe, G., Int. J. Immunopharm. 2, 221, (1980).
10. Goldfarb, R.H., Herberman, R.B., J. of Immunol., 126, 2123, (1981).
11. Hadden, J.W., in "Advances in immunopharmacology" (J. Hadden, L. Chedid, P. Mullen, F. Spreafico Eds.) p. 327, Pergamon Press, Oxford-New-York (1980).
12. Hadden, J.W., in "Advances in immunopharmacology" (J. Hadden, L. Chedid, P. Mullen, F. Spreafico Eds.) p. 457, Pergamon Press, Oxford-New-York (1980).
13. Israel, L., Lagrange, P.H., Salomon, J.C., Cancer Immunology and parasite immunology (L. Israel, Ph. Lagrange, J.C. Salmon Eds.),Vol.97, INSERM Publ. Paris, (1981)
14. Jones, S., Salmon, S., Haskins, C., in "Immunotherapy of cancer : present status of trials in man"(W.D. Terry Ed.) Elsevier/North-Holland, Amsterdam, (1982) In press.
15. Mathe, G., Cancer Immunol. Immunoth. 3, 1, (1977).
16. Mathe, G., Cancer Immunol. Immunoth. 2, 81, (1977).
17. Kleinerman, E.S., Zwelling, L.A., Muchmore, A.V., Cancer Research, 40, 3099, (1980).
18. Kleinerman, E.S., Howser, D., Young, R., Zwelling, L., Barlock, A., Decker, J., Muchmore, A., Lancet ii, 1102, (1980).
19. Patek, P.Q., Collins, J.L., Yogeeswarian, G., Dennert, G., Int. J. Cancer, 24, 624, (1979).
20. Peck, G.L., Yoder, F.W., Lancet, ii, 1172, (1976).
21. Pouillart, P., Palangie, P., Huguenin, P., Morin, P., Gauthier, H., Baron, A., Mathe, G., Lededente, A., Boto, G., Nouv. Presse Med., 7, 265, (1978).
22. Riccardi, C., Santoni, A., Barlozzari, T., Pucceti, P., Herberman, R.B., Int. J. Cancer, 24, 475, (1980).
23. Rosenberg, B., in "Cisplatin Current Status and New Developments" (A.W. Prestayko, S.T. Crooke, S.K. Carter Eds.) p. 9, Academic Press, New-York, (1980).
24. Serrou, B., Dubois, J.B., Acta Cytologica, 20, 577, (1976).
25. Serrou, B., Gauci, L., Caraux, J., Cupissol, D., Thierry, C., Esteve, C., in "Recent Results in Cancer Research" (G. Mathe and F.M. Muggia Eds.) p. 41, Springer-Verlag, Heidelberg, New-York, (1980).
26. Serrou, B., Cupissol, D., Flad, H., Goutner, A., Lang, J. M., Spirzglas, H., Plagne, R., Beltzer, M., Chollet, P., Mathe, G., Inter. J. Immunopharmacol. 2, 168, (1980).
27. Serrou, B., Cupissol, D., Flad, H., Goutner, A., Lang, J. M., Spirtglas , H., Plagne, R., Beltzer, M., Chollet, P., Marneur, M., Mathe, G., in "Immunotherapy of cancer : present status of trials in man (W.D. Terry Ed.) Elsevier/ North-Holland, Amsterdam, (1892) In press.

28. Serrou, B., Cupissol, D., Thierry, C., Caraux, J., Gerber, M., Pioch, Y., Esteve, C., Lodise, R., in "Interferon-properties and clinical use" (A. Khan, N.O. Hill, L.G. Dorn Eds.) p. 621, (1981).
29. Touraine, J.L., Hadden, J.W., Navarro, J., Simon-Lavoine, N., Inter. J. Immunopharmacol., 2, 226, (1980).
30. Umezawa, H., Hoyagi, T., Suda, H., Hamada, M., Takeuchi, T. J. Antibiot., 29, 97, (1976).

EFFECT OF DIETHYLSTILBESTROL AND ESTRAMUSTINE PHOSPHATE (ESTRACYT^R) ON NATURAL KILLER ACTIVITY IN PATIENTS WITH CARCINOMA OF THE PROSTATE

Sven Haukaas[1]
Terje Kalland[1]

Institute of Anatomy
University of Bergen
Bergen, Norway

INTRODUCTION

The effects of estrogen on the immune system are well documented in experimental animals (1). Continous administration to adult mice of 17β - estradiol has been reported to induce a substantial reduction in Natural Killer (NK) cell activity (2), and exposure to the non-steroidal estrogen diethylstilbestrol (DES) resulted in a persistent defect NK function in adult female mice (3). Ablin et al. have provided evidence that both general as well as tumor-associated immunity are impaired after DES therapy (4,5). These observations initiated the present study concerning effects of DES therapy on NK activity in peripheral blood of patients with carcinoma of the prostate.

MATERIALS AND METHODS

Patients

Twelve patients with a confirmed diagnosis of adenocarcinoma of the prostate were included in this study, ranging in age from 61 to 80 years.

The grade of malignancy and clinical stage of the prostatic cancer were evaluated as described (6). Patients with stage I and II, well or moderately differentiated carcinoma

---

1) Present address: Department of Anatomy, Lund, Sweden

ISBN 0-12-341360-5

did not receive any treatment and are not included in this study. Four patients with poorly differentiated carcinoma, representing stage I to IV, received a daily amount of 840 mg estramustine phosphate (EMP, Estracyt^{R}), an estradiol - cytostatic complex, divided in three oral doses. Conventional hormone therapy was given to five patients with stage III to IV, well or moderately differentiated cancer. The conventional hormone regimen consisted of oral treatment with 10 mg DES daily for the first two weeks, thereafter 2 mg DES daily. As a supplement to DES, the patients in this group also received polyestradiol-phosphate (PEP, Estradurin^{R}) at the start of the treatment (80 mg im) and later every 4 weeks.

Three patients with advanced prostatic cancer were selected for high dose DES (Honvan^{R}) treatment intravenously as primary treatment and received 1 000 mg DES daily for a week.

NK activity was determined immediately before the start of treatment as well as one and four weeks later.

Effector cells

Mononuclear cells from peripheral blood were separated on Lymphoprep^{R} (Nyegaard & Co, Oslo, Norway) according to Böyum (7). The cells were washed in phosphate buffered saline (PBS, pH 7.4) and resuspended in RPMI 1640 with 20 % heat inactivated (30 min at 56° C) pooled human AB serum to 10^7 cells/ml.

Target cells

MOLT-4, a T-lymphoblast cell line, originally established from a patient with acute lymphoblast leukemia (8), was maintained as stationary suspension culture in RPMI 1640 medium supplemented with 10 % fetal calf serum (FCS, Bio-Cult, GIBCO).

Cytotoxicity assay

The NK cytotoxicity assay has been described previously (9). In brief, 10^6 effector cells were admixed to 10^4 ^{51}Cr-labelled MOLT-4 target cells in a total volume of 200 µl and incubated at 37° C (humidified incubator, 5 % CO_2, 95 % air) in round bottom microtiter plates (Greiner Original, Greiner & Co., Cambridge, Mass.). After 5 hours, the plates were centrifuged (10 min, 500 G) and 100 µl supernatant aliquots were removed with a micropipette for counting in a Searle 1185 -spectrophotometer.

Spontaneous release was determined by incubation of

labelled target cells in medium only, and varied from 10 - 20 % of total radioactivity. Total radioactivity was determined by counting samples of labelled target cells. Variations between triplicates was below 5 %. Percent cytotoxicity was determined by:

$$\frac{\text{test cpm - spontaneous release cpm}}{\text{total radioactivity}} \times 100$$

The percentage of cytotoxicity prior to therapy ranged from 14 - 42 (mean 31). No correlation with grade or stage of the prostatic cancer could be noted in this small patient group. NK activity was expressed relative to pre-therapy level for the individual patient, each patient thus serving as his own control.

RESULTS

The cytotoxicity of NK cells from patients with prostatic cancer treated with hormone or the estradiol-cytostatic complex EMP before and during treatment is shown in Fig. 1. EMP did not affect Natural Killing as the level of NK activity was unaltered after one and four weeks of therapy.

Conventional hormonal therapy (a combination of DES and PEP) substantially reduced NK activity in peripheral blood of the patients after a treatment period of one week. NK activity was only slightly further lowered three weeks later.

The three patients on high-dose DES treatment had NK activity from 20 - 50 % of pre-therapy levels when the therapy was changed to the conventional regimen after one week.

DISCUSSION

The present study demonstrates differences in susceptibility of NK cells to chemotherapeutic agents in current use for treatment of prostatic carcinoma. EMP is a hormone-cytostatic complex consisting of estradiol linked to the alkylating agent nitrogen mustard. The clinical effects on prostatic cancer are well documented and bone marrow related side effects of EMP treatment are rare (10). Küss et al. found no effects on immunity as judged by the cutaneous response to recall antigens and the total immunoglobulin levels in 30 patients treated with EMP for prostatic cancer

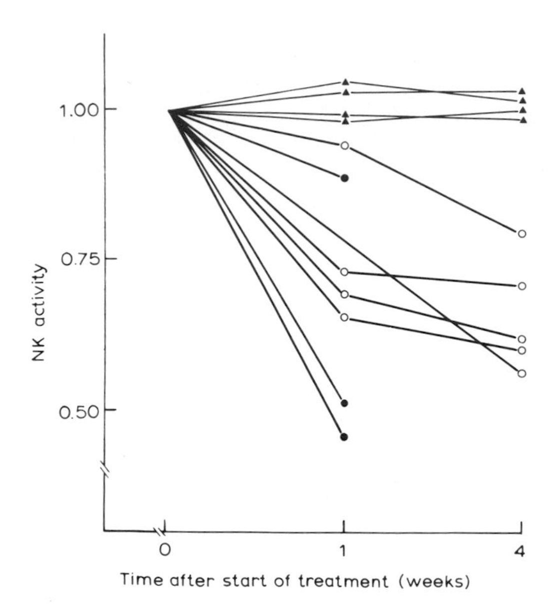

Figure 1. Natural Killer cell activity in peripheral blood from patients treated for cancer of the prostate with EMP (▲), DES-PEP (○) or high doses of DES (●) (see MATERIALS AND METHODS). The activity after 1 and 4 weeks of treatment are related to the pre-therapy level.

(11). In studies from our laboratory, no suppressive effect of EMP treatment was found on mitogen stimulated proliferation of peripheral blood lymphocytes from patients with prostatic cancer (12). Thus, the present results, demonstrating no effect of EMP on NK activity, add further to the evidence that EMP is relatively inert to the immune system.

Administration of DES to patients with prostatic cancer substantially reduced the level of NK activity in peripheral blood within one week. In mice, suppressive activity of DES on NK activity was seen only after 4 - 6 weeks and was probably a result of osteopetrotic bone marrow replacement (2). The rapid effect in patients may indicate a direct effect of DES on NK cells or the bone marrow capacity of these patients is severely limited. Macrophage related suppressor cells regulating NK activity have been described (13), and DES may induce suppressor cells (14). However, neither Seaman et al. (2) nor Kalland (3) were able to demonstrate the involvement of suppressor cells related to the estrogen induced NK impairment in mice. DES has been shown to inhibit lymphocyte functions _in vitro_ (15) and a direct effect of DES on NK cells or their precursors in the bone marrow can not be excluded.

Whether the difference in susceptibility of NK cells to DES and EMP has any biological significance remains to be established. Provided the postulated role of NK cells in the control of metastatic disease holds true, it is possible that EMP might compare favourably to DES, especially in the treatment of tumors with high metastatic potential and in patients with low bone marrow capacity.

REFERENCES

1. Slivic, V.S., Warr, G.V. Period Biol. 75, 231, 1973.
2. Seaman, W.E., Blackman, M.A., Gindhart, T.D. et al. J. Immunol. 121, 2193, 1978.
3. Kalland, T. J. Immunol. 124, 1297, 1980.
4. Ablin, R.J., Bruns, G.R., Al Sheik, H. et al. J. Lab. Clin. Med. 87, 227, 1976.
5. Ablin, R.J., Bhatti, R.A., Guinan, P.D., Cancer Res. 38, 3702, 1978.
6. Höisaeter, P.Å., Haukaas, S., Dahl, D. et al. In Nordiskt Symposium om prostata cancer och dess behandling, p. 83, Helsingborg, Sweden (1980).
7. Böyum, A. Scand. J. Clin. Lab. Invest., Suppl. 97:1, 1968.
8. Minowade, J., Ohnuma, T., Moore, G.E., J.N.C.I. 49, 891, 1972.
9. Kalland, T., Haukaas, S., Invest. Urol. 18, 437, 1980.
10. Jönsson, G., Olsson, A.M., Luttorp, W. et al. In Endocrine Control of the Prostate (P.L.Munson, E. Diczfalusy, J. Glover and R.E. Olsson, eds.) Vitam. Horm. 33, 351, 1975.
11. Küss, R., Khoury, S., Richard, S. et al. Br. J. Urol. 52, 29, 1980.
12. Haukaas, S., Höisaeter, P.Å., Kalland, T. Submitted for publication.
13. Savary, C.A., Lotzova, E. J. Immunol. 120, 239, 1978.
14. Luster, M.T., Boorman, G.A., Dean, J.H. et al. J. Reticuloendothel. Soc. 28, 561, 1980.
15. Harty, J.J., Catalona, W.J., Gomolka, D.M. J. Urol. 116, 484, 1976.

EVIDENCE FOR *IN VIVO* REACTIVITY: AGAINST TRANSPLANTABLE AND PRIMARY TUMORS

Sonoko Habu
Department of Pathology
Tokai University School of Medicine
Isehara, Kanagawa, Japan

Ko Okumura
Department of Immunology
Faculty of Medicine
University of Tokyo
Tokyo, Japan

I. INTRODUCTION

There is accumulating evidence that the rate of transplanted tumor growth inversely carrelate well to the host NK activity (1-3). However, a direct evidence that the NK activity alone is responsible for the regulation of the tumor growth remains to be establish since the strains of mice employed for the different NK activity have different characteristics. Thus, there still is a possibility that other factors associated with each strain may be responsible for the tumor resistance in different ways. Ideally, one way to clarify the possible function of NK cells in a tumor-host system is to utilize mice which have the same genetic background, but differ only with NK cell activity. This can be accomplished with the use of antisera against asialo GM1, since the antisera selectively eliminate asialo GM1 positive ($GA1^+$) cells as well as NK activity *in vivo* and *in vitro* as we have demonstrated in recent years (4,5). We will discuss in this review mainly our recent studies on the regulation of tumor growth by NK cells in nude mice. The use of nude mice offered several advantages; this strain of mice has high NK activity and the contribution of NK cells on the tumor growth

ISBN 0-12-341360-5

can be clearly isolated from that of T cell mediated activity.

II. EFFECT OF DELETION OF ASIALO GM1 POSITIVE CELLS ON TRANSPLANTED TUMOR GROWTH IN NUDE MICE

Repeated injection of anti-asialo GM1 (anti-GA1) every three-days into mice maintain the NK activity at less than 5% level and abrogated $GA1^+$ cells in the spleen (Table I) without elimination of T and B cells. Thus, it was judged adequate for study of the biological role of NK cells in vivo. Two million syngeneic tumor cells, RL♂-1 or two million allogeneic tumor cells, YAC-1, were subcutaneously injected into each BALB/c nude mouse on the same day, and each mouse also received intravenous (i.v.) injection of rabbit anti-GA1 or normal rabbit serum (NRS). For a duration of 24 days, each mouse received repeated injections of the antiserum every three-days and the size of palpable tumors was recorded. The mice which received anti-GA1 resulted in markedly higher incidence of RL♂-1 tumor take as well as the size of tumors was larger than those received the tumor cells and NRS. As seen in Fig. 1, 100 percent palpable tumor of RL♂-1 appeared in anti-GA1 injected mice, whereas the tumors were detectable less than 25% of mice injected with NRS. The difference was more striking in YAC-1 inoculated mice; no palpable tumors appeared in NRS injected mice within 3 weeks. These correlations between the elimination of NK activity and the enhancement of tumor growth are consistent with the other

TABLE I. Effect of Repeated Injection (i.v.) of Anti-GA1 on NK Activity and on Other Cells (CBA/J Mice)

Injected with 20 μl of	% lysis against YAC-1 (E/T=100)	% $GA1^+$ cells	% $Thy\text{-}1^+$ cells	Mitogen response (Δcpm at 48 hr) Con A (2.5 μg/ml)	LPS (100μg/ml)
Anti-GA1					
8 times	2.3	1.8	26.0	42,500	92,200
6 times	3.5	1.0	32.0	49,800	89,300
NRS					
8 times	46.5	14.5	31.5	52,900	100,400
6 times	39.8	13.4	28.3	45,300	95,900

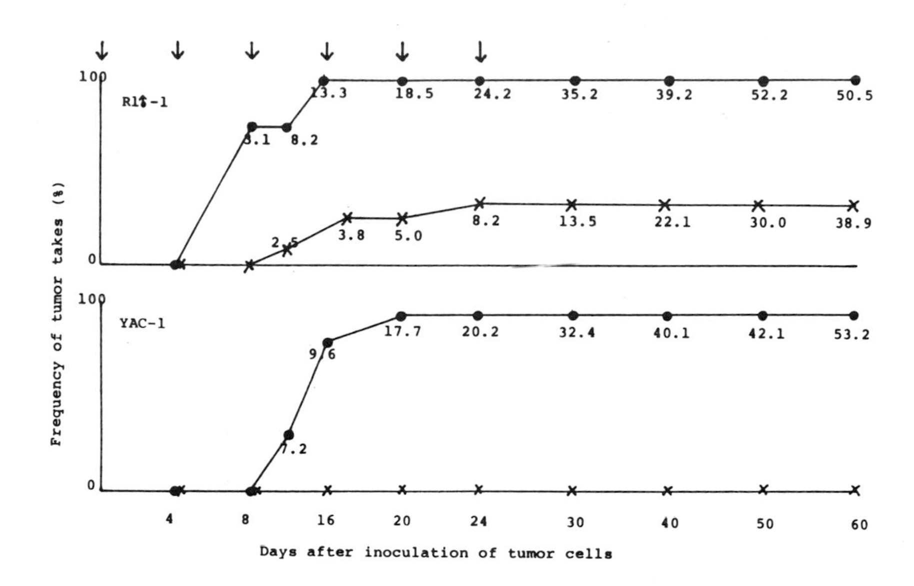

FIGURE 1. Effect of anti-GA1 injection on transplanted tumor growth. 2×10^6 syngeneic (RL♂-1) or allogeneic (YAC-1) tumors were inoculated subcutaneously into BALB/c nude mice receiving anti-GA1 injection (● - ●) or NRS (x - x) on the same day followed by every four-day injection (indicated by arrows on the top of figure). Rates (%) of tumor establishment were compared in 20 anti-GA1 injected mice and 20 control mice each group. The mean tumor sizes were shown with figures indicated by $\sqrt{\text{major axis} \times \text{minor axis}}$ (mm) at every four-day. After interrupting injection on day 24, no newly arising RL♂-1 tumors were detectable but the existing tumors increased their sizes. Of the tumor bearing mice, 30% and 25% died by day 60 in anti-GA1 and NRS injected group, respectively.

observation that beige mice which have low NK cell activity have a low natural resistance to transplanted leukemia (6). The same effect of the antiserum was observed even when heterogeneic tumors, such as human malignant lymphoma and gastric carcinoma, were transplanted into the nude mice. These observations suggest that the surveillance of NK cells extents over the control of the growth of transplanted syngenic, allogeneic and heterogeneic tumors.

III. ROLE OF ASIALO GM1 POSITIVE CELLS IN AUTOCHTHONOUS TUMOR GROWTH AND METASTASIS

Our observations that NK cells participate in suppression of transplanted tumor growth *in vivo* gave rise to further important questions with regard to biological role of NK cells; whether or not NK cells function in surveillance of primary tumor and of the metastasis of the tumors. There is yet no direct evidence which demonstrate the NK cell activity against autochthonous tumor growth, but a fact that nude mice develop similar number of tumors to that of their normal litter mates (7). Since numerous cell types in addition to the NK cells may be involved with regulation of autochthonous tumor growth and metastasis, it is difficult to define the role of one cell population. Such study should be also possible when nude mice injected with anti-GA1 are used. As shown in Fig. 2, tumors induced by methylcholanthrene in the nude mice receiving anti-GA1 were detectable two weeks earlier

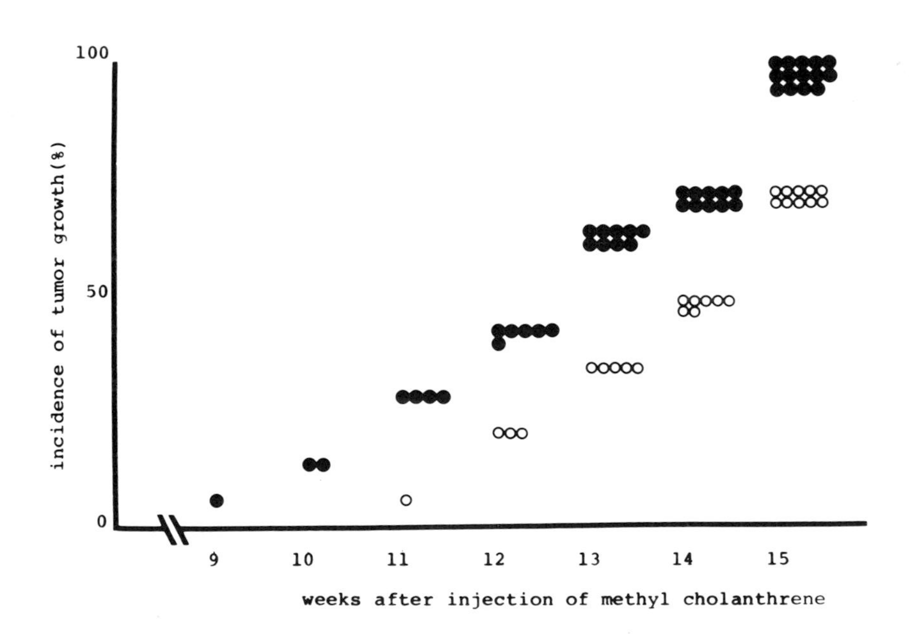

FIGURE 2. Effect of anti-GA1 on methylcholanthrene induced tumor growth. Methylcholanthrene (600 μg) was intramusculally injected into BALB/c nude mice receiving anti-GA1 or NRS injection (i.v.) every 5 days. Each symbol indicates number of mice bearing palpable tumors in 19 anti-GA1 injected mice (●) and 19 NRS injected ones (○) and also rates (%) of the tumor established mice.

than in the control mice: on the 10th week following injection of the carcinogen, tumor was first detectable in 5% of the mice injected with anti-GA1 and on the 16th week tumors appeared in 100% of the mice injected with anti-GA1, whereas in the control mice the tumor first appeared on the 12th week and no palpable tumors were detectable in 20% of mice during 36 week experiment. This is the first evidence which suggest that NK cells act in autochthonous tumor growth. The frequency of tumor appearance in the control mice was low but not zero. The reason for this may have been because of the low sensitivity of the methylcholanthrene induced tumor to NK cell activity. This explanation is consistent with the finding by Talmage et al. that the growth rate of transplanted tumors insensitive to NK cell lysis in beige mice (low NK cell activity) was same as that in C57BL/6 (moderate NK cell activity) (8). They also reported that the frequency of metastasis of syngeneic NK cell sensitive transplanted tumors was markedly more in NK deficient beige mice than in normal C57BL/6 mice. A role of NK cells in metastasis was initially suggested by the observation that transplanted tumors rarely metastasized in nude mice (1). The direct evidence was demonstrated in our system using nude mice injected with anti-GA1. Metastasis of the transplanted human gastric carcinoma was observed in nude mice injected with anti-GA1, whereas no metastasis of the tumors was found in the control mice (no data shown). The above observations together with ours suggested that NK cells have an important function in the host's control of metastasis as well as growth of tumors even in the mice having normal T cell function.

IV. EFFECT OF DELETION OF ASIALO GM1 POSITIVE CELLS ON INTERFERON PRODUCTION

It remained unclear how NK cells recognize and reject tumor cells. However, there is a suggestive evidence that NK activity may be augmented for a short duration following injection of certain tumor cells (9-11). The nonspecific enhancement of NK activity was also observed after the injection of agents which induce the production of interferon (12-14). Indeed, there has been reported a good correlation between NK activity and a level of serum interferon induced by poly I:C, and an injection of anti-interferon into the mice inhibited the augmentation of NK activity induced by poly I:C or transplanted tumors (18,20). More recently, Reid et al. (15) observed that anti-interferon serum suppressed the resistance of nude mice to tumor cells infected with virus. These observations agree with our interpretation that the resistance to tumor growth is suppressed by elimination of NK activity *in vivo*. Because, there is suggestive evidence that NK cells and interferon producing cells belong to the same cell population in human and mice (16). Thus, we examined the ability of interferon production in nude mice which are injected with anti-GA1. As shown in Table II, following intraperitoneal poly I:C (100 μg) injection, the interferon production and NK activity were markedly reduced in the anti-GA1 injected conventional and nude mice. A similar result was obtained in the *in vitro* when spleen cells treated

TABLE II. Effect of Treatment with Anti-GA1 on NK Activity and Interferon Production

Strain	Anti-GA1 injection	% lysis against YAC-1 (E/C=100) poly I:C	PBS	Unit of IF/serum poly I:C	PBS
Experiment 1					
CBA/J	+	2	2	800	510
CBA/J	-	68	46	3,100	640
C57BL/6	+	2	1	60	20
C57BL/6	-	41	27	1,280	20
C57BL/6 nude	+	3	2	620	480
C57BL/6 nude	-	55	42	1,520	20

Strain	Treatment with anti-GA1 + c	% lysis against YAC-1 (E/C=100) poly I:C	PBS	Unit of IF/medium poly I:C	PBS
Experiment 2					
CBA/J	+	1	1	20	44
CBA/J	-	52	43	480	32
C57BL/6	+	1	1	20	20
C57BL/6	-	32	21	600	20
C57BL/6 nude	+	1	1	20	20
C57BL/6 nude	-	54	37	600	20

a. All mice injected (i.v.) with anti-GA1 or NRS on day 0 received 100 μg poly I:C or PBS intraperitoneally on day 1. 4 hr later, their sera were pooled to examine for circulating IF levels.

b. Spleen cells treated with anti-GA1 plus complement were cultured with poly I:C (100 μg/ml) for 12 hr and the supernatant was harvested for examination of IF level.

c. The level of IF in the serum of cultured medium was measured by the ability to inhibit 50% plaque formation in L929 cells.

with anti-GA1 plus complement were incubated with poly I:C for 12 hr. It is uncertain at this time whether interferon produced by the NK cells stimulate the recruitment of new NK cells or interferon produced by the NK cell in turn stimulate further the activity of NK cells. Based on the fact that activation of NK cell activity by interferon is rapid following stimulation and declines rather fast, it is reasonable to assume that NK cells may serve as the first barrier against tumor growth and as a defense system prior to generation of T effector cells.

REFERENCES

1. Herberman, R.B., in "Nude Mouse in Experimental and Clinical Research. p. 135. Academic Press, New York (1978).
2. Petranyi, G.G., Kiessling, R., Povery, S., Klein, G., Herzenberg, L.A., and Wigzell, H., Immunogenetics 3:15 (1976).
3. Kiessling, R., Klein, E., Pross, H., and Wigzell, H., Eur. J. Immunol. 5:117 (1975).
4. Kasai, M., Iwamori, M., Nagai, Y., Okumura, K., and Tada, T., Eur. J. Immunol. 10:175 (1980).
5. Habu, S., Fukui, H., Shimamura, K., Kasai, M., Nagai, Y., Okumura, K., and Tamaoki, N., J. Immunol. 127:34 (1981).
6. Karre, K., Klein, G.O., Kiessling, R., Klein, G., and Roder, J.C., Nature 284:624 (1980).
7. Rygaard, J., and Povlsen, C.O., Transplantation 17:135 (1976).
8. Talmage, J.E., Meyers, K.M., Prieur, D.J., and Starkey, J. R., Nature 284:622 (1980).
9. Herberman, R.B., Nunn, M.E., Holden, H.T., Staal, S., and Djeu, J.Y., Int. J. Cancer 19:555 (1977).
10. Trinchieri, G., and Santoli, D., J. Exp. Med. 147:1314 (1978).
11. Djeu, J.Y., Huang, K., and Herberman, R.B., J. Exp. Med. 151:781 (1980).
12. Gidlund, M., Orn, A., Wigzell, H., Senik, A., and Gresser, I., Nature 273:759 (1978).
13. Wolfe, S.A., Tracey, D.E., and Henney, C.S., Nature 262:584 (1976).
14. Merigan, T.C., and Finkelstein, N.B., Virology 35:363 (1968).
15. Reid, L., Minato, N., Gresser, I., Kadish, A., and Bloom, B.R., Proc. Natl. Acad. Sci. U.S.A. 78:1171 (1981).
16. Trinchieri, G., Santoli, D., Dee, R.R., and Knowles, B.B., J. Exp. Med. 147:1299 (1978).

ACCELERATION OF METASTATIC GROWTH IN ANTI-ASIALO GM1-TREATED MICE

E. Gorelik,[1] R. Wiltrout,[1] K. Okumura,[2]
S. Habu[2] and R.B. Herberman[1]

[1]Laboratory of Immunodiagnosis
National Cancer Institute
Bethesda, Maryland
and
[2]University of Tokyo
Tokyo, Japan

INTRODUCTION

There are several indications that natural killer (NK) cells can be efficient in the destruction and growth suppression of tumor cells *in vivo*.

The levels of this inhibition were shown to correlate with the levels of NK reactivity in the recipient mice and to some extent with the susceptibility of the tumor cells to cytotoxic action of NK cells (Haller *et al*., 1977; Karre *et al*., 1980; Gorelik and Herberman; 1981).

It also has been suggested that NK cells may participate in the control of metastatic spread by the elimination of tumor cells entering the blood stream (Gorelik *et al*., 1979). This hypothesis was based on the finding that 3LL tumor cells from pulmonary metastases in comparison with locally growing 3LL tumors, were more resistant *in vitro* and *in vivo* to the cytotoxic action of normal spleen cells (Gorelik *et al*., 1979). Selection of 3LL tumor cells for resistance of NK cells was accompanied by an increase in the metastatic ability of the selected tumor sublime (Gorelik *et al*., 1982). There are numerous experimental indications that i.v. inoculated tumor cells are eliminated very quickly and only a small fraction of cells survive and may establish tumor colonies in the lungs (Fidler *et al*., 1978). Riccardi *et al*. (1980) have shown that NK cells are primarily responsible for the initial *in vivo* elimination of tumor cells, since suppression or activation of NK activity was

ISBN 0-12-341360-5

paralleled by an increase or decrease in the number of surviving tumor cells in the lungs. In the NK-suppressed or activated conditions, the number of artificial metastases increased or decreased, respectively (Hanna and Fidler, 1980 and 1981).

The participation of NK cells in antimetastatic defenses is further supported by the observation that in nude and beige mice the number of lung metastasis positively correlated with the levels of NK cell activity (Hanna and Fidler, 1980; Hanna, 1980; Talmage *et al.*, 1981).

Kasai *et al.* (1980), have found that rabbit anti-asialo GM1 (anti-asGM1) serum selectively inhibited the cytotoxic activity of NK cells without detectable inhibition of T-cell-mediated immunity. Persistently low levels of NK reactivity in mice could be maintained by reinoculation of this antiserum every 3-5 days. By such treatment, substantial acceleration of the local growth of YAC-1, RL♂1 and human tumor cells was induced in nude mice (Habu *et al.*, 1981).

Specific and prolonged suppression of mouse NK reactivity by anti-asialo GM1 serum provided us with the opportunity to study the role of NK cells in the control of metastatic spread and growth of tumor cells. We studied the development of artificial and spontaneous post-operative metastases in C57BL/6 (B6) mice treated with anti-asialo GM1 serum. In parallel, B6 beige mice were used in these experiments.

RESULTS

In the first series of experiments we studied the influence of anti-asGM1 serum on the ability of mice to eliminate tumor cells inoculated into blood stream.

For these purposes 1 x 10^6 YAC-1 cells labeled with ^{111}InOx (Wiltrout *et al.*, 1981) were inoculated i.v. into mice with different levels of NK reactivity: B6, beige, BALB/c, or BALB/c nude mice (Table 1). Three hours following inoculation of radiolabeled YAC-1 cells, the highest level of tumor cells remained in the lungs of beige mice and the lowest was in nude mice. One i.v. injection of 0.2 ml of anti-asGM1 serum into nude mice dramatically suppressed the natural antitumor resistance of these mice (Table 1). The number of remaining tumor cells in the lungs of treated mice increased to 36-38.4%, in contrast to 1.4% in nude mice treated with normal rabbit serum at the same dilution. It was remarkable that this suppressive effect was observed in mice treated with very small amounts of anti-asGM1 serum (dilution 1:40-1:320). After antiserum treatment, nude mice had levels of NK reactivity as low as that of beige mice.

Table 1. Survival of YAC-1 Tumor Cells in the Lungs of Beige Mice and of Mice Treated with Anti-asGM1 Serum

Mice	Treatment with anti-asGM1 (dilution)	CPM in the lungs (±SE)	% remaining radioactivity
B6 +/+	--	3427 (445)	4.1
B6 beige	--	27014 (2173)	32.3
BALB/c	--	3166 (672)	3.8
BALB/c nude	--	1384 (220)	1.6
BALB/c nude	--	3570 (308)	1.4
	1:40	97904 (12567)	38.4
	1:320	91584 (7482)	36.0

Mice were inoculated i.v. with 1 x 10^6 of YAC-1 cells labeled with ^{111}InOx. Some of nude mice received i.v., 1 day before, anti-asGM1. Radioactivity in lungs was determined 3h after tumor cell inoculation; 5 mice per group.

In the next series of experiments we studied the growth of artificial or spontaneous metastasis in the lungs of C57BL/6 mice treated with anti-asialo GM1 serum.

One day after inoculation of 0.2 ml of anti-asGM1 serum, at a dilution of 1:80, B6 mice received i.v. 4 x 10^4 B16 melanoma cells. Fifteen days later, mice were killed and the number of tumor foci in various organs was determined.

In mice treated with anti-asGM1 serum, the number of tumor foci in the lungs was dramatically increased above that in control mice (Table 2). The entire surface of the lungs was covered with metastases and their precise enumeration was impossible. It is of interest that mice treated with anti-asGM1 also had 70 metastatic foci in the liver, whereas mice treated with normal rabbit serum had no detectable extrapulmonary metastases.

Results similar to those observed in the antiserum treatment were obtained when 2 x 10^4 B16 melanoma cells were inoculated i.v. into beige mice (Table 2). The number of metastatic foci in the lungs of beige mice was about 12 times higher than in normal B6 mice, and beige mice had metastatic tumor in the liver. This increase in the number of metastases can be attributed to a higher number of surviving tumor cells in beige mice. In the same experiment, groups of normal B6 and beige mice were inoculated with 2.5 x 10^5 B16 melanoma cells labeled with ^{111}InOx. Eighteen hours later, 2.9 ± 0.9% the initially inoculated tumor cells

Table 2. Artificial Metastases of B16 in B6 Mice Treated with Anti-asGM1 and in Beige Mice

Mice	Median No. of Metastases[a]	
	Lungs	Liver
B6 treated with normal rabbit serum, d.-1	38	0
B6 treated with anti-asGM1, d.-1	> 300[b]	70[b]
B6	13	0
beige	155[b]	28[b]

a) Determined 15 days following tumor inoculation.
b) Significantly different ($p<0.01$) from control group of mice by Mann-Whitney U test; 8-10 mice per group.

remained in the lungs of normal B6 mice, whereas in the lungs of beige mice, significantly more cells survived ($7 \pm 0.9\%$).

These data support the involvement of NK cells in the control of metastatic spread and growth of tumors. However, these studies were performed using an artificial experimental model, with tumor cells inoculated i.v. into healthy mice. In order to receive direct evidence about the role of NK cells in antimetastatic defenses, we studied the development of spontaneous metastases in mice with suppressed NK cell function. B6 mice were inoculated i.v. with 0.2 ml of anti-asGM1 serum at a dilution of 1:80. The next day, 0.5×10^6 B16 melanoma cells were inoculated into the footpads of mice. Every four days, mice received i.v. 0.2 ml of anti-asGM1 serum (1:150), for a total of six injections.

Control mice received 0.2 ml of normal rabbit serum at the same dilution. As a further comparison 0.5×10^6 B16 melanoma cells were inoculated i.f.p. into beige mice. Local growth of B16 melanoma was similar in control and antiserum-treated mice and also in beige mice. Twenty-five days after tumor cell transplantation, local tumors had reached 7-9 mm in diameter (171-364 mm^3) and were removed. Tumor excisions were performed one day after the last injection of anti-asGM1 serum. Sixteen days later all groups of mice were killed and their lungs were examined.

The number of metastases in mice inoculated with anti-asGM1 serum was significantly higher than in mice treated with normal rabbit serum (Table 3). Beige mice also had more metastases in the lungs than did B6 mice treated with normal rabbit serum. The number of metastases in beige mice and in B6 mice treated with anti-asGM1 were not significantly different, ($p>0.05$). Although some of the metastases in the lungs of mice treated with anti-asGM1 were larger or grew more confluently, especially in mice with a short survival period. Four out of 10 mice from this group died during the period of observation. During the same period, only 1 of 9 beige mice died.

All these results indicate that treatment of mice with anti-asGM1 induced the dramatic acceleration of the growth of both artificial and spontaneous postoperative metastases. This acceleration can be attributed to the fact that anti-asGM1 selectively suppressed, in vitro or in vivo, the cytotoxic activity of NK cells. (Kasai et al., 1980; Habu et al., 1981; Gorelik and Herberman, 1981). Our present data show that in mice treated with anti-asGM1 serum the ability to eliminate tumor cells from the blood stream can be dramatically suppressed. Consequently, more tumor cells could survive and develop metastatic tumors. This conclusion is supported by the similar data obtained in beige mice, which also have very low levels of NK reactivity.

Table 3. Postoperative Pulmonary Metastases of B16 Melanoma in mice, Treated with Anti-asGM1

Mice	Median No. of Metastases (range)
B6 treated with normal rabbit serum	8(4-25)
B6 treated with anti-asGM1 serum	56(15-136)
Beige	37(13-58)

One day after serum inoculation, mice were inoculated i.f.p. with 5 x 10^5 B16 cells. Every 3-4 days mice received additional injections of anti-asGM1 or normal rabbit serum (1:150 dilution). 26 days later, local tumors (177-365 mm^3) were removed. Number of metastases in the lungs was determined 16 days after tumor removal; 8-10 mice per group.

Groups 2 and 3 were significantly different ($p<0.05$) from Group 1. No significant difference ($p>0.05$) between Groups 2 and 3, by Mann-Whitney U test.

From previous studies, on the role of NK cells in resistance to artificial metastases (Hanna and Fidler, 1980; Hanna, 1981), it remained unclear whether NK cells could prevent the development of spontaneous metastases in tumor-bearing or tumor-excised mice. Talmage et al. (1980) found that the number of postoperative metastases in the lungs was higher in bg/bg mice than in bg/+ mice. However, in these experiments local B16 melanomas were removed after 21 days of growth, when local tumors growing in the ear of bg/bg mice reached 93 ± 22 mm^3 in volume, whereas in bg/+ mice tumors were much smaller (4.1 ± 0.9 mm^3). Thus, it was not surprising that in bg/+ mice metastases were not found in the lungs of the bg/+ mice, since the number of postoperative pulmonary metastasis depends on the size of the primary tumor mass (Gorelik et al., 1980). Therefore, the size of the primary tumor at the time of resection needs to be controlled, in order to adequately study metastatic growth in NK-suppressed or activated conditions. In some experiments, we removed local B16 melanomas growing in normal, beige or anti-asGM1 treated B6 mice when local i.f.p. tumors reached the sizes of 62-150 mm^3, and we found few metastatic foci developed in the lungs of these mice. No differences in the number of metastases were observed in these mice (data not shown). In contrast, in the experiment described above (Table 3), local i.f.p. tumors were removed when they were larger (171-364 mm^3).

In experiments with another tumor, Lewis lung carcinoma (3LL), we observed dramatic acceleration of the development of postoperative pulmonary metastases in B6 mice with NK reactivity suppressed by inoculation of anti-asGM1 serum (Gorelik et al., in preparation).

Therefore, our data indicate that inoculation of anti-asGM1 serum suppress the natural antitumor resistance of mice which resulted in the increase in the survival of metastatic cells and development metastatic foci. Thus, NK cells appear to play an important role in the control of metastatic spread and growth of several transplantable tumors.

REFERENCES

Fidler, I., Gerstein, D. and Hart, I. (1978). Adv Cancer Res 28:149.

Gorelik, E., Fogel, M., Segal, S. and Feldman, M. (1979). J Natl Cancer Inst 63:1397.

Gorelik, E., Segal, S. and Feldman, M. (1980). J Natl Cancer Inst 65:1257.

Gorelik, E. and Herberman, R.B. (1981). Int J Cancer 27:709.

Gorelik, E., Segal, S. and Feldman, M. (1982). Cancer Immunol Immunother, in press.
Habu, S. Fukue, H., Shimamura, K., Kasai, M., Nagai, Y., Okumura, K. and Tamaoki, N. (1981). J Immunol 127:34.
Haller, O., Kiessling, R., Orn, A. and Wigzell, H. (1977). J Exp Med 145:1411.
Hanna, N. (1980). Int J Cancer 26:675.
Hanna, N. and Fidler, I. (1980). J Natl Cancer Inst 65:801.
Hanna, N. and Fidler, I. (1981). J Natl Cancer Inst 66:1183.
Karre, K., Klein, G.O., Kiessling, R., Klein, G. and Roder, J. (1980). Int J Cancer 26:789.
Kasai, M., Yoheda, T., Habu, S., Maruyma, Y., Okumura, K. and Tokumaga, R. (1980) Nature 291:334.
Riccardi, C. Santoni, A. Barlozzari, T., Puccetti, P. and Herberman, R.B. (1980). Int J Cancer 25:475.
Talmage, J., Meyers, K., Prieur, D. and Starkey, J. (1980). Nature 284:622.
Wiltrout, R. Taramelli, D. and Holden, H. (1981). J Immunol Meth 43:319.

THE EFFECT OF SELECTIVE NK CELL DEPLETION ON THE GROWTH OF NK SENSITIVE AND RESISTANT LYMPHOMA CELL VARIANTS

Ichiro Kawase
David L. Urdal
Colin G. Brooks
Christopher S. Henney[1]

Program in Basic Immunology
Fred Hutchinson Cancer Research Center
Seattle, Washington

Despite the enormous interest in natural killer (NK) cells in recent years, there is still a paucity of information regarding their function *in vivo* and in particular, in the role that NK cells may play in protection against tumor development. However, recently several reports have strongly suggested that NK cells may play an important role in resistance against tumor development *in vivo*. Haller *et al.* (1) showed that the ability of irradiated and bone-marrow reconstituted mice to reject small inoculae of syngeneic tumor cells was related to the NK responder genotype of the bone-marrow donor. Similarly, Riccardi *et al.* (2) found that the rate of the clearance of intravenously injected tumor cells was often correlated with NK cell activity in the host animals. Furthermore, Ojo (3) reported a close relationship between host NK activity following treatment with *Corynebacterium parvum* and the ability of the treated animals to reject tumors.

The most direct approach towards elucidating a physiological role for NK cells, has been to evaluate tumor growth and development in animals selectively deficient in NK activity. The *bg/bg* mutant of C57BL/6 mice, which has a selective defect in NK function (4), has been a useful model in this regard and both Karre *et al.* (5) and Talmadge *et al.* (6) have documented a lowered resistance to

[1] Supported by grants CA 24537 and AI 15384 from the National Institutes of Health and IM-247 from the American Cancer Society. David L. Urdal is supported by a human cancer-directed fellowship DRG-528 of the Damon Runyon-Walter Winchell Cancer Fund.

ISBN 0-12-341360-5

tumor growth in such mice. However, as this mutation is currently inbred into a very restricted number of mouse strains, its usefulness is somewhat limited. An alternative and more general approach would be to develop a reagent capable of specifically inhibiting the NK system in vivo. Reid et al. (7) have described one such reagent, anti-interferon antibody. Nude mice injected with this material were reported to retain normal NK activity, but failed to mount an augmented NK response following injection of virus-infected tumor cells. Such animals showed a dramatically increased susceptibilty to a variety of xenogeneic tumors, with markedly increased invasiveness and metastatic spread. The general usefulness of anti-interferon antibody as an NK-suppressive reagent is, however, compromised by two problems: (1) tumor cell replication is often directly inhibited by interferon derived from the same species (8) and hence any promoting effect of anti-interferon antibody on the growth of syngeneic tumors could not be unambiguously assigned to its effects on the NK system; (2) such a reagent would not be useful in, for example, studying the role of NK cells in viral infections, where interferon is already established as a major component of host defenses.

Recently, we (9) and others (10) have demonstrated that rabbit antisera against the neutral glycolipid gangliotetraosylceramide (asialo GM1) can, in the presence of complement, eliminate murine NK cell activity in vitro without affecting cytotoxic T cell activity. We therefore examined whether administration of this antiserum in vivo might lead to a selective suppression of NK function.

Rabbit anti-asialo GM1 antiserum was obtained by immunizing a New Zealand white rabbit with an emulsion containing purified asialo GM1, bovine serum albumin (BSA) and complete Freund's adjuvant (11). The immune serum was inactivated at 56°C for 1 hr and adsorbed 3 times with BSA insolubilized with glutaraldehyde.

Intravenous (iv) injection of anti-asialo GM1 anti-serum resulted in significant reduction of splenic NK activity in CBA, BALB/c nu/nu, and DBA/2 mice (Table 1). The maximal suppressive effect was noted by 2 days after antiserum injection. After a single iv injection of anti-asialo GM1 antiserum, NK activity was partially restored by 4 days, and by 7 days normal levels of NK cell activity were observed. Multiple iv injections of the antiserum resulted in a more profound reduction of NK cell activity than did a single injection, with suppression of NK activity typically lasting 2-3 weeks.

As a recent study has referred that cytotoxic T cell precursors may share with NK cells the asialo GM1 marker (12), we were anxious to address the possibility that in vivo generation of cytotoxic T cells might also be suppressed by anti-asialo GM1 therapy. However, DBA/2 mice which had been injected iv on 3 successive days with 150 µl of undiluted anti-serum developed an undiminished cytotoxic T cell response to alloantigen, whereas splenic NK activity of these mice was significantly suppressed. Additionally, it was found that the treatment of mice with anti-asialo GM1 antiserum had no effect on the subsequent ability of the spleen cells from such mice to develop cytotoxic T cell activity in response to allogeneic cells in vitro.

One explanation for this preferential effect of systemic anti-asialo GM1 on the NK cell compartment may be that NK cells have an anatomical distribution which makes them exquisitely susceptible to elimination after coating with specific antibody. Alternatively,

Table I. Effect of IV Injection of Anti-asialo GM1 Antiserum on NK Cell Activity in Mice

Mice	Injected with	% Specific cytolysis[a] cl27v-IC2	YAC-1
CBA[b]	NRS	30.1	55.9
	Anti-asialo GM1	7.0	16.1
BALB/c nu/nu[c]	NRS	29.9	53.3
	Anti-asialo GM1	7.6	20.0
DBA/2[d]	NRS	19.5	28.2
	Anti-asialo GM1	7.5	8.7
DBA/2[e]	RPMI medium	5.1	8.9
	Poly I:C + NRS	40.4	51.7
	Poly I:C + Anti-asialo GM1	15.2	25.0

a) Cytotoxicity test was carried out by ^{51}Cr-release assay. Data represent the mean of 4 mice.

b), c) Mice were injected iv with 150 µl of undiluted normal rabbit serum (NRS) or antiserum, and 4 days later, cytolytic activities of spleen cells were assayed at an E/T ratio of 50/1 for 4 hr.

d) Mice were injected iv with 150 µl of undiluted NRS or anti-serum once a day for 3 consecutive days. Two days after the last injection, cytolytic activities of spleen cells were assayed at an E/T ratio of 100/1 for 8 hr.

e) Mice were injected ip with 100 µg of poly I:C and given three daily iv injections of 150 µl of undiluted NRS or antiserum beginning one day prior to poly I:C injection. One day after the last injection of serum, cytolytic activities of spleen cells were assayed at an E/T ratio of 50/1 for 4 hr.

the amount of antigen on cytotoxic T cell precursors may make them poor targets for systematically administered antibody.

The specificity of the effect of anti-asialo GMI administration on the function of NK cells was also confirmed by experiments to determine the effect of antiserum administration on the activation of NK cells and macrophages in response to polyinosinic-polycytidylic acid (poly I:C) The augmentation of splenic NK activity caused by poly I:C injection was markedly inhibited in mice treated with anti-asialo GMI antiserum whereas the development of cytotoxic macrophages was unaffected. Furthermore, anti-asialo GMI antiserum treatment also failed to inhibit the cytotoxic macrophage response following intraperitoneal injection of Bacille Calmette-Guerin (BCG). These results show that *in vivo* treatment with anti-asialo GMI serum selectively reduces NK cell activity with relatively little effect on the cytotoxic T-cell or macrophage compartments.

Table II. Effect of Anti-asialo GM_1 Antiserum on the Transplantability NK-susceptible and NK-insusceptible Tumor Cells in DBA/2 and BALB/c nu/nu Mice

Mice and Treatment[a]	Tumor	Tumor incidence[b]
DBA/2		
None	cl 27v-1C2 (10^5)[c]	1/15
Anti-asialo GM_1	cl 27v-1C2 (10^5)	10/15
NRS	cl 27v-1C2 (10^5)	0/5
None	cl 27av (10^3)	8/20
Anti-asialo GM_1	cl 27av (10^3)	6/10
BALB/c *nu/nu*		
NRS	cl 27v-1C2 (10^5)	0/7
Anti-asialo GM_1	cl 27v-1C2 (10^5)	5/7
NRS	cl 27av (10^3)	1/7
Anti-asialo GM_1	cl 27av (10^3)	2/7

a) Mice were injected iv with 150 µl of undiluted serum 1 day prior to subcutaneous tumor cell inoculation.

b) Palpable tumor scored 4 weeks after tumor cell injection.

c) No. of tumor cells injected subcutaneously.

It was of interest to us to ascertain whether specific suppression of NK function, by administration of anti-asialo GM1 serum, would influence the growth of syngeneic tumor cells. We have previously established NK-sensitive (cl27v) and NK-resistant (cl27av) cloned lines from L5178Y lymphoma of DBA/2 origin (13) and these were chosen for study. The NK-sensitive cloned cells, cl27v, possessed small amounts of asialo GM1 on their surface. This line was therefore recloned and a variant, cl27v-1C2, was isolated which retained high NK sensitivity but which lacked chemically detectable asialo GM1 (14). Thus, systemic therapy with anti-asialo GM1 should have direct effects on host cells, but not on the tumor.

As described previously (14), the cl27v-1C2 cell line was susceptible to NK cell-mediated lysis, while the cl27av cell line was resistant. The susceptibility to NK cells *in vitro* was reflected in the relative tumorigenicity of the tumor cell lines. One million cl27v-1C2 cells were necessary to produce a consistent tumor growth in DBA/2 mice, whereas as few as 1000 cl27av cells produced a considerable incidence (approximately 40%) of tumor. This difference in transplantability between NK-sensitive cl27v-1C2 and NK-resistant cl27av cells was also seen in BALB/c *nu/nu* mice.

Table II shows the effect of anti-asialo GM1 antiserum on the tumorigeneicity of NK-sensitive and NK-resistant lymphoma cell variants. The tumor incidence of NK-sensitive cl27v-1C2 was significantly increased following antiserum therapy in both DBA/2 and BALB/c *nu/nu* recipient animals. The tumor-bearers invariably died of progressive tumor growth within 6 weeks of tumor inoculation. In contrast, treatment with the anti-serum had no significant effect on the incidence of NK-resistant cl27av tumors in either DBA/2 or BALB/c *nu/nu* mice.

As an extension of these observations, we reasoned that agents which augment NK activity might suppress the growth rate of the NK-sensitive tumor, but be counteracted by anti-asialo GM1 therapy. The results of an experiment of this type are shown in Figure 1. The growth of 10^6 cl27v-1C2 cells was significantly retarded when tumor inoculation was preceded by a single ip injection of poly I:C. Moreover, the growth retarding effects of poly I:C were overcome by the treatment with anti-asialo GM1 therapy. By contrast, these various treatments had no effect on the growth of cl27av cells.

These results, collectively considered, provide compelling evidence that NK cells have a major role in regulating tumor growth, especially, the growth of NK-sensitive lymphoma cells. The present studies and those of others have thus established an important role for NK cells in controlling primary and secondary growth of experimental tumors. A key issue to be resolved is the part, if any, played by NK cells in suppressing growth of spontaneously arising neoplasms. The method of selectively depleting NK cells described here may provide a tool for answering this question, as well as for investigating the involvement of NK cells in other disease states.

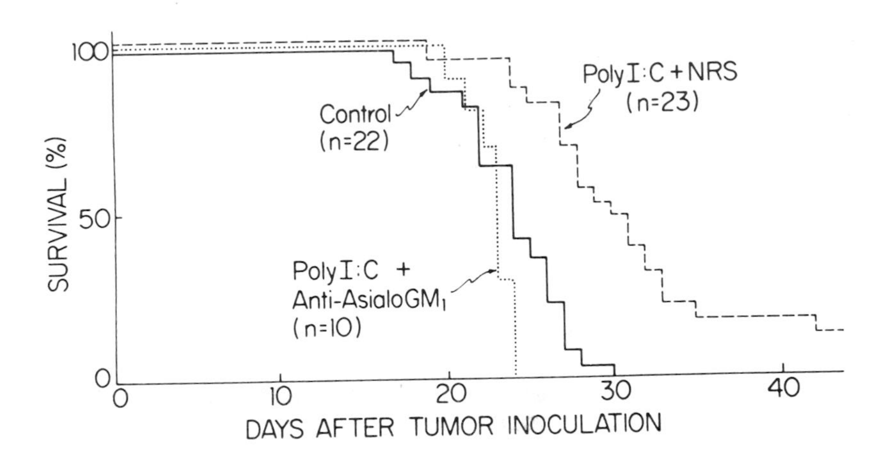

FIGURE 1. Growth of cl27v-1C2 tumor cells in DBA/2 mice treated with poly I:C. DBA/2 mice (10-23 per group) were injected ip with 100 μg poly I:C, and given three daily iv injections of either 150 μl anti-asialo GM1 antiserum (-----) or NRS (— —) beginning one day prior to poly I:C. Mice injected ip with RPMI medium served as control (——). One day after the last injection of serum, mice were inoculated sc with 10^6 cl27v-1C2 tumor cells. Survival rate of mice in the poly I:C plus NRS group was significantly ($P < 0.001$) different from both the control group and from the poly I:C plus anti-asialo GM1 antiserum group.

REFERENCES

1. Haller, O., H. Hanson, R. Kiessling, and H. Wigzell, Nature (London) 270:609 (1977).
2. Riccardi, C., P. Pucceti, A. Santoni, and R.B. Herberman, J. Natl. Cancer Inst. 63:1041 (1979).
3. Ojo, E., Cell. Immunol. 45:182 (1979).
4. Roder, J.C., M.-L. Lohmann-Mathes, W. Domzig and H. Wigzell, J. Immunol. 123:2174 (1979).
5. Karre, K., G.O. Klein, R. Kiessling, G. Klein, and J.C. Roder, Nature (London) 284:624 (1980).
6. Talmadge, J.E., D.M. Myers, D.J. Prieur, and J.R. Starkey, Nature (London) 284:622 (1980).
7. Reid, L.M., N. Minato, I. Gresser, J. Holland, A. Dadish, and B.R. Bloom, Proc. Natl. Acad. Sci. 78:1171 (1981).

8. Burton, R.C. and H.J. Winn, J. Immunol. 126:1985 (1981).
9. Young, W.W., S.-I. Hakomori, J.M. Durdik and C.S. Henney, J. Immunol. 124:199 (1980).
10. Kasai, M., M. Iwamori, Y. Nagai, K. Okumura and T. Tada, Eur. J. Immunol. 10:175 (1980).
11. Hakomori, S.-I., Methods Enzymol. 28:232 (1972).
12. Beck, B.N., S. Gillis, and C.S. Henney, Transplantation, in press (1982).
13. Durdik, J.M., B.N. Beck, E.A. Clark, and C.S. Henney, J. Immunol. 110:1470 (1980).
14. Young, W.W. Jr., J.M. Durdik, D. Urdal, S.-I. Hakomori, and C.S. Henney, J. Immunol. 126:1 (1981).

DIRECT EVIDENCE FOR ANTI-TUMOR ACTIVITY BY NK CELLS *IN VIVO*: GROWTH OF B16 MELANOMA IN ANTI-NK 1.1-TREATED MICE

Sylvia B. Pollack

Department of Biological Structure
University of Washington School of Medicine
Seattle, Washington

I. INTRODUCTION

Previous studies from a number of laboratories had indicated that *in vivo* tumor resistance varies with the level of natural killer (NK) cell activity (cf. review by Herberman and Ortaldo, 1981). We recently developed a model to directly test the role of NK cells *in vivo* by specific depletion of cells: C57Bl/6 (B6) mice were depleted of NK cells by *in vivo* treatment with antisera to the NK-associated alloantigen NK 1.1 (Pollack and Hallenbeck, 1982). When tested *in vitro* in the presence of complement, anti-NK 1.1 kills fewer than 5% of B6 spleen cells (SC), has no effect on cytolytic T cells or plaque forming cells (Pollack *et al.*, 1979; Tam *et al.*, 1980 and unpublished observations). Anti-NK 1.1 alone does not affect NK activity. Injection of B6 mice with 25 μl of anti-NK 1.1 reduced the ability of their SC to lyse YAC-1 *in vitro* by 70% within 2 hr. NK activity gradually returned to control levels but still was significantly depressed at 48 hr. Decreased NK activity was comparable when the antiserum was injected i.p. or i.v.

To test the requirement for NK cells in clearance of circulating lymphoma cells *in vivo*, we pretreated B6 mice with 25 μl anti-NK 1.1 and subsequently injected i.v.

[1]Supported by USPHS CA -18647 and DOE Research Contract 2225 (AT 06-79EV10270).

ISBN 0-12-341360-5

10^6 ^{125}IUdR (5-iodo-2'-deoxyuridine)-labeled YAC-1 or RBL-5 T lymphoma cells. Cpm in lungs, liver and spleen were determined 4 hr later. In a series of four experiments, tumor cell clearance was reduced 2-4 fold in the anti-NK 1.1 treated mice compared to injection controls. These results provided direct evidence that NK cells play an *in vivo* role in the elimination of circulating tumor cells.

We have now extended these studies to test the effect of NK depletion on the growth of a solid tumor, the B16 melanoma, in B6 mice.

II. MATERIALS AND METHODS

A. Antisera

Anti-NK 1.1 serum was produced as previously described (Glimcher *et al.*, 1977; Pollack *et al.*, 1979) in (C3H x BALB/c)F_1 mice immunized with CE SC every 10-14 days for 6-12 immunizations. Individual sera were tested for C-dependent depletion of NK activity and high titered sera were pooled.

B. Mice and Tumor

C57Bl/6J (B6) and C57Bl/6J (bg^J/bg^J) mice were obtained from the Jackson Laboratories (Bar Harbor, ME) and bred in our laboratory to obtain the mice used in these studies.

B16 melanoma was obtained from Drs. Karl Erik and Ingegerd Hellstrom (Fred Hutchinson Cancer Research Center) and maintained *in vivo* by subcutaneous transplantation. Single cell supensions were obtained by trypsinization of nonnecrotic tumor tissue.

C. Experimental Protocol

B6 mice were injected i.p. with 25 μl (0.5 ml of a 1:20 dilution) of either anti-NK 1.1 or preimmune (BALB/c x C3H)F_1 serum, one or two times as indicated in Results. Beige mice were not injected with serum. Each mouse was injected with 5 x 10^4 B16 melanoma cells subcutaneously on each flank. The mice were examined 3 times per week for four weeks. Latent period and tumor growth, assessed by two perpendicular diameters, were recorded. Tumor measurements were done "blind" without knowledge of the experimental treatment.

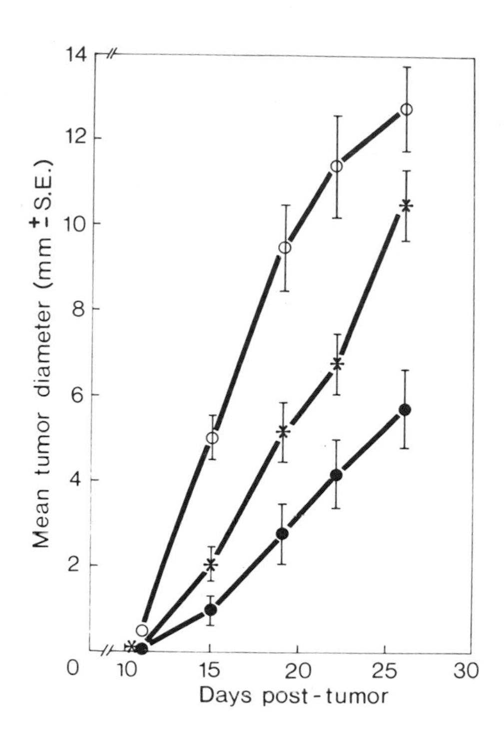

Figure 1. Growth of B16 melanoma in B6 mice injected with 25 µl NMS (●), in B6 mice injected with 25 µl anti-NK 1.1 serum (*), and in B6 $bg^J bg^J$ mice (o); ten mice per group. Sera were injected i.p. 4 hr prior to subcutaneous inoculation of 5 x 10^4 B16 cells in each flank. Differences between the groups were significant as follows:

Day	B6-NMS vs B6-Anti-NK 1.1	B6-anti NK 1.1 vs B6 Beige
15	$p < 0.05$	$p < 0.001$
19	$p < 0.02$	$p < 0.001$
22	$p < 0.02$	$p < 0.002$
26	$p < 0.001$	$p < 0.10$

III. RESULTS

In the first experiment, the effect of one injection of anti-NK 1.1 prior to tumor transplantation was assessed. Three groups of mice were used: a) B6 mice pretreated with 25 µl anti-NK 1.1 serum 4 hr prior to tumor implantation, b) B6 mice pretreated with normal mouse serum (NMS) and c) B6 bg/bg beige mice which are genetically deficient in NK activity. Tumor growth in the anti-NK 1.1 treated mice was significantly greater than in NMS-treated controls but not as great as in beige mice (Fig. 1).

The second experiment tested whether an additional injection of anti-NK 1.1 would have any further effect on tumor growth. B6 mice were pretreated with anti-NK 1.1 at -4 hr as before and given an additional injection of 25 µl of antiserum on day 2 after tumor implantation. Growth of B16 melanoma in these mice was indistinguishable from tumor growth in beige mice (Fig. 2). Both groups had significantly larger tumors than the NMS-treated controls.

IV. DISCUSSION

We have documented that *in vivo* treatment of mice with anti-NK 1.1 serum reduces splenic NK activity to YAC-1 targets *in vitro* with a concomitant reduction in the ability of intact mice to eliminate ^{125}IUdR-labeled lymphoma cells *in vivo* (Pollack and Hallenbeck, 1982). Those data provided direct evidence for a role of NK cells in the elimination of circulating tumor cells *in vivo.*

This model has now been used to test the effects of *in vivo* depletion of NK cells on the growth of a solid tumor, the B16 melanoma. The NK-sensitivity of B16 *in vivo* was strongly implicated by the studies of Talmadge *et al.* (1980) using B6 beige, NK-deficient mice. B16 grew significantly faster in beige than in normal control mice, an observation confirmed in the experiments reported here. Talmadge *et al.* concluded that the enhanced growth of B16 in beige mice was due to the abnormally low NK activity in these mice. Beige mice have other deficits, however, such as a defective B cell response to LPS (E. A. Clark *et al.*, this volume).

The results presented here strengthen the conclusion that the enhanced growth of B16 in beige mice is due to their inherent NK-deficit. Depletion of NK cells in B6 mice with one injection of NK 1.1 antiserum allowed significantly greater tumor growth. Two injections (25 µl each) of anti-NK 1.1 produced mice in which growth of B16 was iden-

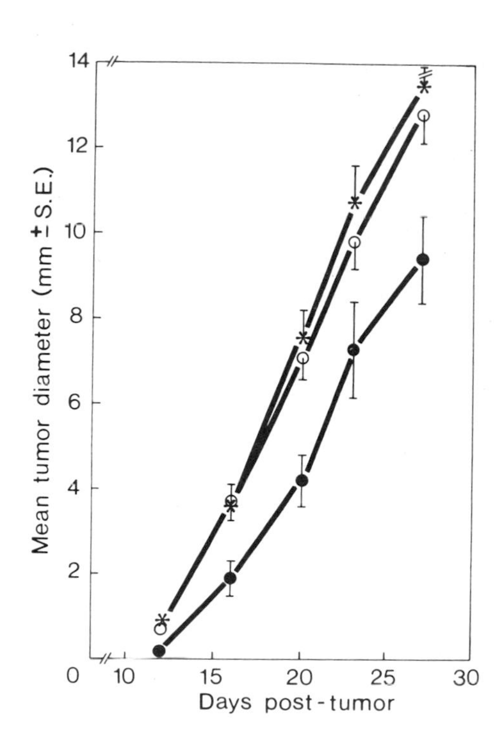

Figure 2. Growth of B16 melanoma in B6 mice injected 2X with NMS (●), or anti-NK 1.1 (*) and in B6 beige mice (○); ten mice per group. Sera injected at -4 hr and +2 days relative to subcutaneous inoculation of 5×10^4 B16 cells into each flank.

Day	B6-NMS vs. B6-anti NK 1.1	B6-NMS vs. B6 beige
16	$p < 0.01$	$p < 0.005$
20	$p < 0.001$	$p < 0.001$
23	$p < 0.02$	$p < 0.05$
27	$p < 0.005$	$p < 0.01$

tical to that in beige mice.

What is not known at this point is whether the effects of NK cells are mediated only by direct lysis of the tumor cells or whether indirect mechanisms also come into play. The intriguing possibility that NK cells play a regulatory role in the immune response remains to be explored. Our model of transient depletion of NK cells in mice will be useful in exploring that possibility.

In addition, other *in vivo* roles which NK cells may play can be directly tested with this model. In another report in this volume, we demonstrate that *in vivo* treatment with anti-NK 1.1 eliminates hybrid resistance of $B6D2F_1$ mice to B6 parental bone marrow and allogenic resistance of B6 mice to BALB/c bone marrow (Lotzova, Pollack and Savary, this volume).

V. ACKNOWLEDGEMENTS

The expert technical assistance of Linda Katzenberger and M. Craig Baily is gratefully acknowledged.

REFERENCES

Glimcher, L., Shen, F. W., and Cantor, H. (1979). J. Exp. Med. 145:1.
Herberman, R., and Ortaldo, J. (1981). Science 214:24.
Pollack, S. B., and Hallenbeck, L. A. (1982). Int. J. Cancer
Pollack, S. B., Tam, M. R., Nowinski, R. C., and Emmons, S. L. (1979). J. Immunol. 123:1818.
Talmadge, K., Meyers, M., Prieur, D. J., and Starkey, J. R. (1980). J. Nat. Cancer Inst. 65:929.
Tam, M. R., Emmons, S. L., and Pollack, S. B. (1980). J.

INVOLVEMENT OF SPONTANEOUS RATHER THAN INDUCED ANTITUMOR MECHANISMS IN RESISTANCE TO PRIMARY FIBROSARCOMA IMPLANTATION AND ITS SECONDARY SPREAD[1]

Robert Keller

Immunobiology Research Group
Institute of Immunology and Virology
University of Zurich
Switzerland

INTRODUCTION

The notion of a central role for immunospecific, T cell-dependent antitumor surveillance has been rigorously challenged by various observations. It is now widely recognized that quite apart from cytotoxic T lymphocytes, so long considered to play a central role, other, broad-range mechanisms capable of responding spontaneously and promptly to foreign cells are likely to be more directly involved than T cells in surveillance against tumors and microbial agents. Notable among these categories of effector cells are mononuclear phagocytes, polymorphonuclear leukocytes and 'natural killer' (NK) cells. These effectors manifest either spontaneous cytotoxicity, exhibiting no clearly defined specificity (but some selectivity for tumor targets) or (via their surface receptors for IgG) interact with target cells opsonized by specific antibody, thereby mediating antibody-dependent cellular cytotoxicity (1-3). Despite a great deal of work, a causal relationship between augmented activity of these natural effectors and resistance to neoplasia remains elusive.

[1] This work was supported by the Swiss National Science Foundation (grants 3.173.77 and 3.609.80) and the Canton of Zurich.

ISBN 0-12-341360-5

Towards establishing the role of these effectors in resistance to neoplasia, we felt that other models should now be explored. Briefly, the rationale for use of the present host-tumor system was as follows: Most cancer deaths are not attributable to the primary tumor but rather to its systemic spread and progressive growth at secondary sites (4-6). There is varied evidence that detachment and invasion of tumor cells into the vascular and/or lymphatic tree is a continuous process (7-9). However, the dissemination of neoplastic cells does not, of itself, invariably lead to metastastic tumor growth. Actually, the likelihood of metastases developing from cells gaining access to the circulation is relatively small. Observations such as these are viewed by us as indirect evidence for the presence and efficient functioning of host defense mechanisms capable of controlling, to some extent, the process of metastasis. These are complex interactions and surely depend on multiple factors pertaining to both host and tumor (3,9).

The present work, based on an experimental metastasizing tumor system, develops further evidence in support of a meaningful role for noninduced, spontaneous defense mechanisms in mediating resistance to the primary tumor inoculum and to secondary tumor cell spread.

Local Tumor Growth

Among the numerous fibrosarcomas induced in DA rats with dimethylbenz(a)anthracene, D-12 cells are unique in their capacity to grow in ascites form and to propagate in vitro for a considerable number of passages (3-9 m). Consequently, host-tumor interactions can be studied comparatively under both in vivo and in vitro conditions. Upon interaction in vitro with C. parvum-induced activated macrophages (initial effector to target cell ratio $\lesssim$2:1), 40 to 80 per cent of the D-12 cells are killed within 36 h; it is noteworthy that the macrophage-mediated tumoricidal process is detectable only after a lag phase of 8-16 h and requires 24-36 h for its full expression. In sharp contrast, D-12 cells resist NK-type killing (10). These in vitro findings, together with the results from interventions interfering with local tumor growth in vivo, are taken as evidence for a role for macrophages rather than NK cells in spontaneous and augmented resistance against this particular tumor. Subcutaneous, intramuscular or intraperitoneal inoculation of D-12 cells into the syngeneic host leads to progressive local tumor growth and the demise of the host. The ascites tumor model is operationally simple and reproducible (10-12,14).

Characteristics of Ascites Tumor Growth

Because of its versatility, the conditions for local tumor growth were studied preferentially in the D-12 ascites tumor model. The principal characteristics of ascites tumor growth are as follows.

1) Intraperitoneal inoculation of 10^2 to 10^3 D-12 cells consistently induces progressive local tumor growth in untreated animals, leading to their demise within 15-25 d. Within this relatively brief period of host survival, tumor growth remains essentially localized to the peritoneal cavity (10-12,14).
2) Spontaneous resistance to the tumor is diminished with age (15).
3) Intraperitoneal inoculation of immunostimulants (C. parvum, BCG) results in temporary enhancement of tumor resistance (10); this augmentation in resistance is operative only locally (14). In rats pretreated with C. parvum (3 mg i.p. on d -7) for example, i.p. inoculation of 10^3 D-12 cells on day 0 seldom led to progressive tumor growth ($<5\%$ of the animals). However, in the absence of special precautions, a considerable proportion of the animals died after an interval of 2-3 months as a consequence of progressive growth of a few tumor cells deposited along the needle injection channel in the subcutaneous tissue (50-70%). These findings indicate that the C. parvum-induced, marked increase in tumor resistance is restricted to the compartment into which it had been inoculated. Rechallenge i.p. of the remaining, tumor-free animals (between 25 and 45%) approximately 3 m after the first challenge consistently led to progressive local tumor growth; it is noteworthy that these animals died within the same time period as animals receiving a primary tumor cell challenge of the same size (10^3; 14).
4) Intraperitoneal inoculation of 72 h supernatants from Con A-stimulated rat spleen cells, disposing of considerable macrophage-activating capability but expressing no T cell growth factor activity, led to enhanced local resistance to the tumor (Keller, unpublished).
5) Adoptive transfer of C. parvum-induced adherent phagocytic cells from an individual rat into another individual,

performed on day -1 or 0, enhanced tumor resistance to an extent comparable to C. parvum (d -7; 14). Similar results were obtained in a Winn-type experiment when C. parvum-induced macrophage-type peritoneal cells from an individual rat were first mixed in vitro with D-12 ascites tumor cells and this mixture then inoculated i.p. into another individual rat. In experiments, in which a cell suspension from inguinal, retroperitoneal and axillary lymph nodes from an individual control rat were similarly mixed with tumor cells in vitro and the mixture inoculated i.p. into another control, tumor growth was measurably enhanced (13).

6) Ablation of spontaneous resistance to the tumor is effected with carrageenan, silica particles, hydrocortisone, dextran sulphate, polyvinyl sulphate or tumor promoters; this array of agents of diverse structure and composition share the capability of diminishing macrophage cytotoxicity in vitro (10,14, Keller, unpublished). Interestingly enough, their depressive effect on local host tumor resistance is attained only when they are given shortly before tumor cell challenge and into the same compartment (11).

These findings are interpreted by us as pointing to a role for local, spontaneous rather than immunospecific mechanisms mediating resistance to the primary tumor inoculum in this system, and in particular to a role for mononuclear phagocytes.

Secondary Tumor Growth

In the present system, the incidence of macroscopic metastases is quite dependent on the site of the primary tumor implant (12). Subcutaneous inoculation of tumor cells into the back seldom results in macroscopic metastases. In sharp contrast, subcutaneous inoculation of tumor cells into the thigh gives rise to spontaneous metastatic tumor growth in the majority of the animals. Surgical intervention involving excision of the primary tumor on the back leads frequently to macroscopic metastases or, when primary tumor inoculum is on the thigh, marked enhancement of the outgrowth of metastatic foci. Histologic and biologic findings affirm that dissemination of neoplastic cells from the primary tumor into draining lymph nodes is a spontaneous, early and continuing process (13). Thus, surgical removal of the primary tumor seems to promote the outgrowth of viable tumor cells already present in lymph nodes. Further experiments have shown that the growth characteristics of tumor cells derived from metastases are not measurably different from those inducing primary tumor growth (13).

It is noteworthy that animals which did not resume either local tumor growth nor develop macroscopic metastases within three months after surgical removal of the primary tumor were not immune (14); upon i.p. rechallenge with D-12 cells, almost all (48/50) died within 15-25 d (two animals died at 40 d). In this respect, animals that had experienced a significant exposure to the primary tumor responded to challenge just as did controls which had never previously experienced the tumor.

In other experiments, total peritoneal cells, spleen cells or a suspension from inguinal, retroperitoneal and axillary lymph nodes from an individual rat which did not resume local tumor growth nor develop macroscopic metastases after surgical removal of the primary tumor, were inoculated i.p. into another control on d -1; in no case was local resistance to the ascites tumor enhanced (Keller, unpublished).

These findings argue against a protective role by conventional, specific immunity in mediating resistance to this tumor.

Thus, the available evidence, though still fragmentary, indicates that a form of natural resistance, which can be selectively enhanced or suppressed, is operative in a syngeneic tumor-host model. This noninduced state of resistance seems decisively to affect the outcome of the early, local interaction with the host as regards the primary tumor implantation and its metastatic spread.

ACKNOWLEDGMENT

The excellent technical assistance of Ms. R. Keist, F. Chytil and R. Werschler is gratefully acknowledged. I thank Dr. M. Landy for reviewing the manuscript.

REFERENCES

1. James, K., McBride, W.H., and Stuart, A. (1977). "The Macrophage and Cancer." Econoprint, Edinburgh.
2. Herberman, R.B. (1980). "Natural Cell-Mediated Immunity Against Tumors." Academic Press, New York.
3. Herberman, R.B., and Ortaldo, J.R. (1981). Science 214:24.
4. Willis, R.A. (1973). "The Spread of Tumours in the Human Body." 3rd ed. Butterworths, London.
5. Baldwin, R.W. (1978). "Secondary Spread of Cancer." Academic Press, London.
6. Roos, E., and Dingemans, K.P. (1980). Biochem. Biophys. Acta 560:135.

7. Van de Velde, C., and Carr, I. (1977). Experientia 33:837.
8. Griffith, J.D., and Salsbury, A.J. (1965). "Circulating Cancer Cells." Thomas, Springfield, Ill.
9. Woodruff, M.F.A. (1980). "The Interaction of Cancer and Host. Its Therapeutic Significance." Grune and Stratton, New York.
10. Keller, R. (1980). In "Natural Cell-Mediated Immunity Against Tumors." (R.B. Herberman, ed.), p. 1219. Academic Press, New York.
11. Keller, R. (1980). In "Mononuclear Phagocytes. Functional Aspects." (R. van Furth, ed.), p. 1725. M. Nijhoff, The Hague.
12. Keller, R. (1981). Invasion and Metastasis 1:136.
13. Keller, R., and Hess, M.W. (submitted)
14. Keller, R. (submitted)
15. Keller, R. 1978. Br. J. Cancer 38:557.

NATURAL KILLER CELL-MEDIATED CYTOTOXICITY AGAINST SOLID TUMORS: *IN VITRO* AND *IN VIVO* STUDIES

Nabil Hanna[1,2]

Cancer Metastasis and Treatment Laboratory
NCI-Frederick Cancer Research Facility
P.O. Box B
Frederick, Maryland 21701

I. INTRODUCTION

The heterogeneity of the effector cells responsible for natural cell-mediated cytotoxicity (NCMC) against tumor target cells *in vitro* has been suggested by several investigators (Burton et al., 1981; Koo et al., 1980; Minato et al., 1981; Stutman et al., 1978). Stutman et al., (1980) provided evidence that the natural cytotoxic (NC) cells that kill solid tumors (Meth-A fibrosarcoma) differ from the natural killer (NK) cells that lyse lymphoma targets (YAC-1 lymphoma) in such characteristics as presence of antigenic markers, target cell specificity, strain and organ distribution, ontogeny, and sensitivity to X-irradiation, cyclophosphamide (Cy) and β-estradiol treatment. However, studies of NCMC against murine ultraviolet (UV)-induced fibrosarcomas and K-1735 and B16 melanomas, both *in vitro* and *in vivo*, have shown that the effector cells active against these adherent solid tumor target cells are NK cells (Hanna and Burton, 1981; Hanna and Fidler, 1980). These findings suggested that the type of effector cells involved in NCMC could not be determined solely by the *in vitro* assay used (4 hours vs. 24 hours) or by the histological classification of the target cells

---

[1]Research sponsored by the National Cancer Institute, DHHS, under Contract NO. NO1-CO-75380 with Litton Bionetics, Inc. The contents of this publication do not necessarily reflect the views or policies of the Department of Health and Human Services, nor does mention of trade names, commercial products or organizations imply endorsement by the U.S. Government.
[2]Present address: Smith Kline and French Laboratories, Dept. of Immunology, 1500 Spring Garden Street, P.O. Box 7929, Philadelphia, Pennsylvania 19101.

ISBN 0-12-341360-5

(lymphomas vs. solid tumors). The definition of the effector mechanisms involved in host-tumor interaction is crucial for the design of a relevant in vivo model for evaluating the role of NK and/or NC cells in the control of the growth and dissemination of malignant neoplasms.

Although the NK and NC systems share several cellular and biological characteristics, the present discussion will be limited to models that allow the two effector mechanisms to be clearly distinguished.

II. ASSAY SYSTEMS

In these studies, NCMC against YAC-1 lymphoma target cells was determined by a short-term (4 hour) ^{51}Cr-release assay and cytotoxicity against [^{3}H]proline-prelabeled adherent fibrosarcoma and melanoma targets by a long-term (20 hour) assay (Hanna and Fidler, 1980). Occasionally, the 18-hour ^{51}Cr-release assay also was used to assess killing of adherent solid tumor cell lines (Hanna and Fidler, 1981a).

III. THE EXPRESSION OF NCMC AGAINST THE UV-2237 FIBROSARCOMA AND THE K-1735 and B16 MELANOMAS IS AGE DEPENDENT.

One of the major differences between NK and NC cells is the effect of the donor's age on the expression of their activity. The level of NK-cell-mediated cytotoxicity was found to be very low or absent in mice younger than 3 weeks old (Herberman et al., 1975), whereas the level of NC-cell-mediated cytotoxicity was already high in newborn mice and did not change with age (Stutman et al., 1978). We tested the NCMC of spleen cells obtained from 2- to 20-week-old syngeneic C3H, HeN/MTV^{-}, C57BL/6N, and allogenic nude (BALB/c or N:NIH (S)) mice against UV-2237 fibrosarcoma and K-1735 and B16 melanoma target cells in vitro. No cytotoxic activity by spleen cells from 2-week-old or younger mice and only marginal cytotoxicity (<10%) by spleen cells from 3-week-old donors could be detected (Hanna, 1980; Hanna and Fidler, 1980). In contrast, mice older than 5 weeks exhibited significant levels of activity. Although marked fluctuations were observed among individual animals, no consistent pattern of an age-dependent decline of cytotoxicity could be detected in mice older than 12 weeks old. The cytotoxic activity of spleen cells from 3-week-old mice, however, could be boosted readily by the administration of bacterial adjuvants or interferon inducers, suggesting that the precursors of the effector cells are present but not yet activated in young mice. The young nude mouse model was used to evaluate the correlation between the levels of NK and/or NC cell activity and the in vivo destruction of circulating fibrosarcoma and melanoma cells and the inhibition of their metastatic spread to distant organs (Hanna and Fidler, 1980). The results of this study can be summarized as follows: 1) the incidence of lung tumor colonies in 3-week-old mice injected with metastatic syngeneic tumor cells (UV-2237, K-1735, B16) was 10- to 100-fold higher than in 8-week-old mice (adult), 2) activation of NK cells in young recipients reduced the incidence of metastases to levels observed in adult mice, and 3) more [^{125}I]iododeoxyuridine-labeled tumor cells survived in the circulation during the first 24 hours after intravenous injection in 3-week-old mice than in adult mice. These findings indicate that the ontogeny of the natural effector mechanism active against UV-induced fibrosarcomas and K-1735 and B16 melanomas, both in vitro and in vivo, is similar to that of NK cells active against lymphoma target cells and different from that of NC cells.

IV. ACTIVITY IN NUDE MICE

In contrast to NC-cell-mediated cytotoxicity, which is present at comparably high levels in both athymic nude mice and normal immunocompetent mice (Stutman et al., 1980), the killing of UV-induced fibrosarcomas and K-1735 and B16 melanomas by spleen cells from adult nude mice was found to be consistently and significantly higher than by spleen cells from age-matched normal mice. The high levels of NK cell activity in adult nude mice was associated with marked resistance to the development of metastases of allogeneic and xenogeneic tumors (Hanna, 1980; Hanna and Fidler, 1981b). Tumor cell variants selected for high metastatic potential in syngeneic mice failed to metastasize in adult nude mice. In contrast, they readily metastasized in 3-week-old nude mice that exhibited low NK-cell-mediated cytotoxicity. Moreover, partial depletion of NK cells by treatment with β-estradiol (Seaman, et al., 1978) or Cy (Mantovani et al., 1978), which does not influence the levels of NC cell activity (Stutman et al., 1980), rendered adult nude mice susceptible to the development of pulmonary metastases of B16 melanomas (Table I).

These findings substantiate the correlation between NK activity and the resistance of nude mice to hematogeneous metastasis of certain solid tumors. A positive correlation was consistently found in tests of more than 30 UV-induced, chemically induced, and spontaneous fibrosarcomas and melanomas of mouse, rat, and human origin.

TABLE I. Correlation between the levels of NK cell-mediated cytotoxicity and the incidence of tumor metastases in nude mice.

Age of nude mice (wks)	Treatment[a]	NK cell activity against: YAC-1 ($LU/10^7$ spleen cells)[b]	NK cell activity against: UV-2237[c] (% cytotoxicity)	Median no. of lung tumor colonies[d] (range)
8	--	145	68	0 (0- 4)
8	Cy	25	22	85 (41-123)
3	--	21	8	168 (98-230)
12	--	83	54	3 (0- 11)
12	β-estradiol	35	31	68 (24- 98)

[a]200 mg/kg Cy injected 4 days before test.
[b]One lytic unit (LU) = number of spleen cells that kill 25% of the target cells.
[c]Percent cytotoxicity at 10^6 effector cells per well.
[d]B16-F10 melanoma cells were injected intravenously (10^5 cells per mouse).

V. ACTIVITY IN BEIGE MICE

The beige mutant of the C57BL/6 mouse showed that it exhibited low levels of NK-cell-mediated cytotoxicity (Roder and Duwe, 1979), but normal levels of NC cell activity (Stutman and Cuttito, 1981). The NK cell activity of spleen cells from untreated and lymphocytic choriomeningitis-virus-treated beige mice against B16 melanoma cell lines *in vitro* was also found to be lower than that of spleen cells from normal C57BL/6 mice (Talmadge et al., 1980). These findings clearly indicate that the beige mouse can be a valuable tool for evaluating the *in vivo* role of NK cells in the resistance against the growth and dissemination of solid tumors. Indeed, recent studies have indicated that after intraveneous or subcutaneous injection, B16 melanoma tumor cells metastasized more readily and at higher frequency in beige mice than in normal age-matched C57BL/6 mice (Hanna and Fidler, 1981a; Talmadge et al., 1980).

VI. SENSITIVITY TO CY

Several studies have shown that NK, but not NC, cells are susceptible to treatment with Cy (Mantovani et al., 1978; Stutman et al., 1980). We found that treatment of C3H or C57BL/6 mice with a single dose of 200 mg/kg Cy four days before *in vitro* testing reduced the expression of NK-cell-mediated cytotoxicity against YAC-1 targets by 70-85% and against UV-2237 and B16 tumors by 40-65%. The effect was more pronounced (>90% inhibition) when the cytotoxic activity was calculated per total spleen cell number which took into consideration the reduced number of mononuclear cells recovered from spleens of Cy-treated mice. Moreover, our studies of the *in vivo* metastatic behavior of UV-induced fibrosarcomas and melanomas in Cy-treated syngeneic mice substantiated further that Cy-sensitive NK cells play a significant role in the destruction of circulating tumor cells and in the inhibition of hematogeneous tumor metastasis (Hanna and Burton, 1981; Hanna and Fidler, 1980). These studies demonstrated that 1) experimental pulmonary metastasis of various lines and clones of the UV-2237, K-1735, and B16 tumors was markedly increased (20- to 200-fold) in mice treated four days earlier with Cy, 2) the survival of radiolabeled tumor cells in the circulation of Cy-treated recipients was enhanced during the first 24-48 hours after intravenous tumor inoculation, and 3) adoptive transfer of normal spleen cells 24 hours before, but not after, tumor cell injection reversed the enhancing effect of Cy. The active lymphoid cell was 1) nylon wool non-adherent, 2) Cy-sensitive, 3) found at low levels in young 3-week-old mice, 4) present in high concentrations in adult nude mice, and 5) positive for NK-1.2 antigen. These cellular characteristics, in particular the sensitivity to treatment with anti-NK-1.2 serum and complement (C), provide conclusive evidence that the effector cell active against these solid tumors, both *in vitro* and *in vivo*, is an NK cell.

VII. SENSITIVITY TO β-ESTRADIOL

Chronic administration of the estrogen β-estradiol into mice produced a significant decrease in anti-YAC-1 NK cell activity activity that could not be boosted by the administration of the interferon inducer polyinosinic-polycytidylic acid (poly I•C)(Seaman et al., 1978, 1979). In contrast, estrogen treatment had no effect on NC-cell-mediated cytotoxicity against Meth-A target cells (Stutman and Cuttito, 1981). Using UV-2237 fibrosarcoma cells as targets, we found 25-50% less cytotoxic activity by spleen cells obtained from β-estradiol-treated mice than by control groups. The NK activity was more profoundly and consistently inhibited, however, when NK cells were activated by poly I•C or *Corynebacterium parvum*. Administration of either agent to β-estradiol-treated mice failed to boost the cytotoxic activity against both YAC-1 lymphoma and UV-2237 fibrosarcoma target cells (Table II).

In vivo studies showed that [^{125}I]iododeoxyuridine-labeled UV-2237 and B16 tumor cells in the circulation of β-estradiol-treated syngeneic mice survived longer than in normal controls. This enhanced survival was associated with an increased incidence of experimental and spontaneous pulmonary metastasis of the three tumor systems studied. Furthermore, treatment with poly I•C failed to inhibit the development of experimental metastases in β-estradiol-treated mice (Table II). This result contrasts sharply with the antimetastatic effect of poly I•C when it is administered to normal mice. In these experiments, treatment with β-estradiol did not inhibit interferon production or the

TABLE II. Effect of treatment of C3H mice with β-estradiol on NCMC against syngeneic UV-2237 fibrosarcoma cells

		NK cell activity against		
Treatment groups	Poly I•C (50 μg/ mouse	YAC-1 (LU/10^7 spleen cells)[a]	UV-2237 percent cytotoxicity[b]	Median no. of lung tumor colonies[c] (range)
Normal controls	-	46	42	78 (48-109)
	+	237	79	4 (0- 11)
β-estradiol	-	12	24	136 (73-194)
	+	18	28	120 (61-173)

[a]One lytic unit (LU) = the number of effector cells that kill 15% of the YAC-1 target cells.
[b]Percent cytotoxicity at 10^6 spleen cells per well.
[c]UV-2237-M3 cell line was injected intravenously (5 X 10^4 cells per mouse) and the number of lung colonies was counted 3 weeks later. Poly I•C was injected intraperitoneally one day before inoculation of tumor cells.

in vivo activation of tumoricidal alveolar or peritoneal macrophages by the injection of poly I·C or *C. parvum*. These findings substantiate the concept that poly I·C-induced prevention of tumor metastasis is mediated primarily by activated NK cells.

VIII. SPECIFIC ANTIGENIC MARKERS

The NK-1 alloantigens (NK-1.1 and NK-1.2) have been reported to be expressed exclusively on NK cells active against lymphoma target cells (Burton and Winn, 1981; Glimcher et al., 1977). Treatment of spleen cells with anti-NK-1.2 serum and C reduced the NCMC against YAC-1 lymphoma cells by 90-95% (Burton et al., 1981). Conversely, NC cells active against Meth-A target cells were NK-1 negative, and their cytotoxic activity was not affected by treatment with anti-NK-1.2 serum and C (Stutman et al., 1980). Using UV-2237 fibrosarcoma or B16 melanoma cells as targets in a 24-hour cytotoxicity assay, we found that treatment of spleen cells with anti-NK-1.2 antibodies and C reduced their NCMC by 75-95% (Hanna and Burton, 1981, Table III). The same treatment, however, reduced the cytotoxic activity against Meth-A cells only 12-18%.

TABLE III. Effect of treatment with anti-NK-1.2 antibodies and C on NCMC against YAC-1 UV-2237, B16, and Meth-A target cells

Treatment	% cytotoxicity			
	YAC-1[a] (50:1)	UV-2237[b] (100:1)	B16[b] (100:1)	Meth-A[b] (100:1)
None, control	43	48	31	72
NMS + C	39	51	33	70
Anti-NK-1.2 + C[c]	2	6	5	57

[a]4-hour ^{51}Cr-release assay.
[b]20-hour [^{3}H]proline cytotoxicity assay.
[c]Effector-cell-to-target-cell ratio.
[d]Anti-NK-1.2 (CE X NZB)F1, anti-CBA was used at 1:100 final dilution. Rabbit C was used at 1:8.

The selectivity and effectiveness of anti-NK-1 antibody in depleting NK cells active against metastatic variant cell lines of UV-2237 and B16 tumors enabled us to provide direct evidence that NK cells are responsible for the reconstitutive antimetastatic capacity of normal spleen cells adoptively transferred to Cy-treated mice (Hanna and Burton, 1981).

The asailo GM1 is another antigenic marker that has been found on NK cells (Young et al., 1980). Treatment with anti-asialo GM1 serum and C reduced the anti-YAC-1 NK cell-mediated cytotoxicity of normal spleen cells by 75-85% and of *C. parvum*-activated NK cells by 45-65%. The same treatment with anti-asialo GM1 antibody reduced the cytotoxic activity against UV-2237 fibrosarcoma target cells by 50%. Since activation of precursors of NK cells may occur during the long-term (24 hour) cytotoxicity assay *in vitro*, the effect of antibody treatment on such precursors should be determined. Recent results indicate that pre-NK cells are NK-1 positive, since spleen cells treated with anti-NK-1.2 and C could not be activated by interferon *in vitro* (Hanna, unpublished observation).

In summary, the effector cells mediating natural cytotoxicity against UV-induced fibrosarcomas and K-1735 and B16 melanomas are nonadherent to nylon wool, present at high levels in nude mice and low levels in beige mice, and sensitive to treatment with Cy, β-estradiol, anti-asialo GM1 and C, and anti-NK-1 sera and C. In addition, the expression of their activity is age dependent. The evidence presented here clearly indicates that NK cells are effective in killing solid tumor cells in a long-term cytotoxicity assay *in vitro* and in destroying circulating tumor cells *in vivo*, thus inhibiting the subsequent establishment of distant organ metastases.

REFERENCES

1. Burton R.C., Bartlett, S.P., Kumar, V., and Winn, H.J. (1981). J. Immunol. 127:1864.
2. Burton, R.C. and Winn, H.J. (1981). J. Immunol. 126:1985.
3. Glimcher, L., Shen, F.W., and Cantor, H. (1977). J. Exp. Med. 145:1.
4. Hanna, N. (1980). Int. J. Cancer 26:675.
5. Hanna, N. and Burton, R.C. (1981). J. Immunol. 127:1754.
6. Hanna, N. and Fidler, I.J. (1980). J. Natl. Cancer Inst. 65:801.
7. Hanna, N. and Fidler, I.J. (1981a). J. Natl. Cancer Inst. 66:1183.
8. Hanna, N. and Fidler, I.J. (1981b). Cancer Res. 41:438.

9. Herberman, R.B., Nunn, M.E. and Laurin, D.H. (1975). Int. J. Cancer 16:216.
10. Koo, G.C., Jacobson, J.B., Hammerling, G.J. and Hammerling, U. (1980). J. Immunol. 125:1003.
11. Mantovani, A. Luini, W., Peri, G., Vecchi, A., and Spreafico, F. (1978). J. Natl. Cancer Inst. 61:1255.
12. Minato, N., Reid, L., and Bloom B.R. (1981). J. Exp. Med. 154:750.
13. Roder, J. and Duwe, A. (1979). Nature 278:451.
14. Seaman, W.E., Blackman, M.A., Gindhart, T.D., Roubinian, J.R., Loeb, J.M. and Talal, N. (1978). J. Immunol. 121:2193.
15. Seaman, W.E., Merigan, T.C., and Talal, N. (1979). J. Immunol. 123:2903.
16. Stutman, O., and Cuttito, M.J. (1981). Nature 290:254.
17. Stutman, O., Figarella, E.F., Paige, C.J. Lattime, E.C. (1980). In "Natural Cell-Medicated Immunity Against Tumors. (R.B. Herberman, ed.), p. 187. Academic Press, New York.
18. Stutman, O., Paige, P.J., and Figarella, E.F. (1978). J. Immunol. 121:1819.
19. Talmadge, J.E., Meyers, K.M., Prieur, D.J., and Starkey, J.R. (1980). Nature 284:622.
20. Young, W.W., Hakomori, S.I., Durdik, J.M., and Henney, C.W. (1980). J. Immunol. 124:199.

THE BEIGE MODEL IN STUDIES OF NATURAL RESISTANCE TO SYNGENEIC, SEMISYNGENEIC, AND PRIMARY TUMORS[1]

Klas Kärre
Gunnar O. Klein
Rolf Kiessling
Shmuel Argov
George Klein

Dept. of Tumor Biology
Karolinska Institute
Stockholm, Sweden

I. INTRODUCTION

The concept of natural surveillance against tumors postulates that the immune system can recognize and eliminate neoplastic cells before they give rise to overt malignancies. The role of NK-cells in this hypothetical defence is sometimes discussed in terms of spontaneous tumor incidence in various inbred strains of mice that show constant differences in their levels of NK-activity. It may for instance be argued that certain strains have a high incidence of spontaneous tumors in spite of their high NK-activity, whereas other strains have low cytolytic levels, and yet do not develop tumors at high frequency. This line of reasoning requires two presumptions: i) the formation of incipient neoplastic cells is a frequent event in all strains of mice, and ii) NK-cells represent the only mechanism under genetic control which can prevent the outgrowth of such cells. Since none of these presumptions are likely to be true (1,2), we consider comparisons of tumor resistance in individuals on common genetic background as a more valid approach to assess the *in vivo* function of NK-cells. Here, we will review some of our own studies along these lines,

[1]Supported by NIH grants CA 14054-06 and CA 26782-02, and by grants from the Swedish Cancer Society.

ISBN 0-12-341360-5

with mice carrying the bg^J (bg) mutation. This model involves littermate animals that differ only at a single genetic locus interfering with NK-function (3,4). We will include already published data as well as some more recent observations, and we will focus the discussion on a comparison between natural resistance as observed against syngeneic, semisyngeneic and primary tumors.

Homozygosity for the bg allele leads to a partial impairment of NK-activity, whereas several monocyte-macrophage as well as T- and B-cell functions remain intact (3). One may however note that macrophage mediated tumor cell lysis showed a delay in beige mice in one experimental system (5). Furthermore, it has been reported that beige mice infected with lymphocytic choriomeningitis virus show a reduced specific antiviral cytotoxicity compared to wild type controls (4). The phenotypic manifestations of the bg mutation, which is thought to primarily affect lysosomal membrane functions, also include abnormal granulocyte morphology and function as well as melanosomal function, leading to a light coat color (6).

Clearly, the bg mutation does not provide an ideal model-the NK-defect it causes is neither absolute, nor selective. However, in comparisons with other mutations that affect NK activity (7), we consider C57Bl.bg/bg mice and their phenotypically normal +/bg littermates as the best available genetic model for studies of NK-function *in vivo*; for instance if no difference in tumor resistance can be demonstrated between beige and wild type mice in an experimental system, this would argue against any NK-mediated defence in that particular situation, irrespective of the fact that the mutation is not selective for NK-function.

II. RESULTS AND DISCUSSION

A. Resistance to Syngeneic Lymphomas

We have mainly worked with small grafts of T-lymphomas as a model for a recently induced and still small tumor mass. Titrated doses of two virally (P-52, Rad-LV; RBL-5, Rauscher virus) and one chemically (EL-4) induced ascites lines were inoculated subcutaneously in B6.+/bg mice (see ref. 8 and 9 for details on animal breeding, tumor lines and the transplantation assay). For each of the tumors, a threshold dose giving tumor takes in 30-45% of the normal syngeneic animals were chosen for further studies. This would provide a situation where the inoculum was large enough to establish a growing tumor, although only in a proportion of the animals, possibly due to an active rejection mechanism. The bg/bg mice were more susceptible to outgrowth of progressively growing tumors at the chosen cell doses. The P-52 (8,9) and EL-4 (fig. 1a) behaved similarly, giving significantly more tumor takes in bg/bg mice than in +/bg littermates. This difference in rejection potential was only marginal with RBL-5 cells (9, compare also with results reported in section B below and fig. 1c). The differences in tumor takes between beige and wild type mice were more impressive in the early phase after tumor inoculation (9), indicating that early acting mechanisms might be responsible for tumor rejection. In agreement with these observations, we found that P-52 and EL-4 cells (^{125}I-IUdR labeled while growing in vivo or in vitro) were eliminated more efficiently from organs of +/bg than bg/bg mice during the first 4-36 h after i.v. injection (8,9). With this assay for natural killing in vivo (10,11) we observed differences in pulmonary as well as in splenic and hepatic radioactivity between beige and wild type mice, suggesting a difference in rapid elimination and not just in organ distribution of the injected lymphoma cells.

We havc also designed two different models where the effect of the bg mutation on natural resistance can be studied in mice with thymus deficiency. In the first approach, we were able obtain "beige-nude" (C57Bl.bg/bg.nu/nu) mice through a breeding scheme with C57Bl.bg/bg.+/+ x C57Bl.+/+.+/nu as the original cross (9). The bg mutation was found to cause a defect in NK function also in nude mice, although the homozygosity for the nu alleles had compensated partly for this by elevating the NK-activity above the very low levels observed in beige mice without the nu defect (9). In a second approach, we constructed thymus deficient bone marrow chimeras, taking advantage of the fact that the beige NK defect can be transferred to wild type recipients by transplantation of beige bone marrow after whole body irradiation (12). Thymectomized and control C57Bl.+/+ mice were irradiated (750 R) and reconstituted with syngeneic (anti-Thy-1 treated) +/bg or bg/bg bone marrow. After 3 weeks the mice were challenged with 5 x 10^2 EL-4 cells, and followed by weekly palpations (fig. 1b). Although the number of mice in these experiments are limited, the data indicate that the mechanism responsible for resistance in +/bg mice which is defective in bg/bg mice is bone marrow derived and does not require an intact thymus to develop. Indeed, thymectomy before irradiation and reconstitution seemed to increase rather than to abrogate the resistance of +/bg bone marrow recipients. This observation is reminiscent of the strong resistance against certain transplantable tumors in nude mice, which is thought to be NK-mediated (13).

Several different mechanisms may probably contribute to natural resistance against tumors (2). However, in the natural surveillance against syngeneic lymphomas observed with the beige model, the differences in in vivo killing of radiolabeled lymphoma cells were paralleled by in vitro cytotoxicity mediated by non-adherent cells (8,9), arguing against monocytes or macrophages as the responsible effector mechanism. Taken together with our observations that the mechanism responsible for resistance is bone marrow dependent and develops in the absence of a thymus, the findings support that NK-cells can contribute to natural defense against syngeneic tumors.

B. Resistance to Lymphoma Grafts in Mice Segregating for H-2 and the Bg Mutation

In spite of the profound NK-defect associated with the bg mutation, the differences in resistance between +/bg and bg/bg mice as discussed in the previous section were small and came out convincingly only after studies of large groups (30-70) of animals. Furthermore, a tenfold increase of the number of grafted cells to 5 x 10^3 or 10^4 resulted in growth in all mice, also of the wild type (9). It is interesting to compare these observations with our previous studies of H-2 linked control of NK-activity and "hybrid resistance" against RBL-5 lymphoma cells in F_1 hybrid and backcross mice (14,15). In that situation, the distribution of NK-activity in $H-2^{b/b}$ mice (NK-low) had partly overlapped with that of $H-2^{d/b}$ heterozygotes (NK-high), and yet the differential in tumor resistance was quite impressive and observed with up to 10^4 grafted cells. This implied that hybrid resistance might depend on other (or additional) factors than the preexisting NK-activity, and the question arose whether the in vitro cytotoxicity or the presence of the H-2 linked gene(s) was the better predictor of the ability of mice to reject tumor transplants. In previous studies this could not be tested, due to the simultaneous presence of both. We therefore constructed the backcross (C57Bl/10.D2♂ x C57Bl/6.bg/bg♀)♀ x C57Bl/6.bg/bg♂, where two loci that influence NK-activity - H-2 and bg - should segregate independently on C57Bl background.

There were four types of offspring in the backcross. Half of the mice were homozygous for $H-2^b$, whereas the other half were $H-2^{d/b}$ heterozygotes, as determined by serological typing on spleen cells. In each of the two groups, half of the mice carried the beige mutation in homozygous form (beige phenotype) and the remaining were heterozygous (wild type, fig. 1c). In spite of the fact that all mice of the beige phenotype were deficient in NK-activity against YAC-1 and RBL-5 targets in the in vitro test (not shown), the $H-2^{d/b}$ bg/bg were more resistant than $H-2^{b/b}$ (+/bg and bg/bg) mice to in vivo challenge with RBL-5 cells, i e H-2 linked resistance was expressed in beige mice (fig. 1c). However, within the H-2 heterozygous group, the bg/bg mice were more susceptible to growth of RBL-5 than the +/bg mice. There are three possible interpretations of these findings:

i) NK-cells do not contribute to H-2 linked resistance, which is mediated by some other mechanism (such as T-cells, macrophages or NC-cells).

ii) NK-cells, or their precursors, present in beige mice, might be able to suppress growth of parental tumor cells in heterozygous mice by other means than direct killing, e.g. by the production and release of interferons. This possibility is raised since the bg mutation does not seem to affect the number of target binding, putative NK-cells, but only their lytic mechanism (3).

iii) The defective NK-cells in beige mice could be triggered to become cytolytic *in vivo* in a process controlled by H-2 linked genes, and thereby contribute to "hybrid resistance". NK-activity can be induced in beige mice by interferon inducers, although the cytolytic levels do not reach those of similarly treated wild type animals (4). Our results would be in line with such an *in vivo* triggering of NK-cells, since the resistance introduced by $H\text{-}2^d$ in bg/bg backcross mice did not reach the level of resistance conferred by this haplotype to +/bg mice in the same offspring (fig. 1c).

Intravenous injection of rabbit anti-asialo-GM-1 antiserum reduces NK-activity of mice dramatically, and it can be maintained at negligible levels by repeated administrations of the antiserum (16). There is evidence that the reduction of cytolytic activity is caused by elimination rather than by functional suppression of NK-cells (16). Hence, it should be possible to exclude alternatives ii) and iii) above if H-2 linked resistance was maintained in mice treated with anti-asialo-GM-1. A part of the +/bg backcross mice received 30 µl of the antiserum (kindly donated by Dr. S. Habu, Tokai University) in i.v. injections every third day, starting one day prior to tumor challenge, and with a total of six administrations. This treatment rendered the majority of $H\text{-}2^{d/b}$.+/bg mice susceptible to outgrowth of RBL-5 grafts (fig. 1c). Note also that the slight delay before tumor appearance in $H\text{-}2^{b/b}$.+/bg (as compared to $H\text{-}2^{b/b}$.bg/bg) mice disappeared when the former were treated with anti-asialo-GM-1 (fig. 1c). These results are consistent with the hypothesis that resistance to H-2 identical as well as to H-2 semisyngeneic tumor cells is mediated by NK-cells, and alternatives ii) and iii) above can not be excluded. However, in order to prove that one of these interpretations is correct, it would be essential to demonstrate that the tumor growth enhancing effect of anti-asialo-GM-1 really depends on the abrogation of the NK-system.

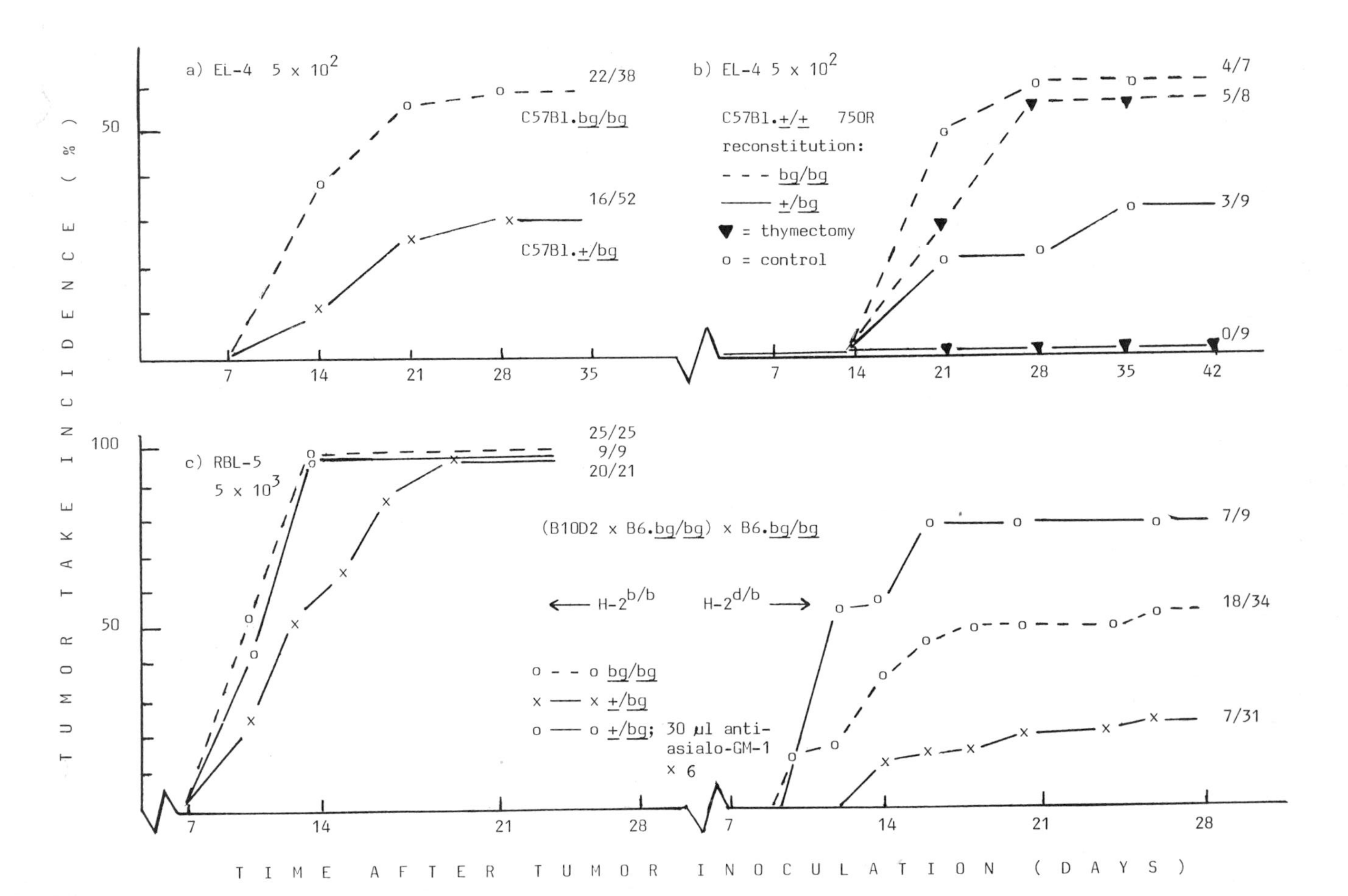

a) EL-4 5 x 10^2
22/38
C57Bl.bg/bg
16/52
C57Bl.+/bg
b) EL-4 5 x 10^2
4/7
5/8
C57Bl.+/+ 750R
reconstitution:
- - - bg/bg
—— +/bg
▼ = thymectomy
o = control
3/9
0/9
c) RBL-5
5 x 10^3
25/25
9/9
20/21
(B10D2 x B6.bg/bg) x B6.bg/bg
H-2^b/b
H-2^d/b
7/9
18/34
7/31
o - - o bg/bg
x —— x +/bg
o —— o +/bg; 30 µl anti-asialo-GM-1 x 6
50
100
7
14
21
28
35
42
TIME AFTER TUMOR INOCULATION (DAYS)
TUMOR TAKE INCIDENCE (%)

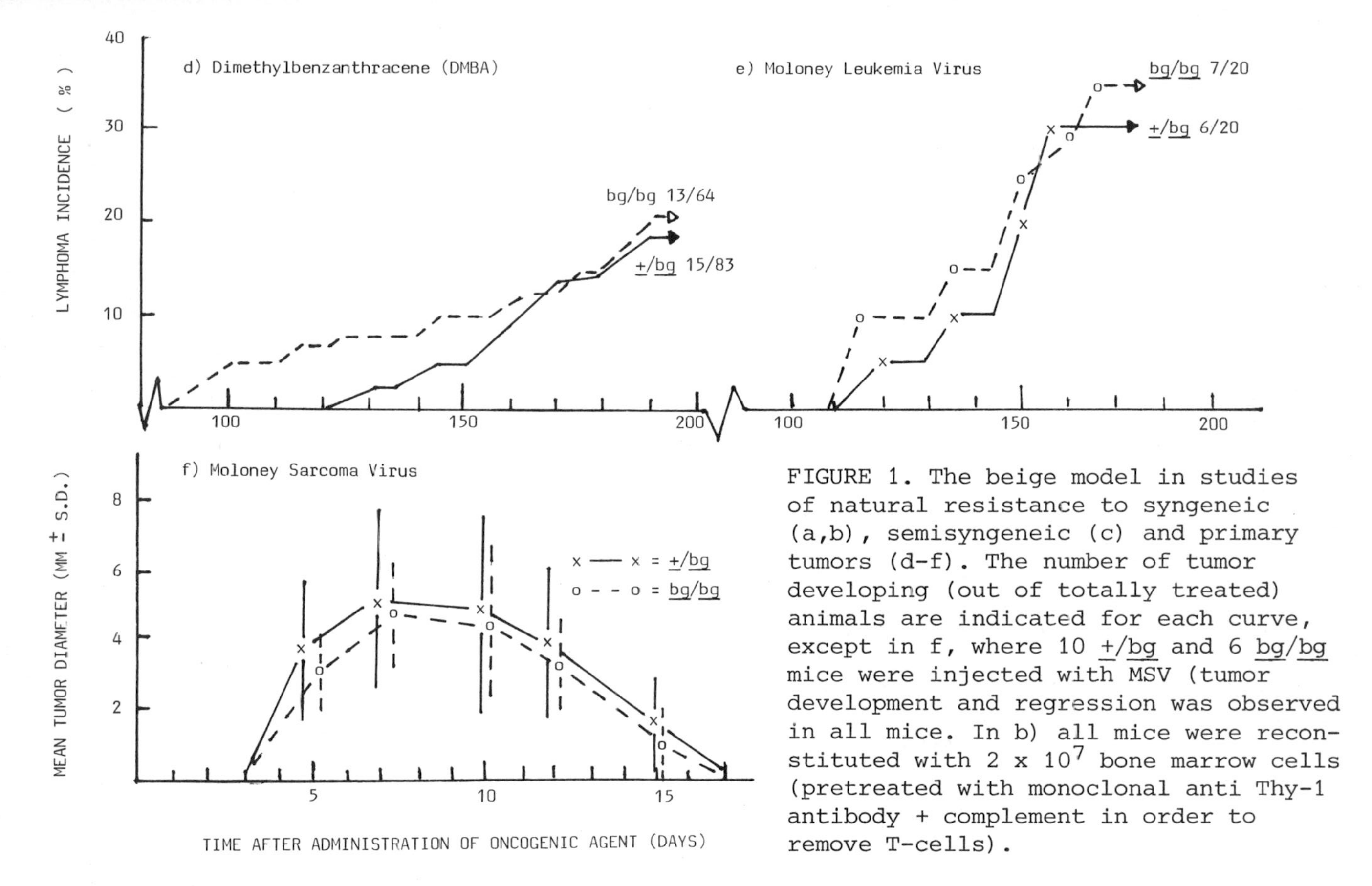

FIGURE 1. The beige model in studies of natural resistance to syngeneic (a,b), semisyngeneic (c) and primary tumors (d-f). The number of tumor developing (out of totally treated) animals are indicated for each curve, except in f, where 10 +/bg and 6 bg/bg mice were injected with MSV (tumor development and regression was observed in all mice. In b) all mice were reconstituted with 2 x 10^7 bone marrow cells (pretreated with monoclonal anti Thy-1 antibody + complement in order to remove T-cells).

C. Resistance to primary tumors

In our first attempt to study whether the bg mutation also affects resistance to induction and outgrowth of primary tumors, a total of 64 C57Bl.bg/bg and 83 +/bg littermates received the chemical carcinogen dimethylbenzanthracene (DMBA) in 5 weekly intragastric doses of 1 mg. By 165 days after the first feeding with DMBA, 18% of the wild type and 31% of the beige mice had developed tumors (17). This difference was due to a higher incidence of cutaneous and subcutaneous tumors in beige mice (20% vs 8% in wild type mice, mostly carcinomas but also including papillomas and non-epithelial tumors). The incidence of lymphoma was similar in the two genotypes, although the lymphomas appeared somewhat earlier in beige than in control mice (see fig. 1d, where the mice have been followed for 200 days). These slight differences between the two genotypes of mice were not statistically significant, nor could we detect any significant differences in tumor incidence in subgroups of 33 +/bg and 27 bg/bg mice that had been monitored for 500 days (17). If NK-cells indeed can contribute to resistance against primary tumor development, the reduced differential between beige and wild type mice in the DMBA system (in comparison with the studies of transplantable tumors) may have been due to partial suppression of NK-activity caused by the carcinogen (17,18). In contrast, tumor transplantation may cause a transient increase in NK-activity, even if the graft contains only 10^3 cells (9).

The oncogenic action of Moloney Leukemia Virus (Mo-LV) is restricted to cells of the lymphoid system. Similar to DMBA, leukemogenesis induced by Mo-LV requires a long latency period, and a large proportion of the resulting lymphomas show trisomy of chromosome 15 (19). Groups of 20 +/bg and 20 bg/bg suckling mice (age < 7 days) were injected i.p. with semipurified Mo-LV (kindly donated by Dr. R. Jaenisch, Heinrich-Pette-Institut, Hamburg) corresponding to 5×10^4 plaque forming units in the XC-plaque assay. Approximately one third of the animals had developed lymphomas within 165 days, with no significant difference in incidence or median latency period between the two groups of mice (fig. 1e). All out of 8 bg/bg and 11 +/bg mice that received the same virus dose at 4-6 weeks of age have remained healthy for 300 days, indicating an age dependent resistance in both genotypes (not shown).

An age dependent resistance has been reported also for tumors induced by Moloney Sarcoma Virus (MSV). This virus causes progressively growing tumors when injected into newborn (or immunodeficient) mice, whereas the tumors induced in normal adult mice regress spontaneously after temporary growth (20). There is evidence to suggest that the tumors do not represent conventional neoplasias consisting of dividing, transformed cells, but rather a granulomatous response dependent on continuous reinfection of cells by the virus (21). Ten +/bg and six bg/bg mice were injected i.m. in a hindleg with 0.2 ml of a MSV preparation (M-MuSV homogenate, University Labs. Inc., Highland Park, N.J.) at 5 weeks of age. The mice were followed with daily palpations and measurements of the appearing tumors. As shown in figure 1f, the patterns of growth and rejection were virtually identical in the two genotypes of mice. This is in contrast to previous comparisons between mice with selective T-cell deficiency and normal controls; the former are extremely susceptible to progressive growth of MSV-tumors, indicating that rejection is T-cell dependent (20).

III. CONCLUDING REMARKS

Figure 1 summarizes our studies of natural resistance to syngeneic, semisyngeneic and primary tumors in the beige model. We observed a weak resistance against syngeneic lymphoma grafts in wild type mice which was defective in beige mice, developed normally in thymectomized bone marrow chimeras, and had the potential to reject $5 \times 10^2 - 10^3$ grafted cells (a,b). We have not been able to demonstrate any significant differences between +/bg and bg/bg mice with regard to resistance to primary tumor development after administration of oncogenic agents (d-f). However, the curves for DMBA-induced lymphomas are consistent with a resistance mechanism that operates early in the preleukemic period in wild type mice, but eventually will "lose the race" with expanding malignant clones. The strongest rejection potential (elimination of 5×10^3 cells) was seen with H-2 heterozygous hosts and H-2 homozygous lymphoma grafts (c). Taken together, these results confirm that NK-cells can contribute to rejection of grafted cells in vivo, although we still do not understand whether the NK-system has evolved as a specific surveillance device against neoplasia. The positive findings with syngeneic transplantable lymphomas involved a boosting of NK-activity as an intrinsic part of the experimental system. On the other hand, the outcome of the DMBA study may have been influenced by an artificial suppression of NK activity caused by the high dose of carcinogen. As to the strong H-2 linked resistance, the experimental system here may not be reflecting specific tumor surveillance, but rather a general defence system against "non-self", geared to rapidly eliminate cells lacking a complete set of histocompatibility antigens expressed by the host (22).

REFERENCES

1. Möller, G., and Möller, E., Transplant. Rev. 28:3 (1976).
2. Chow, D. A., Greene, M. I., and Greenberg, A. H., Int. J. Cancer 23:788 (1979).
3. Roder, J. C., Lohman-Matthes, M. L., Domzig, W., and Wigzell, H., J. Immunol. 123:2174 (1979).
4. Welsh, R., and Kiessling, R., Scand. J. Immunol. 11:363 (1980).
5. Mahoney, K. H., Morse, S. S., and Morahan, P. S., Cancer Res. 40:3939 (1980).
6. Windhorst, D. B., and Padgett, G., J. invest. Dermat. 60:529 (1973).
7. Clark, E. A., Windsor, N. T., Sturge, J. C., and Stanton, T. H., in "Natural Cell Mediated Immunity Against Tumors" (R. B. Herberman, ed.), p. 417, Academic Press, New York, 1980.
8. Kärre, K., Klein, G. O., Kiessling, R., Klein, G., and Roder, J. C., Nature 284:624 (1980).
9. Kärre, K., Klein, G. O., Kiessling, R., Klein, G., and Roder, J. C., Int. J. Cancer 26:789 (1980).
10. Riccardi, C., Pucetti, T., Santoni, A., and Herberman, R. B., J. Natl. Cancer Inst. 63:1041 (1979).
11. Carlson, G. A., Melnychuk, D., and Meeker, M. J., Int. J. Cancer 25:111 (1980).
12. Roder, J. C., J. Immunol. 123:2168 (1979).
13. Warner, N. L., Woodruff, M. F., and Burton, R. C., Int. J. Cancer 20:146 (1977).
14. Klein, G. O., Klein, G., Kiessling, R., and Kärre, K., Immunogenetics 6:561 (1978).
15. Klein, G., Klein, G. O., Kiessling, R., and Kärre, K., Immunogenetics 7:391 (1978).
16. Habu, S., Fukui, H., Shimamura, K., Kasai, M., Nagai, Y., Okumura, K., and Tamaoki, N., J. Immunol. 127:34 (1981).
17. Argov, S., Cochran, A. J., Kärre, K., Klein, G. O., and Klein, G., Int. J. Cancer 28:379 (1981).
18. Ehrlich, R., Efrati, M., Bar-Eyal, A., Wollberg, M., Schiby, G., Ran, M., and Witz, I. P., Int. J. Cancer, 26:315 (1980).
19. Spira, J., Babonits, M., Wiener, F., Ohno, S., Wirshubski, Z., Haran-Ghera, N., and Klein, G., Cancer Res. 40:2609 (1980).
20. Levy, J. P., and Leclerc, J. C., Adv. Cancer Res. 24:1 (1972).
21. Becker, S., and Haskill, J. S., J. Natl. Cancer Inst. 65:469 (1980).
22. Kärre, K. Doctoral thesis, Karolinska Institute, Stockholm, 1981.

NK CELL AND NAb ANTI-TUMOR ACTIVITY IN VIVO[1]

Donna A. Chow
Garth W. Brown
Arnold H. Greenberg

Manitoba Institute of Cell Biology
University of Manitoba
Winnipeg, Manitoba, Canada

Recent efforts to determine whether immunological responses contribute to the surveillance against incipient tumors have focused principally on natural killer (NK) cells. Correlative evidence from ontogenic, genetic and response modulation studies, performed mainly in F1 hybrid mice, suggests that NK cells may indeed contribute to host resistance against NK^{s} targets (Kiessling et al, 1975, 1977; Haller et al, 1977; Kärre et al, 1980a; Riccardi et al, 1980; Gorelik et al, 1981). Furthermore, this hypothesis was supported by the more direct observation that recipients of inocula containing tumor and selected NK cell populations exhibited reduced tumor frequencies (Kasai et al, 1979; Gorelik et al, 1981). Although the growing weight of evidence is strongly in favor of NK cell participation in natural resistance (NR), the correlations between tumor survival and host levels of NK activity, or tumor sensitivity to NK cells are not totally consistant (Kärre et al, 1980b; Riccardi et al, 1980; Collins et al, 1981; Gorelik et al, 1981) and suggest that NR may be a heterogeneous phenomenon. The participation of a number of other effectors, including natural antibody (NAb) (Menard et al, 1977; Wolosin and Greenberg, 1979; Chow et al, 1979, 1981a, 1982) and macrophages (Keller, 1976; Kiessling et al, 1977; Carlson et al, 1980; Greenberg et al, 1980), has been postulated.

In the last few years we have been exploring this question in a syngeneic murine model with emphasis on elucidating the role of NAb-associated processes in NR. This necessitated the parallel examination and delineation of any NK cell contribution. Since there was no definitive means to identify an in vivo NK cell process, this posed a difficult problem. We therefore chose a variety of experimental designs and assays to evalute both phenomena.

---

[1]Supported by the NCI and MRC of Canada.

ISBN 0-12-341360-5

Host-mediated NR was determined in two in vivo assays. Initially we examined the fate of threshold SC tumor inocula producing tumor frequencies of less than 100%, an assay which describes a non-adaptive, thymus-independent phenomenon (Greenberg and Greene, 1976). More recently host resistance was assessed by the rate of elimination of radioisotope from mice in the first few days after injection of 131-I deoxyuridine labeled tumor cells, which Carlson et al (1980) has similarly shown is a measure of T-independent natural resistance.

Evidence in support of a role for NK cells in NR was obtained through examination of clonal variants of the NK-sensitive (NK^s) SL2 lymphoma and interferon (IFN) modulation of the NK target structure of the SL2-5 clone. Our first experiments in which the tumor frequency of threshold SC inocula of two clones from the SL2 correlated inversely with their susceptibility to NK cells in vitro (Chow et al, 1981a) led us to examine the rate of clearance of six 131-I labeled SL2 clones which were selected for their varied NK sensitivity and minimal difference in NAb susceptibility. The SC elimination of these clones was again inversely correlated (Fig 1A), and clearance from the peritoneum exhibited a similar but less striking relationship (Chow et al, submitted). Further evidence of NK participation came from the manipulation of tumor phenotype by IFN. Incubation of the NK^s SL2-5 clone with 10 to 100 U/ml of IFN decreased tumor sensitivity to lysis by NK cells from Poly I:C-stimulated CBA/J and syngeneic DBA/2, and produced a corresponding slowing in the clearance of 131-IUdR labeled cells injected IP into mice of the same strain (Greenberg and Pohajdak, 1982). These observations using a novel NK^s tumor target in a syngeneic system are in agreement with Welsh et al (1981) who assessed the YAC-1 response to IFN in semi-syngeneic recipients, lending further credence to the hypothesis that NK cells participate in NR.

Notwithstanding the argument that NK activity may affect the fate of tumors in vivo, inconsistencies observed in the relative ranking of tumor sensitivities in vivo and in vitro (Riccardi et al, 1980), and the differences in tumorigenicity demonstrated for NK^r tumor lines (Chow et al, 1982), suggest that NK cells are not the singular constituent of NR. Although investigation of the ontogeny of NK cells and NR in F1 hybrids demonstrated positive correlations for activity against NK^s tumors (Haller et al, 1977; Kiessling et al, 1977; Riccardi et al, 1980), the capacity of syngeneic mice to reject or eliminate SC inocula of the NK^s SL2-5 did not decrease with advancing age (Chow et al, 1982) in parallel with the characteristic onset and decline of NK activity. A similar discrepancy

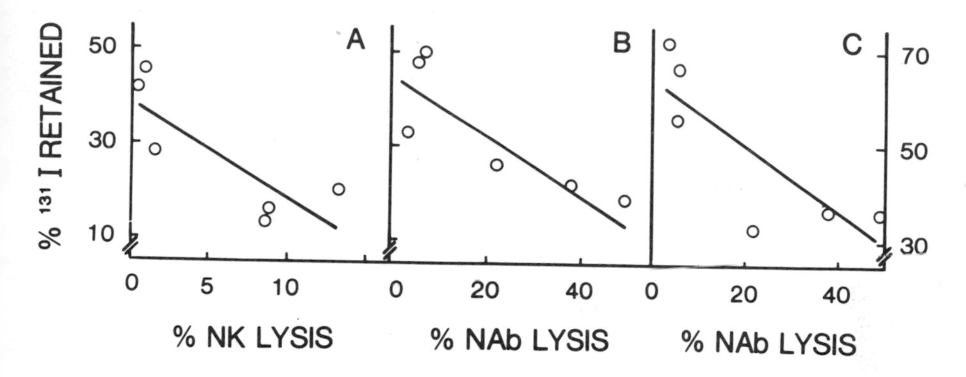

FIGURE 1. Correlations of in vivo and in vitro activity. A) SC elimination of six SL2 clones on day 2 vs. cytolysis by NK cells ($r= -.844$, $p<.05$); B) IP elimination of two clones from each of the L5178Y, P815, and SL2 on day 2 vs. cytolysis by NAb ($r= -.851$, $p<.05$); and C) IV elimination at 4 hours for clones in (B) vs. cytolysis by NAb ($r= -.824$, $p<.05$). Inocula of 5×10^5 cells were employed throughout.

between the ontogeny of NR and that of NK cells has recently been reported for the YAC-1 in an allogenic and a syngeneic system (Gorelik et al, 1981). In addition, the NK deficient Bg/Bg mice exhibited the same degree of resistance as their normal heterozygous littermates using the YAC-1 (Gorelik et al, 1981) and the NK^r L20u.2 (Carlson and Greenberg, 1982). Furthermore, in our more recent studies hybrid resistance in (C57Bl/6 x DBA/2)F1 was not distinguishable for SL2 variants differing 4 fold in their lytic sensitivity in vitro (Carlson and Greenberg, 1982).

Our investigations of anti-tumor NAb from several approaches suggest that this mediator contributes to NR. Through the use of tumor clones which were mainly NK^r, but differed in their sensitivity to NAb, we attempted to minimize the NK contribution. Initial studies showed that the tumorigenicity of threshold SC inocula of two NK^r L5178Y tumor clones correlated inversely with their sensitivity to NAb (Chow et al, 1981a). These studies were extended by examining a panel of six clones derived from three different DBA/2 tumors, following the IV, IP or SC elimination of 131-IUdR labeled cells. A significant inverse correlation was detected between tumor sensitivity to NAb, and the rate of elimination from the peritoneum and clearance following IV inoculation (Fig 1B and C) (Chow et al, 1981a). We have been able to detect NAb most readily in these two sites (Chow et al, submitted). Tumor cells recovered from the peritoneum have detectable membrane bound antibody within 3 hours of inoculation, and up to 70% of the cells can be lysed, with the addition of complement, by 18 hours in vivo (Wolosin and Greenberg, 1979).

Other correlative studies also related NR to NAb activity: (1) The ontogeny of host anti-tumor activity to NK^r and NK^s tumors, measured by both NR assays, improved with aging and more closely followed the development of NAb against these cells (Menard et al, 1977; Chow et al, 1981a, 1982); (2) Modulation experiments employing the non-specific stimulation produced by LPS revealed that this agent could decrease the tumor frequency of threshold inocula in normal and ATxBM mice, while rapidly increasing serum NAb levels against these tumors (Chow et al, 1981a), and (3) the incidence of lymphoma due to endogenous murine leukemia virus in some mouse strains was inversely related to quantitative differences in serum NAb (Aoki et al, 1966; Ihle et al, 1973).

Using a more direct approach to test the in vivo action of NAb, we pretreated the tumors with normal serum in vitro and found that the tumor frequency in vivo was inversely related to the antibody activity of the serum (Chow et al, 1981a).

More recently we have exploited the observation that interferon can modulate the NAb target structure (Greenberg and Pohajdak, 1982). IFN treatment of the NK^r LY-F9 lymphoma in vitro markedly increased the elimination of these cells from the peritoneum of syngeneic mice, in contrast to the slowing of in vivo eliminaton of an NK^s tumor (Chow et al, 1982). IFN enhanced the in vitro binding of NAb to the LY-F9 (Pohajdak and Greenberg, submitted), and in recent experiments, the in vivo acquisition of natural antibody (Greenberg et al, submitted), suggesting that the augmented host resistance may be related to more efficient antibody-tumor interaction produced by IFN.

The hypothesis that both NK cells and NAb contribute to host-mediated resistance leads to the prediction that tumors which survive should exhibit a reduced sensitivity to NR and to these effectors. Other investigators have observed the loss of NK sensitivity in a tumor passaged in vivo (Durdick et al, 1980). Repeated cycles of NK cytolysis in vitro and tumor growth in a high NK F1 hybrid have also produced tumor lines which exhibited a decreased sensitivity to NR and to NK cells in vitro (Grönberg et al, 1981). The growth of small inocula of the LY-F9 for 3 weeks in a syngeneic mouse resulted in a tumor population which upon subsequent testing was eliminated significantly slower than the parental clone (Table 1). This decrease correlated with a reduced sensitivity to syngeneic NAb in vitro. Similarly a decrease in the susceptibility to NR using the NAb^s, NK^s SL2-5 clone grown in vivo corresponded with significant reductions in sensitivity to both syngeneic NAb and splenic NK cells from Poly I:C-stimulated DBA/2. This suggests then, that NAb and NK cell associated mechanisms may direct tumor selection in vivo.

In total the results of our experiments support the hypotheses that both NK cells and NAb contribute to anti-tumor resistance. However, the discrimination of the effects of these two mediators in vivo still presents a serious problem. The chance that some of our in vivo observations with the NK^r tumors may relate to NK activity, or that effects attributed to NK cells may be a result of NAb, cannot be excluded completely. In this regard, all murine tumors we have examined including the NK^s lines, are to some extent, sensitive to syngeneic NAb in vitro (Chow et al, 1981a, submitted) and bind natural antibody very efficiently when inoculated intraperitoneally (Wolosin and Greenberg, 1979). Furthermore, the augmentation of host resistance by adjuvents and IFN inducers not only enhances NK activity, but also natural antibody (Chow et al, 1981b) and macrophages (Kaplan et al, 1974; Shultz et al, 1977), making the interpretation of such experiments with NK^s tumors impossible. The capacity to distinguish NK and NAb associated mechanisms may in addition be somewhat compromised in the YAC-1 system, where evidence suggests that the target structures for NK and NAb activity are coexpressed on clonal variants (Chow et al, 1981b; Greenberg and Pohajdak, 1982). In vivo selection resulting in a decrease in NK sensitivity could, therefore, be attributed to either NK cells or NAb. One must also consider the possibility that these two mediators act cooperatively since NK cells can function through an ADCC mechanism (Ojo and Wigzell, 1978) and NAb may participate in this process (Koide and Takasugi, 1977). NK dependence of host resistance would then not exclude an NAb contribution or vice versa.

The likelihood that the current assays of NR can measure rapidly evoked increases in effector activity adds to the complexity of the problem. Tumor cells themselves have been shown to raise NK levels in vivo (Kärre et al, 1980b) and recent reports indicate that some tumor targets previously considered to be insensitive to NK cells from normal mice were susceptible to the NK cytotoxicity of LCMV stimulated mice (Welsh and

TABLE I. In Vivo Selection for Reduced Sensitivity to NR, NAb and NK Cells

Cells	%131-I ± S.E. Day 2	% Cytotoxicity ± S.E. NAb	NK cells
LY-F9 in vitro	30.6 ± 2.8	9.4 ± 1.6	0.0
LY-F9 3 wks SC	60.6 ± 3.5	6.2 ± 1.6	not done
SL2-5 in vitro	18.4 ± 4.8	27.5 ± 2.8	31.6 ± 2.3
SL2-5 3 wks SC	43.4 ± 4.1	17.7 ± 2.9	27.2 ± 2.1

Differences in the 131-I retention following IP injection of in vitro and 3 week SC cells were significant using t-independent statistics ($p<.001$, df=8). NAb cytolysis of in vitro and 3 week SC cells were significantly different using a paired t-test for the LY-F9 cells ($p<.02$: df=2) and for the SL2-5 cells ($p<.001$, df=6). The sensitivity of the two SL2-5 populations to NK cells were distinguishable using a paired t-test ($p<.04$, df=9).

Kiessling, 1980). Although our NK^r tumors are insensitive to NK lysis of normal and Poly I:C-stimulated syngeneic or CBA mice, it is difficult to know what constitutes absolute resistance, or conversely, what level of NK activity in vitro identifies susceptibility in vivo. Perhaps some extremely low level of sensitivity under conditions of augmented NK activity in mice exposed to tumor cells alone, or in the presence of adjuvants, may result in NK mediated NR. For example, the induction of NK cells by tumor inocula in older animals may at least partially account for the maintenance of NR against some NK^s tumors (Chow et al, 1982). Similarly the idea that NR in vivo is more directly related to rapidly induced anti-tumor activity rather than pre-existing levels of effectors may also be true for NAb. Tumor cells injected into the peritoneum of syngeneic mice were observed to take up markedly more NAb during 18 hours than the maximum amounts absorbed from serum in vitro (Wolosin and Greenberg, 1979). These high levels of NAb acquisition may underlie a potent anti-tumor response. Although the ultimate effect of any mediator will be determined by tumor sensitivity, the ability of the tumor to elicit the effector would also contribute to the observed NR. This could explain the quantitative discrepancy between the degree of selection noted after in vivo passage, and the extent of the modification in tumor phenotype for sensitivity to NAb or NK in vitro (Table I). It may, in general, also account for weakness in the in vivo correlations for a single mediator, NK cells or NAb, in addition to the possibility that both effectors are active.

Several important conclusions can be drawn from our investigations of NR against tumors. The idea that NR is a heterogeneous phenomenon is probably fundamental to its further characterization. A positive correlation between in vitro and in vivo activities can, consequently, provide support for a contribution by an individual effector, but cannot exclude the participation of others. Therefore, a clearer interpretation of these studies would be facilitated by the formulation of experimental designs which minimizes the role of one agent. This could be accomplished through the use of tumor targets that are relatively resistant to an individual mediator, and by studying strains of mice which bear a genetic mutation for one or more effector mechanisms, for example, Xid strains which are deficient in NAb for some tumors (Martin and Martin, 1975) and, of course, the Bg/Bg strains. Furthermore, until the regulation of NK and NAb induction by tumors is clarified, the degree to which they participate in NR will remain uncertain and consequently it will be difficult to assign them a relative role in surveillance.

REFERENCES

Aoki, T., Boyse, E. A., and Old, L. J. (1966). Cancer Res. 26:1415.

Carlson, G. A., and Greenberg, A. H. (1982). In "Natural Cell-mediated Immunity," (R. B. Herberman, ed.), Vol. 2. Academic Press, New York.

Carlson, G. A., Melnychuk, D., and Meeker, M. J. (1980). Int. J. Cancer 25:111.

Chow, D. A., Greene, M. I. and Greenberg, A. H. (1979). Int. J. Cancer 23:788.

Chow D. A., Miller, V. E., Carlson, G. A., Pohajdak, B., and Greenberg, A. H. (1982). Invasion and Metastasis (in press).

Chow, D. A., Wolosin, L. B., and Greenberg, A. H. (1981a). Int. J. Cancer 27:459.

Chow, D. A., Wolosin, L. B., and Greenberg, A. H. (1981b). J. Natl. Cancer Inst. 67:445.

Collins, J. L., Patek, P. Q., and Cohn, M. (1981). J. Exp. Med. 153:89.

Durdick, J. M., Beck, B. N., Clark, E. A., and Henney, C. S. (1980). J. Immunol. 125:683.

Gorelik, E., and Herberman, R. B. (1981). Int. J. Cancer 27:709.
Greenberg, A. H., Chow, D. A., and Wolosin, L. B. (1980). In "Genetic Control of Natural Resistance to Infection and Malignancy," (E. Skamene, ed), p. 455. Academic Press, New York.
Greenberg, A. H., and Greene M. I. (1976). Nature 265:356.
Greenberg, A. H., and Pohajdak, B. (1982). In "Natural Cell-mediated Immunity" (R. B. Herberman, ed.), Academic Press, New York.
Grönberg, A., Kiessling, R., Ericksson, E., and Hansson M. (1981). J. Immunol. 127:1734.
Haller, O., Hansson, M., Kiessling, R., and Wigzell, H. (1977). Nature 270:609.
Ihle, J. N., Yurconic, M., and Hanna, M.G. (1973) J. Exp. Med. 138:194.
Kaplan, A. M., Morahan, P. S., and Regelson, W. (1974). J. Natl. Cancer Inst. 52:1919.
Kärre, K., Klein, G. O., Kiessling, R., Klein, G., and Roder, J. (1980a). Nature 284:624.
Kärre, K., Klein, G. O., Kiessling, R., Klein, G., and Roder, J. (1980b). Int. J. Cancer 26:780.
Kasai, M., Leclerc, J. C., McVay-Boudreau, L., Shen, F. W., and Cantor, H. (1979). J. Exp. Med. 149:1260.
Keller, R. (1976). J. Natl. Cancer Inst. 57:1355.
Kiessling, R., Hochman, P. S., Haller, O., Shearer, G. M., Wigzell, H., and Cudkowicz, G. (1977) Eur. J. Immunol. 7:655.
Kiessling, R., Klein, E., Pross, H., and Wigzell, H. (1975). Eur. J. Immunol. 5:117.
Koide, Y., and Takasugi, M. (1977). J. Natl. Cancer Inst. 59:1099.
Ménard, S., Colnaghi, H. I., and Della Porta, G. (1977). Int. J. Cancer 19:267.
Martin, S. E., and Martin, W. J. (1975). J. Immunol. 115:502.
Ojo, E., and Wigzell, H. (1978). Scand. J. Immunol. 7:297.
Riccardi, C., Santoni, A., Barlozzari, T., Puccetti, P., and Herberman, R. B. (1980). Int. J. Cancer 25:475.
Riesenfeld, I., Orn, A., Gidlund, M., Axberg, I., Alm, G. V., and Wigzell, H. (1980). Int. J. Cancer 25:399.
Shultz, R. M., Papamatheakis, J. D., Luetzeler, J., Ruiz, P. and Chirigos, M. A. (1977). Cancer Res. 37:358.
Welsh, R. M., Kärre, K., Hansson, M., Kunkel, A. L., and Kiessling, R. W. (1981). J. Immunol. 126:219.
Welsh, R. M. and Kiessling, R. W. (1980). In "Natural Cell-mediated Immunity" (R. B. Herberman, ed.), Vol. 1., p. 671. Academic press, New York.
Wolosin, L. B., and Greenberg, A. H. (1979). Int. J. Cancer 23:519.

THE IN VIVO EFFECTS OF INTERFERON (IFN) SUPPRESSION OF THE NK TARGET STRUCTURE[1]

A. H. Greenberg
V. Miller
B. Pohajdak
T. Jablonski

Manitoba Institute of Cell Biology
Univerity of Manitoba
Winnipeg, Manitoba, Canada

The role of interferon as a regulator of anti-tumor host resistance was suggested by Gresser (1972) after demonstrating that mice bearing an IFN-resistant clone of the L1210 lymphoma responded to treatment with exogenous IFN. The ability of the hormone to enhance both natural killer cell (Gidlund et al, 1978) and macrophage (Haller et al, 1979) activity is considered by many investigators to be the mechanisms by which this effect is mediated in vivo. It has also been known for many years that IFN can modulate tumor cell surface antigens, however, the significance of the phenomenon in host resistance has not been fully explored. Trinchieri and Santoli (1978) were the first to point out that IFN may have a paradoxical effect on host-tumor interaction by its ability to suppress the expression of NK target structures (TS) while enhancing NK activity. Welsh et al (1981) have also recently suggested that the TS modulation can occur in vivo and that this may reduce the effectiveness of the hosts' NK response to tumors. We have examined this hypothesis using an NK-sensitive clone of the SL2 lymphoma which proved to be exquisitely responsive to the membrane modulating effects of IFN. Incubation of the tumor with as little as 10 U/ml of L cell IFN (obtained from Dr. C. Ogburn, Dept. of Microbiology, Medical College of Pennsylvania) for 20 hours prior to exposure to NK cells reduced lysis by over 60% of control (Table 1). This inhibition appeared to be a result of loss of the NK target structure since the IFN reduced the capacity of unlabelled tumor to inhibit the lysis of labelled target cells (Table 1).

---

[1]Supported by the NCI and MRC of Canada.

ISBN 0-12-341360-5

TABLE 1 The Effect of IFN on the NK Lysis of the SL2-5 Lymphoma

	TREATMENT				
EXP	TARGET	COLD INHIBITOR++	I:T+	PERCENT LYSIS*	PERCENT INHIBITION
1.	none	--	--	54.2	--
	10 U/ml	--	--	20.1	62.9
	100 U/ml	--	--	13.4	75.3
	2000 U/ml	--	--	8.9	83.6
2.	--	none	--	39.6	--
	--	SL2-5	5:1	15.9	59.8
	--	" (IFN)	5:1	27.7	30.1
	--	"	2.5:1	21.4	46.0
	--	" (IFN)	2.5:1	32.9	16.9
3.	--	--	--	39.6	--
	--	LY-F9	5:1	26.8	32.3
	--	" (IFN)	5:1	30.4	23.2
	--	"	2.5:1	31.0	21.7
	--	" (IFN)	2.5:1	35.3	10.9

* CBA/J spleen cells were taken 24 hours after injection of 100 ug Poly I:C, Effector:Target = 150:1. Similar results were obtained with Poly I:C-stimulated DBA/2.
+ I:T = Inhibitor to target ratio
++ IFN treatment at 100 U/ml for 20 hours

The sensitivity to IFN in vitro was reflected in the behaviour of the tumor in vivo. Significant slowing of the elimination of 131-IUdR labelled SL2 from the peritoneal cavity was observed when tumor was incubated with 10-100 U/ml of IFN for 20 hours in vitro (Fig 1A). At the same time we noted that the reduction in elimination rate could best be demonstrated in young (6-7 week) mice, while older (10-12 week) mice often must showed little or no response. Both of these results were, however, consistent with an NK mediated elimination mechanism. The dose response curve of NK TS modulation in vitro was identical to the in vivo slowing, and the loss of NK mediated host resistance with aging would make it more difficult to detect

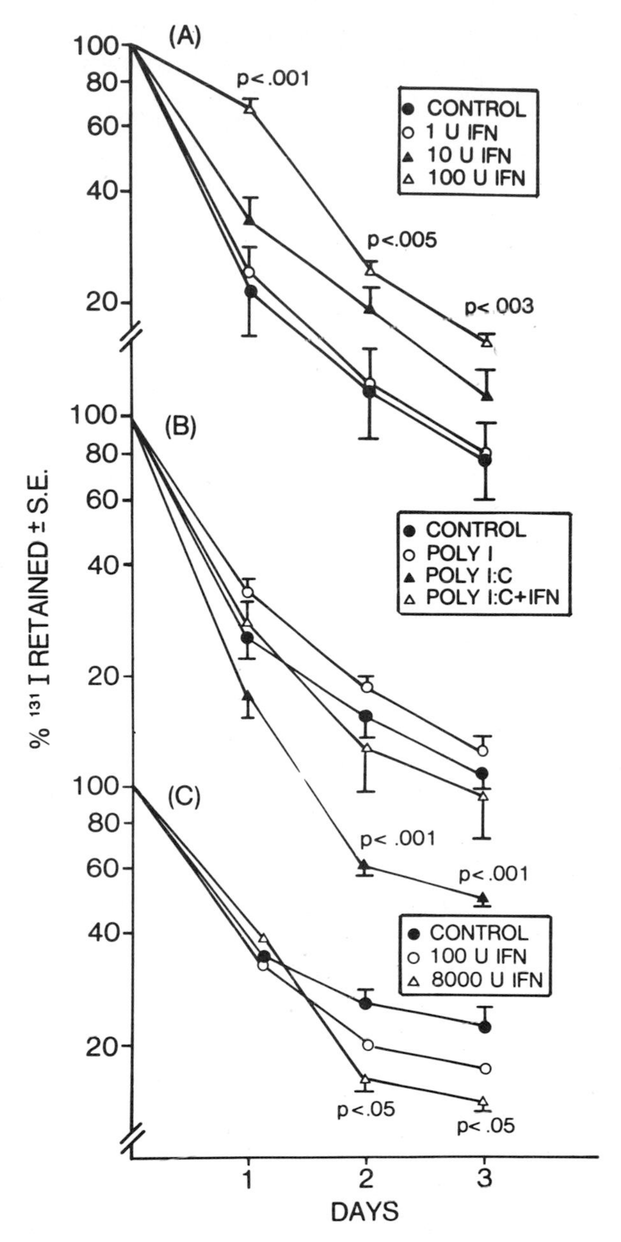

FIGURE 1. The effect of IFN treatment of NK^s and NK^r tumors on their elimination in syngeneic mice. (A) 131-I-labelled SL2-5 (NK^s) was incubated in IFN for 20 hours, washed and injected IP. (B) Mice primed with Poly I:C (or Poly I) 24 hours earlier were inoculated with normal or IFN-treated SL2-5. (C) LY-F9 (NK^r) was treated with IFN, washed and injected IP.

further IFN induced reduction in NK sensitivity. We then re-examined the phenomenon in 10 week old mice after stimulation of NK activity by the intraperitoneal injection of the interferon inducer Poly I:C 24 hours before the tumor. The Poly I:C significantly enhanced the tumor elimination rate over untreated controls while a group receiving Poly I were unaffected (Fig 1B). However, when the tumor was treated with 100 U/ml of IFN, the tumor elimination rate in Poly I:C treated mice returned to control levels (Fig 1B). One could argue from these data that Poly I:C was stimulating the elimination of the tumor as a result of its ability to enhance NK activity. Poly I:C is an interferon inducer, and is thought to stimulate NK activity in vivo via this hormone (Djeu et al, 1979) while Poly I is neither an IFN inducer (Field et al, 1968) nor does it affect NK activity (Greenberg, A. H., unpublished). This view is further supported by the observation that depleting the tumor of an NK TS by exposure to IFN prior to injection reverses the Poly I:C enhancement.

Other interpretations of these data are, however, worth considering. For example, it is known that injected tumors themselves are interferon inducers (Djeu et al, 1980), although what determines their IFN-inducing capacity is not entirely clear. It has been suggested this may be in part dependent on the expression of an NK TS (Ohmori et al, 1979; Djeu et al, 1980). If this is the case, then a decrease in the expression of the TS by pretreatment with IFN will decrease the in vivo induction of IFN and in so doing, NK activation, resulting in net slowing of the elimination of the tumor in normal mice. Poly I:C priming of the mice, however, would overcome this deficiency in tumor IFN and NK induction and tumor elimination rate in Poly I:C stimulated mice then would not be affected by IFN pretreatment of the tumor. Since IFN was observed to slow tumor elimination in the primed mice then reduced NK induction cannot be the only explanation. However, a loss in NK-inducing capacity by IFN-treated tumor, in addition to the suppression of lytic sensitivity, may contribute to the slowing of tumor elimination in unstimulated mice.

Trinchieri et al (1981) argued that the net effects of IFN on host NK induction and TS suppression favors the protection of highly IFN sensitive normal fibroblasts over tumor cells, which he finds respond less well to NK TS modulation. One must then consider why the Poly I:C-induced endogenous IFN does not itself protect the SL2-5 tumor in vivo. Since the NK TS is depressed so readily with small amounts of IFN in vitro, one might predict that the tumor would become rapidly insensitive to NK cells in vivo and show little augmentation of the elimination rate by treatment with Poly I:C. The simplest explanation is that it takes time for the endogenous IFN to modulate the NK TS. By pretreating in vitro with IFN, we are presenting the host with a "fait accomplit". If the important events for effector lysis of the tumor in our assay are completed before the TS modulation occurs, then the effects of endogenous IFN will not be seen. TS suppression of the SL2-5, however, does occur rapidly (Greenberg and Pohajdak, unpublished) and the effector action would have to be completed within 6 to 8 hours of inoculation of the tumor. A second, but more complex, possibility is that our assumption that the IFN is solely affecting NK mediated resistance in vivo is erroneous, and we are detecting other non-NK host effectors whose tumor target structures are also IFN modulated.

Accordingly, we examined the response of the host to the NK-resistant lymphoma L5178Y-F9 (LY-F9) and found that, unlike the SL2-5, treatment of the tumor in vitro with IFN enhanced the elimination rate (Fig 1C). We considered several possible interpretations of this result: (1) IFN increased the expression of the NK TS, and therefore the NK sensitivity of the tumor resulting in an increased elimination rate. The LY-F9, however, was NK resistant at all doses of IFN, and an NK target structure identifiable in "cold" inhibition studies was suppressed, much the same as observed with the SL2-5 (Table 1), (2) IFN reduced the viability of the tumor and therefore promoted its elimination. The tumor cells, however, were nearly one hundred percent viable after in vitro treatment with IFN. In addition, the SL2-5 was growth inhibited as efficiently as the LY-F9, yet slowing of the elimination of this tumor was observed rather than acceleration. Finally, it was demonstrated that the enhanced elimination was dependent on the size of the inoculum, while the elimination of heat killed tumor was significantly greater than untreated tumor at all tumor doses, (3) A third explanation, therefore, which must be considered is that IFN was modifying the recognition of the tumor by another, non-NK, effector mechanism. IFN can, of course, produce a number of other cell membrane changes in addition to its effect on the NK target structure that could influence host-tumor interaction including increased expression of major histocompatibility antigens (Lindahl et al, 1975), carcinoembryonic antigen (Attallah et al, 1979), Fc_{γ} receptors (Fridman et al, 1980), and in recent work in this laboratory the antigens recognized by natural anti-tumor antibody (Greenberg and Pohajdak, in this volume).

Very little is known about the biological consequences of these types of IFN-induced membrane modification, yet the host response during the development of many neoplasms ensures that the tumor cells will be bathed in interferon from their nascence. It is clearly of importance for the understanding of the progression of malignancy to delineate the IFN-induced modifications that can alter the host-tumor interaction. Since IFN can, in the experimental models described here, either enhance or suppress the host's ability to eliminate tumors, having the means to predict which of these effects will predominate may also allow a more rational utilization of interferon as a therapeutic modality.

REFERENCES

Attallah, A., Needy, C., Noguchi, P., and Elisberg, B. (1979). Int. J. Cancer 24:49.

Djeu, J. Y., Heinbaugh, J. A., Holde, H. T., and Herberman, R. B. (1979). J. Immunol 122:175.

Djeu, J. Y., Huang, K.-Y., and Herberman, R. B. (1980). J. Exp. Med. 151:781.

Field, A. K., Tytell, A. A., Lampson, G. P., and Hilleman, M. R., (1968). Proc. Natl. Acad. Sci, USA 61:340.

Fridman, W. H., Gresser, I., Bandu, M. T., Aguet, M., and Neauport-Sautes, C. (1980). J. Immunol. 124:2436.

Gidlund, M., Orn, A., Wigzell, H., Senik, A., and Gresser, I. (1978). Nature 273:759.

Gresser, E., Maury, C., and Brouty-Boye, D. (1972). Nature 239:167.

Haller, O., Arnheiter, H., Gresser, I., and Lindenmann, J. (1979). J. Exp. Med. 149:601.

Lindahl, P., Leary, P., and Gresser, I. (1975). Proc. Nat. Acad. Sci. USA 70:2785.

Ohmori, I., Kawata, M., Okumura, K., Kuwata, T., and Tada, T., (1979). Immun. Letters 1:57.

Trinchieri, G., Granato, D., and Perussia, B. (1981). J. Immunol. 126:335.

Trinchieri, G., and Santoli, D. (1978). J. Exp. Med. 147:1314.

Welsh, R. M., Kärre, K., Hansson, M., Kunkel, L. A., and Kiessling, R. W. (1981). J. Immunol. 126:219.

INTERRELATIONSHIP BETWEEN NK ACTIVITY AND T CELL-MEDIATED IMMUNITY IN SYNGENEIC TUMOR REJECTION[1]

James Urban[2]
Hans Schreiber[3]

La Rabida-University of Chicago Institute,
Committee on Immunology and Department of Pathology,
The University of Chicago
Chicago, Illinois

The evolution of a tumor from initial transformed cells is a multistep process which is characterized by a progressively increased malignant potential of the tumor cells - a process which has been referred to as "neoplastic progression" (Foulds, 1954). This process is apparently based upon the biologic principle of continuous Darwinian selection of variant tumor cells which have an increased potential to grow and survive in the face of homeostatic controls including immunologic defenses of the host. One example is the 1591 fibrosarcoma which has been induced in mice immunosuppressed by ultraviolet light (Kripke, 1977). The original 1591 tumor will ordinarily not grow in syngeneic C3H mice which possess immunocompetent tumor-specific T cells (Flood et al., 1980; Flood et al., 1981); however, during continuous growth in immunosuppressed mice, variant tumor cells arise from the parental tumor which are resistant to the tumor-specific T cell response of normal immunocompetent mice and which are, therefore, selected for when the tumor is finally passed into normal mice (Urban et al., 1982). These selected variant tumors now grow progressively upon further transfer into normal mice.

---

[1]Supported by CA-27326, CA-22677 and CA-19266.
[2]Supported by PHS 5 T32 GMO 7281 and PHS 5 T32 AI-07090.
[3]Supported by Research Career Development Award CA-00432.

ISBN 0-12-341360-5

We were interested in determining the immunologic mechanisms which could restrain the growth of the variant progressor tumor cells and the effector cells to which they were still sensitive. In particular, we have studied changes in the relative capability of these variant tumor cells to induce NK cells and tumor-specific T cells and we have studied changes in their relative sensitivity to these two types of effectors. Such an approach should give some insight into the relative importance and interrelationship of T cell- and NK cell-mediated host defenses against malignant growth.

To analyze these questions we have used direct isolation of effector cells from the site of injection of viable tumor cells, i.e. the peritoneal cavity. With this direct analysis the _in vivo_ relevance of our study seemed to be less questionable since none of the effector cells had to be subjected to any long-term culturing _in vitro_ prior to testing. Thus, mice were injected with 1591 parental regressor tumor cells or with a subtumorigenic dose of 1591 progressor variant tumor cells and 8 days later effector cells were removed from the peritoneal cavities of these animals and tested for cytolytic activity. As shown in Fig. 1, animals injected with the parental 1591 tumor cells developed Lyt-2^+ T cells which lysed the 1591 tumor cells. Neither the antigenically distinct UV-induced 1316 tumor cells, nor seven independently isolated 1591 progressor variant cells lines (data pooled), were lysed. Thus, the variant cells appear to have lost the original tumor-specific transplantaton antigen.

Fig. 1 also shows that Lyt-2^+ tumor-specific reactivity could not be elicited by the injection of progressor variant tumor cells. Nevertheless, injection of progressor variant cells induced high levels of some other type of Lyt-2^- cytolytic effector cell. Further analysis of these cells (Fig. 2) showed that their cytolytic activity was almost totally abrogated by pretreatment with an anti-NK-1.2 antiserum and complement. Thus, these effector cells were NK cells. The partial sensitivity of the effector cells to a monoclonal anti-Thy-1.2 antiserum and complement was consistent with the NK origin of the effector cells (Burton et al., 1981).

The parental tumor cells, exhibiting a strong T cell-recognized antigen, induced tumor-specific cytolytic T cells in normal hosts, but failed to induce NK activity. Interestingly, such tumor cells were capable of inducing NK activity in athymic nude mice. Fig. 3 shows that an injection of 1591 parental regressor cells into nude mice induced high levels of Lyt-2^- effector cells, which upon further analysis proved to be sensitive to anti-NK antiserum (not shown). The level of reactivity was comparable to that produced by normal animals injected with progressor tumor cells (see Fig. 1). Thus, the

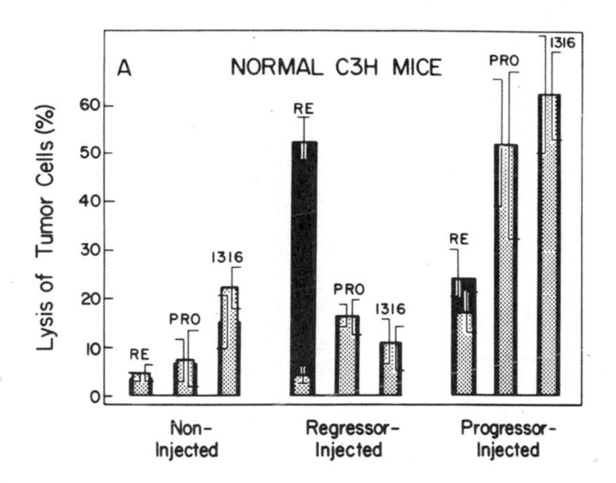

FIGURE 1. Lack of sensitivity of 1591 progressor variant tumor cells (PRO) to Lyt-2^+ 1591 regressor (RE)-specific T cells, and induction of cross-reactive cytolytic immunity by injection of PRO. The vertical bars indicate the percent lysis of target cells after treatment with anti-Lyt-2 and complement (▨) or after treatment with complement alone (■). Vertical lines show the S.E. for 5-11 mice individually tested in 4 independent experiments. E:T 200:1 in a 6-hr ^{51}Cr release assay. (Reproduced from J. Exp. Med. (1982) 155:No. 2 with permission)

structure(s) necessary for induction of NK cells were present on the parental cells, yet the induction of tumor-specific T cells in normal mice somehow interfered with the induction of NK cells. Despite the fact that both types of tumor cells could induce equal levels of NK activity under the appropriate conditions, progressor cells appeared to be somewhat more sensitive than regressor cells to the effects of NK cells, as shown in Fig. 1 and in Fig. 3.

Although a single injection of progressor tumor cells induced comparable levels of NK activity in normal and nude mice, we have found that progressor tumor cells grow faster in nude than in normal hosts (Urban et al., 1982). We therefore thought that a T cell response might be retarding the growth of the progressor tumor in normal mice and such a response might become detectable in normal hosts after repeated injection of the progressor tumor cells. Table I shows that a second injection of these tumor cells in fact induced appreciable levels of such Lyt-2^+ tumor-specific cytolytic effector cells. Interestingly, significant NK activity was no longer detected, in contrast to the high level of NK activity induced by a single injection. In other experiments not presented here, we have found that these Lyt-2^+ effector cells speci-

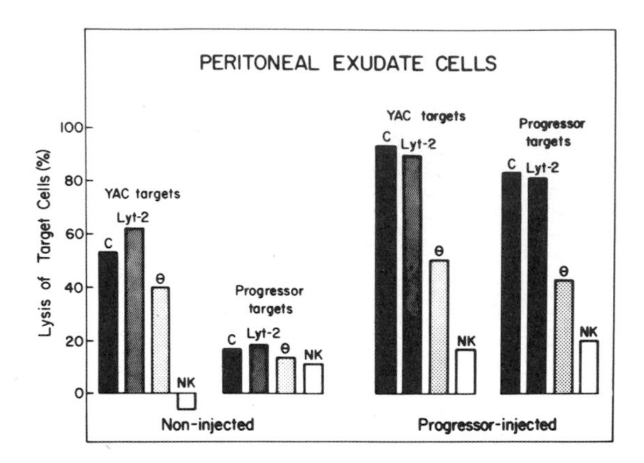

FIGURE 2. Differentiation markers on NK lymphocytes induced by injection of 1591 progressor variant tumor cells. Effector cells were pretreated with antiserum [anti-Lyt-2 (Lyt-2), monoclonal anti-Thy-1.2 (θ), or anti-NK-1.2 (NK)] and/or complement (C). For other details see Fig. 1.

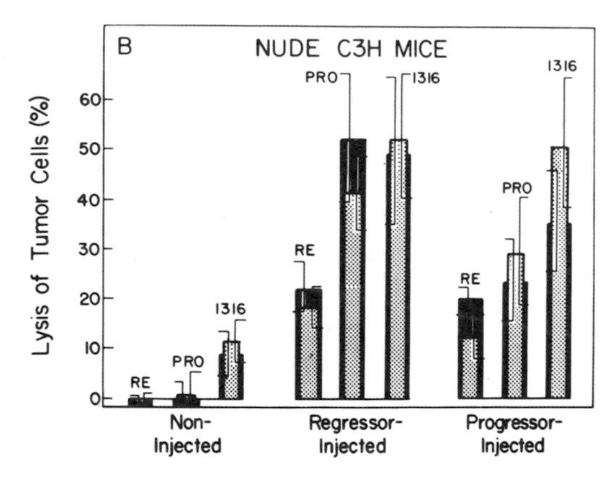

FIGURE 3. Induction of cross-reactive Lyt-2$^-$ cytolytic immunity by injection of 1591 progressor (PRO) or 1591 regressor (RE) tumor cells into nude mice, individually tested in 3 independent experiments. For other details see Fig. 1. (Reproduced from J. Exp. Med. (1982) 155:No. 2 with permission)

fically recognize a second 1591-specific tumor antigen. This antigen had not been lost by the variant tumor cells, possibly because it is less immunogenic and, therefore, less of a target for immune selection by T cells.

NK cells do not seem to be necessary or sufficient for the rejection of the parental 1591 tumor. First, we could find no evidence that significant levels of NK cells were induced after the injection of this tumor into normal mice, which reject the tumor challenge. Second, the parental 1591 tumor grows progressively in nude mice despite the induction of a high level of NK activity. Third, all seven progressor variant tumors tested had lost only the strong tumor-specific transplantation antigen, while no selection against NK-sensitivity occurred. This suggests that only T cell-mediated immunity, and not NK activity, exerted a significant selective pressure upon the parental tumor population. In fact, progressor variant cells were more sensitive than the parental tumor cells to lysis by NK cells and induced NK cells more effectively. However, these phenotypic alterations may allow NK cells to function as a valuable second line of defense against the growth of progressor variants. For example, NK cells may provide a transient suppression of the growth of the progressor variants until a T cell-mediated immunity against these more weakly immunogenic variant tumor cells is generated. However, NK cells alone do not seem to provide absolute protection against tumor growth since nude mice regularly die from the progressive growth of the variant tumors, despite developing high levels of NK activity.

The above data show that 1591 tumor cells which lose a strong T cell-recognized antigen are capable of inducing NK activity in normal mice, as long as no T cell-mediated immunity is elicited to these variant tumors. This finding is compatible with our other observation that the parental tumor cells exhibiting the strong T cell-recognized antigen can induce NK cells in animals that cannot respond with T cells, i.e. in nude mice. At present we do not yet fully understand how the induction of T cell immunity interferes with the induction of NK activity. Although regressor cells have the potential to induce NK activity in nude mice, they cannot do so in normal mice, which develop a strong T cell-mediated immunity. Similarly, normal mice only develop NK cells to progressor variant tumors as long as no T cell-mediated immunity is induced. We know of no published reports showing that high levels of T cell-mediated and NK activity can be induced simultaneously. The interference could be related to some unknown feedback inhibition exerted by T cells on NK cells, or to other mechanisms of competition during the maturation of these two cell types. It would be interesting to know if in other tumor systems increased sensitivity to NK cells and increased induction of NK activity is regularly associated with the loss of a strong tumor-specific T cell antigen. We do know, however, that other UV-induced fibrosarcomas which have a low immunogenicity and progressive growth behavior also seem to be highly sensitive to NK cells (Hanna et al., 1981).

TABLE I. Cytolytic Activity Generated *In Vivo* in Response to One or Two Injections of Progressor Variant Tumor Cells[a]

No. of injections	E:T	Percent lysis of target cells: Progressor variant, C' alone	Progressor variant, C'+αLyt-2	1316, C' alone	1316, C'+αLyt-2
1	100:1	49	54	42	40
	50:1	44	35	28	21
	25:1	25	18	22	13
2	100:1	44	14	8	10
	50:1	43	8	7	9
	25:1	30	8	0	0

[a]C3H mice received 1 or 2 injections of 10^7 progressor tumor cells i.p. Peritoneal exudate cells were removed 8 days after the last injection and treated with anti-Lyt-2 antiserum and/or complement before testing in a 6-hr ^{51}Cr release assay. Values equal the mean for two independent experiments.

Thus NK cells could be a valuable second line of defense acting on more malignant, less immunogenic or less differentiated (Gidlund et al., 1981) tumor cells which develop during tumor progression. Despite this possibility, however, our data suggest that eventual development of effective T cell-mediated immunity is still necessary for ultimate tumor rejection in this tumor system.

REFERENCES

Burton, R. C., Bartlett, S. P., Kumar, V., and Winn, H. J. (1981). J. Immunol. 127:1864.
Flood, P. M., Kripke, M. L., Rowley, D. A., and Schreiber, H. (1980). Proc. Natl. Acad. Sci. U. S. A. 77:2209.
Flood, P. M., Urban, J. L., Kripke, M. L., and Schreiber, H. (1981). J. Exp. Med. 154:275.
Foulds, L. (1954). Cancer Res. 14:327.
Gidlund, M., Orn, A., Pattengale, P. K., Wigzell, H., and Nilsson, K. (1981). Nature 292:848.
Hanna, N., and Burton, R. C. (1981). J. Immunol. 127:1754.
Kripke, M. L. (1977). Cancer Res. 37:1395.
Urban, J. L., Burton, R. C., Holland, J. M., Kripke, M. L., and Schreiber, H. (1982). J. Exp. Med. 155:No. 2.

EVIDENCE FOR *IN VIVO* NK REACTIVITY AGAINST PRIMARY TUMORS

Tina Haliotis*†
John Roder*¶
David Dexter**

*Department of Microbiology and Immunology

**Department of Pathology
Queen's University
Kingston, Ontario, Canada

There is strong evidence that Natural Killer (NK) cells play a role in protection against the growth (Roder, J.C., 1980) and metastasis of transplantable tumors (Talmadge, J.E., 1980). There is, however, little evidence, as yet, that NK cells play a protective role against the development of primary tumors. The most important evidence on this topic can be summarized as follows:

A. In the Murine System

(i) Aging mice (>6 months) have diminished NK function (Kiessling, R., 1975) and also exhibit an increased incidence of spontaneous tumors.

(ii) Raised *in vitro* NK activity against YAC or syngeneic, MCA-induced fibrosarcomas in mice deprived of B cells by chronic anti-μ treatment, correlated well with an increased resistance to induction of primary tumors by MCA (Brodt, P., submitted). An important control in these studies was the observation that the incidence and growth-rate of progressing tumors induced by Moloney sarcoma virus (MSV) did not differ in anti-μ-suppressed or normal control mice that were untreated with normal gammaglobulin (Gordon, unpublished

†Supported by MRC.
¶Supported by NCI.

ISBN 0-12-341360-5

observations). Since host defence in the MSV system is clearly T cell-dependent (Gorczynski, R.M., 1974), this indicates that B cell-deprived mice have normal T function and augmented NK activity does not influence a T cell-dependent rejection system.

(iii) Gorelik and Herberman (1981) presented data indicating a positive correlation between susceptibility of A/J, CBA/J and C57BL/6 mice to urethane-induced lung carcinogenesis and inhibition of in vitro and in vivo natural cell-mediated cytotoxicity, thereby supporting the hypothesis that NK cells play a role in resistance to urethane-induced lung carcinogenesis.

(iv) On the other hand, Argov et al (1981), in a study of the incidence, latency and type of tumors induced in NK cell-deficient C57BL bg/bg mice and NK-intact bg/+ littermates, by oral administration of DMBA, found that the beige mice exhibited only slightly (if at all) increased susceptibility to induction and outgrowth of certain tumor types. Mice with more than one histologically confirmed tumor were slightly more common in the beige group (22% vs 12%). There were small but statistically non-significant differences in the incidence of carcinomas, bile duct adenomas, angiomas and non-thymic lymphomas. Although the latency periods showed wide variations and over-lapping between the two groups, the mean latency periods were slightly shorter (non-significant) in bg/bg mice for all tumor types except thymic lymphoma. In interpreting these data, it is important to consider the recent report that DMBA administered under certain conditions may cause a transient suppression of NK activity (Ehrlich, R., 1980).

(v) In addressing the same question, our group carried out a series of experiments in which was monitored the development of virally-, chemically- or irradiation-induced tumors, in C57BL/6 bg/bg mice compared to C57BL/6 bg/+ mice, as controls. The data, summarized in Table I, on the development of primary tumors in mice treated with carcinogens, show no statistically significant differences between the two groups in any of the experimental conditions.

However, the treatment could have an effect on NK function. In this regard, it should be noted that Gorelik and colleagues, in a study of the depression of NK activity of C57BL/6 mice by a leukemogenic schedule of irradiation, found that NK activity in C57BL/6 +/+ mice was still significantly greater than that of irradiated C57BL/6 bg/bg mice. Furthermore although fractionated doses of irradiation were found to inhibit spontaneous NK activity, the treated mice still had

TABLE I

Mode of Induction	Dose	Percent tumors in dead mice		Prevalent Histological Type
		bg/bg	bg/+	
MLV*	0.01 ml	9/13 (70%)	13/16 (80%)	malignant lymphoma
Benzo(alpha)pyrene, s.c.	1.6%	11/13 (85%)	35/39 (90%)	rhabdomyosarcoma
	0.4%	25/30 (83%)	27/46 (59%)	
	0.1%	14/22 (64%)	23/39 (59%)	
Gamma-irradiation	160 R/wk x 4			malignant lymphoma
expt. 1 (200 days)		6/13 (46%)	7/19 (37%)	
expt. 2 (150 days)		12/38 (31%)	26/51 (50%)	

There were no statistically significant differences in latency or tumor size between the two groups.

*MLV: DMBA-induced murine leukemia virus (type 485-10).

interferon-responsive pre-NK cells, and that after 16 hours of incubation of their spleen cells with interferon, NK activity dramatically increased (Parkinson, D.R., 1981).

With regard to the possible effect of the virus-treatment on NK activity, we cite experiments carried out on the effect of infection with vesicular stomatitis virus (VSV) on NK activity (McKinnon, K.P., 1981). In these experiments, it was found that VSV infection of bg/bg mice induced levels of NK cytolytic activity comparable to that of uninfected normal bg/+ controls but considerably lower than NK activity in VSV-infected bg/+ mice.

(vi) A good way around the problem is the study of tumors arising spontaneously rather than by induction with carcinogens. J.F. Loutit and colleagues, in a short communication to

Nature (Loutit, J.F., 1980) reported an incidence of spontaneously arising disseminated lymphoma in 11 out of 40 (27%) mice of the beige genotype crossed onto a (C3H x 101)F_1 background. However, the only information on controls was that neither C3H nor 101 strains, or the F_1 hybrid were known to be noted for natural lymphomas.

(vii) Along these lines, our group has undertaken a study involving a total of over 200 C57BL/6 bg/bg and bg/+ mice for spontaneously arising tumors. The data collected so far, summarized in Table II, indicate an increased incidence of tumors in the group carrying the beige mutation. We have also observed that the C57BL/6 bg/bg group appears to have a shorter life span. This would be in agreement with an observation made by Goodrick (1977). It is interesting to note

TABLE II

Tumors Arising Spontaneously in bg/bg and bg/+ Mice

Mouse Genotype	Percent tumors in dead mice*	Histological Type
bg/bg	7/15 (46%)	malignant lymphoma
bg/+	3/11 (27%)	malignant lymphoma

*The age profiles were the same for each group.

that thus far the only histological type of spontaneously arising tumors discovered has been malignant lymphoma. Thus, it was found that, as in mink, cattle and man carrying the beige mutation (Padgett, G.A., 1968), the beige mouse may also have a preponderance of spontaneous lymphoproliferative disease, which could be due to any of the phenotypic defects caused by the beige mutation, including reduced NK activity.

B. In the Human System

(i) As we (Roder, J.C., 1980) and others (Klein, G.O., submitted) have shown, Chédiak-Higashi (C-H) patients, the human analogue of the beige mouse, have a severe depletion of NK activity, which is selective within their immune system. It is intriguing to know that, as mentioned above, these patients also have a profound increase in spontaneous lymphoproliferative disease.

(ii) Along these lines, it was found that among the C-H syndrome and three other immunodeficiencies, X-linked lymphoproliferative disease, ataxia telangiectasia and severe combined immunodeficiency, which are also characterized by a profound increase in spontaneous lymphoproliferative disease, the only common denominator with respect to aberrant immune functions was the depression in NK activity (Roder, J.C., 1981).

(iii) Other evidence favouring the hypothesis in question was gathered by T. Morizane and colleagues (Morizane, T., 1980) who investigated the immunological state in patients with liver cirrhosis, in whom it is well known that the incidence of hepatocellular carcinoma is increased. Natural Killer cell activity against HeLa cells was significantly decreased in these patients although the T cell population of peripheral blood lymphocytes (PBL) and blast transformation by phytohemagglutinin were also significantly decreased. This suggests, but does not prove, that the impaired NK activity may be one of the etiological factors in the development of hepatocellular carcinoma in patients with liver cirrhosis.

(iv) Hersey and colleagues, in a study extending up to two years on the possible relationship of cell-mediated cytotoxicity against melanoma cells to prognosis in melanoma patients (Hersey, P., 1978) found that patients who after surgery had low cell-mediated cytotoxicity (CMC) of blood mononuclear cells against cultured human melanoma cells had a significantly higher incidence of recurrence from melanoma than patients with normal or high CMC values, suggesting that this may be an important predisposing factor in the development of recurrent melanoma.

In conclusion, it could be stated that according to the evidence presented to date, it is quite possible that NK cells play a protective role in surveillance against the development of some primary tumors but that further evidence is clearly necessary for a definitive conclusion to be drawn on this issue.

REFERENCES

Argov, S., Cochran, A.J., Karre, K., Klein, G.O. and Klein, G. Int. J. Cancer 28(6):739 (1981).
Brodt, P. and Gordon, J. (submitted for publication).

Ehrlich, R., Efrati, M. and Witz, I.P. In R.B. Herberman (ed) Natural Cell-mediated Immunity against Tumors, pp 47-58, Academic Press, N.Y. (1980).
Goodrick, C.L. Gerontology 23:405 (1977).
Gorczynski, R.M. J. Immunol. 112:533 (1974).
Gorelik, E. and Herberman, R.B. J. Natl. Cancer Inst. 66:543 (1981).
Hersey, P., Edwards, A., Mieton, G.W. and McCarthy, H.W. Br. J. Cancer 37:505 (1978).
Kiessling, R., Klein, E. and Wigzell, H. Eur. J. Immunol. 5: 112 (1975).
Klein, G.O., Ramirez-Duque, P., Merino, F., Forsgren, M. and Henle, W. (submitted for publication).
Loutit, J.F., Townsend, K.M.S. and Knowles, J.F. Nature 285: 66 (May 8, 1980).
McKinnon, K.P., Hale, A.H. and Ruebush, M.J. Infect. Immun. 32(1):204 (1981).
Morizane, T., Watanabe, T., Tsuchimoto, K. and Tsuchiya, M. Gastroenterol. Jpn. 15(3):226 (1980).
Padgett, G.A. Adv. Vet. Sci. 12:240 (1968).
Parkinson, D.R., Brightman, R.P. and Waksal, S.D. J. Immunol. 126(4):1460 (1981).
Roder, J.C. and Haliotis, T. Immunol. Today, Nov. 1980: 96-100.
Roder, J.C., Haliotis, T., Klein, M., Korec, S., Jett, J.R., Ortaldo, J., Herberman, R.B., Katz, P. and Fauci, A.S. Nature 284(5756):553 (Apr. 10, 1980).
Roder, J.C., Laing, L., Haliotis, T. and Kozbor, D. In NK Cells: Fundamental Aspects and Role in Cancer (R.B. Herberman, ed). Human Cancer Immunology, vol. 6, North-Holland Publishers, Amsterdam, 1981.
Talmadge, J.E., Meyers, K.M., Pricur, D.J. and Starkey, J.R. Nature (London) 284:622 (1980).

TRANSFER OF NK ACTIVITY AND LYMPHOMA RESISTANCE TO AKR MICE BY MARROW FROM HIGH NK, LYMPHOMA-RESISTANT (C57xAKR)F_1 MICE

John J. Trentin[1]
Surjit K. Datta
Elizabeth L. Priest
Michael T. Gallagher[2]
Anthony G. Nasrallah[3]

Division of Experimental Biology
Baylor College of Medicine
Houston, Texas

I. INTRODUCTION

In discussing the need for more direct assessment of the in vivo role of NK cells in immune surveillance, Herberman and Ortaldo (1981) have stated that "the most convincing procedure might be to reconstitute animals with depressed NK activity, by adoptive transfer of purified NK cells or their precursors, and determine the effects on carcinogenesis--". The present experiment was designed to transfer high NK activity and hopefully leukemia resistance to the normally low-NK-reactive and high-spontaneous-leukemia AKR mouse by means of bone marrow transplantation from the leukemia resistant (C57xAKR)F_1 hybrid.

In 1973 we pointed out that genetic resistance to parental or allogeneic or xenogeneic marrow transplantation was either non-immunological, or a previously undefined type of immune response (Rauchwerger et al., 1973a; Trentin et al., 1973). We demonstrated a correlation with genetic resistance to leukemogenesis, in that this atypical post-irradiation marrow rejection phenomenon was strongest in the leukemia resistant C57 strain and its hybrids (Rauchwerger et al., 1973b), and

[1]Supported by USPHS grants CA-03367, CA-05021 and K6 CA-14219.
[2]Present address: City of Hope National Medical Center, Duarte, California.
[3]Present address: Cook Children's Hospital, Fort Worth, Texas.

ISBN 0-12-341360-5

leukemogenic fractionated irradiation (225 R four times at weekly intervals) abrogated genetic resistance to both leukemogenesis and to marrow transplantation (Rauchwerger et al., 1976). More directly, lethally irradiated (C57xAKR)F_1 mice, which reject C57 but not AKR parental marrow cells, vigorously rejected spontaneous AKR lymphoma cells, whereas irradiated (C3HxAKR)F_1 mice, which do not reject either parental marrow, do not reject AKR lymphoma cells (Gallagher et al., 1976a; Gallagher et al., 1976b; Trentin et al., 1976), Fig. 1.

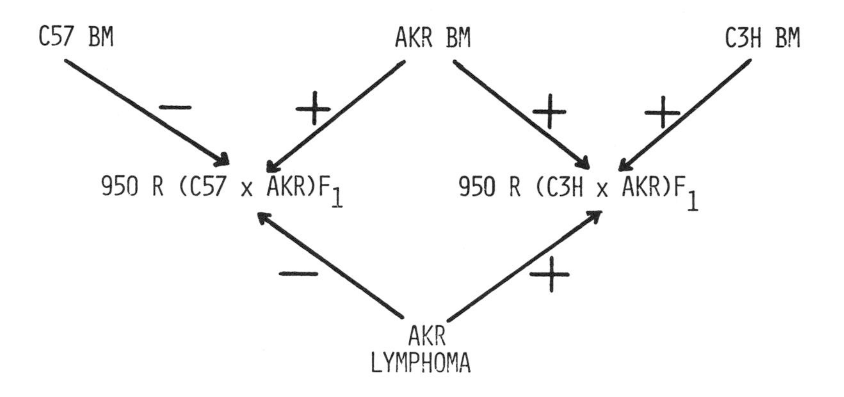

FIGURE 1. Schema of take (+) or rejection (-) of intravenously transfused inbred parental bone marrow (BM) or lymphoma cells in lethally irradiated (950R ^{137}Cs, total body) (C57xAKR)F_1 responder mice and (C3HxAKR)F_1 nonresponder mice.

We and others subsequently demonstrated that hybrid resistance to parental marrow shares the many unique characteristics of NK cell-mediated lysis of lymphoma cells in vitro (Trentin et al., 1977; Kiessling et al., 1977; Datta et al., 1979) and it is now generally agreed that NK cells are the main effector cells of this phenomenon in vivo (Herberman and Ortaldo, 1981). Hybrid resistance to parental marrow transplantation, long recognized as an exception to the laws of tissue transplantation (Cudkowicz and Bennett, 1971), may therefore represent the first recognized but misunderstood in vivo manifestation of NK cell surveillance activity. Natural spleen cell mediated cytolysis of YAC-1 lymphoma cells in vitro was found to be significantly higher for the low leukemia $(C57xAKR)F_1$ than for either the high-leukemia $(C_3HxAKR)F_1$ or AKR strains (Gallagher et al., 1980).

II. METHODS AND RESULTS

AKR mice, aged 4 to 6 weeks, received 950 R whole-body irradiation (^{137}Cs) followed by intravenous transfusion with 10^7 marrow cells from either the AKR strain, or the high NK, low-spontaneous-leukemia $(C57xAKR)F_1$ hybrid, or the low NK, high-spontaneous-leukemia $(C_3HxAKR)F_1$ hybrid.

Age-matched untreated control and chimeric AKR mice were tested at 2 to 4 week intervals for spleen cell mediated cytolysis of YAC-1 lymphoma cells by 4 hour chromium release assay as previously described (Datta et al., 1979). From 2 to 12 weeks post-irradiation and marrow transfusion, spleen cell mediated lysis was higher (in most cases significantly so) in AKR mice rendered chimeric with marrow from the high NK-reactive, low leukemia $(C57xAKR)F_1$ strain than from the low NK-reactive, high-leukemia $(C_3HxAKR)F_1$ or AKR strains, and higher than age control untreated AKR mice (TABLE I).

The incidence of spontaneous thymomas and mean age at death in untreated control mice of the 3 strains, and in the three types of chimeric AKR mice are shown in TABLE II. It is apparent that chimerism with $(C57xAKR)F_1$ marrow conferred not only high NK cell activity but also a high degree of protection against the development of thymomas.

III. DISCUSSION

Under the conditions of these experiments, T_6 marker studies have in the past demonstrated that the chimeric mice should be total, not partial, chimeras (Hellstrom et al., 1973). The question arises therefore of whether the conferred resistance is due to a change of target cells, in addition to or instead of the increased spleen natural cell mediated cytolytic activity. Although resistance to oncogenesis has been shown to reside at the target cell level under certain circumstances, that this is not likely the case here is indicated by the facts that a) target cells of the leukemia-resistant C57 strain are highly susceptible to leukemogenesis if C57 mice received 4 weekly sublethal doses of irradiation, which impairs NK cell activity (the subject of the following paper) and b) unirradiated C57 or $(C57xC_3H)F_1$ thymus transplanted into similarly irradiated $(C57xC_3H)F_1$ mice undergoes indirect "radiation leukemogenesis" of donor cell origin that has been shown to be RAD-LV virus-mediated (Kaplan, 1967).

TABLE I. Spleen Cell Mediated Cytolysis of YAC-1 Lymphoma Cells (Four Hour Assay), % ± S.D.

Spleen Cell Donor	Weeks Post-Irradiation (950R 137 Cs) and Marrow Transfusion				
	2	4	6	8	12
Untreated AKR Controls	3.32 ± 1.1	5.56 ± 2.4	6.13 ± 3.31	3.43 ± 0.44	4.11 ± 0.69
AKR + R + AKR BMC	6.57 ± 1.9	4.62 ± 1.99	7.32 ± 1.02	5.67 ± 0.47	8.6 ± 0.60
AKR + R + (C_3HxAKR)BMC	2.48 ± 0.8	5.96 ± 0.54	5.33 ± 2.64	6.18 ± 0.96	16.21 †
AKR + R + (C57xAKR)BMC	10.76 ± 3.52 ‡	11.43 ± 2.74*	10.47 ± 1.06*	7.62 ± 0.73*	22.18 ± 0.5*

‡ † All groups contain 3 mice except ‡ contains 2 mice, † contains one mouse.

* P ≤ .05 versus AKR + R + AKR BMC

TABLE II. Spontaneous Incidence of Thymoma by Strain and Treatment

No. Dead with Thymoma/Total Dead to Date = % with Thymoma

NORMAL CONTROLS			AKR + 950 R 137 Cs + BMC FROM		
AKR	(C_3HxAKR)F_1	(C57xAKR)F_1	AKR	(C_3HxAKR)F_1	(C_{57}xAKR)F_1
39/39 = 100%	23/23 = 100%	2*/31 = 6%	21/21 = 100%	30/31 = 97%	6*/24 = 25%
314 days †	319 days	561 days	318 days	343 days	362 days

† Mean age at death; all mice were autopsied at spontaneous death, or after sacrifice when moribund, or when showing dyspnea, ruffled hunched posture, or other signs of thymoma, except the (C57xAKR)F_1, all but one of which were sacrificed in apparent good health at an average age of 548 days. One died without thymoma at 955 days. Others are being held for autopsy at spontaneous death.

* The 2 thymomas found in (C57xAKR) controls, and the 6 found in AKR + R + (C57xAKR)BMC were significantly smaller than those found in all other groups.

In the present experiment it is not ruled out that marrow-derived cells other than NK cells may play a part in the observed adoptive transfer of resistance to leukemogenesis. This is the subject of continuing studies.

REFERENCES

Cudkowicz, G., and Bennett, M. (1971). J. Exp. Med. 134:83.
Datta, S.K., Gallagher, M.T., Trentin, J.J., Kiessling, R., and Wigzell, H. (1979). Biomedicine 31:62.
Gallagher, M.T., Datta, S.K., and Trentin, J.J. (1980). Biomedicine 33:73.
Gallagher, M.T., Lotzová, E., and Trentin, J.J. (1976a). Biomedicine 25:1.
Gallagher, M.T., Lotzová, E., and Trentin, J.J. (1976b). In "Immuno-Aspects of the Spleen" (J.R. Battisto and J.W. Streilein, eds.), p. 359. North Holland Publishing Co., Amsterdam.
Hellstrom, I., Hellstrom, K.E., and Trentin, J.J. (1973). Cell. Immunol. 7:73.
Herberman, R.S., and Ortaldo, J.R. (1981). Science 214:24.
Kaplan, H.S. (1967). Cancer Res. 27:1325.
Kiessling, R., Hochman, P.S., Haller, O., Shearer, G.M., Wigzell, H., and Cudkowicz, G. (1977). Eur. J. Immunol. 7:655.
Rauchwerger, J.M., Gallagher, M.T., Monié, H.J., and Trentin, J.J. (1976). Biomedicine 24:20.
Rauchwerger, J.M., Gallagher, M.T., and Trentin, J.J. (1973a). Biomedicine 18:109.
Rauchwerger, J.M., Gallagher, M.T., and Trentin, J.J. (1973b). Proc. Soc. Exp. Biol. Med. 143:145.
Trentin, J.J., Gallagher, M.T., and Lotzová, E. (1976). Transplant. Proc. 8:463.
Trentin, J.J., Kiessling, R., Wigzell, H., Gallagher, M.T., Datta, S.K., and Kulkarni, S.S. (1977). In "Experimental Hematology Today" (S.J. Baum and G.D. Ledney, eds.), p. 179. Springer-Verlag Publishers, New York.
Trentin, J.J., Rauchwerger, J.M., and Gallagher, M.T. (1973). Biomedicine 18:86.

POSSIBLE ROLE OF NK CELLS IN RADIATION LEUKEMOGENESIS: ADOPTIVE REPAIR OF NK DEFICIT OF FRACTIONALLY IRRADIATED MICE BY MARROW TRANSFUSION[1]

Surjit K. Datta
Elizabeth L. Priest
John J. Trentin

Division of Experimental Biology
Baylor College of Medicine
Houston, Texas

I. INTRODUCTION

Kaplan et al. demonstrated that a) fractionated sublethal whole body irradiation of C57BL/6 mice, caused a high incidence of thymic-lymphoma-leukemia, but a post-irradiation transfusion of bone marrow prevented it; b) such irradiation-induced lymphoma is indirectly mediated by a latent virus, RAD-LV (Kaplan, 1967). There is increasing evidence that NK cells play an important role in immune surveillance against lymphoma-leukemia, and against virus infected cells (Herberman and Ortaldo, 1981) and that NK cell activity is adoptively transferable by bone marrow transfusion (Herberman and Ortaldo, 1981; Trentin et al., 1982). To investigate a possible role of natural killer (NK) cells in radiation leukemogenesis, we studied the effect of fractionated irradiation of C57BL/6 mice, with or without marrow transfusion, on NK cell activity and development of thymic lymphoma.

II. METHODS: Over a period of many months, C57BL/6 mice reared in our barrier sustained caesarean derived, SPF mouse colony, aged 4 to 6 weeks, half of each sex, were randomized into unirradiated age control

[1] Supported by USPHS Grants CA-03367 and K6 CA-14219.

ISBN 0-12-341360-5

TABLE I. Effect of Leukemogenic Fractionated Irradiation of C57BL6 Mice, and Post-Irradiation Bone Marrow Transfusion on Spleen Natural Cell Mediated Lysis of YAC-1 Lymphoma Cells[a]

Treatment of Spleen Cell Donor / # of Mice/Point	% Lysis ± S.D. of YAC-1 Cells at Indicated Number of Weeks Post-Irradiation								
	1 Week		2 Weeks		3 Weeks	4 Weeks	5 Weeks	6 Weeks	8 Weeks
Age Control 3,4,3,3,3,6,3,8,6[b]	7.7±1.0	8.3±2.6	14.6±2.6	17.0±5.1	14.6±1.3	11.2±3.2	8.9±1.8	11.1±4.0	11.1±4.1
Age Control + CP[c] 2,3,3,3		18.8±2.7		36.7±9.5	24.2±4.2		17.6±4.6		
Age Control + PIC[d] 6,8,6						26.6±9.3		23.8±5.0	24.8±6.8
R[e] 4,4,3,6,3,9,5	3.4±0.9		6.2±0.7		5.0±3.3	5.4±2.3	5.5±0.7	8.2±2.7	6.5±4.0
R + CP 4,4,3,3		6.9±1.3		3.7±1.0	10.0±5.7		7.7±2.0		
R + PIC 6,9,5						9.7±4.0		13.4±2.4	13.5±4.1
R + BM[f] 4,4,3,6,3,9,6	8.8±2.8		11.6±3.0		12.6±2.3	11.7±3.4	11.7±1.4	14.0±1.7	11.4±3.1
R + BM + CP 4,4,3,3		11.6±3.6		12.3±2.9	19.2±1.2		19.7±1.0		
R + BM + PIC 6,9,6						18.6±2.0		22.4±2.9	27.5±0.5

[a] 4 hour chromium release assay was performed at E:T ratios of 50:1 and 25:1. Only 50:1 data are shown.

[b] number of mice, assayed individually, 2 to 4 per experiments. Points of 6 or 9 mice are pools of 2 or 3 experiments.

[c] 0.25 mg *Corynebacterium parvum* injected i.p. 5 days prior to sacrifice.

[d] 100 μg Poly I:C injected intraperitoneally 1 day before sacrifice.

[e] 4 weekly exposures to 225 R ^{137}Cs.

[f] 10^7 C57 bone marrow cells intravenously 4 to 7 hours after the fourth and last exposure to 225 R.

groups and irradiated groups. The latter received 4 weekly exposures to 225 R ^{137}Cs. Four to seven hours after the last exposure, half of the irradiated mice received 10^7 C57 bone marrow cells intravenously. Marrow-transfused, non-transfused and age-control mice were tested at intervals thereafter for spleen natural cell-mediated lysis of YAC-1 lymphoma target cells, by a 4 hour chromium release method previously described (Datta et al., 1979), with or without 1 or 5 days prior intra-peritoneal injection of the interferon inducers Poly I:C (PIC) or *Coryne-bacterium parvum* (CP), respectively. Mice were housed and all procedures performed within our barrier sustained SPF mouse colony and associated laboratories, with all personnel entering the barrier in sterile coveralls, cap, gloves and mask.

III. RESULTS: Spleen NK cell activity was depressed by fractionated irradiation (Table I), in agreement with the finding of Parkinson et al. (1981). Bone marrow transfusion after the last irradiation significantly increased NK cell activity by 1 week post-transfusion, to essentially normal levels throughout the 8 weeks post-transfusion. NK cell activity was boosted by either PIC or CP, usually by a factor of 2. But at no time did PIC or CP treated fractionally irradiated mice achieve the high NK cell activity of comparably treated control mice. However, fractionally irradiated and marrow transfused mice boosted with either PIC or CP reached PIC or CP treated age control NK levels respectively by 5 weeks in the case of CP, and by 6 and 8 weeks post-irradiation in the case of PIC.

The development of thymic lymphoma and other data are being collected. These results suggest that NK cells may play a role in the control of radiation leukemogenesis.

REFERENCES

Datta, S.K., Gallagher, M.T., Trentin, J.J., Kiessling, R., and Wigzell, H. (1979). Biomedicine 31:62.

Herberman, R.B., and Ortaldo, J.R. (1981). Science 214:24.

Kaplan, H.S. (1967). Cancer Res. 27:1325.

Parkinson, D.R., Brightman, R.P., and Waksal, S.D. (1981). J. Immunol. 126:1460.

Trentin, J.J., Datta, S.K., Priest, E.L., Gallagher, M.T., and Nasrallah, A.G. (1982). In "Natural Cell-Mediated Immunity" (R.B. Herberman, ed.), vol. 2. Academic Press, New York (preceding article, this volume).

ROLE OF NATURAL-CELL-MEDIATED IMMUNITY IN URETHAN-INDUCED LUNG CARCINOGENESIS

Elieser Gorelik[1,2], and Ronald B. Herberman[1,3]

[1]Laboratory of Immunodiagnosis,
National Cancer Institute,
Bethesda, MD.

I. INTRODUCTION

The assumption that NK cells may participate in antitumor immune surveillance is mostly based on the following facts:

a) NK cells are largely responsible for the quick elimination of some transplanted tumor cells from the blood stream or from the local site of growth (Kiessling *et al*, 1975; Riccardi *et al*, 1980; Gorelik and Herberman, 1981a).

b) In T-cell depleted nude or neonatally thymectomized mice, the expected increase in the incidence of spontaneous or induced tumor has not been observed. Since these mice display high levels of NK reactivity, it has been suggested that NK cells may be responsible for the persistent antitumor defense in these mice.

c) Retionic acid has been shown to have both antitumor and NK-augmenting effects and, in contrast, tumor promoters have been shown to depress NK cell function (Goldfarb and Herberman, 1981).

d) Low levels of NK cell activity in some patients (Chediak-Higashi syndrome, X-linked lymphoproliferative disease, patients with allografted kidney) are paralleled

---

[2]Present address: Surgery Branch, National Cancer Institute, Bethesda, Maryland.
[3]Present address: Biological Research and Therapy Branch, National Cancer Institute, Frederick, Maryland.

ISBN 0-12-341360-5

by a high incidence of malignancy, particularly lymphomas (Roder et al, 1980; Sullivan et al, 1980; Lipinsky et al, 1980).

Each of these lines of evidence fits one of the major predictions of the immune surveillance theory, that tumor development would be associated with, and in fact preceded by, depressed immune function. A related prediction of the immune surveillance theory is that carcinogenic agents would cause depressed immune function, thereby impairing the ability of the host to reject the transformed cells. Indeed, numerous experimental data indicate that a variety of carcinogenic factors (irradiation, chemical carcinogens, oncogenic viruses) can be extremely efficient in the suppression of T cell-mediated and humoral immunity (Berenbaum, 1964; Ceglowski and Friedman, 1969). The ability of carcinogenic factors to suppress natural cell-mediated immunity, considered to be a possible first line in the antitumor immune surveillance, is mostly unknown. The involvement of NK cells in antitumor resistance would be supported by indications that carcinogenic agents suppress their activity, and that restoration of function would be accompanied by inhibition of tumor development.

We have investigated the carcinogenic and NK-suppressive effects of urethan. Urethan is carcinogenic in mice, rats, and hamsters, producing lung tumors, lymphomas, hepatomas, melanomas, and vascular tumors. The susceptibility of mice to the urethan-induced carcinogenesis in the lungs depends on their age and genotype (Shimkin, 1958). Therefore we studied the carcinogenic and NK-suppressive effects of urethan in mice of different strains. The carcinogenic effect of urethan was assessed under conditions when NK reactivity of mice was suppressed or restored by transplantation of normal lymphoid cells.

These investigations were performed with A/J, CBA/J, C57BL/6 and C57BL/6 beige mice at 6-8 weeks of age. Urethan (ethyl carbamate) was inoculated i.p. (0.75 - 1.0 mg/g body weight) and at various times thereafter, the cytotoxic activity of spleen cells against YAC-1 cells was tested in a 4 hr. ^{51}Cr-release assay (Gorelik and Herberman, 1981b). In vivo natural antitumor resistance also was assessed by the ability of mice to eliminate i.v. inoculated [^{125}]dUrd-labeled YAC-1 cells from the lungs (Gorelik and Herberman, 1981c).

The carcinogenic effect of urethan in the lungs was determined at 12 weeks following urethan. Lungs were impregnated with India ink (Wexler, 1966), and were easily detected as small white spots on the black surface of the lungs.

As expected, the in vitro NK activity of spleen cells of untreated normal mice was found to vary with their genotype (Table I).

TABLE I. Influence of Urethan on Natural Cell-Mediated Immunity and its Carcinogenic Effect in the Lungs of Mice of Different Genotypes

Strain	Days after urethan[a]	NK (LU_{10}/ spleen[b])	In vivo natural resistance[c]	Lung tumors after 12 wks. +/total mice	Lung tumors after 12 wks. nodules/ lung
A/J	Untreated	38	9838		
	1	6.4[d]	35071[d]	20/20	13
	7	0.2[d]	16701[d]		
CBA/J	Untreated	172.8	2405		
	1	80.0[d]	5448[d]	10/15	2
	7	189.2	2851		
C57BL/6	Untreated	63.2	6080		
	1	43.7	6923	2/20	0.1
	7	76.0	5739		

[a]Seven-week-old mice were injected with urethan (1mg/g).
[b]Cytotoxic activity against YAC-1 cells, 4-hour ^{51}Cr-release assay (Gorelik and Herberman, 1981).
[c]1×10^6 [^{125}I]dUrd-labeled cells inoculated i.v; 3 hours later, remaining radioactivity in lungs was determined.
[d]Different ($p<0.05$) from untreated mice.

The highest level of cytotoxicity of spleen cells was displayed by CBA/J and the lowest by A/J mice. At 1 day following urethan, the cytotoxicity of spleen cells of A/J and CBA/J mice was significantly reduced (Table I). As previously described NK reactivity of A/J and CBA/J mice returned to the normal level 4 days after urethan treatment (Gorelik and Herberman, 1981a,b) but the NK activity of A/J mice became depressed again at around day 7. In contrast, this secondary depression in NK reactivity was not observed in CBA/J mice (Table I). The NK activity of C57BL/6 mice appeared to be completely resistant to suppression by urethan. Spleen cells of normal and urethan-treated C57BL/6 mice at all times between 1 and 8 days following urethan inoculation had similar levels of cytotoxic reactivity.

Rapid clearance of i.v. inoculated tumor cells from the lungs has been shown to be largely mediated by NK cells (Riccardi et al, 1980) and this method provides useful information about *in vivo* NK reactivity of mice. Since urethan is highly efficient in the induction of tumors in the lungs, it was of interest to determine its effects on the rate of elimination of radiolabeled cells from this site. While the number of tumor cells that settled in the lungs at 10 min after tumor cell inoculation was similar in treated and non-treated mice (data not shown), 3 hours later the number of tumor cells remaining in the lungs varied in parallel with their splenic NK reactivity (Table I). Elimination of tumor cells from the lungs of urethan-treated A/J mice was dramatically suppressed. A slight suppression of the tumor cell clearance observed in CBA/J mice, only at 1 day after urethan treatment, and C57BL/6 mice again were entirely resistant to the suppressive action of urethan. Thus, a single injection of urethan was able to suppress natural antitumor resistance of A/J mice, not only in the spleen but also in the lungs, target organ for the predominant, early oncogenic effects of this agent. 12 weeks after urethan treatment, 100% of A/J mice had multiple tumor nodules in the lungs, whereas CBA/J mice had substantially fewer tumors in the lungs and C57BL/6 mice were almost completely resistant.

The participation of NK cells in resistance to urethan-induced carcinogenesis would be further supported if lung carcinogenesis was increased in mice with additional depression of NK reactivity. To assess this possibility, NK reactivity of A/J mice was suppressed by cyclophosphamide (Gorelik and Herberman, 1981c), and also C57BL/6 beige mice which are deficient in NK cell function were studied. Since in the preliminary experiments urethan at the usual dose of 1 mg/g body weight in combination with Cy was lethal for some A/J mice, we decreased the dose of urethan to 0.75 mg/g, which was injected i.p. 2 days after inoculation of Cy (0.15 mg/g) (Table II). Three months later the number of tumors arising in the lungs of A/J mice pretreated with Cy was increased by 54%. However, beige mice treated with urethan had only a low incidence of lung tumors, similar to that of C57BL/6 +/+ mice.

In the next series of experiments, we studied the influence of transplanted normal lymphoid cells on the lung carcinogenesis of A/J mice treated with urethan (Table III). Three months later, the number of visible tumor nodules in the lungs of mice transplanted with normal syngeneic bone marrow or spleen cells was significantly less than in control urethan-treated mice. This inhibition of urethan carcinogenesis was more profound in mice transplanted with bone marrow cells than in mice inoculated with normal spleen cells (Table III). It was of note that spleen cells of donors pretreated with urethan lost their ability to transfer the anticarcinogenic effect. Thymocytes also were unable to influence the

TABLE II. Urethan-Induced Carcinogenesis in Mice with Suppressed NK Reactivity

Strain	Treatment[a]	Mice with tumor/total	No. tumors/lungs
A/J	urethan	35/35	11.4
A/J	Cy + urethan	25/25	17.6
C57BL/6	urethan	2/20	0.1
beige	urethan	3/25	0.1

[a]A/J mice (2 mos. old) were treated i.p. with Cy (0.15 mg/g); 2 days later mice received i.p. urethan (0.75 mg/g). C57BL/6 and C57BL/6 beige mice received i.p. urethan (1 mg/g). After 3 mos., the number of visible tumor nodules in the lungs were counted.

TABLE III. Inhibition of Urethan-Induced Lung Tumors by Transplantation of Normal Spleen or Bone Marrow Cells

Lymphoid cells inoculated[a]	No. of mice per group	No. of tumors/lungs	% inhibition
None	46	10.8	
Normal spleen cells	41	8.1	25
Spleen cells of urethan-treated donors	30	11.6	
Normal bone marrow cells	18	6.4	40
Thymocytes	25	11.6	
Macrophages	20	12.2	

[a]A/J mice (8 weeks old) were inoculated i.p. with urethan (1 mg/g). Two days later, 50×10^6 spleen cells, thymocytes or 25×10^6 bone marrow cells were inoculated i.v. or 20×10^6 thioglycollate-elicited peritoneal macrophages were inoculated i.p. in urethan-treated recipients. Spleen cells of urethan-treated donors were obtained 2 days after urethan-treatment.

carcinogenesis. Peritoneal thioglycollate-elicited macrophages (20×10^6) were inoculated i.p., since i.v. inoculation of more than 5×10^6 cells caused the immediate death of of the recipients and i.v. inoculation of the macrophages also did not change significantly the number of tumors depending in the urethan-treated mice. In contrast, we previously demonstrated that inoculation of spleen or bone marrow cells i.p. into urethan-treated mice inhibited urethan-induced lung carcinogenesis as well as i.v. transfer of these cells (Kraskovsky *et al*, 1973).

Thus our data indicated the possible involvement of NK cells or other natural effector cells in protection against urethan-induced lung carcinogenesis: 1) a carcinogenic dose of urethan suppressed NK reactivity of mice; 2) suppression of NK reactivity by urethan depended on the genotype of the mice and positively correlated with their susceptibility to lung carcinogenesis; 3) young A/J mice, which are more sensitive to the carcinogenic effect of urethan than older A/J mice, also were more sensitive to the NK-suppressive action of urethan (Gorelik and Herberman, 1981b); 4) the incidence of urethan-induced lung tumors was significantly increased in mice pretreated with Cy, which is shown to further suppress NK reactivity; 5) Urethan-induced lung carcinogenesis is suppressed in nude mice. Kaledin *et al* (1978) found one tumor in the lungs of 20% of nude mice, in contrast to multiple tumors in 70% of +/+ or nu/+ urethan-treated littermates; 6) Normal spleen and bone marrow cells but not thymocytes or macrophages inhibited the development of lung tumors in the urethane-treated recipients. In an analogous manner, we have found that transplantation of normal spleen or bone marrow cells can be efficient in restoration of NK reactivity of mice irradiated with small fractionated (179 R x 4) leukemogenic doses of irradiation (Gorelik *et al*, this volume).

In our present experiments we did not find an increase in the incidence of urethan-induced tumors in the lungs of beige mice, which have very low levels of NK cell activity. However, these data do not contradict the possible involvement of NK cells in antitumor defense, since C57BL/6 lung-target cells appear to be inherently resistant to malignant transformation by urethan. The genes responsible for susceptibility or resistance to lung tumorigenesis is mostly located in the target tissue, i.e., the lungs (Heston and Dunn, 1951; Heston, 1976). Therefore, in the absence of malignant cells, any suppression of the function of immune system cannot result in an increased incidence of lung tumors.

Although our data support the hypothesis that NK cells may participate in immune surveillance, they do not argue against the possible involvement of other immune mechanisms.

Indeed, urethan (and other carcinogenic agents) can suppress the function of B- and T-cell immunity (Gorelik et al, 1973). In addition, different immune suppressive procedures may accelerate carcinogenesis (Gatti and Good, 1971; Schnek and Penn, 1971). In order to understand the role of the immune system in antitumor surveillance, the function of both specific and natural immune mechanisms should be analyzed.

REFERENCES

Berenbaum, M. (1964). Brit. Med. Bull. 20:159.
Ceglowski, W., and Friedman, H. (1969). Nature 224:1318.
Gatti, R., and Good, R. (1971). Cancer 28:89.
Golfarb, R., and Herberman, R. (1981). J. Immunol. 126:2129.
Gorelik, E., and Herberman, R. (1981a). Int. J. Cancer 27:709.
Gorelik, E., and Herberman, R. (1981b). J. Natl. Cancer Inst. 66:543.
Gorelik, E., and Herberman, R. (1981c). J. Natl. Cancer Inst. 67:1317.
Gorelik, E., Rosen, B., and Herberman, R. (1982). This volume.
Heston, W. (1976). Adv. Cancer Res. 23:1.
Heston, W., and Dunn, T. (1951). J. Natl. Cancer Inst. 11:1057.
Kaledin, S., Gruntenko, E., and Morozkova, T. (1978). Bull. Exp. Biol. Med. 86:1490.
Kiessling, R., Petranyi, G., Klein, G., and Wigzell, H. (1975). Int. J. Cancer 15:933.
Kraskovsky, G., Gorelik, L., and Kagan, L. (1973). Proc. Acad. Sci. BSSR. 11:1052.
Lipinski, M., Tursz, T., Kreis, M., Finale, Y., and Amiel, J. (1980). Transplant. 29:214.
Riccardi, C., Santoni, A., Barlozzari, T., Puccetti, P., and Herberman, R. (1980). Int. J. Cancer. 25:475.
Rubin, B. (1964). Prog. Exp. Tumor Res. 5:218.
Roder, J., Haliotis, T., Klein, M., Korec, S., Jett, J., Ortaldo, J., Herberman, R., Katz, P., and Fauci, A. (1980). Nature. 284:553.
Schnek, S., and Penn, I. (1971). Lancet. 1:983.
Shimkin, M. (1955). Adv. Cancer Res. 3:223.
Sullivan, J., Byron, K., Brewster, F., and Purtilo, D. (1980). Science 210:543.

DEPRESSION OF NK REACTIVITY IN MICE BY LEUKEMOGENIC DOSES OF IRRADIATION

E. Gorelik[1,2], B. Rosen[1,3] and R.B. Herberman[1,4]

[1]Laboratory of Immunodiagnosis
National Cancer Institute
Bethesda, Maryland

Irradiation has a profound immunosuppressive effect on humoral and T-cell mediated immunity. Investigations on the influence of irradiation on natural-cell mediated immunity have shown that natural killer (NK) cell activity is rather radioresistant, with little or no effect immediately or in some studies, at 1 day after treatment (Kiessling et al, 1977; Hochman et al, 1978; Cudkowicz and Hochman, 1979). However, dramatic suppression of the cytotoxic activity of spleen cells was observed in mice by 2 weeks after sublethal irradiation (Hochman et al, 1978). NK cell activity in irradiated mice could be reconstituted by transfer of normal bone marrow cells and the level of reconstituted NK activity in lethally irradiated mice depended on the genotype of the donors. Bone marrow cells from mice with high or low levels of NK reactivity reconstituted, respectively, high or low levels of natural cell-mediated cytotoxicity in the spleens of the recipients (Haller et al, 1980). It is of interest that bone marrow cells of beige mice were able to reconstitute several immunological functions in lethally irradiated mice, but failed to reconstitute NK activity in the recipients (Roder, 1979).

Most of these studies were performed with lethal or sublethal doses of irradiation. However, it is known that small fractionated doses of irradiation can be extremely efficient in the induction of leukemia in some strains of

[2]Present address: Surgery Branch, National Cancer Institute, Bethesda, Maryland.

[3]Present address: Medical School, Beer-Sheba University, Beer-Sheba, Israel.

[4]Present address: Biological Research and Therapy Branch, National Cancer Institute, Frederick, Maryland.

ISBN 0-12-341360-5

mice, whereas high nonfractionated doses of irradiation had no leukemogenic effect (Kaplan, 1974; Haran-Ghera, 1980). It has been suggested that NK cells may participate in the elimination of malignant cells during the earliest phases of tumor growth (Herberman and Holden, 1979) and it is intriguing to consider the possibility of a role for NK cells in surveillance against irradiation-induced leukemogenesis.

Investigation of the effect of leukemogenic doses of irradiation on the function of natural cell-mediated immunity is needed to begin to understand the possible role of NK cells or other related cells in defense against development of these tumors. Numerous experimental data indicate that normal bone marrow and spleen cells transplanted into mice irradiated with leukemogenic doses of irradiation have substantial antileukemogenic effects (Kaplan, 1974; Haran-Ghara, 1980). It is quite possible that the transplanted normal lymphoid cells may in parallel restore NK reactivity in mice irradiated with small fractionated leukemogenic doses. In addition, the leukemogenic effects of irradiation are age-dependent. When irradiation of mice was started at 1 month of age, the incidence of induced leukemia was higher than in mice irradiated at older ages (Kaplan, 1974; Haran-Ghera, 1980). Although the differences in the susceptibility of C57BL/6 mice to the leukemogenic action of irradiation may be attributed to a variety of mechanisms, natural cell-mediated immunity might contribute to the level of resistance of mice to leukemogenesis.

In order to study the possible involvement of NK cells in irradiation-induced leukemogenesis, we investigated: 1) the influence of small fractionated (179R x 4) doses irradiation on the NK reactivity of C57BL/6 mice of different ages (1-3 months old) to the NK-suppressive action of irradiation or cyclophosphamide.

RESULTS

C57BL/6 +/+ and C57BL/6 bg/bg mice were obtained from Jackson Laboratory, Bar Harbor, Me., at the age of 3.5 weeks. Whole-body irradiation was performed with a ^{137}Caesium source of irradiation, at a dose of 179R (138R/min), 4 times at weekly intervals. One day after the last dose of irradiation, some mice received i.v. 10 x 10^6 bone marrow cells from nonirradiated 2-3 month-old C57BL/6 +/+ or C57BL/6 bg/bg donors. Cytotoxic activity of spleen cells of irradiated and nonirradiated mice was tested in a 4 hr ^{51}Cr-release assay as previously described (Gorelik and Herberman, 1981a).

In vivo natural antitumor resistance was assessed by the ability of mice to eliminate i.v. inoculated ^{51}Cr-labeled YAC-1 cells. The level of radioactivity remaining in the lungs was determined at 3 hours after inoculation of radio-labeled cells (Gorelik and Herberman, 1981b).

The results, presented in Table 1, indicated that 17 days following the last dose of irradiation, the cytotoxic activity of spleen cells of irradiated mice was substantially depressed. Following i.v. inoculation of 10 x 10^6 bone marrow cells of C57BL/6 +/+ mice into irradiated mice, the NK activity of spleen cells was significantly increased, although it did not reach the levels observed in nonirradiated normal C57BL/6 mice. Spleen cells of irradiated and nonirradiated beige mice consistently had very low levels of NK activity. It was of note that transfer of bone marrow cells from beige mice had little if any effect on the NK activity of spleen cells against YAC-1 cells.

This evaluation of the effect of irradiation on NK activity of spleen cells had some limitations, since this activity is age-dependent. Spleen cells of mice develop cytotoxic activity at 3-4 weeks of age, their activity reaches a peak at 6-8 weeks and then declines to low levels (Herberman and Holden, 1978). In our experiments, mice were irradiated between 4 and 8 weeks of age and when NK activity

Table 1. Natural Cell-Mediated Immunity of Irradiated C57BL/6 (+/+) and C57BL/6 Beige (bg/bg) Mice Inoculated with Bone Marrow Cells

Mice	Treatment[a]		NK	% labeled YAC remaining in the lungs
	R	BM	% lysis	
+/+	-	-	8	5
+/+	+	-	1	18
+/+	+	+/+	5	5
+/+	+	bg/bg	2	14
bg/bg	-	-	2	17
bg/bg	+	-	2	32
bg/bg	+	+/+	5	9
bg/bg	+	bg/bg	3	26

[a]Mice were irradiated (R) and 2 days following the last dose, 10x 10^6 bone marrow (BM) cells were inoculated i.v. At 17 days following irradiation, NK activity of spleen cells was tested in a 4 hour ^{51}Cr release assay (E/T,100:1).

[b]1 x 10^6 ^{51}Cr-labeled YAC-1 cells were inoculated i.v. and 3 hours later the level of remaining radioactivity in the lungs was measured.

was to be assessed 2-4 weeks later, the mice were already 10-12 weeks old. This probably explains the relatively low cytotoxicity by spleen cells of untreated C57BL/6 mice, which is slightly increased in a 16 hour cytotoxic assay. Nevertheless, in all experiments performed, the pattern shown in Table 1 was reproducible (Gorelik and Herberman, 1982). In contrast to the decline in splenic NK activity, elimination of radiolabeled cells from the bloodstream remained as efficient as in 6-week-old mice (data not presented). In vivo elimination of tumor cells has been shown to be primarily mediated by NK cells (Riccardi *et al*, 1980). In parallel, it was found that blood lymphocytes have higher levels of NK activity than spleen cells and their activity did not display strong age dependence, with blood lymphocytes of 3-5 month-old mice still having high levels of NK activity (Lanza and Djeu, this volume).

Based on these considerations, an in vivo assay of the ability of mice to eliminate i.v. inoculated tumor cells would be expected to be a more valuable method for analysis of the natural antitumor resistance in mice at $\geq$ 12 weeks of age. Therefore, 17 days after the last dose of irradiation, the ability of C57BL/6 mice to eliminate radiolabeled YAC-1 cells was measured and found to be significantly depressed in irradiated mice.

The number of tumor cells surviving in the lungs of irradiated mice was more than 3 times higher than in the non-irradiated mice. After transfer of bone marrow cells from C57BL/6 mice, the ability to eliminate tumor cells from the lungs was considerably higher than in irradiated mice and was similar to that in normal control C57BL/6 mice. In contrast, bone marrow cells from beige donors failed to restore the antitumor resistance of irradiated C57BL/6 mice (Table 1).

Nonirradiated beige mice had impaired ability to eliminate tumor cells, in comparison with normal C57BL/6 mice and in this respect they were similar to irradiated C57BL/6 +/+ mice. However, as previously reported (Brunda *et al*, 1980), beige mice were not completely deficient in natural cytotoxic function, since after irradiation (179 R x 4) their ability to eliminate tumor cells after i.v. inoculation was further diminished (Table 1). Transfer of bone marrow cells of C57BL/6 +/+ mice into irradiated beige mice reconstituted the natural antitumor resistance of these mice, to levels above those of nonirradiated beige mice, and close to, but not identical with, those in nonirradiated C57BL/6 mice. In contrast, transfer of bone marrow from beige donors did not significantly change the NK reactivity of irradiated C57BL/6 +/+ or bg/bg recipients.

In the next series of experiments we studied the suppressive effect of irradiation (179 R x 4) on the NK reactivity of mice of different ages. C57BL/6 mice at 1, 2 or 3 months of age were subjected to the irradiation protocol. The ability to eliminate i.v. inoculated ^{51}Cr-labeled YAC-1 cells was assessed at 1, 3 and 5 weeks after the last dose of irradiation. Since the irradiation procedure took 4 weeks, at the time of the first test, the mice were more than 2, 3 and 4 months of age, respectively. Nevertheless, all control non-irradiated mice had similar ability to eliminate radiolabeled tumor cells. At 1-3 weeks after irradiation, the antitumor resistance of all the irradiated mice was depressed, with more tumor cells remaining in the lungs of irradiated than in those of nonirradiated control mice. However the level of this suppression was higher in the mice which underwent the irradiation procedure at the age of 1 month. The higher level of sensitivity to irradiation of the younger mice was supported by the observation that 5 weeks after irradiation, the ability to eliminate tumor cells normalized in mice which entered the study at the age of 3 months. In contrast, in the youngest group of mice, natural antitumor resistance was even more suppressed than it was at 1 week after irradiation (Fig. 1).

In order to assess the selectivity of the age-related susceptibility of C57BL/6 mice to depression of natural resistance by irradiation, mice of the various ages were treated with cyclophosphamide (Cy), another NK-suppressive treatment. To determine the minimal NK-suppressive dose, 2-month old C57BL/6 mice were inoculated with Cy (doses 12.5 to 200 mg/kg body weight). One day after i.p. inoculation of Cy, mice received i.v. 7 x 10^6 of ^{51}Cr-labeled tumor cells and the level of radioactivity remaining in the lungs was determined. The clearance of i.v. inoculated tumor cells was suppressed in the mice treated with $\geq$ 100 mg/kg Cy. This dose was chosen to treat C57BL/6 mice of age 1, 2 or 3 months. 4, 7, 11 days after treatment, mice received i.v. radio-labeled YAC-1 cells and the level of remaining radioactive cells was measured 3 hours later (Fig. 2). 4 days after Cy, the level of radioactivity remaining in the lungs of Cy-treated mice was 3 times higher than in untreated control mice. This suppression of tumor cell clearance and its duration was similar in mice of each age group.

DISCUSSION

The prolonged suppression of NK reactivity after irradiation of young mice could be explained by the assumption that fractionated doses of irradiation are able to eliminate stem cells or precursors of NK cells. By transplantation of bone marrow cells from normal syngeneic donors, precursors of NK cells were transferred, resulting in reconstitution of NK

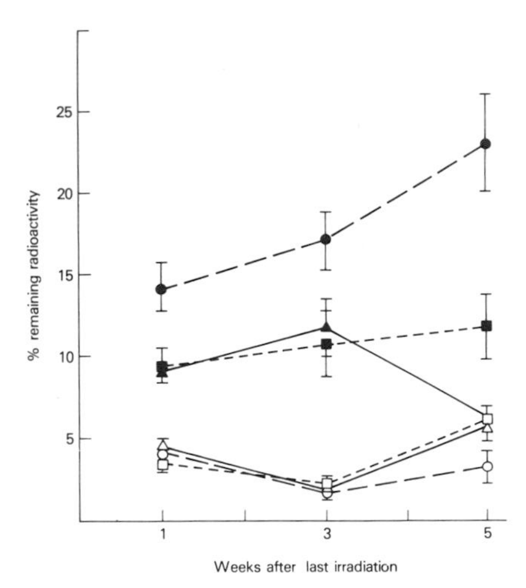

FIGURE 1. Susceptibility to NK suppressive action of irradiation (179 R x 4) of mice at different ages. At various times following the last dose of irradiation, mice were inoculated i.v. with 1 x 10^6 ^{51}Cr-labeled YAC-1 cells. 3 hours later, the level of radioactivity remaining in the lungs was determined. Tested were mice irradiated at 1 (●), 2 (■) or 3 (▲) months of age and nonirradiated control mice at 1 (○), 2 (□) or 3 (△) months of age at the beginning of the protocol.

reactivity. It is of interest that bone marrow cells from beige mice failed to restore the NK reactivity of the irradiated mice.

This is consistent with the previous suggestion that the NK cell deficiency of beige mice is at the level of the stem cells or precursors for NK cells (Roder, 1979). Mature NK cells probably are more radioresistant. Indeed, shortly after lethal or superlethal doses of irradiation the cytotoxic activity of spleen cells of these mice did not decrease and mice still had intact in vivo resistance to transplantation of tumor cells or bone marrow cells. Only later was profound and prolonged suppression of natural cell-mediated cytotoxic activity observed (Hochman et al, 1978; Cudkowicz & Hochman, 1979).

By comparing the susceptibility of C57BL/6 mice of different ages (1-3 months old) to the NK-suppressive action of fractionated irradiation, we have found the highest susceptibility in 1 month old mice. We obtained similar results with A/J mice treated with the chemical carcinogen, urethan (Gorelik & Herberman, 1981a). The suppressive affect of urethan on NK cell functions was more profound and prolonged in young (5-17 days old) than in older mice (6-8 weeks old).

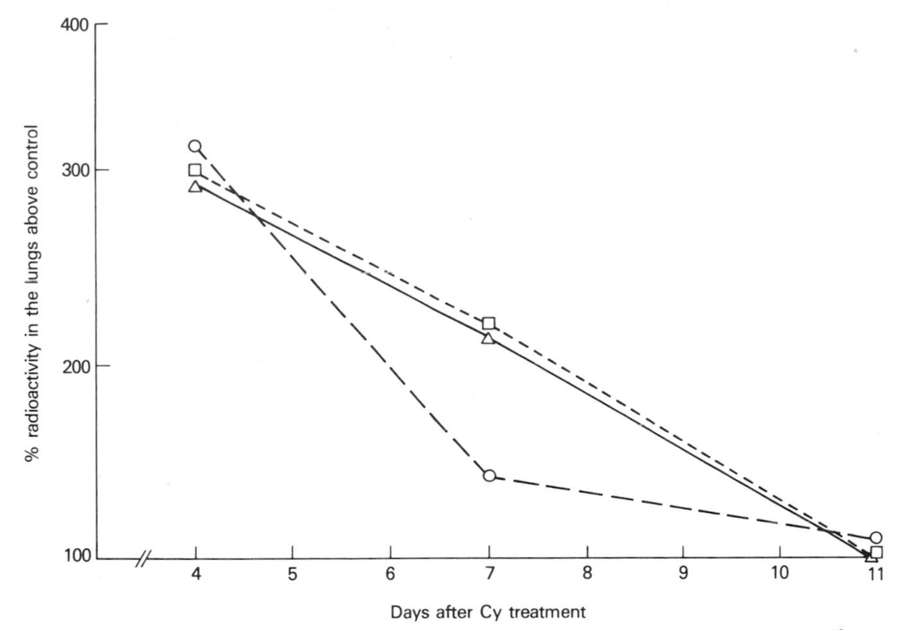

FIGURE 2. NK suppressive effect of Cy in 1 (○), 2 (□), or 3 (△) month old C57BL/6 mice. Cy (100 mg/kg) was injected i.p. and 4, 7, 11 days later, mice were tested for clearance of i.v. inoculated 1 x 10^6 ^{51}Cr-labeled YAC-1 cells. The radioactivity in the lungs of Cy-treated mice was expressed as % of the control nontreated mice:

$$\frac{\text{CPM of Cy-treated mice}}{\text{CPM control mice}} \times 100$$

In contrast, we have shown here that a noncarcinogenic NK-depressive agent, Cy, was equally efficient in the suppression of NK reactivity of mice of various ages. This suppressive effect was transient and 11 days later the NK activity of all treated mice had returned to normal.

Parkinson et al (1981) previously reported that the splenic NK activity of mice treated with leukemogenic doses of irradiation was depressed. However, the differences between irradiated and control mice tended to be small, especially because of the above discussed problem of age-dependent decline of NK activity of spleen cells. The natural antitumor resistance of mice assessed by ability to eliminate in vivo i.v. inoculated tumor cells was found here to be more useful for such studies. It provided the possibility to monitor the natural antitumor resistance of older mice (3-5 months old).

The fractionated doses of irradiation (179 R x 4) which were strongly suppressive for NK cell activity of young

C57BL/6 mice were demonstrated to be strongly leukemogenic. Thus, 93% of C57BL/6 mice irradiated at 1 month of age developed lymphomas within 6 months after irradiation. The protocol for reconstitution of NK reactivity of irradiated mice by transplantation of bone marrow cells has previously been reported to inhibit irradiation-induced leukemogenesis in these mice (Kaplan, 1974; Haran-Ghera, 1980). The higher susceptibility of young mice to the NK-suppressive action of irradiation provided a further correlation to the well-known greater leukemogenicity in younger mice (Kaplan, 1974; Haran-Ghera, 1980). Thus our results are fully consistent with the hypothesis that successful leukemogenesis by irradiation is dependent on depression of natural cell-mediated resistance in addition to direct malignant transformation of T cells or their precursors.

Further experiments will be needed to define more precisely the mechanism of radiation-induced depression of NK activity and particularly to assess more directly the role of NK cells in resistance to radiation-induced leukemogenesis.

REFERENCES

Brunda, M., Holden, H., and Herberman, R. (1980). In "Natural Cell-Mediated Immunity Against Tumors" (R. Herberman, ed.), p. 411. Academic Press, New York.
Cudkowicz, G., and Hochman, P. (1979). Immunol. Rev. 44:13.
Gorelik, E., and Herberman, R. (1981a). J. Natl. Cancer Inst. 66:543.
Gorelik, E., and Herberman, R. (1981b). J. Natl. Cancer Inst. 67:1317.
Gorelik, E., and Herberman, R. (1982). J. Natl. Cancer Inst. In Press.
Haller, O., Orn, A., Gidlund, M., and Wigzell, H. (1980). In "Natural Cell-Mediated Immunity Against Tumors" (R. Herberman, ed.), p. 1219. Academic Press, New York.
Haran-Ghera, N. (1980). In "Viral Oncology" (G. Klein, ed.), p. 161. Raven Press, New York.
Herberman, R., and Holden, H. (1978). Adv. Cancer Res. 27:305.
Hochman, P., Cudkowicz, G., and Dausset, J. (1978). J. Natl. Cancer Inst. 61:265.
Kaplan, M. (1974). Ser. Haematol. 7:2.
Kiessling, R., Hochman, P., Haller, O., Shearer, G., Wigzell, H., and Cudkowicz, G. (1977). Eur. J. Immunol. 7:655.
Parkinson, D., Brightman, R., and Waksal, S. (1981). J. Immunol. 126:1460.
Riccardi, C., Santoni, A., Barlozzari, T., Puccetti, P., Herberman, R. (1980). Int. J. Cancer. 25:475.
Roder, J. (1979). J. Immunol. 123:2168.

THE SUPPRESSION OF NK ACTIVITY BY THE CHEMICAL CARCINOGEN DIMETHYLBENZANTHRACENE AND ITS RESTORATION BY INTERFERON OR POLY I:C[1]

Rachel Ehrlich
Elinor Malatzky
Margalit Efrati
Isaac P. Witz[2]

Department of Microbiology
The George S. Wise Faculty of Life Sciences
Tel Aviv University
Tel Aviv Israel

I. INTRODUCTION

Several studies show that tumor-bearing humans and animals have altered (mostly decreased) NK activities (1,2). Only a few studies dealt with the immediate effects of oncogenic insults on the natural immune system. Goldfarb and Herberman (3) showed that in vitro, the tumor-promotor PMA inhibited both murine and human NK activities. PMA also inhibited interferon mediated augmentation of NK activity. Keller (4) described in vivo and in vitro inhibitory effects of PMA on NK activity in the rat and Seaman et al (5) found that PMA inhibited the NK activity of human PBL. The chemical carcinogen urethan was shown to suppress NK activity only in certain mouse strains (6) and not in others (7). Other oncogenic treatments such as irradiation (8) and certain sex hormones (9) also had a suppressive effect on NK activity.

---

[1]This investigation was supported by Contract NO1-CB74134 awarded by the National Cancer Institute DHEW and by a grant of Concern Foundation in conjunction with the Cohen-Appelbaum-Feldman Families Cancer Research Fund.

[2]Incumbent: The David Furman Chair of Immunobiology of Cancer.

ISBN 0-12-341360-5

Early reports from our laboratory showed that mice bearing mammary tumors induced by the chemical carcinogen dimethylbenzanthracene (DMBA) had a severe decrease in their NK activity. (10). Based on these findings we wished to know when, in relation to carcinogen administration, is this decrease first observed.

In the present report we present results of a time course study on the effects of DMBA on the NK activity of the treated mice. In addition we report on our attempts to augment and reconstitute the DMBA-induced suppression of NK activity using interferon or interferon inducers.

II. RESULTS

Time course study of the effect of DMBA on NK activity

Normal BALB/c mice or pituitary isografted mice were fed with DMBA according to the procedure described previously (10). Tumor incidence was 60% in the former and about 100% in the latter group. Untreated mice which were only given a pituitary isograft did not develop tumors. Mice belonging to these 4 groups were inspected regularly for tumor appearance and tested each week for NK activity.

These assays (results not shown) demonstrated that pituitary isografted and DMBA fed mice or mice treated only with DMBA showed a significant decrease in the number of spleen cells and a significant decrease in the NK activity. The number of lytic units per spleen in these mice was 6-10 times lower than in control mice. This decrease was evident already after the third DMBA feeding and it lasted throughout the entire tumor induction period. The NK activity levels of mice which developed tumors during the 100-day observation period were as low as those of DMBA-fed mice which were still tumor-free. Some of the tumor bearers had no measurable NK activity. The existence of an NK-suppressing activity in the tumor-bearing mice was reported previously (7,10). Pituitary isografted mice did not show any decrease in NK activity.

Augmentation of NK Activity in DMBA-fed Mice

Interferon as well as interferon inducers are known to augment NK activity in vitro and in vivo (11,12). Figure 1 shows that the in vivo administration of poly I:C as well as of interferon to DMBA-treated mice augmented the depressed

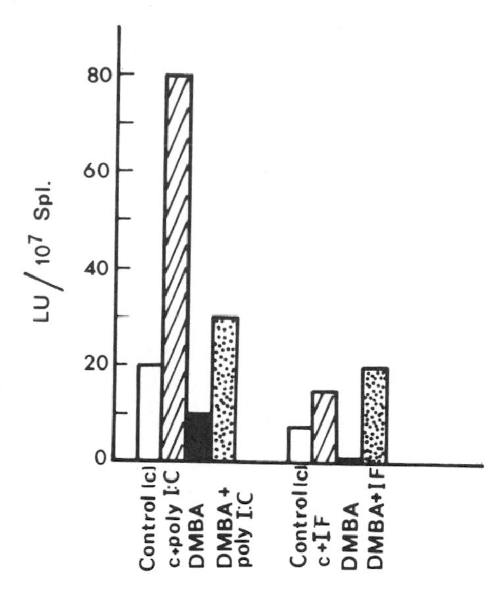

FIGURE 1. The effect of interferon or poly I:C on NK activity

Control (□) and DMBA-fed (■) BALB/c mice were injected i.v. with poly I:C (50μg/mouse) or i.p. with interferon (10^5 units/mouse). NK activity was assayed 24 hours after the administration and calculated as lytic units/10^7 splenocytes.

NK activity of such mice. However, the enhanced activity never reached the level of NK activity mediated by normal splenocytes treated with these materials. Figures 2 and 3 show that the level of NK activity after poly I:C or interferon administration remained high only for 48 hours and then decreased gradually. This pattern was identical to that which was obtained in interferon or poly I:C-treated normal mice (results not shown). Figure 3 shows also that the NK activity of DMBA-treated mice could be restimulated with poly I:C following its decrease to original levels 48 hours after a previous treatment. Consecutive injections of poly I:C (every 24 hours) enabled the maintenance of high levels of NK activity in DMBA-fed mice, throughout the period of treatment(results not shown). Even then, however, the NK activity declined to the original low levels 96 hours after the last poly I:C injection. During this study we observed that the NK activity could not be augmented by poly I:C administration when it was given on the day of DMBA feeding, but only when administered 24 hours later (results not shown).

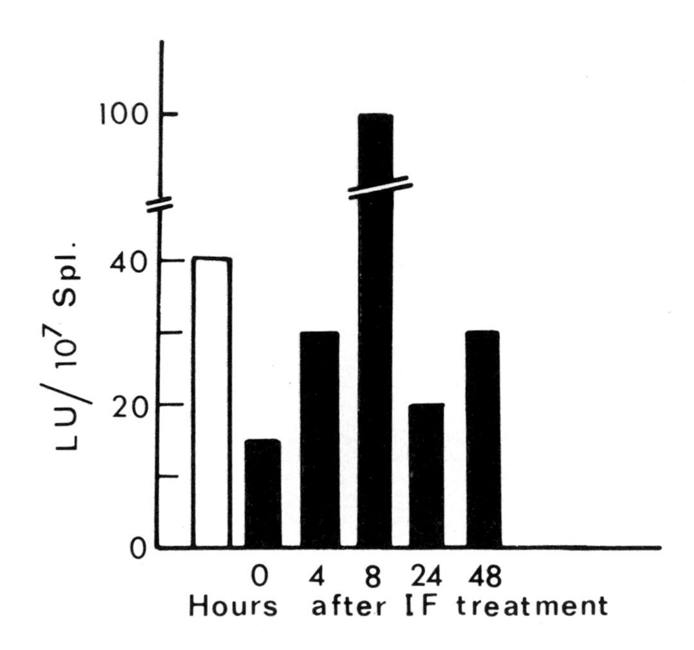

FIGURE 2. The duration of interferon-mediated augmentation of NK activity

DMBA-fed mice were injected i.p. with interferon (IF) (10^5 units/mouse). NK activity of control (□) and DMBA-fed mice (■) were assayed at the indicated intervals (hrs) after IF injection.

III. DISCUSSION

The results of this study show that the chemical carcinogen DMBA suppresses NK activity. Unlike suppression of NK activity induced by carrageenan (13), hydrocortisone acetate (13), aspirin (14) and cyclophosphamide (13), the DMBA-induced NK suppression was not restored spontaneously after ceasing treatment. Such a long-lasting effect was obtained only by treatments affecting bone marrow stem cells, for example, estrogen administered chronically (9) or radioactive strontium (8).

Whether or not the decrease in NK activity is a contributing factor to subsequent development of malignant mammary tumors is still not known. However, indirect evidence does not support such a possibility. Firstly, a pituitary isograft which increases considerably tumor incidence in DMBA-

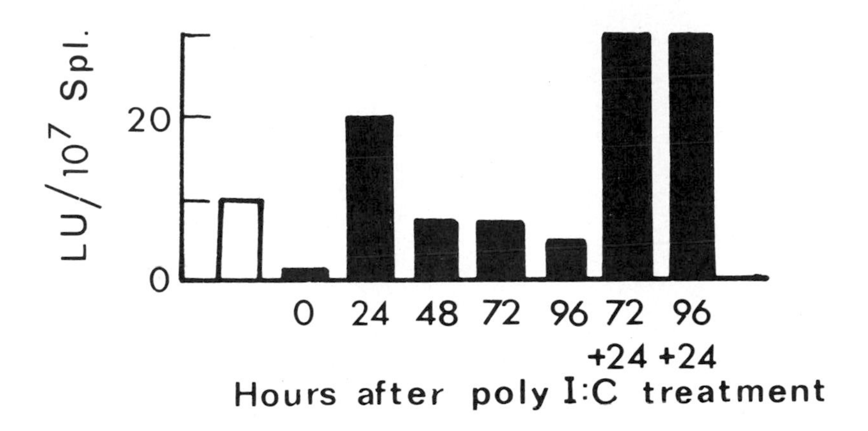

FIGURE 3. The duration of poly I:C mediated augmentation of NK activity

For experimental details see legends of Figures 1 and 2. In addition 2 groups of mice were given 2 poly I:C treatments: the first was treated 72 and 24 hours before the assay and the second was treated 96 and 24 hours before the assay.

treated animals did not suppress NK activity whatsoever. Secondly, a decreased NK activity was already evident after three 1 mg DMBA feedings. Such a carcinogen dose induced only a low incidence of malignant tumors (unpublished data).

Another approach to establish the regulatory role of NK cells (if any) on the progression of DMBA-transformed cells would be to reconstitute the DMBA-induced suppressed NK activity and then to compare tumor incidence in mice whose NK activity was artificially restored with that in mice whose depressed NK activity was not reconstituted. Interferon and poly I:C had an ability to transiently reconstitute NK activity to levels of untreated normal controls. Attempts to restore NK activity over prolonged periods of time by multiple poly I:C treatments revealed the following findings: The animals did not become refractory to multiple treatments and they responded by increased NK activity after each treatment. Moreover, even after multiple treatments the NK activity still decreased to its original low level after 24-48 hrs.

Preliminary results (not shown) demonstrated that 10 consecutive injections of poly I:C during the period of DMBA feeding (from the 3rd to the 6th feeding), although stimulating NK activity during the entire treatment period, did not delay or prevent primary tumor development. It should be noted however, that the 20 day period covered by this treatment was relatively very short compared to the average 100 day tumor induction period.

In any case, the significant decrease in NK activity following a carcinogenic treatment could be one of the first events signalling the occurrence of a malignant transformation much before any tumor load is evident.

REFERENCES

1. Eremin, O. In: "Natural Cell-Mediated Immunity Against Tumors" (Herberman, R.B. ed.), p.1011, Academic Press, New York, 1980.
2. Gerson, J.H. In: "Natural Cell-Mediated Immunity Against Tumors" (Herberman, R.B. ed.) p.1047, Academic Press, New York, 1980.
3. Goldfarb, R.H. and Herberman, R.B. J. Immunol. 126:2129 (1981).
4. Keller, R. Nature 282:729 (1979).
5. Seaman, W.E., Gindhart, T.D., Talal, N. and Werb, Z. Fed. Proc. 39:760 (1980).
6. Gorelik, E. and Herberman, R.B. J. Natl. Cancer Inst. 66:543 (1981).
7. Ehrlich, R., Efrati, M. and Witz, I.P. In: "Natural Cell Mediated Cytotoxicity Against Tumors" (Herberman, R.B. ed.) p.997, Academic Press, New York, 1980.
8. Kumar, V., Ben-Ezra, J., Bennet, M. and Sonnenfeld, G. J. Immunol. 123:1832 (1979).
9. Seaman, W.E., Gindhart, T.D., Greenspan, J.S., Blackman, M.A., and Talal, N. J. Immunol. 122:2541 (1979).
10. Ehrlich, R., Efrati, M., Bar-Eyal, A., Wollberg, M., Schiby, G., Ran, M. and Witz, I.P. Int. J. Cancer 26:315 (1980).
11. Senik, A., Gresser, I., Maury, C., Gidlund, M., Orm, A. and Wigzell, H. Cell Immunol. 44:186 (1979).
12. Ohler, J.R., Lindsay, L.R., Nunn, M.E., Holden, H.T. and Herberman, R.B. Int. J. Cancer 21:210 (1978).
13. Cudkowicz, G. and Hochman, P.S. Immunological Rev. 44:13 (1979).
14. Grohmann, P.H., Porzsolt, F., Quirt, I., Miller, R.G. and Phillips, R.A. Clin. Exp. Immunol. 44:611 (1981).

EFFECT OF DEPRESSION OF NK ACTIVITY BY NEONATAL EXPOSURE TO DIETHYLSTILBESTROL ON SUSCEPTIBILITY TO TRANSPLANTED AND PRIMARY CARCINOGEN-INDUCED TUMORS

Terje Kalland

Institute of Anatomy
University of Bergen
Bergen, Norway

I. INTRODUCTION

Estrogen effects on the immune system are well documented (Slivic & Warr, 1973). Long-lasting treatment of adult animals with high doses of estrogens results in involution of the thymus, induce lymphopenia and impair several aspects of the immune response, including natural killer cell (NK) activity (Luster et al., 1980; Seaman et al., 1981). The effects, however, are reversible and immune competence return to normal some time after cessation of estrogen exposure (Seaman et al., 1981; Haukaas & Kalland, 1982). In contrast, neonatal treatment of female mice with the non-steroidal estrogen diethylstilbestrol (DES) gives rise to alterations in immunological function of persistent nature mainly through its interference with T lymphocyte differentiation (Kalland, 1980; Kalland, 1980).

Later in life, neonatally estrogen-treated female mice develop adenocarcinomas or squamous carcinomas of the vagina and uterine cervix, have an increased insidence of mammary tumors and an increased sensitivity of the mammary gland and stomach to chemical carcinogens (Forsberg & Kalland, 1981; Mori et al., 1976; Shellabarger et al., 1980).

The present study is dealing with long term consequences for NK activity of exposure of the neonatal female mouse to DES and the possible effect on tumor susceptibility.

II. MATERIALS AND METHODS

A. Animals

NMRI mice were from an outbred stock kept at our insti-

ISBN 0-12-341360-5

tute, BALB/c and C57BL/6 mice were obtained from G1. Bomholtgård, Denmark and AKR/J mice from Olac Ltd. (Oxford, England). Newborn female mice were separated into experimental and control groups and treatment started within 24 hrs after birth. Experimental animals were given daily subcutaneous injections of 5 µg DES (Sigma Chemicals, London) in 0.025 ml olive oil for the first 5 days after birth. Control animals received olive oil only. This treatment schedule has earlier been found to induce persistent changes in the immune system and to give rise to malignant changes in the genital tract of old animals (Forsberg & Kalland, 1981).

B. Tumor cells

The T lymphoma cell lines I-51 and I-522 established *in vitro* from spontaneous thymomas of aged AKR/J mice by Riesenfeld et al. (1980) were obtained from Dr. Anders Ørn, Dept. of Immunology, University of Uppsala, Sweden. These cell lines have been shown to have grossly similar growth pattern in low NK-reactive syngeneic mice, as determined by timedependent increase in subcutaneous tumor size, while they differ in NK sensitivity (Riesenfeld et al., 1980).

All cell lines were maintained as serially passaged cell culture lines on RPMI 1640 medium supplemented with 10 % fetal calf serum and antibiotics.

C. *In vitro* NK assay

A standard chromium-release cytotoxicity assay with spleen cells as effector cells was used as described in detail earlier (Kalland, 1980).

D. *In vivo* NK assay

The *in vivo* NK assay was as described by Riccardi et al. (Riccardi et al., 1979).

E. Tumor cell growth *in vivo*

Thousand cells from the I-51 or I-522 lymphoma cell lines in 0.1 ml RPMI 1640 medium were inoculated in the neck of 6 week old control or neonatally DES-treated AKR/J female mice. Tumor cell-inoculated mice were followed every day and the day of animal death was recorded. The experiment was terminated 30 days after the last death of a tumor bearing animal.

F. Induction of sarcomas with methylcholanthrene

At the age of 8-10 weeks, groups of control or neonatally DES treated female NMRI mice were given a single subcutaneous injection of 10, 20, 50 or 100 μg methylcholanthrene (MCA) (Sigma Chemicals, London) in 0.05 ml tricaprylin (Sigma Chemicals, London). The injection site was laterally on the right hind leg. Randomly chosen tumors were prepared for histological study. The experiment was terminated 36 weeks after MCA injection.

III. RESULTS

A. In vitro NK activity

Spleen lymphocytes from the inbred strains C57BL/6, BALB/c and AKR/J as well as the outbred stock NMRI all showed significant cytotoxic activity against the targets YAC-I and MPC-II. When AKR/J mice were tested against the syngeneic lymphomas I-51 and I-522, the earlier reported difference in susceptibility to NK lysis was confirmed (Riesenfeld et al., 1980). I-522 was quite sensitive to NK-mediated lysis, while I-51 was almost totally resistant in the present assay system (Kalland,I980).

All strains tested showed a significant depression of NK activity as a consequence of neonatal exposure to DES. Natural killing by spleen cells was not reduced by the addition of spleen lymphocytes from neonatally DES-exposed animals over a wide range of dilutions, thus no evidence of cellular suppressors of NK activity induced by DES could be detected. Similarly, NK activity of spleen cells was not suppressed when mouse serum from neonatally DES treated animals was substituted for FCS in the assay (Kalland, 1980). Further evidence against the possibility that DES-induced alterations in hormone levels might suppress NK activity was inherent from the finding that pregnancy did not influence NK activity significantly (Kalland, 1980). The possible involvement of defects in interferon stimulation of NK activity was explored by boosting of interferon production with poly I:C or by in vitro pretreatment of effector cells with partially purified mouse interferon. Both procedures increased cytotoxicity of cells from control animals but had no significant effects on effector cells from DES-treated mice (Kalland, 1980).

B. In vivo NK activity

The radioactivity was determined in the spleen and lungs at different times after intravenous injection of ^{125}I-2′deoxyuridine labelled tumor cells. The sensitive cell I-522 was cleared more effectively from both organs in control than in neonatally DES-exposed female mice (Fig. 1).

When the relatively NK-resistant cell line I-51 was injected, no difference in the ability to eliminate the tumor cells was seen between the two animal groups and a substantial amount of radioactivity was found in the lungs and spleen 4 hrs after tumor cell injection.

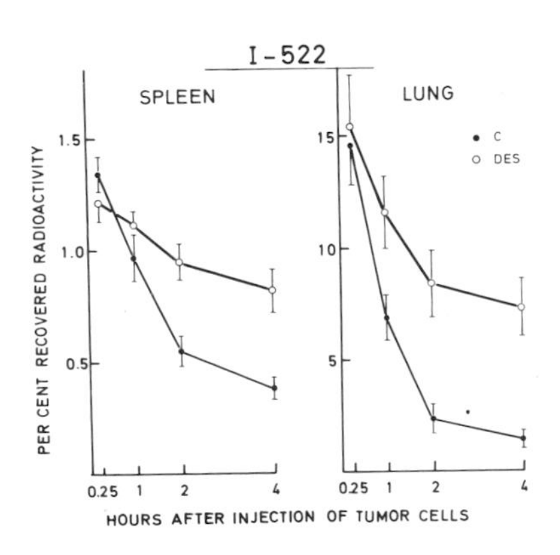

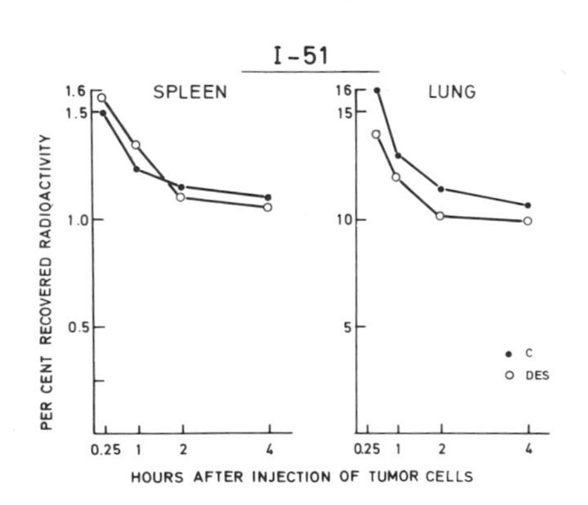

Figure 1. Per cent radioactivity recovered in spleen and lungs at different times after injection of labelled tumor cells.

C. *In vivo* tumor growth

Cumulative death incidence of control and neonatally DES-treated females after transplantation of I-522 and I-51 lymphoma cells into AKR/J mice is shown in Fig. 2. All the animals had large tumors at death. When the relatively NK-resistant tumor I-51 was used, a high percentage of the animals developed tumors, but no difference in tumor-related death incidence between the 2 animal groups were found. Inoculation of cells from the NK sensitive cell line I-522 resulted in a significant higher percentage of deaths among DES-treated animals than controls ($P < 0.01$).

D. Induction of primary tumors with MCA

The incidence of local sarcomas as a function of time after injection of MCA is shown in Fig. 3. After injection of 50 µg or lower doses of MCA, the yield of sarcomas was consistently higher among DES-treated animals than among controls ($P < 0.01$). At the 100 µg dose level, there were only insignificant differences.

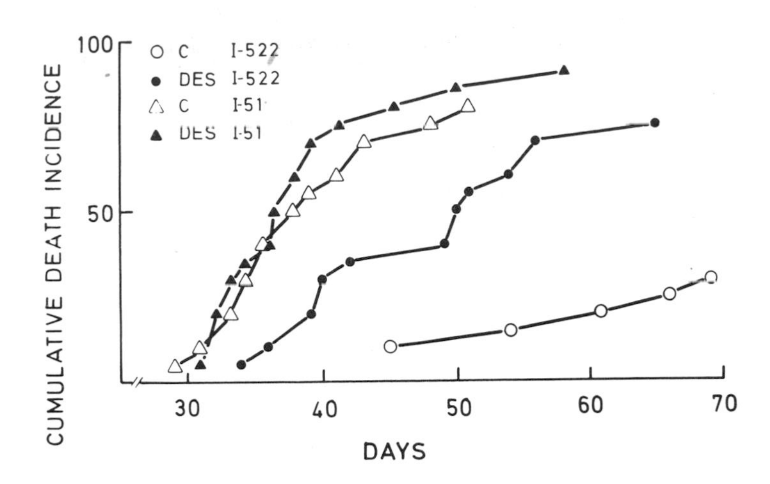

Figure 2. The cumulative death incidence of DES and control females after inoculation with I-522 or I-51 cells. Every group comprises 20 females.

IV. DISCUSSION

The present study demonstrate a significant reduction of NK activity of spleen lymphocytes from adult female mice as a consequence of neonatal exposure to DES. The mechanism of this impairment is not clear, but has not been shown to involve cellular or humoral suppressor factors or impairment in interferon production induced by DES. It could not be excluded, however, that DES may interfere with interferon-mediated maintainance of NK activity (Ørn et al., 1980).

DES disturbs normal T cell differentiation (Kalland, 1980) and suppression of NK activity may be a related phenomena since the NK cell is believed to be in the T cell lineage (Roder et al., 1981). The same treatment that impairs NK activity also results in an increased incidence of genital tract, lung and forestomach tumors in mice, rats and hamsters (Forsberg & Kalland, 1981; Mori et al., 1976; Shellabarger et al., 1980). The increased susceptibility to tumors in various organs, including appearent estrogen non-target tissues, may point to a systemic effect of perinatal estrogen treatment. This study focuses the importance of alterations in NK activity, the branch of the immune system presently believed to be involved in immune surveillance of tumor development (Roder et al., 1981). While the immune system may be of importance at several stages of tumor development, its most important role in surveillance of neoplastic development is probably exerted at a very early step in tumorigenesis against a small number of transformed cells only. In the present study, a very close parallel between the ability to rapidly eliminate intravenously injected tumor cells and in vitro NK activity was found. Moreover, using the cell lines I-522 and I-51, with in vitro as well as in vivo high and low sensitivity to NK lysis, respectively, a close correlation was found between the growth of transplanted tumor cells, their sensitivity to NK and the general NK activity of the host. Most studies ascribing a role for NK cells in tumor surveillance have involved transplantation of established tumor cells and only a limited number of studies have been devoted to the influence of NK cells on primary carcinogenesis. MCA induced carcinogenesis was thus included in the present system. Neonatally DES-treated female mice were significantly more susceptible to MCA-induced carcinogenesis when lower doses of carcinogen were used. MCA has long-lasting immunosuppressive effects at doses higher than 50 μg, and this effect could override the difference between control and neonatally DES-exposed animals. In the beige mouse mutation, known to abrogate NK activity, no effect was seen on induction of sarcomas with 100 μg MCA (Salomon et al., 1979).

Studies on death rate did not indicate that DES-females injected with 10-50 μg MCA are healthier than controls, and thus the difference in tumor incidence could not be explained on a simple "non-immunologic" basis.

In a summary, the present study shows that neonatal exposure of female mice to DES leads to persistent impairment in NK activity and also to an increased susceptibility to transplanted as well as primary carcinogen-induced tumors. Whether this correlation is causal and has any relevance for the development of genital tract tumors of these animals remains to be established.

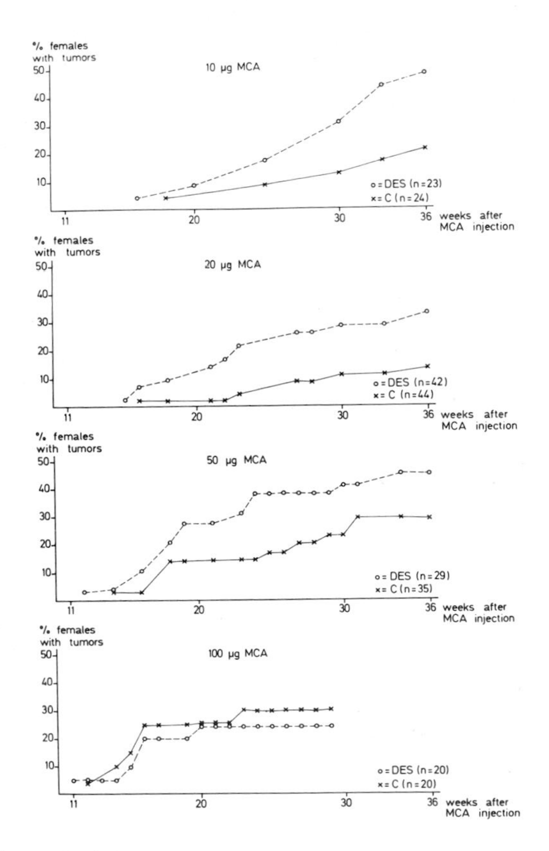

Figure 3. The yield of local sarcomas (in percent of number of females injected) as a function of time after one injection of different doses of MCA.

REFERENCES

Forsberg, J.-G., Kalland, T. Cancer Res. 41, 721, 1981.
Haukaas, S., Kalland, T. Invest. Urol., in press, 1982.
Kalland, T. Cell. Immunol. 51, 55, 1980a.
Kalland, T. J. Immunol. 124, 197, 1980b.
Kalland, T. J. Immunol. 124, 1297, 1980c.
Luster, M.I., Boorman, G.A., Dean, J.H. Reticuloend. Soc. 28, 561, 1980.
Mori, T., Bern, H.A., Mills, K.T. J. Natl. Cancer Inst. 57, 1057, 1976.
Riccardi, C.,Puccetti, C., Santoni, A., Herberman, R.B. J. Natl. Cancer Inst. 63, 1041, 1979.
Riesenfeld, I., Ørn, A., Gidlund, M. Int. J. Cancer, 25, 399, 1980.

Roder, J., Kärre, K., Kiessling, R. Progr. Allergy, 28, 35, 1981.
Salomon, J.-C., Creau-Goldberg, N., Lynch, N.R. Cancer Immunol. Immunother. 8, 67, 1980.
Seaman, W.E., Blackman, M.A., Gindhart, T.D. J. Immunol. 121, 2193, 1981.
Shellabarger, C.J., Knight, B., Stone, J.P. Cancer Res. 40, 1808, 1980.
Slivic, V.S., Warr, G.V. Period Biol. 75, 231, 1973.
Ørn, A., Gidlund, M., Ojo, E. et al. In Natural Cell Mediated Immunity against Tumors, Vol. I (R.B. Herberman, Ed.), Acad. Press, 1980.

THE ROLE OF NK(HSV-1) EFFECTOR CELLS IN RESISTANCE TO HERPESVIRUS INFECTIONS IN MAN

Carlos Lopez
Dahlia Kirkpatrick
Patricia Fitzgerald

Sloan-Kettering Institute for Cancer Research
New York, New York

Host defense against herpesvirus infections undoubtedly depends on a number of humoral factors and cell subpopulations working in concert to resist and clear the invading organism (1). Definition of these factors and cells is required in order to direct the development of new modalities of antiviral therapy based on a prohost concept (2). In order to develop an understanding of the mechanisms involved in host defense against herpesvirus infections, a murine model of genetic resistance against herpes simplex virus-type 1 (HSV-1) was developed (3). Surprisingly, this resistance appeared to be determined by non T-cells since athymic nude mice were no more susceptible than their normal littermates (4,5). Of great interest was our finding that treating mice with strontium-89 (^{89}Sr) abrogated genetic resistance (6). In ^{89}Sr-treated mice, the spleen takes over hematopoiesis and cellular immunity, as gauged by skin graft rejection (7), remains normal. Since ^{89}Sr treated mice showed a loss of virtually all natural killer (NK) activity (8), an NK assay, using HSV-1-infected fibroblasts as targets [NK(HSV-1)], was developed to determine whether this function correlated with genetic resistance to HSV-1 in man (9). Our studies of the effectors of NK(HSV-1) clearly showed that they differ from cells which lyse K562 erythroleukemia cells, the targets most often used to evaluate human NK cell function (10). Not only

Supported in part by NIH grants CA-23766, CA-08748 and CA-17404; NIH training grant CA-09149; grant IM-303 and Faculty Research Award #193 (to CL) from the American Cancer Society.

ISBN 0-12-341360-5

do these effector cells differ with respect to cell surface characteristics (Fitzgerald, et al., this volume), but cold competition studies indicate that the structures recognized on the target cells also differ (11). Because NK(HSV-1) effectors are clearly different than NK(K562) cells and because our continuing interest was in the role of NK cells in host resistance against herpesvirus infections, we felt that an assay with virus infected targets might better reflect this function. In this chapter we present a correlation between low NK(HSV-1) levels and unusual susceptibility to herpesvirus infections in man.

I. METHODS AND MATERIALS

Fresh cord bloods were obtained from uncomplicated deliveries without signs of infection. Heparinized bloods were obtained from patients being evaluated immunologically. All tests were set-up on the same day blood was drawn.

The NK(HSV-1) assay was carried out as described earlier (9). Briefly, HSV-1-infected or uninfected fibroblasts were chromated, washed, and aliquoted. Effector cells were added to yield the desired ratios and the plates were incubated for 14 hours in a humidified CO_2 incubator. The data was analyzed using the Von Krogh equation (12) and the percent lysis at a 50:1 ratio was estimated and used for comparison. Control values were generated with 58 normals. Normal values were considered to be those within 2 standard deviations of the mean.

II. RESULTS

A. Patients Susceptible to Unusually Severe and/or Persistent Herpesvirus Infections. Approximately 75% of all babies infected with Herpes Simplex Virus (HSV) at the time of birth go on to develop an infection disseminated to the viscera, skin, or central nervous system (13). The other 25% of exposed children have localized infections and thus demonstrate some resistance to HSV not found with the other patients. We have studied cord bloods as an indication of the newborns' capacity to generate an NK(HSV-1) response. A total of 22 fresh cord bloods were obtained from uncomplicated, natural deliveries. All bloods were tested on the day obtained and were not incubated overnight. Eight (36%) of these cord bloods demonstrated NK(HSV-1) responses within 2 standard deviations (SD) of the normal response or higher (normal response was 21-69%) and only 5 (23%) demonstrated a response within 1 S.D. of the normal mean or higher. This later figure correlates well with the percent of newborns demonstrating some resistance to HSV infections (13).

Since premature infants are known to be at a 4 times greater risk of severe HSV infections than are full term babies, a study was undertaken of the NK(HSV-1) capacity of their peripheral blood mononuclear cells. Five such patients were studied and all were found to have NK(HSV-1) responses more than 3 S.D. below the normal mean.

Patients with Wiskott-Aldrich syndrome (WAS) also usually demonstrate a marked susceptibility to herpesvirus infection (14). A total of 6 patients with WAS have been evaluated for NK(HSV-1) and 5 were found to have low or low normal responses. The only WAS patient with a clearly normal NK(HSV-1) response is unusual in that he has lived much longer than other WAS patients and has had little difficulty with infections.

Although these data on cord bloods, premature newborns, and WAS patients suggest a relationship between low NK(HSV-1) and susceptibility to herpesvirus infections, these patient groups have been shown to also demonstrate other defects which could also contribute to their exaggerated susceptibility. Because of these other possibilities, it was necessary to carry out a study of individuals without a known cellular immunodeficiency but also susceptible to severe herpesvirus infections. To determine that these individuals were susceptible, they would have to be recognized because of the unusually severe nature of their herpesvirus infection.

B. Patients with Severe Herpesvirus Infections. Five patients have been brought to our attention because of the unusually severe nature of their herpesvirus infections (15). All were hospitalized for their infections and 3 were treated with antiviral drugs. Three of these patients had NK(HSV-1) responses below the normal range and the other two had responses at the bottom end of normal. As a group, these patients were significantly below the normal mean ($p < 0.001$). Cells from two of these patients were incubated with α-interferon prior to evaluating them for NK(HSV-1) activity. One responded with significantly increased lysis of target cells while the other did not.

Studies in Los Angeles and New York City have recently described what appears to be a secondary immunodeficiency state found almost exclusively in young adult homosexual men (16). Although cytomegalovirus and "street" drugs have been considered as possible immunosuppressive agents, the cause of the apparently acquired immunodeficiency has not been determined. Of special interest was the finding that several of these individuals developed severe perianal ulcerative herpesvirus infections (17). The first 5 patients with this clinical syndrome have been evaluated immunologically to attempt to determine the reason for their susceptibility to these infections. Four of the 5 patients demonstrated lymphopenia when first studied (which, for these 4, was relatively late into the infection) as well as skin anergy (17). The last patient was studied earlier in the disease and showed normal lymphoid cell number and function when first studied. The first 4 patients demonstrated depressed NK(HSV-1) function when evaluated on a "per ml of blood" basis; although 2 were low normal on a per cell basis. As the disease progressed in the last patient, he too demonstrated lymphopenia and skin anergy suggesting that these may be late manifestations of this disorder rather than causal of the disease. This patient, however, had low NK(HSV-1) responses throughout this entire course, independent of how it was calculated. Four of these 5 patients have died and one has developed Kaposi's sarcoma.

III. DISCUSSION

Our studies of a mouse model of genetic resistance to HSV-1 indicated that effector cells with many of the properties of NK cells were responsible for the differences between susceptible and resistant strains of mice (4,6). We have carried these observations over to the study of human resistance to HSV infections by developing an NK assay utilizing virus infected fibroblasts as targets (11). The argument for a possible role for NK(HSV-1) effector cells in host resistance to herpesvirus infections would be greatly strengthened if patients unusually susceptible to these infections were found to be deficient in this function.

New born infants are known to be susceptible to severe herpes and other virus infections. About 25% of newborns, however, demonstrate clear indications of resistance since herpesvirus infections in these patients are localized (13). Our finding that 23% of cord bloods demonstrated NK(HSV-1) responses within 1 S.D. of the normal mean or higher correlates well with this finding. In addition, the markedly increased susceptibility of premature infants was also reflected in even lower NK(HSV-1) levels.

Wiskott-Aldrich syndrome (WAS) is a sex-linked, inherited immunodeficiency disorder which has been found to be associated with susceptibility to infections, in general, and with severe herpesvirus infections. This is a heterogeneous disease with some patients having particular problems with infections while others have only a few mild problems. The patients with WAS presented here had low NK(HSV-1) responses and the level of NK(HSV-1) response appeared to correlate with the severity of their problems with infection.

It is difficult to ascribe increased susceptibility to infection to low NK(HSV-1) in newborn and patients with WAS since other deficiencies have been described in these individuals which might also contribute to their unusually severe herpesvirus infections. For example, newborns have been found to be unable to make gamma interferon (18) and WAS patients have been found to have a dysfunction of monocytes (19). Therefore, it was necessary to develop similar correlations in patients without known cellular immunodeficiencies. Our studies of the 5 patients with severe or persistent herpesvirus infections indicate that low NK(HSV-1) responses again correlated with susceptibility to these infections. Our studies of the male homosexuals with ulcerative HSV infections supports these findings and, if the last of the 5 patients is indicative, suggest that low NK(HSV-1) response may antedate the secondary cellular immunodeficiency found later in these patients.

Studies are required to define the deficiencies found in the above described patients. The fact that the cells from one patient with a severe herpesvirus infection responded with significantly elevated kill after preincubation with interferon while the cells from another did not suggests that the nature of the deficiencies can vary among different patients.

In summary, we have presented correlative evidence of an association between low NK(HSV-1) and susceptibility to herpesvirus infections. It is likely that these effector cells play an important defensive role early during an infection. These cells, in concert with macrophages and interferon, probably make up the bulwark of the early defense against herpesvirus infections.

REFERENCES

1. Allison, A.C., Transplant. Rev. 19:3 (1974).
2. Hadden, J.W., Lopez, C., O'Reilly, R.J., and Hadden, E.M. Ann. N.Y. Acad. Sci. 284:139 (1977).
3. Lopez, C., Nature 258:152 (1975).
4. Lopez, C., in "Oncogenesis and Herpesviruses III (G. de The, W. Henle, F. Rapp, eds.) p.775 IARC Scientific Publications, Lyon (1978).
5. Zawatzky, R., Hilfenhaus, J., and Kirchner, H., Cell Immunol. 47:424(1979).
6. Lopez, C., Ryshke, R., and Bennett, M., Infect. Immunity 28:1028 (1980).
7. Bennett, M., J. Immunol. 110:510 (1973).
8. Kiessling, R., Hochman, P.S., Haller, O., Shearer, G.M., Wigzell, H., and Cudkowicz, G., Eur. J. Immunol. 7:663

(1977).
9. Ching, C., and Lopez, C., Infect. Immunity 26:49 (1979).
10. West, W.H., Cannon, G.B., Kay, H.D., Bonnard, G.D., and Herberman, R.B., J. Immunol. 118:355 (1977).
11. Fitzgerald, P., Evans, R., and Lopez, C., in Proc. XIII Leukocyte Culture Conference (K. Resch and H. Kirchner, eds.) p.595, Elsevier/North Holland, New York (1981).
12. Pross, H.F., Barnes, M.G., Ruben, P., Shragge, P., and Patterson, M.S., J. Clin. Immunol. 1:51 (1981).
13. Vinsintine, A., Nahmias, A., Whitley, R., and Alford, C., in The Human Herpesviruses (A.J. Nahmias, W.R. Dowdle, and R.F. Schinazi, eds.) p.599. Elsevier/N. Holland, New York (1981).
14. St. Geme, J.W., Jr., Prince, J.T., Burke, B.A., Good, R. A., and Krivit, W., N. Engl. J. Med. 273:229 (1965).
15. Lopez, C., Kirkpatrick, D., Read, S., Fitzgerald, P., Pitt, J., Pahwa, S., Ching, C., and Smithwick, E.M. (submitted).
16. Durack, D.T., N. Engl. J. Med. 305:1465 (1981).
17. Siegal, F.P., Lopez, C., Hammer, G.S., Brown, A.E., Kornfeld, S.J., Gold, J., Hassett, J., Hirschman, S.Z., Cunningham-Rundles, C., Adelsberg, B.R., Parham, D.M., Siegal, M., Cunningham-Rundles, S., and Armstrong, D., N. Engl. J. Med. 305: 1439 (1981).
18. Winter, H.S., Bryson, Y.J., Gard, S.E., Fischer, T.J., and Stiehm, E.R., Pediat. Res. 12:488 (1978).
19. Blaese, R.M., Strober, W., Brown, R.S., and Waldman, T.A., Lancet i:1056 (1968).

THE ROLE OF NATURAL KILLER CELLS AND INTERFERON IN RESISTANCE TO MURINE CYTOMEGALOVIRUS

Geoffrey R. Shellam[1]
Jane E. Grundy (Chalmer)[2,3]
Jane E. Allan

Department of Microbiology
University of Western Australia
Nedlands, Western Australia 6009

I. INTRODUCTION

The herpesvirus murine cytomegalovirus (MCMV) has been extensively studied in a number of strains of inbred mice as a model of human cytomegalovirus infections. The outcome of infection with MCMV is strongly influenced by the host genotype (Selgrade and Osborn, 1974; Chalmer, et al., 1977) with the H-2^k haplotype as well as non-H-2 genes in C57Bl/10 and C3H mice conferring relative resistance to lethal infection (Grundy [Chalmer], et al., 1981a). The mechanisms of resistance are complex and although not completely understood they must be effective early in the infection since variations in virus titres between mouse strains are detected by the second day (Allan and Shellam, [unpublished observation]) and susceptible mice die as early as the third day of infection (Chalmer, et al., 1977). As discussed previously (Bancroft, et al., 1981) although

---

[1]Supported by grants from the N.H. & M.R.C. of Australia and the Cancer Council and T.V.W. Telethon of Western Australia.
[2]Present address: National Cancer Institute, N.I.H., Bethesda, MD, U.S.A.
[3]Supported by the Netherlands Organization for the Advancement of Pure Research.

ISBN 0-12-341360-5

virus-specific immunity to MCMV is undoubtedly very important for recovery from infection, it is unlikely to be the only determinant of early genotype-related variations in resistance. On the other hand there are a number of essentially non-specific and rapidly activated mechanisms including interferon (IFN), natural killer (NK) cells, macrophages, and complement operating via the alternative pathway, which can provide early protection during virus infections. Here we review our studies on the role of NK cells and interferon in genetically-determined resistance to infection with MCMV.

II. RELATIONSHIP BETWEEN THE ACTIVITY OF NK CELLS AND RESISTANCE TO MCMV

A. Mouse Strain Variation in the Augmentation of Cytotoxicity by MCMV; Correlation with Resistance

In mice a number of viruses have been shown to rapidly stimulate the cytotoxicity of NK cells, the common feature of these infections being the production of IFN which appears to be chiefly responsible for the stimulation of cytotoxicity. Similarly we have demonstrated the stimulation of splenic NK cell activity within 12 hrs of MCMV inoculation towards MCMV-infected Y-1 cells (Bancroft, et al., 1981). This early augmented cytotoxicity is clearly not virus-specific because spleen cells from MCMV-infected mice exhibited greater lysis of N3T3, SC-1 and NZB cell lines infected with MCMV or HSV-1, of L929 cells infected with HSV-1 or encephalomyocarditis virus (Burston and Shellam, unpublished observation) and of RBL-5 and YAC-1 lymphoma cells (Bancroft et al. 1981), than did spleen cells from uninfected mice.

The augmentation of the cytotoxicity of splenic NK cells after MCMV infection was studied in a number of mouse strains to determine whether this was related to resistance to infection (Bancroft, et al., 1981). Mice were injected intraperitoneally (i.p.) with salivary gland-derived virus and 12 to 15 week old animals were used because resistance which is an age-related phenomenon, has stabilized at this time. The virus dose is given in BALB/c LD_{50} units. Cytotoxicity was measured in a standard 4 hr ^{51}Cr release assay using RBL5 target cells, and cell-mediated lysis was expressed in cytotoxic units (c.u.) per 10^6 lymphoid effector cells which is the slope of the curve fitted to

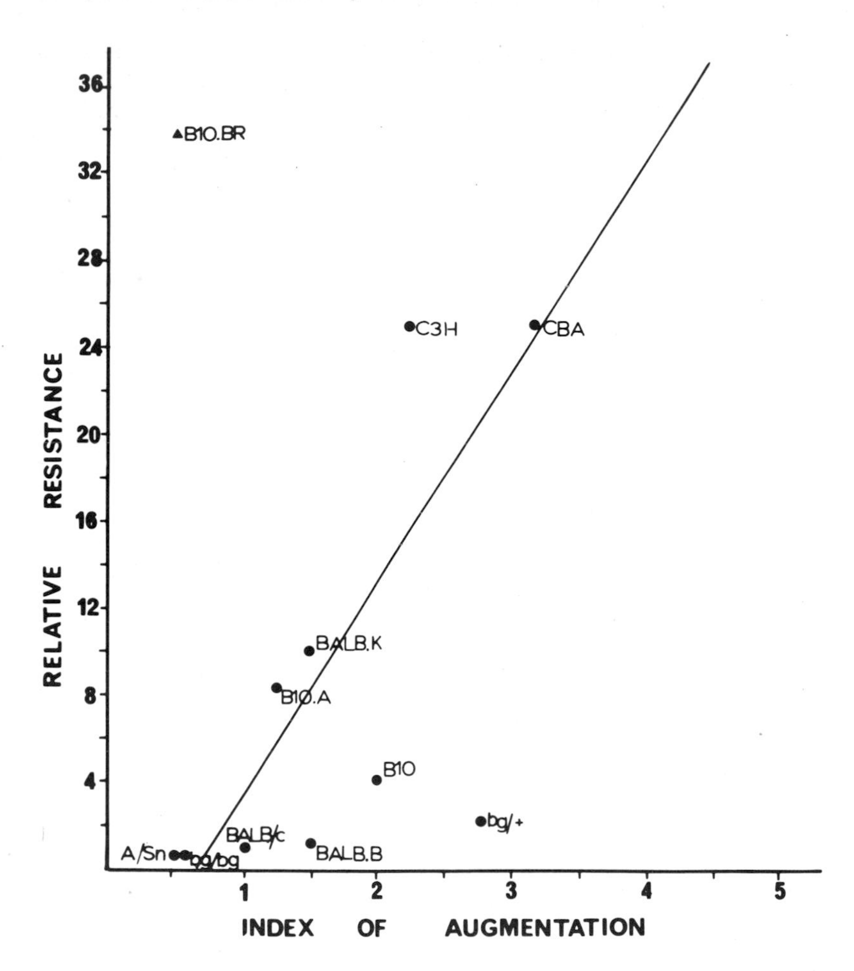

FIGURE 1. Correlation between resistance to MCMV and NK cell augmentation during infection. Augmentation is indexed to the cytotoxic units of BALB/c. Taken from Bancroft et al., 1981.

data obtained using effector:target ratios of 12.5 to 100:1 (Dawkins and Shellam, 1979).

MCMV stimulated cytotoxicity in a dose and time dependent manner and in general greater cytotoxicity was observed in resistant than susceptible mice (Bancroft, et al., 1981, Fig. 1). Thus spleen cells from resistant C3H mice (whose LD_{50} relative to the LD_{50} of BALB/c mice was 25.8) exhibited greater cytotoxicity than cells from BALB/c with doses between 0.1 and $3LD_{50}$ and at times between 12 and 36 hr after infection. A similar result was obtained in a comparison between CBA (relative LD_{50} = 25.8) and the A/WySn strain (relative LD_{50} = 0.5). The augmentation of cytotoxicity 24 hr after infection with $3LD_{50}$ of MCMV was compared in 11 strains and a plot was made of the relative resistance (expressed as relative LD_{50} compared with BALB/c mice) of each strain and the augmentation of cytotoxicity (Fig. 1). A significant positive correlation between resistance and augmentation ($r = 0.68$, $p<0.025$) was observed in 10 of the 11 strains, the exception being the highly resistant B10.BR strain. However, the $H\text{-}2^k$ haplotype did not exert a clear effect as the increase in cytotoxicity was only 1.5 times greater in BALB.K ($H\text{-}2^k$) as compared to BALB/c ($H\text{-}2^d$), and B10.BR ($H\text{-}2^k$) showed less cytotoxicity than B10 ($H\text{-}2^b$). Non-H-2 effects were more pronounced with B10.A ($H\text{-}2^a$) exhibiting greater augmentation than A/WySn ($H\text{-}2^a$; $p<0.005$) and CBA ($H\text{-}2^k$) expressing greater cytotoxicity than BALB.K ($H\text{-}2^k$, $p<0.001$). It is possible that non-H-2 genes influence the production of IFN following MCMV infection as discussed below, or that such genes influence the size of the pool of pre-NK and NK cells capable of responding to IFN. The latter possibility could only be properly evaluated by use of germ-free animals.

B. The Effect of the Beige Mutation on Susceptibility to MCMV

Mice which are homozygous for the beige mutant gene (bg/bg) are homologues of the Chediak-Higashi syndrome and have a defect in NK-mediated cytolysis (Roder and Duwe, 1979). To examine the role of NK cells in resistance to MCMV more directly we have studied the course of the infection in C57BL/6 bg/bg and bg/+mice.

The main findings were that bg/bg mice were significantly more susceptible to lethal and sublethal infection and that the defect appears to be a property of bone marrow-derived cells, presumed to be NK cells (Shellam, et al., 1981). The bg/bg mice were 3.5 times more susceptible to MCMV than bg/+ mice and they developed 33 to 44 fold higher virus titres in the liver, spleen and kidney during a sublethal infection induced by 10^4 plaque forming units (pfu) of virus (Fig. 2), although the kinetics of infection and the organ distribution of the virus were similar in both groups. These results were confirmed by histopathological examination in which greater tissue damage was observed in bg/bg than bg/+ during a lethal or a sublethal infection (Papadimitriou, et al., 1981). These observations did not

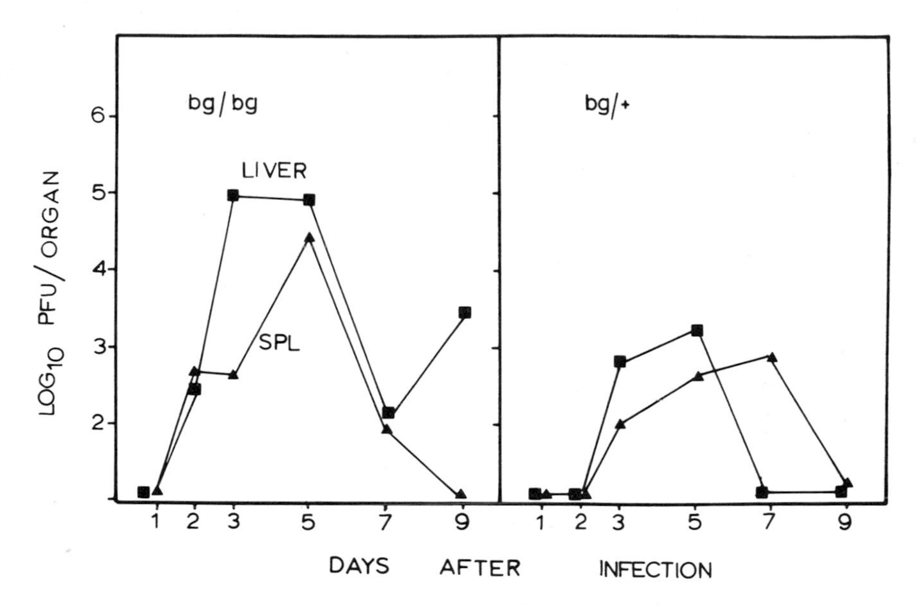

FIGURE 2. Virus titres in bg/bg or bg/+ mice infected with 10^4 pfu MCMV. Liver (·), spleen ( ). Taken from Shellam et al., 1981.

appear to reflect any effect of the mutation on virus replication per se because MCMV replicated equally well in vitro in embryo fibroblasts or respiratory epithelial cells from bg/bg and bg/+ mice (Shellam, et al., 1981). Moreover, our experiments demonstrated that the increased susceptibility of bg/bg was due to a defective host protective response. Firstly, in the liver, which is the major target organ in acute infection of C57BL/6 mice, the inflammatory response was measured histologically in both groups and appeared to be both defective and delayed in bg/bg mice (Papadimitriou, et al., 1981). Secondly, (Fig. 3), spleen cells from bg/bg mice exhibited a lower NK cell response to YAC-1 target cells than bg/+ cells during both a sublethal (10^4 pfu) or a lethal (3 x 10^5 pfu) infection. In addition bone marrow chimeras were constructed in bg/bg and bg/+ mice using bg/bg or bg/+ marrow as shown in Table I. Five to 6 weeks after reconstitution the mice received 10^4 pfu of MCMV i.p. and splenic cytotoxicity and virus titres in the liver and spleen were measured 3 and 5 days later. Irrespective of the recipient genotype mice receiving bg/bg marrow exhibited the more susceptible phenotype and lower levels of cytotoxicity (Shellam, et al., 1981).

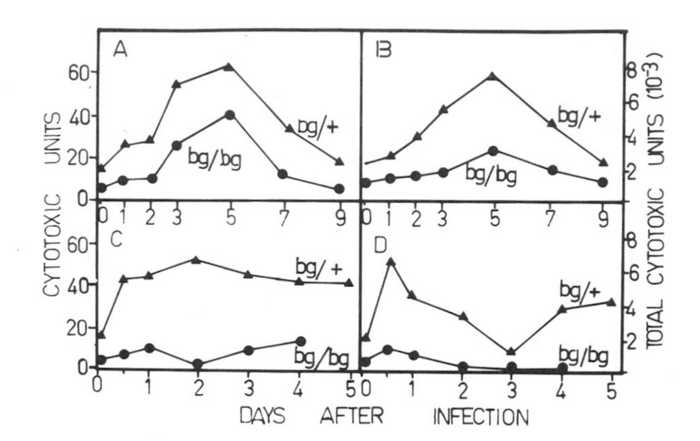

FIGURE 3. Cytotoxicity in bg/bg or bg/+ mice after infection with 10^4 pfu (A, B) or 3 x 10^5 pfu (C, D) of MCMV. The calculation of total cytotoxic units using the spleen cells yields at various times p.i. is described elsewhere (Shellam et al, 1981).

III. ROLE OF TYPE I INTERFERON IN RESISTANCE TO MCMV

Following the i.p. inoculation of a number of mouse strains of different resistance phenotype with 3 BALB/c LD_{50} of MCMV two peaks of plasma type I IFN were detected at 6 or 48 hr p.i. The 48 hr level was similar in all strains (Allan and Shellam, unpublished observations). The 6 hr peak however, varied and appears to be influenced by non-H-2 genes Grundy [Chalmer], et al., 1981b). Thus BALB/c (H-2^d), C57BL/10 (H-2^b) and C3H/HeJ (H-2^k) produce 8.08, 8.90 and 10.55 $\log_2$ units/ml respectively of IFN which can be neutralized by antibody to type I (Grundy [Chalmer], et al., 1981b). Control by non-H-2 genes is suggested by the finding that BALB.K (H-2^k) and B10.BR (H-2^k) have similar IFN titres at 6 hrs p.i. to BALB/c and C57BL/10 respectively Grundy [Chalmer], et al., 1981b). Since non-H-2 genes of C57Bl/10 and C3H mice confer relative resistance to lethal infection in vivo compared with BALB/c background genes (Grundy [Chalmer], et al., 1981a), it is conceivable that this mechanism of genetic control is the regulation of type I IFN production 6 hrs after infection.

Finally, the importance of type I IFN is shown by the observation that treatment of C3H/HeJ and C57BL/10 mice with a single i.v. inoculation of a rabbit-anti-type I antibody 2 hrs before the virus reduced their resistance

TABLE 1. Cytotoxicity of NK Cells and Virus Titres in Chimeric Mice[a]

Organ	Group	Cytotoxicity		Virus titre	
		0 days	3 days	3 days	5 days
Spleen	bg->bg	6.8	14.1	3810	4685
	bg->het	4.4	17.7	95	1701
	het->het	12.3	28.5	140	435
	het->bg	21.3	32.3	6	312
Liver	bg->bg	ND	ND	51296	51279
	bg->het	ND	ND	23408	21605
	het->het	ND	ND	1627	205
	het->bg	ND	ND	956	2597

[a]Lethally irradiated bg/bg (bg) or bg/+ (het) mice received bone marrow i.v. from bg or het donors as indicated and 10^4 pfu of MCMV i.p. 5 weeks later. Cytotoxicity was assayed using YAC-1 target cells. Virus titres are given as pfu/organ. Taken from (Shellam, et al., 1981).

to lethal MCMV infection (Grundy [Chalmer], et al., 1981b). Antibody treatment reduced the resistance of C3H/HeJ mice by 2/3 (p. 0.01, χ^2) and of C57BL/10 mice by 1/2 ($p<0.02$). This treatment also increased the MCMV titre in the plasma, liver and spleen of C3H mice as compared to controls inoculated with normal rabbit serum and MCMV. The 6 hr peak, but not the 48 hr peak, of plasma IFN was abrogated and NK cell cytotoxicity was markedly diminished for 24 hr p.i. (Grundy [Chalmer], et al., 1981b). This strongly suggests that the 6 hr peak of IFN production which is controlled by non-H-2 genes is important in conferring resistance to infection.

IV. CONCLUSION

Our analysis of the role of early mechanisms in protection against MCMV have shown that the stimulation of NK cells and the production of type I IFN are associated with resistance to lethal infection. Both functions are controlled by non-H-2 associated but not necessarily identical genes.

The contribution to resistance of NK cells has not been directly established but resistance is correlated with the early stimulation of cytotoxicity, and NK defective bg/bg mutants of C57BL/6 mice are more susceptible to infection than bg/+ mice. In marrow chimeras inoculated with MCMV, recipients of bg/bg marrow had lower NK cell responses and were more susceptible to infection than were recipients of bg/ + marrow. The production of type I IFN peaks at 6 and 48 hrs post-infection and an association was found between the early IFN peak and resistance in a number of mouse strains. The inoculation of antibody to type I IFN made C3H, and to a lesser extent C57BL/10 mice more susceptible. Also the 6 hr peak of IFN was abrogated and the stimulation of NK cells was diminished. It remains to be resolved to what extent IFN mediates its protective effect directly by inhibiting virus replication or indirectly by the stimulation of NK cells to lyse virus-infected cells.

ACKNOWLEDGEMENTS

We acknowledge the contribution of G. Bancroft, J. Trapman and J. Papadimitriou to this research.

REFERENCES

1. Bancroft, G.J., Shellam, G.R., and Chalmer, J.E. (1981). J. Immunol. 126, 988.
2. Chalmer, J.E., Mackenzie, J.S., and Stanley, N.F. (1977). J. Gen. Virol. 37, 107.
3. Dawkins, H.J.S. and Shellam, G.R. (1979). Int. J. Cancer, 24, 235.
4. Grundy (Chalmer), J.E., Mackenzie, J.S., and Stanley, N.F. (1981a). Infect. Immunity. 32, 277.
5. Grundy (Chalmer), J.E., Trapman, J., Allan, J.E., Shellam, G.R., and Melief, C.J.M. (1981b). Submitted
6. Papadimitriou, J.M., Shellam, G.R., and Allan, J.E. (1981). Submitted.
7. Roder, J.C. and Duwe, A.K. (1979). Nature, 278, 451.
8. Selgrade, M.K. and Osborn, J.E. (1974). Infect. Immun. 10, 1383.
9. Shellam, G.R., Allan, J.E., Papadimitriou, J.M., and Bancroft, G.J. (1981). Proc. Nat. Acad. Sci. (U.S.A.), 78, 5104.

LOSS OF GENETIC RESISTANCE TO MURINE CYTOMEGALOVIRUS INFECTION IN C3H MICE TREATED WITH ^{89}Sr

Aoi Masuda
Michael Bennett

Department of Pathology
Boston University School of Medicine
Boston, Massachusetts

I. INTRODUCTION

Treatment of mice with ^{89}Sr chronically irradiates the bone marrow and forces the spleen to become the source of hemopoietic stem cells. Under these conditions, any effector or immune function that is lost and cannot be explained by the presence of suppressor cells/factors have been interpreted to be marrow-dependent (Bennett, 1973). The natural killer (NK) cells capable of lysing YAC-1, MPC-11, C1-18, or RL♂1 tumor cells are marrow-dependent (M) cells whereas NK cells capable of lysing EL-4, FLD-3, or WEH1-164.1 tumor cells are functional in ^{89}Sr-treated mice and are thus marrow independent NK cells (Haller and Wigzell, 1977; Kumar et al., 1979b; Lust et al., 1981). Recently, a correlation has been demonstrated between NK(YAC-1) function and genetic resistance to the lethal effects of murine cytomegalovirus (MCMV) (Bancroft et al., 1981; Shellam et al., 1981). We sought to determine if the abolition of NK(YAC-1) functon of genetically resistant

[1]Supported by NIH grants AI18811, HL24201, CA31792, CA21401 and BR56RR-5380 and CNPq fellowship 4170(A.M.)
[2]Present address: Department of Parasitology, Faculty of Medicine of Ribeirao Preto, Sao Paulo, S.P.Brazil.
[3]Present address: Department of Pathology, Southwestern Medical School, Dallas, Texas.

ISBN 0-12-341360-5

C3H/St mice would their resistance to MCMV. We observed that ^{89}Sr-treated C3H/St mice were now susceptible to MCMV (Masuda and Bennett, 1981a) but that 'NK' cell function was enhanced, unless a highly virulent MCMV passage was employed. This boosting of NK cell function by MCMV contrasts with the inability of interferon inducers or interferon preparations to boost NK cell function in mice treated with ^{89}Sr (Kumar et al., 1979a; Luevano et al., 1981).

II. MATERIALS AND METHODS

(C57BL/6 x DBA/2)F1 (B6D2F1) and C3H/St mice were purchased from the Jackson Laboratory, Bar Harbor, Maine, and West Seneca Laboratories, West Seneca, New York, respectively. SIM.R mice were bred in our colony. At age 8-10 weeks they were injected i.p. with 100 μCi ^{89}Sr and were infused with 10^7 syngeneic marrow cells 10 days later. The mice (and controls) were infected 3-5 weeks after marrow cell transfer with 10^4, 10^5, or 10^6 plaque forming units (pfu) of the Smith strain of MCMV.

The virus was passaged in SIM.R mice using salivary glands as the source. Second-passage mouse embryo fibroblasts were used to quantitate pfu (Selgrade and Osborn, 1974). Infections center assays were performed using cells from spleens and livers of infected mice.

Survival of groups of 5-8 mice, infections centers in spleen and livers, and serum glutamic pyruvic transaminase (SGPT) were the measures of susceptibility to MCMV.

The assessment of NK cell function was performed exactly as described (Kumar et al., 1979b). Each microtiter well contained $2X10^4$ ^{51}Cr-labeled tumor cells and 25 μl 2% saponin (maximum release), C.07-$1X10^6$ lymphoid cells, or media only (spontaneous release). The optimal times of incubation were used and are: 4 hours-YAC-1 (A strain lymphoma); 18 hours-EL-4 (C57BL strain lymphoma), L1210 (DBA/2 strain lymphoma), C1-18 (C3H strain myeloma), WEHI-164-1 (BALB/c strain fibrosarcoma); 24 hours-FLD-3 (BALB/c erthroleukemia).

III. RESULTS

Whereas all 8 C3H/St age/sex control mice survived infection with 10^6 pfu MCMV, all 8 mice previously treated with ^{89}Sr and infected with 10^6pfu died, 7/8 infected with 10^5pfu died and 4/8 infected with 10^4pfu died within 14 days ($p<0.025$). Genetically susceptible B6D2F1 mice became somewhat more sensitive to MCMV ($p<0.05$).

The infectious center assay performed 3 days after infection revealed an increase (control to ^{89}Sr-treated) from Log^{10} 1.75 to 4.48 in spleens and 3.55 to 4.86 in livers ($p<0.05$) of C3H/St mice.

The SGPT levels (international units/liters) were 43 and 34 in mock-infected and infected control C3H/St mice and were 61 and > 350 in mock-infected and MCMV infected C3H/St mice previously treated with ^{89}Sr. Both control and ^{89}Sr treated B6D2F1 mice had SGPT titers>350 3 days after infection. Histology of the spleens and livers of infected mice were of interest. MCMV inhibited lymphopoiesis and myelopoiesis in spleens of control and ^{89}Sr-treated mice, especially the former. Splenic necrosis with associated mature granulocytes were seen in spleens of control but not of ^{89}Sr-treated mice. The livers of control mice infected with MCMV revealed more leukocytic infiltration and more swelling and vacuolization of liver cells but many less inclusion bodies than livers of mice treated with ^{89}Sr.

NK reactivity against YAC-1, EL-4, L1210, C1-18, WEH1 164.1 and FLD-3 cells was augmented in mice infected with MCMV (Fig. 1).

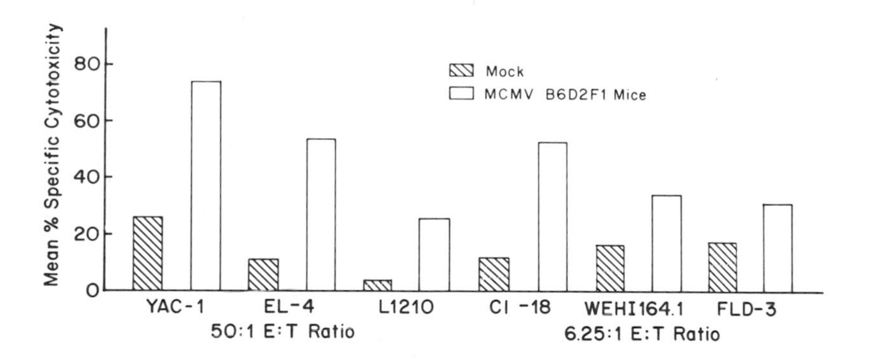

FIGURE 1. NK cell function of B6D2F1 mice 3 days after infection with MCMV (Spleen cells).

Thus both marrow-dependent and independent NK cells were stimulated. NK(YAC-1) cell function was boosted in spleen, bone marrow and peritoneal cell suspensions but not in lymph node or thymus cell suspensions of B6D2F1 mice infected with MCMV 3 days earlier (Fig. 2).

The surprising finding was that MCMV stimulated NK(YAC-1) cell function in ^{89}Sr-treated mice (Fig. 3). The same dose (10^5 pfu) of virus kills the ^{89}Sr-treated mice a few days later. We did note, however, that in later

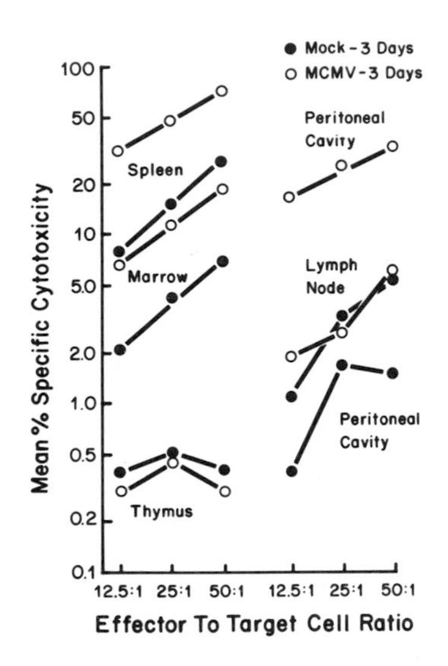

FIGURE 2. NK(YAC-1) cell function 3 days after infection of B6D2F1 mice with MCMV.

passages of the virus, two changes occurred. The virus now killed a small fraction of C3H/St mice at 10^5 pfu and augmented NK(YAC-1) cell function less well in B6D2F1 than in C3H/St mice and less well in ^{89}Sr-treated B6D2F1 or C3H/St mice than in control mice (Masuda and Bennett, 1981a). The NK(YAC-1) cells detected in spleens of MCMV-infected B6D2F1 mice have been characterized. They are resistant to both anti-Thy-1.2+C and anti-NK-1.2+C, do not adhere to nylon wool or plastic and are suppressed by cyclophosphamide treatment of mice and by incubation at 37^0 C for 4 hours. Thus, these effectors resemble NK(YAC-1) cells of non-infected mice except for the resistance to anti-NK-1.2+C (Lust et al., 1981). Preliminary experiments suggest that they are resistant to C.

IV. DISCUSSION

We favored the hypothesis that depletion of M cells by ^{89}Sr would weaken genetic resistance of MCMV by affecting NK cells which are 'activated' during viral infections (Welsh and Zinkernagel, 1977). We were influenced by the observations that NK(YAC-1) activity is low in ^{89} Sr-treated mice and cannot be boosted by interferon or interferon inducers (Kumar et al., 1979b; Luevano et al., 1981) and that degree of boosting of NK(YAC-1) activity correlated with genetic resistance to MCMV (Bancroft et al., 1981;

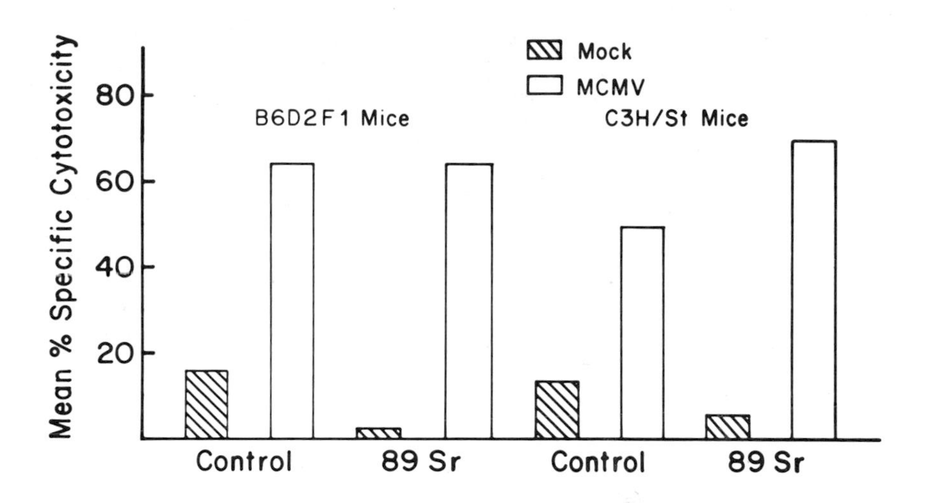

FIGURE 3. Splenic NK(YAC-1) cell function 3 days after infection with MCMV (10^5 pfu).

Shellam et al., 1981). We are forced to consider alternative mechanisms because MCMV augments NK(YAC-1) cell function in ^{89}Sr-treated mice (Fig. 3) and yet kills the mice. Infant mice less than 3 weeks of age are also susceptible to the lethal effects of MCMV but their low NK(YAC-1) activities are boosted by MCMV (Masuda and Bennett, 1981a). In preliminary experiments, we have extended the findings in ^{89}Sr-treated mice to mice treated with extradiol so as to cause osteopetrosis and deplete the animals of M cell functions (Seaman et al., 1979ab). Thus estradiol-treated C3H/St mice lose resistance to MCMV but their NK(YAC-1) cell function is boosted.

The relative lack of inflammatory response to MCMV infection in ^{89}Sr-treated mice is similar to the lack of inflammation detected in the brains and spinal cords of ^{89}Sr-treated mice dying of herpes simplex virus type-1 induced encephalomyelitis (Lopez et al., 1980). We consider two explanations: (i) there may be lack of myelopoiesis in ^{89}Sr-treated mice since the marrow is aplastic and MCMV inhibits splenic myelopoiesis (Masuda and Bennett, 1981a). The marrow of control mice infected with MCMV appears normal. (ii) there appears to be a defect in secretion of lymphokines by mice treatd with ^{89}Sr. For example, Concanavalin A fails to stimulate their lymphocytes to secrete a macrophage activating factor capable of enhancing listeriocidal activity of peritoneal macrophages (Masuda and Bennett, 1981b).

We conclude from these studies that NK cells alone cannot explain the genetic resistance of C3H/St mice to MCMV, although the correlation between genetic resistance virulent passage of MCMV is tested. The relative lack of inflammation observed in ^{89}Sr-treated mice was also observed in athymic mice infected with MCMV (Starr and Allison, 1977). We, therefore, expect that mice depleted of M cells have a defect in lymphokine responses to infection which greatly increase their susceptibility. Indeed, preliminary experiments performed with G. Sonnenfeld indicate that ^{89}Sr-treated mice secrete less interferon than control mice after infection with MCMV. The mechanism of the lymphokine defect remains to be determined.

REFERENCES

Bancroft, G.J., Shellam, G.R. and Chalmer, J.E. J. Immunol. 126.988 (1981).
Bennett, M. J. Immunol. 110:510 (1973).
Haller, O. and Wigzell, H. J. Immunol. 118:1503 (1977).
Kumar, V., Ben-Ezra, J., Bennett, M. and Sonnenfeld, G. J. Immunol. 123.1832 (1979).
Kumar, V., Luevano, E. ad Bennett, M. J. Exp. Med. 150: 531 (1979).
Lopez, C., Ryshke, R. and Bennett, M. Infect. Immunity 28:1025 (1980).
Luevano, E., Kumar, V., and Bennett, M. Scand. J. Immunol 13:563 (1981).
Lust, J.A., Kumar, V., Burton, R.C., Bartlett, S.P. and Bennett, M. J. Exp. Med. 154:306 (1981).
Masuda, A. and Bennett, M. Infect. Immunity 34:970 (1981).
Masuda, A. and Bennett, M. Eur. J. Immunol. 11:556 (1981).
Seaman, W.E., Gindhart, T.D., Greenspan, J.S., Blackman, M.A. and Talal, N.J. Immunol 122:2541 (1979).
Seaman, W.E., Merigan, T.C. and Talal, N. J. Immunol. 123:2903 (1979).
Selgrade, M.K. and Osborn, J.E. Infect. Immunity 10:1383 (1974).
Shellam, G.R., Allan, J.E., Papadimitriou, J.M. and Bancroft, G.J. Proc. Natl. Acad. Sci. USA 78:5104 (1981).
Starr, S.E., and Allison, A.C. Infect. Immunity 17:458 (1977).
Welsh, R.M. and Zinkernagel, R.M. Nature 268:646 (1977).

STUDIES OF NK CELL ACTIVATION AND OF INTERFERON INDUCTION AFTER INJECTION OF MOUSE HEPATITIS VIRUS TYPE 3 (MHV3)

Holger Kirchner
Liesel Schindler

Institute of Virus Research
German Cancer Research Center
Heidelberg, FRG

INTRODUCTION

Injection of viruses is known to induce interferon production in mice (Baron and Buckler, 1963). More recently, it has been found that activation of NK cells occurs *in vivo* in response to virus injection (Armerding and Rossiter, 1981; Bancroft et al., 1981). Interferon is a powerful activator of NK cells (Gidlund et al., 1978). However, not in all instances it is totally clear if the activation of NK cells that occurs after injection of virus is entirely due to interferon. We have found that under certain circumstances very low titers of interferon are found in the peritoneal fluid after injection of Herpes Simplex Virus (HSV) and nonetheless high activation of NK cells is observed (Engler et al., in preparation). However, one may argue that titers of local interferon that are too low to be detectable in the washout fluid are sufficient for the activation of NK cells. Thus, the relationship between virus injection, induction of interferon production and activation of NK cells is as yet not fully understood.

Furthermore, it is of great interest to find out if NK cells or interferon play a role in primary resistance against viral infections. In a model of murine cytomegalovirus infection, a correlation was observed between high titers of NK cell activation after virus injection and high resistance (Bancroft et al., 1981). A similar correlation between high

ISBN 0-12-341360-5

resistance on one side and high titers of early local interferon production and NK cell activation on the other side has been observed in a model of HSV infection of mice (Engler et al., manuscript submitted). We have decided to study if there was such a correlation in an additional viral model in which resistance is genetically controlled.

SOME PERTINENT DATA FROM THE LITERATURE

Primary resistance of mice against MHV3 infection is genetically controlled (Bang and Warwick, 1960). In our studies, for example, we have compared C57BL/6 mice that are highly susceptible with A/J mice that are resistant. The differences in the LD_{50}s between the two mouse strains after ip infection may be as high as a factor of 10^5. Nevertheless, resistance is not absolute, in that very high virus doses also kill A/J mice. It is of interest that in this system C57BL/6 mice are highly susceptible, whereas in many models of primary antiviral resistance this mouse strain is a resistant one. Resistance to MHV3 infection is expressed on the level of the target cell which is the macrophage (see below). Macrophages of resistant mice do not replicate the virus and do not develop the typical "cytopathic" effect which is the development of multi-nucleated macrophages ("giant cells").

A role of NK cells and/or of interferon in primary resistance against MHV3 infection has been deduced from two lines of evidence:

First, Tardieu et al. have developed an adoptive transfer protocol in which cell transfers were performed from adult to newborn A/J mice (1980). In these experiments they have found that a cell type that bore some relation to NK cells was involved in protection. However, in our opinion, this cell type may well have been another type of bone marrow-dependent effector cell. For example, the cell type that produces alpha interferon in response to viruses also appears to be a bone marrow-dependent cell (Kirchner et al., 1980).

Secondly, a role of interferon in the primary resistance of mice against MHV3 infection has been postulated from experiments in which the resistance of A/J mice was broken by the injection of anti-interferon serum (Virelizier and Gresser, 1978). However, this serum certainly was not monospecifically directed against interferon.

Resistance of A/J mice develops with age and up to about 20 days of age A/J mice are susceptible. There is an interesting parallel in that the resistance of C57BL/6 mice to HSV-infection also develop at 20 days of age (Zawatzky et al., manuscript submitted). This observation at first has led us to speculate that mechanisms similar to those that cause the resistance of C57BL/6 mice to HSV play a role in the resistance of A/J mice to MHV3. Thus we wanted to study the potential role of endogenously produced interferon and of activated NK cells in the resistance of mice to MHV3.

BONE MARROW DERIVED MACROPHAGES AS TARGETS FOR THE REPLICATION OF MHV3

In search of suitable target cells for the replication of MHV3 we have adopted the technique described by Klimetzek and Remold (1980) for titration of MHV 3. Bone marrow-derived macrophages can be grown by this technique in reasonably large quantities and MHV3 replicates to high titres in these cultures.

We have found - similarly to a previous observation in cultures of peritoneal macrophages (Schindler et al., 1981) - that addition of C. parvum to cultures of bone marrow-derived macrophages inhibited the replication of MHV3. This observation was of interest since the bone marrow system represents a virtually pure culture of macrophages. It therefore appears that the effect of C. parvum was a direct effect on the macrophages, a conclusion which was not possible from the previous experiments in which peritoneal exudate cells (PEC) were used that contained other cell types besides macrophages.

IN VIVO PROTECTION OF MICE AGAINST MHV3 INFECTION

We have previously shown that injection of C. parvum protected mice against infection with HSV, when given several days before viral infection (Kirchner et al., 1978). No such protection was observed when C. parvum was given simultaneously with HSV. In contrast, protection against MHV3 was observed only when C. parvum was given on the day of viral infection. However, it did not matter if C. parvum was given several hours before or several hours after injection of MHV3. Protection was observed in homozygous nude mice as well as in their heterozygous littermates suggesting that mature T cells did not play a role in the protection exerted by C.parvum.

However, the reasons for this protection are yet unclear. Interestingly, we have recently observed that A/J mice could not be protected against lethal doses of MHV3 by injection of C. parvum (Schindler and Kirchner, unpublished).

Injection of C. parvum induces both interferon and activation of NK cells in mice (Storch et al., manuscript submitted). We tend to believe that these two mechanisms do not play a role in the protection against MHV3. There was a clear-cut dissociation between in vivo antiviral protection on one side and activation of NK cells and interferon induction on the other side. Optimal stimulation of the latter two parameters occured at low doses of C. parvum whereas no antiviral protection was observed after injection of low doses of C. parvum (Schindler et al., 1981). This conclusion is in keeping with the conclusion of our above-mentioned in vivo studies suggesting that the protective effect of C. parvum may be mediated by a direct effect of this compound on macrophages themselves.

To further substantiate this conclusion, we have performed additional in vivo experiments in which we have injected silica before injecting C. parvum. We were expecting to observe that silica might abrogate the protective effect of C. parvum on MHV3 infection.

The results of these experiments unexpectedly have shown that injection of silica itself caused a protection of mice against MHV3 infection. The reasons for this protective effect are not understood. Maybe, silica which is known to destroy macrophages, effectively inactivates the target cells of MHV3 infection in the peritoneal cavity.

TESTING OF INTERFERON TITERS IN MHV3 INFECTED MICE

In many virus systems - and also after injection of certain non-virual inducers, high amounts of interferon are produced within a few hours after injection (Baron and Buckler, 1963). There was however, no interferon detectable in the serum during the first 10 hr after injection of MHV3 (Schindler et al., 1982). Interferon was also not detectable during this period of time in the wash-out fluid from the peritoneal infection site. There was, however, late interferon found 40-60 hr after infection both in the serum and the wash-out fluid.

Unexpectedly, we have observed an inverse correlation between the magnitude of these interferon titers and antiviral resistance. Significantly higher titers of interferon were observed in the peritoneal wash-out fluid and in the serum of susceptible C57BL/6 mice than were observed in susceptible A/J mice.

TESTING OF NK CELL ACTIVITY IN PEC OF MHV3 INJECTED MICE

PEC from untreated C57BL/6 mice in our laboratory show low, if any NK cell activity against YAC-1 lymphoma cells. However, 2-3 days after injection of MHV3 high levels of NK cell activity can be measured in the washed-out PE cell population (Schindler et al., 1982). In contrast, in A/J mice resistant to MHV3 infection, only very low NK cell activity was observed in the PEC after ip infection with MHV3. A deficient NK cell response was seen at all times tested and after injection of various different virus doses in A/J mice. Thus, similarly to the results of interferon testing, our data have shown an inverse correlation between the magnitude of NK cell activity and resistance against MHV3 infection. Quite obviously, there is a link between the interferon data and the NK cell data. Probably, the lack of an NK cell response in A/J mice is caused by a lack of a sufficient interferon response.

CONCLUSIONS

Since we have also shown that there were high titers of MHV3 in the peritoneal cavity of C57BL/6 mice wheras only low virus titers were observed in A/J mice we want to propose the following conclusions: It has been known - and has been confirmed in our laboratory - that adult A/J mice are resistant to MHV3 infection because their macrophages are non-permissive for the virus. Thus, no virus replication occurs in A/J mice there is no interferon induced and pre-NK cells are not activated at this site. In contrast, in C57BL/6 mice high virus titers are produced by peritoneal macrophages which in turn leads to activation of NK cells. At this time, however, neither interferon nor NK cells are able to decisively influence the outcome of the viral infection and the mice die from viral hepatitis within the next two days.

Our data further have shown that there is no early interferon detectable in either strain of mice after injection of MHV3. This is in marked contrast to the situation observed in the mouse model of HSV infection. There, we observe early interferon production at the local site in resistant strains of mice and this early response is obviously caused by the infecting virus. Thus, viruses may differ in their capacity to elicit an interferon response by the primary infecting dose and this differences may cause completely different patterns of resistance mechanisms. Furthermore, endogenous interferon production and associated mechanisms such as activation of NK cells may be relevant for defence in some virus infections but not in others.

Armerding, D., and Rossiter, H. (1981). Immunobiol. 158:369.

Bancroft, G.J., Shellam, G.R., and Chalmer, J.E. (1981). J. Immunol. 126:988.

Bang, F.B., and Warwick, A. (1960). Proc. Natl. Acad. Sci. USA, 46:1065.

Baron, S., and Buckler, C.E. (1963). Science 141:1061.

Engler, H., Zawatzky, R., and Kirchner, H. in preparation.

Engler, H., Zawatzky, R., Kirchner, H., and Armerding, D. manuscript submitted.

Gidlund, N., Oern, A., Wigzell, H., Senik, A., and Gresser, I. (1978) Nature 273:759.

Kirchner, H., Scott, M.T., Hirt, H.M., and Munk, K. (1978). J. Gen. Virol. 41:97.

Kirchner, H., Keyssner, K., Zawatzky, R., and Hilfenhaus, J. (1980). Immunobiol. 157:401.

Klimetzek, V., and Remold, H. (1980). Cell. Immunol. 53:257.

Schindler, L., Streissle, G., and Kirchner, H. (1981). Infect. Immun. 32:1128.

Schindler, L., Engler, H., and Kirchner, H. (1982). Infect. Immun., in press.

Storch, E., Schindler, L., and Kirchner, H., manuscript submitted.

Tardieu, M., Héry, C., and Dupuy, J.M. (1980). J. Immunol. 124:418.

Virelizier, J.L., and Gresser, I. (1978). J. Immunol. 120:1616.

Zawatzky, R., Engler, H., and Kirchner, H., manuscript submitted.

INTERFERON, NATURAL KILLER CELLS AND GENETICALLY DETERMINED RESISTANCE OF MICE AGAINST HERPES SIMPLEX VIRUS

Holger Kirchner
Helmut Engler
Rainer Zawatzky

Institute of Virus Research
German Cancer Research Center
Heidelberg

INTRODUCTION

The idea has been put forward that NK cells besides their potential role in the defense against tumors also may have antiviral effects (Welsh, 1981). Some data to support this concept have been obtained in two models of experimental infections with herpesviruses (Armerding and Rossiter, 1981; Bancroft et al., 1981). Data in two other virus models have argued against a role of NK cells in antiviral resistance (Welsh and Kiessling, 1980; Hirsch, 1981). The situation in regard to Mouse Hepatitis Virus type 3 will be discussed in a separate chapter of ours in this book.

One way of evaluating the potential role of NK cells in resistance is testing animals that differ in genetically determined resistance against experimental virus infections. A correlation between high NK cell activity after infection and high resistance would be supportive for a role of NK cells. Armerding and Rossiter (1981) and ourselves (Kirchner et al., 1980) have indeed observed such a correlation in a mouse model of experimental Herpes Simplex Virus (HSV) infection.

We have also observed a correlation between high titers of virus-induced <u>early</u> interferon and antiviral resistance (Zawatzky et al., 1982).It has been known that interferon induction invariably causes activation of NK cells. We have

ISBN 0-12-341360-5

therefore expressed the view that in our model the primary decisive event in resistance may be the interferon response whereas activation of NK cells represents a secondary effect and therefore may be less relevant in defense. This opinion may be wrong, but we do believe that an investigation of NK cell activation is incomplete without simultaneous investigation of interferon production.

In the following we want to summarize our recent data in the HSV model. The experimental details will not be given, they are contained in papers that are published or in press (Zawatzky et al., 1981; Engler et al., 1981; Zawatzky et al., 1982 a,b). In all experiments, except for those reported in the last paragraph, one strain of HSV type 1 was used, designated HSV-1 WAL, which was prepared as described (Zawatzky et al., 1981). It will be subsequently referred to as HSV.

COMPARISON OF INBRED STRAINS OF MICE

Inbred strains of mice differ in their relative resistance to HSV (Lopez, 1975). Previously we have compared serum interferon titers and splenic NK cell activity in these mice after injection of HSV (Kirchner et al., 1980; Zawatzky et al., 1981). In addition, we wanted to study these two parameters at the infection site, i.e. in the peritoneal cavity. For this pupose we have obtained small samples of cell-free peritoneal fluid at various times after infection and have assayed them for their interferon contents. From parallel groups of mice, the peritoneal exudate cells were recovered and tested for their NK cell activity against YAC-1 lymphoma cells in a 4 hr chromium release assay.

A comparison was performed between 8 different inbred strains of mice that were either resistant, medium resistant or susceptible to HSV infection. Our data have established a clear-cut correlation between resistance and high titers of early (2-4 hr after infection) interferon in the peritoneal fluid. However, when interferon titers were measured between 12 and 24 hr, such a correlation did not exist. In fact, in most experiments, the interferon titers at this time were higher in susceptible mice than in resistance mice. On the other hand, titers of HSV at 24 hr were significantly higher in susceptible than in resistant mice. Additional data have led us to suggest that the late interferon is induced by newly formed virus, whereas the early interferon, that is seen only in resistant mice, is induced by the infecting (input) virus. It will have to be determined, if the early and late interferon may represent different subtypes of interferon, and perhaps are products of different cell types.

In addition, we have measured NK cell activity of peritoneal exudate cells (PEC) after injection of HSV in different inbred strains of mice. Again the time of testing proved to be important (Armerding and Rossiter, 1981). Significant differences between mouse strains were observed when sampling PEC 18-24 hr after virus infection. No such differences were seen at 3 days after infection. There was again a certain correlation in that most resistant strains showed high NK cell activity against YAC-1 lymphoma cells 24 hr after infection with HSV. However, there was one notable exception. Mice of the strain SJL had only very low levels of virus-induced NK cells. This observation is in accordance with the work of Cudkowicz's laboratory that SJL mice are generally defective in their NK system. It is, however of interest that SJL mice are relatively resistant against HSV and that they do produce high titers of early interferon after injection of HSV.

STUDIES OF NEWBORN MICE

Newborn mice are highly susceptible to i.p. infection with HSV, even if they are of the strain C57BL/6 that are resistant as adults. We have confirmed the data of Lopez (1978) that resistance of C57BL/6 develops at about 20 days of age. Newborn C57BL/6 mice, in contrast to adult mice of the same strain, do <u>not</u> produce early interferon when injected with HSV. Neither do they have detectable activity of NK cells at 24 hr post infection.

Newborn mice do show high virus titers in the peritoneal exudate 24 hr after infection (Zawatzky et al., 1982a). It is of interest that at 24 hr high titers of interferon are found in the peritoneal fluid of newborn C57BL/6 mice and that NK cell reactivity is also measurable 3 days after infection. Thus, our data seem to indicate that pre-NK cells are present in newborn mice that are activated by interferon. Again, as in adult mice of susceptible strains, there is no early interferon response and therefore no early NK cell response in newborn C57BL/6 mice. Later on, however, new virus is formed, which induces interferon production and, in turn, causes activation of NK cells.

STUDIES OF NUDE MICE

Homozygous nude mice are not more susceptible to i.p. infection with HSV-1 (WAL) than their heterozygous littermates (Zawatzky et al., 1979). In fact in many experiments, nu/nu mice are somewhat more resistant. When interferon production and NK cell activity were assayed in the peritoneal exudate, nu/nu mice showed higher activity in both of these two parameters (Engler et al., 1982). Thus, a minimal conclusion may be reached that both functions are independent of mature T lymphocytes. Our data reiterate the conclusions of Lopez that primary resistance of mice against HSV is bone-marrow dependent (Lopez, 1978).

STUDIES OF THE VIRUS-INDUCED NK CELLS AND OF THE INTERFERON PRODUCING CELLS

For these studies our experimental protocol had to be somewhat modified. PEC were recovered from virus-injected mice at 24 hr and treated by various methods before performing NK tests. The same cell populations were put in culture for another 18 hr without any further stimulation and tested for interferon production. Similarly to the conclusions reached by Armerding and his coworkers (1981), our data have established that the HSV-induced NK cell shared typical properties of NK cells. However, we are not certain as yet about the nature of the interferon producing cell. It is present in nude mice and it is very radiosensitive. It is not affected by treatment with several monoclonal antisera including anti-theta, anti-Qa 4 and anti-Qa 5 (Engler et al., 1981). The producer cell of HSV-induced interferon is not phagocytic, but it is removed by nylon wool column passage. Thus, collectively the data indicate that the interferon producing cell is a non-T cell. Perhaps, it may be a B cell, but the latter point is far from being proven.

FURTHER DATA TO SUPPORT THE ROLE OF INTERFERON IN RESISTANCE AGAINST HSV

Two sets of data provided additional support for a role of interferon in primary resistance of mice against HSV. First, in our experimental system, we have confirmed a previous observation of Gresser's laboratory (Gresser et al., 1976), that administration of anti-interferon was able to break resistance of adult C57BL/6 mice to HSV infection. Secondly, we were able to protect mice against HSV infection by injection of fibroblast interferon (Zawatzky et al., 1982b). This interferon was provided by E. DeMaeyer and contained all three major species of mouse interferon alpha/beta.

INVESTIGATIONS OF A STRAIN OF HSV-1 THAT IS APATHOGENIC FOR MICE AFTER i.p. INFECTION

We have previously shown that HSV-1 ANG is apathogenic for adult mice of all strains after i.p. infection (Schröder et al., 1981). However, this virus kills mice after intracerebral infection. It kills adult mice treated with cyclophosphamide after i.p. infection and it kills newborn mice. In contrast to the situation observed with HSV-1 WAL in C57BL/6 mice, mice of all strains developed resistance against HSV-1 ANG at 10 days of age. NK cell reactivity in the peritoneal exudate does not differ after injection of HSV-1 ANG and HSV-1 WAL and there is no difference in the interferon response to these two viruses. Initial replication of HSV-1 ANG in the peritoneal cavity is of the same magnitude as the replication of a virus strain that is derived from HSV-1 ANG and that is able to kill mice after i.p. infection (Kümel et al., unpublished data). Preliminary data suggest that homozygous nude mice in contrast to their heterozygous littermates are susceptible to i.p. infection with HSV-1 ANG. Thus, our data seem to indicate that the defense against HSV-1 ANG (i.e. the mechanisms that restrict HSV-1 ANG in the mouse) are of quite different nature than the mechanisms that cause the resistance of adult mice of certain inbred strains against HSV-1 WAL, another strain of HSV-1.

CONCLUSIONS

Our data have strengthened the role of early interferon at the local site after infection in one virus system. In this system a correlation was found between high titers of early interferon and resistance. This was seen both in a comparison between newborn and adult mice and in a comparison between different strains of inbred mice. Such a correlation was not seen when the late interferon response (12-24 hr) was tested. At this time, high virus titers in susceptible mice were accompanied by high interferon titers. There was a limited correlation between high activity of NK cells after virus injection and resistance. However, SJL mice that have a defective NK cell system are nevertheless resistant. SJL mice produce high titers of interferon. Thus, interferon may play a role as an antiviral protein in preventing virus replication, but NK cell activation by interferon may be of less relevance. In another contribution to this volume we will report data in another virus system that show an inverse relationship between resistance and the magnitude of the interferon response (and the NK cell response).

Armerding, D., Simon, M.M., Hämmerling, U., Hämmerling, G., and Rossiter, H., Immunobiol. 158:347 (1981).
Armerding, D., and Rossiter, H., Immunobiol. 158:369 (1981).
Bancroft, G.J., Shellam, G.R., and Chalmer, J.E., J. Immunol. 126:988 (1981).
Engler, H., Zawatzky, R., Goldbach, A., Schröder, C.H., Weyand, C., Hämmerling, G.J., and Kirchner, H. J. Gen. Virol. 55:25 (1981).
Engler, H., Zawatzky, R., Kirchner, H., and Armerding, D. in preparation.
Gresser, I.,Tovey, M.G., Maury, C., and Bandu, M.-T., J. Exp. Med. 144:1316 (1976).
Hirsch, R.L., Immunology 43:81 (1981).
Kirchner, H., Zawatzky, R., Engler, H., and Schröder, C.H., in "Genetic Control of Natural Resistance to Infection and Malignancy" (E. Skamene, P. Kongshavn, and M. Landy, eds.), Academic Press New York - London, 1981.
Kümel, G., Kaerner, H.C., et al., unpublished data.
Lopez, C., Nature (London) 258:152 (1975).
Lopez, C. in "Oncogenesis and Herpesviruses" (G.de Thé, W.Henle, and F. Rapp, eds.), IARC Scientific Publications 24:2, 1978.
Schröder, C.H., Engler, H., and Kirchner, H., J. Gen. Virol. 52:159 (1981).
Welsh, R.M., jr., and Kiessling, R.W., Scand. J. Immunol. 11:363 (1980).
Welsh, R.M., Antiviral Research 1:5 (1981).
Zawatzky, R., Hilfenhaus, J., and Kirchner, H., Cell. Immunol. 47:424 (1979).
Zawatzky, R., Hilfenhaus, J., Marcucci, F., and Kirchner, H., J. Gen. Virol. 53:31 (1981).
Zawatzky, R., Engler, H., and Kirchner, H., J. Gen. Virol. in press (1982a).
Zawatzky, R., Gresser, I., DeMaeyer, E., and Kirchner, H., J. Infect. Dis. in press (1982b).

NATURAL RESISTANCE AGAINST MOUSE HEPATITIS VIRUS 3

Jean-Marie Dupuy[1]

Immunology Research Center
Armand-Frappier Institute
University of Quebec
Laval-des-Rapides, Quebec

I. INTRODUCTION

Natural killer (NK) cell function is thought to represent a major barrier in the resistance toward certain syngeneic or semi-syngeneic virus-induced lymphomas in the mouse (Kiessling *et al.*, 1976). Similarly, there are indications that NK or NK-like cell functions may be important in resistance against microbial infections (Tardieu *et al.*, 1980; Welsh, 1981), but the precise *in vivo* significance of this reactivity is unclear. One important problem encountered by such studies resides in the fact that viral infections induce the production of factors such as interferon which may, in turn, modify the sensitivity of target cells to *in vitro* natural killing (Trinchieri and Santoli, 1978; Welsh, 1981). The observation that NK-like cells may contribute to natural resistance in viral infections would, however, be of a primary importance in linking together a group of effector cells that may be operative in regulatory processes and defense mechanisms such as anti-tumor surveillance (Paige *et al.*, 1978; Stutman *et al.*, 1978; Welsh, 1978; Herberman and Holden, 1979; Herberman *et al.*, 1980), hemopoietic histocompatibility system (Cudkowicz and Bennett, 1971), and defense against infectious agents.

---

[1]This work was supported by a grant from the Medical Research Council of Canada (MA-7241) and INSERM AT 79-5-119-1C.

ISBN 0-12-341360-5

We approached this problem of natural resistance of mice against mouse hepatitis virus 3 (MHV3) infection by reconstitution of immunologically immature neonates with various populations of cells originating from mature, syngeneic animals. We have previously reported that transfer of MHV3 resistance to newborns requires the administration of 3 types of cells: adherent spleen cells, T lymphocytes, and a third population, M cells, which have many characteristics of NK cells (Levy-Leblond and Dupuy, 1978; Tardieu *et al.*, 1980). M cells are present in the non-adherent fraction of the spleen, in peritoneal exudates and in bone marrow but are absent from the thymus. They are age-dependent, non adherent to plastic, sensitive to treatment with puromycin, mitomycin C, silica particles and ^{89}Strontium, and they are enhanced by BCG (Tardieu *et al.*, 1980).

In the study reported here, we have found that M cells are also required in transfer of resistance to X-irradiated, bone marrow reconstituted mice and that spleen cells of ^{89}Strontium-treated mice can inhibit their differentiation.

II. MATERIALS AND METHODS

Animals. A/J strain mice were obtained from The Jackson Laboratory (Bar Harbor, Maine). A/J newborns were bred in our mouse colony. Mice of the A/J strain are resistant to MHV3 infection when tested after 3 weeks of age, whereas 100% of A/J neonates die in 5 to 7 days when infected during the first 2 weeks of life (Le Prévost *et al.*, 1975).

^{89}Strontium Treatment. Mice were injected with one dose of 100 μCi ^{89}Sr (Amersham Searle Co., Oakville, Ontario, Canada) intraperitoneally (i.p.) and were utilized 4 weeks later unless otherwise mentioned.

Preparation of Cell Suspension. All suspensions were obtained from normal or from ^{89}Sr-treated adult A/J mice. Whole spleen cells, adherent spleen cells, nylon wool-separated spleen cells (T lymphocytes) and bone marrow cells were prepared as already described (Tardieu *et al.*, 1980).

Virus. MHV3 was passaged in susceptible DBA mice and on L cells. Virus titer was expressed either in LD50 or in tissue culture infectious dose 50 (T.C.I.D.$_{50}$) (Le Prévost *et al.*, 1975; Krzystyniak and Dupuy, 1981).

III. RESULTS AND DISCUSSION

A/J susceptible newborns were protected against MHV_3 infection with spleen cells from syngeneic adult mice (Tardieu *et al.*, 1980). Transfer of MHV_3 resistance was achieved with either whole spleen cells or with bone marrow cells mixed with adherent spleen cells and T lymphocytes (Table I). Protection was also obtained (5/5) when normal spleen cells were mixed with spleen cells from ^{89}Sr-treated mice. Newborns, however, were not protected against MHV_3 infection by spleen cells of ^{89}Sr-treated mice administered alone (0/7), or mixed with normal cells consisting of bone marrow cells, adherent spleen cells and T lymphocytes (1/7; Table I).

TABLE I. Effect of Cells from ^{89}Sr-Treated Mice on Transfer of Resistance to MHV_3 in A/J Newborns[a]

^{89}Sr-treated mice	Normal mice				Results
Whole spleen cells	Whole spleen cells	BM cells	T Lymphocytes	Adherent spleen cells	No. survivors/ no. tested
-	-	-	-	-	0/15
-	10^7	-	-	-	9/12
2×10^7	-	-	-	-	0/7
2×10^7	10^7	-	-	-	5/5
-	-	10^7	5×10^6	5×10^6	8/10
10^7	-	10^7	5×10^6	5×10^6	1/7

[a]A/J mice were injected i.p. with 100 µCi ^{89}Sr. Four weeks later, spleen cells from treated animals were injected into 4 to 7 day-old A/J recipients, alone or mixed with various populations of cells originating from normal A/J adult mice. Eighteen hours later, newborns were challenged with 100 LD_{50} of MHV_3.

The lack of protection capacities to MHV_3 of spleen cells of ^{89}Sr-treated mice and their inhibitory effect on normal cells was confirmed in a different experimental system. Lethally X-irradiated adult A/J mice, reconstituted with syngeneic bone marrow cells were fully susceptible to MHV_3

infection during the first 3 weeks after reconstitution (Table II). As in newborns, protection was transferred either with whole spleen cells (16/16), or with normal adherent spleen cells mixed with T lymphocytes (8/10). However, when spleen cells of ^{89}Sr-treated mice were injected alone or mixed with normal adherent cells and T lymphocytes, no protection was achieved (2/7 and 1/7, respectively).

TABLE II. Protection of X-Irradiated, Bone Marrow Reconstituted A/J Mice Against MHV_3 Infection[a]

X-irradiated A/J recipients	Results
Cell populations[b]	No. survivors/ no. tested
Marrow cells	0/14
Marrow cells + normal whole spleen cells	16/16
Marrow cells + ^{89}Sr-spleen cells	2/7
Marrow cells + adherent and T spleen cells	8/10
Marrow cells + adherent and T spleen cells + ^{89}Sr-spleen cells	1/7

[a]Adult A/J mice were lethally X-irradiated (850 R) and reconstituted with 20 x 10^6 syngeneic bone marrow cells.

[b]Forty-eight hours after marrow reconstitution, A/J recipients received an i.v. administration of either spleen cells (50 x 10^6) from normal or ^{89}Sr-treated mice, or adherent spleen cells (10^7) and T spleen lymphocytes (10^7) given alone or mixed with ^{89}Sr-spleen cells. Two hours later, recipients were challenged with 100 LD_{50} of MHV_3.

These data indicate that, in newborns as well as in lethally X-irradiated, bone marrow reconstituted mice, the transfer capacity of protection of cells was abrogated by treatment with ^{89}Sr. Whole spleen cells originating from ^{89}Sr-treated mice were devoid of protective capacities against MHV_3. The lack of protection was not restored by normal bone marrow cells or by adherent spleen cells mixed with T lymphocytes but was fully restored when ^{89}Sr-spleen cells were mixed with normal spleen cells. These data strongly suggest that MHV_3 effector M cells may be found at 2 different stages of differentiation: a precursor stage where cells in the marrow are not competent against MHV_3 and a later stage, in the spleen, where M cells are competent. Results also suggest that the differentiation process required for precursor M cells to become fully competent effector cells is inhibited by a cell population present in the spleen of ^{89}Sr-treated mice. Such a differentiation process was corroborated by previous work which showed that M cells in marrow were sensitive to X-irradiation whereas they were radioresistant in the spleen (Tardieu *et al.*, 1980). Administration of ^{89}Sr to mice resulted in a decreased activity of NK cell functions against most tumor cell targets and in the generation of suppressor cells (Bennett, 1973; Kumar *et al.*, 1974; Haller and Wigzell, 1977; Levy *et al.*, 1979; Kumar and Bennett, 1981). It is possible, then, that NK cell differentiation is susceptible to a negative regulation by suppressor cells and that ^{89}Sr-treated mice possess cells capable of suppressing NK cell differentiation.

Since it was observed that M cells (NK-like cells) played a role in MHV_3 resistance and that ^{89}Sr-treatment inhibited their normal differentiation, it was important to study the behaviour of ^{89}Sr-treated A/J mice against MHV_3 infection. Three groups of 5 mice were injected with 100 μCi of ^{89}Sr and were challenged with MHV_3 4 weeks, 6 weeks or 12 weeks later. All animals survived. Destruction of M cells in ^{89}Sr-treated mice was, however, shown by the lack of protection transfer capacities of their spleen cells (Tables I and II) and by the absence of any boosting effect of MHV_3 infection on NK cell functions (data not shown). These results suggest that the anti-MHV_3 resistance related to M cells and natural killing is, in normal conditions, an auxiliary mechanism which can easily be compensated, but plays a major role in circumstances of high susceptibility such as immaturity or immunodeficiencies.

REFERENCES

Bennett, M. (1973). *J. Immunol. 110:*510.

Cudkowicz, G., and Bennett, M. (1971). *J. Exp. Med. 134:*1513.

Haller, O., and Wigzell, H. (1977). *J. Immunol. 118:*1503.

Herberman, R.B., and Holden, H.T. (1979). *J. Natl. Cancer Inst. 62:*441.

Herberman, R.B., Timonen, T., Ortaldo, J.R., Bonnard, G.D., and Gorelik, E. (1980). In "Progress in Immunology IV" (M. Fougereau and J. Dausset, eds.), p. 261. Academic Press, New York.

Kiessling, R., Petranyi, G., Klein, G., and Wigzell, H. (1976) *Int. J. Cancer 17:*7.

Krzystyniak, K., and Dupuy, J.M. (1981). *J. Gen. Virol. 57:* 53.

Kumar, V., and Bennett, M. (1981). *Current Topics in Microbiol. and Immunol. 92:*67.

Kumar, V., Bennett, M., and Eckner, R.J. (1974). *J. Exp. Med. 139:*1093.

Le Prévost, C., Levy-Leblond, E., Virelizier, J.L., and Dupuy, J.M. (1975). *J. Immunol. 114:*221.

Levy, E.M., Bennett, M., Kumar, V., Fitzgerald, P., and Cooperband, S.R. (1979). *Fed. Proc. 38:*5568.

Levy-Leblond, E., and Dupuy, J.M. (1978). *J. Immunol. 118:* 1219.

Paige, C.J., Feo Figarella, E., Cuittito, M.J., Cahan, A., and Stutman, O. (1978). *J. Immunol. 121:*1827.

Stutman, O., Paige, C.J., and Feo Figarella, E. (1978). *J. Immunol. 121:*1819.

Tardieu, M., Héry, C., and Dupuy, J.M. (1980). *J. Immunol. 124:*418.

Trinchieri, G., and Santoli, D. (1978). *J. Exp. Med. 147:*1314.

Welsh, R.M. (1978). *J. Immunol. 121:*1631.

Welsh, R.M. (1981). *Current Topics in Microbiol. and Immunol. 92:*83.

INDUCTION OF NATURAL KILLER CELLS AND INTERFERON PRODUCTION DURING INFECTION OF MICE WITH BABESIA MICROTI OF HUMAN ORIGIN[1]

Mary J. Ruebush[2]
Donald E. Burgess[3]

Department of Microbiology and Immunology
The Bowman Gray School of Medicine
of Wake Forest University
Winston-Salem, North Carolina

I. INTRODUCTION

Babesia microti is an intraerythrocytic protozoan parasite of wild rodents which has been recently found to infect man (Gleason et al., 1970). Natural resistance to this infection in humans and laboratory animal models appears to be affected by many factors, including the age and immunologic status of the host (Ruebush et al., 1977). Because correlations have been found between natural killer (NK) cell levels and resistance to hemoprotozoa in mice (Eugui and Allison, 1980), we examined the possible role of NK cells and interferon (IFN) in resistance to B. microti in mice.

II. EFFECT OF B. MICROTI INFECTION ON SPONTANEOUS CYTOTOXICITY AGAINST YAC CELLS

To examine the effect of B. microti infection on NK activity, groups of female +/+, +/bg, and bg/bg mice were infected i.v. with 10^8 parasites in +/+ blood.

---

[1]Supported by NSF grant SPI-80-09156, NIH grant AI 15785 and ACS grant MV-41.

[2]Present address: Department of Preventive Medicine and Biometrics, Uniformed Services University of the Health Sciences, Bethesda, Maryland.

[3]Present address: Department of Immunology, Walter Reed Army Institute of Research, Washington, D.C.

ISBN 0-12-341360-5

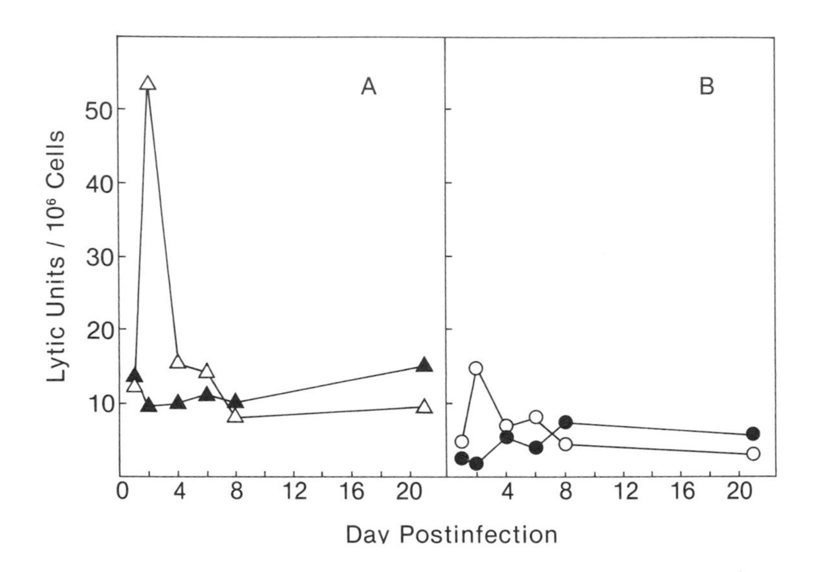

Figure 1. Splenic NK activity of uninfected (closed symbols) or infected (open symbols) beige (B) or wild-type, B6 (A) mice. The difference in lytic units between uninfected and infected mice was significant on days 1, 2, 4 and 6 ($p \leq 0.005$).

Peak induced NK cytotoxicity occurred on day 2 postinfection in each mouse strain (Fig. 1), at levels which were comparable to those observed after inoculation of mice with NK-inducing agents such as poly I:C or VSV (data not shown). Similar results were obtained when the size of the parasite inoculum was reduced (data not shown). On days 4, 6, 8 and 21 postinfection, spontaneous cytotoxicity in _B. microti_-infected mice decreased toward or below control values. The level of NK activity of the infected bg/bg mice on days 1, 2 and 4 was approximately equal to the normal level of NK activity observed in wild-type mice. Thus, the relative increases in NK activity due to _B. microti_ infection were similar in beige and wild-type mice, although the absolute levels of NK activity in the two mouse strains were always significantly different.

Induction of NK-like activity was probably not due to the presence of a virus infection in the mouse colony, experimental mice or B. microti inoculum, since mice and inoculum were repeatedly tested and found to be serologically negative for 11 distinct murine viruses. No significant difference was ever noted between the natural cytotoxicity of uninfected +/+ or +/bg mice. Furthermore, inoculation of uninfected +/+ blood cells had no significant effect on NK activity in mice of any genotype tested.

Depression of NK activity below control levels late in the infection has been previously reported during murine malarial infections (Hunter et al., 1981). Because prostaglandins are known to modulate NK activity, and parasitic infections might induce prostaglandin synthesis, in two experiments, bg/bg and +/bg mice were treated with indomethacin to inhibit prostaglandin synthesis. Results were equivocal since injection of the vehicle alone was sufficient to ablate the NK depression observed (data not shown).

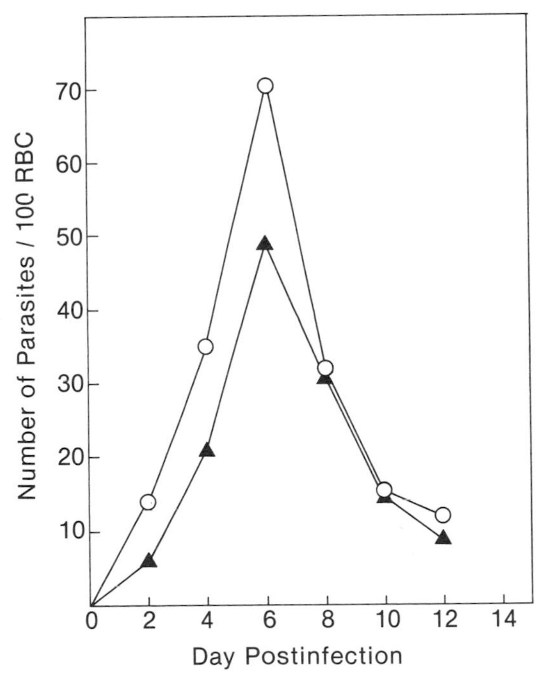

FIGURE 2. Course of B. microti parasitemia in heterozygous (▲) or homozygous (O) beige mice infected i.v. with 10^8 B. microti. Mean parasitemias (Ruebush and Hanson, 1979) were significantly different on day 2 of infection. This experiment was repeated 4 times with similar results.

The cell which was responsible for the spontaneous cytotoxicity induced by *B. microti* infection had the characteristics of the classical NK cell as defined by nonadherence to nylon wool, resistance to treatment with anti-Ly-1 or anti-Ly-2 sera in the presence of complement (C), and susceptibility to 6 hours of preincubation at 37 C and anti-NK-1.2 serum plus C (Burton, 1980).

III. COURSE OF PARASITEMIA IN bg/bg AND +/bg MICE

After i.v. inoculation of 10^8 parasites, beige mice developed significantly higher mean parasitemias (p always ≤ 0.01) than their heterozygous wild-type counterparts on each day postinfection through the day of peak parasitemia (Fig. 2). The difference in parasitemia between these mouse strains was most pronounced on day 2 postinfection and diminished over successive days of infection. Mean parasitemias of +/+ and +/bg mice did not differ significantly. Similar results were obtained when mice were inoculated with 10^7 parasites, but when lower numbers of parasites were introduced (10^6, 10^5, 10^4, 10^3 or 10^2) there were no clear-cut differences between the bg/bg and +/bg groups (data not shown). In these cases, the prepatent period was delayed 6-12 days and peak parasitemias decreased stepwise with inoculum size.

TABLE I. Induction of IFN Production by *B. microti* Infection

Time after Infection	Mouse Strain	IFN Titer[a]
12 hr	+/bg	14.21
24 hr	+/bg	145.07
	bg/bg	130.56
36 hr	+/bg	116.03
48 hr	+/bg	464.21
72 hr	+/bg	26.76
7 day	+/bg	32.64
Uninfected	+/bg	21.76
	bg/bg	16.32

[a]Values shown are the average of 2-3 determinations and were consistent within one serial two-fold dilution. The final dilution which prevented CPE was determined for the serum samples and is expressed in IFN units compared to a reference source of IFN (Calbiochem, La Jolla, CA).

Thus, the defects in the beige mouse appeared to have their most pronounced effects on B. microti parasitemia early in infection and when high numbers of parasites were inoculated. The early depression of parasitemia in +/bg mice coincided temporally with the time of peak NK activity, therefore it seems plausible that differences in parasitemia would not be observed in cases where parasitemia peaked long after NK activity had subsided to or below background levels. Since it is clearly possible to induce NK activity in beige mice, the eventual resolution of parasitemia in these mice could be due to the induction of low numbers of NK cells during infection, or more likely, activation of an alternative protective mechanism later in the infection.

IV. ROLE OF IFN DURING B. MICROTI INFECTION

Using a wide variety of concommitant infections or adjuvants, Clark and his coworkers have hypothesized that the protective activity of a given compound is related to its IFN-inducing ability (Clark, 1979; Clark et al., 1976; Clark et al., 1977a; Herod et al., 1978). Because IFN is known to modulate NK activity (Djeu et al., 1979), it seemed possible that the NK activation observed was simply a result of parasite-induced IFN production and had no relevance to control of disease.

Therefore, we examined the possibility that B. microti might induce NK activity through IFN induction. Infection with B. microti caused +/bg mice to develop IFN titers which were significantly above controls as early as 24 hours after infection (Table I). Thereafter, titers rose to peak near 48 hours and fell to background levels within 72 hours postinfection. Homozygous beige and heterozygous mice had similar titers of IFN before treatment, and developed similar, elevated levels of IFN 24 hrs after infection with B. microti (Table I) or inoculation with IFN-inducing agents such as poly I:C (data not shown). It therefore seems unlikely that differences in IFN induction after B. microti infection in bg/bg vs. +/bg mice might have resulted in the observed differences in NK activity and parasitemia.

Treatment of B. microti-infected +/bg mice with 100 U of IFN i.v. daily beginning one day before infection caused a transient reduction in early parasitemia which was repeatable, but never statistically significant (data not shown). Treatment of mice with 100 μg poly I:C i.p. every other day did not affect parasitemia (data not shown). Infected mice receiving poly I:C or IFN had NK levels which were comparable to those of uninfected mice receiving these agents.

In order to determine whether the transient depression of parasitemia after IFN therapy was due to a direct effect of IFN on parasite viability or to an indirect stimulation of NK activity, infected RBC were pretreated for 6 hr *in vitro* with 10^4 U of IFN/ml or an equivalent volume of PBS. After this time, cells were washed 2 times in PBS and resuspended to the original volume so that mice received 10^9 infected RBC based on determinations made before the beginning of the incubation period. Preincubation of infected RBC either in PBS or IFN caused a marked decrease in parasite infectivity. However, there was no significant difference in the parasitemias of mice receiving PBS- or IFN-pretreated blood on any day tested (data not shown).

V. CONCLUSIONS

It has previously been shown that the mature T lymphocyte is important in natural resistance to *B. microti* infection (Clark and Allison, 1974; Ruebush and Hanson, 1980). The temporal relationship between NK cell activation and early depression of parasitemia demonstrated in the present studies suggests that NK cells may also play a role in control of parasite multiplication. Because beige mice may have other, as yet undefined immunological defects, it is not possible at present to rule out the contributions of other factors, or definitively demonstrate the function of NK cells in this disease. Direct proof of this will only be possible using direct cytotoxic assays with infected RBC as targets.

Because it has been shown that IFN production is stimulated during *B. microti* infection, it is possible that the NK cell activation observed is simply a result of parasite-induced IFN production. However, the fact that no significant difference in natural or induced IFN levels has ever been noted between bg/bg and +/bg mice in this or previous publications (Welsh and Kiessling, 1980) and that it was not possible to affect parasite viability with IFN preparations strongly suggest that the elicited NK response itself is the contributor to the observed immune response. Clearly, the NK cell alone is not sufficient for expression of strong natural resistance, since nude mice which lack mature T cells but possess high degrees of NK activity (Minato et al., 1979) are unable to resolve *B. microti* parasitemia (Clark and Allison, 1974). If NK cells can be demonstrated to have an important effect on *B. microti*, it will probably occur early in infection, before specific immune responses to the parasite are activated.

ACKNOWLEDGMENTS

We would like to thank Dr. R. C. Burton for his gift of anti-NK 1.2 antiserum.

REFERENCES

Burton, R. C. (1980). In: Natural cell-mediated immunity to tumors, R. B. Herberman, ed. Academic Press, NY.

Cerottini, J. C. and Brunner, K. T. (1974). Immunology 18, 67.

Clark, I. A. (1979). Infect. Immun. 24, 319.

Clark, I. A. and Allison, A. C. (1974). Nature 252, 328.

Clark, I. A., Allison, A. C., and Cox, F. E. G. (1976). Nature 259, 309.

Clark, I. A., Cox, F. E. G., and Allison, A. C. (1977a). Parsitology 74, 9.

Djeu, J. Y., Heinbaugh, J. A., Holden, H. T., and Herberman, R. B. (1979). J. Immunol. 122, 175.

Eugui, E. M. and Allison, A. C. (1980). Parasite Immunol. 2, 277.

Gleason, N. N., Healy, G. R., Western, K. A., Benson, G. D., and Schultz, M. G. (1970). J. Parasitol. 56, 125.

Herod, E., Clark, I. A., and Allison, A. C. (1978). Clin. Exp. Immunol. 31, 518.

Hunter, K. W., Folks, T. M., Sayles, P. C., and Strickland, G. T. (1981). Immunol. Lett. 2, 209.

Ruebush, M. J. and Hanson, W. L. (1979). J. Parasitol. 65, 430.

Ruebush, M. J. and Hanson, W. L. (1980). Am. J. Trop. Med. Hyg. 29, 507.

Ruebush, T. K., Cassaday, P. B., Marsh, H. J., Lisker, S. A., Voorhees, D. B., Mahoney, E. B., and Healy, G. R. (1977). Ann. Int. Med. 86, 6.

Welsh, R. M. and Kiessling, R. W. (1980). Scand. J. Immunol. 11, 363.

NATURAL CELL-MEDIATED IMMUNITY AND INTERFERON IN MALARIA AND BABESIA INFECTIONS

Elsie M. Eugui [1]
Anthony C. Allison

Institute of Biological Sciences,
Syntex Research,
Palo Alto, California

I. INTRODUCTION

Natural cell-mediated immunity has been studied mainly in the context of immunity to tumors and viruses. However, marked increases in natural cell-mediated cytotoxicity occur in hemoprotozoan infections. In this paper observations on natural cell-mediated immunity and interferon in malaria and babesiosis are summarized. Although the effector cells have not been rigorously defined, they appear to be natural killer (NK) cells and/or macrophages, and the possibility that these cell types are of the same lineage is considered. The role that they may play in immunity and immunopathology in these infections is discussed.

II. NATURAL CELL MEDIATED IMMUNITY IN HEMOPROTOZOAN INFECTIONS

A. Murine Malaria and Babesiosis

Immune responses in malaria and babesiosis are complex, involving both humoral and cell-mediated components, the relative importance of which varies from one host-parasite combination to another. Antibodies appear to play a major part in the recovery of mice from *Plasmodium berghei* and *P. yoelii*

[1] Present address: Institute of Biological Sciences Syntex Research, 3401 Hillview Ave., Palo Alto, CA 94304

ISBN 0-12-341360-5

infections. Stage-specific monoclonal antibodies have been shown to protect mice from sporozoites of P. berghei (Yoshida et al., 1980) and asexual blood forms of P. yoelii (Freeman et al., 1980). However even in such infections a thymus-dependent, antibody-independent mechanism (or set of mechanisms) can control established infections and prevent re-infection.

Many years ago we demonstrated that neonatally thymectomized rats are more susceptible to P. berghei infections than are their intact littermates (Brown et al., 1968). Later we found that congenitally thymus-deprived nu/nu mice are unable to recover from P. yoelii or Babesia microti infections (Clark & Allison, 1974). Roberts et al.(1977) observed that malaria recrudesced in nu/nu mice after termination of the acute disease with clindamycin.

In mice made B-cell deficient by neonatal injections of anti-μ serum, infections with a normally avirulent strain of P. yoelii produced lethal disease (Weinbaum et al., 1976). However, when the acute disease was terminated by treatment with clindamycin, B-cell deficient mice recovered and were resistant to homologous challenge despite the absence of demonstrable antibodies to plasmodia (Roberts and Weidanz, 1979). This immunity was not sterilizing. With another murine parasite, P. chabaudi adami, infection of B-cell deficient mice was found to activate a T-cell-dependent immune mechanism which terminated the malaria in a manner similar to that seen in immunologically intact mice (Grun & Weidanz, 1981). The immunized B-cell-deficient mice were resistant to homologous challenge as well as to infections with P. vinckei.

Rank and Weidanz (1976) found that in chickens made agammaglobulinemic by chemical bursectomy P. gallinaceum produced fulminating infections. However, when they were rescued by chloroquine therapy they were able to suppress but not eliminate the infection, and they also resisted challenge with the same parasite.

These observations emphasize the importance of a thymus-dependent, antibody-independent mechanism of immunity to re-challenge with all species of malaria parasites so far tested and the ability of this mechanism to bring about recovery from P. chabaudi adami without chemotherapy.

These observations focus attention on the role of T-lymphocytes and cell-mediated immunity in malaria. Apart from helper effects on antibody formation, T-lymphocytes are thought to contribute to recovery from infections in three ways: (1) lysis of parasites or infected cells; (2) Activation of macrophages which then control replication of parasites within them and (3) activation by T-lymphocyte-derived products of macrophages and/or natural killer cells which in turn liberate mediators limiting the replication of parasites within erythrocytes. In the case of mammalian malaria parasites, mechanism (2) is not applicable and there is no evidence that mechanism (1) operates. Hence mechanism (3) is thought to be the most important.

Evidence has accumulated that natural cytotoxicity is increased during the course of malaria and babesia infections and that the resistance of different strains of mice to these infections is positively correlated with their natural cytotoxic activity.

Our investigation began with an analysis of the susceptibility of different mouse strains to *P. chabaudi* and *B. microti* (Eugui and Allison, 1980). It was found that strain A/He mice are highly susceptible to both infections; in this strain *P. chabaudi* is nearly always lethal, and the mice do not recover from *B. microti* infections. Investigations of congenic strains suggested that susceptibility does not correlate with H-2 haplotype (Table 1). The F_1 offspring of susceptible A and resistant B10.A mice are resistant.

TABLE 1. Differences in the susceptibility of various mouse strains to hemoprotozoan infections

Mouse strain	Haplotype	Persistent *B. microti* infection	Lethal *P. chabaudi* infection
A/J Crc	$H\text{-}2^a$	+	+
B10.A/Olac	$H\text{-}2^a$	-	-
CBA/CaCrc	$H\text{-}2^k$	-	-
C57Bl/10ScSACrc	$H\text{-}2^b$	-	-
BALB/cCrc	$H\text{-}2^d$	-	-
$(B10.AxA)F_1$	$H\text{-}2^a$	-	-

Strain A as well as ICR mice have more recently been found defective in their response to non-specific stimulation of immunity against *P. berghei* by *Mycobacterium tuberculosis* BCG (BCG) or *Corynebacterium parvum* (Murphy, 1981).

Strain A mice have two known defects: the capacity of their activated macrophages to kill tumor target cells is decreased (Adams *et al.*, 1981), and the capacity of their non-adherent spleen cells (natural killer of NK cells) to kill tumor or other sensitive target cells is also less than that of other strains (Herberman *et al.*, 1975; Kiessling *et al.*,

1975). In most strains of mice infected with malaria or babesia parasites, there is a rapid increase in the size and number of spleen cells (about four-fold), and in the capacity of their spleen cells to kill tumour cells, expressed in terms of ratio of splenic leucocytes to target cells (Eugui and Allison, 1980; table 2). Thus the total killing capacity of the spleen, a measure of activation of macrophages and NK cells for this property, is considerably augmented. In contrast, strain A mice show little increase in spleen cell number and their capacity to kill tumour cells is slightly, if at all, increased. In murine malaria infections an early increase in natural killer activity followed by suppression has been observed by Hunter and colleagues (1981).

TABLE 2. Changes in the number of nucleated cells in the spleen, and in NK activity, of different strains of mice infected with _P. chabaudi_. Observations on _nu/nu_ mice are also shown.

Mouse strain	Parasitemia* (%)	Spleen cells ($x10^7$)	NK activity** (% specific lysis)
C57Bl	0	5.9(5.8-6.0)	10.9(10.7-11.1)
	13-28	22.4(12.6-32-0)	30.2(26.7-33.0)
A/He	0	12.3(11.3-13.4)	8.2(7.0-9.4)
	56-77	11.0(9.4-14.0)	12.9(9.6 -19.4)
CBA _nu/+_	0	6.8(6.2-8.7)	25.5(23.9-26.6)
CBA _nu/nu_	0	7.3(6.5-8.2)	47.8(45.6-50.1)
CBA _nu/nu_	60-70	10.7(9.0-13.0)	35.3(25.7-43.0)

*Samples were taken at peak parasitemia, 8 or 9 days after infection.

**NK activity was measured using 100 spleen cells to 1 YAC-1 target cell, in a ^{51}Cr release assay. Arithmetic means and ranges (in brackets) are taken from Eugui and Allison (1980). Four animals per group were individually tested.

During the course of hemoprotozoan infections there is a rapid increase in the number of mononuclear cells rising to a peak about the time of maximum parasitemia; thereafter there is a marked increase in the number of erythropoietic precursors (Freeman & Parish, 1978). In _nu/nu_ mice infected with _P. chabaudi_ the increase in spleen cells and capacity to kill tumor cells does not occur (table 2). Hence the recruitment and activation of mononuclear cells in malaria infections are thymus-dependent and under genetic control. Immune spleen cells placed in diffusion chambers with parasitized erythrocytes elicited a marked mononuclear infiltrate around the chamber (Allison et al,,1979). Presumably this is due to the liberation of mononuclear chemotactic and immobilizing factors: Wyler and Gallin (1977) and Coleman et al., (1976) have described the presence in the spleen of mice recovering from malaria of mononuclear chemotactic activity and macrophage migration-inhibition factor, which could explain the accumulation of mononuclear cells in the spleen of infected animals. In vaccinated, P.yoelii-infected mice, mononuclear cells also accumulate in the liver (Dockrell et al., 1980).

The observations which have just been described suggest that sensitized T-cells, reacting with parasite antigens, recruit and activate macrophages and NK cells. These effector cells, may in turn, liberate factors that can inhibit the replication of parasites in circulating erythrocytes, leading to degenerate "crisis forms". Such forms were originally described by Taliaffero and Taliaffero (1944) in Cebus capucinus monkeys recovering from P. brasilianum. In mice recovering from B. microti infections or protected by BCG, such degenerating parasites are observed within circulating erythrocytes (Clark et al., 1977). Similar degenerating intra-erythrocytic parasites are seen in mice recovering from P. chabaudi and P. vinckei infections (personal observations) and in rats recovering from P. berghei infections (Quin & Wyler, 1979).

Mice inoculated with live BCG, killed C. parvum and other stimulators of non-specific immunity are highly resistant to challenge with B. microti or the virulent B. rodhaini (no parsitaemia) and recover from infections with the normally lethal P. vinckei (Clark et al., 1976, 1977). Protection by C. parvum but not by BCG is thymus-independent, since it occurs in nu/nu mice. Evidently the effector cells can be activated through T-lymphocytes, as in normal infections, or through a mechanism bypassing T-lymphocytes, which can be activated by C. parvum. The presence of degenerate parasites within circulating erythrocytes shows that immunity is not effected by interfering with parasite entry or erythrophagocytosis, as was formerly thought. Clark et al. (1981) postulated that macrophages liberate tumor-necrosis factor (TNF) which interferes with parasite growth. Injections of serum containing TNF were found to protect mice against P. vinckei petteri. However, the sera contained mediators other than TNF, and highly purified TNF, or a specific antibody against it, is required to establish unambiguously the role of this mediator in recovery.

It has long been known that BCG increases NK activity (Herberman et al., 1977; Ojo et al., 1978) and a positive correlation has been found between levels of natural cytotoxicity and in vivo resistance to syngeneic tumour transplants as influenced by various routes of administration of C. parvum (Ojo, 1979).

Murine hemoprotozoa can be ranked according to the importance of antibodies in recovery from primary infections on one side and their susceptibility to non-specific immunity on the other side (Table 3).

TABLE 3. Susceptibility of asexual erythrocytic forms of murine hemoprotozoa to non-specific immunity elicited by C. parvum and other agents, and requirement of antibodies for recovery from primary infections

Parasite	Susceptibility to non-specific immunity	Requirement for Antibodies
Babesia microti	+++	?
Babesia rodhaini	+++	?
Plasmodium chabaudi	++	-
Plasmodium vinckei	++	?
Plasmodium yoelii	+	+
Plasmodium berghei	+	+

B. Human Malaria

Whether cellular immune mechanisms are important in human malaria is not at present known. Humoral immunity is thought to be involved (Cohen, 1979), but its role is difficult to establish unambiguously.

Increased NK activity of peripheral blood cells, assayed against K562 target cells, has been observed in children infected with P. falciparum (Ojo-Amaize et al., 1981). The sera of most infected children were found to have antiviral activity, predominantly α-interferon. Addition of exogenous interferon increased NK activity of peripheral blood cells in normal children while having no significant effect on cells from malaria-infected children.

C. Possible Role of Mediators Derived from Effector Cells in Pathogenesis of Malaria

Clark *et al.* (1981) pointed out that many aspects of the illness observed during acute malaria infections in humans and experimental animals are not easily attributable to the presence of protozoan parasites inside circulating erythrocytes. For instance there may be signs of liver damage, hypoglycemia, activation of the blood clotting system to produce a consumptive coagulopathy and retarded division of hemopoietic stem cells in the bone marrow (Clark *et al.*, 1981).

Clark (1978) suggested that something functionally similar to bacterial endotoxin, released from parasitized erythrocytes at schizogony, induces endotoxin-sensitive macrophages to secrete a range of mediators which cause much of the illness which is recognized as malaria. When mice which have been pretreated with BCG, *C. parvum* or zymosan are injected with endotoxin, several functionally important mediators are released, presumably from cells of its macrophage lineage. One of these is tumor-necrosis factor (TNF), already mentioned. Another is Type I (α)-interferon and a third is glucocorticoid-antagonizing factor (GAF), which blocks the induction of several enzymes required for gluconeogenesis, especially phosphenolpyruvate carboxykninase. Presensitized mice injected with endotoxin become hypoglycemic mainly because gluconeogenesis fails. Another mediator liberated under the same conditions is interleukin 1 (otherwise known as lymphocyte-activating factor or LAF) which may contribute to polyclonal lymphocyte activation in malaria. So far indistinguishable from interleukin 1 is endogenous pyrogen, which acts on the hypothalamus to induce fever, and a factor acting on liver cells causing them to produce acute-phase proteins such as serum amyloid-A inducer.

Clark *et al.* (1981) have shown that mice infected with *Plasmodium vinckei petteri* and injected with a small dose of endotoxin release into the serum TNF, interleukin 1 and type 1 interferon. Thus malaria infections can sensitize mice to endotoxin in the same way as BCG, *C. parvum* and zymosan. Clark *et al.* argue that humans with malaria infections might likewise become sensitized, and the presence in the blood of patients of small amounts of material giving endotoxin-like reactivity in the *Limulus* lysate assay has been reported by Tubbs (1980). Whether this material comes from the parasite or the gut of the patients is unknown. Clark *et al.* suggest that material released from parasitized cells at schizogony triggers the release in hosts sensitized by the infection of such mediators as TNF, interferon, interleukin 1 and GAF. These contribute to the pathophysiology of the disease. For example, TNF can inhibit the division of stem cells in the bone marrow.

A relationship between macrophage activation and hypercoagulability has been demonstrated. Human monocytes activated by endotoxin, immune complexes or products of activated lymphocytes release thromboplastin, which initiates blood coagulation (Prydz _et al_., 1979). This may well be happening in patients with malaria, contributing to a consumptive coagulopathy.

Because lipopolysaccharide endotoxins are specialized products of certain Gram-negative bacteria, it seems unlikely that chemically homologous molecules are present in protozoa. Nevertheless it is conceivable that other structures in the parasites, for example, surface-active complex lipids or phosphocarbohydrates may be able to trigger the release of mediators from sensitized mononuclear cells and even show activity in the _Limulus_ lysate test. Alternatively, small amounts of lipopolysaccharide released from intestinal bacteria could have dramatic effects in individuals whose mononuclear cells are presensitized by malaria or other infections. Another possibility is that parasite antigens released at the time of schizogony form immune complexes which activate mediator release by mononuclear cells.

D. Interferon in Malaria

The presence of interferon in the serum of mice infected with _P. berghei_ was demonstrated by Huang _et al_. (1968). Administration of exogenous interferon (Jahiel _et al_., 1970) or interferon inducers (Jahiel _et al_., 1968) has been found to delay the progress of _P. berghei_ infections in mice.

Furthermore, Sauvager and Fauconnier (1978) reported that administration of sheep anti-mouse interferon globulin to mice slightly accelerated _P. berghei_ infections and significantly increased the parasitaemia. These results suggest that α or/β-interferon has some retarding effect on the progress of _P. berghei_ infections in mice, although the infection is usually fatal even in the presence of interferon.

In contrast, Clark _et al_. (1981) were unable to demonstrate any effect of repeated injections of potent anti-interferon serum on the course of _P. vinckei petteri_ infections in mice. Likewise we could find no influence of repeated injections of potent sheep-anti-mouse interferon globulin on *P. chabaudi* infections in male or female BALB/c or B10.A mice (*unpublished* observations, 1978). These results suggest that α and β-interferons are not an important component of the cell-mediated resistance of mice to *P. vinckei* or *P. chabaudi*. Possibly other interferons (such as immune γ-interferon released from activated lymphocytes, see Chun *et al*., 1981) are a link between T-cells, activation of mononuclear cells, increased natural cytotoxicity and protection.

The presence of anti-viral activity, mostly α-interferon, in African children during acute *P. falciparum* infections (Ojo-Amaize *et al*., 1981) has already been mentioned.

E. Possible Relationship Between NK Cells and Cells of the Macrophage Lineage

NK cells are defined by their capacity to bind to, and kill efficiently, sensitive target such as the murine YAC-1 cells and human K562 cells. The lineage of NK cells is disputed (see Herberman, 1980). Lohmann-Matthes *et al*. (1979) reported that promonocytes growing out of murine bone marrow cultures had high NK activity. These cells were lysed with anti-macrophage serum and complement, were non-adherent and non-phagocytic and lacked non-specific esterase activity. Reinhertz *et al*. (1980) observed that human cells bearing receptors for the Fc region of IgG and that mediate NK activity react with a monoclonal antibody OKM1 that is thought to be specific for human monocytes. Holmberg *et al*. (1981) found that a rat anti-mouse macrophage monoclonal antibody (M1/70) recognizes both human monocytes and human NK cells in peripheral blood lysing K562 targets; it also reacted with NK cells in peritoneal exudates of mice that had been injected intraperitoneally with live *Listeria monocytogenes*. The M1/70 positive cells, separated by a fluorescence-activated cell sorter and binding to YAC-1 targets, had a distinctive morphology and contained azurophil granules. Presumably these cells are analogous to the "large granular lymphocytes" described by Herberman's group.

Although these studies do not exclude participation of cells bearing T-lymphocyte markers in NK activity, they suggest that the increased NK activity observed in acute infections may be predominantly due to cells of macrophage lineage. Infections may stimulate the release of macrophage precursors from the bone marrow and their accumulation in the spleen, peritoneal cavity and other sites. Possibly increased proliferation of precursors is also involved. The speculation can be offered that macrophage precursors can adhere to sensitive targets and mediate NK activity, and at an intermediate stage of differentiation can release tumour-necrosis factor and/or other mediators that can limit malaria parasite growth within circulating erythrocytes. On this hypothesis the increased NK activity and macrophage activation observed in malaria infections would be manifestations of the same basic mechanism. The low NK activity and macrophage-mediated cytotoxicity in mice of the A strain, and their high susceptibility to hemoprotozoan infections, would likewise represent associated defects.

ACKNOWLEDGMENTS

The anti-interferon serum used in our experiments was kindly provided by Dr. A. Inglot, Wroclaw, Poland.

REFERENCES

Adams, D.O., Marino, P.A., and Meltzer, M.C. (1981). J. Immunol. 126: 1843.

Allison, A.C., Christensen, J., Clark, I.A., Elford, B.C., and Eugui, E.M. (1979). In: "The Role of the Spleen in the Immunology of Parasitic Diseases" (G. Torrigiani, ed.), p. 151. Schwabe and Co., AG, Basel.

Brown, I.N., Allison, A.C., and Taylor, R.B. (1968). Nature, Lond. 219:292.

Clark, I.A. (1978). Lancet ii:75.

Clark, I.A., and Allison, A.C. (1974). Nature, Lond. 252:328.

Clark, I.A., Allison, A.C., and Cox, F.E.G. (1976). Nature, Lond. 259:309.

Clark, I.A., Richmond, J.E., Wills, E.J., and Allison, A.C. (1977). Parasitology 75: 189.

Clark, I.A., Wills, E.J., Richmond, J.E., and Allison, A.C. (1977). Infect. Immun. 17:430.

Clark, I.A., Virelizier, J-L., Carswell, E.A., and Wood, P.A. (1981). Infect. Immun. 32: 1058.

Cohen, S., McGregor, I.A., and Carrington, S. (1961). Nature, Lond. 192: 733.

Coleman, R.M., Bruce, A., and Rencricca, N.J. (1976). J. Parasitol. 62: 137.

Chun, M., Fernandes, G., and Hoffman, M.K. (1981). J. Immunol. 126:331.

Dockrell, H.M., de Souza, J.B., and Playfair, J.H.L. (1980). Immunology 41: 421.
Eugui, E.M., and Allison, A.C. (1980). Parasite Immunol. 2: 277.
Freeman, R.R., and Parish, C.R. (1978). Immunology 35: 479.
Freeman, R.R., Trejdosiewicz, A.J., and Cross, G.A.M. (1980). Nature, Lond. 284: 366.
Grun, J.I., and Weidanz, W.P. (1981). Nature, Lond. 290: 143.
Herberman, R.B. (1980). In "Natural Cell-Mediated Immunity Against Tumors." Section I, parts A, B and D. Academic Press, New York.
Herberman, R.B., Nunn, M.E., and Lavrin, D.H. (1975). Internat. J. Cancer 16: 216.
Herberman, R.B., Nunn, M.E., Holden, H.T., Staal, S., and Djeu, J.Y. (1977). Internat. J. Cancer 19: 555.
Holmberg, L.A., Springer, T.A., and Ault, K.A. (1981). J. Immunol. 127: 1792.
Huang, K-Y., Schultz, W.W., and Gordon, F.B. (1968). Science 162: 123
Hunter, Jr. K.W., Folks, T.M., Sayles, P.C. and Strickland, G.T. (1981). Immunol. Letters 2:209.
Jahiel, R.I., Vilcek, J., Nussenzweig, R.S., and Vandenberg, J. (1968). Science 161: 802.
Jahiel, R.I., Vilcek, J., and Nussenzweig, R.S. (1970). Nature, Lond. 227: 1350.
Kiessling, R., Petranyi, G., Klein, E., and Wigzell, H. (1975). Internat. J. Cancer 15: 933.
Lohmann-Matthes, M-L., Domzig, W., and Roder, J. (1979). J. Immunol. 123: 1883.
Murphy, J.R. (1981). Infect. Immun. 33: 199.
Ojo, E. (1979). Cell. Immunol. 45: 182.
Ojo, E., Haller, O., Kimura, A., and Wigzell, H. (1978). Internat. J. Cancer 21: 444.
Ojo-Amaize, E.A., Salimonu, L.S., Williams, A.I.O., Akinwolere, O.A.O., Shabo, R., Alm, G.V., and Wigzell, H. (1981). J. Immunol. 127: 2296.
Prydz, H., Lyberg, T., Deteix, P., and Allison, A.C. (1979). Thrombosis Res. 15: 465.
Quin, T.C., and Wyler, P.J. (1979). J. Clin. Invest. 63: 1187.
Rank, R.G., and Wiedanz, W.P. (1976). Proc. Soc. Exp. Biol. Med. 151: 257.
Reinherz, E.L., Moretta, L., Roper, M., Breard, J.M., Mingari, M.C., Cooper, M.D., and Schlossman, S.F. (1980). J. Exp. Med. 151: 969.

Roberts, D.W., Rank, R.G., Weidanz, W.P., and Finerty, J.F. (1977). Infect. Immun. 16: 821.
Roberts, D.W., and Weidanz, W.P. (1979). J. Trop. Med. Hyg. 28: 1.
Sauvager, F., and Fauconnier, B. (1978). Biomedicine 29: 184.
Taliaferro, W.H., and Taliaferro, L.G., (1944). J. Infect. Dis. 75: 1.
Tubbs, H. (1980). Trans. Roy. Soc. Trop. Med. Hyg. 75: 409.
Weinbaum, F.I., Evans, C.B., and Tigelaar, R.E. (1976). J. Immunol. 117: 1999.
Wyler, D.J., and Gallin, J.J. (1977). J. Immunol. 118: 478.
Yoshida, N., Nussenzweig, R.S., Potocnjak, R., Nussenzweig, V., and Aikawa, M. (1980). Science 207: 71.

NATURAL CELL-MEDIATED RESISTANCE IN CRYPTOCOCCOSIS[1]

Juneann Murphy
Department of Botany-Microbiology
University of Oklahoma
Norman, Oklahoma

I. INTRODUCTION

Until recently, natural cellular resistance against bacterial, mycotic, and parasitic infections had been attributed to the polymorphonuclear leucocyte (PMN) and the monocyte/macrophage system of the host. The natural killer (NK) cell system is now emerging as a potential third means of innate cellular resistance against these groups of organisms. It has been fairly well established that NK cells play a role in first line defense against neoplastic cells (Herberman and Holden, 1978; Herberman and Holden, 1979; Herberman et al., 1979; Riccardi et al., 1979; Riccardi et al., 1980a; Riccardi et al., 1980b; Riccardi et al., 1981; Talmage et al., 1980) and against certain virus-infected cells (Bloom et al., 1980; Santoli et al., 1980; Welsh et al., 1979); however, relatively little work has been done looking specifically at the effects of NK cells on other microbial infectious agents. Bennett and Baker (1977) have presented data which implicated NK cells in resistance to *Listeria monocytogenes* during the first 48 hr of infection. Ruebush et al. (in press) have reported that there was a correlation between NK reactivity and resistance to an intracellular protozoan parasite, *Babesia microti*. Data generated by us and others suggest that NK cells may play a role in reducing the numbers of fungal organisms either *in vitro* (Murphy and McDaniel, in press; Tewari et al. 1981) or during the early period after exposure to the etiological agent (Cauley and Murphy, 1979; Cozad and Lindsey, 1974; Cutler, 1976; Graybill and Mitchell, 1978). Recently we have reported on experiments which were designed specifically to assess the potential of NK cells against a yeast-like organism *Cryptococcus neoformans* in an *in vitro* system. We were able to demonstrate that the NK reactivity levels of pools of murine splenic nylon wool nonadherent cells paralleled the ability of the cell pools to inhibit the growth of

1. This study was supported by an NIH Biomedical Research Support Grant No. 2 507 RR07078, a Brown-Hazen Research Corporation Grant and Public Health Service Grants AI-15716 from the National Institutes of Allergy and Infectious Diseases.

ISBN 0-12-341360-5

cryptococci and that the cells responsible for the cryptococci growth inhibition had characteristics of NK cells (Murphy and McDaniel, in press). Other data which support the contention that NK cells can inhibit C. neoformans are presented in another section in this volume. Studies discussed in this section are concerned with the in vivo effectiveness of NK cells against C. neoformans and with the ability of cryptococci to augment NK reactvity in vivo. The approach taken has been to correlate NK reactivity of normal unstimulated mice with their ability to clear cryptococci from various tissues within the first 7 days after infection.

II. NATURAL RESISTANCE OF NUDE MICE AGAINST C. NEOFORMANS

BALB/c nude (nu/nu) and heterozygote (nu/+) mice were injected ip with 10^4 viable C. neoformans cells seven days prior to collecting spleens, livers, lungs, and brains for determinations of the numbers of cyptococci colony forming units (CFU) per organ. The results are presented in Table 1.

TABLE 1. Clearance of C. neoformans from tissues of BALB/c nude (nu/nu) and heterozygotes (nu/+) mice 7 days after an ip infection with 10^4 cells.

Mice	Mean Number ($\log_{10}$) $\pm$ SEM C. neoformans Colony Forming Units Per			
	Spleen	Liver	Lungs	Brain
nu/nu	1.88±1.09	2.3±1.3	1.9±0.8	0.57±0.3
nu/+	3.87±0.25	4.5±0.2	3.18±0.24	1.04±0.3

Nude mice had fewer cryptococci in their tissues than did the heterozygotes. Since this difference appeared early after infection and since the lower CFU counts were in animals which could not mount a protective acquired immune response to C. neoformans (Cauley and Murphy, 1979), innate or natural mechanisms had to be functioning. There are two possible explanations for the reduced numbers of cryptococci in the nude mice. One, nude mice reportedly have an active macrophage system (Cheers and Waller, 1975), and two, they have high levels of NK reactivity (Herberman, 1978). Several groups have studied the ability of the macrophage to phagocytize and kill C. neoformans, and there is disagreement as to the effectiveness of the macrophage against this organism (Bulmer, 1975; Diamond and Bennett, 1973; Diamond et al., 1972; Gentry and Remington, 1971; Griffin, 1981; Karaoui et al., 1977; Kozel and McGaw, 1979; Mitchell and Friedman, 1972; Monga, 1981; Sethi and Pelster, 1974). Our recent studies demonstrating that cells with characteristics of NK cells can inhibit the growth of C. neoformans in vitro plus the lack of strong evidence that the macrophage is effective, have influenced us to favor the possibility that NK cells were responsible, at least in part, for the reduction in numbers of cryptococci in the nude mice. However further supportive data was necessary before this hypothesis could be accepted. And to gain this, a better animal model was required, preferably a model in which the macrophage variable could be held constant while varying the NK reactivity, thus eliminating the possibility of the macrophage as the effector cell responsible for the clearance of cryptococci.

III. ELIMINATION OF C. NEOFORMANS FROM TISSUES OF NK-DEFECTIVE (BEIGE) MICE, HETEROZYGOUS LITTERMATES AND C57BL/6J MICE

The beige (bg/bg) mutant of the C57BL/6 (B6) mouse which has impaired NK cell function (Roder, 1979; Roder and Duwe, 1979) but apparently normal promonocyte, macrophage and T lymphocyte functions (Roder et al. 1979) and heterozygote (bg/+) littermates with normal NK cell reactivity were chosen for in vivo studies on the effectiveness of NK cells in natural resistance against cryptococci. Normal B6 animals were also included for comparison. NK cell reactivities of seven week old animals of each of the three stains were determined using a 4 hr ^{51}Cr release assay with YAC-1 targets and splenic nylon wool nonadherent cells as effectors. The B6 and bg/+ cells had similar cytolytic activities ranging from 12-25%, and the bg/bg cells were very low in activity, ranging from 1-3%. The NK activity in the B6 and bg/+ animals was always significantly higher than the activity in the beige mutants. To study in vivo clearance of C. neoformans, seven week old beige, bg/+ littermates, and B6 mice were injected iv with 2 X 10^4 viable C. neoformans cells, then on days 1, 3, and 7 after giving the infecting dose 3-5 mice from each group were sacrificed to determine the numbers of cryptococci CFU per lungs spleen, and liver. At each time period after infection, the beige mice had higher numbers of cryptococci in their lungs than did the heterozygotes or the B6 animals (Table 2).

TABLE 2. Clearance of C. neoformans from lungs of beige (bg/bg), heterozygote (bg/+) and C57BL6/J(B6) mice.

Day After iv Infection With 2×10^4 C. neoformans	Mean No. C. neoformans CFU Per Lungs ± SEM		
	bg/bg	bg/+	B6
1	195±5	65±17	42±8
3	250±56	133±109	103±52
7	1080±553	125±17	112±42

The number of cryptococci in spleens and livers in general were comparable in all three groups; however, it was noted that at 3 days after infection, spleens of bg/bg mice had significantly higher numbers of cryptococci CFU than did B6 spleens (data not shown). Since large numbers of beige mice were not available, replicate experiments were done at only one autopsy period. Three days post infection was chosen for convenience, and the results of those experiments are shown in Table 3. In all cases the beige mice had greater numbers of cryptococci in their lungs than the heterozygotes or B6 mice. The enhanced abilities of the bg/+ and B6 mice over the bg/bg mice to clear C. neoformans from the lungs could be attributed to the NK cell reactivity of the bg/+ and B6 animals and lack of it in bg/bg mice, since all other variables should

TABLE 3. Clearance of C. neoformans from tissues of beige (bg/bg), heterozygotes (bg/+), and C57BL6/J (B6) mice 3 days after an iv injection of 2 X 10^4 C. neoformans cells.

Group	Mouse	Mean No. C. Neoformans CFU ± SEM		
No.	Stain	Speen	Liver	Lungs
1	bg/bg			
	Exp. 1	1167±117[a]	8917±944	250±56[b]
	Exp. 2	2116±568[b]	10750±501[b]	373±50[c]
	Exp. 3	ND[d]	ND	143±33[e]
2	bg/+			
	Exp. 1	566±145	8900±453	133±109
	Exp. 2	1250±65	12783±2207	120±26
	Exp. 3	ND	ND	36±13
3	B6			
	Exp. 1	966±148	9250±966	103±52
	Exp. 2	966±71	9800±931	60±18
	Exp. 3	ND	ND	43±16

a. P<0.005 when compared to group 2; NS when compared to group 3.
b. NS when compared to group 2; P<0.05 when compared to group 3.
c. P<0.001 when compared to group 2; P<0.0005 when compared to group 3.
d. ND=not done
e. P<0.025 when compared to group 2 or 3.

have been similar in the three groups. It is probably fair to assume that the NK cell levels in the lungs of the mice used in these experiments were comparable to the levels of NK reactivity of the spleens, since Puccetti et al. (1980) demonstrated that splenic and lung cells from mice of the same strain had similar levels of NK reactivity. As in the previous experiments, occasionally we found that bg/bg animals had greater numbers of cryptococci in spleens and livers than heterzygotes or B6, but this was not a consistent observation. The variation may have been due to inherent variation in using small numbers of mice, or it may have been due to overwhelming the cellular resistance of those organs by excessive numbers of the organism, such that the effect of the NK cells was masked. In general, spleens had up to 15 times and livers approximately 100 times the numbers of cryptococci cultured from the lungs. Riccardi et al. (1980b), studying in vivo clearance of tumor cells in mice by NK cells, also noted that usually the NK cell levels in mice correlated best with tumor cell clearance from the lungs, so this is not a unique phenomenon with the cryptococcus system. In spite of the fact that clearance of cryptococci from the spleens and livers were similar in all three groups of mice, the data showing clearance of the organism from the lungs support the hypothesis that NK cells are effective against cryptococci in vivo.

IV. AUGMENTING EFFECTS OF C. NEOFORMANS ON NK REACTIVITY

We have attempted to use a third animal model for in vivo studies and have failed to obtain cryptococci clearance data which paralleled the splenic NK cell levels of the mice. The approach was similar to the previous studies. Briefly, young (7 week old) and old (13-18 week old) (CBA/N X A.TH)F_1 mice were given 2 X 10^4 viable C. neoformans cells iv, then the numbers of cryptococci CFU per organ were determined on day 3 after injection. The levels of splenic NK reactivity of the mice were determined on age-matched uninfected mice. We expected the young mice that had high NK reactivity in their spleens to clear the cryptococci, at least from the lungs, more efficiently than the old animals with low NK activity; however, in three experiments there were no differences in the numbers of cryptococci CFU in the two groups of mice. Herberman (personal communication) has recently observed that a decline in peripheral blood or lung NK activity or a decrease in clearance of tumor cells from lungs does not occur in mice until approximately 25 weeks of age. This could be one explanation for our inability to demonstrate differential clearance of C. neoformans from lungs of young (7 week old) vs old (13-18 week old) mice. Another possibility is that the cryptococci could have augmented

NK reactivity in the old mice, thus enhancing their abilities to clear the cryptococci. Experiments were performed to see if this was a viable suggestion. Old (16-20 week old) B6, CBA/N, and (CBA/N X A.TH)F_1 mice were injected iv with 10^4 or 10^6 heat killed C. neoformans cells, then on days 1 and 3 after injection of cryptococci, a 4 hr ^{51}Cr release assay with YAC-1 targets was performed with splenic nylon wool nonadherent effector cells from each group of injected mice and strain- and age-matched uninjected controls. In mice given 10^6 heat killed cryptococci, significant increases in NK cell reactivity were observed by day 3 in all the mouse strains tested. Animals receiving 10^4 heat killed cryptococci had slightly higher % ^{51}Cr release data than the controls; however, the % release values of the injected mice were not significantly different from the values for the controls. From the limited amount of data available, it appears that NK cell reactivity can be enhanced in mice by C. neoformans and that augmentation of NK activity is dose related. This could also explain why it was impossible to demonstrate differences in clearance of C. neoformans from lungs of 7 week old vs 13-18 week old mice. To demonstrate differential clearance of an infectious agent an animal model must be used in which NK cells are defective and/or their activity can not be significantly augmented by the organism being studied. At this time, the beige mutant model is the best for this type clearance experiments.

V. CONCLUSIONS

The data presented here in conjunction with the data from in vitro experiments (Murphy and McDaniel, in press and included in this volume) strongly suggest a role for NK cells, or a subset of lymphocytes with similar characteristics and coexisting with NK cells, in first line host resistance against C. neoformans. This is not saying that polymorphonuclear leucocytes and macrophages are not also functioning in primary resistance against this pathogen; they most likely are to some degree (Gadebusch, 1972; Monga, 1981; Sethi and Pelster, 1974). Considerably more work is required to determine the relative importance of the roles of these various natural cellular resistance mechanisms in cryptococcosis. It is certainly possible the NK cells may be effective against other nontissue targets, so hopefully the results presented here will encourage others studying host defense mechanisms against bacterial, mycotic or parasitic agents to consider NK cells as potential effector cells in their systems.

ACKNOWLEDGEMENTS

I thank Dr. Olga McDaniel and Ms. Mitra Rahimi for technical assistance and Ms. Beverly Richey for typing of the manuscript.

REFERENCES

Bennett, M. and Baker, E. E. (1977). Cell. Immunol. 33:203
Bloom, B., Minato, N., Neighbour, A., Reid, L., and Marcus, D. (1980). In "Natural Cell-Mediated Immunity Against Tumors" (R. B. Herberman, ed.), p. 505. Academic Press, New York.
Bulmer, G. S., and Tacker, J. R. (1975). Infect. Immun. 11:73.
Cauley, L. K. and Murphy, J. W. (1979). Infect. Immun. 23:644.
Cheers, C. and Waller, R. (1975). J. Immunol. 115:844.
Cozad, G. C. and Lindsey, T. J. (1974). Infect. Immun. 9:261.
Cutler, J. E. (1976). J. Reticuloendothel. Soc. 19:121.
Diamond, R. D., Root, R. K., and Bennett, J. E. (1972). J. Infect. Dis. 125:267.
Diamond, R. D., and Bennett, J. E. (1973). Infect. Immun. 7:231.
Gadebusch, H. H. (1972). Crit. Rev. Microbiol. 1:311.
Gentry, L. L. and Remington, J. S. (1971). J. Infect. Dis. 123:22.
Graybill, J. R. and Mitchell, L. (1978). Infect. Immun. 21:674.
Griffin, F. M. (1981). Proc. Natl. Acad. Sci. 75:3853.
Herberman, R. B. (1978). In "The Nude Mouse in Experimental and Clinical Research" (J. Fogh and G. C. Giovanella, ed.), p. 135. Academic Press, New York.
Herberman, R.B. and Holden, H.T. (1978). In "Advances in Cancer Research" (G. Klein and S. Weinhouse, ed.) p. 305. Academic Press, New York.
Herberman, R. B. and Holden, H. T. (1979). J. Natl. Cancer Inst. 62:441.
Herberman, R. B., Djeu, J. Y., Kay, H. D., Ortaldo, J. R., Riccardi, C., Bonnard, G. D., Holden, H. T., Fagnani, R., Santoni, A., and Puccetti, P. (1979). Immunol. Rev. 44:43.
Karaoui, R. M., Hall, N. K., and Larsh, H. W. (1977). Mykosen 20:409.
Kozel, T. R. and McGaw, T. G. (1979). Infect. Immun. 25:255.
Mitchell, T. G. and Friedman, L. (1972). Infect. Immun. 5:491.
Monga, D. P. (1981). Infect. Immun. 32:975.
Murphy, J. W. and McDaniel, D. O. J. Immunol. in press.
Puccetti, P., Santoni, A., Riccardi, C., and Herberman, R. B. (1980). Int. J. Cancer 25:153.
Riccardi, C., Puccetti, P., Santoni, A., and Herberman, R. B. (1979). J. Natl. Cancer Inst. 63:1041.

Riccardi, C., Santoni, A., Barlozzari, T., Puccetti, A., and Herberman, R. B. (1980a). Int. J. Cancer 25:475.

Riccardi, C., Santoni, A., Barlozzari, T., and Herberman, R. B. (1980b). In "Natural Cell-Mediated Immunity Against Tumors" (R. B. Herberman, ed.), p. 1121. Academic Press, New York.

Riccardi, C., Barlozzari, T., Santoni, A., Herberman, R. B., and Cesarini, C. (1981). J. Immunol. 126:1284.

Roder, J. C. and Duwe, A. (1979). Nature 278:451.

Roder, J. C. (1979). J. Immunol. 123:2168.

Roder, J. C., Lohmann-Matthes, M., Domzig, W., and Wigzell, H. (1979). J. Immunol. 123:2174.

Ruebaush, M. J., Hale, A. H., and McKinnon, K. P. Proc. Am. Soc. Trop. Med. Hyg., in press.

Santoli, D., Perussia, B., and Trinchieri, G. (1980). In "Natural Cell-Mediated Immunity Against Tumors" (R. B. Herberman, ed.), p. 1171. Academic Press, New York.

Sethi, K. K. and Pelster, B. (1974). In "Activation of Macrophages" (W. H. Wagner and H. Holm, ed.), p. 269. Excerpta Medica, Amsterdam.

Talmage, J. E., Meyers, K. M., Prieur, D. J., and Starkey, J. L. (1980). Nature 284:622.

Tewari, R. P., Mkwananzi, J. B., McConnachie, P., Von Behren, L. A., Eagleton, L., Kulkarni, P., and Bartlett, P. C. (1981). ASM Abstracts. p 324.

Welsh, R. M., Zinkernagel, R. M., and Hallenbeck, L. A. (1979). J. Immunol. 122:475.

NATURAL CELL-MEDIATED IMMUNITY AGAINST BACTERIA

Emil Skamene
Mary M. Stevenson
Patricia A.L. Kongshavn

Montreal General Hospital Research Institute
Montreal, Quebec, Canada

The cell-mediated natural resistance to bacterial infections can be best appreciated in those instances where the role of humoral factors of host defense can be excluded with reasonable certainty. An example of such an infection is murine listeriosis. Listeria monocytogenes is a facultative intracellular bacterial parasite that can survive and multiply in mammalian macrophages. Normal or convalescent serum offers no protection against the infection and, in fact, specific antibodies to Listeria cannot be demonstrated in the course of listeriosis. The infected host has at its disposal several lines of defense, each of them being mobilized in successive steps in the natural history of the disease and recovery. It has recently been demonstrated (1,2) that the key mechanisms of the early, non-immunologically induced, resistance to this infection are genetically regulated by a single, dominant, autosomal gene designated **Lr** (3). Mice of various inbred strains possess either the resistant **(Lr^r)** or the susceptible **(Lr^s)** allele of this gene. This chapter provides the evidence that **Lr** gene-controlled natural resistance to Listeria rests with the ability of the infected host's hematopoietic system to promptly mount an efficient macrophage inflammatory response.

COURSE OF INFECTION IN RESISTANT AND SUSCEPTIBLE MICE

When mice of different inbred strains are infected intravenously with Listeria and the course of infection is followed by examining the bacterial burden in liver and spleen, a characteristic pattern can be observed (Fig. 1).

NK CELLS AND OTHER NATURAL EFFECTOR CELLS

ISBN 0-12-341360-5

The liver captures 90% of the infecting inoculum and destroys 50-75% of this load within 6-12 hours. This initial mechanism of defense operates in both Listeria-resistant and Listeria-susceptible animals and it appears that this mechanism represents an innate bactericidal activity of resident Kupffer cells. Surviving organisms start to multiply and the bacterial load increases in the liver and spleen for approximately 2-3 days. The genetically-determined natural resistance to Listeria becomes apparent at this phase of the infection: mice of the **Lr^s** strains, (A,BALB/c,DBA/2,CBA) which are deficient in this facet of host response, fail to prevent bacterial numbers from reaching a lethal level. On the other hand, mice of the **Lr^r** strains (C57BL/6 and substrains, NZB,SJL,C3H/He and substrains) are able to hold bacterial multiplication in check until such time as specific immunity develops, subsequently leading to a rapid sterilization of the organs (from the 3rd-4th day of infection onwards). It can be demonstrated that mice of susceptible strains possess the potential to develop an effective acquired immunity but they succumb to infection before this potential is realized.

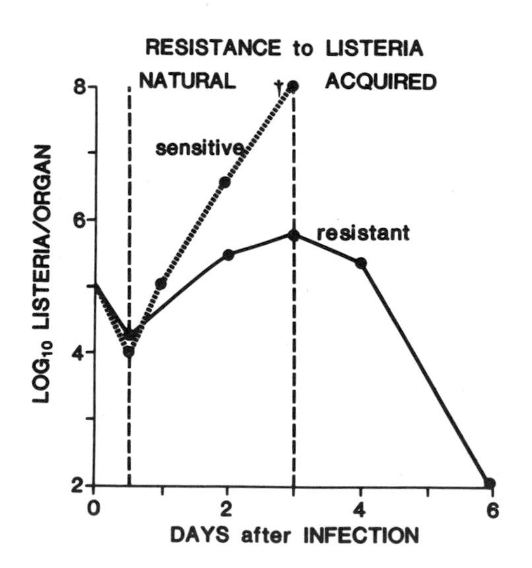

FIGURE 1. Course of listeriosis in **Lr^s** (sensitive) and **Lr^r** (resistant) mice infected with 10^5 organisms intravenously on day 0. †denotes death of all animals.

Early control of bacterial multiplication observed in the organs of **Lr^{r}** mice is not dependent on immune mechanisms since this phase is preserved in either T cell or B cell - deprived animals (4,5). The fact that the strain distribution pattern of **Lr^{r}** and **Lr^{s}** alleles does not correlate with high or low NK-cell activity and the demonstration that beige +/+ mutant mice on a C57BL/6 background are as resistant as their normal +/- littermates argue against the role of NK cells in genetic resistance to Listeria. Whole-body 200-900 R irradiation totally eliminates the genetic advantage of **Lr^{r}** strains, thus indicating that the cell population responsible for natural resistance to Listeria is derived from radiosensitive hemopoietic precursors (6). The bone marrow origin of these cells was confirmed by the ability of isologous bone marrow inoculum to restore the radiation-induced impairment of resistance to Listeria. These findings collectively have led to the indirect conclusion that a radiosensitive cell of macrophage-lineage is responsible for natural resistance to Listeria and that a certain aspect of macrophage response is regulated by the **Lr** gene.

MACROPHAGE RESPONSES OF LISTERIA-RESISTANT AND -SUSCEPTIBLE MOUSE STRAINS

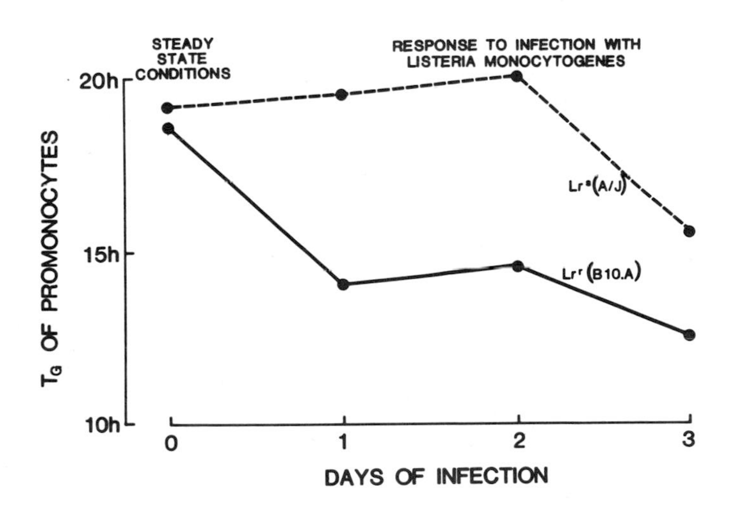

FIGURE 2. Promonocyte generation time of **Lr^{r}** and **Lr^{s}** mice at steady-state conditions and after infection with 10^{4} microorganisms on day 0.

The pattern of anti-listerial resistance of radiation bone-marrow chimeras suggests that the **Lr** gene product influences the microenvironment in which macrophage differentiation from the bone marrow precursors takes place (7). Cytokinetic studies on monocyte production suggest that the **Lr** gene product has the functional characteristics of an inducible monocytopoietin (Fig. 2). It appears to act by promptly decreasing promonocyte generation time (T_G), within hours following listerial infection of the animals of **Lr^r** strains. This response is delayed by at least 48 hours in the **Lr^s** strain animals (8).

The enhanced monocyte production that occurs in infected animals of **Lr^r** strains results in the prompt development of monocytosis and presumably also in the appearance of young, newly arrived macrophages in the foci of infection. Similar changes in the kinetics of monocytopoiesis can be mimicked by stimulation of **Lr^r** animals with an extract of listerial cell walls and the enhanced wave of macrophage response can, in fact, also be demonstrated by stimulation of the peritoneal cavity of such animals with a variety of sterile irritants (9).

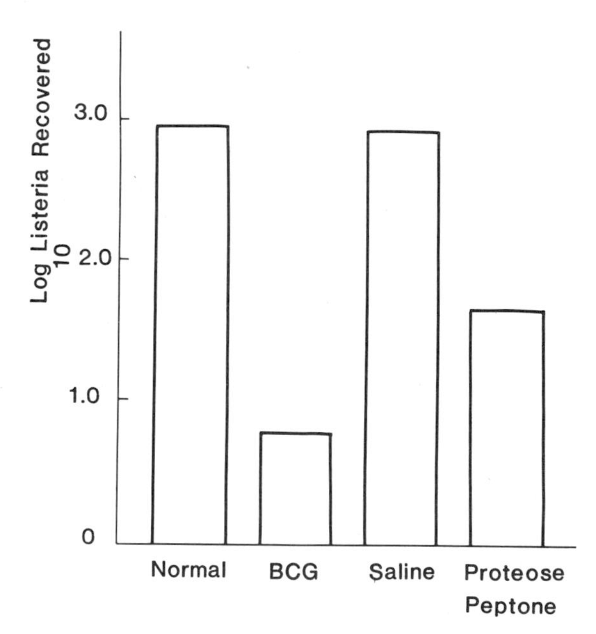

FIGURE 3. Recovery of live Listeria from the peritoneal cavities of **Lr^r** strain normal mice and of mice stimulated by BCG or a sterile irritant (proteose peptone), 4 hours after local infection with 10^5 microorganisms.

Such irritant-induced inflammatory macrophages have enhanced chemotactic ability and are almost as effective in inactivating the locally-injected Listeria as those which are "activated" by immunologically-specific mechanisms (Fig. 3).

Studies on the phenotypic expression of the **Lr** gene that controls natural resistance to Listeria and which were described up to this point were based on the correlation of the resistance with several distinct steps in macrophage responses among mice of selected inbred strains. In other words, animals of $\mathbf{Lr^r}$ strains (C57BL/6 or B10.A in our studies), which are genetically resistant to infection with Listeria, exhibit a prompt macrophage inflammatory response characterized by an enhanced tempo of monocytopoiesis (when induced by infection or by an inflammatory stimulus) and by the appearance of large numbers of young, chemotactically and bactericidally active macrophages at sites of infection or irritation. In order to prove that these correlations are, in fact, the result of a cause-and-effect relationship, a formal proof using genetic analysis as a probe was sought.

LINKAGE ANALYSIS OF NATURAL RESISTANCE TO LISTERIA WITH MACROPHAGE INFLAMMATORY RESPONSE

We reasoned that, if the enhancement of the macrophage inflammatory response were the cause of the high level of natural resistance to Listeria, then these two traits should segregate together in backcross and recombinant mouse populations derived from genetically-resistant and -susceptible progenitors. Two experimental approaches confirmed this assumption. In the one, individual animals of the backcross population derived from $\mathbf{Lr^r}$ (B10.A) and $\mathbf{Lr^s}$ (A) progenitors were first examined for their magnitude of macrophage inflammatory response to local peritoneal irritation with thioglycollate medium. Identical animals were later infected with a typing dose of 10^4 Listeria organisms and identified as Listeria-resistant **($\mathbf{Lr^r}$)** or Listeria-susceptible **($\mathbf{Lr^s}$)**. Among a total of 31 /F_1 (B10.A x A) x A/ backcross animals examined, a statistically highly significant correlation was found between resistance to Listeria and the magnitude of the macrophage inflammatory response, compatible with genetic linkage of these two traits. Such linkage was further confirmed in BXA recombinant inbred (RI) strains derived from C57BL/6 **($\mathbf{Lr^r}$)** and A/J **($\mathbf{Lr^s}$)** progenitors. In this experiment, the two traits (resistance to Listeria and macrophage inflammatory response) were examined independently in groups of 4-6 animals from each of seven BXA recombinant inbred strains

(Fig. 4). We found that in 6 out of 7 BXA RI strains so far examined there was concordance between the magnitude of the macrophage inflammatory response to a sterile irritant injected into the peritoneal cavity and the level of resistance to Listeria. These two approaches thus provide the formal proof that natural resistance to this particular infectious agent (Listeria monocytogenes) rests with the genetically-determined ability of the host's bone marrow to respond promptly to an infectious or sterile inflammatory stimulus.

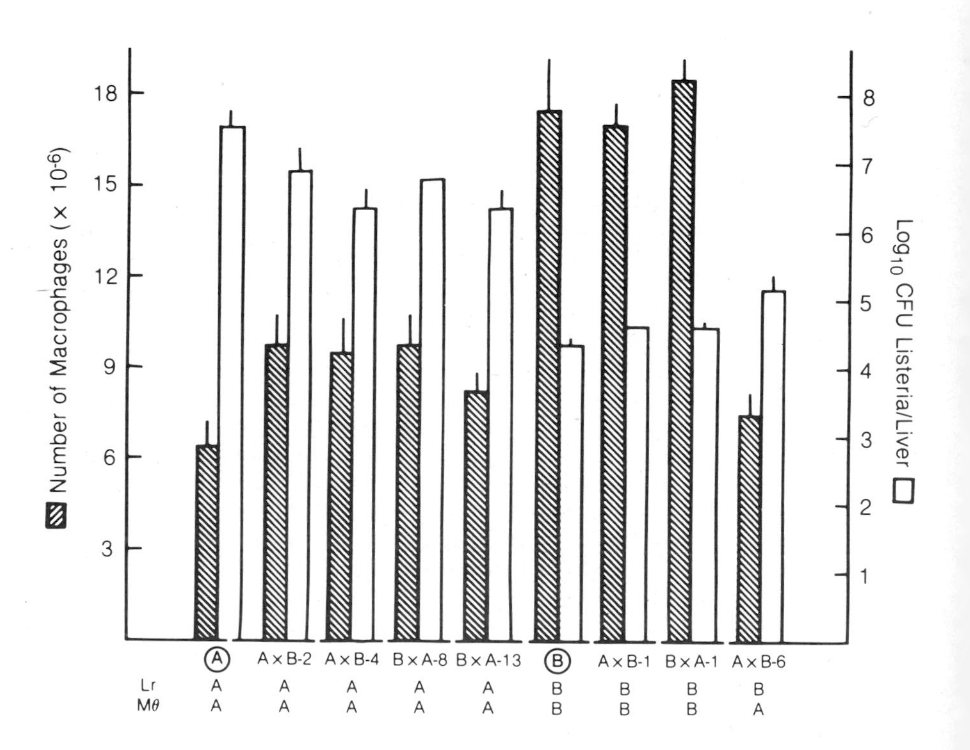

FIGURE 4. Strain distribution pattern of natural resistance to Listeria monocytogenes and macrophage inflammatory response. Natural resistance to Listeria was measured at 48 hours after intravenous infection with 10^4 bacterial, by counting the number of Listeria CFU in liver homogenates. Macrophage inflammatory response (MØ) was measured by the macrophage influx into the peritoneal cavity 3 days after injection of 1 cc thioglycollate medium. The designation of Lr and MØ alleles (A or B) for each strain is based on 95% confidence limit of both typing methods.

GENETIC AND CELLULAR MECHANISMS OF NATURAL RESISTANCE TO INTRACELLULAR BACTERIA[1]

Christina Cheers
Jenny Macgeorge[2]

Department of Microbiology
University of Melbourne
Parkville, Victoria
Australia

INTRODUCTION

With the demonstration that the major histocompatibility complex (MHC) plays a fundamental role in induction of specific immunity, including the immune response to infection has come a burgeoning interested in the role of HLA in human susceptibility to disease. Nevertheless at the experimental level it has been difficult to demonstrate a role for MHC linked genes in disease susceptibility, possible exceptions being susceptibility to LCM virus in mice (1) and induction of splenomegaly by killed Brucella abortus in mice (2). On the other hand there are increasing numbers of reports on non-MHC linked genes governing resistance in experimental animals (3). Many of these determine the growth of organisms during the early phase of infection, when natural resistance mechanisms apply. It is important that such models should be fully investigated, not only for their genetic interest, but also for what they can reveal about the mechanisms of resistance to infection.

We have established a model to investigate the genetics of natural resistance to infection with an intracellular bacterium, Listeria monocytogenes (4). Thirty-four strains

[1]Supported by National Health and Medical Research Council of Australia and the Melbourne University Medical Research Committee Fund.
[2]Present address: Walter and Eliza Hall Institute, Parkville, Victoria, Australia.

ISBN 0-12-341360-5

of mice studied fell clearly into two groups differing in their LD_{50} to L. monocytogenes by 100 fold. The single gene, Lr, controlling the difference between C57Bl and BALB/c mice is a dominant, non-H-2 linked, autosomal gene. It governs the ability to limit bacterial growth during the first 24 to 48 hours, especially in the liver, and also the time of onset of specific cell mediated immunity which may be 1 to 2 days earlier in the resistant mice (5). Once specific immunity is established, there is no detectable deficiency in the susceptible BALB/c mice.

Within hours of infection, Listeria organisms can be seen in the Kupffer cells of the liver, where they initially congregate in similar numbers in both strains. By 24 hours many more foci of infection are visible in the susceptible BALB/c than in resistant C57Bl/10 (6). Each focus comprises a central infected Kupffer cell surrounded by polymorphs.

Early resistance or susceptibility in C57Bl and BALB/c mice does not involve T lymphocytes. Chimeras prepared by reconstituting neonatally thymectomized mice with thymocytes from H-2 congenic resistant or susceptible donors showed that the course of infection was determined by the host, not the T cell donor (7). On the other hand, increased numbers of monocytes appear earlier in the blood of resistant than susceptible mice (6). Others have demonstrated a genetic linkage between this rapid response of monocytes and resistance to Listeria infection (8).

RESULTS AND DISCUSSION

A. Effect of Silica or Carageenan on Listeria Infection

C57Bl/10 or BALB/c mice were injected with 2.5 mg silica intravenously or carageenan (200 mg/kg bodyweight intraperitoneally) 24 hours before infection with a sublethal dose (5×10^2) of L. monocytogenes. Bacterial counts in both strains were markedly exacerbated, presumably because the resident macrophages had been destroyed before infection. Carageenan treated mice were dead by day 3. The effect of silica was less marked in the resistant C57Bl/10 mice (2-10 fold exacerbation on day 2) than in BALB/c mice (40-50 fold exacerbation), consistent with the suggestion that the resistant mice can rapidly recruit other cells to the site of infection. Nevertheless these results suggest that resident macrophages play a fundamental role in both strains.

B. Effect of Irradiation or Vinblastin on Listeria Infection

In order to prevent proliferation of monocyte and polymorph precursors in the bone marrow, mice were given 600 Rads irradiation 48 or 2 hours before infection with 4×10^4 Listeria organisms. Bacteria counts were made 24 hours later (Table 1). When irradiation was given 48 hours before Listeria, infection was exacerbated in the livers of both mouse strains. This radiosensitive function is similar to to that described in nude mice (9), which allows them some degree of control of infection with Listeria in the absence of T lymphocytes. When irradiation was given 2 hours before infection, bacterial numbers were increased in C57Bl/10 mice (Table 1) but were not affected in BALB/c mice until later in infection, after the onset of acquired immunity. It was shown that irradiation did not affect initial localization of the bacteria in liver and spleen, but increased the subsequent growth of the organisms, mostly in the liver. Doses as low as 100 Rads given 2 hours before infection exacerbated infection in the C57Bl/10 mice. This observation agrees with that of Sadarangani et al (10).

C57Bl/10 and BALB/c mice were injected intravenously with 100 μg vinblastin 1 hour before infection with 5×10^2 Listeria or 24 hours later and their bacterial numbers assayed over the following 4 days. Table 2 shows bacterial counts in the livers of treated and untreated mice. In contrast with the effect of irradiation, there was no effect of vinblastin treatment in either strain for the first two days after infection. This suggests that the effect of irradiation on

TABLE 1. Effect of whole body irradiation on numbers of Listeria monocytogenes in C57Bl/10 and BALB/c mice

Time of Irrad.	Log_{10} Listeria/g liver[a]			
	C57Bl/10		BALB/c	
	Unirradiated	600 Rad	Unirradiated	600 Rad
-2 hours	5.06±0.21	6.22±0.27	6.26±0.06	6.51±0.33
-48 hours	5.49±0.12	6.32±0.29	6.23±0.42	7.00±0.13

[a] Mice irradiated 2 or 48 hours previously with 600 Rads were injected intravenously with 4.0×10^4 viable organisms, and viable counts were made 24 hours later on groups of 5 mice. Results expressed as log_{10} geometric mean ± standard deviation.

Table 2. Effect of vinblastin on numbers of *Listeria monocytogenes* in C57Bl/10 and BALB/c mice 2 and 4 days after infection.

Treatment	Log_{10} *Listeria*/liver ± S.D.			
	C57Bl/10		BALB/c	
	Day 2	Day 4	Day 2	Day 4
Control	4.20±0.16	2.59±0.72	6.62±0.43	4.07±0.40
Vinblastin d0	3.90±0.21	3.78±0.58	6.96±0.49	7.85±0.47
Vinblastin d1	3.89±0.19	5.15±0.61	6.43±0.40	7.55±0.48

[a] Groups of 5 mice were injected intravenously with 2×10^2 *Listeria* and with 100 µg vinblastin 1 hour before (d0) or 24 hours after (d1). Bacterial counts were made 1 and 4 days later on groups of 5 mice. Results expressed as $\log_{10}$ geometric mean ± standard deviation.

the day of *Listeria* infection in the Lr^+ C57Bl/10 mice was not on cell division, but possibly on the migration of preformed cells to the site of infection. Vinblastin does not affect cell migration unless cell division is a prerequisite (11). On the other hand, it is known that low doses of irradiation can affect the membrane structure and thus migration of lymphocytes (12), although less seems to be known about radiosensitivity of migration of monocytes and polymorphs in inflammatory foci. Alternatively radiosensitivity of bactericidal activity by phagocytic cells has been demonstrated (13, 14). Such a radiosensitive bactericidal mechanism in the C57Bl/10 could be expressed in either or both the incoming inflammatory cells or the resident macrophages. It should be noted that because of the large numbers of polymorphonuclear cells in the foci of infection (6), the effect of these treatments on monocytes at the site could not be assessed. Irradiation did not increase the number of foci but lead to fewer polymorphs per focus, and to necrosis of many of the polymorphs. However, polymorphs are said not to be important in resistance to *Listeria*, at least in ddN mice (15). By day 4, when untreated mice had developed cell mediated immunity, infection was markedly exacerbated by vinblastin in livers and spleens of both strains. This presumably reflected the requirement for bone marrow derived monocytes to express cell mediated immunity, and possibly also the requirement for proliferating T lymphocytes (16).

C. Listeriosis in Beige Mice

The "beige" mutation in mice, analogous with Chediak-Higashi syndrome in humans, leads to loss of neutral protease in polymorphonuclear leukocytes (17), formation of abnormal lysosomes in monocytes, neutrophils and eosinophils (18), loss of NK activity (19) and susceptibility to infection (20). In different systems, macrophages are either unaffected (21) or show a slower rate of killing of target tumour cells (22). When beige mice, their heterozygous litter mates or the parent strain, C57Bl/10, mice were infected with 5×10^2 L. monocytogenes there was no difference in the bacterial growth curves over 5 days in the three types of mice. In particular there was no difference in liver counts at the critical 24 hour time point when BALB/c mice had tenfold higher counts (Table 3), showing that NK cells were not involved in early resistance, and that the enzymic deficiencies and lysomal abnormalities in the polymorphs and monocytes of these mice were not important in Listeria resistance.

SUMMARY

Early control of Listeria infection in mice is normal in beige mice which lack NK cells. On the other hand, it apparently involves resident macrophages, cells dividing during the 2 days before infection and, in genetically resistant mice only, the recruitment of a highly radio-sensitive mechanism which nevertheless does not require cell division.

Table 3. Numbers of Listeria organisms in livers of beige mice and other strains 24 hours after infection

Strain	Log_{10} Listeria/liver ± S.D.
BALB/c	4.70±0.30
C57Bl/10	3.71±0.42
Beige	3.52±0.70
Heterozygotes	3.31±0.59

[a]Mice infected intravenously with 5×10^2 Listeria organisms and bacterial counts performed 24 hours later in the livers of groups of 5 mice.

ACKNOWLEDGEMENTS

The invaluable assistance of Merilyn Ho, Jacinta Walsh, and Suzanne Gaucci is gratefully acknowledged. We thank Marianne Gierveld for her patient typing of this manuscript.

REFERENCES

1. Oldstone, M.B.A., Dixon, T.J., Mitchell, G.T. and McDevitt, H.O. J. exp. Med. 137: 1201 (1973).
2. Oth, D., Sabalovic, D. and Le Ganec, Y. Folia Biol. Praha. 20: (1974).
3. Skamene, E., Kongshavn, P.A.L. and Landy, H. Genetic control of natural resistance to infection and malignancy. Academic Press, New York (1980).
4. Cheers, C. and McKenzie, I.F.C. Infect. Immun. 19: 755 (1978).
5. Cheers, C., McKenzie, I.F.C., Pavlov, H., Waid, C. and York, J. Infect. Immun. 19: 763 (1978).
6. Cheers, C. and Mandel, T.E. Infect. Immun. 30: 851 (1980).
7. Cheers, C., McKenzie, I.F.C., Mandel, T.E. and Chan, Y.Y. in: Genetic control of natural resistance to infection and malignancy ed E. Skamene, P.A.L. Kongshavn and M. Landy, p.141, Academic Press, New York, (1980).
8. Stevenson, M.U., Kongshavn, P.A.L. and Skamene, E. J. Immunol. 127: 402 (1981).
9. Newborg, M.T. and North, R.J. J. Immunol. 124:571 (1980).
10. Sadarangani, C., Skamene, E. and Kongshavn, P.A.L. Infect. Immun. 28: 381 (1980).
11. McGregor, D.D. and Logie, P.S. J. exp. Med. 137: 660 (1973).
12. Anderson, R.E. and Warner, N.L. Adv. Immun. 24: 215 (1976).
13. Paul, B., Strauss, R., Sbana, A.J. J. Reticuloendothelial Soc. 5: 538 (1968).
14. Nelson, E.L. and Becker, J.R. J. Infect. Dis. 104:13 (1959).
15. Tatsukawa, K., Mitsuyama, M., Takaya, K. and Nomoto, K. J. gen. Microbiol. 115: 161 (1979).
16. North, R.J. J. exp. Med. 132: 535 (1970).
17. Vassalli, V.J., Granelli, A.P., Griscelli, C. and Reich, E. J. exp. Med. 147: 1285 (1978).
18. Oliver, C. and Essmer, E. Lab. Invest. 32: 17 (1975).
19. Clark, E.A., Shulty, L.D. and Pollack, S.B. Immunogenetics 12: 601 (1981).
20. Elin, R.J., Edelin, J.B. and Wolff, S.M. Infect. Immun. 10: 88 (1974).
21. Roder, J. and Duive, A. Nature 278: 451 (1979).
22. Mahoney, K., Morse, S.S. and Morahan, P.S. Cancer Res. 40: 3934 (1980).

THE ROLE OF ASIALO GM1+ (GA1+) CELLS IN THE RESISTANCE TO TRANSPLANTS OF BONE MARROW OR OTHER TISSUES

Ko Okumura

Department of Immunology
Faculty of Medicine
University of Tokyo
Tokyo, Japan

Sonoko Habu
Kazuo Shimamura

Department of Pathology
Tokai University School of Medicine
Isehara, Kanagawa, Japan

I. INTRODUCTION

In addition to the recent postulated role of NK cells in surveillance against neoplastic transformation, it has been suggested that NK-like cells capable of regulating cell differentiation are the effector cells in natural resistance (NR) to transplants of allogeneic and semisyngeneic parental hemopoietic cells (1). Numerous studies suggest that NK-like cells play an important role in natural resistance systems, but because of the lack of specific markers for NK cells, there still remains the question as to whether or not NK cells and effectors of NR to hemopoietic grafts belong to a common cellular lineage.

In the course of studying murine T cell differentiation markers, we found that the neutral glycolipid asialo GM1 (GA1) is a common cell surface markers expressed on both murine early fetal thymocytes and natural killer cells but not on mature T or B cells (2,3). Furthermore, in addition to the potent ability of anti-asialo GM1 antibodies (anti-GA1) to delete NK activity _in vitro_, it was found that _in vivo_

ISBN 0-12-341360-5

injection of microliter amounts of anti-GA1 also abolished NK activity of athymic as well as euthymic mice (4,5). Taking further advantage of this in vivo effect of antibody, experiments were performed to know whether or not the effector mechanisms of F1 hybrid resistance is abolished by anti-GA1 injections. In this paper, we discuss the role of asialo GM1 bearing ($GA1^+$) lymphocytes in murine natural resistance systems.

II. EFFECT OF ANTI-GA1 INJECTIONS ON F1 HYBRID RESISTANCE

According to the experimental systems originally reported by Cudkowicz et al. (6,7), lethaly irradiated (950R)) F1 hybrid mice of C3H x C57BL/6 (C3B6) were reconstituted with 2 x 10^6 bone marrow cells of semisyngeneic C57BL/6 (B6) or syngeneic C3B6 and 5 days later these bone marrow recipients were sacrificed to examine hematopoietic colonies in the spleens. To delete $GA1^+$ NK cells in bone marrow recipients, 0.01 ml of anti-GA1, having a 1:1000 flocculation titer, was injected intraperitonealy into one group of F1 mice immediately after irradiation. As shown in Figure 1, higher numbers of hematopoietic colonies were observed in the spleens of F1 recipients of anti-GA1 and either B6 or F1 bone marrow cells (50 - 70 colonies in the sections). In contrast, only a few small colones (5 - 20 colones) were observed in spleens from F1 mice injected with normal rabbit serum (NRS) and B6 bone marrow. This ability of in vivo injection of anti-GA1 to promote colony expansion in F1 mice was confirmed by incorporation of ^{125}IUDR (data not shown). F1 mice injected with anti-GA1 usually had 5 - 10 fold higher counts in their spleens compared to spleens of the control group that received NRS. To understand how injection of such a minute amount (0.01 ml) of anti-glycolipid antiobdy can affect NK activities (4,5) as well as the effector mechanisms of F1 hybrid resistance, the number of $GA1^+$ cells were counted after membrane fluorescence staining. The number of $GA1^+$ cells were usually about 5% in F1 mice injected with anti-GA1, whereas about 40 - 50% of $GA1^+$ cells were observed in spleen cells of irradiated F1 mice administered NRS. Although it is not clear that this in vivo deletion of $GA1^+$ cells is dependent on the activation of serum complement as is discussed elsewhere in this volume (Kasai and Okumura, Habu and Okumura) the number of $GA1^+$ cells completely correlates to the effector activities of NR. This correlation was also observed in the case of NK activity of athymic nude mice (5).

To avoid the possible effect of anti-GA1 antibodies on radioresistant T cells, athymic nude F1 mice, made by mating

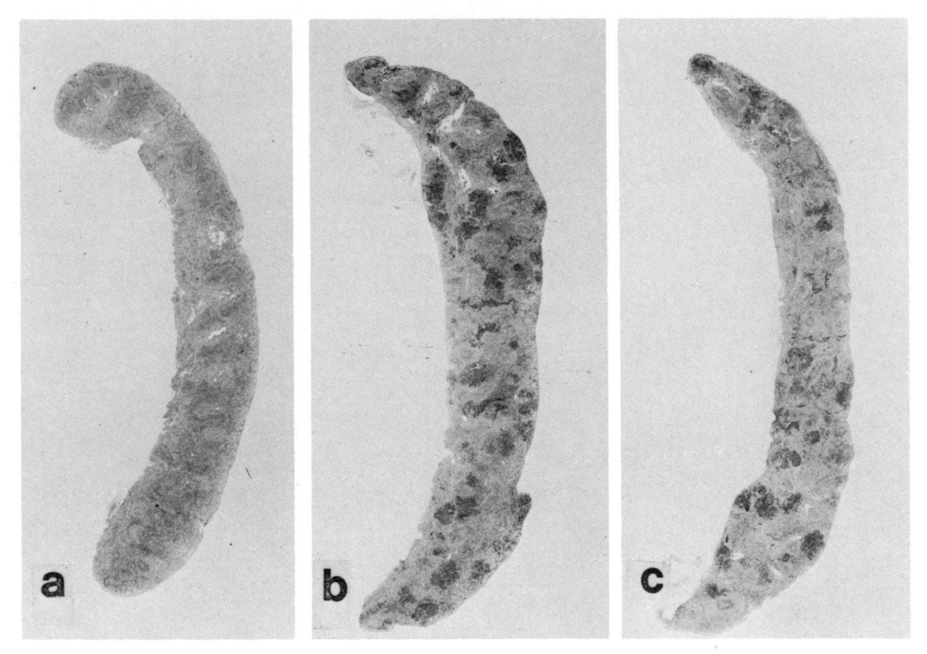

FIGURE 1.

a. Spleen section from irradiated F1 (C3B6) mice injected with 2 x 10^6 B6 bone marrow cells after administraiton of 0.01 ml of normal rabbit serum.

b. Spleen section of irradiated F1 (C3B6) mice injected with 2 x 10^6 bone marrow cells after administration of 0.01 ml of anti-GA1.

c. Spleen section from irradiated F1 (C3B6) mice injection with 2 x 10^6 syngeneic (C3B6) bone marrow cells.

C3H background nude (C3H nu/nu) and C57BL/6 background nude (C57BL/6 nu/nu), were used as bone marrow receipients after 650R irradiation. As shown in Table 1, the effect of anti-GA1 on the number of hematopoietic colonies was more dramatic compared to that in euthymic F1 mice. Only a few small sized colonies were found in spleens of NRS injected F1 nude mice but in the spleen of anti-GA1 injected group, almost same size and same number of colonies were appeared comparing to receipient received bone marrow cells from syngeneic F1 mice mice. Again, depletion of GA1$^+$ cells was confirmed in the mice reccived anti-GA1 injections. The results in Table I, show that natural resistance system of irradiated homozygous nude mice (BALB/c) given allogeneic bone marrow graft was also abrogated by anti-GA1 injections, although the resistance is not as potent as in F1 nude mice.

TABLE I. Effect of Anti-GA1 on Hybrid Resistance of Nude Mice

Donors and recipients[a]	Administration of anti-GA1 (0.01 ml)[b]	Number of colonies[c]	Number of GA1 cell(%)[d]
C57BL/6 ⟶ F1	+	++ ~ +++	<5
C57BL/6 ⟶ F1	-	-	60
C57BL/6 ⟶ BALB/c	+	++	50
C57BL/6 ⟶ BALB/c	-	+	<5
F1 ⟶ F1	+	+++	
F1 ⟶ F1	-	+++	

a. Nude F1 (C57BL/6 x BALB/c) and nude BALB/c were used as recipients. 2×10^6 of bone marrow cells were transferred immediately after irradiation (650R).

b. Antibody was injected intraperitoneally. Same amount (0.01 ml) of NRS was injected to the control group showed as -.

c. Number of colonies were counted by microscope after making section. The number of colonies in median sections of spleens were scored as +++:>50, ++: 50 ~ 30, +: 30 ~ 10, -: <10. Four recipients for one group were examined.

d. The number of GA1+ cell were counted under fluorescence microscope after the indirect cell membrane staining using anti-GA1 and FITC conjugated goat-anti-rabbit Igs.

To know the target specificity of this natural resistant mechanism, the resistance or rejection mechanism against cell of the B cell lineage was investigated by utilizing homozygous nude mice and Ig allotype congeneic strain. Irradiated BALB/c nu/nu (Igh^a) were injected with 1 x 10^6 bone marrow cells from CB-20 (Igh^b) which is an Igh congenic strain with the BALB/c background. Immunoglobulin of the $Igh\text{-}1^b$ allotype was undetected in the sera of the BALB/c recipients of CB-20 bone marrow throughout experiments (more than 8 weeks). In contrast, sera of BALB/c nu/nu given 0.01 ml of anti-GA1 before bone marrow transfer, contained detectable levels of Igh-1 allotype in 2 weeks after the cell transfer and, by 4 weeks, these recipients became completely chimeric for immunoglobulin producing cells (B-cells) (Habu et al, in preparation). We could not examine natural resistance against cells of the T cell lineage with these experimental systems but our results demonstrate that deletion of $GA1^+$ cells in nude mice also deleted natural resistance mechanisms against cells of the B cell lineage, in addition to hemopoietic stem cells.

The biological meaning of natural resistance has been extended by the recent work. Still, one of the most vague parts of the phenomenon is the specificity effector cells of NR or NK cells for their targets. If the sensitivity of the target is associated with a differentiation marker, the experimental system mentioned above may provide a tool to investigate the target specificity of NR effectors as well as NK cells.

III. CHARACTERIZATION OF ASIALO GM1 BEARING ($GA1^+$) CELLS

The preceeding experiments show that in vivo administration of anti-GA1 affects the natural resistance system, as well as the number of $GA1^+$ cell in the spleen. The next question asked was what kinds of lymphocytes express GA1. We have already reported that anti-GA1 does not influence T cell functions such as killer and helper activities (2,8). Furthermore, concanavalin A (Con A) responding T cells as well as Ia-bearing macrophage like cells which are essential for Con A responses are unaffected by anti-GA1 (5). $GA1^+$ cells are radiation resistant, so $GA1^+$ cell and NK activities are enriched in the spleens of lethaly irradiated mice. In those $GA1^+$ enriched cell populations, we could morphologically distinguish mainly two kinds of cells, namely large and small lymphocytes, but not which lymphocytes were responsible for NK activities. Immunoelectromicroscopy of NK enriched cell populations from nude mice, indicated that the cells with

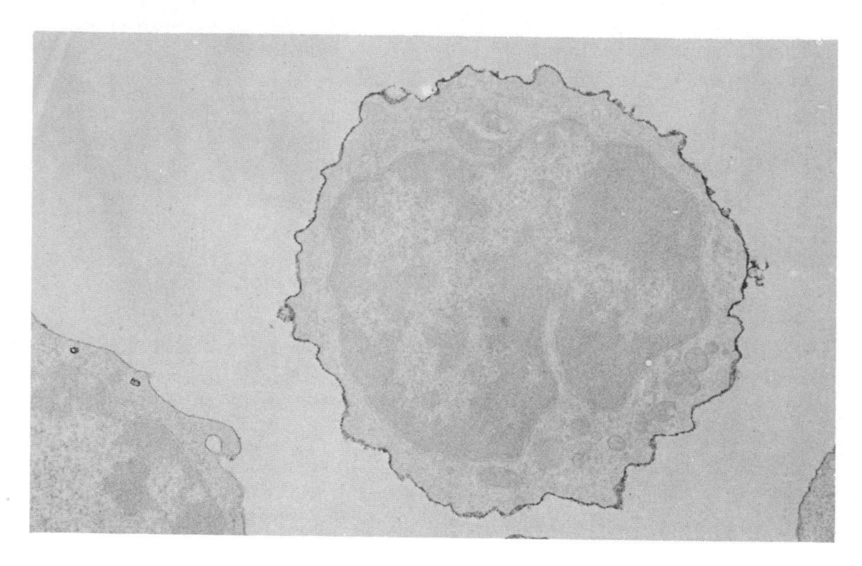

FIGURE 2. Immunoelectron micrograph of an GA1+ lymphocytes. Reaction products with anti-GA1 antibody are noted on the surface of the small lymphocytes (x 10,000)

activities against YAC-1 target are small lymphocytes (Fig. 2), whereas larger GA1+lymphocytes, predominately monocytes do not participate in cytolysis during short time (4 - 6 hr) incubation period (9). In vivo administration of anti-GA1 causes a decrease of both large and small GA1+ cells, it is difficult to know whether the participation of GA1+ monocytes in NR effector system is essential. We have also observed by cell membrane staining that GA1+ monocyte, representing less than 30% of total splenic monocytes, do not bear detectable amounts of Ia antigen on them (unpublished data). Thus it may be concluded that Ia+ monocytes are not involved in NR systems.

As mentioned elsewhere in this volume, anti-GA1 is able to react with fetal thymocytes as well as some immature cells in fetal liver which migrate into thymus (Habu and Okumura), then we can postulate that GA1+ small lymphocytes of adult spleens with NK activity may belong to the T cell lineage. In any case, as GA1+ is a marker common to both NK cells and effectors of NR, it may further provide useful approach to understand the murine natural immune systems.

REFERENCES

1. Cudkowicz, G., and Hochman, P.S., Immunol. Rev. 44:13 (1979).
2. Kasai, M., Iwamori, M., Nagai, Y., Okumura, K., and Tada, T., Eur. J. Immunol. 10:175 (1980).
3. Habu, S., Kasai, M., Nagai, Y., Tamaoki, T., Tada, T., Herzenberg, L.A., and Okumura, K., J. Immunol. 125:2284 (1980).
4. Kasai, M., Yoneda, Y., Habu, S., Maruyama, Y., Okumura, K., and Tokunaga, T., Nature 291:334 (1981).
5. Habu, S., Fukui, H., Shimamura, K., Kasai, M., Nagai, Y., Okumura, K., and Tamaoki, N., J. Immunol. 127:34 (1981).
6. Cudkowicz, G., and Benett, M., J. Exp. Med. 134:83 (1971).
7. Cudkowicz, G., and Benett, M., J. Exp. Med. 134:1513 (1971).
8. Habu, S., Hayakawa, K., Okumura, K., and Tada, T., Eur. J. Immunol. 9:938 (1979).
9. Shimamura, K., Habu, S., Fukui, H., Akatsuka, A., Okumura, K., and Tamaoki, N., J.N.C.I. in press.

DIRECT EVIDENCE FOR THE INVOLVEMENT OF NATURAL KILLER CELLS IN BONE MARROW TRANSPLANTATION

Eva Lotzová[1,3], S.B. Pollack[2,3], and C.A. Savary[1,4]

[1]M.D. Anderson Hospital and Tumor Institute
Houston, Texas and [2]University of Washington
Seattle, Washington

I. INTRODUCTION

Murine natural killer (NK) cells have been implicated in resistance to allogeneic and parental bone marrow (BM) transplants (Kiessling et al, 1977; Lotzová and Savary, 1977; Clark and Harmon, 1980; Lotzová, 1980b). This implication rests primarily on the striking similarities between NK cells and the effector cells involved in rejection of histoincompatible BM transplants (BM effector - BME cells) in mice. Both NK and BME cells are naturally occurring and rapidly reactive, do not appear to have immunological memory, are independent of thymic influences, but most likely are dependent on BM for their origin and function, are radioresistant and do not function before 3 weeks of life (Kiessling et al, 1977; Lotzová and Savary, 1977; Lotzová, 1980b). Furthermore, NK and BME cell reactivity is sensitive to the same cytocidal and immunomodulating agents; for instance, when administered in vivo, cyclophosphamide (Lotzová and Savary, 1977), silica (Kiessling et al, 1977; Lotzová and Savary, 1977), glucan (Lotzová and Gutterman, 1979), radioactive strontium-89 (Kiessling et al, 1977), Corynebacterium parvum [under certain experimental conditions] (Lotzová and Savary, 1977; Lotzová and McCredie, 1978a,b; Savary and Lotzová, 1978; Lotzová, 1980a), and cortisone (Lotzová and Savary, 1981) diminished NK cell cytotoxicity as well as BM rejection potential. Conversely, agents that augment NK cell cytotoxicity, e.g., poly I:C (Herberman et al, 1978; Djeu et al, 1979), also increased resistance to BM transplants (Lotzová, unpublished observations). Even though these observations are highly suggestive of NK cell and BME cell similarity (or perhaps iden-

[3]Supported by NIH grants CA 21062, CA 14528 and CA 18647.
[4]Recipient of Rosalie B. Hite Postdoctoral Fellowship.

ISBN 0-12-341360-5

tity), they were based on comparative studies only. In this study we present data demonstrating abrogation of resistance to BM grafts by antiserum directed against a NK cell specific cell surface antigen, NK 1.1. These data demonstrate involvement of NK cells in BM transplantation.

II. MATERIALS AND METHODS

The following female mice were used at 11-16 weeks of age: (C57BL/6 x DBA/2)F_1 (B6D2F_1), C57BL/6 (B6), C3H/Anf (C3H), and BALB/c.

Anti-NK 1.1 serum was raised as described previously (Glimcher et al, 1977; Pollack et al, 1979). The NK 1.1 antiserum killed fewer than 5% of B6 or B6D2F_1 splenocytes in complement dependent assays in vitro.

NK cell activity was assessed on YAC-1 target cells labelled with ^{51}Cr, as described previously (Pollack et al, 1979; Lotzová and Savary, 1981).

Recipient mice were exposed to total body irradiation from a ^{60}Co source; B6 mice received 875R, C3H mice 950R, and B6D2F_1 mice 1100R. Twenty-four hrs after irradiation the mice were transplanted with 10^6 BM cells of various donors, as indicated in Section IV. The growth of transplanted BM cells was evaluated 9 days after transplantation by ^{125}IUdR technique (Lotzová, 1977), or spleen colony-forming technique (Till and McCulloch, 1961). The ^{125}IUdR technique measures all proliferating cell populations present in the marrow graft, whereas the spleen colony-forming technique measures only proliferation of hemopoietic stem cells (CFU-S).

III. REDUCTION OF SPLENIC NK CELL CYTOTOXICITY BY IN VIVO TREATMENT WITH ANTI-NK 1.1 SERUM

It has been demonstrated that in vivo treatment of B6 mice with as little as 25 µl of anti-NK 1.1 serum depletes splenic NK cell activity in vitro to YAC-1 within 2 hrs (Pollack and Hallenbeck, 1982). Such NK cell-depleted mice have a reduced capacity to eliminate ^{125}IUdR labelled lymphoma cells (Pollack and Hallenbeck, 1982), and additionally, an enhanced growth of B16 melanoma (Pollack, this volume). As shown in Table I, B6D2F_1 mice injected intravenously (i.v.) with 0.4 ml of NK 1.1 antiserum also had significantly less splenic NK cell activity than did age and sex matched controls. Splenocytes were obtained 6 hrs after injection of antiserum and tested on YAC-1 target cells.

Table I. Effect of NK 1.1 Antiserum on Splenic NK Cell Cytotoxicity of $B6D2F_1$ Mice to T Cell Lymphoma, YAC-1

Treatment of effector cells[a]	Percent of NK cell cytotoxicity[b] mean (±SE)[n]	
	Target-to-effector cell ratio	
	1:25	1:50
None	14.5 (1.3)[5]	24.0 (1.9)[5]
NK 1.1	8.0 (0.8)[6]	14.9 (0.9)[6]

[a]Six hrs after i.v. administration of 0.4 ml of NK 1.1 antiserum (1:7 or 1:8 dilution), splenocytes were tested for cytotoxicity to YAC-1 target cells.

[b]The differences between untreated and NK 1.1 antiserum-treated groups were evaluated statistically with the Student's t-test analysis and the probability (P) was calculated. $P<0.01$ for both 1:25 and 1:50 target-to-effector cell ratios.

IV. ABROGATION OF ALLOGENEIC AND HYBRID RESISTANCE TO BONE MARROW TRANSPLANTS BY ANTI-NK 1.1 SERUM

Since NK cells have been implicated as effectors in allogeneic and hybrid resistance, we tested whether treatment with NK 1.1 antiserum would also affect allogeneic and parental BM transplant rejection potential in vivo. In the first experiment, B6 mice, resistant to allogeneic BALB/c BM cells, were irradiated and 18 hrs later injected i.v. with 0.4 ml of NK 1.1 antiserum at a dilution of 1:7. Six hrs thereafter, antiserum-treated B6 mice were transplanted with 10^6 BALB/c BM cells.

The results of a typical allogeneic transplant experiment are illustrated in Table II. BALB/c BM cells grew without impairment in C3H mice, which do not exhibit allogeneic resistance (Lotzová et al, 1977); this was evident by significant ^{125}IUdR uptake, as well as the colony-forming potential. On the contrary, there was no proliferation of BALB/c BM cells in untreated B6 recipients that exhibit allogeneic resistance. When, however, BALB/c marrow cells were transplanted into anti-NK 1.1 pretreated B6 recipients, significant uptake of ^{125}IUdR as well as high colony-forming activity was observed.

Table II. Abrogation of Allogeneic Resistance to Bone Marrow Transplants by NK 1.1

Donor → recipient combination	Recipient treatment	Mean splenic uptake of ^{125}IUdR (±SE) [n]	Mean CFU-S formation (±SE) [n]
BALB/c → C3H[a]	None	0.43(0.07) [6]	>20 [6]
BALB/c → B6	None	0.06(0.01) [6]	0.4(0.2) [6]
BALB/c → B6	NK 1.1	0.32(0.07) [5]	>20 [2] 17(1.0) [3]
B6 - R.C.[b]	None	0.04(0.01) [3]	None [3]
B6 - R.C.	NK 1.1	0.05(0.01) [2]	None [2]

[a]C3H mice, which fully support the growth of BALB/c BM cells (Lotzová et al, 1977) were used as positive controls.

[b]R.C. - radiation controls were mice that were irradiated, but not transplanted with BM cells.

The antiserum did not affect endogenous colony formation or endogenous ^{125}IUdR uptake, since NK 1.1-treated radiation controls did not exhibit any spleen colony-formation potential or significant ^{125}IUdR uptake.

The effect of NK 1.1 antiserum on hybrid resistance also was tested (Table III). Hybrid resistance of $B6D2F_1$ mice to

Table III. Abrogation of Hybrid Resistance to B6 Bone Marrow Transplants by NK 1.1

Recipient	Recipient treatment	Mean splenic uptake of ^{125}IUdR (±SE) [n]	Mean CFU-S formation (±SE) [n]
B6	None	0.86(0.09) [6]	>20 [6]
B6	NK 1.1	0.65(0.11) [6]	>20 [6]
$B6D2F_1$	None	0.16(0.03) [6]	6.8(1.9) [6]
$B6D2F_1$	NK 1.1	0.60(0.08) [6]	>20 [6]
	Radiation controls[a]		
$B6D2F_1$	None	0.03(0.01) [6]	None [6]
$B6D2F_1$	NK 1.1	0.04(0.02) [4]	None [4]

[a]Radiation controls were mice that were irradiated but not transplanted with BM cells.

B6 BM cells was abrogated by pretreatment with NK 1.1 antiserum. This was illustrated by the growth of parental BM cells (evaluated by ^{125}IUdR uptake) and formation of spleen colonies in B6D2F$_1$ hybrids. NK 1.1 antiserum did not affect in any way growth of BM cells in syngeneic mice, since there was a comparable ^{125}IUdR uptake in syngeneic untreated and NK 1.1-pretreated mice.

V. CONCLUDING REMARKS

Our experiments have shown that in vivo treatment with antiserum directed against the NK cell specific cell surface antigen, NK 1.1 (Pollack et al, 1979) abrogated resistance to allogeneic and parental BM transplants. These observations provided the first direct evidence of the involvement of NK cells in BM transplantation. That the effect of antiserum was indeed directed against NK cells was supported by evidence that NK 1.1 antiserum does not decrease the cytolytic activity of alloimmune T cells (Pollack et al, 1979) and that monoclonal Thy-1.2 antibodies do not abrogate allogeneic resistance to BM grafts (Lotzová et al, manuscript in preparation). While our results implicate NK cells as one of the components of the BM transplantation reaction, they do not necessarily imply that NK cells are the only cells involved in marrow graft rejection. In fact, our previous studies suggested that macrophages may also play a part in anti-BM immunity (Lotzová and Cudkowicz, 1973; Lotzová et al, 1975; Lotzová, 1977).

VI. ACKNOWLEDGMENTS

The expert technical assistance of Abbas Khan and skillful secretarial assistance of Geraldine Harrison is greatly appreciated.

REFERENCES

Clark, E.A. and Harmon, R.C. (1980). *Adv. Cancer Res. 31*:227.

Djeu, J.Y., Heinbaugh, J.A., Holden, H.T. and Herberman, R.B. (1979). *J. Immunol. 122*:182.

Glimcher, L., Shen, F.W. and Cantor, H. (1977). *J. Exp. Med. 145*:1.

Herberman, R.B., Djeu, J.Y., Ortaldo, J.R., Holden, H.T., West, W.H. and Bonnard, G.D. (1978). *Cancer Treatment Reports 62:*1893.

Kiessling, R., Hochman, P.S., Haller, O., Shearer, G.M., Wigzell, H. and Cudkowicz, G. (1977). *Eur. J. Immunol. 7:*655.

Lotzová, E. (1977). *Exp. Hematol. 5:*215.

Lotzová, E. (1980a). In "Natural Cell-Mediated Immunity Against Tumors" (R.B. Herberman, ed.) p. 735. Academic Press, New York.

Lotzová, E. (1980b). In "Natural Cell-Mediated Immunity Against Tumors" (R.B. Herberman, ed.), p. 1117. Academic Press, New York.

Lotzová, E. and Cudkowicz, G. (1973). *J. Immunol. 110:*791.

Lotzová, E., Dicke, K.A., Trentin, J.J. and Gallagher, M.T. (1977). *Transplantation Proc. IX:*289.

Lotzová, E., Gallagher, M.T. and Trentin, J.J. (1975). *Biomedicine 22:*387.

Lotzová, E. and Gutterman, J.U. (1979). *J. Immunol. 123:*607.

Lotzová, E. and McCredie, K.B. (1978a). *Cancer Immunol. and Immunother. 4:*215.

Lotzová, E. and McCredie, K.B. (1978b). *Proc. Am. Assoc. Cancer Res. 19:*52 (abstract).

Lotzová, E. and Savary, C.A. (1977). *Biomedicine 27:*341.

Lotzová, E. and Savary, C.A. (1981). *Exp. Hematol. 9:*766.

Pollack, S.B. and Hallenbeck, L.A. (1982). *Int. J. Cancer,* in press.

Pollack, S.B., Tam, M.R., Nowinski, R.C. and Emmons, S.L. (1979). *J. Immunol. 123:*1818.

Savary, C.A. and Lotzová, E. (1978). *J. Immunol. 120:*239.

Till, J.E. and McCulloch, E.A. (1961). *Radiat. Res. 14:*213.

IN VIVO ROLE OF NK(HSV-1) IN THE INDUCTION OF GRAFT VERSUS HOST DISEASE IN BONE MARROW TRANSPLANT RECIPIENTS

Carlos Lopez
Patricia Fitzgerald
Dahlia Kirkpatrick

Sloan-Kettering Institute for Cancer Research
New York, New York.

I. INTRODUCTION

Graft-versus-host disease (GvHD) remains one of the major difficulties associated with bone marrow transplantation in man (1). This allogeneic reactivity has been found in 70% of recipients of donor marrows matched with respect to the serologically defined (HLA-A,B, and C) and lymphocyte determined (HLA-D) antigens indicating that genes outside of this region also contribute to this reaction. GvHD is associated with a prolonged immunodeficiency state and an associated increased susceptibility to severe infections, especially pneumonitis caused by cytomegalovirus.

Studies of GvHD induced in lethally irradiated mice with allogeneic marrow transplants demonstrated that donor lymphoid cells are responsible for mediating this reactivity (2). However, other studies, with chick embryos also showed that the induction of GvHD was absolutely dependent on a host cellular function which matures late in ontogeny (3). In a parallel but different set of studies, Cudkowicz and Bennett (4) showed that the capacity of lethally irradiated mice to resist allogeneic bone marrow grafts was dependent on a cellular function which matures at about 3 weeks of age. This function, termed allogeneic resistance, was found to be derived from bone marrow stem cells since engraftment of stem

---

Supported in part by NIH grants CA-23766, CA-08748, and CA-17404: NIH training grant CA-09149; grant IM-303 and Faculty Research Award #193 (to C.L.) from the American Cancer Society.

ISBN 0-12-341360-5

cells from a resistant to a susceptible strain of mouse resulted in a resistant chimera. Further, it was found to be dependent on an intact bone marrow since treatment of mice with strontium-89 (^{89}Sr), which causes a severe marrow aplasia, abrogated this genetic resistance (5). In ^{89}Sr-treated mice the spleen takes over hematopoiesis and cellular immunity, as gauged by skin graft rejection, remains normal (5). Since murine natural killer cells which lyse YAC targets [NK(YAC)] also matured after 3 weeks of age, were bone marrow derived, and were dependent on a normal bone marrow environment, these effector cells were through to play an important role in allogeneic resistance (6).

Because of our studies of a mouse model which indicated that effector cells with many of the properties of NK cells probably play an important role in genetic resistance to HSV-1 (7), we developed an assay using HSV-1 infected fibroblasts as targets [NK(HSV-1)] to determine whether a similar role could be defined for resistance to herpesvirus infections in man. In another chapter of this book we summarize evidence to suggest that low NK(HSV-1) is associated with increased susceptibility to severe infections (Lopez et al., this volume). In addition, our studies of bone marrow transplant recipients prior to transplantation indicated that normal levels of NK(HSV-1) were predictive of the patients who would go on to develop GvHD in the post transplant period (8). In this chapter, we present an up-date on this correlation as well as evidence to suggest that NK(HSV-1) function is derived from bone marrow stem cells, and is dependent on a normal marrow microenvironment, properties which have been used to define the effectors of allogeneic resistance and NK(YAC) in the mouse.

II. METHODS AND MATERIALS

Patients undergoing bone marrow or fetal liver transplantation to restore marrow function were studied prospectively (8). The treatments in preparation for transplantation were as described (8). GvHD was graded by the criteria of Gluksberg et al (9).

NK(HSV-1) assay was carried out as described earlier (10) with minor modifications (Lopez et al., elsewhere in this book). The estimated lysis at a 50:1 ratio was used for comparisons. Normal control values were used to develop a geometric mean and normal range (± 2 standard deviations).

III. RESULTS

Table 1 summarizes our results of pretransplant NK(HSV-1)

and development of GvHD in the post-transplant period. Clearly, there was a highly significant ($p < 0.0005$) correlation between normal pretransplant NK(HSV-1) and the development of GvHD in the period after transplantation. This association held for each of the major patient groups when evaluated individually (aplastic anemia, leukemia, and severe combined immunodeficiency disease). Interestingly, there was no correlation between the grade of GvHD and the NK(HSV-1) high NK(HSV-1) response only predicted that the patient would have GvHD and did not suggest that it would be severe.

TABLE I.

		Pretransplant NK(HSV-1)	
		Normal	Low
Development of GvHD	Yes	17	3
	No	0	17

More recently we have evaluated the NK(HSV-1) responses of the bone marrow donors to determine whether these might also show a correlation with the development of GvHD. Fourteen donors have been evaluated to date and no correlation was found with their NK(HSV-1) response and the development of GvHD in the recipient ($p > .1$). In this same group there was a significant correlation between NK(HSV-1) of the recipient prior to transplant and his development of GvHD afterward ($p < 0.01$).

Studies were also undertaken to determine whether, like the effector cells of allogeneic resistance, NK(HSV-1) cells also were derived from bone marrow stem cells. A total of 6 patients with very low NK(HSV-1) responses prior to transplantation have been followed after engraftment and found to develop normal levels of response. Each of these patients was shown to be a chimera indicating that the stem cells for the NK(HSV-1) effector cells probably were of donor origin.

Our studies of patients with aplastic anemia (AA) and osteopetrosis also indicate that NK(HSV-1) function is dependent on an intact, normal bone marrow microenvironment. Twelve AA were studied who demonstrated complete marrow failure (no evidence of monocytoid/myeloid precursors) and 10 of these also showed very low NK(HSV-1) responses. Ten AA patients had some bone marrow precursors and 8 of these had normal NK(HSV-1) responses.

Osteopetrosis is an inherited disorder leading to obliteration of the marrow cavity by bone (11). Four of the 5 patients studied with this disorder demonstrated low NK (HSV-1) responses. Disease in the other patient was not far advanced (as evidenced by the lack of anemia) and NK(HSV-1) was normal.

Studies of patients with severe combined immunodeficiency disease (SCID) indicate that the effectors of NK(HSV-1) are clearly not T-Cells, B-Cells, or pre-T-Cells. Three of the 10 SCID patients studied had normal NK(HSV-1) responses even though lymphoid cell numbers and function were grossly abnormal. In addition, those SCID patients who had pre-T-cells (12) in their marrow usually (3/4) failed to demonstrate normal NK(HSV-1).

IV. DISCUSSION

As has been detailed elsewhere in this book (Fitzgerald, et al.), the effectors of NK(HSV-1) are clearly different than the cells which lyse K562 targets. This fact has been most convincingly made by the results of our study of adult patients with aplastic anemia: Four of these patients demonstrated normal kill of K562 targets but abnormally low NK(HSV-1) responses. This heterogeneity of effector cells is a very important consideration when one attempts to correlate NK function with a biologic activity. For example, Livnat et al (13) evaluated NK cell function of bone marrow transplant recipients using K562 target cells and were unable to demonstrate a correlation between pretransplant levels and GvHD in the post-transplant period. Clearly, these two target cells detect different effector populations and a biologic correlation with one does not necessarily apply to the other.

The association between pretransplant NK(HSV-1) and GvHD after engraftment is even stronger than when first described (8). One possibility which we raised earlier was that since most marrow transplant donors were matched siblings, perhaps the correlation was with family NK(HSV-1) levels and our finding was only fortuitous. We have recently started evaluating donor as well as recipient NK(HSV-1) levels and only the latter has correlated with GvHD. These data strongly indicate that NK(HSV-1) function reflects the capacity of the recipient to stimulate GvHD.

Our data on the characteristics of the (HSV-1) indicate that they share many properties with the effectors of allogeneic resistance and NK(YAC) in the mouse. Clearly, NK(HSV-1) effector cells are not classical T-Cells, B-Cells, or pre-T-Cells. In addition, these effector cells are derived from bone marrow stem cells and, most importantly, appear to be dependent on a normal marrow microenvironment.

This latter characteristic has been used to distinguish the effectors of allogeneic resistance from the cells which mediate skin graft rejection (5) as well as to distinguish the various subpopulations of murine NK cells (14). It is interesting to note that the four AA patients with normal NK(K562) [but low NK(HSV-1)] all demonstrated complete marrow failure, suggesting that the effectors of NK(K562) are probably not dependent on a normal marrow.

Allogeneic interactions require a "stimulator" cell which is relatively resistant to acute irradiation, matures late in ontogeny, and induces relatively nonspecific reactions (2). Our studies suggest that effectors of NK(HSV-1) may be "stimulator" cells since they are required for the induction of GvHD in bone marrow transplant recipients, they appeared to be resistant to irradiation in these patients, and our studies of newborns (Lopez et al., this volume) indicate that these effectors mature late in ontogeny. Since the relative strength of the NK(HSV-1) response was not predictive of the grade of GvHD obtained, induction of these reactions is probably an "all or none" phenomenon and the clinical expression of disease is probably dependent on other disparities as well as the interaction of lymphoid cells.

If allogeneic resistance (or host versus graft reaction, HvGR) is mediated by NK cells, as suggested by the mouse studies (6), and GvHD is induced by recipient NK cells as proposed by us in human transplant recipients, then it is possible that HvGR and GvHD are slightly different expressions of the same basic reaction. They may only be at different ends of the same continuum. It is also possible that chronic GvHD also falls on this same continuum.

The mixed leukocyte reaction (MLR) is an example of an allogeneic response requiring a stimulator cell function (2). Low MLR reactivity (matching at the lymphocyte determined antigens) between donor and recipient is probably required to avoid graft rejection in human marrow transplantation. Unfortunately, the MLR is not predictive of GvHD in marrow transplantation, suggesting that genes outside of the major histocompatibility region also are involved. Since GvHD requires a stimulator cell for induction, the NK(HSV-1) assay probably only determines that such a system is intact in the recipient and is thus poised for inducing this reaction. Patients with low NK(HSV-1) lack such a system and thus fail to induce GvHD.

REFERENCES

1. Thomas, E.D., Storb, R., Clift, R.A., Fefer, A., Johnson, F.L. Neiman, P.E., Lerner, K.G., Glucksberg, H., and Buckner, C.E.: N. Engl. J. Med. 292:832, 895 (1975).

2. Cerottini, J.C., Nordin, A.A., Brunner, K.T., J. Exp. Med. 134:83 (1971).
3. Lafferty, K.J., and Talmage, D.W. Transplant. Proc. 8: 349 (1976).
4. Cudkowicz, G., and Bennett, M. J. Exp. Med. 134:83(1971).
5. Bennett, M. J. Immunol. 110:510 (1973).
6. Kiessling, R., Hochman, P.S., Haller, O., Shearer, G., Wigzell, H., and Cudkowicz, G. Eur. J. Immunol. 7:663 (1977).
7. Lopez, C., Ryshke, R., Bennett, M. Infect. Immunity 28:1028 (1980).
8. Lopez, C., Kirkpatrick, D., Sorell, M., O'Reilly, R.J., and Ching, C. Lancet ii:1103 (1979).
9. Glucksberg, H., Storb, R., Fefer, A., et al. Transplantation 18:295 (1974).
10. Ching, C., and Lopez, C. Infect. Immunity 26:49 (1979).
11. Sorell, M., Kapoor, N., Kirkpatrick, D., Rosen, J.F., Chaganti, R.S.K., Lopez, C., Dupont, B., Pollack, M.S., Terrin, B.N., Harris, M.B., Vine, D., Rose, J.S., Goossen, C., Lane, J., Good, R.A., and O'Reilly, R.J. Amer. J. Med. 70:1280 (1981).
12. Pahwa, R., Pahwa, S., Good, R.A. J. Clin. Invest. 64: 1632 (1979).
13. Livnat, S., Seigneuret, M., Storb, R., and Prentice, R. L. J. Immunol. 124:481 (1980).
14. Kumar, V., Leuvano, E., and Bennett, M. J. Exp. Med. 150:513 (1979).

Index

A

ABM, *see* Anti-beta-2 microglobulin
N-Acetylglucosamine, inhibition of K562 natural killing by, 972–973
Activated murine macrophages, augmented binding of tumor cells by, 1003–1008
Acute myeloid leukemia patients, *Corynebacterium parvum* effects on cell-mediated immunity in, 1285–1290
ADCC (antibody-dependent cellular cytotoxicity)
  human mononuclear phagocyte populations and, 170
  leukemia blasts as effectors in, 1141–1146
  NK cells and, 119–122, 824
  vs. NK porcine activity, 345
  ontogenic development of, 342–343
Age-independent natural killer cell activity, in murine peripheral blood, 335–339, *see also* Natural killer activity
Aging mice, spontaneous tumors in, 1399
Alcohol ingestion, NK activity and, 651–655
Alloantibody augmentation, of mouse NK activity, 449–463
Alloantigens, for natural killer cells, 105–111
Alloantisera, reactivity to NK cell, 212–213
Alloantiserum NK augmentation, characteristics of, 449–453
Allogeneic cells, lymphocyte stimulation with, 791–796
AMC, *see* Autologous mixed culture
Anti-asialo GM1-treated mice, metastatic growth in, 1331–1336
Anti-BAT antiserum, 125–126
Anti-beta-2-microglobulin antisera
  human NK cells and monocytes in relation to, 251–256
  monocyte natural cytotoxicity and, 254–255
Antibodies
  monoclonal, *see* Monoclonal antibodies
  natural, *see* Natural antibody
  in normal rabbit sera, 758–763
Antibody-coated HPBL, natural cytotoxic activity of, 251–253
Antibody-dependent killer cells, 31
Anti-GA1 antibody
  interferon and, 127–128
  NK activity and, 126–127
Antigen-specific cloned TCGF-dependent cells, 894
Anti-HLA microglobulin antisera, human NK cells and, 251–256
Anti-interferon-α, and herpesvirus fibroblasts, 389–390
Anti-K562 cytotoxic activity
  of HPBL, 253
  following yellow fever vaccination, 499–502
Anti-K562 killer cells, enrichment of, 96
Anti-macrophage antibody plus complement, NK activity elimination with, 243–250
Anti-macrophage monoclonals, pretreatment with, 248–249
Anti-myelomonocytic antibodies, NK and K cell reactivities with, 43
Anti-NK-1.1 serum, reactivity of, 325
Anti-NK-2.1, as activity of NZB and BALB/c serum, 113–118
Anti-rat monoclonal antibodies, rat leukocytes and, 141
Antisera survey, of H-2 compatible strain combinations, 108

Anti-tumor activity, of NK cells, 1339–1344, 1347–1352
Anti-tumor host resistance, interferon as regulator of, 1387
Anti-tumor lytic potential, of lymphocytes from solid tumors, 1061–1067
Anti-tumor mechanisms, spontaneous vs. induced, 1353–1357
Anti-tumor reactivity, induction of by IL-2, 859–870
Arachidonic acid metabolism, interferon and, 406
Ascites tumor growth, 1355–1356, *see also* Tumor growth
Asialo GM1, as marker of immature cells of T-cell lineage, 210–212
Asialo GM1-bearing cells, characterization of, 1531–1532
Asialo-GM1 deletion, 1324–1327
  interferon production and, 1327–1329
Asialo GM1$^+$ (GA1$^+$) cells, in resistance to bone marrow and other transplants, 1527–1532
Atopic dermatitis, peripheral blood natural-cell-mediated cytotoxicity in, 1249–1252
Augmented binding
  capacity for, 1006–1007
  cell biology and, 1005–1006
  cytolysis and, 1006
  of tumor cells, 1003–1008
Autochthonous tumor cells, *see also* Tumors
  asialo GM1 positive cells in, 1326–1329
  interferon and, 1071–1072
Autoimmune disease, beige gene influences in development of in SB/Le male mice, 301–306
Autologous allogeneic and xenogeneic NK insensitive target cells, PHA-induced susceptibility of, 463–472
Autologous lymphoblastoid cell lines, lymphocyte stimulation with, 791
Autologous mixed culture, lytic potential generation in, 799–805
Autologous tumor cells
  lysis of, 693
  T-cell lytic effect against, 829–833

B

Babesia microti, NK cells and interferon production during infection of mice with, 1483–1488
Babesia microti parasitemia, 1485–1487
Babesiosis, murine malaria and, 1491–1496
Bacillus Calmette-Guerin, 396, 1004
Bacteria, natural cell-mediated immunity against, 1513–1525
BALB/c mice, anti-GA1 and, 126–127
BCG, *see* Bacillus Galmette-Guerin
Bcg gene, mapping of, 316–317
Bcg typing, of inbred mouse strains, 314
Beige gene, in autoimmune disease in SB/Le male mice, 301–306
Beige model studies, in natural resistance to syngeneic, semisyngeneic, and primary tumors, 1369–1377
Beige mutants
  C. neoformans and, 1505–1507
  listeriosis in, 1525
  MCMV susceptibility and, 1454
  murine NK activity and, 645–650
Bestatin, 1310
Biological response modifiers
  defined, 1291
  and NK activity in cancer patients, 1309–1314
  objective markers in study of, 511–517
Blood lymphocyte subsets, interferon and PHA in cytotoxic activity of, 475–482
Blood mononuclear cells, increased anti-K562 cytoxicity following yellow fever vaccination, 499–502
Bone marrow cells, grafting of into SR-89-treated mice, 331
Bone marrow dependent NK cells, 329–334
Bone marrow mononuclear cells, 1085–1090
Bone marrow transfusion, NK defect repair by, 1411–1414
Bone marrow transplantation
  asialo GM$^+$(GA1$^+$) in resistance to, 1527–1532
  lymphoma resistance transfer through, 1405–1409
  natural killer cells in, 1535–1539
  NK (HSV-1) assay in, 1542–1543
  recovery of NK cell activity following, 1253–1256
Bone marrow transplant recipients, NK (HSV-1) role in induction of GVHD in, 1541–1545
Bone-seeking isotope, mouse NK cells and, 329–334
Brain neocortex, murine NK activity control by, 639–642

Breast-cancer patients, NK and K cell activity in, 1189–1194, 1317–1321
Breast cancer regional lymph node, natural cytotoxicity in, 1133–1140
Broncho-alveolar macrophages, suppressive activity of, 581–587
Brucella abortus, 396–397, 1521

C

C3H/St mice, genetic resistance to MCMV in, 1459–1464
Calcium ions, cytotoxic T lymphocytes and, 989–990
CAMP levels, in inhibitory target-cell recognition by human NK cells, 607–613
Cancer, 1069–1071, 1167–1173, 1183–1187, 1194–1200, 1309–1314, 1353–1357
  of lung, 1415–1421
  NK cells and, 47, 1380–1382, 1439
Cancer immunotherapy, tumor antigens and, 1309
Cancer patients, *see also* Primary tumors; Solid tumors; Tumors
  interferon treatment for, 1265–1270, 1273–1283
  NK and K cell activity assessment in, 1175–1181, 1189–1194, 1265–1270, 1303–1308, 1317–1321
  NK suppression by pleural effusions in, 589–594
  OK432 augmentation of NK cell activity in, 1303–1314
Carageenan
  in *Listeria* infection, 1522
  in NK suppression, 202
Carbohydrate receptors, in natural cell-mediated cytotoxicity, 977–982
Cell-mediated cytotoxicity, *see also* Natural cell-mediated cytotoxicity
  in human NK-T cell clones, 888–890
  mechanism of, 989
Cell surface thiols, in human natural cell-mediated cytotoxicity, 1021–1026
Cellular cytotoxicity, *see also* Natural cell-mediated cytotoxicity
  inhibition of with mono- and oligosaccharides, 771–776
  spontaneous and antibody-dependent, 772–776
  sugars and, 774
Cellular secretion, human NK cell activity and, 983–987
Central nervous system, immune functions and, 639
Cerebral cortex, surgical lesions of, 640
Cervical cancer patients, NK and K-cell activity in, 1189–1194
CF, *see* Cytolytic factor
Chemotherapeutic agents, and NK activity in cancer patients, 1309
Chicken erythrocytes, lymphocyte cytotoxicity in, 157
Cholera toxin, NK reactivity inhibition of, 595–599
Chromium release data, conversion to relative NK, 1178
Chromosome 1 locus, in natural resistance to intracellular pathogens, 313–318
Chronic disease, human NK cell function and, 1241–1247
Chronic lymphocytic leukemia cells
  increased NK cell sensitivity in, 734
  after TPA treatment, 735
Cis-platinum, in murine tumor inhibition, 1311
CLL, *see* Chronic lymphocytic leukemia
Cloned cell lines
  cytotoxic activities of, 895
  with natural killer activity, 843–849
Cloned murine cell lines, with high cytolytic activity against YAC-1 targets, 851–858
Colon carcinoma, natural effector function in, 1127–1128
"Concomitant immunity," 237
Contaminating endotoxin, mononuclear phagocyte cytotoxicity and, 165–171
Cord blood lymphocytes, surface markers and cytotoxicity of, 156
Corynebacterium parvus, 396–397
  host NK activity following treatment with, 1339–1340, 1364
  in vivo effects of in natural-cell-mediated immunity in leukemia patients, 1285–1290
  murine NK activity and, 527, 1493
  Percoll gradient fractionation of spleen cells and, 1106–1108
Corynebacterium parvum-induced peritoneal cells, 245
Corynebacterium parvum-induced suppressor cells, for mouse NK activity, 519–525
CRBC-ADCC activity, NK cells and, 121–122

Cryptococcosis, natural cell-mediated resistance in, 1503–1509
Cryptococcus neoformans
- augmenting effects on NK reactivity, 1508–1509
- natural resistance of nude mice to, 1504–1505
- NK cell reactivity to, 1105–1111, 1503

CTL, *see* Cytotoxic T lymphocytes
Cultured natural killer cells, 807–813, *see also* Natural killer cells
Cultured spleen cells, cytotoxic activity of, 875, *see also* Spleen cells
Culture-induced killer cells, "unstimulated," 810
Cyclic adenosine 3′,5′-monophosphate, *see* cAMP
Cyclomunine, *in vitro* lymphocyte effects of, 1311
Cytolysis, stages in, 977
Cytolytic activity, potentiation of by lymphokines and interferon, 161–162
Cytolytic factor
- effects on target selectivity, 951–952
- macrophage activation and, 951
- in tumor cytotoxicity, 952–953

Cytolytic protease secretion, 949–953
Cytomegalovirus, murine, 1451–1464
Cytostasis
- LPS and, 204–205
- tumor-host immune relations and, 208

Cytostatic activity, murine splenocytes in mediation of, 201–208
Cytostatic cells, characterization of, 206–207
Cytotoxic activity, against NK-sensitive cultured tumor cell lines, 1047–1053
Cytotoxic assays, outlines of, 239
Cytotoxic capacity, *in vitro* amplification or stimulation of, 861
Cytotoxic cells, surface markers of, 44
Cytotoxic effector cells, generation of, 808–810
Cytotoxicity
- proteolytic enzymes in, 940–942
- selective vs. nonselective, 456

Cytotoxic leukocytes
- natural killing function of, 786–787
- vs. NK-like cells, 793–794

Cytotoxic T lymphocytes
- vs. natural killer cells, 239
- in NK cell lysis, 61, 989–990
- specificity of fresh and activated forms of, 691–697

D

Daudi tumor, prevention of growth of, 868
Dermatitis, atopic, 1249–1252
Diethylstilbestrol
- in depression of NK activity, 1437
- neonatal exposure to, 1437
- NK activity in prostate carcinoma, 1317–1321

Differentiation antigen, of murine NK cells, 125–129
Dimethylbenzanthracene, NK suppression by, 1431–1436
Diterpene esters, macrophage differentiation induction by, 601–605
DMBA, *see* Dimethylbenzanthracene
DTC, *see* Sodium diethyldithiocarbamate

E

Effector cells, *see also* Natural cell-mediated cytotoxicity
- bone marrow mononuclears as, 1087
- *Cryptococcus neoformans* and, 1109–1111
- function analysis of, 194
- generation of, 808–810
- incubation of at 37 °C, 622
- interferon pretreatment of, 376
- as large granular lymphocytes, 676
- in lymphocyte cytotoxicity enhancement, 153–154
- lysis of two targets bound simultaneously to, 150
- NCMC modulation and, 461, 1359
- NKCF release by, 963
- reactivity against primary tumor targets, 1059
- in resistance to herpesvirus infections in man, 1445–1449
- serum treatment and, 459
- specific cytotoxicity scores for, 686
- tissue distribution of, for NK and ADCC, 343–344

Effector populations
- characterization of, 196
- separation of, 194

Endotoxin contamination, of tissue culture reagents, 165–170
Epstein-Barr virus
- B cell transformation of, 791
- LGL and, 1047–1049, 1052–1053

E-rosette cells, in NK and ADCC deficiencies, 1195–1200

Estracyt®, in prostate cancer, 1317–1321
β-Estradiol, NCMC and, 1363–1364
Ethanol, natural killer activity and, 651–655
Experimental allergic encephalomyelitis, 1219
Experimental allergic neuritis, 1219

F

F1 hybrid mice, anti-GA1 injections in, 1528–1531
F344 rats, LGL leukemias in, 1161–1164
FACS, *see* Fluorescence-activated cell sorting
Fcγ⁻ receptor expression, lymphocyte subset cytotoxic activity and, 97
Feedback growth control, and differentiation of proliferating cells, 235
Ficoll-Hypaque centrifugation, 676
Fluorescence-activate sorting, in human NK cell size distribution and β_2-microglobulin content, 99–103
Free oxygen radicals, in NK cytolytic pathway activation, 1011–1019
FTS, *see* Serum thymic factor
Functional bone marrow, NK cells and, 329–334, *see also* Bone marrow

G

GA1⁺ lymphocytes, immunoelectron micrography of, 1532, *see also* Asialo GM1
Genetic analysis, of natural resistance *in vivo*, 291–298
Genetic control, of NC activity in mouse, 281–288
Glycophorin A, 727
Gnotobiotic miniature swine system, NK and K cell activities in, 381–385
Golgi apparatus, morphology of, 985
Graft-versus-host disease, NK(HSV-1) *in vivo* role of, 1541–1545
Granular lymphocytes, *see* Large granular lymphocytes
GVHD, *see* Graft-versus-host disease

H

H-2 associated natural resistance. NK cells and, 292-295
Hamsters, NK cells in, 1207–1210
Hematopoietic cells, natural killing of, 1077–1083
Hemin, K562 cell NK sensitivity and, 727-732
Hemoprotozoan infections, natural cell-mediated immunity in, 1491–1500
Herpesvirus saimiri, marmoset T-cell transformation by, 835-842
Herpesvirus ateles, marmoset T-cell transformation by, 835-842
Herpesvirus-infected fibroblasts, interferon in natural killing of, 387-392
Herpesvirus infections
  interferon in, 387–392, 1474–1475
  NK(HSV-1) effector cells in, 1445–1449
Herpesvirus murine cytomegalovirus, NK cells and interferon in, 1451–1458, *see also* Murine cytomegalovirus infection
Herpesvirus-transformed marmoset T cell lines, natural killer cell-like cytotoxicity mediated by, 835–842
High-risk tumor and cancer patients, NK cell activity in, 1183–1187, *see also* Cancer patients
HLA system, multiple sclerosis and, 1227–1232
HNK-1 antigen, nature of, 32
HNK-1⁺ cells
  distribution of in human lymphoid tissues, 34
  expression of other surface antigens on, 33
  morphology and cytotoxic function of, 32
Hodgkin's disease
  cell-mediated immune defects in, 1201
  natural killing in, 1201–1205
  T-cell mediated NK deficiency in, 1205
HPBL, *see* Human peripheral blood leukocytes
Human blood lymphocyte subsets
  E-rosette negative and positive subsets of, 92–95
  monoclonal reagents in characterization of, 91–98
Human cytotoxic lymphocytes, *see also* Cytotoxic T lymphocytes
  clones of, 864–865
  specificity of fresh and activated forms of, 691–697
  surface markers of, 866–867
Human interferons, in porcine NK and ADCC, 383–385, *see also* Interferon
Human large granular lymphocytes, *see also* Large granular lymphocytes

Human large granular lymphocytes *Continued*
identification criteria for, 258
lineage of, 261–262
morphology and cytotoxicity of, 1–7
as new cell lineage, 257–262
T-cell lineage and, 259–260
Human leukocytes, human NK cell augmentation with, 349–353
Human lymphocytes, recognition structures for NK cells in, 683–688
Human lymphoid tissues, HNK-1+ cell distribution in, 34
Human malaria, immunity in, 1496, *see also* Malaria
Human monocytes
in endotoxin-free conditions, 166–167
natural cytotoxicity of, 159–164
Human mononuclear phagocytes, natural cytotoxicity of, 165–171, 581–587
Human natural cell-mediated cytotoxicity, *see also* Human natural killer activity; Natural killer activity
augmentation by α- and γ-interferon inducers, 355–360
cell surface thiols in, 1021–1026
modification by MVE-2, 1291–1295
proteases with specificity of cleavage at aromatic amino acids in, 923–929
surface-associated proteases in, 939–946
Human natural killer activation, interferon and, 361–366
Human natural killer activity, *see also* Natural killer activity
cellular secretion associated with, 983–987
vs. human primary tumors with LGL, 1055–1059
negative regulation by monomeric IgG, 621–627
OK432 augmentation of, 431–436
regulation against tumor target cells by monocyte subpopulation, 657–667
suppressor lymphocytes of, 575–580
Human natural killer cells, *see also* Natural killer cells
analysis by monoclonal antibodies against myelomonocytic and lymphocytic antigens, 47–52
augmentation of with human leukocyte and human recombinant leukocyte interferon, 349–353
cAMP inhibition of target-cell recognition by, 607–613
cell surface phenotypes of, 69
characteristics of, 77
clonal analysis of, 879–884
cloned populations of, 883–884
as complex and heterogeneous system, 71
differentiation pathway of, 35–37
FcR for IgG and IgM in, 631–636
Fc_GR and Fc_MR distribution in, 632
as homogeneous and discrete cell subset, 39–45
identification by monoclonal HNK-L (Leu-7) antibody, 31–37
interferon and, 361–366, 657–667, 882–883
interleukin-2 and, 409–414
vs. lectin-dependent effector cells, 145–151
limiting dilution analysis of, 879–882
monoclonal antibodies in analysis of, 67–71
phenotypes of, 59–64
purified cultures of, 821–826
size distribution and β_2 microglobulin content of, 99–103
surface markers on, 53–58, 85
tumor cell lysis by, 931–936
Human natural killer cytolytic activity, *see also* Cytolytic factor
lymphocyte-target cells and, 1041–1046
protease secretion and, 949–953
Human natural killer effector cells, heterogeneity of, 73–78, *see also* Natural killer cells
Human natural killer function
in chronic diseases, 1241–1247
monoclonal antibody in analysis of, 777–784
Human natural killer T-cell clones, continuous culture of, 887–892
Human ovarian tumors, natural killer activity in, 1119–1125
Human peripheral blood leukocytes
lineage of, 253–254
natural cytotoxic activity of, 251–252
Human pulmonary tumors, NK activity of tumor infiltrating and lymph node lymphocytes in, 1113–1118
Human recombinant leukocyte interferon, augmentation of human NK cells with, 349–353
HVA, *see Herpesvirus ateles*
HVS, *see Herpesvirus saimiri*
Hydrocortisone acetate, in NK suppression, 202
Hypophysectomy, murine NK activity and, 645–650

I

IF, IFN, *see* Interferon
IL-2, *see* Interleukin-2
Immune complexes, in NK activity, 633–636
Immunoglobulin G, as inhibitor of NK activity, 623–624
Immunomodulatory agents, 1279–1301
Interferon
in advanced cancer patients, 1265–1270, 1273–1283
and anti-Daudi lytic potential of cultured cells, 803
anti-GA1 and, 127–128
arachidonic acid metabolism and, 406
in *B. microti* infection, 1487–1488
in cancer patients, 1265–1270, 1273–1283
in chronic diseases, 1242–1243
in cytotoxic activity of blood lymphocyte subsets, 475–482
in cytotoxic reactivity regulation of NK cells, 615–620
defective function of, 1245–1246
defined, 1279
differential effects on NK and ADCC, 345
differential effects on porcine NK and K-cell activities, 381–385
effector lymphocytes and, 1048
and E-rosette cells from leukemia patients, 1195–1200
extravascular natural effectors and, 1128
herpesvirus infection and, 1474–1475
human fibroplastic, 1310
human NK activity vs. tumor target cells and, 657–667
in human NK cell activation, 361–366, 882–883
human NK cytotoxicity augmentation with, 355–360
immunomodulatory effects and, 882–883, 1297–1301
inhibition of by pretreatment with 17.2, 999
in *in vitro* modulation of NK activity, 395–400
and K562 killing of lymphocytes from different sites, 1129
as major modifier of NK cytolytic activity, 615
in malaria, 1498–1499
modulation of macrophage cytolysis by, 401–407
mouse strains and, 1472
in multiple sclerosis, 1223–1224, 1229–1230, 1233–1239
in murine cytomegalovirus, 1451–1458
and murine hepatitis virus type 3, 1465–1476
in natural killing of herpesvirus-infected fibroblasts, 387–392
NK activity and, 44, 109, 127, 132–134, 350, 369–373, 431, 621, 995–996, 1431–1436
in NK-cell-mediated cytotoxicity, 10–15
NK defective function and, 1245–1246
in NK lysis of SL2-5 lymphoma, 1388–1391
NK trophoblast cell sensitivity and, 752
NK-1$^+$ cell proportion to, 327
pleiotropic effects on augmentation of rat NK activity, 375–379
potentiation of cytolytic activity by, 161–163
in pretreatment of effector cells, 376
production by human NK cells, 669–674
production by peripheral blood mononuclear cells, 371–372
protective effect on target cells, 372–373
in RA patients, 1246
vs. reactivity against autochthonous tumors, 1071–1072
recombinant leukocyte, 1279–1283
in regulation of IL-2-dependent growth of NK progenitor cells, 909–915
as regulator of anti-tumor host response, 1387
TPA and, 753
in trophoblast cell sensitivity to NK cell activity, 751–756
tumor promoter TPA and, 751–756
wart virus and, 1301
Interferon-activated LGL, phenotypic characterization of, 81
Interferon activity, testing of serum for, 458
Interferon-α
enhanced NK activity and, 1259–1262
human NK cytotoxicity augmentation with, 355–360
lytic potential modification by, 799–805
Interferon-γ, inducers of, 355–360
Interferon induced NK cells
porcine alveolar macrophage interferon and, 382
surface antigens and, 119–121
Interferon induction, NK cell activation of, 179–185, 1465–1470
Interferon-stimulated NK cells, compared with TPA-stimulated cytotoxic cells, 438

Interferon pretreatment, NK cell growth frequency and, 911
Interferon-producing cells, virus-induced NK cells and, 1474
Interferon production
asialo GM-1 positive cells and, 1327–1329
in Hodgkin's disease, 1205
by human lymphocytes fractionated by Percoll density gradient centrifugation, 675–680
independence of, 389
T-cell stimulation in, 673
Interferon suppression, *in vivo* effects of, 1387–1392
Interferon-treated melanoma patients, NK cytotoxicity in, 1265–1270
Interleukin-1, human NK activity and, 657–667
Interleukin-2
discovery of, 821
human NK cells and, 409–414
induction of NK-like anti-tumor reactivity by, 859–870
natural killer cells and, 133–134
NCMC modulation by, 415–420
Interleukin-2-dependent allospecific lymphoid populations, NK activity in, 903–908
Interleukin-2-induced augmentation, of NK cell activity, 421–426
Interleukin-3, natural cytotoxic activity of mouse cell cultures maintained with, 917–920
Intestine NK effector cell, serological characterization of, 29
Intracellular bacteria, natural resistance to, 1521–1525
Intracellular levels of cAMP, in inhibition of target-cell recognition by human NK cells, 607–613
In vivo reactivity, of NK cells, 549–555
Irradiation doses, in NK reactivity depression in mice, 1423–1430
^{125}IUDR, in B16-F10 melanoma cells, 202

K

K562 tumor cells
inhibition of natural killing by saccharides *in vitro*, 972–973
interaction with LGL, 1038–1040
interferon in killing of, 1129
lytic effect of T-cell cultures against, 829–833
NK sensitivity of, 727–732
Killer cells, *see also* Human natural killer cells; Natural killer cells
cell lineages of, 346–347
interferon and, 44, 109, 127, 132–134, 350, 355–366, 369–373, 381–383, 431, 621, 657–667, 882–883, 995–996, 1431–1432
monoclonal antibodies used as, 40
natural killer cells and, 343–346, *see also* Natural killer cells
ontogenic development of in gnotobiotic miniature swine, 341
reactivity with anti-myelomonocytic antibodies, 43

L

Lactobacillus sp., 342
Large granular lymphocytes
allogeneic tumor biopsy cells and, 1047–1053
and anti-myelomonocytic and anti-lymphocytic monoclonal antibodies, 48–50
characteristics of, 319
cytotoxic activity of, 825
defined, 319
effector cells and, 25–30, 1072–1073, *see also* Effector cells
Epstein-Barr virus and, 1047–1049, 1052–1053
in F344 rats, 1161–1164
functional inhibition by monoclonal antibodies, 82–83
granule formation in, 4–6
growth curves for, 822
as heterogeneous population, 96
human NK activity vs. human primary tumors with, 1055–1059
identification of, 1036
K562 tumor cells and, 1038–1040
lysosomal system and, 2–4
lytic capacity of, 14
metastatic paraganglioma and, 1300
monoclonal antibodies and, 265–267
morphology and cytotoxicity of, 1–7, 823
natural killer activity and, 9–15, 26–28, 39, 79, 614–615, 703
NCMC effector cells as, 676

Large granular lymphocytes, *Continued*
origin of, 47
Percoll fractions and, 880
phenotypic characterization of, 81
proteolytic activity by, 934–935
and rat NK cells, 20–22, 139–142
as separate lineage, 269
strain and age variation in frequency of, 323
surface markers and, 824
tumor cell lines and, 1047–1053
ultrastructure of, 2
uropod as integral and specialized structure in, 1035–1040
LCLs, *see* Lymphoblastoid cell lines
LDCC, *see* Lectin-dependent cellular cytotoxicity
Lectin-dependent cellular cytotoxicity
defined, 145
NK effector cell as, 148
Lectin-dependent effector cells, vs. human NK cells, 145–151
Lectin-dependent NK cellular cytotoxicity, in mice, 483–490
Lectin-induced cellular cytotoxicity, NK cells and, 824
Lentinus edodes, 1297
Leukemia blasts, as effectors in natural killing and antibody-dependent cytotoxicity assay, 1141–1146
Leukemia patients, *C. parvum* and, 1285–1290
Leukemia-resistant mice, marrow transfer of NK activity from, 1405–1409
Leukemia cells, anti-GA1 reactivity and, 128–129
Leukemogeneic irradiation doses, NK reactivity depression in mice by means of, 1423–1430
Leukocyte K562 cells, sensitivity to NK lysis, 727, *see also* K562 tumor cells
Leukocytes, human, 349–353
Leukocytoid lymphocytes, 1
LGL, *see* Large granular lymphocytes
LICC, *see* Lectin-induced cellular cytotoxicity
Ligand binding, radioactive labeling in, 1004
Limiting dilution analysis, of NK progenitor cell growth, 910–915
Lipopolysaccharides, splenocytes and, 202–204
Listeria monocytogenes, 1499, 1503
natural resistance to, 1513–1524
LT, *see* Lymphotoxins
Lumbricus terrestris, 262
Lung cancer, *see also* Cancer patients
natural-cell-mediated immunity in, 1415–1421
urethan-induced, 1415–1421
Lymph node lymphocytes, suppressor cell activity and, 1116–1118
Lymphoblastoid cell lines, lymphocyte stimulation with, 791–796
Lymphocyte cytotoxicity, to chicken erthrocytes, 157
Lymphocyte fractions, NK and UDCC of, 155
Lymphocyte-mediated cytotoxic effects, nomenclature of, 239–242
Lymphocyte populations, T-cell division of, 799
Lymphocytes
anti-tumor lytic potential separated from, 1061–1067
cytotoxicity enhancement for, 153
E-rosetting population in, 85–88
granular, 1–7
interferon pretreatment of, 659
virus-dependent natural cytotoxicity of, 153–158
Lymphocyte stimulation, with allogeneic normal cells and autologous lymphoblastoid cell lines, 791–796
Lymphocyte-target cell intensities, morphological characterization of, 1041–1046
Lymphohemapoietic cells
natural killer cells from, 109
subset of, 105
Lymphoid cells, lytic capacity of, 12
Lymphokines
and NK cell proliferation and cytolytic activity, 844–846
potentiation of cytolytic activity of, 161–163
Lymphoma resistance, transfer from AKR mice by marrow from high NK lymphoma-resistant mice, 1405–1409
Lymphoproliferative syndrome, NK deficiency in, 1211–1217
Lymphotoxins
defined, 969
generation of, 970–971
in natural cytotoxicity, 969–975
Lysis
induction and enhancement of, 995–1001
NK, *see* Natural killer cell lysis

Lytic capacity, enhancement of, 1000–1001
Lytic efficiency, enhancement of, 995–1001
Lytic events, natural vs. immune, 242
Lytic potential, generation of in mixed cultures, 799–805

M

Macrophage activation analysis, objective markers in, 511–517
Macrophage activation stages, 512
Macrophage cytolysis, modulation by interferon, 401–407
Macrophage cytotoxicity, against protozoa, 1099–1103
Macrophage-mediated tumor cytotoxicity, and secretion of novel cytolytic protease, 949–953
Macrophages
broncho-alveolar, 581–587
spontaneous cytotoxicity of, 165
Macrophage tumoricide activity, 601–605
Malaria
effector cells in pathogenesis of, 1497–1498
human, 1496
interferon in, 1498–1499
murine, 1491–1492
natural cell-mediated immunity and interferon in, 1491–1500
Maleic anhydride and divinyl ether, as biological response modifiers, 1291
Mammary tumors, *see* Breast cancer
Mannose-6-phosphate receptor, on target cell, 977–978
Marmoset T-cell transformation, by HVA and HVS, 835–842
Marrow transfusion, 1411–1414
Marrow transplantation, *see* Bone marrow transplantation
MCA, *see* Methylcholanthrene
MCMV, *see* Herpesvirus murine cytomegalovirus
Melanoma, in anti-NK treated mice, 1347–1352
Melanoma patients, interferon treatment for, 1265–1270, *see also* Cancer patients
Membrane perturbations, NK reactivity and, 721
Methylcholanthrene, in sarcoma induction, 1439
Mice, *see also* Mouse; Murine
genetic strains vs. NK activity in, 528
lack of suppressor cell activity for NK cells in, 557–561
lectin-dependent NK cellular cytotoxicity in, 483–490
NK activity absence in at birth, 528
protection against MHV3 infection, 1467–1468
radioactive stimulation treatment for, 563–568
spontaneously cytotoxic cells in, 131
suppressor cell activity regulation in, 557–572
Micrococsis lysodeikticus, 1295
Microflora, NK and ADCC effects of, 344
Mitochondrial respiration, macrophage-induced inhibition of, 1028–1030
Mitogen-induced cellular toxicity, NK cells and, 632
Mitogens, and interferon production by human NK cells, 669–672
Mixed leukocyte cultures
NK activity and, 61
stimulated lymphocytes in, 791–794
Mixed lymphocyte cultures
lytic potential generation and modification in, 799–805
NK-like cells and, 785–790
NL-like cytotoxicity in, 787–788
MLC, *see* Mixed lymphocyte cultures
MNAb, *see* Murine natural antibody
Moloney sarcoma virus, 1399
Monensin, and human NK cell-mediated cytolysis, 985
Monoclonal antibodies
anti-K562 cytotoxicity and, 57
cytotoxic activity reduction with, 59–64
human NK cell analysis by, 67–71, 777–784
inhibition of LGL by, 82–83, 265–267
from mice immunized with LGL, 50–51
natural killer cytotoxicity and, 70, 75, 85–90, 139, 244, 355–360
phenotype and functional characterization of NK cells of, 79–84
production of, 31–32
purified null or T cells and, 87–88
reactivities of, 48–49
surface markers and, 42
T-CLL cells and, 55
T lymphocytes vs. NK-like cells and, 791–796

Monoclonal cells, fluorescent assay of, 80, *see* Monoclonal antibodies
Monoclonal HNK-2 (Leu-7) antibody, human NK cell identification by, 31–37
Monoclonal reagents
human blood lymphocyte subsets characterized by, 91–98
recoveries of cells defined with, 94–95
Monoclonal specific anti-macrophage antibody plus complement, 243–250
Monocyte-mediated cytotoxicity, spontaneous, 815–819
Monocyte subpopulations, human NK antibody vs. tumor target cells in, 657–667
Monocytic/granulocytic antigens, natural cytotoxicity and, 56–57
Monocytoid lymphocytes, 1
Monomeric IgG, negative regulation of human NK activity by, 621–627
Mononuclear cells
characterization of, 197
titration of NK cell activity in, 87
Mononuclear phagocytes
activation of, 511
ADCC and, 170
NK activity modulation by, 581–587
Monosaccharides, cellular cytotoxicity and, 772–776
Mouse, genetic control of NC activity in, 281–288, *see also* Mice; Murine
Mouse bone marrow, NK cell dependence on, 329–334, *see also* Bone marrow
Mouse hepatitis virus 3, 1465–1470, 1477–1481
Mouse inbred strains
Bcg typing in, 314
linkage analysis in, 315–316
Mouse mammary tumors, 1069–1075
Mouse NK activity, *see* Murine natural killer activity
Multiple sclerosis
ADCC and, 1221
HLA system and, 1227–1232
impaired NK function in, 1227–1232
interferon and, 1223–1224, 1229–1230, 1233–1239, 1350
monozygotic twin studies and, 1233–1239
NK cell activity and, 1219–1224, 1233–1239
Murine cell clones
with NK-like activities, 893–900
permanently growing, 893–900
Murine cell lines, with high cytolytic activity against YAC-1 targets, 851–858
Murine cultured lymphoid cell, surface markers for, 866
Murine cultures, in IL-2 containing media, 862
Murine cytomegalovirus infection, loss of genetic resistance to, 1451–1464
Murine gut mucosal lymphoid cells, NK activity and, 27
Murine lectin-dependent NK cellular cytoxicity, 483–490
Murine leukemia, tunicamycin-induced differentiation in, 736
Murine leukemic cells, anti-GA1 reactivity to, 128–129
Murine mammary tumors, 1069–1071, *see also* Breast cancer; Cancer
Murine natural killer activity, *see also* Natural killer activity
age and genetic factors in, 528–531
alloantibody augmentation in, 449–453
cloned lines of, 873–877
control in, 281–288
C. parvum-induced suppressor cells for, 519–525
heterogeneity of, 175–176
hypophysectomy and, 645–650
immunoregulation by serum thymic factor, 215–223
natural suppressor cells for, 527–532
phorbol esters vs. interferon-induced activation of, 437–441
starvation effects in, 645–650
suppression of by normal peritoneal macrophages, 535–540
tumor growth and, 645–650
Murine natural killer antibody, 719–725
Murine natural killer cell-mediated cytotoxicity, 977
Murine natural killer cells, differentiation antigen and, 125–129, *see also* Natural killer cells
Murine peripheral blood, age-independent natural killer activity in, 335–339
Murine peritoneal exudate cell preparations, 193–196
Murine resident peritoneal cells, 193
Murine skin tumors, diterpene esters and, 601–603
Murine spleen cells
alcohol and, 653–654
interferon and, 327
in mediation of cytostatic activity, 201–208

natural cytotoxic activity of when maintained with IL-3, 917–920
Murine *Trypanosoma cruzi* infection, NK activity during, 503–507
MVE-2, *see* Maleic anhydride and divinyl ether
Mycobacterium bovis, 1519
Mycobacterium tuberculosis, mouse susceptibility to, 1493
Myelomonocytic origin, of NK cells, 47

N

NAb, *see* Natural antibody
Natural antibody
in host-mediated resistance, 1382
and NK cell activity in syngeneic murine model, 1379–1385
Natural antibody anti-tumor activity, NK cells and, 1379–1385
Natural antibody target structures, NK cells and, 719–725
Natural cell-mediated cytolysis, inhibitor sites and, 991–993
Natural cell-mediated cytotoxicity, *see also* Effector cells
carbohydrate receptors in, 977–982
against different target cells, 339
effector cells and, 713, 962–963, 1359
enhancement of, 1000–1001
inhibition of, 979–980
modulation of by interleukin-2, 415–420
murine natural killer cells and, 179, 187–191
NKCF role in, 961–967
NK-like effector cells and, 962–963
poly-1C augmented, 182
in solid tumors, 1359–1366
specific and nonspecific modulation by serum components, 455–462
vs. spontaneous mouse mammary tumors, 1069–1075
in tumor development in animals, 395
Natural cell-mediated immunity
against bacteria, 1513–1519
in urethan-induced lung carcinogenesis, 1415–1421
Natural cell-mediated lysis, and sialic acid of tumor cells, 765–770
Natural cell-mediated resistance, in cryptococcosis, 1503–1509
Natural cytotoxic cells, *see also* Natural killer cytotoxicity
chromosome 17 and, 281–288
in murine peritoneal exudate cell preparations, 193–196
target cell recognition by, 713–717
Natural cytotoxic effectors, in human tumors and tumor-draining nodes, 1127–1132
Natural killer activation, mechanisms of, 995–1001
Natural killer activity, *see also* Natural killer cells
adherent cancer cells in suppression of, 589–594
age-independent type in murine peripheral blood, 335–339
in aging mice, 1399
alcohol and, 151–155
assessment of in cancer patients, 1175–1181
augmentation by retinoic acid, 427–429
and biological response modifiers in cancer patients, 1309–1314
following bone marrow transplantation, 1253–1256
boosting conditions and, 183
in breast and cervical cancer patients, 1189–1194, 1303–1308
carrageenan suppression of, 202
change in related to interferon dose, 1281–1282
chromium release data in, 1178
chromosome 17 and, 277
clinical variables affecting, 1179
cloned cell lines with, 843–849
in cloned IL-2-dependent allospecific lymphoid populations, 903–908
by *C. parvum*-induced suppressor cells, 519–524
^{51}Cr-release assay for, 660
detection of after antisera treatment, 338
diamide and qBBR inhibition in, 1024–1025
diethylstilbestrol and, 1437–1442
dimethylbenzanthracene suppression of, 1431–1436
early maturation of, 342
FTS direct action on, 222
genetic variation in, 319–324
in high-risk tumor and cancer patients, 1183–1187
in Hodgkin's disease, 1201–1205
in human ovarian tumors, 1119–1125

Natural killer activity, *Continued*
hydrocortisone acetate in suppression of, 202
IgG as inhibitor of, 623–624
immune complexes in, 633–636
interferon and, *see* Interferon
interleukin-2 and, 421–426
in vitro modulation of, 395–400
large granular lymphocytes and, 9, 26, 79
Lentinian administration and, 1298–1299
lipopolysaccharide and, 202–205
in mast cell-deficient W/W^v mice, 273–274
in melanoma patients, 1167–1173
metabolic requirements for augmentation of, 398
modulation by human mononuclear phagocytes, 581–587
modulation during murine *Trypanosoma cruzi* infection, 503–508
monoclonal antibodies and, 62–63, 139, 244
in multiple sclerosis, 1233–1239
murine, *see* Murine natural killer activity
"natural" antibodies and, 344, 1379–1385
against primary tumors *in vivo,* 1399–1403
in prostate cancer, 1317–1321
recombinant leukocyte interferon in modulation of, 1279–1283
reduction of by anti-LGL antibody, 142
in SM/J mice, 307–311
in spleen and PBL strains, 336–337
in spontaneous and transplanted reticulum cell neoplasms, 1147–1152
strain and variations in, 323–324
suppression of, 202, 519–524, 535–540, 589–594, 1431–1436
and T-cell mediated immunity in syngeneic tumor rejection, 1393–1398
transfer of to AKR mice by marrow from high NK lymphoma-resistant mice, 1405–1409
against transplantable and primary tumors, 1323–1329
trophoblast cell sensitivity to, 751–756
against *Trypanosoma cruzi* forms, 1091–1096
tumor-related changes and prognosis in melanoma patients, 1167–1173, 1380–1382
against YAC-1 cells, 275–280, 294
Natural killer activity inhibitory cells, characterization of, 544
Natural killer-associated antigens
natural cells and, 187–191
NZB anti-BALB/c serum and, 114
Natural killer B cells, advantages of, 109
Natural killer cell activity, *see* Natural killer activity
Natural killer cell boosting conditions, 183
Natural killer cell depression, in multiple sclerosis, 1219–1224
Natural killer cell-like cytotoxicity, mediation by herpesvirus-transformed marmoset T-cell lines, 835–842, *see also* Natural killer cytotoxicity
Natural killer cell lysis
cellular secretion role in, 987
mechanism of, 989–994
"Natural killer cell memory," 236
Natural killer cell population, postnatal expansion of, 34
Natural killer cells, *see also* Natural killer activity
abnormality in, 34–35
"activated" designation for, 177, 240, 369–371
ADCC cells and, 121, 824
alloantigens specific for, 105–111
alloantisera reactive to, 212–213
anti-asialo GM1 antiserum for, 210
anti-tumor activity of, 1339–1344, 1347–1353
C-dependent effects of, 114
in bone marrow transplantation, 1253–1256, 1535–1539
cell lineage of, 209–213, 346–347
cell surface markers of, 73–78, 848, *see also* Surface markers
cloned populations of, 884–885
CRBC-ADCC activity and, 121–122
vs. *Cryptococcus neoformans,* 1105–1111, 1508–1509
CTL cell proliferation and, 231–234
culture-activated, 187–191, 807–813
cytotoxic T lymphocyte activity and, 61, 239, 519
defective function of, 1244–1245
as defense against cancer and viral infections, 47
defined, 903
depletion of, 225–226
development of, 224–237
distinction from analogues, 35
effector-population mediation of, 193–196
elimination by monoclonal specific antimacrophage antibody plus complement, 243–250

Natural killer cells, *Continued*
enhancement by E-rosette cells from lymphocytic leukemia patients, 1195–1200
fractionation of on target-cell monolayers, 699–704
frequencies of, 9–10
genetic factors in, 319
and H-2 associated natural resistance, 292–295
in hamsters, 1207–1210
heterogeneity of, 176
in host resistance against tumors, 1303
human, *see* Human natural killer cells
immunopharmacological manipulation of activity in, 519
as immunosurveillance mechanism against tumor growth, 1077, 1085
interferon and, 132–134, 184, 350, 369–373, *see also* Interferon
interferon plus immunomodulatory agents in, 1297–1301
interleukin-2 and, 133–134
in vivo reactivity of, 549–555
kinetics of, 785–786
and lack of suppressor cell ability in mice, 557–561
as large granular lymphocytes, 20, 39, 139, 1244
as LDCC effector cells, 148
vs. lectin-dependent killer cells, 145–151
leukemic blasts and, 1141–1146
as lymphohemapoietic cells, 109
lymphoma cell growth and, 1339–1344
lysing capacity of, 903
lysis induction and enhancement in, 995–1001
marrow dependence of, 329–334
MNAb and RNAb in relation to, 719–725
monoclonal antibodies and, 40, 79–84
of mouse, *see* Murine natural killer activity
in murine cytomegalovirus resistance, 1451–1458
myelomonocytic origin of, 47
and NAb anti-tumor activity *in vivo*, 1379–1385
natural antibody target structures and, 719–725
"natural" designation in, 240
NCMC mediation by, 455
new subpopulation of, 483–490
NK-associated cell surface antigens and, 187–191
"null" lymphocytes and, 39, 209
ontogenic development of, 342–343
ontogeny of NK-1$^+$ strain in, 325–328
original functional definition of, 177
origin and specificity of, 224–237
oxygen radicals and, 1012–1014
phenotypic and functional characterization of, 79–84
phenotypic heterogeneity of, 63–64
precursors of in ^{89}Sr-treated B6D2F1 mice, 330
pre-T cells and, 332–333
proliferation and role of during viral infection, 493–498
in radiation leukogenesis, 1411–1414
in rats, 17–23
RCN-associated origin of, 1150–1152
reactivity with anti-myelomonocyte antibodies, 43
recognition structures for, 683–688
recruitment of pre-NK to active NK type, 996–997
regulation of cytotoxic reactivity by interferon and PGE, 615–620
as regulators of hematopoietic cells, 1077–1083
as resistance to semi-syngeneic virus-induced lymphomas, 1477–1481
as separate lineage, 265–270
solid tumors killed by, 179–185
specificity of, 743–749, 846–848
in splenic PNA$^-$ fraction, 226, *see also* Splenic natural killer cells
standardized formula for, 1280
surface markers in, 73–78, 848, *see also* Surface markers
surveillance, self-recognition, and growth control related to, 234–238
target cell recognition by, 713–717
target cell structures and, 743–749
T cells and, 209–213, 229–237, 268, 411–413
tumor cells and, 291, 387, 1326–1329
tumor development and, 733–739, 1339–1344, 1347–1352
Natural killer cell specific alloantigens, 106–108
Natural killer cell subsets, 173–178
Natural killer cell susceptibility, malic acid and, 768
Natural killer cell system
biological significance of, 136
heterogeneity and regulation of, 131–137

Natural killer cell tumor-target sensitivity changes, 733–739
Natural killer cytotoxicity, *see also* Natural cytotoxic cells
  characteristics of, 964–965
  effector cells and, 25–30
  genetic control in mouse, 281–288
  immune complexes and, 631–636
  inhibition of by tumor promoters and cholera toxin, 595–599
  interferon production and, 675–680
  in interferon-treated melanoma patients, 1265–1270
  lymphotoxin role in, 969–975
  monoclonal antibody effects in, 70
  in NK cell-mediated cytotoxicity, 961–967
  in regional lymph node in breast cancer, 1133–1140
Natural killer cytolytic pathway, 1011–1019
Natural killer deficiency, in X-linked lymphoproliferative syndrome, 1211–1217
Natural killer depression, in chronic diseases, 1242–1244
Natural killer effector cells, 25–30, 148, 1445–1449
Natural-killer-like cells, 785–790
Natural-killer patterned binding, 705–710
Natural killer progenitor cells
  interferon and T cells in regulation of, 909–915
  limiting dilution analysis of, 909–915
Natural killer reactivity, *see also* Natural killer activity
  against primary tumors *in vivo,* 1399–1403
  regulation by suppressor cells, 541–547
Natural killer sensitive and resistant lymphoma cell variants, growth of, 1339–1344
Natural killer sensitive cell lines, lysis of, 695
Natural killer sensitivity, surface sialic acid and, 768
Natural killer strains, cytotoxic activity of F_1 offspring from, 322
Natural killer susceptibility, target cell H-2 expression and, 714–715
Natural cell target interaction, competitive inhibition of, 744–745
Natural killer target structures, serological approaches to elucidation of, 757–763
Natural killing, human bone marrow mononuclear cells as effectors and targets of, 1085–1090
Natural resistance, genetic analysis of, 291–298
Natural suppressor cells, for murine NK activity, 527–532
NCMC, *see* Natural cell-mediated cytotoxicity
Neuraminidase, LGL and, 12–13
Neutral serine proteases, in tumor cell lysis by NK cells, 931–936
Newcastle disease virus, 1242
New Zealand B/J mice, immunizing with BALB/c spleen cells, 113
NK cells, *see* Natural killer cells
NK_A and NK_B cells, in culture, 810–812
NKCF, *see* Natural killer cytotoxic factors
NKCMC, *see* Natural killer cell-mediated cytotoxicity
NK (HSV-1) assay, in bone marrow transplantation, 1542
NK-like anti-tumor resistivity, induction of by IL-2, 859–870
NK-like cells, 483–490, 788–790
$NK\text{-}1^+$ cells, 326
Non-NK leukocytes, 705–710
Non-specific host defense, 369–373
Novel cytolytic protease secretion, macrophage-mediated tumor cytotoxicity and, 949–953
NPT 15 392, 1311
Null cells, 39, 85, 209
Nylon wool passed spleen cells, binding to various tumor targets, 707–708

O

OK432 (heat- and penicillin-treated lyophilized powder)
  human NK activity augmentation by, 431–436
  in NK cell activity in cancer patients, 1303–1308
Oligosaccharides, cellular cytotoxicity and, 77–2776
Ontogenic development, of porcine NK and K cells, 341–348
Ovarian tumors, NK activity in, 1119–1125
Oxygen intermediates, modulation of release of, 405–406
Oxygen radicals
  generation mechanisms for, 1015–1017
  NK cells and, 1012–1014

P

PBL, *see* Peripheral blood lymphocytes
PBMC, *see* Peripheral blood mononuclear cells
PEC, *see* Murine resident peritoneal cells
Percoll density gradient centrifugation or fractionation
 interferon production and, 675–680
 of rat NK cells, 18, 79
 of rat splenic cells, 1106–1108
 suppressor cells and, 543
Peripheral blood large granular lymphocytes, ultrastructure of, 2–3, *see also* Large granular lymphocytes
Peripheral blood leukocytes, reactivity against primary tumors, 1055–1058
Peripheral blood lines, NK activity of, 700–702
Peripheral blood lymphocytes
 fractionation of, 699–704
 NK activity and, 1114
Peripheral blood monocytes, *see* Peripheral blood mononuclear cells
Peripheral blood mononuclear cells
 interferon production by, 371–372
 natural cytotoxicity of, 160–161
 natural killer cells and, 1242
Peripheral blood natural-cell-mediated cytotoxicity, 1249–1252
Peritoneal exudate cells, 852
Peritoneal macrophages, suppression of murine NK activity by, 535–540
PHA, *see* Phytohemagglutinin
Phorbol esters
 and induction of macrophage differentiation in bone marrow precursors, 602–603
 vs. interferon-induced activation of mouse NK cells, 437–441
 tumor-promoting capacity of, 438
Phospholipase A_2 activation, in recognition and lysis, 955–958
Phospholipid metabolism, in NK cell activity, 955–958
Photohemagglutinin
 and cytotoxic activity of blood lymphocyte subsets, 475–482
 LDCC and, 145
 and susceptibility induced in NK cell lysis, 463–472
Plasmodium berghei, 1491–1492, 1498
Plasmodium chabaudi adami, 1492
Plasmodium yoelii, 1491–1492
Poly I-C, NK activity and, 1431–1436
Polymorphonuclear leukocytes
 in anti-tumor surveillance, 1353
 rat LGL and, 139
Porcine alveolar macrophage interferon, 382
Porcine NK and K cells
 interferon effects on, 381–385
 ontogenic development of, 341–348
Pre-natural killer cells, recruitment to active-NK cells, 996–999
Pre-T cells, dissociation from NK-A cells, 332–333
Primary fibrosarcoma, spontaneous vs. induced anti-tumor mechanisms in resistance to, 1353–1357
Primary mammary tumors, normal lymphoid cell resistivity against, 1069–1071
Primary tumors
 beige model in studies of natural resistance to, 1369–1377
 diethylstilbestrol and, 1437–1442
 effector cell reactivities against, 1059, 1072–1073
 in vivo NK reactivity against, 1323–1329, 1399–1403
 PBL reactivity against, 1055–1058
Prostaglandin(s), modulation of release of, 405–406
Prostaglandin E_2
 in activated NK cell suppression, 618
 endogenous, 618
 in NK cell activity, 621
 in regulation of NK cell cytotoxic reactivity, 615–620
Prostate cancer, diethylstilbestrol and estramustine phosphate in, 1317–1321
Proteolytic enzymes, in cytotoxicity, 940–942
Prothymocytes, vs. splenic natural killer cells, 225–227
Protozoa, natural macrophage cytotoxicity against, 1099–1103
Purified human T cells, natural cytotoxic activity of, 85–90, *see also* T cells
Purified natural human killer cells, cultures of, 821–826
Purified null cells, monoclonal antibodies and, 87

Q

Qa-5 + C effector population, poly-IC augmented NCMC and, 182

R

RA, *see* Retinoic acid; Rheumatoid arthritis
Rabbit natural antibody, 719–725
Rabbit sera, antibodies in, 758–763
Radiation leukemogenesis, NK cells in, 1411–1414
Radioactive strontium, in murine suppressor cell activity, 330, 1459–1464, 1478–1489
Rat anti-mouse macrophage monoclonal antibodies, characteristics of, 250
Rat large granular lymphocytes, cell surface antigenic characteristics of, 20–22, 139–142, *see also* Large granular lymphocytes
Rat leukocytes, reactivity of with other antibodies, 141
Rat natural killer activity, *see also* Natural killer activity
  genetic variation in, 319–324
  interferon pleiotropic effects on, 375–379
Rat natural killer cells, isolation of, 17, 139, *see also* Natural killer cells
Rat strains, NK activity in, 319–324
RCN, *see* Reticulum cell neoplasms
RCS, *see* Reticulum cell sarcomas
Recognition structures, for NK cells on human lymphocytes, 683–688
Recombinant human leukocyte interferon, 350–352, 1279–1283
Reticulum cell neoplasms, NK cell activity in, 1147–1152
Reticulum cell sarcomas, in SJL/J mice, 1153–1159
Retinoic acid
  anti-tumor immunity and, 428
  NK activity augmentation by, 427–429
Reverse ADCC reaction, in alloantiserum NK augmentation, 449–453, *see also* ADCC
Rheumatoid arthritis, NK function in, 1245–1246
RNAb, *see* Rabbit natural antibody
RO 10 9359 retinoic derivative, 1310–1311

S

Sarcomas, methylcholanthrene induction of, 1439
Sb/Le male mice, beige gene in development of autoimmune disease in, 301–306
Secondary tumor growth, 1356–1357, *see also* Tumor growth; Tumors
Secretory pathway, blocking of, 985
Serum components, NCMC modulation of, 455–462
Serum thymic factor, in immunoregulation of mouse NK activity, 215–223
Sham-operated mice, NK activity levels in, 641
Silica, in *Listeria* infection, 1522–1523
SJL/J mice, reticulum cell neoplasms in, 1147–1159
SJL/J spleen cells, proliferation of, 874
SM/J mice, NK cell activation in, 307–311
Sodium diethyldithiocarbamate
  CNS and, 639
  mouse NK activity and, 443–447
Sodium butyrate, K562 cell NK sensitivity and, 727–732
Solid tumors, *see also* Tumors
  anti-tumor lytic potentials of lymphocytes separated from, 1061–1067
  natural killer cell-mediated cytotoxicity against, 179–185, 1359–1366
Spleen cells
  C. parvum-induced adherent, in NK activity suppression, 520
  LPS and, 202–204
  phorbol esters in cytotoxicity of, 438
  target binding, 326
Splenic natural killer cells, vs. prothymocytes, 225–227
Spontaneous macrophage cytotoxicity, murine peritoneal exudate cell preparations and, 193–196
Spontaneous monocyte-mediated cytotoxicity, T helper activity in, 815–819
Spontaneous mouse mammary tumors, natural cell-mediated cytotoxicity against, 1069–1075
Staphylococcus aureus, 1297
Starvation, murine NK activity in, 645–650
Streptococcus sp., 342, 431
Strontium-89, in murine suppressor cell activity, 563–568, 1459–1464, 1478–1479
Suppressor cells
  marrow-dependent cell regulation of, 569–572
  in mice, 557–582
  natural killer reactivity regulation of, 541–547, 575–580

Suppressor cells, *Continued*
Percoll fractionation and, 543
Surface antigens
of ADCC cells, 121–122
of interferon-induced NK cells, 119–121
of murine NK cells, 125
NK and NC cells in expression of, 188–190
Surface-associated proteases, in human NK cell-mediated cytotoxicity, 939–946
Surface enzymes, exposure of during cytolysis, 939–946
Surface malic acid, of tumor cells, 765–770
Surface markers
of cord blood lymphocytes, 156
for HNK cells, 53–58, 85
for human and murine CLC, 866–867
on interferon-induced natural killer cells, 119–122
LGL receptors and, 824
on monoclonal antibodies, 42
murine antibody-dependent cellular cytotoxicity cells and, 119–122
on NK cells, 53–58, 73–78, 85, 119–122, 848
for permanently growing murine cell clones, 897–900
in VDCC of human lymphocytes, 153–154
in XLP syndrome, 1214
Syngeneic murine model, NK cell and NAb anti-tumor activity in, 1379–1385
Syngeneic tumor rejection, NK activity and T-cell-mediated immunity in, 1393–1398
Syngeneic virus-induced mouse lymphomas, NK cell function in, 1477–1481
Systemic lupus erythematosus, NK cell and regulating indices of PBMC in, 1243

T

Target-cell binding
exposure of concealed cytolytic proteases following, 943–946
lytic function and, 1044
NK activity and, 632
Target cell monolayers, fractionation of NK cells on, 699–704
Target cell-specific IgG, NK cell activation with, 361–366
Target cell structures, NK cells and, 743–749
T-cell cultures
cytotoxic activity of, 825
growth curves for, 822
lytic effect of against tumor biopsy cells and K562, 829–830
morphology of, 823
T-cell growth factor, 821, 893–894, *see also* Interleukin-2
T-cell lineage
asialo GM1 and, 210–214
LGL and, 259–260
of NK cells, 209–213, 268
T-cell lines, of marmoset, 835–842
T-cell mediated events, DTC and, 443
T-cell mediated immunity, and NK activity in syngeneic tumor rejection, 1393–1395
T-cell proliferation, MHC Ia antigens and, 231
T cells, *see also* Lymphocytes; T lymphocytes; Target cells
activity in VDCC, 154
antigen-sensitive, 155
in depression effect of interferon on frequency of cytotoxic NK cells, 912–913
in increased anti-K562 cytotoxicity following yellow fever vaccination, 499–502
lytic function and, 241
NK and K cell reactivities with antibodies specific for, 41–43, 53–54
regulation of IL-2 dependent growth of NK progenitor cells, 909–915
TCGF, *see* T-cell growth factor
T-CLL cells, NK and ADCC activity in, 54–55
T helper activity, in spontaneous monocyte-mediated cytotoxicity in man, 815–819
Thymic antigens, natural cytotoxicity and, 56–57
Thymopoietin sensitive precursor, 332
TIL, *see* Tumor infiltrating lymphocytes
Tissue culture reagents, endotoxin contamination of, 170
T lymphocytes, *see also* T cells
in anti-tumor surveillance, 1353
NK-like cells and, 791–796
TPA (12-0-tetradecanoylphorbol-13-acetate)
CLL cell sensitivity after treatment with, 735
interferon and, 753
NK cell activity and, 751–756
surface sialic acid and, 768
TPA-induced activation, macrophage removal and, 441

TPA-treated trophoblast, tumorogenicity and, 754–755
Transformed target cells, activated macrophage-mediated cytotoxicity for, 1027–1033
Transmethylation, phospholipid metabolism and, 955–958
Transplantable tumors
  diethylstilbestrol and, 1437–1442
  in vivo reactivity against, 1323–1329
  NK reactivity against, 1399–1403
Trichomonas vaginalis, natural cytotoxicity against, 1099–1103
Trypanosoma cruzi infection, NK activity during, 503–507, 1091–1096
Tumor-associated lymphoid cells, NK activity and, 1119–1125
Tumor biopsy cells, T-cell lytic effect against, 829–833
Tumor cell lines, LGL cytotoxic activity against, 1050
Tumor cell lysis
  by human NK cells, 931–936
  neutral serine proteases in, 931–936
Tumor cells
  augmented binding of by activated murine macrophages, 1003–1008
  and leukomogenic doses of radiation in NK depression, 1423–1430
  NK activity and, 291, 387, 1326–1329
  surface salic acid of, vs. NK cell-mediated lysis, 765–770
  natural resistance against, 295–297
  transplantable, *see* Transplantable tumors
Tumor cytotoxicity, activated murine macrophages and, 1003–1008
Tumor draining nodes, natural cytotoxic effectors in, 1127–1132
Tumor growth
  in hamsters, 1207–1210
  monocytic cytotoxicity and, 162
  murine NK activity and, 645–650
  NK cell augmentation and suppression in, 1207–1210
  NK cells as immunosurveillance mechanism in, 1077, 1085
Tumoricidal activity, functional heterogeneity and expression of, 167–169
Tumor-infiltrating lymphocytes
  NK activity of in human pulmonary tumors, 1113–1118
  suppressor cell activity and, 1116–1118
Tumor patients, high-risk, *see* High-risk tumor and cancer patients; *see also* Cancer; Cancer patients
Tumor promoters, NK cell cytotoxic reactivity inhibition by, 595–599
Tumor-promoting diterpene esters, in macrophage differentiation, 601–605
Tumors, *see also* Cancer
  NK cells in host resistance against, 1303
  primary, *see* Primary tumors
  secondary, 1356–1357
  solid, *see* Solid tumors
  transplantable, 291–298, 1323–1329, 1399–1403, 1437–1442
  urethan-induced, 1416–1420
Tumor target cells, human peripheral blood monocytes and, 159, *see also* Target cells
Tumor target structures, natural antibody and NK cells in relation to, 719–725
Tumor transplantation, 1323–1329, 1399–1403, 1437–1442
  natural resistance to, 291–298
Tunicamycin-induced differentiation, in murine leukemia, 736

U

Urethan, carcinogenic effects of, 1416
Uropod, as integral and specialized structure of LGL, 1035–1040
UV-2237 fibrosarcoma, NCMC against, 1360–1361

V

VCN, *see* Neuraminidase
VDCC, *see* Virus dependent natural cytotoxicity
VEP8-VEP9 monoclonal antibodies, 48–50
Vesicular stomatitis virus, 751
Vinblastin, in *Listeria* infection, 1523–1524
Virus-dependent natural cytotoxicity, 153–158
Viruses, and interferon production by human NK cells, 669–672
Virus-dependent natural cytotoxicity, 153–158
Virus-infected cells, natural immunity against, 495–497
Virus infection, NK cell proliferation and role in, 493–498

W

Warts, interferon and, 1301

X

X-linked lymphoproliferative syndrome, NK deficiency in, 1211–1217

Y

YAC-1 (murine lymphoma) cells
Babesia microti infection and, 1483–1484
cloned murine cell lines and, 851–858
development of resistance in NKCMC and NKCF, 965–966
murine NK cells in lysing of, 1542
NCMC cytotoxicity against, 1360
YAC-1 tumor, cold target inhibition of lysis of, 1073–1074
Yellow fever vaccination, increased anti-K562 cytotoxicity following, 499–502